FOOD SAFETY MANAGEMENT

FOOD SAFETY MANAGEMENT

A PRACTICAL GUIDE FOR THE FOOD INDUSTRY

Edited by

YASMINE MOTARJEMI
Nyon, Switzerland

HUUB LELIEVELD
Global Harmonization Initiative, Bilthoven, The Netherlands

ELSEVIER

AMSTERDAM • BOSTON • HEIDELBERG • LONDON • NEW YORK
OXFORD • PARIS • SAN DIEGO • SAN FRANCISCO
SINGAPORE • SYDNEY • TOKYO
Academic Press is an imprint of Elsevier

Academic Press is an imprint of Elsevier
32 Jamestown Road, London NW1 7BY, UK
225 Wyman Street, Waltham, MA 02451, USA
525 B Street, Suite 1800, San Diego, CA 92101-4495, USA

British Library Cataloguing-in-Publication Data
A catalogue record for this book is available from the British Library

Library of Congress Cataloging-in-Publication Data
A catalog record for this book is available from the Library of Congress

ISBN: 978-0-12-381504-0

For information on all Academic Press publications visit our website at www.store.elsevier.com

Typeset by MPS Limited, Chennai, India
www.adi-mps.com

Printed and bound in United States of America

14 15 16 17 10 9 8 7 6 5 4 3 2 1

Contents

6 Meat and Meat Products

JOHN N. SOFOS

7 Poultry and Eggs

SHAI BARBUT, IRWIN PRONK

8 Seafood

SANJA VIDAČEK

9 Fruits and Vegetables (including Herbs)

VLASTA PILIŽOTA

10 Coffee, Cocoa and Derived Products (e.g. Chocolate)

ANETT WINKLER

11 Honey, Confectionery and Bakery Products

SUCHART CHAVEN

12 Edible Nuts, Oilseeds and Legumes

W.C.A. GELDERBLOM, G.S. SHEPHARD, J.P. RHEEDER, S.K. SATHE, A. GHIASI, Y. MOTARJEMI

13 Oils and Fats

GERRIT VAN DUIJN

14 Bottled and Drinking Water

ARNOLD F. DIJKSTRA, ANA MARIA DE RODA HUSMAN

15 Pet Food

PABLO A. CARRIÓN, LARRY J. THOMPSON

16 Food Contact Materials

CARLOS DE LA CRUZ GARCIA, GLORIA SÁNCHEZ MORAGAS, DAVID NORDQVIST

II

TECHNOLOGIES AND FOOD SAFETY

17 Thermal Treatment

TIBOR DEAK

18 Non-thermal Processing Technologies

OLGA MARTÍN-BELLOSO, ROBERT SOLIVA-FORTUNY, PEDRO ELEZ-MARTÍNEZ, A. ROBERT MARSELLÉS-FONTANET, HUMBERTO VEGA-MERCADO

19 Acids and Fermentation

MARTIN R. ADAMS

20 Chilling and Freezing

STEPHEN J. JAMES, CHRISTIAN JAMES

21 Detection of Physical Hazards

GILLES DEMAUREX, LAURENT SALLÉ

III FOOD SAFETY ASSURANCE SYSTEMS

22 Principles and Systems for Quality and Food Safety Management

PETER OVERBOSCH, SARAH BLANCHARD

23 Hygiene in Primary Production

GISELA KOPPER, SLAVKO MIRECKI, IGOR S. KLJUJEV, VERA B. RAICEVIC, BLAZO T. LALEVIC, JELENA PETROVIC-JOVICIC, STOJMIR STOJANOVSKI, DIJANA BLAZEKOVIC-DIMOVSKA

24 Hygiene in Food Processing and Manufacturing

JOHN HOLAH

25 Site Selection, Site Layout, Building Design

HUUB LELIEVELD

26 Hygienic Design and Maintenance of Equipment

FRANK MOERMAN, JACQUES KASTELEIN

27 Development of a Comprehensive Cleaning and Sanitizing Program for Food Production Facilities

ROBERT RYTHER

28 Personal Hygiene and Health

EWEN C.D. TODD

29 Pest Management

CHRISTOPHER H. BELL

30 Safe Handling of Food in Homes and Food Services

PATRICIA DESMARCHELIER

31 Hazard Analysis and Critical Control Point System (HACCP)

YASMINE MOTARJEMI

32 HACCP Misconceptions

YASMINE MOTARJEMI, CAROL WALLACE, SARA MORTIMORE

33 Management of Microbiological Hazards: Role of Testing as Verification

TIM JACKSON

34 Management of Chemical Contaminants

YASMINE MOTARJEMI

35 Food Defense

NED MITENIUS, SHAUN P. KENNEDY, FRANK F. BUSTA

IV

SUSTAINABILITY AND ETHICS

List of Contributors

Tanja Pajk Žontar International Consumer Research Institute, Ljubljana, Slovenia

Martin R. Adams University of Surrey, Guildford, Surrey, UK

Lucia E. Anelich Anelich Consulting, Pretoria, South Africa

Jean-Michel Antoine Danone, Paris, France

Diána Bánáti International Life Sciences Institute (ILSI) Europe, Brussels, Belgium. Former Central Food Research Institute, Hungary

Shai Barbut University of Guelph, Guelph, Canada

Christopher H. Bell The Food and Environment Research Agency, Sand Hutton, York, UK

Sarah Blanchard Metro AG, Duesseldorf, Germany

Dijana Blazekovic-Dimovska Faculty of Biotechnical Sciences, University of Bitola, Bitola, Macedonia

Arnaud Bouxin FEFAC, Brussels, Belgium

Frank F. Busta National Center for Food Protection and Defense, University of Minnesota, St. Paul, MN, USA

Pablo A. Carrión Nestlé Purina PetCare, Product Technology Center, St. Louis, MO, USA

Suchart Chaven Pepsi-Cola International Ltd., Dubai, UAE

Stella Cochrane Unilever, Bedford, UK

Anne Constable Nestlé Research Center, Lausanne, Switzerland

René Crevel Unilever, Bedford, UK

Carlos de la Cruz Garcia Nestec SA, Nestlé System Technology Center, Orbe, Switzerland

Tibor Deak† Corvinus University, Budapest, Hungary

Gilles Demaurex Nestle Product Technology Centre, Orbe, Switzerland

Patricia Desmarchelier Food Safety Consultant, Brisbane, Australia

Arnold F. Dijkstra Dijkstra Food Consultancy, Leeuwarden, Netherlands

Pedro Elez-Martínez University of Lleida, Lleida, Spain

W.C.A. Gelderblom Department of Biochemistry, Medical Research Council, Tygerberg, and Stellenbosch University, Stellenbosch, South Africa

A. Ghiasi Food Safety Consultant, Tehran

John Holah Campden BRI, UK

Serge Imboden Haute Ecole de Sion, Sion, Switzerland

Tim Jackson Nestlé USA and Nestlé Canada

Christian James The Grimsby Institute of Further & Higher Education (GIFHE), North East Lincolnshire, UK

Stephen J. James The Grimsby Institute of Further & Higher Education (GIFHE), North East Lincolnshire, UK

P.J. Jooste Colorado State University, Fort Collins, CO, USA

Jacques Kastelein TNO, Zeist, The Netherlands

Shaun P. Kennedy Veterinary Population Medicine, University of Minnesota, St. Paul, MN, USA

Igor S. Kljujev University of Belgrade, Serbia

†To the regret of the editors and the publisher, the author of chapter 17, Prof. Tibor Deak, passed away after completion of his chapter. It is a loss to the scientific community.

Gisela Kopper Latin American and Caribbean Association of Food Science and Technology, San Jose, Costa Rica

Blazo T. Lalevic University of Belgrade, Serbia

Huub Lelieveld Formerly Unilever, Vlaardingen The Netherlands

Olga Martín-Belloso University of Lleida, Lleida, Spain

Slavko Mirecki University of Montenegro, Podgorica, Montenegro

Ned Mitenius Periscope Consulting, Grapevine, TX, USA

Frank Moerman Catholic University of Leuven - KU Leuven, Leuven, Belgium

Sara Mortimore Land O'Lakes, Inc., St. Paul, MN, USA

Yasmine Motarjemi Food Safety Management Consultant, Nyon, Switzerland

G. Moy Food Safety Consultants International, Geneva, Switzerland

David Nordqvist Nestec SA, Nestlé System Technology Center, Orbe, Switzerland

Thomas Ohlsson Thoohls, Molndal, Sweden, Previously at SIK – The Swedish Institute for Food and Biotechnology, Gothenburg, Sweden

Peter Overbosch Metro AG, Duesseldorf, Germany

Irène Perrin Nestlé Research Center, Lausanne, Switzerland

Marjana Peterman Slovene Consumers' Association, Ljubljana, Slovenia

Jelena Petrovic-Jovicic University of Belgrade, Serbia

Vlasta Piližota University of Josip Juraj Strossmayer in Osijek, Osijek, Croatia

Irwin Pronk University of Guelph, Guelph, Canada

Vera B. Raicevic University of Belgrade, Serbia

J.P. Rheeder Medical Research Council, Tygerberg, South Africa

A. Robert Marsellés-Fontanet University of Lleida, Lleida, Spain

Robert Ryther ECOLAB, Eagan, MN, USA

Laurent Sallé Nestle Product Technology Centre, Orbe, Switzerland

Gloria Sánchez Moragas Institute of Agrochemistry and Food Technology, Paterna, Valencia, Spain

S.K. Sathe College of Human Sciences, Florida State University, Tallahassee, FL, USA

Benoît Schilter Nestlé Research Center, Lausanne, Switzerland

F. Tracy Schonrock Schonrock Consulting, formerly with the United States Department of Agriculture (USDA)

G.S. Shephard Medical Research Council, Tygerberg, South Africa

John N. Sofos Colorado State University, Fort Collins, CO, USA

Robert Soliva-Fortuny University of Lleida, Lleida, Spain

Stojmir Stojanovski Hydrobiological Institute, Ohrid, Macedonia

Larry J. Thomson Nestlé Purina PetCare, Product Technology Center, St. Louis, MO, USA

Ewen C.D. Todd Ewen Todd Consulting LLC, Okemos, MI, USA

Nina Veflen Olsen Nofima, Ås, Norway

Humberto Vega-Mercado Merck & Co., West Point, PA, USA

Sanja Vidaček Faculty of Food Technology & Biotechnology, Zagreb, Croatia

Carol A. Wallace University of Central Lancashire, Preston, UK

Anett Winkler Kraft Foods R&D Inc., Munich, Germany

Ana Maria de Roda Husman National Institute for Public Health and the Environment (RIVM), Bilthoven, The Netherlands

Gerrit van Duijn Oil Processing Consultant formerly Unilever R&D, Vlaardingen, The Netherlands

Foreword

It is a pleasure to write this foreword, because of the importance of food safety for the food industry, governments and consumers alike. The frequent food safety incidents occurring globally illustrate that food safety management in the industry is a subject that badly needs attention. In recent years, we have seen many incidents that have made people suffer or even lose their lives; the victims also included children. Although in some cases this has been due to emerging threats, such as new pathogenic bacteria, in many cases these incidents result from the reoccurrence of previous failures.

With professional management of food safety, incidents and certainly their recurrence can be prevented. In the case of emerging threats, adequate management should also limit the impact of incidents.

Food Safety Management: A Practical Guide for the Food Industry is a unique book and a reference for the future. For the first time, it gathers all essential and basic information that managers and professionals need to know about the management of food safety in the food industry and other related topics such as leadership, management of people, ethics and sustainability. It relays past experience to novice mangers. In line with the modern approach to management of food safety, it examines food safety management from the perspective of the entire food chain from "farm to fork." Farm standing for the primary food production, i.e. animal husbandry, agriculture and fishery, discussing the hazards and risks during the very first stages of the production of food. The book also emphasizes that skills and training in hard sciences on their own are not sufficient; it is the integration of knowledge, skills and the attitude and mindset of all involved, including top management, that make the management of food safety possible.

The guidance given in this book is applicable and relevant to all parts of the world and the book will be an invaluable resource and manual for training all present and future food safety managers or regulatory officials supervising food operations.

The editors are the most outstanding food safety experts in the world, and I congratulate them for the successful completion of this remarkable undertaking. I recommend this book with full confidence to colleagues all over the world.

Ping-fan Rao Prof. Dr.

Director, Institute of Biotechnology, Fuzhou University, China,
Vice-president, Chinese Institute of Food Science and Technology,
President, International Union of Food Science and Technology

Foreword

I am very pleased to be invited to write a foreword for this book because it brings together important information regarding food safety risks and practical approaches for managing them across the supply chain. From a manufacture's perspective, the safety and quality of our products is of the highest importance. The relationship we have with our customers and consumers is built on the confidence and trust they have in us and the food we make. For us – food producers, processors and manufacturers – food safety is an essential part of our culture and we work hard to maintain that trust. Food safety incidents shake that trust and erode consumer confidence. As an industry, we must act together to understand what has happened in these situations, share what we learn and determine how to prevent such situations in the future. It is essential that we help each other advance our programs and drive safety improvements into our product designs.

We often speak of controlling hazards and managing food safety. While these are extremely important, the contribution of a successful design in products, processes and packaging safety should not be overlooked. A good food safety program identifies hazards that are reasonably likely to occur and eliminates them from the design. The desired outcome is product and packaging that delivers safety during intended shelf-life and consumer use; controllable processes that ensure elimination or reduction of hazards to acceptable levels; and an environment that prevents recontamination, both by the physical parameters as well as by the presence of an educated workforce that understands and cares about food safety. Only when we are unable to fully eliminate a hazard in the design should we proceed to manage it using appropriate preventive controls.

The basis for all product, process and equipment design as well as food safety programs is the ability to identify and understand the risks that are reasonably likely to occur. This is a critical step needed in order to conduct an effective hazard analysis. Knowledge of the risks associated with the different aspects of the supply chain is extremely important. This publication shares examples of real incidents and their root causes as well as the various risks associated with the different sectors of the production process. It also shares possible methods that may be used to control these hazards. Having this type of end-to-end information is important to the understanding of the supply chain and will lead to more informed hazard analysis.

Finally, over and above technical aspects, the book underlines the importance of company culture, leadership, people management, crisis management and communication, ethics and sustainability, all of which are important aspects of today's food operations.

In this book you will be provided with guidance from some of the leading food safety figures in our industry. I am sure you will benefit from their experience and knowledge and hope their perspective inspires you to enhance and strengthen your own food safety programs and activities.

Matilda Freund Dr.
Senior Director, Quality, Mondelēz International

Preface

As stated by the Persian poet Ferdowsi, in the 10th century knowledge is power.*

Knowledge of food safety is a *conditio sine qua non* for professionals of food safety to be able to meet their responsibilities. However, at the outset it should be said that knowledge is not enough and as explained in Chapters 37 (Human Factors in Food Safety Management) and 47 (Training and Education) a number of other factors come into play.

In relation to knowledge, the questions that come to mind are what should food safety managers in the food industry, or governmental, or non-governmental officials overseeing their operations, know, and what should students of food safety courses be taught at minimum before starting their work?

There are many books on food safety. For the most part, these books address specific aspects of food safety in depth such as foodborne pathogens, chemical contaminants, quality assurance systems or sanitation procedures. Such books are often for specialists on a given subject. The aim with this book is to give the essentials that food safety professionals from any discipline should know about food safety management in the industrial setting, taking into consideration that food safety is a multidisciplinary subject and not all professionals have the same scientific and technical background.

The book intends to give an overview and an integrated perspective of food safety management, including risk and control measures for various categories of foods. It describes the elements of the food safety assurance systems in the food industry and provides guidance for their practical implementation.

As such, together with the *Encyclopedia on Food Safety* (Elsevier, 2014), it aims to be a practical resource for the education and training of present and future food safety professionals working in the food industry[1] or in governmental roles, such as food inspectors and auditors verifying food operations.

More specifically, the book attempts to:

- Consolidate essential knowledge for the management of food safety and facilitate its application, using practical examples and cases studies.
- Use practical examples to create awareness of pitfalls and past incidents, their cause(s) and lessons learned.
- Provide observations on what works and what does not (dos and don'ts) and

*Although in his work *Religious Meditations, Of Heresies* (1597) Sir Francis Bacon (1561–1626) has used the quote "Knowledge is Power," the adage is first attributed to the Persian poet Ferdowsi (pseudonym of Abū al-Qasem Man ūr (born c. 940, near ūs, Iran – died c. 1020). Ferdowsi is the author of the Persian national epic the *Shāh-nāmeh* ("Book of Kings"), which is the historical and mythological tale of Persian kings and the heroes of Iran. Overall, Ferdowsi spent 35 years of his life composing the book, which originally contained 60,000 couplets. Another translation of this adage is "A learned human is a powerful one too; the old hearts grow young through knowledge."

[1]For the purpose of this chapter, the term "food industry" is defined as all relevant sectors associated with the production, storage and handling of food, from primary production to retail and the food service level.

on issues to be aware of in food safety management.
- Bridge past experiences to state-of-the-art food safety assurance systems, and anticipate potential future risks and mitigation steps.

The educational objectives of this book are to enable food safety professionals to:

- Identify hazards and controls at various stages of the food chain.
- Understand food processing technologies in order to be able to determine which parameters need to be controlled, how they must be monitored and which limits (or critical limits) must be observed.
- Get an integrated perspective on food safety and quality assurance in the food industry and the interactions and interrelation of the elements of food safety assurance systems (e.g. how suppliers should be audited in the light of hazards identified in the HACCP study).
- Get an overview of food safety management in society, i.e. what are the respective functions of governmental and regulatory authorities, industries and other stakeholders and how these need to interact with each other.
- Understand the role and responsibilities of each sector in the food chain.
- Bring specific issues to the attention of auditors and inspectors.
- Enable food safety professionals to anticipate and respond to future challenges.

The target audience of the book is food safety managers working in different food sectors (including pet food), from primary production to processing, transport, retail and distribution, as well as the food service sector. Secondary target groups of the book are students and future food safety professionals as well as food inspectors, auditors, trainers and food safety consultants.

Part I of the book reviews risks and control measures in specific food sectors along the food chain providing food safety managers with awareness of the major risks with their raw material and operations. Part II describes technologies that may be used to ensure the safety of food, and explains pitfalls and other factors that would be important for their application (i.e. control and monitoring parameters, critical limits, validation). Part III describes elements of food safety assurance systems in the food industry and presents an overview of information on the role of various sectors in the management of safety of the food supply. Part IV discusses the more topical issues of sustainability and ethics as well as food safety trends in modern society. The book finishes with some final words on the training and education itself, which we hope are "words of wisdom."

Unfortunately, official reports and data on foodborne illnesses from developing countries are scarce or anectodal. Therefore, most of the examples are from industrialized countries. However, the know-how presented in the book is still valid for developing countries. It also shows what is on the horizon in terms of best practices for less developed businesses as well as the improvements that can be brought to the management of food safety of companies of all sizes.

For a detailed description of individual pathogens and/or chemical contaminants we recommend readers to refer to the *Encyclopedia of Food Safety* or other suggested reading material mentioned at the end of each chapter of this book. Readers will also find information on other products (e.g. spices, cereal products, beverages,) as well as many other subjects (e.g. risk analysis process) in these sources. To ease the use of the relevant sections of the book, at the risk of repetition, each chapter is written in a self-standing manner.

This book focuses on food safety; however, at the outset of this book we would like to deplore the abuse that animals and food animals are undergoing and we call on the conscience of all individuals in society to unite in putting a stop to such abuse and mistreatment, as well as overexploitation of the environment and work for the safeguard of the planet. At the same time, we appeal to the sense of responsibility of all individuals working in the food industry, in particular managers, to challenge company policies and practices which are favoring shortsighted benefits to the detriments of ethics, the good of the larger society, the planet and humankind.

Together with contributors, we have worked diligently to provide a valuable resource for food safety professionals and to share our vision, knowledge and experience. As Ferdowsi expresses it for Persia, with this book we hope to have contributed to spreading the seeds of knowledge. Comments from the readers are welcome for improving future editions of the book and in this way contributing to better practices.

Much hard labor have I done for thirty years
In the end I have revived Persia through this Persian verse.
I shall not pass away since I will remain alive
Through the seeds of this language I have spread everywhere.

Ferdowsi

Nomenclature

ABBREVIATIONS OF IMPORTANT TECHNICAL TERMS

(This is a non-exhaustive list of commonly used abbreviations in the area of food safety)

ADI	Acceptable Daily Intake
ADME	Absorption, Distribution, Metabolism and Excretion
AI	Adequate Intake
ALARA	As Low As Reasonably Achievable
ALOP	Appropriate Level of Protection
ARfD	Acute Reference Dose
BMD	Benchmark Dose
BMDL	Benchmark Dose at Lower Confidence Limit
CCP	Critical Control Point
CFR	Case-Fatality Rate
CFU	Colony Forming Unit
CIP	Cleaning in Place
DALY	Disability-Adjusted Life Year
DGGE	Denaturing Gradient Gel Electrophoresis
DNA	Deoxyribonucleic acid
EAR	Estimated Average Requirement
ED_{50}	Effective Dose 50%
ELISA	Enzyme-Linked Immunosorbent Assay
EMRL	Extraneous Maximum Residue Limit
FSO	Food Safety Objective
GAHP	Good Animal Husbandry Practice
GAP	Good Agricultural Practice
GHP	Good Hygienic Practice
GAqP	Good Aquacultural Practice
GC	Gas Chromatography
GC-MS	Gas Chromatography-Mass Spectrometry
GHP	Good Hygienic Practice
GLP	Good Laboratory Practice
GM	Genetically Modified
GMOs	Genetically-Modified Organisms
GMP	Good Manufacturing Practice
GPVD	Good Practice in the use of Veterinary Drugs
GRAS	Generally Recognized As Safe

HABs	Harmful Algal Blooms
HACCP	Hazard Analysis and Critical Control Point
HPLC	High Performance Liquid Chromatography
HPLC-MS	High Performance Liquid Chromatography-Mass Spectrometry
HPP	High Pressure Processing
HTST	High Temperature Short Time
HUS	Hemolytic Uremic Syndrome
IEDI	International Estimated Daily Intake
IESTI	International Estimated Short-Term Intake
LD_{50}	Lethal Dose 50%
LOAEL	Lowest-Observed-Adverse-Effect Level
LOD	Limit of Detection
LOQ	Limit of Quantitation
MFFB	Moisture on a Fat-Free Bases
ML	Maximum Level
MLST	Multi-Locus Sequence Typing
MLVA	Multiple-Locus Variable-Number Tandem Repeat Analysis
MOE	Margin of Exposure
MRL	Maximum Residue Limit
mRNA	Messenger Ribonucleic Acid
MS	Mass Spectrometry
NEDI	National Estimated Daily Intake
NOAEL	No-Observed-Adverse-Effect Level
NOEL	No-Observed-Effect Level
OPRP	Operational Prerequisite Programme
PC	Performance Criterion
PCR	Polymerase Chain Reaction
PDCA	Plan-Do-Check-Act
PEF	Pulsed Electric Fields
PFGE	Pulsed Field Gel Electrophoresis
PMTDI	Provisional Maximum Tolerable Daily Intake
PO	Performance Objective
PRP	Prerequisite Program
PrP	Protease Resistant Protein
PTMI	Provisional Tolerable Monthly Intake
PTWI	Provisional Tolerable Weekly Intake
QPS	Qualified Presumption of Safety
RDA	Recommended Dietary Allowance
RNA	Ribonucleic acid
SMEs	Small- and Medium-sized Enterprises
SOP	Standard Operating Procedure
SPS Agreement	Agreement on the Application of Sanitary and Phytosanitary Measures
TBT Agreement	Agreement on Technical Barriers to Trade
TDI	Tolerable Daily Intake
TDS	Total Diet Study

TEF	Toxic Equivalency Factor
TEQ	Toxic Equivalence
TMDI	Theoretical Maximum Daily Intake
TSE	Transmissible Spongiform Encephalopathy
UHT	Ultra High Temperature
UL	Upper Limit
UV	Ultra Violet

ABBREVIATION OF SELECTED ORGANIZATIONS INVOLVED IN FOOD SAFETY

(This is a non- exhaustive list of some of the frequently mentioned abbreviation in the Encyclopedia of Food Safety)

CAC	Codex Alimentarius Commission
CDC	Centers for Disease Control and Prevention (USA)
CI	Consumers International
EFSA	European Food Safety Authority
EHEDG	European Hygienic Engineering and Design Group
EPA	Environmental Protection Agency (USA)
EC	European Commission (European Union)
FAO	Food and Agricultural Organization of the United Nations
FDA	Food and Drug Administration (USA)
IARC	International Agency for Research on Cancer
ICMSF	International Commission on Microbiological Specifications for Foods
IFIC	The International Food Information Council
ILSI	International Life Sciences Institute
ISO	International Standards Organization
IUFoST	International Union of Food Science and Technology
JECFA	Joint FAO/WHO Expert Committee on Food Additives
JEMRA	Joint FAO/WHO Expert Meetings on Microbiological Risk Assessment
JMPR	Joint FAO/WHO Meetings on Pesticide Residues
OIE	World Organisation for Animal Health
PAHO	Pan American Health Organization
RASFF	Rapid Alert System for Food and Feed
UNEP	United Nations Environment Programme
USDA	United States Department of Agriculture
WHO	World Health Organization
WTO	World Trade Organization

CHAPTER

1

Fundamentals in Management of Food Safety in the Industrial Setting: Challenges and Outlook of the 21st Century

Yasmine Motarjemi[1] *and Huub Lelieveld*[2]

[1]Food Safety Management Consultant, Nyon, Switzerland,
[2]Global Harmonization Initiative, Bilthoven, The Netherlands

OUTLINE

Food Safety Management.
DOI: http://dx.doi.org/10.1016/B978-0-12-381504-0.00001-9

CONSUMER TRUST: THE CORNER STONE OF A FOOD BUSINESS

Successful and sustainable businesses are those that give priority to consumers and are able to meet their expectations. Three fundamental expectations of consumers are to be able to trust the food businesses from which they buy their food products, to be able to rely on their ability to ensure the safety of their products, and to have confidence that, in the event of a mishap, they will take the necessary measures to protect them and will act in a truthful and transparent manner. Therefore, it cannot be stressed enough that the trust of consumers (and customers) is one of the most important assets of a food business and that food safety is the foundation of this trust.

THE 21ST CENTURY: A NEW ERA IN FOOD SAFETY

The end of the 20th century was marked by a drastic increase in the incidence of foodborne illnesses, large-scale outbreaks and the emergence of new foodborne pathogens and chemical hazards. An alarming number of food safety incidents and crises, widely reported by the media, also fueled the feeling of insecurity among consumers. A combination of different factors is believed to have contributed to this trend, among which:

- The industrialization of agricultural production, mass production and increase in the number of food service establishments;
- The liberalization of trade and the growing number of imported foodstuffs;
- Tourism, urbanization with subsequent changes in lifestyle, food consumption pattern and food preparation practices.

Additionally, the increased availability of and accessibility to information and its rapid communication through mass and social media further amplified the feeling of insecurity. The consumer perception and the trend of requiring better quality, fresher food and more ethical food production practices has also weighed in the decision-making process. These developments have been the impetus for major changes in the management of food safety and the development of new procedures and principles for decision-making, changes in systems and requirements for food production and processing, and for the strengthening of the infrastructure for food safety management (Table 1.1).

Hence, the 21st century sets the beginning of a new era in food safety. Reviewing the history of food safety from prehistoric times, we can divide it in three major eras:

- A time where consumers were directly managing the safety of products by consuming a food and judging the safety by its impact on their health;
- A period where governments were managing food safety by testing products and removing contaminated or non-compliant products from the market; in general, food was considered safe unless people became ill or tests would indicate otherwise; and
- The present era where food businesses have become responsible for providing evidence that they have taken necessary measures to prevent contamination of foods. This means that foods are considered safe when there are proofs that the safeguard measures have been taken and the hygienic conditions of production, processing, transport and distribution or preparation have been observed.

TABLE 1.1 Key Guiding Principles that Gained Prominence in the 1990s for the Management of Food Safety (Motarjemi, in Press)

1. Integrated approach, i.e. consideration of the risks and control measures along the entire food chain, from primary production up to the point of consumption.
2. Shared responsibility, which is the recognition that all sectors, including consumers, have a responsibility in ensuring food safety.
3. Multi-disciplinary approach, which comes from the understanding that ensuring food safety requires different types of scientific and operational expertise.
4. Evidence-based and risk-based decision-making to ensure objectivity and the most efficient use of resources in food safety management. This principle is important to assure stakeholders or trading partners that measures are based on scientific and technical evidence, and are effective and commensurate to the degree of risk. The principle also facilitates the implementation of the WTO/SPS article on equivalence as it allows countries to deviate from the requirements of importing countries, if they can demonstrate the equivalence of measures on a scientific basis.
5. Transparency, uncertainty and precautionary principles. Transparency is an obvious consequence of the above-mentioned principles on the evidence-based decision-making process. However, it gains particular importance when there is uncertainty in data or when data are lacking. The value of transparency is that, in absence of full scientific information or variation in the degree of risk, the uncertainty and variability are declared, and evidence of the adequacy of protective measures is provided. Transparency also increases trust in stakeholders and trading partners. The precautionary principle states that if a product, an action or a policy has a suspected risk of causing harm to the public or to the environment, protective action should be supported before there is complete scientific proof of a risk.
6. Structured approach: i.e. while risk managers and risk assessors should maintain an active interaction, there should be a functional separation between risk assessment and risk management to ensure objective and unbiased decision-making, balancing scientific consideration with societal values and economic interests, as well as considering the risk perception of consumers.
7. Harmonization of food standards which is a goal as well as a consequence of the WTO/SPS Agreement.
8. Continuous improvement. As in any quality management system, a Plan, Check, Act and Do/Review process should be applied to food safety management. Such activities will continuously improve the safety of foods by reducing risks to a level that is as low as technically/reasonably achievable.[a]

[a]*This principle does not apply to foods which present an immediate and/or an unacceptable risk to consumers' health and where a crisis management procedure should be implemented. The principle applies both in governmental functions, which should progressively drive the contamination of food supply and incidence of illnesses to as low a level as technically and reasonably achievable (ALARA principle), and to industry where it is expected to have a yearly objective for improving the food safety assurance system.*

From the above, it follows that food safety management in the industry is not first a question of addressing food safety problems, but essentially one of taking the necessary measures to prevent them, including the necessary research and tests to confirm that the control measures are effective (validation) and properly implemented (verification). Since the introduction of the HACCP (hazard analysis and critical control point) system, the role of governments has shifted from identifying potentially unsafe food or unsafe practices to supervising and verifying the implementation of food safety management systems by industry.

THE CONCEPT OF FOOD SAFETY AND ITS DEFINITION

Today, the subject of food safety has become a discipline in its own right and a formal definition was elaborated by the Codex Alimentarius Commission[1] (CAC) in 1997.

[1]Codex Alimentarius Commission is an intergovernmental body, operating under the auspices of the World Health Organization and the Food and Agriculture Organization of the United Nations.

According to the CAC, "food safety is the assurance that food *will not cause harm* to the consumer when it is prepared and/or eaten according to its intended use."

This definition embodies several important notions:

1. The notion *of harm*, which separates safety aspects of food from other quality aspects that make food unfit for human consumption without necessarily presenting a danger to health. The aspects of food which make it unfit for human consumption, even though it is safe, are referred to by the CAC as *food suitability*.
2. The concept of *assurance*, i.e. food safety and its management should be based on measures that are in place to provide assurance that food is safe. In other words, food safety depends on the conditions in which food is produced and prepared, and not on the results of the end-product testing, which for many contaminants cannot be a reliable method for food safety assurance. The conditions for ensuring both safety and suitability are referred by the CAC as "food hygiene."
3. *Preparation and/or use of a food product should be considered in product design*. A food product is considered safe if it is prepared and/or used according to its intended use. Subsequently, the intended use should be considered by the manufacturer in the design of the product as well as in their information conveyed to the consumer. The consumer must also follow on-pack instructions as provided by the manufacturer.

ELEMENTS OF FOOD SAFETY MANAGEMENT

The modern approach to food safety management recognizes the need for cooperation of different sectors and a role and a responsibility for each sector. Figure 1.1 illustrates the functions of the different sectors as described below.

Government

Public health and food control authorities have the leading role in managing food safety and have the responsibility of overseeing the safety of the food supply, from primary production up to the point of consumption. With this responsibility, they have to do the following:

- Foresee all infrastructures and public health services that are necessary for a good food safety management, such as public health laboratories, water supply and sanitation, etc.;
- Promulgate laws and regulations, which give priority to public health but which can also meet other societal and environmental factors;
- Enforce legislation through the provision of advice to trade and the commercial sector, inspection and monitoring of food supply, and, where necessary, prosecuting offenders;
- Provide education to caregivers, consumers, travelers, health professionals and the public at large.

Today, decisions on measures required to manage risks are taken in the context of the *risk analysis process*. There are different types of models for describing the risk analysis process. Figure 1.2 depicts the process of risk analysis according to Codex Alimentarius. The process includes: risk assessment, risk management and risk communication.

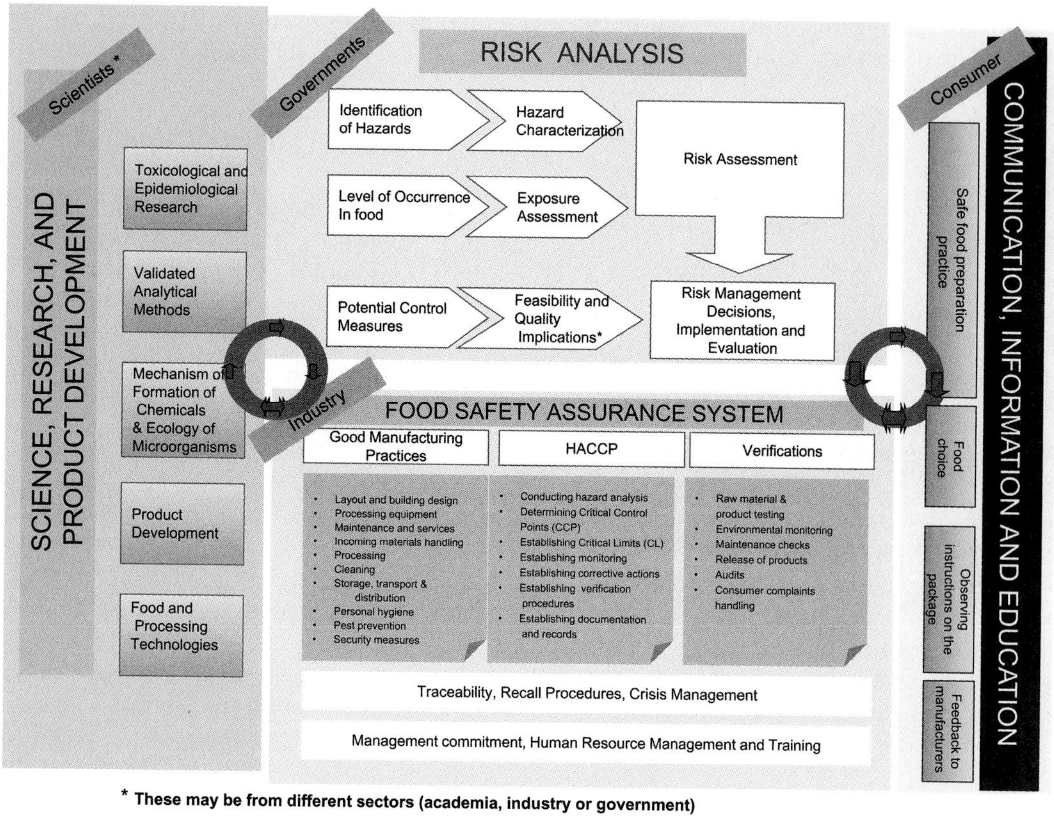

FIGURE 1.1 General overview of the organization of food safety management in society.

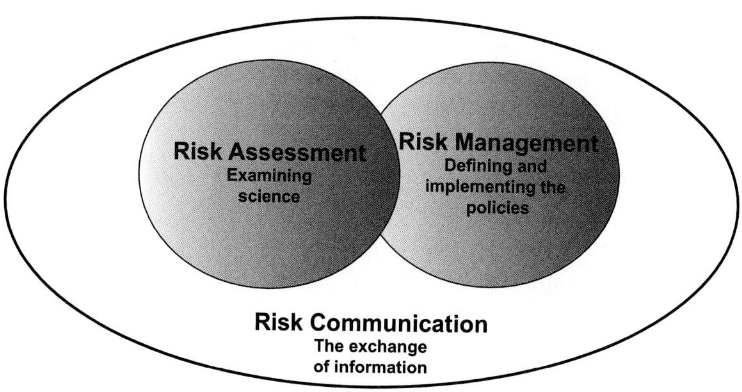

FIGURE 1.2 Risk analysis process.

As risk managers, regulatory authorities are, among others, responsible for (1) driving the risk analysis process, (2) setting public health goals and (3) deciding on risk management priorities.

The risk management process itself comprises a number of steps which are briefly discussed here. For a more in-depth review, the reader is referred to Moy (in press) and Gorris (in press).

The first step is referred to as *preliminary activities*. As part of this, governmental risk managers will commission a preliminary risk profile for a given hazard or hazard/food. Based on the outcome of this and in the light of existing data, they will decide if a risk assessment is required, or if it is possible to evaluate the various control options. Should risk managers find that a risk assessment is justified, bearing in mind the resource and time investment, they may decide to commission a qualitative or a quantitative risk assessment. In this case, they are responsible for elaborating a *risk assessment policy* in consultation with risk assessors and other interested parties. A risk assessment policy is a guidance to risk assessors, outlining information such as:

- The purpose and scope of the risk assessment, e.g. sector of the food chain, types of food and products to consider;
- Target populations or subpopulations;
- Key scientific judgments, particularly when there is a high degree of uncertainty in existing data or in data gaps;
- The type and sources of data to be considered;
- How the data should be presented, in particular the types of assumptions and uncertainties.

The process of risk assessment and risk management follows an iterative interaction between risk assessors and risk managers, during which these need to foster mutual understanding and refine the risk assessment so that it responds as closely as possible to the questions posed by risk managers. When deciding on the appropriate control measures, risk managers need to take into consideration a number of other factors, sometimes also referred to as "other legitimate factors." These factors vary according to the nature of the hazard under consideration and can include costs, feasibility, benefits, other risks (e.g. environmental or nutritional), consumer preferences and societal values such as animal welfare. At times, a risk assessment may be required to advise on the efficiency of the control measures, to develop an understanding of the public health outcome according to different levels of contamination, to have an estimation of the risk of various foods/hazards combination, etc.

In managing a risk, depending on the nature and degree of the risk and on other factors mentioned above, risk managers have different options at hand. These range from taking a regulatory action, such as those listed below, to taking no action.

- Compliance with certain standards (e.g. setting a norm for a chemical hazard or a food safety objective or microbiological criteria for a microbiological hazard);
- Labeling;
- Testing and/or certification of foods;
- A specific processing of foods to inactivate pathogens;
- Application of a code;
- Recalling a product in case of an incident.

Alternatively, they may decide to manage the risk by providing education to consumers or requiring the training of food handlers in food service establishments. It can also happen that they decide not to take any action (e.g. if the risk is low or negligible). In any event, the food safety authorities have the responsibility to communicate and explain their decision to the stakeholders.

To identify possible food safety problems and to review the implementation of the risk management decisions and to evaluate the need for any revision in decisions or implementation, the collection of various types of data need be considered. Examples are:

- Inspection reports and evaluation of implementation of risk management decisions by the food industry;
- Monitoring of chemical contaminants;
- Surveillance of foodborne diseases (data from different types of surveillance methods need to be considered);
- Consumer complaints;
- Trade rejections;
- Public recalls, withdrawals and/or incidents; and
- Applied research based on defined indicators (knowledge, gaps).

Other types of information may also be required for planning improvement or preventive actions. Examples are trends in incidents and alerts, be they occurring in a country or outside the national boundaries, adequacy of resources, as well as various changes in the society such as changes in climate, demography, international trade and travel, or emergence of new pathogenic agents.

Industry

The food industry is responsible for ensuring that the food that it puts on the marketplace or that is served in food service establishments is safe, fit for human consumption and meets regulatory requirements of the country where it is marketed. They have to consider the regulatory norms for hazards as food safety standards and ensure that their products are not violating these limits. To meet these responsibilities, the food industry is required to have an integrated food safety assurance system.

A model for this system consists in combining three sets of measures according to the three lines of defense (Figure 1.1).

The first line of defense is the implementation of codes of good practices. These are a set of general principles and measures which have been identified through past experience as necessary to ensure the safety and wholesomeness of the foods produced; with some adaptation, they are generally applicable to all categories of foods and products and/or establishments regardless of location, specific conditions and type of business. Depending on the sector, they are referred to as Codes of Good Agriculture Practice, Codes of Animal Husbandry, Codes of Good Manufacturing Practice, Codes of Good Transport or Storage Practice, etc. Very often, such codes are voluntary, but at times they are legally established by regulatory authorities. However, where they do not exist or are not stringent enough, the industry may also develop such codes. The Codex Alimentarius Commission has developed a large number of codes of practice. The recommended *International Code of Practice –General*

Principles of Food Hygiene is one of the "horizontal" codes that has wide application in the food industry. For given categories of products, there are also product-specific codes where guidance is provided for the handling of that particular group of products.

The second line of defense is the application of the HACCP system. During this process, hazards specific to a food and/or process are proactively identified and control measures specific to the hazards in question are determined. Concerning steps that are considered critical for ensuring the safety of the food product, monitoring parameters characterizing the control measures and critical limits for the monitoring parameters are established and the steps are monitored to ensure that the critical limits are respected at all times. Additionally, any regulatory requirements (such as codes of practices or national standards, food safety objectives, sampling plans, etc.) or customer requirements, e.g. specifications, performance criteria for intermediary processes, need to be considered during product/process design and respected during operations. Needless to say that during the development of an HACCP plan, measures identified for controlling the hazards and the parameters as well as limits to be respected need to be validated, short of which the HACCP study will become a simple paper exercise. HACCP also has other elements such as corrective actions in case the process is going out of control, and, as explained below, verification and documentation.

A strategy that has been used by some governments to assist small or less developed businesses in applying the HACCP system is to develop HACCP-based codes of practice for specific categories of food products. Such an approach is important for small or less developed businesses as these often lack expertise in food safety, and unless they are assisted by a trade organization, they may not be in a position to carry out an HACCP study by themselves. A HACCP-based code of practice for a specific sector combines both the general principles of food hygiene and the considerations and requirements specific to a given food sector.

Frequently, the question is raised about the difference between the code approach to food safety assurance versus the HACCP system, and their respective benefits. Originally, a code approach was viewed as a general and prescriptive system of management of food safety in a business. Subsequently, HACCP was recommended by public health authorities to promote a preventive approach based on the analysis of hazards in foods or processes, before these lead to an incident. When applying the HACCP system, hazards specific to a particular food product, process and to the conditions in which the food is prepared are identified and control measures specific to the hazard in question are devised. In this way, as opposed to codes that are general guidance, through the HACCP system control measures are targeted to hazards specific to the product (raw material or conditions of production). However, with experience, it became evident that both approaches have their respective values, and that HACCP would be more efficient if some basic hygienic conditions and preventive measures were in place. Today, these are referred to as *prerequisites* in food safety assurance systems of the food industry, and it is recognized that it is by combining both approaches that the optimum conditions of food safety management are attained.

Very often, the documentation required as part of HACCP has given the HACCP system the negative image of being burdened by paperwork. However, records and documentation are essential as support material for communication between members of the HACCP team and/or with time, for the maintenance of the plan, i.e. for the HACCP team to be able to consider the need for any change in the plan and thus ensure that the system remains valid and up to date. Also, records are required to provide evidence to customers and/or inspectors of the adequacy of measures.

The third and last line of defense is verification activities. These are also part of the HACCP application, but to delineate between measures implemented for prevention and those required for verifying that preventive measures are effective and performing correctly, these are presented separately.

As for the governmental evaluation process, verification activities include all tests and other data collected to verify that preventive measures achieve the objectives set. Verification should not be mistaken for validation, which is a process to ensure that control measures are effective to achieve the objectives desired. The validation process is usually implemented during the product and process design stages, or when a change has been made in product design or during its manufacturing. If verification data indicate that a product is not meeting a set standard, even though the plan has been implemented, the validation of control measures may need to be questioned.

In principle, where codes of good practice and the HACCP system are optimally implemented, a high degree of safety can be assured. Nevertheless, verification measures are important to detect any dysfunction in the system. They also provide evidence of compliance with the food safety standard and should not be stopped on the grounds that data on contamination are negative, as data are needed for proving the performance of the food safety assurance system at all times. Examples of verification measures are:

- Raw material and end-product testing;
- Environmental monitoring;
- Calibration and other maintenance checks;
- Release of products;
- Audits;
- Consumer complaints handling.

Should verification data indicate non-compliance, the adequacy of the implementation of the HACCP system and the prerequisites must be examined in the first place. In absence of any failure in implementation, the validation of the elements of the HACCP study can then be questioned.

At times, in spite of all measures, it can happen that a raw material used in a product is contaminated or a product that is contaminated is marketed. Through a traceability system, i.e. information on the source of raw materials or on the customers who have received the product, a contaminated product can be traced and recalled. Regulatory authorities in some countries require that the traceability system of an establishment ensure that information on the source of a raw material or destination of a final product be available for one step up or one step down. With the globalization of the food supply and the passing of food ingredients through various traders, it is sometimes difficult to ensure precise or valid information on the condition of the production of raw materials. Where information on traceability is lacking, the investigation of outbreaks and the identification of the implicated food become more difficult as observed in an outbreak of *Salmonella* Saintpaul the United States in 2008 (CDC 2008). Originally, the outbreak was attributed to tomatoes until it was discovered that the main vehicle was jalapeño and serrano peppers. The outbreak lasted from April to August and some reported 1442 persons fell ill. The weaker the traceability, the larger the scale of the outbreak or product loss. This was demonstrated in the dioxin incident in Ireland where a full product recall was conducted for pork meat, whereas for beef meat it was possible to limit the recall to the contaminated product because after the BSE crisis, a traceability system was established for beef products (Casey et al. 2010). Similarly, in the food manufacturing industry, the finer the traceability, e.g. indicating the date and time the product was produced, the smaller the quantity of product wasted in case of recall.

Finally, the entire food safety assurance system should be supported by a well-performing crisis management system to protect consumers from exposure to contaminated products.

Fundamental to all these systems are the training and education of the staff as well as the management's commitment. Therefore, fostering a culture of food safety, from training of the staff to motivating them and appreciating their constraints, constitutes one of the most important pillars of food safety management in industry and in governmental functions. The importance of organization culture cannot be emphasized enough. Reporting any non-compliance or a risk-prone situation at an early stage can contribute to preventing outbreaks before they occur.

Consumers and the Informal Sector

Consumers at large and domestic and professional food handlers in particular also have an equally important role in food safety. These include, but are not limited to:

- Observation of good hygienic practice in the preparation of food;
- Reading information (e.g. "use by date" of products, product storage, possible presence of allergens, target consumer) on the labels of products and observing the instruction for the preparation and storage of products;
- Reporting defective (unsafe) products to the public health authorities and/or manufacturer;
- Being discriminatory in the selection of products, brands and establishments (incl. restaurants, caterers) to exclude those that may present a risk for health, do not respect food hygiene, do not meet regulatory requirements or have unethical practices.

To enable consumers to assume their responsibility in the hygienic handling of food as well as to judge potential risks with certain products, practices or establishments, consumer information and education are key. This is best carried out by professionals who are both trusted by the general public and who also are in dialogue with the pubic in their every-day work. Examples are representative of consumer organizations, health professionals and school teachers.

Academia

Scientists in general, whether they work in academic institutions, in government or in industry, also have an important function. With the trend in evidence-based decision-making and taking science into consideration, be it life or social sciences, the role of this sector in the risk analysis process has increased during recent years. Their integrity, excellence and relevance make them ideal communicators for managers (e.g. report of their results, articles) or for the general public (e.g. interviews in the mass/social media). As such, they play an important role in both the management of food safety (in particular risk assessment and risk communication) and the management of a crisis.

On the technical aspects, scientists contribute to food safety management by providing different types of scientific data and their assessment, which is necessary for making decisions. Examples are:

- Toxicological information, mechanisms of contamination of foods with chemicals, or their formation;
- Ecology of microorganisms and epidemiology of foodborne diseases;
- Validated analytical methods;
- Process and technologies to control hazards;
- Consumer perception, beliefs and practices.

In industry, scientists can minimize risks associated with products and processes by designing out risks during product development and defining necessary control measures for managing the operational risks during the production or manufacturing of foods.

Additionally, scientists can further contribute to the management of food safety by creating tools to make information on food safety easily accessible to all stakeholders in society.

CHALLENGES IN MANAGEMENT OF FOOD SAFETY AND OUTLOOK

In spite of measures implemented during the last decades and advances in science and technology (Table 1.2), managing food safety remains today a daunting task. Many factors

TABLE 1.2 Some Major Developments in Food Safety Management in the Last Two to Three Decades

1. Increased general awareness about food safety.
2. Research on pathogens, chemical contaminants and technologies and increased scientific and technical know-how.
3. Development and emergence of high-performing food technologies and analytical methods.
4. Increased availability of epidemiological and scientific data on foodborne pathogens and chemical contaminants.
5. Improvement in the procedures for risk assessment and risk management.
6. Strengthening of national legislation (standards, codes of practices) and its enforcement (inspection, monitoring).
7. Strengthening of the international requirements (Codex Alimentarius, Agreement of the Sanitary and Phytosanitary Measures of the World Trade Organization, ISO 22000 refers).
8. Increased preventive measures at the primary industry.
9. Improvements in quality assurance, including application of the HACCP system.
10. Strengthening the foodborne disease and food contamination surveillance systems, alerts, traceability and incident management.
11. Increased training of professionals involved in food safety (governments, food industry and food service sector).
12. Recognition of the importance of risk perception and good risk communication.
13. Educational campaigns for consumers and the general public.
14. Improved waste management, protection of the environment and of water and sanitation facilities.

contribute to this; understanding and recognizing these factors is important for managing food safety better and foreseeing the infrastructure, procedures, systems and resources that are required to this end.

For the food industry, a first challenge is to be able to ensure the safety of its products, and at the same time meet consumers' expectations in terms of quality. As will be seen later, managing food safety is by itself very complex; to achieve this as well as to provide high quality products is very challenging and at times quite difficult as some requirements for food safety do not necessarily go in the same direction as the perceived quality.

With the progress in the industrialization process and advances in science and technology, consumers' expectations have also increased. Today, for most consumers, food is not only a source of nutrition, but it is also a source of pleasure and an "emotional experience." Moreover, in recent years, with the fortification of food and the development of functional foods, some consumers see foods as a means to alleviate their health risks. In modern societies, where the lifestyle and the structure of family have changed, consumers also need food products to be more convenient in terms of accessibility, transportation, storage, preparation and use (e.g. easy opening). Many consumers also attach importance to the attractiveness of products, e.g. color of the product or its packaging. Worldwide, price is also an important determining factor and many consumers seek foods that offer the best value for price. Food businesses also have to respect the cultural and traditional values of the societies in which they operate, for instance they must observe the kosher and halal rules. Other factors which may also impact consumers' decisions with regard to their preference for one brand rather than another are issues related to the environment, animal welfare, ethical practices and in general their perception of the responsible behavior of a company. Thus, over and above safety, a successful business needs to meet a broad range of criteria that varies with lifestyle, values, culture, the level of education and the perception of consumers.

It should also be mentioned that food is also a means of subsistence for many people. The food industry, from the primary, manufacturing, retail to the service industry is by far the biggest industry in the world; thus it provides job opportunities to a considerable proportion of the world population. According to the International Labour Organization, in countries that have official statistics, the food manufacturing industry alone employs 22 million people. Besides providing job opportunities, the food industry is an engine for economic development as it provides food for the world population and supports international trade and food export, which is a source of foreign exchange.

Thus, as part of their social responsibility, food industries also have obligations towards their employees, their job security and the economic role that they have in the community where they are established. Overcoming certain food safety issues, in particular where food safety standards are too stringent compared to what is possible, or where legislation is not feasible, can be at the cost of compromising the livelihood of many people and crippling the frail economies of certain countries. On the other hand, in businesses that fail to ensure the safety of their products and are forced to close a factory or go out of business, employees may also run the risk of losing their jobs.

With regard to the management of food safety, there are many factors which contribute to this complexity and present challenges.

Complexity of the Subject

In food safety, we are facing a mind-boggling number of hazards of biological, chemical, physical and other nature, not to mention the unexpected chemicals that may find their way into the product due to accidental contamination, tampering or sabotage. With developments in science and technologies, increases in our knowledge and analytical capabilities, the list of hazards is becoming longer. Chemical hazards alone group countless chemical agents, e.g. hundreds of different types of pesticides, antibiotics, food additives, environmental contaminants from heavy metals to PCBs and radionuclides, naturally occurring toxins, including a number of mycotoxins, as well as processing and packaging contaminants. Microbial hazards are also numerous, but their greatest challenge lies in their multifaceted nature. For instance, they vary in their:

- Conditions for growth (pH, water activity, aerobic versus anaerobic conditions);
- Mechanism of pathogenicity and ability to produce toxins with different sensitivity to heat;
- Virulence and in their opportunistic nature, i.e. some pathogens target mainly vulnerable population groups;
- Dose–response relationship, which also depends on the food matrix and the target person;
- Resistance to various control measures, e.g. heat, acidity, chlorination, etc.;
- Ecology and vehicle of transmission; and
- Health consequences.

Managing this technical and scientific complexity and communicating this complex set of information to decision-makers or other actors in the food chain are not always easy. A particular difficulty lies in communicating with food handlers/caterers in the food service industry, who have a generally low level of academic background, or with the general public in a convincing manner; yet this knowledge is fundamental to the decision-making process and a prerequisite to good practices. This communication becomes an intricate task when food safety measures are intertwined with economic factors, ingrained cultural habits and beliefs, or simply food preferences. An example is the consumption of raw milk or cheese made with raw milk versus the pasteurization of milk for health protection.

In the food industry, even in the most resourceful companies where there is access to technical expertise, the complexity of food safety is an issue in designing food safety control systems, particularly when other conflicting quality criteria have to be met. An example where this technical complexity has led to a mistake in decision-making can be seen in an outbreak of salmonellosis in the United Kingdom in 2006 (Carroll 2009). In this incident, the company in question undermined the consequences of low level of salmonella in chocolate. It assumed that at the level salmonellae were present, it did not present a concern for public health while a number of previous outbreaks provided evidence of the risks. Another example is found in an outbreak of *Staphylococcus aureus* where it was believed that by reheating milk which was subject to time–temperature, the milk could be rendered safe, while toxins *of S. aureus* are heat stable. In the area of chemicals, the contamination of soft drinks with benzene shows the global difficulty in keeping track of the scientific and technical know-how and transferring it to a new generation of professionals. In 2006, it was found that some soft drinks, where a combination of sodium benzoate and citric or ascorbic acid was used, contained unacceptable levels of benzene due to the interaction between these ingredients. This interaction and potential formation of benzene was known already in the 1990s, but presumably, with time, this knowledge had faded away from the institutional memory of the scientific and technical organizations. A rigorous implementation of HACCP can prevent many such technical errors; however, it requires that HACCP studies be carried out by a team of competent experts and be duly validated. This is still not the case in many businesses.

Thus, communicating the science of food safety to all stakeholders of the food chain, commensurate to their role and in a responsible manner, will be an important task in the 21st century. This is a huge but important challenge, as this means basically educating the entire world population in food safety since everyone is a potential food handler and some may have a professional life in the food sector. This can be achieved only if food safety is taught systematically in schools, starting from primary schools to academia, be it food science and technology or public health and medicine. Making this science accessible to every individual will promote common understanding. For the food industry personnel, training of the professionals is fundamental; while human error can be forgiven by consumers, ignorance or negligence cannot.

Complexity of Food Operations

A second factor that undermines the efficiency of control measures is the complexity and variability of food operations, a situation that makes employees of food businesses more prone to error. In the food industry, food operations can be very complicated by the number and variability of ingredients, recipes, processes and standards to meet, in particular if products are to be marketed to different countries with different regulatory requirements. Changes in the various aspects of operations add to this complexity: product formulation, raw materials, packaging, production and processing scheme, construction and maintenance work, shift of personnel (including temporary personnel) and markets where the product is sold. Export of food to different markets with different regulations, consumer practices or climatic conditions requires that these factors are considered in the design of the products. Such a complexity necessitates a very well organized and managed logistic infrastructure, planning and discipline in execution. Short of this, the situation becomes conducive to human error.

The example of allergen management illustrates this point. Where processing or manufacturing of various products is shared on the same line of production, the management of allergens may present a greater risk versus the use of dedicated lines. However, the latter is not always possible particularly when small quantities of products are manufactured. Sharing the production lines would require a careful scheduling of operations, dismantling and effective cleaning of equipment and careful labeling of allergens, sometimes in several languages and according to the legislation of the different countries where products are marketed. A slight change in any of the above parameters, if not managed, can lead to an error in labeling.

At the agricultural level, the trend to use food for purposes other than consumption, e.g. fuel, will add a new dimension to the complexity of the food chain and to the management of food safety, as this type of food may contain substances that are not appropriate for human consumption. Segregation of these foods from crops destined for human consumption will require additional logistic infrastructure and thus again become a potential for human error.

Thus, a slight change in any of the production parameters requires a thorough examination of the consequences of these for the management of food safety and, if necessary, changes in the control measures, and these changes may impact on other hazards or quality parameters. Such situations often necessitate a re-examination of the HACCP study, and, if necessary, a change in the plan and its communication to operators and other relevant persons in the business. A case in point shows that a change of supplier has been the source of many incidents: in Germany, from April to September 1993, a nationwide outbreak of salmonellosis associated with paprika potato chips affected an estimated 1000 children below the age of 14. In this outbreak, the trader changed the supplier of paprika for a totally unknown supplier without informing the customer (Lehmacher et al., 1995). An accidental breakdown in infrastructure, resulting in temporary change in conditions of processing and manufacturing foods, has also been the cause of food safety problems and foodborne disease outbreaks; in an extensive outbreak of staphylococcal foodborne intoxication associated with low-fat milk in Osaka, Japan, as many as 13,420 persons became ill. The incident was due to a power cut and storage of the milk in time–temperature conditions allowing growth of *S. aureus* and formation of toxins (Asao 2003). Reconstruction and maintenance work is a frequent cause of post-process contamination of products. From August 1998 to February 1999, a large multi-state outbreak of listeriosis occurred in the United States (CDC 1999). Investigators documented more than 100 illnesses in 22 states. A total of 21 deaths including 15 adults and six stillbirths/miscarriages were reported. CDC (Centers for Disease Control) and state and local health departments identified the vehicle for transmission as hot dogs and possibly deli meats produced under many brand names by one manufacturer. It is believed that dust kicked up during summer maintenance of the air-conditioning system at the plant.

In the food service industry or at the household level, preparation of food requires a multitude of tasks; a minor change in the ingredient, quantity, conditions or procedure can make a difference in the safety of products. Without a specific knowledge of the consequences of the change, there is a risk for an incident. For instance, one of the frequent occasions for foodborne illness is festivity and/or other occasions where food is prepared in large quantities. In such occasions, the normal procedure may not be appropriate as the refrigerator may not have the capacity for rapid cooling of the food; conditions of storage of food become then favorable for pathogen growth and/or production of toxins.

A key measure for overcoming this factor is to carefully map the requirements for managing food safety, define the roles and responsibilities, the processes and principles for the management, in particular the change management, and, last but not least, provide training and education commensurate to the role and responsibilities.

Complexity of the Food Supply and External Environment

While in the past foods were produced locally, today many ingredients are imported from distant countries, and produced under different legislative and social conditions. The liberalization of trade and also the tendency to provide consumers with a varied and sometimes exotic food supply has of course encouraged this change. Food businesses often have difficulty in foreseeing the hazards that may be associated with a raw material produced elsewhere. Testing the raw material is not always an option or effective as it is difficult to test for all types of potential hazards and view the large number of agents which may be present, considering that some of these may be unpredictable. The incapacity to foresee hazards, particularly when there are fraudulent practices, has been the cause of numerous small- or large-scale incidents. Examples are the import of honey contaminated with the prohibited antibiotic chloramphenicol in Europe in 2001, import of wheat gluten adulterated with melamine from China to the USA in 2006, import of guar gum contaminated with pentachlorophenol (PCP) and dioxins from India to Europe in 2007, sunflower oil contaminated with mineral oil imported from Ukraine to Europe in 2007, etc. Emergence of foodborne hazards is also another factor which aggravates this unpredictability. Import of raspberries with *Cyclospora cayetanensis* in USA, fenugreek seeds contaminated with shiga-producing *Escherichia coli* O104: H4 (Germany and France 2011) are some examples of outbreaks of microbiological origin (Herwaldt et al. 1997; EFSA 2011; Motarjemi 2011).

Urbanization and industrialization have caused large-scale incidents of foodborne disease and, consequently, they make negative publicity in the media. Perhaps the largest crisis in the history of food safety was the BSE epidemic, which affected many countries worldwide and led to a total reconsideration of the system of food safety management. For regulatory authorities, additional complexity comes from the fact that the size of food businesses varies from a few persons to a few hundreds or thousands of persons; thus the resources and conditions for processing and handling foods are not equivalent. Therefore, devising regulations which are applicable to all the different sizes and types of business, and yet are specific enough to be effective, is often difficult. At the international level, this becomes even more complex as the resources, infrastructure, environmental and climatic conditions, lifestyle and sociocultural values differ among nations. Therefore, at times, harmonizing food safety legislation, although a necessity in the light of the international trade in food and feed, becomes particularly cumbersome.

With today's globalized food supply, there is a need for worldwide strengthening of regulatory control, including promulgating the necessary legislation and supervision of their implementation. Considering the limited resources for control, it is important to reinforce the accountability and responsibility of food businesses. A continued effort for harmonizing food safety regulations, and a reinforcement of international collaboration for sharing best practices, will remain essential for a more efficient management of food safety in the global food market.

Human Factor

People, be they managers of food businesses, farmers, workers on the production line, professionals working in the regulatory agencies or consumers, all play an important role in the management of food safety. Their knowledge, perception and attitude as well as their motivation and commitment, and most importantly their capabilities to meet their responsibilities, all impact on the safety of the food supply.

Therefore, a first challenge is to ensure management commitment. Most companies or governmental organizations have policies that pledge for food safety; however, many fail to

TABLE 1.3 Some Common Misperceptions Observed in Management of Food Businesses

Misperceptions	Correction
Food safety management is in conflict with economic interests.	1. There is no business without food safety 2. A good management of food safety can promote the business.
Food safety management is addressing food safety problems.	Food safety management is taking necessary measures to prevent food safety problems, including confirmation that the measures are effective (validation) and implemented (verification).
Our products are safe, as we have never had any incident.	A past record of safety is no guarantee for the future.
Our products are safe because the tests were negative.	End-product testing is not an evidence of microbial safety, but a confirmation of the efficiency of the food safety management system.
Regulatory requirements are impediments to the business.	1. Regulatory requirements and their enforcement will: a. Facilitate fair trade and a healthy competitive environment; b. Ensure that all stakeholders in the food chain fulfill their role; this decreases potential risks with suppliers and their raw material; c. Provide guidance to businesses, in particular small and less developed businesses, on matters related to food safety, such as norms needed in designing and validating food safety assurance systems; d. Increase the confidence of consumers in the food supply and reassure consumers that commercial products are safe and meet the nationally and/or internationally agreed safety and quality standards.

implement their policies and to provide optimum infrastructure, processes, organizational culture and required resources. One reason is that leaders do not always appreciate the magnitude of the effort that is needed to ensure food safety in a sustainable manner. Some consider safety as a granted attribute of a food product. While they are aware of the complexity of their operations (as described above), however, they do not realize how this complexity impacts on food safety and its management. Food safety needs to be considered in the implication of any business decision at the outset, such as the decision to make a new acquisition, or develop or change a product or a process, implications for using contract manufacturer or subcontractors for specific services, or any other cost-cutting decision. Table 1.3 describes some frequently observed misperceptions. The financial competitiveness and the drive to increase profits are another reason for the gap between policies and actual practices. A possible consequence of this attitude is an approach to risk management and organizational culture that oscillates between a protective approach with a reasonable safety margin and a more productive approach with a high level of risk taking. As Figure 1.3 shows, following each incident there is a shift towards a more conservative approach to decision-making; later, the drive for increased profit swings the organizational culture and decision-making principles in the other direction, until another accident of greater magnitude occurs. Eventually, a catastrophic situation ensues. Often when the food manager does a good job, there are no incidents and hence the management wonders why they should spend funds on safety. A past record of safety is no guarantee for the future.

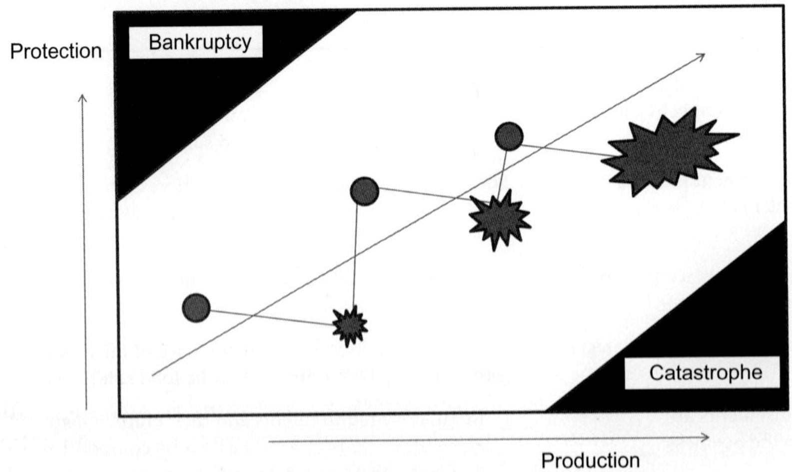

FIGURE 1.3 Changes in organizational culture leading to incidents. *Adapted from Reason (1997).*

As to the technical staff or workers in food production, their role in food safety management is key. None of the food safety systems and control measures can be effective without the intervention or supervision of qualified, skillful and motivated staff. This is true in any organization, be it private or public. Yet, very often the importance of a professional human resource management for food safety is overlooked.

Over and above providing necessary training and education, a proficient human resource management should consider the needs of the employees to meet their responsibilities. These needs may be (1) material, such as equipment, tools, optimum physical conditions, or (2) managerial support, such as time for performing their asks, clarification of authority or other prerogatives, supporting policies, etc. – in other words, all that is needed to make the implementation of measures humanly feasible and enable employees to meet their responsibilities. Short of this, the situation can lead to human error, or the employees are forced to take shortcuts or violate the rules to perform their task.

Finally, as alluded to above, the organizational culture will set the environment and the context in which employees work. This will have a major influence on the attitude and motivation of employees. An organizational culture may seem to be a vague concept, but it basically boils down to the set of values, written and unwritten rules that leaders "practice" and reward or penalize their staff for.

The role of leaders is essential for organizational culture. Leaders must have an exemplary behavior. Where there are discrepancies between the written and unwritten rules, or where managers preach values that they do not follow themselves, the staff will suffer from stress and they are more likely to be complacent or non-compliant. The most detrimental factor in food safety management is an organizational culture which breeds fear. In a fear culture, employees are discouraged from reporting potential problems; they may cover up gaps and increase opportunities for incidents.

CONCLUSIONS

Incidents in food safety in the past few decades have eroded the trust of consumers and have created misperception on the subject, although among experts there is a general consensus that the food supply has never been safer. The new technologies that could contribute to the enhancement of food safety, such as food irradiation and biotechnology, carry the burden of this mistrust.

Regaining the trust of consumers and developing an international consensus among stakeholders on the acceptable level of risks and the safety measures for effectively addressing these risks remain an important challenge for the 21st century.

Finally, for an effective management of food safety over and above science, systems, equipment and procedures, consideration of the human factor is essential. This ranges from factors underlying consumer choice and practices, to commitment and motivation of managers in the food industry in providing adequate infrastructure and organizational culture conducive to professional food safety management. Today, there is a wealth of scientific and technical know-how and an array of technologies and systems available to ensure a safe food supply. The challenge is to facilitate the access of food safety professionals and the general population to this know-how. The present book *Food Safety Management: A Practical Guide for the Food Industry* is developed with this perspective.

References

Asao, T., et al., 2003. An extensive outbreak of staphylococcal food poisoning due to low-fat milk in Japan: estimation of enterotoxin A in the incriminated milk and powdered skim milk. Epidemiol. Infect. 130 (1), 33–40.

Carroll, C., 2009. Defying a reputational crisis – Cadbury's salmonella scare: why are customers willing to forgive and forget? Corp. Reputation. Rev. 12, 64–82.

Casey, D.K., James, S.L., Wall, P.G., 2010. A tale of two crises: the Belgian and Irish dioxin contamination incidents. Brit. Food J. 112 (10), 1077–1091.

CDC, 2008. Outbreak of salmonella serotype Saintpaul infections associated with multiple raw produce items – United States. MMWR 57 (34), 929–934.

CDC., Press release 17 March 1999. <http://www.cdc.gov/od/oc/media/pressrel/r990114.htm>.

Codex Alimentarius Commission. Basic Text, Food and Agriculture Organization. Rome, Italy.

EFSA (European Food Safety Authority), 2011. Technical report of EFSA. Tracing seeds, in particular fenugreek (Trigonella foenum-graecum) seeds, in relation to the Shiga toxin-producing *E. coli* (STEC) O104:H4 2011. Outbreaks in Germany and France. European Food Safety Authority, European Food Safety Authority (EFSA), Parma, Italy.

Gorris, L., Encyclopedia of Food Safety. Elsevier (in press).

Herwaldt, B.L., Ackers, M.-L., the Cyclospora Working Group, 1997. An outbreak in 1996 of cyclosporiasis associated with imported raspberries. N. Engl. J. Med. 336, 1548–1556.

Lehmacher, A., Bockemühl, J., Aleksic, S., 1995. Nationwide outbreak of human salmonellosis in Germany due to contaminated paprika and paprika-powdered potato chips. Epidemiol. Infect. 115 (3), 501–511.

Motarjemi, Y., 2011. Food safety: what is the role for gastroenterologists? World Gastroenterol News 16 (3), 3–4. (1).

Motarjemi, Y., Modern approach to food safety management. In: Encyclopedia of Food Safety. Elsevier (in press).

Moy, G., Encyclopedia of Food Safety. Elsevier (in press).

Read, K., 2007. "Corporate pathos": new approaches to quell hostile publics. J. Commun. Manage. 11 (4), 332–347.

Reason, J.T., 1997. Managing the Risks of Organizational Accidents. Ashgate, Aldershot (UK).

Further Reading

Käferstein, F., Motarjemi, Y., Moy, G., 2009. Food safety – its role in health and development: the problems related to our food supply. In: Medical Sciences (eds. Yola, L., Verhasselt, G., Mansourian, B.P., Wojtczak, A.M., McArthur Sayers, B., Szczerban, J., et al.), in Encyclopedia of Life Support Systems (EOLSS), Developed under the Auspices of the UNESCO. Eolss Publishers: Oxford, UK.

Käferstein, F., Abdussalam, M., 1999. Food safety in the 21st century. Bull. World Health Organ. 77 (4), 347–351.

Motarjemi, Y., 2009 Management of food safety in the industrial setting. In: Medical Sciences (eds. Mansourian, B.P., Wojtczak, A., Mahfouz, S.M., Arata, A.A., Aluwihare, A.P.R., Brauer, G.W., et al.), in Encyclopedia of Life Support Systems (EOLSS), Developed under the Auspices of the UNESCO. Eolss Publishers: Oxford, UK. <http://www.eolss.net> (retrieved April 14, 2011).

Motarjemi, Y., 2010. Food regulations: global harmonization. Encyclopedia Biotechnol. Agric. Food 1 (1), 262–268. (Francis and Taylor)

Wallace, C., Sperber, W., Mortimore, S.E., 2012. Food Safety for the 21st Century: Managing HACCP and Food Safety throughout the Global Supply Chain. Wiley-Blackwell.

SECTION I

RISKS AND CONTROLS IN THE FOOD SUPPLY CHAIN

CHAPTER

2

Management of Safety in the Feed Chain

Arnaud Bouxin
FEFAC, Brussels, Belgium

OUTLINE

Food Safety Management.
DOI: http://dx.doi.org/10.1016/B978-0-12-381504-0.00002-0

A prerequisite for the production of safe food of animal origin is safe feed. Over the last decades, a series of food safety incidents occurred, taking their origin in feed (BSE agent in meat and bone meal, PCBs/dioxins in fats and bread meal, melamine in protein concentrates, etc.). An effective feed safety management system is key to ensure the safety of products of animal origin. However, feed safety is not just about preserving the safety of animal products: any feed safety management system should address both human food safety and animal health and welfare. In certain countries such as those in the EU, operators are also required to take into account in their risk management system the potential negative impact of contaminants on zootechnical performance. On top of these safety considerations, market specifications excluding certain ingredients (e.g. non-GMO feed or organic feed) impose additional constraints on operators in the feed chain. These are not addressed in the present chapter.

OVERVIEW OF THE FEED CHAIN

The feed chain is extremely complex as it involves interactions with many sectors: feed is a main input to the production of food but the food chain generates itself at different processing stages. Co-products can be found in the feed chain, and even part of foodstuffs, which are withdrawn from the food market for logistical reasons or because they have exceeded the use-by date (so-called former foodstuffs), may also be used for feed purposes. The feed chain cannot therefore be represented linearly (see Figure 2.1). When it comes to safety management, the impact of this iterative process on, e.g., concentration of contaminants is essential. The feed chain interacts also with a number of other sectors whose core business (and level of interest/consideration) is often not the feed outlet (biofuels industry, chemical industry, mining companies, etc.). This also has to be taken into account when considering risk management in the feed chain. The feed market is global, in particular as regards unprocessed cereals and co-products from the biofuel and crushing industry.

Operators involved in the feed chain may be classified in five main categories:

- Producers of feed ingredients: these may be farmers producing crops or processors of vegetable products, produce from the biofuel industry or producers of feed additives, etc.;
- Traders, transporters, warehouses;
- Premix manufacturers, mixing certain feed additives on a carrier;
- Compound feed manufacturers mixing feed ingredients with premixes to produce a complete or complementary feed;
- Livestock holders who produce their own feed (home mixers) and/or deliver feed to animals.

CHARACTERISTICS OF THE FEED CHAIN

An understanding of certain features of the feed chain is important for managing its safety and realizing the extent of the challenges. Feed is defined by Codex as "Any single or multiple materials, whether processed, semi-processed or raw, which is intended to be fed directly to food producing animals."

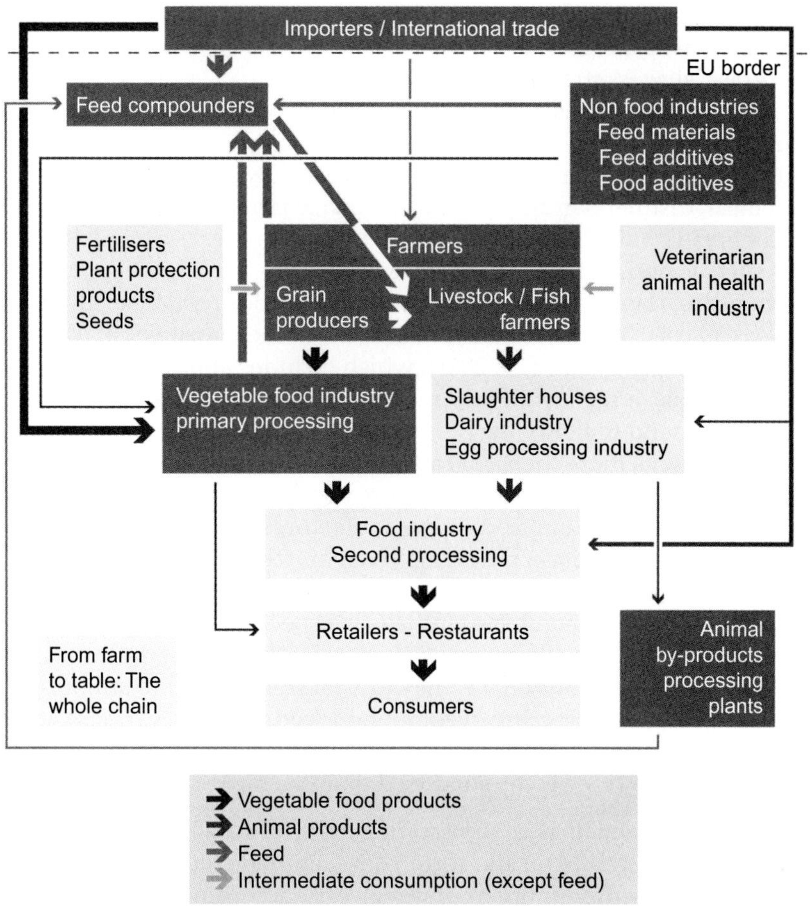

FIGURE 2.1 Functioning of the feed chain (FEFAC 2009).

There are four extensive categories of feed with different risk profiles:

1. Feed additives defined by Codex as "Any intentionally added ingredients not normally consumed as feed by themselves, whether or not they have nutritional value, which affect the characteristics of feed or animal products": these are chemically well-defined substances which are added to the diet to exert a specific function, whether nutritional (vitamins, etc.), technological (binders, etc.), sensorial (flavorings, etc.) or zootechnical (enzymes, etc.). In several countries these feed additives are subject to an authorization procedure based on a risk assessment and are often subject to restrictions in terms of dosage or target species. The EU register of feed additives includes more than 2800 substances.
2. Premixes: these are uniform mixtures of micro-ingredients and feed additives on a carrier to facilitate their even distribution in a larger mix. Premixes are often dedicated to a given target species.

3. Feed materials: these are feed from vegetable, animal or mineral origin, which can be classified in four main categories:
   a. Forages (grass, silages, straw);
   b. Unprocessed feed materials, i.e. feed crops such as peas or feed wheat or the surplus of food crops;
   c. By-products of the food, drinks and biofuel processing industries (e.g. bran, beet pulp, rapeseed meal, distiller's grains, soya meal, fish trimmings, etc.) or former foodstuffs (surplus of bread);
   d. Minerals such as phosphates, limestone, etc.

   In several countries, inventories of feed materials have been established, such as in the USA (AAFCO list). In 2011, the EU published a non-exhaustive catalogue of feed materials (Regulation (EC) No. 575/2011), which includes almost 600 feed ingredients. Next to this catalogue, a register of feed material (www.feedmaterialsregister.eu), where EU operators are bound to notify the placing on the market of any feed material not listed in the catalogue, counts more than 2000 additional feed materials.
4. Compound feed: a compound feed is a mixture of several feed materials, whether or not with feed additives. Compound feeds are manufactured to meet specifications prepared by specialists in animal nutrition providing the required nutritional needs according to the particular species of animal and its growth stage or position in the production cycle. The compound feed manufacturer may be a specialized company or the farmer himself. Compound feed may be complete, i.e. sufficient to meet the animal's needs, or complementary (or concentrate), i.e. they must be distributed to animals together with other feed. The manufacturing of compound feed involves various categories of operators, i.e. the producers of the feed ingredients (feed additives or feed materials), the premix manufacturer and the compound feed manufacturer.

The diet of ruminant animals is composed in general of forages, completed by other feed materials whether or not mixed in the form of a complementary feed. The proportion of feed other than forages in the ruminant diet will vary according to the level of intensification of the production system. Monogastrics (poultry, pigs, etc.) do not get forages.

The key features of the feed chain from a feed safety management point of view are the following:

- A huge number of feed ingredients and variety of risk profiles (see above): not all feed ingredients are used at the same time. A compound feed usually contains between three ingredients for the simple mixtures and up to 30 ingredients for elaborate compound feed (half being feed additives). An average composition of a compound feed in the EU is given in Figure 2.2. The most important feed ingredients are cereals incorporated at more than 50% and oilseed meals. However, the type of ingredients, their number and their incorporation rates in compound feed vary significantly depending on the species of destination and also on the availability and the quotations of the feed ingredients, which depend on the location of the compound feed manufacturing plant.
- A vast number of feed chain operators and origins: the risk profile of a feed may be affected at any stage of its life cycle: primary production (i.e. mining company, crop producer or chemical company), transport, handling, storage, intermediate processing (food or biofuel industry), mixture with other feed ingredients and distribution to animals. The risk at the

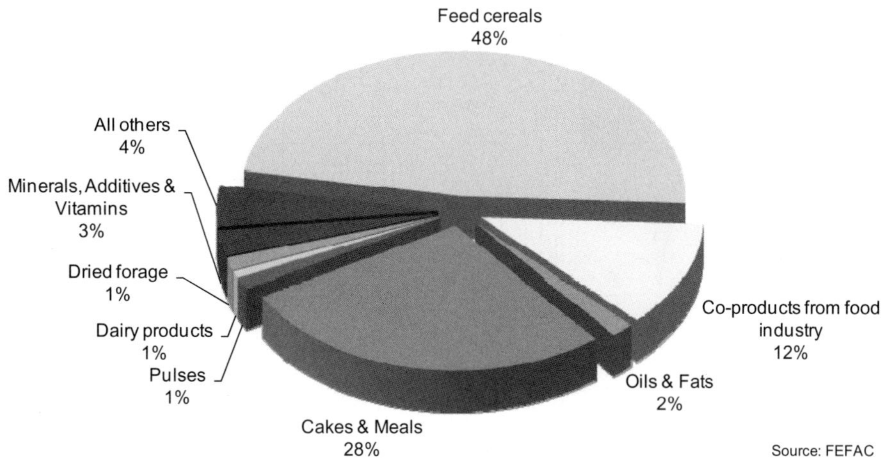

FIGURE 2.2 Average composition of a compound feed produced by industrial compound feed manufacturers in the EU in 2011.

first stages of the chain (feed additives and feed materials production) is the introduction of hazards in the feed chain (e.g. contaminants such as dioxins) whereas the risk at the premix and compound feed stages is mistakes in the formulation of the feed or cross-contamination. This means that an efficient risk management strategy in the feed chain should aim at preventing the introduction of hazards at the first stages of the chain, whereas controls at the subsequence stages should focus on the control of formulation and cross-contamination. It should also be stressed that each stage of the chain has the potential to extend the scope of an incident through the multiplication of operators involved, meaning that an incident occurring at the beginning of the chain has a potentially larger impact than if occurring at the end. A large proportion of certain feed materials such as soybean meal are subject to international trade, and the number of potential geographical origins is also an important dimension to be considered for feed safety management.

- The different animal species: the toxicity of contaminants is not the same for all animal species and the transfer of contaminants from feed to products of animal origin is also species specific. As compound feed manufacturers often produce feed for several categories of animals, the animal species is an additional dimension to the risk analysis. Furthermore, certain feed ingredients may be restricted for use by certain species only. The risk of cross-contamination during transport or storage of feed ingredients or within the feed mill between batches of compound feed destined to different animal species must also be taken into account by the feed businesses in their feed safety management procedures (Figure 2.3). The same applies for those feed manufacturers involved in the manufacturing and delivery of medicated feed.

It can be concluded that, in some respects, feed management of feed safety is more complex than often realized and presents challenges of a different nature than food safety. This means that, for products which may be used either as food or feed, the results of a food safety assessment, although useful information, are not sufficient to guarantee the safety

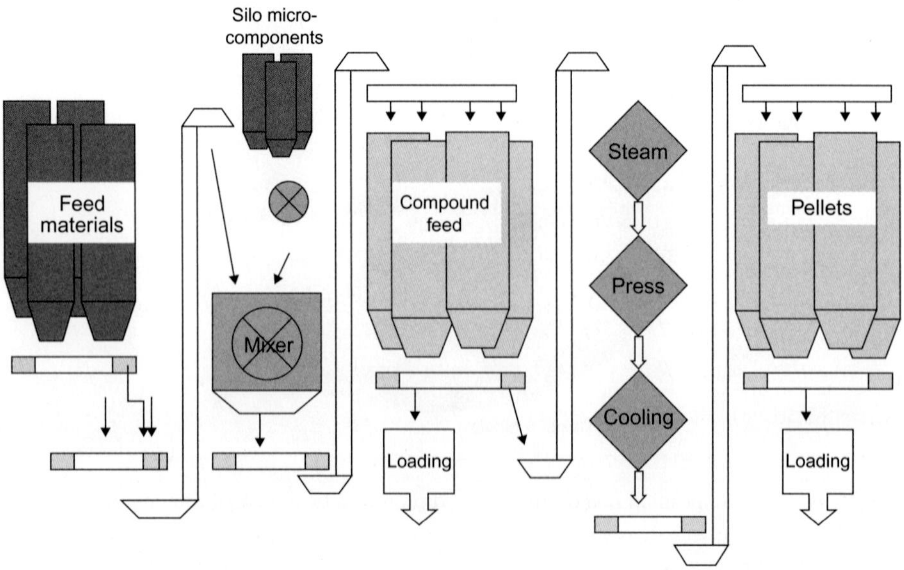

FIGURE 2.3 Schematic representation of the compound feed manufacturing process.

of the product as feed. However, the principles and approaches used to assess and prevent potential hazards in feed are similar to those prevailing in the management of food safety.

POTENTIAL HAZARDS

A proper risk management requires a solid risk assessment, which starts with hazard identification.

Feed hazards associated with human or animal health issues may be of biological (prion, pathogenic microorganisms, parasites), chemical (heavy metals, dioxins, mycotoxins, glucosinolates, excessive levels of pesticides, veterinary medicinal substances or additives, adulterants) or physical origin. Products of their biotransformation in edible products shall also be considered (e.g. aflatoxin B1 in feed transformed into aflatoxin M1 in milk). These hazards may be introduced with source materials or via carry-over or cross-contamination during handling, storage, transport and manufacturing. The presence of these hazards may be natural (development of mycotoxins in crops), or due to inadequate process control (dioxin formation during heating, carry-over), bad hygiene practice (*Salmonella*, use of contaminated raw materials or processing aids) or fraud (deliberate adulteration with products not destined for feed). All these hazards are potentially harmful for animals and/or consumers of animal products. Certain mycotoxins are harmful to animals but are not transferred to animal tissues and do not pose a risk to human health. On the other hand, some hazards such as specific *Salmonella* serotypes may not be harmful for certain animal species but may be so for human health if present in animal products.

The Codex Task Force on Animal Feed started establishing guidance for governments in prioritizing their national feed hazards. They base their work on the IFIF/FAO manual of *Good Practices for Animal Feeding* (covering activities by the feed industry, grazing and forages, home mixing and distribution of feed to farm animals), which establishes a current non-exhaustive list of hazards of importance for feed safety, based on the following criteria:

- Relevance of the hazard to public health;
- Extent of occurrence in feed for food producing animals and food of animal origin;
- Potential impact on international trade in feed and food.

This list has been completed to take into account hazards of relevance for animal health as well and has been sorted by categories.

Biological Hazards

Bacteria

Pathogenic microorganisms in feed may transfer to food-producing animals and then to animal products. They may be introduced into pastures, forages/silages (*Clostridium* spp., *Brucella*) or may be present in feed from animal origin, e.g. dairy products, animal meals (*Salmonella*), and/or may be introduced to feed by cross-contamination or carry-over during processing, transport and storage.

Endoparasites

Some animal endoparasites, such as *Echinococcus*, *Toxoplasma gondii* and *Cysticercus* and *Trichinella*, are human health hazards and may contaminate pasture and forages.

Prions

Prions are responsible for the transmission of transmissible spongiform encephalopathies (TSEs). They may be present in ruminant protein meals and are extremely resistant to denaturation by chemical and physical agents including heat. The Codex Code of Good Practice for Animal Feeding recommends that animal products that could be a source of BSE should not be used for feeding directly to, or for feed manufacturing for, ruminants.

Chemical Hazards

Elements

Elements which have a relatively long chemical or biological half-life will tend to accumulate in edible products after repeated exposure. The following are non-exhaustive examples:

- Arsenic, found typically in minerals in inorganic form and in fishmeal in the less toxic organic form.
- Cadmium, in particular in minerals (such as phosphate and zinc sources) and in forages. The risk of contamination is greater in crops produced on soil where contaminated manure, sewage, sludge or phosphate fertilizers have been spread.
- Lead may occur in grain or forage grown on contaminated soil and also as a contaminant in minerals.

- Fluorine can be found in particular in feed ingredients of marine origin.
- Radionuclides including caesium-134, caesium-137, strontium-90 and iodine-131, when present in animal feed and forages, may transfer to edible products. They may arise from water or windborne environmental contamination.

Mycotoxins

Mycotoxins are found most commonly in cereals (especially wheat, sorghum and maize) but can also be found in oilseed meals and cakes, and silage. The most significant mycotoxin from a food safety point of view is aflatoxin B1, which is transferred into milk in the form of aflatoxin M1. Its presence on the outer part of the grains means that most food processes tend to concentrate the mycotoxins in the co-products such as brans or middlings.

Mycotoxin contamination in feed may occur on the field or during storage. Transfer from feed to edible products has been demonstrated for aflatoxin and, to a lesser extent, ochratoxin A. Other mycotoxins such as zearalenone, deoxynivalenol and fumonisin have no or limited transfer to food but can pose serious risk for animal health, animal welfare, in particular for young animals such as piglets, or affect significantly animal performance.

Terrestrial Plant Toxins

Toxin-producing plants may occur in grasslands used for forage or in certain crops. Toxins can include pyrrizolidine alkaloids, ergot alkaloids and other alkaloids (e.g. atropine, caffeine, cocaine, ephedrine, morphine, nicotine, solanine), terpenes (e.g. camphor, menthol, pinene), tetrahydrocannabinol, gossypol, isoflavones and glycosides (e.g. glucosinolates, cyanogenic glycosides, digitalis).

Bacterial Toxins

Toxins produced by bacteria such as *Bacillus* spp., *Clostridium botulinum*, *C. perfringens* or *Staphylococcus aureus* are acutely toxic to food-producing animals when ingested with feed. Transfer of toxin to edible products is therefore unlikely.

Organic Chemicals

Many organic chemical contaminants that are present in the environment may contaminate feed. The lipophilic compounds such as dioxins and some organic chemicals such as organochlorine pesticides (e.g. aldrin, dieldrin, DDT) have the greatest tendency to accumulate in the environment and in edible products of food-producing animals, in particular milk and fats. Some of these substances are classified as persistent organic pollutants and subject to prohibition of use by international agreements such as the Stockholm Agreement.

The duration of exposure is an important element to take into account in the risk analysis in case of contamination.

Dioxins are the most emblematic group of hazardous chemicals including polychlorinated dibenzodioxins (PCDD), dibenzofurans (PCDF) and dioxin-like polychlorinated biphenyls (DL-PCBs). The different congeners hold different levels of toxicity.

A number of cases of contamination of animal products with dioxins with a feed origin have been reported over the last 20 years (see Table 2.1). The reason for this is the multiplicity of contamination sources, e.g. by direct contamination due to bad practices (e.g. from use of wood tainted with dioxin-containing preservatives as a carrier of premixture), fraud (recycling

TABLE 2.1 Non-exhaustive List of Cases of Contamination of Feed with Dioxins in the EU over the Last 15 Years

Year	Product	Origin
1998	Dioxin in citrus pulp pellets	Process (contaminated limestone)
1999	Feed fats	Fraud (disposal of waste oil)
1999	Kaolinitic clays	Natural (prehistoric fire)
2000	Choline chloride	Process (use of treated wood as carrier)
2002	Carbosan copper	Process
2003	Dried fodder	Process (direct drying with treated wood)
2004	Potato pulp	Process (use of contaminated clay)
2008	Bread meal	Process (drying using contaminated fuel oil)
2010/2011	Feed fats	Mixing with technical fatty acids (under investigation)

of hazardous waste such as mineral oils in the feed chain) or from combustion sources (e.g. waste incineration plants, fossil fuel power stations, bush fires, exhaust gases). Dioxins may also be present in mineral sources, such as clays (due to prehistoric forest fires), recovered copper sulfate and zinc oxide. Fish meal and fish oils may present high levels of dioxins depending on the origins or certain types of fish (e.g. blue whiting from the North Atlantic).

The Code of Practice for the Prevention and Reduction of Dioxin and Dioxin-like PCB Contamination in Food and Feeds (CAC/RCP 62-2006) provides guidance on the occurrence, reduction and prevention of dioxin contamination.

The presence of chemicals in feed ingredients may also result from:

- the use of substances as pesticides: residues of pesticides may be present in feed ingredients as a result of their use on crops; non-intentional presence of pesticide residues in crops may result from the uptake of residues present as a result of treating a previous crop with pesticides or from spray-drift, volatilization and/or runoff;
- the intentional use of substances in feed: substances used as feed additives may be toxic for animals and/or humans above certain levels; in certain countries, maximum permitted levels are set in animal diets by official regulators (EU register of feed additives);
- the use of processing aids in biofuels or food manufacturing processes which may end up in by-products (e.g. antibiotics used to control microbiological contamination may concentrate in yeast cultures used for ethanol production and be sold as a dehydrated protein source and in distiller's dried grains with solubles after their use in fermentation for ethanol production);
- the presence in former foodstuffs of food additives or contaminants (theobromine) which are not of concern for human health but may be for animal health/welfare;
- the presence of residues of veterinary medicinal products in feed from animal origin (whether approved or unapproved such as nitrofurans in shrimps, chloramphenicol in milk powder);
- deliberate adulteration of feed materials (e.g. incorporation of melamine in vegetable protein concentrates to increase the nitrogen level);
- packaging residues resulting from mechanical unwrapping of former foodstuffs.

An overview of the potential biological and chemical hazards in feed and their potential for transfer to animal products is presented in Table 2.2.

Physical Hazards

These are mostly bones and pieces of metal, plastic or glass. They are potentially harmful to animals as they can provoke severe animal health problems (e.g. gut injuries, sudden death). However, they are unlikely to transfer to and impair the safety of animal products. The origin of physical hazards is often their presence in the environment (pollution) or poorly designed or maintained facilities and equipment or improper employee practices. They may also originate from former foodstuffs whose packaging has not been effectively removed.

This list of hazards is indicative and must be adapted to the specific situation of any feed operator for their own risk assessment, taking into account hazard characterization, exposure assessment and risk characterization. This indeed depends in particular on the position of the operator in the chain (i.e. supplier of unprocessed feed, trader/transporter, manufacturer of feed ingredients, manufacturer of premixes and/or compound feed, farmer),

TABLE 2.2 Non-exhaustive List of Potential Biological and Chemical Hazards in Feed and Potential Transfer into Animal Products

Hazard	Potential sources	Animal product
Bacteria (e.g. *Salmonella*, *Brucella*, *Listeria*)	Pasture, forages, animal meals, oilseed meals	Eggs, poultry, milk and milk products
Endoparasites (e.g. echinococcus, toxoplasma, trichinella)	Pasture, forages, compound feed	Various tissues containing infective cysts
Prions	Ruminant proteins	Specified risk materials (e.g. nervous system tissues, distal ileum)
Radionucleides: 90Sr, 131I, 137Cs	Pasture, forages, crops	Milk, meat
Heavy metals (As, Cd, Pb, Hg, Ni, ...)	Sea plants, micro and macro minerals, soil, etc.	Higher: fish, kidney, liver Lower: meat and milk
Mycotoxins (fusarium trycothecens, etc.)	Grains, co-products from grain processing	Milk (aflatoxin) (limited transfer for most other toxins)
Plant toxins (tremetone, alkaloids)	Botanical impurities in forages and crops	Milk, meat
Dioxins, PCBs	Natural presence; environmental contamination; heat processes	Fat (in milk, meat, egg yolk)
Organochlorine pesticides	Environmental contamination	Fat
Veterinary drugs, pesticides, processing aid residues	Feed produced from treated animals/crops; use of antibiotics in fermentation processes	Meat, milk, eggs
Adulterants (melamine, etc.)	Deliberate adulteration of feed	Milk, meat

the type of raw material used and its geographical origin, and the type of animal species of destination of the feed. Some of the above hazards may therefore not be relevant for all feed operators, whereas some hazards not listed above may have to be taken into account. Additional source of information for hazard identification may be the Rapid Alert System for Food and Feed established in the EU to ensure rapid information and coordination of risk management in case of contamination. Figures 2.4 and 2.5 provide an illustration of the number of feed-related notifications to the RASFF in 2011. The absolute figures shall be handled with care as there is still a lack of harmonization in the procedures used by national authorities to notify contamination cases to the RASFF but the scenarios attached to them are more interesting to guide the hazard management at the level of operators.

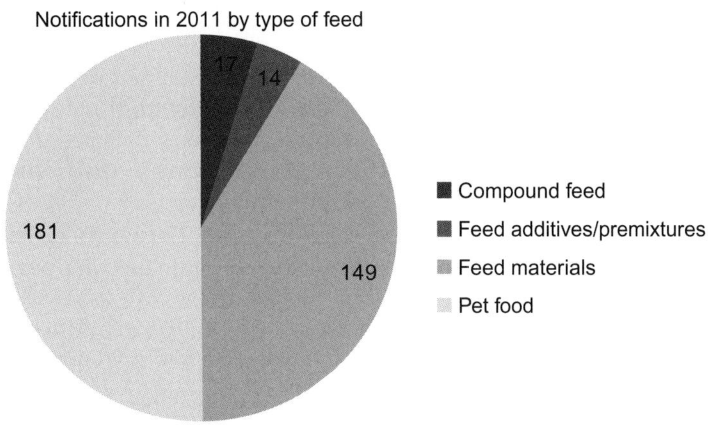

FIGURE 2.4 Notifications of feed safety contamination to the RASFF in 2011 by type of feed in the EU.

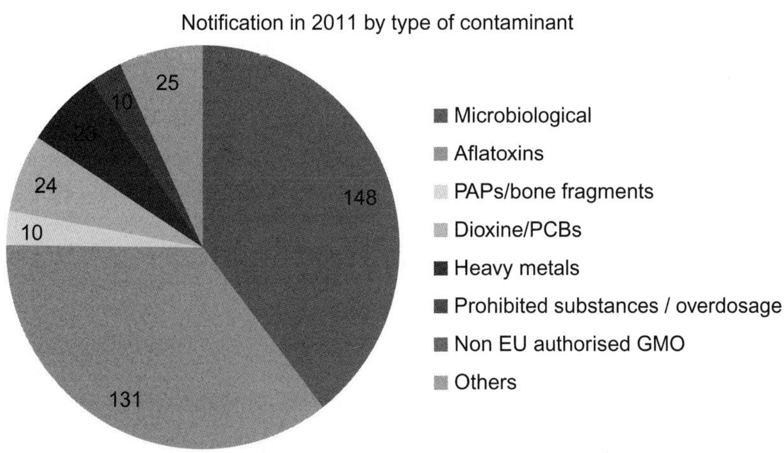

FIGURE 2.5 Notifications of feed safety contamination to the RASFF in 2011 by type of contaminant in the EU.

A number of factors may also turn a negligible hazard into relevant risk because of an increase of its prevalence or the emergence of new hazards. The development of new processing methods for crops (e.g. biofuels production) generating new co-products involves using new processing aids with potentially harmful residues. Global warming may also imply a change in the geographic presence of, e.g., fusarium.

GOOD HYGIENE PRACTICES IN THE FEED SECTOR

The proper management of feed safety lies in the identification and implementation of good practices. The tools developed for the management of food safety such as ISO 22000, FSSC 22000, or PAS220 can also be used for the management of feed safety. A number of tools exist to help operators implement the hygienic practices that are relevant to their activities, namely:

- Codex Code of Good Animal Feeding (2004): this document establishes basic principles for the management of feed safety as well as minimum good practices for animal feeding at any step of the feed chain from feed ingredients producers to distribution to animals. It addresses human health risk only.
- IFIF/FAO Manual for Industrial Feed (2009): this document provides practical guidance on how to implement the Codex Code of Practice by commercial compound feed manufacturers and farmers producing feed on their farms. It addresses human health risk.
- PAS222 (2011): the prerequisite program for food safety in the manufacturing of food and feed was developed by the British Standards Institute for any operator of the feed chain and addresses hazards that may adversely affect both animal and human health.

In addition to these tools, a number of professional standards were developed at national/regional level and by specific sectors of the feed chain to meet the requirements relevant to their activities and to their national legal requirements. In particular, EU Regulation (EC) No. 183/2005 on Feed Hygiene encourages organizations of the feed chain to develop such sectoral guides to good hygiene practice and foresees an assessment of the relevance of such guides by authorities. These tools qualify either as codes/guides to good practice or as feed safety management systems and may be linked to a certification scheme. EU schemes are used to integrate animal health and even animal performance in their scope. A non-exhaustive list of such schemes is provided in Table 2.3.

Feed Safety Management Principles

All the above mentioned codes/guides are based on three essential principles:

1. Responsibility of each operator for the safety of the feed it places on the market or uses. This principle supposes:
   a. The implementation of a feed safety management system including a hazard analysis (HACCP recommended or imposed);
   b. The commitment of all staff from CEO to the operational management to the implementation of the feed safety management system;

   c. A definition of the scope of the good practice (human health/animal health, type of feed, production sites, type of operations, definition of terms) and establishment of feed safety objectives;
   d. An audit of the performance of the feed safety management system;
   e. A review of the feed safety management system at defined intervals in particular when major or significant changes to plant or products occur, to ensure its suitability and effectiveness of changes and improvements;
   f. Proper internal communication flow and adequate training/qualification of all staff members.
2. Traceability of products one step back/one step forward. This principle supposes:
   a. A system of documentation to ensure traceability, which identifies (1) suppliers and intermediaries of purchased materials, and (2) to whom these incoming feeds have been supplied once processed into finished feeds;
   b. Records of the details of all suppliers/intermediaries of purchased feed and batch numbers of all purchased batches as well as the nature and quantity of outgoing feed with their manufacturing date and the name and address of the customer to whom each batch is delivered;
   c. Keeping records in such a way as to be easily accessible and allowing prompt identification of potentially contaminated products in case of incident and, if needed, withdrawal/recall of contaminated products further to a risk analysis.
3. Cooperation along the chain to ensure a proper handling of the risk (e.g. instructions for use on labels) and with public authorities in case of contamination. This principle supposes:
   a. Proper information regarding the nature of the product and its intended use: this should include in particular the following elements as appropriate:
      - For all feed: a clear denomination of the feed in a manner that should not mislead the user of the feed as to its real nature, the identity of the supplier and the lot identification;
      - In addition for feed additives, premixes and compound feed: the manufacturing date, the shelf-life and instructions for safe handling and use;
      - In addition for compound feed and premixes: information about the species or category of animals for which the feed is intended; the purpose for which the feed is intended; a list of feed ingredients, including additives.

      In principle, national legislation establishes such labeling requirements.
   b. An external communication policy towards customers so that, in case of non-conformity of a safety nature affecting feed and triggering product withdrawal/recall, the feed supplier effectively, accurately and in a transparent way informs users of the feed at stake of the reason for its withdrawal/recall;
   c. A communication policy towards authorities whereby operators, when deemed necessary, inform competent authorities if they consider that a feed or feed ingredient does not satisfy the feed safety requirements and statutory standards. The information should be as detailed as possible and should at least contain a description of the nature of the problem, a description of the feed or feed ingredients, the species for which it is intended, the lot identifier, the name of the manufacturer and the place of origin.

Prerequisite Programs

A number of basic principles shall be applied to assist in controlling feed hazards and address in particular:

- The location of the site and the surroundings;
- Processes and workspaces;
- Supply of air, water, energy and light;
- Waste management;
- Equipment suitability, cleaning and maintenance;
- Management of ingredients;
- Prevention and management of contamination/carry-over;
- Pest control;
- Management of internal/external returns;
- Product withdrawal procedures.

Below are some elements of these prerequisite programs which are extremely relevant for the management of feed safety.

Management of Ingredients

- Incoming products must be delivered by suppliers assessed on a regular basis by the purchaser (and prior to any delivery, in the case of a new supplier) or participating in a feed safety assurance system, subject to certification by a third party, and recognized by the purchaser. The purpose of the supplier assessment is in particular to check that there is an effective feed safety control system in place and to appraise the outcome of the monitoring program implemented by the supplier.
- Each batch of incoming ingredient shall be visually inspected. Documentation shall be checked to verify the integrity of the material.
- Ingredients should be stored separately from each other and from finished products.
- Procedures should be established to keep to a minimum the proportion of out-of-date stocks (e.g. first-in-first-out principle) by applying a careful stock rotation. Materials must be stored in such a way that they are clearly identifiable, and that their intake identification is easily visible.

Prevention and Management of Carry-over

- Contamination may arise from traces of products of a previous run that cannot be completely cleaned from the product line due to technical limitation: this type of contamination is called carry-over. Controlling carry-over is essential in particular in multispecies compound feed and premix plants handling substances prohibited or subject to restrictions of use for certain animal species (veterinary medicinal substances, feed additives or feed ingredients).
- Several factors may influence the level of carry-over of a substance in a feed mill: the facilities themselves (the equipment of the facilities), the substance itself, the feed matrix and the measures that are taken to control carry-over.
- Feed operators shall in the first place measure their level of carry-over in order to identify and apply the adequate measures, taking into account the statutory standards regarding carry-over.
- Feed operators producing feed for several species must draw up production schedules derived from the HACCP study taking into account the premise-bound carry-over, the

characteristics of the substances (depending on adhesive strength, electrostatic properties and the size and density of the particles) and the species for which they are authorized. This may include scheduling exclusions.
- In order to establish this schedule, the company must define for each substance regarded as at-risk further to the HACCP study the number of batches to be produced between a batch containing a given active substance (additive including coccidiostats and histomonostats or veterinary medicinal substances) and a batch for a non-target species or for withdrawal feed or for continuous food-producing animals (dairy cows, laying hens). This number of batches will be defined for each animal species, taking into account the level of carry-over of the plant, the physical characteristics of the substance and the level of risk for animal and public health.
- Where necessary, the equipment must be flushed to avoid carry-over between batches. Flushing must be done using a specified amount of wheat feed or other suitable material, proven to purge the system adequately.

Management of Internal/External Returns

- The production of finished feed must be organized, both on an internal and external level, with an eye to limit possible returns to a minimum.
- External returns (from customers) should be avoided. When occuring, they must be assessed and, if needed, placed in separate adequately segregated storage to prevent contamination of other feed.
- Internal returns, other than flushing or cleaning material, must, whenever possible, be reincorporated into their original batch or "run."
- Procedural rules must lay down in which feed formulation returns may be incorporated and the maximum percentage of returns in the respective feed type. In no case should a product containing an ingredient subject to restrictions of use be reprocessed into a batch designed for a species for which this material is prohibited.

More elements regarding prerequisite programs in the feed sector can be found in the standards listed in Table 2.3, in particular as regards transport, storage, etc.

Hazard Analysis and Monitoring Plans

The prerequisite programs shall be completed by a hazard analysis specific to the situation of the individual company. The Codex Code of Practice for Animal Feeding specifies that, where applicable, HACCP principles should be followed. HACCP in its full extent requires expertise and resources that are not always available, in particular to small operators, especially small farms.

A number of publications provide guidance on how to perform HACCP. The Codex guidelines remain the internationally accepted reference (ftp://ftp.fao.org/codex/Publications/Booklets/Hygiene/FoodHygiene_2003e.pdf). Although there are currently no such Codex guidelines for feed, the approach for food is applicable to feed, while taking into account the specificities of the feed sector as specified above, i.e.:

- The number of operators, in particular the number of operators whose core business is not feed but food or non-feed/food industry (e.g. biofuels); some of these operators do not even know that their by-products are used in the feed sector and some even refuse to be regarded as feed operators for image purposes;

TABLE 2.3 Type, Origin and Scope of Feed Safety Management Documents

Name	Scope FM	FA	PR	CF	TR	Country of origin	Geographical area	Type of tool
SFSF	x	x	x	x		USA	USA	Certifiable scheme
EFMC			x	x		EU	EU	Code
GMP+ International	x	x	x	x	x	NL	Global	Certifiable scheme
QS	x	x	x	x	x	DE	DE	Certifiable scheme
FEMAS	x	x				UK	Global	Certifiable scheme
UFAS			x	x		UK	UK	Certifiable scheme
OQUALIM			x	x		FR	FR	Certifiable scheme
OVOCOM	x	x	x	x	x	BE	BE	Certifiable scheme
EFISC	x					EU	EU	Certifiable scheme
GTP					x	EU	Global	Certifiable scheme
FAMI-QS		x				EU	Global	Certifiable scheme
Feed & Food Safety	x	x	x	x		BR	BR	Certifiable scheme
GLOBAL G.A.P.				x	x	Global	Global	Certifiable scheme

FM: Feed material; FA: Feed additives; PR: Premixes; CF: Compound feed; TR: Trade

- The type of risk addressed, i.e. first the combination of human health, animal health/ welfare, and, in certain cases, animal performance;
- The specificity of the human health risk assessment involving the biological interference of the animal;
- The number of animal species involved, with different sensitivity to hazards;
- The number of feed ingredients.

The feed safety standards listed in Table 2.3 may also provide guidance on how to perform HACCP in the feed sector. The new Codex Task Force on animal feed is currently developing specific risk assessment guidance for national governments on feed safety impacts on food safety using the present food HACCP guidance as reference.

A number of tools are available to operators to carry out their hazard analysis and help set monitoring plans and establish critical limits:

- The national statutory standards: national legislation may establish maximum limits for contaminants in feed ingredients and/or compound feed: in the EU, Directive 2002/32/ EC establishes maximum limits for chemical contaminants such as heavy metals, mycotoxins, certain pesticides, dioxins, etc. based on human health, animal health and animal performance as well as on the ALARA (As Low As Reasonably Achievable) principle; likewise, MRLs may be established in national legislation for pesticides and/or veterinary medicinal products in feed ingredients of vegetable or animal origin;

- The Codex standards: in case of absence of national standards, Codex MRLs for pesticides in unprocessed plant products is the reference for feed business operators (Codex Pesticides Residues in Food Online Database);
- Results of monitoring programs: certain national authorities are used to publish the outcome of their monitoring programs on certain contaminants. Certain food safety authorities such as EFSA may also publish reports on the occurrence of contaminants in food and feed, e.g. for *Salmonella;*
- Rapid alert systems, e.g. the RASFF (see Chapter 22).

From Good Practices to Certified Feed Safety Assurance Schemes

Assessment of suppliers is an essential element for feed safety risk management. Third party certification of compliance with good practices enables reduction of the number of audits while preserving know-how.

A number of bodies having developed codes of good practices established also a third party certification scheme (see Table 2.3). Such schemes cover either part of the chain or embrace the feed chain as a whole, and may also be integrated in broader schemes covering the whole feed and food chain. Alternatively, they may be also cross-references with downstream assurance schemes run by the livestock industry or retailers such as GlobalGAP. In certain countries such as the UK, Belgium or the Netherlands, certification against the national feed safety assurance scheme is a prerequisite to market access.

Several Feed Safety Assurance Schemes such as GMP+ International, OVOCOM or QS have also established collective monitoring programs. This allows reduction of the need for analytical checks while improving the knowledge on the occurrence of contaminants.

EXAMPLES OF FEED SAFETY INCIDENTS AND WHAT LESSONS CAN BE LEARNED

MPA in Glucose Syrup in 2002

Background

Medroxyprogesterone acetate (MPA) is a synthetic hormone having progestogen activity and is used in human and veterinary medicine. It is no longer permitted for use as growth promoters in the EU.

Fertility problems occurred in three pig farms in the Netherlands on 20 May 2002. Animals were fed with wet feed containing a high concentration of contaminated glucose syrup sourced directly by farmers from a Belgian company, which happened also to be a waste processor. The first step of the investigation carried out by Dutch authorities as regards the origin of the contamination was completed on 20 June 2002 and notified to the EU Commission and other member states through the Rapid Alert System Food and Feed. Further investigations enabled the identification of two other contamination tracks. The whole tracking and tracing operation was completed on 24 July 2002.

The MPA contamination found its origin in Ireland in the illegal mixture of pharmaceutical non-hazardous waste with hazardous waste containing MPA at some point in the waste disposal chain. From September 2000 to June 2002, 1850 kg of MPA was illegally classified as non-hazardous waste and shipped without notification to the Belgian waste processor. This company then mixed up the waste with glucose syrup.

The contaminated syrup was first sold to soft drink producers up until December 2001. Then, the Belgian company supplied its glucose syrup to (1) wet feed companies which resold the product directly to home mixers and to a compound feed manufacturer who mixed up the glucose syrup with molasses, the mixture then being included in compound feed distributed to farmers and (2) a molasses trader who mixed up the glucose syrup with molasses, the mixture being sold to feed manufacturers for inclusion in compound feed. As a consequence, the spread of the contamination involved a large number of feed business operators in the Netherlands and Belgium and almost half of the Dutch livestock farms were subject to temporary blockage.

A product recall for contaminated glucose syrup, molasses and feed was undertaken according to a procedure approved by the EU authorities in Belgium, Germany and the Netherlands, and, to a lesser extent, in other member states. Pig market prices fell in mid-July 2002. The cost of the MPA contamination was estimated between EUR 107 and 132 million.

What did not Work?

- The main shortcoming in this incident was that there was no clear physical separation between the hazardous waste stream and the feed chain.
- Although the Dutch feed/food chain has developed a comprehensive risk management-oriented quality assurance system (GMP+) subject to external audits and imposing the sourcing of feed materials from GMP+-approved suppliers only, the Belgian waste processor, which did not handle a Dutch GMP+ approval, managed to sell its products to GMP+-approved home mixers and a molasses trader.

Lessons to be Learned

- As a matter of principle, the waste management streams should be kept physically separated from the feed chain.
- Any introduction of a new material or modification of the composition of a material or change in the manufacturing process should be subject to a hazard analysis by the supplier.
- Suppliers who are not audited by a third party for their feed safety risk management should make available to their customers the results of their own hazard assessment.
- Audits systems from certified Feed Safety Assurance Schemes should be more efficient in the detection of non-compliances.

Contamination of Bread Meal with Dioxins in Ireland in 2008

Background

The dioxin contamination incident originated in the detection of elevated levels of PCBs in a pork fat sample, and feed analyzed on a pig farm was also positive for PCBs. The contaminated feed was traced back to a company specialized in the processing of bread crumb not used for human consumption. Samples of fats and feed taken in other pig farms that were identified as having received similar feed from this company also proved positive. Considering that those 10 farms represented more than 10% of the Irish pig slaughters, the Irish government ordered a full recall of pork and pork products manufactured from all pigs slaughtered in Ireland within the last 3 months before the identification of the contamination. This decision was due to the traceability system in place in the pork chain but was not sufficient to allow for a targeted recall, considering the high degree of commingling of product in secondary processing.

All of the evidence available suggested that the incident occurred as a result of contaminated fuel being used in an oil-fired burner (direct flame drying system) that generated the heat to dry the feed at the bread crumb recovery operation. Laboratory tests showed that the oil used as a fuel in the burner at the plant was contaminated with PCBs.

What did not Work?

- The feed business processing the bread crumb should have identified the risks associated with the direct drying process in its feed safety management system based on HACCP principles.
- The authorities failed to verify that the feed business was complying with the legislation in that the HACCP plan was not fit for purpose, and the inspection of the premises was inadequate. The company was classified as "low risk" by the authorities simply because it was not using animal by-products with no consideration of direct drying as a "high-risk" process.

Lessons to be Learned

- Direct drying should be regarded as a high-risk operation.
- There is a need to ensure that any feed operation is placed under the supervision of the feed safety authorities.

Dioxins in Feed Fats in Germany in 2010/2011

Background

On 21 December 2010 a feed mill located in Niedersachsen detected contamination of compound feed for laying hens with dioxins above the maximum permitted levels as part of its own checks.

The German authorities having been informed by the company started investigations and discovered that a fat processor located in Schleswig-Holstein had purchased several consignments of fatty acids for the purpose of producing fat. This fat processor produced both feed fat and fat for industrial uses in separate lines. However, the processor subcontracted the production of feed fat containing the fatty acids to another fat processor located in Niedersachsen.

The fatty acids had been purchased from a biodiesel plant, from which the material was delivered directly to the fat processor in Niedersachsen, via a Dutch trader that handled both fat for the production of feed and fat for industrial uses. The fat processor in Niedersachsen subsequently used these fatty acids for the production of feed fat that was directly dispatched to several manufacturers of compound feed as feed fats.

The mixed fatty acids were confirmed to be contaminated by dioxins above the EU maximum permitted levels. The likely source of the contamination was the use of contaminated raw materials or technical processing aids for the production of biodiesel, which through the process were concentrated in the fatty acids.

Investigations from the authorities at the level of the two fat producers in Schleswig-Holstein and Niedersachsen led to the identification of eight deliveries of potentially contaminated fatty acids representing a volume of 206t. These 206t where mixed up with other fat products and sold as feed fats (2256t) to 25 compound feed manufacturers. The compound feed was then delivered to almost 4500 farms that were subsequently blocked. Most compound feed was already used at the time of the blocking of the farms. Random testing of the samples of compound feed kept by the manufacturers showed levels of contamination below the maximum permitted for compound, which is logical given the low inclusion rates of feed fats in compound feed and the contamination load of the feed fats. Farms were unblocked based on a risk assessment including estimation of the theoretical highest contamination level of compound feed; further analysis was then performed. Only a few farms have shown results on eggs and pig lard above the maximum permitted levels for animal products.

The fat processor in Schleswig-Holstein benefited from a good reputation and was GMP+ and QS certified; feed companies believed that this certificate also covered the activities of the Nierdersachsen plant.

What did not Work?

- The mixed fatty acids were identified in the contract between the biodiesel producers and the Dutch trader as "Mixed fatty acids from cooking oils – not intended for food or feed purposes." But in the contract between the Dutch trader and the fat processor in Schleswig-Holstein, the fatty acids were identified as "Technical mixed fatty acids." As a matter of principle, there was no legal requirement in the EU prohibiting the use of technical fatty acids in the production of feed providing they meet all feed safety criteria and unless it is explicitly mentioned that these are not to be used for this purpose, which was the case for the contract between the biodiesel plant and the feed fat company from Niedersachsen.
- The fat processor in Niedersachsen was not registered by public authorities as a producer of feed fat, only as a transporter of feed.

Lessons to be Learned

- Traceability procedures when well implemented allow the quick identification of potentially contaminated batches of fatty acids and the farms that received compound feed containing the feed fats produced with the contaminated fatty acids.
- There is a need to secure the identity of the products along the chain and in particular their suitability/non-suitability for feed use.
- Confidence in suppliers should not replace regular checks and audits.
- A proper risk assessment is required at all stages of the chain.
- Traders shall feel responsible for the safety of the products they place on the market.
- Although monitoring by users of feed materials enabled the detection of the contamination in this case, the detection at the level of the supplier is by far the most effective way to detect contaminations at an early stage and prevent their propagation ("top of the pyramid" principle).

CONCLUSIONS

Feed safety management is particularly complex in that it involves a larger area in terms of type of risk (human health, animal health and welfare and even livestock performance), type of assessment (direct impact on animal but also transfer/biotransformation from feed to products of animal origin) and the variety of animal species concerned. The involvement of all operators in the chain is essential to ensure cost-effective feed safety management. Contaminations should be identified at the earliest stage possible, taking into account the structure of the feed chain and the rapid turnover at premixes and compound feed mills. To this end, it is essential that operators are made aware that (part of) their product is destined for use in feed. Leaving the responsibility to control feed safety to the last stages of the chain is not only costly but also inefficient. Collaboration among all operators in the chain is therefore essential and the safety management tools developed at different levels of the chain have contributed to a significant improvement in the management of feed safety over the past years.

References

AFIA (American Feed Industry Association). Safe Feed/Safe Food Certification Program. USA. <www.safefeed-safefood.org/main/home.cfm>.

AIC (Agricultural Industries Confederation). Universal Feed Assurance Scheme (UFAS) – Feed Materials Assurance Scheme (FEMAS). UK. <www.agindustries.org.uk/content.template/28/28/Trade%20Assurance/Trade%20Assurance/Trade%20Assurance%20Homepage.mspx>.

BSI (2011) PAS 222:2011 – Prerequisite programmes for food safety in the manufacture of food and feed for animals. <http://www.bsigroup.com/en/sectorsandservices/Forms/PAS-2222011-free-download>.

COCERAL. European Code of Good Trading Practices (GTP). Europe. <www.gtpcode.eu>.

Codex Alimentarius., 2004 Code of Practice on Good Animal Feeding. <http://www.codexalimentarius.org/standards/list-of-standards/en/>.

EFIP., European code on good practice for the industrial manufacturing of safe feed materials (EFISC). Europe. <www.efisc.eu>.

FAO and IFIF., 2010. Good practices for the feed industry – Implementing the Codex Alimentarius Code of Practice on Good Animal Feeding. FAO Animal Production and Health Manual No. 9. Rome. <http://www.fao.org/docrep/012/i1379e/i1379e.pdf>.

FEFAC., 2009. European Feed Manufacturers' Guide – EFMC. Europe. <www.fefac.eu/file.pdf?FileID=25125>.

FEFANA., Quality and Safety System for Specialty Feed Ingredients and their Mixtures. Europe (FAMI-QS). <www.fami-qs.org/documents.htm>.

GMP+ International. GMP+ Feed Safety Assurance Scheme. The Netherlands. <www.gmpplus.org/en/>.

OQUALIM., Guide de Bonnes Pratiques de la Fabrication des Aliments Composés pour Animaux. France. <www.oqualim.fr/>.

OVOCOM (Consultation Platform for the Animal Feed Chain). Good Manufacturing Practice – GMP. Belgium. <www.ovocom.be>.

QS. Leitfaden für die Futtermittelwirtschaft. Germany. <www.q-s.de/dc_feed_sector.html>.

Sindiraçoes – Feed and Food Safety Program. Brazil. <www.sindiracoes.org.br/programa-feed-food/o-programa>.

CHAPTER

3

Naturally Occurring Toxicants of Plant Origin

Benoît Schilter, Anne Constable and Irène Perrin

Nestlé Research Center, Lausanne, Switzerland

OUTLINE

INTRODUCTION

With advances in science and technology, the quality of food has continuously improved. Significant developments have been achieved on various sensorial, safety and nutritional attributes. In addition, scientific efforts are now increasingly devoted to the design of functional food products providing health benefits beyond basic nutrition.

Parallel to this evolution, food has also been found to be a source of public health issues. Illnesses may be caused by infectious (microbiological) or toxic (chemical) agents entering

Food Safety Management.
DOI: http://dx.doi.org/10.1016/B978-0-12-381504-0.00003-2

the body with the ingestion of food. Because of the globalization of food trade, intensive agriculture and environmental pollution, this hidden aspect of food quality has become an increasing concern for consumers, regulatory authorities and the food industry.

Chemical food safety has mainly focused on synthetic chemicals such as additives, residues from agricultural and veterinary practices (e.g. pesticides, antibiotics) and environmental contaminants of industrial (e.g. dioxins, heavy metals) or natural (mycotoxins, heavy metals) origin. In most countries, such chemicals are the subject of legislation, which governs the establishment of limits in foods. This allows food manufacturers to develop and implement quality management strategies ensuring the safety of food products. Food safety is defined as "the assurance that food will not harm the consumer when it is prepared and/or eaten according to its intended use."

In this context, it is interesting to note that up to 99.9% of the non-nutrient chemicals that humans ingest with their normal diet are actually natural and not synthetic (Ames and Gold, 1997). Food crops produce not only nutrients but also a vast array of non-nutrient secondary metabolites (Dolan et al., 2010; Gry et al., 2007; Essers et al., 1998). Importantly, it is considered that dietary exposure to these naturally occurring non-nutrient chemicals can greatly exceed exposures to any types of man-made chemicals occurring in food (Mattsson, 2007; Paustenbach and Galbraith, 2006). Since some of these natural chemicals have been shown to induce severe toxicity, the questions of their significance for consumer safety and of the need for implementing management options aimed at keeping exposures under control are raised.

To address the public health significance of naturally occurring chemicals is not straightforward for a number of reasons. First, for most of them adequate toxicological and exposure data are missing, which prevents an accurate risk assessment. Second, it is increasingly acknowledged that some naturally occurring chemicals in food plants may induce biological effects with beneficial health impact. Finally, interactions with other constituents of food may significantly modulate the expression of the toxic and/or beneficial effects of naturally occurring chemicals (matrix effect).

To prioritize resources allocated to food safety management, it is essential to have a sound basis for evaluating the health risk of the many different chemical hazards. The present chapter provides a brief overview on the chemical diversity and toxic properties of substances occurring naturally in food plants. In addition, it brings some insights on how such chemicals can be managed in order to ensure food safety.

SCOPE AND DEFINITIONS

In terms of chemical composition, plant-derived foods can be assumed as mixtures of chemicals, which can be grouped in two broad categories: intrinsic components which are inherent constituents of the plants and extrinsic components which are chemicals of both natural or industrial origin, reaching the food either by direct addition (food additives), by contamination (e.g. pollutants, mycotoxins, packaging migrants) or indirectly as a result of agricultural practices (e.g. pesticide residues). Intrinsic components encompass a wide range of chemicals with various potential health impacts:

- Macro- (proteins, lipids, sugars) and micro- (e.g. vitamins) nutrients that determine the nutritional value of the plant food.

- Anti-nutrients, which may reduce the nutritional value of the plant food (e.g. protease inhibitors blocking protein digestion, phytate inhibiting absorption of minerals such as iron).
- Inherent plant toxicants, which are non-nutrient secondary plant metabolites identified because of their potential to produce toxicity in humans. Glycoalkaloids in potatoes and cyanogenic glucosides are well-known inherent toxicants.

The scope of the present chapter is on the safety aspects of inherent plant toxicants. There is no official definition of inherent plant toxicants. The EU-AIR-NETTOX project adopted the following definition: "inherent plant toxicants are plant constituents, which might give rise to adverse effects in humans when the plant or plant products are ingested" (Gry et al., 2007; Essers et al., 1998).

INHERENT PLANT TOXICANTS: CHEMICAL DIVERSITY AND ROLES IN THE PLANTS

Plants produce a vast array of secondary metabolites of highly diverse chemical structures, ranging from relatively simple organic chemicals to complex molecules such as proteins. Table 3.1 shows some well-documented examples. The actual function in the plant of inherent toxicants is often not known. They are thought to primarily play a significant role in the defense of plants against bacteria, fungi, viruses and insects (Mattsson, 2007; Lattanzio et al., 2006; Essers et al., 1998; Ames and Gold, 1997; Beier, 1990). Therefore they are sometimes called "natural pesticides." The example of glycoalkaloids in potatoes illustrates this hypothesis: highly pathogen-resistant potato cultivars selected through breeding programs are usually high in glycoalkaloids (Speijers et al., 2010). Defoliation of potato plants by insects was found to induce the production of toxic glycoalkaloids in the tubers, while manual defoliation did not have any effects (Speijers et al., 2010; Pariera-Dinkins et al., 2008). Because some inherent plant toxicants have a strong bitter taste, a role in preventing feeding by mammals has also been advocated (Essers et al., 1998). Other roles more related to plant physiology are also documented. For example, evidence is available for a role of plant phenolics as internal physiological regulators or chemical messengers within the intact plants, with involvement in phototropism and plant growth (Lattanzio et al., 2006).

TOXICOLOGICAL AND BIOLOGICAL CONSIDERATIONS

Whereas inherent plant toxicants have traditionally been identified because of their toxic properties and effects in humans, in recent years significant interest has focused on bioactivities of non-nutrient plant chemicals that are compatible with beneficial health effects. Secondary plant metabolites have indeed been associated with outbreaks of adverse health effects in humans but they are also thought to largely explain the epidemiological evidence for the health benefits derived from a diet rich in fruits and vegetables (Walter, 2003). To fully understand the net health impact of inherent plant toxicants, and more generally of inherent non-nutrient plant constituents, both potential toxicological and beneficial effects need to be considered.

TABLE 3.1 Inherent Plant Toxicants

Inherent Toxicants	Typical Food Plants	Reported Effects in Human	Mechanism Reported	Reference
α-Solanine	Potato	*Gastrointestinal effects:* diarrhea, vomiting, abdominal pain *Neurological effects (at higher dose):* drowsiness, apathy, confusion, vision disturbances, death	Cholinesterase inhibition, disruption of cell membrane	Kuiper-Goodman and Nawrot (1992)
Glycyrrhizic acid	Licorice	Hypokalemia, sodium retention, cardiac arrhythmia, hypertension	Suppression of the rennin–angiotensin–aldosterone system through inhibition of 11-beta-hydrosteroid dehydrogenase in liver and kidney	Van Gelderen et al. (2000)
Linamarin	Cassava	Mediated by hydrogen cyanide effects: *Acute high dose:* nausea, vomiting, giddiness, headache, hyperpnea, dyspnea, convulsion, death *Moderate dose:* neurological effects (konzo)	Cyanide binding to cytochrome oxidase resulting in reduced oxygen utilization and anoxia	Speijers (1992)
Genistein	Soybean	Various hormonal effects which may be interpreted either as adverse or beneficial	Interaction with estrogen receptor beta, various interferences with thyroid hormone system	BfR (2007)
8-methoxypsoralen	Celery	In combination with sunlight or UVA light, acute phototoxicity and skin burns. medium-term exposure may increase skin cancer	Intercalation between base pairs of DNA to form a non-covalent DNA complex. With UVA, formation of photoadducts from this complex. Modification of protein. Lipid peroxidation. Lysosome damage	SKLM (2006)
α-Thujone	Wormwood oil, absinthe	Seizure, coma	Modulation of GABA type A receptor	SCF (2003)

Toxic Properties

Any chemical has the potential to produce adverse health effects, i.e. toxicity. In this context toxicity is defined as the "inherent property of a chemical agent to cause an adverse biological effect." Toxicity depends on the chemical structure of chemicals and is therefore highly substance specific. Because of the broad structural diversity of inherent plant toxicants, a broad range of toxic effects and modes of action is expected and actually observed. This is illustrated by the examples provided in Table 3.1. Knowledge is generally more limited for inherent non-nutrient plant chemicals than for food additives or pesticides. Because many inherent plant toxicants have been identified as a result of outbreaks or individual cases of intoxication, often limited human data are available. They mostly deal with acute or subacute/subchronic exposures. Classical examples below provide some insight into the toxicological properties of inherent plant toxicants.

The common potato (*Solanum tuberosum*) produces several glycoalkaloids (mainly α-solanine and α-chaconine) from cholesterol. They are located mostly in the peel and their levels depend on a number of factors such as cultivar, storage conditions and sprouting. There have been many case reports of human poisoning resulting from consumption of potatoes rich in glycoalkaloids. The acute symptoms of low-grade glycoalkaloid poisoning are acute gastrointestinal upset with diarrhea, vomiting and abdominal pain. In severe cases, neurological symptoms, including drowsiness and apathy, confusion, weakness and vision disturbances, followed by unconsciousness and in some cases death, have been reported (Kuiper-Goodman and Nawrot, 1992). Years ago, the hypothesis of a link between potato glycoalkaloid exposure and the incidence of neural tube defects was raised. Although some animal data may provide some support, this hypothesis has not been substantiated by epidemiological data. Experimental data in animal models indicate that potato glycoalkaloids are not genotoxic *in vivo* but are embryotoxic and teratogenic, producing central nervous system abnormalities (Kuiper-Goodman and Nawrot, 1992). Although significant data are available on the toxicity of potato glycoalkaloids no official safe level of exposure (such as an acceptable daily intake) could be established. For example, the WHO/FAO Joint Expert Committee on Food Additives and Contaminants (JECFA) concluded that the data in animals and humans did not permit the establishment of a safe level of exposure. However, it was recognized that based on a large experience of potato consumption, normal glycoalkaloid levels (20–100 mg/kg potato) were not of concern (Kuiper-Goodman and Nawrot, 1992). Others have considered levels up to 200 mg/kg as tolerable for human consumption (Essers et al., 1998).

Cassava is a tropical shrub, the root of which is an important staple food in Africa and Asia. Cassava roots produce two cyanogenic glucosides, linamarin and lotaustralin (Speijers, 1992; Speijers et al., 2010). The acute toxicity of cyanogenic glucosides *per se* is low. They are biotransformed into thiocyanate during detoxification in the mammalian body. Significant toxicological concern occurs when the glucosides come into contact with linamarase during food processing. Linamarase is an inherent glucosidase localized in various tissue compartments of the same plant. Disruption of the cell structure during food processing with release and mixing of the intracellular chemicals results in linamarase hydrolyzing the glucosides to cyanohydrins, which degrade into hydrocyanic acid (HCN). Several adverse health effects have been attributed to HCN exposure from inappropriately processed cassava. Acute, fatal poisonings have been documented. Chronic dietary consumption of insufficiently processed cassava has been strongly associated with a central nervous system syndrome named Konzo (Nyirenda et al., 2011; Speijers et al., 2010; Speijers, 1992). JECFA has recently established health-based guidance values for cyanogenic glucosides (expressed as cyanide equivalents): an Acute Reference Dose (ARfD) of 0.09 mg/kg bw/d for episodic short-term consumption and a Provisional Maximal Tolerable Daily Intake (PMTDI) of 0.02 mg/kg bw/d for regular chronic consumption. According to JECFA, a content at the level of the Codex Maximum Limit (ML) of 10 mg/kg HCN in cassava flour does not lead to exposures above ARfD or PMTDI (JECFA, 2011).

Pyrrolizidine alkaloids (PAs) form a group of >350 individual heterocyclic compounds of broad structural diversity (Edgar et al., 2011). They share as the basic structure the four necine bases, retronecine (most frequent), platynecine, heliotridine and otonecine. The toxic potency of individual compounds depends on their chemical structure. PAs are secondary plant metabolites commonly found at variable levels (up to 180 g/kg) in many (about 6000) flowering plant species worldwide. They often occur in mixtures. After ingestion PAs are rapidly absorbed from the gastrointestinal tract and can be activated to toxic metabolites in the liver through the action of various cytochrome P450 enzymes. In humans, the characteristic lesion of PA toxicity is veno-occlusive disease (VOD) in the liver resulting from endothelial cell damage, fibrin deposition and hemorrhage in the centrolobular area, leading to fibrotic occlusion of the central and sublobular veins with progression to cirrhosis (Edgar et al., 2011; COT, 2008). Typically, liver lesions continue to progress well after the elimination of PAs from single or repeat exposure (delayed toxicity). In animals, acute and chronic hepatotoxicity is also observed. In rodent bioassays, individual PAs induced various types of tumors, mostly of the liver. There is no evidence for PA-related cancer formation in humans. The main sources of human PA exposure and the only source of reported poisoning are staple food crops (e.g. grains) contaminated with seeds or dust from PA-producing plants and the intentional ingestion of PA-containing herbs, teas and dietary supplements (Edgar et al., 2011; COT, 2008). Another documented source of human exposure is honey, especially if derived predominantly from a single flower species (unifloral). The actual health significance of PA exposure from honey is not clear but acute effects are unlikely (Edgar et al., 2011; COT, 2008). Several approaches have been applied to establish health-based guidance values for PAs. Using limited human data, the Food Standards Australia New Zealand (FSANZ) established a Provisional Tolerable Daily Intake (PTDI) of 1 μg/kg bw/d heliotrine (FSANZ, 2001). The UK Committee on Toxicity (COT) concluded that the available reports on human VOD did not provide sufficiently reliable data on exposures and established a Tolerable Daily Intake (TDI) of 0.1 μg/kg bw/d based on 2-year rat and mice studies with riddelline (COT, 2008). Such an exposure is considered unlikely to cause any non-neoplastic toxic effects. It was considered that the ratio of LD50 values can be used to convert other PAs to riddelline equivalents for comparison with this dose. However, it was also acknowledged that PAs should be considered as genotoxic carcinogens. Based on animal data on lasiocarpine, an exposure <0.007 μg/kg bw/d was considered unlikely to increase cancer risk. Assuming equal genotoxic and carcinogenic potency for all PAs, this value could be used to establish a level of carcinogenic concern for any other, less characterized PAs (COT, 2008).

Bracken fern (*Pteridium aquilinum*) grows worldwide preferentially on recently deforested areas, poorly managed pastures and abandoned farmland. Cattle eat bracken fern only in absence of alternatives (draught, overgrazed or heavily infested pastures). Bracken fern is carcinogenic in several animal species (Alonso-Amelot et al., 1993, 2002; IARC, 1987). In cattle it causes bovine enzootic hematuria, an ultimately lethal disease with hemorrhage from the urinary tract and multiple bladder tumors. The major toxic and carcinogenic principle in bracken fern is believed to be ptaquiloside (PT). Since most experiments were carried out with whole bracken fern, toxicological data on isolated PT are very limited. PT is excreted into cow's milk at 1.2–8.6% of the ingested dose. Contaminated milk has been shown to be carcinogenic in rodents (Alonso-Amelot et al., 1993, 2002; IARC, 1987). Humans can be exposed to PT either directly from consumption of bracken fern (especially Japan and Brazil) or indirectly via milk. The latter route of exposure is probably restricted to populations consuming the milk of local cows grazing on pastures heavily infested with bracken fern (especially northern Wales and Central America). There is some epidemiological evidence suggesting a correlation between bracken fern intake and esophageal carcinoma in Japan and between prevalence of bovine enzootic hematuria or bracken fern infestation of pastures and gastric and esophageal carcinoma in North Wales, Costa Rica and Venezuela (Alonso-Amelot et al., 2002). In its most recent evaluation the International Agency for Research on Cancer (IARC) has classified bracken fern as *possibly carcinogenic to humans* (group 2B) (IARC, 1987).

Proteins may also fit the definition of inherent plant toxicants. For example, lectins are proteins that bind specifically to carbohydrates. They are present in many food raw materials, at particularly high levels in legumes. Phytohemaglutinin (PHA) is found in significant amounts in red kidney beans. It is a lectin known for its ability to agglutinate mammalian red blood cells. PHA has been assumed to be the agent responsible for the adverse effects associated with the consumption of undercooked food products based on red kidney beans (Noah et al., 1980). The time from consumption to onset of symptoms is short (1–3 hours). Symptoms include extreme nausea followed by vomiting, which can be severe. Diarrhea develops somewhat later. Usually, there is rapid spontaneous recovery (3–4 hours after onset of symptoms). Outbreaks of intoxication are well documented. The severity of reported symptoms is directly related to the levels of PHA in the incriminated foods. PHA levels are measured using the red blood cell agglutination test. The unit is the hemagglutinating units (hau). Raw red kidney beans contain up to 70,000 hau/g dry weight, while properly processed beans contain 200–400 hau/g dry weight (FDA, 2011; Noah et al., 1980).

Modulation of Toxic Effects

Exposure to pure, isolated inherent plant toxicants is very unlikely. Instead, exposure is through the consumption of food where they are embedded in a complex matrix together with other natural chemicals, which themselves can potentially produce adverse or beneficial health effects. Theoretically this leaves multiple possibilities of interactions between different molecules that may result in increased or decreased net toxicity of the food plant as a whole. Several different cases have been described.

The toxicological concerns associated with cyanogenic compounds in cassava illustrate the matrix effect. As discussed above, cyanogens are present in cassava as glucosides. Because of their relatively high stability in the human body, the toxicity of these parent compounds is not considered of high acute concern. However, their contact during food processing with glucosidase co-occurring in the same plant induces the formation of free cyanogens, which degrade into the highly neurotoxic HCN. Interestingly, the physical form in which the product is consumed may also be of high relevance for safety. When cassava flour is prepared into a stiff paste, it forms an elastic ball in the stomach, which reduces in size very slowly. This causes a slow release of the toxicant, which can then be detoxified more effectively (Schultz, 1984). Thus, the food matrix may play an important role in the release of the ultimate toxicant.

Other possibilities for interactions relate to mixture effects. These occur when one constituent affects the toxicokinetics (absorption, distribution, metabolism and elimination) or toxicodynamics (action at target cells) of other constituents present in the same plant. Interactions can lead to increased or decreased biological activities (adverse or beneficial) upon combined exposure. The most important toxicokinetic interaction is thought to occur through effects on xenobiotic metabolism (Schilter et al., 2003), either at the level of enzymatic activity (e.g. inhibition) or of modulation of enzyme gene expression (induction, repression). Examples include furocoumarins in grapefruit, which inhibit cytochrome P450 CYP3A4 (He et al., 1998), or suforaphane in broccoli (Zhang et al., 1994), which is documented to activate Nrf2, a transcription factor responsible for the regulation of enzymes involved in detoxification and cell protection. Toxicodynamic interactions are most important for chemicals that share the same target organ and/or mode of action (Schilter et al., 2003). This may be of particular importance for inherent plant toxicants since they often consist of a family of similar substances that may share common sites or modes of action. Examples include pyrrolizidine alkaloids and potato glycoalkaloids discussed in the previous sections.

RISK ASSESSMENT CONSIDERATIONS

The general paradigm of risk assessment, namely (1) hazard identification, (2) hazard characterization, (3) exposure assessment, and (4) risk characterization, has been extensively and successfully applied to food additives and other man-made chemicals in food such as pesticide residues. For these chemicals, extensive toxicological databases are generated in animal models, and health-based guidance values such as acceptable daily intakes (ADI) are established through the application of uncertainty factors to the most conservative no observed adverse effect levels (NOAEL) obtained in the most relevant toxicological feeding studies. Safety is then managed by ensuring that the total human exposure from all dietary sources does not exceed the specific ADI. Although considered as a good basis for safety evaluation, the direct application of the general paradigm of risk assessment to inherent toxicants is not straightforward. For most of them, the available toxicological data are insufficient to establish an ADI. The data often refer to acute effects and very little is generally available on potential chronic toxicity and carcinogenicity. In addition, the limited number of appropriate toxicological studies have often been performed with isolated chemicals and do not provide any relevant information on potential matrix and mixture effects and hence on the safety of the whole food as consumed. Information on outbreaks of human poisoning usually does not include reliable data on intake of the toxicant and clinical course for individual subjects. This precludes the estimation of safe or minimal toxic intake levels even for short-term exposure.

An important question is whether similar or different toxicological considerations should be applied to inherent plant toxicants and synthetic chemicals. Several authors have argued that many inherent plant toxicants actually should raise more safety concern than man-made chemicals because of their toxic potency together with likely high exposure levels

(Mattsson, 2007; Essers et al., 1998; Ames and Gold, 1997). For many inherent plant toxicants, the margin of exposure between the actual exposure and the level documented to produce adverse effects in humans is very low. However, their elimination is often difficult because of their probable physiological function in the plant. Therefore, the application of standard uncertainty factors as usually applied in risk assessment would severely limit the intake of the plant itself and could theoretically lead to prohibition. The actual public health impact of limiting plant consumption must take into account other aspects such as nutritional value and food security. Furthermore, as suggested above, the safety of a food cannot be assessed just based on the toxicology of its constituents tested in isolation. Coffee is a good illustrating example (Ames and Gold, 1997). More than 1000 chemicals have been described in coffee, 28 have been tested and 19 are rodent carcinogens. However, animal carcinogenicity studies with whole coffee (Schilter et al., 2001) and the extensive epidemiological data (Arab, 2010; Nkondjock, 2009; George et al., 2008; Schilter et al., 2001) do not indicate any carcinogenicity in humans associated with coffee consumption. On the contrary, the most striking data available in both animal models and in human epidemiological studies are actually compatible with protective, cancer-preventive effects (Arab, 2010; Nkondjock, 2009; George et al., 2008; Schilter et al., 2001).

It appears obvious that although the standard risk assessment provides a good basis to evaluate inherent plant toxicants and to set up management priorities, improvements are strongly warranted. This will be possible through a better application of the principle of history of (safe) use (Schilter et al., 2003; Essers et al., 1998). In addition, optimized paradigms for toxicity testing, applying biomarkers of effects together with toxicokinetic modeling in both animal models and humans will in the future allow the replacement of large default uncertainty factors by science-based specific factors (Schilter et al., 2003; Essers et al., 1998). A method for integrating risks and benefits based on the application of the Disability Adjusted Life Year (DALY) concept has recently been developed for food (Hoekstra et al., 2010) and will certainly provide additional information relevant for public health management. However, such an approach requires comprehensive and high-quality data. It will not be widely applicable without generating additional experimental information.

RISK MANAGEMENT OF INHERENT PLANT TOXICANTS

General Context of Risk Management

Risk management is defined as "the process of weighing policy alternatives in light of the result of a risk assessment and other relevant evaluations (feasibility, cost–benefit) and, if required, selecting and implementing appropriate control options."

Because of the limitations of risk assessment, risk management is not straightforward. Indeed risk assessment in general suffers from significant weaknesses and uncertainties such as often insufficient information on exposure, difficulties to extrapolate from high-dose laboratory animal studies to actual low-level human exposure, and limited data on both origin and extent of variability in sensitivity to toxicity within the human population. As mentioned above, most of these uncertainties are dealt with through the application of conservative default assumptions.

Risk management is further complicated by the fact that a zero risk level cannot be attained or even expected. In consequence, risk management implies the difficult task of defining levels of risks that are acceptable for the society. Accepting a risk is a matter of perception. Experts and the general public often differ in their perception of a particular risk. Experts base their judgment on a risk assessment derived from scientific findings, the public on personal experience, beliefs, culture and values such as equity and fairness. Understanding the risk perception of consumers and addressing their concerns are essential for effective food safety management and communication with consumers.

In this context, perception of inherent plant toxicants differs from that of synthetic chemicals. Although margins of exposure are likely to be low for many inherent plant toxicants, relatively little effort is actually undertaken to generate toxicological and exposure data in order to improve risk assessment and management. This may be explained by the fact that natural chemicals are often perceived as being of less health concern than synthetic ones. It is widely thought that natural toxicants and humans have coexisted since the beginning of time and therefore humans have developed resistance to them. Examples highlighted above provide arguments that such ideas about inherent toxicants are misconceptions. It is also sometimes advocated that feasible options to reduce these chemicals are limited because they play important physiological and/or defense roles in the plants. In principle, from a safety perspective, chemical hazards to be managed are those which could constitute a health risk. Several inherent plant toxicants have been associated with toxic effects in humans and therefore must be managed appropriately. Some examples of management options are provided below.

Selective Breeding and New Cultivar Development

It is possible to obtain plant cultivars low in inherent toxicants by selective breeding. This has been applied to potatoes to reduce glycoalkaloid content. Importantly, this may not be compatible with other desired attributes of potatoes such as resistance to pathogens. Indeed, since they play an important role in plant protection, glycoalkaloids are usually present in high levels in resistant cultivars (Speijers et al., 2010; Kuiper-Goodman and Nawrot, 1992).

A similar approach can be applied to obtain cassava cultivars low in cyanogenic glucosides. However, because linamarin and linamarase may greatly vary not only between cultivars but also between plants of the same cultivar and between roots of the same plants, development of cultivars consistently low in cyanogenic glucosides is challenging. Levels of cyanogenic glucosides in cassava roots are known to depend on several factors including linamarin biosynthesis, catabolism and transport from the leaves to the roots (Nambisan, 2011; Speijers et al., 2010). Modulation of these factors through genetic engineering has been considered a promising avenue for reducing cyanogenic glucosides in cassava in the long term (Nambisan, 2011).

Agricultural, Storage and Handling Practices

Since many inherent plant toxicants are part of the plant's self-defense system and can be induced in response to various crop damages and stresses, agricultural, handling and storage practices are expected to have an impact on their occurrence in the edible parts. This is illustrated by glycoalkaloids in potatoes: pre-planting fungicide treatment of soils together with insecticide and fungicide applications during the growth period are common practices to prevent diseases known to increase glycoalkaloid levels in potatoes and are anticipated to reduce their occurrence in tubers (Speijers et al., 2010). A slight increase in glycoalkaloid was reported in organically versus conventionally grown potatoes (Hajslova et al., 2005) suggesting only a relatively small influence of agricultural practices. However, such results are difficult to interpret since total glycoalkaloid levels vary significantly according to potato variety and geographical location where the field trials were conducted.

Physical injury and exposure to light of potatoes are well known to induce biosynthetic pathways resulting in increased levels of glycoalkaloids (Speijers et al., 2010). This clearly indicates a role for handling and storage practices in order to keep glycoalkaloid levels in tubers under control. This is of particular importance since most potatoes are stored prior to consumption. For long-term post-harvest storage, appropriate conditions involving low exposure to light, humidity and temperature have to be applied to minimize glycoalkaloid levels in tubers.

Processing

There are a number of examples documenting the impact of food processing and preparation on the levels of inherent toxicants in finished foods as eaten. Processing may allow either removing or inactivating toxicants, or both. Optimizing processing in terms of detoxification requires a full understanding of the behavior of the toxicants to be controlled in the relevant processing conditions.

Potato glycoalkaloids are heat stable and therefore most traditional food processes do not significantly impact their levels in finished products as consumed. However, glycoalkaloids are mainly concentrated at the surface of the tuber, and therefore peeling of potatoes is widely known as an efficient way to eliminate glycoalkaloids from food as consumed.

Another classical example of the importance of processing is detoxification of bitter cassava containing high levels of cyanogenic glucosides (Nambisan, 2011; Speijers et al., 2010). Many procedures are applied that result in significant reduction of cyanogen contents. Crushing presoaked roots followed by sun-drying is a traditional way to process cassava before consumption. Most procedures include a step that results in volatilization of the toxic HCN formed by the interaction of cyanogenic glucosides with linamarase.

Lectins such as PHA are usually heat sensitive and can be destroyed by adequate processing. For red kidney beans, it is recommended to soak them in water for at least 5 hours prior to cooking, then to discard the water and boil them in fresh water for at least 10 minutes. Temperature must be high enough. At equivalent original glycogen levels, beans heated to 80°C were reported to be more toxic than non-treated raw beans (FDA, 2011; Noah et al., 1980).

CONCLUSIONS

- Most non-nutritional chemicals that humans ingest with their regular diet are natural constituents of food plants.
- Quantitatively, dietary exposures to some of these inherent plant constituents may significantly exceed exposures to most of the man-made chemicals occurring in food.

- Some inherent plant constituents are highly toxic (inherent plant toxicants) and have to be managed to ensure food safety.
- The principal management solutions currently applied include development of cultivars low in inherent toxicants, specific agricultural and storage practices and appropriate specific food processing.
- Currently, major resources are devoted to the safety assessment of synthetic chemicals. From a public health perspective, focusing efforts on specific inherent plant toxicants may be more important.
- More research on the actual health significance of inherent toxicants in food is needed. This requires the generation of toxicological data on specific chemicals identified as of potential concern. In addition, refining the methods for health assessment of such chemicals is critical.
- A holistic approach is required including not only standard risk assessment but also understanding of the health benefits associated with the consumption of secondary plant metabolites and of the pertinent food plants as such.
- Improvement of the methodology to assess the health impact of food plants containing inherent toxicants will allow optimizing priority setting for research and ensure that the limited resources are devoted to issues of real public health concern.

References

Alonso-Amelot, M.E., Castillo, U., De Jongh, F., 1993. Passage of the bracken fern carcinogen ptaquiloside into bovine milk. Lait 73 (3), 323–332.

Alonso-Amelot, M.E., Avendaño, M., 2002. Human carcinogenesis and bracken fern: a review of the evidence. Cur. Med. Chem. 9 (6), 675–686.

Ames, B.N., Gold, L.S., 1997. Environmental pollution, pesticides, and the prevention of cancer. FASEB J. 11, 1041–1052.

Arab, L., 2010. Epidemiologic evidence on coffee and cancer. Nutr. Cancer 62, 271–283.

Beier, R.C., 1990. Natural pesticides and bioactive components in foods. Rev. Environ. Contam. Toxicol. 113, 48–137.

BfR, 2007. Isolated isoflavones are not without risk. BfR expert opinion No. 039/2007.

COT, Committee on toxicity of chemicals in food, consumer products and the environment, 2008. COT statement on pyrrolizidine alkaloids in food. <http://cot.food.uk/cotstements/cotstementsyears/2008>.

Dolan, L.C., Matulka, R.A., Burdock, G.A., 2010. Naturally occurring food toxins. Toxins 2, 2289–2332.

Edgar, J.A., Colegate, S.M., Boppré, M., Molyneux, R.J., 2011. Pyrrolizidine alkaloids in food: a spectrum of potential health consequences. Food Add. Contam. 28 (3), 308–324.

Essers, A.J.A., Alink, G.M., Speijers, G.J.A., Alexander, J., Bouwmeister, P.-J., van den Brandt, P.A., et al., 1998. Food plant toxicants and safety risk assessment and regulation of inherent toxicants in plant foods. Environ. Toxicol. Pharmacol. 5, 155–172.

FDA, Food and Drug Administration, 2011. Phytohaemagglutinin. In: Bad Bug Book. US Department of Health & Human Services. <http://www.fda.gov.food/foodsafety/foodbornillness/foodborneillnesspathogen/phytoaemagglutinin>.

FSANZ, Food Standards Australia New Zealand, 2001. Pyrrolizidine alkaloids in food. A toxicological review and risk assessment. Technical Report Series No. 2.

George, S.E., Ramalakshmi, K., Mohan Rao, L.J., 2008. A perception on health benefits of coffee. Crit. Rev. Food Sci. Nutr. 48, 464–486.

Gry, J., Black, L., Folmer, D.E., Pilegaard, K., Plumb, J., Rhodes, M., et al., 2007. EuroFIR-BASIS – a combined composition and biological activity database for bioactive compounds in plant-based foods. Trends Fd Sci. Technol. 18, 434–444.

Hajslova, J., Schulzova, V., Slanina, P., Janné, K., Hellenäs, K.E., Andersson, C., 2005. Quality of organically and conventionally grown potatoes: four-year study of micronutrients, metals, secondary metabolites, enzymatic browning and organoleptic properties. Food Add. Contam. 22, 514–534.

He, K., Iyer, K.R., Hayes, R.N., Sinz, M.W., Woolf, T.F., Hollenberg, P.F., 1998. Inactivation of cytochrome P450 by bergamotin, a component of grapefruit juice. Chem. Res. Toxicol. 11, 252–259.

Hoekstra, J., Hart, A., Boobis, A., Claupein, E., Cockburn, A., Hunt, A., et al., 2010. BRAFO tiered approach for benefit–risk assessment of foods. Food Chem. Toxocol. doi: 10.1016/j.fct.2010.05.049.

IARC, International Agency for Cancer Research, 1987. Bracken fern. Summaries & Evaluations, Supp. 7. <http://www.inchem.org/documents/iarc/suppl7/brackenfern.html>.

JECFA, Joint Expert Committee on Food Additives and Contaminants, 2011. Cynogenic glycosides. In: Summary Report of the 74th Meeting of the Joint FAO/WHO Expert Committee on Food Additives (14–23 June 2011). <ftp://ftp.fao.org/ag/agn/jecfa/JECFA_74_Summary_Report_4July2011.pdf>.

Kuiper-Goodman, T., Nawrot, P.S., 1992. Solanine and chaconine. In: WHO Food Additives Series 30. World Health Organization, Geneva. <http://www.inchem.org/documents/jecfa/jecfamono/v30je19.htm>.

Lattanzio, V., Lattanzio, V.M.T., Cardinali, A., 2006. Role of phenolics in the resistance mechanisms of plants against fungal pathogens and insects. In: Imperato, F. (Ed.), Phytochemistry: Advances in Research, 37/661 (2), pp. 23–67. Research Signpost, Fort P.O., Trivandrum-695 023, Kerala, India.

Mattsson, J.L., 2007. Mixtures in the real world: the importance of plant self-defense toxicants, mycotoxins, and the human diet. Toxicol. Appl. Pharmacol. 223, 125–132.

Nambisan, B., 2011. Strategies for elimination of cyanogens from cassava for reducing toxicity and improving food safety. Food Chem. Toxicol. 48, 690–693.

Nkondjock, A., 2009. Coffee consumption and the risk of cancer: an overview. Cancer Lett. 277, 121–125.

Noah, N.D., Bender, A.E., Reaidi, G.B., Gilbert, R.J., 1980. Food poisoning from raw kidney beans. Brit. Med. J. 19 July, 236–237.

Nyirenda, D.B., Chiwona-Karltun, L., Chitundu, M., Haggblade, S., Brimer, L., 2011. Chemical safety of cassava products in regions adopting cassava production and processing – experience from southern Africa. Food Chem. Toxicol. 49, 607–612.

Pariera-Dinkins, C.L., Peterson, R.K.D., Gibson, J.E., Hu, Q., Weaver, D.K., 2008. Glycoalkaloid responses of potato to Colorado potato beetle defoliation. Food Chem. Toxicol. 46, 2832–2836.

Paustenbach, D., Galbraith, D., 2006. Biomonitoring: is body burden relevant to public health? Regul. Toxicol. Pharmacol. 44 (3), 249–261.

SCF, 2003. Opinion of the Scientific Committee on Food on thujone. Opinion 6 February 2003. SCF/CS/FLAVOUR/23 ADD2 Final.

Schilter, B., Cavin, C., Tritscher, A., Constable, A., 2001. Coffee: health and safety considerations. In: Clarke, R.J., Vitzhum, O.G. (Eds.), Coffee: Recent Developments Blackwell Sci., London, pp. 165–183.

Schilter, B., Andersson, C., Anton, R., Constable, A., Kleiner, J., O'Brien, J., et al., 2003. Guidance for the safety assessment of botanicals and botanical preparations for use in food and food supplements. Food Chem. Toxicol. 41, 1625–1649.

Schultz, V., 1984. Clinical pharmacokinetics of nitroprusside, cyanide, thiosulfate and thiocyanate. Clin. Pharmacokinet. 9, 239–251.

SKLM, 2006. Toxicological assessment of furocoumarins in foodstuffs. Opinion of 24 September 2004.

Speijers, G., 1992. Cyanogenic glycosides. In: WHO Food Additives Series 30. World Health Organization, Geneva. <http://www.inchem.org/documents/jecfa/jecfamono/v30je18.htm>.

Speijers, G., Alink, G., de Saeger, S., Hardy, A., Magan, N., Pilegaard, K., et al., 2010. Evaluation of agronomic practices for mitigation of natural toxins. ILSI Europe Report Series. ILSI Europe: Brussels.

Van Gelderen, C.E., Bijlsma, J.A., van Dokkum, W., Savelkoul, T.J., 2000. Glycyrrhizic acid: the assessment of a no effect level. Hum. Exp. Toxicol. 19 (8), 434–439.

Walter, P., 2003. Why are fruits and vegetables so healthy? Int. J. Vit. Nutr. Res. 72 (3), 52–53.

Zhang, Y., Kensler, T.W., Cho, C.G., Posner, G.H., Talalay, P., 1994. Anticarcinogenic activities of sulphorafane and structurally related synthetic norbornyl isothicyanates. Proc. Natl. Acad. Sci. USA 91 (8), 3147–3150.

CHAPTER

4

Allergens

René Crevel and Stella Cochrane

Unilever, Bedford, UK

OUTLINE

Food Safety Management.
DOI: http://dx.doi.org/10.1016/B978-0-12-381504-0.00004-4

INTRODUCTION

Food allergy has been long recognized as a clinical phenomenon, with numerous reports in the 20th century medical literature (Prausnitz and Küstner, 1921; Loveless, 1950). However, while it was known that patients could suffer extremely severe and sometimes fatal reactions following ingestion of minute amounts of the offending food, food allergy was perceived as a problem for the individual sufferers alone and their clinicians. In the last decade of the 20th century this perception changed and food allergy is now recognized as an important public health problem. A major factor in this increased concern is probably the rise in the prevalence of atopic disease (Lewis et al., 1996) of which IgE-mediated food allergy can be considered a manifestation. The prevalence and incidence of food allergy and the number of severe reactions (Venter and Arshad, 2011) appears be increasing, although the lack of sound baseline epidemiological data precludes firm conclusions. The new perception of food allergy has been accompanied by the recognition that the solution to the problem lies with collaboration between all the stakeholders, including those with a food allergy and those who look after them, clinicians, public authorities and the food industry.

Many factors influence the development of allergy to common foods. However, these are outside the scope of this chapter, which is concerned with the elicitation of reactions in people who already have an allergy. In this context, the ultimate aim for all stakeholders is to prevent people with food allergy reacting to the allergens to which they are sensitized. This can be achieved in two ways. One is to ensure accurate allergen declaration through labeling, so that sufferers can avoid the relevant foods. The other is to ensure that where a specific allergen is present inadvertently, for instance through cross-contact, the product does not contain it in an amount that would pose a risk and food allergy sufferers can assume it is safe for them. Both these requirements can only be fulfilled by detailed knowledge of the composition of products. The process of food manufacture is extremely complex. This complexity derives from several factors including material sourcing, processing, efficient use of equipment and other resources, and product formulation. Managing allergen risks requires an integrated approach, which takes into account all these factors throughout the supply chain, from ingredient suppliers through to retailers, and ultimately the consumer (see Chapters 1 and 39).

Total avoidance of cross-contact and therefore absence of specific allergens from products where they are not part of the formulation is often not practicable. Managing allergen risks therefore requires an analysis of the risk arising from residual allergen, and subsequently a thorough and, wherever possible, quantitative assessment of risks. Although knowledge of minimum provoking doses for many allergens is inadequate, knowing how much allergen is present in a product is a key element in this assessment, and the subsequent management of the allergen risk. This chapter will focus on how to translate the requirement for safe products into practical allergen management.

FOOD ALLERGY: A PUBLIC HEALTH PROBLEM

Food Allergy and Food Intolerance

Food allergy forms part of a wide spectrum of adverse reactions to foods, which also includes microbial and chemical toxicity, pharmacological effects and those due to errors

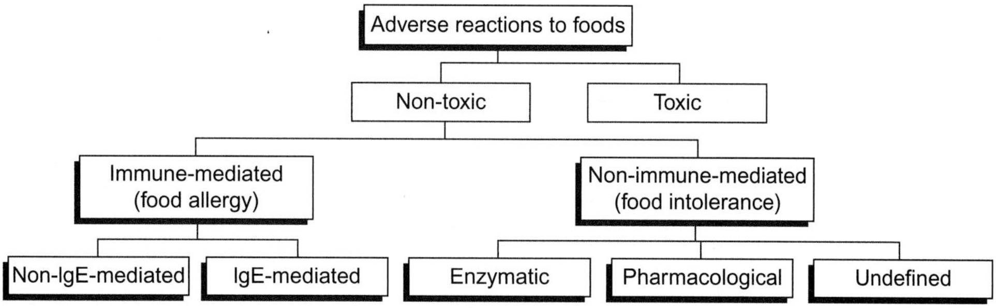

FIGURE 4.1 Classification of adverse reactions to foods by the European Academy of Allergy and Clinical Immunology.

of metabolism, as well as idiosyncratic reactions (European Academy of Allergy & Clinical Immunology classification) (Figure 4.1) (Bruijnzeel-Coomen et al., 1995). Reactions which are attributable neither to toxic mechanisms nor allergy are often referred to as food intolerance, but are also frequently confused, possibly because the symptoms can often be the same. These intolerances cover a range of mechanisms, such as lactase deficiency in lactose intolerance or inborn errors of metabolism such as in galactosemia.

Food allergy refers to a condition where an individual has generated an immune response to a food, and a subsequent encounter with the same food provokes an adverse (allergic) reaction. Foods can produce many different types of immune and allergic responses, but the public health concern lies largely with those in which formation of IgE antibodies to proteins in the food occurs, which are then implicated in immediate-type reactions on subsequent exposure. Allergic reactions mediated by IgE can vary from very slight, indeed barely perceptible to severe and occasionally fatal, depending on the dose, the individual and other factors. Data on the number of allergic reactions to food, and more importantly their severity, are scarce. Sampson (2005) cites a figure of 200 deaths and 30,000 emergency room (ER) visits for the USA, while a recent survey of ER in a representative sample of US hospitals estimated 125,000 reactions per annum, of which approximately 14,000 were due to anaphylaxis, the most severe and potentially lethal manifestation of food allergy (Ross et al., 2008). Allergenic foods most frequently responsible for severe and fatal reactions include peanuts, tree nuts, cows' milk and hens' eggs in all regions where such data are collected (Worm et al., 2010).

Food allergy affects a higher proportion of children than adults (Sampson, 2005) and reactivity to some allergenic foods, such as milk and egg, tends to be largely outgrown, while allergy to others, such as peanuts, generally persists (Venter and Arshad, 2011). Little is known about why allergy to certain foods develops, although exposure and its pattern, the characteristics of the implicated proteins, but also individual characteristics, such as atopy, all play a role. The range of minimum doses required to elicit a reaction in allergic people spans at least six orders of magnitude. Until recently, the distribution of these doses remained uncharacterized, making risk assessments arduous and fraught with uncertainty (Taylor et al., 2002; EFSA, 2004; Threshold Working Group, 2008; Crevel et al., 2008). However, recent work analyzing results from double-blind placebo-controlled food

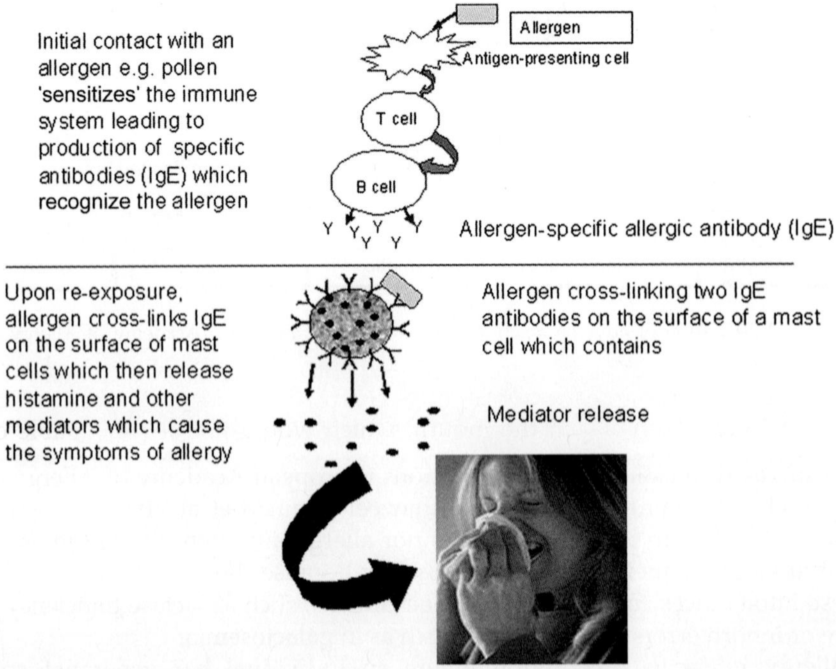

FIGURE 4.2 Mechanisms of food allergy.

challenges (DBPCFC) conducted under well-defined conditions demonstrates that sufficient data are available to characterize the response to some allergenic foods, such as peanut (Taylor et al., 2009, 2010).

Mechanisms of IgE-mediated Food Allergy

All allergic responses are characterized by two phases: sensitization and elicitation. During the sensitization phase, the immune system recognizes a component of the allergenic food (almost invariably a protein) as foreign, resulting in a series of events culminating in the production of circulating IgE antibodies and their distribution around the body. These IgE antibodies do not all remain in the blood and lymph, but attach themselves via a specialized receptor (denoted FcεR1) to specific types of cell, in particular mast cells. During the elicitation phase, allergenic protein cross-links IgE antibodies bound to mast cells, resulting in the release of chemical mediators which then cause the symptoms of an allergic reaction (Figure 4.2).

Celiac disease is grouped with food allergy for purposes of allergen management although clinically it is classed as an auto-immune disease rather than an allergy. People with celiac disease are unable to tolerate in their diet the proteins known as gluten found in wheat and related cereals. In susceptible individuals, exposure to gluten results in the formation of auto-antibodies against certain endogenous proteins, and the ensuing reaction ultimately leads to atrophy of the lining of the small intestine (villus atrophy) which greatly reduces its ability to absorb nutrients.

Food intolerances are managed in the same way as allergens, with labeling of the ingredient as the main measure. Some ingredients responsible for intolerances, such as lactose, are indeed included on some regulatory allergen lists.

Symptoms of Food Allergy

The symptoms of an IgE-mediated reaction reflect directly the inflammatory response to the chemicals released from cells such as mast cells. They can affect one or more organ systems, including the skin, the gastrointestinal tract, the respiratory and cardiovascular system, with skin reactions being among the most frequently implicated. Symptoms range from pruritis or tingling in the mouth, which would not be perceptible other than to the allergic person, through eczema and rashes, angioedema, shortness of breath to the drop in blood pressure and cardiovascular collapse characteristic of anaphylactic shock. Gastrointestinal symptoms include stomach cramps, nausea, vomiting and diarrhea. Severe allergic reactions to foods can be fatal, but information about the doses implicated in such reactions is very limited and problematic to interpret because of the circumstances under which it is generated. Recently Wainstein et al. (2010) showed that a dose of peanut as small as 20 mg (5 mg peanut protein) could result in anaphylaxis, but this occurred in the clinic under controlled conditions.

While celiac disease may mimic some of the symptoms of food allergy, the underlying mechanisms are very different, as is the timing of reactions after consumption of gluten. Symptoms thus include diarrhea, bloating, abdominal pain, weight loss, failure to grow at the expected rate and malnutrition. In adults, fatigue is common. The speed with which symptoms occur after ingestion depends to some extent on the dose, but reactions are never of the rapid and catastrophic type like anaphylaxis that are associated with IgE-mediated allergies.

Prevalence of Food Allergy

One reason why food allergens need to be managed is that they constitute a threat to public health. One aspect of this threat is the potential severity of reactions and consequences for the quality of life of sufferers, but another is its prevalence in populations (Figure 4.3). Until recently, estimates of the prevalence of food allergy as a whole and allergy to individual foods were scarce and provided an inadequate basis for risk assessment and management (Rona et al., 2007). One particular problem was the considerable overestimate arising from self-reporting compared to formal diagnosis by food challenge. However, recent studies in several regions and continents, including Europe, the United States and Australia, have provided high-quality data. Thus a cross-sectional study in over 40,000 children (up to 18 years) by Gupta et al. (2011) in the USA indicated an overall prevalence of 8%, of which about 40% reported having experienced a severe reaction. Peanut, milk and shellfish were the foods implicated most frequently. Osborne et al. (2011) demonstrated that over 10% of infants up to 1 year old in Australia suffered from a challenge-verified food

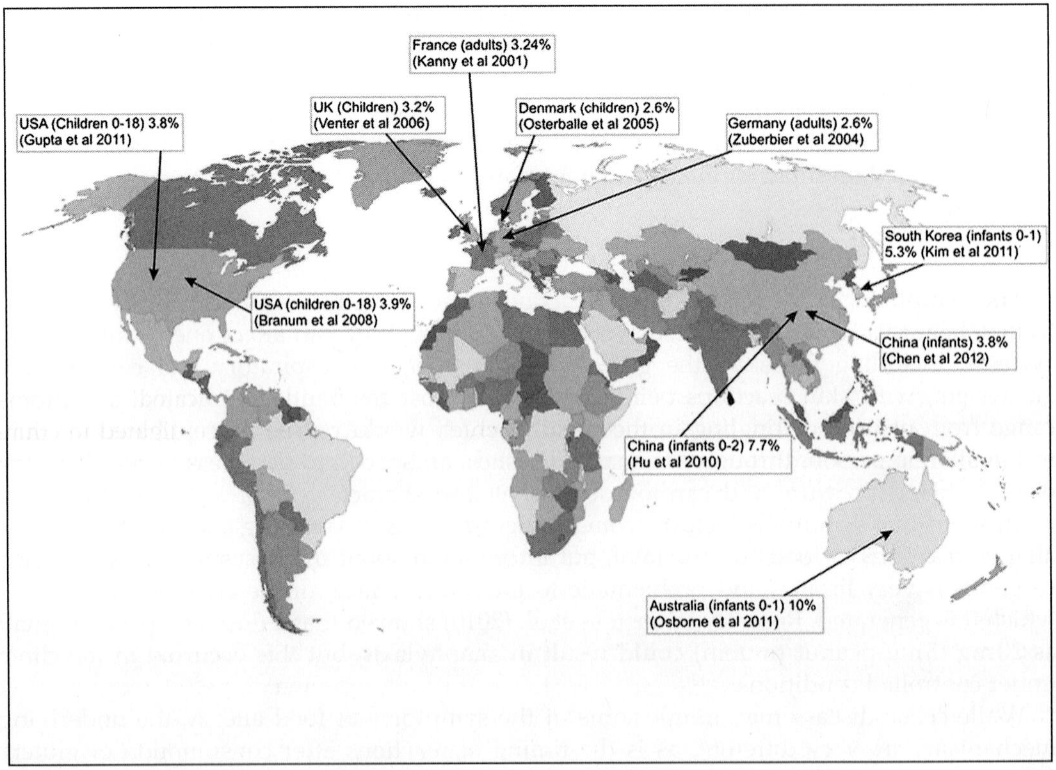

FIGURE 4.3 Prevalence of food allergy across the world.

allergy. Countries with emerging economies also show similar trends. Kim et al. (2011) estimated that 5.3% of a birth cohort of Korean infants suffered from a food allergy, while Hu et al. (2010) observed a rise from 3.5 to 7.7% in challenge-verified food allergy from 1999 to 2009 in cross-sectional studies of infants up to 2 years old in Chongqing (China). Thus, as the social and environmental changes seen in Europe and the USA spread to other parts of the world, they will likely start to experience similar increases in the prevalence of food allergies, as seen for instance in Hong Kong and Singapore. Despite the recent promising news about specific immunotherapy for food allergens, avoidance remains the primary means whereby allergic consumers protect themselves. This requires that they know that the allergen is present (labeling) or its presence must be reduced to the point where it poses a negligible risk, hence the importance of defining minimum eliciting doses and their distribution in populations.

Celiac disease was long thought to be rather rare, but recent studies indicate that it may affect over 1% of the population (Bingley et al., 2004; Lamireau and Clouzeau, 2011), but much of it is undiagnosed.

ALLERGENIC FOODS OF PUBLIC HEALTH IMPORTANCE

Evolution of Regulatory Allergen Lists across the World

Over 160 foods have been reported to provoke allergic reactions (Hefle et al., 1996), but far fewer are considered to be of sufficient public health importance that they must be specifically managed.

Prioritization of allergenic foods as a function of their public health importance began in earnest with the FAO-WHO Expert Consultation on Food Allergies of 1995 (FAO-WHO, 1995), which identified eight major foods or food groups associated with the vast majority of allergic reactions (over 90%). This list was adopted in 1999 into the Codex General Standard on Labelling. Although the participants had rather scarce data upon which to base their conclusions, this Codex list still remains the foundation of most national and supranational regulatory allergen lists. Since then, allergen lists have been promulgated in countries covering over half the world population (Table 4.1). Identification of new allergenic foods of public health importance continues, notably through systematic epidemiological studies in projects such as Europrevall, so the lists are likely to get longer with time.

Mandatory declaration of allergenic ingredients required by labeling legislation is, however, but the first and perhaps most visible consequence of priority allergen lists. More significant from an industry point of view is the implication that these priority allergens need to be actively managed to ensure that people suffering from allergies to them are not placed at risk. Inclusion of foods on such lists and additions to them therefore requires careful consideration of the benefits in terms of public health and needs to be based on sound scientific criteria, as discussed by Bjorksten et al. (2008) and van Bilsen et al. (2011). Such an approach ensures that the prioritization reflected by the lists is not diluted by inclusion of foods which pose only a relatively insignificant risk to public health.

Legal/Regulatory Aspects

Management of allergens starts with compliance with the regulatory requirements of the country where the product is sold, but at minimum the key allergens identified by Codex (Table 4.1). The requirement to manage allergens falls within the general ambit of food safety, which has been defined by recent standards and regulations. For instance, the Codex Alimentarius defines food safety as the concept that food will not cause harm to the consumer when it is prepared and/or eaten according to its intended use. In this definition, it notes that food safety refers specifically to the occurrence of food hazards and does not include adverse effects that may result from nutritional considerations, in other words nutritional imbalances. According to the Codex Alimentarius Commission (2003), "food safety refers to all those hazards, whether chronic or acute, that may make food injurious to the health of the consumer" and "is not negotiable." The European Union's Food Law (Regulation (EC) 178/2002) elaborates the concept further. Under the Regulation, food is deemed unsafe if it is either injurious to human health or unfit for human consumption. In line with the EU White Paper on food safety, it adopts a risk-based approach, recognizing that safety is not an absolute condition. This concept mirrors the criterion of "reasonable

TABLE 4.1 Main Regulatory Allergen Lists across the World

Allergenic Food or Food Group, Including Derived Products	Codex	USA	Canada	Australia/ New Zealand	South Africa	China (PRC from 2012)	Switzerland	EU Annex IIIa	Japan
Cereals containing gluten, i.e. wheat, rye, barley, oats, spelt, kamut or their hybridized strains	✓	✓[a]	✓	✓	✓	✓	✓	✓	✓[a]
Crustaceans	✓	✓	✓	✓	✓	✓	✓	✓	✓[b]
Mollusks			✓	✓	✓	✓		✓	✓[b]
Eggs	✓	✓	✓	✓	✓	✓	✓	✓	✓
Fish	✓	✓	✓	✓	✓	✓	✓	✓	✓[b]
Peanuts	✓	✓	✓	✓	✓	✓	✓	✓	✓
Soybeans	✓	✓	✓	✓	✓	✓	✓	✓	✓[b]
Milk and dairy products	✓	✓	✓	✓	✓	✓	✓	✓	✓
Tree nuts	✓	✓	✓	✓	✓	✓	✓	✓	✓[b]
Sesame seeds			✓	✓	✓		✓	✓	
Mustard			✓					✓	
Celery							✓	✓	
Lupin								✓	
Buckwheat									✓
Beef									✓[b]
Chicken (poultry)									✓[b]
Pork									✓[b]
Mushrooms									✓[b]
Apples									✓[b]
Kiwi fruit									✓[b]
Oranges									✓[b]
Peaches									✓[b]
Yams									✓[b]
Sulfites >10 mg/kg	✓	✓	✓	✓	✓	✓	✓	✓	

[a]*Wheat only*
[b]*Recommended by notice*

certainty of no harm" used in US law (FQPA, 1996). Thus, in determining whether a food is unsafe, the Regulation requires that two key characteristics are taken into account: the normal conditions of use of the food by food operators as well as by the final consumer, and the information provided about the food, for instance by labeling, but also more generally.

The regulation also provides guidance on how to determine whether any food is injurious to health. This includes consideration of the probable immediate and/or short-term and/or long-term effects of that food on the health of a person consuming it, but also the particular health sensitivities of a specific category of consumers where the food is intended for that category of consumers. The guidance agreed by the EU's Standing Committee on the Food Chain and Animal Health indicates that the presence of traces of an allergen (for instance, by cross-contact) does not automatically make a food injurious to health, unless that food had specifically been made for consumers with allergies.

Contemporary approaches to food safety emphasize the need for a comprehensive integrated process (Codex Alimentarius Commission, 2003; Regulation (EC) 178/2002) both in individual food businesses and along the whole food chain. This approach follows logically from the observation that food hazards may arise at any point along that chain and highlights the need for good communication at all levels, as well as traceability.

MANAGEMENT OF FOOD ALLERGENS

Allergens continue to form a major cause of alerts and recalls, despite legislation being now well established. For instance, in 2011, 114 allergen incidents generated 59 alerts by the UK Food Standards Agency out of a total of 105 alerts issued. Major contributors to those alerts were incorrect labeling and cross-contact (Figure 4.4). Labeling issues related either to the mandatory information (e.g. allergen not listed in ingredient list despite being deliberately added) or to incorrect precautionary labeling, highlighting the fact that if such voluntary labeling is used, it must be correct and not misleading. Cross-contact issues illustrate that minimizing the unintended presence of allergenic constituents remains a challenge for the food industry.

That allergens pose a threat to public health and must therefore be managed is now beyond argument. Clearly, it is important that in addressing the risks arising from allergens, new risks are not created, which may affect even more people. Thus a key aspect of allergen management is the need to integrate it into general food safety management. An integrated system is likely to be inherently more efficient, but it is also absolutely required because the measures required to deal with one safety hazard, e.g. microbiological, may conflict with those needed to mitigate another, e.g. allergens. Thus wet cleaning is generally extremely effective in reducing allergen contamination, but can lead to severe microbiological problems in dry mix systems. However, more fundamentally, allergens differ from other contaminants with consequences for their management. Unlike those contaminants, allergens can generally be consumed safely by the vast majority of the population in any reasonable quantity and many are also important sources of nutrients, and may also have important functional attributes.

The Practice of Allergen Management

Allergen management implies actively dealing with allergens when making food products so that allergic consumers can make safe choices. This goes well beyond just avoiding

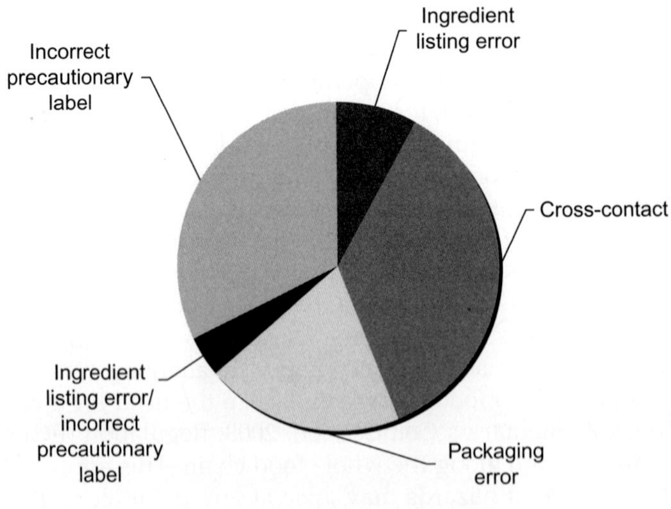

FIGURE 4.4 UK Food Standards Agency Allergen alerts by category in 2011.

the use of allergens or telling the consumer that a product may contain a particular allergen or allergens. Rather, it is about knowing where and what allergens are present throughout the food manufacturing process, deliberately or, perhaps even more importantly, unintentionally. It is also about assessing the residual risk if an unintended allergen cannot be completely removed from a product and communicating clearly and accurately that risk to consumers, neither exaggerating it nor playing it down, or requiring them to assess the risk themselves. Allergen management thus concerns the whole supply chain from the farm to the final consumer and requires accurate and comprehensive information about allergens from all those stages. Implementation of allergen management demands significant resources and therefore requires engagement of senior management within companies, as recognized by the Food Safety Management Standard ISO 22000:2005.

Underlying allergen management and food safety generally are prerequisite programs which describe the basic conditions considered necessary to assure safe food production. These include considerations of premises design, hygiene, etc. and will not be described here as they are covered in other chapters.

Allergen management requires, first, identification of all sources of the allergen risks, then assessment of those risks and subsequently their management. This will be an iterative process, since, having identified a risk and determined that it is significant, the first step will be to look at ways of reducing it. Allergen risks can occur at all stages of the food manufacturing process, which can be summarized as Design, Sourcing, Manufacture and Delivery (Table 4.2).

At the Design stage, key considerations include product composition and ingredient specification. Typically the allergen's contribution to product functionality should be critically assessed and the feasibility of substitution considered. For instance, if an allergen is present as a flavor carrier, does an alternative exist which does not use that allergen or uses

TABLE 4.2 Allergen Management Considerations at Different Stages of Food Production

Design	Sourcing	Manufacture	Delivery
1. Need for allergen **2.** Alternative to allergen **3.** Ingredient specifications Traceability Cross-contact Claims **4.** Manufacturing Implications for existing operations	**1.** Ingredient specifications **2.** Suppliers' allergen management proficiency **3.** Change control Ingredient formulation/ specifications Process	**1.** HACCP studies **2.** Knowing factory and equipment **3.** Separation of allergens and non-allergens Time (scheduling) Space (dedicated storage and equipment) **4.** Sanitation Visually clean standard Analytical assessment Validation and verification	**1.** Labeling Clear Accurate Comprehensive **2.** Precautionary labeling Based on risk assessment

an allergen already present in the formulation? Ingredient specifications are also critical, particularly in respect of unintentional presence of allergens. For instance, a "gluten-free" claim would require assurance from suppliers of absence of gluten, supported by evidence that their ingredients meet appropriate specifications. Another consideration during development is what measures will be needed to manufacture the product so that no additional allergen risks are created. This requires consideration of whether a particular manufacturing site already handles a specific allergen, or at the limit, even whether the product should be made at all, since the measures needed to control allergen risks could easily make a low-volume product economically unviable.

At the Sourcing stage, the critical consideration is obtaining comprehensive, accurate and reliable information about the ingredients, ensuring specifications are appropriate. Supplier questionnaires should provide information about their allergen management, including the extent to which they understand and apply processes such as HACCP (hazard analysis and critical control points). Absence of management thresholds for allergens has led many suppliers to use disclaimers or "may contain" assertions about possible allergen presence by cross-contact. Scrutiny of such disclaimers and statements may be needed to understand better the resulting risk, and quantitative information may need to be sought to permit a quantitative risk assessment. Periodic audits, either by the company's own auditors or by auditors accredited under the major standards (e.g. BRC, IFS, GFSI) should support this process. Suppliers must also understand that they cannot change a formulation or specification without agreement or without informing their customer so that the necessary information can be conveyed to the final consumer through labeling. Inclusion of provision of appropriate (sufficient to make sound decisions about management and precautionary labeling) allergen information in contractual terms is strongly recommended.

The Manufacture stage is the one over which the manufacturer has the greatest control, but possibly also the most complex. Detailed knowledge of the design and operation of the plant are imperative to successful management of allergens, particularly where it was designed before allergens were considered a food safety issue. Critical elements include identifying where the risk of allergen cross-contact arises and devising systems to minimize it. Parts of the manufacturing system through which the ingredients do not obviously flow can seriously challenge attempts to manage allergens as well as validation studies. The filters used to protect the machinery in vacuum/pneumatic material transport systems are a good example. Product residues will build up on them and unless they are replaced or cleaned at appropriate intervals, they can act as reservoirs which will release material, including allergenic ingredients at random intervals into product which should not contain it. As a result, the quantitative analysis of risk from allergens based on validation studies can be totally negated. Measures to minimize cross-contact include allergen segregation in both space and time, including careful design of storage areas to minimize potential contamination in the event of spills, dedicated equipment and, occasionally, whole lines or facilities. Production scheduling affects separation in time and is a powerful measure in allergen management. However, the complexity involved should not be underestimated, since allergens are not the only variable that needs to be taken into account, with flavor and color among two other important considerations. Of course each allergen needs to be considered individually too.

Cleaning also separates allergens from other components and each other, both in time and space. Indeed cleaning can also be considered as a control measure in a HACCP plan. The cleaning process, in particular if the step in the operation is a critical control point, should be subject to validation, monitoring and verification. This means that prior to manufacturing, the procedure and method of cleaning need to be validated to ensure that they are indeed effective for adequately removing allergens to the degree required. During the manufacturing process, the cleaning process needs to be monitored periodically to ensure that it is implemented as planned. Further data collected through various verification procedures, e.g. periodic testing of final products, audits or any consumer or customer complaints, can serve to verify if the management plan is effective and implemented. Clearly, any non-compliance should prompt an investigation of the verification of the plan (i.e. is the plan implemented as planned?). If this is the case, the validation of the cleaning procedure would need to be questioned and re-evaluated (see Chapters 24 and 31).

Finally, measures which can only be implemented as part of a longer-term plan include equipment and factory design. Of course, any new factory or new operations must take account of allergen management requirements at the design stage and in particular the need to minimize the presence of unintended allergen through cross-contact. Key considerations in the design are that equipment can be easily cleaned in place or that it can be dismantled. It should also avoid corners, crevices or dead-ends where material can build up and effectively form a reservoir from which it can be released into products of which it does not form part, thereby giving rise to sporadic and unpredictable cross-contact.

The Delivery stage is the one at which the product is brought to the consumer. It is difficult to minimize the importance of allergen considerations at this stage and failures at this point account for a significant proportion of allergen alerts (see for example, UK Food Standards Agency reports). Critical attention to artwork is needed to ensure that the correct packaging has been used and that all allergens are listed and clear to the consumer or purchaser. At this stage the packaging should be checked for incompatible elements, such as a "dairy-free" logo, but with milk in the ingredients. In line with Codex guidelines on validating control measures, it is important to validate that the information for consumers is presented in a clear and understandable manner. Indeed, the European Union's Regulation,

1169/2011 on Food Information for Consumers prescribes that allergens shall always be declared by reference to their common name, as defined in the Regulation, as well as being emphasized. Furthermore, the Regulation prescribes a minimum font size to improve legibility (which also applies to other ingredients).

The packaging is also the vehicle for any precautionary labeling that a hazard analysis and an evaluation of the risk have shown to be required for the product; it is critical to remember in this regard that precautionary labeling can never be a substitute for good allergen management measures and does not, of itself, exonerate the manufacturer from any legal liability. If an allergen box is used, then this also needs to accord with the ingredients list – again allergy alerts, not to mention allergic reactions in consumers, have occurred as a result of discrepancies in this area.

Training

Underpinning all components of allergen management is training appropriate to the level of responsibility and role within the food business. Personnel included should range from senior management to the operatives directly involved in production and other activities. Senior management requires an understanding of the impact of food allergy on the consumer as well as on the business, and of their accountability for food safety. In contrast, those involved more directly with production and food handling activities will need guidance and instruction on procedures for minimizing cross-contact and best practice for sanitation. However, these will be most effectively conveyed if they are also placed in the wider context of food allergy. Training in allergen management is also crucial for personnel in food service establishments/catering establishments, particularly since these are well known to represent a high risk to allergic consumers. This should cover understanding and interpretation of allergen information provided by suppliers, including any precautionary labeling, as well as avoidance of cross-contact during food preparation and providing accurate information to consumers. In the European Union, provision of allergen information for non-prepacked foods, including food served by catering establishments will become mandatory under Regulation, 1169/2011.

Allergen Control Plans

Allergen control plans summarize all the necessary elements that must be checked in order to determine the allergen status of a specific facility and define the control measures that may be needed. It can thus be developed as part of the more general HACCP plan, considering the flow of materials through the factory. The allergen control plan can therefore follow the schema outlined above and should cover the following questions.

Raw Material Sourcing

- Are auditors being briefed to pay particular attention to allergen management at the supplier?
- Is the specification of raw materials and semi-finished ingredients accurate and comprehensive with regard to allergens?

- Does the specification provide enough information to assess the allergen risks accurately, given the use of the raw material?
- Have all allergenic materials that are used at the facility been identified and taken into account?

Raw Material Receipt and Storage

- Do appropriate procedures exist to assure integrity of the separation between raw materials during transport (i.e. no cross-contact during this stage)?
- Do appropriate procedures exist to ensure that raw materials are correctly assigned for storage location?
- Is storage designed to ensure segregation of allergens from other raw materials and each other and maintain it in case of failure to contain them (e.g. damage to containers)?

Manufacturing Operations

- Are material flows comprehensively described and understood, so that all possibilities for cross-contact have been identified? This should include possible reservoirs where materials can be held up and subsequently released, as well as shared pipework, etc.
- Have all operations where cross-contact can take place been identified?
- Are dedicated utensils provided where necessary for products containing food allergens?
- Have the opportunities for scheduling (e.g. non-allergen before allergen) been explored and implemented?
- Do positive measures exist to ensure that formulations are correctly made up, in particular to avoid an allergen being added by mistake?
- Is work in progress properly labeled?
- Are procedures in place to ensure that rework of products containing allergens is controlled?
- Do procedures exist to avoid mispackaging, with resulting incorrect allergen declaration?
- Do protocols exist for all cleaning operations and have they been validated?
- What measures exist to verify cleaning operations?
- Has a study to validate allergen management at the facility been conducted and documented?
- Are there procedures to avoid inadvertent introduction of allergens into manufacturing areas (e.g. on clothing, tools, etc.)?

Personnel and Training

- Have all personnel (from top management, to workers and auditors), including part-time and temporary staff, undergone training in aspects of allergen management to a level appropriate to their role?
- Is basic allergen training included in staff induction procedures appropriate to each role?

Assessing the Risk from Food Allergens

Allergy was long thought to be an area where the conventional risk assessment paradigm could not be applied. In a publication on the Threshold of Toxicological Concern, Kroes et al. (2002) acknowledge that "a particular challenge is the evaluation of food allergens and components causing other forms of intolerances, and how to determine the levels present and actual intakes vs. the limited knowledge of amounts needed for induction or elicitation of a response." The authors in fact decided to exclude consideration of this issue from their paper.

Recent work also changed this perception and demonstrated that quantitative assessment of the risk from allergens was possible. Consideration of the risk assessment paradigm revealed that the most striking gap was in the characterization of the relationship between the dose of allergen, the proportion of the allergic population that reacted to that amount and the nature of those reactions. Statistical modeling of dose distributions using data on minimum eliciting doses from food challenges has proved very successful in filling this gap for several allergens, while avoiding the difficulties of defining an absolute (population) threshold or no observed effect level (NOAEL) experimentally. The principle of this approach consists in defining the dose interval within which an individual reacts during a food challenge with the relevant allergen (Taylor et al., 2009). This interval contains this individual's minimum eliciting dose (threshold). Thresholds from a range of individuals allergic to the same allergen are plotted against the dose as a cumulative distribution, which can then be fitted with different models for the shape of curve produced. All distributions so far have been sigmoidal when frequency of response is plotted against the logarithm of the dose and have produced good fits with the lognormal, loglogistic and Weibull models. Being able to define the distribution in this way has permitted derivation of eliciting doses corresponding to amounts of allergen predicted to cause reactions in small proportions of the allergic population (5% or less). This methodology makes use of all the available data rather than a single point and can be used to generate quantitative estimates of risk, when information about exposure is available.

One of the outcomes of risk assessment can be a decision to apply a precautionary label. Under ideal circumstances, a precautionary label would result in avoidance of the product by the relevant allergic individuals. In practice and particularly in the current circumstances of extensive use of precautionary labeling, observance of the warning is far from absolute and, indeed, reaches in some cases quite low values (<50%). The reasons for this are complex, but they include the overuse previously mentioned as well as consumer confusion over the message, no doubt exacerbated by the large number of different precautionary statements (Pieretti et al., 2009). The statistical modeling approach can help to define quantitative action levels, since information about the extent of consumer compliance with precautionary labeling as a function of its prevalence can be factored in as an additional quantitative factor.

Practical Aspects of Assessing the Risk from Allergenic Ingredients

The protein component of allergenic ingredients is the determinant of allergenic risk and, as discussed above, allergic individuals react to the amount consumed on any one eating occasion (i.e. meal, snack, etc.). For any given allergenic ingredient, therefore, the starting point for assessing the risk is the protein content and the amount that will be present in a portion (or amount eaten on any one occasion). The protein content of different types of ingredient should be available from the general specification provided by the supplier, both for intended and unintended allergens. However, in the event that the supplier cannot readily provide this information, generic information is available from a variety of sources on food compositions. Taylor and colleagues summarized generally available data on several allergenic foods some years ago (Taylor et al., 2002) and the values have also been used by the US FDA Threshold Working Group (2008). Values are also available for derived ingredients such as oils derived from allergenic sources (Crevel et al., 2000).

Where allergenic constituents are used as ingredients in foods (i.e. deliberately added), they are required by law to be declared irrespective of the amount present in accordance with Codex Alimentarius or regulatory requirements. The focus of risk assessment is therefore on unintended allergens present by cross-contact or otherwise. Thus the next step is to consider how much can be present by cross-contact, usually in a worst-case scenario. A typical worst-case scenario would be a product without the allergen being made on the same equipment immediately after one with a high concentration of allergen, with any already established cleaning protocol between products being used. As previously discussed, the worst-case carry-over may be subject to other constraints than allergens, such as taste, color, etc. Depending on the process and equipment, the proportion of the previous product carried over may be measured by collecting and weighing residual product in the equipment, or the allergen itself can be assayed in the following product. In some cases the proportion carried over will have been previously established or it may be sufficient initially to make a reasonable assumption based on other factors such as those already mentioned. Once a value is available for the proportion carried over, the allergenic protein concentration in the following product can be calculated, as can be the amount in a portion of the product. The amount of allergenic protein can then be compared to the amounts reported to cause reactions and a conclusion drawn about the risk posed by cross-contact.

Although, at the time of writing, no generally agreed reference amounts have been published, dose-distribution data can be used to derive the likely proportion of allergic consumers reacting to any particular amount of allergenic protein to decide whether the residual risk can be considered low enough. This decision can be based on one of a number of approaches, such as safety assessment, margin of exposure or probabilistic modeling, as described in Madsen et al. (2009). Alternatively a system such as the Australian Allergen Bureau's VITAL (Voluntary Incidental Trace Allergen Labelling) (Allergen Bureau, 2008) system can be used. This system comprises a grid of reference amounts for the most common allergenic constituents at and below which the risk to the vast majority of allergic individuals is considered negligible, based on an expert analysis of the available clinical data. The principles described above can also be applied when deciding whether a product containing an undeclared allergen needs to be recalled to minimize the risk to public health. Variables impacting on this decision include, beside the amount of allergen per portion and whether it varies across the batch of product, the number of units of product on the market (reflecting exposure).

The website of the European Trade Association for edible oils and fats (FEDIOL) also shows a publicly available example of a risk assessment which led to the conclusion that precautionary labeling of edible oils, because of the possible presence of peanut proteins from highly refined peanut oil, was unwarranted.

The initial risk assessment thus leads to a conclusion as to whether the unintended allergen poses a significant risk in the process as it currently operates. If that risk assessment is favorable, then all that is required is periodic verification. If the risk assessment is unfavorable, then the next step depends on the initial analysis. If the latter was based on theoretical calculations, then a quantitative analysis is required to establish whether the theoretical scenario accurately reflected reality. If an initial quantitative analysis formed the basis of the conclusion, then the next step would be to review the sources of unintended allergen with a view to reducing the proportion of carry-over. Only after establishing that further measures to mitigate the risk to a negligible level could not reasonably be put in place should precautionary labeling be applied to the product.

When using precautionary labeling, the wording should be carefully considered. Recent studies show that allergic consumers can readily misinterpret labels (Barnett et al., 2011). A simple phrase such as "may contain" or "may be present" (VITAL Allergen Management guide) is to be preferred to those which may imply a lesser degree of risk, such as "may contain traces of" or which require food allergic consumers to try to evaluate the risk associated with a particular production facility or process, such as "made in a factory which also produces" and "made on a line which also makes."

ANALYTICAL ASPECTS OF ALLERGEN MANAGEMENT

Validation and Verification

As discussed in the preceding section, the risk arising from the unintended presence of allergens needs to be assessed. The first step in this analysis is to establish the extent of cross-contact and, if necessary, investigate different measures to reduce it.

Validation is the process of checking whether or not current allergen management procedures, particularly cleaning procedures, control allergen cross-contact to an acceptable level. *Verification* is checking and recording that validated procedures are being implemented.

Allergen Detection Methods

A variety of methods are used for allergen cleaning validation studies and the most common are listed in Table 4.3 along with their major advantages and disadvantages.

ELISAs are currently the most commonly used allergen detection tests and although they have a number of advantages, they also suffer from a number of disadvantages that need to be mitigated to assure the validity of the results they generate. These are covered in more detail below:

- ELISAs rely on an antibody reaction with a protein(s) and proteins exist in different forms and relative abundances in different foodstuffs. Thus the antibodies in an ELISA may have been raised against a different mixture of proteins to those present in the potential contaminating material. Target protein(s) can also differ between ELISA kits that have the same purpose, e.g. ELISAs for milk may detect beta-lactoglobulin *or* casein *or* a mixture of milk proteins. Therefore, knowledge of the protein composition of the allergen source is required in order to ensure the correct ELISA kit is chosen to detect it and for the interpretation of the data. It is also important to understand the reporting units of the chosen ELISA (e.g. for milk, ppm beta-lactoglobulin or ppm skimmed milk powder).
  - Example: if the source of potential milk protein carry-over is a whey concentrate then beta-lactoglobulin would be a suitable choice of target protein; however, if the source

TABLE 4.3 Comparison of Analytical Methods for Allergens

Method	Target	Main Advantages	Main Disadvantages
Non-specific methods 1. Visual check 2. Adenosine triphosphate (ATP) 3. Total protein	Not applicable ATP Protein	Rapid and cheap	Visual inspection only applies to accessible areas Relationship between visually clean and allergen levels is unknown and will depend upon the surface, allergen(s) and matrix. ATP and total protein assays are non-specific and positive results difficult to interpret
DNA detection methods 1. Polymerase chain reaction (PCR)	Species specific DNA sequences	Very specific. DNA is stable and less affected by processing. Can be used as a confirmatory technique	Measuring DNA is not measuring the allergen (i.e. the protein): provides an indication of potential presence/absence of an allergen. Useful confirmatory technique but should only be used with caution where other methods are unavailable. Difficult to use as basis for quantitative risk assessment
Antibody-based detection methods/immunochemistry 1. Enzyme-linked immunosorbent assay (ELISA) 2. Lateral flow/dipstick devices	Both detect allergenic/antigenic protein	Relatively fast and measure specific proteins. ELISAs are sensitive and quantitative and lateral flow devices are cheap and simple to use	ELISAs are not available for all allergens and are affected by factors that can affect extraction of target proteins from samples and/or ability to detect target proteins. There is also a need to understand what protein(s) an ELISA has been developed to detect and reporting units to ensure correct selection and interpretation. Lateral flow devices are only qualitative (semi-quantitative at best)
Mass spectrometry (MS)	Allergenic protein, through peptide mass fingerprinting +/− peptide sequence analysis	Highly specific, can detect multiple allergens in a sample and is much less affected by processing and food matrices than ELISAs	Currently only commercially available as a qualitative screen for some allergens. Not currently as sensitive as some ELISAs and only validated in a limited number of matrices

is skimmed milk powder then the dominant protein present would be casein. Also in the former case, reporting units in beta-lactoglobulin would be required but in the latter results could be in ppm casein or ppm skimmed milk powder.

- Food processing can alter the ability to detect an allergen, due to changes in the protein such that the antibody used in the ELISA no longer recognizes the target protein or

because proteins associate with other components of the formulation and become more difficult to extract. ELISAs thus may have difficulty recognizing and produce a false negative result/reduced quantification for:
  - Heated products
  - Fermented products
  - Hydrolyzed products
- ELISAs require the extraction of the protein into an aqueous environment prior to analysis. The efficiency of this extraction depends on the solubility of the protein(s) of interest and the formulation of the food they are to be extracted from, e.g. high fat matrices or recipes rich in polyphenols can affect extraction. To check extraction efficiency for a given sample matrix a "spike and recovery test" is recommended, e.g. if the aim is to detect skimmed milk powder in a milk-free product, then a known milk-free sample of product (e.g. prepared in the QA kitchen) can be "spiked" with a known amount of skimmed milk powder and the level of extraction quantified. The food matrix can also affect ELISAs directly, e.g. some ingredients could cross-react with the antibodies in the ELISA to give a false positive reading and others may produce colored backgrounds that need to be controlled for. Thus provision of a known allergen-free sample as a control has further value.

Whether ELISA or PCR is used, the ability to provide a reliable service will depend on the experience and expertise of the analytical laboratory with the individual allergen and tests. An experienced operator should be aware of and control for any sources of potential contamination, while carrying out the test. A certified laboratory will also ensure that all equipment is calibrated and accurate. A good laboratory should offer a confidential service and welcome, indeed even request, early discussion of the validation study providing advice on correct test selection and study design.

Design of Validation Studies

Starting with a qualitative risk assessment and then moving onto a semi-quantitative one is recommended in order to determine whether or not an analytical-based validation study is required or applicable. For example, it is sometimes possible to estimate levels of allergen carry-over from one production run to another by "worst-case scenario calculations," i.e. measuring how much material is left behind in a process (e.g. based on film thickness on equipment or weighing brushed-out residual), what the levels of such material would be after dilution with the next product (or in the next process step), what amount of the material is allergen and therefore allergen levels in the final product that could be consumed. An example of such a calculation is described in Table 4.4.

If an analytical study is required, accurate and robust analytical results are only useful if the samples analyzed have been taken as part of a correctly designed study. The aim of any validation study should be clearly defined and understood, so that the sampling procedures and subsequent analyses are correctly designed or selected and implemented.

For a food product, development of a scientifically sound sampling plan includes a statistical analysis of the probability that all allergens are detected and ensures that any allergens present are accurately measured. Important sampling questions that need to be considered

TABLE 4.4 Example of a Worst-case Scenario Calculation

Aim of study	To determine whether or not carry-over of soy lecithin into non-soy containing products made on a shared line presents a risk to consumers
Information required	The maximum potential concentration of soy protein in any recipe made on the line, based on: **1.** the maximum amount of soy lecithin in any product recipe **2.** the amount of soy protein in the soy lecithin The worst-case potential carry-over for each stage of the process, based on the maximum level of residual product left in tanks/lines, etc.
Calculation	Maximum amount of soy lecithin in any product made on the line = 25 mg/kg Amount of soy protein in the soy lecithin used in the factory (maximum) = 1000 mg/kg Therefore the maximum amount of soy protein in any recipe made on the line = 0.025 mg/kg Worst-case potential carry-over from one recipe to another on the line = 3% Therefore the maximum potential carry-over of soy protein into a non-soy recipe is 0.00075 mg/kg

include whether the allergen is likely to be evenly distributed within the batch,[1] the number of samples per batch that should be tested, which batches should be tested, which portion of a run should be tested, and how to obtain a specific degree of confidence (e.g. 95% confidence) that no allergen is present.

The six main stages in the design of an analytical validation study are summarized in Table 4.5 and an example of an analytical study is provided in Table 4.6.

Verification

Cleaning processes should be periodically (e.g. yearly) verified to confirm that they remain effective and when changes are made that might impact allergen management, e.g. if there are design alterations to a process line/equipment a revalidation should be performed.

[1]For example is the potential carry-over from skimmed milk powder or small pieces of nut? In the former case an analytically based cleaning validation study may be suitable as the potential carry-over should be evenly spread throughout the product. However, such an approach cannot be used for sporadic contamination, as the probability of actually detecting the contamination by analytical techniques is very small and therefore non-detection only offers false reassurance. For particulate allergens a visual inspection should take place after cleaning to ensure that no particulates are left. The build-up of allergenic material on process line/equipment (e.g. heat exchange plates, vacuum equipment filters, etc.) also needs to be assessed (e.g. through regular inspection and swabs) as this can be a source of spot contamination that is very difficult to prevent and detect.

TABLE 4.5 Design of a Validation Study

Stage	Description
1. Procedure	Define and document procedure to be validated, e.g. details of cleaning procedures, define the worst-case scenario
2. Contaminant	Define and understand what is the potential "contaminating" material is to ensure analytical methods are available and correctly chosen
3. Samples	Define what to sample There are three main types of sampling that can be carried out to assess the presence of allergen after cleaning: 1. *Direct surface sampling* – swab sampling can be used to identify contaminated surfaces, but cannot be used in inaccessible areas and results are difficult to quantitate when positive. Approach not suitable for complex mixing and filling lines 2. *Sampling of rinse/push materials* – whole process equipment can be assessed. Method assumes that if the rinse material is clean then the equipment is clean. However, allergenic material insoluble in rinse material or physically trapped may invalidate results. Allergen assays can be affected by high levels of alkali and acid so assayed rinse water should be neutral. All parts of production processes need to be considered such as loops, bulk dead ends and filler heads or nozzles that require special cleaning. Volume of rinse material needs to be compared with product volume for risk assessment calculations 3. *Final products* – most relevant to consumer exposure and therefore assessment of consumer risk. However, food matrix may affect allergen assays and require controls. Legal issues may arise if a product could be deemed to pose a significant risk
4. Controls	Appropriate controls (positive and negative) must be included to ensure accuracy, validity and interpretability of results. These include: 1. A sample known to be free from the allergenic ingredient (a negative control) 2. A sample known to be free from the allergenic ingredient for "spike and recovery" to assess extraction efficiency Additional standards based on the cross-contact allergenic ingredient(s): e.g. an allergen kit for casein, supplied with a pure casein standard may not give a true representation of the level of contamination with cheese powder, which will contain a mixture of milk proteins including casein
5. Protocol	Clearly define in the study protocol how to take, label and store samples to avoid contamination, sample leakage, confusion over results and microbial spoilage
6. Results	Finally the results of the sampling need to be evaluated and a risk assessment needs to be performed to establish whether levels are acceptable. When the validation has been finalized the work should be documented in the Quality Management System

Interpretation of Validation Studies

Currently there are no agreed clinical thresholds for food allergens, although there have been some attempts in different areas of the world to provide labeling guidance. In Switzerland an action limit for labeling of 1 g/kg (one part per thousand) was defined in 2001 and, as previously mentioned, in Australia and New Zealand the Allergen Bureau (an initiative of the Australian Food and Grocery Council) developed the voluntary incidental

TABLE 4.6 Example of an Analytically-based Cleaning Validation

Aim of study	To determine whether or not current CIP protocols are sufficient to control carry-over of milk into non-milk products made on a shared line
Information required/steps taken	**1.** HACCP assessment suggested a minimal risk of cross-contamination but analyses requested to confirm **2.** Experienced, certified laboratory with a good reputation and experience with allergen measurement was identified and approached and the validation study discussed with analytical staff **3.** The potential source of carry-over of milk allergens (ingredient of concern) was a whey powder and therefore an ELISA measuring beta-lactoglobulin, the dominant protein in whey, was recommended **4.** To fully validate the process CIP flush water samples were taken from all loops **5.** Final product was also sampled in addition to flush water, as the laboratory had extensive experience of measuring allergens in the final product type and therefore had validated ELISAs for use in the product matrix
Calculation	**1.** Rinse water samples from one CIP loop gave results above the limits of detection for the ELISAs: 2 mg/kg beta-lactoglobulin **2.** ~50–60% of a whey powder protein is beta-lactoglobulin and therefore 2 mg/kg of beta-lactoglobulin is equivalent to ~3–4 mg/kg of total whey powder proteins **3.** Therefore the amount of total milk protein in the flush water sample is taken to be 3–4 mg/kg. As the amount of water in the final CIP flush is 300 l, but a 1500 l batch of product is usually passed through the line, the amount of milk protein in a batch of product would be expected to be diluted four times, i.e. ~1 mg/kg of product **4.** Furthermore the 1500 l of mixture went into 6 kg of final product and therefore the milk protein content would be diluted further to ~0.25 mg/kg **5.** This was supported by a lack of detectable milk protein in the final product, i.e. <1 mg/kg beta-lactoglobulin, the limit of detection of the ELISA used

trace allergen labeling (VITAL) system, which includes a set of action levels that specify whether or not a precautionary label is required based on the level of cross-contact identified. It is clear that there is a need for agreed, acceptable limits for the labeling of non-deliberately added allergens in foods and indeed there is a great deal of time and effort currently being invested in addressing this challenge.

SUMMARY

Food allergy is now recognized as an important public health issue, requiring collaboration between multiple stakeholders, including the food industry, to be effectively addressed. Allergen management in food production has the ultimate aim of assuring the safety of vulnerable allergic consumers. This is achieved by ensuring accurate allergen declaration through labeling and ensuring where allergens are present inadvertently either products do not contain amounts that pose an unacceptable risk to food allergy sufferers or meaningful precautionary labeling is applied. To ensure this aim is met food allergen risk management must be integrated into general food safety management and allergen risks considered at all stages of the food manufacturing process, i.e. design, source, make and deliver. With recent advances in understanding of minimum eliciting doses for some allergens it is now possible to conduct quantitative risk assessments for cross-contact. However, validation studies and the use of analytical methods such as ELISAs to quantify levels of carryover are subject to many potential pitfalls and benefit from a rigorous approach to design, conduct and interpretation.

References

Allergen Bureau, 2008. Voluntary Incidental Trace Allergen Labelling (VITAL). <http://www.allergenbureau.net/allergen-guide/vital/>.

Barnett, J., Leftwich, J., Muncer, K., Grimshaw, K., Shepherd, R., Raats, M.M., et al., 2011. How do peanut and nut-allergic consumers use information on the packaging to avoid allergens? Allergy 66, 969–978.

Bingley, P.J., Williams, A.J.K., Norcross, A.J., et al., 2004. Undiagnosed coeliac disease at age seven: population based prospective birth cohort study. BMJ 328, 322–323.

Bjorksten, B., Crevel, R., Hischenhuber, C., Lovik, M., Samuels, F., Strobel, S., et al., 2008. Criteria for identifying allergenic foods of public health importance. Regul. Toxicol. Pharmacol. 51, 42–52.

Branum, A.M., Lukacs, S.L., 2008. NCHS data brief. Food allergy among U.S. children: trends in prevalence and hospitalizations (10), pp. 1–8.

Bruijnzeel-Coomen, C., Ortolani, C., Aas, K., et al., 1995. EAACI position paper. Adverse reactions to food. Allergy 50, 623–635.

Chen, J., Liao, Y., Zhang, H.Z., Zhao, H., Chen, J., Li, H.Q., et al., 2012. Jan; Prevalence of food allergy in children under 2 years of age in three cities in China.[Article in Chinese] 50 (1), pp. 5–9.

Codex Alimentarius Commission. 2003. Principles for the risk analysis of foods derived from modern biotechnology. CAC/GL44-2003. Codex Alimentarius Commission, Joint FAO/WHO Food Standards Programme, Food and Agriculture Organization, Rome. (available at: <http://www.codexalimentarius.net/web/standard_list.do?lang=en>).

Crevel, R.W., Kerkhoff, M.A., Koning, M.M., 2000. Allergenicity of refined vegetable oils. Food Chem. Toxicol. 38 (4), 385–393.

Crevel, R.W., Ballmer-Weber, B.K., Holzhauser, T., Hourihane, J.O., Knulst, A.C., Mackie, A.R., et al., 2008. Thresholds for food allergens and their value to different stakeholders. Allergy 63 (5), 597–609.

EFSA, 2004. European food safety authority. opinion of the scientific panel on dietitic products, nutrition and allergies on a request from the commission relating to the evaluation of allergenic foods for labelling purposes. EFSA J. 32, 1–197.

FAO, 1995. Report of the FAO Technical Consultation on Food Allergies. Rome.

Food Quality Protection Act (FQPA). 1996. Public Law 104–170 – Aug. 3, 1996, 110 STAT. 1489.

Gupta, R.S., Springston, E.E., Warrier, M.R., Smith, B., Kumar, R., Pongracic, J., et al., 2011. The prevalence, severity, and distribution of childhood food allergy in the United States. Pediatrics 128 (1), e9–e17.

Hefle, S.L., Nordlee, J.A., Taylor, S.L., 1996. Allergenic foods. Crit. Rev. Food Sci. Nutr. 36 (Suppl), S69–S89.

Hu, Y., Chen, J., Li, H., 2010. Comparison of food allergy prevalence among Chinese infants in Chongqing, 2009 versus 1999. Pediatr. Int. 52, 820–824.

Kanny, G., Moneret-Vautrin, D.A., Flabbee, J., Beaudouin, E., Morisset, M., Thevenin, F.J., 2001. Population study of food allergy in France. Allergy Clin. Immunol. 108 (1), 133–140.

Kim, J., Chang, E., Han, Y., Ahn, K., Lee, S.I., 2011. The incidence and risk factors of immediate type food allergy during the first year of life in Korean infants: a birth cohort study. Pediatr. Allergy Immunol. 22 (7), 715–719.

Kroes, R., Müller, D., Lambe, J., Löwik, M.R.H., van Klaveren, J., Kleiner, J., et al., 2002. Assessment of intake from the diet. Food Chem. Toxicol. 40, 327–385.

Lamireau, T., Clouzeau, H., 2011. Epidemiology of celiac disease. Pathol. Biol. (Paris) May 24. [Epub ahead of print].

Lewis, S., Butland, B., Strachan, D., Bynner, J., Richards, D., Butler, N., et al., 1996. Study of the aetiology of wheezing illness at age 16 in two national British birth cohorts. Thorax 51, 670–676.

Loveless, M.H., 1950. Milk allergy: a survey of its incidence; experiments with a masked ingestion test. J. Allergy 21, 489.

Madsen, C.B., Hattersley, S., Buck, J., Gendel, S.M., Houben, G.F., Hourihane, J., et al., 2009. Approaches to risk assessment in food allergy: report from a workshop "developing a framework for assessing the risk from allergenic foods. Food Chem. Toxicol. 47 (2), 480–489.

Osborne, N.J., Koplin, J.J., Martin, P.E., Gurrin, L.C., Lowe, A.J., Matheson, M.C., et al., 2011. Prevalence of challenge-proven IgE-mediated food allergy using population-based sampling and predetermined challenge criteria in infants. J. Allergy Clin. Immunol. 127 (3), 668–676.

Osterballe, M., Hansen, T.K., Mortz, C.G., Høst, A., Bindslev-Jensen, C., 2005. The prevalence of food hypersensitivity in an unselected population of children and adults. Pediatr. Allergy Immunol. 16 (7), 567–573.

Pieretti, M.M., Chung, D., Pacenza, R., Slotkin, T., Sicherer, S.H., 2009. Audit of manufactured products: use of allergen advisory labels and identification of labeling ambiguities. J. Allergy Clin. Immunol. 124, 337–341.

Prausnitz, C., Küstner, H., 1921. Studien über die Uberempfindlichkeit (Study of hypersensitivity). Zentralbl. Bakteriol. Mikrobiol. Hyg. Abt. Orig. 86, 160–169.

Regulation (EU) 1169/2011 (EC) of the European Parliament and of the Council of 25 October 2011 on the provision of food information to consumers. Official Journal of the European Union L304/18–L304/63.

Rona, R.J., Keil, T., Summers, C., Gislason, D., Zuidmeer, L., Sodergren, E., et al., 2007. The prevalence of food allergy: a meta-analysis. J. Allergy Clin. Immunol. 120, 638–646.

Ross, M.P., Ferguson, M., Street, D., Klontz, K., Schroeder, T., Luccioli, S., 2008. Analysis of food-allergic and anaphylactic events in the National Electronic Injury Surveillance System. J. Allergy Clin. Immunol. 121 (1), 166–171.

Sampson, H.A., 2005. Food allergy – accurately identifying clinical reactivity. Allergy 60 (Suppl. 79), 19–24.

Taylor, S.L., Hefle, S.L., Bindslev-Jensen, C., Bock, S.A., Burks, A.W., Christie, L., et al., 2002. Factors affecting the determination of threshold doses for allergenic foods: how much is too much? J. Allergy Clin. Immunol. 109, 24–30.

Taylor, S.L., Crevel, R.W.R., Sheffield, D., Kabourek, J., Baumert, J., 2009. Threshold dose for peanut: risk characterization based upon published results from challenges of peanut-allergic individuals. Food Chem. Toxicol. 47, 1198–1204.

Taylor, S.L., Moneret-Vautrin, D.A., Crevel, R.W., Sheffield, D., Morisset, M., Dumont, P., et al., 2010. Threshold dose for peanut: risk characterization based upon diagnostic oral challenge of a series of 286 peanut-allergic individuals. Food Chem. Toxicol. 48, 814–819.

Threshold Working Group, 2008. Approaches to establish thresholds for major food allergens and for gluten in foods. J. Food Prot. 71, 1043–1088.

van Bilsen, J.H., Ronsmans, S., Crevel, R.W., Rona, R.J., Przyrembel, H., Penninks, A.H., et al., 2011. Evaluation of scientific criteria for identifying allergenic foods of public health importance. Regul. Toxicol. Pharmacol. 60 (3), 281–289.

Venter, C., Arshad, S.H., 2011. Epidemiology of food allergy. Pediatr. Clin. N. Am. 58, 327–349.

Venter, C., Pereira, B., Grundy, J., et al., 2006. Incidence of parentally reported and clinically diagnosed food hypersensitivity in the first year of life. J. Allergy Clin. Immunol. 117, 1118–1124.

Wainstein, B.K., Studdert, J., Ziegler, M., Ziegler, J.B., 2010. Prediction of anaphylaxis during peanut food challenge: usefulness of the peanut skin prick test (SPT) and specific IgE level. Pediatr. Allergy Immunol. 21, 603–611.

Worm, M., Timmermans, F., Moneret-Vautrin, A., Muraro, A., Malmheden Yman, I., Lovik, M., et al., 2010. Towards a European registry of severe allergic reactions: current status of national registries and future needs. Allergy 65, 671–680.

Zuberbier, T., Edenharter, G., Worm, M., Ehlers, I., Reimann, S., Hantke, T., et al. 2004. Prevalence of adverse reactions to food in Germany – a population study. Allergy 59 (3), 338–345.

CHAPTER

5

Milk and Dairy Products

Y. Motarjemi[1], G.G. Moy[2], P.J. Jooste[3] and L.E. Anelich[4]

[1]Food Safety Management Consultant, Nyon, Switzerland, [2]Food Safety Consultants International, Geneva, Switzerland, [3]Colorado State University, Fort Collins, CO, USA, [4]Anelich Consulting, Pretoria, South Africa

OUTLINE

Food Safety Management.
DOI: http://dx.doi.org/10.1016/B978-0-12-381504-0.00005-6

INTRODUCTION

Milk is a fluid secreted by the female of all mammals for the nutrition of their offspring. The Codex Alimentarius Commission (CAC) defines milk as *the normal mammary secretion of milking animals obtained from one or more milkings without either addition to it or extraction from it, intended for consumption as liquid milk or for further processing* (CAC, 1999). Other food safety authorities have different definitions. For example, the US definition mentions that milk should be essentially free from colostrums.[1] While the main components of milk are common for most mammals, the quantities of the components will vary as shown in Table 5.1. On the other hand, many of the biologically active components of milk are species specific. Therefore, cows' milk should never be substituted for human milk.

Approximately 85, 11, 2, and 2% of the global milk production is obtained from cows, water buffaloes, sheep, and goats, respectively. Milk from camel, yak, reindeer, horse, and donkey is important in certain regions, but is insignificant in global trade (Fox, 2011). Because it is highly perishable, milk has traditionally been processed into a broad range of more stable products. The removal of water from some of these products also facilitates their transport. Dairy products with long shelf-lives, such as cheeses, also provide sources of food over many months. Table 5.2 shows some of the diverse products and technologies used in the dairy sector. Milk can also be fractioned into its principal constituents, e.g. lactose, milk fat fractions, milk protein products (casein, caseinates), and whey protein concentrates.

In many countries of the world, the dairy industry is one of the most important food sectors and it has, by and large, been very successful in providing safe products. Nevertheless, the concern for the safety of these products remains high on the agenda of public health authorities. There are several reasons for this:

1. Milk is particularly rich in nutrients and provides an ideal environment for growth of many microorganisms;
2. Contamination of these products can occur at different points in the food chain through often complex pathways; and
3. These products have been the source of foodborne outbreaks caused by a broad range of microbial and chemical hazards.

TABLE 5.1 Composition (%) of the Milk in Selected Species (Adapted from Fox, 2011)

Species	Total Solids	Fat	Protein	Lactose	Ash
Human	12.2	3.8	1	7.0	0.2
Cow	12.7	3.7	3.4	4.8	0.7
Buffalo	16.8	7.4	3.8	4.8	0.8
Goat	12.3	4.5	2.9	4.1	0.8
Sheep	19.3	7.4	4.5	4.8	1.0

[1]The Code of Federal Regulations (CFR) defines milk as "the lacteal secretion, practically free from colostrums, obtained by the complete milking of one or more healthy cows."

TABLE 5.2 Diversity of Dairy Products (Fox, 2011)

Process Primary	Product	Further Products
Centrifugal separation	Cream	Butter, anhydrous milk fat, ghee; creams: various fat content (10 ± 50%) (pasteurized, ultra-heat treated sterilized, in-container stabilized), cream cheeses
	Skim milk	Skim milk powders, casein, cheese, milk protein concentrates
Concentration by thermal evaporation or ultrafiltration		In-container or ultra-heat treated-sterilized concentrated milks; sweetened condensed milk
Concentration and drying		Whole-milk powders; infant formulae; dietary products
Enzymatic coagulation	Cheese	Numerous varieties; further products, e.g. processed cheese, cheese-based ingredients
	Rennet casein cheese analogues	Whey, whey powders, demineralized whey powders, whey protein concentrates, whey protein isolates, individual whey proteins, whey protein hydrolysates, nutraceuticals
		Lactose and lactose derivatives
Acid coagulation	Cheese	Fresh cheeses and cheese-based products
	Acid casein/caseinates	Functional applications, e.g. coffee creamers, meat extenders; nutritional applications
	Whey	As for rennet casein
Fermentation		Various fermented milk products, e.g. yoghurt, buttermilk, acidophilus milk, bioyoghurt
Freezing	Ice cream	numerous types and formulations

HISTORICAL PERSPECTIVE

A few milestones in the history of the safety of milk and dairy products are helpful in understanding the present state of affairs. In the earliest days of human civilization, the main concern was the adulteration of milk, which presents both nutritional and microbiological concerns. Through the ages, many cultures had laws to prevent the adulteration of milk. Some developed rules for the handling of milk based on religious precepts, such as kosher foods. In many Middle Eastern countries, boiling milk for consumption is a traditional practice that continues to the present day.

However, it was not until the 1860s that Louis Pasteur firmly established the germ theory of disease and demonstrated many of its practical applications in medicine and food technology. With the heat treatment of wine and beer to destroy competing microorganisms, Pasteur laid the foundation for the most significant advance in microbial safety of milk and dairy products. In the 1890s, milk pasteurization was officially introduced in a number of US cities and in some European countries. Some of the earliest international standards were those for milk and dairy products, which were elaborated by the International Dairy Federation in 1903. In 1953, the World Health Organization (WHO), jointly with the Food and Agriculture Organization (FAO), published its first recommendations on milk pasteurization (WHO, 1953). While the original intent of milk pasteurization was to prevent transmission of *Mycobacterium tuberculosis* var. *bovis*, in the late 1950s, pasteurization parameters were later adjusted to destroy the most heat-resistant pathogen in milk, namely *Coxiella burnetii* which causes Q fever. By 1963, standards and codes of practice on the safety of milk and dairy products were being developed by the FAO/WHO Codex Alimentarius Commission and its Committee on Milk and Milk Products hosted by New Zealand.

Consequently, the safety of milk and milk products has improved dramatically and these products are consistently been shown to be some of the safest foods on the market in most countries. In recent times, however, outbreaks involving new pathogenic organisms, such as *Listeria monoycotogenes*, enterohemorrhagic *E. coli*, *Campylobacter*, *Cronobacter sakzakii*, and *Staphylococcus aureus* as well as toxic substances that have been deliberately added to milk, such as melamine, have drawn the attention of the dairy industry, governments and consumers to re-examine the safety of milk and dairy products.

FOODBORNE DISEASE OUTBREAKS

Up until the mid-20th century, consumption of raw milk was common and, not surprisingly, so was milkborne disease. This is unfortunately still the case in some developing areas of the world and among certain groups of people where raw milk is consumed. Improvements in sanitary practices, milk pasteurization, and animal health have all had significant impact on the prevention and reduction of milkborne illnesses. Today, such diseases remain a problem only in those places where raw milk and products made from it are still consumed, either legally or through ignorance of the health consequences. From 1998 to 2005, data in the USA indicate 39 outbreaks causing 831 cases with 66 hospitalizations and one death and these were related to the consumption of raw milk. Other sources of illness were homemade ice cream, soft unripened cheese made from raw milk, and rarely butter and milk powder.

In Europe and the USA, milk and milk products are implicated in 2–6% of all bacterial foodborne outbreaks. In the industrialized countries, most outbreaks are related to fresh, soft or semi-soft cheese made from raw ewes' or goats' milk, often produced using artisanal methods. Although little data are available, the prevalence of milkborne diseases may be higher in the developing countries, where sanitary infrastructure and refrigeration are less available. Reconstitution of dried milk powder or infant formula can be a frequent cause of contamination of products and infections. Official reports on the cause of outbreaks and factors leading to contamination of products are scarce or anecdotal. Most available reports are from countries which have well-developed surveillance and outbreak investigation systems. Even then, many reports fail to provide an in-depth explanation of the errors or shortcomings that have led to the primary contamination of the product. Consequently, the examples cited here are from the rare reports which have shed light on the root cause of incidents and that have provided lessons for risk management. A selection of reported outbreaks of illness implicating milk dairy products is given in Table 5.3.

TABLE 5.3 Selected Reported Outbreaks or Incidents of Foodborne Illnesses Associated with Dairy Products

Year	Country	Illness/Agent	Cases/Impact	Implicated Food
2013	Australia	Listeriosis	21 (3 deaths); Recall of 100 cheese products	Soft cheese
2012	USA	Listeriosis	22 (4 deaths)	Imported Italian ricotta cheese
2012	Spain	Listeriosis	2	Latin-style fresh cheese
2012	Israel	Botulism (bovine)	Destroying 10 tankers of milk over 250,000 liters of milk	Raw milk
2012	Finland	Enterohemorrhagic *E. coli* infection	1 (child)	Unpasteurized milk
2012	USA	*E. coli* O157 infection *Cryptosporidium, Campylobacter* infections	21	Raw milk
2012	USA	Campylobacteriosis	18	Unpasteurized milk
2012	Russia	Salmonellosis (*S. Oranienburg*)	16 (infants)	Infant formula
2012	USA	Enterohemorrhagic *E. coli* infection	5 (children)	Organically produced milk
2011	China	Aflatoxin	Recall of contaminated milk (1.2 microg/kg)	Milk
2011	Belgium	Listeriosis	12	Hard cheese made with pasteurized milk
2011	USA	Yersiniosis	14	Pasteurized milk – ice cream
2011	USA	Q fever	3	Raw milk
2011	USA	Campylobacteriosis	8	Unpasteurized milk
2011	Russia	Tick borne encephalitis	1 (child)	Goats' milk
2011	China	Nitrite	38 (3 deaths – children)	Milk
2010	USA	Campylobacteriosis, *E. coli* O157 infection	30 (2 children)	Unpasteurized goats' milk
2010	China	Melamine	Seizure of 76 tons	Dairy ingredients
2010	USA	*E. coli* O157 infection	5	Unpasteurized milk
2010	USA	Campylobacteriosis	15	Unpasteurized milk
2010	USA	Listerioisis	14	Hog head cheese
2009	Russia	Botulism	4	Cheese (prepacked)
2009	USA	Salmonellosis (*S.* Newport)	2100 (estimated)	Unpasteurized cheese

(Continued)

TABLE 5.3 (Continued)

Year	Country	Illness/Agent	Cases/Impact	Implicated Food
2009	Mexico	Brucellosis	48	Unpasteurized cheese
2009	USA	Campylobacteriosis	12	Unpasteurized milk
2009	USA	Listeriosis	8	Pasteurized milk cheese
2009	Canada	Listeriosis	38 (16 maternal – neonatal)	Pasteurized milk cheese
2008	Chile	Listeriosis	91 (5 deaths)	Cheese (Brie)
2008	Canada	*E. coli* infection	16	Raw milk cheese
2008	China	Melamine	294,000 infants (6 deaths)	Milk, infant formula
2008	Spain	Salmonellosis (*S. Kedougou*)	21 (19 infants)	(Infant formula)
2008	Canada	Salmonellosis (*S. Enteritidis*)	87 (1 death)	Cheese
2008	USA (ex Mexico)	*Mycobacterium bovis*	1 child	Cheese (quesco fresco)
2008	Italy	Dioxins	Recall	Buffalo mozzarella cheese
2007	Austria	*Staphylococcus aureus* intoxication	40 school children	Milk products (vanilla and chocolate milk)
2007	USA	Campylobacteriosis	87 (2 outbreaks)	Unpasteurized milk
2007	USA	Campylobacteriosis	3	Unpasteurized milk sold as pet food
2007	USA	Campylobacteriosis	19	Unpasteurized milk/fresh cheese
2007	USA	Campylobacteriosis	7	Unpasteurized milk
2007	Canada	Yersiniosis, listeriosis	2 (children)	Unpasteurized milk/cheese
2007	Paraguay	Foodborne illness	300	Pasteurized milk
2007	USA	Salmonellosis (*S. Typhimurium*)	29	Raw milk or raw milk products
2007	The Netherlands	Campylobacteriois	16	Raw milk
2006	USA	*E. coli* O157 infection	2 (children)	Unpasteurized milk
2006	USA	Food poisoning (jail)	40	Reconstituted powder milk
2006	Finland	*Streptococcus equi* infection	3	Fresh goats' cheese
2006	Netherlands	Salmonellosis (*S. Typhimurium*)	200	Hard cheese made from raw milk
2006	France	Salmonellosis (*S. Montevideo*)	23	Raw milk cheese

(Continued)

TABLE 5.3 (Continued)

Year	Country	Illness/Agent	Cases/Impact	Implicated Food
2006–2007	USA	Salmonellosis (*S. Newport*)	67	Mexican-style cheese
2005	Netherlands	*Campylobacter*	22	Raw milk
2005	Europe (Italy, France, Portugal, Spain)	Chemical contamination with isopropylthioxantone (ITX)	Recall of 2 million liters	Liquid infant formula
2005	France	Salmonellosis (*S. Worthington*)	49	Milk powder
2005	Kenya	Food poisoning (possibly botulism)	43 (4 deaths – children)	Fermented milk (colostrums)
2005	USA	*Mycobacterium bovis*	1 death (35 from 2001–2004)	Cheese (e.g. queso fresco from Mexico)
2005	Canada	*E. coli* O157 infection	17 (2 children)	Unpasteurized milk
2005–2006	USA	*E. coli* O157 infection	18	Unpasteurized milk
2004	Lithuania	Shigellosis	5	Unpasteurized milk curd
2004	France	*E. coli* O157 infection	3	Unpasteurized goats' cheese
2003	Denmark	*E. coli* O157 infection	25 (18 children)	Milk (organic)
2003	USA	Salmonellosis (*S. Typhimurium*)	62	Raw milk
2003	Israel	Thiamine deficiency	Several infants (3 deaths)	Infant formula
2003	Spain	Campylobacteriosis	81	Cross-contamination of a custard milk with chicken made during the preparation
2002	Canada	Botulism	Recall	Blue cheese
2002	Canada	*E. coli* O157 infection	13	Unpasteurized Gouda cheese
2002–2003	USA	Samonellosis (*S. Typhimurium*)	2 children	Unpasteurized milk
2001	USA	Campylobacteriosis	5	Unpasteurized milk
2001	France	Salmonellosis (*S. Enteritidis*)	2 outbreaks (190 and 23)	Cantal cheese made with raw milk
2000	Japan	*S. aureus* intoxication	13,420	Pasteurized milk
2000–2001	USA	Listerioisis	12	Mexican-style cheese
1999	USA	Salmonellosis (*S. Hadar*)	15	Swiss cheese
1999	Brazil	*S. aureus* intoxication	328	Cheese
1998	UK	*E. coli* O157 infection	7	Unpasteurized milk/cream

(Continued)

TABLE 5.3 (Continued)

Year	Country	Illness/Agent	Cases/Impact	Implicated Food
1998	USA	*E. coli* O157	8	Fresh cheese curd
1998	USA	Samonellosis (*S. Typhimurium*)		Unpasteurized milk
1997	France	Salmonellosis (*S. Typhimurium*)	113	Raw milk, soft cheese
1997	USA	Samonellosis (*S. Typhimurium*)	54	Raw milk cheese
1997	Iran	Botulism	27 (1)	Cheese
1996	Italy	Botulism	8 (1 death)	Mascarpone cheese
1993	France	Samonellosis (*S. Paratyphi*)	273	Goats' milk cheese
1985	USA	Campylobacteriosis	250	Raw milk
1985	USA	Salmonellosis (*S. Enteritidis*)	160,000–190,000 (estimated)	Pasteurized milk

RISK AND CONTROLS

Hazards associated with milk and dairy products are numerous and varied. These hazards are usually grouped under three categories, namely biological, chemical and physical (see Table 5.4). These hazards may contaminate milk and dairy products at multiple points in the food chain. This chapter highlights a selected number of these. While not exhaustive, the information presented here should be an aid when conducting a hazard analysis of milk and dairy products.

The major potential hazards that can be found in milk are discussed in the following sections which are organized according to the milk and dairy production chain. Note, however, that certain hazards are potential contaminants at more than one point in the production chain. Outbreaks can also involve several pathogens.

Feed

Hazards

Animal feed plays an important role in the health of food-producing animals and the safety of products derived from them, namely milk and dairy products. Animal feed can be a source of infections in food animals, with various pathogens, e.g. viruses, bacteria and parasites, which subsequently may lead to the contamination of milk. For instance, improperly fermented silage ($pH > 4.5$) can transmit *Listeria monocytogenes* to ruminants. Infected or healthy asymptomatic carriers can excrete high numbers of *L. monocytogenes* in their feces and contaminate the environment and ultimately the milk (Ryser, 2011).

Mycotoxins in feed are also a problem. In particular, aflatoxin B is a known human carcinogen present in maize, peanuts and other crops and their fodders. It is metabolized and transferred to milk in the form of aflatoxin M1. Proper drying and storage of the feed

TABLE 5.4 Potential Hazards in Milk and Dairy Products

Biological Hazards	Chemical Hazards	Physical Hazards
Bacillus cereus	Cleaning agents/sanitizers	Metal fragments, screws and rivets
Brucella spp.	Antimicrobials	Machine filings
Campylobacter jejuni	Pesticides	Glass shards
Coxiella burnetii	Hormones	Wood splinters
Cronobacter sakazakii	Dioxins and PCBs	Jewelry
Cryptosporidium parvum	Aflatoxin M1	Stones
Pathogenic *E. coli*	Heavy metals	Insulation/paint
Enterohemorrhagic *E. coli*	Radionuclides	Plastic fragments
Listeria monocytogenes	Processing contaminants	Personal effects – such as jewelry, buttons, nail fragments, nail varnish, dressings
Leptospira	Packaging contaminants (bisphenol A)	Hair, dust and dirt
Mycobacterium bovis	Melamine and other adulterants	Insect parts/fragments
Mycobacterium paratuberculosis		
Salmonella (non-typhi)		
Shigella spp.		
Staphylococcus aureus (enterotoxins)		
Yersinia enterocolitica		
Fecal–orally transmitted pathogens such as hepatitis A, *Salmonella typhi* and *paratyphi*, pathogenic *E. coli*		

are important measures for preventing growth of the toxigenic molds such as *Aspergillus flavus*, which is mainly responsible for aflatoxin formation. Other mycotoxins such as ochratoxin, T-2 toxins, deoxynivalenol and zearalenone may also be carried over into milk. However, they are more of concern for animal health rather than milk safety. Once feed has been contaminated, processing the feed by means of heat treatment has little effect in eliminating most mycotoxins. Therefore, preventing contamination at its source is the most effective method of reducing the risk of mycotoxin contamination. Preventive measures need to be applied during crop production, handling, storage and processing. Proper drying of the feed is particularly important. Continuous monitoring of feed can ensure that contamination does not exceed tolerable levels.

Animal feed is also a potential source of exposure of farm animals to environmental contaminants such radionuclides, polychlorinated biphenyls (PCBs) and dioxins. During the Chernobyl accident in 1986, a cloud of the airborne emissions passed over several European countries depositing radionuclides on pastures and resulting in the contamination of milk and dairy products. In 1999, feedstuff contaminated with a mixture of PCBs and dioxins was found to be responsible for the intoxication of laying hens in Belgium. Subsequently, the tainted animal feed was found to have contaminated pork, milk and dairy products in Belgium and the Netherlands. This incident was reportedly poorly managed and resulted in enormous conflicts, economic losses and significant loss of consumer confidence in the safety of the food supply.

The nutritional composition of the feed can have an impact on the composition of milk and should be taken into account in the formulation of infant products. For instance, a major recall of infant formula due to excessive amounts of iodine in these products in China in 2005 was attributed possibly to the excessive fortification of feed with iodine and its subsequent appearance in milk.

Bovine spongiform encephalopathy (BSE) has been attributed to the practice of feeding meat and bone meal from infected animals to cattle in the UK. To control the BSE epidemic, in 2000 a total ban of meat and bone meal in animal feed was implemented in the affected countries.

In 2008, an investigation of the melamine incident in China led to the discovery that, in addition to the direct addition of melamine to milk (see below), melamine was also added to animal feed to falsify the protein content and was possibly a source of trace amounts of melamine in milk.

Possible Procedures to Minimize the Risks of Feed and Milk Contamination

The production, processing, storage, transport and distribution of safe and suitable feed and feed ingredients are the responsibilities of all stakeholders along the food chain, including farmers, feed ingredient manufacturers, feed compounders, transport contractors, etc. Each sector is responsible for the activities under its direct control, including compliance with applicable statutory requirements. The Dutch animal feed sector has opted for a quality assurance system based on the hazard analysis and critical control point (HACCP) system this is widely applied by the European food industry. In particular, the animal feed industry and the preceding ingredient suppliers are now seen as essential parts of the food safety assurance chain.

- Production of feed and feed ingredients on the farm
- Adherence to good agricultural practices (GAP) is encouraged in the management of natural, improved and cultivated pastures, and in the production of forage and cereal grain crops used as feed or feed ingredients for food-producing animals. This is particularly important for plant materials that may contain residues of pesticides. Following GAP prescriptions will minimize the risk of biological, chemical or physical contaminants from entering the food chain.
- Crop residuals and stubbles used for grazing after harvest should also be considered as livestock feed. The same applies to livestock bedding since most livestock will consume a portion of their bedding. Straw or wood shavings should therefore be managed in the same manner as animal feed ingredients. Rational grazing and dispersion of manure should be applied in such way as to reduce cross-contamination among animals.
- Other factors that should be taken into consideration are the proximity of the agricultural land to industrial operations where effluents or air emissions can lead to feed contamination and chemical fertilizers, manure, pesticides and other agricultural chemicals which should be stored, managed and disposed of correctly.
- Monitoring and identification of health hazards
- When purchasing feed ingredients from suppliers, such suppliers should be able to demonstrably guarantee product safety. Audit procedures can include inspection, sampling and analysis for undesirable substances. Feed ingredients should meet acceptable and, if applicable, statutory standards for levels of pathogens, mycotoxins, pesticides and other undesirable substances that may constitute a health hazard for the consumer. Any feed or feed ingredient unsuitable for animal feed should be disposed of

properly. Traceability of feed and feed ingredients, including additives, should be enabled by proper labeling and record-keeping at all stages of production and distribution.

- Processing, storage and distribution of feeds and feed ingredients
- The effective implementation of GMPs and where applicable HACCP-based approaches should ensure that the following areas are addressed as a minimum:
  - Premises
  - Buildings and equipment should be constructed to permit ease of operation, maintenance and cleaning. Water should be of a suitable quality and effluent should be adequately disposed of.
  - Receiving, storage and transportation
  - Feed and feed ingredients should be stored separately from fertilizers, pesticides and other potential toxic materials. Processed material should also be stored separately from unprocessed ingredients. The presence of undesirable substances should be monitored and controlled. Finished products should be delivered and used as quickly as possible on a first-in, first-out basis. During storage, precautions should be taken to restrict microbial growth in feedstuffs and ingredients.
  - Management of medication and other additives
  - Medications and other additives that are to be added to the feed should be managed and stored correctly in order to avoid cross-contamination and potential overdosage.
  - Personnel training
  - Personnel should be adequately trained and aware of their role and responsibilities in protecting food safety.

Farm: Milk and Animal Health

Pathogenic Organisms

As shown in Table 5.4, milk can harbor a wide range of microorganisms; therefore raw milk can be a direct source of many types of foodborne infections. The contamination of milk may follow many different pathways. Some organisms are directly shed into the milk, particularly if the dairy herd suffers from mastitis or other infections, such as bovine tuberculosis, brucellosis and Q fever. Several bacteria can cause mastitis, including *Staphylococcus aureus*, *Streptococcus* spp. and *Corynebacterium bovis*. Other bacteria may contaminate the milk during the milking process through contact of the milk with the hide, udder or milking equipment. Animal feces and the environment are important sources of microorganisms, particularly when hygiene is poor. Farm workers can also be a source of contamination if personal hygiene is not respected. Therefore, contamination of milk goes beyond pathogens which are typically transmitted from the animal reservoir and can include any fecally transmitted pathogen of human origin, including viruses, parasites (e.g. *Cryptosporidium*) and bacteria such as *Shigella* and *Salmonella typhi*.

ENTEROBACTERIACEAE

The family Enterobacteriaceae includes a large number of organisms (*Escherichia*, *Salmonella*, *Shigella*, *Yersinia*, *Klebsiella*, *Enterobacter*, *Serratia*, *Citrobacter*, *Proteus*, *Edwardsiella*,

Erwinia, *Morganella* and *Providencia*). The presence of any member of the Enterobacteriaceae family is undesirable in pasteurized milk and dairy products. This is due to:

1. The inherent spoilage capacity of many genera in this family;
2. The fact that the presence of certain genera in water and food may be indicative of fecal contamination; and
3. The serious food safety implications that the presence in food or water of the many pathogens in this family may have.

Some are of specific concern to the milk and dairy industry:

- *Salmonella* spp. Salmonellae are a frequent contaminant of milk and, as illustrated throughout this chapter, it is a source of many outbreaks involving milk and other dairy products. Various surveys in the USA show that some 10–21% of the dairy farms are positive for salmonellae. In 2003, 32 of 678 (4.7%) raw bulk milk samples were reported to test positive for salmonellae in the USA. The serovars and the level of contamination, however, differ in different regions of the world.
- *Salmonella* is shed from infected and clinically ill animals, in particular in the febrile condition. However, asymptomatic carrier animals may also intermittently excrete this organism for prolonged periods, even months or years. Infected cows are a major source of contamination of the environment and other neighboring animals in the stable. *Salmonella* can contaminate milk from milking equipment and the environment (bedding, manure, aerosols and fecal matter). Infected farmers and handlers, pets and other farm animals as well as wild animals, such as birds, may also be the source of contamination of the environment. Cattle can be contaminated by drinking contaminated water or grazing on pasture fertilized with human sewage.
- *Campylobacter* spp. *C. jejuni* and *C. coli* are the common strains of *Campylobacter* and are a frequent cause of infections involving raw milk. A survey done in England in 1988 showed that 5.9% of the raw milk samples tested were positive for *C. jejuni* and that there was a significant association between the presence of *E. coli* in milk and that of *C. jejuni*. These organisms are also intermittently shed by subclinically infected bovines and other animals. Milk can also be contaminated during the milking process by organisms that colonize the teat canal as well as from fecal contamination on the outside of the udder in which case milk may contain up to 10^3–10^4 cfu ml^{-1} of these organisms.
- *Yersinia*. Of the various strains of *Yersinia*, only *Y. enterocolitica* and *Y. pseudotuberculosis* are viewed as foodborne. The main reservoirs of *Y. enterocolitica* are pigs and rodents. Therefore contamination of milk is mainly environmental (e.g. through feces or polluted water) although farm workers may also introduce the organism. *Y. enterocolitica* is regarded as an unusual cause of milkborne illness because of the low incidence of human pathogenic strains in the raw milk supply and the high susceptibility of the organism to pasteurization. The role of milk and dairy products in the transmission of *Yersinia* is a subject of debate. Raw milk and inadequately pasteurized milk and dairy products have, nevertheless, been implicated in the transmission of *Y. enterocolitica* infections to humans. In 2005, the first recorded food-associated outbreak of yersiniosis occurred in New York, where more than 220 individuals were stricken with acute intestinal illness after consumption of contaminated milk. In addition, epidemiological studies have

revealed that refrigerated food stored over prolonged periods of time poses an additional risk since *Y. enterocolitica*, as a psychrotrophic microbe, is able to grow at temperatures as low as 0°C.

- *Escherichia coli*. This organism is currently the best known indicator of fecal contamination, primarily of water, but also of raw food products. Its recovery from fresh dairy products consequently suggests that other organisms of fecal origin, including pathogens, may be present. *E. coli* strains are commonly associated with the normal facultative anaerobic microflora found in the intestinal tracts of humans and animals. Although many of these strains are harmless commensals, various *E. coli* strains have acquired virulence genes that render them pathogenic for both humans and animals.
- Although enterohemorrhagic *E. coli* (EHEC) is often the primary concern in industrialized countries, other pathogenic *E. coli*, e.g. enteroaggregative *E. coli* (EAEC) and enteropathogenic *E. coli* (EPEC), have also been associated with milk and dairy products. The pathways of contamination of milk of these strains are similar to other fecal–orally transmitted pathogens. However, in the case of EHEC, the organism may colonize the intestinal tract of healthy asymptomatic dairy cattle and thus provide multiple opportunities for the contamination of milk.
- Milk can be contaminated in various ways. Dairy cattle can directly shed the organism in the milk. Up to one-third of dairy herds and 4–10% of cattle within the herd may be excreting organisms at any time. The organism can also be present in milk through direct contact with the hide of animals, manure and feed or indirectly through contact with contaminated bedding, soil, dust, water and equipment. In addition, farm personnel can also be a source of contamination.
- *Cronobacter sakazakii* (previously *Enterobacter sakazakii*). This organism is a member of the genus *Cronobacter* and is widely dispersed in the environment, including soil, water, food processing plants and the home. The main pathogenic organism in this group has been associated with cases of sepsis, meningitis and necrotizing enterocolitis in infants. The case-fatality rate can be important and in one case a case-fatality rate of 50% has been reported. Powdered infant formula (PIF) has been identified as a major route of infection, and therefore these organisms are of particular concern to manufacturers of infant formula and suppliers of milk powder and derivatives to this industry. In the farm environment, *Cronobacter* has been found in dried pellets of animal feed, but not in cattle feces.

Other organisms in the family, e.g. *Proteus* spp. and *Hafnia alvei*, can also be present in the environment and contaminate milk.

LISTERIA MONOCYTOGENES

This organism is a pathogen of major concern in the milk and dairy industries. It is commonly encountered in the dairy farm environment and prevalent in dairy animals producing clinical and subclinical mastitis. *L. monocytogenes* is psychotropic and can grow in milk at low temperatures. It has the ability to survive adverse environmental conditions. Infected dairy cattle, sheep and goats can intermittently shed *L. monocytogenes* in their milk at levels of up to 10^4 cfu ml^{-1}. Healthy animals can also shed *L. monocytogenes* in their milk for a long period. However, a major source of contamination of raw bulk-tank milk is environmental, with fecal sources and manure playing major roles. The organism can be transmitted to cows via feed, such as improperly fermented silage and other feedstuffs, causing infection in the animals. Incidences of *L. monocytogenes* of 4.2, 2.2 and 2.6% have been reported in farm milk samples of bovine, ovine and caprine origin, respectively.

Refrigeration of milk right after milking is important for controlling growth. However, in view of the ability of *L. monocytogenes* to survive adverse environmental conditions, particularly low temperatures, the collected milk should be processed as soon as possible. In the USA and Europe, 2.5–5% of the raw milk is contaminated with the organism. *L. monocytogenes* populations in naturally contaminated raw milk can increase 1000-fold after 4 days' storage at 10°C or after 10 days' storage at 4°C.

The variability in virulence of *L. monocytogenes* strains is gaining wider recognition and acceptance. Throughout the world, three serotypes (i.e. 4b, 1/2a and 1/2b) account for 89–96% of cases of human listeriosis, providing evidence that certain strains are more likely to cause illness. Although it is difficult to entirely eliminate the presence of *Listeria* in raw milk, prevention must begin at the farm with good animal husbandry, in particular with attention to the quality of feed, good sanitation, proper milking practices and animal health (Ryser, 2011).

BRUCELLA SPP.

Brucella abortus and *B. melitensis* cause infections in cattle but also in humans and in a range of other animal species, including buffalo, camel, deer, dog, goat, horse, pig and sheep. Infections in these animals cause abortion or premature births. Infected animals can shed the organism in their milk.

Brucellosis eradication programs in many countries have resulted in a decrease of outbreaks. In many parts of the world this eradication has been so successful that the organism no longer poses a hazard to human health. Where the disease still occurs, pasteurization of milk has minimized the number of human outbreaks by these microorganisms. However, brucellosis is regarded as a re-emerging disease in certain regions of the world and closer attention is warranted.

COXIELLA BURNETTI

Coxiella burnetti is the causative agent of Q fever and is an organism that may infect the udder, probably by the hematogenous route. Consumption of, or contact with, contaminated milk can lead to human infection. This organism was found to be more heat resistant than *Mycobacterium tuberculosis* and in 1956 the recommended vat pasteurization temperature was raised from 61.7°C to 63°C (holding time 30 minutes) to ensure destruction of the organism.

MYCOBACTERIUM SPP.

There are some 100 different species of mycobacteria, most of which are non-pathogenic to humans and are found mainly in soil and water. Both *M. tuberculosis* and *M. bovis* are pathogenic to humans and cattle. The primary source of *M. bovis* is cattle species, but other animals (deer, goats, pigs, dogs, cats) may also be infected. The organism can be shed in the milk. *Mycobacterium avium* subspecies *paratuberculosis* (MAP) is the causative agent of Johne's disease, which is primarily a disease of ruminants. The disease is a slow developing colitis in which the intestinal macrophages are infected. In the process inflammatory reactions are induced in the host gut. This affects the ability of the gut to absorb protein from the diet resulting in clinical features that include diarrhea and chronic weight loss.

Due to the similarity of symptoms between Johne's disease and Crohn's disease in humans, MAP has been purported to be associated with Crohn's disease. Although the nature of this relationship is still debated, there are sufficient reasons to eradicate Johne's disease in cattle and to prevent MAP from entering the human food supply. In any case, Johne's disease is itself devastating for animals and a cause of reduced milk productivity.

MAP can be passed from mother to calf through both colostrum and milk. *In utero* transmission of MAP has also been observed. MAP bacteria are shed intermittently in high numbers in the feces of infected cattle, contaminating bedding, pasture and water sources. Milk may be contaminated by the natural shedding of infected macrophages or by fecal contamination. While the infectious dose is reported to be as low as 1000 organisms, clinically affected animals may shed up to 5×10^{12} mycobacterial cells per day. Occurrence of the pathogen in milk-producing animals is consequently a challenge to animal health, milk quality and the safety of the milk supply. As the organism shows patterns of heat resistance and appears to survive pasteurization, the most efficient control measure is the eradication of the illness in cattle and the sourcing of milk from healthy animals.

STAPHYLOCOCCUS AUREUS

S. aureus is found on the skin, teats and mucous membranes of animals. It is also present in the infected mammary glands of dairy cattle and other animals. Throughout the world, it is a significant cause of mastitis in dairy cows. Shedding of the organism is irregular and depends on the age of the animals. Various other studies have implicated bulk-tank milk as a potential source of enterotoxigenic *S. aureus* in milk and milk products.

Humans (nose, pharynx, skin, infected wounds and lesions) are also an important source of *S. aureus* and may contaminate the milk during the milking process. However, *S. aureus* can also have an environmental origin. In Europe, milk and dairy products constitute 1–9% (mean 4.8%) of all *S. aureus* outbreaks (EC, 2003). Numbers in excess of $10^6\,g^{-1}$ are necessary to produce enterotoxins.

BACILLUS CEREUS

B. cereus is ubiquitous in the environment that includes the soil, bedding material and milking equipment, as well as pastures. The spores are present in soil from $10^2\,cfu\,g^{-1}$ and up to more than $10^5\,cfu\,g^{-1}$. It is thus a frequent contaminant of raw milk. There is a seasonal variation in the number of *B. cereus*, with higher numbers during the grazing period as the teats become contaminated by the soil. Vegetative *B. cereus* cells are found in raw milk at $<10\,cfu\,ml^{-1}$ to a few hundred per ml. These cells are killed by pasteurization. Spores are found at much lower numbers, i.e. from $<10\,l^{-1}$ to a few thousand per liter of milk.

CLOSTRIDIA SPP. (E.G. *CLOSTRIDIUM BOTULINUM, CL. PERFRINGENS*)

Cl. botulinum and *Cl. perfringens* are widely distributed in soil, dust, water, sediments, sewage and vegetation. Feeds and especially silage can also be contaminated. Under favorable conditions of water activity (a_w), pH and temperature, they can grow and contribute to the spread of spores. Raw milk can become contaminated during the milking process. Numbers of *Cl. botulinum* spores in raw milk are generally very low; in cheese production, during the centrifugation and filtration steps, they can increase to $10\,g^{-1}$ or more in cheese. However, in favorable conditions of growth, even low numbers of *Cl. botulinum* spores can be dangerous. Most outbreaks of *Cl. botulinum* intoxication are associated with proteolytic strains as they are more heat stable and acid resistant than the non-proteolytic ones. Outbreaks are often associated with pasteurized and heat-treated milk where competitive flora are killed. Outbreaks have been reported with cheese and yoghurt due to the addition of contaminated ingredients. Contamination of infant formula with *Cl. botulinum* B spores has also been suspected to be a cause of infant botulism. Another potential concern is the consumption of milk derived from cattle or other animals affected by botulism. High-speed centrifugation (bactofugation) removes most spore-forming organisms from milk and does not affect the composition, flavor and nutritional value of the milk.

CRYPTOSPORIDIUM SPP.

Among the various protozoa that can find their way into milk and dairy products, *Cryptosporidium* (in particular *Cr. muris* and *Cr. parvum*) is the most significant for milk and dairy products. The reservoirs for *Cryptosporidium* spp. are various animal species (cattle, pigs, sheep, mice, rodents, cats, mice) as well as humans. The oocysts of this organism can also be present in water and the environment and can survive for many weeks under cool and moist conditions. *Cryptosporidium parvum* cannot survive pasteurization of milk and 100% inactivation is achieved by heating milk to 71.7°C for 5 seconds. *Cr. parvum* has shown 0–5% viability after 48 hours in ice cream stored at −20°C. Prolonged storage of contaminated yoghurt for up to 240 hours has not been sufficient to destroy *Cr. parvum*. A decrease in viability of the organisms from 83% at time 0 to 61% after 240 hours has been noted.

Cryptosporidium can cause clinical disease and death in young animals. Young farm animals are very commonly infected with *C. parvum* and milk becomes contaminated through feces. Several outbreaks of cryptosporidiosis have been associated with milk. In 1985 an outbreak of 22 cases of cryptosporidiosis occurred in Mexico in which contaminated milk was suspected. In 1995 50 cases of cryptosporidiosis were confirmed in the United Kingdom. Junior-level schoolchildren were infected after drinking milk that was distributed to the school by a small-scale local producer. The on-farm pasteurizer was found to be faulty and hence the milk was not adequately pasteurized. In 1984 a mother and her 1-year-old child were infected with *Cryptosporidium* after drinking unpasteurized goats' milk that had been purchased locally in Australia.

Fecal–Orally Transmitted Pathogens

Many fecal–orally transmitted human pathogens (including bacteria, viruses and protozoa) can be transmitted via milk and dairy products by contamination during the milking process at the farm. This happens when an ill or subclinically infected farm worker is involved in the milking process. For instance, *Shigella* spp. can contaminate milk and remain viable for at least 72 hours at 4°C in raw milk. As the infective dose is very low (10 to $100\,cfu\,ml^{-1}$), even without proliferation of the organism, human illness can occur. Analysis of bulk cows' milk indicates that in some developing countries, a significant proportion of the samples, e.g. 20%, contain *Shigella* spp. in the range of 1×10^6 cfu/ml.

VIRUSES (E.G. *ROTAVIRUS*, HEPATITIS A VIRUS, *POLIOVIRUS*)

It is reported that the first enteric virus associated with milk was the poliovirus in the period before the Second World War. Most foodborne viruses are transmitted via the fecal–oral route and can be infective at very low doses. Milk can be contaminated by the farm worker, polluted water, or directly by fecal contamination. In experiments where pasteurized and boiled milk were artificially inoculated with poliovirus and coxsackievirus B, these viruses survived for at least 90 days at 4°C and for up to 30 days at 25°C. Similarly, yoghurt stored at 4°C supported the survival of poliovirus and coxsackievirus B5 for 90 days. On the other hand, both viruses are readily inactivated by pasteurization.

In 1993, seven people were infected with the tickborne encephalitis (TBE) virus after drinking raw goats' milk. Previous cases of alimentary TBE were recorded in 1984 (four cases) and in 1989 (two cases). Both of the latter outbreaks were associated with the consumption of unpasteurized goats' milk. TBE belongs to the flavivirus family and is the only enveloped virus known to be associated with foodborne infections. The virus infects dairy animals via the tick vector and infected animals shed the virus in their milk, which if ingested without pasteurization may infect humans.

Finally, in the 1990s, prions, the agent of transmissible spongiform encephalopathies (TSEs), emerged as a new potential foodborne pathogen. Up to the present, however, except for the hypothetical case of bovine spongiform encephalopathy in small ruminants, there is no scientific evidence to suggest that milk and dairy products can pose any risk to human consumers.

Control of Microbial Hazards

Prevention and control of microorganisms in dairy products starts at the farm level by placing different hurdles to minimize opportunities for contamination. Maintaining animal health and welfare is critical to producing high-quality milk and minimizing contamination as many of the foodborne pathogens are shed in milk. To prevent spread of diseases and contamination of the environment, every effort should be made to identify and separate sick animals as soon as possible. As an indicator for animal health, the somatic cell count (SCC) is often used. An uninfected udder will typically have an SCC less than 100,000 cells^{-1}. Generally, the lower the SCC, the better the animal health. The upper accepted limit varies among countries. The European Union (EU), New Zealand, Australia, Switzerland and Norway have all set an upper limit of 400,000 cells/ml. Coliform bacteria are indicators of poor udder preparation or unhygienic handling of the milking machines. Generally, coliform counts should be less than 100 cfu ml^{-1} for milk intended to be pasteurized and less than 10 cfu ml^{-1} if milk will be consumed raw. The European Union has also established microbiological criteria for milk that will be consumed raw or used for cheese production (EC, 2005).

Where applicable, vaccination can be used to prevent infections. Under certain circumstances, testing and culling of chronically infected animals may be undertaken. These measures have led to the effective eradication of brucellosis, Q fever and bovine tuberculosis in certain regions of the world.

The milking process is an important step for the safety and quality of milk and dairy products. Milking should take place on animals in good health and with consideration of the withdrawal period if a medication is administered. Unsanitary udders have significantly higher prevalence of infection with mastitis pathogens. Good hygiene practices can effectively reduce the bacterial contamination in milk. Before milking, the udder and the teat need to be thoroughly washed, sanitized and dried. The collected milk should be cooled as soon as possible after milking, and any suspect milk (off-color or off-odor) should be separated.

Environmental hygiene is essential for prevention of contamination after the milking process as many of the above pathogens are transmitted through direct or indirect contact with animal feces, contaminated water and soil. Bedding material can itself be highly contaminated and contain up to 10^{10} cfu g^{-1} of pathogenic bacteria. Rodents, pets and birds are reservoirs for *Leptospira*, *Campylobacter* and *Salmonella*, respectively, and can be sources of contamination in the housing area of the cows. Dairy equipment (bulk tanks, heating or cooling units, milking equipment) can also be contaminated and must be cleaned and disinfected after use. Water can be the source of many pathogens, particularly *Cryptosporidium* which is relatively resistant to simple chlorination. Farm workers must respect personal hygiene and abstain from milking if ill, as they may shed human pathogens.

Veterinary Drugs

ANTIMICROBIALS

The most commonly used veterinary drugs associated with milk are antibiotics, employed to combat mastitis-causing pathogens in the dairy cow. National surveys in developed countries show that between 0.1 and 0.5% of tanker milk samples test positive for antibiotic residues. The occurrence of antibiotic residues in milk may have economic, technological and even human health implications. In the first place, such residues can lead to partial or complete inhibition of acid production by cheese starter cultures. This can lead to inadequate ripening and maturation of the cheese, resulting in flavor or texture defects and substantial financial loss for the dairy industry. There has also been increasing public concern over the possible links between antimicrobial drug residues in milk and the transfer of antibiotic-resistant organisms to humans as a result of veterinary and prophylactic use of antibiotics in food animals. A third concern is that sensitive individuals could exhibit allergic reactions to drug residues or their metabolites, especially in the case of beta-lactam antibiotics.

As from 1990, maximum residue limits (MRLs) have been set in Europe for veterinary drugs in foodstuffs of animal origin like milk. Most dairy companies also use rapid tests to monitor all incoming milk for the presence of beta-lactam antibiotics. Some of these companies are claiming compensation from the responsible farmer for the costs of disposing of the milk of a contaminated tanker load.

The solution to the problem of drug residues in milk lies in the application of the general principles of "Good Farming Practice." These include the following principles:

- Good farm management should in the first place be directed toward the prevention of infectious diseases, such as clinical and subclinical mastitis, in order to limit the use of veterinary drugs;
- In the process, the farmer must keep his animals in sound physical condition by ensuring proper hygiene and good housekeeping practices and implementing sound farm management;

- In preventing mastitis, the use of properly functioning milking machines is of primary importance. The use of veterinary drugs, nevertheless, remains necessary, but this option should only be exercised after a correct diagnosis by a veterinarian. Only registered pharmaceutical products with known depletion patterns should be used;
- Correctly administering veterinary drugs is also very important in terms of prescribed dose, frequency, route of administration and withdrawal period;
- Keeping reliable records of such drug use is also essential. It remains the responsibility of the milk producer to respect the prescribed withdrawal period. In the process, the treated animal needs to be marked clearly to allow for correct identification (e.g. by taping a hind leg);
- Treated cows need to be milked last during their withholding period so that the milk can be discarded in the proper way. The milking equipment should also be cleaned properly after contact with the contaminated milk;
- Special care should be taken with milk from cows that have been treated with long-acting dry cow products or with milk from cows that have been recently purchased; and
- Good communication is also important. Everyone on the dairy farm should be informed of any treatment. To facilitate this, the number of people authorized to administer antibiotic and other drugs should be limited. If there is any doubt, the milk should be tested or discarded.

OTHER VETERINARY DRUGS

Many human drugs are contraindicated during lactation and a similar problem occurs in dairy animals. The treatment of animals with ectoparasiticides and endoparasiticides can result in residues in milk if the withholding period before returning to milk production is not observed. Endoparasiticides used to control helminths (including tapeworms, roundworms and flukes) may be administered as feed additives, by injection or cutaneously. The most commonly used compounds in the past were levamisole and the benzimidazoles, but have now been largely supplanted by ivermectin. Studies on the excretion rate of these drugs indicate that a withdrawal time of 5 days is adequate after therapeutic treatment. Cutaneous treatment of animals against ectoparasites includes a wide variety of pesticides that are evaluated as veterinary drugs. In most cases, a 2-day withdrawal period is adequate for assuring that residues are within the safe limits. Consequently, for these types of drugs, the safety of the milk of treated animals depends on their proper use. However, the use of pesticides to control environmental problems, such as insects and rodents, also needs to be handled with care to avoid contamination.

HORMONES

A number of hormones are often used in relation to dairy animals, such as oxytocin and prostaglandins. However, one of the more controversial is bovine somatotropin (BST) (sometimes referred to as bovine growth hormone) and its genetically engineered counterpart recombinant BST (rBST). Adopted by a contentious vote in 2012, the use of rBST to "freshen" cows is recommended by the Codex Alimentarius Commission and is approved for use in the USA and in other countries such as South Africa, but not in the EU because of animal health reasons. However, the milk from rBST-treated cows may be traded without restrictions. This is because both BST and rBST are not hormonally active in humans and if ingested, they are rapidly digested because they are protein hormones.

Mycotoxins

As mentioned above, dairy cattle and animals are exposed to mycotoxins through feed ingredients such as maize and peanuts. Mycotoxins pose risks to both animal and human health and, depending on the mycotoxin, may have carcinogenic, estrogenic, neurotoxic, dermonecrotic or immunosuppressive effects. Fungal species of greatest concern in the dairy industry are *Aspergillus flavus, A. parasiticus* and *A. nomius.* These species produce aflatoxin B1 and related toxins under favorable conditions of temperature, water activity and nutrient availability, which are common in subtropical climates. In recent years concern has been expressed about the presence of aflatoxin M1 in milk and milk products, which is an animal metabolite of aflatoxin B1.

The exposure of humans to aflatoxins has resulted in liver damage and cancer. Aflatoxin M1 levels in dairy products are regulated in many countries. Codex has adopted a maximum limit (ML) of 0.5 μg/kg of liquid milk, but has not set an ML for its precursor aflatoxin B1 in animal feed. The EU has establish MLs for feed commodities that vary between 0.05 and 0.005 mg/kg. Provided that these MLs for aflatoxin B1 (and other mycotoxins) in feeds are observed, there should be no health problem from residues in milk.

Mycotoxins produced by fungal species other than *Aspergillus* and possibly *Penicillium* are of minor concern for dairy products. Nevertheless contamination of feed and forage with zearalenone (a mycotoxin of *Fusarium* spp.) has been shown to result in residues of zeranol in forage fed to cattle. Hydrogenation of alfa-zearalenol, probably in the rumen, is responsible for the formation of zeranol. Zeranol is approved in some countries as a hormonal growth promoter, but is specifically prohibited from use in food animals in the EU. This finding of a "natural" zeranol source in cattle has complicated control measures and makes it necessary to differentiate zeranol arising from feed and forage contamination from its deliberate use as a growth promoter.

Industrial and Environmental Contaminants

PESTICIDE RESIDUES

Pesticides include insecticides, herbicides and fungicides. The most common concern is insecticides including organochlorines, the organophosphates and carbamates. Organochlorine pesticides enter the food chain as a result of their lipophilic properties, in this way biomagnifying in the food chain and bioaccumulating in individuals. Milk is considered as one of the more convenient indicators for measuring the extent of these persistent residues that originate in contaminated animal feed. The main route of human exposure to many organochlorine pesticides is through foods of animal origin. Typical contaminants of milk are the persistent fat soluble organochlorine pesticides such as DDT (dichlorodiphenyltrichloroethane) and other organochlorine pesticides. However, most of these pesticides have been withdrawn from use, but remain as environmental contaminants as they are quite stable in soil. Cows grazing in contaminated pastures can ingest soil with these pesticides, particularly under drought conditions.

Organophosphate and carbamate pesticides are the most widely used insecticides today, but degrade rapidly in the environment and are further metabolized by animals. Codex routinely establishes MRLs for these and other pesticides on animal fodders as a result of their use on crops. In some cases, MRLs are also established for residues in milk and in other animal products, such as eggs. For example, the use of the insecticide cyromazine and its metabolite melamine can appear as residues in milk.

DIOXINS AND POLYCHLORINATED BIPHENYLS (PCBS)

The term "dioxins" covers a group of 75 polychlorinated dibenzo-p-dioxin (PCDD) and 135 polychlorinated dibenzofuran (PCDF) congeners of which 17 are of toxicological concern. The most toxic congener is 2,3,7,8-tetrachloro dibenzo-p-dioxin (TCDD) which is a known human carcinogen, but possesses other toxic properties. Adverse effects of dioxins were considered not to occur in humans at levels below certain thresholds, but this is being challenged by epidemiological studies showing effects at low doses. The maximum level set in EU regulations for dioxins in milk and milk products, including butter fat, is 3 pg World Health Organization Toxic Equivalents (WHO-TEQ)/g fat.

The polychlorinated biphenyls (PCBs) are a group of 209 congeners which can be divided into two groups according to their toxicological properties. Twelve congeners exhibit toxicological properties similar to dioxins and are therefore termed "dioxin-like PCBs." The other PCBs have a different toxicological profile.

Dioxins and PCBs are extremely resistant to chemical and biological degradation and therefore persist in the environment and accumulate in the feed and food chain. Dioxins arise during the production of chloro-organics and in emissions of municipal incineration and other pyrolytic processes, such as forest fires. PCB production ceased many years ago, but PCBs remain as environmental pollutants. Contamination of animal feed occurs via particle-bound distribution on grass and other fodder plants. The accumulation of dioxins in animals is mainly from these contaminated feeding stuffs. Human foods of animal origin in turn contribute to approximately 80% of the overall human exposure to dioxins and PCBs.

For these reasons feeding stuffs and in some cases soil raise concerns as potential sources of dioxins. Like the organochlorine pesticides, dioxins and PCBs are fat soluble. Case studies that have involved these contaminants include the Belgian PCB/dioxin incident in 1999. Feedstuffs produced from a contaminated source were sent to over 2500 farms and subsequently appeared in nearly every category of animal-derived food (pork, milk, chicken and eggs).

Based on 13,797 food samples taken between 1995 and 2010, the European Food Safety Agency (EFSA) has estimated that the percentage of individuals exposed to dioxin and dioxin-like PCBs above the tolerable weekly intake (TWI) of 14 pg WHO-TEQ/kg bodyweight was between 1.0 and 52.9%. The major contributor to total exposure was the food category of milk and dairy products for almost all groups of infants and toddlers, whereas for most of the groups of adolescents, adults, elderly and very elderly, fish and shellfish were first, followed by milk and dairy products. For the total dioxins and dioxin-like PCBs, the upper bound estimate for 1422 samples of milk and milk products was 1.92 pg WHO-TEQ/kg. The corresponding EU maximum limit for milk and dairy products is 5.5 pg WHO-TEQ/kg. Of all samples tested, 0.5% exceeded this limit by a small margin. As the EU has implemented source-directed measures, the levels in food have been falling for many years.

HEAVY METALS

Heavy metals is a general term that applies to a group of metals and metalloids, including elements such as cadmium (Cd), chromium (Cr), nickel (Ni), copper (Cu), zinc (Zn), mercury (Hg) and lead (Pb). This sometimes includes arsenic (As) because of its toxicity.

Because of their widespread distribution in the environment and their many adverse health effects, heavy metals have become a global health concern. Pb is one of the most common heavy metal pollutants and tolerable levels have fallen as the adverse effects of Pb, especially on the development of cognitive brain function in children, have become recognized. Pb is readily transmitted to milk as it is associated with calcium metabolism. However, in a 1999 study, the highest metal concentrations in dairy cattle feeds were for Zn and Cu. Mineral supplements contained higher concentrations of Ni, Pb, Cd, As and Cr than did other feed components.

Cleaning Agents and Sanitizers

Cleaning and sanitizing agents are essential components of any good manufacturing practice in the food industry, but are particularly important for milk and milk products to remove any bacteria from food contact surfaces. However, it is important that residues of cleaning agents and sanitizers are also removed to avoid contamination. At the farm level, maintaining hygiene of the udder is of critical importance to prevent microbial contamination of milk and various disinfectants have been developed to clean the udder before milking. At the plant level, the use of cleaning-in-place (CIP) methods requires careful cleaning, sanitizing, draining and rinsing procedures. Some sanitizers, such as certain iodophores, do not require rinsing and therefore occur in milk and dairy products at very low concentrations as indirect food additives. The most commonly used sanitizers, including iodophores, chlorhexidine and hypochlorites, contain iodine or chlorine as the active agent. In addition, hydrogen peroxide and quaternary ammonium compounds are also used.

Generally, only limited information is available on the toxicology of these compounds and their occurrence in food, with the exception of iodine. Iodine levels in milk have been increasing in a number of countries. On the other hand, milk was found to be an important dietary source of iodine in New Zealand and as the industry moved to quaternary ammonium compounds, the government found it necessary to initiate the fortification of bread with iodine to prevent deficiencies of this micronutrient in their population.

Other Potential Chemical Hazards

RADIONUCLIDES

Other contaminants that may arise in milk and milk products include radioactive isotopes whose sources are both anthropogenic and non-anthropogenic. For instance, background radiation in milk may vary from 40 Bq/l for potassium-40 to below 0.1 Bq/l for cesium-137. Radioactivity of food may increase as the result of certain human activities such as weapons testing and nuclear accidents. The latter presents the greatest source of contamination as demonstrated by the Chernobyl accident in 1986 and the Fukushima disaster in 2011. The Codex Alimentarius Commission has developed derived intervention levels for milk and other foods for various radionuclides following such accidental releases.

FRAUD AND ECONOMIC ADULTERANTS

The addition of water to milk is probably one of the oldest forms of economic adulteration of food. However, other materials, such as chalk, were often added to mask this fraud. The most recent variation of this practice was the addition of melamine to milk in China. This was done to avoid detection by the standard analysis of crude protein in milk, sometimes called total protein, which is used to monitor and control milk quality in the dairy industry. The Kjeldahl analysis measures the total nitrogen content of milk, which is then simply multiplied by 6.38 to express the result on a protein equivalent basis. Melamine is a chemical substance rich in nitrogen, inexpensive and widely available as it is used in the manufacture of many laminates, plastics, coatings, glues and kitchenware. In 2008, infant formula made with melamine-adulterated milk resulted in illness in a reported 300,000 infants, including over 50,000 hospitalizations and six deaths. A range of other products including liquid milk, ice cream and yoghurt were also contaminated. Because melamine can be present in milk and other products as a result of other sources of contamination, the Codex Alimentarius Commission has recommended MLs to accommodate these situations at 1 $mg\,kg^{-1}$ in infant formula and 2.5 $mg\,kg^{-1}$ in other foods and animal feed.

Physical Hazards

Physical hazards, or more generally foreign bodies, include glass, metal, stones, wood, plastic, dust, dirt, hair and insect fragments. Although some technologies are available for the verification of any incidental contamination, effective removal of physical hazards at later stages of the processing and manufacturing is difficult. A more effective strategy is to prevent such contamination all along the food chain starting at the farm, with policies and programs such as pest management, a glass-free policy where feasible, protective clothing and good housekeeping practices.

Transportation

Collection and transportation of milk is a point in the food chain where recontamination of milk with chemical or microbial hazards and/or growth of microorganisms can occur. Therefore, the bulk tanks and vessels used for milk transportation need to be cleaned and disinfected. Care should be taken that these are not used for any other purpose, especially the transport of potentially toxic materials. To minimize growth of microorganisms, milk should be chilled to a temperature of 6°C or below and processed as soon as possible. Good transportation practices, in particular maintaining the cold chain, need to be observed, in particular for sensitive products. In a major outbreak of salmonellosis in 1994 in the USA, an estimated 224,000 persons became ill after consuming contaminated ice cream. This outbreak of salmonellosis was most likely due to contamination of the pasteurized ice cream premix during transport in tanker trailers that had previously carried non-pasteurized liquid eggs containing *Salmonella Enteritidis*. This incident highlights the importance of observing hygienic measures during transportation.

Transportation is also a point of vulnerability in the food chain where actions of sabotage or tampering may occur. Addition of *Cl. botulinum* toxin or other chemical hazards in milk have been considered as potential risks for bioterrorism or sabotage. Therefore, measures should be taken to secure the transport vehicle, e.g. sealing tanks with tamper-proof tags.

Processing and Manufacturing

Milk

In the dairy plant, processing of milk involves several unit operations: storage, clarification, preheating, separation, standardization for fat content, homogenization, pasteurization, cooling and packing. The raw milk is stored in silos for a limited time (in the USA, it is up to 72 hours) at below 7°C or preferably 4°C according to local regulatory requirements. Refrigeration is essential for limiting growth of organisms.

Of various unit operations, pasteurization is the most critical for safety as raw milk, in spite of all efforts made at the farm level, may still contain foodborne pathogens. Vegetative cells of foodborne pathogens are sensitive to heat and are readily killed by the pasteurization process.[2] During the process, spoilage microorganisms and undesirable enzymes (lipases and protease) are also reduced. Hence pasteurization both ensures safety and prolongs shelf-life with minimal changes to flavor and nutritional quality of the product. Three different heat treatment conditions are usually applied, namely:

- 63°C for at least 30 minutes is the low temperature–long time (LTLT) method, which is often used for batch pasteurization. In this process, the milk is stirred regularly to ensure that all particles receive adequate heat treatment;
- 71.7°C for at least 15 seconds is the high temperature–short time (HTST) pasteurization method and is applied in heat exchangers that process milk continuously. The method provides higher energy efficiency; and
- 135°C for 1 second is the ultra-high temperature (UHT) process. When this process is combined with aseptic packaging, the unopened product is shelf-stable and can be kept unrefrigerated.

Pasteurization is designed to destroy the most heat-resistant pathogens: *C. burnetii*, *M. tuberculosis* and *L. monocytogenes*. The efficacy of pasteurization depends on the initial bacterial load. This is generally about 10^4 or 10^5 cfu ml^{-1} of milk. Other pathogens such as *Brucella*, *Campylobacter*, *E. coli*, *Salmonella*, *S. aureus* and *Yersinia* are all killed during the process. However, thermophilic and some mesophilic organisms as well as spore-forming bacteria (e.g. *Bacillus* and *Clostridium*) can survive the heat process and contribute to the spoilage of milk. To demonstrate that milk has been adequately pasteurized, the alkaline phosphatase test can be carried out. This is based on detecting the presence of a temperature-sensitive enzyme in milk (slightly more heat resistant than *C. burnetti*) that is inactivated during pasteurization.

Occasionally, there have been reports of foodborne disease outbreaks associated with pasteurized milk. These usually occur as a result of failures in the pasteurization process or are due to post-process recontamination. For instance, in 1985, the post-process contamination of pasteurized milk with *Salmonella Typhimurium* led to a major outbreak of salmonellosis affecting an estimated 168,791 to 197,581 persons. The likely cause of contamination was an error in equipment design leading to a cross-connection between a tank containing raw milk and a tank with pasteurized skim milk. Other outbreaks have occurred as a result of failures in the heating process, use of the wrong thermocouple and failing to monitor time–temperature parameters. In a major outbreak of *S. aureus* intoxication, where over

[2]Pasteurization is a microbiocidal heat treatment aimed at reducing the number of any pathogenic microorganisms in milk and liquid milk products, if present, to a level at which they do not constitute a significant health hazard. Pasteurization conditions are designed to effectively destroy the organisms *Mycobacterium tuberculosis* and *Coxiella burnettii* (CAC, 1999).

13,800 school children were affected in Japan in 2000 after consuming contaminated skim milk, an electricity cut led to time–temperature abuse of the product. The company then applied an inappropriate corrective action by attempting to reheat the milk without recognizing that the *S. aureus* enterotoxins are heat stable once formed. In an outbreak of yersiniosis in the USA caused by chocolate milk, chocolate syrup contaminated with Yersinia was added to milk after it had been pasteurized.

Pasteurization does not eliminate spores of bacteria, and many spoilage bacteria are resistant to pasteurization temperatures. To destroy endospores, higher heat treatment such as UHT must be applied. Therefore, to prevent growth of microorganisms surviving or incidentally recontaminating milk, pasteurized milk products need to be refrigerated as soon as possible and maintained cold. As seen above, this is important to prevent growth of *S. aureus* and its production of heat stable enterotoxins.

In developing countries where facilities for pasteurization and cooling are not available, the lactoperoxidase (LP) system can be used to minimize microbial growth and extend the shelf-life of milk. The LP system exploits the antimicrobial system naturally present in milk by increasing the concentrations of two components or activators (thiocyanate and hydrogen peroxide) reacting with each other. This reaction is catalyzed by the enzyme lactoperoxidase which is naturally present in milk and leads to the formation of antibacterial compounds (CAC, 1991).

One of the frequent problems associated with food processing and manufacturing is foreign objects, some of which are health hazards and pose risks of injury or choking. Examples are glass, stones, wood, plastic fragments and metal (or metal particles resulting from friction between metal parts).

Preventive measures should be put in place to protect products. These include:

- Hygienic design of equipment and preventive maintenance to prevent loose parts falling in the products and friction between metal parts;
- Using shatterproof light covers to prevent glass contamination from taking place; and
- Prohibiting jewelry, glass (glass-free policy) and wooden pallets in the processing area.

During the processing of milk, it is invariably subjected to procedures that will remove any physical contaminant. Centrifugal clarifiers are standard equipment in any commercial milk processing operation and filters are employed in many places. To further reduce risk, sieving milk powder or using magnets for incidental presence can be used. As a final verification measure, products can be passed through metal detectors or X-ray equipment (important if glass jars or bottles are used) to confirm that preventive measures are effective or as a corrective measure in case of failure.

Cheese

Cheese is the product of casein coagulation in the milk followed by separation and removal of the whey from the curd. There are many types of cheese: hard, soft, semi-soft or semi-hard cheese, as well as cheese made from pasteurized milk, or cheese made from unpasteurized milk (Muehleman, 2014).

In relation to cheese, a broad range of organisms are of concern. These include *S. aureus*, *Bacillus* spp., *Cl. botulinum*, *L. monocytogenes*, pathogenic *E. coli*, *Salmonella*, *Streptococcus* groups A and C, *B. abortus*, *M. bovis* and *C. burnetti*.

Production of cheese made from raw milk requires high-quality raw milk, minimal microbiological contamination and prevention of growth at all levels of the production chain. Some pathogens, e.g. most viruses, *Shigella*, enterohemorrhagic *E. coli*, *Campylobacter* and *Cryptosporidium*, have low infective doses and even a slight contamination of products can be the source of infection. In high fat products such as cheese, even pathogens usually requiring large numbers to cause infection can be infective in low doses. This implies that the presence of low numbers of pathogens in milk destined for the production of raw milk cheeses can constitute a threat to the consumer.

To ensure safety, strict hygienic control based on the concept of hurdle technology must be applied (Muehleman, 2014). This consists of:

- Sourcing milk of high microbial quality from healthy animals;
- Rapid cooling and processing of the milk;
- Rapid acidification of the cheese and salting of the cheese during its ripening and maturation to reduce the water activity; and
- Hygiene controls during its aging.

Outbreaks related to cheeses made from unpasteurized milk are often related to one or a combination of the following factors:

- Animals shedding pathogenic bacteria;
- Improper storage of milk prior to cheese production, e.g. no refrigeration for several days prior to manufacture;
- Poor starter activity with consequent production of cheese with a too high pH;
- Poor starter activity due to inhibition of acid production by phage and/or antibiotic residues in the milk;
- Poor plant hygiene, gross environmental contamination;
- Faulty pasteurization; and
- Shedding of the causative organisms by plant personnel.

In industrial processes, safety is based on the pasteurization of milk, use of acid-producing starter culture to produce a rapid decrease in pH, and prevention of post-process contamination. Phage contamination and/or presence of antibiotic residues in the milk can slow down the acidification process and create pH conditions that are favorable for growth of pathogens during the first hours of production. Some starter cultures also produce bacteriocins. These have the ability to inhibit other bacteria. Additionally, during the ripening process in hard and semi-hard cheeses, most pathogens die off. Several factors such as moisture, salt, nitrate, pH and temperature control these processes. The effect of these conditions on pathogens varies according to the ecology of the pathogen. On the other hand, soft and semi-soft cheese (e.g. Camembert and Brie) contain a high level of moisture and the pH of the surface increases during ripening due to yeast activity. This will provide optimum conditions for growth of pathogens and, should the product be contaminated with *L. monocytogenes*, pose a particular risk as the organism may establish itself in the processing environment, proliferate at prevailing low temperatures and contaminate other products.

Outbreaks related to fresh, soft or semi-soft cheeses made from pasteurized milk can usually be linked to failures in pasteurization and inadequate post-pasteurization hygiene. For example, the investigation of an outbreak of salmonellosis due to *Salmonella Heidelberg* in

Denver and Pueblo, Colorado, in 1976, showed that the raw milk used to make the cheddar cheese contained more than 3 million bacteria per ml. The raw milk was stored for 1–3 days in insulated, but unrefrigerated, holding tanks (Poppe, 2011).

Another large outbreak of salmonellosis occurred in Canada in 1984 and affected some 2700 people who had consumed cheddar cheese contaminated with *S. Typhimurium* PT10.

An investigation into the outbreak showed one of the cows in the herd was intermittently shedding *S. Typhimurium* PT10, although clinically healthy. Additionally, the manual turning of an electronic flow diversion valve in the plant allowed some raw milk to flow into vats used for cheese-making. The faulty pasteurization process was confirmed by the phosphatase test. The investigation further showed that in a high fat matrix such as cheese only a few *Salmonella* bacteria might cause infection in consumers and that *Salmonella* can survive refrigerated storage for more than 40 weeks (Poppe, 2011).

Utensils used for cheese-making may also be a source of contamination. In an outbreak of salmonellosis caused by *Salmonella Berta*, the buckets used for manufacture of the unpasteurized soft cheese had previously been used for soaking chicken carcasses. *S. Berta* was also isolated from the chicken carcasses (Poppe, 2011).

Salmonellosis has also been associated with cheese made with raw goats' milk. In 1993 in an outbreak of salmonellosis in France, consumption of goats' milk cheese made from unpasteurized milk caused a large number of consumers to be infected with *Salmonella* Paratyphi B. The organism was isolated from milk at the plant and was found in the milk from one of the farms supplying milk.

L. monocytogenes is of particular concern with soft cheese made from raw milk. The first and largest known outbreak of listeriosis implicating cheese occurred in Los Angeles in 1985 and was implicated in an estimated 300 cases and led to 48 deaths. The outbreak was due to consumption of California-made Jalisco-brand Mexican-style cheese. Factory records indicated that raw milk may have been intentionally added to pasteurized milk used in making the cheese.

Another *Listeria* outbreak involving cheese occurred in Switzerland that was associated with consumption of Vacherin Mont d'Or. From 1983 to 1987, the outbreak claimed 122 victims and 34 deaths. The strain implicated in the outbreak was isolated from the surface of the cheese at levels of 10^4–10^6 cfu g^{-1}. The wooden shelves and brushes in cheese-ripening cellars were also contaminated. The outbreak was brought under control after cleaning and disinfection of the ripening room and replacing the wooden ripening shelves with metal shelves.

In general, *L. monocytogenes* is quite well adapted to dairy factory environments. When strains of *L. monocytogenes* become established in a food-processing facility, they can remain members of the resident microbial flora for many years. Soft and semi-hard cheeses that were improperly prepared have periodically been implicated in outbreaks of listerioisis.

In processing plants, the primary source of *Listeria* spp. are most likely drains and floors, particularly areas which are wet, cool and inaccessible to cleaning and sanitation. Cooling waters should also be considered as a possible source of contamination.

To control *L. monocytogenes* in the processing area, the following measures need to be considered (Joost and Anelich, 2008; Kozak et al., 1996):

- Hygienic zoning, i.e. separation of the milk-receiving area and from the processing and packaging area and preventing any raw product from entering the processing area. The zoning plan should include the flow of people, pallets, raw material and equipment used for the raw material (e.g. forklifts).

- Hygienic design of premises, drains and equipment to prevent build-up of organic matter, formation of biofilms and contamination through aerosols, and also to facilitate the effective cleaning and sanitization of the premises and equipment. In this regard, drains should not be located under or near filling and packaging equipment. Areas that have the potential for pooling of product should be eliminated. Outside air should be filtered and be free of condensate. All conditions for airborne contamination should be avoided.
- Effective pasteurization is the key step to control *Listeria* during processing. In the case of HTST pasteurization of milk, a minimum of 72°C for 15 seconds is essential. Products containing higher fat or sugar levels require higher temperatures to ensure effective destruction of *Listeria* spp., such as 75°C for 15 seconds to be safe.
- Cleaning and sanitizing programs are vital in ensuring that post-pasteurization contamination does not occur. Absorbent items such as rags and sponges should be eliminated to reduce potential harborage and spread of the organisms. Separate brushes should be used for product contact and non-product contact surfaces.
- An environmental monitoring program should ensure that environmental hygiene is under control. In this regard, particular attention should be given to niches where *L. monocytogenes* is likely to establish itself, e.g. hollow rollers on conveyors, cracked tubular support rods on equipment, the space between close-fitting, metal-to-metal or metal-to-plastic parts, worn or cracked rubber seals around doors, on–off valves and switches for equipment as well as saturated insulation material, and cool and moist areas.
- In the case of a positive sample, there should be rapid and effective corrective action.

Various outbreaks of botulism have also been reported in connection with cheese. In 1997, in Italy, mascarpone cheese contaminated with *Cl. botulinum* type A led to an outbreak affecting eight persons and causing one death.

Cheese can be a source of biogenic amines, notably histamine and tyramine, both of which cause intense, but transient intoxications. They are produced by spoilage organisms that contain enzymes that decarboxylate the corresponding amino acids to the amine. In an outbreak of histamine intoxication in Canada, the amount of histamine in the cheddar cheese was $40\,mg\,100\,g^{-1}$. In another outbreak in the Netherlands, Gouda contained $5\,mg\,100\,g^{-1}$ histamine.

Other contaminants that can be of concern are nitrate and nitrite that cows are exposed to during grazing and through drinking water. Milk can be contaminated by either secretory or post-secretory processes. However, the level of nitrate in the diet of dairy cows does not seem to affect milk composition. The nitrate content can, however, increase as a result of intentional addition during cheese production or as a consequence of residues of sanitizers. Another source of nitrate can be contamination with incoming wash-water. However, except for situations where nitrates are added intentionally or accidentally, at the levels they usually occur in milk and dairy products, they do not present a major public health concern (Indyk and Woollard, 2011). However, in 2011, 38 persons were reported to suffer from nitrite poisoning following deliberate addition of nitrite to milk and three children died.

A range of other chemical contaminants might also occur at the processing stage. However, unless introduced by other ingredients, milk and dairy products normally contain very low levels of these substances.

Yoghurt, Ice Cream and Dairy Desserts

Industrially produced dairy desserts generally include a microbiocidal step such as pasteurization, retort sterilization, or ultra-high-temperature–short-time processing, which confers product safety. Under conditions where there is no post-process contamination, these products have a good safety record.

Post-process contamination has been a concern with *L. monocytogenes* in ice cream. However, due to the inability of the organism to grow at freezing temperatures, the public health of risk associated with ice cream is viewed as low.

The main concern with dairy desserts, yoghurt and ice cream is with pathogens that are introduced with other ingredients, in particular eggs, or by the food handler where these products are prepared artisanally. For instance, in 2007, an outbreak of Shiga-like toxin-producing *E. coli* (STEC) O145 and *E. coli* O26 occurred in Belgium. The outbreak was caused by ice cream produced and sold at a farm in the province. Five children aged between 2 and 11 years developed hemolytic uremic syndrome (HUS) and seven other co-exposed persons contracted severe diarrhea. The investigation showed that the ice cream was made from pasteurized milk but was most likely contaminated by one of the food handlers (De Schrijver et al., 2008).

In 1989, in the United Kingdom, hazelnut yoghurt was implicated in an outbreak of botulism. The outbreak affected some 27 persons and led to one death. The source of contamination was hazelnut conserve sweetened with aspartame rather than sugar. The investigation indicated that the processing of the hazelnut conserve was inadequate to destroy *Cl. botulinum* spores (O'Mahony et al., 1990).

Although underreported and poorly documented, cream-filled bakery products or ice cream made in an artisanal way are also susceptible to causing intoxication by *S. aureus* entorotoxin, resulting from a combination of contamination of the product and time–temperature abuse. In 2002, ice cream of a major company was implicated in intoxication with *S. aureus* toxin. Over 20 persons were reported ill, but it is estimated that up to a few hundred persons may have been affected. The failure was in the maintenance of the dispensing machine, of which the circuit of cleaning and disinfection of the system did not function properly and the pasteurization system also failed to function. The owner of the shop was not aware of the importance of the pasteurization step.

E. coli O157 shows patterns of acid resistance. Therefore its survival in acid curd cheeses, yoghurt and other fermented dairy products for long periods needs to be taken into consideration.

Dried Milk Powder

Dried milk powder (including infant formulae) has occasionally been implicated in outbreaks of salmonellosis or other infections. The presence of the bacteria is usually as a result of post-process contamination. For instance, in an outbreak of *Salmonella Ealing* in the United Kingdom in 1985, all of the infected infants had been fed with a dried milk product from one manufacturer. The product was contaminated at a low frequency, i.e. four positives out of 267 sealed packets. This shows the difficulty and limitation of end-product testing for ensuring safety. The source of infection was traced to the factory spray-drier, which had a hole in its inner lining and allowed the escape of powder and its return after it was contaminated by insulation material. Following the incident, it was recommended that raw milk and whey, which frequently contain salmonellae, should not be allowed onto the site of milk-drying plants (Rowe et al., 1987). In 1973, a large-scale outbreak of salmonellosis occurred in Trinidad in which nearly 3000 infants were infected with *Salmonella Derby*. The investigation traced the illness to seven brands of powdered milk packed in a single processing plant. In 1977, another major outbreak of *Salmonella Bredeney* infection occurred in Australia and was linked to contamination of powdered milk-based infant formulas during manufacture. Investigation of the manufacturing conditions revealed that contamination occurred in the spray-driers (Cahill et al., 2008).

With regard to manufacturing infant formula, the two pathogens *Salmonella* and *Cronobacter sakazakii* are of particular concern. Although *C. sakazakii* is destroyed by heat treatment of milk, it has been implicated in several cases of infection or outbreaks in Belgium, France, Iceland, Israel, New Zealand and the USA. The major routes of contamination are believed to be post-process contamination, contaminated ingredients in dry mix processes, or contamination by caregivers during preparation. In infant formula manufacturing, the organism may establish itself in the processing environment and contaminate the products at various steps during or following the drying process. To prevent post-process contamination, strict hygiene (including zoning and dry cleaning) of the processing area is essential (see Chapter 33). In 2008, the CAC developed specific microbiological criteria for *Salmonella* and *C. sakazakii* in infant formula (CAC, 2008).

In some instances, it was shown that upon the reconstitution of the infant formula, hygienic principles had not been respected. Such errors are critical with respect to infectious agents as well as spore forming bacteria such as *Bacillus cereus*. If the time-temperature of storage of the product is not respected, spores which survive the heat treatment can grow upon reconstitution of the PIF and produce toxins. In one incident infant formula was kept warm in bottle warmers for several hours. Therefore during preparation and reconstitution of PIF, adhering to hygienic principles is of utmost importance. To this end, use of water at a temperature above 70°C and thereafter rapid refrigeration must be applied (WHO, 2007).

Over and above microbial and chemical hazards that may enter the product through milk, nutritional composition of infant formula should be considered in managing its safety as micro- or macronutrients in either excess or deficiency can be harmful to the health of infants. In 2003, infant formula deficient in vitamin B1 (thiamine) was the cause of a major outbreak in Israel causing three deaths and dangerously affecting the health of 23 other babies (see Chapter 40).

Packaging

Packaging protects milk from UV light, bacterial contamination and tampering. Glass bottles are less used as they pose a glass hazard and are heavy. Paperboard packaging with polyethylene coating offers a low-cost packaging option. Another option is the high density polyethylene (HDPE) packaging. The safety of packaging in general needs further study as this is an area that has evolved rapidly without full consideration of the safety of components and their interaction. In 2005, liquid infant formula of a major food company had to be recalled due to migration of isopropylthioxanthone (ITX), a photoinitiator used in printing inks on the product carton. Recently, bisphenol A (BPA) has raised safety concerns as it is a suspected endocrine disrupter with possible adverse health effects on the developing fetus. BPA is used in several food contact materials, including bottles used for infant formula feeding. Several countries including Canada, EU and recently the USA and South Africa have banned its use in baby bottles as a precautionary measure.

Milk powder must be packed in containers that adequately protect it from moisture, light, oxygen and foreign bodies. For the safety and stability of dried products, water activity is the most important parameter. The packaging should ensure that a safe range of water activity is maintained throughout the shelf-life of the product, and designed in such a way that temperature changes during transport and distribution will not affect the water activity of the product. The type of packaging of milk and dairy products can be of many varieties. The more common type are multilayered, consisting of paper or cardboard with polyethylene lining, or, alternatively, metal barrels lined with polyethylene, or cans sealed with aluminum foil. Packing is also carried out in an atmosphere of inert gas or under partial vacuum to prevent oxidation of the product (Alvarez and Pascall, 2011; Hotchkiss and Meunier-Goddik, 2002).

Warehouse

A potential source of contamination of dairy products in warehouses is pests such as birds, rodents, cockroaches, flies and other insects. Pests are the reservoir for many pathogens; hence, through their droppings and urine, they are direct and dangerous sources of contamination of food. Contamination of products can also occur indirectly through the environment. Pest management is thus important both from a general hygiene perspective as well as for control of specific pathogens. Good warehouse management is also important to prevent any accidental contamination of the products or raw material with industrial chemicals.

Labeling

It is the responsibility of food manufacturers to determine the shelf-life of products, and through labeling provide information on the durability and storage conditions so that safety and suitability (i.e. fitness for consumption) of the product is maintained throughout the period specified. Reasonably anticipated temperature abuse by consumers should be considered in establishing shelf-life as well as in the design of products. Also, as products such as dried milk or infant formula require final preparation or reconstitution, the instruction on the package should be clearly, visibly and unambiguously presented to inform consumers of potential safety issues and to provide the necessary information for the safe use of the product. The communication should take into account the risk perception of users and potential misuse of products. Very often consumers have the misperception that dried shelf-stable products are sterile and may ignore potential risks associated with the growth of any surviving organisms.

Soft cheese made from raw milk or other potentially hazardous products may present greater risks for vulnerable groups, such as pregnant women. Another example is a dairy dessert made with raw eggs that should be properly labeled to inform vulnerable population subgroups of the potential risk such a product may hold for them.

Another important consideration during labeling is allergens. Many consumers may be allergic to milk protein (casein) or may have intolerance for lactose. Cows' milk allergy is encountered with a prevalence of 2–3% in infants and 0.5–3% in adults, but this varies among different populations. In some cases, exposure to milk proteins can be life-threatening. About 75% of adults are estimated to experience a decrease in lactase activity during adulthood, although this varies by region. For example, it is only 5% in northern Europe where milk is consumed throughout life, but is 90% in Asia and Africa. In many countries, appropriate legislation for labeling of allergens exists and most of these contain a requirement for labeling of milk as an allergen. Nevertheless, manufacturers should be vigilant to include milk on the label as one of the ingredients of the product in which it is used; and as far as possible prevent cross-contamination through use of dedicated production lines, scheduling products and proper cleaning to minimize the need for precautionary labeling of cross-contact traces.

Preparation and Consumption

Worldwide, consumption of raw milk, soft cheeses made from raw milk, or desserts containing raw eggs has been the source of infections and outbreaks of foodborne illness. In spite of good animal husbandry at the farm and observation of hygienic practices in milk and cheese manufacturing plants, such products still present a residual risk for consumers. Therefore, consumers, particularly vulnerable groups (e.g. pregnant women, the young, immunocompromised and the elderly), should be informed of potential risks for their health of consuming such products and be advised to avoid them.

In developing countries, where access to a safe water supply is limited and infrastructure for hygienic preparation of food is rudimentary, the reconstitution of milk from powder or infant feeds from PIF can lead to the product becoming contaminated. The product may also be subjected to time–temperature abuse allowing growth of foodborne pathogens. Most of these cases or outbreaks are not reported or published officially. For instance, in 2004, an unpublished massive outbreak of foodborne illness occurred in Indonesia in which several hundred school children fell ill after consuming reconstituted milk, ironically during a campaign for promotion of milk. There are similar concerns with infant formula that have been recognized as the source of infant diarrhea in developing countries. The importance of education of caregivers and hospital personnel in the risks and safe preparation of infant formula cannot be overemphasized and the WHO has produced specific guidelines for this purpose (WHO, 2007).

CONCLUSION

In a large part of the world, milk and dairy products are important components of the daily diet. Ensuring the safety of these products requires the careful management of a wide range of microbial and chemical hazards. It is now generally accepted that this is only possible if basic rules of hygiene are respected all along the food chain from farm to final preparation of the product for consumption. Key elements include ensuring good animal health, a hygienic milking process, pasteurization of milk (or the equivalent) and prevention of post-process contamination. Together with other prerequisite programs, the application of the HACCP system can further enhance the safety of milk and dairy products by ensuring that the critical measures are implemented correctly.

In regard to human factors, education and outreach to dairy farmers is important in assuring the safe production and handling of raw milk. Training of personnel working in food processing and manufacturing is essential for ensuring that pasteurization is properly conducted and post-process contamination does not occur. Equal attention should be given to education both through labeling as well as in various settings, such as health clinics and schools. In all cases, the risk perceptions of the target groups must be considered in order to effectively influence their attitudes and change their behavior.

References

Alvarez, V.B., Pascall, M.A., 2011. Packaging. In: Encyclopedia of Dairy Sciences, second ed. Elsevier., (pp. 16–23).

CAC, 1991. Codex Guidelines for the Preservation of Raw Milk by Use of the Lactoperoxidase System was adopted [CAC/GL 13–1991] Codex Alimentarius.

CAC, 1999. General Standard for the Use of Dairy Terms (CODEX STAN 206–1999). Joint FAO/WHO Food Standard Programme, Rome, FAO.

CAC, 2008. Code of Hygienic Practice for Powdered Formulae for Infants and Young Children CAC/RCP 66–2008. Joint FAO/WHO Food Standard Programme, Rome, FAO.

Cahill, S.A., Wachsmuth, K., Costarrica, M.L., Ben Embarek, P.K., 2008. Powdered infant formula as a source of Salmonella infection in infants. Clin. Infect. Dis. 46 (2), 268–273.

De Schrijver, K., Buvens, G., Possé, B., Van den Branden, D., Oosterlynck, O., De Zutte, L., et al., 2008. Outbreak of verocytotoxin-producing *E. coli* O145 and O26 infections associated with the consumption of ice cream produced at a farm, Belgium, 2007. Eurosurveillance 13 (1–3) <www.eurosurveillance.org>.

EC-European Commission, 2003. Opinion on Staphylococcal Enterotoxins in milk products, particularly cheeses (adopted on 26–27 March 2003) <http://ec.europa.eu/food/fs/sc/scv/out61_en.pdf>.

EC-European Commission, 2005. Commission Regulation (EC) No. 2073/2005 of 15 November 2005 on microbiological criteria for foodstuffs. European Union, Brussels.

Fox, P.F., 2011. Milk Encyclopedia of Dairy Sciences, second ed. Elsevier., (pp. 458–466).

Hotchkiss, J.H., Meunier-Goddik, L., 2002. Encyclopedia of Dairy Sciences. Elsevier., (pp. 2201–2206).

Indyk, H.E., Woollard, D.C., 2011. Nitrates and nitrites as contaminants. In: Encyclopedia of Dairy Sciences, second ed.. Elsevier., (pp. 906–911).

Jooste Peter, J., Anelich Lucia, E.C.M., 2008. "Safety and Quality of Dairy Products" *Safety and Quality of Dairy Products* In: Trevor, J.B. Richard, K.R. (Eds.), Advanced Dairy Science and Technology, 2008 Blackwell Publishing Ltd, New Jersey, pp. 153–182.

Kozak, J., Balmer, T., Byrne, R., Fisher, K., 1996. Prevalence of Listeria monocytogenes in foods: incidence in dairy products. Food Control 7 (4/5), 215–221.

Muehleman, M., 2014. Safety of cheese. In: Motarjemi, Y., Moy, G., Todd, E. (Eds.), Encyclopedia of Food Safety. Elsevier.

O'Mahony, M., Mitchell, E., Gilbert, R.J., Hutchinson, D.N., Begg, N.T., Rodhouse, J.C., et al., 1990. An outbreak of foodborne botulism associated with contaminated hazelnut yoghurt. Epidemiol. Infect. 104 (3), 389–395.

Poppe, C., 2011. Salmonellosis Encyclopedia of Dairy Sciences, second ed. 2011, pp. 190–194.

Rowe, B., Begg, N.T., Hutchinson, D.N., Dawkins, H.C., Gilbert, R.J., Jacob, M., et al., 1987. *Salmonella Ealing* infections associated with consumption of infant dried milk. Lancet 2 (8564), 900–903.

Ryser, E.T., 2011. Listeria monocytogenes. In: Encyclopedia of Dairy Sciences. Elsevier., (pp. 81–86).

WHO, 1953. Milk Pasteurisation. World Health Organization, Geneva, (World Health Organization Monograph Series No. 14).

WHO, 2007. Guidelines for the Safe Preparation, Storage and Handling of Powdered Infant Formula. World Health Organisation, Geneva.

Further Reading

ADASC., 1999. Australian Manual for Control of Listeria in the Dairy Industry. Australian Dairy Authorities' Standards Committee, July 1999.

Asperger, H., 1994. Staphylococcus aureus Monograph on the Significance of Pathogenic Micro Organisms in Raw Milk. International Dairy Federation, Brussels. (pp. 24–42).

Bad Bug Book., 2012. Foodborne Pathogenic Microorganisms and Natural Toxins Handbook, second ed. USA: FDA (available at: <http://www.fda.gov/Food/FoodSafety/FoodborneIllness/FoodborneIllnessFoodbornePathogensNaturalToxins/BadBugBook/default.htm>.

Bindels, J.G., Hoijer, M., 2000. Allergens: latest developments, newest techniques. In: Safety in Dairy Products. Bulletin of the International Dairy Federation 351/2000, pp. 31–2.

Britz, T.J., Robinson, R.K., 2008. Advanced Dairy Science and Technology. Blackwell Publishing, Oxford.

CAC, 2004. Code of Hygienic Practice for Milk and Milk Products CAC/RCP 57-2004. Joint FAO/WHO Food Standard Programme, Rome, FAO.

Codex Alimentarius Commission (various standards) <http://www.codexalimentarius.org/>.

Dairy Food Safety Victoria, 2002. Code of Practice for Dairy Food Safety <www.dairysafe.vic.gov.au> (accessed December 2005).

Encyclopedia of Dairy Sciences 2011. Elsevier (various papers) <http://drsani.iauq.ac.ir/imagesMasterPage/Files/drsani/file/Advanced%20Dairy%20Science%20and%20Technology,%201405136189.pdf> (accessed 13.02.13).

ICMSF, 1996. Microorganisms in Foods 5: Characteristics of Microbial Pathogens. Springer, London. (Blackie Academic & Professional. ISBN: 041247350X).

ICMSF, 2002. Microorganisms in Foods 7: Microbiological Testing in Food Safety Management. Kluwer Academic/Plenum Publishers, (ISBN: 0306472627. Available from Springer).

ICMSF, 2005. Microorganisms in Foods 6: Microbial Ecology of Food Commodities, second ed. Kluwer Academic/Plenum Publishers, NY. (ISBN: 0-306-48675-X. Springer).

ICMSF, 2011. Microorganisms in Foods 8: Use of Data for Assessing Process Control and Product Acceptance. Springer, (ISBN: 978-1-4419-9373-1).

Kodba, Z.C., 2000. Organochlorine pesticide residues in dairy milk. In: Bulletin of the International Dairy Federation 351/2000 "Safety in Dairy Products." IDF: Brussels, pp. 34.

Lee Wong, A.C., 1998. Biofilms in food processing environments. J. Dairy. Sci. 81, 2765–2770.

McEvoy, J.D.G., 2002. Contamination of animal feedingstuffs as a cause of residues in food: a review of regulatory aspects, incidence and control. Analitica. Chemica. Acta. 473, 3–26.

Michel, A.L., McCrindle, C.M., 2004. An overview of zoonoses with importance to the dairy industry. In: A Farm to Table Approach for Emerging and Developed Dairy Countries: Proceedings of the IDF/FAO International Symposium on Dairy Safety and Hygiene, Cape Town, South Africa, 2–5 March 2004, pp. 47–51.

Mostert, J.F., Jooste, P.J., 2002. Quality control in the dairy industry. In: Robinson, R.K. (Ed.), Dairy Microbiology Handbook (third ed.). John Wiley & Sons, Inc., pp. 655–736.

Nicholson, F.A., Chambers, B.J., Williams, J.R., Unwin, R.J., 1999. Heavy metal contents of livestock feeds and animal manures in England and Wales. Bioresource. Technol. 70, 23–31.

Oliver, S.P., Jayarao, B.M., Almeida, R.A., 2005. Foodborne pathogens in milk and the dairy farm environment: food safety and public health implications. Foodborne Path. Dis. 2 (2), 115–129.

Reybroeck, W., 2003. Role of the farmer in preventing residues of antibiotics in farm milk. In: Conference of Quality Management at Farm Level. IDF World Dairy Summit and Centenary, 7–12 September 2003, Bruges, Belgium, pp. 239–40.

Robinson, R.K., Tamime, A.Y., Wsolek, M., 2002. Microbiology of fermented milks. In: Robinson, R.K. (Ed.), Dairy Microbiology Handbook (third ed.). John Wiley & Sons, Inc., pp. 367–430.

Ryser, E.T., Marth, E.H. (Eds.), 2007. Listeria, Listeriosis and Food Safety (third ed.). Taylor & Francis Group, NY

Tompkin, R.B., 2002. Control of Listeria monocytogenes in the food processing environment. J. Food Protect 65 (4), 709–725.

Websites

Codex Alimentarius Commission (various standards)
http://www.codexalimentarius.org/
Food and Agriculture Organization
http://www.fao.org/food/food-safety-quality/home-page/en/
European Dairy Federation
http://www.fooddrinkeurope.eu/member/eda/
International Dairy Federation
http://www.alphagalileo.org/Organisations/Default.aspx?OrganisationId=8375
World Organization for Animal Health
http://www.oie.int/fr/
World Health Organization
http://www.who.int/foodsafety/en/

CHAPTER

6

Meat and Meat Products

John N. Sofos

Colorado State University, Fort Collins, CO, USA

OUTLINE

Food Safety Management.
DOI: http://dx.doi.org/10.1016/B978-0-12-381504-0.00006-8

INTRODUCTION

The safety of meat products has had a prominent position among societal concerns in recent years. Consumer health hazards associated with meat products are of a physical, chemical or biological nature. Besides allergens, which affect sensitive persons, the most serious meat safety issues resulting in immediate consumer health problems and recalls from the marketplace of potentially contaminated products are associated with hazards of a biological nature, especially pathogenic bacteria. Microbial pathogens cause mild, severe, brief or chronic gastrointestinal or invasive human illness, or death. Viral pathogens cause the highest numbers of foodborne illness cases and are a major concern at food service, while parasitic agents become problematic under specific circumstances and in certain geographical areas. Spoilage microorganisms result in loss of quality and shortening of shelf-life, which lead to reduced food supplies and economic losses (Sofos, 2013; Sofos et al., 2013).

The microbiological quality and safety of meat products is compromised by system failures or abuses during food animal production, product processing and distribution, and preparation for consumption, as well as by consumption habits. Despite continuous improvements in meat processing, documented disease episodes and concern about meat products acting as vehicles of hazards are increasing rather than diminishing. The reasons for this trend are multiple and may include: changes in animal production, product processing and distribution; increased worldwide meat consumption; increased international meat trade; changing consumer needs, such as increased preference for minimally processed products; increasing numbers of consumers at risk for infection; emerging pathogens of increased virulence and resistance to control or clinical treatment; advances in microbial detection methodologies; inadequate food worker and consumer education and training in proper food handling; and increased interest, awareness and scrutiny of foodborne illness episodes by consumers, news media and consumer organizations (Sofos et al., 2013).

Sources contributing microbial contamination to animals and meat include animal feces, soil, water, air, feed, hides, intestines, lymph nodes, processing equipment, utensils and humans. Identifying the sources and modes of meat product contamination is important for proper hazard control and enhancement of meat safety. Proper sanitation and hygienic practices are essential in keeping contamination at low levels. However, as some level of contamination is unavoidable, and because products support microbial growth, they should be handled and preserved properly throughout the supply chain in order to maintain quality and safety. A comprehensive strategy for controlling microbial problems in meat products should involve implementation of interventions or procedures that: (1) prevent or minimize access and transfer of microorganisms to the product; (2) reduce initial contamination by removal or inactivation of microorganisms which have gained access; (3) inactivate or kill microorganisms on products; and (4) prevent, delay or slow down growth of microorganisms, which have gained access and have not been inactivated. Proper implementation and management of such an approach ensures the safety of meat products or at least reduces the incidence of microbial meatborne illness. It should be mentioned that modification of food service and consumer habits and behavior should also contribute to improved safety of meat and meat products. Microbial meatborne illness may be associated with intentional (e.g. steak tartar) or accidental (i.e. Jack-in-the Box *Escherichia coli* O157:H7 outbreak through consumption of undercooked ground beef in the United States) consumption of raw or undercooked products.

Overall, there is a need for better understanding of hazard behavior and approaches for effective control in meat and meat products. The objectives of this chapter are to: identify and briefly describe hazards associated with meat and meat products; summarize data on contamination levels, outbreaks and incidence of disease; describe control of risks at different stages of the meat product chain/from primary production, processing, distribution, etc., including packaging; and overview the implementation of the hazard analysis critical control point (HACCP) process management system for pathogen control in meat products.

HAZARDS ASSOCIATED WITH MEAT AND MEAT PRODUCTS

General

Hazards that compromise the safety of meat and meat products are of a physical, chemical or biological nature. Physical hazards originate from the environment and the raw materials, and include bone chips and foreign bodies such as metal, glass, wood, plastic, stones, etc. Chemical hazards include natural and synthetic environmental contaminants such as residues of animal drugs and pesticides, or industrial chemicals present in the animal or processing environment, or resulting from excessive use of food additives in processed products. Biological hazards include bacteria, viruses, parasites and abnormal prions (agents causing transmissible spongiform encephalopathies (TSE) such as bovine spongiform encephalopathy (BSE) commonly known as mad cow disease).

Hazards of a chemical nature that may be present in meat products include mycotoxins resulting from feeding animals with moldy feeds or due to uncontrolled mold growth in certain meat products aged improperly and for long periods of time. Use of chemicals (e.g. antibiotics, hormones, growth promoting agents) during meat animal production, for better growth, feed efficiency and disease control, as well as accidental contamination with industrial chemicals (e.g. dioxins), is of concern to consumers, and causes trade conflicts between certain countries. A portion of the consuming public continues being concerned about the safety of residues of food additives (e.g. common salt, nitrate, nitrite, lactates, phosphates and other compounds) in processed meat products. These materials play important functions in the processing, quality, shelf-life and microbial safety of processed meat products. Overall, the contribution of additives and their residues to the overall food safety concerns in meat products is considered small, in particular in industrialized countries where there is extensive legislation for their control. Nevertheless, concerns for potential presence of residues of certain toxic chemicals, such as dioxins, in meat products are well founded as it has occurred and has caused major consumer scares from time to time (Sofos, 2013). In general, as consumer concerns and food safety issues associated with chemical residues continue, the search for "natural" products to replace synthetic ones will also continue. Chemicals, acting as allergens in sensitive consumers are often causes of product recalls. They become a problem when they accidentally contaminate a product or have not been declared on the product label. It should be recognized that food allergen issues are important and should be addressed through labeling or production of additive-free alternative products when feasible.

The debate over the positive and negative aspects of meat in the human diet also re-emerges from time to time. Studies are needed to better elucidate the issues and allow prudent recommendations concerning these debates. Overall, meat, even in reduced portions, plays an important role in the human diet and health and will remain a main component of human diets. In contrast, currently, biological pathogens, which cause problems of immediate and obvious human health concern, will continue receiving most of the public attention (Sofos et al., 2013).

In general, control of physical and chemical hazards is accomplished through good production, manufacturing and hygienic practices, including proper facilities and equipment designs, as well as other prerequisite programs, such as letters of warrantee from suppliers, validation and verification of the need and amounts used, etc.

Major challenges related to microbial pathogens include foodborne illness outbreaks and associated deaths, recalls from the marketplace of potentially contaminated products and regulatory compliance problems. The most severe biological hazards in foods are pathogenic bacteria, which may cause illness as direct agents (infections), through production of various toxins (intoxications) or both (toxicoinfections). Typical clinical symptoms of foodborne bacterial and viral diseases include acute diarrhea, abdominal cramps, vomiting or some other manifestation in the gastrointestinal system. In addition, in their invasive severe form, some bacterial pathogens are associated with syndromes affecting the central nervous system or various organs, as well as being involved in various chronic sequelae. Individuals with suppressed or compromised immune systems are more susceptible to severe foodborne microbial illness. Biological hazard-associated issues become more challenging with emergence of pathogens with increased virulence or of low infectious doses, or of resistance to antibiotics used in animal production and to antimicrobial additives used in meat processing. In general, the most important food safety challenge of current concern in meat products is the need to control pathogenic microorganisms, especially bacteria and viruses, and in certain regions parasites.

Additional challenges are associated with animal manure (large quantities generated in intense animal production settings) disposal needs, which result in cross-contamination of water and foods of plant origin with enteric pathogens. Manure treatment and proper disposal during food animal production, and the associated need for development and implementation of pathogen control programs at the farm, are major issues. Further areas of interest are: animal identification and traceability needs; development of improved and rapid pathogen detection methodologies for laboratory and field use; presence of food additives and chemical residues in meat products; regulatory inspection harmonization at the national and international level; delineation of responsibilities for zoonotic diseases between animal health and regulatory public health agencies; international harmonization through the establishment of risk assessment-based food safety targets; and complete and routine implementation of HACCP principles at the production and processing level and on the basis of food handler training and consumer education. The above and related issues have been reviewed and discussed extensively in publications included in the list of references or listed under "Further Reading."

Microbial Contamination of Meat Products

Live and healthy animals could be described as consisting of sterile tissue surrounded by surfaces heavily contaminated with microorganisms. Contamination is present on external animal surfaces such as hides and the gastrointestinal tract, as well as sporadically, at low levels and transiently, inside organ tissues such as lymph nodes. Microbial contamination is easily introduced from the environment, sometimes at high levels, and may include spoilage causing organisms, pathogens and species involved in production of desirable fermented meat products. Meat animal carcasses are contaminated during slaughter and dressing immediately as the hide is incised for its removal and the underlying tissue is exposed to the environment. Overall, the cycle of contamination involves animal feces and manure, soil, decaying matter, air, pastures and other animal feeds, and animals and their products, as well as water and other foods or the environment. Additional vehicles of biological hazards within this continuum may include rodents, birds, insects, animal transportation crates and vehicles, and other equipment and utensils, which contribute to cross-contamination.

The primary source of contamination for meat is the animals' hide, which carries high levels of microorganisms originating from animal feces, soil and water during animal production and during animal transportation and holding before slaughter. Additional sources of contamination include the processing environment, equipment, utensils and humans contributing through cross-contamination. During carcass deboning and cutting, contamination on external carcass and equipment surfaces is spread and distributed on equipment and other environmental surfaces as well as other carcasses or meat cuts and leads to a cycle of contamination spreading and cross-contamination. Initial contamination levels may be lower than 10^2 or exceed 10^7 colony forming units (cfu) per cm^2 depending on processing operation and carcass site (Koutsoumanis et al., 2006).

Animal manure serves as a source of contamination of water used for drinking or to irrigate or wash plant crops resulting in contamination spreading or cross-contamination of other foods, as demonstrated by the increased occurrence of foodborne outbreaks of enteric pathogens associated with vegetable consumption. Since sources of contamination are diverse, and facilities and practices of slaughtering and processing operations are variable, there may be variation in the types of microorganisms introduced and, especially, in the extent (prevalence and concentration) of contamination in meat. Contamination may also vary depending on the characteristics of each animal, geographic origin and season of the year. The extent of microbial transfer from the above sources to meat and other foods depends on sanitation and hygienic practices, product handling, processing, preparation and serving procedures, and conditions of storage and distribution. In general, meat carcass surfaces become easily contaminated during animal slaughter and carcass dressing.

Contamination may be spread to meat during cutting, processing, storage, merchandizing, preparation and serving, if manufacturing and hygienic practices allow. Then, if not properly handled, processed, preserved and prepared for consumption, meat supports growth of various microorganisms leading to spoilage and foodborne illness. The habit of consuming steak tartar (raw meat) in parts of Europe is not immune of the potential to lead to foodborne illness, similar to any raw meat, especially when handled or served unhygienically or by subclinically infected pathogen carrying employees. Reasons that unacceptable consumption habits are not being linked frequently to major outbreaks may be that the meat is kept, handled and prepared under strict hygienic and sanitary conditions, and that it is prepared on demand in limited servings at the point and time of consumption. It is also important to notice that some processes are not always sufficient to eliminate some hazards (e.g. fermented pork meat may not destroy parasites).

Spoilage Microorganisms in Meat Products

Microorganisms, such as micrococci and Gram-negative rods, are of concern because they are responsible for spoilage and loss of acceptable eating quality in meat products. They include *Pseudomonas* (*P. fragi*, *P. fluorescens*, *P. putida* and *P. lundensis*), *Shewanella putrefaciens*, *Photobacterium phosphereum*, *Brochothrix thermosphacta*, cold-tolerant Enterobacteriaceae (e.g. *Hafnia alvei*, *Serratia liquefaciens* and *Enterobacter agglomerans*), *Acinetobacter* spp., *Alcaligenes* spp., *Moraxella* spp., *Flavobacterium* spp., *Staphylococcus* spp., *Micrococcus* spp., coryneforms, fecal streptococci, lactic acid bacteria (LAB), etc. (Koutsoumanis et al., 2006; Sofos et al., 2013).

Spoilage changes vary with types of microorganisms that dominate, meat product composition and properties (e.g. pH, enzymatic activity, sugar and lipid content, additives, etc.), storage conditions (e.g. temperature, packaging and gas atmosphere, etc.) and time–length of storage. Spoilage differs among cooked, cured, heat processed, fermented or dried products of varying water activity and pH. Microorganisms commonly involved in spoilage occurring in aerobically stored meat at cold temperatures rely on oxidative metabolism. Microorganisms dominating in processed meat products may include micrococci, streptococci, lactobacilli, and *B. thermosphacta*. LAB lead to sour spoilage of meat when exposed to restricted oxygen environments (Koutsoumanis et al., 2006; Sofos et al., 2013).

Adverse health effects may also develop in consumers from consumption of stored meat products due to the presence of biogenic amines (e.g. histamine, putrescine, spermidine, etc.). Biogenic amine production is associated with growth of microorganisms such as Enterobacteriaceae and *Lactobacillus*. Control of growth and reduction of biogenic amine production is achieved through proper sanitation, proper storage temperature and storage time limitation (Sofos et al., 2013).

Meat Fermentations

Certain bacterial species, mostly LAB, are involved in useful meat fermentations through metabolic processes that convert substrates into desirable food products (e.g. fermented sausages) or ingredients (e.g. vitamins and enzymes). A variety of fermented meat products are found throughout the world. In addition, LAB produce microbial antagonists, such as bacteriocins which are active against pathogens, including *Listeria monocytogenes*. Bacteriocins have been researched for over 30 years but they are not approved and used in foods, with the exception of nisin. Interest has also been shown in the development of fermented meats with probiotic microorganisms. However, knowledge is still preliminary to commercialize fermented meats with probiotic health benefits (Sofos et al., 2013).

Biological Hazards in Meat Products

General

There is a long list of biological hazards, including bacteria, viruses, parasites and prions, that may be present in animals and meat products. Some of these hazards are transmitted to humans through handling and consumption of meat products. Some hazards present in animals are associated with non-food transmission routes such as aerosols and direct contact with animals or diseased tissues, while others have not been documented as being transmitted to humans through meats. Hazards of most concern in terms of deaths or severity of illness are bacterial, while viruses cause large numbers of usually mild gastrointestinal illness as a consequence of poor sanitation and unhygienic practices; parasites have lesser involvement in meatborne illness in developed countries, but they may be of major concern in certain regions (Table 6.1).

Meat processed under sanitary and hygienic conditions should generally be contaminated infrequently and with low concentrations of pathogens compared to those of spoilage causing bacteria. Levels of pathogens on meat carcasses can vary from 1 to >30 most probable number (MPN)/cm^2, and prevalence may differ within an animal species (e.g. cow and bull compared to steer and heifer carcasses). Pathogen prevalence on carcasses may also be affected by season of the year (e.g. higher prevalence of *Escherichia coli* O157 on cattle in the summer and early fall than winter months). Since contamination is unpredictable, any raw, unprocessed and uncooked meat product should be considered as potentially contaminated with pathogens (Koutsoumanis et al., 2006; Sofos et al., 2013).

Ground, comminuted, or in general non-intact, meat products contain higher contamination levels than carcass or intact product surfaces, due to cross-contamination from grinders and utensils, spreading of contamination and greater surface area. Further, in these products, contamination is entrapped within the tissue, which makes it more difficult to kill during cooking. Fresh meat products classified as non-intact include intact meat cuts that are mechanically tenderized by cubing, frenching or pounding devices, blades, solid- or hollow-needle injectors, used to inject solutions for marinating, flavoring, moisture enhancement or tenderizing. Beef trimmings destined for processing into formed items such as gyros as well as any chopped, ground, flaked or minced product are also considered non-intact. Processing increases tenderness, juiciness and flavor, and makes use of such products desirable in institutional settings. In the United States, the total volume of such products may exceed 70% of total beef carcass volume including, especially, those of lower tenderness (Sofos et al., 2013).

Spreading and entrapment of contamination throughout the interior of non-intact products, compared to intact products that are contaminated only on external surfaces,

TABLE 6.1 Biological Hazards, their Origin, Meat Products Affected, Type of Illness, Need for Growth in Food before Consumption, and Control Approaches

Biological Agent	Type of Meat	Type of Illness	Sources	Growth in Food	Control Approach
BACTERIA					
Bacillus cereus	Beef, pork, lamb, poultry	Toxicoinfection (diarrheal) in meat products; intoxication (emetic) in other foods (e.g. rice)	Soil, animals	Needed (some grow in the cold)	Inactivate spores (heat); temperature control (>60°C or <10°C) to prevent spore germination and growth; pH or a_w control to prevent growth; cook foods when needed for consumption; reheating, including stir-frying, does not destroy heat-resistant emetic toxins
Campylobacter spp. (thermophilic)	Beef, pork, lamb, poultry	Invasive infection	Animals	Not needed	Inactivate cells (pasteurization, cooking); hygienic slaughter and processing procedures; irradiation of meat and poultry; treatment of water; prevention of cross-contamination of contact surfaces; personal hygiene in food preparation (hand washing after contact with animals); keeping pets away from food-handling areas; avoid eating raw or partially cooked poultry and drinking raw milk
Clostridium botulinum	Beef, pork, lamb, poultry	Intoxication (toxicoinfection, infant botulism)	Soil, water, animals	Needed	Inactivate spores (canning/sterilization); control cell growth (refrigeration); boiling to destroy toxins; nitrites in pasteurized meat products; acid-preservation (pH < 4.6); thorough cooking of home-canned food (boiling or stirring for 15 minutes); discard swollen cans
Clostridium perfringens	Beef, pork, lamb, poultry	Toxicoinfection	Soil, water, animals	Needed	Inactivate spores (heating or cooking); control growth (refrigeration)
Escherichia coli (STEC/EHEC and other pathogenic groups)	Beef, pork, lamb	Toxicoinfection	Animals	Not needed	Prevent fecal contamination of food and water; inactivate cells (pasteurization, cooking); drinking water treatment; proper sewage disposal; good personal hygiene; irradiation of meat; separation of raw and cooked foods; hand washing before food preparation; avoid eating raw or partially cooked meat and poultry; refrigeration

Listeria monocytogenes	Ready-to-eat beef, pork, lamb, poultry	Invasive infection	Processing environment, soil, water, animals	Needed (grows in the cold)	Inactivate cells (pasteurization, cooking); control growth (freezing, no long-term refrigerated storage); avoid cross-contamination of ready-to-eat high-risk processed foods; thorough reheating before consumption; sensitive populations (e.g. pregnant, immunocompromised, aged) avoid high-risk foods (e.g. soft cheese from unpasteurized milk, ready-to-eat meat such as paté, raw meat, raw milk, pre-prepared salads, cold, smoked or raw seafood)
Salmonella enterica	Poultry, beef, pork, lamb	Invasive infection	Animals	Needed for meat products. Not needed for certain strain/food combinations	Inactivate cells (pasteurization, cooking); control growth (refrigeration); irradiation of meat and poultry; reheating of food; prevention of cross-contamination; cleaning and sanitation of food preparation surfaces; exclusion of pets and other animals from food-handling areas; vulnerable consumers should avoid raw and undercooked meat and poultry, and other animal origin foods
Shigella		Infection	Water; mostly human (fecal–oral) transmission	Not needed	Inactivate cells (pasteurization, cooking); hand washing; treatment of drinking water; effective sewage disposal; safe food preparation practices; thorough cooking and reheating of food; sanitation of food preparation surfaces
Staphylococcus aureus	Ready-to-eat beef, pork, lamb, poultry	Intoxication (heat-resistant toxin)	Humans, processing environment, animals	Needed	Exclusion of infected food handlers; inactivate cells (pasteurization, cooking); toxins are heat-resistant; control food handlers (skin lesions, boils, cuts, etc.); good personal hygiene of workers; no time–temperature abuse of cooked/ready-to-eat foods (refrigeration)
Yersinia enterocolitica	Pork, poultry	Invasive infection	Animals, soil, water	Needed (grows in the cold)	Inactivate cells (pasteurization, cooking); control growth (freezing); prevention of cross-contamination
PARASITES					
Cryptosporidium parvum	Beef, poultry	Invasive infection	Meat, milk, water, human (fecal–oral transmission)	Not needed	Hand washing; pasteurization/cooking; irradiation; filtration and disinfection of water; sanitary disposal of excreta, sewage and wastewater; boiling water when unsafe; boiling of milk if not pasteurized
Giardia duodenalis	Beef	Invasive infection	Meat, water, human (fecal–oral transmission)	Not needed	Hand washing; pasteurization/cooking; thorough washing of fruit and vegetables; irradiation; filtration and disinfection of water; sanitary disposal of excreta, sewage water; treatment of irrigation water; good hand hygiene; campers should avoid drinking surface water; boil or filter water

(*Continued*)

TABLE 6.1 (Continued)

Biological Agent	Type of Meat	Type of Illness	Sources	Growth in Food	Control Approach
Sarcocystis spp.	Beef, pork	Invasive infection	Meat	Not needed	Salting; pasteurization; cooking; irradiation
Taenia spp. (cysticercosis, taeniasis)	Beef, pork	Invasive infection	Meat	Not needed	Prevention of fecal contamination; safe sewage disposal; no sewage water for irrigation; heating (pasteurization or cooking); freezing; irradiation; early diagnosis and treatment
Toxoplasma gondii	Beef, pork, poultry (raised outdoors)	Invasive infection	Meat	Not needed	Heating (pasteurization, cooking); irradiation; good personal hygiene after contact with cats and before food preparation; safe disposal of cat feces; pregnant women to avoid raw or undercooked meat
Trichinella spp.	Pork, game	Invasive infection	Meat	Not needed	Freezing; pasteurization/cooking; irradiation
VIRUSES					
Hepatitis E	Pork	Invasive infection	Animals, water, human (fecal–oral transmission)	Not needed	Hand washing; sanitation; cooking
Other foodborne viruses (e.g. hepatitis A, norovirus)	Ready-to-eat meats and other foods	Infection	Water, human (fecal–oral transmission)	Not needed	Hand washing; sanitation; cooking; water treatment; safe sewage disposal; good personal hygiene; thorough cooking of shellfish; vaccination of professional food handlers and travelers for hepatitis A; abstinence from handling food when ill (e.g. diarrhea)
PRIONS (ENCEPHALOPATHIES)					
Bovine spongiform encephalopathy (BSE)	Beef	Invasive infection	Animals	Not needed	Special controls during animal growth (e.g. control of safety of animal feed) and slaughter (e.g. control of animal health and removal of specified risk material)

Modified from Adams and Motarjemi, 1999; Sofos, 2013.

constitutes a public health concern because cells of pathogens such as *E. coli* O157:H7 may survive cooking if the product is intentionally or accidentally undercooked, especially if consumers perceive non-intact products as intact. The United States Department of Agriculture Food Safety and Inspection Service (USDA/FSIS) considers non-intact meat products, together with raw ground beef (the ultimate non-intact product), as adulterated if samples are contaminated with *E. coli* O157:H7.

The risks associated with non-intact meat products may be controlled through implementation of effective meat decontamination interventions, application of approved and effective antimicrobial treatments to subprimal meat cuts before tenderization, proper chilling and rotation of injection solutions, potential use of antimicrobials in injection brines, effective sanitation and temperature controls, and cooking procedures that are selected based on type of cooking method and product characteristics. Inactivation of *E. coli* O157:H7 in these products may be influenced by antimicrobials in brine solutions or in products, and cooking procedures (Sofos et al., 2013).

Bacterial Hazards

Pathogens that have been documented as transmitted by meat products include pathogenic *Bacillus* spp. (e.g. *Bacillus cereus*), *Campylobacter* spp. (thermophilic), *Clostridium perfringens*, *Clostridium botulinum*, pathogenic *E. coli* serotypes, especially shigatoxin-producing/verotoxigenic (STEC/VTEC) enterohemorrhagic (EHEC) strains, *L. monocytogenes*, *Salmonella enterica* serotypes, *Staphylococcus aureus* and *Yersinia enterocolitica.* Other pathogenic bacteria, which may be or have been suggested as potentially transmitted through meat, include *Aeromonas*, *Arcobacter* (previously mesophilic *Campylobacter*), *Bacillus anthracis*, *Brucella*, *Clostridium difficile*, *Enterobacter*, *Helicobacter*, *Mycobacterium*, *Plesiomonas* and *Shigella* (EFSA, 2011; EFSA/ECDC, 2011; Sofos, 2013).

Some pathogens may be of no major concern as meatborne in developed world regions such as the United States and Europe (EFSA, 2011). For example, *B. anthracis* may be responsible for endemic disease in Africa and Asia through direct contact with infected animals or carcasses. Extensive inhalation of aerosolized spores during processing of hides and wool in enclosed facilities may also lead to pulmonary anthrax, while cutaneous anthrax may be acquired by handling contaminated hides or wool. Gastrointestinal anthrax may be linked to consumption of raw or undercooked meat from infected animals. *Bacillus cereus* causes intoxication (emetic form, mostly from starchy foods) or toxicoinfection (diarrheal form, including from meat products) after spore germination and multiplication in foods (Table 6.1).

Spores of *Clostridium* are found in soil, dust and water as well as in the intestines of animals from where they may contaminate various foods including meat products. Common species associated with food, including meatborne transmission, are *C. botulinum* and *C. perfringens.* They may become problematic when the food is temperature abused and allows growth of the pathogen. Specifically, spores of *C. perfringens* become activated by heat treatment and upon time–temperature abuse they germinate, multiply and produce enterotoxin, which is released during sporulation in the intestine, leading to illness. Spores of *C. botulinum*, after activation and germination, grow in anaerobic or micro-aerobic conditions and produce deadly neurotoxins in the food. Therefore, control measures are different. *Clostridium difficile* is an emerging pathogen with increased incidence in nosocomial infections. Potential sources of the organism include farm (cows, pigs, horses) and domestic (dogs, cats) animals. Although it has been isolated from meat and poultry products, the available information is inadequate to establish any role of meat in human epidemiology (Sofos et al., 2013).

Bacterial hazards such as *Coxiella burnetti* are transmitted to humans mostly through aerosols, direct contact or consumption of foods such as milk, but not meat. Similar modes of transmission exist for *Mycobacterium avium, M. avium* subsp. *paratuberculosis* and *Mycobacterium bovis.* Evidence for the presence of *M. bovis* in meat is limited and inconclusive, and transmission through meat consumption has not been verified; confirmed modes of transmission include aerosols and consumption of unpasteurized milk (EFSA, 2011). *Mycobacterium paratuberculosis* causes Johne's disease, a chronic enteritis in cattle and other ruminants. Humans also develop a similar chronic inflammatory condition in the intestine (Crohn's disease); however, any relationship of *M. paratuberculosis* with Crohn's disease is unverified. Samples of blood, liver, kidney, lymph nodes and muscle tissue obtained from carcasses of cows with advanced Johne's disease as well as samples of cooked muscle tissues and cooked hamburger patties that contained chopped mesenteric lymph nodes were tested and *M. paratuberculosis* was recovered from samples of mesenteric lymph nodes, and raw and cooked meat. Therefore, human exposure to *M. paratuberculosis* should be controlled (Sofos et al., 2013).

Shigella may be introduced by humans (e.g. through poor hand hygiene) in ready-to-eat products. *Staphylococcus aureus* may be present in raw meat but it is outcompeted by other bacteria. It may become a problem in processed meat products, where it is introduced usually by humans, and competition by spoilage organisms is limited. Then, if the product is temperature abused, the pathogen is able to produce heat-resistant enterotoxins. *Yersinia enterocolitica* includes pathogenic serotypes, which may be transmitted with foods contaminated through water, including pork products (Sofos, 2008, 2013; Sofos et al., 2013).

Campylobacter spp. are common enteric pathogens in developed countries; *C. jejuni*, and to a lesser extent *C. coli*, are the most common species in foods. Although most common in poultry, they are found in all food-producing animals. Foods commonly implicated in infections include milk, eggs, meats, poultry and water. Most cases are attributed to handling and consumption of broiler meat, with cross-contamination being very important.

Campylobacteriosis, the infection caused by *Campylobacter*, is one of the most frequently reported bacterial foodborne illnesses in the United States and European countries, and may result from ingestion of as few as 500 cells. Symptoms of acute colitis, fever, malaise, abdominal pain, headache, watery or sticky diarrhea with traces of blood (occult), inflammation of the lamina propria and abscesses appear within 2–5 days and may persist for up to 10 days (Bacon and Sofos, 2003).

The cells of *Campylobacter* are curved, slender, Gram-negative, nonspore-forming rods of corkscrew-type motility. Being microaerophilic, they grow optimally at 2.0–5.0% oxygen and 5.0–10.0% carbon dioxide, but not in normal air of 21% oxygen. The optimum temperature range for growth of *Campylobacter* is 37–42°C, while under favorable nutritional and environmental (e.g. atmospheric) conditions growth occurs at 30–45°C. Growth occurs at pH values of 4.9–8.0, but the range 6.5–7.5 is preferred. Cells are sensitive to cooking. Being sensitive to drying, *Campylobacter* requires water activities above 0.91 (Bacon and Sofos, 2003).

Most strains of *E. coli* are harmless inhabitants of the gastrointestinal tract of humans and other warm-blooded animals. Strains of certain serotypes cause diarrheal-type illness.

Diarrheagenic, enterovirulent or pathogenic *E. coli* serotypes are Gram-negative, facultatively anaerobic, nonspore-forming, motile rods. From the gastrointestinal tract, they contaminate animal exteriors, soil and water, and consequently meat products and foods of plant origin. Of the various disease-causing *E. coli* groups, enterohemorrhagic *E. coli* (EHEC) strains of shigatoxin or verotoxigenic *E. coli* (STEC/VTEC) serotypes are of most concern in undercooked meat products, especially non-intact meat products such as ground beef, and a variety of other foods (Sofos et al., 2013).

STEC/VTEC serotypes cause mild to severe bloody diarrhea (hemorrhagic colitis), or in some cases hemolytic uremic syndrome (HUS), which is characterized by microangiopathic hemolytic anemia, thrombocytopenia and acute renal failure. The symptoms appear 3–9 days following ingestion of >10 cells, and last for 2–9 days. Approximately 6% of infected individuals develop HUS, which is associated with 80% of serotype *E. coli* O157:H7 cases in North America. As the STEC/VTEC is of most concern, *E. coli* O157:H7 has been declared as an adulterant for raw ground beef and other non-intact beef products in the United States, since 1994 and 1999, respectively. Additional pathogenic STEC/VTEC serogroups (i.e. O26, O45, O103, O111, O121 and O145) have also been placed under similar regulatory action in the United States since 2012. Foods other than meat implicated in infection are fermented meat products such as salami, unpasteurized milk and cheese, fruit juice, sprouts, lettuce, spinach, cantaloupe and mushrooms (Bacon and Sofos, 2003).

Certain animals shed *E. coli* O157:H7 cells for longer periods of time and because of that they are characterized as "persistent shedders," while some animals may be considered as "super-shedders" because they shed large numbers of pathogen cells. Thus, such animals may be the major source of contamination for the environment and foods. Identifying the factors that result in persistent- or super-shedding animals could contribute to the development of interventions for better control of the pathogen (Sofos et al., 2013).

Escherichia coli O157:H7 can grow at temperatures as low as 7–8°C and as high as 44–46°C, with an optimum in the range 35–40°C, and a minimum required water activity of 0.95. Their optimum pH for growth is 6.0–7.0, but they can grow in the range 4.4–9.0 and can tolerate acid more than other pathogens. Normal cooking temperatures easily inactivate cells of these mesophilic nonspore-forming pathogens (Bacon and Sofos, 2003).

Listeria monocytogenes causes a severe invasive infection in sensitive individuals such as the elderly, immunocompromised and the unborn, where it exhibits a case-fatality rate of 20–30%. Invasive listeriosis is characterized by serious syndromes of the central nervous system such as meningitis and meningoencephalitis, while in its mild form the infection is a typical gastrointestinal foodborne illness. While pregnant women may develop a flu-like illness, the infected fetus develops meningitis, neonatal septicemia, stillbirth or spontaneous abortion. The incubation period of listeriosis may vary from a few days to 2–3 months, which complicates attribution and identification of outbreaks. The infection is usually associated with ready-to-eat meat, poultry and other foods (e.g. deli-type foods, soft cheeses, seafood and unpasteurized dairy products) contaminated after processing. Growth occurs during prolonged storage even at refrigeration temperatures. The infectious dose should be greater than 100 cells/g; however, the possibility of lower infectious doses has not been excluded, especially for sensitive individuals (Bacon and Sofos, 2003).

Listeria includes the species *L. monocytogenes*, *L. innocua*, *L. ivanovii*, *L. seeligeri*, *L. welshimeri* and *L. grayi*, which now includes *L. murrayi*, and the new species *L. roucourtii* and

L. marthii. They are nonspore-forming, psychrotrophic, aerobic, microaerophilic or facultatively anaerobic, Gram-positive rods that are motile at 28°C. Of the 13 known *L. monocytogenes* serovars, the only human pathogen species, 1/2a, 1/2b and 4b, account for 95% of human isolates, with serovar 4b strains involved in 33–50% of human cases worldwide. Strains show major differences in virulence; however, no major correlation has been established between virulence and origin (human, animal, food, etc.) or characteristics (serovar, genotype, etc.). At this time, all strains of *L. monocytogenes* are considered capable of causing listeriosis (Bacon and Sofos, 2003).

Listeria is ubiquitous in the environment and may be found in many animals. It is of major concern because it grows under adverse conditions and limited nutrition. It is found in floors, walls, drains, condensed and standing water, and food residues on equipment in meat processing environments. The pathogen is sensitive to normal cooking but it may contaminate products after heating, when exposed to the environment during cutting, slicing and repackaging. Growth occurs in the range −0.4°C to 45°C, with optimum growth at 30–37°C. The optimum pH for growth is 7.0 with a range of 4.39–9.40. The pathogen survives at sodium chloride levels of up to 25% but requires water activity above 0.92 for growth (Bacon and Sofos, 2003).

Strains of *Salmonella enterica* subspecies *enterica* serotypes are causes of foodborne illness throughout the world. This fecal organism is widely distributed in the environment. Thus, foods involved in human illness include animal products, as well as any other food. Meat and poultry products are considered major sources since the main habitat of *Salmonella* is the gastrointestinal tract of food-producing animals. Salmonellosis can also be caused through contact with animals or infected humans.

As a member of the Enterobacteriaceae family, cells of *Salmonella* are Gram-negative, facultatively anaerobic, nonspore-forming rods. The approximately 2500 serotypes of *Salmonella* grow at temperatures in the range 5.2–46.2°C with optimum at 35–43°C. The pH range allowing growth is 3.8–9.5, with an optimum at 6.5–7.5, while the minimum water activity is at 0.93. The incubation period is 5h to 5 days, and the symptoms of the infection include diarrhea, nausea, mild fever, chills, and vomiting and abdominal cramps. The duration of the symptoms is 1–2 days, but may last longer and lead to severe chronic sequelae. The infectious dose in foods may be as low as 10–100 cells, depending on serotype, food type and the immune state of the individual (Bacon and Sofos, 2003).

Meat may be contaminated with *Salmonella* throughout the slaughtering, dressing and boning process, starting with the carcass during knife incision for hide removal. The USDA/FSIS has established microbiological criteria for *Salmonella* in animal carcasses and ground products as a measure for verification of pathogen reduction since the implementation of HACCP programs. Approaches controlling STEC/VTEC should also be effective against *Salmonella* strains and similar enteric vegetative pathogens.

Other Biological Hazards and Concerns in Meat

Parasitic agents that may be transmitted with pork include *Taenia solium, Trichinella spiralis, Sarcocystis suihominis* and *Toxoplasma gondii*, resulting in taeniasis, trichinosis, sarcocystosis and toxoplasmosis, respectively (Table 6.1). Beef may be the source of tapeworms (*Taenia saginata* cysticercus) and *Sarcocystis hominis*, and through fecally contaminated water may serve as an indirect vector for transmission of *Giardia duodenalis* (or *lamblia*) and

Cryptosporidium parvum. Poultry (mainly when raised outdoors) may transmit *Cryptosporidium* and *Toxoplasma gondii*. *Trichinella* may also be transmitted through game meat. Inactivation of parasites is achieved through proper cooking, freezing, salting, chemical treatments and ionizing radiation of meat and meat products (EFSA, 2011, 2012; Sofos et al., 2013).

Viral agents, including norovirus, hepatitis A and enteroviruses, cause the highest number of mild foodborne gastroenteritis cases. Transmission is mostly associated with poor sanitation, inadequate cooking or cross-contamination before consumption. Similar to parasites, viruses are unable to grow in foods and are generally sensitive to cooking. Their control in ready-to-eat foods should be through proper sanitation and hygienic practices of food service workers.

Transmissible spongiform encephalopathies (TSE) or prion diseases, and especially BSE, emerged as a major animal health problem in the 1990s. The concern was high because of potential association with human TSEs such as the new variant Creutzfeldt–Jakob disease (vCJD). Preventive controls, including feed bans and control of "specified risk materials" (brain, skull, eyes, spinal cord, small intestines, etc.), during slaughter of all cattle (small intestines) or those exceeding 30 months of age; increasing process controls for material obtained with "advanced meat recovery" systems; banning use in food products of "mechanically separated meat"; and banning use of the above materials in dietary supplements and cosmetics, have greatly reduced spreading and apparently resulted in containment of BSE. It is important to recognize emerging challenges, such as BSE, early in their development, in order to properly contain them through well-coordinated worldwide efforts (Sofos, 2008; Sofos et al., 2013).

Animal Health, Welfare and Humane Treatment

Animal health pandemics, such as avian influenza and foot-and-mouth disease, may cause major economic losses to local, domestic or international markets. In addition, they may become technical, economic, political or diplomatic or trade issues among countries. Worldwide health authorities should cooperate to achieve early detection and diagnosis for their prevention or containment and eradication (Sofos, 2008).

The issue of humane treatment and welfare of food animals deserves increased attention worldwide. Evidence suggests that animal stressing may damage meat quality and lead to contamination shedding and cross-contamination. However, irrespective of whether good animal husbandry practices make meat products safer or of better quality, humane treatment of animals is ethically essential and should be practiced by all involved in animal handling.

Animal and Meat Traceability

Maintenance of custody of the identity of animals and their products, from production to retail, requires effective traceability programs. Traceability is useful in the protection of animal and public health, and its use is a consequence of the BSE epidemic in Europe. European Union legislation requires animal food traceability systems, based on product labeling. Traceability can play a major role in management of food safety risks and in product authentication, and is necessary in food product recalls. It should be noted, however, that traceability of composite products, such as ground beef, is complicated and difficult, if not impossible, to apply in commerce. Animal identification and traceability systems must be complete and mandatory, and must be based on technologies that identify animals and their origin as well as origin of feed. In general, the overall concept of biotracing is gaining prominence in food safety (Sofos, 2008; Sofos et al., 2013).

Pathogen Resistance

Strains of bacteria may be resistant to control procedures, survive better in their hosts, be more virulent at lower doses, or exhibit resistance to drugs used for disease treatment. Stress adaptation and development of resistance by bacteria to antibiotics and potentially to traditional food preservation barriers, such as low pH, heat, cold temperatures, dryness or low water activity and chemical additives, is an issue of emerging concern. A common sense approach for control of antimicrobial resistance is prudent use and avoidance of overuse, abuse or misuse of antimicrobials. Control approaches should be based on risk analysis and examination of all issues related to a situation (Sofos, 2008).

Environmental Contamination Issues

Pathogens of enteric origin, such as *E. coli* O157:H7, have in recent years been involved in human illness through consumption of vegetable and fruit products, such as apple juice, salad seed sprouts, watermelon, spinach, lettuce and onions. *Salmonella* outbreaks have been associated with consumption of cantaloupes, watermelon, sprouts, tomatoes, peppers, chocolate, peanut products and dry breakfast cereal. These events confirm the role of environmental cross-contamination with enteric pathogens from animal feces or manure to a variety of food products of non-animal origin.

If not properly handled, composted and processed, manure leads to environmental and water contamination with pathogens of concern to humans. Natural water runoff or use of contaminated water to irrigate food crops or to wash plant food products, as well as wildlife movement in the field, results in cross-contamination. The origin of the pathogens may be the gastrointestinal system of food-producing or wildlife animals in proximity to plant food-producing farms. Irrespective of whether birds, wild animals or human negligence are the source of the problem, the meat animal industry should contribute to efforts needed to address this issue (Sofos, 2008; Sofos et al., 2013).

Additional issues related to meat safety include the safety and quality of organically and biologically grown products, the need for and development of improved and rapid testing and pathogen detection methodologies for laboratory and field use, regulatory inspection harmonization issues at the national and international level, establishment of risk assessment-based food safety objectives, and routine implementation of the HACCP system at the production and processing level on the basis of food handler training and consumer education, which is discussed in subsequent sections (Sofos, 2008).

CONTAMINATION FREQUENCY AND INCIDENCE OF DISEASE

Contamination Frequency

Although prevalence and levels of pathogens in animals, raw meat and meat products are generally infrequent and low because their occurrence is random (but not statistically randomly distributed) it is impossible to predict. Therefore, all raw meat products should be considered as potentially carrying pathogens and, thus, treated and handled accordingly. A comprehensive review by Rhoades et al. (2009) found that prevalence in cattle and beef is usually affected by pathogen type, animal type, animal age, season, feed, housing and general meat production practices. Mean prevalence rates (and ranges of means from individual surveys) of *E. coli* O157 were 6.2% (0.0–57%), 44% (7.3–76%), 0.3% (0.0–0.5%) and 1.2% (0.0–17%) for feces, hides, chilled carcasses and raw beef products, respectively. Corresponding prevalence rates for *Salmonella* were 2.9% (0.0–5.5%), 60% (15–71%), 1.3% (0.2–6.0%) and 3.8% (0.0–7.5%), while for *L. monocytogenes* the mean prevalence rates were 19% (4.8–29%), 12% (10–13%) and 10% (1.6–24%) for feces, hides and raw beef products, respectively.

In the United States, *E. coli* O157 in beef cattle had prevalence rates of 0.3–19.7% in feedlots and 0.7–27.3% on pasture, as summarized in a review by Hussein (2007); corresponding prevalence rates of non-O157 STEC/VTEC were 4.6–55.9% and 4.7–44.8%. Prevalence of O157 and non-O157 STEC/VTEC was 0.01–43.4% and 1.7–58.0% on whole carcasses, 0.1–54.2% and 2.4–30.0% in ground beef, 0.1–4.4% and 17.0–49.2% in sausage, and 1.1–36.0% and 11.4–49.6% in various retail cuts, respectively. Hussein (2007) also found that of 162 STEC/VTEC serotypes isolated from beef products, 43 were also detected in HUS patients and 36 are known to cause other human illnesses. Of 373 STEC serotypes isolated from cattle feces or hides, 65 were detected in HUS patients and 62 are known to cause other human illnesses. Fratamico et al. (2008) reported that 58 STEC/VTEC serotypes were isolated from swine feces and 13, 6, 80, 21, 6.4, 4.6, 42.9, 11.4 and 0.46% of the isolates carried the stx_1, stx_2, stx_{2e}, *estIa*, *estIb*, *fedA*, *astA*, hly_{933} and *cdt-III* genes, respectively; none of the strains possessed the *elt*, *bfp*, *faeG*, *fanA*, *fasA*, $fimF_{41a}$, *cnf-1*, *cnf-2*, *eae*, *cdt-I* or *cdt-IV* genes (Sofos et al., 2013).

A survey under the United States National Antimicrobial Resistance Monitoring System (NARMS) indicated that, in the period 2002 to 2007, of a total of 7258 *E. coli* isolates collected from retail meats, 16 ground beef and one pork chop isolates were positive for *stx* genes; specifically, five were positive for stx_1 and stx_2, two for stx_1 and 10 for stx_2. None of the isolates carried *eae*, while seven carried the *hlyA* (enterohemorrhagic *E. coli*; EHEC gene); 16 of the isolates were toxic against Vero cells. The 17 strains belonged to serotypes O83:H8, O8:H16, O15:H16, O15:H17, O88:H38, ONT:H51, ONT:H2, ONT:H10, ONT:H7 and ONT:H46, while subtyping by pulsed-field gel electrophoresis (PFGE) yielded 14 distinct restriction patterns (Xia et al., 2010). These findings led to the conclusion that diverse STEC/VTEC strains may contaminate retail meats (Sofos et al., 2013).

Data of the USDA/FSIS (www.fsis.usda.gov), summarizing beef product recalls (Table 6.2) for potential contamination with *E. coli* O157:H7 (Sofos, 2013), indicate that during the period 1994–1997, when concern over *E. coli* O157:H7 was increasing and procedures for its control and detection methodologies were evolving, recall numbers ranged from 2 to 6 per year. Then, as detection methods improved and scrutiny increased for the years 1998, 1999, 2000, 2001 and 2002, recall numbers increased to 12, 9, 30, 29 and 34, respectively. In following years, potentially due to better contamination control procedures during slaughter, recall numbers decreased to 8, 7, 5 and 8 for the years 2003, 2004, 2005 and 2006, respectively. For the years 2007, 2008, 2009, 2010 and 2011 they were 18, 9, 13, 7 and 10, respectively. Amounts of fresh beef recalled in each incident during this period ranged from 74,180 to 13,522,505 kg (Table 6.2).

TABLE 6.2 Numbers of USDA/FSIS Recalls due to *E. coli* O157:H7 in Ground Beef Products and Amount of Product Recalled in the United States (1994–2011)

Year	Number of Recalls	Amount of Product Recalled (kg)
1994	2	391,609
1995	5	430,439
1996	2	74,180
1997	6	11,618,305
1998	12	926,837
1999	9	325,882
2000	30	1,228,908
2001	29	987,520
2002	34	10,879,970
2003	8	472,734
2004	7	543,676
2005	5	566,287
2006	8	82,508
2007	18	13,522,505
2008	9	199,856
2009	13	316,945
2010	7	903,849
2011	10	304,531

Data of the USDA/FSIS raw ground beef testing program in the United States indicate (Table 6.3) that *E. coli* O157:H7 prevalence rates ranged between 0.00 and 0.86% for the period 1994–2011. *Salmonella* prevalence, as tested by USDA/FSIS under the HACCP Pathogen Reduction Regulation, is shown in Figure 6.1. Based on samples taken by USDA/FSIS from United States plants *Salmonella* positive samples (Figure 6.1) of ground beef were 1.6, 1.1, 2.0, 2.7, 2.4, 1.9 and 2.2% in 2004, 2005, 2006, 2007, 2008, 2009 and 2010, respectively, compared to a baseline of 7.5% in 1994. Corresponding data for ground chicken were 25.5, 32.4, 45.0, 26.3, 25.5, 18.2 and 18.8%, with a baseline of 44.6%. Overall, as shown in Figure 6.1, positive sample numbers declined for all meat products tested during the period of operation under the HACCP system.

Meatborne Illness Episodes

Meat and meat products are important vehicles in the transfer of foodborne hazards to humans. However, considering the enormous quantities of meat products consumed on

TABLE 6.3 USDA/FSIS Results from Analysis of Raw Ground Beef Samples for *E. coli* O157:H7

Year	Number Analyzed	Number Positive	Percent Positive
1994	891	0	0.00
1995	5,407	3	0.06
1996	5,703	4	0.07
1997[a]	6,065	4	0.07
1998	8,080	14	0.17
1999[b]	7,785	32	0.41
2000	6,375	55	0.86
2001	7,010	59	0.84
2002	7,025	55	0.78
2003	6,584	20	0.30
2004	8,010	14	0.17
2005[c]	10,976	19	0.17
2006	11,779	20	0.17
2007	12,292	29	0.24
2008[d]	11,630	54	0.46
2009	12,797	41	0.32
2010	12,590	30	0.24
2011	13,455	10	0.07

[a]*During October 1997, the amount of sample analyzed was increased from 25 g to 325 g to provide increased detection sensitivity.*
[b]*On 3 September 1999, a new selection and detection method was introduced to further increase test sensitivity.*
[c]*During October 2005, a new screening method was introduced to reduce the number of screen positives that do not confirm positive.*
[d]*Beginning with CY 2008, annual microbiological sample results were posted according to the date the sample was collected. Prior to CY 2008, annual posting of microbiological data results was based on the sample analysis completion date. For this reason, data from CY 2008 cannot be directly compared to data from prior years. In addition to the change in date criterion, target sampling that incorporates production volume and results history was introduced as well as a change in the laboratory testing method.*

a daily basis worldwide, and the resulting number of servings to be contaminated, their safety record should be considered, overall, as high. Nevertheless, any amount of product found potentially contaminated and recalled from the marketplace, as well as any illness, and especially even a single death, through consumption of food, including meat, is unacceptable. Indicative data and reasons for product recalls in the United States are presented in Table 6.4. Data showing numbers of outbreaks and cases of illness associated with various meat products in the European Union and the United States have been summarized and presented by Sofos (2013). The latest United States Centers for Disease Control and Prevention (CDC) estimates for pathogens (and some foods) responsible for most foodborne illness are *Campylobacter* (poultry), *E. coli* O157 (ground beef, leafy greens, raw milk), *L. monocytogenes* (delicatessen meats, unpasteurized soft cheeses), *Salmonella* (eggs, poultry,

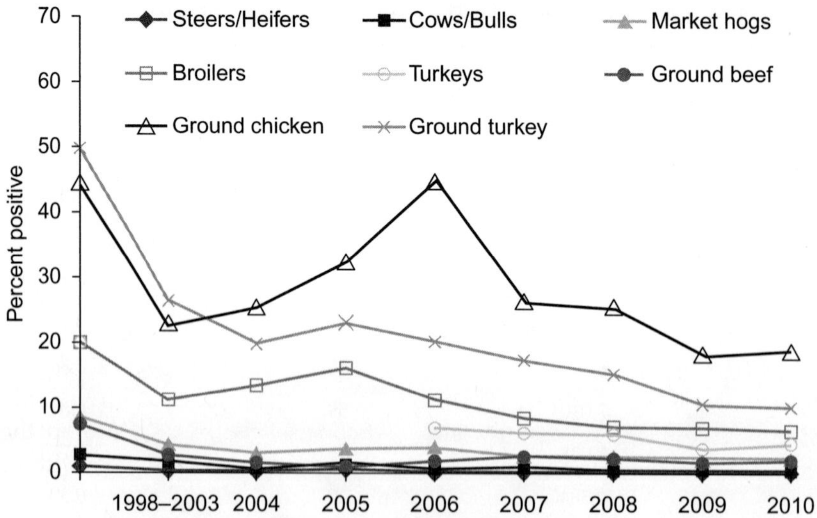

FIGURE 6.1 USDA/FSIS nationwide monitoring data indicating percent positive *Salmonella* samples of various meat products in United States plants as part of testing under HACCP.

TABLE 6.4 Summary of Number of USDA/FSIS Recalls and Amount of Product Recalled in the United States (2005–2011)

	Number of Recalls (Amount of Product Recalled; ×1,000,000 kg)						
Agent/Product	**2005 (53)**[a]	**2006 (34)**	**2007 (58)**	**2008 (54)**	**2009 (69)**	**2010 (70)**	**2011 (103)**
E. coli O157:H7	5 (0.6)	8 (0.1)	22 (15.2)	17 (3.2)	16 (0.6)	11 (1.0)	13 (0.5)
Salmonella	0	1 (<0.1)	1 (38.1)	0	6 (1.0)	7 (4.5)	10 (16.4)
L. monocytogenes	30 (1.6)	6 (<0.1)	11 (1.4)	15 (0.2)	8 (<0.1)	8 (0.2)	11 (0.2)
Undeclared allergen	9 (0.2)	9 (0.5)	12 (0.1)	7 (0.1)	13 (0.4)	18 (0.3)	40 (0.6)
Other	9 (0.5)	10 (2.0)	12 (10.1)	15 (66.7)	26 (2.3)	26 (9.5)	29 (0.3)
Beef	12 (0.8)	15 (0.3)	26 (14.0)	24 (68.9)	34 (1.8)	28 (10.4)	35 (0.6)
Pork	9 (<0.1)	4 (<0.1)	9 (0.1)	9 (0.1)	9 (1.2)	16 (0.7)	14 (0.3)
Poultry	14 (0.3)	8 (0.1)	9 (1.3)	15 (1.0)	13 (0.9)	17 (3.3)	31 (16.7)
Buffalo	–	1 (<0.1)	–	–	–	1 (<0.1)	–
Mixed meats	18 (1.8)	6 (2.3)	14 (49.5)	6 (0.2)	13 (0.4)	8 (1.0)	23 (0.4)

[a]*Total number of recalls.*

meat), *Vibrio* (raw oysters), norovirus (many foods; e.g. sandwiches, salads) and *Toxoplasma* (meats). Further, viral pathogens are of major concern in food service, while bacterial pathogens, such as *E. coli* O157:H7 and other STEC/VTEC, *Salmonella* and *Campylobacter* continue to be of major importance in the safety of raw meat and poultry, and *L. monocytogenes* in ready-to-eat processed products (www.cdc.gov/foodsafety).

Examples of meatborne outbreaks, demonstrating the diversity of causes and reasons for their occurrence, were presented by Sofos (2013) and include: the first United States confirmed incidence of listeriosis due to a meat product in 1988, involving a female cancer patient who ate microwave-heated turkey frankfurters of the same brand every day; the highly publicized *E. coli* O157:H7 outbreak associated with consumption of undercooked hamburgers from fast food restaurants in the Pacific Northwest of the United States that killed four children and sickened 700 others in 1992–1993; the outbreak of *E. coli* O157:H7 that occurred in central Scotland in 1996, involving 496 cases and 21 deaths of elderly persons due to cross-contamination between raw and cooked meat at a butcher's shop; the deadly (14 deaths and four miscarriages or stillbirths) listeriosis outbreak associated with consumption of delicatessen meats in 24 states of the United States in 1998–1999; the *S. aureus* intoxication outbreak at a company picnic in an amusement park in Georgia, United States, in 2000, attributed to consumption of pork barbecue kept in a cooler; the 2008 outbreak of listeriosis in Toronto, Canada, linked to cold meat cuts that led to 22 deaths and a total of 57 confirmed cases; the hemolytic uremic syndrome outbreak in Bordeaux, France, associated with the death of three children in 2012; and numerous others (Sofos, 2013).

Active surveillance data collected by CDC (FoodNet) in the United States for cases of *E. coli* O157:H7 per 100,000 population were 1.73, 1.1, 0.9, 1.06, 1.31, 1.2, 1.12, 0.99 and 0.9 for the years 2002, 2003, 2004, 2005, 2006, 2007, 2008, 2009 and 2010, respectively, compared to 2.7 in 1996, while corresponding data for non-O157 STEC were 0.33, 0.46, 0.57, 0.45, 0.57 and 1.0 for 2005, 2006, 2007, 2008, 2009 and 2010, respectively (Sofos et al., 2013). Data from CDC (www.cdc.gov) indicate that for 2008, beef and poultry were implicated in 31 and 32, respectively, of 218 outbreaks with confirmed food source. It is noteworthy that a food vehicle was reported for 481 (47%) of the total outbreaks. European Union data indicate that in 2009, pork, beef, broiler and other meat products were responsible for 7.8, 2.5, 3.6 and 3.4%, respectively, of verified outbreaks (EFSA/ECDC, 2011). Pathogens involved in verified pork product outbreaks included *Trichinella*, *Clostridium*, *Salmonella*, *Staphylococcus*, *E. coli*, *Bacillus*, viruses and other agents at 39.5, 22.4, 15.8, 6.6, 2.6, 2.6, 2.6 and 2.6%, respectively.

Efforts to collect data on the extent of foodborne illness outbreaks and episodes, as well as on causative agents and on food vehicles involved in transmission need to be improved worldwide. Attribution of illness episodes to the implicated food will be improved with better tracing of pathogens. Pathogen tracing can be accomplished through proper surveillance activities, microbial source tracking, and use of phenotypic and genotypic methods. The United States CDC has increased emphasis in tracking foodborne illnesses through more intense epidemiological and molecular investigations conducted by FoodNet and PulseNet. The data generated provide estimates for trends in the food safety burden, and as they become better, it will be easier to link changes in foodborne illness prevalence with regulatory and industrial pathogen control activities for specific food industry sectors, such as meat products. The data also provide a better picture of food safety problems, causes of problems and progress in their control, and they are necessary in microbial risk assessments (Sofos, 2008; Sofos et al., 2013).

CONTROL OF HAZARDS AT DIFFERENT STAGES OF THE MEAT CHAIN

Microbial Control Strategy

Meat and meat products support rapid and extensive microbial growth because they are moist and rich in nutrients, while microbial types exist that are able to proliferate and dominate in all types of meat products. Therefore, meat products need to be preserved in order to maintain quality, delay spoilage and assure safety. The strategy for microbial control in meat products involves implementation of measures that: (1) prevent or minimize introduction of microorganisms into the product; (2) reduce contamination by removal (decontamination) or inactivation (killing) of microorganisms which have gained access; and (3) prevent, delay or slow down growth of viable microorganisms, which have been introduced and have not been inactivated. Implementation of this approach also needs proper management of the interventions in order to at least reduce the potential for microbial meatborne illness (Koutsoumanis et al., 2006).

Hurdles applied for pathogen control in foods may be of a physical, physicochemical or biological nature. Physical treatments include thermal (low or high temperature) and non-thermal (ionizing radiations such as gamma rays, X-rays, electron beams or high hydrostatic pressure, pulsed electric fields, sonication, ultrasonic waves, ultraviolet light, pulsed UV light and microwaves), as well as packaging methods including modified atmospheres such as vacuum, high oxygen, low oxygen and oxygen-free. Newer packaging approaches are termed as active packaging, smart packaging, coatings or antimicrobial edible films, etc. Acidity or low pH, low water activity, modified oxidation-reduction potential (E_h) and application of chemical antimicrobials as ingredients or externally in the form of solutions or preparations are hurdles of a physicochemical nature. Biological interventions include addition of microbial starter cultures, mostly LAB, or use of their metabolites as antimicrobial preparations (Sofos et al., 2013).

The specific goals of the strategy for microbial control are accomplished through implementation of adequate cleaning, good sanitation, proper hygiene and effective antimicrobial intervention technologies in order to: (1) harvest and ship for slaughter and processing food animals with reduced contamination levels; (2) reduce potential for transfer of microorganisms to carcasses and meat from live animals, water and the environment; (3) apply safe and effective decontamination interventions, when approved and needed, for reduction of microbial levels on carcasses or meat; (4) apply processes (e.g. heat, high pressure, irradiation, etc., when approved and useful) in order to reduce or eliminate, by killing, microbial contamination on processed or cooked products; (5) avoid or minimize cross-contamination at all stages of the chain, from production, slaughter, processing and preparation to consumption; and (6) maintain the cold chain by keeping products under low temperature and packaging conditions that inhibit growth of surviving microorganisms. Following the HACCP principles for proper design, validation, implementation, monitoring, verification and documentation, this common sense approach is the best strategy available for assuring meat safety and quality (Koutsoumanis et al., 2006; Sofos et al., 2013) and is discussed in subsequent paragraphs.

Keeping Contamination Low

General

While complete prevention of raw meat product contamination is impossible, restriction of sources and control of access or transfer of contamination is possible. This is possible only if processing facilities and equipment are of proper design and through implementation of effective and documented cleaning, sanitation and hygienic procedures. Packaging also serves as a barrier to contamination of processed products.

Minimizing levels of contamination should be the number one priority of HACCP prerequisite programs because it reduces the probability that errors at subsequent points of the food chain (e.g. food processing and preparation) will result in foodborne illness. In addition, low initial contamination levels help processors meet specifications of contractual agreements and regulatory standards such as the zero tolerance requirement for *E. coli* O157:H7 and *L. monocytogenes* in raw and cooked products, respectively. Thus, products consumed with minimal or no further processing for pathogen destruction should be safer. Further, processes designed to kill target populations of pathogens will not fail due to excessive initial contamination levels. In addition, low initial contamination reduces risks from pathogens of high infectious doses; improves the safety of products contaminated with pathogens of low infectious doses; and limits cross-contamination risks at all stages of food processing including preparation for consumption and serving (Sofos, 2008).

Cleaning and Sanitation

Proper cleaning and sanitation procedures for removal of food residues and contaminants, and prevention or control of biofilm formation in the food environment, are needed to maintain low product contamination levels. Biofilms consist of microbial cell clusters attached to surfaces where they multiply forming a cell mass of microcolonies encapsulated and held together by an exopolysaccharide matrix. The hydrated matrix has channels and pores throughout its structure, allowing transport of oxygen, nutrients and waste materials, and increases resistance of biofilm cells to sanitizers. Food processing areas prone to biofilm formation include floors, walls, pipes and drains. Biofilms are formed, as mono- or mixed cultures, by bacteria such as *Pseudomonas*, *Listeria*, *Salmonella*, *Campylobacter*, *E. coli* and LAB. They may be established on materials such as stainless steel, aluminum, nylon, Teflon, rubber, plastic, glass, etc., which are used in non- or food-contact surfaces of pasteurizers, conveyor belts, gaskets, crevices and dead spaces. Biofilms are especially a problem in areas that are hard to clean and sanitize. A concern is that biofilms may be much more resistant to chemical sanitizers than free-flowing planktonic cells of the same species. Also, the concentration of sanitizers and exposure time need to be increased in order to effectively destroy cells in biofilms, compared to free-flowing cells. Summarizing, to control contamination and enhance meat safety through improved hygienic practices, it is crucial to prevent formation or to remove and inactivate biofilms if formed. Biofilms can be removed through application of procedures that dissolve organic material, including physical force through scrubbing. Effective cleaning and sanitation programs are very important, while frequent rotation of sanitizing agents and thorough drying of equipment provide hurdles against microorganisms attempting to establish biofilms. Inadequate cleaning allows residual organic matter to be present and react with the sanitizer, reducing its antimicrobial activity. Cleaning and sanitation programs should be well developed, validated, verified and documented (Sofos et al., 2013).

Contamination Control at the Pre-harvest Level

Reduction of microbial contamination on the farm has been researched in recent years in relation to animal and plant food production for safer consumer products. The reasons for this include the need for reduction of pathogen sources and levels in order to reduce animal product contamination, direct animal-to-human transmission of pathogens, as well as water contamination and subsequent cross-contamination of vegetables through animal feces. Research for pre-harvest control of animal contamination has been directed toward animal diet modifications, feed additives or supplements, vaccination, bacteriophage application, antibiotic treatments, competitive exclusion, prebiotics or probiotics and good production management practices, including pen management, clean feed, chlorinated water, clean and unstressful transportation to slaughter, clean lairage and animal cleaning before slaughter. With the exception of good production management practices, and to some extent feeding of probiotics, the remaining approaches are still of limited or no use.

Carcass and Raw Meat Decontamination

The notable *E. coli* O157:H7 outbreak linked to consumption of undercooked contaminated ground beef in 1992–1993 in the United States renewed interest in carcass washing and chemical or thermal decontamination. Such treatments were first proposed in the early 1970s without noticeable commercial use. Currently, the meat processing industry in the United States, Canada and Australia is using carcass or meat decontamination interventions extensively. Goals are to provide safer products for consumers, help in compliance with regulatory requirements and meet trade specifications for raw materials (e.g. raw fresh meat trimmings). The target of carcass and meat decontamination is to reduce prevalence and numbers of pathogenic bacteria, especially STEC/VTEC, *Salmonella*, *Campylobacter*, etc. Specific interventions applied include: total or partial external animal washing or hair removal; removal by knife-trimming and/or steam-vacuuming of soiled spots in order to meet the "zero tolerance" requirement for visible contamination before carcass washing; decontamination with organic acid solutions and/or hot water applied immediately after hide removal but before evisceration (pre-evisceration washing); spray-washing with water after carcass splitting, removal of specified risk materials and zero tolerance inspection; spraying, dipping, deluging, etc. with hot water or steam (thermal pasteurization), and/or rinsing with chemical solutions (e.g. lactic acid, acidified sodium chlorite, peroxyacetic acid-based preparations, etc.); dry- or spray-chilling of carcasses; chemical spraying of chilled carcass sides before deboning and of carcass cuts or trimmings at packaging or before grinding or processing into ground or other non-intact or brine-injected products. Decontamination treatments are applied in sequence (e.g. trimming, washing, chemical, chilling, etc.) and some simultaneously (e.g. warm acid solutions, steam and vacuum) following the multiple hurdle concept (Sofos, 2005; Sofos et al., 2013).

Decontamination treatments assist plants in meeting regulatory criteria and industry specifications, but they are inadequate for complete microbial removal or inactivation (reductions of 1 to 3 log units) because they are of short duration and their intensity is mild.

However, they help in limiting cross-contamination and improving product quality, and should increase product safety by reducing the probability of illness from consumption of undercooked product (Sofos, 2008; Sofos et al., 2013). The effectiveness of a decontamination treatment is variable as it is affected by water pressure, temperature, chemicals and their concentration, method of application, time or stage of application during slaughter and processing, time duration of exposure (which varies with speed of slaughter and length of the application cabinet), etc. It needs to be emphasized that decontamination must be only one component of an integrated pathogen control system that is applied only when all necessary prerequisite programs, including proper and sanitary designs for facilities and equipment, good manufacturing and hygienic practices and proper management based on HACCP principles, are followed (Sofos et al., 2013).

Although organic acids are commonly used, other chemical decontamination solutions proposed, evaluated or used in certain situations include chlorine or chlorine dioxide, trisodium phosphate, cetylpyridinium chloride, hydrogen peroxide, ozone, etc. Decontamination treatments are approved by USDA/FSIS for use in meat, even without labeling, if it is demonstrated that: (1) they are Generally Recognized as Safe (GRAS); (2) they do not result in product adulteration; (3) they are considered as a processing aid; and (4) scientific data demonstrate their effectiveness. European Union regulations, in principle, allow application of water, including hot water, for decontamination of carcasses. Approval of chemical decontaminants in the European Union is based on evaluation of: (1) the toxicological safety of the material; (2) the efficacy of decontamination; (3) the potential for emergence of reduced susceptibility to biocides and/or resistance to therapeutic antimicrobials; and (4) potential risks associated with release of effluents into the environment.

Universal considerations in approval and use of decontamination treatments for use in meat include the: (1) safety of workers and consumers; (2) potential effects on product quality; and (3) potential for spreading of bacterial cells over the carcass surface or penetration into the tissue. These concerns can be addressed through proper selection and approval of safe substances and application in properly designed and operating equipment. Other issues deserving consideration include potential for injury or stress resistance in bacterial cells, selection or accumulation of bacterial spores and changes in the microbial ecology of the environment and product, such as elimination or inhibition of normal Gram-negative spoilage bacteria, and selection of yeasts or LAB by acid treatments or of Gram-negative bacteria by water/steam treatments (Sofos, 2008; Sofos et al., 2013).

Destruction or Inhibition of Contamination

General

Unless frozen, fresh raw meat is highly perishable by nature. Thus, in addition to being cooked for consumption, it may be processed into products of longer shelf-life and desirable eating characteristics. A variety of technologies, based on physical (e.g. refrigeration, freezing, heating, drying, irradiation, smoking and packaging), chemical (e.g. curing agents such as nitrite or salt, lactate, acetate, diacetate, citrate, propionate, sorbate, benzoate, as well as acidifying agents like acetic, lactic and citric acid) and biological (e.g. LAB and their products including low pH, organic acids, hydrogen peroxide, bacteriocins, etc.) factors are used to process meat products for assurance of quality, stability and safety. This is achieved through inactivation or inhibition of growth of spoilage and pathogenic microorganisms. As indicated, combinations of such technologies, applied at individually sublethal levels (hurdle technology), also result in stable and safe products of better quality (Koutsoumanis et al., 2006; Sofos, 2008).

Based on the specific interventions applied, the intensity of their application and the combinations used, inactivation of microbial contamination of importance in meat safety and shelf-life may be complete (commercial sterility) or adequate for long-term or partial delay, or complete inhibition (freezing) of growth (longer shelf-life meat products; usually pasteurized, fermented or acidified, dried and refrigerated). Such strategies lead to production of processed meat products, either shelf-stable or of longer shelf-life under refrigeration, most of which are ready-to-eat, indicating absence of pathogens (Koutsoumanis et al., 2006).

Bacterial Destruction

Microbial inactivation in meat products is achieved by exposure to adequate levels of physical treatments such as thermal processing, ionizing radiation or high hydrostatic pressure. Inactivation of both cells and spores of foodborne pathogens is usually achieved by heating at temperatures above boiling of water (canning), resulting in commercially sterile meat products that are shelf-stable for long periods of time provided that the integrity of the container has remained intact during and after the lethal treatment and no recontamination has occurred. This is achieved either through aseptic processing and filling of containers (applicable mostly to liquid or flowing foods) or through processing in the final container (more applicable to solid meat products). Destruction of bacterial cells but not spores (pasteurization) is achieved by heating at milder temperatures (below water boiling) and in certain products and countries by exposure to irradiation or high hydrostatic pressure treatments; irradiation and high hydrostatic pressure treatments of intensities adequate to kill bacterial spores are not used in meat processing as they damage product quality (Koutsoumanis et al., 2006).

Inhibition of Bacterial Growth

Technologies that delay or stop growth of microorganisms rely on control of factors affecting microbial survival and growth, such as type and extent of initial contamination as well as factors intrinsic or extrinsic to the product; the most important of which are moisture expressed as water activity, pH, antimicrobials, storage temperature and gas atmosphere. Thus, major preservation technologies, based on inhibition of microbial growth, include temperature control (freezing, refrigeration, pasteurization), decreased water activity (drying), acidification (direct addition of acid) or fermentation–biopreservation (acid and bacteriocins such as nisin), addition of chemical preservatives (curing), and modified atmosphere, including vacuum, packaging (Koutsoumanis et al., 2006). It should be restated that, in addition to the extreme modification of a single factor, inhibition of microbial growth is often achieved with a combination of technologies at individually sublethal levels that yield the "multiple hurdle" effect.

A characteristic example of controlling a pathogen through reduction of contamination levels and inhibition or delay of growth when inactivation is not complete is *L. monocytogenes* in ready-to-eat meat products that support growth before consumption.

When this highly fatal pathogen was involved in major deadly outbreaks through consumption of post-lethality treatment-contaminated products, the need for intervention became obvious. Although thermal processing of meat products inactivates cells of the pathogen in the formulation, contamination may be reintroduced during product slicing and repackaging. Recontamination is sometimes difficult to avoid through sanitation because this pathogen survives and persists in harsh conditions and is able to grow under refrigeration. Therefore, some processors may need additional control interventions for better assurance of pathogen control. Because of these concerns the USDA/FSIS established a regulation for control of *L. monocytogenes* in ready-to-eat meat and poultry products that may be contaminated after processing, and that allow growth of the pathogen during distribution and storage before consumption. According to this regulation processors should select and implement one of three alternatives for control of *L. monocytogenes* in their products. The three alternatives are: (1) application of a post-lethality treatment (it may be an antimicrobial agent) that reduces or eliminates microbial contamination on the product in combination with an antimicrobial agent or process that controls growth; (2) application of either one of the above; or (3) application of a validated, verified and documented sanitation program in combination with a microbiological testing program for food-contact surfaces, including holding of product when results of testing are positive. The zero tolerance (absence in two 25 g samples) for *L. monocytogenes* in ready-to-eat meat and poultry products in the United States is still in effect, and product found contaminated is recalled and destroyed. This regulation offers the industry alternatives that they can put in place in order to resume production when there is a zero tolerance failure, provided that the selected alternative is validated and introduced in their HACCP plan or prerequisite programs (Sofos et al., 2013).

Post-lethality physical treatments evaluated or introduced as *L. monocytogenes* control alternatives include radiant heating, flash steam heating and steam pasteurization or hot water immersion. In addition, several chemicals, including potassium and sodium lactate, diacetate, acetate, acetic and lactic acid, acidic calcium sulfate, lauric arginate, pediocin and cetylpyridinium chloride, have been evaluated for antilisterial effects in ready-to-eat meat products. Solutions of acetic or lactic acid, nisin, benzoate, sorbate and their combinations applied post-processing by immersion of frankfurters, bologna, ham, smoked sausage and turkey breast, formulated with or without antimicrobials (e.g. lactate and diacetate), caused reductions in *L. monocytogenes* counts and inhibited survivors during storage. Antimicrobial activity against survivors during product storage varied with type or combination of antimicrobials, type of product, concentration of antimicrobial, length of exposure time and sequence of exposure. It should be emphasized that processors need to validate formulations and treatments that fit their product specifications and expectations (Koutsoumanis et al., 2006; Sofos et al., 2013).

Non-thermal Processing Treatments

In efforts to satisfy the increasing number of consumers who prefer foods that are minimally processed, without synthetic or with natural additives, of high quality, nutritious, fresh in appearance, convenient, with natural flavor and taste, and of an extended shelf-life without compromising safety, the industry is considering alternative preservation technologies such as high hydrostatic pressure. Other non-thermal technologies evaluated or used to some extent include ionizing radiation, pulsed X-rays, ultrasound, pulsed light and pulsed electric fields, high-voltage arc discharge, magnetic fields, dense phase carbon dioxide light pulses, natural biopreservatives and active packaging systems (Koutsoumanis et al., 2006). Levels of these processes considered for use in foods are adequate for inactivation of vegetative cells, but not bacterial spores. Antimicrobial activity may increase if non-thermal and thermal technologies are used in milder combinations following the multiple hurdle concept.

Ionizing radiation treatments have been scientifically established as safe and effective, and have been approved for use in raw meat and poultry in the United States since 1992 and 1997, respectively. Application may be with gamma rays or high-energy electrons and X-rays which kill microorganisms, extending the shelf-life and enhancing food safety. Maximum irradiation doses applied depend on type of product and whether the product is treated in the refrigerated or frozen state. For raw/chilled meat, raw/frozen meat and fresh or frozen poultry approved maximum doses are 4.5, 7.0 and 3.0 kGy, respectively, in the United States. Irradiation finds only limited commercial application in meat products because of consumer resistance and conflicting reports of potential undesirable effects on meat color and odor. Interventions proposed to preserve color and flavor during irradiation treatment include feeding food animals dietary vitamin E or conjugated linoleic acid, treatment of fresh products with antioxidants such as tocopherol, reduced fat content and double packaging (Sofos et al., 2013).

Atmospheric pressures of up to 1000 MPa kill microbial cells and extend product shelf-life without major changes in nutrients and sensory quality of foods. Microbial cells are killed through protein, including enzyme, denaturation, cell membrane damage and solute loss. High-pressure treatments combined with mild heat allow for improved synergistic effects. In the United States, high-pressure processing is used as a natural alternative to chemical antimicrobials for control of *L. monocytogenes* in commercially processed meat products such as cooked ham. High-pressure processing could kill STEC/VTEC and improve dry fermented sausage safety without the undesirable effects of heat or other treatments on product quality. The USDA/FSIS requires a 3–5 log unit reduction of *E. coli* O157:H7 in these products because they were implicated in outbreaks (Sofos et al., 2013).

Meat Packaging

Meat products are packaged to control moisture and weight loss, prevent introduction of additional contamination or cross-contamination, and delay chemical and microbial spoilage. Meat packaging systems include simple overwrapping with paper or air-permeable films for short-term chilled storage and/or retail display. More complex systems involve a large variety of modified atmospheres including vacuum packaging, bulk-gas flushing or systems using up to 100% carbon dioxide or other gas mixtures for longer chilled storage.

Packaging innovations include use of high barrier films, active or smart packaging systems, or use of low levels of carbon monoxide for better bright red color retention in fresh meat (Koutsoumanis et al., 2006; Sofos et al., 2013). The cherry-red color of fresh red meat, which is the result of the reaction of oxygen with myoglobin and residual hemoglobin to form oxymyoglobin and oxyhemoglobin, respectively, is preferred by consumers, and is assured by packaging aerobically on Styrofoam trays overwrapped with polyvinyl chloride (PVC) film. This color turns into brown within 5–7 days due to oxidation of these pigments to form metmyoglobin. When at less than 0.15%, oxygen prevents browning, while at 0.15–2.0% oxygen enhances browning. Bacterial growth, usually pseudomonads, is faster under PVC overwrap compared to vacuum packaging conditions, which favor growth of LAB such as *Lactobacillus.*

Modified atmosphere packaging involves placement of product under mixtures of non-toxic gases, such as oxygen, nitrogen and carbon dioxide, in proportions different than air. Such an approach is more acceptable for light-colored cuts of pork and poultry meat. Absence of oxygen inhibits lipid oxidation and discoloration in meat products. Case-ready packaging is a form of modified atmosphere packaging done at a centralized location before transportation for display at retail stores. The ready-for-display individual portions of meat are wrapped in PVC and placed in a master pack, which is sealed under the selected modified atmosphere. The master pack is shipped, and when desired at the store the individual PVC packages are removed and placed in display for red color blooming (Sofos et al., 2013).

Carbon dioxide is used at levels above 20% because it acts as an antimicrobial especially at lower temperatures of storage which increase solubility. In recent years carbon monoxide has found use in packaging of fresh meat in the United States because at low levels it maintains the desirable red color, inhibits oxidized flavors and reduces microbial growth for 28–35 days. Consumers have expressed concerns that carbon monoxide is potentially hazardous and that the extended fresh meat appearance may mask spoilage and high bacterial counts.

A concern associated with modified atmosphere packaging and delay of meat spoilage is potential creation of an environment favoring anaerobic or facultative pathogen growth before adequate evidence of spoilage. Thus, it is recommended to store such products under proper refrigeration and for a defined length of time for pathogen control. Products not processed to inactivate bacterial spores (e.g. mostly seafood products) must be stored below 3.3°C for the duration of shelf-life if packaged under reduced oxygen in order to prevent growth of and toxin production by non-proteolytic *C. botulinum*. Use of time–temperature integrators, antimicrobial agents or freezing may provide additional antimicrobial hurdles (Sofos et al., 2013).

Incorporation of additives into packaging systems, either within the package, attached to the inside or incorporated into the packaging material, are known as active packaging systems. Additives that may be used include oxygen scavengers, carbon dioxide scavengers and emitters, moisture controlling agents and antimicrobials. Recent packaging development efforts address the use of natural, disposable, potentially biodegradable or recyclable materials, or the use of edible coatings as packaging films. Such advances in packaging approaches are presented as smart or intelligent packaging techniques. Intelligent packaging also uses sensors, indicators (e.g. integrity, time–temperature indicators (TTI)) and radio frequency identification (RFID) systems to monitor conditions within the package and to provide information related to product quality. Additional advances are needed, however, for practical use of such systems in meat products (Koutsoumanis et al., 2006; Sofos et al., 2013).

Optimization of Sublethal Multiple Hurdles

As indicated, control of pathogenic bacteria in various processed meat products is accomplished through application of combinations of individually sublethal antimicrobial treatments in the form of multiple sequential or simultaneous interventions (hurdle technology concept). This approach meets consumer demands for safe foods that are of good quality, wholesome, nutritional and economically affordable, but with no additives, are convenient to use and have been exposed to only minimal processing. A concern expressed for this approach is that, if not properly designed and applied, the individually sublethal hurdles may lead to adaptation or selection of resistant bacteria which may express multiple resistances or cross-protection to food-related sublethal stresses such as sanitizers, decontamination interventions, acids, cold, heat, drying, anaerobiosis, etc. Stress-adapted pathogens may be more difficult to control, leading to failure of preservation systems.

Recent progress in predictive mathematical modeling has contributed to better selection of hurdle levels and combinations. However, there is still opportunity for optimization of hurdles to achieve maximum product shelf-life and safety without selection of strains with compromised sensitivity to hurdles. The hurdles should be selected and applied in appropriate combinations, intensities or concentrations, and in a sequence leading to optimal synergistic antimicrobial effects in complex food systems. The goal should be to achieve cell death or inhibition of growth through exhaustion of cells trying to repair injuries caused by the hurdles. Design of such intelligent meat preservation systems requires better knowledge of cell functions as well as modes and mechanisms of antimicrobial activity of different hurdles (Sofos et al., 2013).

MEAT SAFETY PROCESS MANAGEMENT

Regulatory Requirements

The interest in meat safety generated after the 1992–1993 undercooked ground beef *E. coli* O157:H7 outbreak led to regulatory inspection changes in the United States. Changes established by USDA/FSIS included: (1) reinforcement of the "zero tolerance" policy requiring removal of visible soil from carcasses by knife-trimming before washing and decontamination (in 1993); (2) declaration of *E. coli* O157:H7 as an "adulterant" in ground beef (in 1994) and in all other non-intact beef products (in 1999); and (3) implementation of formally inspected sanitation standard operating procedures (in 1996; SSOP), implementation of HACCP (in 1998–2000) as a process management system and compliance with performance criteria for *E. coli* biotype I to verify process control (done by meat processors) and *Salmonella* as a verification of HACCP and for tracking pathogen reduction (done by USDA/FSIS). In addition, USDA/FSIS has issued various Directives (www.fsis.usda.gov) to industry, such as the need for re-evaluation of HACCP plans and testing of ground beef and ground beef raw materials, including trimmings derived from steaks/roasts ("bench trim"), for *E. coli* O157:H7. In 2003, the USDA/FSIS issued another regulation addressing control of *L. monocytogenes* in ready-to-eat meat and poultry products that may be contaminated after processing (e.g. during slicing and repackaging) and allow growth during distribution and storage even at refrigeration temperatures. Operation under the principles of HACCP and other related requirements have also been established through legislation in various countries and regions, including the European Union, Canada, Australia, Japan and New Zealand.

In addition to regulatory microbiological performance criteria, the United States meat industry, especially the ground beef sector, has imposed contractual microbial specifications and related microbial control requirements on their raw meat suppliers. The reasons why the industry sought pathogen control improvements in their products were the importance of providing safer products for consumers and the negative publicity and other consequences associated with illness outbreaks and product recalls. Consequently, the scientific sector has responded through research that has contributed to improved pathogen control through development and validation of various antimicrobial intervention technologies discussed in previous paragraphs. As indicated, an effective pathogen control program should include activities employed pre-harvest or in the field, post-harvest or during processing, at retail, food service and at home. Methods, treatments, processes, interventions and hurdles applied for pathogen control should be managed properly based on the HACCP principles and spirit.

Prerequisite Programs and HACCP

HACCP is a proactive strategy designed to anticipate food safety hazards and solve problems in advance in order to prevent production and consumption of unsafe products (see Chapter 31). As a pathogen control management system, HACCP is based on the concept of establishing and managing controls following a complete hazard analysis. For improved effectiveness the concept of HACCP should be applied throughout the food chain. Further, HACCP implementation should be based on strong prerequisite programs (PP), including Good Manufacturing Practices (GMP) and Good Hygiene Practices (GHP), which are its essential foundation. Proper hazard identification, selection of effective controls and detection of a deviation as it occurs are the basis for the success of HACCP in hazard control through prevention. Inclusion of steps to be taken when deviations occur assures quick and effective re-establishment of process control, and proper disposal of potentially hazardous product. For proper implementation and success of HACCP there is a need for commitment by top management, while effectiveness is based on proper employee education and training through appropriate standard operating procedures (SOP), i.e. job instructions. Governments, trade groups and the industry universally have accepted the HACCP principles. Benefits of proper HACCP implementation include enhanced food safety assurance, better use of resources, timely response to problems and compliance with regulations, customer specifications and consumer demands (Sofos, 2008).

HACCP is defined as a systematic approach for the identification, evaluation and control of food safety hazards based on seven principles: hazard analysis; critical control points (CCP); critical limits (CL); monitoring procedures; corrective actions; verification procedures; and record-keeping/documentation procedures. As indicated, HACCP needs to be established and implemented on a solid foundation of PP, which may also be known as "control points." PP are defined as universal steps or procedures applied to control the environmental and operational conditions in a food establishment for the production of safe food. Common PP include: Good Agricultural Practices (GAP) or Good Production Practices (GPP); GMP; and GHP. Important prerequisites deal with facility premises and land history, including location and structure (design, construction, maintenance and working environment such as lighting, temperature, humidity, etc.); facilities, equipment and instrument design and standards, including maintenance and calibration services; foreign material control; water, ice and air quality and control; cleaning and sanitation programs and SSOP, including sanitary services, disposal of waste materials and provision of electricity, water, refrigeration and steam; personnel training in hygiene principles and task accomplishment; specifications for raw materials, including live animals, food ingredients, chemicals and packaging; product traceability and recall plans; documentation and maintenance of records; etc. GHP are defined as operations involved in providing a clean sanitary environment in product processing, preparation, handling and storage. GHP include: cleaning of plant and equipment; personnel health and cleanliness; cleanliness of live animals and other raw materials; and proper packaging, labeling, specifications and storage. Additional PP include: purchasing requirements; supplier selection, certification and approval; product specifications; product storage and transportation; rework control; receiving, storing and controlling ingredients; pest control; allergen control; chemical control; personnel facilities; product identification, labeling, tracing and recall; and record and document control (Sofos, 2005).

HACCP Implementation through SOP

Proper and effective implementation and management of food safety assurance processes needs adequate and proper implementation of validated and verified HACCP plans. As indicated, HACCP programs should be applied throughout the food chain and should be based on a foundation consisting of effective PP, including GMP and GHP. This can be accomplished through development and implementation of SOPs or job instructions. SOPs include procedures for each step in a process, procedures describing how each GMP and GHP is to be carried out, and procedures to be followed at each CCP. In other words, SOPs describe in detail, at least, how each activity is done. A complete SOP should address the following: describe the task to be accomplished; who is responsible for the job; when and how frequently the task is performed; the importance of the task; the steps involved in accomplishing the task; and provide guidance as to what should be done if a deviation or other problem develops. SOPs should be written in a way that is understandable to workers and should be used for personnel training in order to maintain consistency in HACCP implementation. Of course, they should be revised when the need arises (Sofos, 2005, 2008).

Validation of CCP and CL

Validation is a necessary component of HACCP, which may be considered as a form of verification. It is used to ensure that the CLs at each CCP of a HACCP plan are achieved or achieve their targets. Simply, HACCP plans and CCPs and CLs are validated to determine whether they are working as intended for the prevention, elimination or reduction of food safety hazards. Validation may be based on scientific literature and/or historical data, regulatory requirements, or sometimes it may be necessary to evaluate CCPs as implemented within a processing operation in order to determine efficacy of an intervention through a validation study involving microbial challenge testing. Initial validation should be repeated when changes in processing occur or problems arise as determined by monitoring and verification. For more detail on validation, refer to Chapter 31 on Hazard Analysis and Critical Control Point System.

Monitoring of CCPs and CLs

The effectiveness of a HACCP plan in controlling food safety hazards depends on development of proper monitoring systems, statistically valid monitoring frequencies, proper training of personnel conducting monitoring, validation and verification activities, and continuous evaluation of production processes. Monitoring facilitates continuous tracking of the operation at CCPs, produces written documentation for use in verification, determines lack or loss of control and facilitates proper corrective actions. Effective monitoring of CCPs is often based on the use of statistical process control (SPC) to summarize monitoring data on charts, providing a pictorial, more realistic and continuous assessment of process performance. This allows a better and early warning of tendencies for potential upcoming process failures. SPC is based on the continuous analysis of process monitoring data, which is assessed against predetermined criteria. The system may be completely automated, semi-automated or manual, depending on type of monitoring parameters, which may include pH, microbiological and temperature data, or visual observations, etc. (Sofos, 2005).

Verification of HACCP

According to the United States NACMCF, the HACCP verification principle is defined as activities, other than monitoring, that determine the validity of the HACCP plan and that the system is operating according to the plan. Designing proper verification procedures, as well as hazard analysis, and identification of CCPs and CLs, are HACCP principles that require complete knowledge of the process as well as the hazards and their characteristics within each process. While validation covers the scientific and technical quality elements of the HACCP plan, verification covers procedures that determine compliance with the HACCP plan. As listed in the United States HACCP regulation, verification procedures include: calibration of process monitoring equipment; direct observations of monitoring activities and corrective actions; and review of records generated and maintained within HACCP. Additional procedures used for verification in the United States and the European Union involve determination of bacterial counts, such as Enterobacteriaceae or *E. coli* counts, as indicators of process hygiene. Verification should be designed to randomly determine that a meat processing plant is producing carcasses or meat within set microbiological criteria.

Verification activities are less frequent than monitoring and should be based on process capability studies; stable and capable processes are verified less frequently than those that are less stable. In addition to routine verification, verification activities are also important in situations of emerging product safety concerns; to confirm that changes in processing or the HACCP plan have no adverse effect; and to determine whether the HACCP plan needs modification due to changes in the process, equipment, product formulation, etc. A reassessment of the HACCP plan is a periodic comprehensive verification and should include a thorough technical evaluation of all HACCP plan elements by unbiased, independent, internal or external experts (Sofos, 2005).

Microbial Testing in Meat Safety Assurance

It is important to stress that, by nature, microbial testing is slow in producing results and therefore not effective in monitoring CCPs, which is optimal when continuous and instantaneous if possible. Thus, microbial testing should not be relied upon as a routine means of HACCP monitoring or as a final determinant of product safety. The difficulty in assuring the microbiological safety of meat through testing alone actually led to the development and widespread adoption and implementation of preventive HACCP principles. However, it is impossible to operate an effective system for control of microbiological contamination without reference to densities or prevalence of bacteria as criteria for verification and as to whether the system is working properly. Thus, microbiological testing should continue, but with correct goals and objectives. Microbial testing is essential for determination of hazards and risks, finding pathogen niches and sources, and development, establishment, implementation, validation, maintenance and verification of effective safety assurance systems, such as HACCP.

Microbiological criteria, determined through microbial testing, do not assure product safety but are needed in monitoring the performance of meat processing operations designed to produce safe products. Thus, regulators in the United States, European Union and other countries have established microbiological performance criteria for meat and poultry products that need to be met by plants operating under the principles of HACCP. The criteria include numbers of microorganisms, mostly indicators, recovered from meat products or the frequency of their recovery. This is a routine use of microbial testing in meat safety assurance through verification of CCPs and other pathogen control activities. As indicated, the United States meat inspection regulation requires testing carcasses for *E. coli* counts by slaughter plants to determine compliance with process performance criteria and for prevention or reduction of fecal contamination on carcasses; testing for presence of *Salmonella* was selected as an indicator of pathogen reduction through HACCP implementation. The European Union has also established testing requirements and criteria for process hygiene control.

Since HACCP is an effective system based on sufficient and properly validated controls, it requires little end-product testing. Thus, processors and regulators should emphasize frequent reviews of HACCP plans, verification that the HACCP plan is being followed correctly, and review of CCPs monitoring and corrective action records, instead of relying on end-product testing for food safety assurance. However, microbiological testing is useful in verifying that the CLs at each of the CCPs are achieving the purpose of reducing, preventing and/or eliminating food safety hazards. The frequency of such process control verification testing should be based on the level of process control achieved in the operation, through the use of control charts and capability studies. In cases of significant variations in the control parameters, it may be necessary to conduct microbiological studies under operational conditions within the establishment to verify the efficacy of the CCPs (Sofos, 2005).

In conclusion, microbial testing should not be implemented with the sole objective of determining the safety of a product lot before shipment after processing. Frequency of pathogen occurrence in meat products is naturally low, unpredictable and non-statistically random. Thus, statistical sampling plans for detection of pathogen positive samples, at a reasonable probability rate, require testing of very high, sometimes unrealistic, numbers of samples. The results are usually negative and this may provide a false sense of security that the whole product lot is safe. If the test is positive, testing provides a valuable service, but the probabilities for this occurring increase only when the lot is highly contaminated which should have been prevented by proper management of effective control processes. Thus, reliance on process monitoring is preferred.

Education and Training

Correct, complete and routine implementation of HACCP is expected to occur only when there is proper, complete and routine education and training of management and employees in the importance of controlling foodborne hazards, the goals of HACCP, and the proper and continuous application of its principles. If company management appreciates the function of HACCP, it will be more willing to provide materials, equipment and adequate time for training and education of employees in the implementation of HACCP. Implementation of HACCP is not complete until there is adequate training of food handling employees. Food workers should be trained through proper SOPs to effectively perform their work activities, especially those associated with CCPs, GMPs and GHPs. Education and training in food safety principles and proper food handling procedures is also needed for consumers. They also handle food and may introduce hazards or cause their spreading or proliferation through their actions. It is necessary to teach consumers the basics of proper cooking of animal foods, thorough washing of raw vegetables, separation of uncooked from ready-to-eat foods, and washing of hands, cutting boards, knives, etc. At-risk individuals should be instructed to avoid or cook risky foods, and to avoid raw or unpasteurized foods (Sofos, 2008). Further, it is important to emphasize the value of developing a food safety culture throughout our society, and especially within each food operation.

Non-intervention HACCP

In contrast to the United States, where use of decontamination interventions has become common, in other countries or regions, such as the European Union, animal slaughter is conducted and inspected under non-intervention HACCP. As described by Bolton et al. (2001), the non-intervention HACCP relies on GMPs and GHPs to control contamination, and on hygiene audits to assess performance of operations in controlling microbial loads of carcasses. Success of contamination control under this system is based on identification of poor online practices through inspection of carcasses for the presence of visible fecal contamination as a CCP (zero tolerance). This approach requires continuous monitoring of operations needing hygiene control, such as removal of the hide, evisceration, spinal cord removal and carcass chilling. According to Bolton et al. (2001), specific examples of monitoring targets include: equipment sanitization with 82°C water; zero residual spinal cord tissue residue and critical limits for visible contamination rate (%) defects; rodding or sealing of the esophagus with crocodile clips, plastic rings or potato starch cones; sanitization of knives and rodding applicator; bagging and tying of bung with plastic bag, and knife sanitization; removal for rendering or incineration of BSE-specified risk material; and criteria for air temperature, relative humidity, air flow/velocity and carcass spacing in the chiller. In advanced facilities, some of these criteria may be monitored online and involve electronic activation of computerized systems in the office or visual displays in processing areas visible by employees, optional activation of alarms, and photo-eye counts of carcasses processed and with defects. Corrective actions include knife-trimming, retraining and determination of cause for future prevention.

HACCP with Interventions

Intervention HACCP is applied in the United States and allows application of specific online decontamination interventions to reduce levels of bacterial contamination on carcasses but after enforcement of zero tolerance for visible contamination. A number of such

TABLE 6.5 Fresh Meat Decontamination Interventions Applied in the United States

Before hide removal:
Animal cleaning
Hair removal
After hide removal:
Knife-trimming
Steam-vacuuming
Before evisceration:
Hot water spraying/rinsing
Organic acid (e.g. lactic) spraying/rinsing
After zero tolerance inspection for visible soil:
Water washing
Pressurized steam
Hot water spraying
Chemical rinsing (e.g. organic acid)
Carcass chilling
Deboning/packaging/ grinding:
Chemical rinsing (e.g. organic acid, etc.)

interventions, which may be applied sequentially or simultaneously, are currently in use (Table 6.5). It is possible to produce hygienically acceptable meat animal carcasses with both intervention and non-intervention HACCP. Factors that may make use of interventions needed probably include fast slaughter line speeds, old and outdated facilities, not well-trained workers, dirtier animals, etc.

Examples of generic process flowcharts for HACCP plans in the meat industry, operating with decontamination interventions, are shown for beef slaughter in Figure 6.2, and for heat treated but not fully cooked, not shelf-stable meat products such as smoked sausage, partially cooked chicken patties, etc. in Figure 6.3. Activities within such simplified flowcharts are multiple and sometimes confidential by each company; especially in high output plans, and involve specific tasks by each of many employees, sometimes exceeding 100. For example, in animal slaughter, some additional steps (not comprehensive) within a flowchart are: hock removal; first incision for hide removal; securing the hide at the flank after opening; head removal; first incision for opening of the abdomen for evisceration; kidney fat removal, etc. Employees visually inspect carcasses online to detect soiled carcass spots, which may have been introduced during the process. Detected feces or fecal stains are removed immediately by knife-trimming or steam-vacuuming. Associated additional

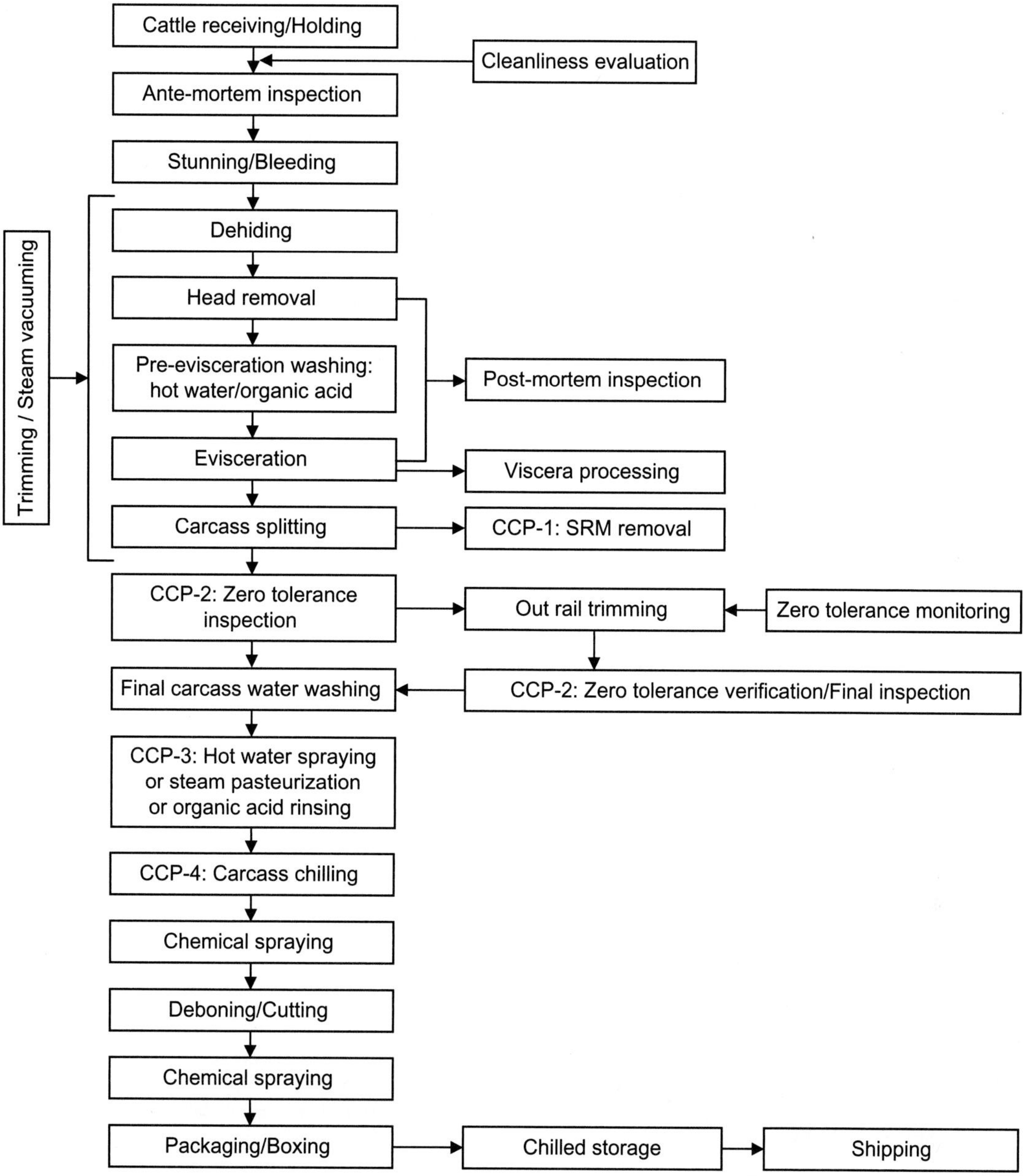

FIGURE 6.2 Cattle slaughter processing flowchart.

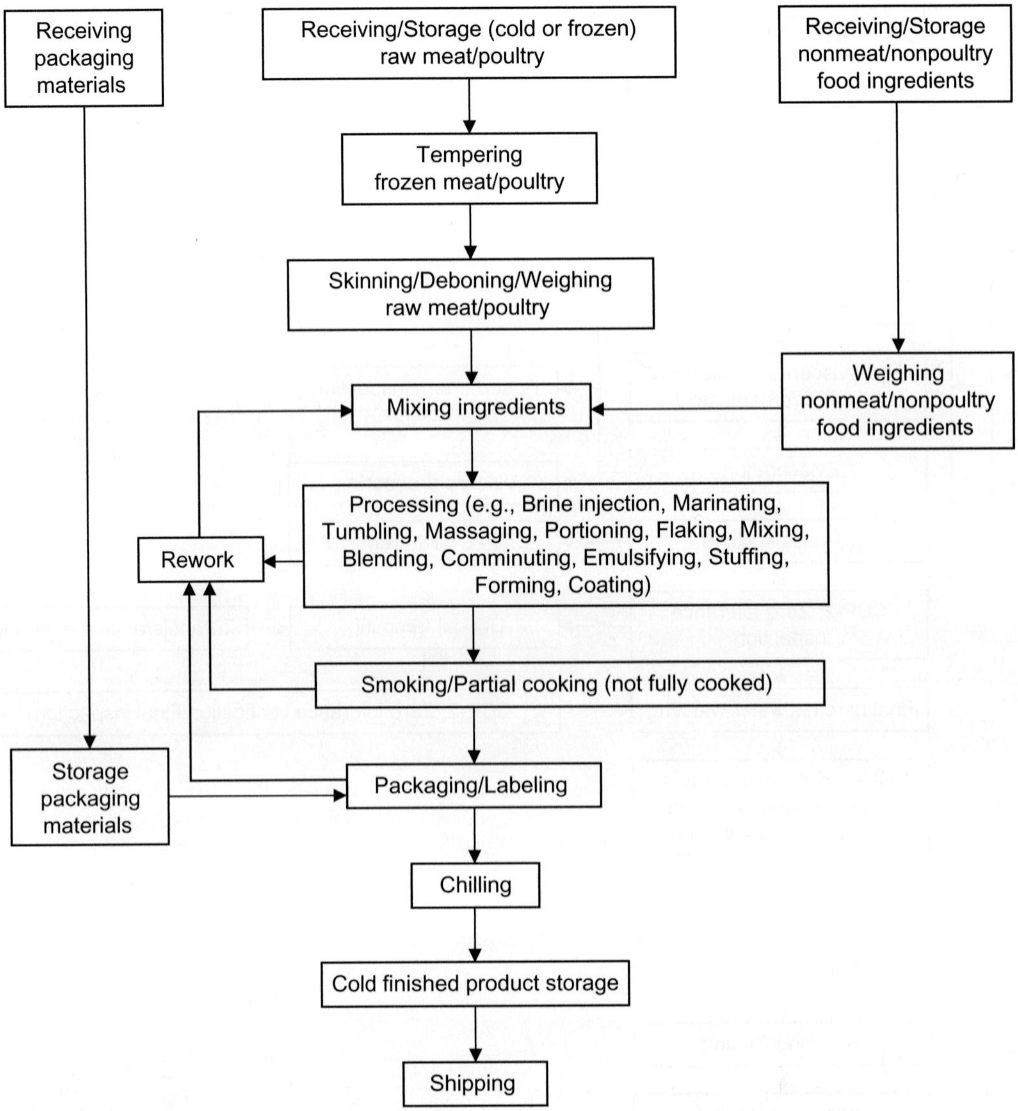

FIGURE 6.3 Processing flowchart for heat-treated but not fully cooked, not shelf-stable products such as smoked sausage, partially cooked chicken patties, etc.

corrective actions required may include retraining or replacing the person performing the operation, replacement of knives, steels and scabbards, checking the temperature of hot water sanitizers for knives, etc.; improved operations may rely on alternating use of two knives by each employee, one being sanitized while the other is in use.

Common CCPs (in some operations these may be PPs or GMPs) in United States slaughtering, deboning or meat cutting, and further processing operations (with associated CLs,

TABLE 6.6 Common Critical Control Points (in Some Operations these may be GMP) and Associated Critical Limits, Monitoring Procedures, Corrective Actions, and Verification Activities in United States Slaughtering, Meat Deboning or Cutting, and Further Processing Operations

CCP: Visible soil zero tolerance enforcement by knife-trimming (required before carcass washing):

Critical limits: No visible feces, ingesta, milk, hair

Monitoring: Visual observation, continuous for each carcass

Corrective action: Retrimming followed by reinspection

Verification: Visual observation by designated supervisor at selected frequency

At least three failures: Re-evaluation of HACCP plan

CCP: Hot water carcass spraying:

Critical limits: Water of 75–85°C applied at 15–25 psi for 5–12 sec

Monitoring: Water temperature, continuous recording

Corrective action: Reprocess carcasses; reset water application requirements

Verification: Check instruments and recordings at frequency selected by supervisor

CCP: Organic acid solution rinsing of carcasses before chilling:

Critical limits: 2.0–5.0%; solution of pH 2.0–3.0 applied at 25–55°C and 15–25 psi for 5–12 sec

Monitoring: Measure acidity or pH of solution; once per hour or continuously

Corrective action: Reprocess carcasses; reset process parameters and materials

Verification: Check instruments, continuous, automated or at selected intervals

CCP: Carcass chilling:

Critical limits: Carcass surface ≤4°C before deboning; 36–48 h after slaughter

Monitoring: Continuously; record air temperature, air velocity, relative humidity, carcass spacing and load of carcasses in chiller

Corrective action: Extend chilling time

Verification: Measurements and visual checks at specified intervals; set warning alarms

CCP: Cooking of pasteurized refrigerated meat product:

Critical limits: Internal temperature of 71°C to be reached within 4 h

Monitoring: Thermocouple temperature recording; oven relative humidity; oven load limits; product composition (fat/moisture); product item spacing in oven

Corrective action: Recook, rework or destroy; reset oven parameters; check product and load

Verification: Supervisor check parameters at specified intervals

monitoring procedures, corrective actions and verification activities), are shown in Table 6.6 (these are only examples; those selected by an operation need validation before use). Tasks performed at each step of Table 6.6 are described in corresponding SOPs used for worker training, completion of activity, verification of performance and auditing or inspection by outside experts or regulators.

CONCLUSIONS

It is natural for raw meat to be contaminated with spoilage and pathogenic microorganisms. Pathogens may cause problems ranging from mild gastrointestinal discomfort to severe, acute or chronic illness, or death. Contamination is introduced during animal growth, harvesting, storage, further processing, distribution, retailing, preparation and consumption, and it originates from soil, decaying material and animal fecal waste, which contaminate water, air, animals, plants, processing facilities, equipment and humans, leading to a complete contamination cycle. Although it is impossible to produce meat products free of contamination, efforts should continue to minimize prevalence and levels of microbial pathogens on raw meat, and to control contamination through inhibition of growth or destruction.

Prevalence, extent and type of contamination on meat are influenced by sanitary, hygienic and processing conditions during handling at all stages of the chain. Control of pathogens and management of food safety risks should be based on an integrated effort and approach that applies to all sectors, from producer through processor, distributor, packer, retailer, food service worker and consumer. Pathogen control at the pre-harvest level is difficult because knowledge is still limited relative to pathogen reservoirs, methodology limitations, ubiquitous presence of some pathogens, numerous and complicating variables involved, and cost. However, reduction of pathogen prevalence on animals pre-harvest may lead to a reduced probability that errors occurring in subsequent parts of the food chain will lead to foodborne illness; this should also reduce pathogen problems associated with water and foods of plant origin. Interventions applied during meat processing include sanitation, decontamination, heating, chilling, freezing, drying, fermentation, use of acidulants or antimicrobials, packaging, proper storage and distribution, and appropriate handling and preparation for consumption.

Proper application of controls results in products that should be safe for consumption following proper cooking and serving, provided that they are managed properly under the principles of HACCP. Foods should be stored and handled under conditions that minimize cross-contamination (clean and sanitary environment), are cooked properly (e.g. ground beef to 70°C), and are stored or held at the correct temperatures (cold: under 4°C; hot: above 60°C) and for the indicated length of time.

In general, the safety of meat products can be maintained and continuously improved if they are produced in facilities and with equipment that are properly designed for adequate cleaning and sanitation; follow good manufacturing and good hygiene practices and other prerequisite programs; apply processes by properly trained employees following necessary standard operating procedures; are managed by a properly designed HACCP plan in a company environment of food safety culture. In addition, there should always be efforts for improvement or update of operations and modification when necessary. Even then, the need for consumer education in food safety and proper food handling should not be ignored. Consumers should be advised to handle and prepare all foods, including meat products, properly, and to follow labeling instructions.

References

Adams, M., Motarjemi, Y., 1999. Basic Food Safety for Health Workers. World Health Organization, Geneva, Switzerland.

Bacon, R.T., Sofos, J.N., 2003. Characteristics of biological hazards. In: Schmidt, R.H., Rodrick, G.E. (Eds.), Food Safety Handbook. Wiley-Interscience, Hoboken, New Jersey, USA, pp. 157–195.

Bolton, D.J., Doherty, A.M., Sheridan, J.J., 2001. Beef HACCP: intervention and non-intervention systems. Int. J. Food Microbiol. 66, 119–129.

EFSA (European Food Safety Authority), 2011. Scientific opinion on the public health hazards to be covered by inspection of meat (swine). EFSA J. 9 (10), 2351, pp. 198. Available online: <http://www.efsa.europa.eu/en/efsajournal/doc/2351.pdf>.

EFSA/ECDC (European Food Safety Authority, European Centre for Disease Prevention and Control), 2011. The European Union summary report on trends and sources of zoonoses, zoonotic agents and food-borne outbreaks in 2009. EFSA J. 9 (3), 2090, pp. 378. Available online: <http://www.efsa.europa.eu/efsajournal>.

EFSA (European Food Safety Authority), 2012. Scientific opinion on the public health hazards to be covered by inspection of meat (poultry). EFSA J. 10 (6), 2741, pp. 179. Available online: <http://www.efsa.europa.eu/en/efsajournal/doc/2741.pdf>.

Fratamico, P.M., Bhagwat, A.A., Injaian, L., Fedorka-Cray, P.J., 2008. Characterization of Shiga toxin-producing *Escherichia coli* strains isolated from swine feces. Foodborne Pathog. Dis. 5, 827–838.

Hussein, H.S., 2007. Prevalence and pathogenicity of Shiga toxin-producing *Escherichia coli* in beef cattle and their products. J. Anim. Sci. 85 (E. Suppl.), E63–E72.

Koutsoumanis, K.P., Geornaras, I., Sofos, J.N., 2006. Microbiology of land muscle food. In: Hui, Y.H. (Ed.), Handbook of Food Science, vol. 1. CRC Press Taylor & Francis Group, Boca Raton, FL, pp. 52.1–52.43.

Rhoades, J.R., Duffy, G., Koutsoumanis, K., 2009. Prevalence and concentration of verocytotoxigenic *Escherichia coli, Salmonella enterica* and *Listeria monocytogenes* in the beef production chain: a review. Food Microbiol. 26, 357–376.

Sofos, J.N., 2005. Improving the Safety of Fresh Meat. CRC/Woodhead Publishing, Limited, Cambridge, UK.

Sofos, J.N., 2008. Challenges to meat safety in the 21st century. Meat Sci. 78, 3–13.

Sofos, J.N., 2013. Meat and meat products. In: Motarjemi, Y., Moy, G., Todd, E. (Eds.), Encyclopedia of Food Safety, Elsevier.

Sofos, J.N., Flick, G., Nychas, G.-J., O'Bryan, C.A., Ricke, S.C., Crandall, P.G., 2013. Meat, poultry and seafood. In: Doyle, M.P., Buchanan, R.L. (Eds.), Food Microbiology: Fundamentals and Frontiers, fourth ed. ASM Press, Washington, DC, pp. 109–165 (Chapter 6).

Xia, X., Meng, J., McDermott, P.F., Ayers, S., Blickenstaff, K., Tran, T., et al., 2010. Presence and characterization of Shiga toxin-producing *Escherichia coli* and other potentially diarrheagenic *E. coli* strains in retail meats. Appl. Environm. Microbiol. 76, 1709–1717.

Further Reading

Arthur, T.M., Brichta-Harhay, D.M., Bosilevac, J.M., Kalchayanand, N., Shackelford, S.D., Wheeler, T.L., et al., 2010. Super shedding of *Escherichia coli* O157:H7 by cattle and the impact on beef carcass contamination. Meat Sci. 86, 32–37.

Byelashov, O.A., Sofos, J.N., 2009. Strategies for on-line decontamination of carcasses. In: Toldrá, F. (Ed.), Safety of Meat and Processed Meat. Springer, New York, NY, pp. 149–182.

Carlson, B.A., Nightingale, K.K., Mason, G.L., Ruby, J.R., Choat, W.T., Loneragan, G.H., et al., 2009. *Escherichia coli* O157:H7 strains that persist in feedlot cattle are genetically related and demonstrate an enhanced ability to adhere to intestinal epithelial cells. Appl. Environ. Microbiol. 75, 5927–5937.

Cocolin, L., Dolci, P., Rantsiou, K., 2008. Molecular methods for identification of microorganisms in traditional meat products. In: Toldrá, F. (Ed.), Meat Biotechnology. Springer, New York, NY, pp. 91–127 (Chapter 4).

Codex Alimentarious, 1997. Codex Committee on Food Hygiene: Recommended International Code of Practice, General Principles of Food Hygiene, CAC/RCP 1-1969, Rev. 3, 1997 in Codex Alimentarius Commission Food Hygiene Basic Texts, Food and Agriculture. Organization of the United Nations, World Health Organization, Rome.

Cornforth, D.P., Hunt, M.C., 2008. Low-oxygen packaging of fresh meat with carbon monoxide: meat quality, microbiology, and safety. American Meat Science Association White Paper Series, January. Amer Meat Sci Assoc, Savoy, IL.

Davidson, P.M., Sofos, J.N., Branen, A.L. (Eds.), 2005. Antimicrobials in Foods, third ed. CRC/Taylor & Francis, Boca Raton, FL.

Davies, A., Board, R., 1998. The Microbiology of Meat and Poultry. Blackie Academic and Professional, London, UK.

De Boer, E., Zwartkruis-Nahuis, A., Heuvelink, A.E., Harmanus, C., Kuijper, E.J., 2011. Prevalence of *Clostridium difficile* in retailed meat in The Netherlands. Int. J. Food Microbiol. 144, 561–564.

De Vuyst, L., Falony, G., Leroy, F., 2008. Probiotics in fermented sausages. Meat Sci. 80, 75–78.

Doyle, M.P., Erickson, M.C., 2006. Emerging microbiological food safety issues related to meat. Meat Sci. 74, 98–112.

Eblen, D.R., 2007. Public health importance of non-O157 Shiga toxin-producing *Escherichia coli* (non-O157 STEC) in the US food supply. USDA, FSIS, OPHS. Available from: <http://www.fsis.usda.gov/PDF/STEC_101207.pdf> (accessed 24.02.2011.).

EFSA (European Food Safety Authority), 2009. Scientific opinion on food safety aspects of dairy cow housing and husbandry systems. EFSA J. 1189, 1–27. Available online: <http://www.efsa.europa.eu/en/efsajournal/doc/1189.pdf>.

EFSA (European Food Safety Authority), 2010. Revision of the joint AFC/BIOHAZ guidance document on the submission of data for the evaluation of the safety and efficacy of substances for the removal of microbial surface contamination of foods of animal origin intended for human consumption. EFSA J. 8 (4), 1544. pp. 32. Available online: <http://www.efsa.europa.eu/en/efsajournal/doc/1544.pdf>.

EFSA (European Food Safety Authority), 2012. Scientific opinion on Norovirus (NoV) in oysters: methods, limits and control options. EFSA J. 10 (1), 2500. pp. 39. Available online: <http://www.efsa.europa.eu/en/efsajournal/doc/2500.pdf>.

Eltholth, M.M., Marsh, V.R., Van Winden, S., Guitian, F.J., 2009. Contamination of food products with *Mycobacterium avium paratuberculosis*: a systematic review. J. Appl. Microbiol. 107, 1061–1071.

FSIS (Food Safety and Inspection Service), 1996. Pathogen reduction; Hazard Analysis and Critical Control Point (HACCP) systems: final rule, 9CFR Part 304, et al. Fed. Reg. 61, 38805–38989.

FSIS (Food Safety and Inspection Service), 1999a. FSIS policy on non-intact raw beef products contaminated with *E. coli* O157:H7. Available at: <http://www.fsis.usda.gov/OA/background/O157policy.htm> (accessed 31.03.2010.).

FSIS (Food Safety and Inspection Service), 1999b. Irradiation of meat and meat products. Fed. Regist. 4, 9089–9105.

FSIS (Food Safety and Inspection Service), 2002. Comparative risk assessment for intact (non-tenderized) and non-intact (tenderized) beef: technical report. Available at: <http://www.fsis.usda.gov/PDF/Beef_Risk_Assess_Report_Mar2002.pdf> (accessed 31.03.2010.).

FSIS (Food Safety and Inspection Service), 2003. Control of *Listeria monocytogenes* in ready-to-eat meat and poultry products; final rule. Fed. Regist. 68, 34208–34254.

Grant, M.A., Hedberg, C., Johnson, R., Harris, J., Logue, C.M., Meng, J., et al., 2011. Executive summary: the significance of non-O157 Shiga toxin-producing *Escherichia coli* in food. Food Prot. Trends 31, 33–45.

Hoorfar, J., Wagner, M., Jordan, K., Boquin, S.L., Skiby, J., 2011. Towards biotracing in food chains. Int. J. Food Microbiol. 145, S1–S4.

Hugas, M., Tsagarida, E., 2008. Pros and cons of carcass decontamination: the role of the European Food Safety Authority. Meat Sci. 78, 43–52.

Hunt, M.C., Mancini, R.A., Haachmeister, K.A., Kropf, D.H., Merriman, M., Delduca, G., et al., 2004. Carbon monoxide in modified atmosphere packaging affects color, shelf life, and microorganisms of beef steaks and ground beef. J. Food Sci. 69, FCT45–FCT52.

ICMSF (International Commission for Microbiological Specifications in Foods), 1996. Microorganisms in Foods 5: Characteristics of Microbial Pathogens. Blackie Academic & Professional, London, UK.

ICMSF (International Commission for Microbiological Specifications in Foods), 1998. Microorganisms in Foods 6: Microbial Ecology of Food Commodities. Blackie Academic & Professional, London, UK.

ICMSF (International Commission for Microbiological Specifications in Foods), 2002. Microorganisms in Foods 7: Microbiological Testing in Food Safety Management. Kluwer Academic/Plenum Publishers, New York, NY.

IFT (Institute of Food Technologists), 2006. Antimicrobial Resistance – Implications for the Food System. Chicago, IL.

Jöbstl, M., Heuberger, S., Indra, A., Nepf, R., Köfer, J., Wagner, M., 2010. *Clostridium difficile* in raw products of animal origin. Int. J. Food Microbiol. 138, 172–175.

Juneja, V.K., Sofos, J.N., 2002. Control of Foodborne Microorganisms. Marcel Dekker, Inc., New York, NY.

Juneja, V.K., Sofos, J.N., 2009. Pathogens and Toxins in Foods: Challenges and Interventions. ASM Press, Washington, DC.

Kalac, P., 2006. Biologically active polyamines in beef, pork and meat products: a review. Meat Sci. 73, 1–11.

Koutsoumanis, K.P., Sofos, J.N., 2004. Microbial contamination of carcasses and cuts. In: Jensens, W.K. (Ed.), Encyclopedia of Meat Sciences. Elsevier Academic Press, Amsterdam, The Netherlands, pp. 727–737.

Leistner, L., Gould, G., 2002. Hurdle Technologies. Kluwer Academic/Plenum Publishers, New York, NY, USA.

Loretz, M., Stephan, R., Rand, C., Zweifel, C., 2010. Antimicrobial activity of decontamination treatments for poultry carcasses: a literature survey. Food Control 21, 791–804.

Loretz, M., Stephan, R., Rand, C., Zweifel, C., 2011. Antibacterial activity of decontamination treatments for cattle hides and beef carcasses. Food Control 22, 347–359.

McDowell, D.A., Sheridan, J.J., Bolton, D.J., 2005. HACCP in slaughter operations. In: Sofos, J.N. (Ed.), Improving the Safety of Fresh Meat. Woodhead Publishing, Ltd., Cambridge, UK, pp. 696–730 (Chapter 29).

McMillin, K.W., 2008. Where is MAP going? A review and future potential of modified atmosphere packaging for meat. Meat Sci. 80, 43–65.

Mutharia, L.M., Klassen, M.D., Fairles, J., Barbut, S., Gill, C.O., 2010. *Mycobacterium avium* subsp. *paratuberculosis* in muscle, lymphatic and organ tissues from cows with advanced Johne's disease. Int. J. Food Microbiol. 136, 340–344.

NACMCF (National Advisory Committee on Microbiological Criteria for Foods), 1998. Hazard analysis and critical control point principles and application guidelines. J. Food Prot. 61, 62–775.

NACMCF (National Advisory Committee on Microbiological Criteria for Foods), 2002. Final Report: *E. coli* O157:H7 in Blade Tenderized Beef. Available at: <http://www.fsis.usda.gov/OPHS/NACMCF/2002/rep_blade1.htm>.

NACMCF (National Advisory Committee on Microbiological Criteria for Foods), 2010. Assessment of food as a source of exposure to *Mycobacterium avium* subspecies *paratuberculosis* (MAP). J. Food Prot. 73, 1357–1397.

Nørrung, B., Andersen, J.K., Buncic, S., 2009. Main concerns of pathogenic microorganisms in meat. In: Toldrá, F. (Ed.), Safety of Meat and Processed Meat. Springer, New York, NY, pp. 3–29 (Chapter 1).

Nychas, G.-J.E., Marshall, D.L., Sofos, J.N., 2007. Meat, poultry, and seafood. In: Doyle, M.P., Beuchat, L.R. (Eds.), Food Microbiology, Fundamentals and Frontiers, third ed. ASM Press, Washington, DC, pp. 105–140 (Chapter 6).

Rantsiou, K., Cocolin, L., 2008. Fermented meat products. In: Cocolin, L., Ercolini, D. (Eds.), Molecular Techniques in the Microbial Ecology of Fermented Foods. Springer, New York, NY, pp. 91–118 (Chapter 3).

Samelis, J., Sofos, J.N., 2003. Strategies to control stress-adapted pathogens. In: Yousef, A.E., Juneja, V.K. (Eds.), Microbial Stress Adaptation and Food Safety. CRC Press, Boca Raton, FL, pp. 303–351.

Scallan, E., Griffin, P.M., Angulo, F.J., Tauxe, R.V., Hoekstra, R.M., 2011. Foodborne illness acquired in the United States – unspecified agents. Emerg. Infect. Dis. 17, 16–22.

Scallan, E., Hoekstra, R.M., Angulo, F.J., Tauxe, R.V., Widdowson, M.-A., Roy, S.L., et al., 2011. Foodborne illness acquired in the United States – major pathogens. Emerg. Infect. Dis. 17, 7–15.

Sebranek, J., 2008. The political saga of carbon monoxide packaging. Scientists Speak. April 20, Marco Island Marriott Beach Resort Annual Convention & Expo. IARW-WFLO-IRTA. Available at: <http://www.WFLO.org>.

Skandamis, P.N., Nychas, G.-J.E., Sofos, J.N., 2010. Meat decontamination. In: Toldrá, F. (Ed.), Handbook of Meat Processing. Wiley-Blackwell, Ames, IA, pp. 43–85.

Smith, G.C., Tatum, J.D., Belk, K.E., Scanga, J.A., Grandin, T., Sofos, J.N., 2005. Traceabiltity from a US perspective. Meat Sci. 71, 174–193.

Sofos, J.N., 1994. Microbial growth and its control in meat, poultry and fish. In: Pearson, A.M., Dutson, T.R. (Eds.), Quality Attributes and Their Measurements in Meat, Poultry and Fish Products. Blackie Academic and Professional, Glasgow, UK, pp. 359–403.

Sofos, J.N., 2002. Approaches to pre-harvest food safety assurance. In: Smulders, F.J.M., Collins, J.D. (Eds.), Food Safety Assurance and Veterinary Public Health, Food Safety Assurance in the Pre-harvest Phase, Vol. 1. Wageningen Academic Publishers, Wageningen, The Netherlands, pp. 23–48.

Sofos, J.N., 2004a. Pathogens in animal products: major biological hazards. In: Pond, W., Bell, A. (Eds.), Encyclopedia of Animal Science. Marcel Dekker, Inc., New York, NY, USA, pp. 698–700.

Sofos, J.N., 2004b. Pathogens in animal products: sources and control. In: Pond, W., Bell, A. (Eds.), Encyclopedia of Animal Science. Marcel Dekker, Inc., New York, NY, USA, pp. 701–703.

Sofos, J.N., 2006. Field data availability and needs for use in microbiological risk assessment. In: Smulders, F.J.M. (Ed.), Food Safety Assurance and Veterinary Public Health. Towards a Risk-based Chain Control, Vol. 4. Wageningen Academic Publishers, Wageningen, The Netherlands, pp. 57–74.

Sofos, J.N., 2009a. ASAS centennial paper: developments and future outlook for postslaughter food safety. J. An. Sci. 87, 2448–2457.

Sofos, J.N., 2009b. Biofilms: our constant enemies. Food Saf. Mag. 38, 40–41.

Sofos, J.N., 2009c. Preparing safe nonintact meat products – a review. Fleischwirtschaft Int. 24, 68–73.

Sofos, J.N., 2010. Biofilms – importance and need for control. Fleischwirtschaft Int. 25, 27–30.

Sofos, J.N., Smith, G.C., 1998. Non-acid meat decontamination technologies: model studies and commercial applications. Int. J. Food Microbiol. 44, 171–188.

Sofos, J.N., Geornaras, I., 2010. Overview of current meat hygiene and safety risks and summary of recent studies on biofilms, and control of *Escherichia coli* O157:H7 in nonintact, and *Listeria monocytogenes* in ready-to-eat, meat products. Meat Sci. 86, 2–14.

Sofos, J.N., Geornaras, I., Belk, K.E., Smith, G.C., 2008. Nonintact whole muscle food safety: the problem and research needs. In Proceedings of the 61st Reciprocal Meat Conference, American Meat Science Association, June 22–25, Gainesville, FL. p. 7. American Meat Science Association, Champaign, IL.

Sofos, J.N., Beuchat, L.R., Davidson, P.M., Johnson, E.A., 1998. Naturally Occurring Antimicrobials in Food. Council for Agricultural Science and Technology, Ames, IA (ISBN 1-887383-12-3).

Songer, J.G., Trinh, H.T., Killgore, G.E., Thompson, A.D., McDonald, L.C., Limbago, B.M., 2009. *Clostridium difficile* in retail meat products, USA, 2007. Emerg. Infect. Dis. 15, 819–821.

Sperber, W.H., Stevenson, K.E., Bernard, D.T., Deibel, K.E., Moberg, L.J., Hontz, L.R., et al., 1998. The role of prerequisite programs in managing a HACCP system. Dairy Food Environ. Sanit. 18, 418–423.

Stopforth, J.D., Sofos, J.N., 2006. Recent advances in pre- and post-slaughter intervention strategies for control of meat contamination. In: Juneja, V.K., Cherry, J.P., Tunick, M.H. (Eds.), Advances in Microbial Food Safety, ACS Symposium 931. Recent Advances in Intervention Strategies to Improve Food Safety. American Chemical Society, Oxford University Press, Washington, DC, USA, pp. 66–86.

Stopforth, J.D., Sofos, J.N., Taylor, S.L., Baumert, J.L., 2012. Food safety issues in animal source foods related to animal health and welfare. In: Pond, W.G., Bazer, F.W., Rollin, B.E. (Eds.), Animal Welfare in Animal Agriculture: Husbandry, Stewardship, and Sustainability in Animal Production. CRC Press/Taylor & Francis Group, Boca Raton, FL, pp. 205–231.

Thippareddi, E.A., Boyle, E., Burson, D.E., 2005. Monitoring, validating and veifying the effectiveness of HACCP systems. In: Sofos, J.N. (Ed.), Improving the Safety of Fresh Meat. Woodhead Publishing, Ltd., Cambridge, UK, pp. 731–766 (Chapter 30).

Toldrá, F. (Ed.), 2009. Safety of Meat and Processed Meat. Springer, New York, NY.

Toldrá, F. (Ed.), 2010. Handbook of Meat Processing. Wiley-Blackwell, Ames, IA.

Tompkin, R.B., 2002. Control of *Listeria monocytogenes* in the food-processing environment. J. Food Prot. 65, 709–725.

Wallace, C., Williams, T., 2001. Pre-requisites: a help or a hindrance to HACCP? Food Control 12, 235–240.

Weese, J.S., Avery, B.P., Rousseau, J., Reid-Smith, R.J., 2009. Detection and enumeration of *Clostridium difficile* spores in retail beef and pork. Appl. Environ. Microbiol. 75, 5009–5011.

Weese, J.S., Reid-Smith, R.J., Avery, B.P., Rousseau, J., 2010. Detection and characterization of *Clostridium difficile* in retail chicken. Lett. Appl. Microbiol. 50, 362–365.

Yiannas, F., 2010. Food Safety Culture: Creating a Behavior-based Safety Management System. Springer.

Zhou, G.H., Xu, X.L., Liu, Y., 2010. Preservation technologies for fresh meat – a review. Meat Sci. 86, 119–128.

CHAPTER

7

Poultry and Eggs

Shai Barbut and Irwin Pronk
University of Guelph, Guelph, Canada

OUTLINE

INTRODUCTION

The poultry industry has seen significant changes in the methods used to harvest poultry meat over the past half a century. Some of the major points include an over fourfold increase in line speed (new plants are designed to process 13,500 broilers per hour), a large increase in the proportion of cut up and deboned meat produced, as well as substantial improvements in sanitation. It should be realized that in 1970 a high-speed line was running at 3000 broilers per hour. Quadrupling line speed has presented many challenges in terms of equipment design, reducing cross-contamination, inspection and product quality. As will be discussed below, the introduction of scientific-based hazard analysis and critical control points (HACCP) programs has been very effective in improving sanitation standards in modern poultry processing plants, and improving food safety. When it comes to heavier birds, such

Food Safety Management.
DOI: http://dx.doi.org/10.1016/B978-0-12-381504-0.00007-X

as turkeys, line speed is lower (3500 per hour) but the amount of meat processed per line is usually higher than in a high-speed broiler line (Barbut, 2010).

The main steps included in a poultry (e.g. broilers, turkeys, ducks, ratite) primary processing plant are:

- Live bird receiving
- Stunning (electrical/gas/no stun)
- Bleeding
- Defeathering
- Electrical stimulation (optional, but used in many high-speed lines)
- Evisceration
- Inspection
- Chilling
- Aging (optional)
- Portioning and cutting
- Packaging and distribution

Later in the chapter you will find a detailed flow diagram of the whole process, but the list above should provide a general concept to begin with.

The process starts by bringing the birds, in cages, to the processing plant. It is important to remember that the feathers and skin are covered with a natural microflora and sometimes even dry manure. Microbial counts of skin and feather surfaces can show numbers of 10^4 to 10^9 microorganisms per cm^2 or per gram. In addition, the internal viscera (stomach, gut) are populated with a high number of microorganisms that also represent a challenge when removing the viscera while preventing/minimizing cross-contamination – this is similar in other meat-producing animals such as beef and pork. The challenge therefore is to process the meat with minimal transfer of bacteria and obtain products with the lowest possible microbial counts.

Due to the factors mentioned above, the process of hazard analysis is extremely important (see also previous chapters). While there are many examples to support this (governments and industry associations) it is crucial that each plant carefully analyzes the hazards in their own plant, using their own data and understanding their own unique plant culture. When the hazard analysis is done well, the HACCP team is more likely to determine the correct critical control points (CCPs), therefore making clearer and more applicable the prerequisite programs since they also should be designed to address the issues raised in the hazard analysis. The entire process must be integrated with the HACCP plan at its heart, making it the central repository of everything we know about food safety in the plant.

A brief discussion concerning the three types of hazards (i.e. microbial, chemical and physical) found in every HACCP program is provided below, prior to presenting an actual HACCP model.

MICROBIAL HAZARDS

Some important background information from the Centers for Disease Control indicates that foodborne agents cause an estimated 48 million illnesses annually in the United States alone. This includes 9.4 million illnesses from known pathogens, and the top commodities to which outbreaks have been attributed are poultry (15%), beef (14%) and finfish (14%) (CDC, 2011). It has also been shown that about 65% of chickens on retail sale in the UK are contaminated with *Campylobacter*, which has been shown to be the leading cause of food-borne illness and is responsible for about 30% of cases in the UK – 371,000 estimated cases in England and Wales in 2009, resulting in more than 17,500 hospitalizations and 88 deaths.

Disease in humans is mainly caused by zoonotic pathogens such as *Salmonella* and *Campylobacter*. Due to the *Campylobacter jejuni*'s widespread occurrence in the environment, its epidemiology remains poorly understood. It is generally accepted, however, that chickens are a natural host for human pathogens such as *C. jejuni*, and for *Salmonella*; these pathogens are even considered by some as part of the normal microbial flora of poultry. In general, colonized broiler chicks are an important vector for transmitting these pathogens to humans. However, despite the increasing evidence that the chicken reservoir is the number one risk factor for the disease in humans, no effective strategy exists to reduce *Campylobacter* prevalence in poultry flocks. This can in part be explained by the incomplete understanding of the epidemiology of *C. jejuni* in broiler flocks. As a result, the number of human campylobacteriosis cases associated with the chicken vector remains high. On farms, the current emphasis appears to be on *Salmonella*, and specifically *Salmonella Enteriditis*, where more progress has been achieved. In the USA, for example, regulatory sampling prevalence of *Salmonella* for the years 1998–2002, as compared to the performance standard established in the PR/HACCP rule, showed that broiler prevalence was reduced to 10.9% compared to a standard of 20%, and ground chicken to 19.8% compared to a standard of 44.6% (Anonymous, 2003). Positive *Salmonella* samples from very small broiler plants showed the greatest decrease; i.e. from 37.2% in 2001 to 8.4% in 2002. Fries (2002) suggested a number of different strategies that might be used by the poultry industry to reduce the rate of transferring *Salmonella* from live birds to the product. Later, Franchin et al. (2010) have more closely looked at methods associated with rinsing poultry carcasses in different stages during primary processing and integrating HACCP as measures to reduce *Salmonella*. Overall, some of the strategies that work have been implemented successfully in various industrialized countries; however, problems do still persist in other parts of the world. Another very important aspect today is the role of consumer education in the prevention of cross-contamination (at home/restaurant kitchens) and proper cooking methods which have been shown to be another barrier in reducing food-borne diseases. It is now also customary in numerous countries to include labels explaining proper handling and cooking instructions to help at the consumer level.

In the case of *Campylobacter*, the European Food Safety Authority (EFSA, 2010a,b) has published an evaluation of factors that may contribute to its spread in live chickens and chicken carcasses in the European Union (EU). The EFSA panel states that batches of chickens infected with *Campylobacter* are 30 times more likely to produce carcasses contaminated with this bacterium. Also, EFSA experts say that measures before slaughter could reduce the risk by up to 50%, although this figure is expected to vary considerably between EU member states. Such measures focus mostly on preventing the bacteria from entering the housing in which the chickens are kept and on reducing the number of *Campylobacter* in the intestines of chickens sent to slaughter. Using fly screens, reducing the age at which chickens are sent to slaughter and discontinuing thinning practices (as humans entering chicken housing may carry bacteria from outside) have been suggested as effective measures.

Corry and Atabay (2001) indicated that thermophilic *Campylobacter* is not generally thought to be transmitted vertically via eggs, nor via feed or litter, provided rearing houses are cleaned and disinfected between flocks, and litter renewed. Flocks usually become infected at about 3 weeks of age. Every bird is usually rapidly colonized, with high levels (10^6–10^7 cfu/g) in the cecal contents. The source of infection can be via unchlorinated water, but in situations where the water supply is safe, the precise source of infection is seldom identified. Infection could be via wild birds, rodents, or from farm operatives via boots or clothing. Infection has sometimes been associated with "thinning" of flocks about a week prior to slaughter. Avoidance of infection during rearing therefore relies mostly on careful attention to hygiene, exclusion of vermin and a clean water supply.

EFSA (2011) has published a scientific opinion assessing the public health impact of control measures that could be used to reduce the occurrence of *Campylobacter* in chickens and chicken meat. Possible measures for risk reduction in the meat production chain include, for instance: cooking on an industrial scale or irradiating the meat, which are both likely to destroy all *Campylobacter* that may be present on the meat; and freezing carcasses for 2 to 3 weeks, which would reduce the risk by more than 90% (note: each of the examples can serve as a critical control point in an integrated operation). Freezing carcasses for short periods of time (2–3 days) or treating chicken carcasses with hot water (at 80°C for 20 seconds) or with chemicals, such as lactic acid, was estimated to reduce the risk by between 50 and 90%.

During transport, slaughter and dressing, *Campylobacter*-negative flocks can readily be contaminated from positive flocks. Contamination can be reduced by improved disinfection of transport crates, slaughter of uninfected flocks prior to infected flocks, and by careful attention to major points of cross-contamination on the line. A more effective measure would be to use a terminal decontamination step, such as trisodium phosphate, lactic acid, reduced pressure steam or gamma irradiation.

As discussed above, prevention of pathogens is taking a number of forms. An example of a very new approach is the study of maternal antibodies that are passed from hens to their chicks. Udakis (2012) indicated that: "these antibodies protect chicks from becoming colonized by *Campylobacter* in the first week of life…our group has now identified the bacterial molecules that these antibodies attack, which has given us a starting point for a vaccine against *Campylobacter*…we have already found that chickens injected with these specific molecules – found on the surface of *C. jejuni* – produce antibodies against the bacterium. This response partially protects them from colonization…preventing contamination of poultry at slaughter has not been effective at reducing illness in humans."

CHEMICAL HAZARDS

An example of a chemical hazard would be antibiotic residues, which have been in the news for several years and are a source of health concern for consumers. This concern is one of the driving factors in the growth of the organic foods market. Consumers are particularly alarmed by information such as researchers at the Johns Hopkins Bloomberg School of Public Health and Arizona State University who found evidence suggesting that a class of antibiotics previously banned by the US government for poultry production is still in use. Love (2012) looked for drugs and other residues in feather meal, a common additive to chicken, swine, cattle and fish feed. The most important drugs found in the study were fluoroquinolones – broad spectrum antibiotics used to treat serious bacterial infections in people, particularly those infections that have become resistant to older antibiotic classes. The banned drugs were found in eight of 12 samples of feather meal in a multistate study. The findings were a surprise to scientists because fluoroquinolone use in US poultry production was banned by the US Food and Drug Administration in 2005. In the US, antibiotics can be used at low levels as feed additives, primarily to make poultry grow faster, rather than to treat a specific disease. An estimated 13.2 million kg of antibiotics were sold in 2009 to the US poultry and livestock industries. Residues in feather meal/meat are considered a chemical hazard, as are residues of cleaning compound, pesticides, etc. Another example is the Belgian dioxin incident. The incident occurred in January 1999 when a mixture of polychlorinated biphenyls (PCBs) contaminated with dioxins was accidentally added to a stock of recycled fat used in the production of animal feeds (Bernard et al., 2002). It impacted more than 2500 farms and resulted in a major food crisis, which rapidly extended to the whole country and could be resolved only by the implementation of a large PCB/dioxin food monitoring program. The Belgian PCB incident was due to a single source of PCB oil introduced into the food chain. The total amount of PCBs added to recycled fats was estimated at 50 kg which corresponds to about 100 liters of PCB oil. The highest concentrations of PCBs and dioxins were found in poultry and especially in the reproduction animals (hens), which showed the classical manifestations of chick edema disease. Pigs were also affected but to a lesser extent and no sign of intoxication was observed.

PHYSICAL HAZARDS

In this category foreign materials ranging from plastic to metal and glass, which present a hazard to the consumer, should be eliminated from the product (Mortimore and Wallace, 1995). Inspection of incoming non-meat raw materials is extremely important as well as the final raw meat product itself for potential presence of metal parts (e.g. screws falling from machinery, broken injection needle), glass (e.g. pieces of shattered light bulb; ideally light bulbs should be covered or coated to prevent such a problem), plastic (e.g. broken piece of a conveyor belt), etc. These hazards must first of all be addressed by prerequisite programs aimed at prevention, e.g. preventive maintenance. Additional measures such as use of metal detectors and X-ray can be implemented to confirm that prerequisite programs are effective and to enhance safety assurance.

HACCP GENERIC MODEL

An example of a model developed by the Canadian Food Inspection Agency for chilled, ready-to-cook, whole chicken (CFIA, 2012) will be used to demonstrate the construction of a tailor-made in-house program. The model has a lot in common with the USDA model, but it elaborates on a few more issues. The Poultry Slaughter Model encompasses the process starting from receiving the live bird (model written for chickens, but can be applied to turkeys, ducks, etc.) through packaging the whole bird. The illustration will take the reader step by step, demonstrating the potential hazards and ways to control them. While generic models are helpful for the industry they do not always reflect the real issues in a specific processing plant. Government inspectors reviewing the processor's HACCP plan will typically be using the generic model as a reference so any deviations from the list of CCPs will require a great deal of sound, scientific data.

The process flow diagram (Figure 7.1) lists the different raw materials and processes involved in obtaining the final product. The general product description (part of the document required by the CFIA which is also responsible for approving HACCP plans in Canada) includes the following:

- Product name: raw whole chicken (bone-in and deboned);
- Important product characteristics (pH, a_w): none;
- Intended use: carcasses, portions, giblets and paws – ready to cook or for further processing;
- Packaging: Styrofoam trays, absorbent pads, plastic films, cardboard boxes;
- Shelf-life: X days at ≤4°C or less;
- Where it will be sold: retail, restaurants, institutions, and further processors;
- Labeling instructions: keep refrigerated, keep frozen, best before date, safe handling instructions (recommended);
- Special distribution conditions: ≤4°C at all times, or maintain in a frozen state.

The meat is obtained from live birds transported to the processing plant. In some countries there are also HACCP programs for growing the birds (i.e. dealing with the live bird), but this part will not be discussed here.

The non-meat components coming into the plant (Figure 7.1) include water, ice and processing aids such as CO_2 for modified atmosphere packaging, ingredients such as antimicrobial agents, and salt. Note: most of these items are commonly covered by the prerequisite program.

As can be seen in Figure 7.1, provisions have been made for different processing schemes, such as using manual versus automatic carcass transfer (Steps 18 and 31), and for plants equipped with new evisceration machines (e.g. physically separating viscera from carcass prior to inspection). Overall, any new/proposed interventions should be approved by the local authorities within each specific HACCP plan. The figure outlines possible critical control points and numbers them in a sequential order. For example, the first critical control point is suggested for viscera defect detection and is identified as CCP 1B, meaning that this is the first point that represents potential intervention to reduce biological (B) hazards.

Table 7.1 is an example of the listing provided in the HACCP generic model related to the first CCP. It is important to identify all the potential problems so adequate measures can be taken to eliminate/minimize the hazards.

Table 7.2 lists the different biological hazards that could be controlled within the chilling operation, and ways they can be addressed.

There are various conditions that can contribute to microbial and chemical contaminations, as well as different physical hazards that can affect the products' safety. Examples of a few potential hazards that might be a problem from previous steps of the process include:

- Incomplete documentation from the farm;
- No evidence of feed withdrawal or withdrawal of medications;

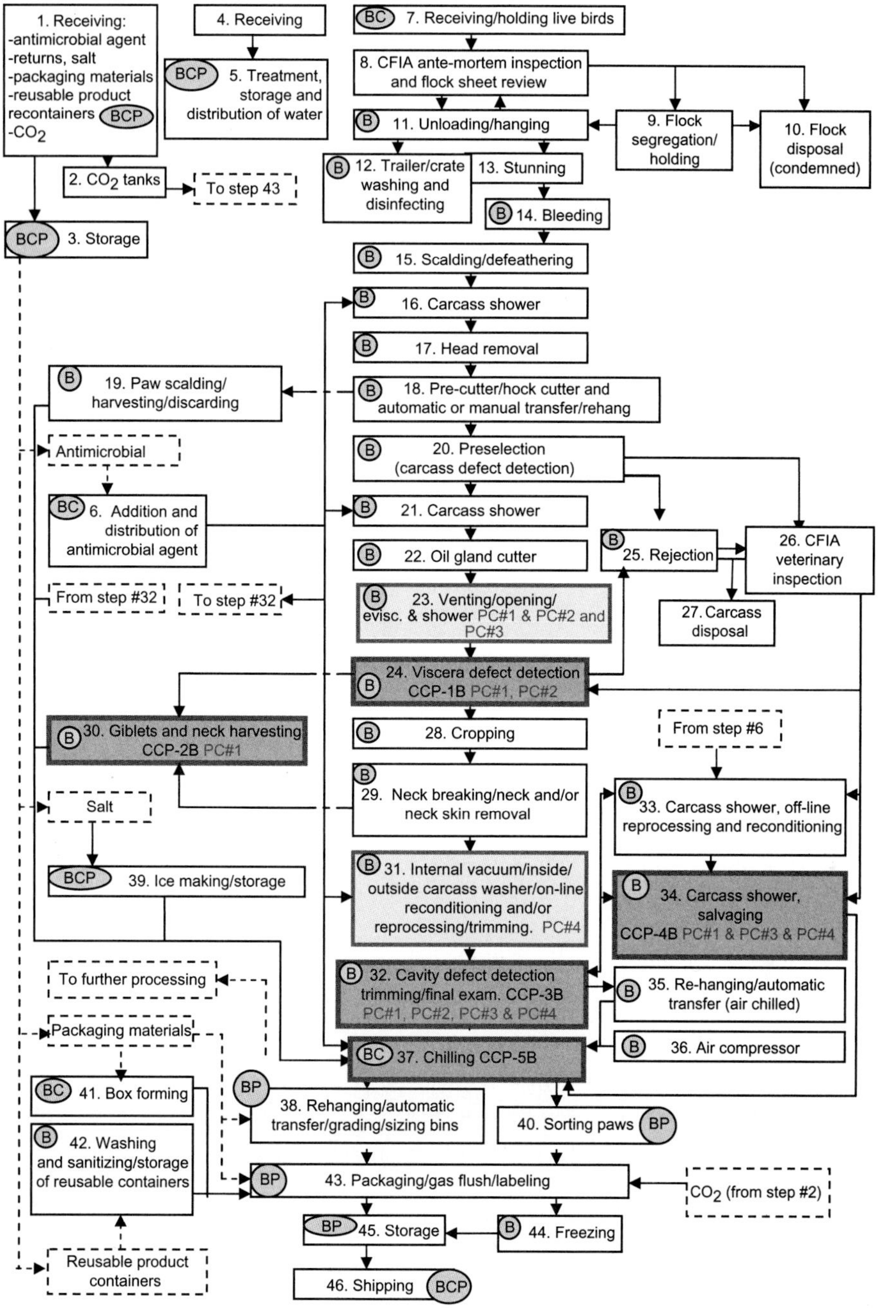

FIGURE 7.1 Process flow diagram for poultry primary processing, including potential hazards (B = biological, C = chemical, P = physical), and suggested critical control points (CCP). *From the HACCP Generic Model by the CFIA (2012).*

TABLE 7.1 Example Describing the Critical Control Point (CCP) Involved in Viscera Defect Detection. See Figure 7.1 for Process Flow Diagram. Part from an HACCP Generic Model (CFIA, 2012)

Form #10

Critical Control Points
Product Name: Poultry Slaughter

Process Steps	CCP/Hazard Number	Hazard Description	Critical Limits	Monitoring Procedures	Deviation Procedures	Verification Procedures	HACCP Records
24. Viscera Defect Detection	CCP-1B	Presence of pathogens in or on viscera due to failure to detect visible fecal and/or ingesta contamination and/or failure to detect visceral pathological conditions and/or improper removal (e.g. Septicaemia/Toxaemia or Hepatitis).	As per MOP (Manual of Procedures 19.6.2.4) viscera defect group as per DDS program.	Randomly, once per hour, "CCP-1B monitor" will visually monitor "X" number of randomly selected viscera on the line after the viscera helper, for fecal, ingesta and/or pathological defects as per DDS program. Records observations and date\time and signs on "CCP-1B Form". Note: see Company Random Selection Procedures.	If lot is rejected. "CCP-1B monitor" will contact maintenance to find and correct the cause of deviation. "CCP-1B monitor" will contact Supervisor to conduct a food safety assessment and either add additional employees or slow down the line. "CCP-1B monitor" will conduct a retest. If the re-test also fails, product since last successful test will be held and the DDS decision tree will be followed as per MOP 19.6.2.5.2.10. If the lot is rejected for Septicaemia/Toxaemia, immediate carcass and viscera post chill verification is required as per DDS decision tree MOP 19.6.2.5.2.10. The following information is documented on deviation CCP-1B Form:	The "CCP-1B verifier" observes the "CCP-1B monitor" once every "Y" (validated frequency) to ensure he/she is performing his/her task as per written program. The "CCP-1B verifier" also examines "X" day(s) worth of "CCP-1B Forms" and "Defect Log" once per "Y" days to ensure monitoring is performed as specified by written procedures and forms are completed and appropriate corrective and preventative measures were taken as required. Also to ensure Pre-shipment review is completed as per MOP 11, USA, Annex Q, Q.1.1b).	"CCP-1B Form" "CCP 1B Verification Form" "Defect Detection Standards Defects Log Post Chill Product Verification" record.

1. A description of the deviation and its cause 2. Action(s) taken to control affected product 3. Corrective action(s) taken to restore control of the CCP 4. Measures taken to prevent reoccurrence of the deviation The following information is documented on the "Defect Detection Standards Defects Log Post Chill Product Verification" record. 1. Verification of effectiveness of corrective and preventative actions taken (re-tests) Both forms must include initials, date and exact time an entry is made Any deviation will require an evaluation at the supporting PC#1 and PC#2.	If deficiencies are noted during verification procedures, a root cause analysis and food safety assessment will be performed. Corrective actions/preventative measures may include retraining of "CCP -1B monitor" and/or employees and/or re-evaluation of monitoring/deviation procedures. Verification observations, verifier's signature and date\time are recorded on "CCP-1B Verification Form".

TABLE 7.2 Example Describing the Critical Control Point (CCP) Involved in the Chilling Operation. See Figure 7.1 for Process Flow Diagram. From the CFIA HACCP Generic Model (2012)

Form #10

Critical Control Points
Product Name: Poultry Slaughter

Process Steps	CCP/ Hazard Number	Hazard Description	Critical Limits	Monitoring Procedures	Deviation Procedures	Verification Procedures	HACCP Records
37. Chilling	CCP-5B	Pathogen growth due to inadequate chilling resulting from time/temperature abuse.	As per "dressed poultry carcasses and parts" of the MOP (19.8.2.4.1 & 19.8.4.1) Portions/necks/giblets - chilled to 4°C or lower within 2 hours after evisceration (salvaged turkey breasts, breast fillets, legs, drumsticks and thighs shall be chilled to 4°C or lower within 4 hours after evisceration) as per MOP 19.8.2.4.2	Every "Y" hour(s) for "X" number of carcasses, CCP-5B Monitor inserts a calibrated digital thermometer in the deepest part of the breast and records product temperature at the time/location specified within the validated chilling procedure. Every "Y" hour(s) for "X" number of portions/necks/giblets, CCP-5B Monitor inserts a calibrated digital thermometer into each and records product temperature (s) at the time/location specified within the validated chilling procedure.	If product temperature in carcasses and portions is not being brought down according to the prescribed temperature according to the operator's chilling protocol, the Supervisor is contacted and appropriate measures must be readily inflated to bring down the product temperature within the timeframe specified in the MH-MOP. Whenever a deviation is noticed in the chilling of carcasses or parts, the product could be either cooked or if kept fresh, the shelf life/best before date must be re-evaluated. If the violations result in the spoilage of the product, then the product must be disposed of to prevent its entry in the human food chain. In the case of portions/necks/giblets. If a deviation is noticed, then the product must be disposed of to prevent its entry in the human food chain.	The "CCP-5B verifier" observes the "CCP-5B monitor" once every "Y" days to ensure he/she is performing his/her task as per written program. The "CCP-5B verifier" also examines "X" day(s) worth of "CCP-5B Forms" once per "Y" days to ensure monitoring is performed as specified by written procedures and forms are completed and appropriate corrective and preventative measures were taken as required. Also to ensure Pre-shipment review is completed as per MOP 11, USA, Annex Q, Q.1.1b).	"CCP-5B Form" "CCP-5B Verfication Form"

Dressed poultry carcasses can be shipped to another registered establishment provided the product surface temperature has reached 7°C or lower before being shipped.	It is highly recommended that the monitoring frequencies established by the operator must allow for the opportunity to effectively further cool the product prior to exceeding the regulatory time frame CCP-5B Monitor records observations and date/time and signs on "CCP-5B Form".	Dressed poultry carcasses to be shipped to another registered establishment must reach product surface temperature of 7°C or lower before being shipped, otherwise the product must stay in the approved continuous chilling process until appropriate temperature is reached. The following information is documented on deviation CCP-5B Form: 1. A description of the deviation and its cause 2. Acton(s) taken to control affected product 3. Corrective action(s) taken to restore control of the CCP 4. Verification of effectiveness of corrective measures 5. Measures taken to prevent reoccurrence of the deviation 6. Verification of effectiveness of preventative measures CCP-5B Form must include initials, date and exact time an entry is made	If deficiencies are noted during verification procedures, a root cause analysis and food safety assessment will be performed. Corrective actions/preventative measures may include retraining of "CCP-5B monitor" and/or employees and/or re-evaluation of monitoring/deviation procedures. Verification observations, verifier's signature and date/time are recorded on "CCP-5B Verification Form".

- Scalding all the birds in a common bath without continued water change;
- Showers with not enough water, incomplete coverage or too little pressure;
- Cross-contamination between carcasses from contaminated machinery or workers' gloves by pathogens such as *Salmonella* and *Campylobacter*.

In some countries antimicrobial agents may be used during the process. They include chemicals such as: chlorine, sodium chloride, phosphate and lactic acid.

It is recognized that the different steps involved cannot eliminate all pathogenic bacteria; however, research findings have indicated that certain procedures can reduce the bacterial load on ready-to-cook poultry meat. Some of the most important preventive measures include the following:

- Adequate feed withdrawal to reduce contamination due to evisceration accidents;
- Effective reprocessing or decontamination procedures to reduce visible and bacterial contamination;
- Rapid chilling and maintaining the temperature ≤4°C to minimize bacterial growth;
- Use of countercurrent water flow in hot water scalders, or the use of steam only in the new generation of scalders;
- Effective spray washing of the carcasses with sufficient water (volume and pressure);
- Incorporation of a sanitizing agent such as chlorine (where permitted), hot water or lactic acid, during whole-carcass spray wash;
- Treating product transfer belts and automatic evisceration equipment with water sprays for cleaning during continuous operation;
- Operating counter-flow water chillers (see discussion below);
- Adding bactericidal agents such as chlorine or phosphate to water chillers and/or use as a spray.

Through the slaughter process there are a number of contamination sources as well as a number of steps where decontamination takes place. Again, it is exceedingly important that each plant uses the process data to determine the extent of hazards in their process as well as the effectiveness of their control measures.

Keener (2004) and others reported that contamination occurs both on the farm and in poultry slaughter plants. Routine procedures on the farm such as feed withdrawal, poultry handling, and transportation practices have a documented effect on *Campylobacter* levels at the processing plant. At the plant, defeathering, evisceration, and carcass chillers have been documented to cross-contaminate poultry carcasses. Carcass washings and the application of processing aids have been shown to reduce populations of *Campylobacter* on carcasses by 0.5 to 1.5 log; however, populations of *Campylobacter* have been shown to enter a poultry processing plant at levels of up to 10^5 colony-forming units (CFU)/mL of carcass rinse. This contamination can be easily spread from carcass to carcass during processing if certain measures are not employed. Further effort is needed to design more efficient and effective washing systems. It is estimated that an average poultry processing plant spends $500,000 to $1 million per year on water for washing chicken carcasses; however, sometimes with minimal reductions in bacteria. A large portion of the bacterial reduction can occur by the application of processing aids such as TSP and ASC. Although *Campylobacter* contamination is reduced during processing, they can still be present on the carcass after processing at levels of 10^2 to 10^4 organisms. It is important to mention that the industry is constantly working on equipment/procedures to reduce contamination. An example is the 2012 introduction of the Aero-scalder in which steam is used for the defeathering process. This means that the birds are not submerged in a common hot water bath, and chances of cross-contamination are reduced.

Carcass washing is an example of an intervention that is used to remove blood residues and gut spills (Step 33; Figure 7.1). The results can be affected by the volume, pressure and coverage of the spray. Sometimes processors use city water, with a low residual chlorine level, but in other cases high levels of chlorine, TPP, lactic acid, etc. are used. Laboratory and field studies evaluated lactic acid efficacy as a chlorine alternative. Considering laboratory and field studies, lactic acid produced greater reductions in *Salmonella*, APC, and coliforms, validating its effectiveness as a chlorine alternative in mobile poultry slaughter operations (Killinger et al., 2010).

Visual inspection (see Step 26; Figure 7.1) of each finished carcass is commonly used by plant and government inspectors. However, it should be noted that there is some controversy about this. An example is a report by a Washington, DC-based Food & Water Watch which says that it found evidence that inspectors (referring to company staff rather than government inspectors) at poultry companies regularly miss quality defects on the line. The group released an analysis of more than 5000 documents that it says support its claim that company inspectors often miss defects at facilities that operate under the voluntary HACCP-based Inspection Models Project (HIMP). The group requested sampling results from USDA from 1 March 2011 through 31 August 2011. It received records for 11 of the 20 broiler plants in the HIMP pilot and three of the four turkey plants. Its analysis established that, on average, company inspectors found that the majority of non-compliance records filed for the 14 plants under the pilot was for "fecal contamination found on the carcasses" and that 90% of the non-compliance records were the result of visible fecal contamination (Fielding, 2012). The news comes as the USDA's Food Safety and Inspection Service (FSIS) prepares to all but abandon the food inspection system under which federal inspectors examine chicken and turkey carcasses on the slaughter line by sight, touch and smell and move to a modernized system stressing offline quality assurance.

IMPORTANCE OF EQUIPMENT/PROCESS SELECTION

Paying attention to equipment design, ease of cleaning, and minimizing cross-contamination is an important process in obtaining a product with low microbial counts. Over the years the poultry industry has evolved (e.g. increased line speed) but also come up with ways to improve the hygienic standards of poultry meat. The importance of selecting the right equipment and procedure will be demonstrated by using the chilling operation. Currently there are two main approaches that include water and air chilling; some combination of the two can also be found. Choosing one method over the other depends on availability and cost of water/electricity, yield desired, previous scalding temperature used, market requirements, etc. Water chilling has been traditionally used for poultry where a cold water bath sometimes supplemented with ice is employed (Figure 7.2). This is also an example of a potential critical control point where all the birds pass through a common

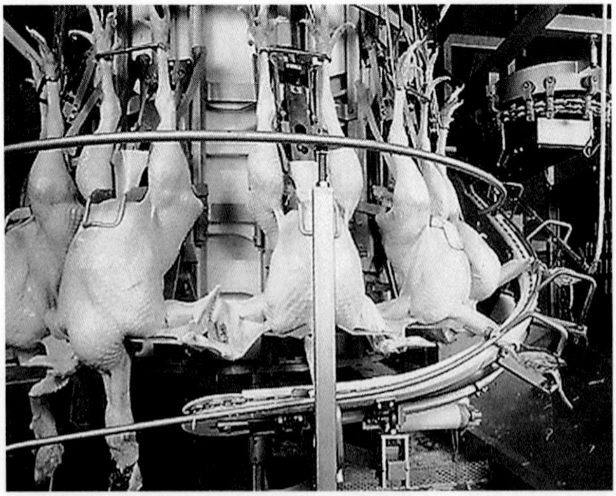

FIGURE 7.2 Automatic equipment used for poultry evisceration. A high-speed broiler line processes a few thousand birds per hour. *Courtesy of Marel – Stork Poultry Processing.*

bath and hazards can be minimized if the processor is careful about the water quality, overflow (i.e. different countries have certain regulations for the amount of fresh water needed for each incoming bird), water temperature, flow rate, etc. Later developments focused on employing a counter-flow design (Figure 7.3b) where clean cold water is introduced at the exit end of a long, relatively narrow, water tank. The birds are moved by an auger or large paddles while clean cold water is being pumped from the exit end towards the entry point. By doing so, the industry has achieved higher hygienic standards as the clean water helps to remove more microorganisms. There have been numerous publications describing the process and reduction of microorganisms (Allen et al., 2000). Later developments have resulted in the introduction of a multistage chiller (two to three separate tanks) where birds are moved through the tanks and the water from the previous tank is allowed to drip prior to placing the birds in the next tank. It should be mentioned that this approach has also been used in the scalding tank (Step 15; Figure 7.1) that uses hot water to loosen the feathers (not discussed in detail in this review).

Large commercial air chilling is another development that started to appear on the market about 25 years ago, and today is very popular in Europe and other places where the cost of water is relatively high. Air chilling is done while the birds are either moving on a shackle line (Figure 7.3a; see right side) or put on carts that are placed in an air-chilling room. Overall, air chilling is considered by some to be a more hygienic process since there is no common bath and potentially less cross-contamination among the birds. However, in the scientific literature there is still some controversy about the microbial quality of the birds, as some researchers have shown an improvement by using air chill while others have shown an advantage of using the washing effect of the cold water (Huezo et al., 2007; Barbut et al., 2009).

Overall, while immersion chilling is still the predominant method in the USA, the popularity of air chilling is growing. Many of the advocates of air chilling also speak of improved flavor and color. Northcutt and Smith (2008) compared air-chilling and immersion-chilling

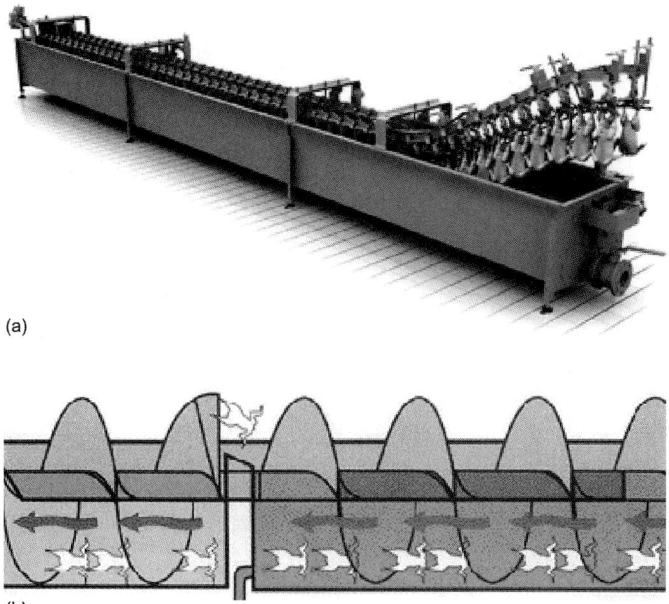

FIGURE 7.3 Chilling operation of eviscerated poultry. The carcasses can stay on the line and be submerged in cold water (a), or stay on a similar line and going through a cold air chilling tunnel (not shown here). A more common way of water chilling is a screw-type chiller (b) where carcasses are dropped into a long chiller with a stream of counter-flow cold water coming from the exit end. In the case shown here, carcasses move through two tanks. See text for additional explanation. *Courtesy of Stork Poultry Processing.*

methods to ascertain the best method for ensuring meat quality, food safety and water management. They concluded that air-chilled chicken can be processed quicker while maintaining meat quality, while also providing higher cooked meat yields and using less water than immersion-chilled poultry. However, the researchers said that they found no significant difference in bacterial pathogen levels between the two chilling methods. At the same time, a USDA-sponsored study by the University of Nebraska found that 350 air-chilled chickens had about 20% fewer bacteria (such as *Salmonella* and *Campylobacter*) than on the same number of water-cooled poultry. The study, however, examined only one air-chilling plant and one water-immersion plant.

ADVANTAGES OF IMPLEMENTING HACCP

Advantages can be clearly illustrated by the results of a very large-scale program in the USA when the Food Safety and Inspection Service (FSIS) issued the Pathogen Reduction/ Hazard Analysis and Critical Control Point – Final Rule (the PR/HACCP) on 25 July 1996. The FSIS got the mandate to also verify that industry PR/HACCP systems are effective in controlling the contamination of raw meat and poultry products with human disease-causing bacteria, as this rule sets product-specific *Salmonella* performance standards that must be met by slaughter establishments and establishments producing raw ground products. The performance standards are based on the prevalence of *Salmonella*, as determined from the FSIS's nationwide microbial baseline studies, and are expressed in terms of the maximum number of *Salmonella*-positive samples that are allowed in a given sample set. From January 1998 through December 2000, federal inspectors collected 98,204 samples and 1502 completed sample sets for *Salmonella* analysis from large, small, and very small establishments that produced at least one of seven raw meat and poultry products: broilers, market hogs, cows and bulls, steers and heifers, ground beef, ground chicken, and ground turkey. *Salmonella* prevalence in most of the product categories was lower after the implementation of PR/HACCP than in pre-PR/HACCP baseline studies surveys conducted by the FSIS. According to Rose et al. (2002), the results of 3 years of testing at establishments of all sizes combined show that >80% of the sample sets met the *Salmonella* prevalence performance standards (e.g. 20.0% for broilers, 8.7% for market hogs, 2.7% for cows and bulls). The decreased *Salmonella* prevalence was partly reflected by industry improvements, such as improving process control, incorporation of antimicrobial interventions, and increased microbial monitoring; all in conjunction with PR/HACCP implementation. A later follow-up in 2003 revealed that 81% of establishments never had a failed test. In establishments that did experience set failure(s), the failed sets were generally collected early in the establishment testing history. Small establishments were more likely to have experienced a set failure than large or very small establishments (Eblen et al., 2006). The FSIS response to failed *Salmonella* sample sets in the form of in-depth verification reviews and related establishment-initiated corrective actions have likely contributed to declines in the number of establishments that failed sets. The authors mentioned that focusing on food safety measures in small establishments should further reduce the number of sample sets that fail to meet the *Salmonella* performance standard. Fletcher (2006) mentioned that implementing an HACCP-based inspection program has been credited with reducing the incidence of *Salmonella*-positive carcasses from approximately 20 to 10%. He also stressed the importance of using standardized sampling procedures for reporting *Salmonella*.

Overall, the introduction of various interventions should be validated to determine the efficacy (Codex, 1993; Barbut, 2002). A few examples will be provided below but they are by no means a comprehensive list of all possible intervention procedures. Stopforth et al. (2007) looked at changes in aerobic plate counts (APC), total coliform counts (TCC), *Escherichia coli* counts (ECC), and *Salmonella* incidences on poultry carcasses and parts as well as poultry processing water. They examined samples before and after individual interventions and after poultry carcasses were exposed to multiple-sequential interventions at various stages during the slaughter process in three different plants. Interventions included post-evisceration wash, inside-outside bird washes, chlorine dioxide wash, chlorine dioxide wash plus chlorine chiller, chiller exit spray, post-chiller wash, and a tri-sodium phosphate wash at two of the plants. The majority of individual interventions effectively or significantly ($P < 0.05$) reduced microbial populations on or in carcasses, carcass parts, and processing water. Reductions in microbial counts ranged from 0 to 1.2 log CFU/ml of sample rinse. Multiple-sequential interventions resulted in significant reductions in APC, TCC, ECC, and *Salmonella* incidence of 2.4, 2.8, and 2.9 log CFU/ml and 79%, respectively, at the first plant; 1.8, 1.7, and 1.6 log CFU/ml and 91%, respectively, at second plant; and 0.8, 1.1, and 0.9 log CFU/ml and 40%, respectively, at the third plant. These results enabled validation of in-plant poultry processing interventions and provided a source of information to help the industry in its selection of antimicrobial strategies. Gill et al. (2006) looked at different groups of bacteria after various processes applied during broiler processing at an HACCP-approved plant. The log mean numbers of aerobes, coliforms, *E. coli* and presumptive staphylococci plus listerias on carcasses after scalding at 58°C and plucking were about 4.4, 2.5, 2.2 and 1.4 log cfu/cm^2, respectively. The numbers of bacteria on eviscerated carcasses were similar. After the series of operations for removing the crop, lungs, kidneys and neck, the numbers of aerobes were about 1 log unit less than on eviscerated carcasses, but the numbers of the other bacteria were not substantially reduced. After cooling in water, the numbers of coliforms and *E. coli* were about 1 log unit less and the numbers of presumptive staphylococci plus listerias were about 0.5 log unit less than the numbers on dressed carcasses, but the numbers of aerobes were not reduced.

Another consideration is the emerging issue of the avian influenza virus (AIV) and more specifically the H5N1 highly pathogenic strain. The possible human health threat that it poses has raised concerns over the food safety implications of this virus infecting poultry. The European Food Safety Agency and the US Department of Agriculture's Animal and Plant Health Inspection Service have identified legal and illegal importations of infected poultry commodities as reviewed by Beato et al. (2009). The authors indicate that AIVs may be recovered from a variety of poultry products. However, the presence of AIVs in poultry products is influenced by the characteristics of the viral strain, particularly its pathogenicity and thus the ability to cause systemic infection. As a consequence the host also influences the likelihood of the virus being present. Data are still fragmentary and further studies should be carried out in a more extensive and coordinated manner in order to establish proper risk assessments on the spread of infection to a given area and/or host by poultry products. Although only limited studies have been published, it is reassuring that heat and pressure treatments have been shown to inactivate, to acceptable levels, any viable viruses in selected commodities (Beato et al., 2009).

The machinery used also plays an important role in maintaining the operation clean and reducing problems with cross-contamination. An example is the evisceration process (Figure 7.2) where the first step is a machine used to make a cut and open the abdominal cavity. It is important to realize that the machine needs to work at high speed (e.g. 13,500 broilers per hour) with no/minimum damage to the carcass as well as prevent damaging the intestines so no gut spills occur later on (i.e. bacterial count of gut content is about 10^8 to 10^9 per gram). The length of the opening cut should be adjusted to correspond to the size of the carcasses processed. It is very important that adjustments can be done quickly and easily during production as there is very little time between flocks. Overall, equipment design is related to the prerequisite program (but is mentioned here) as hygiene-focused design helps keep the machine clean during operation. Features such as sloped surfaces (to prevent water/debris accumulation) and no blind spots enable the machine to stay clean during operation while water spray can be used to remove any material falling on it. As indicated before, the point of consumer education and providing adequate instructions cannot be overlooked. An example of the need for very clear cooking instructions is the 2007 case concerning the recalling of frozen chicken and turkey pot pies which were undercooked by some customers in the USA (Anonymous, 2007). This led to 152 cases of *Salmonella* poisoning in 31 states and 20 people hospitalized. The company responded by (1) asking

FIGURE 7.4 High-speed egg breaking machine (http://www.sparboe.com/eggs-as-ingredients.html).

customers to return suspected products, (2) reminding consumers that these products were not ready to eat, and must always be thoroughly cooked, and (3) most importantly, revising cooking instructions for future marketing.

EGG BREAKING OPERATIONS

In large egg processing operations eggs are brought into a processing plant for either washing, grading and sorting prior to selling as fresh table eggs, or for washing followed by breaking (Figure 7.4) and separating the components (egg yolks, whites, liquid whole eggs) to be used as an ingredient in other food products (e.g. noodles, cake mixes, mayonnaise). The main focus in this section will be on the breaking operation where a more elaborate HACCP program is used (Figure 7.5).

Eggs are usually picked up from the producer three to four times a week. Often, eggs arrive at the breaking plant on the same day they are laid. Eggs are typically gathered on automatic belts and refrigerated in large producing farms. All eggs are washed at the breaking plant under closely controlled temperature conditions using approved detergent sanitizers. Washing and disinfecting procedures have greatly enhanced the retention of egg quality and reduced the incidence of bacterial spoilage (Froning et al., 2001). It is interesting to note that using very fresh eggs in the breaking operation has led to some pasteurization concerns, especially for egg white, as eggs reaching the breaking plant often have an egg white with a lower pH. *Salmonella* is somewhat more heat resistant in egg white at a lower pH. Over time, egg white pH will increase from 7.6 to 9.4, depending on temperature of storage; this increase may take 7 to 10 days. Obviously, the egg processing industry must maintain high albumen quality goals and pasteurization guidelines must work within today's improved quality assurance programs (Anonymous, 1990; Sugihara et al., 1996).

As indicated above, the process starts with bringing the eggs to the grading station. Some of the eggs can have visible dirt (e.g. manure, small feathers attached, stain), while others

FORM #3: PROCESS FLOW DIAGRAM

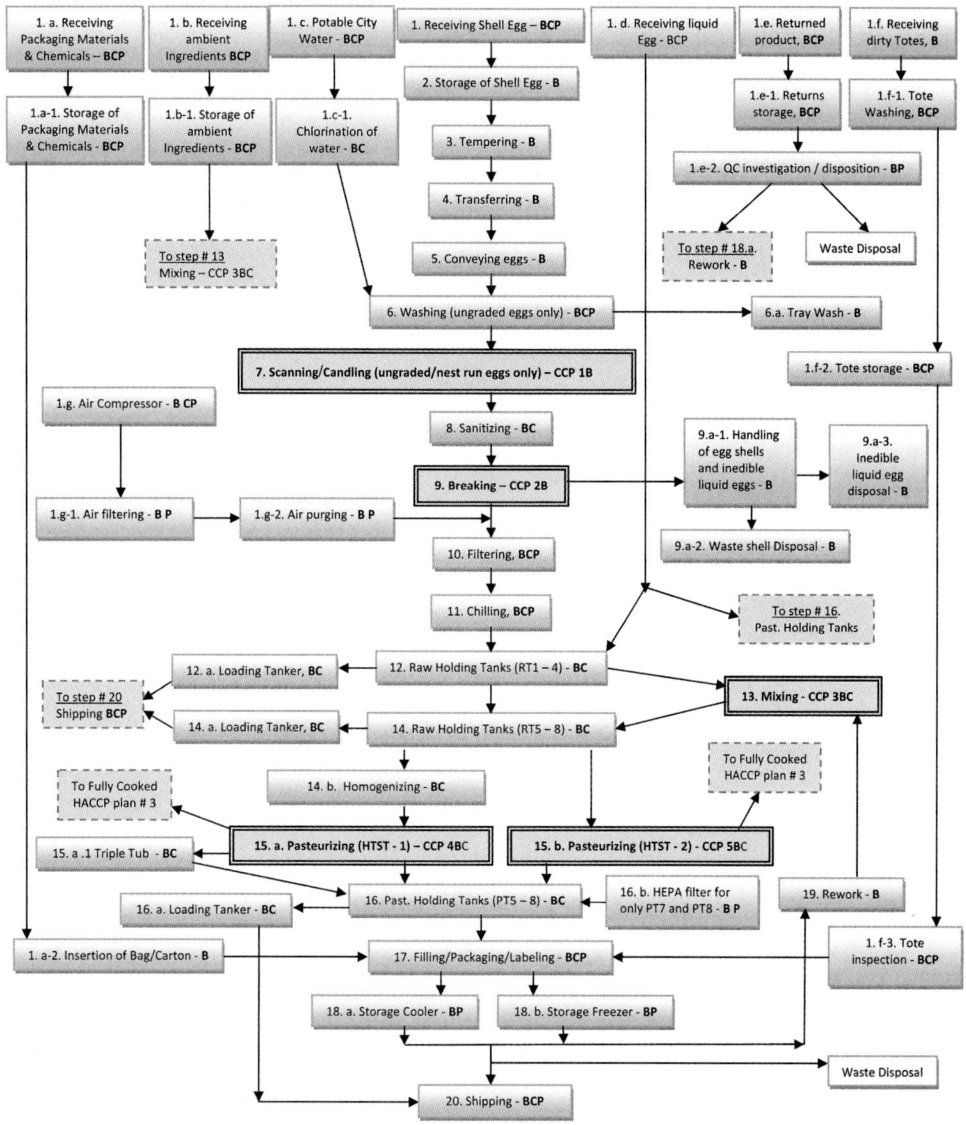

	Date	HACCP Coordinator	Page 1 of 1
Effective:			
Supersedes:			

FIGURE 7.5 An example of a flow diagram of an egg breaking operation with suggested critical control points and potential hazards indicated in each step.

appear visibly clean. However, that does not mean that they are microbiologically clean, as the process by which they are obtained is not sterile and actually involves a normal farm environment. The first step when they arrive at the station is a cooling operation (if not cooled before), followed by candling, where a strong light is used to check for cracks in the shell/leaking egg content (i.e. eggs that will not withstand the following washing operation). The candling step is also used to check for things like blood spots and double yolk inside the egg itself. The next step is washing the eggs where a warm solution containing a detergent is used. Attention should be given to the water temperature and its relation to the egg temperature, as a cold solution will result in shrinkage of the egg content and absorption of some cleaning solution through the porous shell (i.e. shell is very porous to allow the embryo to breath). At this point the eggs can be graded for the table egg market or cracked open by a high-speed breaking machine (Figure 7.4), where the egg white and yolk are separated while the shell is removed. As one can see there is a chance for cross-contamination and the liquid egg content has to be pasteurized immediately after the breaking operation.

In North America, for example, the first comprehensive egg pasteurization manual was produced by the USDA in 1969. It reviewed research available at that time to develop present pasteurization requirements. This later led to the Egg Products Inspection Act of 1970, established by the FDA (Froning et al., 2001). Although egg pasteurization was first utilized by the egg products industry in the 1930s, this Act required that all egg products be *Salmonella* free through use of approved pasteurization methods. Later Cunningham (1995) provided an extensive review of pasteurization methods used by the egg industry.

A flow diagram showing a typical process used to produce liquid eggs, with an overlay of a generic HACCP plan, is shown in Figure 7.5. The diagram is part of an HACCP document used by a large egg producer in North America and serves here as an example to review the different steps, potential hazards and suggested CCP. The raw materials coming to the plant comprise graded/ungraded shell eggs, as well as liquid eggs which can arrive in a pasteurized or unpasteurized form (see top row). The raw materials can potentially carry microbial hazards (e.g. *Salmonella*), chemical hazards (e.g. antibiotics), and physical hazard (e.g. pieces of plastic in the liquid eggs). The first CCP in this scheme is the scanning/candling operation (as indicated before, shell egg passing above a strong source of light) where dirty, cracked or leaking eggs are identified and removed. The eggs are then washed and sanitized where attention should be given to chemical concentrations, and water temperature (e.g. water temperature below the egg content temperature can cause shrinking of the egg content and sucking of cleaning solution into the air cell space). The following step is the breaking operation (see equipment in Figure 7.4). In this HACCP model it is designated as the second CCP. At this point, unacceptable products are removed and this can also include defects such as blood spots.

The eggs can be separated into whites and yolks which are filtered to remove any residual pieces of shell that might have entered the process. At this point the eggs can be mixed with ingredients such as salt, sugar, and citric acid (depending on end use and customers' requirements). The following pasteurization step is one of the key operations to assure pathogen-free product and therefore designated as CCP. Time–temperature values approved by government regulators must be met (see example in Table 7.3 and additional discussion later in the chapter). Records of operation conditions should be available to the government inspector at all times, and usually should be kept on site for a period of 1–5 years

TABLE 7.3 Liquid Egg Pasteurization Guidelines (Froning et al., 2001)

	Temp.	Time
Liquid Egg Product	**°C (°F)**	**Min***
Egg white with pH adjusted	56.7 (134)	4.3
Egg white without pH adjusted	57.7 (136)	6.3
Egg white pH 8.6 with hydrogen peroxide (Standard Brands process)	54.4 (130)	3.5
USDA scrambled egg mix (30% solids)	62.2 (144)	2.0
Scrambled egg mix (22% solids)	60.0 (140)	2.4
Fortified whole egg "Tex" product (32% solids)	62.2 (144)	2.0
Fortified egg yolk "Tex" product (49% solids)	63.3 (146)	3.5
Imitation egg product	56.7 (134)	4.6
Salted yolk (10%)	63.3 (146)	4.5
Salted whole egg (10%) without storage	63.3 (146)	5.7
Salted whole egg (10%) + 96 hour storage	63.3 (146)	3.5
Sugared yolk (10%)	63.3 (146)	3.5
Sugared whole egg (10%)	61.1 (142)	3.5
Plain yolk	61.1 (142)	3.5
	60.0 (140)	6.2
Plain whole egg	60.0 (140)	3.5

**Based on a 5 log reduction.*

(depending on the country). Plant personnel are also responsible to check and validate the effectiveness of the pasteurization process; similar to validation of all other CCPs. In this case samples are routinely taken for microbial analysis to verify that no pathogens (e.g. *E. coli*, *Salmonella*) are present. The resulting liquid eggs are going through several more steps as outlined in the figure, and can then be dried (e.g. egg powder), filled into cartons/plastic bags and then stored fresh, or frozen for later distribution.

As indicated above, pasteurization requirements for various liquid egg products are shown in Table 7.3. These USDA requirements provide minimum temperatures and holding times (Froning et al., 2001). As with other heat processing procedures, pasteurization time and temperature combinations must be validated to ensure they are able to sufficiently reduce the number of pathogenic bacteria to a safe level.

Examples for whole egg pasteurization in other countries are: 63.3°C for 2.5 minutes in China; 62°C for 2.5 minutes in Australia; 64.4°C for 2.5 minutes in Great Britain; and 65°C for 90 to 180 seconds in Denmark (Cunningham, 1995).

Microbial inactivation of pathogens without heat can also be done in heat-sensitive food ingredients such as egg whites. Hydrogen peroxide has been shown to eliminate *Salmonella* in egg white at room temperature (Ayres and Slosberg, 1949). After treatment, catalase is added to decompose hydrogen peroxide to water and oxygen. The Armour Company developed a patent which utilized this technology in which heating to 51.7°C and holding for 1.5 minutes is used to inactivate the natural catalase in the egg white. Overall, hydrogen peroxide (10% solution) is then metered into the holding tube at a level of 0.5 kg per 100 kg of egg white. The mixture is held at 51.7°C for 2 minutes, after which the pasteurized product is cooled to 7°C and catalase added to remove residual hydrogen peroxide.

The importance of pasteurization can be illustrated by the massive problem created by mistakenly mixing non-pasteurized egg material within an ice cream premix. In September 1994, the Minnesota Department of Health detected an increase in the number of reports of *Salmonella Enteritidis* infections. After a case–control study implicated a nationally distributed major brand of ice cream in the outbreak, the product was recalled. It was estimated that *S. Enteritidis* gastroenteritis developed in 224,000 persons in the USA after they ate the ice cream (Hennessy et al., 1996). Ice cream associated with infection contained a higher percentage of premix that had been transported by tanker which had carried non-pasteurized eggs immediately before. To prevent further outbreaks, food products not destined for repasteurization should be transported in dedicated containers.

Radiation pasteurization has also been studied extensively. Many of the earlier research efforts in the 1950s and 1960s have been reviewed in the 1968 *International Egg Pasteurization Manual*; and later in the 2001 manual (Froning et al., 2001). Gamma irradiation was emphasized at that time, as it has excellent penetration, particularly in frozen egg products. Yolk-containing egg products were noted to have off-flavors which were largely volatized during spray drying. Egg white was less prone to off-flavor development during gamma irradiation. Kijowski et al. (1994) observed that gamma irradiation of frozen whole egg at 2.5 kGy did not adversely affect functional or sensory properties (D value of 0.39 kGy).

Processes using ultra-high pasteurization temperature are also allowed for producing egg products on continuous flow, high temperature, short-time pasteurization equipment "to provide liquid whole egg products for refrigerated distribution which have greatly reduced levels of spoilage microorganisms, while still having good functional properties." Some patents in this area include Swartzel et al. (1991a,b).

Distributing contaminated eggs can be a risky operation. The USDA Food Safety Inspection Service, the *Salmonella Enteritidis* Risk Assessment Team created a baseline model for shell eggs (FSIS, 2005). In their simulation of average production of 46.8 billion shell eggs per year in the USA, 2.3 million contain *Salmonella Enteritidis.* The consumption of these eggs results in a mean of 661,633 human illnesses ranging from 126,374 to 1.7 million cases per year (5th and 95th percentiles). It is estimated that about 94% of these cases recover without medical care, 5% visit a physician, an additional 0.5% are hospitalized, and 0.05% of the cases result in death.

Overall there has been a decline in the consumption of shell eggs in recent years in industrialized countries, though the demand for liquid egg products (i.e. mostly pasteurized prior to leaving the egg grading station) in the food industry has been increasing. Several hundred million pounds of frozen, pasteurized egg products are produced by the US egg industry every year. Increases in the number of manufactured food items using egg products has resulted in a 30% growth in the use of liquid, frozen and dried egg over the past few years. Some of the applications for pasteurized egg products include liquid whole eggs used for custards and cakes, and salted yolks for salad dressings and mayonnaise. In addition, liquid egg products are used for the manufacture of pet foods, culture media for growth of microorganisms, vaccine production, cosmetics and hair shampoos (Anonymous, 1990). Since 1966, US food laws require that all commercial eggs broken out of the shell for manufacturing must be pasteurized (to destroy *Salmonella*).

A decreasing trend in the notification rate of salmonellosis cases in the EU, particularly those caused by *S. Enteritidis*, has been seen over recent years (EFSA, 2010a,b). This has largely been attributed to the implementation of *Salmonella* national control programs in the laying flocks (e.g. culling of infected grandparents flocks, and sometimes production flocks, as well as using competitive exclusion). Nevertheless, most of the reported foodborne outbreaks reported in the EU are still caused by *Salmonella*, with the most important food source being eggs and egg products.

A recent massive recall of table eggs due to *Salmonella* can be used to illustrate some of the challenges in the field. Overall, the US Food and Drug Administration conducts hundreds of inspections of egg producers to ensure they adhere to newly implemented safety rules. During the recall the FDA Commissioner Margaret Hamburg said it was an "unfortunate irony" that the new rules (i.e. procedure to check for *Salmonella*) took effect from 9 July 2012, after the *Salmonella* outbreak likely began. If the rules had been in place, she said, the FDA "very likely" could have prevented the recent recall of a half-billion eggs. The incidence involved two Iowa egg producers – Wright County Egg Co., one of the largest producers in the US, and Hillandale Farms of Iowa Inc., which shares some egg suppliers with Wright – both of whom announced voluntary recalls of eggs. Wright recalled 380 million eggs and Hillandale 170 million. The FDA says they don't think any other farms were involved in the outbreak. Nearly 2000 people were sickened with *Salmonella* from May through July, three times the normal rate, according to the Centers for Disease Control and Prevention. No deaths have been reported. CDC officials said they expected to record more cases of *Salmonella* from that timeframe in coming weeks because of the lag time between exposure and formal reports. FDA officials also said that the two companies will not be able to ship products again until the farms have presented detailed plans on how they will avoid contamination risks. "The FDA would expect the firms to correct any violations and will continue to work to ensure that eggs are not sold to consumers from these farms until we know they can be shipped safely," an FDA spokeswoman said. The two companies are still producing eggs, but they are being diverted to a "breaking" facility for processing or pasteurization.

As indicated above, the typical process is that eggs from the farms are transported to grading stations. The highest quality eggs are then sent to the retail and wholesale trade while the cracked and dirty eggs are sent to the breaking plant. It is the pasteurization step of the cracked eggs that reduces the risk of pathogens. The two US companies involved in the *Salmonella* recall had their farms' eggs diverted to breaking plants as a precaution. All of this stresses the importance of the pasteurization step, where feasible (Martin and Mundy, 2010).

According to the risk assessment for *Salmonella Enteritidis* in shell eggs and *Salmonella* spp. in Egg Products Report in the USA (FSIS, 1998) pasteurization was predicted to be effective for reducing illnesses from *Salmonella Enteritidis* in shell eggs. If all eggs produced in the USA were pasteurized for a 3-$\log_{10}$ reduction of *Salmonella Enteritidis*, the annual number of illnesses would be reduced from 130,000 to 41,000. A 5-$\log_{10}$ reduction would reduce the annual number of illnesses to 19,000.

Novel methods of pasteurization are being studied. One example is disinfection of liquid egg products by using UV light. While UV is not commonly used, Atilgan (2007) has investigated the possibilities and limitations and showed some promising results.

In summary, there are still opportunities in both the poultry and egg processing industries to improve food safety. While a great deal of research has been done, and improvements made to poultry and egg processing, there is still a significant risk to consumers, microbiological in particular; efforts should be made at all levels. This should start at the breeder flocks (usually today operating under very high bio security standards in most countries), to hatcheries, growing farms for both egg and meat producing poultry, feed mills, meat and egg processing plants, and obviously the consumer (e.g. proper handling and cooking as discussed in other chapters).

References

Allen, V.M., Corry, J.E.L., Burton, C.H., Whyte, R.T., Mead, G.C., 2000. Hygiene aspects of modern poultry chilling. Int. J. Food Microbiol. 58, 39–48.

Anonymous, 1990. Ready eggs. Technical Bulletin. APV Crepaco, Wisconsin.

Anonymous, 2003. USDA data show incidence of Salmonella reduction in raw meat and poultry. <http://www.usda.gov/wps/portal/usda/usdahome?contentidonly=true&contentid=2003/04/0127.html> (accessed Nov 2012.).

Anonymous, 2007. Product incidents. <http://en.wikipedia.org/wiki/Banquet_Foods> (accessed Nov 2012.).

Atilgan, M.R., 2007. Disinfection of liquid egg products by using UV light. PhD thesis. Izmir Institute of Technology.

Ayres, J.C., Slosberg, H.M., 1949. Destruction of Salmonella in egg albumen. Food Technol. 3, 180–183.

Barbut, S., 2002. Hazard analysis critical control point. In Poultry Products Processing – An Industry Guide. CRC Press, New York, NY, pp. 379–428.

Barbut, S., 2010. Past and future of poultry meat harvesting technologies. Worlds Poult. Sci. 66, 399–410.

Barbut, S., Moza, L., Natteress, F., Dilts, B., Gill, C.O., 2009. The microbiological condition of air- or water-chilled carcasses produced at the same poultry packing plant. J. App. Poult. Res. 18, 501–507.

Beato, M.S., Capua, I., Alexander, D.J., 2009. Avian influenza viruses in poultry products: a review. Avian Path. 38, 193–200.

Bernard, A., Broeckaert, F., De Poorter, G., De Cock, A., Hermans, C., Saegerman, C., et al., 2002. The Belgian PCB/dioxin incident: analysis of the food chain contamination and health risk evaluation. Environ. Res. 88 (1), 1–18.

CDC, 2011. Centers for Disease Control, USA. Surveillance for foodborne disease outbreaks. MMWR 60 (35), 1197–1202.

CFIA, 2012. HACCP Generic Model. Poultry Slaughter – Chilled Ready to Cook Whole Chicken. Canadian Food Inspection Agency, Ottawa, Canada.

Codex, 1993. Guidelines for application of Hazard Analysis Critical Control Point (HACCP) system. Codex Alimentarius Commission, ALIMORM 95/13, Appendix II, Rome.

Corry, J.E.L., Atabay, H.I., 2001. Poultry as a source of Campylobacter and related organisms. J. Appl. Micro. 90 (56), 96S–114S.

Cunningham, F.E., 1995. In: Stadelman, W.J., Cotterill, O.J. (Eds.), Egg Science and Technology, fourth ed. Haworth Press, Binghamton, NY, pp. 289–321.

Eblen, D.R., Barlow, K.E., Naugle, A.L., 2006. U.S. Food Safety and Inspection Service testing for Salmonella in selected raw meat and poultry products in the United States, 1998 through 2003: an establishment-level analysis. J. Food Prot. 69, 2600–2606.

EFSA, 2010a. European food safety authority. The community summary report on trends and sources of zoonoses, zoonotic agents and food-borne outbreaks in the European Union in 2008. EFSA J. 8 (1), 1496.

European Food Safety Authority, 2010b. EFSA evaluates factors contributing to Campylobacter in chicken. EFSA J. 9 (4), 2105–2246.

European Food Safety Authority, 2011. BIOHAZ panel: EFSA issues advice on reduction of Campylobacter in chickens. EFSA.

Fielding, M., 2012. Analysis finds that company inspectors often miss defects, Meating Place. <http://www.mtgplace.com/meatingplace/DailyNews/>.

Fletcher, D.L., 2006. Influence of sampling methodology on reported incidence of Salmonella in poultry. J. AOAC Int. 89, 512–516.

Franchin, P.R., Battistella, P.D., Vieira, C., 2010. Evaluation of multi-sequential interventions with water to reduce microbial loading as applied to chicken carcasses during slaughtering – a review. Worlds Poult. Sci. 66, 203–213.

Fries, R., 2002. Reducing Salmonella transfer during industrial poultry meat production. Worlds Poult. Sci. 58, 527–538.

Froning, G.W., Peters, D., Muriana, P., Eskridge, K., Travnicek, D., Sumner, S., 2001. International Egg Pasteurization Manual. American Egg Board, Park Ridge, IL. <http://www.aeb.org/images/website/documents/food-manufacturers/order-aeb-resources/Pasteurization_Manual.pdf>.

FSIS, 1998. Food Safety and Inspection Service. *Salmonella enteritidis* risk in shell eggs and egg products. Prepared by the Assessment Team. FSIS, Washington, DC.

FSIS, 2005. Risk assessments of *Salmonella enteritidis* in shell eggs and *Salmonella* spp. in egg products. Report by the United States Department of Agriculture. Washington, DC. <http://www.fsis.usda.gov/science/risk_assessments/index.asp#eggs>.

Gill, C.O., Moza, L.F., Badoni, M., Barbut, S., 2006. The effects on the microbiological condition of product of carcass dressing, cooling, and portioning processes at a poultry packing plant. Int. J. Food Micro. 110, 187–193.

Hennessy, T.W., Hedberg, C.W., Slutsker, L., White, K.E., Besser-Wiek, J.M., Moen, M.E., et al., 1996. A national outbreak of *Salmonella enteritidis* infections from ice cream. The Investigation Team. New Engl. J. Med. 334 (20), 1281–1286.

Huezo, R., Northcutt, J.K., Smith, D.P., Fletcher, D.L., 2007. Effect of chilling method and deboning time on broiler breast fillet quality. J. Appl. Poult. Res. 16 (4), 537–545.

Keener, K.M., Bashor, M.P., Curtis, P.A., Sheldon, B.W., Katharious, S., 2004. Comprehensive review of Campylobacter and poultry processing. Compr. Rev. Food Sci. Food Saf. 3 (2), 105–116.

Kijowski, J., Lesnierowski, G., Zabielski, J., Fizer, W., Magnuski, T., 1994. In: Sim, J.S., Nakai, S. (Eds.), Egg Uses and Processing Technologies. CAB International Press, pp. 340–348.

Killinger, K.M., Kannan, A., Bery, A.I., Cugger, C.G., 2010. Validation of a 2 percent lactic acid antimicrobial rinse for mobile poultry slaughter operations. J. Food Prot. 73 (11), 2079–2083.

Love, D., 2012. Researchers Find Evidence of Banned Antibiotics in Poultry Products. Johns Hopkins University Bloomberg School of Public Health, Environmental Science & Technology.

Martin, T.W., Mundy, A., 2010. FDA's new rules signal closer look at egg farms. Wall St. J. (8-24-10).

Mortimore, S., Wallace, C., 1995. HACCP – A Practical Approach. Chapman and Hall Publ., New York, NY.

Northcutt, J., Smith., D., 2008. ARS studies best method to chill chickens. World Poult. Mag. <http://www.worldpoultry.net/news/ars-studies-best-method-to-chill-chickens-2425.html>.

Rose, B.E., Hill, W.E., Umholtz, R., Ransom, G.M., James, W.O., 2002. Testing for Salmonella in raw meat and poultry products collected at federally inspected establishments in the United States, 1998 through 2000. J. Food Prot. 65, 937–947.

Stopforth, J.D., O'Connor, R., Lopes, M., Kottapalli, B., Hill, W.E., Samadpour, M., 2007. Validation of individual and multiple-sequential interventions for reduction of microbial populations during processing of poultry carcasses and parts. J. Food Prot. 70, 1393–1401.

Sugihara, T.F., Ijichi, K., Kline, L., 1996. Heat pasteurization of liquid whole egg. Food Technol. 8, 100–107.

Swartzel, K.R., Ball Jr., H.R., Hamid-Samimi, M.-H., 1991a. US Patent 4,994,291.

Swartzel, K.R., Ball Jr., H.R., Hamid-Samimi, M.-H., 1991b. US Patent 5,019,408.

Udakis, L., 2012. Vaccinating could prevent foodborne illness. Soc. Gen. Microbiol. <http://www.eurekalert.org/pub_releases/2012-03/sfgm-vcc032312.php>.

CHAPTER

8

Seafood

Sanja Vidaček

Faculty of Food Technology & Biotechnology, Zagreb, Croatia

OUTLINE

Food Safety Management.
DOI: http://dx.doi.org/10.1016/B978-0-12-381504-0.00008-1

INTRODUCTION

Fish, mollusks and crustaceans are valuable sources of proteins, important fatty acids, minerals and vitamins to humans. In some developing countries they represent the main protein source in nutrition. In recent years, the nutritional benefits of aquatic organisms have mainly been associated with their exceptionally advantageous fatty acids profile rich in polyunsaturated fatty acids (PUFAs), particularly eicosapentaenoic acid (EPA) and docosahexaenoic acid (DHA). These well-known nutritional benefits combined with its exceptional gastronomic value and diversity of species, make this food category very attractive to consumers in developed countries over the world.

Fish, mollusks and crustaceans are also highly perishable foods. Delicate structures and characteristic properties make them prone to fast bacterial and enzymatic changes *post mortem* which result in short shelf-life.

Global total production of fish, crustaceans and mollusks has been continuously increasing from the beginning of the century, mostly as a result of strong growth in the aquaculture sector. At the same time, the international fish trade has undergone tremendous expansion during the last three decades, increasing from US$8 billion in 1976 to a record export value of US$102.5 billion in 2010 (FAO, 2012). While becoming more important in a global food chain, seafood is continuously being associated with a number of foodborne outbreaks.

Food safety challenges differ by region, due to differences in income level, diets, local conditions and government infrastructures. Food safety concerns in developing countries typically include the use of untreated or partially treated wastewater, the use of sewage or animal manure on crops and fish ponds, the inappropriate use of agricultural chemicals, the absence of food inspection, a lack of infrastructure (such as adequate refrigeration) and poor hygiene including a lack of clean water supplies. Infections with *Vibrio cholerae*, which have been controlled in many parts of the world, have been a major health concern in many developing countries from the southeast Asian region (particularly in Bangladesh and India), Central and Latin America and Africa. Cholera outbreaks generally are linked to contaminated water, but transmission can occur through contaminated foods, often seafood. Data regarding foodborne diseases from some parts of the world (the African region in particular) are extremely scarce (CSPI, 2005).

Data from developed countries show that histamine poisoning is the highest food safety concern from fish consumption and it seems to be underreported in many developing countries. In Europe in 2010, the majority of outbreaks caused by fish were the result of histamine intoxication (EFSA, 2012a). In the USA, common causes of fishborne outbreaks are histamine and ciguatera toxin (Lynch et al., 2006; Dickey and Plakas, 2009). In Australia in 2009, 6% of all outbreaks were due to or suspected to be due to fish or seafood dishes. Toxin-mediated outbreaks comprised 9% of all foodborne outbreaks, with 33% of these due to fish toxins (ciguatera and histamine poisoning) (OzFoodNet, 2009). In Japan, for the period from 1998 to 2008, there were 89 incidents of histamine poisoning with 1577 cases reported (Toda et al., 2009). Having in mind that histamine production can be completely avoided, as its occurrence in fish is a consequence of inadequate time–temperature conditions of fish sometime post-catch, it is obvious that the basic food safety principles are not always properly applied. Ciguatera outbreaks are also caused by a toxin, but unlike histamine, it is accumulated in fish pre-mortal. Ciguatera fish poisoning is a disease endemic to tropical and subtropical coral reef regions of the world, but over time has become a hazard in nonendemic regions because of expanding international trade in seafood from tropical fisheries (Dickey and Plakas, 2009).

An even greater risk than fish is the consumption of shellfish. The majority of outbreaks in Europe caused by seafood other than fish in 2010 were related to calicivirus followed by marine biotoxins (EFSA, 2012a), bivalve mollusks being the main source. In Japan in 2009, the most frequent implicated foods in terms of number of foodborne incidents were the products of fish and shellfish, with infections by shellfish being the most often reported. There were 33 viral incidents with shellfish caused by calicivirus with 401 cases reported (Japan Food Poisoning Statistics, 2009). Bivalve mollusks are a well-documented source of infections and intoxications because of their way of feeding, but more important because they are often consumed raw. Raw mollusks are also often the source of infections by *Vibrio* spp. In Japan in 2009 *V. parahaemolyticus* was the causative agent of two bacterial infections reported that year for shellfish (Japan Food Poisoning Statistics, 2009). The changing of consumers' preferences to fresh or lightly preserved products is also a trigger for the illnesses because many of the pathogenic organisms are easily destroyed by heat.

The contamination of aquatic animals by environmental chemicals, like dioxins, polychlorinated biphenyls and mercury, is also a public health concern, particularly in the EU and USA (CSPI, 2005). Inappropriate use of aquaculture chemicals and antibiotics has been a food safety issue in aquaculture of some developing countries, but with the export of the products to other areas, this hazard has become a global problem.

It is obvious that food safety remains a major concern facing the seafood industry and according to the FAO (2012) it is a critical component in ensuring food and nutrition security worldwide.

PRODUCTION OF SAFE SEAFOOD – PREREQUISITE PROGRAMS AND HACCP

The seafood industry, as any other food industry, has to assure that the product it is producing is safe for consumers. Moreover, in light of significant growth in international fish trade, it is clear that safety is not limited just to production, but also to distribution and storage. It means that all the players in the seafood business need to have a food safety plan which is implemented in an efficient way. For the processing industry, it means an HACCP-based approach as a means to enhance food safety, preceded by a prerequisite program. The Codex Alimentarius Commission, established by FAO and WHO in 1963, has been developing harmonized international food standards, guidelines and codes of practice to protect the health of consumers and ensure fair trade practices in the food trade. The principal document published by Codex Alimentarius on general principles of food hygiene (CA, 1969) lays a foundation for ensuring food hygiene and the code of practice for fish and fishery products (CA, 2003) is intended for all those engaged in the handling, production, storage, distribution, export, import and sale of fish and fishery products.

Prerequisite programs are practices and/or conditions which are an essential part of the overall food safety plan. Prerequisite programs include hygienic design and control of conditions of facilities and production equipment; avoiding cross-contamination from

TABLE 8.1 Seafood-Related Hazards

Hazards	Seafood-Related Hazards
Biological	Bacteria: *Salmonella* spp., *Shigella* spp., *Escherichia coli*, *Aeromonas hydrophila*, *Clostridium botulinum*, *Clostridium perfringens*, *Bacillus* spp., *Staphylococcus aureus*, *Listeria monocytogenes*, *Vibrio cholerae*, *Vibrio vulnificus*, *Vibrio parahaemolyticus*, *Yersinia enterocolitica*, *Campylobacter* spp. Viruses: calicivirus, hepatitis A Parasites: nematodes, trematodes, cestodes
Chemical	Histamine, toxins produced by pathogenic bacteria, biotoxins, aquaculture drugs, heavy metals (Hg, Cd, As, Pb) and other environmental chemicals, additives, allergens, processing hazards
Physical	Foreign bodies: metal parts from the equipment or tools, plastic from the packaging, nails from personnel, pests Fishbone

personnel, materials or inadequate structure of the facilities; supplier control; water and ice control; toxic chemicals control; cleaning and sanitation; pest control; personal hygiene and training; storage and transportation control; management of waste; and traceability and recall procedures (see Chapter 1). An example of the importance of the efficient prerequisite program is control of *Listeria monocytogenes* during the processing of cold smoked or ready-to-eat products. Its presence on the product is mainly related to the processing environment and its control to cleaning, disinfection and "cleanability" of equipment.

When properly designed and implemented, the science-based HACCP system significantly reduces the risk of biological, physical or chemical hazards reaching the consumer. HACCP has been the gold standard of food safety since the 1990s. There are seven well-known principles of HACCP (hazard analysis, critical control points determination, establishing critical limits, establishing monitoring systems, establishing corrective actions, establishing verification procedures, establishing record keeping and documentation) and 12 steps required to develop an HACCP plan. These 12 steps are designed to ensure that the seven principles are applied correctly. In short, after establishing the HACCP team, describing all the products and their intended use, a flow diagram should be drawn and confirmed in practice. After these steps, the HACCP team should identify and analyze the hazards (see Chapter 31).

Effective hazard identification and hazard analysis are the keys to a successful HACCP plan. Hazard identification is the HACCP principal task in which the HACCP team reviews all processes, procedures and ingredients to compile a list of hazards. The hazards in seafood are presented in Table 8.1. The key to managing and getting control over a number of possible hazards is to identify only those hazards likely to occur.

The seafood-borne pathogenic bacteria is the most numerous category of hazards. They can be categorized in three groups based on their origin (Huss et al., 2003):

- Bacteria indigenous to the aquatic environment that belong to the natural microflora of fish (*Clostridium botulinum;* non-proteolytic types B, E, F, pathogenic *Vibrio* spp., *Aeromonas hydrophila*);
- Bacteria from the animal/human reservoir (*Salmonella* spp., *Shigella* spp., *Escherichia coli*, *Campylobacter jejuni*, *Yersinia enterocolitica*, *Staphylococcus aureus*);

- Bacteria from the general environment (*Clostridium botulinum* proteolytic type A, B, *Clostridum perfringens*, *Bacillus* spp., *Listeria monocytogenes*).

Similar categorization is given by Lyhs (2009). In his categorization, the second group involves only bacteria which are present in seafood as a result of fecal contamination from animal/human. Categorized like this, *Staphylococcus aureus* is not a part of this group as the main source of this pathogen in fish are human nasal passages. The third group involves bacteria that contaminate the product during processing, storage or preparation for consumption.

Pathogenic bacteria can cause intoxications (species *Clostridium*, *Bacillus*, *Staphylococcus*) or infections (the others). The probability of seafood-borne pathogenic bacteria causing illness depends on a number of bacteria and, in the case of bacteria causing infections, on their minimum infective dose (the number of viable bacterial cells necessary to cause disease). Most often intoxications require that the toxin-producing bacteria have grown to high numbers (10^5–10^8 cfu/g) in the food before it is eaten (Huss et al., 2003). Pathogenic bacteria that cause infections will continue to grow inside the person and cause illness. Minimum infective dose varies considerably among these bacterial species and does not always mean that infections are caused by bacteria that have a low infective dose. According to Huss et al., (2003):

- The number of pathogenic bacteria indigenous to the aquatic environment that cause infections is in general low in fish and they have a high minimum infective dose.
- The number of pathogenic bacteria from the general environment is also in general low in fish; most of them are bacteria-causing intoxications meaning that they need to grow in the product in order to cause illness; the minimum infective dose has not been determined for *Listeria monocytogenes* mainly because the same dose does not cause the same effects in different populations.
- The number of bacteria from the human/animal reservoir is different in fish and they have different minimal infective doses.

The preventive measure for bacteria-causing intoxications and those with high minimum infective dose is control of growth. The growth of bacteria is slow at low temperature and therefore the most important preventive measure is chilling the fish soon after harvest and maintaining the low temperature during distribution and storage. Other measures against growth of bacteria include control of water activity, pH or oxygen, or temperature during processing as bacteria have different limiting conditions for growth (see FDA, 2011). Prevention of illness from bacteria with low minimal infective dose is control of contamination, applying the requirements set by the prerequisite program. These bacteria are easily destroyed by cooking.

HAZARDS ASSOCIATED WITH SEAFOOD

Owing to the variety of seafood species (with differences in composition and structure), the different environments they come from and different handling and processing practices, food safety hazards are numerous for this food category. Some types of hazards are species

TABLE 8.2 Significant Hazards in Seafood

Hazards	Originate from the Marine (Aquatic) Environment or Naturally Occur and are Present at the Time of Catch	Originate from the Processing Environment or Occur as a Result of Inadequate Processing or Handling
BIOLOGICAL		
Bacteria	+ especially raw mollusks	+
Viruses	+ especially raw mollusks	+
Parasites	+ especially fish	–
CHEMICAL		
Biotoxins	+	–
Histamine	–	+ only certain species of marine fish
Toxins produced by pathogenic bacteria	–	+
Aquaculture drugs (veterinary residues)	+ only farmed species	–
Environmental chemicals	+ only farmed species and coastal areas	–
Additives and allergens (phosphates, sulfites, nitrites, intrinsic allergens)	–	+
Processing hazards (PAHs, nitrosamines)	–	+ traditionally smoked products
PHYSICAL		
Foreign bodies Fishbone	–	+

related. Potential species-related hazards for many species of fish, mollusks and crustaceans are listed in a guide for fish and fishery products published by FDA in the USA (FDA, 2011). With the purpose of easier hazard analysis at different stages of the seafood chain, Table 8.2 shows significant hazards in seafood, grouped as the ones that may be present at the time of catch or that may occur during handling and processing.

Hazards that Originate from the Marine (Aquatic) Environment or Naturally Occur and are Present at the Time of Catch

Bacteria and Viruses

Seafood-related bacterial and viral hazards that originate from the aquatic environment or are naturally present in animals involve pathogens with a human reservoir that can occur when growing areas are contaminated with human sewage or pathogens naturally occurring in the aquatic environment. These pathogenic bacteria and viruses are destroyed by cooking. Raw bivalve mollusks (oysters) are especially risky organisms as they feed by filtering large volumes of water and are consumed raw. Fish can also be infected by these pathogenic organisms, but they can be found on the skin and the gut, and the edible part – muscle – is considered sterile at the time of catch. Bacteria can be transferred to the muscles by mishandling the fish. These microorganisms are generally a major source of foodborne illnesses.

The shellfish-borne bacterial infections from sewage waters may include infections with pathogenic bacteria such as *Salmonella* spp., *Shigella* spp. or *Escherichia coli*. Hazards also present *Yersinia enterocolitica* and *Campylobacter* spp. In the EU in 2010, the sources of the highest number of notifications reported for *Escherichia coli* were live bivalve mollusks (RASFF, 2011). To control the risk of developing seafood-borne infections from sewage waters, government authorities have monitoring programs for classifying the waters where shellfish are harvested according to a number of bacteria. When the number of bacteria exceeds the set criteria, depuration of bivalve mollusks is required prior to marketing to ensure that they are safe for consumption (Huss et al., 2003).

Besides bacteria, shellfish can also accumulate large numbers of viruses from sewage waters. The most important are noroviruses and hepatitis A. The most significant outbreak of hepatitis A infection occurred in Shanghai, China, in 1988, in which almost 300,000 cases were caused by consumption of clams harvested from a sewage-polluted area. The genus *Norovirus*, belonging to the family Caliciviridae, is considered the leading cause of non-bacterial human gastroenteritis in developed countries (Croci et al., 2012). As stated above, government authorities have programs for monitoring fecal pollution in water, but it is not a reliable means of determining the extent of viral contamination of shellfish. Current treatment regimens for placing live mollusks on the market (depuration and relaying) as is commonly practiced do not effectively reduce noroviruses. The most effective measure to control infection by norovirus from raw mollusk consumption is to produce them in areas which are not fecally contaminated (EFSA, 2012b).

Seafood-borne diseases caused by the genus *Vibrio* are primarily caused by *Vibrio parahaemolyticus*, *Vibrio vulnificus* and *Vibrio cholerae*. *Vibrio vulnificus* and *Vibrio parahaemolyticus* are ubiquitous bacterial pathogens found naturally in marine and estuarine waters. *V. cholerae*, unlike most other vibrios, can survive in freshwater environments (CA, 2010). Infections with *Vibrio cholerae* have still been a major health concern in many developing countries. Infections by non-cholerae *Vibrio* species have recently attracted much attention because these infections have started occurring in some new geographical areas, probably as a result of a climate change (Martinez-Urtaza et al., 2010). The incidents and levels of vibrios present on marine organisms are greatly affected by water temperature, as they multiply rapidly between 20°C and 40°C. Due to the halophilic nature and the marine source of these pathogens, raw seafood is naturally contaminated and is the main food responsible for infection. *V. parahaemolyticus* and *V. vulnificus* occupy a similar ecological niche but have different disease symptoms, growth temperatures and salt tolerancies. *V. vulnificus* does not tolerate low temperatures or high salinity (FAO/WHO, 2005; Martinez-Urtaza et al., 2010).

Similar to noroviruses, monitoring waters for fecal bacteria is not an effective way of controlling vibrios in bivalve mollusks. Risk assessments for *V. vulnificus* in oysters and *V. parahaemolyticus* in seafood were conducted by the Food and Agricultural Organization and World Health Organization (FAO/WHO, 2005; FAO/WHO, 2011a). As there has been an increase in reported outbreaks and cases of foodborne disease attributed to pathogenic *Vibrio* species in seafood, the Codex Alimentarius published guidelines for their control in 2010 (CA, 2010). There is currently a lack of detailed surveillance information regarding non-cholerae *Vibrio* infections in Europe (Baker-Austin et al., 2010).

Parasites

A large number of parasites infect fish but only a few cause illnesses in humans: Opisthorchiidae and Heterophyidae (Class *Trematodea*, Subclass *Digenea*), Anisakidae and Gnathostomidae (Phylum *Nematoda*) and Diphyllobothridae (Class *Cestoda*) (Lima dos Santos and Howgate, 2011). Humans acquire the fishborne parasitic diseases through the consumption of infected raw, undercooked or inadequately preserved fish. Although many aquatic organisms may carry these parasites, human infections are mainly related to fish consumption.

These parasites have different life cycles, with different organisms as primary, intermediate and definitive hosts. Therefore, routes of human infections by these parasites are different. Infections by nematodes are caused by consumption of marine fish and cephalopods mainly from open marine waters (only one case of farmed salmon infected by *Anisakis* has been reported). Infections by trematodes are related to consumption of farmed freshwater fish and crustaceans. More than 100 species of freshwater fishes belonging to 13 families, especially the Cyprinidae, and three species of freshwater shrimp can serve as a second intermediate host of liver flukes, which are trematodes of the highest public health concern. For cestodes, wild and farmed freshwater and marine fish living in cold water habitats can be intermediate hosts.

In the worldwide picture of fishborne parasitoses, cestodiasis (diphyllobothriasis) is considered a mild disease. Nematodiasis is also not as severe as trematodiases – the illnesses that can be severe, and the incidences which are high in endemic areas (Lima dos Santos and Howgate, 2011).

The number of people currently infected with fishborne trematodes exceeds 18 million (WHO, 1995), but worldwide the number of people at risk, including those in developed countries, is more than half a billion. Infections by the liver flukes (trematodes *Clonorchis* and *Opisthorchis*) are a major health concern specifically in the Far East, Eastern Europe and Southeast Asia. The public health significance of these diseases is increasing because of intensification of aquaculture, environmental damage, a lack of appropriate tools for control, links with poor sewage treatment and poverty (90% of world aquaculture is situated in Asia), and cultural traditions of eating raw or minimally processed fishery products (WHO, 1995, 2004).

In Europe, most of the parasitic infestations of fish are related to nematodes in marine fish. In 2010, a 41% increase of parasitic infestation with *Anisakis* of fish (one case of squid) has been reported in the EU (compared to 2009). The reports were mostly for chilled fish and, in some cases, for frozen fish (RASFF, 2011). The other health-related issue regarding *Anisakis* is allergy. *Anisakis* is the only parasite known to cause allergic-type reactions to sensitive individuals. The responsibility of the industry is to provide fish that have no visible parasitic larvae and to ensure that the fish do not pose a health risk to humans. Control strategies for industry to reduce the risk of helminthic infections include visually inspecting fish, by means such as candling, for parasites that are large enough to be detected visually (Huss et al., 2003).

Freezing or heat treatments are the most effective processes used for the killing of parasitic larvae. However, only for nematodes temperature–time conditions of freezing and heating are well defined. The only data available for trematodes would seem to indicate a higher heat resistance of trematodes compared to nematodes. More research is also required on the survival of trematodes in edible fish tissues during traditional processing and preparation (WHO, 1999). For the killing of *A. simplex* larvae, requirements include freezing at −20°C for not less than 24 hours at the core of the fishery products (or treatments which provide an equivalent level of health protection, like freezing at −35°C for at least 15 hours or at −15°C for at least 96 hours) and heat treatment at >60°C for at least 1 minute (EFSA, 2010). The freezing of fish to be consumed either raw or after mild processing (cold smoking, marinating) is compulsory in many European countries (EU, 2004).

For prevention and control of liver flukes, education campaigns are important for communicating the risk to consumers, who should be advised to consume only cooked fish. Environmental sanitation is also important because efficient control should only produce parasite-free fish. More recently, mass chemotherapy of people at risk in endemic areas was recommended as the most effective control strategy (WHO, 2004).

Biotoxins

Marine biotoxins are mainly produced by algae or phytoplankton. In the case of shellfish biotoxins, these toxins can accumulate in the digestive gland (hepatopancreas) of filter-feeding molluskan shellfish and pose a health risk to humans if contaminated shellfish are consumed. The fish toxin causing ciguatera accumulates in all tissues but particularly in the head, roe, liver and other viscera of fish that have eaten the dinoflagella and moved up the food chain to larger fish, and subsequently to humans – the last host in the food chain.

Most of the algal toxins associated with seafood poisoning are heat stable and are not inactivated by cooking. It is also not possible to visually distinguish toxic from non-toxic fish and shellfish.

Common classifications of shellfish biotoxins are based on the symptoms experienced by humans following consumption of contaminated shellfish. Four categories are distinguished: PSP (paralytic shellfish poisoning), NSP (neurotoxic shellfish poisoning), DSP (diarrheic shellfish poisoning) and ASP (amnesic shellfish poisoning). To control these hazards, monitoring programs for marine biotoxins have been established by governments in many countries (Lawrence et al., 2011). When toxin levels in bivalve or cell numbers of toxic algae exceed the accepted limit, harvesting areas are closed or some sort of restriction of harvesting is imposed. Procedures to reopen closed areas include increased sampling from the area and adjacent open areas (Huss et al., 2003).

The fish toxin causing ciguatera is produced by the dinoflagellate *Gambierdiscus toxicus*, which is widely distributed on coral reefs and in lagoons. There are other toxins in fish, but ciguatera poisoning is the most common nonbacterial, fishborne poisoning in the USA and Australia (Lynch et al., 2006; OzFoodNet, 2009) and has been a significant concern in tropical areas for centuries. With the growth in international trade, the risk of ciguatera poisoning is becoming worldwide. Few specific regulations for the control of ciguatera toxin exist. Although the most suspected are carnivorous fish, many marine species may be ciguatera carrying. The list is given in FDA (2011). The toxin is thermally stable and the contaminated fish can remain toxic for years. The most widespread measure applied for the prevention of ciguatera and other fish toxins is the prohibition of the sale of fish species known to be potentially toxic, or for which some ciguatera outbreaks have been reported (FAO, 2004).

Aquaculture Drugs

Aquaculture is a fast-growing sector of the world food economy and accounts for nearly half of all seafood production worldwide (FAO, 2012). Veterinary drug residues in fishery products have become a hazard associated only with farmed species. Aquaculture drugs must be approved by the national authorities, but the regulation about permitted aquaculture drugs is not equally set around the world. As a safeguard to human health, authorities have set the acceptable limits to concentrations of approved drug residues in farmed fish. During the last decade there were a number of cases where the veterinary drug residues were found to exceed the maximum level or cases where forbidden drugs have been found. In 2009 in the EU, the majority of cases reported on noncompliance with the regulations regarding veterinary drug residues in foods was for imported crustaceans containing nitrofuran metabolites (semicarbazide) (RASFF, 2010). However, only a few cases were reported on this topic in 2010 (RASFF, 2011). Seafood farmers have an obligation to use approved chemicals and to assure that the product they sell has the level of the residue beneath the maximum limits. Some countries have monitoring programs that aim to detect the presence of unapproved chemicals in aquaculture products. The application of good aquaculture practices is an important preventive approach to control the misuse and use of unapproved chemicals (FAO, 2009). Currently China's output from aquaculture accounts for about 67% of the world's total production. The dominant export species of aquatic products are shrimp, shellfish, eel, tilapia and large yellow croaker. However, the safety issues, especially using chemicals and antibiotics, have become a very serious problem in China's aquaculture. Nowadays the Chinese government is making substantial efforts to improve food safety and quality. The training of farmers has been highlighted, a traceability system was built, and testing and monitor techniques have become the basis of food safety control. The government has planned to develop new technologies and healthy culture. The Food Safety Law of the People's Republic of China, which was adopted on 28 February 2009, was promulgated and came into force on 1 June 2009 (NSBO, 2010).

Chemicals from the Environment

Heavy metals, such as mercury, cadmium, lead and arsenic, polychlorinated biphenyls (PCBs), especially the so-called "dioxin-like" PCBs and dioxins, represent a group of highly toxic substances accumulating in the tissues of marine organisms and being conveyed through the food chain to humans (Llobet et al., 2003; Storelli, 2008). Most of the contaminants are present in environment through natural occurrence and from industrial and agricultural sources (except the presence of the PCBs – anthropometric activity). Unlike aquaculture drugs, chemical contaminants are hazards from both wild and farmed species; however, more risky is seafood harvested from coastal and estuarine areas and from contaminated fresh water, than from fish harvested from the open seas. As a control measure, guidelines and regulations stipulating maximum permissible levels of contaminants in foods have been set to limit human dietary exposure. The control strategies are monitoring programs set by government authorities and the closure of harvesting areas that pose a risk (FAO, 2009).

The most often reported chemical from the environment related to fish consumption is mercury. Mercury can be found in different forms with methylmercury being the most toxic. People are exposed to methylmercury mainly through their diet, especially through the consumption of freshwater and marine fish and of other animals that consume fish (such as marine mammals) (WHO, 2008). The first well-documented outbreak of acute methylmercury (MeHg) poisoning by consumption of contaminated fish occurred in Minamata, Japan, in 1953 (which had been officially recognized in 2001) with 2265 victims (1784 of whom had died).

The critical target for methylmercury toxicity is the nervous system, especially during its developmental stage. Therefore, most research on methylmercury hazards has studied the effects of a mother's fish consumption during pregnancy on a baby's developing brain, and consequently government recommendations have been focused on women of childbearing age. In humans, the indices of neurotoxicity include neurobehavioral deficits, neuronal loss, ataxia, visual disturbances, impaired hearing, paralysis and death (WHO, 2008). At the international level, for the Joint FAO/WHO Expert Committee on Food Additives (JECFA) the hazard characterization is expressed as the Provisional Weekly Tolerable Intake (PTWI) and is currently established at 1.6 μg of methylmercury per kg bodyweight (FAO/WHO, 2007). The highest levels of mercury are found in fish that are apical predators of older age (such as king mackerel, pike, shark, swordfish, walleye, barracuda, large tuna, scabbard and marlin) and fish-consuming mammals (such as seals and toothed whales).

Moderate consumption of a variety of fish is not likely to result in exposures of concern. However, people who consume large amounts of contaminated fish or marine mammals may be highly exposed to methylmercury and therefore could be at risk (UNEP, 2002). This was also confirmed recently by results of 22 cases involving 24 individual patients in the USA who acquired methylmercury poisoning from eating fish (Groth, 2008). The patients ate fish more than three times a week, sometimes every day. The patients' symptoms closely matched the symptoms observed in Japan (Minamata) and included pains in extremities, fatigue, impairments of speech and hearing, stinging or needle-like sensations in the extremities and mouth, and loss of coordination. The blood mercury levels associated with symptoms in several cases were below the level regarded as safe which suggested that sensitive individuals may experience some adverse effects at low dose levels. When patients stopped eating high-mercury fish, their blood mercury level decreased and symptoms largely or completely dissapeared. The patients acquired their mercury doses from just six fish: tuna, swordfish, sea bass, halibut, yellowtail and king mackerel, with the prevalence of tuna in 86% of cases. The cases documented the need to expand government advisories on fish consumption to include advice for people who eat a lot of fish.

Shellfish are often reported as sources of cadmium. High concentrations of cadmium are present in mollusks and crustaceans such as oysters and other bivalve mollusks, cephalopods and crabs. Maximum levels for heavy metal contaminants have been established in many countries so it is important to be aware of the legislative limits which apply if exporting.

Finally, the Fukushima nuclear disaster that occurred on 11 March 2011 in Japan deserves to be mentioned. The Tohoku earthquake and tsunami caused extensive damage to the Fukushima Daiichi nuclear power plant. Radioactive contamination of the Pacific Ocean following the nuclear incident has raised public concerns about seafood safety. The impact of the incident on seafood safety in Japan, on migratory fish from Japanese waters and on seafood in other parts of the world is given in FAO/WHO (2011b).

Hazards Originating from the Processing Environment/Originating from the Fish as a Result of Mishandling the Fish or Inadequate Processing Practice

Bacteria and Viruses

The bacteria from the Enterobacteriaceae family (*Salmonella* spp., *Shigella* spp., *Escherichia coli*, *Yersinia enterocolitica*) are important pathogenic bacteria principally occurring in the gastrointestinal tracts of humans/animals and environments polluted with human or animal excreta (Huss et al., 2003). Fish and shellfish can acquire these bacteria from polluted waters and/or can become contaminated with them during storage and processing. As previously mentioned, unlike fresh-caught bivalve mollusks that can be contaminated by polluted water, these bacteria should not be present on fresh-caught fish. The contamination of fish with these pathogenic bacteria probably occurs during handling of fish and during the production process.

From 1973 to 2006, in the USA, *Salmonella* spp. and *Shigella* spp. each were associated with about 10% of the reported illnesses associated with seafood. An outbreak in the USA in March 2012 was caused by *Salmonella* in frozen tuna. It has been shown that *Salmonella* was the most common contaminant of imported fish and fishery products in the USA in 2001. Most *Salmonella* contamination problems in fishery products were associated with shrimp (Allshouse et al., 2004). In the EU, *Salmonella* was the main causative agent of foodborne outbreaks, responsible for 35.4% of all reported outbreaks in 2008; fish and fish products were the source in 1.4%, while crustaceans, shellfish, mollusks and products thereof accounted for 1% (EFSA, 2010). Various aspects associated with the microbiological risks posed by the presence of *Salmonella* in seafood, data of incidence on a global level and some prevention and control strategies are presented by Amagliani et al. (2012).

Unlike raw bivalve mollusks, there is no indication that fish is an important source of *E. coli* infection. *Yersinia enterocolitica* is also not often associated with infections caused by fish.

As mentioned above, these bacteria have different minimal infective doses. Infection in fish can be prevented by effective prerequisite programs, good personal hygiene and health education of food handlers, and chilling at low temperature. These bacteria are also easily destroyed by thermal treatment, so adequate cooking is also a control method for this hazard.

Besides the bacteria from the family Enterobacteriaceae, *Campylobacter* (Family Campylobacteriaceae) can be also isolated from fish but the most important source of the bacterium are raw mollusks. In Japan in 2009, there was an outbreak after consumption of fish paste caused by *Campylobacter* with 65 patients involved (Japan Food Poisoning Statistics, 2009).

Although it has to be mentioned as the pathogen that can contaminate fish, mollusks and crustaceans post-catch, infections of *Vibrio* spp. are also mostly related to raw oysters and are the result of the harvest environment. Foods associated with illnesses due to consumption of *V. parahaemolyticus* include also fish-balls, boiled clams, fried mackerel and steamed/boiled crabmeat. Hygiene, low temperature of handling and cooking post-catch are preventive measures against infections from vibrios during processing (CA, 2010).

Seafood can be infected with viruses during handling and processing as a consequence of poor hygiene of food handlers infected with them. Viruses do not multiply outside the host, and thus their numbers will not increase on fish after the initial contamination event (Huss et al., 2003). As already mentioned, the major source of viral infections is raw shellfish that become contaminated through polluted harvest waters.

L. monocytogenes is ubiquitous in nature. Different kinds of fish, squid and crustaceans from water environments containing a high organic load have been found to contain *L. monocytogenes* (Miettinen, 2006). However, unpolluted water is free from this organism, and it could not be detected in fish from such locations. Therefore, although *L. monocytogenes* may be part of the natural microflora of fish, there are strong indications that the raw material is not the primary source for contamination of the final product with *L. monocytogenes* (Huss et al., 2000). *L. monocytogenes* can grow in vacuum- and gas-packaged products at refrigeration temperatures. Ready-to-eat (RTE) food products which are stored at refrigeration temperature for more than 10–15 days and are consumed without sufficient heating to kill living bacterial cells belong to the risk foodstuffs causing listeriosis (Gudbjörnsdóttir et al., 2004). Effective cleaning was found to be an essential preventive measure in reducing the amount of *L. monocytogenes* contamination in fish processing. Often the procedures used for cleaning and disinfection were, however, insufficient in removing persistent *L. monocytogenes* contamination in fish processing factories (Miettinen, 2006). In the EU, there is an increasing trend toward smoked fish being infected with *L. monocytogenes* (RASSF, 2011). A combination of control measures set by the prerequisite programs is the efficient way to control this pathogen. Efficient cleaning and disinfection mean not just good cleaning agents and programs, but also skilled and educated workers as well as hygienically designed equipment correctly installed within the plant. Sanitary principles for food equipment manufacture are available through EHEDG guidelines (European Hygienic Engineering & Design Group) or 3-A standards in the USA.

Histamine

Histamine fish poisoning (or scombrotoxin fish poisoning) is an allergy-like form of food poisoning that continues to be a major problem in food safety. It is caused by ingestion of certain species of marine fish that contain high levels of histamine and possibly other biogenic amines. Histamine is produced from free histidine due to the action of bacterial histamine decarboxylase following time–temperature abuse. Some fish species have higher concentrations of histidine in their tissues (tuna, mackerel, mahi-mahi, sardines, anchovies) and these fish cause histamine fish poisoning in the majority of intoxications (with salmon and swordfish being exceptions) (Hungerford, 2010). Fish species that have been associated with histamine fish poisoning or elevated levels of free histidine are listed in FAO/WHO (2012). Unlike previously mentioned toxins, the histamine formation is completely related to the post-catch period and therefore histamine poisoning should be an entirely preventable condition. Histamine-forming bacteria are mainly indigenous bacteria of fish capable of growing and producing histamine over a wide temperature range; however, growth is more rapid at high temperature. Other biogenic amines produced during bacterial growth in fish may potentiate histamine's effect. Control of histamine formation includes freezing of fish, but a histamine decarboxylase will be active again after thawing. The efficient way is cooking, as thermal treatment destroys both the bacteria and the enzyme. However, histamine, if it is formed before, cannot be eliminated or destroyed by heating or any other processing technology. Histamine development will therefore be most likely formed in raw, unfrozen fish as a result of time–temperature abuse. An incident of foodborne poisoning due to ingestion of fried fish cubes occurred in Taiwan in June 2007. The incident caused 347 victims to become ill. They all suffered from allergy-like symptoms, including rash, nausea, diarrhea and flushing, but all recovered within 24 h. The use of poor quality raw fish for cooking contributed to the presence of high histamine levels in fried fish cubes and resulted in foodborne poisoning (Chen et al., 2010).

Control of temperature of fish during processing, storage and transport is the main preventive measure to assure the control of histamine production. The critical limit for histamine production is time at a given temperature. The practical recommendations on maximal time after catch before chilling (depending on the temperature of fish) can be found in an FDA guide (FDA, 2011). These recommendations are very suitable for fresh/frozen fish processing. However, some of the histamine-producing bacteria grow at lower pH, higher salt concentrations and at reduced oxygen level, which means that histamine formation can occur during processing of certain types of products. Processors should take into account the total time from catch and storage before processing, and are advised to follow careful monitoring procedures on histamine specifically at the receiving step to the plant that include checking the temperature of raw material, sensory assessment and testing histamine levels. Processors should set up critical histamine levels according to the type of processing to be applied (Köse, 2010). The risk from histamine poisoning is best controlled by applying an effective prerequisite program and, where feasible, an HACCP system. Appropriate sampling plans and testing for histamine should be used to validate the HACCP systems, verify the effectiveness of control measures and detect failures in the system (FAO/WHO, 2012). Several of the existing standards include maximum levels for histamine in different fish and fishery products. In the EU, the critical levels of histamine are different according to whether the products have undergone enzyme maturation treatment in brine or not. For the enzyme-matured products, the critical concentration of histamine is 200 mg/kg, and for simple fish products it is 100 mg/kg (EU, 2005). In the USA (FDA, 2011) a critical level of histamine at 50 mg/kg is used.

Toxins Produced by Pathogenic Bacteria

As stated before, bacteria causing intoxications need to be present in a sufficient number before producing toxins and therefore they do not present risks at the time of catch. Growth conditions of these bacteria are different and can be found in Huss et al. (2003) and FDA (2011). The main preventive measure is control of growth.

Spore-forming bacteria that produce toxins and are associated with seafood outbreaks – *Clostridium* spp. and *Bacillus* – are commonly found in soil, involved in organic matter decay and are natural inhabitants of the gastrointestinal tract of insects and many warm-blooded animal species. Most of them are bacteria from the general environment (except some types of *C. botulinum*) (Huss et al., 2003) that contaminate products during processing.

Clostridium botulinum is ubiquitous in nature and its spores are naturally present in soil and water. The bacterium produces a neurotoxin – botulin – under anaerobic, low-acid conditions. The types of bacteria pathogenic to humans (types A, B, E and F) can be divided into two groups. Group I strains (proteolytic types A, B and F), the spores of which are highly heat resistant, mesophilic, NaCl tolerant and have the general environment as the natural habitat, are frequently related to insufficiently processed home-preserved foods such as canned vegetables and cured meats. Group II strains (non-proteolytic types B, E and F) are heat sensitive, NaCl sensitive and have the aquatic environment as their natural habitat, and owing to their ability to grow at refrigerated temperatures they are a safety risk in modern industrially processed foods. These foods are processed with mild heat treatments that may allow the survival of group II spores (Huss et al., 2003; Lindström et al., 2006).

Most seafood-associated botulism cases in the USA are caused by toxin type E. Implicated seafood has been fermented under anaerobic conditions that favor the germination of *C. botulinum* (Iwamoto et al., 2010). The main preventive measures are control of growth of the bacteria by controlling the temperature, pH, oxygen or salt, or by adding preservatives. Unlike biotoxins and histamine, botulism toxin is sensitive to heat, so cooking for a sufficient time can inhibit the toxin, which can be an added measure to ensure safety.

Clostridium perfringens and *Bacillus cereus* are also spore-forming bacteria and are ubiquitous. *C. perfringens* is an anaerobe commonly found in mammalian feces and soil. *B. cereus* is an aerobic bacterium that is commonly found in soil, on vegetables and in many raw and processed foods. Spores may survive cooking, and rapid growth may occur if the food is not chilled promptly. Outbreaks are usually associated with food left at inappropriate temperatures for prolonged periods, allowing multiplication of the bacteria. Only a few reports of illness due to the presence of these microorganisms in seafood have been published (Iwamoto et al., 2010). An incident in 2007 in Spain was caused by *B. cereus* in ready-to-eat tuna (Doménech-Sánchez et al., 2011): several vomiting episodes were reported a few hours after the tuna fish consumption in a beach club. Microbiological analyses detected high bacterial levels of *B. cereus* in ready-to-eat fish samples, indicating inappropriate cooking procedures.

The important non-spore-forming, toxin-producing pathogenic bacterium in seafood is *Staphylococcus aureus.* Although *S. aureus* is a ubiquitous organism, the largest reservoir of enterotoxin-producing staphylococci is human nasal passages, but they are also found on skin, hands, wounds and cutaneous abscesses. The presence of staphylococci in cooked or processed foods can serve to indicate poor hygiene among food handlers. Freshly caught fish is generally free from this bacterium. A recent study conducted during 2008 and 2009 showed high incidence of *S. aureus* (~25%) in fish products in Spain. The incidence was highest in fresh (43%) and frozen (30%) products, but it was high in salted and smoked fish, ready-to-cook products and ready-to-eat products (Vázquez-Sánchez et al., 2012). Unlike botulin, enterotoxins produced by *S. aureus* are heat resistant. The main preventive measure to control *S. aureus* during processing is an effective prerequisite program (hygienic handling).

Additives and Allergens

The fish processing industry does not use many additives when compared to other food processing industries.

Phosphates have been used for many years by fishery products processors to reduce the loss of natural moisture in products during processing. Typical products are shrimps, frozen fish fillets or surimi. Products containing added phosphates must be declared. The EU allows processors to add up to 0.5% of phosphates in seafood products.

Nitrite is used as a preservative and color fixative in the processing of fish. Potassium nitrite inhibits the growth of *Clostridium botulinum*. It is often used in cured fish, especially in smoked products. Its level in a finished product has also been regulated.

Certain food and color additives can cause hypersensitivity reactions, or food intolerances, in some consumers. Examples of such food and color additives that are used in fish and fishery products include sulfiting agents and FD&C Yellow No. 5 (FDA, 2011). The addition of sulfite to raw prawns/shrimps is a way to control the development of a black spot (melanosis) and it has been a worldwide practice for many years. Because sulfites have become implicated in causing allergic-type reactions (hypersensitivity reactions principally in asthmatics), their concentration in food products has been limited and therefore the industry has to control their use (Hardisson et al., 2002). Control measures for allergic-type reactions that can result from the presence of sulfites or other additives are declaring their presence on the finished product. When added, sulfites must be declared on product labels at levels above 10 mg/kg. When the processor uses shrimps or prawns as its raw material, it should ask to receive a supplier's certification of the lack of sulfiting agent used and make its own test on sulfites. Yellow No. 5 is sometimes added to smoked fish to impart color. Its use should be declared on the product label (FDA, 2011).

Except for these additives, a number of foods contain allergenic proteins. Allergenic proteins are natural constituents of the food that can pose a health risk to certain sensitive individuals. There are eight foods that are defined as major allergens. These are milk, eggs, fish (cod, bass, flounder), crustacean shellfish (crab, lobster, shrimp), tree nuts, peanuts, wheat and soybeans).

Seafood allergies are among the most common types of food allergies on a worldwide basis. Tropomyosins and parvalbumins are two of the largest animal food allergen families. Tropomyosins are the major allergens of crustaceans and mollusks, and parvalbumins are abundant in the white muscle of many fish species. Allergic reactions to shellfish are more common than they are to fish. The overall prevalence of shellfish allergy in the western world (USA, Canada and Europe) is about 0.6%, ranging between 0 and 10%. Of the shellfish, prawns are most frequently implicated (62% of shellfish allergy), followed by crab, lobster and then the molluskan species. In Asia, a similar pattern of shellfish allergy is seen (Lee et al., 2012). In Europe, allergy to fish and shellfish prevails in Scandinavia and Northern Europe. In the USA, seafood allergy is reported by 2.3% of the general population, mostly to shellfish (2%) (Sicherer et al., 2004). Nevertheless, studies are lacking from some regions with high fish and seafood consumption.

Processors of foods that contain major allergens or contain proteins derived from eight major allergens must label their use. For example, the seafood industry should list the individual fish species on the food ingredient label to protect individuals sensitive to fish allergens. Major allergens must be listed on the product label even if it is used as a minor ingredient such as a flavoring. Mollusks (clams, mussels, oysters, scallops and squid) are not considered major allergens although they possess the allergenic proteins as well, so their presence in the product is not mandatory to be labeled (Taylor, 2008). Hazard analysis and HACCP plans for major food allergens and additives causing allergic-type reactions are presented in FDA (2011).

Processing Hazards (PAH, Nitrosamines)

The traditional smoking process can result in the presence of carcinogenic polycyclic aromatic hydrocarbons (PAHs) in the final product. Prevention involves modification of traditional technology. Filtered smoke and mild conditions of smoking can reduce the contamination of the products with carcinogenic PAHs (Stołyhwo and Sikorski, 2005).

Traditional smoked, fermented, salted and salt-dried products are sometimes associated with the presence of carcinogenic nitrosamines. Nitrosamines are generally formed through reactions between secondary and tertiary amines and nitrite under certain conditions, but there are many factors influencing their formation or degradation in food. To control the formation of nitrosamines, it is advisable to use good quality fish and water, limited amounts of nitrite, pure salt and an efficient prerequisite program (Köse, 2010).

Physical Hazards

Physical hazards are foreign materials which may be unintentionally introduced to fish products (like metal fragments in minced products) or naturally occurring objects (like bones in fish). Typical hazards include metal parts from the equipment, parts of packaging (glass or plastic) or hazards that are introduced to the product by inadequate (unhygienic) handling of fish by workers. These latter are prevented by the prerequisite program and the former by the implementation of the HACCP system (metal detectors installed on the processing lines are efficient in monitoring the metal parts in finished products).

RISKS AT DIFFERENT STAGES OF THE FOOD CHAIN

Aquaculture

In primary production (aquaculture), farmers should control the hazards that originate from the environment or the ones that may naturally occur in the animal at the time of catch (Table 8.2). Products from aquaculture have sometimes been associated with certain food safety issues, as the risk of contamination of products by chemical and biological agents is greater in freshwater and coastal ecosystems than in the open seas. On the other hand, some hazards are not highly probable in most farmed seafood, the presence of *Anisakis* spp., for example, and consequently it is not likely that it presents a high risk for consumers. Similar to prerequisite programs and good manufacturing or hygienic practice in processing, good aquaculture practices are a series of considerations, procedures and protocols designed to achieve efficient aquaculture production and to help ensure final product safety and environmental sustainability. Food safety risks from the products from aquaculture will of course differ from region to region and from species to species, and will vary according to the method of production, management practices and environmental conditions. In general, fish and shellfish farmers are responsible to assure the safety of their products and exclude the possibility of having bacteria (*Salmonella* spp., *Shigella* spp., *Escherichia coli*, *Vibrio* spp.; especially in mollusks to be eaten raw), parasites (trematodes and cestodes), viruses (calicivirus, hepatitis A), biotoxins, aquaculture drugs or environmental chemicals in fish/shellfish. The code of practice for fish and fishery products (CA, 2003) offers a list of potential hazards and defects as well as technical considerations for operations in aquaculture with the aim of production of safe fish/shellfish.

Processing Industry

The processing industry (secondary production) should control the farmers' or suppliers' documentation regarding hazards that originate from the environment or those that may naturally occur in the animal at the time of catch. The processing industry should focus on the control of the hazards that originate from their processing environment or may occur as a result of inadequate processing conditions or fish handling (Table 8.2). These hazards include pathogenic bacteria (*Salmonella* spp., *Shigella* spp., *Escherichia coli*, *Listeria monocytogenes*) and toxins from pathogenic bacteria (*Clostridium* spp., *Bacillus* spp., *Staphylococcus aureus*), viruses, additives, processing hazards, histamine (when processing tuna or other histamine-forming species) and physical hazards (Table 8.2). The level of risk from these hazards will depend on species and type of product. The most diverse group of hazards is pathogenic bacteria, and their multiplication or survival will depend on the pH, oxygen, water activity, temperature and presence of preservatives. The basic requirement is to chill the fish soon after catch and maintain low temperature during processing. Limiting conditions for pathogen growth are listed in a guide for fish and fishery products (FDA, 2011). The code of practice for fish and fishery products (CA, 2003) is intended as a guide to the seafood processing industry as well as to set up an HACCP program. It offers flowcharts of the production processes, a list of potential hazards and defects and technical guidance of every step of the production process. A detailed review of the hazards causing public health concerns in fish and fish products and detailed HACCP plans for different products are presented in Huss et al. (2003).

Fresh Seafood

About 47% of the fish destined for human consumption in the world today is in live and fresh form (FAO, 2012). Fresh fish is distributed as whole, whole eviscerated, filleted or sliced and transported on ice. It can be packed in vacuum or in a modified atmosphere. Processing into fillets or steaks can be automated, semi-automated or manual. Shellfish is also often traded as just harvested, sorted, washed and packed. Fresh fish and other seafood are considered highly perishable products. Basic requirements include low-temperature processing and hygienic working practice and environment.

The hazards to be controlled in mollusk processing are:

- Biological and chemical hazards originating from the marine (aquatic) environment or naturally occurring, and are present at the time of catch: bacteria, viruses, biotoxins, environmental chemicals;
- Biological hazards originating from the processing environment: bacteria and viruses.

As already stated, the hazards originating from the marine (aquatic) environment or that naturally occur in the shellfish are controlled by monitoring the harvesting area or by checking the suppliers of the mollusks. Bacteria and viruses originating from the processing environment are controlled through the requests set by the prerequisite program. Growth of pathogens is temperature dependent so the temperature during processing and storage should be controlled.

The hazards to be controlled in fresh fish processing are:

- Biological and chemical hazards originating from the marine (aquatic) environment or naturally occurring, and are present at the time of catch: parasites, biotoxins, aquaculture drugs (for farmed fish), environmental chemicals (for farmed fish and fish from coastal waters);
- Biological, chemical and physical hazards originating from the processing environment or occurring as a result of inadequate processing or handling (bacteria, viruses, histamine-only in certain fish species, metal parts, etc.).

The hazards originating from the marine (aquatic) environment or that naturally occur in fish are controlled by different mechanisms by the industry. The control of the aquaculture chemicals of farmed fish is achieved through monitoring the harvesting area and the control of the environmental chemicals in coastal-caught fish through the government-controlled monitoring programs. The freezing of fish is a control measure for the risks of parasites and if a fish has a record of causing ciguatera, it should be avoided.

The risks are different for fish that are to be eaten raw and those that are to be cooked before consumption. For example, parasites, bacteria and viruses originating from the processing environment are a lesser risk for fish to be cooked before consumption, as these pathogenic organisms are easily destroyed by elevated temperatures during cooking. The hazards originating from the processing environment or occur as a result of inadequate processing or handling are controlled through the requests set by the prerequisite program (for control of bacteria and viruses; if the fish are to be eaten raw) and by control of temperature (histamine). As stated before, the bacteria that belong to the natural microflora of fish need to grow to a great number and these are not a risk for fresh fish (Huss et al., 2003). Technical guidance for the production of safe fresh fish is also given in CA (2003).

Packaging conditions that reduce the amount of oxygen present in the package or eliminate it (e.g. vacuum packaging or modified atmosphere packaging) extend the shelf-life of a product by inhibiting the growth of aerobic spoilage bacteria. In these products, toxin formation by *C. botulinum* is a significant hazard as there is the increased potential for growth of anaerobic bacteria.

Frozen Seafood

Freezing represents the main method for preservation of seafood. Frozen products account for around 29% of total global fishery production (FAO, 2012). Freezing prevents bacterial growth and destroys parasites, and therefore is a critical step in the production of frozen seafood. However, the freezing process does not kill bacteria, so if pathogenic bacteria are present on the seafood before freezing, they may cause illness after thawing. Therefore, the raw material for freezing has to be fresh and needs to be handled in a hygienic manner. The freezing process and subsequent frozen storage are critical steps where growth of bacteria may occur. Control measures include measurements of temperature and time of freezing, and temperature and duration of frozen storage. Sometimes the technology of freezing includes addition of phosphates or sulfites, so these are also hazards that need to be controlled.

The outbreak of *Salmonella* infection associated with tuna is a recent example of an incidence with frozen fish. The incident occurred in March 2012 in the USA. Until June a total of 390 persons infected with *Salmonella* had been reported. The product was sold through distributors to restaurants; it was not available for sale to individual consumers and was to be cooked before consumption. Many of those who became ill reported eating raw tuna in sushi as "spicy tuna." A month after the first cases of illness, an inspection was conducted at the product manufacturer. Based on the initial tour of the facility, inspectors identified several seafood HACCP deficiencies such as lack of controls for histamine at receipt of product, lack of controls for *Clostridium botulinum* at storage, and an ineffective prerequisite program regarding safety of water and condition and cleanliness of food surface areas. Based on the results of the inspection of the facility and the product, the importer had to recall the product from the market.

Cured Products

Cured products are a diverse group of dried, smoked, salted, marinated and fermented foods. Their technology differs significantly among countries. The shelf-life of these products is prolonged by reduction of pH (marinated, sometimes fermented products), water activity (dried, salted products) or a combination of these factors (smoked fish). Salt content is often expressed as Water Phase Salt (WPS). The potential for microbial growth decreases as WPS increases. Water activity is also related to the microbial (and enzymatic stability) of a product and can be significantly reduced by drying or salting. In these technologies, raw material is not just the fish but also salt, sugar, acid, nitrites or different spices that should be controlled. Also, when the smoking is performed in a traditional way, there is a possibility of having elevated levels of processing hazards. Hazard analysis is more demanding in these products as they again involve the control of fish/shellfish but also all the biological, chemical and physical hazards during production. As stated before, different pathogens need different conditions for growth. As cured products have different preservative this means that they will have growth of different pathogens as significant hazards. From the food safety perspective, according to Huss et al. (2003) these products can be divided into:

- Lightly preserved fish – WPS < 6% and pH > 5.0, preservatives possible (examples: lightly salted, marinated and cold smoked products)
- Fermented fish – products which contain a carbohydrate source and in which WPS < 8% (examples: typically southeast Asian products)
- Semi-preserved fish – WPS > 6% or pH < 5.0, preservatives possible (examples: salted and marinated fish, fermented fish and caviar products)
- Dried, smoke-dried, heavily-salted fish – > 10% WPS and/or a very low water activity (a_w 0.85) (example: stock fish)

In these products, it is essential to achieve targeted levels of WPS, pH and water activity, because these are limiting conditions for growth of bacteria. Therefore, salting, marinating and drying are very critical processing steps. In the processing of all types of cured products, the significant hazards include: growth of bacteria and viruses (the species depend on the type of the product), parasites (not a significant hazard, only for dried, smoke-dried, heavily-salted fish), biotoxins (ciguatera, certain fish species), histamine (certain fish species), toxins produced by pathogenic bacteria, aquaculture drugs (farmed fish) and chemicals (for farmed fish and fish from coastal waters). HACCP plans for every type of these products are given in Huss et al. (2003). Technical guidance for the production of safe cured products can be found in CA (2003).

Cooked and Canned Products

Cooked seafood products are very often readily available. This group includes pasteurized or cooked and breaded fish fillets (which need to be cooked before consumption) or cooked shrimps and crabmeat (ready to eat). In the processing of cooked products, besides the ones which originate from the marine (aquatic) environment or naturally occur and are present at the time of catch, the significant hazards include pathogen survival during cooking/heating and pathogen growth during storage: these significant hazards that must be included in the HACCP plan. Details are given in Huss et al. (2003) and CA (2003).

Canning is a technology where the aim is to obtain a commercially sterile product stored at an ambient temperature. In the processing of canned products, the significant hazards include: survival of bacterial spores during the sterilization process, recontamination during cooling the cans, contamination during post-process handling, biotoxins (ciguatera), histamine (certain fish species), toxins produced by pathogenic bacteria, aquaculture drugs (farmed fish) and chemicals (for farmed fish and fish from coastal waters). Tuna and many other histamine-producing fish species (sardine, mackerel) are often processed to a canned product. In these productions, histamine control is very important.

Transportation and Storage

Fish and fishery products must be handled and transported by highly efficient distribution channels that can ensure that the integrity of the produce is maintained. Improvements in packaging help in preserving the quality of products. In the last few decades, major innovations in refrigeration, ice-making and transportation have also allowed the distribution of fish in fresh and other forms (FAO, 2012). The method of handling temperature during transport and storage and hygiene of a transport vehicle are of the greatest importance. Appropriate measures should be applied to minimize damage to products and also their packaging. Fish, shellfish and their products should be adequately protected against contamination from dust. Frozen products should be maintained at −18°C or below and fresh fish, shellfish and their products should be kept at a temperature as close as possible to 0°C. Potential hazards and technological guidelines that can be used to develop control measures and corrective action during transportation are given in CA (2003). Transportation vehicles should be clean and sanitized to avoid cross-contamination from the vehicle to the product.

CONCLUSION

The seafood business has been changing a great deal over the last decades in terms of continuous growth of aquaculture production, marked international trade of the products and the shifts in the consumers' preferences in developed countries which absorb most of the products. Processing is becoming more intensive, geographically concentrated, vertically integrated and linked with global supply chains (FAO, 2012). These changes pose new food safety challenges to all the parties in the seafood business.

All through the food chain, from farming (catching), processing, storing, distributing, to selling and serving, the hygiene principles should be applied. As seafood products are traded internationally, harmonization of regulations regarding hygienic requirements is needed. The requirements for the use of aquaculture drugs, for example, complicate the international trade and make the control by the authorities challenging.

There are many existing guidelines on hygienic handling and processing of seafood, but they are not always implemented in the right way by the industry. Histamine poisoning is a good example of lack of implementation of the food safety guidelines.

Another important source of seafood-borne infections particularly in recent years is *Vibrio*, a pathogen that has started to be a problem in certain geographical areas as a result of climate change. The reasons for seafood-related food safety problems are obviously different and, therefore, providing safe seafood continues to be a common challenge for the producers, authorities and researchers.

References

Allshouse, J., Buzby, J., Harvey, D., Zorn, D., 2004. United States Department of Agriculture Economic Research Service, Seafood Safety and Trade. Agriculture Information Bulletin (Number 789-7).

Amagliani, G., Brandi, G., Schiavano, G.F., 2012. Incidence and role of *Salmonella* in seafood safety. Food Res. Int. 45, 780–788.

Baker-Austin, C., Stockley, L., Rangdale, R., Martinez-Urtaza, J., 2010. Environmental occurrence and clinical impact of *Vibrio vulnificus* and *Vibrio parahaemolyticus*: a European perspective. Environ. Microbiol. Rep. 2 (1), 7–18.

CA, Codex Alimentarius, 2003. Code of practice for fish and fishery products. CAC/RCP 52-2003.

CA, Codex Alimentarius, 2010. Guidelines on the application of general principles of food hygiene to the control of pathogenic *Vibrio* species in seafood. CAC/GL 73-2010.

CA, Codex Alimentatius, 1969. Recommended international code of practice. General principles of food hygiene. CAC/RCP 1-1969.

Chen, H.-C., Huang, Y.-R., Hsu, H.-H., Lin, C.-S., Chen, W.-C., Lin, C.-M., et al., 2010. Determination of histamine and biogenic amines in fish cubes (*Tetrapturus angustirostris*) implicated in a food-borne poisoning. Food Control 21, 13–18.

Croci, L., Suffredini, E., Di Pasquale, S., Cozzi, L., 2012. Detection of *Norovirus* and *Feline calicivirus* in spiked molluscs subjected to heat treatments. Food Control 25 (1), 17–22.

CSPI, 2005. Center for Science in the Public Interest. Global and Local: Food Safety Around the World. Washington, DC. Available at: <http://cspinet.org/new/pdf/global.pdf>.

Dickey, R.W., Plakas, S.M., 2009. Ciguatera: a public health perspective. Toxicon doi:10.1016/j.toxicon.2009.09.008.

Doménech-Sánchez, A., Laso, E., Pérez, M.J., Berrocal, C.I., 2011. Emetic disease caused by *Bacillus cereus* after consumption of tuna fish in a beach club. Foodborne Pathog. Dis. 8 (7), 835–837.

EFSA, 2010. Scientific opinion on risk assessment of parasites in fishery products. EFSA J. 8 (4), 1543.

EFSA, 2012a. The European Union summary report on trends and sources of zoonoses, zoonotic agents and food-borne outbreaks in 2010. EFSA J. 10 (3), 2597.

EFSA, 2012b. Scientific Opinion Scientific Opinion on Norovirus (NoV) in oysters: methods, limits and control options. EFSA J. 10 (1), 2500.

EU, 2004. Corrigendum to Regulation (EC) No. 853/2004 of the European Parliament and of the Council of 29 April 2004 laying down specific hygiene rules for food of animal origin. Off. J. Eur. Union L 226, 22–82.

EU, 2005. Commission regulation (EC) No. 2073/2005 of 15 November 2005 on microbiological criteria for foodstuffs. Off. J. Eur. Union 338, 1–25.

FAO, Food and Agriculture Organization, 2004. Marine biotoxins. FAO Food and Nutrition Paper 80, Rome.

FAO, Food and Agriculture Organization, 2009. Guidelines for risk-based fish inspection. FAO Food and Nutrition Paper 90, Rome.

FAO, Food and Agriculture Organization, 2012. The state of world fisheries and aquaculture. ISBN 978-92-5-107225-7, Rome.

FAO/WHO, 2005. Risk assessment of *Vibrio vulnificus* in raw oysters: interpretative summary and technical report. Microbiological Risk Assessment Series No. 8. Rome, Italy: FAO.

FAO/WHO, 2007. Evaluation of certain food additives and contaminants. Sixty-seventh report of the Joint FAO/WHO Expert Committee on Food Additives. WHO Technical Report Series No. 940. Rome, Italy: FAO.

FAO/WHO, 2011a. Risk assessment of *Vibrio parahaemolyticus* in seafood: interpretative summary and technical report. Microbiological Risk Assessment Series No. 16. Rome, Italy: FAO.

FAO/WHO, 2011b. Impact on seafood safety of the nuclear accident in Japan. (Available at: <http://www.iaea.org/newscenter/focus/fukushima/seafoodsafety0511.pdf>.

FAO/WHO, 2012. Joint FAO/WHO expert meeting on the public health risks of histamine and other biogenic amines from fish and fishery products. Rome, Italy: FAO. (Available at: <http://www.fao.org/fileadmin/user_upload/agns/news_events/1_FAO-WHO_Expert_Meeting_Histamine.pdf>).

FDA, 2011. Fish and Fishery Products Hazards and Controls Guidance, fourth ed. US Department of Health and Human Services Food and Drug Administration.

Groth, E., 2008. Eating too much high-mercury fish. Mercury Policy Project. (Available at: <www.mercurypolicy.org>).

Gudbjörnsdóttir, B., Suihko, M.-L., Gustavsson, P., Thorkelsson, G., Salo, S., Sjöberg, A.-M., et al., 2004. The incidence of *Listeria monocytogenes* in meat, poultry and seafood plants in the Nordic countries. Food Microbiol. 21 (2), 217–225.

Hardisson, A., Rubio, C., Frías, I., Rodríguez, I., Reguera, J.I., 2002. Content of sulphite in frozen prawns and shrimps. Food Control 13 (4–5), 275–279.

Hungerford, J.M., 2010. Scombroid poisoning: a review. Toxicon 56 (2), 231–243.

Huss, H.H., Ababouch, L., Gram, L., 2003. Assessment and Management of Seafood Safety and Quality. FAO Fisheries Technical Paper 444. Rome, Italy: FAO.

Huss, H.H., Jørgensen, L.V., Vogel, B.F., 2000. Control options for *Listeria monocytogenes* in seafoods. Int. J. Food Microbiol. 62 (3), 267–274.

Iwamoto, M., Ayers, T., Mahon, B.E., Swerdlow, D.L., 2010. Epidemiology of seafood-associated infections in the United States. Clin. Microbiol. Rev. 23 (2), 399–411.

Japan Food Poisoning Statistics, 2009. Inspection and Safety Division, Department of Food Safety, Pharmaceutical and Food Safety Bureau, Ministry of Health Labour and Welfare. (Available at: <http://www.mhlw.go.jp/english/topics/foodsafety/poisoning/dl/Food_Poisoning_Statistics_2009.pdf>.

Köse, S., 2010. Evaluation of seafood safety health hazards for traditional fish products: preventive measures and monitoring issues. Turk. J. Fish. Aquat. Sc. 10, 139–160.

Lawrence, J., Loreal, H., Toyofuku, H., Hess, P., Karunasagar, I., Ababouch, L., 2011. Assessment and management of biotoxin risks in bivalve molluscs. FAO Fisheries and Aquaculture Technical Paper 551. Rome.

Lee, A.J., Gerez, I., Pei-Chi Shek, L., Lee, B.W., 2012. Shellfish allergy – an Asia-Pacific perspective. Asian. Pac. J. Allergy. Immunol. 30, 3–10.

Lima dos Santos, C.A.M., Howgate, P., 2011. Fishborne zoonotic parasites and aquaculture: a review. Aquaculture 318, 253–261.

Lindström, M., Kiviniemi, K., Korkeala, H., 2006. Hazard and control of group II (non-proteolytic) Clostridium botulinum in modern food processing. Int. J. Food Microbiol. 108 (1), 92–104.

Llobet, J.M., Falcò, G., Casas, C., Teixidò, A., Domingo, J.L., 2003. Concentrations of arsenic, cadmium, mercury and lead in common foods and estimated daily intake by children, adolescents, adults and seniors of Catalonia, Spain. J. Agric. Food Chem. 51, 838–842.

Lyhs, U., 2009. Microbiological methods. In: Rehbein, H., Oehlenschläger, J. (Eds.), Fishery Products: Quality, Safety and Authenticity. Wiley-Blackwell, Oxford, UK, pp. 318–348.

Lynch, M., Painter, J., Woodruff, R., Braden, C., 2006. Surveillance for foodborne-disease outbreaks – United States, 1998–2002. MMWR Surveill Summ 55 (10), 1–42.

Martinez-Urtaza, J., Bowers, J.C., Trinanes, J., DePaola, A., 2010. Climate anomalies and the increasing risk of *Vibrio parahaemolyticus* and *Vibrio vulnificus* illnesses. Food Res. Int. 43, 1780–1790.

Miettinen, H., 2006. *Listeria Monocytogenes* in Fish Farming and Processing. Dissertation, University of Helsinki, Finland.

NSBO, 2010. Netherlands Business Support Office (NBSO). An overview of China's aquaculture in Dalian. (Available at: <http://www.zakendoeninchina.org/binaries/content/assets/postenweb/c/china/zaken-doen-in-china/import/kansen_en_sectoren/agrofood/rapporten_over_agro_food/an-overview-of-chinas-aquaculture>.

OzFoodNet, 2009. Monitoring the incidence and causes of diseases potentially transmitted by food in Australia: annual report of the ozfoodnet network. The OzFoodNet Working Group. (Available at: <https://www.health.gov.au/internet/main/publishing.nsf/content/cda-cdi3404-pdf-cnt.htm/$FILE/cdi3404b.pdf>.

RASFF, The Rapid Alert System for Food and Feed, 2010. Annual Report 2009, Health and Consumers Directorate-General, European Commission.

RASFF, The Rapid Alert System for Food and Feed, 2011. Annual Report 2010, Health and Consumers Directorate-General, European Commission.

Sicherer, S.H., Muñoz-Furlong, A., Sampson, H.A., 2004. Prevalence of seafood allergy in the United States determined by a random telephone survey. J. Allergy. Clin. Immunol. 114 (1), 159–165.

Stołyhwo, A., Sikorski, Z.E., 2005. Polycyclic aromatic hydrocarbons in smoked fish – a critical review. Food Chem. 91, 303–311.

Storelli, M.M., 2008. Potential human health risks from metals (Hg, Cd, and Pb) and polychlorinated biphenyls (PCBs) via seafood consumption: estimation of target hazard quotients (THQs) and toxic equivalents (TEQs). Food Chem. Toxicol. 46 (8), 2782–2788.

Taylor, S.L., 2008. Food toxicology. In: Metcalfe, D.D., Sampson, H.A., Simon, R.A. (Eds.), Food Allergy: Adverse Reactions to Foods and Food Additives, fourth ed. Blackwell Publishing Ltd, Oxford, UK, pp. 498–507.

Toda, M., Yamamoto, M., Uneyama, C., Morikawa, K., 2009. Histamine food poisonings in Japan and other countries. Kokuritsu Iyakuhin Shokuhin Eisei Kenkyusho Hokoku 127, 31–88.

UNEP, 2002. United Nations Environment Programme (UNEP), Global Mercury Assessment. UNEP Chemicals Mercury Programme. (Available at: <www.chem.unep.ch/mercury/Report/Final%20Assessment%20report.htm>.

Vázquez-Sánchez, D., López-Cabo, M., Saá-Ibusquiza, P., Rodríguez-Herrera, J.J., 2012. Incidence and characterization of *Staphylococcus aureus* in fishery products marketed in Galicia (Northwest Spain). Int. J. Food Microbiol. 157 (2), 286–296.

WHO, 1995. World Health Organization. Control of foodborne trematode infections. WHO Tech. Rep. Ser., 849.

WHO, 1999. World Health Organization. Food safety issues associated with products from aquaculture. WHO Technol. Rep. Ser., 883.

WHO, 2004. World Health Organization. Report of Joint WHO/FAO workshop on food-borne trematode infections in Asia. Ha Noi, Vietnam, 26–28 November, WHO, WPRO, 1–58.

WHO, 2008. World Health Organization. Guidance for Identifying Populations at Risk from Mercury Exposure. WHO, Geneva.

CHAPTER

9

Fruits and Vegetables (including Herbs)

Vlasta Pilizota

University of Josip Juraj Strossmayer in Osijek, Osijek, Croatia

OUTLINE

Food Safety Management.
DOI: http://dx.doi.org/10.1016/B978-0-12-381504-0.00009-3

INTRODUCTION

Among the agricultural products for which there is a continuing and expanding demand by markets are fruits, vegetables and herbs.

Fruits, vegetables and, at a certain level, herbs play a significant role in human nutrition by supplying nutrients such as vitamins, minerals and dietary fiber. Plant foods, fruits, vegetables and herbs contain an immense variety of biologically active, non-nutritive secondary metabolites known as phytochemicals that have disease-fighting properties (Jongen, 2002). Some of those phytochemicals, such as polyphenols, pigments (e.g. carotenoids) and glucosinolates, may have nutritional value. While many fruits and vegetables are consumed primarily in their fresh state many produce are also consumed to a significant degree in their processed state.

Herbs, the leafy plant parts (e.g. parsley, basil, oregano, mint), usually referred to as herbs in European and North American cuisines (Raghavan, 2007), are seasonings of vegetative origin, commonly used as a food ingredient in a fresh and dried form, in both the commercial and domestic setting.

Fruit, vegetable and herb production and processing involve a complex supply chain from the farm to the point of consumption. From many points of view it is of great importance to strengthen each link in the chain and improve the integration of the supply chain as a whole if high quality and safety of produce have to be maintained (Jongen, 2002).

International trade of fresh fruits, vegetables and herbs is a billion dollar business that has significantly increased in the last decade. This trade is important also from a safety viewpoint, and faces (primary) producers, manufacturers, distributors and retailers who place a great deal of value on their reputation for marketing naturally delicious and nutritious products.

This chapter presents both a very short description of the main hazards that could contaminate fruits, vegetables and herbs, and an overview of their risk and the possible measures to control them. It focuses on their management using the HACCP approach.

The main objective of this chapter is to provide a problem-oriented look at fruit, vegetable and herb contamination, its avoidance, and how to manage food safety in the whole food chain.

FOOD SAFETY

Food safety is an assurance that food will not cause harm to the consumer when it is prepared and/or eaten according to its intended use (CAC, 2003a). Safety is one of the most important specifications of all products. Safety of fruits and vegetables (including herbs) starts with agriculture and primary production. One major factor leading to food contamination during food preparation and storage is time–temperature abuse, which results in the survival, growth and production of toxins by pathogens.

Each year in many countries, foodborne illnesses cause sickness, death and accompanying economic costs that can cripple companies and erode public trust in the safety of the food supply.

In recent decades, microbial safety has become a concern and a series of large foodborne outbreaks has occurred around the world. The situation is more alarming due to a number of population explosion, urbanization and changes in lifestyle, consumption of, so-called, minimal processed ready-to-eat foods, international trade in food and animal feed, international tourism and immigration, and a short supply of potable drinking water (Varadaraj, 2010).

HAZARDS ASSOCIATED WITH FRUITS AND VEGETABLES (INCLUDING HERBS)

Consumption of raw fruits, vegetables ("fresh produce") and herbs has increased worldwide due to nutritional awareness and promotions by national dietary health programs in many countries.

Fruit, vegetable and herb contamination problems can occur in the growing environment, after harvest, during preparation for storage and processing, shipping to the market and in the home.

Types of Hazards

In recent years, there has been a growing awareness and concern about the potential risk of all kinds of hazards. Many hazards may enter the food supply, making the food potentially harmful when consumed. They can cause injury or illness in the absence of their control. These foodborne hazards could be divided into three categories: biological, chemical or physical. Conditions of food handling or processing can also be the source of hazards as they may lead to the survival or growth of pathogens. As food products could be produced in different geographical locations and climates, throughout the world, they may be exposed during the growth to hazardous environmental contaminants present in the local region.

Biological Hazards

Biological hazards include pathogenic fungi, bacteria, viruses, prions, protozoans and helminthic parasites, namely certain trematodes. These could cause foodborne illness due to pathogen–host interaction. Fruits and vegetables normally carry a non-pathogenic epiphytic microflora. However, there are certain sources/factors that contribute to the microbiological contamination of these products with pathogens, all of which must be controlled (McDowell et al., 2007). Contamination can arise as a consequence of treating soil with untreated manure and sewage sludge and from irrigation water. Additionally, handling and the application of technologies such as cutting, slicing, skinning and shredding (in the case of minimally processed fruits and vegetables) will remove the natural protective barriers of the intact plant and open the possibility for providing a suitable medium for the growth of contaminating microorganisms (EU EC, 2002).

Pathogens most commonly associated with fruits and vegetables include *Salmonella, Shigella, Escherichia coli* O157:H7, *Listeria, Campylobacter, Cryptosporidia*, viruses such as hepatitis A and parasites such as *Entomoeba histolytica, Giardia lamblia, Cyclospora cayetanensis, Cryptosporidium parvum, Toxoplasma gondii*, and certain trematodes (*Fasciola hepatica* associated with watercress). In general, it should be reckoned that any microbial agent transmitted through the fecal–oral route, such as some of those mentioned above or *V. cholerae* or rotavirus, is also of relevance to fruits and vegetables. The numbers of bacteria present in fruits and vegetables will vary depending on seasonal and climatic variation and may range from 10^3 to 10^8 per gram. Shredding and slicing were found to increase counts of mesophilic bacteria from 10^3–10^4 to 10^5–10^6 CFU g^{-1} for a range of vegetables (Francis et al., 1999). Accelerated growth and spoilage occurs due to increased nutrient availability and larger surface areas for microbial growth. In contrast with bacteria, many different yeast species of comparable quantitative importance have been identified in minimally processed vegetables, including species of *Candida, Cryptococcus, Rhodotorula, Trichosporon, Pichia* and *Torulaspora* (FAO, 2007). In the case of herbs, most problems are connected with mold, high moisture contents and aflatoxin contents (FAO, 2008).

Chemical Hazards

Chemical hazards may appear in food products either by natural occurrence (e.g. naturally occurring toxin in certain mushrooms, solanine in potatoes) in a raw material or by deliberate or unintentional addition during primary production and/or processing. They could result from a number of sources: the environment such as heavy metals or radionuclides; agricultural chemicals such as insecticides, fungicides; packaging materials; cleaning/sanitizing agents; certain toxins; and misuse of food chemicals (e.g. additives) (Wallace et al., 2011).

Fruits and vegetables (including herbs) are prone to chemical contamination that can occur under growing, harvesting or post-harvest conditions and can result from deliberate exposures, such as pesticide application, or unintentional exposures, such as those resulting from fungal contamination (McDowell et al., 2007). Control of chemical hazards along the food chain is of primary importance as the consumer has no opportunity to reduce them substantially during food preparation.

Physical Hazards

Physical hazards are commonly called "foreign materials" or "foreign bodies" because their presence in food is unnatural. They include: inadvertent field matter (stones, wood, metal, pieces of bone, insect fragments, etc.); inadvertent processing residues (glass, metal fragments, pieces of plastic, personal objects, etc.); intentional materials (employee sabotage) and miscellaneous particulates and fragments (Schmidt et al., 2008); or food itself (pits, stones, stems) – these may enter foods at almost every point of the food supply. Some foreign matters may *per se* not be a hazard but their finding in food may be indicative of poor hygienic practice and a distressing event for consumers.

FACTORS AFFECTING BIOLOGICAL CONTAMINATION

Fresh fruits, vegetables and herbs normally carry a non-pathogenic epiphytic microflora but may be contaminated with foodborne pathogens, which makes their microbial safety of prime importance. Regardless of the production system used, they are grown in environments that have a wide range of accidental or intentional inputs that are potential sources of microbial foodborne hazards and may lead to contaminated produce. An increase in their number at any stage of post-harvest operations will lead to exposure of consumers, and therefore prevention of contamination throughout the supply chain is essential. Contamination can be transiently present on the surface of fresh produce, and/or microorganisms can become internalized in fruits and vegetables by penetrating deeper tissues through damaged sites on the surface of these products.

Foodborne Diseases

Factors contributing to outbreaks of human infections associated with consumption of raw and minimal processed fruits and vegetables (and at a certain level, herbs) may include changes in agronomic and processing practices, an increase in per capita consumption of raw or minimally processed fruits and vegetables, increased international trade and distribution, and an increase in the number of immunocompromised consumers (Beuchat, 2002). The presence of unwanted contaminants increases the risk of illness for those consuming the produce.

The incidence of foodborne diseases around the world has been recorded (Table 9.1) in both developing and developed countries likely due to the globalization of food supply and

TABLE 9.1 Reported Food Poisoning Incidents (Lynch et al., 2009; Wallace et al., 2011; Marler, 2012)

Year	Pathogen	Cases	Regions	Food
2011	*E. coli* O26	14	North America	Raw clover sprouts
2011	*E. coli* O104:H4	4000+	Europe	Fenugreek seeds
2008	*Salmonella*	1442	North America	Fresh peppers, tomatoes
2007	*Salmonella*	51	Europe, North America	Fresh basil
2007	*Shigella*	175	Australia, Europe	Alfalfa sprouts
2007	*Salmonella*	45	Europe	Alfalfa sprouts
2006	*E. coli* O157:H7	206	North America	Fresh spinach
2006	*Salmonella*	20+	Europe	Arugula

trade. Trends toward greater geographic distribution of fruits, vegetables and herbs from central processing facilities and subsequent storage and handling practices in food preparation areas may also be contributing to an increased frequency of produce-associated infections. Many large outbreaks involving widely consumed commodities such as apple cider, cantaloupe, raspberries, bagged lettuce and spinach, tomatoes, green onions and sprouts have been reported during the past decade (Beuchat, 2002). Most of the reported outbreaks have been associated with bacterial contamination, particularly members of the Enterobacteriaceae. Of these, *Salmonella* and *Escherichia coli* O157 in sprouted seeds, tomatoes and fruit juices are of particular concern. The viruses involved in outbreaks have a human reservoir (e.g. Norwalk-like and hepatitis A) and can be associated with intact products grown in contact with the soil and/or water. Outbreaks linked to protozoa (e.g. *Cryptosporidium, Cyclospora, Giardia*) have been associated more with fruits than with vegetables. Protozoa and viruses are most often associated with contaminated water or food handlers. The natural microbial load depends to a great extent on the type of commodity, environmental considerations, seasonality and the conditions under which a particular fruit, vegetable or herb is grown (EU EC, 2002).

It is estimated that foodborne diseases cause about 76 million illnesses per year in the United States alone. Known pathogens account for an estimated 14 million illnesses, 60,000 hospitalizations and 1800 deaths. More than 75% of these are caused by known pathogens, while unknown agents account for the remaining 62 million illnesses (Novak et al. 2003). In the USA, since 1999, 80% of leafy green outbreaks and 98% of illnesses have been from fresh-cut products (Harris, 2010).

Surface Characteristic

The surfaces of fruits and vegetables show a large diversity in structure and composition and present a variety of surfaces to which a bacterium may bind. Those surfaces provide a habitat for a variety of microorganisms including bacteria, yeast and molds (Mendonca, 2005). The epidermis is covered by an epicuticular wax on aerial organs (leaves, stem, flowers and fruits) or periderm on roots and tubers. Stomata, lenticels, broken trichomes and scars from detached organs represent natural ways of entry for microorganisms. Since cracks in the surface of fruits and vegetables, as well as in herbs, may occur in certain growing conditions and as post-harvest handling may cause injuries and bruising, microorganisms transferred to fresh produce can enter areas of pre-existing damage (Ukuku et al., 2005). Damage to the cuticular layer can permit microbial proliferation in cellular fluids and moisture released from the damaged sites. Sugars in released juices from damaged tissue attract insects, which can further injure fresh produce and facilitate entry of microorganisms (Ukuku et al., 2005).

Bacterial Attachment

The mechanism of attachment of bacterial cells to plant surfaces has been studied most extensively for plant pathogens. According to Fletcher (1996) bacterial adhesion occurs in three steps: reversible adsorption, primary adhesion and colonization. During the reversible adsorption phase, the bacterium is at a distance of greater than 50 nm and is affected by van der Waals interactions with the substratum. This means that the bacteria can be easily washed off at this stage. At the primary adhesion stage, the distance between the bacteria and the substratum ranges from 10 to 20 nm and the type of force affecting adhesion is electrostatic unless the opposing surface has a net surface charge, then attractive forces will come into play. The colonization step is the final phase and biofilm may be formed (Ukuku et al., 2005).

Irregularities such as roughness, crevices and pits have been shown to increase bacterial adherence by increasing cell attachment and reducing the ability to remove cells. However, preventive mechanisms should be geared towards physical or chemical treatments to prevent bacterial transfer from the surfaces of the produce to the interior flesh. The effectiveness of chlorination of wash water in reducing the population of bacteria on produce is dependent on the interval between contamination and application of the washing treatment (Sapers et al., 1999, 2001; Ukuku et al., 2001).

Bacterial Infiltration and Internalization

Bacterial adhesion to or contact with damaged or intact plant surfaces precedes entry of these organisms into fresh produce (Mendonca, 2005).

Most microorganisms on the surface of intact fresh produce are prevented from entering subsurface tissues by the cuticular layer that covers the epidermis of leaves, stems and fruits (Nguyen-The and Carlin, 2000). However, natural openings on the surface of fruits and vegetables can provide channels through which bacteria can enter these products, as well as through scars from detached organs or cracks in the surface of vegetables and fruits (including herbs) that occur in certain growing conditions (i.e. via infiltration with contaminated water in the fields), as well as during harvesting and post-harvest handling. Internalized bacteria can increase post-harvest losses or compromise the microbial safety of fruits and vegetables (Ukuku et al., 2005).

Bacterial Biofilm Formation

The ability of bacteria to produce extracellular polysaccharides on surfaces, which results in the formation of biofilm, enhance bacterial colonization and survival on plant surface and increase their resistance to cleaning and to antimicrobial agents, is well known (Ukuku et al., 2005). The biofilms appeared to originate on the cuticle in distinct micro-environments such as in the natural depression of the stomata, or in the intercellular junction. Bacteria also adhered to and developed biofilm colonies within an hour of contact and with clean stainless steel surfaces (Carmichael et al., 1998). Containers used to harvest, transport and display raw fruits and vegetables are often not effectively cleaned and sanitized, which can lead to the development of biofilms. Even single-use containers may hold produce for a sufficient time to allow the formation of biofilms. Contamination of fresh produce with pathogens may result from contact with surfaces harboring these biofilms. If pathogens attach to biofilms during transport or processing, their survival and growth may be enhanced (Ukuku et al., 2005).

CONTAMINATION ALONG THE FOOD CHAIN

Food safety is a growing concern for consumers and professionals in the food and food-service sectors. The risk of contamination can occur at any stage of crop production and processing. For that reason all actors in the food chain should be aware that hazards need to be controlled and minimized.

Pre-harvest, Harvest and Post-harvest Measures

It is in the interest of the grower, and the industry, to produce a high-quality product that will attract a premium market price. Variety selection, climatic conditions and growing practices at harvest will greatly affect the quality of fresh produce (Cantwell, 2007).

Pre-harvest operations involve the preparation of the facilities for the harvest material, which will ensure the crop is stored and processed quickly under hygienic conditions. It is of great importance that all personnel working along the whole food chain (at each step in cultivation and preparation for market) have full knowledge of good hygienic and agricultural practices and be aware of their role and responsibility in maintaining the hygiene, quality and innocuousness of the products in order to improve product safety (FAO, 2002).

Training and education of all employees with direct access (processing, storage and transport workers) and indirect access (equipment operators, buyers, pest control operators) to the production areas of fresh fruits, vegetables and herbs should be considered as a primary preventive control measure, or risk mitigation strategy (FAO, 2008; FDA, 2008).

General requirements for training are outlined in the Codex Code of Hygienic Practice for Fresh Fruits and Vegetables (CAC, 2003b, Sections 10.1 and 10.2) and are:

- Good health and hygiene for personal health and food safety.
- Hand washing for food safety and proper hand-washing techniques.
- Using sanitary facilities to reduce the potential for contaminating field, produce, other workers and water supplies.
- Techniques for hygienic handling and storage of fruits, vegetables and herbs by transporters, distributors, storage handlers and consumers.
- Shared responsibility among stakeholders: agricultural workers, government, NGOs and the media.

Pre-harvest Factors

Pre-harvest factors affecting fruits, vegetables and herbs can influence their post-harvest quality and safety (Crisosto and Michell, 2002).

Within each commodity, there is a range of genotypic variation in composition, quality and post-harvest life potential (Kader, 2002). By choosing the correct genotype for given environmental conditions the incidence and severity of decay, insect damage and physiological disorders in commodity can be reduced. Primary producers face challenges in utilizing technologies for producing high-quality crops. Farmers, scientists, extension specialists and market personnel must work together to provide knowledge, best practices and enabling tools for growers to ensure pre-harvest conditions optimized for production of high-quality horticultural crops that satisfy and reward discerning consumers (Hewett, 2006).

During growing periods sources of hazards could be: feces, soil, irrigation water, water used to apply pesticides, foliar treatments, growth hormones, inadequately composted manure, air (dust), wild and domestic animals (including fowl and reptiles), insects and human handling.

SOIL

Soil is a rich reservoir for a variety of chemical and physical hazards and non-pathogen and pathogen microorganisms (such as *Bacillus cereus*, *Clostridium botulinum*, *Clostridium perfringens*, *Listeria monocytogenes*). Other pathogenic organisms from the human/animal reservoir can be found in the soil due to irrigation and fertilization with manure and sludge or due to droppings of animals in the farming area. The contamination rate and survival of bacteria in soil appears to be dependent on several factors including soil type, prior use of land, moisture content, ultraviolet light exposure, temperature and presence or absence of a ground crop.

WATER

Water is used for irrigation of plants, chemicals application, produce washing and cooling systems. Water of inadequate quality has the potential to be a direct source of contamination as well as a vehicle for spreading localized contamination in the field and in facilities used for post-harvest processes. Water quality will vary depending on its source. Groundwater, surface water and human waste water are commonly used for irrigation. The transfer of foodborne pathogenic microorganisms from irrigation water to fruits, vegetables and herbs will depend on the irrigation technique and on the nature of the produce (NACMCF, 1999). Surface water from streams and lakes may be contaminated with pathogenic protozoa, bacteria and viruses. Wherever water comes into contact with fresh produce, its quality may directly determine the potential for persistent pathogen contamination.

The use of waste water for agricultural irrigation has been practiced for centuries, especially in arid regions with limited water resources. Using contaminated or waste water in irrigation is associated with some health risks due to the possibility of the presence of a wide spectrum of pathogens. Consequently, ensuring proper quality (extensive waste water treatment and improvement) of crop production water on site is the key to safe production of fresh fruits and vegetables. Although standards for the use of reclaimed waste water exist for food eaten raw, irrigation using reclaimed water for crop irrigation is seldom practiced. (The United States Environmental Protection Agency guidelines for surface water recommend fewer than 1000 fecal coliforms/100 ml of surface water, including river water, for irrigation of crops.) Growers should identify the sources of water used for a particular purpose and minimize contamination from livestock, run-off, heavy rainfall and excess irrigation. It is also recommended that the microbial and chemical quality of water is tested at appropriate intervals.

FERTILIZERS

Sewage, animal manure, slurry, sludge and compost of human and animal origin may be used as fertilizers for fruit, vegetable and herb production particularly in organic production systems. Considering the source of these fertilizers, there is potential for contamination with fecal-orally transmitted pathogens.

Pathogens associated with manure (e.g. *L. monocytogenes*, members of the Enterobacteriaceae like *Salmonella*, *Shigella*, *Yersinia*, *E. coli*, as well as *Campylobacter*) may survive for extended periods, and while there has been a substantial amount of research in this area, uncertainties regarding pathogen behavior remain.

Potential risks can be significantly reduced by treatment procedures in order to reduce the potential pathogenic microbial load in manure or biosolid waste. Measures that can help kill pathogens that are present in manures and slurries include: exposure to sunlight and ultraviolet rays, high temperatures (above 55°C), low acid or high alkaline conditions (use of quick lime or slaked lime to raise pH levels), drying and the passage of time (bacteria such as *E. coli* can survive in soil for several months) (EU EC, 2002). Also, potential risks can be significantly reduced by prevention of direct or indirect contact between organic fertilizers and produce.

PLANT PROTECTION PRODUCTS

Chemical biocides (herbicides, nematicides, insecticides and pesticides) are in general used for protection of plants against pests and plant diseases. Even though substances authorized for this purpose have undergone extensive safety evaluations, there is consumer concern about their need and safety, which has stimulated the development of alternative control methods.

The use of microorganisms for biocontrol is a scientific field where limited knowledge exists on the potential risk for the consumers at the time of consumption. A wide range of microorganisms are used in biological control. Also, bioactive crystalline protein of some strains of microorganisms has been used for the control of insects. In some countries certain antibiotic substances are used for plant protection (EU EC, 2002).

The metabolism of agrochemicals in plants has never been greater. In a world where food safety and environmental concerns are increasing, knowledge of the metabolic processes within plants and the terminal residues of agrochemicals in food crops are invaluable.

Residues of agrochemicals in foodstuffs, water (and animal feeds) are regulated by the establishment of maximum residual levels (MRLs) or tolerances. These levels represent the maximum level of an active substance and relevant metabolites that can legally be present in or on the food, water (or animal feed). MRLs are set on the basis of supervised trials in which GAP is observed and must not pose an unacceptable risk to human health. In practice, these levels are also subject to the requirements of international trade with agricultural products.

International regulations on MRLs, e.g. of pesticides in food, now cover hundreds of individual components (there are currently over 1000 recognized pesticides utilized that can be grouped into more than 40 classes of chemical families) at levels down to parts per billion (ppb). Legislation should stipulate that the buyer (i.e. the producer) is held responsible for each gram or milliliter of the product. The purpose of the legislation is to protect people (including the producer) and the environment against pollution by agrochemicals. National and international legislation (e.g. EC Regulation No. 1107/2009) on uptake and metabolism of agrochemicals in crops and farm animals is a basis for the assessment of consumer risks.

Distribution of plant protection products in plants is not uniform but depends on the kind of active substance and the route of penetration into the plant. Types and amounts of metabolites are influenced by the uptake, distribution and time of persistence of the product in the plant. Plants generally metabolize plant protection products (in most cases in several stages) to water-soluble conjugate compounds (down to non-extractable residues) and bound residues which stay in the plant. Complete degradation pathways are known for only a few of the active substances currently in use (EU EC, 2011).

Harvesting

Harvest, at the mature stage of commodity, marks the end of the growing period and the commencement of market preparation or conditioning for fresh or processed products (FAO, 2002). Harvesting can be performed by hand or mechanically, and involves a number of other activities undertaken in the field. This includes those of commercial interest: pre-sorting, removal of foliage and other non-edible parts, and others. Mechanical harvesting is recommended for produce that can readily withstand physical handling (i.e. carrots, potatoes and radishes). It is generally used to harvest produce destined for the processing industry. For commodities destined for the fresh market (lettuce, berries, grapes, peppers, apples, etc.), which can be damaged easily, integrity and appearance are important. Therefore, manual harvesting is widely used for these products. With manual harvesting, personal hygiene is particularly important since there is a great deal of handling that could lead to contamination of the product. Proper hygiene during harvesting (e.g. handling of tools) is also critical to product safety (EU EC, 2002).

Fruits, vegetables and herbs can become contaminated with pathogenic microorganisms through fecal material, human handling, harvesting equipment, transport containers, wild and domestic animals, air, transport vehicles, ice or water (Beuchat, 1995).

The first washing of vegetables at harvest that removes much of the adhering soil and dirt, however, could be a source of microbial contamination. Whenever produce is dumped into water or washed with recirculated water that is not maintained properly, there is a good chance that contamination will occur. It is useful to confirm the absence of pathogens in processing water. Clean, well-designed and maintained equipment is less likely to cause damage to fresh produce and to introduce spoilage and pathogenic microorganisms (Brackett, 1992). Dirty storage facilities and the presence of rodents, birds and insects may increase the risk of contamination with foodborne pathogens (FDA, 1998). Harvesting at the appropriate time and keeping the harvested product under controlled environmental conditions (cool storage) will help retard growth of post-harvest spoilage and pathogenic microorganisms (Brackett, 1992). It is important that hygienic practices are followed throughout the processing of fresh produce and that raw materials and finished product are stored and handled in such a manner as to prevent contamination and damage which may lead to internalization of organisms. It is also critical that the temperature of processing is controlled to prevent product spoilage and also to prevent the growth of pathogens.

Post-harvest Factors

Post-harvest operations can be very varied, from simple open-air packing to cleaning, trimming, in some cases washing, drying, waxing, packing, transportation and storage. During these practices conditions may arise which lead to cross-contamination of the produce from other agricultural materials, equipment, facilities or from the workers. Environmental conditions and transportation time will also influence the quality and safety of the produce prior to processing or consumption. The potential for contamination may be enhanced when fruits or vegetables have fallen from the plant to the ground and are picked and placed into the handling and processing chain. Animals are the primary reservoir for the pathogenic organisms associated with outbreaks of *E. coli* O157:H7 infection and cryptosporidiosis. In particular, cattle, deer and sheep can asymptomatically carry *E. coli* O157:H7 and *Cryptosporidium*, and many animals, including cattle, chickens and pigs (when inadvertently entering fields), can asymptomatically carry *Salmonella*.

Also, contamination may occur when improperly composted manure has been applied as a fertilizer. Because contaminated manure may become airborne dust particles it is possible that fruits on trees and vines may become contaminated. These mechanisms of contamination are somewhat speculative at present and must be thoroughly investigated before appropriate interventions can be introduced to reduce the risk.

Poor handling (during sorting, packaging and transport) can damage fresh produce, rendering the product susceptible to the growth/survival of spoilage and pathogenic microorganisms. The presence of cut and damaged surfaces provides an opportunity for contamination and growth of microorganisms and internalization into plant tissues (Francis et al., 1999).

Patulin, a toxic fungal metabolite (mycotoxin) produced by certain molds of the genera *Penicillium*, *Aspergillus* and *Byssochlamys* growing on certain food commodities, especially fruit (e.g. apples and pears), is of concern from a food safety perspective. Patulin exhibits a number of toxic effects and its presence in food is undesirable. Patulin occurs most often in apples that have been spoiled by mold growth (e.g. after storage at room temperature and at 1°C or if stored at room temperature following 4–5 months' storage at 4°C), or in products made from spoiled apples, such as apple juice, pies and conserves. Contaminated apple juice usually contains patulin at levels below 50 μg/l, but much higher levels (up to 4000 μg/L) have been reported occasionally. At harvest, damaged and rotten fruits should be discarded, as these are much more likely to contain patulin. It has also been found in pears and grapes, as well as in vegetables. Although patulin is now considered to be a less significant food safety hazard than previously, a number of countries have introduced regulations specifying maximum permitted levels in susceptible products. (The Codex Alimentarius Commission has also set a recommended upper limit of 50 μg/kg for patulin in apple juice and apple ingredients in other beverages.)

Storage and Handling

To ensure safety and to obtain the maximum benefit from fresh or processed fruit, vegetables and herbs, careful storage, handling and preparation are necessary. The aim of proper post-harvest storage is to extend and ensure shelf-life of raw materials. Storage rooms can be grouped accordingly as those requiring refrigeration and those that do not (FAO, 2002). Since fresh produce is alive and respiring (i.e. enzymatically converting sugars and acids in the presence of oxygen to carbon dioxide and heat) there is a need for it to be cooled (maintenance of the cold chain minimizes the growth of bacteria). The desired environment can be obtained in facilities where temperature, air circulation, relative humidity and sometimes atmospheric composition can be controlled. The selected method of cooling will depend greatly on the anticipated storage life of the commodity.

After harvesting, and before cooling, fruits and vegetables are precooled (temperature is reduced) ready for cold storage or safe transport. Precooling may be done with cold forced air, cold water (hydrocooling), direct contact with ice or by evaporation of water from the product under a partial vacuum (vacuum cooling). A combination of cooled air and water in the form of a mist called hydraircooling is an innovation in cooling vegetables.

Water quality is important in reducing contamination during post-harvest cooling, washing and disinfection operations. Water used for post-harvest operations should be potable and free from disease-causing organisms. Post-harvest water can become contaminated easily and it quickly becomes saturated with organic matter (e.g. soil, materials leaching from the fruit, etc.), therefore, procedures to ensure good water quality are critical. These include frequent filtering, regular changing of wash water and the use of disinfectants (FDA, 1998).

It is important for most fruits and vegetables to keep them within their optimal ranges of temperature and relative humidity, air circulation, under conventional, controlled (CA) or modified atmosphere (MA), to maintain their quality and safety and minimize post-harvest losses. Although a selected low temperature (depending on the product) should be maintained throughout shelf-life, fruits and vegetables can still spoil, as a consequence of fungal attacks.

The distribution chain rarely has the facilities to store each commodity under ideal conditions and requires handlers to make compromises as to the choices of temperature and relative humidity. These choices can lead to physiological stress and loss of shelf-life and quality. For long-distance distribution the use of refrigerated transport or the use of other coolants is necessary. The weakest two links in the post-harvest handling cold chain of fresh produce are the retail and home handling systems (FAO, 2002).

In the case of herbs the extent of post-harvest handling depends on the demands of the market. Some crops are sold fresh, and maintenance of the cold chain is important, while others require various levels of processing. The most common method used to protect the herbs from spoilage is drying.

Packing

The purpose of food packaging is to protect against food pathogens, spoilage organisms, pests, tampering, damage, etc. In some cases, the raw agricultural product is completely prepared for the market in the field. Sometimes, preparatory treatments include cleaning, disinfecting, waxing and adding color (even brand name stamping on individual fruits) to improve appearance and maintain quality (FAO, 2002). After sorting and classifying, produce should be carefully packed to achieve uniformity and to prevent damage (compression, scrapes, etc.) which causes decay and inferior quality. Good hygienic practices should be followed in handling containers and packing materials to prevent product contamination.

The demand for year-round supplies at ever higher quality standards by both the processing industry and retail sector is driving the development of new technical and managerial strategies. Although refrigeration throughout the cool chain is likely to remain the most important technology for maintaining product quality, a broader range of approaches are increasingly in use, such as MA during transport, storage and in individual produce packages.

Risks with fruits and vegetables lie with initial contamination. Growth of pathogens is of lesser concern as normally by the time pathogenic organisms can develop fruits and vegetables will spoil and will become undesirable.

There is, however, evidence that sealing fresh fruits, vegetables and herbs in modified atmosphere packaging (MAP) may extend shelf-life, while still allowing the growth of pathogenic bacteria, in particular *Listeria* spp. and *Escherichia coli* O157 (Phillips, 1996).

Processing

Growth in demand has led to increased marketing of fresh horticultural products in lightly processed form (Cantwell, 2002). While many fruits and vegetables are consumed primarily in their fresh state (minimally processed, ready-to-eat produce), many products are also consumed to a significant degree in their processed state.

Risk associated with preservation or failures in processing are treated in other parts of the book (e.g. thermal treatment processing).

Minimal Processing

Whereas most food processing techniques stabilize the products and lengthen their storage and shelf-life, light processing of fruits and vegetables increases their perishability (Cantwell, 2002).

The term "minimally processed" refers to raw fruits and vegetables that have been lightly processed. Early terminology referred to "minimally processing," which was described as handling, preparation, packaging and distribution of agricultural commodities in a fresh-like state (Shewfelt, 1987). "Fresh-cut" refers to raw fruits and vegetables that have been cut, shredded, peeled, abraded or otherwise prepared to produce ready-to-eat or ready-to-cook portions. However, the key criteria for fruit and vegetable products to be considered "fresh-cut" are that they consist of 100% usable material and that the tissue is in a living, respiring, physiological stage (IFPA, 2001). The value of fresh-cut produce lies in the primary characteristics of freshness and convenience. Food safety, nutrition and sensory quality are required while providing extended shelf-life and freshness. Fresh-cut produce is a safe, wholesome food when produced under GAPs, GMPs and sanitation procedures.

CONTROL MEASURES IN PREVENTING CONTAMINATION

Knowledge of the nature of fruits and vegetables (including herbs) as they relate to pre- and post-harvest handling, processing, packaging and storage are essential for ensuring their wholesomeness and nutritional value, and for developing the most effective procedures and innovative technologies for maintaining their quality and safety (Lamikanra, 2002).

Cleaning and Washing

Cleaning and washing are often the only preservation treatments applied to raw agricultural commodities and at a certain stage minimally processed fruits, vegetables and herbs. As the first step in processing, cleaning is a form of separation concerned with removal of foreign materials like twigs, stalks, dirt, sand, soil, insects, pesticides and fertilizer residues from raw material, as well as from containers and equipment. The cleaning process also involves separation of light from heavy materials via gravity, flotation, picking, screening, dewatering and others (Wiley, 1997). However, many existing methods of cleaning and disinfecting fresh produce are incapable of achieving reductions in pathogen levels sufficient to ensure product safety. The method of treatment for cleaning depends on the ability of produce to tolerate water. Soft tissue and delicate commodities with large water-adhering surface areas such as berries and grapes do not tolerate water. They are "dry cleaned" using air blowers or vacuum methods (FAO, 2002).

Sanitation of whole fruits is conducted generally with an initial rinse in tap water to eliminate pesticide residues, plant debris and other possible contamination, followed by a dip in chlorinated water (50 to 150 ppm of added free chlorine are commonly used, and at pH below 8, usually at 6.0 to 7.5 for effective disinfection without damaging equipment surfaces) to reduce effectively the microbial loads on the raw material surface. Chlorine is normally used to disinfect the fruit surface by adding sodium hypochlorite (NaOCl) to the wash water. However, there are some health concerns related to the use of chlorine in disinfection of fruits and vegetables because of its potential reaction with some organic compounds and formation of potentially mutagenic or carcinogenic reaction products.

The effectiveness of most chlorine-based sanitizers is influenced by several factors, such as pH, temperature, exposure time, type of pathogen and surface morphology. Many sanitizers were used and/or are in use for washing raw agricultural material but not one sanitizer is effective for all possible pathogens and products. Furthermore, parasites and viruses generally exhibit higher resistance to chlorine than bacteria.

The efficiency of disinfectants is limited by the neutralization effect of fruit and vegetable tissue components on the surface and also by inaccessibility of disinfectant to the microbial cells in creases, crevices, pockets and natural openings in the skin. Thus, there is a need to examine the factors that limit the efficacy of washing in reducing microbial populations on produce and to devise means of overcoming such limitations (Sapers, 2003).

The minimally processed fruit and vegetable product is immersed in a bath in which bubbling is maintained by a jet of air. This turbulence permits one to eliminate practically all traces of air and foreign matter without bruising the product (FAO, 2002). If bacterial attachment occurs more than 24 hours prior to washing, detachment or inactivation using chlorine or hydrogen peroxide treatments is shown to be less effective. It is likely that the limited ability of washing to remove established bacterial populations from the surface of fresh produce is due in part to biofilm formation, microbial infiltration and internalization.

Conventional washing and sanitizing methods, even using sanitizing agents such as chlorine, chlorine dioxide, hydrogen peroxide, ozone and peroxyacetic acid, are not capable of reducing microbial populations by more than 90 or 99%. Although such reductions represent a large decrease in the numbers of microorganisms present on the commodity and may result in significant improvements in product quality and shelf-life, they are not equivalent to surface pasteurization and may be inadequate to ensure product safety (Sapers, 2003).

At the retail level or at food establishments, produce is usually washed only using potable water, and the fresh-cut pieces may not always be prepared using clean and sanitized utensils. Thus, fresh-cut fruits and vegetables may not be adequately sanitized and protected from cross-contamination. However, because the time of contamination is not generally known and may precede washing by many days, more effective means of decontaminating produce are needed (Sapers et al., 2001; Ukuku et al., 2005).

Transport and Storage

Following harvest, raw foodstuffs are normally transported to holding, shipping or processing facilities. Transport conveyances are thus a part of the food chain where contamination can occur. Most produce must be harvested and shipped within 12 to 72 hours so it can be received in distribution centers with approximately 10 days of shelf-life remaining (Whitaker, 2010).

Temperature is known to be important in produce production for quality and safety reasons. It is generally accepted that most foods need to be maintained at cold temperatures from harvest to consumption. Improper cold holding of food is the most frequent temperature violation for nearly all facility types.

Packing

Raw commodities may be packed at the field (grapes, strawberries, etc.) to prevent additional handling, water loss, possible damage and contamination during shipping and/or storage. Before packaging, certain commodities (parsley, carrot, etc.) may be trimmed to remove nonedible parts. All packaging materials should be made of food contact grade materials to ensure that toxic compounds in the packaging materials do not leach out of the package and into the produce. Other things to consider in packaging are temperature and shipping. Precooling before packing and cooling after packing are essential.

Modified Atmosphere Packaging

MAP is used in fruit and vegetable storage to extend shelf-life by decreasing the metabolic activity of the product and the growth of microorganisms. MAP involves the creation of a modified atmosphere by altering the normal composition of air (78% nitrogen, 21% oxygen, 0.03% carbon dioxide and traces of noble gases) to provide an optimum atmosphere for increasing the storage length and quality of food (Phillips, 1996). Atmospheric modification can be achieved by using controlled atmosphere storage and/or active or passive MAP. Active modification creates a slight vacuum inside the package that is then replaced by a desired mixture of gases. Passive modification occurs when the product is packaged using a selected film type, and a desired atmosphere develops naturally as a consequence of the product's respiration and the diffusion of gases through the film (IFT, 2000). The design of packaging systems and the selection of materials (gas diffusion rates vary greatly among films) have an effect on the risk of foodborne pathogens in fresh produce. Therefore it is important to apply sound packaging technology knowledge in order to select the correct materials and package design.

MAP could be considered as one of the technologies used to extend shelf-life of fresh produce to protect it from pathogens and damage. But MAP is not without safety concerns and alone is not sufficient to prevent pathogen growth. Chilling produce at 5°C or less is essential.

Processing and Preservation Techniques

Consumers are becoming more aware of the importance of eating fresh and fresh-cut fruits and vegetables (Castell-Perez and Moreira, 2011). However, these types of produce have repeatedly become a source of foodborne illnesses all over the world. Most of the commercially used interventions to protect produce from contamination employ chemical agents, such as washing with 2% chlorinated water, which cannot wash these pathogens off the produce or inactivate them. Another side effect of this treatment is detrimental effects on the organoleptic properties of the food.

Because thermal processing of fresh produce is not an option, new techniques for maintaining quality and inhibiting undesired microbial growth are the only means of including a lethality step in the processing and handling of fresh produce in the distribution chain (Castell-Perez and Moreira, 2011).

Chemical-based Washing Treatments

CHLORINE DIOXIDE

Chlorine dioxide (ClO_2) has been recognized as a strong oxidizing agent with a broad biocidal effectiveness, due to its high oxidation capacity being about 2.5 times greater than chlorine. It does not react with nitrogen-containing compounds or ammonia to form dangerous compounds. Many studies have demonstrated its antimicrobial activity and its use was allowed in washing fruits and vegetables by the FDA (1998).

It can reduce microbial populations in dump tanks and wash water, but tests with cucumbers resulted in less than a 1-log population reduction on product surfaces. In tests conducted in an apple packinghouse, addition of chlorine dioxide (3–5 mg/ml) to dump tank water reduced the population of filamentous fungi. Treatment of pears inoculated with *Botrytis cinerea*, *Mucor piriformis* or *P. expansum* with 10 mg/ml chlorine dioxide for 10 min suppressed decay, but addition of 0.5 mg/ml of chlorine dioxide to flume water did not reduce decay of inoculated fruit. Chlorine dioxide reduced the population of *E. coli* O157:H7 on inoculated apples by only 2.5 logs at 80 mg/ml (Sapers, 2003).

ORGANIC ACIDS

Organic acids (e.g. lactic acid, citric acid, acetic acid, tartaric acid) have been described as strong antimicrobial agents against psychrophilic and mesophilic microorganisms in fresh-cut fruit and vegetables. Acetic acid has been tested as an antimicrobial agent for apples. In one study, a 5% acetic acid wash was reported to reduce the population of *E. coli* O157:H7 on inoculated apples by about 3 logs. However, these apples were inoculated only 30 min prior to treatment, probably providing insufficient time for strong bacterial attachment and possible biofilm formation. In another study, apples that had been inoculated with *E. coli* O157:H7 and air dried for 30 min were treated with 5% acetic acid at 55°C for as long as 25 min. Although the *E. coli* population was greatly reduced in the apple skin and stem areas, as many as 3 to 4 logs survived in the calyx tissue (Sapers, 2003).

HYDROGEN PEROXIDE

Hydrogen peroxide possesses bactericidal and inhibitory activity due to its properties as an oxidant (Sapers, 2004), but is less active against fungi. Dilute hydrogen peroxide solutions are effective in washing mushrooms, controlling post-harvest decay of vegetables, extending the shelf-life of fresh-cut vegetables and melons, and decontaminating apples containing nonpathogenic *E. coli*. Studies have shown that 5% hydrogen peroxide solutions, alone or combined with commercial surfactants, can achieve substantially higher log reductions for inoculated apples than 200 ppm of chlorine. When applied at a temperature of 50 to 60°C, reductions as great as 3 to 4 $\log_{10}$ CFU/g have been obtained (Sapers et al., 1999). A 5% hydrogen peroxide wash applied at 50 or 60°C to the whole cantaloupe melon prior to rind removal was superior to chlorine in extending the shelf-life of fresh-cut melon cubes. Visual observations of spoilage were consistent with the microbiological data showing suppression of bacterial growth following the peroxide treatment, perhaps indicative of injury to spoilage-causing bacteria (Sapers et al., 2001).

Hydrogen peroxide vapor treatments have been used to inhibit post-harvest decay in some commodities. However, vapor treatments tend to be slow and can cause injury to some commodities such as mushrooms, raspberries and strawberries.

COMBINATION OF DIFFERENT DISINFECTANT AGENTS

Combination of several disinfectant agents such as lactic acid, chlorinated water, thyme essential oil solution, sodium lactate, citric acid, hydrogen peroxide, ozone and peroxyacetic acid, has been already widely report (Ukuku et al., 2005).

An acidified surfactant treatment, applied in a brush washer to maximize soil removal, might be followed by hydrogen peroxide treatment, applied by immersing the commodity in a dip tank. In general, combinations of chemical disinfectants maintain better sensory and microbial quality of the product.

Other examples of treatment combinations with the potential for synergism include an acidified surfactant wash treatment combined with surface pasteurization, vacuum infiltration of hydrogen peroxide or ozone solution, or with vapor-phase application of a sanitizer vapor. Such innovations might not be capable of achieving greater than 5-log reductions in pathogen populations possible with true pasteurization treatments, but they might bring about large improvements in the microbiological quality and safety of minimally processed fruits and vegetables. Therefore, for the future, more studies should be carried out to determine the synergistic effects of combining technologies (Sapers, 2003).

Physical Treatments

Recently, many studies have demonstrated the effectiveness of surface decontamination techniques to reduce the microbial risk involved with the consumption of fresh fruits and vegetables (Erkan et al., 2001; Allende et al., 2006).

UV-C LIGHT

UV-C (in the range of 240–260 nm) could be very effective in inactivating common enteric human foodborne bacterial pathogens and maintaining fruit quality during post-UV storage. Exposing packaged watermelon cubes to UV-C light at 4.1 $kJ\,m^{-2}$ produced >1 log reduction in microbial populations by the end of the product's shelf-life without affecting juice leakage, color and overall visual quality.

Non-ionizing, artificial ultraviolet-C (UV-C) radiation is extensively used in a broad range of antimicrobial applications including disinfection of water, air, food preparation surfaces and food containers, and has also been combined with other post-harvest treatments such as mild thermal treatments. Many researchers have already tested the synergistic effects of combining UV-C light with chemical disinfection and/or MAP on vegetable produce (Allende et al., 2006).

Also, UV-C light can catalyze oxidative changes in certain products that lead to rancidity and discoloration.

ULTRASOUND

Power ultrasound, as used for cleaning in the electronics industry, has a potential application for decontamination. Ultrasound treatment at 38.5–40.5 kHz can enhance the effectiveness of chemical sanitizers in killing pathogens that are able to grow in raw agricultural produce (Allende et al., 2006).

PULSED ENERGY

Pulsed energy processing is based on the concept of applying any energy (electric, magnetic or light) that has been stored for a long time in a very short amount of time. This results in huge power generation and causes microbial death.

Pulsed energy has been applied in three forms to food, namely, pulsed electric field (PEF), pulsed light (PL) and pulsed magnetic field (PMF) (Tewari, 2003).

LIGHT PULSES

Light pulses have been used successfully as a new technique for the inactivation of bacteria and fungi on the surface of food products when the major composition of the emitted spectrum is UV light. Very little information is available about the efficacy of light pulses to inhibit microbial growth and prolong shelf-life of fresh-cut fruits, vegetables and herbs (Tewari, 2003). Some studies have focused on the microbial and sensory quality of fresh-cut vegetables using intense light pulses combined with MAP. Microbial reductions up to 2.04 log have been reported by the combination of both techniques, although the shelf-life of the product was not always extended. Additionally, combination of pulsed light with mild heat treatments prolonged the shelf-life of some raw commodities without visible fungal growth for a few days compared to the control (Allende et al., 2006).

PULSED MAGNETIC FIELD

Similar to PEF, PMF can be used to inactivate microorganisms. The technological advantages of PMF include minimal thermal denaturation of nutritional and organoleptic properties, reduced energy requirement for sufficient processing, and potential treatment of foods inside a flexible film package. Additional work is required to correlate the inactivation of microorganisms in food to PMF strength, PMF to the denaturation of nutritional characteristics of food, and the energy efficiency of PMF to the extended shelf-life of food. Little information is available on spore-forming pathogens' inactivation using PMF, which makes this technology several years away for possible commercialization for shelf-stable low-acid foods. However, there seems to be a potential for producing foods with minimal nutritional loss using PMF (Tewari, 2003).

IRRADIATION

Low-dose gamma irradiation is very effective for reducing bacterial, parasitic and protozoan pathogens in raw foods. Irradiation was approved by the FDA for use on fruits and vegetables at a maximum level of 1.0 kGy. Irradiation can be accomplished using gamma rays, X-rays and high-energy electrons (e-beams). Treatment of fresh produce using ionizing radiation has a significant strategic importance for the future of food safety worldwide. This is simply because it is the most researched non-thermal food process technology and has been proven that it is safe, when done properly (Castell-Perez and Moreira, 2011). Irradiation has the potential to reduce internalized *E. coli* O157:H7 in leafy greens by 3 to 4 logs, and has been combined with conventional disinfection methods such as chlorinated water or preservation technologies by using MAP. Treatment of fresh-cut lettuce with low-dose irradiation of about 0.20–0.35 kGy, combined with a chlorine (80–100 ppm NaOCl) wash and MAP, increased the microbiological shelf-life without adversely affecting the visual quality or flavor of the product.

HIGH-PRESSURE PROCESSING (HPP)

HPP is nonthermal processing that has been used for the last 15 years. It has been explored extensively in the food industry and related research institutions due to its lethal effects on food microorganisms without losing nutritional and sensory characteristics of food and "fresh" taste (Tewari, 2003).

HPP in the range of 200–900 MPa for several minutes inactivates the vegetative cells of microorganisms, compared to the heat pasteurization, without damaging the low molecular weight components. The extent of microbial inactivation of HPP is not only species dependent but also influenced by the physicochemical environment such as water activity and pH. However, the high resistance of bacterial spores to HPP is still a major outstanding issue.

HURDLE TECHNOLOGY

Hurdle technology is the combination of different preservation techniques as a preservation strategy. There are more than 60 potential hurdles for foods that improve the stability and/or quality of minimally processed products. The most important hurdles commonly used in food preservation are based on controlling temperature, water activity, acidity, redox potential and the use of preservatives, modified atmosphere and competitive microorganisms (e.g. lactic acid bacteria). By combining hurdles, the intensity of the individual preservation techniques can be kept comparatively low, minimizing the loss of quality, while the overall impact on microbial growth may remain high. Examples of hurdle technologies are natural preservatives, which are used as hurdles in food deterioration (Rico et al., 2007).

APPLICATION OF THE HACCP SYSTEM

Using the CAC definition HACCP is a system that identifies, evaluates and controls hazards which are significant for food safety. HACCP aims at preventing identified potential problems from occurring. Today, the concept of HACCP is considered to be the reference method for food safety assurance worldwide. It has become a requirement for international food trade.

The HACCP system and its validation and maintenance will be explained in more detail in Chapter 31.

A CASE STUDY ON THE APPLICATION OF THE HACCP APPROACH FOR THE MANAGEMENT OF THE PREPARATION OF FRESH TOMATOES FOR STORAGE AND SALE

This case study is an example of food safety management for the preparation of fresh (salad) tomatoes for storage and sale, developed for the purpose of illustrating the application of the HACCP approach for the management of possible contaminants in fruits and vegetables.

Scope of the Study

This HACCP study will cover the operations from crop (tomato) production, harvesting, receiving raw agricultural material, through steps before storage (washing, sorting, waxing, degreening), packing and distribution to market, and evaluates the potential points of contamination.

Description of the Product

The tomato is a very popular crop. Tomatoes can be grown in the field (outside) or in a greenhouse (inside). In greenhouse production, modeling is focused on yield prediction, optimization of climate and fertigation control and evaluation of strategies of crop management. In field production, focus is more on the prediction of harvest dates and the estimation of water and nutrient requirements.

Tomato fruits are usually picked when fully ripe, and are therefore very susceptible to cracking, bruising and consequently decay. Fruit at the breaker stage is mixed with the fully ripe and a certain proportion of unripe fruits. Fully ripe and green tomatoes could be sorted at the growing site and/or later in the packing/processing facility. After harvest tomatoes are loaded onto trucks in cardboard boxes and transported to the storing and/or packing facilities in a manner that will minimize damage.

Tomatoes can be consumed either fresh or as the main ingredient in a range of processed products. The fresh market emphasizes visual appearance and shelf-life duration, whereas the processing industry gives more value to the dry matter concentration and composition. In both cases, producers have to control their production process to reach the standards defined by their customers. For the market the tomatoes should be ripe, red and firm to soft, free of all mold growth, stems, leaves, dirt and other soils. High quality "salad" tomatoes have the highest value when sold fresh and in good condition. Shriveling percentage can be high since fruits are often exposed to the sun. Adequate receiving into the facility and storage may reduce subsequent losses (Table 9.2). In some cases the storage life of fresh tomatoes can be greatly extended by leaving part of the stalk attached to the fruits at harvest time.

Because of their soft texture tomatoes should be handled gently to minimize bruising and breaking of the skin. After harvesting, tomatoes may be packed after receiving in the packing facility or after a certain time in storage. Fresh tomatoes may be packed in plastic pouches and/or cardboard boxes, and distributed to the market at adequate temperature.

TABLE 9.2 Recommended Temperature and Relative Humidity, and Approximate Storage Life for Tomatoes

	Temperature		Relative Humidity	
Product	°C	°F	(%)	Approximate Storage Life
Tomatoes, mature-green	13 to 18	55 to 65	85 to 90	2 to 3 weeks
Tomatoes, pink	10 to 13	50 to 55	85 to 90	7 to 10 days
Tomatoes, ripe	7 to 10	45 to 50	90 to 95	3 to 5 days

Operational records regarding products and practices can be helpful to companies. Such records help ensure consistency of production, packing and processing operations and end-product quality and safety.

The flow diagram presented in Figure 9.1 describes operations from field production and handling procedures for preparation of tomatoes for storage and sale. Also, the diagram presents possible CCP at which there is a high or medium risk that control may be absent, and at which control must be applied. It is important to identify the market because some markets require certain varieties of produce.

A flow process is constructed for fresh tomatoes to detail how the product is created from raw material, through post-harvesting and handling procedures, storage, packing and distribution to distribution center, retail, foodservice and consumers. The purpose of the flow diagram is to identify specific areas where hazards could occur. Once completed, the flow diagram should be verified by a supervisor for completeness and accuracy.

Distribution and Intended Use

Fresh tomatoes are widely available (year-round basis) and generally provide exceptional nutritional benefits for all groups of consumers. Tomatoes are predominantly eaten raw (intact or peeled and/or cut, minimally processed), within a shelf-life, but may be also processed in industry in many different products or preserved at home.

Eaten raw without washing before consumption they may be harmful if contaminated with injurious substances (pathogens from improperly used manure or fecal-contaminated irrigation water, high levels of pesticides or other toxic agrochemical compounds). Fully ripe tomatoes have to be held refrigerated before consumption for a short time (for less than 5 days) to delay softening. Unripe tomatoes could be held at room temperature until they ripen, usually in a day or two. When tomatoes are sold to consumers for direct consumption or the preparation of meals the responsibility lies with the producer and retailers to ensure that the produce is free from hazard.

HACCP Study

Due to the tomato production process (outside or inside), various harvest and post-harvest operations, and crop management, certain hazards may be present.

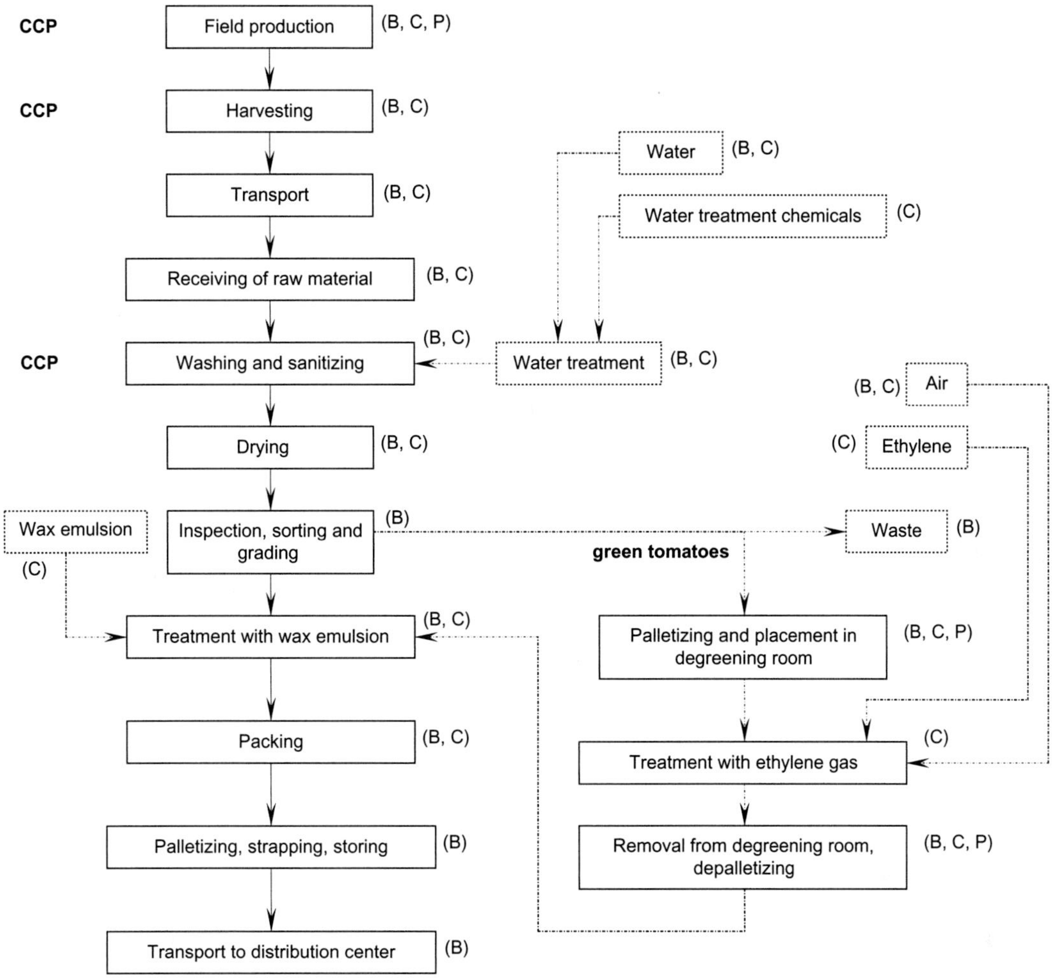

FIGURE 9.1 Flow diagram of operations and handling procedures in the fresh tomato HACCP case study.

Hazard Analysis

Each step of tomato preparation for storage and distribution is taken into consideration individually in terms of food safety. Possible biological, chemical and physical hazards and risks determined in each step, and a plan for the management of contaminants for fresh tomato preparation, CCP, critical limits, monitoring process (what, how, frequency, who), corrective actions, verification and records kept, are suggested and presented in Table 9.3.

FIELD PRODUCTION

Field production may be the first CCP in tomato production due to biological contamination. Production systems for tomatoes fall into two categories – open field and protected

TABLE 9.3 A Plan for the Management of Biological, Chemical and Physical Contaminants in Tomatoes in Line with the HACCP Model

Steps	Hazard	Control Measure	CCP	Limits	Monitoring	Corrective action	Verification
Field production	*Biological*: Seeds contaminated with pathogenic microorganisms such as *Salmonella*, *Shigella*, *E. coli*. Contamination of the soil with manure, wastewater, etc. Contamination of water used for irrigation	Advising producer on good agriculture practices. Supplier management: provide specifications for the seeds, select suppliers able to meet the specifications. Protection of the soil from animal manure. Use of treated safe water for irrigation. Use of clean equipment	Yes	Use of audited and approved source; supplier certificate. Use of treated water and complying with specifications	Monitor source of the seeds; monitor suppliers for any non-compliance. Monitor the microbiolgical quality of irrigation water	Reject seeds if not accompanied by supplier guarantee	Verify that audits have been performed as planned and the suppliers are able to comply with specifications. Verify that the auditors were adequately competent and trained for the job. Periodically test seeds for pathogens and agrochemicals
	Chemical: Contamination of soil due to presence of chemicals (heavy metals). Misuse of agrochemicals such as pesticides, herbicides or insecticides	Advising producers on application of agrochemicals according to regulatory requirements	No				
	Physical: Seeds may contain different foreign matters; however, it is viewed unlikely that these became a danger to public health		No				
Harvesting	*Biological*: Contamination of tomatoes by the field workers with fecal–orally transmitted pathogens (examples are *E. coli*, *Shigella*, hepatitis A) (if the harvest is carried out manually, this step will be of greater risk than if carried out mechanically) Contamination of boxes used for collecting harvested tomatoes	Ensure workers' hygiene and good health conditions, i.e. train workers in respecting hand and personal hygiene, reporting any health potential risk. Ensure access of workers to hand-washing facility Use clean and dedicated boxes	Yes	Unwashed hands of workers before picking tomatoes; health status	Monitoring worker health and hygiene	Reject the tomatoes for fresh use by consumers; use for further processing	Verify the training and attitude of workers towards personal hygiene Test tomatoes for pathogens and agrochemicals
	Chemical: Contamination of boxes used for collecting harvested tomatoes, with agrochemicals	Use clean and dedicated boxes	No				

Transport	*Biological*: Contamination from unclean transport vehicles *Chemical*: Contamination with agrochemicals or allergenic foods or contamination with pathogens from unclean transport vehicles	Good transportation practice, including clean and dedicated transportation vehicles	No				Verify the transportation conditions
Receiving of raw material	*Biological*: Contamination of raw material with pathogen microorganisms from producing site or transport vehicles *Chemical*: Chemical residues from producing site, cleaners and sanitizers, or toxic compounds (from packing material, and/or toxins produced by molds)	Ensure good hygiene practice at receiving point and workers' personal hygiene	No				Verify conditions at receiving point
Washing and sanitizing	*Biological*: Contamination of water with rinsing water *Chemical*: High chlorine level may result in chlorine residue on product	Use of potable water treated with chlorine. Washing of tomatoes (see subsequent steps)	Yes	Active chlorine 100–150 ppm, pH 6.5–7.5. Rising with potable wash water with a residual chlorine of 0.2–0.3 ppm. There should be no pathogens present. Chlorine is major compound used for disinfection of produce. Other substances could be used including organic acids, chlorine dioxide, hydrogen peroxide and ozone	Monitor the amounts of chlorine used for antimicrobiological activity. Monitor the quality of water and the residual chlorine of water	Rewash the tomatoes with treated safe water. Consider use of tomatoes other than for fresh use by consumers	Verify the water treatment process. Test tomatoes for microbial pathogens as mentioned above

(*Continued*)

TABLE 9.3 (Continued)

Steps	Hazard	Control Measure	CCP	Limits	Monitoring	Corrective action	Verification
Drying	*Biological*: Pathogens from dirty drying machine may contaminate product. *Chemical*: Dripping oil may contaminate product.	Filter air used for drying process. Equipment maintenance	No				Verify the conditions of the equipment on a regular basis
Inspection, sorting and grading	*Biological*: Contamination with equipment. Contamination of tomatoes by the workers with focally transmitted pathogens (if the intervention of workers is limited this risk is not very significant)	Ensure workers' hygiene and good health conditions, i.e. train workers in observing hand and personal hygiene, reporting any health potential risk. Ensure access of workers to hand-washing facility. Use clean equipment and sanitary conditions, maintenance of equipment	No				Verify workers knowledge and attitude towards hand and personal hygiene. Verify the maintenance of equipment
Treatment with wax emulsion	*Biological*: Contamination with equipment. *Chemical*: Contamination of tomato due to use of non-food grade wax.	Ensure adequate equipment. Ensure adequate wax emulsion	No	Use of audited and approved supplier			Review the microbial count reports of raw material
Packing	*Biological*: Contamination due to poor hygiene and handling practice. Dirty environment is harborage for pathogen. Contamination by dirty reused packaging material	Train workers in hygienic practice inclusive of hand hygiene and good health conditions. Ensure access of workers to hand-washing facility	No	Workers' personal hygiene/health status		Adapt optimal conditions of packaging. Use of food grade packaging material	Review the microbial count reports of raw material
	Chemical: Non-food grade packaging materials/glue may introduce chemical contamination	Use of adequate food grade packaging materials. Ensure hygienic conditions for packaging to avoid cross-contamination	No				

Palletizing, strapping and storing	*Biological*: Contamination due to poor hygiene and handling practice	Train workers in good hygienic practice. Ensure clean environment and pest management	No	Audit the palletizing and storage conditions for good storage practice and pest management
Palletizing and placement in degreening room	*Biological*: Growth of pathogenic microorganisms due to high temperature, prolonged storage, survival of environmental contaminants, cross-contamination. Raw material with biological contamination mistakenly passes inspection due to poor documentation. Contamination due to poor hygiene and handling practice *Chemical*: Cross-contamination from pallets and equipment *Physical*: Breaking of equipment	Optimize palletizing, strapping and storing conditions. Conduct palletizing, strapping and storing of the product at adequate temperature and humidity. Ensure hygienic conditions of room and equipment to avoid cross-contamination. Ensure workers' hygiene	No	Review the microbial count reports of raw material
Treatment with ethylene gas	*Chemical*: Excessive ethylene concentration may result in residue	Optimize ethylene concentration (100 ppm)	No	Verify supplier of ethylene gas

(*Continued*)

TABLE 9.3 (Continued)

Steps	Hazard	Control Measure	CCP	Limits	Monitoring	Corrective action	Verification
Removal from degreening room, depalletizing	*Biological*: Growth of pathogenic microorganisms due to high temperature, prolonged storage, survival of environmental contaminants, cross-contamination. Raw material with biological contamination mistakenly passes inspection due to poor documentation. Contamination due to poor hygiene and handling practice *Chemical*: Cross-contamination from pallets and equipment *Physical*: Breaking of equipment	Optimize palletizing, strapping and storing conditions. Conduct palletizing, strapping and storing of the product at adequate temperature and humidity. Ensure hygienic conditions of room and equipment to avoid cross-contamination. Ensure workers' hygiene	No				Review the microbial count reports of raw material
Transport to distribution center	*Biological*: Contamination due to poor hygiene and handling		No				Periodic audit of distributors by the manufacturer to verify the implementation of good hygienic practice

culture systems – with a wide variation in terms of inputs, size, location, environmental conditions, productivity and target markets. There are a number of production factors that affect losses and contamination of raw commodity and these should be utilized as much as possible. For that reason reduction of losses and the risk of hazards start at the production site.

According to Table 9.3, biological hazards during production may come from seeds contaminated with pathogenic microorganisms (*Salmonella*, *Shigella*, *E. coli*), from soil contaminated with untreated manure, waste water, etc., and from water used for irrigation, etc. Chemical hazards may come from contaminated soil (chemical residues from primary production and heavy metals). Also, misuse of agrochemicals (pesticides, herbicides or insecticides) may cause chemical contamination. Critical limits for these hazards should be identified in a supplier certificate. Possible physical hazards may come with contaminated seeds and those present in the soil (glass, plastic, wood chips, stones, hard plant material, metal pieces from equipment used in soil management); however, it is unlikely that they become a danger to consumers' health.

Knowledge of the production system (environment) and knowledge of what constitutes a hazard mean that the capacity to identify hazards within a production system is critical for identifying and applying relevant and effective mitigations. Permanent recording systems of the commodity and agronomic activities carried out on the growing site must be established.

HARVESTING

Harvesting may be a CCP since biological hazards may be present. During harvesting tomatoes may be contaminated with pathogenic microorganisms through fecal material present in the soil (from wild and domestic animals, and manure), air, plant waste, human handling, harvesting equipment, field and/or transport containers and vehicles. If the harvest is carried out manually tomatoes could be contaminated by the field workers with fecal–orally transmitted pathogens (*Shigella*, *E. coli*, hepatitis A). In mechanized harvesting this risk could be avoided. In many cases, workers use an inspection line for primary selection on the field.

Agricultural produce routinely comes into contact with harvesting equipment (harvesters, knives, clippers and scissors) and containers (bins, boxes, buckets, trailers and trucks) used for collecting the produce. Equipment (such as tables, conveyor belts, flumes, washing or cooling bins) and containers may retain pathogenic microorganisms from the adhered soil.

Containers (boxes) used for collecting harvested tomatoes could be the source of chemical contamination too. Preventive or control measures for the significant hazards are field sanitation, auditing crop protection chemical application, crop handling and use of clean and adequate containers.

Raw material may be contaminated with hard vegetative material, wood, glass, metal fragments and pieces of plastic, and can be removed by workers at the growing site during harvesting, inspection, sorting and grading, but they are unlikely to become a danger to consumers' health. Metal particles in raw material that cannot be removed by inspection, or during cleaning and/or sorting and grading, should be checked by using a metal detector.

For those involved in harvest and post-harvest processing hygiene is an important consideration due to the widespread use of human hands as part of the process.

Human hands touch tomatoes, for example, at almost every step of the process (see Chapter 28).

TRANSPORT

Truck (or other vehicle for raw material transportation) sanitation faces challenges similar to those encountered in the sanitation of equipment and containers during harvest. Biological and chemical contamination may occur due to unclear transport vehicles. In some instances, commodities may be protected by the containers in which they are packed; however, in most cases, tomatoes grown in the field are loaded directly onto the truck. Maintaining clean transport vehicles (applying good transportation practice) on a frequent and scheduled basis can ensure raw material safety.

RECEIVING OF RAW MATERIAL

After transporting to sorting and/or storing facilities, raw material could be contaminated but it is unlikely to become a danger to consumers' health if steps such as washing and sanitizing are followed.

Nevertheless, it is of great importance to avoid mechanical injury of the produce during all operations, since the disturbance of the commodity's physical barrier would greatly increase the opportunities for pathogen survival and growth, if contamination occurs. Also, it should not contain chemical residues from farms, cleaners and sanitizers, toxic compounds leaked from packing material, and/or toxins produced by molds. To ensure crop safety, growers should be able to provide evidence of all chemical residues testing by an accredited laboratory.

To monitor raw material income, the receiving manager/staff should ensure that a supplier guarantee exists for each incoming shipment and a supplier certificate should be visually confirmed by the personnel. This should be done for each incoming shipment for all raw materials (which has to be accompanied by a supplier guarantee). If quality of raw material is inadequate, raw material should be rejected. Suppliers should be periodically audited and the tomatoes also periodically tested for foodborne pathogens as a means of verifying the hazard analysis. Supplier guarantee records and audit reports (as a verification of the supplier quality assurance) should be kept and archived.

As a general practice, it is important that firms that produce and harvest tomatoes maintain documentation and records related to operational information about the product and practices, as well as tracing information about the product.

WASHING AND SANITIZING

Washing tomatoes is an important step in processing in terms of removal of biological, chemical and physical contaminants. Biological hazards, due to contaminated water, and/or due to high levels of sanitizing compound (e.g. chlorine or other sanitizers), make washing a CCP. Water comes most often from wells and municipal water supplies. It is used for washing (via baths and/or sprays), cooling (through cold water), and conveying produce between points (as when a flume is used). Water used for washing should contain chlorine (generally used as an antimicrobial agent) at adequate concentration (100–150 ppm or higher, active chlorine at pH between 6 and 7.5) for eliminating the microbiological risks of vegetables, and to prevent potential cross-contamination. The temperature of the chlorinated water should be at least 10°C colder than that of the tomatoes to achieve a positive differential, thereby minimizing the uptake of wash water, and entry of microorganisms, through stem or blossom and open areas in the skin (due to mechanical damage). Following the soak, a thorough rinsing with fresh water using pressurized spray nozzles is necessary. During washing and rinsing, high chemical and microbiological quality potable water should be used.

Control of the sanitary quality of water is technologically feasible but requires strict management of operating practices. Periodic water analyses should be performed in accredited laboratories and chlorine used in the process should be purchased with a compliance certificate showing the purity of the chemical. The amount of residual chlorine on tomatoes should be under the upper limits (0.2–0.3 ppm) given in specifications. Treatment at an improper washing facility, instead of the removal of contaminants, could cause tomato contamination if the microbiological and chemical quality of water used is poor.

DRYING

Excess moisture on the surface of the fresh produce can cause deterioration by microorganisms. To remove excess water from the surface, the produce should be dried (by forced air drying). Pathogens from a dirty drying machine (and blower) may contaminate produce. The condition of the equipment should be checked on a regular basis to prevent chemical hazard (e.g. from dripping oil). Also, metal (chips) from moving parts may contaminate produce. These can be identified using metal detection equipment. In many conveyor lines, the line will be stopped because small pieces of metal are detected within it.

INSPECTION, SORTING AND GRADING

After harvest, some handling procedures such as inspection, sorting (by color and size), grading, washing (by water sanitizing procedures) and cooling are carried out to remove off-color and defective fruit which may reduce the occurrence of different hazards. But there is no guarantee of hazard reduction to acceptable levels, or elimination. Inspection, sorting and grading, if conducted at lower temperatures, could reduce certain amounts of biological environmental contaminates. Possible chemical and physical hazards could be prevented by assuring hygienic conditions of the sorting and grading facilities, equipment and workers.

Before storage, vines, twigs, stems, leaves, etc. that might rub on other tomatoes and cause damage during the ripening process should always be removed (trimmed). Trimming raw materials could be applied at the growing site, too.

Since these handling procedures are usually accomplished, mechanically broken pieces of equipment could be a serious physical hazard.

Sorting by size can be carried out before or after sorting by color, and should be always carried out before grading. This is because it is easier to identify tomato fruits with defects on a uniform product, either in terms of size or color. During these procedures biological contamination could occur as a consequence of cross-contamination due to poor hygiene and handling practice (poorly maintained equipment, contaminated facility and employees' personal hygiene). Because of their soft texture, tomatoes should be handled gently to minimize bruising and breaking of the skin which can provide channels through which microorganisms can enter products.

It is of great importance that employees are trained to follow good personal hygiene/health practices, including the use of proper hand-washing techniques, wearing clean clothes and any additional outer coverings (e.g. hairnets and beard covers, disposable gloves, aprons). Employees should be trained consistently in line with the level of complexity of their jobs and additional training should be provided as needed to ensure current knowledge of equipment and process technology.

WASTE

As a result of certain handling and operational procedures, waste could accumulate and become a source of contamination, especially biological. Regular removal of waste from the processing site could prevent contamination.

WAXING

Waxing the surface of tomatoes is a treatment used to retard the rate of moisture loss, and has a cosmetic effect; the wax imparts a gloss to the skin and gives the produce a more shiny appearance than the unwaxed commodity. Waxing of tomatoes can also extend the storage life. To prevent biological contamination equipment should be in a clean and sanitary condition. It is very important that waxes are approved for human consumption. Problems might arise if unregistered formulations are used, or if the skin is eaten by humans. Ingredients in the wax mixtures have to be classified as GRAS (generally recognized as safe) to avoid chemical contamination. Water used for adding waxes has to be of the same quality as water for commodity washing.

PACKAGING

Packaging is a very important step in terms of biological, chemical or physical contamination. The risk of biological contamination exists due to poor hygiene/health condition and handling practices (see Chapter 28).

Produce can be packed in containers at the field, depending on the commodity, or be temporarily placed in bulk bins, baskets or bags which will be transported to the packing/storing facilities. In a packing house products can be prepared continuously for 24 hours regardless of the weather.

The design of packaging systems and the selection of materials have an effect on the risk of foodborne pathogens in tomatoes; it is therefore important to apply sound packaging technology knowledge in order to select the correct materials and package design. Tomatoes may be packed in a variety of containers, depending on the intended market. New, clean and quality materials must be used on the inside to avoid any internal or external damage or biological and/or chemical contamination. All packaging material should be made of food contact grade materials to ensure that toxic compounds in the packing material do not leach into the produce or out of the package. MAP could be used for fresh produce.

Packing of fresh tomatoes at the packing facility should be performed under suitable packing conditions (i.e. temperature, humidity, clean environment, etc.) to avoid the growth of pathogens.

PALLETIZING, STRAPPING AND STORING

During palletizing, strapping for degreening and/or storing of tomatoes, biological contamination could occur due to poor hygiene, damaged equipment and dirty environment.

Storage of produce is carried out in warehouses and specialized storage facilities and also can be an entry point for pathogenic microorganisms or permit the growth of pathogens if present. Precooling of warm product after harvest is favorable before cold storage to prevent spoilage. These facilities are subject to abuse at several stages, and temperature abuse can contribute to the growth of pathogens. Humidity is also an important factor for cold storage and indoor ripening. Too much humidity can encourage rotting (and the dreaded fruit flies); too little humidity can lead to dehydrated tomatoes.

Cold storage facilities (walls, floors, air cooling fans, refrigeration drip pans and coils) should be cleaned and sanitized on a frequent and regular basis. Also, contamination due to refrigerant leak in the facility, or grease and oil from equipment, can cause chemical contamination.

Temperatures and relative humidity of the storage should be monitored, and changes in these parameters should be reset. Adequate storage involves proper regulation of temperature, humidity, air circulation, stacking pattern, regular inspection, and prompt produce disposal as soon as maximum storage life has been attained.

During storage a certain amount of fresh tomatoes could be packed before transport. During storage cross-contamination could happen between contaminated and uncontaminated tomatoes, from dirty packaging material, other products, poor worker hygiene, moldy walls or due to contamination to outside storage. Hygienic barriers might be used, stores should be cleaned and sanitized periodically, and records should be kept for archive and audits.

Physical and chemical hazards should not be present in raw material or in facilities if all precautions for good storage practice are put in place.

DEGREENING (TREATMENT WITH ETHYLENE GAS)

Tomatoes are harvested within a range of maturity and they should be separated by colors before packing. Commercial growers/shippers keep the concentration of the ethylene gas either low to inhibit ripening, or high to promote ripening. As the market requires tomatoes presented at a uniform color stage they could be treated by ethylene (a naturally occurring, odorless, tasteless gas produced by many types of produce) to have fruit mature faster. Mature green tomato fruits exhibit accelerated ripening in the presence of ethylene in controlled conditions (conc. 100–150 ppm; temp. 20–25°C; 24–48 hours). Common storage of mature green tomatoes with ripe tomatoes should be avoided because the ethylene could hasten the ripening of tomatoes located nearby. After degreening, waxing may be applied.

REMOVAL FROM STORAGE, DEGREENING ROOM, DEPALLETIZING

If fresh produce is going to market removal from storage or degreening room unpacking could be needed. If all precautions (due to handling and hygiene practice, control of storage conditions) have not been taken in account, biological, chemical and physical contamination could occur. All equipment and pallets have to be clean and in good condition. Also, the hygiene/health of those involved in this stage is very important.

TRANSPORT TO DISTRIBUTION CENTER (LOADING IN TRANSPORT VEHICLE)

The contents of each packaging unit must be uniform and contain only the same origin, variety, quality and maturity index products. Visible contents must represent the whole. Vehicles used for transporting other than tomatoes should never be used. Vehicles used during transportation should be cleaned and sanitized on a regular basis. Loading temperature and sanitation conditions of the vehicle should be controlled. The best strategy to prevent growth and development of pathogens is to keep produce at recommended transport conditions, particularly temperature. Principles of hygienic practice in transport should be applied to avoid any kind of hazard.

Fresh produce can take many routes to the end user and for that reason each step of the route must be managed to reduce, control or eliminate the risk of contamination.

In need of any support and periodical maintenance of equipment in process for preparation of vegetables for storage, processing and/or sale, maintenance service should be called. The Quality Control/Quality Assurance (QC/QA) department should ensure that potential hazards in all steps of the process are avoided by stating preventive and corrective actions. Effectiveness of the HACCP system can be confirmed by verification.

All activities taking place in the HACCP system should be recorded and archived for periodic internal and external audits. Audits are performed by internal audit and government officials dealing with food safety.

Corrective Actions

When a deviation from the prescribed limits for any identified CCP at any step occurs corrective action must be taken to eliminate the potential contamination. Such corrective action may be: revise data gathering procedures to ensure that appropriate regimes are applied for fresh tomato handling procedures for preparation of tomatoes for storage and sale; restore sanitary conditions; strengthen worker training; and prevent a recurrence of the contamination of product. In addition, when corrective actions are needed, the HACCP team should re-evaluate the Standard Operating Procedures and make appropriate modifications or appropriate improvements in their implementation.

Verifications

Monitoring activities and tests (such as quality of raw material, water quality, quality of produce, the efficiency of cleaning and sanitation of equipment, vehicle and processing environment, the efficiency of temperature and humidity during cold storage, competitiveness of auditors, knowledge and attitude of workers towards personal hygiene, etc.) need "verification" to confirm proper implementation of the HACCP plan. The HACCP plan needs periodic review including review of CCP records, to ensure that the HACCP program is implemented and functioning properly.

Records

Records must be maintained to provide verification that all appropriate prerequisite programs included in the HACCP plan are being followed in accordance with the goals and defined requirements. All papers related to the HACCP system must be kept in order and be accessible.

Those records include:

- All critical control points monitoring records
- Supplier's specifications

- Storage records
  - Temperature and humidity monitoring records
- Analytical results
- Microbial and chemical records (e.g. food contact surfaces, equipment)
- Water quality and supply records
- Equipment monitoring and maintenance records
- Sanitation records
- Corrective action records
- Loading place and distribution records
  - Inspecting delivery vehicles
- Employee training records

Implementation of the HACCP Study

The outcome of the HACCP study is summarized in the HACCP plan (Table 9.3). The implementation consists of:

- Providing guidance and specifications to suppliers of the raw material, as well as distributors in handling the raw material and finished products;
- Training employees to identify the appropriate corrective actions and to be able to understand all steps in the production of fresh or fresh-cut produce;
- Training of auditors, laboratory and personnel engaged in verification measures;
- Monitoring controlling measures as identified in the HACCP plan;
- Implementing the verification measures.

Maintenance of HACCP Plan and Continuous Improvement

The maintenance activities for HACCP are based on keeping the HACCP plan current and suitable for control of all relevant significant tomato safety hazards. The HACCP plan has to be revised if: the raw material supplier changes (where the tomato is sourced); the production environment changes (e.g. if potential source of contamination of soil and irrigation water exists, due to flood, etc.); the field, facility and transport sanitation changes; there is change of any operation for tomato preparation for market; there are personnel changes; the end user (intended target consumer/market) changes.

The HACCP plan has to be continuously reviewed on a quarterly basis. Continuous reviewing, monitoring and verifying of data and potential gaps at all implementations of the HACCP plan could be used to improve the application of the HACCP system.

CONCLUSION

The production of high quality (salad) tomatoes and the maintenance and enhancement of this quality in post-harvest and distribution operations are associated with careful incorporation of technologies applied throughout the production, harvesting and post-harvesting stages. These technologies are crucial to ensure quality and safety, as well as efficient handling of the produce throughout the whole food chain.

Acknowledgment

The author thanks Dr. Mirela Kopjar (Faculty of Food Technology, University Josip Juraj Strossmayer in Osijek, Croatia) for her valuable contribution in reviewing this chapter and technical help.

References

Allende, A., Tomás-Barberán, F.A., Gil, M.I., 2006. Minimal processing for healthy traditional foods. Trends Food Sci. Technol. 17, 513–519.

Beuchat, L.R., 1995. Pathogenic microorganisms associated with fresh produce. J. Food Prot. 59, 204–216.

Beuchat, L.R., 2002. Ecological factors influencing survival and growth of human pathogens on raw fruits and vegetables. Microbes Infect. 4, 413–423.

Brackett, R.E., 1992. Shelf stability and safety of fresh produce as influenced by sanitation and disinfection. J. Food Prot. 55, 808–814.

CAC, 2003a. Codex Alimentarius Commission. Recommended International Code of Practice. General Principles of Food Hygiene. CAC/RCP 1-1969, Rev. 4–2003 <www.codexalimentarius.net/download/standards/23/cxp_001e.pdf>.

CAC, 2003b. Code of hygienic practice for fresh fruits and vegetables, Doc. CAC/RCP 53–2003.

Cantwell, M., 2002. Summary table of optimal handling conditions for fresh produce. In: Kader, A.A. (Ed.), Postharvest Technology of Horticultural Crops. University of California, pp. 511–518.

Cantwell, M., 2007. Postharvest handling of horticultural products: keeping principles in perspective. Small Farm News 1, 1–3. <http://postharvest.ucdavis.edu>.

Carmichael, I., Harper, I., Coventry, M., Taylor, P., Wan, J., Hickey, M.W., 1998. Bacterial colonization and biofilm development on minimally processed vegetables. J. Appl. Microbiol. 85, 45S–51S.

Castell-Perez, E., Moreira, R.G., 2011. An engineering approach to ensuring the safety of fresh and fresh-cut fruits and vegetables. Food Saf. Mag. (February/March).

Crisosto, H.C., Michell, P.J., 2002. Preharvest factors affecting fruit and vegetable quality. In: Kader, A.A. (Ed.), Postharvest Technology of Horticulture Crops. Postharvest Technology Research and Information Center, UC Davis, pp. 49–55.

Erkan, M., Wang, C.Y., Krizek, D.T., 2001. UV-C radiation reduces microbial populations and deterioration in *Cucurbita pepo* fruit tissue. Environ. Exp. Bot. 45, 1–9.

EU EC, 2002. European Commission, health and consumer protection directorate-general, Directorate C – Scientific Opinions, Scientific Committee on Food; C2 – Management of scientific committees II; scientific co-operation and networks. Risk Profile on the Microbiological Contamination of Fruits and Vegetables Eaten Raw, Report of the Scientific Committee on Food; Brussels – Belgium.

EU EC, 2011. Guidelines on comparability, extrapolation, group tolerances and data requirements for setting MRLs – Chemicals, Contaminants, Pesticides (SANCO 7525/VI/95 – rev. 9).

FAO, 2002. Handling and preservation of fruits and vegetables by combined methods for rural areas. FAO Agricultural Services Bulletin 149.

FAO, 2007. Safety and Quality of Fresh Fruit and Vegetables. A training manual for trainers.

FAO, 2008. Microbiological hazards in fresh leafy vegetables and herbs. Microbiological Risk Assessment Series 14.

FDA, 1998. Guide to minimise microbial food safety hazards for fresh fruits and vegetables. <www.cfsan.fda.gov/~dms/prodguid.html>.

FDA, 2008. Guidance for Industry: Guide to Minimize Microbial Food Safety Hazards of Fresh-cut Fruits and Vegetables.

Fletcher, M., 1996. Bacterial attachment in aquatic environments: a diversity of surfaces and adhesion strategies. In: Fletcher, M. (Ed.), Bacterial Adhesion: Molecular and Ecological Diversity John Wiley & Sons, New York, p. 1.

Francis, G.A., Thomas, C., O'Beirne, D., 1999. The microbiological safety of minimally processed vegetables. Int. J. Food Sci. Technol. 34, 1–22.

Harris, L.J., 2010. Preharvest and Harvest Food Safety. Department of Food Science and Technology Western Institute for Food Safety and Security, Western Center for Food Safety, UC Davis Freshcut Workshop.

Hewett, E.W., 2006. An overview of preharvest factors influencing postharvest quality of horticultural products. Int. J. Postharvest Technol. Innov. 1, 4–15.

IFPA, 2001. International Fresh-cut Produce Association. <www.fresh-cuts.org>.

IFT, 2000. Expert Report on Emerging Microbiological Food Safety Issues. Implications for Control in the 21st Century, pp. 6–7. S. Lowry/Univ. Ulster/Stone.

Jongen, W., 2002. Fruit and Vegetable Processing-Improving Quality. Woodhead Publishing.

Kader, A.A., 2002. Quality parameters of fresh-cut fruit and vegetable products. In: Lamikanra, O. (Ed.), Fresh-cut Fruits and Vegetables: Science, Technology, and Market CRC Press, LLC, Boca Raton, London, New York, Washington, DC.

Lamikanra, O., 2002. Fresh-cut Fruits and Vegetables: Science, Technology, and Market. CRC Press LLC, Boca Raton, London, New York, Washington, DC.

Lynch, M.F., Tauxe, R.V., Hedberg, C.W., 2009. Food poisoning from raw fruit and vegetables. Epidemiol. Infect. 137, 307–315.

Marler, B., 2012. <http://www.marlerblog.com/>.

McDowell, D., Maloney, M., Swan, L., Erwin, P., Gormley, R. McKee, R., et al., 2007. A Review of the Fruit and Vegetable Food Chain. Consumer focused review of the fruit and vegetable food chain. <http://www.safefood.eu/Global/Publications/Researchreports/Fruit_and_Veg_CFR_Final.pdf?epslanguage=en>.

Mendonca, A., 2005. Bacterial infiltration and internalization in fruits and vegetables. In: Lamikanra, O., Imam, S., Ukuku, D. (Eds.), CRC Press LLC, pp. 442–457.

NACMCF, 1999. National advisory committee on microbiological criteria for foods. Microbiological safety evaluations and recommendations on sprouted seeds. Int. J. Food Microbiol. 52, 123–153.

Nguyen-The, C., Carlin, F., 2000. Fresh and processed vegetables In: Lund, B.M. Baird-Parker, T.C. Gould, G.W. (Eds.), The Microbiological Safety and Quality of Food, vol. 1 Aspen Publishing, Gaithersburg, MD, pp. 620–684.

Novak, J.S., Sapers, G.M., Juneja, V.K., 2003. Microbial Safety of Minimally Processed Foods, CRC Press LLC.

Phillips, C.A., 1996. Review: modified atmosphere packaging and its effect on the microbiological quality and safety of produce. Int. J. Food Sci. Technol. 31, 463–479.

Raghavan, S., 2007. Handbook of Spices, Seasonings and Flavorings. CRC Press, Taylor and Francis.

Rico, D., Martín-Diana, A.B., Barat, J.M., Barry-Ryan, C., 2007. Extending and measuring the quality of fresh-cut fruit and vegetables: a review. Trends Food Sci. Technol. 18, 373–386.

Sapers, G.M., 2003. Washing and sanitizing raw materials for minimally processed fruit and vegetable products. In: Novak, J.S., Sapers, G.M., Juneja, V.K. (Eds.), Microbial Safety of Minimally Processed Foods CRC Press LLC, pp. 221–255.

Sapers, G.M., 2004. Effect of hot water and hydrogen peroxide treatments on survival of *Salmonella* and microbial quality of whole and fresh-cut cantaloupe. J. Food Prot. 67, 432–437.

Sapers, G.M., Miller, R.L., Mattrazzo, A.M., 1999. Effectiveness of sanitizing agents in inactivating *Escherichia coli* in Golden Delicious apples. J. Food Sci. 64, 734–737.

Sapers, G.M., Miller, R.L., Pilizota, V., Mattrazzo, A.M., 2001. Antimicrobial treatments for minimally processed cantaloupe melon. J. Food Sci. 66, 345–349.

Schmidt, R.H., Sims, C.A., Parish, M.E., Pao, S., Ismail, M.A., 2008. A model HACCP plan for small-scale, fresh-squeezed (not pasteurized) citrus juice operations. Cooperative Extension Service, Institute of Food and Agricultural Science, University of Florida.

Shewfelt, R.L., 1987. Quality of minimally processed fruits and vegetables. J. Food Qual. 10, 143–156.

Tewari, G., 2003. Microbial safety during nonthermal preservation of foods. In: Novak, J.S., Sapers, G.M., Juneja, V.K. (Eds.), Microbial Safety of Minimally Processed CRC Press, pp. 190.

Ukuku, D.O., Liao, C.-H., Gembeh, S., 2005. Attachment of bacterial human pathogens on fruit and vegetable surfaces. In: Lamikanra, O., Imam, S., Ukuku, D. (Eds.), *Produce Degradation – Pathways and Prevention*. Taylor & Francis imprint by CRC Press LLC, pp. 421–441.

Ukuku, D.O., Pilizota, V., Sapers, G.M., 2001. Bioluminescence AT assay for estimating total plate counts of surface microflora of whole cantaloupe and determining efficacy of washing treatments. J. Food Prot. 64, 813.

Varadaraj, M.C., 2010. Capacity building: building analytical capacity for microbial food safety. In: Boisrobert, C., Stjepanovic, A., Oh, S., Lelieveld, H. (Eds.), In Ensuring Global Food Safety – Exploring Global Harmonization Elsevier Inc., London, UK, pp. 151–176.

Wallace, C.A., Sperber, W.H., Mortimore, S.E., 2011. Food Safety for the 21st Century. Managing HACCP and Food Safety throughout the Global Supply Chain. Wiley-Blackwell: a John Wiley & Sons, Ltd. publication.

Whitaker, R.E., 2010. Confronting food safety challenges head-on in produce. Food Saf. Mag., 12–15.

Wiley, R.C., 1997. Preservation methods for minimally processed refrigerated foods and vegetables. In: Wiley, R.C. (Ed.), Minimally Processed Refrigerated Fruits & Vegetables Chapman and Hall, New York, pp. 66–134.

CHAPTER

10

Coffee, Cocoa and Derived Products (e.g. Chocolate)

Anett Winkler
Kraft Foods R&D Inc., Munich, Germany

OUTLINE

Food Safety Management.
DOI: http://dx.doi.org/10.1016/B978-0-12-381504-0.00010-X

GENERAL CONSIDERATIONS

Effective food safety management requires a holistic approach, and, in that regard, could be compared to living organisms: continuously adapting – or being adapted – to new environments. The basis for effective food safety management is provided by applying (prerequisite) programs that assure good manufacturing/hygiene practices and hygienic design, and by applying HACCP principles. Numerous references, guidelines and literature referring to those programs are available (ISO 22000, Codex Alimentarius, ILSI monograph and report series, specific guidance for certain food types), and the EU Food legislation (EC regulation 178/2002 and 852/2004) refers to application of the HACCP principles as well. The chapter on Hazard Analysis and Critical Control Point System (Chapter 31) in this book can also be analyzed for more details on HACCP. In practice, they need to be applied to very different processes/productions – and cultural environments. One of the main factors in successful food safety management is the people involved. A 2006 study in the USA revealed that the top reason for failures in food safety management was untrained/unaware employees (Sertkaya et al., 2006). Therefore, training and retraining (as one of the prerequisite programs) in an understandable manner should be one of the focus areas to ensure food safety.

As new facts emerge, they need to be evaluated for their relevance to food safety management. In order to do that, available information sources need to be screened and their relevance discussed by specialists in their fields. This includes, but is not limited to, rapid alert systems (RASFF in the EU); outbreak data, especially in view of potential root cause analyses; scientific literature; and governmental activities – which may be reactions to recent outbreaks or even searching for emerging hazards (Robinson, 2011). Based on the outcome of discussions, adaptations in food safety management may be needed, and should then subsequently be implemented.

The following sections are meant to provide specific guidance/information on the assumption that the basic knowledge/theories are known by the reader. In addition, since food quality and food safety are often linked, some quality aspects will also be highlighted here.

Coffee

Among about the 40 known different species of the genus *Coffea*, only two are of major importance for worldwide commercial coffee production: *Coffea arabica* L. and *Coffea canephora* var. *robusta*. Around 60–70% of worldwide coffee production is represented by *Coffea arabica*, the major remaining part by the Robusta variety.

Coffee fruits reach their maturity within an average of 9 months. The fruits are normally picked by hand, or mechanically on large farms. After harvest the beans need to be separated from the pulp, which is achieved in two different ways: dry or wet processing. The simpler and cheaper dry processing is mainly used for Robusta and Arabica in Brazil. The overmature fruits are spread out in the sun or put in mechanical dryers, and allowed to dry to a moisture content of 9–13%. The wet processing is mainly used for Arabica, and both steps involve a fermentation step or mechanical removal of the pulp. Following wet processing the beans are then dried to at least 14% moisture, which is equivalent to a water activity of 0.75. From a microbiological point of view a moisture content of <13% is preferred for later storage. An increase of the moisture content >14% results in growth of molds. When stored in rooms with a relative humidity at 80%, visible mold spoilage may be noticed after only 2 weeks of storage (Betancourt and Frank, 1983a).

Next, the dried beans are processed to remove the husk and the parchment, and then sorted for different criteria before being packed in bulk or bags.

More detailed information about all different aspects of coffee and its primary production can be found elsewhere (Wrigley, 1988; Rothfos, 1985, 1986; Müller, 1997; Rotzoll and Müller, 2006).

Roast and Ground, and Instant/Soluble Coffees

Information on microbiological aspects of coffee is rather limited, and refers mainly to the fermentation processes mentioned above. In one study total aerobic mesophilic counts of around 10E5 cfu/g, <10 cfu/g coliforms and <100 cfu/yeast and molds were detected in green coffee (Mohr, 1971). However, the roasting process to achieve the desired color and flavor requires temperatures of 180/190°C or higher. Such roasting decreased the microbiological load of the beans to less than 100 cfu/g. Due to further moisture loss during roasting the resulting water activity of around 0.36 prevents any further microbial growth. In general, coffee is brewed with hot or boiling water before final usage, adding another microbiological safety margin. However, there are some recipes for preparation of "cold brewed coffee" by mixing ground beans with water and letting it stand overnight. Longer storage (>2–3 days) of liquid coffee could lead to significant changes in taste due to growth and metabolic by-products of microorganisms, mainly Bacilli and Lactobacilli (Mohr, 1971).

The mycotoxin ochratoxin A (OTA) is the most significant microbiological hazard related to coffee and its products. OTA-producing molds, mainly *Aspergillus* and *Penicillium* species, are found on green coffee beans, where they can grow under favorable conditions. Critical stages in processing are drying of the green beans, as well as improper storage and transport conditions of the beans before roasting – water activity values of more than 0.8, and temperatures of above 10°C, with an optimum at 35°C allowed for growth and toxigenesis (Suárez-Quiroz et al., 2004). Coffee fermentation seems to be less likely as a main source of OTA formation due to the low pH during that process (Wimmer et al., 2006).

Other mycotoxins are rarely found in coffee, although potential toxin-producing molds are also found on raw beans. However, caffeine and chlorogenic acids are known inhibitors of the biosynthesis of mycotoxins to different degrees; whereas the biosynthesis of sterigmatocysteine is strongly inhibited by them, biosynthesis of OTA is only slightly affected (Betancourt and Frank, 1983b).

Although the molds themselves are killed during roasting, complete degradation of mycotoxins during that process cannot be relied upon as a control step. Depending on the roasting profile, potential OTA content can be reduced by 60–70%. Since mycotoxins are water soluble, they are then transferred to the final beverage being consumed. Due to this known hazard there are several regulations/guidelines referring to maximum values in coffee (EC, FAO/WHO). In view of food safety management this hazard is most suitably controlled by having supplier control programs in place and by operational prerequisite programs through establishing relevant sampling plans of incoming raw materials.

Other potential chemical hazards related to green coffee are dioxins, PCBs, heavy metals, pesticides and fumigants. Therefore a monitoring plan based on risk levels should be implemented. The chapter on management of chemical contaminants can be consulted for setting a risk-based monitoring plan. Since sampling and analyses will only find problems when they exist, i.e. in the raw material, a more proactive approach towards such issues would include establishing close relationships for coffee-producing countries and coffee farmer organizations, as well as continuous monitoring of rapid alert systems.

Pesticides applied to the coffee trees are less of a hazard to the coffee beans, since these would primarily be in contact with the whole coffee fruits and removed during separation of the beans from the pulp. Attention should be drawn towards the potential cross-contamination of pesticides on the bags used to transport the beans. Such contamination could have originated from other goods transported in the bags, as was found to be the case in 2008 in Ethiopia. As a consequence, all old bags were burnt and those concerned in Ethiopia were trained in how to use pesticide detection systems; additionally, transportation of coffee in old bags was also banned.

The roasting itself needs to be looked at with respect to other contaminants that could be formed during the processing step, like acrylamide and furan. Furan is formed during roasting, and it was shown that longer roasts lead to a tendency of higher furan levels being formed. These levels are considered a major contributor to furan exposure in the adolescent population (EFSA 2011, furan).

However, formation from acrylamide in coffee is significantly different from other food products. It is formed at the beginning of roasting, and then subsequently destroyed during further roasting. Therefore, in general, stronger roasted variants like espresso have lower values than light roasted ones (Lahmann, 2010). More guidance for an evaluation of processes with respect to acrylamide formation can be found in the Food Drink Acrylamide Toolbox 2011. This document looks at intervention steps in different productions/processes to reduce or prevent acrylamide formation, and is a useful tool to assess these possibilities. The toolbox also includes considerations towards methods and detection limits used, as well as consumer guidance for the products. In the case of coffee it is stated: "Typical brewing equipment transfers AA almost completely into the beverage. The cup/beverage concentrations for roast coffee and soluble coffee are similar. Espresso brewing may however show lower transfer rates due to specific extraction conditions." EFSA is publishing a survey on the acrylamide levels at regular intervals, with the latest one being published in 2011 (EFSA 2011, acrylamide). For processing contaminants it is recommended to read Motarjemi et al. (2009), or other chapters of the book as most relevant.

Physical hazards, like foreign matters potentially passing through the subsequent sieving/sorting steps during production, are in general not significant in coffee due to the method of final preparation (brewing and fine sieving). A general scheme of coffee processing is provided in Figure 10.1.

Soluble/instant coffee is produced by extraction of roast and ground coffee with hot water/steam (ca. 160°C), the extract is then rapidly cooled and either freeze-dried or spray-dried. Due to its preparation with hot water/steam, the final product can be looked at in the same way as roast and ground coffee with respect to food safety criteria.

Decaffeination of coffee can be performed by different means. Before roasting, caffeine is removed either by using a water-solvent partition or by using supercritical CO_2 under pressure. In both cases raw beans are usually steam treated under pressure before decaffeination to swell the beans and allow for easier removal of caffeine. When solvents (dichloromethane

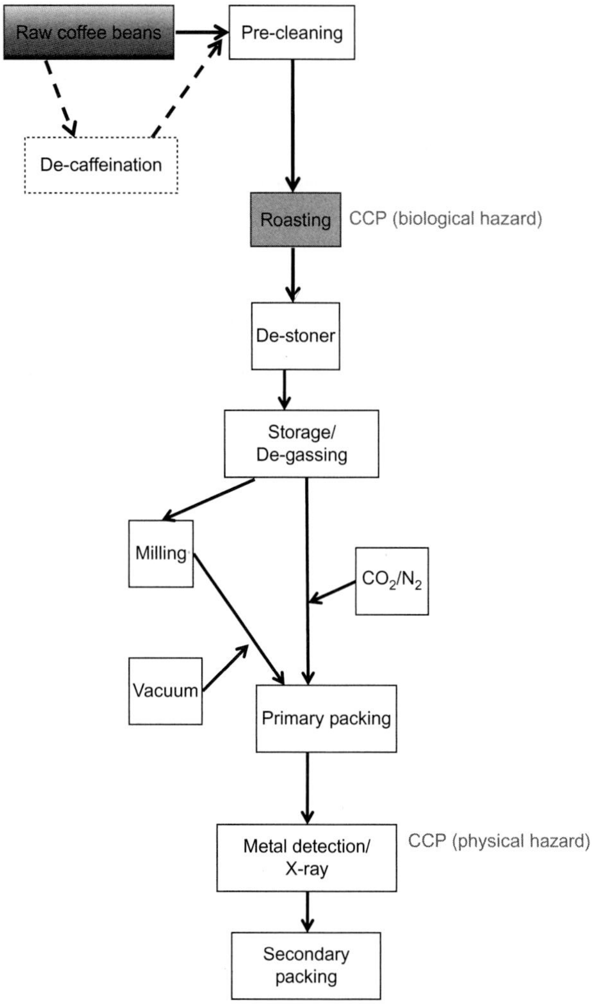

FIGURE 10.1 General scheme of coffee processing. Note: other potential options/orders of processing are outlined in the text.

or ethyl acetate) are used in the process then legal limits with respect to the working environment and residuals have to be adhered to (WHO, IARC, 1999).

For special usages flavors and/or sugar can be added to the coffee, but due to hot water preparation, in general, those would not influence the risks related to coffee consumption. In certain countries coffee is flavored with spice mixes which should be looked at individually with respect to hazards related to those spices and respective control mechanisms. Reliance on final preparation with hot water and immediate consumption may not always be the case in such countries.

Glass used for packaging needs special attention. Soluble coffee is the variety that is often packed in coffee jars. Such jars could contain dust or other extraneous matters when delivered. Therefore, care should be taken to remove these either by blowing or turning them before filling, and preventing contamination later on. Additionally, glass breakage has to be considered and a detailed procedure put in place to trace/find all glass pieces in case such breakage occurs on the production line. The procedure could be handled as "oPrP" as outlined in ISO 22000 (2005). Although X-ray technology can be used to find non-metallic foreign matter, it cannot be relied upon as the only control measure. An example of a potential critical control point (CCP) for foreign matter control is included in Example 10.1, below.

EXAMPLE 10.1 CCP IN COFFEE PROCESSING – FOREIGN MATTER CONTROL

Critical Control Point I

Extraneous Material Detection

Process Step: Metal detector or in line X-ray unit.

Hazard

Physical (extraneous) material of the size and shape to pose a health hazard in finished product, e.g. metal, glass, stones, wood, hard and/or sharp plastic.

Critical Limit

Lines and/or processes identified as reasonably likely to pose a potential extraneous material hazard to finished product shall be equipped with a functioning extraneous material detection device (metal detector or in-line X-ray unit).

and

The Plant HACCP team shall identify in the HACCP plan both the quantity of product (packages, pounds, pieces) as X, and the timeframe or length of production time as Y to establish the critical limit parameters. Once determined, the process shall be managed as follows:

- The process is considered to be in control when there are less than or equal to X confirmed contaminated packages/pieces/pounds of product in Y production hours.
- The process is considered to be out of control, and a deviation to the critical limit has occurred, when more than X packages/pieces/pounds of product are confirmed contaminated in Y production hours.

NOTE: When determining process control status, the X "confirmed contaminated" packages/pieces/pounds of product can also be considered "number of diverted" packages/pieces/pounds of product in cases where on-line confirmation of contamination is not feasible.

NOTE:

- Determination for critical limit values for X and Y should be made based on product history.
- Use a standard set of terms to describe the findings (metal fines, nut, bolt, glass fragment, stone, etc.).
- Hazardous extraneous matter is material which is sharp and hard.
- Photographs of acceptable/unacceptable amounts and types of material are recommended.

Monitoring Activity/Frequency: Continuous: All packages/product shall pass through the operating extraneous material detection device. Rejected packs/pieces/pounds shall be evaluated to determine cause for rejection where feasible, based on the nature of the finished product. Based on the product evaluation, the number of packages/pieces/pounds of product rejected due to confirmed contamination event shall be recorded to identify when a critical limit deviation has occurred.

If the nature of the product precludes inspection, each rejection event will be assumed to be a confirmed contamination event. In this case, the number of packages/pieces/pound of product diverted shall be either automatically or manually recorded at defined frequencies sufficient to identify when a critical limit deviation has occurred.

Corrective Action Activity: If a critical limit deviation occurs, stop the process, place all product (packaged, unpackaged, rework, or other) produced during the suspect timeframe on Category II hold. Notify designated responsible person to determine disposition.

The suspect timeframe is the Y hour run time if the deviation is associated with X, either:

- confirmed contaminated packages/pieces/pounds of product

or

- total rejected packages/pieces/pounds of product in cases where on-line confirmation of contamination is not feasible.

An investigation shall be conducted to identify the root cause of the deviation, including attempts to isolate and identify the actual contaminants. Part of the investigation may include efforts to evaluate the potential of false rejects that are associated with the operation of the extraneous matter detection device instead of actual contamination events. This may include re-running rejected material through the device two or more times to check for additional rejection, or sieving of powdered product to determine the presence of any foreign material. If false rejection is initially suspected, this part of the investigation can occur prior to stopping the line and placing product on hold. Based on the investigation results, those additional actions will occur as needed.

After investigation, product determined to be contaminated with extraneous material of the size, shape, and/or nature to pose a food safety risk shall be placed on Category I hold. Notify Designated Quality Function for product disposition.

If investigation reveals the nature of the contaminant does not meet the criteria to pose a food safety risk, but rather is considered a product quality concern (e.g. metal dust, soft plastic packaging material, etc.), retain product on Category II hold. Based on internal risk assessment of the Quality team the product can be released. In case of uncertainty notify Designated Quality Function for product disposition.

If a detection device is not working at its design limit, stop the line and repair or replace the device. The suspect timeframe is back to the last acceptable equipment verification event if the detection device is found not to be working at its design limit during a verification check. Place the product produced since the last time the device was verified to be operating at its design limit on Category II hold. Re-run the held product through a properly operating device. (If a detection device cannot be repaired or replaced, the line can continue to run if the product produced is placed on Category II hold and run through a properly operating detection device later, or disposition of product can be determined using an alternate method as documented in the HACCP plan.)

Hold/Release documentation is required.
Corrective action must be documented.
RESPONSIBILITY: (Monitoring and Corrective Action)
Designated trained quality, production, and/or maintenance employees.
*RECORD/LOCATION:
Extraneous Material Detector Log
Hold and Release Records
Corrective Action Records
Verification Records

Minimum CCP Verification Activities: The extraneous material detection device is verified to be:

1. Set at the detection limit as defined during the validation (i.e. specific for the finished (i.e. converted) product being run). HACCP plans must list the size and type of target extraneous material that the detector will detect (mm).
2. Operating at the design limit by passing the required test pieces through the detection device in the manner and at the frequencies designated.
3. Operating with functional reject mechanism, verified at regular intervals.

NOTE: Ideally the reject mechanism should automatically divert the test pieces and attached product/package to an isolated identified bin or area to prevent reentry into the product flow. Where automated diversion of rejected test pieces and/or product is not feasible due to the nature of the product or equipment, at minimum the rejection mechanism shall stop the production line when the test pieces and attached product/package are passed through the aperture, with a trained operator responsible to remove the test pieces and attached product/package from the production line.

Verification activities shall be documented and included as part of the records review indicated below.

Designated responsible employee (usually the Supervisor) reviews and signs extraneous material detector records and verification records at least daily.

Scientific Basis (as used for validation of this specific CCP)

Moberg, L.J., 1992. Establishing critical limits for critical control points. In: Pierson, M.D., Corlett Jr., D.A. (Eds.), HACCP Principle and Applications. Van Nostrand Reinhold, NY, pp. 50–61.

Lewis, D.F., 1993. A tutorial and comprehensive bibliography on the identification of foreign bodies found in food. Food Struct. 12, 365–378.

Hyman, F.N., Klontz, K.C., Tollefson, L., 1993. Food and Drug Administration surveillance of the role of foreign objects in foodborne injuries. Public Health Rep. 108 (1), 54–59.

Hyman, F.N., Klontz, K.C., Tollefson, L., 1993. Eating as a hazard to health: preventing, treating dental injuries caused by foreign objects in food. JAM Dental. Assoc. 124, 65–69.

Lehman, A.J., 1958. Quarterly report to the editor on topics of current interest: glass and metal fragments in foods and beverages. Assoc. Food Drug Off. Q Bull. 22 (1), 24–26.

Olsen, A.R., 1998. Regulatory Action Criteria for Filth and Other Extraneous Materials. Regul. Toxicol. Pharmacol. 28, 181–189.

US Food and Drug Administration, 2005. Compliance Guidelines CPG Sec. 555.425. Foods, Adulteration Involving Hard or Sharp Foreign Objects.

* designate the location of each record

Dry Coffee Mixes

Dry coffee mixes refers mainly to mixes of soluble coffee with dairy components, sugar and flavors. All components are usually mixed in a dry stage, and then packed.

Attention needs to be paid to allergens (e.g. milk) and their control, which can be introduced in manufacturing areas at that stage.

Final preparation can be with either hot or cold water, depending on composition. From a food safety point of view the hazards brought in by the individual components should be looked at, and then appropriate controls established. Since the mixes are dry, no microbiological growth occurs over shelf-life.

Ready-to-drink Coffee-based Beverages

There are many different coffee-based beverages on the market; the range covers some containing milk and/or sugar, and/or flavors, as well as pure coffee concentrates. Shelf-stable beverages are either retorted or aseptically processed. For these the same considerations as for production of other commercially sterile products apply. This includes the choice – and thereby contributing microbial load – of the individual components to the final beverage, validation of the production process (retort or UHT and aseptic packaging), packaging integrity, as well as the distribution and final market, especially in view of temperature conditions. Climates with higher (tropical) temperatures would generally require higher heat treatments to achieve commercial sterility.

Special attention should be given to in-processing steps, where intermediate or final solutions are stored for certain times before final processing. Those steps need to be assessed towards growth potential of microorganisms for food safety and quality reasons. Some microorganisms can form heat-stable toxins which would be able to survive the severe heat applied in such processes. Examples are toxins formed by *Staphylococcus aureus* and *Bacillus cereus*.

Detailed information about thermal processing is provided by several other authors/references (ILSI Reports on Thermal Processing, Aseptic Processing, 2011, Deak and Farkas, 2013), and even regulations in certain countries (USA – FDA). It is also referred to in the chapter on thermal treatment (Chapter 17).

Green Coffee Beverages

In the last 10 years beverages produced either with green coffee beans or slightly heated and fermented coffee have entered the market. Some of them are dry powders (soluble) to be prepared with warm water, like "Ryslim." Others are ready to drink and fall in the category of energy drinks, like "Mondicina." At the moment, despite the use of green coffee beans, no other or special hazards are known compared to other beverages in the respective categories.

COCOA AND DERIVED PRODUCTS

Before discussing cocoa in more detail, some information is provided about the currently recognized main microbiological hazard – *Salmonella* – related to cocoa and chocolate products. This information could help to explain some of the challenges/difficulties in controlling this hazard, but also provide some guidance towards areas to focus on.

Salmonella in Cocoa/Chocolate Production (Low Moisture Products)

Until the 1970s cocoa and its products (chocolate, dry powders) had been regarded as a microbiologically safe product, because processing conditions were thought to be severe enough to destroy any pathogenic microorganisms brought in by untreated materials, and the low water activity (<0.6) does not allow for microbial growth. In addition to the low water activity, other antimicrobial parts of the ingredients had been thought to contribute to the microbiological safety of those products (Busta and Speck, 1968).

At the beginning of the 1970s the first outbreak of *Salmonella* could be traced back to cocoa powder and chocolate (Gästrin, 1972). Since then several outbreaks related to chocolate/chocolate products occurred. In all cases *Salmonella* was the microorganism causing the disease, thus making it the main pathogen of concern for low moisture cocoa products. Further analyses and investigations of the involved foods revealed that very minute amounts of living *Salmonella* seem to be sufficient to cause illnesses (see Table 10.1). Besides low moisture another common feature of all those products involved in such outbreaks is their high amount of fat. It had been assumed that because the cells are embedded in fat, they were able to pass through the stomach and infect the intestine (Blaser and Newman, 1982). Another aspect to consider is the long-term survival of the *Salmonella* cells in dry confectionery raw materials, as well as in finished products (chocolate) (Tamminga et al., 1976; Komitopoulou and Penaloza, 2009). Interestingly, serotypes involved in outbreaks (*S. Montevideo, S. Oranienburg*) seem to survive better in cocoa matrixes than others (Juven et al., 1984; Komitopoulou and Penaloza, 2009). Inhibitory/antibacterial properties of cocoa can be overcome by the addition of milk, especially casein (Busta and Speck, 1968; Zapatka et al., 1977). This could also be one factor why *Salmonella* tends to survive longer in cocoa products with milk components (Rieschel and Schenkel, 1971).

Furthermore, Salmonellae showed a much higher heat resistance in low moisture products compared to that in high moisture foods. It became clear that high fat and high sugar

TABLE 10.1 Outbreak Data Related to Chocolate and Cocoa Products

Year	Country	Serotype	Dose Detected	Reference
1970/71	Sweden	*S. Durham*	Unknown	Gästrin et al. (1972)
1973/74	USA/Canada	*S. Eastbourne*	Average 2.5 cfu/g	Craven et al. (1975)
1982	Italy	*S. Napoli*	1.6 cfu/g	Gill et al. (1983)
1985/86	Canada	*S. Nima*	4.3–24 cfu/100 g	Hockin et al. (1989)
1987	Norway	*S. Typhimurium*	≤10 cfu/100 g	Kapperud et al. (1990)
2001	Germany (and others)	*S. Oranienburg*	1.1–2.8 cfu/g	Werber et al. (2005)
2006	UK	*S. Montevideo*	0.3–10/100 g	Cadbury Statement (personal communication).

provided a favorable matrix which protected *Salmonella* from heat damage/destruction (Goepfert and Biggie, 1968; Sumner et al., 1991). The reasons for this are still not completely understood, but it is assumed that the absence of water and a dormant cell state play a role. It is also noteworthy that significant differences in the heat resistance in low moisture foods can be observed between *Salmonella* strains of the same serotype (Ma et al., 2009).

The main considerations for an effective *Salmonella* control in cocoa and cocoa products could be summarized as follows:

1. Appropriate validation studies should be performed to ensure that (heat) treatments, meant to control microbiological hazards, are effective. General guidance for validation studies is provided by a Codex Alimentarius document published in 2008. Before starting a validation, prevalence data of target microorganisms (e.g. *Salmonella*) in the matrix/raw material should be looked for, and then the level of reduction to be targeted by the treatment could be defined. Special attention should be paid towards the choice of target microorganism(s), since even strains of the same subspecies can be differently affected by treatments. In general, validation studies can be performed in different ways:
   a. In a laboratory scale experiments could be designed to mimic industrial conditions/processing parameters to evaluate their effectiveness against the target microorganism(s). In a next step the implementation of required parameters (e.g. time, temperature, moisture required) in production has to be validated as "equipment/processing" validation.
   b. Inoculated samples can be processed in the equipment, and reduction in microbial loads measured after processing. For that purpose surrogate organisms (i.e. non-pathogenic strains showing at least the resistance towards the treatment as the target organism(s)) are commonly used.

   Both validation options have several advantages/disadvantages and require a team of engineers and microbiologists to ensure the validation is effective for the process in question.
2. The long survival and persistence of *Salmonella* in dry environments makes it difficult to completely remove the contamination once it has found a way into production. Rigorous actions and long-term continuous verification activities are needed in such cases, and even then successful removal is not always guaranteed. Such actions could go as far as exchanging floors, removing superficial material layers, whole room disinfections, disassembling of lines and steam treatments of all equipment.
3. In recognition of the potential low infective doses still able to cause illnesses, statistical sampling plans have been developed by different organizations (Foster, 1971; ICMSF, 1986). However, even sampling according to such plans relies on certain statistical distributions of microorganisms in the food, and "spot" – inhomogeneous – contaminations in foods could be missed (Habraken et al., 1986). A more in-depth discussion of sampling plans and their use in food safety management can be found in different publications/guidelines (ICMSF, 2002; Bassett, 2010). Furthermore, there is no known clear correlation between the level of other microorganisms that could be used as "indicators" or even "index" organisms for *Salmonella* in such low moisture products and dry environments (Stadhouders et al., 1982).

Therefore, *Salmonella* sampling and testing of materials/environment cannot be relied upon to ensure the safety of the goods, but have their role in verifying the effectiveness of control measures in upstream processes. Furthermore, due to the potential inhomogeneous distributions, a single confirmed positive result would render the whole lot/batch adulterated, even with repeated negative results afterwards. To allow for appropriate root-cause analyses in such cases, serotyping and even further genotyping of the isolate(s) should be performed.

Much more detailed documents with respect to the control of *Salmonella* in low moisture foods have been published recently (Scott et al., 2009; Chen et al., 2009a). Also risk factors and sources of contamination related to past outbreaks have been reviewed recently (Podolak et al., 2010). Common contributing factors were inadequate processing and segregation of areas, contaminated raw materials, poor sanitation/GMP practices, employee awareness and not investigating potential root-causes in case of pathogen findings. For effective food safety management the last point is of utmost importance, since without knowledge of the root-cause appropriate corrective actions and prevention of recurrence is not possible.

Understanding of outbreaks from current cocoa processing have led – and still lead – to significant changes with respect to infrastructure (strict segregation of raw beans), raw material controls (ongoing supplier control programs), hygiene programs (hygiene sluices verified by regular test programs), involvement of the whole manufacturing environment (to prevent contamination from areas adjacent to production) and even surroundings of the plant including water sourcing.

Raw Cocoa Beans

Cocoa beans are the seeds of the small tree *Theobroma cacao* L. which originates from South/Central America. Today there are three main growing areas in the world: West Africa, South America and Southeast Asia. Early on a cocoa variety known as Criollo had been introduced to Asia. Due to its low yield and susceptibility to diseases this type is no longer widespread. The main variety of cocoa currently cultivated is called Forastero, or Amelonado, which is essentially a variant of Forastero. There is a third variety known as Trinitario, a hybrid between Forastero and Criollo. Nacional is another variety only grown in Ecuador and producing beans with so-called "Arriba" flavor, mainly used in the production of dark chocolates. Around 95% of the world's cocoa production is classified as "bulk cocoa" originating from Forastero-type cocoa trees. The rest of the production is classified as "fine" or "flavor" cocoa, and is used mainly in dark chocolates where they provide the special flavors and/or colors to these products.

The majority of cocoa is produced by small farmers who then sell their product (dried fermented cocoa beans) to intermediate traders who sell it to larger traders. From their warehouses it is then shipped to or processed in larger cities in the countries of origin.

After 5–6 months the cocoa pods ripen and are cut by hand. In this regard cropping can either be continuous like every 2–4 weeks or rather there are 1–2 peak harvesting periods per year. In West Africa the main harvest period is from the beginning of October until December.

The pods are then opened and the beans separated from the placenta. Subsequent fermentation of the beans is an important step for the development of flavor precursors – and therefore cocoa quality. During fermentation the development of cocoa flavor precursors depends on the initiation of germination, the death of the embryo liberating enzymes and substrates, and the hydrolysis of storage proteins into oligopeptides and amino acids. Fermentation is equally important for the production of a stable, non-perishable commodity, which can be traded over long distances and times. Generally, the fermentation process involves three stages: anaerobic yeasts in first 24–36h, then as yeasts become inhibited by ethanol, increasing pH and oxygen, lactic acid bacteria becomes more dominant at around 48–96h. They convert available sugar into lactic acid. As aeration is even more increased, acetic acid bacteria become more dominant, converting ethanol into acetic acid. Since this is an exothermic reaction the temperature rises to 50°C. At the end of the fermentation process numbers of spore-forming bacteria, especially thermophilic Bacilli, increase.

Following fermentation the beans are dried either in the sun by spreading them on any suitable horizontal surface or by mechanical means. Beans are then hand sorted to remove debris, broken beans and other foreign matters. All these steps are susceptible to contamination with Enterobacteriaceae, including *Salmonella*, by animal droppings, pests (insects), and also by manual handling.

The final moisture of the beans must not exceed 8% in order to effectively prevent mold growth. In addition, beans should be stored away from any other odorous substances, since cocoa beans easily absorb such odors making them unpalatable for further usage.

A more detailed description of these processes has been discussed elsewhere (Fowler, 2009).

Raw cocoa beans entering processing facilities can have quite high microbial loads, around 10E6cfu/g, where half of which consists of Enterobacteriaceae, and the majority of the rest consists of thermophilic Bacilli spores. Therefore, incoming microbiological testing of raw cocoa has no relevance, and even the presence of pathogenic bacteria (*Salmonella*) is no reason for rejection. Due to their microbial load raw cocoa beans and their handling should be well separated from the processed side of production.

The potential of mold growth, caused either by insufficient drying or by improper storage, has to be considered for two main reasons. Mycotoxins may be formed; however, cotyledons contain large amounts of inhibitors against aflatoxins, mainly methylxanthines (Buchanan and Fletcher, 1978). The shells, which are more likely to be contaminated, are removed during further processing. On the other hand, ochratoxin A (OTA) formation may not be inhibited to the same extent, and its occurrence has been reported in cocoa (Bisbal et al., 2009).

Another cause of mold growth is lipolytic spoilage, especially by molds from the genus *Aspergillus* (ICMSF, Volume 6, 2005). Monitoring programs with respect to mycotoxins and general mold spoilage should be in place.

Other chemical contaminants, like heavy metals, should also be included in such monitoring programs. It is noteworthy that heavy metal content, especially cadmium, derives partly from the soil and uptake by the plants, and is, therefore, dependent on origin (farms).

Such monitoring programs are most suitably handled as part of the prerequisite programs at processing facilities. Neither mycotoxins nor heavy metals are significantly reduced during further processing. They can potentially even be concentrated in certain products based on their solubility in fat or water.

Attention should also be paid to pest infestation of raw beans. Due to their nutritional value various insects breed on them. Of special importance is the "tropical warehouse moth," which leaves the bean when fully grown to find another location to pupate. This particular behavior enables the moths to infest other areas of a processing plant, when raw bean reception and storage are not strictly controlled. In this regard, effective pest control programs – as with all other prerequisite programs – form an important part of food safety management, since pests can function as "carriers" of pathogens.

A detailed overview of cocoa processing/chocolate manufacturing technologies can be found in the book *Industrial Chocolate Manufacture and Use* (Beckett, 2009).

Cocoa Mass/Liquor Production

Further processing of the raw beans then starts with different cleaning steps to remove extraneous matters by screening, air currents and magnets. Consideration should be given to potential damage of whole beans during cleaning operations, e.g. by dropping them on metal surfaces. Such damage could have a negative impact on separation of nibs and shells later in the process, and, furthermore, increase the level of contamination of the nibs.

After cleaning there are many different ways of processing as outlined in Figure 10.2. The choice of the process step as a control for microbial hazards will determine the zoning concept of the whole plant (see "Chocolate").

Roasting of the Beans

Cocoa beans are dry roasted primarily to achieve a certain flavor, but not to control microbial hazards related to raw beans. Due to the required temperatures to produce the

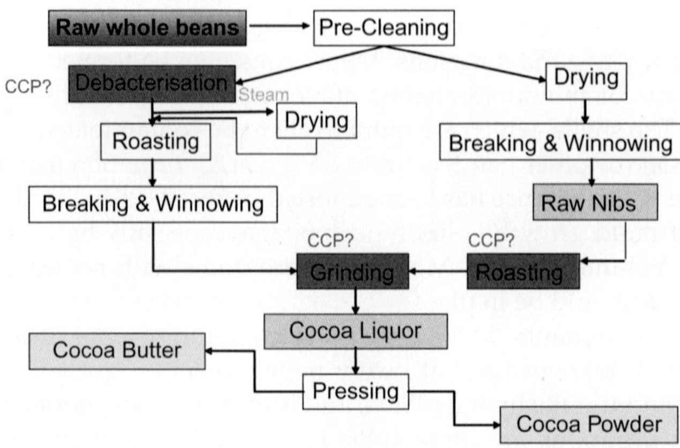

FIGURE 10.2 General scheme of cocoa processing. Note: other potential options/orders of processing are outlined in the text.

desired flavors (110–140°C at exit of roasting), the microbial load of the beans is significantly reduced (Penaloza-Izurieta et al., 2008). However, to be established as a control point for microbial hazards, a consistent, validated reduction of pathogenic microorganisms has to be achieved. This is a big challenge in dry roasting operations where many variables influence the kill effect during processing: fluctuating incoming moisture of the beans, heating transfer by air flow, partially broken beans and residence times in roaster. There are other processes in cocoa processing more suitable as control points for microbial hazards, as outlined below.

Steam Debacterization (Predominantly used for Cocoa Beans)

Originally, different heat treatments have been introduced to allow for easier and better separation of shells and nibs. In view of necessary process controls in place to control microbial hazards, steam treatment under pressure has been recognized as a suitable control step in that regard. By using pressurized steam all beans/particles in a vessel are treated adequately, and the introduced moisture significantly increases the lethality of the process. Depending on the parameters used steam debacterization can also be used to significantly reduce the thermophilic spore load of the cocoa beans (Stehli et al., 2002). Following steam debacterization the cocoa beans have to be dried again to a certain moisture content in order to allow a high quality roast.

Depending on the process flow, cocoa beans can also undergo debacterization after roasting. In such cases the treatment is set in such a way that moisture is not significantly increased by the treatment and beans can be broken afterwards. Moisture uptake by the beans is also minimized because they already enter the vessel with high temperatures (90–100°C) coming from the roasters.

Validation of critical process parameters should especially include tests at start-up of the equipment after longer stoppages. Procedures at start-up normally include warming up the equipment to a certain temperature before loading the product. It should then be ensured that there is no condensation of the steam, which would significantly decrease its effectiveness. In order to prevent this some equipment has a so-called superheater installed before the steam enters the reaction chamber.

Old models may face the problem of having only one door – used for loading of raw beans and unloading of processed ones. In such cases the risk of cross-contamination needs to be assessed and controlled.

In cases of processing failures, required corrective actions have to be taken. This could include reprocessing until process parameters are achieved, or unloading and reprocessing or disposal of the products in question. The unloading operation needs attention with respect to any part of line equipment coming in contact with that product – and its subsequent necessity to clean and disinfect it before reverting to standard operation.

In high-throughput operations a continuous steam debacterization of the nibs can be used. Once again, the adequacy of the start-up operation has to be assessed, and corrective actions in case of process failures carefully looked at. In some operations water is added to the process, but this water is mainly used to remove potential off-flavors.

With respect to chemical hazards the steam in direct contact with the product needs to be of culinary quality. Absence of turbidity, off-flavors and particles should be ensured and verified by routine checks of the condensate.

Breaking and Winnowing

Before breaking, beans can pass through an infrared (IR) heater in order to improve the separation of shells and nibs (as already mentioned for the steam treatment above). Such an IR treatment also decreases the microbial load of the beans/shells to a certain extent, but could not be used as a control step for microbial hazards, since an equally sufficient heat treatment of all, including broken, beans is not achieved.

The most commonly used equipment to break cocoa beans is disc breakers which operate with centrifugal force. After hitting the impact plate broken beans fall down and are conveyed to the winnower for sorting. Separation of shell and nib is achieved by using different sieves and air flows in several sections.

Potential contaminations by extraneous matter, especially by worn impact plates and/or broken sieves, should be taken into account when setting up routine checks and preventive maintenance regimes. Engines should be installed in a way to prevent lubricants to drop into products, and preferably only food grade lubricants should be used.

Nib Roasting/Alkalization

Historically, nib roasting was introduced to cocoa processing when it was recognized that alkalization of the nibs before roasting improved some characteristics of the cocoa powder. Around 1800 the so-called "Dutching process," better known as "alkalization," had been developed in the Netherlands. Alkalization is used to achieve a distinct dark color by roasting the nibs or treating the cocoa mass with alkali solutions. The three main chemicals used in this process are sodium hydroxide and calcium or kalium carbonate. Alkalized cocoa is mainly used for cocoa powder production.

Currently, alkalization is predominantly performed as part of the nib roasting process. Commonly used equipment is batch type roasters (e.g. Barth roasters) which allow for steam/water and alkali mixes to be injected into the nibs that are being rotated to avoid localized overheating. The three main steps in such a process are preheating of the nibs, injecting the liquid and roasting of the nibs. Following roasting the nibs are transferred to an external cooler. The air used to cool the nibs should be adequately filtered so that it does not become a source of contamination. For guidance with respect to air quality, reference is made to Brown (2005).

Validation of that process for control of microbial hazards requires specific knowledge of all roasting profiles used at the line. The least favorable one in terms of bactericidal effects should be chosen for the validation of the process. Main parameters influencing the bactericidal effect consist of moisture of the nibs during processing (or at the end of the process before cooling), amount of water addition, temperature profile and processing time (see CCP example as outlined in Example 10.2).

EXAMPLE 10.2 A CCP IN COCOA PROCESSING

Cocoa Nib Roasting/Grinding

Critical Control Point ID: Heat treatment (grinding and roasting) of cocoa nibs.

Control Step: Time, temperature and moisture to reduce 6 logs of *Salmonella Eastbourne* and *Salmonella Napoli*.

Hazard: Biological (*Salmonella* – strains involved in outbreaks related to dry products, chocolate).

Critical Limits: Time/temperature/moisture as follows:

If moisture 1.0–2.5 %, then:

Min. Temperature	Min. Time
120°C (248°F)	11.70 min.
130°C (266°F)	4.93 min.
140°C (284°F)	2.08 min.
z = 26.67°C (48.08°F)	

If moisture above 2.5%, then:

Min. Temperature	Min. Time
110°C (230°F)	4.40 min.
120°C (248°F)	2.04 min.
125°C (257°F)	1.39 min.
130°C (266°F)	0.95 min.
140°C (284°F)	0.44 min.
z = 30.03°C (54.06°F)	

(lowest applicable temperature is 90°C/194°F)
Note: Applicable moisture has to be measured after roasting, before cooling.
Equivalent time–temperature parameters can be calculated using z values.

Monitoring Activity/Frequency: **Temperature (processes without holding tube):** Temperature of the product at the coldest spot shall be continuously recorded on a temperature chart.

Temperature (processes with holding tube): Temperature of the product at the end of the holding tube* shall be continuously recorded on a temperature chart. If the required holding time is instantaneous (0.5 sec or less), then the temperature sensor can be located after the heat exchanger.

Time: Flow rate shall be recorded continuously or pump setting is recorded once per shift and after speed changes or the pump seal integrity (sealed by authority or plant) is recorded daily or it is technically not possible to exceed the time requirements (this must be documented).

NOTE: The correlation flow rate/holding time for the fastest particle must be documented and filed with the HACCP plan.

Moisture: Water addition is recorded continuously or moisture measurement minimum once per shift.

Corrective Action Activity: Underprocessed product shall be automatically diverted and retreated or post processed product will be identified, put on Cat 1 hold and the designated Quality Function will be contacted to determine next steps or product will be discarded. The system divert shall be reflected by the frequency pen marking on the temperature chart.

Hold/Release documentation is required.

* if the holding tube is heated, the temperature has to be recorded at the coldest spot of the tube.

Corrective action must be documented.
RESPONSIBILITY: (Monitoring and Corrective Action)
Designated trained employee
*RECORD/LOCATION:
Temperature Charts
Moisture Measurements
Measuring Equipment Logs
Hold and Release and Corrective Action Records
Verification Records

Minimum CCP Verification Activities: Designated responsible employee (usually the Supervisor) reviews and signs processing records at least daily.

All measuring devices used to monitor critical control parameters shall be calibrated at a frequency sufficient to demonstrate control (minimum every 6 months).

See also in HACCP standard Section 11 "HACCP System Verification" point 1.1 Individual CCP Verification Activities (1st level).

Scientific Basis (as used for validation of this specific CCP)

Leatherhead Food Research Association, 1990. Effect of moisture level on the heat resistance of *Salmonella* in cocoa liquor, Research Report No. 666, April 1990.

Krapf, T., Gantenbein-Demarchi, C., 2010. Thermal inactivation of *Salmonella* spp. during conching. LWT – Food Sci. Technol. 43, 720–723.

NOTE: In confectionery products *Salmonella* has been recognized as the main pathogen of concern. (ICMSF "Microorganisms in Food 2: Sampling for Microbiological Analysis" 2nd edition, 1986).

* designate the location of each record.

The water used in the process should be of potable quality.

Grinding of the Nibs/Cocoa Liquor Production

Various mills are used to grind the cocoa nibs, and usage depends on the final required characteristics of the cocoa liquor (particle sizes). In general, grinding is a two-stage process: coarse grinding carried out by hammer/disc mills, and a second stage where ball mills are used. Since ball mills consist of many small metal balls, where the pre-grounded nibs are pushed through, abrasion and potential losses of the balls should be considered and adequately controlled. Due to the heat produced by friction in ball mills, these are normally double-jacketed and cooled with water. This water should be included in routine control programs and monitored to ensure adequate quality.

In some processes alkalization is not performed at the stage of the nibs, but with the cocoa liquor. This is normally performed in special vessels, allowing for addition of the alkali solution to the liquor, processing for a certain time, and then removal of the excess water by applying a vacuum. This process step could potentially be used to control microbial hazards. However, the distribution of moisture (as a key factor in achieving bactericidal effects), as well as the coldest point in the reaction vessel, need to be determined and taken into account as least favorable conditions when performing a validation.

There have been a few studies done towards the bactericidal effect of heat treating cocoa liquor at different moisture levels (Davies et al., 1990; Krapf and Gantenbein-Demarchi, 2010). These studies showed that heat resistance of *Salmonella* in cocoa liquor is in the same range as found in chocolate. Bactericidal effects in a timeframe suitable for process controls are only achieved at higher temperatures (>90°C).

Theoretically, several processing steps could be combined to achieve a certain bactericidal effect. However, from a control point of view this is difficult to manage due to the many variables to be included in determining critical limits, and due to diversion/retreatment in the case of processing failures.

Cocoa liquor can be stored/transported in liquid or solid form (normally blocks of 25 kg) under appropriate storage conditions. Storage of the liquid cocoa liquor requires warm temperatures of around 50°C, and is achieved by using double-jacketed equipment.

Cocoa Butter/Cocoa Powder

Cocoa liquor can then be separated by hydraulic pressing into cocoa butter and cocoa cakes, which are then ground to powder. Another rarely used option of separation consists in solvent extraction. Here, the risks of residual chemical residues have to be considered.

Hydraulic pressing is an open process where removed cocoa butter freely flows down in channels towards storage tanks, and the cocoa cakes drop onto a conveyor. The position of equipment parts which need lubrication, as well as the kind of lubricants used, need to be assessed with respect to potential product contact.

In order to prevent environmental contamination at that stage, strict zoning/separation as mentioned above should be implemented. The effectiveness of such measures should be verified at regular intervals by environmental sampling (for more details see below in "Chocolate").

Cocoa cakes can either be intermediately stored or immediately processed further. Intermediate storage needs to be clearly defined in order not to negatively influence the flavor and microbiological quality of the cocoa. Although the low moisture of the cake would not allow for microbial growth, storage in high humidity conditions or conditions allowing for condensation could increase the amount of available water to allow for potential growth.

Cakes are further processed by first breaking them into smaller pieces, usually using a tunnel with a screw inside. During this process excessive heat is produced, and therefore this kind of equipment is usually cooled by water. The water, as mentioned before, should be adequately controlled. Attention should be paid to the prevention of condensation inside the equipment, which could lead to microbial problems, and quality problems with respect to sticking of the powder to the walls. The small pieces are then further ground in mills where often forced air is used as a cooling agent. The air used in this process should be adequately filtered to prevent contamination of the product. Later on in the process sieves and magnets are used to control extraneous matters; the sieves themselves need to be considered as sources of extraneous matter when they break.

The cocoa powder can then either be packed in big bags or small containers/packs or agglomerated when used in soluble cocoa drinks. Agglomeration changes the surface structure of the powder by applying steam in a continuous process to free-flowing powder. The steam is used in direct product contact and should therefore be of culinary quality. The potential of condensation in the equipment should be evaluated and prevented.

Cocoa butter coming from the press is normally filtered to remove small solid particles. This can be done by using paper filters which should be regularly exchanged and monitored for findings. Then the product is stored/used or processed further by deodorization. This process removes odors/flavors by heating to high temperatures and using culinary steam under vacuum.

Cocoa butter, similar to cocoa liquor, can be stored and transported in liquid or solid forms.

Chocolate

Chocolate making involves many steps which can be quite diverse depending on the final products. Nevertheless, there are general points which need to be considered in all chocolate productions.

Raw Materials

All incoming raw materials should be grouped, based on known potential risks, available historical literature, outbreak data, as well as the knowledge of currently used processing technologies. Most well-documented examples of processed materials with increased risk of microbiological contamination (*Salmonella*) are dairy powders, egg and nut products, and cocoa itself (for cocoa see previous section). Detailed discussions about the other raw materials mentioned above can be found in other chapters of this book.

In general, the final steps of chocolate production (from mixing of ingredients until molding) do not include a control step for microbial hazards. Therefore, the safety and quality of the final product would depend on the raw materials used. Any contamination present in a raw material would be processed and not only end up in the respective finished product, but also has the potential to contaminate production lines leading to a more widespread contamination issue in the plant. It is for these reasons that production processes at suppliers should be looked at and controlled as part of internal food safety management. This could include supplier audit programs and audits performed by specially trained and qualified auditors for different material groups.

Fillings and inclusions used in chocolates are very diverse, and all different kinds of raw materials have been/are used. This growing specialty segment also includes dried fruits, spices, herbs and other materials which do not necessarily undergo an effective control step for microbial hazards at the supplier site. Also, other microbiological hazards, like viruses and parasites, need to be considered here. Before using such ingredients a thorough risk assessment is necessary and any potential processing options of such materials at the supplier or internally (before being added to chocolate mass) should be evaluated.

Since all of those individual raw materials are discussed in more detail in other chapters of this book, the reader is referred to these chapters, which detail similar chemical and potential physical hazards.

The effectiveness of supplier control programs can be verified by regular sampling and testing of incoming raw materials. The testing regimes should be set up based on the risks associated with the raw materials, and any finding should be followed up with the respective supplier(s). The limitations and statistical considerations of testing regimes have been discussed at the beginning of this section.

Appropriate zoning of raw material storage and handling is necessary, where raw agricultural commodities with high contamination risks (e.g. raw cocoa, raw nuts, flour) are being processed in chocolate plants.

Additionally, internal processes should be assessed and, if possible, validated to establish control steps for biological hazards. Typical processes to be looked at would be heating steps during production, e.g. cocoa bean/nut treatments, cooking of pre-solutions, caramel production, baking.

Current challenges are related to effective processing techniques in minimally processed raw materials, effective prevention of recontamination, as well as traceability to the primary origin of all ingredients used.

It should also be kept in mind that the continuing globalization of supply and trade can introduce new risks, and also change the significance of hazards, in the raw material supply. This refers as well to physical and chemical hazards associated with raw materials used. However, only allergens will be discussed in more detail below (see subsection on good manufacturing practices (GMP)), due to some specifics in allergen handling and control in chocolate plants.

Individual Steps in Chocolate Manufacturing

As mentioned above, there are many different ways of producing chocolate. The main paths are summarized in Figure 10.3.

Chocolate Crumb

Historically, crumb making started in Switzerland with the first milk chocolates, came to England where it was further developed, and is now being used in different parts of the world. There is not just one crumb-making process, but very different ones, depending on raw materials used as well as throughputs required.

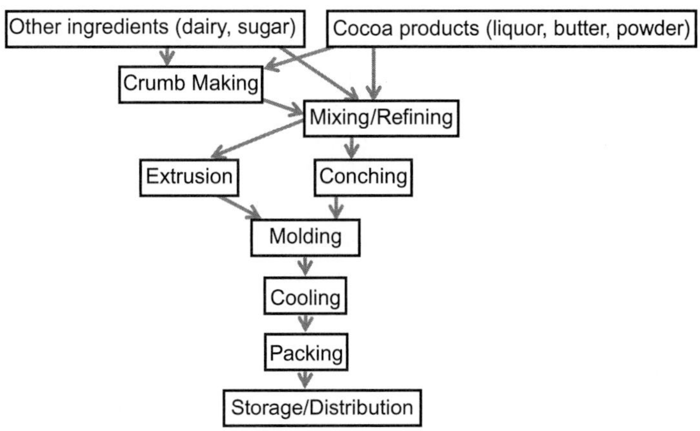

FIGURE 10.3 Chocolate-making processes.

From a food safety point of view, the main difference – and things to consider – is in the raw materials used. Whereas some crumb processes start with only processed raw materials, like milk powders and sweetened condensed milk, others start the process with raw milk being delivered to the plant.

Where raw milk is being used, reference is made to the chapter on milk, since the same considerations apply until the milk is being pasteurized. Following pasteurization the milk is concentrated before being mixed with sugar to produce sweetened condensed milk used in crumb making. Until the production of the crumb each process step has to be evaluated with respect to its growth potential during storage times. Concentrated milk does not provide protection against microbial growth. Such growth can only be controlled by applying defined temperature–time restrictions, and implementing adequate cleaning regimes. This also includes any potential circulation loops, reclaim liquids, as well as defining cleaning requirements in case of production interruptions.

As for processed raw materials used in crumb production the same considerations apply as for the other raw materials used in chocolate production.

All ingredients are mixed together and either water is added at that stage or shortly later during extrusion. The actual crumb making involves heating of all ingredients in order to allow also for certain caramelization of the sugar (to develop specific flavors), and then drying to remove the excess water used in the process. Drying is normally done in vacuum dryers, where attention should be paid towards maintenance of such systems.

Crumb making can be performed in closed vessels or by using extrusion technology. The heat generated during these processes could allow for control of microbial hazards, but this would require detailed knowledge of the behavior of microorganisms in the matrix under the given processing conditions. Despite this, monitoring of the process in order to ensure control and diversion in case of process failures can be difficult to manage.

Due to the low water content the final crumbs can be stored for prolonged periods of time (which was one of the advantages of crumbs).

Pre-mixing/Refining

Dry ingredients for production of chocolate are pre-mixed and then refined to reduce the particle size to the required stage at different refiners. Transportation of ingredients could involve transport by air, and that air should not be a source of contamination. Therefore, appropriate filtration should be installed and the system maintained.

Refiners consists of three or five metal rolls with adjusted distances in between where the sticky mass of mixed ingredients is pushed through from bottom to top. Due to the friction during refining the rolls need to be cooled, which is commonly achieved by water flowing inside the rolls. That water should be controlled in the same way as other water used in double-jacketed equipment (see p. 276, Double-Jacketed Equipment). Abrasion and extraneous matter contamination should be evaluated and controlled by appropriate means, e.g. preventive maintenance and magnets installed.

Conching

Conching is considered an important step for flavor development in chocolate. The refined powder mass is added into the conche, where mixing starts and over time other ingredients like cocoa butter, emulsifiers, flavors, nut pastes will be added to the conche.

This process usually takes several hours, and is performed at a wide range of temperatures (50–90°C). The temperatures used depend on the kind of chocolate produced, where milk chocolate is usually treated at lower temperatures than dark chocolate. From a food safety perspective, this treatment is not sufficient to control microbial hazards, especially *Salmonella* in that low moisture material (Krapf Gantenbein-Demarchi, 2010).

Conches are also double-jacketed and the water used for heating should be controlled correspondingly.

Extrusion

This type of extrusion does not lead to extensive generation of heat like the extrusion used for crumb making. The pre-mixed and refined chocolate is extruded and emulsifiers and flavors are added during the short process. Extrusion times are in the range of minutes.

Attention should be paid to the use of lubricants and prevention of leakages that come into product contact.

Filling Preparation

Many different kinds of fillings are used in chocolate. From a food safety point of view two major groups can be differentiated: fat based and water/sugar based. Fillings with a water activity <0.6 would not allow for microbial growth. However, higher water activity values could allow for osmophilic yeasts to grow and spoil the finished products during the long shelf-life. Since heat treatments severe enough to destroy such osmophilic yeasts are not commonly used at this stage, raw materials should be carefully controlled, and cleaning of the line appropriately done and verified.

The fat-based fillings refer to nougats, nut pastes, marzipan and creams of all kinds. They can be received ready to use in chocolate plants, and only melted and tempered there, or they are prepared in mixing tanks in the plant.

Water/sugar-based fillings contain a high amount of sugar in order to lower water activity (also with a view to minimizing water migration between the different components in chocolate). With respect to osmophilic yeasts the same considerations apply as above. To increase microbial stability of such fillings, preservatives can be added. The most common one is ethanol, which is at the same time the most effective one when used to a concentration of 15–20% in the watery phase. Besides usage of preservatives adequate (heat) treatments in conjunction with prevention of recontamination of such fillings can also be an option to increase microbial stability.

Caramel fillings are also widely used in chocolate products. Caramel is produced by mixing dairy ingredients (or just water) with sugar, and then heating the mixture to high temperatures (>100°C) to ensure caramelization of the sugar. When dairy ingredients are used, storage conditions and cleaning cycles of any liquid dairy storages/processes should be carefully controlled to prevent microbial growth. Although the heat processing of the caramel would most likely destroy vegetative bacterial cells, some toxins formed by them would survive the heat treatment. Examples are toxins formed by *Staphylococcus aureus* and *Bacillus cereus*.

Marshmallows and other forms of aerated fillings based on egg whites are also produced and enrobed in chocolate. In their production the most critical step for microbial growth is the dissolution of the egg white and its further processing, since in that form it could also give rise to bacterial growth and potential toxin formation. In this case there is no heat processing involved, so that any bacteria dissolved and grown in the egg solution would contaminate the finished products.

Molding

The obtained liquid chocolate, after conching/extrusion, is stored in double-jacketed tanks; all transport systems are double-jacketed including pumps. Also, tempering machines and depositors at the molding lines are kept warm the same way. Therefore, preventive maintenance and control of water should be an essential part of the prerequisite programs in chocolate production (see above).

Attention should also be paid to the use of lubricants and prevention of leakages that come into product contact.

The molds used are usually made of plastic – and can break during continuous operation. To prevent contamination with extraneous matter, molds should be controlled for damage and exchanged accordingly. Before depositing chocolate, molds are preheated which can be done by air, where the air should not give rise to contamination.

Deposited molds are vibrated, and cooled in long cooling tunnels. The air used for cooling should be adequately filtered to prevent contamination, and cooling tunnel temperature and humidity have to be controlled to prevent condensation. Temperature-controlled rooms may be required in certain countries to prevent condensation on the chocolate when leaving the cooling tunnels. In climates where outside temperatures are too high, water would rapidly condense on the surface of the (comparably) cold chocolate when leaving the cooling tunnels. Regular visual checks should ensure that no condensation occurs on the products during/after cooling.

For filled chocolates, several molding/filling/cooling steps can be used one after the other, before the final product is completed.

Following demolding chocolates are packed, weight checked and run through metal detection. When metalized foils are used for packaging, metal detection is performed before packaging.

The molds used in production are normally wet cleaned in cleaning machines. Most of these machines include a drying section at the end, and if not, molds have to be dried separately before being reused. In order to ensure the dryness of the molds, a visual control should be implemented after drying operations.

Storage/distribution

Chocolate products should be stored at defined temperatures and humidities. Rapid changes in temperature could cause condensation on the surface of the products in the packaging, leading to mold growth at these spots.

Furthermore, as already mentioned for cocoa, chocolate should be stored away from any distinct odors/flavors, since it readily absorbs these leading to taste defects when reaching consumers.

Chilled Chocolate Products

Recently some chilled chocolate products have appeared on the market. Some of them may be cooled purely because of taste reasons, but some contain dairy fillings which require cooling for safety and quality reasons. These products usually have a significantly shorter shelf-life than traditional chocolates, and should be treated like dairy-based desserts. Food safety considerations for the dairy fillings would be essentially the same as for dairy products in that range, and from a quality point of view yeast and mold growth over their shelf-life needs to be controlled.

Specifics of Cocoa as Ingredient in Other Products

When using cocoa as an ingredient the predominance of thermophilic *Bacillus* spores due to cocoa fermentation needs to be taken into account. These spores could cause problems in products due to

1. their high heat resistance, which could allow them to survive standard heat treatments; and
2. later outgrowth in final products with high water activity (e.g. milk products) and distributed in hot climates.

Therefore, special cocoa powders are available on the market where the spore load is significantly reduced.

Some specific examples of cocoa-derived products are mentioned below.

Shelf-stable cocoa-based drinks and desserts are predominantly milk-based retorted/aseptically processed products, where considerations for these products would apply.

Also available are *chocolate sauces* used as toppings or concentrates to prepare drinks at home or in vending machines. Such sauces are commonly heat processed and packed to prevent growth of osmophilic yeast/molds, but their water activity is low enough to prevent growth of other microorganisms.

Dry instant cocoa beverages are also widely known. These can be produced in two different ways:

1. One process involves dry mixing of all ingredients which are then packed accordingly. Here again, there is no control step in the process, and relevant prerequisite programs need to ensure safety and quality of the finished products.
2. Another process starts with the production of a thick viscous mass consisting mainly of malt extract, cocoa and dairy ingredients. These are then heated in batch ovens or in a continuous process, followed by a vacuum drying step to remove water. Due to the high amount of solids and sugar in the mass, the use of the heating step as a control of microbial hazards is difficult to validate. The solid mass is then broken into pieces, and finally grinded into powder before being packed. The same general considerations would apply as for the dry mixing process.

Good Manufacturing Practices/Hygiene Requirements in Cocoa/Chocolate Production

Allergens

Due to the diversity of the finished products in chocolate manufacturing, different allergens or ingredients containing allergens (e.g. milk, egg, nuts and seeds) are used in production. Most commonly production lines are not dedicated to single specific products, but are used for certain groups of finished products, like plain chocolates, chocolate with inclusions and with fillings, pralines and countlines (small bars). In such cases products containing different allergens are being processed on the lines. Although complete removal of previously used allergens is possible, this requires thorough validation and ongoing verification activities. Therefore, in most cases the aim is to minimize allergen carry-over to the next production by cleaning (see the paragraph on cleaning, below) or flushing. It is recommended to manage allergens risks in the plant by appropriate control measures like:

1. A plant layout should indicate where allergens are stored, transported and handled in order to identify and control potential cross-contamination points within the plant (similar to assessing cross-contamination by microorganisms).
2. For each line all finished products and their raw materials should be evaluated with respect to their allergen profiles (containment or potential carry-over from supplier processes).
3. All shared equipment should be included in the assessment, and cleaning procedures put in place to minimize allergen carry-over.
4. Any rework/flushing masses/overflow used in the plant should be stored, handled and reused in a way to minimize allergen carry-over.
5. Zoning can be used to prevent and minimize allergen cross-contaminations.

Management of allergens is becoming increasingly important, and as work is being initiated towards quantitative thresholds for allergens, more focus on that subject, in terms of detection methods and intervention strategies, can be expected in the near future.

Rework

Rework is created at different stages in manufacturing, and should be assessed in the same way as other raw materials entering the process. The handling, storage and any potential pre-processing of rework need to be part of that assessment, too. The point(s) of rework addition(s) need to be clearly known and evaluated for their adequacy of the operation. The more the upstream rework is added in the process the more it is being spread to different areas, potentially making some points, like allergen minimization or traceability, difficult to handle. Preferably, rework is only added to the same kind of finished products (also called like-to-like addition) at the final stages of chocolate molding.

Any detected issues in the finished products from where the rework is created should lead to an immediate re-evaluation of the respective rework and determination of adequate corrective actions. In case of microbiological problems, sampling and testing are appropriate to verify the controls in place, but cannot be used as a control on their own.

Double-Jacketed Equipment

In order to keep products in a liquid state double-jacketed equipment is used at all stages from cocoa liquor to chocolate. This is not only limited to tanks, but includes all parts of the lines, and even the pumps.

The water used should be microbiologically monitored and preventive maintenance put in place to detect leakages into the tanks/pipes. In general, major leakages would be noticed due to the significant increase in product viscosity, but micro-leakages could potentially go undetected.

Water cycles can vary from large cycles feeding many systems down to individual equipment pieces with closed cycles. It is very important that a current map of all water cycles is available – and is updated accordingly in order to allow for appropriate testing/control regimes. The water used to feed these cycles, as well as its treatment, should be looked at to ensure that the in-feed water is of potable quality.

In other cases, where water is used for cooling, chemicals may be added to the water, and these should be assessed for use in indirect product contact, and for their potential bactericidal properties. An example is propylenglycol, where a bactericidal effect is achieved at concentrations >30%.

Zoning (Separation) of Areas to Prevent Microbial Cross-contamination

Zoning (separation) of areas is necessary to prevent cross-contamination between raw, contaminated areas (e.g. raw cocoa beans) and processed areas in a plant. For effective zoning all incoming raw materials need to be assessed with respect to their microbial load and further processing. For example, chocolate plants may also produce wafers where flour is used as an ingredient in dough. Flour, due to its potential for carrying pathogens, should also be readily separated from other raw materials without any control step for microbial hazards.

Furthermore, all areas of the plant including offices and warehouses need to be taken into account when setting up the zoning of a plant.

Measures to control identified sources of cross-contamination risks should be documented, identified in a risk assessment and implemented. Such control measures can include the use of closed systems (e.g. tanks and pipes with pneumatic transport), structural separation of the area by design (e.g. separate building, walls), restricted and controlled traffic patterns of people, materials and equipment, air pressure differentials (e.g. negative air pressure in raw, contaminated areas, so that air would not flow to processed areas in the plant), use of a sluice as entrance and exit with personnel hygiene and changing measures, use of designated and/or coded tools and equipment, and separation of effluent and waste water drains (e.g. flowing from zones with potentially higher risk levels of contamination to zones with lower risk levels of contamination).

Special attention should be paid to instruct and control any contractors working in different areas in the plant, especially during ongoing construction works. During installations and repairs special attention should be paid to putting in place zoning measures to prevent cross-contamination of other areas. This could include intermediate walls, restricted/changed traffic patterns and increased hygiene measures (sluices, hand washing, clothing/shoe changes, cleaning activities). In such circumstances verification measures should be increased appropriately, e.g. by increasing environmental sampling in those areas.

Environmental Monitoring Programs

The effectiveness of zoning measures should be verified by environmental monitoring programs (see also Chen et al., 2009b). As mentioned above, the main microbiological concern in chocolate production is *Salmonella*. Since there are no known reliable indicators for *Salmonella* in dry productions, sampling needs to concentrate on that pathogen.

Sampling points should concentrate on areas where contamination could most likely occur first, in order to prevent further spread and adapt zoning measures accordingly.

Examples of such points are high traffic areas of people and materials, especially potential crossing points, such as common entrances, lifts and hallways. Since materials are often transported by forklifts or movable containers their wheels are also worth sampling. The brushes of machines/tools used to clean surfaces/floors could also provide indications of potential contamination.

Taking into consideration that in chocolate production the only missing factor to allow for microbial growth is water, any area where water is used (e.g. hand washing, equipment washing, filling productions) should be included in the sampling plan. A plant layout map can be very helpful so that any of these areas are not overlooked; preferably all drains would be indicated on such a map as well. Of equal importance is a trending of results in order to take appropriate actions in case of repeated positive findings.

It should also be kept in mind that single cells will not be detected by sampling due to the detection limits of the methods. Cell concentrations of around 10E3–10E4 cells give rise to positive results, which should trigger immediate actions. Such activities include immediate cleaning and disinfection of the area, review of any unusual observations, traffic restrictions, existing zoning measures, and resampling at increased frequency for a defined period in time.

Besides increased sampling during installations/repairs, etc. emergency actions should also trigger increased awareness for potential cross-contamination. Water usage from hydrants in production should trigger increased sampling of the water, environment and products (when resuming production) to ensure food safety and quality.

For hygiene purposes it is advisable to sample areas, where wet cleaning/water are used, for other hygiene indicators, like Enterobacteriaceae. In general, no Enterobacteriaceae would be expected after effective cleaning activities.

Cleaning

Most of the equipment used in current chocolate manufacturing has not been designed for wet cleaning (with water). Due to the nature of chocolate (low moisture, high fat) not allowing for microbial growth, historically cleaning was focused on removal of product residues for quality reasons (product changes) in the lines. This has changed in the last few years towards increased attention to appropriate cleaning regimes to control microbial as well as allergen issues.

Small equipment can be dismantled and cleaned/disinfected in designated and separated washing stations. Newly designed equipment may be even capable of being cleaned in place (CIP). Most importantly any residual water needs to be removed by drying before putting the equipment back into operation. Drying may be enhanced by heating up the equipment, if indirect heating devices are part of the installation or by means of hot air that is being blown through the line.

Special attention should be paid to points not easily accessible for inspection – and therefore drying. These need to be specifically included in verification by visual checks before reassembling.

Wet (water) cleaning activities should be carefully assessed with respect to the compatibility of the equipment, the effectiveness of drying and appropriate verification activities afterwards. The latter not only include visual checks, but also sampling and microbiological analysis of first products/materials processed on those parts of the lines. Any unusual findings, like increased microbial counts, should trigger immediate action towards cleaning controls and product disposal.

Pipes can be pre-cleaned by using "pigging," where a cylinder usually with a rubber head (the "pig") is pushed through the pipes by air or product, and then stored in a separate chamber at the line. The air used for pigging should be adequately filtered, and the pig stations regularly monitored for build-up of product residues.

Tanks, tempering machines and other equipment, which are difficult to access to clean manually, are commonly cleaned by flushing next product through. Flushing volumes should be defined, and their addition to the product streams carefully controlled in view of allergen carry-over.

Due to their current design tempering machines cannot be cleaned with water.

Flushing with hot oil (>60°C) can also be effective in removing product residues in the lines, and thereby any microorganisms embedded in them. However, such flushing is not regarded as disinfection step due to the increased heat resistance of bacteria in oil.

Other cleaning techniques sometimes used include dry steam cleaning, or cleaning with dry ice. Before using these methods on a regular basis, they need to be proven effective for the purposes for which they are used.

Disinfection of open/accessible parts of the lines is commonly performed by wiping or spraying with disinfection liquids like ethanol-based solutions. From an efficiency point of view it should be remembered that 70% ethanol is more efficient to remove vegetative bacteria than is 96% ethanol.

The efficacy of the cleaning operations, and especially the effectiveness of the drying step at the end, should be verified. For this purpose, visual inspections should be carried out each time, and samples can be taken from the line after cleaning and analyzed for Enterobacteriaceae. Expected acceptable results would be close to the detection limits of the methods used. Since not all parts of the line/equipment might be easily accessible for such activities, the first productions are also analyzed for quality parameters in addition to the regularly used testing regimes. Analyses for total viable count and Enterobacteriaceae are widely used in such cases, and an increase in these parameters would lead to further investigations and the related finished products being withheld. Depending on the results of the investigation further disposition of the finished products would be defined.

In that regard it is noteworthy that the method used for detection of Enterobacteriaceae should be capable of detecting injured cells in the products, i.e. methods using resuscitation steps would be most appropriate here.

Transportation

Liquid products (e.g. cocoa liquor, chocolate masses) are transported in bulk/tankers or mobile containers, depending on customer demands. Procedures should be defined with respect to regular checks on cleanliness of the vessels before loading, as well as towards the previous loads allowed before transporting cocoa products. In general, they should only be used for transporting food products, cleaning certificates should be available and openings should be sealed. Cleaning checks would include visual checks – also in view of residual water residues from cleaning operations, and sometimes rapid hygiene or microbiological sampling. Transportation of certain products, either because of their potential microbial load, risk of residual water, or because of their potential impact on flavor, could require two cleaning cycles. Attention should also be paid to the connecting pipes used to load/unload products with respect to their cleanliness. For more detailed information with respect to transportation reference is made to other chapters in this book.

References

Bassett, J., 2010. Impact of microbial distributions on food safety. ILSI Europe Report. <http://www.ilsi.org/Europe/Pages/ReportSeries.aspx>.

Beckett, S.T. (Ed.), 2009. Industrial Chocolate Manufacture and Use, fourth ed. Wiley-Blackwell, UK.

Betancourt, L.E., Frank, H.K., 1983a. Bedingungen des mikrobiellen Verderbs von grünem Kaffee. 1. Mitteilung: Sorptionsverhalten und Schimmelbildung. Deut. Lebensm.-Rundschau 79 (11), 366–369.

Betancourt, L.E., Frank, H.K., 1983b. Bedingungen des mikrobiellen Verderbs von grünem Kaffee. 2. Mitteilung: Bildung von Mycotoxinen. Deut. Lebensm.-Rundschau 79 (12), 404–407.

Bisbal, F., Gil, J.V., Ramón, D., Martínez-Culebras, P.V., 2009. ITS-RFLP characterization of black Aspergillus isolates responsible for ochratoxin. A contamination in cocoa beans. Eur. Food Res. Technol. 229, 751–755.

Blaser, M.J., Newman, L.S., 1982. A review of human salmonellosis: I. Infective dose. Rev. Infect. Diseases. 4 (6), 1096–1104.

Brown, K.L., 2005. Guidelines on Air Quality Standards for the Food Industry, second ed. CCFRA.

Buchanan, R.L., Fletcher, A.M., 1978. Methylxanthine inhibition of aflatoxin production. J. Food Sci. 43, 654–655.

Busta, F.F., Speck, M.L., 1968. Antimicrobial effect of cocoa on Salmonellae. Appl. Microbiol. 16, 424–425.

Chen, Y., Scott, V.N., Freier, T.A., Kuehm, J., Moorman, M., Meyer, J., et al., 2009a. Control of *Salmonella* in low-moisture foods II: hygiene practices to minimize *Salmonella* contamination and growth. Food Prot. Trends. July, 435–445.

Chen, Y., Scott, V.N., Freier, T.A., Kuehm, J., Moorman, M., Meyer, J., et al., 2009b. Control of *Salmonella* in low-moisture foods III: process validation and environmental monitoring. Food Prot. Trends. August, 493–508.

Codex, 2008. CAC/GL 69: Guideline for the validation of Food Safety Control. Codex Alimentarius Commission, Rome, Italy. List of Codex Alimentarius Standards. Available online: <http://www.codexalimentarius.net/web/standard_list.do?lang=en> (accessed 17.06.11).

Craven, P.C., Mackel, D.C., Baine, W.B., Barker, W.H., Gangarosa, E.J., 1975. International outbreak of *Salmonella* Eastbourne infection traced to contaminated chocolate. Lancet 1, 788–792.

Davies, A.R., Blood, R.M., Gibbs, P.A., 1990. Effect of moisture level on the heat resistance of Salmonellae in cocoa liquor. Research Reports. Leatherhead Food R.A. (contact for this report: Dr. Evangelia Komitopoulou; email: EKomitopoulou@LeatherheadFood.com).

Deak, T., Farkas, J., 2013. Microbiology of Thermally Preserved Foods: Canning and Novel Physical Methods. DEStech Publication, Pensylvania, USA.

EC Regulation 178/2002 and 852/2004. <http://eur-lex.europa.eu> (accessed 17.06.11).

EFSA, 2011. Update on furan levels in food from monitoring years 2004–2010 and exposure assessment. EFSA J. 9 (9), 2347.

Food Drink Acrylamide Toolbox, 2011. <http://www.fooddrinkeurope.eu/uploads/publications_documents/Toolboxfinal260911.pdf> (accessed 12.01.12).

Foster, E.M., 1971. The control of Salmonellae in processed foods: a classification system and sampling plan. J. Assoc. Off. Anal. Chem. 54, 259–266.

Fowler, M.S., 2009. Cocoa beans: from tree to factory. In: Beckett, S.T. (Ed.), Industrial Chocolate Manufacture and Use, fourth ed. Wiley-Blackwell, pp. 10–47.

Gästrin, B., Kaempe, A., Nystroem, K.G., 1972. *Salmonella* Durham outbreak caused by cocoa powder. Laekartidingen 69, 5335–5338.

Gill, O.N., Sockett, P.N., Bartlett, C.L., Vaile, M.S., Rowe, B., Gilbert, R.J., et al. 1983. Outbreak of *Salmonella* napoli infection caused by contaminated chocolate bars. Lancet 12, 574–577.

Goepfert, J.M., Biggie, R.A., 1968. Heat resistance of *Salmonella Typhimurium* and *Salmonella Senftenberg* 775W in milk chocolate. Appl. Microbiol. 16, 1939–1940.

Habraken, C.J.M., Mossel, D.A.A., van den Reek, S., 1986. Management of *Salmonella* risks in the production of powdered milk products. Neth. Milk Dairy J. 40, 99–116.

Hockin, J.C., D'Aoust, J.-Y., Bowering, D., Jessop, J.H., Khanna, B., Lior, H., et al. 1989. An international outbreak of *Salmonella* Nima from imported chocolate. J. Food Prot. 52, 51–54.

ICMSF (International Commission on Microbiological Specifications for Foods), 1986. Microorganisms in Food 2: Sampling for microbiological analysis: Principles and specific applications. Blackwell Sci. Publ. ISBN 0-632-01567-5.

ICMSF (International Commission on Microbiological Specifications for Foods), 2002. Microorganisms in Food 7: Microbiological testing in food safety management. Kluwer Acad. ISBN 0-306-47262-7.

ICMSF (International Commission on Microbiological Specifications for Foods), 2005. Microorganisms in Food 6: Microbiological Ecology of Food Commodities. Kluwer Acad. ISBN 0-306-48676-8.

ILSI report and Monograph series. <http://www.ilsi.org/EUROPE/Pages/Publications.aspx> (accessed 17.06.11).

ISO, 2005. ISO 22000: Food Safety Management Systems – Requirements for any organization in the Food Industry. International Organization of Standardization: Geneva, Switzerland.

Juven, B.J., Cox, N.A., Bailey, J.S., Thomson, J.E., Charles, O.W., Shutze, J.V., 1984. Survival of *Salmonella* in dry food and feed. J. Food Prot. 47 (6), 445–448.

Kapperud, G., Lassen, J., Dommarsnes, K., Kristiansen, B.E., Caugant, D.A., Ask, E., et al. 1990. Outbreak of *Salmonella* Typhimurium infection traced to contaminated chocolate and caused by a strain lacking the 60-megadalton virulence plasmid. J. Clin. Microbiol. 28, 2597–2601.

Krapf, T., Gantenbein-Demarchi, C., 2010. Thermal inactivation of Salmonella spp. during conching. LWT – Food Sci. Technol. 43, 720–723.

Komitopoulou, E., Penaloza, W., 2009. Fate of *Salmonella* in dry confectionery raw materials. J. Appl. Microbiol. 106, 1892–1900.

Lahmann, 2010. Handbuch. Lebensmittelhygiene. Behr's Verlag, Chapter 10: Kaffee und Tee.

Ma, L., Zhang, G., Gerner-Smidt, P., Mantripragada, V., Ezeoke, I., Doyle, M.P., 2009. Thermal inactivation of *Salmonella* in peanut butter. J. Food Prot. 72 (8), 1596–1601.

Mohr, E., 1971. Mikrobiologische Untersuchungen an Rohkaffee, Röstkaffee, Flüssigkaffee und Extraktkaffee. Kaffee und Tee Markt. 21, 4–5.

Motarjemi, Y., et al., 2009. Application of HACCP for the management of processing contaminants. In: Stadler, R.H., Lineback, D.R. (Eds.), Process-induced Food Toxicants: Occurrence, Formation, Mitigation, and Health Risks. John Wiley & Sons, Inc.

Müller, G., 1997. Kaffee, Kakao, Tee, Vanille, Tabak. In: Müller, G., Holzapfel, W., Weber, H. (Eds.), Lebensmittel pflanzlicher Herkunft. Behr's Verlag, Hamburg.

Penaloza-Izurieta, W., Krapf, T., Diep, B., Benz, K., Gantenbein-Demarchi, C., 2008. Reduktion von Salmonellen bei unterschiedlichen Röstparametern. Lebensmittel-Technologie 6/08, 10–12.

Podolak, R., Enache, E., Stone, W., Black, G.B., Elliott, P.H., 2010. Sources and risk factors for contamination, survival, persistence, and heat resistance of salmonella in low-moisture foods. J. Food Prot. 73 (10), 1919–1936.

RASFF online <http://ec.europa.eu/food/food/rapidalert/index_en.htm> (accessed 17.06.11).

Rieschel, H., Schenkel, J., 1971. Das Verhalten von Mikroorganismen, speziell Salmonellen in Schokoladenwaren. Alimenta 10, 57–66.

Robinson, T., 2011. EFSA's perspective on emerging risks. European IAFP Symposium, Ede-Wageningen. <http://www.foodprotection.org/events/european-symposia/11Ede/Robinson.pdf> (accessed 17.06.11).

Rothfos, B., 1985. Coffee – Production. Gordian-Max Rieck GmbH, Hamburg.

Rothfos, B., 1986. Coffee – Consumption. Gordian-Max Rieck GmbH, Hamburg.

Rotzoll, F., Müller, H.-G., 2004. Kaffeewissen: Vom Anbau bis zum Endproduckt. Deutscher Kaffeeverband e.V., Hamburg 9 (12), 30–35, 101.

Scott, V.N., Chen, Y., Freier, T.A., Kuehm, J., Moorman, M., Meyer, J., et al., 2009. Control of *Salmonella* in low-moisture foods. I: Minimizing entry of *Salmonella* into a processing facility. Food Prot. Trends. June, 342–353.

Sertkaya, A., Berlind, A., Lange, R., Zink, D.L., 2006. Top ten food safety problems in the United States food processing industry. Food Prot. Trends. 26 (5), 310–315.

Stadhouders, J., Hup, G., Hassing, F., 1982. The conceptions index and indicator organisms discussed on the basis of the bacteriology of spray-dried milk powder. Neth. Milk Dairy. J. 36, 231–260.

Stehli, Ch., Gantenbein-Demarchi, C., Benz, C., 2002. Sporeninaktivierung auf Kakaobohnen. Lebensmittel-Technologie 35, 7–8.

Sumner, S.S., Sandros, T.M., Harmon, M.C., Scott, V.N., Bernard, D.T., 1991. Heat resistance of *Salmonella* Typhimurium and *Listeria monocytogenes* in sucrose solutions of various water activities. J. Food Sci. 56 (6), 1741–1743.

Suárez-Quiroz, M.L., González-Rios, O., Barel, M., Guyot, B., Schorr-Galindo, S., Guiraud, J.P., 2004. Effect of chemical and environmental factors on Aspergillus ochraceus growth and toxigenesis in green coffee. Food Microbiol. 21, 629–634.

Tamminga, S.K., Beumer, R.R., Kampelmacher, E.H., van Leusden, F.M., 1976. Survival of *Salmonella eastbourne* and *Salmonella* Typhimurium in chocolate. J. Hyg. 76, 41–47.

Werber, D., Dreesman, J., Feil, F., van Treeck, U., Fell, G., Ethelberg, S., et al. 2005. International outbreak of *Salmonella* Oranienburg due to German chocolate. BMC Infect. Dis DOI: 10.1186/1471-2334-5-7. Online access: <http://www.biomedcentral.com/1471-2334/5/7> (accessed 06.10.13).

WHO, IARC monographs vol. 71, 1999. <http://monographs.iarc.fr/ENG/Monographs/vol71/mono71-10.pdf>.

Wimmer, M., Färber, P., Holzapfel, W.H., 2006. Determination of exact physiological parameters determining ochratoxin A-biosynthesis by *Aspergillus ochraceus* growing on green coffee. Poster. *Tagung DGHM*. Suhl: Germany.

Wrigley, G., 1988. Coffee. Longman Scientific & Technical, Harlow, Essex, England.

Zapatka, F.A., Varney, G.W., Sioskey, A.J., 1977. Neutralization of the bactericidal effect of cocoa powder on Salmonella by casein. J. Appl. Bacteriol. 42, 21–25.

Further Reading

Brochures available from CAOBISCO (Association of Chocolate, Biscuit and Confectionery Industry of the EU). <http://www.caobisco.com/page.asp?p=220> (accessed 02.06.11).

European Food Safety Authority, 2011. Results on acrylamide levels in food from monitoring years 2007–2009 and Exposure Assessment. EFSA J. 9 (4), 2133, pp. 48. dx.doi.org/10.2903/j.efsa.2011.2133. Available online: <www.efsa.europa.eu/efsajournal>.

Kornacki, J.L., 2006. Microbiological sampling in the dry foods processing environment. Food Safety. Mag., pp.66.

Kvenberg, J.E., Schwalm, D.J., 2000. Use of microbial data for hazard analysis and critical control point verification – Food and Drug Administration perspective. J. Food Prot. 63 (6), 810–814.

Swanson, K.M.J., Anderson, J.E., 2000. Industry perspective on the use of microbial data for hazard analysis and critical control point validation and verification. J. Food Prot. 63 (6), 815–818.

CHAPTER

11

Honey, Confectionery and Bakery Products

Suchart Chaven

Pepsi-Cola International Ltd., Dubai, UAE

OUTLINE

Food Safety Management.
DOI: http://dx.doi.org/10.1016/B978-0-12-381504-0.00011-1

INTRODUCTION

Confectionery and bakery products are historically one of mankind's oldest food staples; both as sustenance and enjoyment. This chapter provides food safety guidance and considerations for the manufacturing of confectionery and bakery products, including honey.

HONEY

One of the first sweeteners that mankind discovered and used before sugar was honey. Ancient cultures such as those of the Greeks, the Romans, the Egyptians, the Mesopotamians and Chinese used honey to coat fruits, seeds and stems of plants to preserve them or to be used as a kind of confectionery. Honey was also thought to have medicinal properties in many cultures.

Of the approximately 20,000 known species of bees, the honey bee of genus *Apis mellifica* is the primary producer of the sweet product that man collects and consumes as honey. The honey bee collects nectar from flowering plants and transforms the nectar into honey by a process of regurgitation and evaporation which occurs naturally when stored as a food source in wax honeycombs in the beehive.

Processing (Figure 11.1)

Honey is classified by its floral source from which the nectar is collected by the honey bee. Monofloral honey is made primarily from the nectar of one type of flower in which the bees have access to a common flower source. Different monofloral honeys have different flavor and color characteristics, and this largely depends on the flower source. Polyfloral honey, or wildflower honey, is derived from nectar of many types of flowers. As such, there may be differences in color, depending on the flora and seasonal variations.

Early forms of honey harvesting entailed the destruction of the beehive and colony. With the advent of the movable frame hives in the 19th century, the honey can be harvested from the hive without destroying the colony. Honey processing generally consists of removing a frame segment with honeycombs from the hive by beekeepers while using smoke to calm the bees. The honeycomb wax cells are uncapped with a heated knife and the extracted honey is collected and packaged. Additional processing may include melting, centrifugation and straining to further extract honey from the wax capping.

Honey can be classified by packaging and processing:

- Crystallized honey is honey in which the glucose content has crystallized, also called "granulated honey" or "candied honey."
- Pasteurized honey is honey that has been heated or pasteurized at 161°F (71.7°C) or higher.
- Raw honey is honey harvested from the beehive without processing or by extraction or straining without heat processing.
- Strained honey is honey that has been passed through a mesh material to remove particulate material.

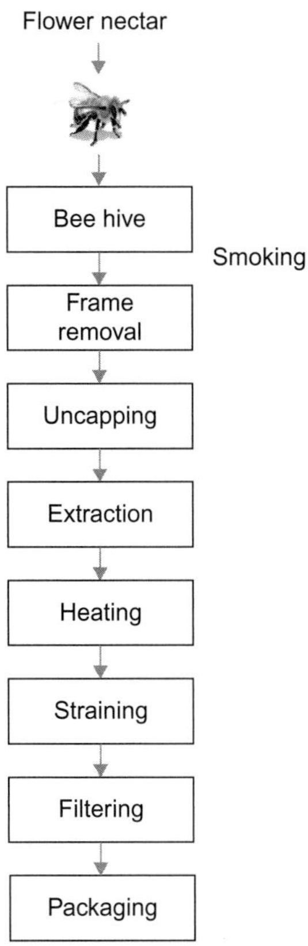

FIGURE 11.1 The processing of honey from source to end product.

- Filtered honey is honey that has been heated and filtered to remove all or most of the fine particles, pollen grains, air bubbles or other materials.
- Ultrasonicated honey is honey that has been processed by ultrasonication to eliminate yeast cells and to inhibit honey crystallization.
- Whipped honey is honey that has been churned or processed to control crystallization and has a smooth consistency.
- Dried honey is honey in which the moisture has been removed to form granules.
- Comb honey is honey in the natural state of the honey comb. Another variant of comb exists where large pieces or chunks of comb honey are packaged with extracted liquid honey.

Intrinsic Properties

Honey is primarily composed of fructose, glucose, maltose and water. Other minor components such as minerals, proteins and enzymes can be found. Compositional variations may occur mainly due to the nectar source and seasonal variation. The water content of

TABLE 11.1 Approximate Equilibrium Points between the Relative Humidity and the Percentage Water Content of Liquid Clover Honey

Water Content (%)	ERH
16.1	52
17.4	58
21.5	66
28.9	76
33.9	81

honey is around 18% with the approximate corresponding environmental relative humidity (ERH) or water activity of 0.60. Martin (1958) demonstrated the equilibrium moisture content of honey exposed to various atmospheres (Table 11.1):

Honey contains a number of acids which include amino acids (0.05–0.1%) and organic acids (0.57%, range: 0.17–1.17%). The average pH of honey is 3.9 (with a typical range of 3.4 to 6.1) (USDA, 1962).

Hazard Analysis

Biological

The predominant microorganisms found in honey are derived from the nectar and the honey bee. Although a wide variety of yeast may be recovered from unprocessed honey, it is predominantly osmophilic yeast varieties such as *Zygosaccharomyces rouxii* or *Zygosaccharomyces bailii*, which are of relevant concern for honey processing. These organisms, under the right environment, may grow and ferment honey resulting in the formation of alcohol and carbon dioxide and a sour taste from the breakdown of alcohol to acetic acid and water. However, these spoilage conditions should be considered rare as Lochhead (1933) demonstrated that honeys with less than 17.1% water will not ferment within a year, no matter what the yeast count may be. At between 17.1 and 18% of moisture, honey with 1000 yeast spores or less per gram will remain stable. However, if the moisture is between 18.1 and 19%, it would be expected that yeast would be able to grow and ferment honey. A heating step in the extraction of honey provides additional controls in minimizing spoilage risks (Table 11.2). Therefore, with proper extraction and heat treatment, honey should not ferment under normal storage conditions.

The low water activity and osmotic environment are not conducive to bacterial growth. However, under rare conditions, *Clostridium botulinum* from honey has been implicated with infant botulism in the United States. Infant botulism is the infectious (intestinal) form of botulism, which results in the spores of *Clostridium botulinum* colonizing infants' large intestine and producing botulinum toxin. A case–control study performed by the California Department of Health Services in 1976–1978 indicated that infants with type B botulism were more likely to have been fed honey and with type B spores were identified

TABLE 11.2 Heating Step in the Extraction of Honey: Additional Controls in Minimizing Spoilage Risks

Temperature (°F)	Equivalent in °C	Heating time (minutes)
128	53.3	470
130	54.4	170
135	57.2	60
140	60.0	42
145	62.8	7.5
150	65.6	2.8
155	68.3	1.0
160	71.1	0.4

See Townsend (1961).

in implicated honey samples (Arnon et al., 1979). The typical heat treatment in the processing of honey is insufficient for mitigating the spores of *Clostridium botulinum*. The American Academy of Pediatrics has advised that honey should not be added to food, water or formula that is fed to infants younger than 12 months of age.

Chemical

GRAYANOTOXIN

In history there have been reported cases of so-called "toxic honey"; one in which the Roman army in the first century BC under Pompey the Great became poisoned after consuming "maddening" honey and lost the war with the Heptakometes in Turkey. Supposedly, the honey was produced from the nectar of the plant *Rhododendron ponticum* which contains alkaloids that are poisonous to humans, but not to bees.

Honey intoxication, or grayanotoxin poisoning, is caused by the consumption of honey produced from the nectar of rhododendrons. The intoxication is rarely fatal, generally lasts for no more than 24 hours and is mainly associated with symptoms of dizziness, weakness, excessive perspiration, nausea and vomiting shortly after the toxic honey is ingested (FDA, 2012). Grayanotoxin poisoning in humans is relatively rare. However, cases of honey intoxication should be anticipated where there is availability of these plants within the area of close proximity to the hive. From 1984 to 1986, cases of honey intoxication were continuously reported in Turkey.

HYDROXYMETHYLFURFURAL

Hydroxymethylfurfural (HMF), or 5-hydroxymethylfurfural, is an organic compound formed as an intermediate in the Maillard reaction, and from hydration of sugars has been known to be associated with thermally processed foods such as milk, fruit juices, jams, jellies and may also be naturally found honey. While HMF is potentially carcinogenic to humans or might be metabolized by humans to potentially carcinogenic compounds

(Capuano and Fogliano, 2011), the concentration in natural honey is found to be several magnitudes lower than many thermally processed foods and the food industry has taken the levels of HMF in honey as a quality measure for excessive heat during extraction, storage changes as HMF increases over time, or for possible adulteration with sugars and syrup. HMF levels in honey are regulated in many countries and are available in international standards such as the Codex Alimentarius Standard for Honey (Codex Stan 12-1981). The hydroxymethylfurfural content of honey after processing and/or blending shall not be more than 40 mg/kg. However, in the case of honey of declared origin from countries or regions with tropical ambient temperatures, and blends of these honeys, the HMF content shall not be more than 80 mg/kg.

ANTIBIOTICS

Bee colonies are subjected to a variety of diseases, one of which is the American Foulbrood (AFB), caused by the spore-forming *Paenibacillus larvae* species. The larvae ingest spores of *Paenibacillus larvae* that are present in their food and die as the bacterial spores germinate and multiply in the gut of the larva. Bee larvae less than 24 hours old are most susceptible to infection. The disease not only affects the bee larvae, but is highly infectious and deadly to bee brood. With the European Foulbrood (EFB), the bacterium *Melissococcus plutonius* multiplies in the gut of the larvae, competes for food in the gut and causes starvation.

Beekeepers utilize routine hive inspection programs, isolation and destruction of infected hives in order to control the spread of the disease. In some countries, antibiotics and antimicrobial agents such as erythromycin, sulfonamides, streptomycin, tetracycline and tylosin may be used, but are subject to local regulatory approval. In the European Union, there have not been established acceptable MRLs for antimicrobial/antibiotic for honey and therefore these products are not authorized for use.

Antibiotics such as chloramphenicol have been banned from being used in many countries due to concerns related to potential carcinogenicity and genotoxicity as well as the potential to cause antimicrobial resistance. In March 2002, during routine testing for drug residues in honey, the Canadian Food Inspection Agency (CFIA) discovered the presence of chloramphenicol in a shipment of honey from China. Health Canada provided the results of a Health Risk Assessment, advising that the honey posed a low, but serious, health risk. It was recommended that the product detained should not be allowed to be sold in Canada and advice was issued of the risk of consuming the contaminated products (Health Canada, 2004). Findings were also reported from the CFIA in December 2006 for imported honey from Ukraine. Similarly, the US Food and Drug Administration reported the detention of honey containing unapproved fluoroquinolone antibiotics from several countries (FDA, 2012).

PESTICIDES AND HEAVY METALS

The primary source of lead contamination in honey is improper use of containers and lead-bearing equipment. Honey which is slightly acidic can react with surfaces containing lead, allowing lead to be absorbed into the honey. Lead-bearing equipment may include galvanized equipment such as extractors and tanks, soldered equipment with lead soldered seams and some bronze and brass fittings, or older equipment.

More recently, the effect of environmental contaminants may also have direct correlation to honey.

TABLE 11.3 Control Measures

Process Step	Potential Hazards	Control Measures
Beehive management	Unapproved antibiotics	Alternative controls without antibiotics
Beehive management	Pesticides for mites and moths	Alternative controls without pesticides
Beehive management	Toxic honey	Beehive location
Honeycomb frame removal	Repellants	Approved repellants/smoker
Uncapping	Metal sharps	Approved knives and tools
Extraction	Toxic metals in vessels	Food grade equipment
Filtering and filtering	Foreign bodies	Appropriate mesh
Packaging and labeling	Glass and brittles	GMP – glass and brittles

Honey contains pollen grains derived from the foraging environment of honey bees as well as the location of the beehive. The variability and exposure of the honey to environmental contaminant types are well documented. Bibi et al. in 2008 analyzed honey samples from selected countries, including Austria, Australia, Canada, Germany, Pakistan, Saudi Arabia and the United States and recommended that further studies should be undertaken to help in finding out possible sources of heavy metal pollution in the vegetation of the area from where the honey originated (Bibi et al., 2008). Porrini et al. suggested that bees and their products may be used as indicators of environmental pollution for heavy metals and pesticides (Porrini et al., 2003). Codreanu et al. concluded similar findings citing that sources of the honey's contamination may be from environmental residuum from auto emissions and recommended that beehives should be located a minimum of 3 km distance (approximately the bees' fly area) from any roads with intense traffic (Codreanu et al., 2009).

Physical

Physical hazard risks in honey production are primarily minimized through straining of honey and in glass breakage procedures where glass packaging is used (see Table 11.3).

CONFECTIONERY

Confectionery is a general term that describes the food products category that has sugar as the primary ingredient; it is also referred to as confections. Other terminologies for confections are "candy" in North America and "sweet" in the United Kingdom. Confectionery products are very diverse, varying in size, shape, flavors, color and hardness. Some well-known representatives of confectionery products include hard candies, candy cane, cotton candy, chewing gums, gummies, lollipops, fondants, marsh mellow, rock candy and taffy. Cocoa-based or chocolate confections are described in Chapter 10.

The sweetness of confectioneries is primarily derived from sugar or sucrose. Most of the world's production of sucrose is manufactured from either sugar cane (*Saccharum*

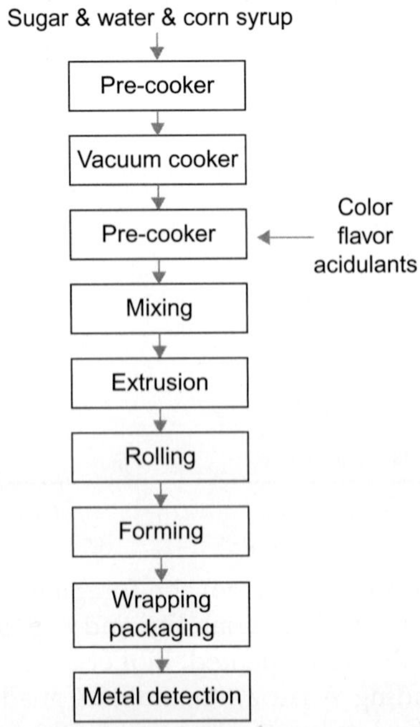

FIGURE 11.2 Example of hard candy processing.

officinarum) or sugar beet (*Beta vulgaris*). Sugar cane is grown in subtropical and tropical regions of the world. Sugar beets are grown in moderate and continental climates. High fructose corn syrup (HFCS), another source of sweetening agent that may be used as a component of candy bars, is derived from the processing of corn starch with heat, caustic soda and/or hydrochloric acid plus the conversion by enzymatic activity (alpha-amylase, glucoamylase and glucose isomerase) to yield HFCS products. Other sugars and sugar substitutes such as dextrose, fructose, manitol, sorbitol and xylitol may also be used.

Processing

The diversity of processing of confectionery varies and is reflected by the diversity of confectionery products. The following are some examples of confectionery processing.

Hard Candy Processing (Figure 11.2)

Sugar is dissolved in hot water, corn syrup is added and steam heated. The moisture content of the sugar syrup is further reduced under vacuum in a cooker. After cooking and before mixing, color, flavor and acidulants such as citric or malic acid are added. After mixing, the semisolid candy is extruded, rolled, formed and wrapped.

Gummy Candy

Gummy candy is a confectionery composed of gelatin, sugar, flavorings and colorings. Similarly to hard candy processing, the compound solution is cooked and pumped to a starch bed depositing machine called a Mogul. The Mogul is a continuous forming machine that is used to form candy shape impressions in a starch tray; it is also used to remove cooled candy shapes from the starch tray. Starch trays are used to dry, cool and set the candy.

Chewing Gums

The primary component of chewing gum is the gum base made of synthetic polymers. The gum base is mixed with sweeteners, softeners, colors and flavors, and rolled and sheeted. The sheeted rolls are scored (for example, into stick shapes) and are stored in conditioning rooms with controlled humidity and temperature. The final product is then wrapped and packaged.

Intrinsic Properties

The stability and safety of confectionery products is primarily defined by both the high temperatures utilized to dissolve the compound syrup to form the candy, and the low water activity that is inhibitory to microbiological growth. During the 1960s, it was common to find hard candy as a common component of US Civil Defense carbohydrate supplement, due to its inherent stability properties when kept under dry conditions.

Hazard Analysis

Biological

Raw materials: cane sugar, beet sugar, and high fructose corn syrup.

Cane sugar is processed into raw sugar by a sequence of operations: harvesting, cutting, crushing, extraction of juice, clarification, evaporation, crystallization, centrifugation and refining.

Beet sugar processing involves harvesting, slicing, extraction of juice, carbonization, evaporation, crystallization and refining.

The heating process used for these processes and the resulting low water activity of the end products greatly decreases the initial microflora with the remaining organisms consisting primarily of heat-resistant spores (Owen, 1977) such as those from *Bacillus* and *Clostridium* species. While spores of *Clostridium botulinum* have been detected in sugar, they are of little relevance due to the low water activity of the confectionery products. For high fructose corn syrup, which utilizes enzymatic processing, the primary organisms of concern are spoilage organisms such as osmophilic yeast like *Zygosaccharomyces rouxii* and *Saccharomyces cerevisiae*, which are not relevant to hard candy confectionery processing.

Chemical

Chemical contaminants of confectionery products are limited to the low risks associated with the primary ingredients and the flavoring and coloring agent.

PESTICIDES

As with other agricultural commodities, sugar cane, beet crops and their products may be subjected to residual contamination by the misuse application of pesticides. Pesticide use and application are controlled by local regulatory and international standards and often with maximum residual limits (MRLs) defined. In 2003, a non-governmental organization in New Delhi alleged that pesticides could be detected in carbonated beverages produced by large multinational companies, implicating sugar as one of the primary raw materials in use. This claim has been mainly refuted by independent analysis for the multinational companies.

HEAVY METALS

Since the California's Department of Public Health began testing confectionery products for lead in 2007, the department has issued numerous warnings to consumers for imported products mainly from Mexico, Malaysia, China and India. While the source has not been defined, products containing tamarind, chili powder or salt that is mined from certain parts of the world may have a higher likelihood of elevated levels of lead. Lead may also be introduced into the candy through improper drying, storing or grinding of the ingredients (California Public Heath, 2012). Similarly, the US Food and Drug Administration (FDA) has issued warnings for potential for children to be exposed to lead from candy imported from Mexico and to develop tighter guidelines for manufacturers, importers and distributors of imported candy. Lead has been found in the wrappers of some imported candies; the ink used on these plastic or paper wrappers may contain lead that leaches into the candy (CDC, 2009).

Physical

Candy is often recognized as one of the leading causes of food choking hazard in children under the age of 3. The physical characteristics of candy such as bite size, shape and texture were analyzed and found to demonstrate a relationship to the severity of clinical outcomes (Altkom et al., 2008).

In 2002, the US Food and Drug Administration announced recalls from an Asian company producing candies in small sealed plastic cups that contained konjac gelatin with or without a chunk of fruit. Due to the packaging, shape, slipperiness and consistency, the candy posed a potentially serious choking risk, particularly to infants, children and the elderly. There had been reported deaths from choking associated with this type of jelly candy throughout the United States as well as in other countries. The packaging of these types of candies now contains a note to consumers, advising them to cut the product into smaller pieces before serving it to small children. Food safety education can help parents to supervise and select for appropriate candies to improve safety for this highest-risk population.

Foreign body (metal pieces or shavings) contamination can occur from equipment used for intensive processing of candies. Metal detection systems remain the primary control measure for these hazards.

BAKERY

Cereal-based products constitute a major food staple in the world with grains principally derived from wheat, rice, maize, barley, sorghum, millet, oat and rye. Common to cereal-based bakery products is that dry heat by convection, typically using an oven, is used to prepare "baked products" such as bread, cakes, pastries, pies, tarts, cookies and crackers. For the most part, bakery products have had a good history with regard to product safety. Where there have been reported food safety issues, they are usually associated with high moisture filling rather than cereal components.

Intrinsic Properties

For high moisture products such as bread with the water activity typically greater than 0.94, post-bake spoilage contamination with mold reduces the ambient shelf-life to days. Preservatives such as calcium propionate may be used to extend the shelf-life as well as prevent the growth of spore formers or "rope bacteria," mucoid variants of *Bacillus subtilis* with spores that can survive the heating process. For cookies and crackers with low moisture and water activity typically less than 0.6, the shelf-life is dependent on quality aspects rather than on microbiological stability and can vary from months to weeks.

Intermediate Bakery Products

Microbiological risks of intermediate bakery products, such as high moisture batter or mixes such as for muffins and cake, should be considered. These intermediates are of high water activity and the microflora reflect the potential growth of innate organisms as well as pathogens such as *Staphylococcus aureus* or *Bacillus cereus* with heat stable enterotoxins that can withstand the baking process. The traditional baking process or batch process may be of lower risk than large-scale industrial processes with extended processing, holding time and recycled batters. Considerations should be made for temperature controls for holding and processing as well as defining the frequency required for cleaning and sanitation.

Processing (Figure 11.3)

The baking process results in complex chemical and physical changes to include the partial gelatinization of starches, coagulation of proteins and the caramelization of sugars and Maillard reaction, which results in the browning of the outer surfaces and providing the product with an attractive appearance, aroma, texture and taste.

Hazard Analysis

Biological

Cereal in the field is exposed to various microorganisms in the soil, birds, animals and other plants. As such, the microorganisms found in flour and milled grains originate from the material from which they are milled. Additional to the environment, the grains are exposed to potential contamination from harvesting, transportation, storage environment and milling operation. Studies have shown that the milling process has little effect on the microbiology of wheat flour other than removing the outer bran of the wheat kernel (Richter et al., 1993). Microbiological loads as high as 10^5 to 10^7 can be found in wheat at harvest (Seiler, 1978).

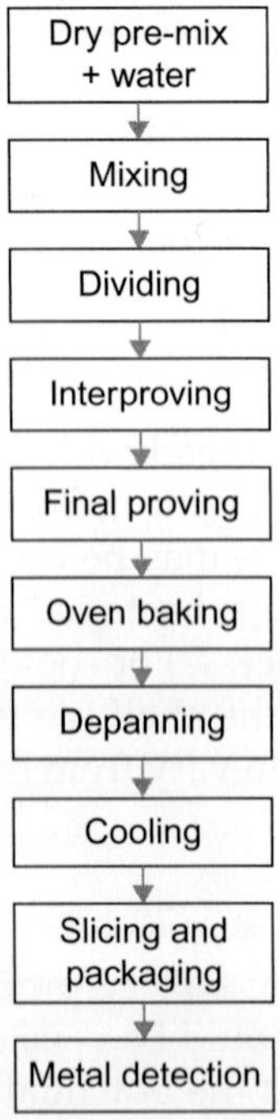

FIGURE 11.3 Example of bread processing.

Low levels of pathogens such as *Salmonella* and *Escherichia coli* have been detected in low numbers and on occasion have resulted in product contamination and recalls. Microbiological surveys of milled cereal from 2003 to 2005, using routine data from North American dry-milling operations, indicated the presence of *Salmonella* in 0.14% of wheat flour ($n=4358$) (Sperber and the North American Millers' Association Microbiology Working Group, 2007).

In 2002, a major ready-to-bake frozen pie company in the United States conducted a voluntary recall as the product was manufactured with flour contaminated with *Salmonella*. Subsequently, the flour milling company also conducted a recall.

In 2005 an ice cream franchise company initiated a voluntary nationwide recall for a cake batter ice cream after the Minnesota Department of Health notified the federal Centers for Disease Control and Prevention (CDC) that four cases of *Salmonella Typhimurium* were linked to a common pulsed field gel electrophoresis (PFGE) pattern. The company had been using a dry cake mix that was not intended for use without a consumer cooking step. In total, 25 people had reported illness.

In 2008, a joint investigation by the New Zealand Ministry of Health and the New Zealand Food Safety Authority (NZFSA) had linked outbreaks of *Salmonella Typhimurium* to a common food source – flour – and surmised that people with the infection, especially children, were more likely to have eaten uncooked flour, for example in home-made play dough, raw cake and batter mixes. Sixty-six cases of reported illness with eight people requiring hospitalization were associated with the incident. In 2009, a company manufacturing pre-packaged frozen cookie dough conducted a recall due to possible linkage of *Escherichia coli* O157:H7 with contaminated flour. Seventy-seven people reported illness, with 35 requiring hospitalization, and 10 developing life-threatening hemolyticuremic syndrome. After the incident, the CDC suggested that foods that contain raw flour "should be considered as possible vehicles of infection of future outbreaks of STEC and that food processors should consider the use of pasteurized flour in ready-to-cook or ready-to-bake foods that are likely to be consumed without cooking or baking, even though label statements may warn against consuming uncooked product" (Neil et al., 2011).

The primary biological control measure is the baking time and temperature to produce the product. In commercial ovens, the internal core product temperature remains close to 97°C while the outer surface temperature increases as the moisture is driven off to form an outer crust.

The validation of internal time and temperature of product bake can be achieved by use of data logging equipment or sensors for thermal profiling. Thermal profiling consists of the use of multiple sensors that measure the internal temperature as well as the oven air temperature as the product is transversed in the oven. Thermal profiling in combination with literature review, or use of relevant pathogen surrogates, may be used to verify that the baking process applied can meet or exceed minimal requirements for controlling vegetative pathogens (Figure 11.4).

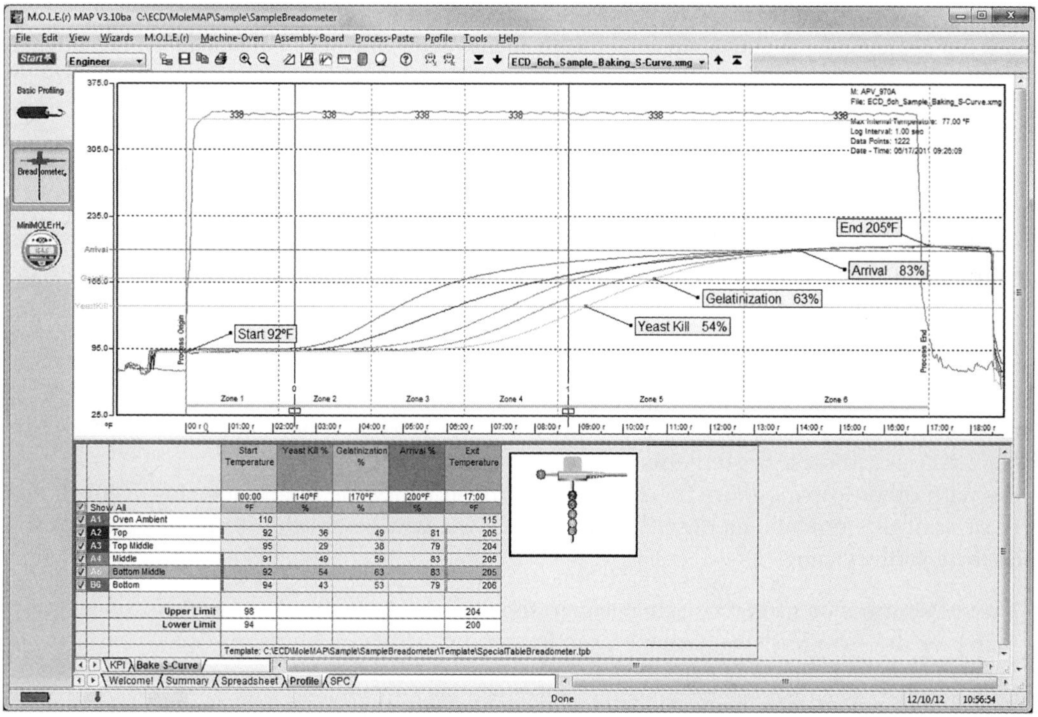

FIGURE 11.4 Example of bread thermal process profiling from Electronic Controls Design Inc. (ECD).

Additional Considerations

Sensitive Ingredients and Inclusions

Sensitive ingredients may consist of raw or unprocessed commodities such as flour, which has been previously described as a potential vector for *Salmonella*, with time and temperature for baking as an adequate control measure.

Sensitive ingredients, such as unpasteurized cream or eggs, should be discouraged for use in post-thermal process applications such as filling, icing or frosting or in products such as tiramisu, where there have been reported cases of illness linked to contaminated eggs. Vegetative pathogen potentials from these unprocessed ingredients may be seen as direct hazards to consumers as there are no mitigation steps.

Sensitive ingredients may also include processed ingredients such as chocolate chips for cookies, chocolate for enrobing of bakery products, or peanuts and tree nuts used as inclusions or as toppings.

Chocolate and cocoa-derived products are considered sensitive ingredients, primarily due to historical associations with food industry *Salmonella* outbreaks. *Salmonella* contamination potential may be linked to various components of processing from cross-contamination between raw and roasted beans, environmental cross-contamination from inadequate separation between clean and unclean process zones, to other vectors such as roof and water leaks in an open processing environment. Common associations of *Salmonella* and chocolate products are that they can survive in contaminated products for years due to the protective effect of fat, have increased heat resistance in the low water activity (a_w) environment, and have a very low infective dose in the product. The control of *Salmonella* in manufacturing chocolate product includes appropriate bean roasting, strict environmental control and monitoring and pathogen verification testing.

Peanuts and tree nuts are also considered to be sensitive ingredients and have been associated in various *Salmonella* outbreaks. As with chocolate products, long survival time in a low moisture environment and low infective dose have been reported. An in-depth review for control of *Salmonella* in low moisture foods has been provided by the Grocery Manufacturer Association (GMA, 2009). Further details on hazards and control measures of nuts and chocolate are described in Chapters 13 and 17.

For bake products that use nuts and chocolate as inclusions, such as chocolate chips in cookies, or as chocolate for enrobing or icing, the primary control measures are with the suppliers of the sensitive ingredients, with verifications by certificates of analysis (COA) for *Salmonella* as the time and temperature for baking is usually insufficient for mitigating *Salmonella* risks in these ingredients.

As with other low moisture foods products, the appropriate manufacturing control measures or control elements, as described by the GMA, can be applied to baked goods for an industrial setting:

1. Prevent ingress or spread of *Salmonella* in the processing facility.
2. Enhance the stringency of hygiene practices and controls in the primary *Salmonella* control area.
3. Apply hygienic design principles to building and equipment design.
4. Prevent or minimize growth of *Salmonella* within the facility.

5. Establish a raw materials/ingredients control program.
6. Validate control measures to inactivate *Salmonella*.
7. Establish procedures for verification of *Salmonella* controls and corrective actions.

In an artisanal bakery setting, where the emphasis is on a quality product and with potentially less understanding of requirements for food safety and cross-contamination, higher risk may be more probable than on an industrial scale, but may be limited to the fewer number of consumers.

High Moisture and Perishable Fillings

Bakery products with high a_w and low pH fillings such as custards and creams that can support microbial growth have been associated with outbreaks. Pathogens such as *Staphylococcus aureus* and *Bacillus cereus* have been associated with contaminated fillings. High hygiene, short shelf-life and refrigerated or frozen storage conditions are preventive measures for these perishable products.

Chemical

The microflora of grains is very diverse. Many aerobic mesophiles, yeasts and molds are indigenous to the cereal plant. Generally, the growth of and colonization of the indigenous microflora is usually not of significance for baked products. However, where there are opportunities for growth, such as under high moisture storage or harvest conditions, pathogenic fungi may be of concern. *Claviceps purpurea* is historically well described for the ergot alkaloid epidemic poisonings during the Middle Ages, also known as St. Anthony's Fire, as well as for the possible linkage to hysterical symptoms of women from the consumption of ergot contaminated rye that led to the Salem witch trials. The growth of *Aspergillus flavus* on corn, wheat or other grains may result in contamination with the highly carcinogenic aflatoxin. *Fusarium graminearum* growth with production of deoxynivalenol (DON) or vomitoxin may render the grain unsuitable for consumption.

Some controls that are applied to reduce the potential growth of these organisms include the type and manner of tillage, crop rotation, type of seed, drying efficiency and condition in storage facilities. For baking manufacturers that rely on the primary ingredients, a COA or internal ELISA test kits, where available, can be used for verification.

Allergen cross-contamination, as described in Chapter 4, remains a high concern for bakery manufacturing. Much of the industry's baking equipment such as mixers, blenders, conveyor belts and rollers have been designed for general cleaning, but not to the extent required for detailed allergen cleaning – whereas other major components such as ovens have not been designed for dismantling, inspection or cleaning. Segregation or dedication of process for production of products of the same allergen profile along with allergen profile scheduling are some of the strategies used to minimize extensive downtime associated with allergen cleaning.

Physical

Hard biscuits and cookies may become a choking hazard for infants. Consumer awareness and education for parents is essential to minimize the hazard (e.g. cookies may be softened with milk or only baby biscuits specifically designed to dissolve and not break should be provided to infants).

Preventive measures such as flour sifting can be used to minimize foreign objects that may come from flour. Evaluation of the sifter tailing (material that has not passed through the screens) for foreign objects as well as for insect contamination provides evidence of the quality of the ingredient. Occasionally, the sifter itself, when damaged, may cause a bakery product recall due to the presence of metal pieces, which depending on the size, length and configuration may not have been detected by a metal detector for the finished product.

Magnets can be used as a quality indicator of tramp metal in free-flowing material.

Routine production equipment inspection and preventive maintenance programs further help to minimize foreign object risks.

In addition to sieving the flour, metal detectors can be used as a verification of the efficiency of preventive measures.

CONCLUDING REMARKS

Honey, confectionery and bakery products belong to mankind's success stories. They constitute a significant food sector for sustenance and enjoyment for people throughout the world. It is also of high significance that historically these products are considered safe and should remain safe as long as the food industry does not take this fact for granted. The application of GMP, GHP and HACCP are still as relevant in honey, confectionery and bakery production as they are in any other products.

References

Altkom, R., et al., July 2008. Int. J. Pediatr. Otorhinolaryngol. 72 (7), 1041–1046.

Arnon, S.S., Midura, T.F., Damus, K., Thompson, B., Wood, R.M., Chin, J., 1979. Honey and other environmental risk factors for infant botulism. J. Pediatr. 94, 331–336.

Bibi, S., Zahoor, S., Malik, R., 2008. Pollen analysis and heavy metal detection in honey samples from seven select countries. Pakistani J. Bot. 40 (2), 507–516.

California Public Heath, CA.GOV. Frequently Asked Questions (FAQ) about Lead and Lead-Contaminated Products. November 2012. <http://www.cdph.ca.gov/programs/Pages/LeadFAQ.aspx>.

Capuano, E., Fogliano, V., 2011. Acrylamide and 5-hydroxymethylfurfural (HMF): a review on metabolism, toxicity, occurrence in food and mitigation strategies. LWT – Food Sci. Technol. 44, 793–810.

Centers for Disease Control and Prevention. CDC – Lead – Tips – Sources of Lead – Candy. June 2009. <http://www.cdc.gov/nceh/lead/tips/candy.htm>.

Codreanu, I., et al., 2009. Correlations between the lead residues in honey and the apiary location. Splaiul Independentei No. 105, 050097, Bucharest, Romania.

Food and Drug Administration. Bad Bug Book, Foodborne Pathogenic Microorganisms and Natural Toxins. Second Edition, Natural Toxins, pp. 294–253, 2012.

Grocery Manufacturer Association (GMA), 2009. Control of *Salmonella* in low moisture foods.

Health Canada, Veterinary Drugs, Factsheets & FAQ, Chloramphenicol, April 2004, <http://www.hc-sc.gc.ca/dhp-mps/vet/faq/faq_chloramphenicol_honey-miel-eng.php>.

Lochhead, A.G., Heron, D.A., 1929. Microbiological studies of honey. Can. Dept. Agr. Bull. 116 (N.S.).

Martin, E.C., 1958. Some aspects of hygroscopic properties and fermentation of honey. Bee World 39, 165–178.

Neil, K.P., Biggerstaff, G., MacDonald, J.K., Trees, E., Medus, C., Musser, K.A., et al., 2011. A novel vehicle for transmission of *Escherichia coli* O157:H7 infections associated with consumption of ready-to-bake commercial prepackaged cookie dough – United States, 2009. Clin. Infect. Dis. 54 (4), 511–518.

Owen, W.L., 1977. Microbiology of cane sugar manufacture and refining. In: Meade, G.P., Chen, J.C.P. (Eds.), Cane Sugar Handbook, tenth ed. Wiley, New York, pp. 405–422.

Porrini, C., et al., 2003. Honey bees and bee products as monitors for environmental contamination. APIACTA 38, 63–70.
Richter, K.S., Dorneanu, E., Eskridge, K.M., Rao, C., 1993. Microbiological quality of flours. Cereal Foods World 38, 367–369.
Seiler, D.A.L., 1978. The microbial content of wheat and flour. FMBRA Bull. February, 26–36.
Sperber, W.H., The North American Millers' Association Microbiology Working Group, 2007. Role of microbiological guidelines in the production and commercial use of milled cereal grains: a practical approach for the 21st century. J. Food Prot. 70, 1041–1053.
Townsend, G.F., 1961. Preparation of Honey for Market. Ontario Department of Agriculture Publication., 544. pp. 24.
US FDA Import Alert 36-04, May 2013, Detention without Physical Examination of Honey and Blended Syrup Due to Presence of Fluoroquinolones.
USDA. 1962. White Jr., J.W., et al., Composition of American Honeys. Tech. Bull. 1261. Agricultural Research Service, USDA, Washington, DC.

Further Reading

Lochhead, A.G., Heron, D.A., 1929. Microbiological studies of honey. Can. Dept. Agr. Bull. 116 (N.S.).
US FDA Import Alert 36-04, May 2013, Detention without Physical Examination of Honey and Blended Syrup Due to Presence of Fluoroquinolones.

CHAPTER

12

Edible Nuts, Oilseeds and Legumes

W.C.A. Gelderblom[1,2], *G.S. Shephard*[1], *J.P. Rheeder*[1], *S.K. Sathe*[3], *A. Ghiasi*[4] *and Y. Motarjemi*[5]

[1]Medical Research Council, Tygerberg, South Africa, [2]Department of Biochemistry, Stellenbosch University, Stellenbosch, South Africa, [3]College of Human Sciences, Florida State University, Tallahassee, FL, USA, [4]Food Safety Consultant, Tehran, [5]Iranian Pistachio Research Institute, Rafsanjan, Iran

OUTLINE

Food Safety Management.
DOI: http://dx.doi.org/10.1016/B978-0-12-381504-0.00012-3

PART 1: PERSPECTIVES ON MYCOTOXINS

INTRODUCTION

Nuts, oilseeds and legumes form part of traditional diets worldwide and are a rich source of dietary proteins, fiber, polyunsaturated fatty acids and phytochemicals. These food commodities are also well known for their potential as functional foods in the fight against chronic diseases of lifestyle (King et al. 2007; Messina, 1999). After cereals (wheat, rice, maize and barley), the former commodities also provide the next highest source of calories in human nutrition, forming part of staple diets (legumes) or are functional ingredients in many processed foods. Legumes contain two to three times more protein than cereals and are the principal source of dietary protein for vegetarians (Deshpande et al., 2000).

Global production of edible nuts is dominated by the United States of America (USA), followed by Indonesia and China; for oilseed mainly by China; and for legumes by Algeria, China and Iraq (data for 2009 at http://faostat.fao.org/site//339/default.aspx). The USA is the largest producer of almonds, followed by Spain (Campbell et al., 2003; ABC, 2010), and although China is by far the largest producer of walnuts, the USA is the largest exporter (Bayman et al., 2002).Turkey is the leading hazel nut producer and exporter, followed by Italy and the USA (Basaran and Ozcan, 2009), while Iran is the largest producer and exporter of pistachios, followed by the USA and Turkey. Brazil nuts originate from the Amazon regions and are mostly collected in the northern part of Brazil and neighboring countries (e.g. Bolivia and Peru) which are the major exporters (Freitas-Silva and Venancio, 2011). Globally, China, India and Nigeria are the top three producers of groundnuts (with shells).

HAZARD ANALYSIS

During harvesting, transport and handling of nuts, oilseeds and legumes, like with other produce, there are many potential hazards. Several of these hazards are discussed in other chapters (Chapters 4, 9 and 34). Although not addressed here, the reader should be also alerted to the risk of *Salmonella* in nuts. Many incidents of *Salmonella* reported from USA have been associated with peanut butter. This chapter focuses on one of the most important hazards: mycotoxins.

It is estimated that 25% of the world's food crops are contaminated with mycotoxins, resulting in an estimated global loss of foodstuffs in the range of 1000 million tonnes per year (WHO, 1991; http://www.fao.org/ag/agn/agns/chemicals_mycotoxins_en.asp). Apart from direct economic losses, mycotoxins are also associated with many animal and human diseases. Mycotoxins, ubiquitous in their occurrence, are produced by a variety of fungi that grow on the foods used by humans and animals alike (Sharma and Salunkhe, 1991). In developed countries susceptible subjects are generally well protected by the high standards of the major food suppliers and retailers, and the regulatory controls that deter the importation of seriously contaminated products (van Egmond et al., 2007) while in developing countries such protections are typically lacking. Stringent mycotoxin standards on exported food crops imply that developing countries export their best-quality food products. However, due to the lack or even the absence of regulatory measures, highly contaminated foods may be utilized domestically which, in addition to significant economic losses, increases the risk of mycotoxin exposure (Wu, 2004).

MYCOTOXIGENIC FUNGI AND MYCOTOXINS

Fungal contamination of food and feed is a worldwide phenomenon with *Aspergillus*, *Penicillium* and *Fusarium* fungal spp. and are considered the most significant toxigenic genera. They produce a diverse group of secondary metabolites of mycotoxins that exhibit a wide range of toxic effects in animals and humans (Table 12.1). Exposure of humans mainly occurs via the food chain and the ingestion of contaminated agricultural commodities such as cereal grains, nuts seeds, oil seeds and food products derived from these sources. Although acute outbreaks of mycotoxicosis in humans are rare, the chronic low-dose exposure is a concern especially as to their modulating role in human disease. This becomes more relevant in developing countries where the lack of mycotoxin regulations, food scarcity or insecurity and diversity, poor infrastructure and malnourishment will impact on the development of certain chronic diseases (Wu, 2004). Major mycotoxins commonly occurring

TABLE 12.1 Major Mycotoxigenic Fungal Species and Secondary Metabolites Associated with Nut Infections

Genus	Fungal Species	Major Mycotoxins
Aspergillus	*A. flavus*	Aflatoxins
	A. parasiticus	
	A. carbonarius	Ochratoxin A
	A. ochraceus	
	A. niger	**Fumonisins**
Penicilium	*P. chrysogenum*	Ochratoxin A
	P. commune	
	P. nordicum	
	P. purpurrescens	
	P. rugulosum	
	P. verrucosum	
Fusarium	*F. verticillioides*	Fumonisins
	F. proliferatum	
	F. graminerum Schwabe	Deoxynivalenol
	F. culmorum	Nivalenol
	F. crookwellense	Zearalenone
Alternaria	***A. alternata***	Alternariol
		Alternariol methyl ether
		Altertoxin-I
		Tenuazonic acid
		Fumonisins

Fungal species and mycotoxins in bold refer to recent discoveries of fungi with the ability to produce fumonisins.

on agricultural products include the aflatoxins, ochratoxins, zearalenone, trichothecenes (deoxynivalenol and nivalenol) and the fumonisins, which are produced by *Aspergillus* and *Fusarium* (Molyneux, 2007).

Mycotoxins produced by *Aspergillus* species, specifically *A. flavus* and *A. parasiticus*, are the major contaminants of peanuts and products derived from them (Magnoli et al., 2007). The major fungi contaminating tree nuts such as almonds, pistachios and walnuts are *Aspergillus*, *Penicillium*, *Rhizopus*, *Alternaria*, *Fusarium* and *Trichoderma* (Jiménes et al., 1991; Bayman et al., 2002). In chestnuts, *Alternaria*, *Fusarium*, *Penicillium* and *Aspergillus* dominate (Overy et al., 2003). In Brazil nuts a similar fungal mycobiotica including *Aspergillus*, *Penicillium* and *Fusarium* exist with *A. flavus* that dominates (Frietas-Silva and Venâncio, 2011). Each nut species, however, has a distinct mycoflora with Aspergilli that normally co-occur, while the presence of *Penicillium* negatively correlates with *Aspergillus*. It would appear that the harvest and post-harvest treatment protocols do have a major influence on the mycoflora and hence mycotoxin contamination (Bayman et al., 2002). The contamination of edible oils normally reflected the level of the relevant mycotoxins in the parent or base material (Schollenberger et al., 2008). The contamination of sunflower and soybean by *A. alternata* has been reported, whereas *F. verticillioides* also frequently contaminated sunflower grains (Pozzi et al., 2005) and *F. proliferatum* has been reported to occur on peas and soybean (Ivić et al., 2009). Insect infestations and damage play a major role in fungal infection and mycotoxin contamination. Insect control either through pest control, breeding or genetic engineering of resistant cultivars and/or biological control through the application of non-toxigenic strains is a promising tool to reduce mycotoxin contamination (Campbell et al., 2003).

Although mycotoxin contamination is not a serious problem in developed countries with respect to human health, economic losses are significant. In contrast mycotoxins cause numerous health problems in developing countries, apart from economic losses (Khlangwiset and Wu, 2010). AFB1 and AFG1 are pro-carcinogens that, upon metabolism, give rise to the toxic, mutagenic and carcinogenic metabolites, disrupting cellular oxidative homeostasis and intercalating with DNA. The aflatoxins are the major mycotoxin contaminants of peanuts, hazel nuts, pistachio nuts, almonds, Brazil nuts, walnuts and therefore the most important mycotoxins entering the human food chain upon consumption. *A. flavus* and *A. parasiticus* are the major producers, and like many secondary metabolites the aflatoxins are a family of closely related compounds which includes aflatoxin B1, B2, G1 and G2, with AFB1 not only occurring at the highest levels but also being the most toxic. It can cause a variety of adverse effects in various animal species while the acute toxicity differs between species and sexes of the same species (Moss, 2002). Based on the acute aflatoxin poisoning in India (Krishnamachari et al., 1975) an LD50 of approximately 5 mg/kg body weight has been proposed in humans. Although acute exposure is rare and seems not to be a concern in relation to tree nuts if not a major component of the diet, the chronic exposure and the carcinogenic properties of AFB1 and AFG1 is a major concern internationally (Molyneux et al., 2007). In Benin and Togo the frequency of aflatoxin exposure in children was much higher in the maize consuming agro-ecological zones as compared to peanuts. The impact of aflatoxin exposure due to groundnut consumption is limited, although it was more prevalent in high socioeconomic strata in certain agro-ecological zones due to a higher consumption frequency (Egal et al., 2005). Other African countries such as the Democratic Republic of

Congo (DRC), however, have to face far worse scenarios, where exposure via the consumption of peanut and peanut oil by far exceeds the regulatory limits set by the EU and the WHO (Kamika and Takoy, 2011). This is of particular interest as the DRC is classified as one of the African countries with a high risk of developing primary liver cancer.

Ochratoxin A (OTA) is produced by *P. verrucosum* and *P. nordicum* and a few *Aspergillus* species including *A. carbonarius* and *A. niger*. The mycotoxin occurs on a wide variety of food products including coffee, grapes, beans, chickpeas and nut seeds such as pecans and pistachios. OTA exhibits immunosuppressive, nephrotoxic, nephrocarcinogenic and terato-carcinogenic effects. The formation of DNA adducts and the induction of oxidative stress have been proposed as possible mechanisms involved in OTA nephrocarcinogenicity, which was classified as a group 2B carcinogen or possibly carcinogenic in humans (IARC, 1993). Involvement of OTA in the development of chronic renal disease and kidney and urinary tumors in the Balkan countries has not been established as yet. However, high concentrations of OTA were recorded in the blood of humans from endemic villages as compared to non-endemic villages (FAO/WHO, 2011).

Deoxynivalenol (DON) is one of the major trichothecene mycotoxins produced mainly by *Fusarium graminearum*, *F. culmorum* and *F. crookwellense* which infect the small grains such as wheat and barley as well as maize, millet, sorghum and rice. A recent report indicated the presence of *Fusarium* species and DON in only a few samples of soybeans (Barros et al., 2011). The major acute toxic effect of DON is related to feed refusal, vomiting and severe gastrointestinal toxicity in animals. Other effects include teratogenicity, cardiotoxicity and disruption of the immune system. DON caused similar acute toxic effects in humans upon the consumption of contaminated cereals. Several outbreaks of acute human illness have been reported upon the consumption of scabby wheat and moldy maize contaminated with high levels of DON which resulted in characteristic symptoms such as nausea, vomiting, abdominal pain and diarrhoea (FAO/WHO, 2011). Zearalenone (ZEA) normally co-occurs with DON and exhibits its activity by binding to estrogen receptors altering the estrogen-responsive elements in the nucleus. ZEA also interferes with steroid metabolism and hence could be involved in the disruption of the endocrine system and has been shown to increase liver cell and pituitary tumors in mice. ZEA, α-zeAranol and the type B trichothecene, 15-acetyl DON, are consistently detected in soybean oil (Schollenberger et al., 2008).

The fumonisins, mainly produced by *Fusarium verticilioides* and *F. proliferatum* and recently also by *A. niger*, commonly occur on maize and to some extent in sorghum and millet. Fumonisins have not been reported to occur on nuts although the major fungal producers, *F. verticillioides* and *F. proliferatum*, have been shown to occur on pistachios (Fernane et al., 2010). Fumonisins cause a wide variety of toxic syndromes in animals, and depending on the animal species could affect the liver, kidneys, lungs and brain. They have been associated with the development of liver and esophageal cancer and neural tube defects in humans. Fumonisins have been classified as apparent non-genotoxic carcinogens that exhibited their mode of action via the disruption of lipid biosynthesis and hence the structure and function of cellular membranes (FAO/WHO, 2001). Reports on the contamination on edible nuts oilseeds and legumes are scarce. Fumonisin-producing *Fusarium* species were isolated from cowpea seeds while fumonisin B1 was found to occur at concentrations ranging between 0.12 and 0.61 μg/kg in cowpea cultivars from South Africa (Kritzinger et al., 2003).

Alternaria toxins alternariol and alternariol monomethyl ether were detected in sunflower meal with levels well below the LD50 necessary to cause toxicity in animals (Pozzi et al., 2005). Olive oil produced from highly *Alternaria*-infected olive samples in the laboratory provided evidence of alternariol transfer, although no human exposure to contaminated oil has been recorded. *Alternaria* toxins are not considered to be major toxic principles; however, the mutagenic potential appears to be a major human health concern. Recently alternariol has reported to exhibit estrogenic potential, inhibited cell proliferation and genotoxic effects in cultured mammalian cells. The most toxic compound produced by *Alternaria* is tenuazonic acid causing a wide variety of toxic lesions in animals and pre-cancerous lesions in the esophagus of mice. It has also been implicated in the etiology of onyalai, a human hematological disorder in Africa manifested as hemorrhagic bullae on the mucosa of the oronasopharynx (Hesseling, 1992; Barkai-Golan, 2008).

CONTROL MEASURES

Control measures for mycotoxins are multifaceted and involve a range of interventions based on a farm to fork approach, which acknowledges that mycotoxins are the result of fungal contamination occurring in the food supply chain. The minimum requirements for mycotoxin control applicable to all the important toxins are the application of good agricultural practices (GAP), for example:

- proper sanitation in tree nut orchards to reduce the presence of *A. flavus* inoculum on decaying leaves and fruit;
- stacking of proper wind-rows at peanut harvest to allow for sufficient drying of the pods; good storage practices (GSP):
- storage of commodities in dry, well-ventilated and preferably temperature-controlled environments;
- refraining from storing good quality food lots with moldy or substandard lots; and good manufacturing practices (GMP):
- regular inspection of processing equipment to ensure cleanliness and no build-up of moldy material in inaccessible areas;
- rapid processing, drying and packaging of foodstuffs to reduce the length of time raw commodities need to be stored before processing.

In addition, utilizing a hazard analysis critical control point (HACCP) system, in conjunction with these practices, is crucial when handling and processing foodstuffs. This latter system is dealt with in detail later in this chapter. Apart from these general principles, a variety of measures have been developed to address control issues for specific mycotoxins. It should be noted that the mixing of good and contaminated lots to reduce the overall contamination level is illegal in many countries, but is still prominent in the food industry. The role of risk management therefore is to identify the problem of mycotoxin contamination, commission a science-based risk assessment, consider the result of the assessment and possible risk management options, implement an intervention strategy if required and then monitor and review the results of any intervention. In this manner, risk management involves the consideration and implementation of food policy options, while taking cognizance of scientific, cultural, economic, social and ethical issues, which will differ between countries.

Aflatoxins

The major naturally occurring aflatoxins that have been evaluated by JECFA (Joint FAO/WHO Expert Committee on Food Additives) are AFB1 in food commodities and AFM1, a metabolic product from the ingestion of AFB1, in milk. The International Agency for Research on Cancer (IARC) has evaluated AFB1 and natural mixtures of aflatoxins as group 1 carcinogens, i.e. there is sufficient evidence to characterize AFB1 as a known human carcinogen, whereas AFM1 was classed as group 2B, possibly carcinogenic to humans (IARC, 1993). JECFA evaluated AFB1 as a genotoxic carcinogen with an absence of a toxicological threshold (WHO, 1998). In deriving a suitable potency estimate, JECFA recognized a synergy between AFB1 exposure and hepatitis B infection as causative factors in liver cancer. Using human epidemiological data, the JECFA chose a potency of 0.01 cancers/year per 100,000 population per ng aflatoxin/kg bodyweight per day for hepatitis B surface antigen negative individuals. For positive individuals, a cancer rate 30 times this level was selected. A comparative toxicological study in rats indicated that AFM1 has a potency approximately an order of magnitude less than AFB1 and in its evaluation, JECFA applied the same factor (one-tenth) to derive a potency for AFM1 (FAO/WHO, 2001). Based on these AFB1 potencies, hepatitis prevalence and aflatoxin exposure data, it has recently been estimated that aflatoxin may play a causative role in 4.6–28.2% of all global primary liver cancer cases (Liu and Wu, 2010). Besides its role in liver cancer and incidences in Kenya of deaths due to acute aflatoxicosis from high exposure rates, chronic exposure to aflatoxins has been associated with the risk of stunting in children and immune suppression, as well as a putative role in kwashiorkor (Shephard, 2008).

Given its substantial adverse effects on human health, the reduction and control of aflatoxin contamination has been the subject of active research for many years. It is generally recognized that no single measure will be adequate and that a package of control measures will be necessary to effectively reduce human exposure, especially in developing countries, where exposure levels are very high. In developed market economies, implementation of legislated maximum tolerated levels (MTLs) with adequate enforcement is effective in controlling consumer exposure. However, in developing countries with poor enforcement infrastructure, alternative food supply chains involving informal markets and with significant subsistence farming communities, legislation is inadequate. Control measures for aflatoxin in these countries can be categorized as pre-harvest, post-harvest, or individual-based strategies.

The pre-harvest control of aflatoxins is addressed by agricultural research for resistant varieties of groundnuts and maize and by the application of GAP, including control of insects and reduction of crop stress by irrigation. The most significant advance in pre-harvest control of aflatoxins has been the field application to the soil of non-toxigenic (atoxigenic) strains of the main aflatoxin-producing fungus *Aspergillus flavus*. The principle behind this strategy is that the applied atoxigenic strains will outcompete the natural aflatoxin-producing strains in the soil and pre- and post-harvest on agricultural crops (Dorner, 2008). As in all post-harvest fungal problems, the rapid and efficient drying of an agricultural crop is an essential component to reduce fungal contamination. In the case of sun drying of groundnuts, pods need to be adequately exposed for maximum effect. Similarly, storage needs to be governed by good storage practices (GSP) that keep the crop dry and free of insect activity. An intervention study conducted in Guinea, West Africa, has demonstrated how a package of post-harvest measures addressing issues such as sorting, drying and storage of groundnuts can have a positive impact on aflatoxin exposure in a rural village (Turner et al., 2005).

Management of aflatoxins in groundnuts can also be achieved by a range of industrial food-processing techniques, including segregation of lots, screening and color sorting (either with or without blanching) (Dorner, 2008). The sorting of poor quality groundnuts can also be undertaken manually in subsistence farming environments. Although the aflatoxins are regarded as heat stable mycotoxins, roasting of peanuts can reduce AFB1 by 50–80%, whereas home cooking of various African fermented maize dishes can reduce AFB1 by 40–80% (WHO, 1998).

A number of strategies to reduce exposure to aflatoxins are targeted at an individual level, but require a large degree of public awareness. These measures encourage a more varied diet, a change from maize back to traditional African cereals like sorghum and millet and the use of enterosorbents such as Novasil clay, which has been shown to chemisorb AFB1 in the clay interlayers and thus exclude it from the gut lumen (Phillips et al., 2008). Although specific chemo-prevention measures have been investigated, such as the use of oltipraz, these are generally regarded as too expensive for general use in developing countries or not proven to be culturally acceptable. Alternatively, although it does not address exposure, the vaccination of the general population against hepatitis B does mean that the cancer potency of AFB1 is significantly reduced due to the synergistic interaction of these two risk factors for liver cancer.

Fumonisin, Deoxynivalenol and Ochratoxin A

Of the major mycotoxins other than aflatoxin affecting human health, fumonisins, deoxynivalenol and ochratoxin A occur only sporadically in nuts, oilseeds and legumes. The risk assessments of these mycotoxins have also been performed by JECFA (FAO/WHO, 2001). Unlike for aflatoxin, JECFA employed a threshold-safety factor approach to generate a provisional maximum tolerable daily (or weekly for ochratoxin A) intake for fumonisins, deoxynivalenol and ochratoxin A. For fumonisins, JECFA used toxicity studies in rodents to set the no observable effects limit (NOEL) for renal toxicity at 0.2 mg/kg bodyweight/day, which combined with a safety factor of 100 gave a provisional maximum tolerable daily intake (PMTDI) of 2 μg/kg bodyweight. This was applied to the three most abundant fumonisin analogues (FB1, FB2 and FB3) either alone or in combination. In a similar manner, the PMTDI for deoxynivalenol was set at 1 μg/kg bodyweight, based on an NOEL of 0.1 mg/kg bodyweight per day for the absence of toxicological effects and significant bodyweight loss in a 2-year study in mice. A more recent evaluation confirmed the PMTDI and extended it to a group PMTDI to include the 3- and 15-acetyldeoxynivalenol analogues, based on evidence that these are converted to deoxynivalenol in the gut (FAO/WHO, 2011).

The acute effect of a high dose of deoxynivalenol is the induction of emesis. Based on this effect in pigs, JECFA derived an acute reference dose for deoxynivalenol of 8 μg/kg bodyweight. Ochratoxin A is largely bound to protein and hence the evaluation by JECFA established a provisional tolerable weekly intake (PTWI) of 100 ng/kg bodyweight (FAO/WHO, 2001). The main sources of human exposure to these three mycotoxins are through the consumption of cereals (mainly wheat and maize). Since these mycotoxins are not considered a major problem in nuts, oilseeds and legumes, specific control measures have not been developed. Hence the control of fumonisins, deoxynivalenol and ochratoxin A in these commodities relies on generic methods such as maintenance of GAP, GSP, GMP and the sorting of contaminated lots.

CONTAMINATION LEVELS, INCIDENCE OF DISEASES

Acute toxic outbreaks in humans consuming mycotoxin-contaminated nuts have not been reported although the major concern regarding mycotoxins such as AFB is the hepatocarcinogenicity, especially in hepatitis B-infected individuals. Due to the deleterious effects of mycotoxins in humans and animals, most countries establish a maximum tolerable level (MTL) to safeguard the health of humans. Due to strict regulations, such as the 2 μg/kg for AFB1 set by the European Commission, mycotoxin contamination of nuts has been studied worldwide, especially in countries that are major producers and exporters of these crops. The natural occurrence in a variety of nuts has been studied in various countries reporting a wide range of contamination in recent surveys (Table 12.2). Although aflatoxins occur widely in a variety of foodstuffs, tree nuts and oilseeds, pistachios having the highest risk of being contaminated. A recent study indicated that the level of AFB1 contamination in pistachio nuts imported into Spain from Iran showed a large variation ranging from no contamination to levels exceeding the EU regulation of an MTL of 2 μg/kg (Ariño et al., 2009). Several samples imported from Turkey and the USA including samples from Spain tested negative. A detailed study on the contamination levels during the 2003/4 harvest season indicated that 11.8% of samples contained AFB1 levels above the MTL of 5 μg/kg set by Iran (Cheraghali et al., 2007). Available pistachios from retail shops also showed high frequency of contamination with 36 and 29% of the nuts that exceeded the MTL set for AFB1 (5 μg/kg) and total aflatoxin (15 μg/kg) by the EU, respectively (Sarhang Pour 2009).

Although OTA is occasionally detected in a few pistachio samples, a recent study showed that *A. carbonarius*, and to some extent *A. niger* isolated from pistachio nuts, produced high amounts of OTA (Marín et al. 2008). The presence of low levels of aflatoxins (0.4 to 0.7 μg/kg) and OTA (170 μg/kg) were also detected in pistachios from various retail outlets in Algeria (Fernane et al., 2010). OTA was also found to occur in 50% of the peanut samples from Argentina during storage with mean levels of 5.6 to 130 μg/kg (Mangoli et al., 2007). *Aspergillus* section *Nigri* was found to be the main producer of OTA in culture as compared to *A. carbonarius*. The findings suggested that humans and animals are frequently being exposed to OTA upon the consumption of peanuts and peanut-derived products.

Contamination of peanuts by aflatoxins has been the focus of a recent study in Kinshasa in the Democratic Republic of Congo where it is utilized as food and oilseed (Kamika and Takoy, 2011). High contamination levels were recorded, specifically in the rainy season

TABLE 12.2 Occurrence and Level of Mycotoxin Contamination on a Variety of Nuts Produced in Different Countries of Origin

Country of Origin	Commodity	Mycotoxin(s)	Level (μg/kg)	Remarks	Reference
Togo and Benin	Groundnut	AFB1	Northern: 12.5 Southern–Northern: 362.8–528.3	Average groundnut consumption frequency: 1.4–4.8 days/week	Egal et al. (2005)
Niger	Groundnut	AFB	>30	Severe water stress high AFB production	Rauford et al. (2006)
Western Kenya	Groundnut	AFB	Busia district: 0–2687.2 Homabay: 0–7525	87% levels <4 μg/kg 5.45% >5–20 μg/kg 7.55 >20 μg/kg	Mutegi et al. (2009)
Democratic Republic of Congo	Groundnut	AFB1	Dry season: 12–937 Rainy season: 15–390	72% of samples positive 70% exceeded maximum level of 5 μg/kg	Kamika and Takoy (2011)
Morocco	Nuts	AFB1 (AFBtotal)	Groundnut: 0.17 (0.3) Walnut: 360 (730) Pistachios: 158 (163)	5 and 20% of pistachio and walnut exceed EU (4 μg AFT/kg) regulation, respectively	Juan et al. (2008)
Algeria	Pistachio	AFBtotal OTA	AFBtotal: 0.4–0.7 OTA: 170	AFBtotal: 6.5% of samples OTA: 3.3%	Fernane et al. (2010)
Malaysia	Nuts and nutty products	AFBtotal	Groundnut shelled: 17.8–711 Walnut: 17.2 Coated nut product: 113–514 Peanut butter: 16.6–67.3 Peanut cake: 61.9–84 Confectionary: 17–21.4	High levels occurred in food from tropical and subtropical regions	Leong et al. (2010)
Qatar	Nuts	AFBtotal	Pistachio: 7.3–289 Almonds: nd Cashew: nd Walnut: 2.8 Hazel: 6 Groundnut: nd	Pistachios without cells highly contaminated. Poor storage and handling the key determinants	Abdulkadar et al. (2000)
Jordan	Nuts	AFBtotal Ochratoxin A	OTA in nuts: 2.75–7.42 OTA in sunflower seeds: 4.34 AFBtotal in walnut: 9.62	Nuts mainly imported	Salem and Ahmad (2010)

Pakistan	Nuts	AFBtotal	Almonds: 2.13 Walnuts (30%): 3.43 Peanuts (20%): 5.2 Pistachios: 6.34 Pine nuts: 3.25	Nuts without cells. Lower levels with cells: walnuts – 30%; peanuts – 20%; pistachios – 20% above legal levels set by EU	Luttfullah and Hussain (2010)
Pakistan	Beans, peas, soybean	AFBtotal	Red kidney beans: 5 Split pea: 4.1 Chick pea: 2.5 Cow pea: 2.2 Soybean: 6.4	Red kidney beans 10% above limit Split peas 6% above limit Soybean 10% above limit	Luttfullah and Hussain (2010)
China	Peanuts	AFB1	1040 samples collected during 2099 to 2010 in 4 zones: 25% of samples contained 0.01–720 μg AFB1/kg	95% of samples <1.0 μg AFB1/kg 1% samples <20 μgAFB1/kg 3.75% exceed EU regulation of <2 μg AFB1/kg	Ding et al. (2012)
Iran	Pistachios	AFBtotal	AFB1: 5.9 ± 41.7 AFBtotal: 7.3 ± 53.2	11.8 and 7.5% samples above the MTL of AFB1 (5 μg/kg) and AFBtotal (15 μ/kg)	Cheraghali et al. (2007)
Iran	Pistachio	AFBtotal	AFB1: 185.89 AFBtotal: 215.05	36 and 29% of samples exceeded the MTL set by the EU	Sarhang Pour et al. (2010)
Spain	Pistachios (Iran/USA/Turkey/Spain)	AFB1	AFB1: 0.12–0.29	All positive samples from Iran. None of the samples exceeded EU regulations	Ariño et al. (2009)
Turkey	Commercialized nuts (hazel, pistachios, peanuts)	AFB1	<5 μg/kg (14.28%) >5 μg/kg (1.84%)	Samples randomly collected from local markets and retail stores	Basaran and Ozcan (2009)
Argentina	Peanuts	Ochratoxin A	Storage: 1st mouth – 30 ng/g 2nd month – 6.5 ng/g 3rd month – 13 ng/g	OTA levels decreased with storage	Mangoli et al. (2007)

where a mean level of 205.7 μg/kg was recorded, with 70% of the samples exceeding the maximum limit of 5 μg/kg set by the WHO. A large survey in China also reported aflatoxin contamination from four agro-ecological zones throughout 12 provinces indicating levels ranging from 0.01 to 720 μg/kg in 25% of the samples (Ding et al., 2012). Despite these high levels, peanuts are not the principal diet and aflatoxin exposure from corn and corn products seems to be the major source. Of interest is that the risk to humans to develop liver cancer was 10 times higher when considering peanut oil with levels reportedly ranging between 8 and 68.8 μg/kg.

Contamination of a variety of nuts, including almonds, walnuts, pistachios, peanuts, hazel and cashew nuts and/or food products derived from these nuts by aflatoxins and, in some occasions, ochratoxin and/or combinations thereof, have been reported (Salem and Ahmad, 2010; Luttfullah and Hussain, 2010; Leong et al., 2010; Abdulkadar et al., 2000; Juan et al., 2008; Mutegi et al., 2009). Regional and seasonal variations of aflatoxin contamination occur while soil moisture content and temperatures optimal for the growth of *A. flavus* play an important role in the prediction of the level of contamination (Rauford et al., 2006).

HACCP CASE STUDIES

The hazard analysis and critical control point (HACCP) system identifies, evaluates and controls hazards (including mycotoxins) that pose significant health risks in terms of food safety (see Chapter 31). It is a structured, systematic approach throughout the commodity chain and is a key element in total quality management (TQM), together with good agricultural practices (GAP), good storage practices (GSP) and good manufacturing practices (GMP). Although some authors argue that HACCP is applicable to the whole food chain, from the producer to the consumer, others maintain that its implementation on farm and from the product distributor to the final consumer is not practical. These latter steps form part of GAP and GSP within each food chain, rather than part of the HACCP plan (Sperber, 2005).

Though structured, each HACCP case study or model is based on some subjectiveness when determining the critical control points (CCPs). Although several case studies might deal with the same commodity being processed, this does not mean that the individual HACCP plans are identical. Each HACCP plan must be tailored according to a very specific commodity chain and it is here that the difference between hazard identification and risk analysis must be clearly stated for each separate case or production line (CAC, 1999).

PART 2: PISTACHIO NUT PROCESSING HACCP STUDY

INTRODUCTION

The pistachio nut is the fruit stone of the tree *Pistaciavera* L. Each fruit has a single stone, which consists of a kernel covered by a testa and enclosed in a shell. The shell itself is enclosed in a protective hull. Pistachio nuts are grown commercially in Iran, the USA, Turkey, Syria, Greece and China (FAO, 2011). In most countries, except in Turkey, the nuts are usually de-hulled soon after harvest and the nuts in their shells are then processed, dried and stored. In Turkey, however, the nuts are usually stored in-hull after harvest, sometimes for many months before any processing takes place. Early de-hulling has the advantage of avoiding staining of the shell, but has the disadvantage of exposing the split nuts at an early stage to contamination by *Aspergillus flavus* and *A. parasiticus* spores, which have the potential to infect the nuts and produce aflatoxin.

One month or more before maturity, the shell partially splits within the hull. The hull should remain intact, but sometimes it also splits naturally prior to harvest and these "early splits" are particularly susceptible to aflatoxin contamination. Early splitting allows invasion by insects, particularly the navel orange worm [*Amyeloistransitella* (Walker)] and insect-damaged nuts are associated with a high risk of aflatoxin contamination.

In Iran and Syria varieties are grown that tend to have large nuts with hulls, which are relatively prone to early splitting, although this is also influenced by environmental and climatic factors. In Turkey and the USA the pistachio varieties tend to yield smaller nuts with greener kernels and these have hulls that are not very susceptible to early splitting.

Collecting a representative sample of pistachio nuts for aflatoxin testing is particularly difficult as it has been established that the incidence of highly contaminated nuts is usually very low in sorted export lots, in the order of 1 in 25,000 nuts (Sommer et al., 1986). A single pistachio with an aflatoxin concentration of 60,000 μg/kg can contaminate an otherwise aflatoxin-free 4.5 kg lot (approximately 3000 nuts) at the US-FDA action level of 20 μg/kg for total aflatoxin.

DESCRIPTION OF THE PRODUCT

Dried pistachios should have a moisture content of <6% and a water activity (a_W) of less than 0.70 for long-term storage. The preferred bulk packaging is in propylene bags with a mass of 50 kg. These can be preserved for at least 1 year at 60–65% relative humidity and at a temperature of between 10 and 20°C. Product labeling should include the product name, product type, net weight, name and trade brand of the exporter, serial number or identification code, country of origin and processing, and the recommended expiry date. Transporting pistachios for export should be done in dry, cool conditions in proper containers to prevent contamination and mold growth within the nuts.

DISTRIBUTION AND INTENDED USE OF THE PRODUCT

Pistachios are mostly used in the confectionery and snack food industries as dried nuts or packaged as roasted, salted nuts in plastic foil and vacuum sealed. The main export markets are the EU and the USA, as well as Russia and the Far East.

PISTACHIO NUT PROCESSING – COMMODITY FLOW DIAGRAM (CFD)

An example of a CFD for pistachio nut processing, as found mainly in the Middle East, is illustrated in Figure 12.1.

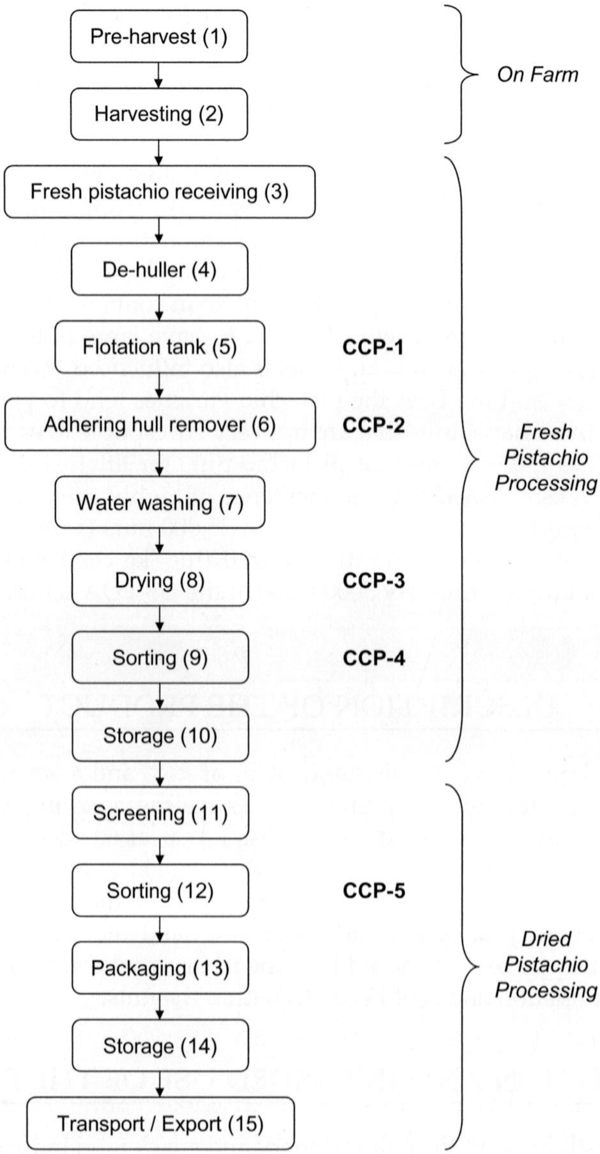

FIGURE 12.1 An example of a commodity flow diagram (CFD) for pistachio nut processing.

HAZARD IDENTIFICATION AND RISK ANALYSIS

According to the Codex Alimentarius Commission (1999), a hazard is defined as a biological, chemical or physical agent that might cause ill-health when consumed, and risk is the probability of occurrence of this hazard. The interaction between hazard and risk results in the severity or seriousness of the hazard to the health of the consumers (Thomas and John, 2002; Bertolini et al., 2007).

Risk analysis, when referring to hazard, should be performed according to the following aspects: quantification, probability of occurrence, severity, reduction, increasing or modification trends along the process. For aflatoxin, the severity is always high due to its carcinogenic potential. It is possible to correlate the hazard risk and severity by a risk classification model to assess the significance of a risk (CAC, 1997).

Identification of the Mycotoxin Hazard

Aflatoxin is the predominant mycotoxin hazard for which the EU, the USA and many other countries have instituted regulatory limits for edible nuts and it is, therefore, the only mycotoxin considered in this case study.

Identification of Aflatoxin Risks in the Commodity Flow Diagram (CFD) and Suitable Control Measures

Step 1: On Farm, Pre-harvest

Risk: Most aflatoxin contamination occurs in the orchard and is associated with damage caused to the fruit's hull, mainly early splitting, prior to harvesting. The exposed nut becomes susceptible to infestation by *A. flavus* spores, leading to aflatoxin accumulation. Subsequent invasion of early-splitters by insects, particularly the navel orangeworm, compounds the problem.

Controls: Cultivation of pistachio varieties, which are not susceptible to early splitting. Carry out early harvesting to reduce the levels of early-split fruit. Pre-harvest aflatoxin contamination can be significantly reduced by applying Integrated Phytosanitary Management (IPSM), which aims to minimize the fungal spore counts and navel orangeworm levels in the orchard. Removal or burial of tree litter is highly recommended.

Step 2: On Farm, Harvesting

Risk: Most aflatoxin contamination in the orchard is associated with damage caused to the fruit's hull, mainly early splitting, prior to harvesting. The exposed nut becomes susceptible to infestation by *A. flavus* spores, leading to aflatoxin accumulation. Infestation of early-split nuts by insects, particularly the navel orangeworm, compounds the problem. This is generally considered, especially in the Middle East, a period of high risk equivalent to a CCP in the processing phase.

Controls: Planned early harvesting of pistachios is the most important practice to reduce the levels of early-split nuts exposed to *A. flavus* contamination. Nut contact with the soil during harvesting should be avoided. Harvested nuts should be transported to the processing plant as soon after harvest as possible.

Step 3: Fresh Pistachio Receiving

No aflatoxin contamination is likely at this step, provided that de-hulling is not delayed after receiving the fresh pistachios at the processing plant.

Step 4: De-huller

Risk: The process of de-hulling can predispose healthy nuts to subsequent fungal contamination due to the release of large volumes of *A. flavus* spores from the plant and fruit debris.

Control: Direct airflow away from the fresh pistachios and through a chlorinated water tank to eliminate the fungal spores.

Step 5: Floatation Tank

Risk: Incorrect management of water flotation can lead to further *A. flavus* contamination. The addition and continuous circulation of contaminated water within the flotation tank system can further contaminate the nuts. Leaving the sorted nuts in the flotation tank for too long can also lead to excessive fungal contamination.

Controls: Use chlorinated water in the flotation system and replace dirty water on a regular basis. Chlorinated dirty water can then be used to flood irrigate the orchards without the threat of spreading more fungal inoculum in the orchard environment.

Step 6: Adhering Hull Remover

No aflatoxin contamination at this step.

Step 7: Washing under Sprayers

Risk: The circulation of contaminated, dirty water from the flotation tank to this washing step will increase aflatoxin levels by infecting healthy nuts.

Controls: Same as Step 5.

Step 8: Drying (Mechanical/Solar)

Risk: Delayed drying of the wet nuts can lead to the development of *A. flavus* on the nuts and subsequent aflatoxin contamination.

Controls: Rapid drying of the pistachios to remove excess water is essential to prevent growth of *A. flavus* on the wet nuts.

Step 9: Sorting (by Hand or Electronic Eye)

Unsplit pistachios are removed. No aflatoxin contamination is likely at this step.

Step 10: Storage

Risk: No aflatoxin contamination is likely at this step, provided that drying to a water activity (a_W) of less than 0.70 has been achieved.

Controls: Storage of pistachios in clean jute bags at a suitable temperature and relative humidity will prevent any aflatoxin accumulation.

Step 11: Screening by Size (Gravity Separator)

Levels of aflatoxin may be reduced at this step due to the mechanical removal of small, shriveled nuts, which are more likely to be contaminated with aflatoxin.

Step 12: Sorting (by Hand or Electronic Eye) (Schatzki and Pan, 1996)

Levels of aflatoxin will be significantly reduced at this step due to the removal of shell-stained, discolored and defective pistachios, which contain high levels of aflatoxin.

TABLE 12.3 Example of an HACCP Study Worksheet for Pistachio Nut Processing

Steps	Hazards	Control Measures	Control/CCP	Critical Limits	Monitoring	Corrective Actions	Verification
ON FARM							
1. Pre-harvest	Fungal contamination	Select varieties which are not prone to early splitting; control deleterious insects; apply IPSM in orchard to reduce fungal spores in soil and air	GAP	Reduce (plow in) >95% orchard litter	Visual inspection	Bury excess orchard litter; control navel orangeworm and other pests	Control of orchard operations and records kept by farmer
2. Harvesting	Fungal and aflatoxin contamination	Early harvesting to reduce level of early-splits; avoid nut contact with soil; transport directly to processing plant	GAP	Max level of early-split nuts: ≤5%	Visual inspection	Plan time of harvest	Control of orchard operations and records kept by farmer
FRESH PISTACHIO PROCESSING							
3. Fresh pistachio receiving	None likely	Commence processing as soon as possible	GMP	None	Management of incoming batches	Process nuts according to sequence of receiving	Control of operations and records by quality assurance manager and HACCP team
4. De-huller	Fungal and aflatoxin contamination	Avoid damage to healthy nuts; control airflow away from the nuts and through chlorinated water tank	GMP	None	Visual inspection	Revise processing procedures	Control of operations and records by quality assurance manager and HACCP team
5. Flotation tank	Increased fungal contamination; insufficient separation of low weight contaminated nuts	Use clean or chlorinated water to reduce fungal spread; optimum efficiency of nut separation with residing time in the water, temperature and pistachio to water ratio	**CCP-1**	Residing time of at least 5 min in the tank; water temperature of 20°C; 1:4 nut to water ratio	Visual inspection	Modify/adjust processing equipment to meet the required standards	Control of operations and records by quality assurance manager and HACCP team

(*Continued*)

TABLE 12.3 (*Continued*)

Steps	Hazards	Control Measures	Control/CCP	Critical Limits	Monitoring	Corrective Actions	Verification
6. Adhering hull remover	Fungal and aflatoxin contamination	Avoid damage to healthy nuts; sufficient removal of adhering debris will deprive fungus of substrate from which to contaminate healthy nuts	**CCP-2**	No adhering debris or other foreign material to spread contamination	Visual inspection	Revise processing procedures	Control of operations and records by quality assurance manager and HACCP team
7. Washing (sprayers)	Fungal contamination	Use clean or chlorinated water to reduce fungal spread; adjust spray nozzles and flow rate for effective washing of nuts	GMP	None	Visual inspection	Revise processing procedures	Control of operations and records by quality assurance manager and HACCP team
8. Drying (mechanical and/or solar)	Aflatoxin contamination	Rapid drying of pistachios to prevent the production of aflatoxin in the wet nuts	**CCP-3**	Drying of nuts to a water activity (a_w) of ≤ 0.70 within 48 h of start of processing	Controlling and recording of drying temperature and period; lab testing to examine product's water activity	Keeping nuts in dryer until desired water activity is achieved	Control of operations and records by quality assurance manager and HACCP team
9. Sorting (hand or electronic)	Aflatoxin contamination	Removal of shell-stained, discolored and defective pistachios by means of hand sorting and/or electronic UV scanning	**CCP-4**	Achieve a total aflatoxin level of $\leq 15\,\mu g/kg$	Chemical analysis according to approved international standards	Further sorting is required if critical limit is not achieved; adequate training of sorters	Control of operations and records by quality assurance manager and HACCP team

10. Storage	None likely	Storage of nuts in clean jute bags at a suitable temperature and relative humidity, provided the nuts have been dried properly in Step 8	GSP	None	Visual inspection of storage facility and ongoing monitoring	Maintain optimum storage conditions	Control of operations and records by quality assurance manager and HACCP team
DRIED PISTACHIO PROCESSING (NOT NECESSARILY THE SAME PROCESSOR AS WITH FRESH PISTACHIOS)							
11. Screening (gravity separator)	Aflatoxin contamination	Mechanical removal of small, shriveled and defective nuts	GMP	None	Visual inspection	Modify processing procedures	Control of operations and records by quality assurance manager and HACCP team
12. Sorting (hand or electronic)	Aflatoxin contamination	Removal of shell-stained, discolored and defective pistachios by means of hand sorting and/or electronic UV scanning	**CCP-5**	Achieve a total aflatoxin level of $\leq 10\,\mu g/kg$ for the final product (local and export quality)	Chemical analysis according to approved international standards	Further sorting is required if critical limit is not achieved; adequate training of sorters	Control of operations and records by quality assurance manager and HACCP team
13. Packaging	None likely	Dry and hermetically sealed packing materials must be used, to ensure an adequate shelf-life for the product	GMP	None	Good management of processing step and visual inspection	Modify processing procedures	Control of operations and records by quality assurance manager and HACCP team
14. Storage	None likely	Storage of nuts in packaging at a suitable temperature and relative humidity	GSP	None	Visual inspection of storage facility and ongoing monitoring	Maintain optimum storage conditions	Control of operations and records by quality assurance manager and HACCP team

(*Continued*)

TABLE 12.3 (*Continued*)

Steps	Hazards	Control Measures	Control/CCP	Critical Limits	Monitoring	Corrective Actions	Verification
15. Transportation and export	None likely	Transportation and storage of nuts at appropriate and homogeneous temperature and relative humidity; prevent any condensation on product when transporting by ocean or between different climatic regions. Prevent temperature differences in a batch (e.g. by using thermally insulated storage areas (walls, bottoms, lids))	GMP/GSP	None	Visual inspection of transport/ storage containers	Maintain optimum storage conditions	Control of export operations and records kept by exporter

TABLE 12.4 Example of an HACCP Plan (CCPs) for Pistachio Nut Processing

CCPs	Hazards	Control Measures	Critical Limits	Monitoring	Corrective Actions	Verification
CCP-1 Flotation tank	Increased fungal contamination; insufficient separation of low weight contaminated nuts	Use clean or chlorinated water to reduce fungal spread; optimum efficiency of nut separation with residing time in the water, temperature and pistachio to water ratio	Residing time of at least 5 min in the tank; water temperature of 20°C; 1:4 nut to water ratio	Visual inspection	Modify/adjust processing equipment to meet the required standards	Control of operations and records by quality assurance manager and HACCP team
CCP-2 Adhering hull remover	Fungal and aflatoxin contamination	Avoid damage to healthy nuts; sufficient removal of adhering debris will deprive fungus of substrate from which to contaminate healthy nuts	No adhering debris or other foreign material to spread contamination	Visual inspection	Revise processing procedures	Control of operations and records by quality assurance manager and HACCP team
CCP-3 Drying	Aflatoxin contamination	Rapid drying of pistachios to prevent the production aflatoxin in the wet nuts	Drying of nuts to a water activity (a_w) of ≤0.70 within 48 h of start of processing	Controlling and recording of drying temperature and period; lab testing to examine product's water activity	Keeping nuts in dryer until desired water activity is achieved	Control of operations and records by quality assurance manager and HACCP team
CCP-4 Sorting (fresh processing)	Aflatoxin contamination	Removal of shell-stained, discolored and defective pistachios by means of hand sorting and/or electronic UV scanning	Achieve a total aflatoxin level of ≤15 μg/kg	Chemical analysis according to approved international standards	Further sorting is required if critical limit is not achieved; adequate training of sorters	Control of operations and records by quality assurance manager and HACCP team
CCP-5 Sorting (dried processing)	Aflatoxin contamination	Removal of shell-stained, discolored and defective pistachios by means of hand sorting and/or electronic UV scanning	Achieve a total aflatoxin level of ≤10 μg/kg for the final product (local and export quality)	Chemical analysis according to approved international standards	Further sorting is required if critical limit is not achieved; adequate training of sorters	Control of operations and records by quality assurance manager and HACCP team

Step 13: Packaging

No risk of aflatoxin contamination, but inappropriate packing may make the nuts susceptible to future contamination if re-wetting occurs.

Step 14: Storage

No aflatoxin contamination is likely, provided that GSP procedures are adhered to (i.e. dry, well-ventilated and preferably temperature-controlled environment).

Step 15: Transportation and Export

No aflatoxin contamination is likely at this stage, provided that transportation and export conditions protect the nuts from excessive heat and fluctuations in moisture. It is also important to select lots that meet the customer's aflatoxin specification.

HACCP STUDY WORKSHEET AND HACCP PLAN (CCPS)

The HACCP study consists of two worksheets: (1) the HACCP Study Worksheet (see Table 12.3) which is a detailed description of the whole commodity process as illustrated in the CFD; (2) the HACCP Plan (see Table 12.4) which lists only the identified CCPs.

VERIFICATION, VALIDATION AND IMPLEMENTATION OF THE HACCP PLAN

Once the HACCP plan has been drawn up by the HACCP team, the full process then needs to be verified and validated to ensure the plan's effectiveness (see Chapter 31 for details). Only then can the HACCP plan be implemented, with ongoing monitoring taking place to check that the plan remains relevant to the specific processes and that all critical limits are met for each batch or lot of processed commodity. These procedures ensure that should any lot not meet the recommended limits for any CCP in the study, it should preferably be subjected to the necessary "Corrective Actions" as outlined in the HACCP plan. A consolidated list of documents and analytical test records for the whole HACCP study will also facilitate traceability in the case of troubleshooting and lot rejections during export.

Acknowledgments

The HACCP case study was part of an FAO project in the Islamic Republic of Iran, carried out by the authors, in cooperation with H. Feizi, H. Hokmabadi, S.A. Mohammadi, M. Moradi and M. Kenny.

References

ABC, 2010. Almond Board of California, Almond Almanac. Available from: <http://www.almondboard.com>.

Abdulkadar, A.H.W., Al-Ali, A., Al-Jedah, J., 2000. Aflatoxin contamination in edible nuts imported. Qatar Food Control 11, 157–160.

Ariño, A., Herrera, M., Estopañan, G, Rota, M.C., Carramiñana, J.J., Juan, T., et al., 2009. Aflatoxins in bulk and prepacked pistachios sold in Spain and effect of sorting. Food Control 20, 811–814.

Barkai-Golan, R., 2008. Alternaria mycotoxins. In: Barkai-Golan, R., Paster, N. (Eds.), Mycotoxins in Fruits and Vegetables. Elsevier Inc., San Diego.

Barros, G., Zanon, M.S., Abod, A., Oviedo, M.S., Ramirez, M.L., Reynoso, M.M., et al., 2011. Natural deoxynivalenol occurrence and genotype and chemotype determination of a field population of the *Fusarium grameniarum* complex associated with soybean in Argentina. Food Add Contam. Part A Jan, 1–11. (Epub).

Basaran, P., Ozcan, M., 2009. Occurrence of aflatoxins in various nuts commercialised in Turkey. J. Food Saf. 29, 95–105.

Bayman, P., Baker, J.L., Mahoney, N.E., 2002. *Aspergillus* on tree nuts: incidence and associations. Mycopathologia 155, 161–169.

Bertolini, M., Rizzi, A., Bevilacqua, M., 2007. An alternative approach to HACCP system implementation. J. Food Engineer. 79, 1322–1328.

Campbell, B.C., Molyneux, R.J., Schatzki, T.F., 2003. Current research on reducing pre- and post-harvest aflatoxin contamination of U.S. almond, pistachio, and walnut. Toxin Rev. 22, 225–266.

Cheraghali, A.M., Yazdanpanah, H., Doraki, N., Abouhossain, G., Hassibi, M., et al., 2007. Incidence of aflatoxin in Iran pistachio nuts. Food Chem. Toxicol. 45, 812–816.

Codex Alimentarius Commission (CAC), 1997. Hazard analysis and critical control point (HACCP) system and guidelines for its application. Annex to CAC/RCP 1-1969. In Basic Texts on Food Hygiene, second ed. Rome. <ftp.fao.org/docrep/fao/005/Y1579e/Y1579e.pdf>.

Codex Alimentarius Commission (CAC), 1999. Recommended methods of analysis and sampling. CODEX STAN 234, Rome. <www.codexalimentarius.net>.

Deshpande, S.S., Salunkhe, D.K., Oyewole, O.B., Azam-Ali, S., Battcock, M., Bressani, R., 2000. Fermented grain legumes, seeds and nuts: a global perspective. FAO Agricultural Services Bulletin 142. Rome, Italy.

Ding, X., Li, P., Bai, Y., Zhou, H., 2012. Aflatoxin B_1 in post-harvest peanuts and dietary risk in China. Food Control 23, 143–148.

Dorner, J.W., 2008. Management and prevention of mycotoxins in peanuts. Food Addit. Contam. 25, 203–208.

Egal, S., Hounsa, A.S., Gong, Y.Y., Turner, P.C., Wild, C.P., Hall, A.J., et al., 2005. Dietary exposure to aflatoxin from maize and groundnut in young children from Benin and Togo, West Africa. Int. J. Food Microbiol. 104, 215–224.

FAO/WHO, 2001. Food and Agriculture Organization of the United Nations/World Health Organization. Safety Evaluation of Certain Mycotoxins in Food. FAO Food and Nutrition Paper 74, WHO Food Additives Series 47. WHO, Geneva, Switzerland.

Fernane, F., Sanchis, V., Marin, S., Ramos, A.J., 2010. First report on mould and mycotoxin contamination of pistachios in Algeria. Mycopathologia 170, 423–429.

Freitas-Silva, O., Venancio, A., 2011. Brazil nuts: benefits and risk associated with contamination by fungi and mycotoxins. Food Res. Internat. 44, 1434–1440.

Hesseling, P.B., 1992. Onyalai. Baillieres Clin. Haematol. 5, 457–473.

IARC, 1993. International Agency for Research on Cancer. WHO IARC Monographs on the Evaluation of Carcinogenic Risks to Humans. Some Naturally Occurring Substances: Food Items and Constituents, Heterocyclic Aromatic Amines and Mycotoxins. Aflatoxins, vol. 56, pp. 245–395. IARC, Lyon, France.

Ivić, D., Domijan, A.M., Peraica, M., Milicević, M., Cvjetković, B., 2009. *Fusarium* spp. Contamination of wheat, maize, soybean, and pea grain in Croatia. Arh. Hig. Rada Toksikol. 60, 435–442.

Jimenez, M., Mateo, R., Querol, A., Huerta, T., Hernandez, E., 1991. Mycotoxins and mycotoxigenic moulds in nuts and sunflower seed for human consumption. Mycopathologia 115, 121–127.

Juan, C., Zinedine, A., Moltó, J.C., Idrissi, J.C., Mańes, J., 2008. Aflatoxins levels in dried fruits and nuts from Rabat-Sale area, Morocco. Food Control 19, 849–853.

Kamika, I., Takoy, L.L., 2011. Natural occurrence of aflatoxin B_1 in peanut collected from Kinshasa, Democratic Republic of Congo. Food Control 22, 1760–1764.

Khlangwiset, P., Wu, F., 2010. Cost and efficacy of public health interventions to reduce aflatoxin-induced human disease. Food Add. Cont. 27, 998–1014.

Krishnamachari, K.A., Bhat, R.V., Nagarajan, V., Tilak, T.B., 1975. Hepatitis due to aflatoxicosis. An outbreak in Western India. Lancet 1 (7915), 1061–1063. 10.

Kritzinger, Q., Aveling., T.A.S., Marasas, W.F.O., Rheeder, J.P., Van der Westhuizen, L., Shephard, G.S., 2003. Mycoflora and fumonisin mycotoxins associated with cowpea (Vigna unguiculata (L) Walp) seeds. J. Agric. Food Chem. 51, 2188–2192.

Leong, Y.-H., Ismail, N., Latif, A.A., Ahmad, R., 2010. Aflatoxin occurrence in nuts and commercial nutty products in Malaysia. Food Control 21, 334–338.

Liu, Y., Wu, F., 2010. Global burden of aflatoxin-induced hepatocellular carcinoma: a risk assessment. Environ. Health Perspect. 118, 818–824.

Luttfullah, G., Hussain, A., 2010. Studies on contamination level of aflatoxins in some dried fruits and nuts of Pakistan. Food Control 23, 32–36.

Magnoli, C., Astoreca, A., Ponsone, M.L., Fernandez-Juri, M.G., Barberis, C., Dalcero, A.M., 2007. Ochratoxin A and Aspergillus section Nigri in peanut seeds at different months of storage in Cordoba, Argentina. Int. J. Food Microbiol. 119, 213–218.

Marín, S., Hodzic, I., Ramos, A.J., Sanchis, V., Predicting the growth/no-growth boundary and ochratoxin A production by *Aspergillus carbonarius* in pistachio nuts. Food Microbiol. 25, 683–689.

Messina, M.J., 1999. Legumes and sobeans: overview of their nutritional profiles and health effects. Am. J. Clin. Nutr. 70, 439S–450S.

Molyneux, R.J., Mahoney, N., Kim, J.H., Campbell, C., 2007. Mycotoxins in edible nuts. Int. J. Food Microbiol. 119, 72–78.

Moss, M.O., 2002. Risk assessment for aflatoxin in foodstuffs. Internat. Biodeter. Biodeg. 50, 137–142.

Mutegi, C.K., Ngugi, H.K., Hendriks, S.L., Jones, R.B., 2009. Prevalence and factors associated with aflatoxin contamination of peanuts from Western Kenya. Int. J. Food Microbiol. 130, 27–34.

Overy, D.P., Seifert, K.A., Savard, M.E., Frisvad, J.C., 2003. Spoilage fungi and their mycotoxins in commercially marketed chestnuts. Int. J. Food Microbiol. 88, 69–77.

Phillips, T.D., Afriyie-Gyawu, E., Williams, J., Huebner, H., Ankrah, N.-A., Ofori-Adjei, D., et al., 2008. Reducing human exposure to aflatoxin through the use of clay: a review. Food Addit. Contam. 25, 134–145.

Pozzi, C.R., Braghini, R., Arcaro, J.R., Zorzete, P., Israel, A.L., Pozar, I.O., et al., 2005. Mycoflora and occurrence of alternariol and alternariol monomethyl ether in Brazilian sunflower from sowing to harvest. J. Agric. Food Chem. 53, 5824–5828.

Rauford, P.Q., Prasad, P.V.V., Waliyar, F., Taheri, A., 2006. Drought, pod yield, pre-harvest *Aspergillus* infection and aflatoxin contamination on peanut in Niger. Field Crops Res. 98, 20–29.

Salem, N., Ahmad, R., 2010. Mycotoxins in food from Jordan: preliminary survey. Food Control 21, 1099–1103.

Sarhang Pour, R., Rasti, M., Zighamian, H., DaraeiGarmakhani, A.D., 2010. Occurrence of aflatoxins in pistachio nuts in Esfahan Province of Iran. J. Food Saf. 30, 330–340.

Schatzki, T.F., Pan, J.L., 1996. Distribution of aflatoxin in pistachios. 3. Distribution in pistachio process streams. J. Agric. Food Chem. 44, 1076–1084.

Schollenberger, M., Müller, H.-M., Rüflee, M., Drochner, W., 2008. Natural occurrence of 16 *Fusarium* toxins in edible oil marketed in Germany. Food Control 19, 475–482.

Sharma, R.P., Salunkhe, D.K., 1991. Occurrence of mycotoxins in foods and feeds. In: Sharma, R.P., Salunkhe, D.K. (Eds.), Mycotoxins and Phytoalexins CRC Press, Inc., Boca Raton, FL, pp. 13–32.

Shephard, G.S., 2008. Risk assessment of aflatoxins in food in Africa. Food Addit. Contam. 25, 1246–1256.

Sommer, N.F., Buchanan, J.R., Fortlage, R.J., 1986. Relation of early splitting and tattering of pistachio nuts to aflatoxin in orchards. Phytopathology 76, 692–694.

Sperber, W.H., 2005. HACCP does not work from farm to table. Food Control 16, 511–514.

Thomas, R., John, S., 2002. A simple, spreadsheet-based, food safety risk assessment tool. Int. J. Food Microbiol. 77, 39–53.

Turner, P.C., Sylla, A., Gong, Y.Y., Diallo, M.S., Sutcliffe, A.E., Hall, A.J., et al., 2005. Reduction in exposure to carcinogenic aflatoxins by postharvest intervention measures in West Africa: a community-based intervention study. Lancet 365 (9475), 1950–1956.

VanEgmond, H.P., Schothorst, R.C., Jonker, M.A., 2007. Regulations relating to mycotoxins in food: perspectives in a global and European context. Anal. Bioanal. Chem. 389, 147–157.

WHO, 1991. Mycotoxins. Fact Sheet No. 5. Basic Food Safety for Health Workers.

WHO. 1998. World Health Organization. Safety Evaluation of Certain Food Additives and Contaminants. Aflatoxins. WHO Food Additives Series, vol. 40, pp. 359–468. WHO, Geneva, Switzerland.

Wu, F., 2004. Mycotoxin risk assessment for the purpose of setting international regulatory standards. Environ. Sci. Tech. 38, 4049–4055.

Further Reading

FAO/WHO, 2011. Food and Agriculture Organization of the United Nations/World Health Organization. Safety Evaluation of Certain Mycotoxins in Food. FAO JECFA Monographs 8, WHO Food Additives Series 63. WHO, Geneva, Switzerland.

King, J.C., Blumberg, J., Ingwersen, L., Jenab, M., Tucker, K.L., 2008. Tree nuts and peanuts as components of a healthy diet. J. Nutr. 138, 1736S–1740S.

CHAPTER

13

Oils and Fats

Gerrit van Duijn

Oil Processing Consultant formerly Unilever R&D, Vlaardingen, The Netherlands

OUTLINE

Food Safety Management.
DOI: http://dx.doi.org/10.1016/B978-0-12-381504-0.00013-5

INTRODUCTION

Functionality

Oils and fats have been used from ancient times for food preparation as well as in non-food applications like lamp oil, lubricant, soap manufacturing and skin care. This chapter will only deal with the use of oils and fats for food preparation or as ingredients in food.

Oils and fats provide functionality in food preparation and use as well as nutritional benefits. They serve as a heat transfer medium at elevated temperatures (e.g. frying), improve taste sensation (spreads and dressings), give texture and flavor to a wide range of foodstuffs, supply a concentrated source of energy, deliver critical building elements for the body and act as a carrier for essential minor components like vitamins A and D. A balanced intake of oils and fats is essential for human health.

Supply Chain

Oils and fats originate from plant and animal sources. The plant-based oils and fats dominate in current food applications and will be the main focus of this chapter.

The supply chains of vegetable oils and fats consist of:

- The growing of oil seeds, fruits, kernels or nuts.
- Oil extraction to recover the oil, the by-product meal is mostly used as animal feed.
- Purification and modification processes to optimize the properties of oils.
- All transport from grower to end user.

Until the industrial revolution in the 19th century, rapeseed, linseed, olives and nuts were the main sources of vegetable oils. Today, the world market is dominated by palm and soybean oil, followed by rapeseed and sunflower oil. This has led to a change in the extraction and purification/modification processes. Originally, the oil extraction process consisted of cleaning, crushing, heating and pressing. From 1900 onwards solvent extraction was applied to recover the residual oil from the pressed cake or to replace the pressing process completely (e.g. for soybean oil). At more or less the same time, the oil purification process changed from a simple decanting and filtration to a combination of neutralization with caustic, bleaching with active clay and deodorization at high temperature under vacuum with steam. Using this refining process, minor components were reduced to improve taste and appearance while the removal of processing residues was required after the introduction of solvent extraction and nickel catalyzed hydrogenation.

Later, the introduction of improved analytical techniques showed that the refining process also reduces the levels of many of the contaminants present in the crude (extracted) oil. The refining process may also introduce side reaction products; some of these products may affect health. This chapter gives an overview of the most important contaminants in the crude oil, the validation of the refining process for the removal of these components, and the formation of potential hazardous by-products during refining.

CONTAMINANTS IN CRUDE OILS AND FATS

Crude Oil Risk Assessment

The presence and levels of contaminants in crude oils depend on:

- Agricultural practices.
- Procedures of oil crop storage, drying and handling.
- Oil extraction practices.
- Contamination and degradation during crude oil transport.

Risk of contamination and the type of contaminant will differ per oil type and origin. Information on the presence of contaminants in various oil types from different origins has been collected in three ways:

1. Visits of all steps of the crude oil supply chains.
2. Analyses of crude oils bought for further processing.
3. Sharing of contaminant information in industry organizations and during conferences.

The following contaminants were found at detectable levels: pesticide residues, polycyclic aromatic hydrocarbons, hydrocarbons of mineral origin and mycotoxins. The following sections give the risk assessments for presence of these contaminants in crude vegetable oils.

Pesticide Residues

Plant protection products or pesticides can be used during the cultivation and storage of oil seeds, fruits, kernels and nuts, to protect the crop during growing, to reduce weeds and to protect seeds during storage and transport. The EU introduced limits for the residual levels of pesticides in the harvested crops – the so-called maximum residue limits (MRL). These limits are requested by pesticide manufacturers based on residues found after pesticide use according to good agricultural practices (GAP). These MRLs are crop and pesticide specific and are much lower than the harmful toxicological thresholds (see Figure 13.1). For crop/pesticide combinations for which an MRL has not been requested or the request has not been granted, the pesticide level in the crop has to be below the level of determination (LOD). The LOD of the individual pesticides is indicated in the EU directives (see EU pesticide database: http://ec.europa.eu/sanco_pesticides/public/index.cfm). Pesticides are considered to be contaminants if the level in crops exceeds the MRL of the pesticide/crop combination.

Depending on their physical/chemical properties, the pesticide residues present in the oil crop will concentrate differently in the products of the oil extraction process:

- They will concentrate in the crude oil if they are oil or hexane (in the case of solvent extraction) soluble. The maximum concentration factor, $X(\text{max})$, of oil- or hexane-soluble pesticides from oil crop to crude oil is $X(\text{max}) = 100\%/C_{\text{oil}}$, where C_{oil} is the fraction of oil in the oil crop (%).
- In palm oil extraction, they will concentrate in the sludge if they are water soluble.

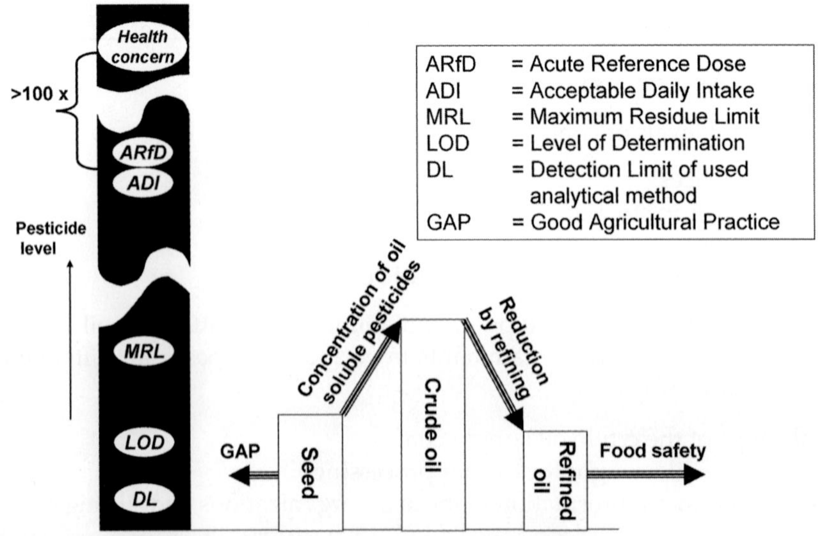

FIGURE 13.1 A qualitative picture of the change of a pesticide level by oil extraction and by subsequent refining. The vertical column shows a relationship with the different health levels, legal limit and detection limit.

- The concentrations in oil and meal fractions are both equal to the level in the crop if the pesticides are not soluble in oil or solvent and equally distributed in oil and meal.

Hence, oil extraction may result in a pesticide level in the crude oil that is higher than the MRL of the oil crop.

The pesticides used in the seed oil supply chain are mainly organophosphorus insecticides. These are applied to protect oil seeds during storage and transport after harvesting (post-harvest treatment). The following pesticides were found at detectable levels in crude rapeseed, crude sunflower and crude soybean oil samples (van Duijn, 2008): fenitrothion, malathion, pirimiphos-methyl, parathion-methyl, dichlorvos, chlorpyriphos, chlorpyriphos-methyl and endosulfan. The highest pesticide levels and frequency of samples with a level >MRL were found in the crude sunflower oil samples. Crude rapeseed and crude soybean oil samples showed much lower levels and frequencies.

Oil palm fruits are processed within a few days and preferably within 24 hours after harvesting for quality reasons. Post-harvest treatment of palm fruits is therefore not required and detectable pesticide levels in the crude palm oil samples were never found.

In the supply chains of palm kernels and coconuts, chemical crop protection is not applied, resulting in non-detectable levels in the crude oils.

Polycyclic Aromatic Hydrocarbons

Polycyclic aromatic hydrocarbons (PAHs) constitute a large class of organic compounds that are composed of two or more fused aromatic rings. They are primarily formed by incomplete combustion or pyrolysis of organic matter. PAHs generally occur in complex mixtures that may consist of hundreds of compounds (Alexander et al., 2008). Humans are exposed to PAHs by inhalation if they smoke and/or consume contaminated food. Oil crop can be contaminated with PAHs by absorption of these components from exhaust gases, when these gases are in direct contact with the crop during drying.

Oil mills set specifications for the moisture levels of oil seeds. At too high moisture levels, oil seeds need to be dried either before arrival at the oil mill or at the mill itself. Drying by direct contact with exhaust gases has been observed for soybeans in wood-fired packed bed dryers and for sunflower seeds in diesel-fired counter-current dryers. Indirect dryers are used in the EU and the USA. In these dryers hot air is generated via a heat exchanger; this excludes contact between exhaust gases and the product.

In the coconut supply chain drying is an essential operation, as it avoids aflatoxin formation and releases the copra (coconut meal) from the shell. In the most commonly used method in the main producing country, the Philippines, halved coconuts are dried upside down on a grid over an open fire, burning coconut shells. Thousands of these drying installations are operated by small farmers.

Palm kernels are washed and dried after removal of the shells in the palm oil mill by cracking. A majority of oil mills dry the kernels in indirect dryers; however, a minority use direct dryers. Oil palm fruit is not dried since it is processed shortly after harvesting while the fruit itself contains around 50% humidity.

Benz(a) pyrene (BaP), a PAH with five aromatic rings, is generally used as a marker for the presence of PAH in crude oils. Figure 13.2 shows the results of BaP analyses in various crude oils (van Duijn and den Dekker, 2010). This graph confirms that the contamination level and frequency of crude coconut oil is very high (maximum BaP level: 70 μg/kg, fraction of samples with BaP >1 μg/kg (frequency) was almost 80%). Crude sunflower oil was both high in maximum (40 μg/kg) and in frequency (12%), followed by crude rapeseed oil (maximum 10 μg/kg, frequency 9%), crude palm kernel oil (maximum 6 μg/kg, frequency 6%) and water degummed soybean oil (maximum 2 μg/kg, frequency 7%). Other oils that may contain PAHs are grape seed oil and cotton seed oil.

Hydrocarbons of Mineral Origin

Hydrocarbons of mineral origin consist mainly of alkanes of different chain length. They are manufactured from crude mineral oils in various refining steps such as distillation, extraction and crystallization followed by purification. Mineral oil products can be divided in product groups on the basis of their carbon number and/or viscosity. Of importance for this section are the following product groups (classified on the basis of carbon number):

C6: Hexane, used as solvent for vegetable oil extraction.
C6–C10: Gasoline.
C10–C24: Mid fraction, consisting of kerosene, diesel and light fuel oil.
C20–C55: Medium and high viscosity oils such as grease oil, hydraulic oils, etc.
Carbon number >56: solids.

Based on toxicity studies, the Joint FAO/WHO Expert Committee on Food Additives (JECFA) accepted different acceptable daily intakes (ADI) for mineral oil with high viscosity (ADI of max 20 mg/kg bodyweight) and for mineral oil with medium and low viscosity

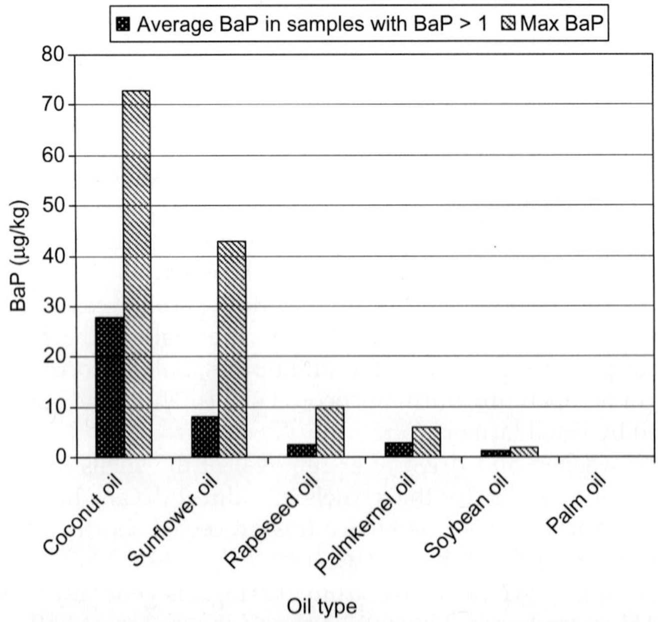

FIGURE 13.2 Results of BaP (benz(a) pyrene) analyses in crude oils. This graph shows the average of the samples containing more than 1 µg/kg BaP and the maximum observed levels.

(ADI of max 10 mg/kg bodyweight for class I and max. 0.01 mg/kg body weight for classes II and III).

Mineral oil products can be present in crude edible oils due to contamination during processing (lubricants and hydraulic oils), as residues from previous cargoes during transport and storage, and by fraudulent addition. However, their presence can also be the result of an allowed use as a processing aid, e.g. as hexane in solvent extraction, as solvent for pesticides and as anti-dusting agent in oil seed storage. It should be noted that long chain alkanes are synthesized by a large number of edible plants and animals, resulting in considerable levels of naturally occurring alkanes in crude edible oils (e.g. max. 160 mg/kg in sunflower oil). Natural alkanes are characterized by a strong predominance of odd carbon numbers.

Oils and fats have to be free from contamination with hydrocarbons from mineral origin. This can be ensured by supply chain auditing and by setting analytical limits. These limits should take into account the presence of mineral oil products from allowed practices, the presence of "natural" alkanes, and the analytical detection limit. The following limits are industry standards based on good agricultural and manufacturing practices.

- Short chain hydrocarbons (shorter than C10) are volatile and are contractually limited by the flashpoint (temperature at which escaping gases can be detected by flashing). The flashpoint has been introduced to exclude explosion risk during transport and storage. The contractual limit is minimum 121°C.

- After an incident with diesel contamination of crude palm oil, the Dutch, Malaysian and Indonesian governments agreed on a limit of 25 mg/kg diesel (expressed as C10–C24) in crude palm oil and palm products.
- In 2008, imported crude sunflower oil from Ukraine had been contaminated with high viscosity mineral oil. The acceptable limit for imported crude sunflower was set at 50 mg/kg presence of total hydrocarbons (C10–C56). This is after correction for known amounts of natural alkanes. These known amounts follow from a historical analytical database of non-contaminated samples.
- For all other vegetable oils an action limit was set at 300 mg/kg of total hydrocarbons (C10–C56). This limit includes an unknown level of natural alkanes and the allowed use of mineral oils as processing aids. A further study of the type of contamination is required if the analysis results in a level higher than the action limit.

Mycotoxins

Aflatoxin

Aflatoxins are mycotoxins that are produced by strains from the Aspergillus family (molds). They are found as contaminants in human and animal food as a result of fungal contamination during growing, and usually to a larger extent, post-harvest storage. Carcinogenic effects of aflatoxins to humans are no longer doubted and legal limits for aflatoxins in foodstuff are very low. Aflatoxins are most commonly associated with groundnuts (peanuts), dried fruit, tree nuts (such as almonds, pecans, walnuts, pistachio and brazil nuts), spices, figs, crude vegetable oils (peanut oil, coconut oil), cocoa beans and a range of agricultural products, the most important being maize, rice, cottonseed and copra. The aflatoxins that may appear in oil seeds and vegetable oils are aflatoxin B1, G1, B2 and G2 of which B1 and G1 are the most common. In general, no more than 10% of the aflatoxins present in seeds, peanuts and copra are transferred to the crude oil after pressing and extraction. This is due to the fact that aflatoxins are mainly protein bound.

An inventory of aflatoxin levels by Unilever in the mid-1990s demonstrated the frequent occurrence of aflatoxins in crude coconut and peanut oil. The average aflatoxin B1 concentration for coconut oil was 14 μg/kg, with a maximum of 75 μg/kg. The average value for the peanut oil was 10 μg/kg, with a maximum of 34 μg/kg. Aflatoxin G1 levels in these oils were about 30% of the aflatoxin B1 levels; aflatoxin B2 and G2 were not detected (<1.0 μg/kg). Crude rapeseed oil showed no aflatoxin contamination (<1.0 μg/kg) (see Table 13.1).

Zearalenone in Crude Maize Germ Oil

Fungi (*Fusarium* species) producing the toxin zearalenone (ZEN) are common soil fungi which mainly develop during flowering. The *Fusarium* fungi are usually found on cereals grown in the temperate regions of America, Europe and Asia. The weather conditions during the 2-week flowering period have a determining effect on the toxin levels. High ZEN levels are found at relatively low temperatures and high humidity. Various studies reported a negative effect of ZEN on the fertility of pigs.

In the maize milling process, where the germs are separated from the rest, ZEN is concentrated in the germs. The maize producer's industry organization (AAF) estimates that

TABLE 13.1 Aflatoxin Analyses in Crude Oil Deliveries to Unilever

Aflatoxin B1		Coconut Oil	Groundnut Oil	Rapeseed Oil
No of samples	–	42	11	3
Min.	μg/kg	1	4	ND
Max.	μg/kg	75	34	ND
Average	μg/kg	14	10	ND

Samples taken from deliveries 1992–1994
***Aflatoxin G1** Around 30 % of B1 level*
***Aflatoxin B2, G2** Not Detectable*

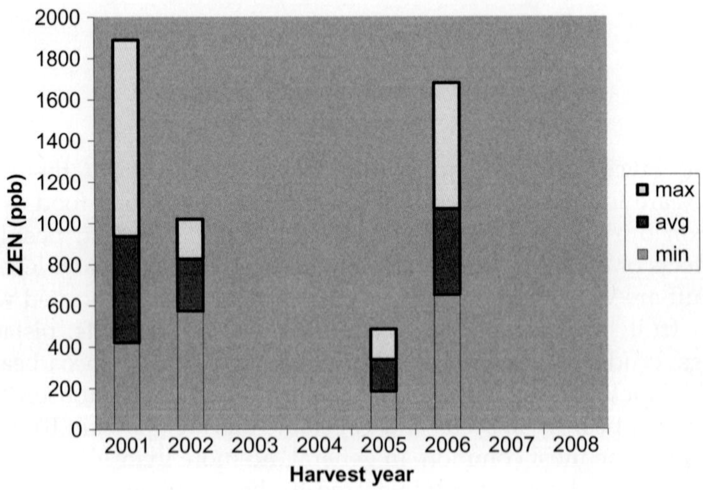

FIGURE 13.3 Results of zearalenone analyses in crude maize oil. The x-axis shows the year of harvesting. The samples were taken from the crude oils produced from the harvested crop during the year following harvesting.

the concentration factor is between 3 and 5. In the last 10-year period, high ZEN levels were found in the European harvests of years 2001, 2002, 2005 and 2006. Other years showed very low ZEN levels (see Figure 13.3). The flowering period of summer 2006 was specifically unfavorable for ZEN formation (cold and wet). Maize producers found an average ZEN level of 370 μg/kg in 236 samples of unprocessed maize. With a concentration factor of 3 to 5, this results in an average level of 1100–1850 μg/kg in germs. Oil producers' data showed that the level in crude maize oil was almost equal to that in maize germs. Levels up to 1810 μg/kg were found in crude maize oil produced in the first quarter of 2007.

Residues of Previous Cargoes

The general principle is that transport of both crude and fully refined oils and fats is only permitted in conveyances that are dedicated to foodstuffs. Conveyances include containers, road tankers, rail tank cars, river barges, coastal ships, deep sea vessels, land tanks, direct pipelines and other handling facilities that may come into contact with the oils and fats. This general principle is applied in the EU and in most other countries.

Intercontinental transport of oils and fats in bulk is carried out in sea-going vessels with a capacity of 30,000 to 70,000 tons. This bulk transport of oils and fats represents around 30% of the total bulk transport in this type of vessel. A restriction to foodstuff-dedicated transport would result in an insufficient availability of ships to serve the overseas oils and fats trade. Therefore, the EU and the international trade have accepted a derogation of the general principle of foodstuff-dedicated transport for this type of transport.

In the EU, this derogation was based on the following criteria:

1. It should not introduce toxicological concerns for which a threshold is difficult to establish (genotoxic or carcinogenic potential).
2. It is based on efficient procedures to clean ship tanks between cargoes.
3. The residue of the previous cargo after cleaning is diluted in the transported quantity of oil or fat.
4. The validated removal of the previous cargo residue by refining after unloading is considered a prerequisite for crude and semi-refined oils.
5. It assumes the availability of analytical methods to verify the absence of residues of previous cargoes in the refined oils and fats.

These criteria led to the following set of rules for bulk transport in sea going vessels (EU Commission 2004):

1. For crude and semi-processed oils and fats which are to be further processed before being used for human consumption (further processing has to be refining according to industry standards):
    a. The immediate previous cargo transported in that tank shall have been a foodstuff or a cargo from the list of acceptable previous cargoes, if the oil or fat is transported in a stainless steel tank, or a tank with epoxy resin coating. This list of acceptable previous cargoes is published by the EU and regularly updated.
    b. If the oil or fat is transported in a tank of a different material than those mentioned under a., then the three previous cargoes transported in that tank shall have been a foodstuff or a cargo from the list of acceptable previous cargoes.
    c. The buyer must obtain access to written information on the three previous cargoes carried in the relevant tanks.
2. For fully refined oils that are not further processed before being used for human consumption:
    a. If the ship's tank is stainless steel or epoxy resin coated, the three previous cargoes transported in that tank shall have been foodstuffs.
    b. In all other cases, the transport must be dedicated to foodstuffs only.

The international trade of oils and fats is using standard contracts. In particular the standard contracts issued by FOSFA (Federation of Oils, Seeds and Fats Associations Ltd, London, UK) contain conditions similar to those mentioned above.

Heavy Metals and Dioxins

Presence of heavy metals in crude oils and fats may originate from the oil crop due to uptake from the soil. Contamination risk during processing and transport of oil crop and during transport and handling of crude oil is very limited since heavy metals are not used in contact materials in this supply chain. Crude oil analyses confirmed that heavy metals are seldom present at detectable levels. Metals like iron and copper, however, are commonly present in crude oils and fats. These metals may affect quality (they are catalysts for oxidation) but not health.

Monitoring programs for dioxins, furans and dioxin-like PCBs showed levels well below the allowed level for oils and fats intended for direct human consumption. Only crude fish oil may contain relatively high dioxin levels due to concentration of dioxin in the fish feed chain.

Crude Oil Risk Matrix

The crude oil risk matrix, shown in Figure 13.4, gives the risk classification (high, medium or low) for the presence of a contaminant in a crude oil, in case the origin of this

	Pesticides	PAH	Mineral oil in edible oil imported in EU	Previous Cargoes in sea going vessels	Dioxins and PCBs	Aflatoxins	Zearalenone
LIMIT	MRL or LOD		FEDIOL Code of Practice	EC/4/2004			
Soybean oil	M	M	R	R	L		
Sunflower oil	H	H	R	R	L		
Rapeseed oil	M	M	R	R	L		
Corn oil	M	M	R	R	L		H
Palm oil	L	L	R	R	L		
Palm kernel oil	L	M	R	R	L		
Coconut oil	L	H	R	R	L	M	
Groundnut oil	L	L	R	R	L	H	
Fish oil	L	M			H		
Linseed oil	M	M			L		
Cottonseed	M	M	R	R	L		
Grape seed	L	H			L		
Olive	M	M			L		

		Occurence	Monitoring frequency:
high riks	H	Regularly (> once a year)	Every batch
medium risk	M	Occasionally (every 1-5 years)	Minimum once per quarter
low risk	L	Seldom (< once every 5 years)	Maximum once per quarter
regulated	R	Not applicable	Every batch

Figure shows the risk classification for contaminant presence in a crude oil, in case the origin of this oil is unknown. The matrix can be used to determine the frequency of analyses.

FIGURE 13.4 Crude Oil Risk Matrix.

crude oil is unknown. Knowledge of practices or procedures in dedicated supply chains may further reduce the risk classification in case these practices reduce contamination risk. The crude oil risk matrix can be used to determine the frequency of contaminant analyses in crude oils. The proposed frequencies are:

- High risk→check every delivery
- Medium risk→quarterly monitoring
- Low risk→annual monitoring

Crude oils and fats limits are set for pesticides, hydrocarbons of mineral origin and previous cargoes.

- The pesticide level in the crude oil should not exceed the MRL for the pesticide/oil seed combination after correction by the concentration factor occurring due to oil extraction (to be confirmed by updated EU regulation).
- The level of hydrocarbons of mineral origin should not exceed the limits defined by the industry (see "Hydrocarbons of Mineral Origin," above).
- Previous cargoes are checked by comparing the previous cargo from the ship's logbook with the EU or FOSFA positive list of allowed previous cargoes, taking into account the construction material of the ship tanks. This activity is normally performed by an independent superintendent.

The other contaminants have no legal or industry limits in crude oil, but are regulated in the fully refined product. The crude oil analytical results are therefore the input for the refining process validation for contaminant removal.

REFINING PROCESS VALIDATION FOR CONTAMINANT REMOVAL

The Refining Process

The refining process is a combination of the following process steps (toolbox); for more process details see Bockisch (1998):

- *Degumming:* A pretreatment process applied to seed oils to reduce the phosphorus content. It is a two-step process with addition of water and/or acid to hydrate phospholipids. The phospholipids are subsequently removed by centrifugation.
- *Neutralization:* The purpose of neutralization is to reduce the concentration of free fatty acids to a maximum of 0.10% with the use of a diluted alkali solution, typically sodium hydroxide. This process can be applied batch-wise in stirred vessels and continuously by means of centrifuges. After alkali treatment, the oil is washed with hot water or treated with silica to reduce the residual soap level in the neutralized oil.
- *Bleaching:* The main purpose is to remove residual soap, pigments and oxidized components. In this process, bleaching earth (activated clay and/or silica) is added to the oil as absorbent. The earth and absorbed impurities are subsequently removed by

filtration. Addition of activated carbon in the bleaching process will also reduce the polycyclic aromatic hydrocarbon level. An acid pretreatment before bleaching earth addition will improve the removal of phosphorous (max 30 ppm) and/or metals during the bleaching process.
- *Deodorization:* Under high vacuum the oil is heated to 180–240°C and brought in contact with stripping steam to remove volatile components and to create an odorless oil with a bland taste and increased storage stability. Also free fatty acids can be removed during deodorization, at increased temperatures (220–270°C).

The process sequence of combined degumming/neutralization followed by bleaching and deodorization is called chemical refining, referring to the chemical removal of free fatty acids. The process sequence of degumming followed by bleaching with acid pretreatment and deodorization at high temperature is called physical refining, referring to the physical removal of free fatty acids (stripping). The physical refining process is generally preferred for low phosphorous oils (acid degummed seed oils and tropical oils) due to lower oil losses and less liquid effluent production.

Refining Process Validation

The levels of most contaminants are regulated for fully refined oils but not for crude oils (EC Commission 2006). The refining process validation will assure that the contaminant level in the fully refined oil is below the regulated limit, even for the crude oil feedstock with the highest observed contaminant level. The validation process is as follows:

1. The refinery is informed in the case of a crude oil delivery with a contaminant level higher than the refined oil limit (or the highest level used in previous process validations) and the contaminated lot is blocked.
2. A minimum batch of the blocked contaminated oil is processed in the refinery using the standard refining recipe. The contaminant levels are analyzed in deodorized end product (and preferably also after the intermediate refining steps).
3. If the contaminant level in the deodorized oil is below the regulated limit, the crude oil is de-blocked and the whole lot can be processed. The validation process needs to be repeated with modified process conditions if the contaminant level in the deodorized oil is still too high. Alternatively, the crude oil can be sold for non-food application (feed or bio-fuel) in case removal is technically or economically not feasible.
4. This validation process needs to be repeated for every delivery of crude oil with a contaminant level higher than the levels used in previous process validations.

Results of pilot plant refining validation trials are shown in Pages et al. (2010).

Pesticide Residues Removal

Pesticide concentrations will reduce during the refining steps. The effect of each step on residual levels depends on the physical/chemical properties of the pesticide.

Water-soluble pesticides dissolve in the alkaline solution during neutralization and are removed with the soap stock.

Some pesticides (e.g. pirimiphos-methyl) are absorbed by acid–base interactions onto the bleaching earth.

Volatile pesticides are removed with the steam during the deodorization process. All organophosphorus insecticides have a higher volatility than free fatty acids and will be removed at increased temperatures (220–270°C).

Figure 13.5 gives an example of a process validation experiment for pesticide removal.

Polycyclic Aromatic Hydrocarbon Removal

Polycyclic aromatic hydrocarbons (PAHs) are removed by active carbon dosing in the bleaching process. Volatile PAH will be additionally reduced during high temperature deodorization. The volatility depends on the number of aromatic groups in the PAH compound; four or less are called light PAH, while five or more are called heavy PAH. The tracer compound benz(a)pyrene (BaP) has five aromatic groups and is a heavy PAH. The current EU regulation sets a limit for BaP of 2 ppb and a total of 10 ppb for 4 selected PAH, for oils and fats intended for direct human consumption or used as ingredient in food.

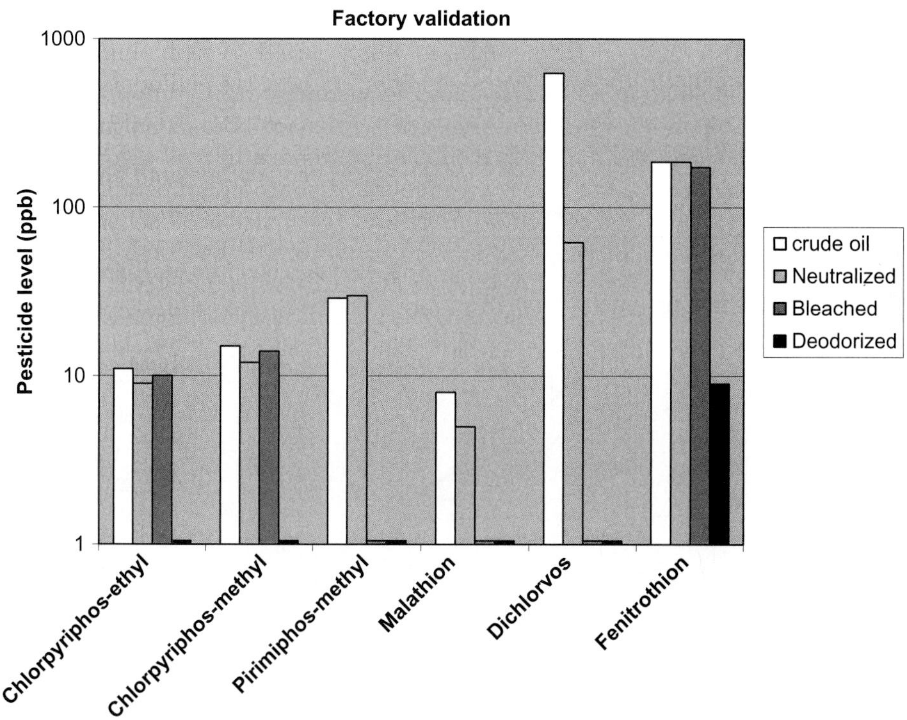

FIGURE 13.5 An example of process validation of pesticide removal. The graph shows the levels of the pesticides present in the crude oil and the levels after neutralization, bleaching and deodorization.

Light PAH will be partly removed during deodorization, the degree of reduction depends on deodorization temperature (around 50% reduction at 180°C, up to 90% reduction at 240°C).

There are two ways for determining the required active carbon dosage for PAH removal:

1. Measure the BaP content in every incoming parcel of oil and add active carbon based on experience with previous process validations. Figure 13.6 shows the results of process validations performed in Unilever refineries.
2. If BaP analysis is not available (on time), the dosage should be based on a realistic worst case BaP level in the crude oil. The standard active carbon dosage will then be the dosage of the successful process validation performed with this worst-case crude oil.

Removal of Hydrocarbons of Mineral Origin

Hexane (C6) is partly removed by the vacuum systems of neutralization and/or bleaching while the remainder is removed in the deodorizer to a level below the detection limit (0.1 mg/kg). EC regulation sets an upper limit to hexane residue in oil of 1 mg/kg.

Gasoline has never been detected as a contaminant of crude oils and fats. Therefore, the refining process has never been validated for gasoline removal. However, experiences with compounds of similar or even lower volatility indicate a complete removal during deodorization.

The upper limit for the presence of mineral oil mid fraction (C10–C24, including kerosene and diesel) in crude palm oil is 25 mg/kg. Process validation has shown that this level can be reduced to below the detection limit (10 mg/kg).

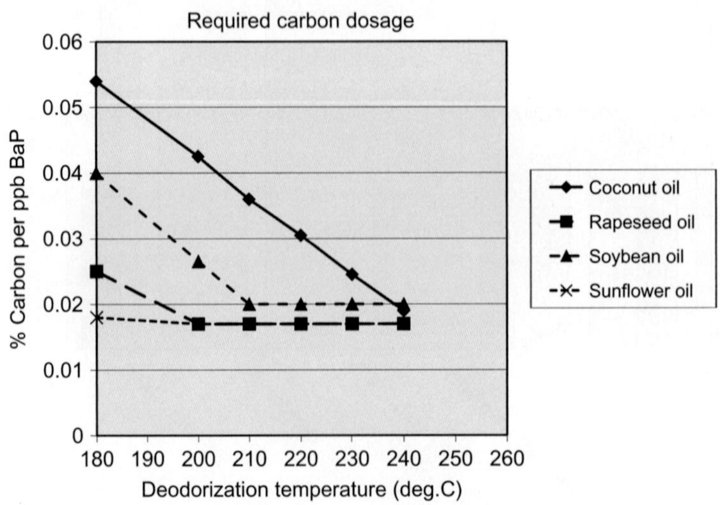

FIGURE 13.6 Process validation for BaP removal. The graph shows the percentage of active carbon, per percentage of BaP in the crude oil, needed to reduce BaP below the legal limit of 2.0 μg/kg and the total of 13 tested PAH below 25 μg/kg. The horizontal axis gives the deodorization temperature; this mainly influences the light PAH.

Mineral oil fractions with a carbon number above C24 will barely be removed in the refining process. The physical and chemical properties of these components differ insufficiently from the properties of edible oils and fats to enable separation.

Mycotoxins Removal

Aflatoxin Removal

Both the chemical and physical process sequence will reduce the aflatoxin levels to below the detection limit (1.0 µg/kg). The processes that are responsible for aflatoxin reduction are neutralization with lye (in chemical refining) and bleaching. The activated bleaching earth is responsible for the removal during bleaching; addition of active carbon will barely increase the separation efficiency. Deodorization, even at high temperature, gives only a modest contribution to aflatoxin removal. The EU limits for aflatoxin in oils and fats intended for direct human consumption or as ingredient in food are: max 2.0 for aflatoxin B1 and max 4.0 for the sum of aflatoxins B1, B2, G1 and G2.

Zearalenone (ZEN) Removal in Maize Oil

The refining process will largely reduce the ZEN content in maize oil, but a complete removal will not be obtained under standard refining conditions. Additionally, the removal efficiency depends on the refining process used (chemical or physical). Chemical refining will remove 80–98% of ZEN while the removal efficiency of physical refining varies between 70 and 80%. The EU limit for refined maize oil is 400 µg/kg based on the ALARA (as low as reasonably achievable) principle.

Other Contaminants

Residues of Previous Cargoes

The validated removal of substances by a standard refining process is one of the criteria for acceptance of these substances as allowed previous cargo. Further refining process validation is therefore not required for the substances on the allowed (positive) list. Oils and fats cannot be accepted for food use (even after refining) when the previous cargo is not on the positive list.

Heavy Metals

Heavy metals are seldom present at detectable levels in crude oils and fats. Therefore, the refining process cannot be validated for heavy metal removal except in exceptional cases where heavy metals are present in the crude oil. Metals like iron and copper are effectively removed by neutralization and bleaching with acid pretreatment. It is assumed also that heavy metal levels will be reduced by these processes. The EU has only regulated the level of lead in oils and fats (max 0.1 mg/kg).

Dioxins

The dioxin level in fish oil is reduced by active carbon addition during bleaching followed by deodorization at moderate temperature (max 190°C to limit isomerization). The

Contaminants	Hydrocarbons < C20	Hydrocarbons > C20	PAH (BaP)	Pesticides	Aflatoxin B1	Zearalenone
Legal or industry limit in refined oil	LOD	LOD	2.0 µg/kg	MRL or LOD	2.0 µg/kg	400 µg/kg
Crude oil reception						
Degumming						
Neutralization						93%
Bleaching						77%
Deodorization						

= *Chemical refining*
= *Physical refining*
= *Chemical and Physical*

Figure summarizes process validation experience of contaminant removal.

FIGURE 13.7 The Refining Link Table.

EU sets limits for the levels of dioxins and the sum of dioxins and dioxin-like PCBs, taking into account the toxicity equivalents of the individual compounds. The limit for the sum of dioxins in vegetable oils and fats is 0.75 pg/g (WHO PCDD/F-TEQ) and 1.5 pg/g (WHO PCDD/F-TEQ) for the sum of dioxins and dioxin-like PCBs. The limits for fish oil are higher: sum of dioxins is 2.0 pg/g (WHO PCDD/F-TEQ) and 10.0 pg/g (WHO PCDD/F-TEQ) for the sum of dioxins and dioxin-like PCBs.

The Refining Link Tables

The refining process validation experience can be summarized in a refining link table (see Figure 13.7). This link table shows the contaminants, the regulated or industry limits in refined oil, and the process step that reduces the concentration of this contaminant in the product. This link table gives a quick reference for process optimization and troubleshooting. It can also be the basis for an HACCP analysis of the refining process.

Contaminant levels below the legal limit or industry standard can be assured by combination of crude oil analyses and refining process validations. This system should be regularly checked by internal and third party (customer) audits (once every 1–3 years) and by monitoring of the contaminant levels in the refined product (at least once every quarter). Crude oil supply chain risk assessments should be periodically repeated (once every 5 years) by combination of supply chain visits and analyses of crude oils for a wide range of potential oil soluble contaminants.

BY-PRODUCTS FORMED DURING OIL REFINING

Cis-trans Isomerization

The previous sections showed that the refining process reduces many of the contaminants present in crude edible oils. However, the high temperature during deodorization may also cause modification of the triacylglycerols and reactions of the triacylglycerols with other components. The best-known example of a triacylglycerol modification is the cis-trans isomerization of the double bonds in the fatty acid chains. Nutritional research published since the early 1990s showed a negative effect of trans fatty acids on blood cholesterol. This section will only deal with the unintended presence of trans fatty acids (TFA) in deodorized oils and not the intended presence of TFA in partially hydrogenated fats.

Cis-trans isomerization mainly occurs in oils containing high levels of polyunsaturated fatty acids at elevated deodorization temperatures (>220°C). The kinetics of trans formation can be predicted by the following relation (van Duijn et al., 2006): $C_{tr} = C_o(1 - e^{-kt})$, in which C_{tr} is the trans isomer concentration of a fatty acid after exposure at high temperature, C_o is the original cis concentration of that fatty acid and t is the exposure time in minutes. The rate constant k in this model is fatty acid and temperature dependent (temperature T in kelvin):

- For linolenic acid (C18:3): $k = 6.3 \times 10^{11} \times e^{-145/RT}$ (min)$^{-1}$
- For linoleic acid (C18:2): $k = 8 \times 10^{8} \times e^{-128/RT}$ (min)$^{-1}$
- For oleic acid (C18:1): a fixed additional contribution of 0.1–0.2%.

This model has been used to predict the TFA levels of the main seed oils after 30 and 60 minutes of deodorization at temperatures ranging from 200 to 260°C. The results are given in Table 13.2: TFA levels above 1% will occur in sunflower oil only at long deodorization time and high temperature, for soybean and rapeseed oil at high temperature and short deodorization time and also at 240°C and long deodorization time.

Reducing TFA in deodorization will be achieved by reducing deodorization time and/or temperature. However, long deodorization time and/or high temperature may be required for the following reasons:

TABLE 13.2 Predicted Trans Fatty Acid Levels (in %) in Different Oils, Deodorized at Different Temperatures and Times

Deodorization temperature	Sunflower Oil		Soybean/Rapeseed Oil	
	30 minutes	60 minutes	30 minutes	60 minutes
200	0.3	0.3	0.3	0.3
220	0.4	0.4	0.4	0.5
240	0.5	0.7	0.7	1.1
250	0.6	1.0	1.0	1.7
260	0.9	1.4	1.6	2.8

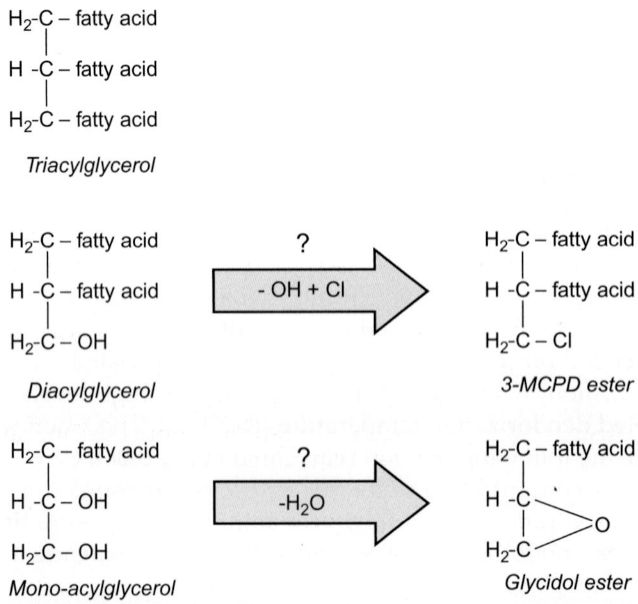

FIGURE 13.8 The chemical formula of 3-MCPD and glycidyl ester and their relation to diacylglycerol and mono-acylglycerol.

- Removal of free fatty acids in physical refining.
- Reduction of pesticide residues and light polycyclic aromatic hydrocarbon levels.
- Required minimum deodorization time and temperature to obtain a bland tasting deodorized product.
- Decomposition of red color at high temperature.

The process window of operational parameters should be defined for each deodorizer/oil combination. The deodorization temperature and time within this process window should be high enough to deliver a product that is in specification for minor components and contaminants, and low enough to limit TFA formation.

3-MCPD and Glycidyl Esters

The formation of 3-monochloropropane-1,2diol (3-MCPD) and glycidyl (or glycidol) esters has recently been reported as undesirable side reactions of refining. Figure 13.8 gives the chemical formula of 3-MCPD and glycidyl ester and their relation with the various acylglycerol compositions. The formation mechanisms of both 3-MCPD and glycidyl ester are under investigation.

The health effects of the low-level presence of 3-MCPD and glycidyl esters in edible oils are so far unknown. Toxicological studies were started but results are so far not yet available. The current assumption is that 3-MCPD esters will hydrolyze in the gastrointestinal tract to free 3-MCPD (Larsen, 2009). Exposures to levels well above tolerable daily intake (TDI) may occur if a 100% hydrolysis is assumed. This exposure reduces to below TDI assuming that only 3-MCPD esters with the chlorine at the 1,3 position are hydrolyzed to free MCPD.

There are no toxicological data on glycidyl esters but glycidol is a known genotoxic and carcinogenic compound.

It has been confirmed by several studies that both 3-MCPD and glycidyl esters are formed mainly during the deodorization step of the refining process. The level of 3-MCPD esters seems to be independent of deodorization conditions if the deodorization temperature is >120°C (Hrncirik and van Duijn, 2011). Also the pretreatment before deodorization (neutralization + bleaching or bleaching with acid pretreatment) and the pretreatment conditions have no significant effect on the 3-MCPD level after deodorization.

The level of glycidyl ester will increase at increasing deodorization temperature and/or time (Hrncirik and van Duijn, 2011). Limiting deodorization temperature and/or time to mitigate glycidyl ester formation could be an additional condition in determining the operational process window for deodorization (see "Cis-trans Isomerization," above).

HACCP

Each food production site needs to perform a hazard analysis critical control points or HACCP to secure the food safety of their products. Such an analysis needs to be based on the seven HACCP principles:

1. Conduct a hazard analysis.
2. Identify critical control points (CCP).
3. Establish critical limits.
4. Establish a monitoring system for each CCP.
5. Establish corrective actions.
6. Establish procedures for verification.
7. Establish procedures documentation and record keeping.

Management commitment and the use of good manufacturing practices are prerequisites of the HACCP system. Scientific research is needed to deliver the basics for the risk analysis (Motarjemi et al., 2009):

- Toxicological assessment.
- Mechanism of contaminant formation.
- Validated analytical methods.
- Likelihood of occurrence.
- Acceptable contaminant level.

Table 13.3 gives an overview of the HACCP-based monitoring plan for crude oil contaminants and refining by-products. This plan is developed based on the information given in the previous chapters.

The following contaminants in crude oil will result in a rejection of the delivery for food use:

- Previous cargoes: based on a paper check (logbook), rejection of the crude oil if not on the positive EU or FOSFA list.

TABLE 13.3 Overview of the HACCP-based Monitoring Plan for Crude Oil Contaminants and Refining By-products

Hazard	Critical Limit	Monitoring	Corrective Action
CRUDE OIL INTAKE			
Not allowed previous cargo	Cargo not on EU or FOSFA positive list	Check ship logbook	Block and reject for food use
Mineral oil contamination	25–300 ppm (see page 331)	Analyze crude oil	Block and reject for food use
Residue of not allowed pesticide	LOD = level of determination	Analyze crude oil	Block and reject for food use
Too high residue of allowed pesticide	MRL = maximum residue limit	Analyze crude oil	Block and reject for food use
PROCESS VALIDATION			
Poly aromatic hydrocarbons	Legal limit in refined oil	Check level in crude oil	Apply validated reduction process
Hexane residue in solvent extracted oils	LOD = level of determination	Check crude oil flashpoint	Apply validated reduction process
Aflatoxin in coconut or groundnut oil	Legal limit in refined oil	Check level in crude oil	Apply validated reduction process
Zearalenone in maize germ oil	Legal limit in refined maize germ oil	Check level in crude oil if ZEN is detected in harvest samples	Apply validated reduction process or reject for food use if removal is not feasible
PROCESS CONTAMINANTS			
Trans fatty acids	Local legislation or industry standard	Check level in deodorized oil	Reduce deodorization temperature

- Mineral oil: only presence of hexane in seed oils is allowed. High viscosity mineral oil cannot be removed by refining, while unknown impurities and additives may be present in all mineral oil.
- Pesticides: crude oils should not contain detectable levels of not allowed pesticides or allowed pesticides above MRL, even when these pesticides can be removed by refining.

The following contaminants have limits in oils designated for direct food use, which are mostly refined oils:

- Polycyclic aromatic hydrocarbons: a too high level in crude oil can be reduced by a combination of active carbon treatment and high temperature deodorization.
- Hexane residues in solvent extracted oils: removal to below detectable level under vacuum at elevated temperature. Mostly removed during the first drying step, completely removed during deodorization.
- Aflatoxin: reduced during pretreatment before deodorization.
- Zearalenone (ZEN) in maize germ oil: level in crude oil depends on weather conditions during flowering. Analysis of representative harvest samples of maize seeds indicates the

risk for presence of ZEN in the harvest of that year from the tested region. Regular crude oil analyses are only required if ZEN is found in the harvest samples of the supply region or for oils of unknown origin.

Only trans fatty acid is shown as a process contaminant. A number of countries have a legal limit for trans fatty acids; other regions and product categories (like the EU margarine industry) have an industry standard. The 3-MCPD and glycidyl esters are not mentioned in Table 3.3. Essential scientific research, required to identify critical control points for these esters, is not yet completed. Further research is needed to establish the toxic effect, the formation mechanism, a validated method of analysis and acceptable contaminant levels.

A site HACCP may result in additional hazards related to actual processing procedures and equipment, like a risk of lubricant oil contamination, foreign bodies in final product, etc. These should lead to appropriate corrective actions following the HACCP principles.

References

Alexander, J., Benford, D., Cockburn, A., et al., 2008. Scientific opinion of the panel on contaminants in the food chain on a request from the European Commission on polycyclic aromatic hydrocarbons in Food. EFSA J. 724, 1–114.

Bockisch, M., 1998. Fats and Oils Handbook. AOCS Press, Campaign.

EC Commission Regulation No. 1881, 2006. Setting maximum levels for certain contaminants in foodstuffs. Off. J. Eur. Union L364, 5–24.

EU Commission Directive 2004/4/EC, 2004. Amending Directive 96/3/EC granting a derogation from certain provisions of council directive 93/43/EEC on the hygiene of foodstuffs as regards the transport of bulk liquid oils and fats by sea. Off. J. Eur. Union L15, 25–31.

EU Pesticide Database, <http://ec.europa.eu/sanco_pesticides/public/index.cfm>.

Hrncirik, K., van Duijn, G., 2011. An initial study on the formation of 3-MCPD esters during oil refining. Eur. J. Lipid Sci. Technol. 113, 374–379.

Larsen, J.C., 2009. 3-MCPD esters in food products. ILSI Eur. Rep. Ser.

Motarjemi, Y., Stadler, R.H., Studer, A., 2009. Application of the HACCP approach for the management of processing contaminants. In: Stadler, R.H., Lineback, D. (Eds.), Process Induced Food Toxicants and Health Risks. John Wiley & Sons Inc., New Jersey, pp. 565–620.

Pages, X., Morin, O., Birot, C., et al., 2010. Raffinage des huiles et des corps gras et élimination des contaminants. OCL 17, 86–99.

van Duijn, G., 2008. Industrial experiences with pesticide removal during edible oil refining. Eur. J. Lipid Sci. Technol. 110, 982–989.

van Duijn, G., den Dekker, G., 2010. Unilever food safety assurance system for refined vegetable oils and fats. OCL 17, 100–103.

van Duijn, G., Dumelin, E.E., Trautwein, E.A., 2006. Virtually trans free oils and modified fats. In: Williams, C., Buttriss, J. (Eds.), Improving the Fat Content of Foods. Woodhead Publishing Limited, Cambridge, pp. 490–507.

CHAPTER

14

Bottled and Drinking Water

Arnold F. Dijkstra[1] *and Ana Maria de Roda Husman*[2,3]

[1]Dijkstra Food Consultancy, Leeuwarden, Netherlands, [2]National Institute for Public Health and the Environment (RIVM), Bilthoven, The Netherlands, [3]Utrecht University, Utrecht, The Netherlands

OUTLINE

Food Safety Management.
DOI: http://dx.doi.org/10.1016/B978-0-12-381504-0.00014-7

WATER AND THE FOOD INDUSTRY

In the food industry, water can be an end product, such as bottled water, or an ingredient in a wide range of commodities. In addition, water may be used as a means to produce the food, such as irrigation water and shellfish growing waters, and in food processing, such as for washing produce and/or the materials for food production/processing. Also, water may be used as a transport mechanism. In each of these cases, the consumer is subjected to possible human health hazards from water. This chapter focuses on the different types of water used for the preparation of drinking water and potential hazards related to water intended for direct use by the consumer (bottled water, tap water, ice cubes), or indirectly as an ingredient of any food commodity that is consumed without further processing for safety. Practical cases are presented for the determination of safe water, processing for safety water treatment systems, water reuse in the food industry and bottled water safety.

DEFINITIONS FOR WATER

Terms used to designate types of water are diverse and diversely used. Here, we exemplify the different terms for water, their origin and their definition.

- **Bottled water (packaged)** addresses natural mineral water, spring water and all other drinking water, according to the European Union.
- **Packaged water:** Packaged drinking water means all water that is sealed in bottles, packages, or other containers and offered for sale for human consumption, including bottled mineral water, with no added ingredients (FDA, 2012).
- **Natural mineral waters** are waters derived from a natural mineral water spring, which:
  - have been extracted from the ground of a member state and are recognized by the responsible authority of that member state as satisfying the provisions of Schedule 1, Part 1 of S.I. No. 225 of 2007, or
  - have been extracted from the ground of a third country and imported into the Community, and have been recognized by the responsible authority of a member state pursuant to certification in the third country, and are intended to be placed on the market in a member state in bottles or containers, according to the EC.
- "**Other waters**" are those waters which are intended for human consumption, are not natural mineral waters as defined in S.I. No. 225 of 2007, are not spring waters as defined in S.I. No. 225 of 2007, and are intended to be placed on the market in a member state in bottles or containers, according to the European Union.
- **Drinking water:** Water that is intended for human consumption and suitable for all usual domestic uses, complying with the requirements of the WHO Guidelines for Drinking-water Quality or appropriate national standards established by the regulating authority (WHO, 2006). Water that meets or exceeds all applicable federal/provincial/local requirements concerning safety. Also known as **potable (drinkable) water** (Symons et al., 2000; WHO The Health and Environment Lexicon, 2012);
- **Tap water** (running water, city water, municipal water, etc.) is potable water supplied to a tap (valve) inside the household or workplace (Wikipedia).

- **Safe water:** see Drinking water.
- **Clean water:** Water that is clean and is acceptable to the consumer with respect to taste, odor and appearance (WHO, 2011).

LEGISLATION

Legally different categories of waters intended for human consumption supplied as bottled water or municipal drinking water are distinguished.

Bottled Water

Bottled water is covered by European Commission (EC) Regulation S.I. No. 225 of 2007 for natural mineral waters, spring waters and other waters in bottles or containers. This legislation covers the definition of natural mineral water, spring water and "other water," their exploitation, treatment, microbiological criteria, chemical contaminants, sales description, labeling and packaging. Spring waters and "other waters" must also comply with EC Regulation S.I. No. 278 of 2007 for drinking water.

Directive 2009/54/EC defines the provisions applicable to the marketing and exploitation of natural mineral waters. Commission Directive 2003/40/EC of 16 May 2003 establishes the list, concentration limits and labeling requirements for the constituents of natural mineral waters and the conditions for using ozone-enriched air for the treatment of natural mineral waters and spring waters. Natural mineral waters are subject to an authorization procedure carried out by the competent authorities of the EU member states or by European Economic Area (EEA) countries.

The lists of natural mineral waters officially recognized by the member states of the EU and of the EEA (Iceland) and (Norway) are published by the European Commission in the *Official Journal of the European Union*. These lists are regularly updated on http://ec.europa.eu/food/food/labellingnutrition/water/index_en.htm.

Natural mineral waters and spring waters may be treated at source to remove unstable elements and some undesirable constituents in compliance with the provisions laid down in Article 4 of Directive 2009/54/EC. Treatments other than filtration with possible oxygenation have to be assessed and authorized at EU level prior to their use by industry. Commission Regulation (EU) No. 115/2010 of 9 February 2010 lays down the conditions for use of activated alumina for the removal of fluoride from natural mineral waters and spring waters.

Municipal Drinking Water

The European Drinking-water Directive (DWD), Council Directive 98/83/EC as amended by Regulations 1882/2003/EC and 596/2009/EC, concerns the quality of water intended for human consumption and forms part of the regulation of water supply and sanitation in the European Union.

The Directive is intended to protect human health by laying down healthiness and purity requirements which must be met by drinking water within the Community. It applies to all water intended for human consumption apart from natural mineral waters (see §18.3.1) and waters which are medicinal products.

Member states shall ensure that such drinking water:

- does not contain any concentration of microorganisms, parasites or any other substance which constitutes a potential human health risk;
- meets the minimum requirements (microbiological and chemical parameters and those relating to radioactivity) laid down by the directive.

Member states will also take any other action needed in order to guarantee the healthiness and purity of water intended for human consumption.

On a global scale, the WHO provides guidelines for the safety of water, the so-called Guidelines for Drinking-water Quality (WHO, 2011). In these guidelines, the use of water safety plans (WSPs; see also "Risk Assessment and Risk Management," on page 366) is suggested as a comprehensive risk assessment and risk management approach that encompasses all steps in the water supply from catchment to consumer to be able to consistently ensure the safety of a drinking-water supply. The WSP approach draws on many of the principles and concepts from other risk management approaches, in particular the multiple-barrier approach and hazard analysis and critical control point (as used in the food industry).

For verification that the WSP has been put into place, minimum requirements for safe drinking water have been set as laid down in the WHO guidelines (Table 14.1). Microbiological testing shall always be regarded as a verification tool due to the retrospective nature of the methodology. Preferably microbial parameters are integrated with physical and chemical parameters such as temperature and disinfectant concentration, parameters that can be measured in real time and therefore are suitable for continuous monitoring.

SOURCES OF WATER

Earth harbors huge amounts of water, in total approximately 1.4 billion cubic kilometers (Table 14.2 and Figure 14.1). Only 3% of this volume consists of fresh water which is mainly employed for consumption. Increasingly, other sources such as brackish and saline waters are considered for the production of drinking water.

Drinking water may be produced from groundwater, surface water, rainwater and/or recycled water. Depending on the quality of such water and the required quality for its application, source waters may need to be treated prior to use.

Groundwater may originate from shallow and/or deep, (un)confined aquifers and is generally considered safe with respect to contamination with microbial, chemical and radiological hazards (discussed in detail in "Hazards Associated with Drinking Water," page 358). Natural mineral waters are characterized by their purity at source and their constant level of minerals.

Surface waters may include rivers, lakes, delta and groundwater areas with brackish water or a sea or ocean. If water is not fresh but brackish or salt then desalination processes are in order for the production of water suitable for consumption.

Alternatively, drinking water may be produced from rainwater and/or recycled water.

TABLE 14.1 Microbiological Monitoring and Verification of Various Water Types within a Factory

Water	Target Organisms	Method	Guideline Value	Frequency of Monitoring
Potable, municipal drinking water at intake point. Water (municipal or well) after treatment	Coliforms presence/absence test using membrane filtration	ISO 9308	ND[a] in 100 ml	As determined by HACCP
	Escherichia coli	ISO 9308	ND in 100 ml	
	Enterococci	ISO 7899	ND in 100 ml	
	Total plate count 22°C	ISO 6222	≤100/ml	
	Total plate count 37°C	ISO 6222	≤10/ml	
Product make up water	Depended on processing	–	–	As determined by HACCP
Chilled water circuits (closed), unpreserved	Coliforms plate count	ISO 9308	≤1/ml	As determined by HACCP
	Total plate count 22°C	ISO 6222	≤1000/ml	
Chilled water circuits (closed), preserved	Coliforms plate count	ISO 9308	≤1/ml	As determined by HACCP. Check preservative concentration continuous or weekly
	Total plate count 22°C	ISO 6222	≤1000/ml	
Hot water circuits	None	–	–	Check temperature storage (60°C) and distribution (56°C) continuously
Final rinse water	Depended on processing	–	For aseptic processes sterility is required	As determined by HACCP
Cooling water for canning	Coliforms plate count.	ISO 9308	≤1/ml	As determined by HACCP. Check chlorine concentration continuous or daily
	Total plate count 22°C	ISO 6222	≤100/ml	
	Chlorination	ISO 7393	2–10 mg/l	
Bottled water	*Escherichia coli*	ISO 9308	ND in 250 ml	As determined by HACCP
	Enterococci	ISO 7899	ND in 250 ml	
	Pseudomonas aeruginosa	ISO 16266	ND in 250 ml	
	Total plate count 22°C	ISO 6222	≤100/ml	
	Total plate count 37°C	ISO 6222	≤20/ml	

Adapted from ILSI, 2008, WHO, 2011 and EC 2012
[a]ND = Not detectable in the defined volumes.

Groundwater

Groundwater is water contained beneath the surface in rocks and soil, and which accumulates underground in aquifers (WHO, 2006). Groundwater constitutes 30% of the global freshwater pool (Figure 14.1). In many parts of the world groundwater sources are the

TABLE 14.2 Estimation of the Global Water Distribution (Gleick, 1996)

Water Source	Water Volume, in Cubic Miles	Water Volume, in Cubic Kilometers	Fresh Water Percentage	Total Water Percentage
Oceans, seas, and bays	321,000,000	1,338,000,000	–	96.5
Ice caps, glaciers and permanent snow	5,773,000	24,064,000	68.7	1.74
Groundwater	5,614,000	23,400,000	–	1.7
Fresh	2,526,000	10,530,000	30.1	0.76
Saline	3,088,000	12,870,000	–	0.94
Soil moisture	3959	16,500	0.05	0.001
Ground ice and permafrost	71,970	300,000	0.86	0.022
Lakes	42,320	176,400	–	0.013
Fresh	21,830	91,000	0.26	0.007
Saline	20,490	85,400	–	0.006
Atmosphere	3095	12,900	0.04	0.001
Swamp water	2752	11,470	0.03	0.0008
Rivers	509	2120	0.006	0.0002
Biological water	269	1120	0.003	0.0001
Total	332,500,000	1,386,000,000	–	100

single most important supply for the production of drinking water, particularly in areas with limited or polluted surface water sources. Groundwater is typically of more stable quality and better microbial quality than surface waters. Groundwater quality from small suppliers suffers more from a lack of information, risk assessment and risk management. Groundwater often requires little or no treatment to be suitable for drinking. There are many examples of groundwater being distributed without treatment. However, groundwater quality may be corrupted by nearby sources of hazards if the groundwater well is insufficiently confined and/or well integrity is compromised. Viruses are considered to be the most critical pathogens for groundwater contamination, because of their ability to travel through the subsurface and their high infectivity (Schijven et al., 2010). Flooding of groundwater wells due to extreme precipitation and unnatural threats to its quality should be recognized (Schijven and de Roda Husman, 2005). It is vital therefore that the quality of groundwater is protected if public health is not to be compromised.

Surface Water

Surface water may consist of fresh or saline water, or a combination of semi-saline water called brackish water. Brackish and saline waters are present in our oceans, seas and river

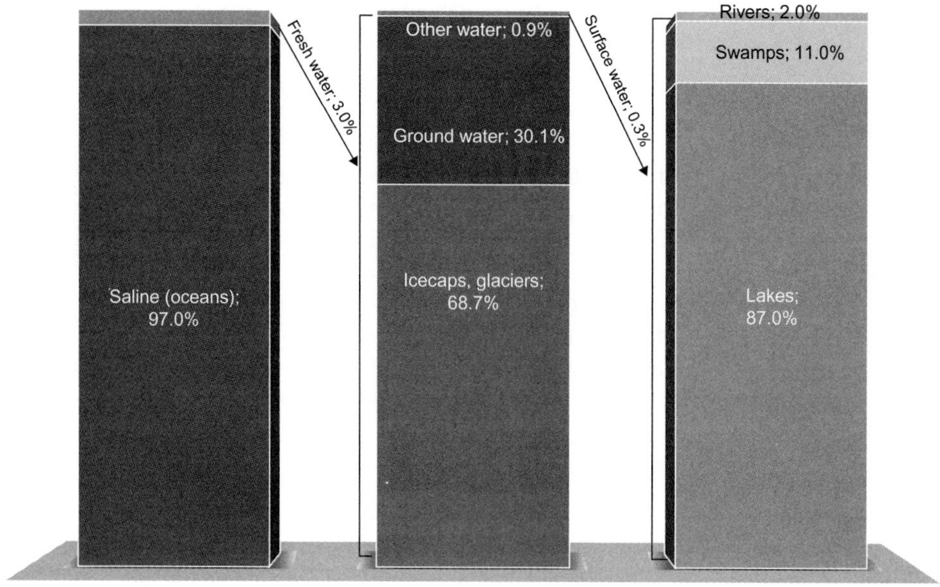

FIGURE 14.1 Freshwater sources.

delta areas. Fresh surface waters include rivers, lakes, swamps and groundwater (Figure 14.1). The larger part of fresh water (two-thirds) is, however, frozen and encapsulated in icecaps and glaciers. Groundwater also constitutes a large part: one-third. Of the remaining part, fresh water mostly includes lakes, swamps and rivers. Surface waters are largely under the influence of contamination from human and animal activities but also from the environment itself, which may compromise public health. The range of human activities in the catchment that may cause pollution of surface waters with microbiological, chemical and radiological hazards includes agricultural activities, sanitation practices, industry, mining, military sites, waste disposal and traffic. As compared with groundwater, surface waters generally need to be treated, often extensively.

Rainwater

Rainwater is initially free of contamination, except for air pollutants (Lye, 2009; Schets et al., 2010; WHO, 2011). However, the quality of rainwater may subsequently deteriorate during harvesting, storage and household use. When collected from rooftops or otherwise, it may become contaminated by animals and humans directly or indirectly from their waste or, alternatively, chemicals may dissolve from collecting and storage devices and human pathogens may grow in stored rainwater. Well-designed rainwater harvesting systems with clean catchments, covered cisterns and storage tanks, and treatment, as appropriate, supported by good hygiene at point of use, can offer drinking water with very low health risks. Rainwater can provide an important source of drinking water in some circumstances as well as be a useful source of water for blending with other sources to reduce the levels of contaminants that may cause health concerns, such as arsenic and fluoride.

Saline Water

In light of climate issues and population growth, it may be difficult to provide sufficient water supply to the world including meeting industry needs. In this respect, desalination of ocean and sea water has been explored. Desalination facilities exist all over the world, particularly in the eastern Mediterranean region, with use increasing on all continents. Desalination is used to remove salts from brackish or saline surface water and groundwater in order to render it acceptable for human consumption or other uses such as in the food industry. Some of the desalination processes used (especially distillation and reverse osmosis) are highly effective in removing microbial and chemical hazards facilitating the use of these processes as single-stage treatments.

Recycled Water

After use in the food industry (ILSI, 2012), water may be of sufficient quality for use in a similar process, e.g. washing. Depending on the contact, both in time and surface, between the water and food ingredient or end product, the recycled water needs to meet quality requirements. If these are not met, there is a range of available treatment options to improve water quality (WHO, 2011). See also "Water Reuse in Food Processing," on page 372.

DRINKING-WATER APPLICATIONS IN THE FOOD INDUSTRY

Water is used widely in the food industry. It is used to move products, to produce and/or wash vegetables, fruits, fish and poultry, and to clean and refresh raw vegetables after harvesting and during distribution. Water, or steam made from it, is used for cleaning, disinfection and heating purposes. Finally, water can be a consumer end product and/or an ingredient in food. Virtually all frozen foods carry a glaze of ice which is often derived from process water, and for certain frozen foods (such as fish and shellfish) a glaze is added as a protective measure.

The amount of water used to produce food commodities is sometimes impressive. The so-called water footprint is defined as the total volume of fresh water that is used to produce the goods and services consumed by an individual or community or produced by a business. The production of 1 kilogram of beef requires 15,415 liters of water, an average of 1,600 liters are needed for 1 kg of bread and 27 liters are needed for a cup of tea (250 ml). (www.waterfootprint.org, accessed June 2013).

Water as End Product

Water is delivered to the consumer either as tap water from a piped distribution system or packaged in bottles, cartons or other containers. The food industry delivers packaged water only.

The following types of water are produced as an end product, packaged in plastic or glass bottles, cartons, water coolers, water dispensers and so on. Sizes range from small single serving PET bottles to large carboys for water coolers.

- **Spring water:** Bottled water derived from an underground formation from which water flows naturally to the surface of the earth. Spring water must be collected only at the spring or through a borehole tapping the underground formation feeding the spring.
- **Purified water:** Water that has been produced by distillation, deionization, reverse osmosis or other suitable processes while meeting the definition of purified water in the United States Pharmacopoeia.
- **Mineral water:** Bottled water containing not less than 250 parts per million total dissolved solids.
- **Sparkling bottled water:** Water that, after treatment and possible replacement with carbon dioxide, contains the same amount of carbon dioxide that it had as it emerged from the source.
- **Artesian water/Artesian well water:** Bottled water from a well that taps a confined aquifer (a water-bearing underground layer of rock or sand) in which the water level stands at some height above the top of the aquifer.
- **Well water:** Bottled water from a hole bored, drilled or otherwise constructed in the ground, which taps the water aquifer.

The amount of bottled water consumed per year was estimated in 2011 to be 262 billion liters worldwide. This represents an average of 37 liters per capita. (http://www.zenithinternational.com/accessed june 2013). However, some countries show extremely higher figures. Table 14.3 shows the 20 countries where most bottled water is consumed, compared to the average global consumption.

The market is forecasted to grow to over 400 billion liters in 2020.

Bottled water has come under criticism in recent years for the environmental impacts of groundwater extraction, the energy and environmental costs of the plastic packaging and transportation costs and concerns about water quality and the validity of some marketing claims. One criticism of bottled water concerns the packaging. Bottled water commonly is packaged in polyethylene terephthalate (PET), which requires a significant amount of energy to produce. While PET is recyclable, only a fraction of plastic bottles made from PET are actually recycled. For example, in the United States, according to a NAPCOR (National Association for PET Container Resources) study, water bottles account for 50% of all the PET bottles and containers collected by curb side recycling, and the recycling rate for water bottles was 28% in 2009. However, bales of PET collected for recycling often contain materials such as polypropylene caps, labels and glue, and other contaminants, which are then weighed and included in the PET recycling rate. The percentage of "clean PET flake" yielded once the contaminants have been removed was 21% in 2009 and is a more accurate depiction of how much PET actually gets recycled. European recycling rates tend to be somewhat higher. In the United States, plastic used to create bottles uses an estimated 15 million barrels of oil annually (data on recycling from http://www.container-recycling.org, accessed September 2012).

TABLE 14.3 Per Capita Bottled Water Consumption by Top Countries 2000 to 2010

	Per Capita Bottled Water Consumption (liters)		
Countries	2000	2005	2010
Mexico	124	179	243
Italy	160	191	187
United Arab Emirates	114	181	153
Belgium–Luxembourg	118	160	148
Germany	102	128	134
France	126	139	132
Spain	105	146	124
Lebanon	77	107	121
Thailand	70	76	114
Hungary	39	70	111
Switzerland	90	104	108
United States	67	99	107
Slovenia	56	81	107
Croatia	47	78	101
Cyprus	72	98	98
Qatar	–	79	95
Saudi Arabia	80	93	95
China, Hong Kong SAR	–	68	95
Czech Republic	68	90	92
Austria	75	81	91

Data from Beverage Marketing Corporation and http://www.worldwater.org – accessed September 2012.

Water as Ingredient

The importance of water quality cannot be underestimated by food manufacturers. It plays a vital role, both as a critical ingredient in ensuring food quality and as a key to efficient production. It provides appropriate water content in the final product. For example, canned soups and vegetables contain a high percentage of added water once they have been cooked and packaged. Another important function of water is to dissolve ingredients. Especially when used as ingredient, it is important that the water produces no hazards, flavors or smells which might affect the quality or consistency of the final product. "Determination of Water

Safety," on page 367, presents a simple safety classification of water "fit for purpose" and an easy-to-use decision tree for assessing the suitability of water for its intended use.

Producers of food products that need to be rehydrated before consumption should be aware of the safety of water used for this purpose. In particular products that will not, or will not sufficiently, be reheated need attention from both the water supplier as well as the supplier of the dried food. The water supplier obviously needs to provide safe water, either bottled or tap. The food supplier is responsible for pathogen-free products, maximal intrinsic security (like water content or capacity, acidity, temperature, packaging) and shall instruct customers in how to safely rehydrate the product – in particular if they sell their products to areas where safe water is not commonly available.

Dried foods intended for babies, such as infant formula, require specifically safe water as infants' immune system is not fully developed and they are particularly vulnerable. Of course, the product itself shall be free from any harmful chemical, microbiological or physical hazards; however, consumers should be alerted that dried foods are not sterile. Risks associated with the rehydration of products and their final preparation for consumption, including recontamination during storage, should be considered in the products' HACCP plan, and validated safety instructions should be provided to consumers. The Codex Guidelines on Validation of Control Measures provide guidance on the validation process for preparation of consumer information (see Codex Alimentarius. Guidelines for the Validation of Control Measures).

The World Health Organization has issued guidelines for safe preparation, storage and handling of powdered infant formula (WHO, 2007b). These guidelines are valuable for consumers and producers of infant formula. A few relevant subjects for producers of infant formula are summarized:

- **Formula preparation:** In most cases, it is safe to mix formula using ordinary cold tap water that has been brought to boil and then boiled for 1 minute and cooled. According to the World Health Organization, recent studies suggest that mixing powdered formula with water at a temperature of at least 70°C (158°F) will eliminate the bacterium *Cronobacter sakazakii* (previously *Enterobacter sakazakii*) and other pathogenic (micro) organisms. Remember that formula made with hot water needs to be cooled quickly to body temperature if it is being fed to the baby immediately. If the formula is not being fed immediately, refrigerate it right away and keep refrigerated until feeding.
- **Water:** Use the exact amount of water recommended on the label. Under-diluted formula can cause problems related to dehydration. Over-diluted formula will not provide adequate nutrition, and, if fed for an extended period of time, may result in slower growth.
- **Bottled water:** If consumers use non-sterile bottled water for formula preparation, they should follow the same directions as described for tap water above. If the water is marketed by the manufacturer as sterile and for infants, it must meet general requirements for commercial sterility.
- **"Use by" or "expiry" date:** This is the date after which a package or container of infant formula should not be fed to infants.
- **Storage:** Manufacturers must include instructions on infant formula packaging for before and after the container is opened. They must also include information on the storage and disposal of prepared formula.

Water for Processing

During food production, water is widely used as a processing aid without the aim of serving it as an ingredient. Examples of demands for water during food processing are (not exhaustive):

- Washing or cleaning of (raw) products.
- Transport of products.
- Treatment of the product (e.g. alteration, separation).
- Cooling processes: for example, fish is typically shipped in ice; poultry may be cooled in water and slush ice and transported in ice.
- Steam generation for heating, directly or indirectly.
- Cleaning or rinsing of equipment.
- Abnormal incidents (like fire protection).
- Sanitation.

Increasingly process water is recycled. This subject is discussed in "Water Treatment Technologies for Safe Water Production," on page 367.

Water at Household Level

To obtain and maintain safe water at the household level, integrated planning, combined with effective monitoring and evaluation, is critical. An estimated 780 million people drink water from unimproved sources, and millions more drink contaminated water from improved sources (WHO and Unicef, 2012). Until safe, reliable, piped-in water is available to every household, interim measures, such as household water treatment and safe storage (HWTS) to prevent contamination during collection, transport and use in the home, are needed to reduce the burden of diarrheal disease. While a growing body of evidence demonstrates that the use of HWTS methods improves the microbial quality of household drinking water and reduces the burden of diarrheal disease in users, there is also increasing evidence that inconsistent and/or incorrect use may be a major challenge in realizing the full potential from HWTS. In order to develop effective mechanisms to encourage and sustain correct use of HWTS, there is a need to monitor and evaluate uptake. Recently, WHO (2012) has provided a toolkit including process monitoring to assess program implementation and quantitative analysis through surveys, direct observation and water quality monitoring. As part of this toolkit, a set of indicators pays attention to reported and observed use; correct, consistent use and storage; knowledge and behavior; other environmental health interventions; and water quality.

HAZARDS ASSOCIATED WITH DRINKING WATER

Microbial, chemical and radiological hazards may compromise water quality and confer public health risks by human consumption of food and water. The great majority of evident water-related health problems are the result of microbial (bacterial, viral, protozoan or other biological) contamination.

An appreciable number of serious health concerns may occur as a result of the chemical contamination of drinking water. Adverse health effects due to exposure to microbial hazards will be acute and may be chronic as opposed to exposure to most chemical hazards that are rarely acute.

Microbial Hazards

Bacteria, viruses, protozoan parasites, algae, amoebae and helminths are known microbial hazards associated with drinking water. Some of these organisms, such as a few bacterial species, algae and helminths, can multiply independently in the aquatic environment whereas other, so-called enteric pathogens are completely dependent on their warm-blooded host, animals and/or humans, for their multiplication (Table 14.4). In case the enteric bacteria, viruses or parasites can be transmitted from animals to humans, whether or not waterborne, these are called zoonotic. Examples are the protozoan parasite *Cryptosporidium*, the hepatitis E virus and the bacterium *Campylobacter*. Viruses (20–300 nm) in general are much smaller than bacteria (approx. 1 μm) which are turn smaller than protozoan parasites (10 μm or larger). The different sizes affect their fate and transport in the aquatic environment as well as their removal and inactivation efficacy by treatment.

Infection with waterborne pathogens may pass without symptoms, or lead to mild disease, severe disease or death. Young children are especially vulnerable for contracting water-related infections and diseases mainly involving diarrhea, and if not properly treated these could be life threatening. On a global level, the UN and partners estimate that child mortality has declined by 41% since 1990, from 12 million deaths per year to 6.9 million in 2011 (data from WHO, accessed November 2012). However, many countries, especially in sub-Saharan Africa, are still far off-target in reducing child deaths. Contaminated water is an important cause of the catastrophe (see also Motarjemi et al., 1993, 2012): worldwide, an estimated 780 million people lacked safe drinking water in 2010 involving 1.8 million diarrheal disease deaths, mostly children, every year (WHO, 2011).

In low income regions exposure to *Vibrio* bacteria causes large cholera outbreaks with many thousands becoming ill, resulting in countless deaths (Mandal et al., 2011). These outbreaks may follow natural disasters such as floods, as in Haiti in 2010. However, outbreaks are ongoing in sub-Saharan Africa. Hepatitis E virus has caused numerous outbreaks among displaced people in Chad and Sudan in 2004 resulting in more than 45 deaths, mostly pregnant women (Boccia et al., 2006). In 2007, in northern Uganda, the virus demanded 160 deaths with more than 10,196 persons diseased (Teshale et al., 2010). Although the problem with unsafe drinking water is strongly related to low income countries, high income countries may also suffer from major outbreaks. One of the largest recorded outbreaks of waterborne disease took place in Milwaukee (USA) in 1993. Over 400,000 people were infected with *Cryptosporidium parvum*. This outbreak was probably caused by polluted water from Lake Michigan, the source of the drinking water. In May 2000 approximately 2300 people became seriously ill and seven died from exposure to contaminated drinking water in the town of Walkerton, Ontario (Canada). A combination of extreme weather, lack of appropriate control systems and human failure resulted in water being contaminated with *E. coli* O157:H7 and *Campylobacter jejuni*. These cases illustrate that the complexity of (tap)water systems in developed countries may be alive to technical and/or human failure.

TABLE 14.4 Sources of Water-related Pathogens

	Source		
Pathogen	**Human**	**Animal**	**Environmental**
Acanthamoeba	–	–	+
Adenoviruses	+	+	–
Aeromonas	+	+	+
Campylobacter	+	+	–
Cryptosporidium	+	+	–
Cyanobacteria	–	–	+
Pathogenic *E. coli*	+	+	–
Enteroviruses	+	+	–
Giardia	+	+	–
Hepatitis A virus	+	–	–
Hepatitis E virus	+	+	–
Legionella	–	–	+
Leptospira	–	+	–
Mycobacterium (nontuberculous mycobacteria)	–	–	+
Naegleria fowleri	–	–	+
Noroviruses	+	–	–
Pseudomonas aeruginosa	–	–	+
Rotavirus	+	–	–
Salmonella (para)typhi	+	–	–
Salmonella nontyphi	+	+	–
Shigella	+	+	–
Staphylococcus aureus	+	–	–
Toxoplasma	+	+	+
Vibrio	+	+	+

(Updated from Schets et al., 2010)

Disease outbreaks may be directly associated with drinking-water consumption but also to more indirect exposure. After heavy rainfall, 60% of cruise participants reported gastroenteritis with stools positive for *Shigella sonnei*, *Giardia* and *Cryptosporidium* after consumption of ice produced from potable water contaminated with lake water (Serdarevic et al., 2012).

Chemical Hazards

Water may contain many different chemicals, usually in low to very low concentrations; however, spills may be extensive. An example is the Minamata disease in Japan (1956). It was caused by the release of methyl mercury in the industrial wastewater from the Chisso Corporation's chemical factory, which continued from 1932 to 1968 (Wikipedia, accessed December 2012). In Bangladesh it is estimated that a major part of the population is at risk of poisoning because groundwater used for drinking has been contaminated with naturally occurring inorganic arsenic (Smith et al., 2000). Due to labor and technical limitations only part of these chemicals are monitored; beyond this the focus is addressed to well-known substances or groups of chemicals (and not the individual elements). The main sources of chemical hazards are (WHO, 2011):

- Naturally occurring: rocks, soils and the effects of the geological setting and climate.
- Industrial sources and human dwellings: mining (extractive industries) and manufacturing and processing industries, sewage, solid wastes, urban runoff, fuel leakages, pharmaceuticals, hormones, personal care products.
- Agricultural activities: manures, fertilizers, intensive animal practices and pesticides.
- Water treatment or materials in contact with drinking water: coagulants, DBPs, piping materials.
- Pesticides used in water for public health: larvicides used in the control of insect vectors of disease.
- Cyanobacteria producing unwanted metabolites: eutrophic water bodies.

The effect of chemical contaminants may be categorized as follows:

- Toxic to live stock.
- Toxic to fish, shellfish or crustaceans, in particular in aquaculture.
- Toxic to crops (phytotoxic).
- Accumulation in fish, livestock, plants and products derived from them.
- Toxic to humans, either directly or indirectly.

Chemical hazards are usually not related to acute toxicity while concentrations are usually very low. Of concern, however, is exposure to very low concentrations with effects that are only evident after a very long period of time.

An extensive overview of chemical hazards has been described by the WHO. Without being complete this list includes the following categories.

Inorganic

This group of potential hazards includes metals and metalloids (like lead, iron, nickel, zinc, mercury, arsenic, boron, cadmium and molybdenum), salts (sodium, chloride, potassium, calcium, manganese and magnesium), nitrate and nitrite and the parameter total hardness.

Some of the inorganic substances are derived from soil and/or rocks, but some metals are potentially released from pipeline systems. An epidemiological study on the extent of lead exposure via tap water in Hamburg (Germany) showed that people with lead in tap water above 5 mg/L showed significantly higher blood lead levels compared to those with no detectable lead in the tap water. Elevated levels of lead (and other metals) may cause adverse health effects after prolonged periods of exposure (Cidu, 2011).

In the UK an incident affecting five children attending a summer camp was related to consumption of "blue colored" drinking water. The contamination occurred in an old building which was being used for the first time after a few months. Because the stored water had been left standing for many months it had become blue tinged due to the copper pipes and tanks. The children's symptoms were consistent with excessive copper ingestion. After the system had been completely flushed through, the water returned to its natural colorless state and the levels of copper were confirmed to be below the (UK) guideline values (Paranthaman, 2010).

These examples elucidate the urge to analyze risks associated with the tap water distribution system.

Nitrate toxicosis can occur through metabolism of nitrate to nitrite, which in turn oxidizes the iron atoms in hemoglobin from ferrous iron (2+) to ferric iron (3+), rendering it unable to carry oxygen. This process can lead to generalized lack of oxygen in organ tissue and a dangerous condition called methemoglobinemia. Methemoglobinemia in infants is known as blue baby syndrome. Although nitrates in drinking water were once thought to be a contributing factor, there are now significant scientific doubts as to whether there is a causal link to disease (Wikipedia, accessed 29 August 2012).

Although salts are necessary for the human body and physiology, excessive salt concentrations may be hazardous. Fresh water normally has a salt concentration $<0.05\%$. Drinking water with elevated amounts of salt can have unfavorable effects on blood pressure and heart rate, and produce physiological changes (headache, dizziness, nausea, blood-stained stools, vomiting). In extreme cases, the increased salt content of drinking water may cause severe illness and even death.

In conclusion, inorganic hazards may cause severe illness in humans. However, the chance that the threshold concentrations end up in tap or bottled drinking water can be controlled relatively easy with an appropriate HACCP system.

Organic

Organic pollutants are a comprehensive group of chemicals that include (Wikipedia, accessed August 2012; WHO, 2011):

- Detergents.
- Disinfection by-products found in chemically disinfected drinking water, such as chloroform.
- Food processing waste, which can include oxygen-demanding substances, fats and grease.
- Insecticides and herbicides, a wide range of organohalides and other chemical compounds.
- Petroleum hydrocarbons, including fuels (gasoline, diesel fuel, jet fuels and fuel oil), lubricants (motor oil) and fuel combustion by-products from storm water runoff.
- Tree and bush debris from logging operations.
- Volatile organic compounds (VOCs), such as industrial solvents, from improper storage.
- Chlorinated solvents that may fall to the bottom of reservoirs, since they do not mix well with water and are denser.

- Polychlorinated biphenyl (PCBs).
- Trichloroethylene.
- Perchlorate (both a naturally occurring and man-made chemical that is used to produce rocket fuel, fireworks, flares and explosives). Perchlorate can also be present in bleach and in some fertilizers (http://water.epa.gov, accessed December 2012).
- Various chemical compounds found in personal hygiene and cosmetic products.

The WHO guideline for drinking-water quality provides detailed information on many of the organic hazards and proposes methods to prevent and control them (WHO, 2011).

Disinfectants

Disinfectants commonly used in the food, drink and catering industries include the following:

- Surface active agents (surfactants). These include the amphoterics (based on amyl alkyl glycines), the cationics (quaternary ammonium compounds – known as QACs or quats) and biguanides/diguanides. Many of the amphoterics and cationics are classified as skin, eye and respiratory irritants. Biguanides/diguanides are of low toxicity and irritancy and are useful skin disinfectants.
- Alcohols. These are used as skin cleaners as well as a transport medium for other active ingredients, but nevertheless are irritating to eyes, nose and throat at high airborne concentrations and can be a fire risk.
- Aldehydes. Glutaraldehyde is classified as a skin and respiratory sensitizer. Formaldehyde is a strong respiratory irritant and is also classified as a category 3 carcinogen.
- Peracetic acid is a powerful oxidizing agent used in the food and drink industries and is also extremely corrosive.
- Hypochlorite and organic chlorine-releasing compounds are corrosive in their concentrated form and are classified as eye and skin irritants in their dilute form (5–10%).

Most disinfectants are used to disinfect equipment or the premises. But drinking water used for food production may also contain disinfectants which are added by water suppliers to control pathogens (see "Water Treatment Technologies for Safe Water Production," on page 367). Disinfectants themselves can react with naturally-occurring materials in the drinking water to form by-products, such as trihalomethanes and haloacetic acids, which may pose health risks. The challenge for water suppliers is to control and limit the risks from pathogens and disinfection by-products as well as health risks to customers from disinfection by-products. For actual information on allowed disinfectants and maximum residual disinfectant levels, food companies shall address to local suppliers and legislation.

The food industry may also be at risk directly because chemicals are used for cleaning and disinfection. Residues may come in contact with the product(s) causing hazards, e.g. as with the supplied drinking water, and HACCP plans should cover these risks appropriately.

Pharmaceuticals, Hormones and Drugs

As a consequence of strong increases in human (and animal) healthcare more and more pharmaceuticals, hormones and drugs are prescribed. Part of the substances themselves or their metabolites are excreted and may reach water sources. In particular substances that are designed to be active in the human body at low levels are of concern. For other substances small quantities mean that effects are only evident after a long period of time. Therefore most standards for drinking water are based on risk assessments for long-term exposure.

A study in 2010 reviewed various QPhRA (quantitative pharmaceutical risk assessment) studies to identify potential threads (Kumar, 2010). In general, for low concentrations of APIs (active pharmaceutical ingredients), none of the QPhRA studies has identified any human health risks via exposure to drinking water, but uncertainties related to the QPhRA still exist and warrant consideration. In particular, knowledge about chronic effects and mixture effects of pharmaceuticals is very limited and requires further study.

Radiological Hazards

Radiation may originate from a number of naturally-occurring and human-made sources. Natural materials like uranium, thorium and potassium-40 can be found in diverse environments. Radioactive constituents of drinking water can result from:

- Naturally-occurring radioactive substances.
- Technological processes from which radioactive materials are released (like mining, processing of mineral sands or phosphate fertilizer production).
- Radionuclides discharged from nuclear fuel recycle facilities.
- Manufactured radionuclides (e.g. for medical and industrial use) that are not properly discharged.
- Past releases of radionuclides into the environment, including water sources (nuclear research programs and tests).

Radiation risks are limited, in particular when water is supplied by reliable suppliers. Food companies using natural sources should analyze the possible radiological risks by assessing the environment and if necessary testing for contaminants.

Greater concerns are related to nuclear disasters.

In 1986 the Chernobyl accident contaminated 125,000 square miles of land in Belarus, Russia and Ukraine with radio nucleotides including cesium-137, strontium-90 and plutonium-239. It is interesting that the water supply is not nearly as contaminated as the soil. Levels in water bodies fell rapidly during the weeks after fallout through dilution, physical decay and absorption of radionuclides to catchment soils. Bed sediments are an important long-term sink for radioactivity. Aquatic habitats also tend to be more tolerant of radioactive contamination (http://environmentalchemistry.com, accessed August 2012 + Chernobyl's Legacy: Health, Environmental and Socio-economic Impacts and Recommendations to the Governments of Belarus, the Russian Federation and Ukraine. The Chernobyl Forum: 2003–2005).

The Fukushima-Daiichi nuclear plant disaster after the earthquake and tsunami that struck Japan on 11 March 2011 again illustrated the risk of radiological contamination of water. Most of the radioactive material ended up in the sea and will be strongly diluted and therefore will not cause concern for the drinking water, as illustrated by Yasuhiro Sonoda (MP) drinking a glass of decontaminated water taken from puddles inside the housing of the reactors. However, some scientists fear that deep water fish, fish at the top of the food chain, mollusks and other filtrating sea life are most sensitive to nuclear contamination/ concentration.

For the food industry radiological hazards from (drinking) water may be relevant only if the company is located near a disaster area or when water is imported from these areas. A thorough risk analysis and monitoring program is required under these conditions.

Organoleptic (Taste, Odor, Appearance) Hazards

Taste and odor in drinking water are two of the most widespread causes of customer complaints. Although there are in general no associated health effects, the importance for the food industry is significant while organoleptic problems may influence product quality.

Since taste and odor work together it is often difficult to distinguish the two. Common organoleptic deviations include (http://extoxnet.orst.edu, accessed September 2012):

- **Strong chlorine taste or smell:** Generally this occurs when the water is treated at the water treatment plant by disinfection (see "Chemical Hazards Associated with Drinking Water," above).
- **Metallic taste:** Some water systems have a high mineral concentration causing a salty or soda taste. In the case of iron and manganese, a strong metallic taste is readily detected.
- **Rotten egg odor:** This is usually a result of decaying organic deposits underground. As water flows through these areas, hydrogen sulfide gas is picked up, and when this water reaches the surface or comes out of the tap, the gas is released into the air. Hydrogen sulfide gas produces the rotten egg odor, can be corrosive to plumbing at high concentrations and can tarnish silver rapidly. As little as 0.5 ppm (parts per million) can be tasted in drinking water.
- **Musty or unnatural smells:** These smells are normally a result of, even low amounts of, organic matter or even some pesticides in the water supply.
- **Turpentine taste or odor:** This smell can be a result of methyl tert-butyl ether (MTBE) contamination. MTBE is a gasoline additive, used as an oxygenate to raise the octane number. The odor threshold of MTBE is fairly low, so many people can smell it.
- **Red or brown color:** A red, brown or rusty color is generally indicative of iron or manganese in the water. It may cause stains in sinks, or discolored laundry.
- **Yellow color:** This coloration occurs in regions where the water has passed through marshlands and then moved through peat soils. It is more commonly found in surface water supplies and shallow wells. Although the yellow color may be displeasing, it presents no health hazard, as it is only small particles suspended in the water.
- **Blue or green color:** A green or blue color is generally a result of copper in the water supply, or copper pipes and corrosive water. Copper has a taste threshold of around 5 ppm. Copper can become a problem if the concentration is higher than 30 ppm. Effects at this dose are vomiting, diarrhea and general gastrointestinal distress.
- **Cloudy white or foamy water:** Cloudy water is usually due to turbidity. Turbidity is caused by finely divided particles in the water. When light hits the water, it is scattered, giving a cloudy look to the water. The particles may be of either organic or inorganic nature. Neither one causes any harmful effects to the body, although they can cause abrasions to pipes, or possible staining of sinks.

When water is used for food production, or may be in contact with food, organoleptic hazards are part of the HACCP plan. If necessary, appropriate measures shall be taken to mitigate aberrant characteristics of the water.

Miscellaneous Hazards

To ensure the safety of water in the food industry, apart from environmental and processing care to produce safe water, the role of the staff in the food industry also needs to be considered. The workers need to be aware that they may be asymptomatic carriers of pathogens and therefore need to exercise optimal (hand) hygiene after defecation. For instance, cruise ships are regularly involved in large-scale gastroenteritis outbreaks associated with norovirus often due to insufficient hygiene of kitchen workers. Water has been epidemiologically identified as one of the risk factors (Verhoef et al., 2008). Prerequisite programs on ship sanitation such as by the WHO and CDC should cover this. In 2011 the WHO launched the third edition of the guide to ship sanitation with global reference on health requirements for ship construction and operation. And the Vessel Sanitation Program (VSP) at the CDC assists the cruise ship industry to prevent and control the introduction, transmission and spread of gastrointestinal illnesses on cruise ships.

The design and maintenance of the entire water distribution system (tanks, boilers, piping) shall be as optimal as possible. Dead ends shall be removed and long setting times must be followed by adequate flushing with hot water (or disinfectant). In particular care should be taken to avoid growth of *Legionella*. The WHO has issued an extensive document on *Legionella* and the prevention of legionellosis (WHO, 2007a). This WHO document has separate chapters on potable water and in-building distribution systems and on cooling towers and evaporative condensers.

Drinking water is also a potential vehicle for the deliberate use of microbial pathogens, microbe-derived products or chemicals that cause harm to humans, livestock or agricultural crops. Food companies shall conduct assessments of their vulnerabilities to terrorist attack or sabotage and set up preventive programs or systems to provide a safe and reliable supply of drinking water (for further information see Chapter 35).

RISK ASSESSMENT AND RISK MANAGEMENT

The most effective means of consistently ensuring the safety of a drinking-water supply is through the use of a comprehensive risk assessment and risk management approach that encompasses all steps in the water supply from catchment to consumer. The WHO has proposed such a water safety framework and the implementation of comprehensive water safety plans (WSPs) to consistently ensure drinking-water safety and thereby protect public health (WHO, 2011). Failure to ensure drinking-water safety may expose the community to the risk of outbreaks of intestinal and other infectious diseases. Outbreaks of waterborne disease are particularly to be avoided because of their capacity to result in the simultaneous infection of a large number of persons and potentially a high proportion of the community.

Water safety plans (WHO, 2009, 2011) are suggested to comprise of:

- A system assessment to determine whether the drinking-water supply (from source through treatment to the point of consumption) as a whole can deliver water of a quality that meets the health-based targets.
- Operational monitoring of the control measures in the drinking-water supply that are of particular importance in securing drinking-water safety.
- Management plans documenting the system assessment and monitoring plans and describing actions to be taken in normal operation and incident conditions, including upgrade and improvement, documentation and communication.

HACCP CASE STUDIES

Determination of Water Safety

As with any hazard, radiological, chemical and (micro)biological hazards in drinking water should be assessed using principles of HACCP (see Chapter 31).

The first step is to establish the intended use of the water. Questions that should be answered: does the water come in contact with the product and, if so, at what stages? Are consumers exposed to the water and, if so, in what form (drinking water, adherent water, ice, steam)? A simple classification for the "fit for purpose" is (adapted from ILSI, 2008):

	Chemically potable	**Chemically non-potable**
(Micro)biologically potable	Class 1	Class 3
(Micro)biologically non-potable	Class 2	Class 4

For each application the right category shall be chosen. For products with little or no further processing for safety, class 1 water shall be used as ingredient. If only class 2 water is available, an appropriate pretreatment shall be applied, or the processing itself contains appropriate steps to eliminate microbiological risks. Classes 3 and 4 will normally not be suitable for water as ingredient, but may be used as processing water that will not be in direct contact with the product itself.

Treatment of the water may change the class. Heat treatment may change class 2 water to class 1 water. Ultrafiltration and additional chemical treatment may even change class 4 water to class 1 water.

To establish whether the water is safe for the intended use, a decision tree was published by ILSI (adapted from ILSI, 2008 – see Figure 14.2).

Water Treatment Technologies for Safe Water Production

An increasing number of technologies are developed to process water for safety. Typical industrial wastewater treatment consists of a combination of physical, biological and chemical processes to remove solids and organic matter, and, if necessary, pathogens, metals and

1. Is the water potentially contaminated with either radiological, chemical or (micro)biological hazards at concentrations which are significant for human health? The fit for purpose classification.	→ NO (Class 1 water)	"SAFE" WATER
→ YES (Class 2, 3 or 4 water)		
2. Will the water be consumed without further treatment or come into contact with products that will be consumed without further treatment?	→ NO (Class 2, 3 or 4 water)	
→ YES (Class 2, 3 or 4 water)		
3. Is the water treated to eliminate potential hazards before consumption or contact with the product that will be consumed?	→ YES (Class 1 water)	
→ NO (Class 2, 3 or 4 water)		
4. Will subsequent treatment of the product for consumption, either in the factory or at home by consumers, eliminate the hazard?	→ YES (Class 1 water)	
→ NO (Class 2, 3 or 4 water)		
UNSAFE WATER		

Question 1 defines the fit for purpose class and refers to knowledge of the potential hazards and criteria set in water guidelines and (inter)national regulations. If no criteria are available a full risk analysis is necessary to establish potential hazards and judgment of chance and impact/severity.

Question 2 refers to the intended use of the water and whether a potential hazard may be in contact with the consumer either directly or indirectly. Therefore this question involves an evaluation of exposure and risk. Is exposure of the hazard to the product or consumer likely (the chance)? If so, how much and how long and what will be the potential consequence (severity)? Interestingly water not fit for use (Classes 2, 3 and 4) can be considered safe when not used for indirect or direct consumption.

Question 3 refers to existing steps in the process that will (un)intentionally act as mitigation step(s) to potential hazards and risks? Steps can involve, for example, heating, filtration, chemical treatment, UV treatment, ozone treatment.

Question 4 addresses the additional mitigation steps, either at the consumers' home or at the producers' factory. In the latter case this usually involves process steps that are not intended to reduce the risks but as side effect do so.

FIGURE 14.2 Water safety decision tree. *Adapted from ILSI, 2008.*

nutrients from wastewater. Table 14.5 summarizes some water treatment alternatives for given challenges.

The goal in designing a processing system to obtain safe water is to develop an integrated cost-effective scheme that is capable of reliably meeting water quality and safety objectives. The degree of treatment required in individual water treatment facilities varies according to the specific (re)use application and associated water quality requirements

Filtration

Filtration involves porous material (filter) to separate (suspended) solids from the water. Most applied systems are granular filtration and require the use of filter cartridges (EHEDG, 2004). Granular filtration uses a filter bed consisting of one or more layers of sand and anthracite. Factors that influence effectiveness are the size, form and nature of the particles, the strength, the porosity, the filtration rate and the bed height. Filter cartridges are usually placed in a pressure vessel. Effectiveness is determined by the right pore size and fouling. Pressure drop over the filter indicates saturation with solids. In time, replacement of the

TABLE 14.5 Water Treatment Alternatives

Challenge	Treatment Option	Advantage	Concern
Microbiological hazards (bacteria, viruses, protozoa)	Chlorination	Easy to handle, effective to most bacteria	Most protozoa are resistant and some viruses are not eliminated. Chemical by-products. Elevated turbidity reduces effectiveness
	Ozone	Very effective against most bacteria and viruses. Viruses generally more resistant than bacteria, effective to *Cryptosporidium*	Complex technology, bromate formation, some viruses are not eliminated
	UV	Easy to handle, effective to *Cryptosporidium*	TSS, turbidity and color may render it inefficient
	Membranes (ultra-filtration, nano-filtration)	No by-products, no smell, no taste	Costs, fouling
	Heating (sterilization)	Very effective, no smell, no taste	Costs (energy)
Suspended solids	Granular media, filters	Low cost, readily available, simple and effective. Large volume, low pressure	Require regular maintenance
	Screen filters	Widely available in specialized materials	Relatively coarse separation. Not suited to heavy loads, clogging
	Tubular screen filters	Robust and offer repeated use	Selection of screen material must match process conditions
	Membrane (micro-filtration, ultra-filtration)	No by-products, no smell, no taste	Higher operating costs, fouling
Organic matter	Advanced biological treatment (e.g. bio-filtration).	Low cost	Only for biodegradable substances
	Adsorption (PAC, GAC)	Very effective for non-polar substances	Costly, residuals (spent carbon)
	AOP (advanced oxidation processes)	No residuals produced	Formation of unknown (biodegradable) compounds
Inorganic compounds: heavy metals	Flocculation/precipitation		Chemicals used increase salinity
Inorganic compounds: salinity	Ion-exchange	Effective	Cost, salt increase
	Reverse osmosis	Effective	Residuals to be disposed may need to be treated to reduce corrosivity

(Adapted from ILSI, 2008 and WHO, 2011).

TABLE 14.6 Types of Membrane Filtration (EHEDG 2004)

Type of Membrane Technology	Pressure Applied (bar)	Porosity (cut-off value)	Retention
Micro-filtration (MF)	1–2	20–1000 nm	Solid particles, bacteria, yeasts, protozoa, colloids
Ultra-filtration (UF)	1–5	20–200 nm	Above + polysaccharides, proteins
Nano-filtration (NF)	5	1–10 nm	Above + sugars, amino acids, hardness (calcium salts), multiple charged ions (e.g. sulfates, phosphates), viruses
Reversed osmosis	15–50	<2 nm	Above + salts

filters or back-washing (water flow in the opposite direction) is necessary. Drawbacks are long running times, insufficient frequency of back-washing or filter replacement and installation of non-compliant cartridges in pressure vessels. For filtration of small solids, soluble materials and microorganisms, membrane filtration is necessary.

Membrane Filtration

Membrane filtration is a pressure-driven technology. Depending on the pore sizes particles are retained (Table 14.6).

The choice of a filtration system is complex and requires specific knowledge of available materials (organic polymers, ceramic and stainless steel), membrane geometry (spiral, tubular, capillary, hollow fiber) and the application involved (temperature, pH, particles in the fluid, cleaning methods/chemicals). Like any filtration technology, membrane filtration is susceptible for fouling and systematic cleaning (or replacement) is required. Leakage of membranes due to chemical and mechanical damage induces risk of post-filtration contamination. Reversed osmosis water may have corrosive properties due to removal of minerals. Remineralization may be required in certain applications.

Chlorination

Chlorination is one of the most used disinfection systems for potable and utility water.

Sodium hypochlorite (N_aOCl) is the predominant chemical used for chlorination. The main reasons are availability, simplicity of the application, cost effectiveness and, if properly used, reliability. Chlorine is effective at inactivating bacteria and viruses, and under certain circumstances parasites like *Giardia*. However, chlorine has little impact on the parasite *Cryptosporidium* at typical water treatment concentrations (up to 5 mg/l). Chlorine's general disinfection capability with respect to microorganisms can be illustrated in the following way from most effective to least effective: bacteria > viruses > *Giardia* cysts > *Cryptosporidium* oocysts (USPHC, 2006).

Even higher numbers of bacteria are generally killed in minutes. This is particularly true for Gram-negative bacteria like *E. coli* and *Salmonella*. Gram-positive bacteria, especially spore-forming species like *Bacillus* and *Clostridium*, tend to be less sensitive but can still be eliminated at appropriate concentrations of chlorine and contact times.

TABLE 14.7 The Various Forms of Chlorine (CAWST, 2012)

Product	Strength	Remarks
High test hypochlorite (HTH) (calcium hypochlorite)	65% – 70%	Usually in granular form. Stable (approximately 2% active chlorine loss per year)
Chlorinated lime, aka bleaching powder	30%	Usually in powder form. Not stable.
Household bleach (sodium hypochlorite)	2.5–10%	Liquid form. Not stable; only use if manufactured recently (<3 months) and stored away from heat and light
Sodium dichloro-isocyanurate (NaDCC), used in products such as "Aquatabs"	50–60% as granules. 5 mg to >5 g active chlorine per tablet	Usually in tablet form, also available in granular form. Tablets pre-dosed for water treatment. Very stable (shelf-life approximately 5 years)

Chlorine has been shown to be a highly effective viricide. Most viruses are killed very effectively after exposure to chlorine within minutes. The most resistant virus was a poliovirus, requiring more than 60 minutes for 4-log removal.

Chlorine has been shown to have limited success inactivating protozoa. An important indicator, *Giardia lamblia*, requires prolonged contact times (30–60 minutes) at chlorine residual concentration (2–3 mg/l) to achieve 99.9% (3-log) inactivation.

The parasite *Cryptosporidium*, however, is very resistant and requires high chlorine concentrations and extreme long exposure times to eliminate cells and oocysts. One *Cryptosporidium* study reported that 80 mg/l of free chlorine required 90 minutes to achieve only a 1-log (90%) inactivation of oocysts! These exposure times and concentrations are generally not feasible and therefore chlorination is not an option to control protozoa.

Chlorine kills bacteria and viruses by interfering with chemical bonds and in particular inactivation of enzymes.

Chlorination for the control of microbiological contamination of drinking and processing water involves the following parameters (WHO, 2011):

- Residual concentration of free chlorine minimal 0.5 mg/l, typical 2–3 mg/l and maximum 5 mg/l.
- Contact time at least 30 min at pH <8.0 (optimum pH 5.5–pH 7.5).
- The contact time is valid at 18–20°C and above. For every 10°C drop in temperature the efficiency of disinfection reduces by 50–60% (at close to 0°C disinfection efficiency is very poor).

Chlorine is available in several forms (see Table 14.7).

Despite the benefits, some disadvantages must be addressed (EHEDG, 2005):

- Reduced effectiveness at pH >8.0 and lower temperatures.
- Reacts with nitrogenous compounds forming chloramines (unpleasant odors and health concerns). Also reactive with several organic materials forming compounds with possible health impacts.

- Easily quenched by organic matter and turbidity in the water.
- Highly corrosive.

An alternative to chlorine is the use of chlorine dioxide, a highly reactive compound that cannot be stored in its active form. Therefore it is generated on site, close to the point of use. Compared to chlorine it mitigates most of the disadvantages; however, the costs and the necessity to generate it at the point of use makes it a less interesting option for smaller companies.

Filtration and Chlorination

While chlorination alone is not (always) effective against protozoa, a dual approach may be applied. Several (household) water treatment systems incorporate both a physical filtration step for particle removal and a chlorination step for disinfection. Alternatively particles, including protozoa, may be removed by flocculation prior to chlorination, using coagulants. Aluminum coagulants include aluminum sulfate, aluminum chloride and sodium aluminate. Iron coagulants include ferric sulfate, ferrous sulfate, ferric chloride and ferric chloride sulfate. Other chemicals used as coagulants include hydrated lime and magnesium carbonate. Overall *Giardia* and *Cryptosporidium* removals after coagulation and filtration may be approximately 5-log (for further reading: www.iwawaterwiki.org, assessed January 2013).

Water Reuse in Food Processing

Fresh water resources are globally subjected to increasing pressure in the form of consumptive water use and pollution. On national and international levels awareness is growing that water resources should be protected both qualitatively and quantitatively. The food industry is in general regarded as a major water consumer resulting in relatively high water footprints. Apart from increased efficiency, reuse of water is a way to reduce fresh water exploitation. When applying reused water it is necessary to identify whether the reused water will be in contact with the product(s) or not. Typical applications of reused water are indirect cooling or the generation of steam that will not be in contact with the product. Direct contact applications may include washing and/or transport of raw products (like fruit or vegetables that will be processed) or cleaning of equipment.

Any food industry considering the application of reused water should ask the following questions (ILSI, 2008)[1] :

- What is the proposed reuse? Will the water come into contact with food or will it be used as a noncontact processing aid (e.g. coolant)?
- What are the regulatory, consumer safety and technical requirements for the water in the proposed reuse application?
- What is the starting quality of the intended reuse criteria and what treatments or controls can be applied so that it meets the criteria defined in the previous question?

[1]ILSI is working on a Water Recovery Guideline which is expected to be released in 2013 (www.ilsi.org).

- What monitoring procedures need to be put in place to adequately monitor the performance of the treatments and/or controls?
- What procedures need to be put in place to overcome existing technical difficulties, such as chemical or biological fouling (e.g. biofilms)?
- What measures need to be taken if a deviation from the required quality is detected?
- What changes to availability or cost are likely in the future and may alter the current situation (e.g. proposals in Brazil to charge industry for water abstracted from either groundwater or rivers)?
- What changes to water supply quality are likely in the future (e.g. salination of groundwater)?
- What treatments will be required to ensure that the water meets the necessary standards?
- What modifications could be incorporated into either existing or new equipment (e.g. appropriate filters on bottle washers) or existing or new process lines to maximize the opportunities for water reuse?
- What regulatory conditions encourage (or discourage) optimized water use?

Example: Recycled Hot Water as a Decontamination Technique for Meat Carcasses

The European Food Safety Authority has delivered a scientific opinion on safety and efficacy of using recycled hot water as a decontamination technique for meat carcasses (EFSA, 2012). At the moment (2013) only the use of potable water is allowed in the EU for carcass decontamination purposes. However, recycled water (i.e. reusing water after reheating) is used for carcass decontamination in some countries (e.g. Canada, Denmark). Environmental care and energy-preserving motives are driving forces for recycling. The EFSA study has considered potential microbiological and abiotic risks for carcasses associated with recycled hot water decontamination and related control options.

From the study it is concluded that the decontamination efficacy of recycled hot water does not differ significantly from that of hot potable water.

By ensuring proper heating regimes of recycled water, vegetative bacterial cells and protozoan parasites are controlled. Microbial toxins are not significantly inactivated in the recycling process, but production of these toxins during the first round of carcass decontamination and prior to heating is not relevant.

According to the EFSA study, only microbiological risks associated with heat-resistant bacterial spores (*C. botulinum*, *C. perfringens*, *C. difficile* and *B. cereus*) are relevant for recycled hot water. These risks can be controlled by ensuring that recycled hot water is verifiably subjected to appropriate reheating and frequency of renewal regimes. These regimes shall ensure that the microbiological risk in recycled water is not higher than in hot potable water. For abiotic risks, the only concern with recycled hot water derives from the potential presence and accumulation of residues of veterinary drugs and other chemical contaminants in the water for decontamination of poultry carcasses.

As with any process recycling of water for decontamination of carcasses shall be subjected to HACCP. Important criteria for efficacy and control of possible risks include minimal heating temperature, time regime and frequency of renewal of recycled water. These criteria shall ensure compliance with existing microbiological criteria for potable water and prevent accumulation of heat-resistant spores. Recycling procedures shall be microbiologically validated, continuously monitored by instrumental measurements, verified periodically by microbiological testing of water and documented. Compliance with the chemical criteria for potable water needs to be verified for recycled hot water by periodic chemical analysis of the water and documented. The absence of residues of veterinary medicinal products in recycled hot water used for decontamination of poultry carcasses has to be verified by periodical testing and be documented.

Finally, the application of recycled hot water applied on carcasses (temperatures, application techniques and related parameters) shall be subject to risk analysis in the same way as with hot potable water decontamination.

Bottled Water Safety

Aside from adhering to the various industry regulations, the best way to minimize the risk of contaminated bottled water is to have a good HACCP system in place. The seven principles of a HACCP system (see Chapter 31) provide the basis for safe production and will help to satisfy business owners and their customers that products are safe in an efficient, reliable and cost-effective way. It is achieved by focusing on hazard prevention throughout the product life cycle rather than relying on end-product testing.

An example from a multinational company that produces bottled water shows a typical production process and accompanying quality assurance and control measures (Figure 14.3).

1. **Source receiving and inspection**
   Water is carefully collected from the source, which may either be a well or municipal supply. Common method of receiving water is through stainless steel pipeline. Water from the source shall be tested prior to internal processing on microbiological and chemical aspects.
2. **Activated carbon filtration (municipal water only)**
   Activated carbon may be necessary to remove substances like chlorine and trihalomethanes. This filtration process should be monitored and tested regularly.
3. **Pretreatment**
   Water softener may be used to reduce water hardness.
4. **Demineralization process**
   Demineralization is the use of cation – and anion resin beds to remove minerals. Technologies include:
   a. Reverse osmosis: Use of high-pressure pump and special membranes, called semi-permeable membranes, to reverse the natural phenomenon of osmosis.
   b. Distillation: A process that boils the water and collects the condensate for bottling.
5. **Water storage and monitoring**
   Water is received into storage tanks. Storage environment and water carefully monitored daily.
6. **Micro-filtration**
   Using micro-filters, usually pharmaceutical grade, particles as small as 0.2 micron in diameter are removed. The pore size guarantees removal of microbiological contaminants.

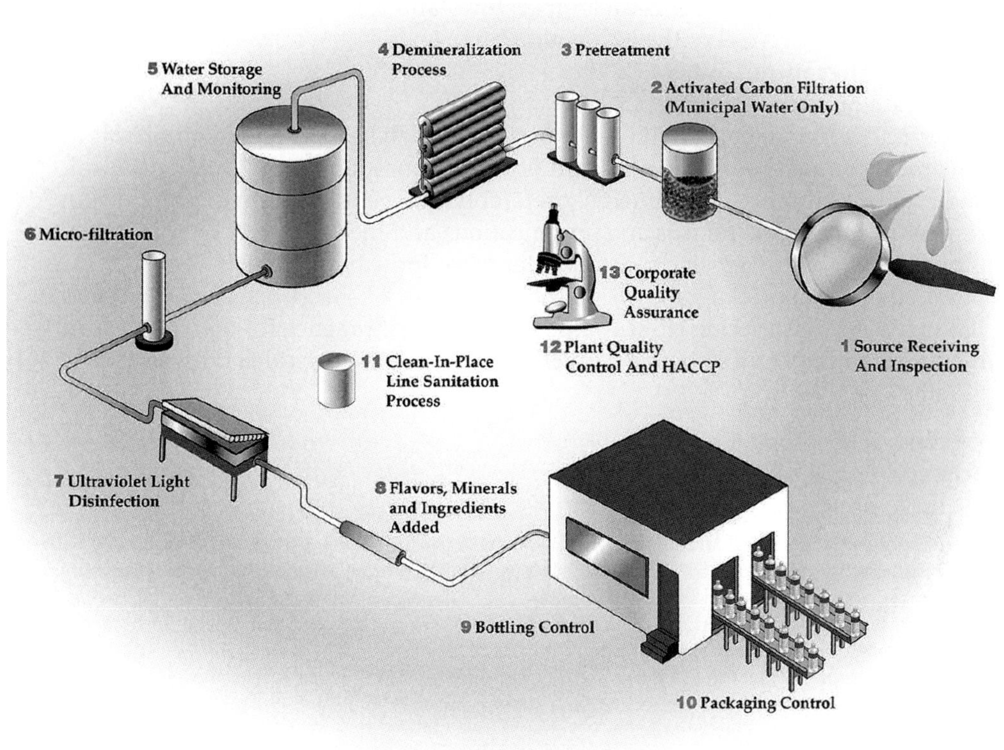

FIGURE 14.3 Bottled water production (http://www.nestle-waters.com/brands/water-quality/Pages/purified-water.aspx: assessed August 2012).

7. **Ultraviolet light disinfection**
   Application of ultraviolet light provides added assurance of product disinfection and safety. As with ultra-filtration this process should be continually monitored by instrumentation.
8. **Flavors, minerals and ingredients added**
9. **Bottling control**
   Bottling should be conducted under highly controlled conditions using state-of-the-art equipment. Each bottle shall be marked with a code that identifies the plant (location), bottling line and time produced. Filling room and environment are subject to high sanitary conditions.
10. **Packaging control**
    Packaging materials not meeting (internal) standards should be rejected before using them. Bottles, caps and labels should be controlled and monitored by lot.
11. **Clean-in-place line sanitation process**
    Line sanitation practices include preferably internal pipe and equipment cleaning methods (cleaning in place – CIP). Such processes should circulate detergent and

sanitizing solutions at the precise temperatures and time to affect total control and maximum effectiveness of the line sanitation process.

12. **Plant quality control and HACCP**
13. **Corporate quality assurance**
    Water, packaging materials and plant processes shall be carefully monitored to ensure they meet company specifications and (inter)national standards. Quality control and quality assurance departments, preferably independent from production, are responsible for the standards and specifications and monitoring of the plant quality programs. A comprehensive set of standards for industries active in bottled water production to ensure safety and quality has been published by the US International Bottled Water Association (IBWA, 2012). This code of practice for bottled water offers monitoring matrices for chemical, microbiological, radiological and organoleptic parameters.

References

Boccia, D., Guthmann, J.P., Klovstad, H., Hamid, N., Tatay, M., Ciglenecki, I., et al., 2006. High mortality associated with an outbreak of hepatitis E among displaced persons in Darfur, Sudan. Clin. Infect. Dis. 42 (12), 1679–1684.

Centre for Affordable Water and Sanitation Technology, 2012. Chlorine Disinfection. Available online: <www.cawst.org>.

Cidu, R., Frau, F., Tore, P., 2011. Drinking water quality: comparing inorganic components in bottled water and Italian tap water. J. Food Comp. Anal. 24, 184–193.

EFSA Panel on Biological Hazards (BIOHAZ), 2012. Scientific Opinion on the safety and efficacy of using recycled hot water as a decontamination technique for meat carcasses. EFSA J. 8 (9), 1827, pp. 69. Available online: <www.efsa.europa.eu/efsajournal.htm>.

EHEDG, 2005. Safe and hygienic water treatment in food factories. Trends Food Sci. Technol. 16, 568–573.

Food and Drug Administration FDA, 2012. Code of Federal Regulations, title 21: Food and Drugs; Part 129 – processing and bottling of bottled drinking water.

Gleick, P.H., 1996. Water resources. In: Schneider, S.H. (Ed.), Encyclopedia of Climate and Weather, vol. 2. Oxford University Press, New York, pp. 817–823.

International Bottled Water Association, 2012. Bottled Water Code of Practice. Available at <www.bottledwater.org>.

International Life Sciences Institute ILSI, 2008. Considering Water Quality for Use in the Food Industry. Aavailable at <http://www.ilsi.org/Europe/Pages/HomePage.aspx>.

International Life Sciences Institute ILSI, 2012. Guidelines for Water Reuse in the Bottling Industry.

Kumar, A., Chang, B., Xagoraraki, I., 2010. Human health risk assessment of pharmaceuticals in water: issues and challenges ahead. Int. J. Environ. Res. Public Health 7, 3929–3954.

Lye, D.J., 2009. Rooftop runoff as a source of contamination: a review. Sci. Total Environ. 407 (21), 5429–5434.

Mandal, S., Mandal, M.D., Pal, N.K., 2011. Cholera: a great global concern. Asian Pac. J. Trop. Med. 4 (7), 573–580.

Motarjemi, Y., Kaferstein, F., Moy, G., Quevedo, F., 1993. Contaminated weaning food: a major risk factor for diarrhoea and associated malnutrition. Bull World Health Org. 71 (1), 79–92. Available at: <http://www.gastrojournal.org/article/S0016-5085(12)01014-1/fulltext>.

Motarjemi, Y., Steffen, R., Binder, H.J., 2012. Preventive Strategy Against Infectious Diarrhea—A Holistic Approach. Gastroenterology 143 (3), 516–519.

Paranthaman, K., Harrison, H., 2010. Drinking-water incidents due to chemical contamination in England and Wales, 2006–2008. J. Water Health 8 (4), 735–740.

Schets, F.M., Italiaander, R., van den Berg, H.H., de Roda Husman, A.M., 2010. Rainwater harvesting: quality assessment and utilization in The Netherlands. J. Water Health 8 (2), 224–235.

Schijven, J.F., de Roda Husman, A.M., 2005. Effect of climate changes on waterborne disease in The Netherlands. Water Sci. Technol. 51 (5), 79–87.

Schijven, J.F., Hassanizadeh, S.M., de Roda Husman, A.M., 2010. Vulnerability of unconfined aquifers to virus contamination. Water Res. 44 (4), 1170–1181.

Serdarevic, F., Jones, R.C., Weaver, K.N., Black, S.R., Ritger, K.A., Guichard, F., et al., 2012. Multi-pathogen waterborne disease outbreak associated with a dinner cruise on Lake Michigan. Epidemiol. Infect. 140 (4), 621–625.
Smith, A.H., Lingas, E.O., Rhaman, M., 2000. Contamination of drinking-water by arsenic in Bangladesh: a public health emergency. Bull World Health Org. 78 (9), 1093–1103.
Symons, J.M., Bradley Jr., L.C., Cleveland, T.C., 2000. The Drinking Water Dictionary. American Water Works Association.
Teshale, E.H., Howard, C.M., Grytdal, S.P., Handzel, T.R., Barry, V., Kamili, S., 2010. Hepatitis E epidemic, Uganda. Emerg. Infect. Dis. 1. Available from: <http://wwwnc.cdc.gov/eid/article/16/1/09-0764.htm>.
US Army Public Health Command – USPHC, 2006. Chlorine Disinfection in the Use of Individual Water Purification Devices. Technical Information Paper #31-002-0306. Available at: <http://phc.amedd.army.mil> (accessed November 2012).
Verhoef, L., Boxman, I.L., Duizer, E., Rutjes, S.A., Vennema, H., Friesema, I.H., et al., 2008. Multiple exposures during a norovirus outbreak on a river-cruise sailing through Europe, 2006. Euro Surveill. 13 (24), pii: 18899.
WHO, 2006. Protecting Groundwater for Health. ISBN: 9241546689, pp. 697. Geneva. Aavailable at: <http://www.who.int>.
WHO, 2007a. Legionella and the Prevention of Legionellosis. Bartram, J., Chartier, Y., Lee, J.V., Pond, K., Surman-Lee, D. (Eds.). ISBN: 92 4 156297 8. Geneva. Available at: <http://www.who.int>.
WHO, 2007b. Safe Preparation, Storage and Handling of Powdered Infant Formula; Guidelines. World Health Organization in collaboration with Food and Agriculture Organization of the United Nations. ISBN: 978 92 4 159541 4. Geneva. Available at: <http://www.who.int>.
WHO, 2009. Water Safety Plan Manual. ISBN: 978 92 4 156263 8, pp. 102. Geneva. Available at: <http://www.who.int>.
WHO, 2011. Guidelines for Drinking-Water Quality. ISBN: 978 92 4 154815 1. Geneva. Available at: <http://www.who.int>.
WHO, 2012. A Toolkit for Monitoring and Evaluating Household Water Treatment and Safe Storage Programmes. ISBN 978 92 4 150462 1, pp. 62. Geneva. Available at: <http://www.who.int>.
WHO and Unicef (2012) Rapid assessment of drinking-water quality: a handbook for implementation. 1. Drinking water – standards. 2. Water supply – standards. 3. Water quality. 4. Environmental monitoring. 5. Handbooks. ISBN 978 92 4 150468 3.

Further Reading

An extensive amount of information is available on the safety of water. Many documents are accessible (or can be downloaded) from websites. For further reading we recommend the following websites and documents (without the intention to be complete, many other sources may be valuable):

The European Hygienic Engineering & Design Group (EHEDG) issues guidelines which are regularly updated and complemented by new documents in various language versions (www.ehedg.org).

The International Life Sciences Institute (ILSI) disseminates science by publishing articles on original research, literature reviews and gap analyses, and meeting proceedings in peer-reviewed journals. ILSI Europe also publishes books, monographs, white papers and other reports through ILSI Press (www.ilsi.org).

The World Health Organization and Unicef have published many relevant documents on water safety. Most of them are freely available at www.who.int. In particular we recommend:

WHO (2011) Guidelines for Drinking-Water Quality. World Health Organisation, 4e edition, Geneva.

WHO (2007) Safe preparation, storage and handling of powdered infant formula; Guidelines.

WHO and Unicef (2012) A toolkit for monitoring and evaluating household water treatment and safe storage programmes. ISBN 978 92 4 150462 1.

The International Bottled Water Association produced a Bottled Water Code of Practice in 2012.

The European Food Safety Authority (EFSA) regularly publishes documents (opinions) on food and water safety. These documents can be retrieved from http://www.efsa.europa.eu. European legislation is available at http://europa.eu/eu-law/index_en.htm.

CHAPTER

15

Pet Food

Pablo A. Carrión and Larry J. Thompson

Nestlé Purina PetCare, Product Technology Center, St. Louis, MO, USA

OUTLINE

INTRODUCTION

Dog and cat pet ownership is popular throughout the world and pets are increasingly treated as members of the family. The pet food industry started in England in 1860, when the first commercial dog biscuits were marketed. Today sales of pet food in the USA alone exceed 18 billon US dollars a year (APPA, 2012). There are three main types of commercial pet food products: dry and semi-moist shelf-stable extruded food; thermally processed low acid canned products; and a variety of product forms sold as treats. With the exception of some treats, most products are formulated to be nutritionally complete and balanced. Thus,

Food Safety Management.
DOI: http://dx.doi.org/10.1016/B978-0-12-381504-0.00015-9

the modern pet food industry provides an essential service to pet owners by making nutritious and palatable pet food convenient to acquire and feed.

The pet food industry utilizes the same ingredient streams as those of the human food supply making use of many of the by-products and co-products. Therefore, the food safety hazards potentially present in pet food ingredients are the same as the ones facing the food industry in general. There is, however, a difference in the severity of health effects of these hazards to cats, dogs and humans. Pets tend to be very resistant to the clinical effects of infection by human food pathogens. On the other hand, they may be very sensitive to certain natural toxins or food components (e.g. alkaloids, caffeine, etc.) as well as veterinary drugs and feed additives.

The most significant historical pet food safety incidents in terms of frequency of occurrence and severity are related to aflatoxins, veterinary drug contamination, *Salmonella* recontamination and, more recently, adulterated ingredients. Together, these hazards account for the vast majority of safety incidents where pets were severely affected. With the exception of the food pathogen *Salmonella*, most other food safety hazards are ingredients or formulations based and have no effective control measures in the manufacturing process itself (Table 15.1). Potential HACCP control strategies to address these food safety threats will be discussed in this chapter.

BIOLOGICAL HAZARDS

Salmonella Contamination of Dry Pet Foods and Treats

Salmonella is a Gram-negative, non-spore-forming, rod-shaped bacterium belonging to the Family Enterobacteriaceae. This genus includes about 2400 different serovars. Non-typhoid strains of salmonellae are a common cause of gastroenteritis and septicemia in humans and pets. Domestic and wild animals are often intestinal carriers of this pathogen. *Salmonella* is widespread in nature and has been found to survive for weeks in water and for several years in soil. In food ingredients, *Salmonella* can contaminate eggs, raw meats, poultry, fish and their by-products (Wareing and Fernandes, 2007). *Salmonella* is one of the leading causes of human gastroenteritis worldwide. In the USA there are an estimated 1.4 million cases a year and some 400 deaths (Voetsch et al., 2004). Salmonellosis remains the second most often reported zoonotic disease of humans in the European Union with 99,020 cases reported in 2010 (EFSA, 2012). Vulnerable populations include people with compromised immune systems, infants and the elderly. The enteric infection has an incubation time of 8–72 hours with symptoms that include nausea, vomiting, abdominal cramps, diarrhea, fever and headache. The symptoms can last from 2 to 5 days (Wareing and Fernandes, 2007).

Salmonella-contaminated feed may cause salmonellosis in animals. Generally, young animals are the most susceptible to an enteric-type infection but in more severe cases the infection may become systemic. In adult animals the infection is more likely to be asymptomatic. Prevalence of *Salmonella* carriage rates have been reported as high as 36% in healthy dogs, and 18% in healthy cats (Leonard et al., 2010; Sanchez et al., 2002). Dogs infected with *Salmonella* often carry multiple strains at a time. Most infections are asymptomatic or mild and are commonly not identified. Prolonged and sporadic fecal shedding of *Salmonella* is

TABLE 15.1 Most Common Hazards Associated with Pet Food Safety Incidents and their Control

Hazard	Type	Root Cause	Control
Salmonella	Biological	Post-CCP cross-contamination from contaminated factory surface, environment or ingredient. Potential sources of contamination include: birds (feces, feathers) entering via air currents or water leaks. Presence of raw materials past CCP due to poor dust tightness, zoning or traffic patterns. Pests	Good manufacturing practices (GMP): e.g. ingredient quality measures, hygiene practices, hygienic design and process validation and verification procedures (GMA, 2009)
Ionophore toxicity	Chemical	Cross-contamination of feed ingredient with antibiotics via shared production lines with medicated feed or labeling errors of medicated feeds or vitamin premixes	Procurement of ingredients from suppliers that do not manufacture medicated products on the same production line
Adulteration (e.g. melamine)	Chemical	Fraud	"Trust but verify" ingredient supplier quality assurance and traceability programs
Nutrient toxicity or deficiency	Chemical	Misformulation or mixing error at batching	Careful accounting of the ingredient usage rate during batching. Vendor assurance measures, including validated mixing processes. Premix monitoring
Mycotoxin toxicity (e.g. aflatoxins and DON)	Chemical	Contaminated cereals (contamination may occur in the field and/or storage at supplier)	A cereal sampling and testing operational prerequisite program is required. Depending on prevalence of aflatoxin and DON, potentially all cereal deliveries to a factory must be sampled and tested before use. Good silo storage practices are required if grain is to be stored at the factory for any length of time
Metal and other hard bodies	Physical	Metal contamination from ingredients or equipment	GMP-based foreign material control programs including inspection, line magnets and metal detection of packaged product (verification)

a well-documented phenomenon (Morse et al., 1976). When symptomatic infections occur, clinical signs in young animals can include fever, anorexia, vomiting, intermittent diarrhea and bloody stools (Carter and Quinn, 2000). Infected dogs in the household pose a documented elevated risk of infection to their owners (Morse et al., 1976). Salmonellosis in cats is relatively rare, with subclinical infections and carriage rates among healthy cats reported to be very low. Nevertheless, cases of symptomatic infection, chronic carriage and transmission to humans have been documented (Van Immerseel et al., 2004).

TABLE 15.2 Recent North American Human Outbreaks of Salmonellosis Linked to Pet Food (FDA, 2012)

Country	Pathogen	Product	Date
Canada	*Salmonella* Infantis	Pig-ear dog treats	1999
USA	*Salmonella* Newport	Beefsteak-patty dog treats	2002
Canada/USA	*Salmonella* Thompson	Pet treats	2005
USA	*Salmonella* Schwarzengrund	Dry pet food	2006–2007

Pet food products contaminated with *Salmonella* pose a risk of infection to pet owners (Morse et al., 1976). Infection can occur via contaminated fomites or from ingestion of contaminated pet food (e.g. by children) (Behravesh et al., 2010; Morse et al., 1976). Numerous incidents in the USA have occurred where pet foods were found to be contaminated with *Salmonella* resulting in at least 13 recalls since 2006 (FDA, 2010a). Several human *Salmonella* infections and outbreaks have been linked to commercial pet food products (Table 15.2). One such outbreak of salmonellosis in the USA during 2007 was thoroughly investigated by the Centers for Disease Control and Prevention (CDC) and illustrates clearly the zoonotic potential of contaminated pet foods (CDC, 2008). Young children were found to be at a greater risk of infection than other family members. The specific family practices involved in the transmission of *Salmonella* to consumers included feeding the pet in the kitchen (Behravesh et al., 2010).

Dry pet foods are considered high fat, low moisture and low water activity (a_w) products. When formulated without humectants or preservatives, these products have an a_w of 0.65 or lower, corresponding to a moisture content of 12% or less. These are typically coated with fat (tallow, poultry fat) for enhanced palatability (Crane et al., 2000). At these low a_w levels, dry pet foods are shelf-stable because bacteria, molds and mites are unable to grow and spoil the food (FDA, 2012). Despite the inability of *Salmonella* to typically grow on low moisture foods, some cells have been shown to survive on pet foods and in pet food manufacturing environments for an extended period of time (GMA, 2009). The ability of some cells to survive on manufacturing surfaces can lead to the persistent contamination of processing areas, including air handling systems, floors and production equipment. The capacity to survive in a desiccated state is further enhanced by the presence of fat on product contact surfaces. Environmental moisture originating from cleaning and other sources can allow the multiplication of *Salmonella* in the factory (GMA, 2010a). Some factors that contribute to the possibility of cross-contamination include the existence of environmental conditions within the factory that generate microenvironments where *Salmonella* can grow in the proximity of the product stream. These include: condensation of moisture on production surfaces, poor hygienic practices (e.g. wet cleaning), poor equipment design, inadequate maintenance of equipment and inadequate zoning (e.g. incomplete segregation of pre- and post-extrusion environments and materials) (GMA, 2009). Important contributing factors for ineffective zoning include complex traffic patterns, poor dust control, uncontrolled ingress of external air and water, and the presence of pests and wild birds in and around the factory (GMA, 2010a). Contaminated ingredients used as post-extrusion flavor coatings can also be a source of *Salmonella* contamination.

Many typical pet food ingredients are potentially contaminated with *Salmonella*; these include meat and poultry by-product meals, raw meats and even cereal grains. HACCP studies of typical pet food manufacturing processes identify extrusion cooking as the only effective critical control point (CCP) for the elimination of *Salmonella*. Given the temperature profiles of subsequent unit operations, it is unlikely that any of the post-extrusion processing unit operations (e.g. kibble drying, flavor coating, cooling, intermediate storage and packaging) are consistently effective in reducing or eliminating *Salmonella*. This indicates that the presence of *Salmonella* on pet foods is the result of a cross-contamination event caused by direct inoculation of the kibble by a contaminated material (Behravesh et al., 2010). To minimize the potential for post-extrusion product cross-contamination, the manufacturer must implement a comprehensive food safety system encompassing good manufacturing practices (GMPs) and HACCP principles. The Grocery Manufacturers Association (GMA) describes in detail seven GMPs and HACCP elements that must be emphasized for the control of *Salmonella* in low moisture foods when additional processing occurs after a heat inactivation control process, as is the case in pet food factories. The seven elements include ingredient quality measures, hygiene practices, hygienic design and process validation and verification procedures (GMA, 2009).

Other Potential Significant Biological Hazards

There have been near incidents and some speculation about the possible contamination risk of commercial pet foods with pathogens other than *Salmonella*. In September of 2007, the FDA issued a recall notice for a frozen chicken blend raw food product contaminated with *Listeria*. In 2001 and 2006 ProMED-mail posts (http://www.promedmail.org: accessed 25 April 2012) discussed the possible transmission of *Escherichia coli* O157 from a dog to a child in the UK and the carriage of this organism by healthy dogs. No clear link was made to commercial pet food. The recent trend towards innovation in the industry for less processed and "fresher" product concepts has led to the introduction of raw, chilled and frozen pet foods. Given the high incidence of microbial pathogens in raw meats, it seems unlikely that products with minimal or no heat treatments can succeed without significant attention to pathogen control strategies in their manufacture. Invariably the search for shelf-stable "fresh" product forms will lead the industry toward emerging processing technologies such as ultra-high hydrostatic pressure (UHP or HHP) pasteurization, among others.

During the mostly European epidemic of bovine spongiform encephalitis (BSE), some 100 cases of feline spongiform encephalitis (FSE) were reported from 1986 to 2001 among domestic cats and exotic zoo felines, mainly in Europe. Commercial cat food was clearly implicated in some instances and the sporadic cases in zoos were probably caused by infected bovine offal. The disease is characterized by progressive neurological signs, behavioral changes and death. The properties of FSE are identical to BSE and the variant Creutzfeldt–Jakob agent. Fortunately the measures taken across Europe to prevent the inclusion of BSE-suspect material in animal feeds, feed materials and pet foods were very successful in preventing new cases. No additional cases of FSE have been reported in cats since 2001 (http://archive.defra.gov.uk/foodfarm/farmanimal/diseases/atoz/bse/othertses, accessed on 25.02.2013). Even though the outbreak is now controlled and no new cases of TSE have appeared in domestic cats, it is important that control measures such as the strict observance of the legally required controls on the disposal and feeding of specified risk materials be observed to prevent its re-emergence.

MYCOTOXICOSIS

Mycotoxins are toxic secondary metabolites produced by various molds (Richard, 2007). Mycotoxins are considered an important group of unavoidable chemical food safety hazards prevalent in many pet food ingredients. Mycotoxins commonly reported in pet food products include aflatoxins, ochratoxin A and the *Fusarium* mycotoxins such as fumonisins, deoxynivalenol (DON), T-2/HT-2 and zearalenone. Of these, only aflatoxins and DON have a significant history of pet food-related incidents. Fumonisins and zearelanone are frequently reported to contaminate pet foods in various concentrations but have not been directly implicated in commercial pet food safety incidents (Leung et al., 2006; Boermans and Leung, 2007). The toxicity of ochratoxin A (Szczech et al., 1973; Kitchen et al., 1977) and zearelanone (Gajecka et al., 2004) have been described for dogs. There is very little toxicological information with respect to cats.

Most mycotoxins are not reduced to an acceptable level or eliminated by typical pet food manufacturing processes. Thus, control of this hazard can only be realized through procurement of commodities with consistently low contamination rates. The sometimes poor track record of the pet food industry in managing this hazard is partly explained by the difficulty of routine and effective upstream supplier quality assurance strategies for agricultural commodities like cereal grains. For example, maize is generally harvested by a myriad of small to large producers and storage occurs in regional silos where the grain is comingled with that from an entire region. This situation combined with the seasonal variation and geographic incidence of various mycotoxins demands careful monitoring of each harvest and frequent verification of these levels in bulk deliveries to the factory. The factory monitoring programs must be based on statistically valid sampling plans and procedures (FAO, 2001). Care must be taken with local bulk storage of grains at the factory as unfavorable storage conditions may lead to molding and mycotoxin development in storage (Codex, 2003). Fortunately, rapid factory-friendly analytical methods, mainly ELISA-based assays, are available commercially to test most ingredients for many mycotoxins (GIPSA, 2013).

The sensitivity of cats and dogs to some prevalent mycotoxins, though not completely understood in all cases, is clearly a significant food safety hazard. In the following section, the specific cases of aflatoxins and DON contamination of pet food are discussed.

Aflatoxins

Aflatoxins are mycotoxins produced by the molds *Aspergillus flavus* and *A. parasiticus* as they grow on foodstuffs either under field conditions or during storage. The major types of aflatoxins are designated B1, B2, G1 and G2 with their main metabolites designated M1 and M2 (CAST, 2003). Aflatoxins are considered unavoidable natural contaminants of various pet food ingredients, especially maize (Table 15.3). The potential for significant aflatoxin contamination of susceptible ingredients varies due to seasonal and regional climatic conditions and local agricultural practices.

TABLE 15.3 Examples of Ingredients Known to be Potentially Contaminated with Aflatoxins

Cereals	Oilseeds/Nuts	Spices/Tubers
Maize (corn)	Peanut	Chili peppers
Corn gluten meal	Soybean	Black pepper
Corn gluten feed	Sunflower	Coriander
Dried distiller's grains (DDGS)	Cotton seed	Turmeric
Sorghum	Almond	Ginger
Millet	Pistachio	Tapioca (yuca, manioc)
Rice	Walnut	
Wheat	Brazil nuts	

Aflatoxins are rapidly and extensively absorbed from the gut and metabolized in the liver to toxic epoxides which bind to and damage essential cell components such as DNA, RNA and protein enzymes. In all animal species studied, the primary clinical effect of aflatoxin ingestion is related to liver damage. Different animal species will have different sensitivities to aflatoxin and young animals are more susceptible than adults (Bohm 2005). Dogs given a single dose of 100 μg/kgbw of aflatoxin B1 have been shown to excrete both the aflatoxin metabolites M1 and Q1 in their urine with 90% of a single dose excreted in 12 hours (Bingham et al., 2004).

Tragic incidents involving aflatoxin-contaminated commercial pet food have been reported in several areas of the world. Table 15.4 lists results of either market surveillance or reports following outbreaks of aflatoxicosis. The US dog food recall that occurred in 2005–2006 had reports of aflatoxin concentrations of 223–598 ppb (Newman et al., 2007; Stenske 2006). Affected animals showed the following progression of clinical signs: feed refusal, lethargy, vomiting, jaundice, diarrhea, peripheral edema with final onset of bleeding disorders and seizures leading to death (Dereszynski 2008). Experimental work has shown that aflatoxins given to dogs at 500 μg/kgbw can kill the dogs in as little as two doses and dogs fed for 10 weeks at 20 μg/kgbw/day (approx. 360 ppb in the diet) developed classic liver lesions (Armbrecht et al., 1971). Dogs fed 5 μg/kgbw/day for 10 weeks (approx. 90 ppb in the diet) did not have clinical changes but calculated projections indicated this level could result in serious problems, including sudden death if fed chronically. Dogs fed at 1 μg/kgbw/day and below for 10 weeks (approx. 20 ppb in the diet and below) showed no adverse effects and were expected to have no chronic adverse effects.

Aflatoxins are stable under conventional pet food manufacturing conditions including extrusion cooking, baking and retorting and are therefore not reduced during manufacturing of pet foods (IARC, 2002). Because there are no critical control points (CCP) for this hazard in the manufacturing process, it is imperative that ingredients used to manufacture pet foods have low levels of contamination within regulatory constraints. Regulatory limits for pet food are set at or below 20 ppb in most countries (Leung et al., 2006). The burden of sourcing low aflatoxin-containing ingredients is especially significant for maize and

TABLE 15.4 Examples of Reports of Aflatoxin-contaminated Commercial Dry Dog Food Products and Home Rations

Location	Year	AFLA (ppb)	Reference
United States	1986	250–450	Liggett et al. (1986)
South Africa	1987	100–300	Bastianello et al. (1987)
United Kingdom	1997	2.1 and 370	Scudamore et al. (1997)
United States	2001	150–300	Garland and Reagor (2001)
Mexico	2001	mean 5 and 8	Sharma and Marquez (2001)
Turkey	2002	1.75–20	Gunsen and Yaroglu (2002)
Portugal	2003	not detected	Martins et al. (2003)
Brazil	2004	mean 19 and 16	Maia and Pereira Bastos de Siqueira (2002)
United States	2006	579	Stenske et al. (2006)
United States	2007	223–579	Newman et al. (2007)
United States	2008	40–800	Dereszynski et al. (2008)
Argentina	2009	2–167	Juri et al. (2009)

its by-products (e.g. corn gluten feed and meal) given its high usage rate in the pet food industry.

Deoxynivalenol

Deoxynivalenol (DON), also known as vomitoxin, is a common and unavoidable mycotoxin contaminant of cereals in temperate climates, especially maize and wheat. DON contamination has been reported in commercial pet food (Table 15.5). In 1995 a product recall occurred in the USA after a commercial dog food containing wheat had been associated with feed refusal and vomiting, with other more severe clinical signs reported but not confirmed (Hughes et al., 1999).

DON is most commonly produced by molds in the genus *Fusarium*. DON-producing *Fusarium* strains are ubiquitous in temperate regions. Plant infections with *Fusarium* molds and DON production occurs mainly in the field during the flowering period which are favored by humid and cool weather. DON contamination affects predominantly maize, wheat and barley, and less often oats, rice, rye, sorghum and triticale. DON can be found in combination with other fusarial mycotoxins such as zearalenone, as well as the trichothecene mycotoxins nivalenol, T-2 and HT-2 toxins. Closely related metabolites of DON include 15-acetyl DON and 3-acetyl DON. Carry-over of DON to food products from animal origin does not appear to be of concern due to the rapid elimination of the compound from the body (meat) and the very low transfer rates to milk and eggs (EFSA, 2007).

TABLE 15.5 Case Reports of DON Levels in Commercial Pet Foods

Country	DON Concentration	Reference
US	7–23 ppm	Hughes et al. (1999)
Germany	22–1837 ppb	Songsermsakul et al. (2007)
Portugal	100–130 ppb	Martins et al. (2003)
Austria	0–1386 ppb	Bohm and Razzai-Fazeli (2005)

TABLE 15.6 Observed Effects of Dietary DON in Cats and Dogs (Data from Hughes et al., 1999)

	Feed Refusal	Vomiting	
	NOAEL[a] ppm diet	LOAEL[b] ppm diet	NOAEL ppm diet
Dog	4.5	8	6
Cat	7.7	10	8

[a]*NOAEL – no observed adverse effect level.*
[b]*LOAEL – lowest observed adverse effect level.*

Cats and dogs are sensitive to the toxic effects of DON, but the variability between individuals is high with low levels associated with feed refusal, vomiting and gastrointestinal upset. DON is rapidly and extensively absorbed from the gut. It is rapidly metabolized and excreted and does not accumulate in the body. It has been shown to inhibit the synthesis of DNA, RNA and protein. Acute DON toxicity appears as vomiting (hence the name vomitoxin) and diarrhea within 1 hour of ingestion. At levels below those leading to acute effects, anorexia (feed refusal) and the associated subsequent altered nutritional efficiency and reduced weight gain have been observed (Table 15.6). These effects are rapidly reversible with removal of DON from the diet. DON is also reported to be immunotoxic *in vitro*. Dogs previously exposed to DON-contaminated food preferentially select non-contaminated food if given the choice (Hughes et al., 1999).

Levels of DON contamination of cereals can exhibit wide annual variation due to regional or local growing conditions. DON is not reduced by milling, and is concentrated by dry milling in the grain by-products, such as wheat midds, fiber or hulls and dry distiller's grains (DDGs). DON is stable under conventional pet food processing conditions and will not be reduced by extrusion cooking, baking or retorting (EFSA, 2007). As with aflatoxin and all other mycotoxins, control of this hazard requires the procurement of consistently low contaminated grain. Routine factory verification of DON levels in the "at-risk" materials remains the core preventive strategy.

TABLE 15.7 Veterinary Drug Residues in Pet Food Ingredients

Ingredients	Origin	Veterinary Drug	Reference
Molasses yeast from ethanol fermentations Dry distiller's grains (DDGS)	Ethanol fermentations	Penicillin Virginamycin Erythromycin Tylosin Ionophores Others?	RG-6 Regulatory Guidance: Ethanol Distiller's Grains for Livestock Feed. Canadian Food Inspection Agency, 2013
Bovine, swine and poultry: Meat Lung Liver Kidney Viscera	Illegal use in farm animals	Clenbuterol Ractopamine	Chan (1998) Salleras et al. (1995) Sporano et al. (1998)
Fish Shrimp	Aquaculture	Chloramphenicol Malachite green Furazolidone	Ellis and Turner (2007)

TOXICITIES CAUSED BY MEDICATED FEED CARRY-OVER INTO PET FOOD RAW MATERIALS

Veterinary drugs added to feeds can be toxic to dogs and cats. Pets may be exposed to a variety of pharmacologically active compounds through ingredient residues resulting from farm or industrial practices, with some of these being illegal (Table 15.7). Nevertheless, the most devastating incidents of toxicities have been associated with cross-contamination of feed ingredients with medicated feeds during feed or premix processing, handling or delivery. The GMP requirements for medicated feed producers (European Union, EC No. 183/2005 and USA, 21 CFR 225.10) cannot completely eliminate the possibility of cross-contamination of medicated residues in subsequent batches. Significant carry-over can occur even after multiple sweeper batches of unmedicated product have passed through the system. The factors that can influence the degree of carry-over include: strength of feed/drug/carrier adhesion to line surfaces, particle size and density and electrostatic properties of the materials (EFSA, 2008). Polyether ionophore antibiotic cross-contamination of pet foods is an example of the potential magnitude of this veterinary drug hazard. In 1996 a very tragic incident involving paralysis and death of several hundred cats occurred in the Netherlands (Van der Linde-Sipman et al., 1999).

Ionophore antibiotics include salinomycin, lasalocid, monensin sodium and narasin, among others. These commercially available feed additives are administered to poultry for control of coccidiosis and to beef cattle and swine for improved feed efficiency and meat production. Ionophores form lipid-soluble complexes with monovalent cations (Na^+, K^+)

and facilitate specific ionic transport across biological membranes. These result in changes in transmembranous ion gradients and electrical potentials. Salinomycin also increases the release of catecholamines (adrenalin, noradrenalin). The primary target organs of ionophore toxicity are cardiac and skeletal muscles and peripheral nerves. Dietary no observed effect levels (NOELs) of 1–2.5 mg/kgbw/d of salinomycin, lasalocid, narasin and monensin have been reported for dogs. However, toxicity has been observed in dogs after ingestion of canned pet food containing 10–15 mg/kg (ppm) of lasalocid. Assuming a 10-kg dog and a food energy content of 1.2 kcal/g, this would correspond to 0.6–0.9 mg/kgbw/d of lasalocid (i.e. slightly below the reported NOEL) (Oehme and Pickrell, 1999; Van der Linde-Sipman et al., 1999). In cats toxicity has been observed after ingestion of dry pet food containing 16–21 ppm of salinomycin. Assuming a food consumption of 16 g/kgbw/d, this would correspond to an intake of 0.26–0.34 mg/kgbw/d of salinomycin. In cats and dogs clinical signs appear as skeletal muscle paresis (incomplete paralysis). Usually the hind limbs are affected first, with more severe cases progressing to complete paralysis, dysphonia (altered voice production), respiratory distress and even death (Espino et al., 2003; Van der Linde-Sipman et al., 1999).

Because a drug may not be destroyed during the pet food manufacturing process, as is the case for ionophores, the most effective preventive strategy for this hazard is eliminating it all together. Pet food ingredient suppliers must be completely drug free. When this is not possible, exacting manufacturing quality control procedures and customer-managed verification programs must be in place.

ADULTERATION FOR PROFIT, THE MELAMINE CASE

The FDA defines an adulterated food as that containing "any poisonous or deleterious substances, such as chemical contaminants, which may or ordinarily render it harmful to health" and includes in this definition unavoidable contaminants that are either naturally present in agricultural commodities (e.g. mycotoxins and heavy metals) or are the result of industrial processing (e.g. dioxins and acrylamide) (FDA, 2010b). Another category of adulteration encompasses the criminal and willful substitution of a higher value ingredient with an ingredient of lesser cost. This type of fraud is defined by the GMA as "the intentional fraudulent modification of an ingredient for economic gain through the following methods: unapproved enhancements; dilution with a lesser value ingredient; concealment of damage or contamination; mislabeling of product or ingredient; substitution of a lesser value ingredient; or failing to disclose required product information" (GMA, 2012b). Food adulteration for profit has existed from ancient times and with today's globalized trade in foodstuff, it can impact any country. The range of recent food adulterations reported by the press actually shocks and disappoints, some recently reported incidents include: fake baby milk formulas, soy sauce made from human hair, fish soaked in ink for color, and eels fed contraceptive pills for enhanced growth (Barbosa and Barrionuevo, 2007).

Ruminants can obtain protein from non-protein nitrogen (NPN) through fermentation by their rumen bacteria and NPN is often added to their diet to supplement protein.

Melamine and cyanuric acid have been used as an NPN in cattle, along with urea, ammonium nitrate and biuret. Nevertheless, melamine is not considered a good NPN because its hydrolysis in cattle is slow and less complete than other NPNs (Newton and Utley, 1978). Melamine is used in a wide range of industrial applications including the production of plastic by combining it with formaldehyde. It is a major component of countertops, fabrics, glues, flame retardants, colorants for plastics, fertilizers and derivatives of some drugs. Cyanuric acid is a structural analogue of melamine and is often found as an impurity of melamine.

Pets and other non-ruminant mammals cannot utilize inorganic nitrogen in the food. Adulteration of protein-rich feed ingredients and feeds has always been a problem in the industry and buyers have routinely screened for NPNs. The use of melamine to adulterate pet food ingredients was unexpected (Dobson et al., 2008). In 2007, fake wheat gluten (a thickening agent and protein supplement), made by blending wheat flour and scrap melamine contaminated with cyanuric acid, caused the deaths of several hundred animals and significant kidney disease in thousands more. The mixture was formulated to match the apparent protein content of wheat gluten as measured by the commonly used Kjeldahl method for total nitrogen content (Rovner, 2008). Smaller amounts of corn gluten and rice protein concentrate were also implicated in other cases. The adulterated materials were all imported from China via a number of middleman transactions that obscured completely the identity of the original manufacturers. A series of canned pet food product recalls followed encompassing over 5300 lots, affecting over 1100 products and brands in North America (Nestle 2008). Another important development in this saga came with publications that identified melamine in tissues of animals that had died in 2004/2005 of kidney disease associated with a pet food recall in Southeast Asia; therefore the industry had been victim of this fraud once before (Brown et al., 2007)! Incredibly, once the pet food feed ingredient stream was no longer available to the counterfeiters, they turned their attention to the human milk industry. In late 2008, melamine was found in China as a contaminant in milk, milk products, infant formula and eggs, resulting in the deaths of several children and causing kidney stones in thousands more (Barbosa, 2009).

Melamine and cyanuric acid alone proved to be remarkably non-toxic, even in large concentrations. Melamine alone when fed to dogs at 3% of diet for 1 year had no adverse effect on general health and produced no histopathological changes (Hodge et al., 1965; Lipschitz and Stokey, 1945). Cats fed melamine alone at up to 1% of wet diet for 11 days (181 mg/kgbw/d) showed no adverse health effects. On the other hand, the combination of melamine and cyanuric acid proved toxic. Cats with a single oral exposure to a mixture of melamine and cyanuric acid at 0.2% of diet (32 mg/kgbw of each) developed depression, vomiting and feed refusal approximately 12 hours after ingestion. The melamine and cyanuric acid were excreted in the kidney where they combined to form crystals which blocked the kidney tubules and resulted in kidney disease or failure. Kidney function was impaired by 36 hours and animals were euthanized at 48 hours because of acute renal failure. Histopathological changes, including crystal formation in the kidney, were similar if not identical to those found in clinical cases of animals ingesting tainted pet food (Puschner et al., 2007).

The HACCP implications of this tragic situation are clear and include: "a trust but verify approach" throughout the supply chain (Henry, 2009), including frequent audits of suppliers. The implementation of routine product identity verification in addition to the standard quality control tests which can be fooled by an able counterfeiter. Reliance on early warning information is useful in allocating risk levels, for example a major fluctuation in ingredient prices can signal an attractive target for fraud. Most countries have now set regulatory limits on melamine and cyanuric acid. Although testing requirements and limits vary, the most common regulatory limit is 1.0 ppm melamine in infant formulas and 2.5 ppm melamine in other foods.

TOXICITIES CAUSED BY NUTRIENT MISFORMULATION

Essential nutrients such as vitamins, minerals and amino acids are many times added to commercial pet foods to assure that they are nutritionally complete and balanced as per trade or regulatory requirements (e.g. AAFCO 2012 Official Publication, http://www.aafco.org). Over- or under-supplementation of nutrients into the product can lead to regulatory non-compliance, risk of toxicity or risk of nutritional deficiencies. The risk of severe nutritional deficiencies exists because a given commercial diet may be the only food a pet animal consumes. A review of the product recall reports in the USA over the last decade shows an interesting pattern of multiple reports of excessive vitamin D_3 incidents involving dog foods and insufficient thiamine incidents involving cat products (Table 15.8). One report exists for excessive methionine in a dog product. Invariably, nutrient misformulation into diets can be traced to industrial accidents either at the pet food manufacturer or at the vitamin premix supplier, often due to formulation errors or improper mixing of the premix ingredients (Bischoff and Rumbeiha, 2012).

Control of this hazard is linked to GMPs at both the vendor of the ingredients and at the pet food manufacturer. Critical GMPs include mixing validation and process capability studies, careful reconciliation of ingredient use to assure proper formulation and ingredient monitoring. Interestingly, the case of vitamin D toxicosis reported in 2010 which involved the carry-over of a vitamin D supplement (25-hydroxy vitamin D) used in other feed products into a correctly formulated pet food premix points to the risks of additive carry-over into products manufactured on the same manufacturing lines as other feed products. This type of sequence error on shared manufacturing lines has also resulted in the carry-over of antibiotics with disastrous consequences (see "Toxicities Caused by Medicated Feed Carry-over into Pet Food Raw Materials," above).

CONCLUSION

Complete and balanced pet food products are formulated to be the single source of nutrition for a pet. Most pets are sustained mainly through feeding of a reduced range of commercial products and a limited number of production batches for a prolonged amount of time. The impact of the diet and therefore food safety hazards on the health of the pet is more like that of a human infant than an older person eating a varied diet. A careful review

TABLE 15.8 Nutritional Toxicities and Deficiencies

Year	Nutrient	Exposure	Root Cause	Number Affected	Reference
1999	Excessive vitamin D_3	14.65 mg/kg BW	Feed-mixing error	Toxicity or death reported in at least 25 dogs	Rumbeiha and Morrison (2011)
2000	Excessive methionine	1.60–2.75%		Anorexia or vomiting was reported in 21 dogs	
2006	Excessive vitamin D_3	Up to 2664 IU/1000 kcal (ME)	Misformulated vitamin premix containing up to 284,700 IU vitamin D_3/kg	Toxicity or death reported in six dogs and five cats	Rumbeiha and Morrison (2011)
2009	Insufficient thiamin	Canned cat food. 1.5 ppm in the product	Misformulated vitamin premix	13 to 20 cats with reversible neurological symptoms including limb ataxia, rigid paralysis, flaccid neck, blindness, circling behavior, seizures, nystagmus and vomiting	Pet Food Recall 2009 – presentation by Karyn Bischoff Assistant Professor Animal Health Diagnostic Center College of Veterinary Medicine Cornell University Ithaca, New York 14853
2009	Excessive vitamin A	Feline research diet	Misformulation	Hypervitaminosis in cats	Bischoff and Rumbeiha (2012)
2010	Insufficient thiamin	Canned cat food			https://www.avma.org/News/Issues/recalls-alerts/Pages/pet-food-safety-recalls-alerts.aspx
2010	Excessive Vitamin D_3	Dry dog food	Scheduling error by Vitamin D supplier allowed for carry-over of vitamin D supplement (25-hydroxy vitamin D) into pet ingredient	16 dogs in eight states hypercalcemia, increased thirst and urination, weight loss, anorexia or azotemia	Hypervitaminosis D in Dogs Associated with Diet – Kent R. Refsal, DVM, PhD Diagnostic Center for Population & Animal Health | 4125 Beaumont Road, Lansing, MI 48910-8104 | PH: 517.353.1683 FX: 517.353.5096 | www.animalhealth.msu.edu WEBCD.GEN.REF.026.01 Issue Date: 10/8/2010
2011	Insufficient thiamin	Canned cat food "less than adequate levels of thiamine"		One consumer complaint received by the FDA	https://www.avma.org/News/Issues/recalls-alerts/Pages/pet-food-safety-recalls-alerts.aspx

of the industry record with regards to pet food safety reveals issues with the control of a small number of food hazards that account for the vast majority of incidents, these are:

- Aflatoxin.
- *Salmonella.*
- Sporadic adulteration of ingredients with veterinary drugs, inorganic nitrogen sources, specific risk materials (BSE) and heavy meals.
- Nutritional misformulation.

Most of these hazards originate in the raw material supply and have no effective control points in the process. Thus their control relies on food safety management practices by the raw material suppliers and a "trust but verify" vendor management program. All raw materials must be risk assessed via a comprehensive HACCP program and all potential hazards defined and controlled. Factories making low moisture pet foods need specific programs aimed at *Salmonella* control in the environment.

References

APPA – American Pet Products Association, Inc., 2012. Industry Statistics & Trends. Information available in the web at: <http://www.americanpetproducts.org/press_industrytrends.asp> (accessed 25.05.2013.).

Armbrecht, B.H., Geleta, J.N., Shalkop, W.T., 1971. A subacute exposure of beagle dogs to aflatoxin. Toxicol. Appl. Pharmacol. 18 (3), 579–585.

Barbosa, D., 2009. Death sentences in Chinese milk case. NY Times (January 22).

Barbosa, D., Barrionuevo, A., 2007. In China, additive to animals' food is an open secret. NY Times (30 April).

Bastianello, S.S., Nesbit, J.W., Williams, M.C., Lange, A.L., 1987. Pathological findings in a natural outbreak of aflatoxicosis in dogs. J. Vet. Res. 54 (4), 635–640.

Behravesh, C.B., Ferraro, A., Deasy, M., Dato, V., Moll, M., Sandt, C., et al., 2010. Human *Salmonella* infections linked to contaminated dry dog and cat food, 2006–2008. Pediatr. 126 (3), 477–483.

Bingham, A.K., Huebner, H.J., Phillips, T.D., Bauer, J.E., 2004. Identification and reduction of urinary aflatoxin metabolites in dogs. Food Chem. Toxicol. 42, 1851–1858.

Bischoff, K., Rumbeiha, W.K., 2012. Pet food recalls and pet food contaminants in small animals. Vet. Clin. Small Anim. 42 (2), 237–250.

Boermans, H.J., Leung, M.C., 2007. Mycotoxins and the pet food industry: toxicological evidence and risk assessment. Int. J. Food Microbiol. 119 (1–2), 95–102.

Bohm, J., Razzai-Fazeli, E., 2005.. In: Diaz, D. (Ed.), The Mycotoxin Blue Book Nottingham University Press, Nottingham, UK, pp. 77–91.

Brown, C.A., Jeong, K.S., Poppenga, R.H., Puschner, B., Miller, D.M., Ellis, A.E., et al., 2007. Outbreaks of renal failure associated with melamine and cyanuric acid in dogs and cats in 2004 and 2007. J. Vet. Diagn. Invest. 19 (5), 525–531.

Carter, M.E., Quinn, P.J., 2000. *Salmonella* infection in dogs and cats. In: Wray, A., Wray, C. (Eds.), *Salmonella* in Domestic Animals CABI Publishing. CAB International (Chapter 14).

CAST – Council on Agriculture Science and Technology, 2003. Mycotoxins: Risks in Plant, Animal, and Human Systems. CAST, Ames, IA.

CDC – Centers for Disease Control and Prevention, 2008. Update: recall of dry dog and cat food products associated with human *Salmonella* Schwarzengrund infections – United States. Morb. Mortal. Wkly Rep. 57 (44), 1200–1202.

Chan, T.Y.K., 1998. Food poisoning following the consumption of clenbuterol-treated livestock. Hong Kong Pract. 20 (7), 366–369.

Codex, 2003. Codex Alimentarius. Code of practice for the prevention and reduction of mycotoxin contamination in cereals, including annexes on ochratoxin A, zearalenone, fumonisins and tricothecenes CAC/RCP 51-2003. Available online at: <http://www.codexalimentarius.org/input/download/standards/406/CXP_051e.pdf> (accessed 25.05.2013.).

Crane, S.W., Griffin, R.W., Messent, P.R., 2000. Introduction to commercial pet foods. In: Hand, M.S., Thatcher, C.D., Remillard, R.L., Roudebush, P. (Eds.), Small Animal Clinical Nutrition, fourth ed. Mark Morris Institute, Topeka, KS. (Chapter 3).

Dobson, R.L., Motlagh, S., Quijano, M., Cambron, R.T., Baker, T.R., Pullen, A.M., et al., 2008. Identification and characterization of toxicity of contaminants in pet food leading to an outbreak of renal toxicity in cats and dogs. Toxicol. Sci. 106 (1), 251–262.

EFSA – European Food Safety Authority. 2007. Opinion of the scientific panel on contaminants in the food chain on a request from the commission related to deoxynivalenol (DON) as undesirable substance in animal feed. EFSA J. (2004) 73, 1–42. Updated 2 February 2007. <http://www.efsa.europa.eu/cs/BlobServer/Scientific_Opinion/contam_op_ej73_DON.pdf> (accessed 05.12.2013.).

EFSA – European Food Safety Authority, 2008. Opinion of the scientific panel on contaminants in the food chain on a request from the European Commission on cross-contamination of non-target feedingstuffs by monensin authorized for use as a feed additive. EFSA J. 592, 1–40.

EFSA – European Food Safety Authority, 2010. The Community Summary Report on Trends and Sources of Zoonoses, Zoonotic Agents and Food Borne Outbreaks in the European Union in 2010. EFSA J. 10(3), 2597.

Ellis, L.J., Turner, J.L., 2007. China Environment Series, Issue 9. The Woodrow Wilson International Center for Scholars.

Espino, L., Suarez, M.L., Miño, N., Goicoa, A., Fidalgo, L.E., Santamarina, G., 2003. Suspected lasalocid poisoning in three dogs. Vet. Hum. Toxicol. 45 (5), 241–242.

FAO, 2001. Food and Agriculture Organization of the United Nations. Food and Nutrition Paper 73: Manual on the Application of the HACCP System in Mycotoxin Prevention and Control. ISBN 92-5-104611-5.

FDA, 2010a. Compliance Policy Guide Sec. 690.800 *Salmonella* in Animal Feed Draft Guidance. Available online at: <http://www.fda.gov/.../ComplianceManuals/CompliancePolicyGuidanceManual/UCM220418.doc> <www.fda.gov/.../ComplianceManuals/CompliancePolicyGuidanceManual/>.

FDA, 2010b. Food Contaminants & Adulteration. Available online at: <http://www.fda.gov/food/foodsafety/foodcontaminantsadulteration/default.htm> (accessed 23.10.10.).

FDA, 2012. Bad Bug Book: Handbook of Foodborne Pathogenic Microorganisms and Natural Toxins. Available online at: <http://www.fda.gov/Food/FoodborneIllnessContaminants/CausesOfIllnessBadBugBook> (accessed 25.05.2013.).

FDA – US Food and Drug Administration, 2010. Compliance Policy Guide Sec. 690.800 *Salmonella* in Animal Feed Draft Guidance. Available online at: <www.fda.gov/.../ComplianceManuals/CompliancePolicyGuidanceManual/UCM220418.doc>. Food Contaminants & Adulteration. Available online at: <http://www.fda.gov/food/foodsafety/foodcontaminantsadulteration/default.htm> (accessed 23.10.2010.). Bad Bug Book: Introduction Foodborne Pathogenic Microorganisms and Natural Toxins Handbook. Available online at: <http://www.fda.gov/Food/FoodSafety/FoodborneIllness/FoodborneIllnessFoodbornePathogensNaturalToxins/BadBugBook/default.htm> (accessed 24.08.2010.). Interim Safety and Risk Assessment of Melamine and its Analogues in Food for Humans, <http://www.cfsan.fda.gov/~dms/melamra3.html> (accessed 24.8.2010.).

Gajecka, M., Jakimiuk, E., Skorska-Wyszynska, E., Zielonka, L., Polak, M., Paluszewski, A., et al., 2004. Influence of zearalenone mycotoxicosis on selected immunological, haematological and biochemical indexes of blood plasma in bitches. Pol. J. Vet. Sci. 7, 175–180.

Garland, T., Reagor, J., 2001. Mycotoxins and Phycotoxins in Perspective at the Turn of the Millennium. Ponsen & Looyen, Wageningen, pp. 231–236.

GIPSA – Grain Inspection, Packers and Stockyards Administration. 2010. GIPSA Performance Verified Rapid Test Kits for Analysis of Mycotoxins. Available online at: <http://www.gipsa.usda.gov/fgis/tech-servsup/metheqp/GIPSA_Approved_Mycotoxin_Rapid_Test_Kits.pdf> (accessed 25.05.2013.).

GMA – Grocery Manufacturer's Association. 2009. Control of *Salmonella* in low-moisture foods (minor corrections 16 March 2009. Available online at: <http://www.gmaonline.org/downloads/wygwam/SalmonellaControlGuidance.pdf> (accessed 25.05.2013.).

GMA 2010a. Annex to Control of *Salmonella* in low-moisture foods. Available online at: <http://www.gmaonline.org/science/Salmonellaguidanceannex.pdf> (accessed 20.08.2010); Consumer Product Fraud Deterrence and Detection. Available online at: <http://www.gmaonline.org/downloads/wygwam/Salmonellaguidanceannex.pdf> (accessed 25.05.2013.).

GMA, 2010b. Consumer Product Fraud Deterrence and Detection. Available online at: < http://www.gmaonline.org/downloads/wygwam/consumerproductfraud.pdf > (accessed 23.05.13.).

Gunsen, U., Yaroglu, T., 2002. Aflatoxin in dog and horse feeds in turkey. Vet. Hum. Toxicol. 44 (2), 113–114.
Henry, C.W., 2009. Economically Motivated Adulteration. FDA Public Meeting May 1, 2009. Craig W. Henry Ph.D. Senior Vice President and Chief Operating Officer, Scientific and Regulatory Affairs Grocery Manufacturers Association, Washington. Available online at: <www.fda.gov/downloads/NewsEvents/.../UCM163651.ppt> (accessed 23.08.2010.).
Hodge, H.C., Panner, B.J., Downs, W.L., Maynard, E.A., 1965. Toxicity of sodium cyanurate. Tox. Appl. Pharm. 7, 667–674.
Hughes, D.M., Gahl, M.J., Graham, C.H., Grieb, S.L., 1999. Overt signs of toxicity to dogs and cats of dietary deoxynivalenol. J. Anim. Sci. 77, 693–700.
IARC (International Agency for Research on Cancer), 1999. Melamine. IARC Summary and Evaluation 73:329 <http://www.inchem.org/documents/iarc/vol73/73-12.html> (accessed December 2008).
IARC (International Agency for Research on Cancer), 2002. IARC Monographs on the Evaluation of Carcinogenic Risks to Humans. Volume 82: Some traditional herbal medicines, some mycotoxins, naphthalene and styrene <http://monographs.iarc.fr/ENG/Monographs/vol82/mono82-7A.pdf> (accessed 31.12.09).
Juri, M.G.F., Bressán, F., Astoreca, A.L., Barberis, C.L., Cavaglieri, L.R., Dalcero, A.M., et al. 2009. Aflatoxins, fumonisins and toxigenic fungi in raw materials and ready dry dog food in Argentina. Rev. Bras. Med. Vet. 31 (2), 109–117.
Kitchen, D., Carlton, W.W., Tuite, J., 1977. Ochratoxin A and citrinin induced nephrosis in beagle dogs. I. Clinical and clinicopathological features. Vet. Pathol. 4, 154–172.
Leonard, E.K., Pearl, D.L., Finley, R.L., Janecko, N., Peregrine, A.S., Reid-Smith, R.J., et al., 2010. Evaluation of pet-related management factors and the risk of *Salmonella* spp. Carriage in pet dogs from volunteer households in Ontario (2005–2006). Zoonoses Public Health 58 (2), 140–149.
Leung, M., Diaz-Llano, G., Smith, T.K., 2006. Mycotoxins in pet food: a review on worldwide prevalence and preventative strategies. J. Agric. Food Chem. 54, 9623–9635.
Leung, M., Smith, T.K., Karrow, N.A., Boermans, H.J., 2007. Effects of foodborne Fusarium mycotoxins with and without a polymeric glucomannan mycotoxin adsorbent on food intake and nutrient digestibility, body weight, and physical and clinicopathologic variables of mature dogs. Amer. J. Vet. Res. 68, 1122–1129.
Liggett, A.D., et al., 1986. Canine aflatoxicosis: a continuing problem. Vet. Hum. Toxicol. 28 (5), 428–430.
Lipschitz, W.L., Stokey, E., 1945. The mode of action of three new diuretics: melamine, adenine and formoguanamine. J. Pharm. Exp. Ther. 83 (4), 235–249.
Martins, M.L., et al., 2003. Fungal flora and mycotoxins detection in commercial pet food. Rev. Port Ciencias Vet. 98 (548), 179–183.
Melnick, R.L., Boorman, G.A., Haseman, J.K., Montali, R.J., Huff, J., 1984. Urolithiasis and bladder carcinogenicity of melamine in rodents. Toxicol. Appl. Pharmacol. 72 (2), 292–303.
Morse, E.V., Duncan, M.A., Estep, D.A., Riggs, W.A., Blackburn, B.O., 1976. Canine salmonellosis: a review and report of dog to child transmission of *Salmonella* Enteritidis. Am. J. Public Health 66 (1), 82–83.
Nestle, M., 1978. The Chihuahua in the Coal Mine. University of California Press.
Newman, S.J., Smith, J.R., Stenske, K.A., Newman, L.B., Dunlap, J.R., Imerman, P.M., et al., 2007. Aflatoxicosis in nine dogs after exposure to contaminated commercial dog food. J. Vet. Diagn. Invest. 19 (2), 168–175.
Newton, G.L., Utley, P.R., 1978. Melamine as a dietary nitrogen source for ruminants. J. Anim. Sci. 47, 1338–1344.
OECD SIDS (Organisation for Economic Co-Operation and Development, Screening Information Data Sets), 2002. Melamine <http://www.inchem.org/documents/sids/sids/108781.pdf> (accessed December 2008).
Oehme, F.W., Pickrell, J.A., 1999. An analysis of the chronic oral toxicity of polyether ionophore antibiotics in animals. Vet. Hum. Toxicol. 41 (4), 251–257.
Puschner, B., Poppenga, R.H., Lowenstine, L.J., Filigenzi, M.S., Pesavento, P.A., 2007. Assessment of melamine and cyanuric acid toxicity in cats. J. Vet. Diagn. Invest. 19 (6), 616–624.
Razzazi, E., Böhm, J., Grajewski, J., Szczepaniak, K., Kübber-Heiss, A.J., Iben, C.H., 2001. Residues of ochratoxin A in pet foods, canine and feline kidneys. J. Anim. Physiol. Anim. Nutr. (Berl) 85 (7–8), 212–216.
RG-6 Regulatory Guidance: Ethanol Distiller's Grains for Livestock Feed. Canadian Food Inspection Agency. Available online at: <http://www.inspection.gc.ca/animals/feeds/regulatory-guidance/rg-6/eng/1329275341920/1329275491608> (accessed 15.05.2013.).
Richard, J.L., 2007. Some major mycotoxins and their mycotoxicoses – an overview. Int. J. Food Microbiol. 119 (1–2), 3–10.
Rovner, S.L., 2008. Anatomy of a pet food catastrophe. Chem. Eng. News 86 (18), 41–43. <http://pubs.acs.org/cen/science/86/8619sci3.html> (accessed December 2008).

Rumbeiha, W., Morrison, J., 2011. A review of class I and class II pet food recalls involving chemical contaminants from 1996 to 2008. J. Med. Toxicol 7 (1), 60–66.

Safran, N., Aizenberg, I., Bark, H., 1993. Paralytic syndrome attributed to lasalocid residues in a commercial ration fed to dogs. *J. Am. Vet. Med. Assoc.* 202 (8), 1273–1275.

Salleras, L., et al., 1995. Epidemiologic study of an outbreak of clenbuterol poisoning in Catalonia, Spain. Public Health Rep. 110 (3), 338–342.

Sanchez, S., Hofacre, C.L., Lee, M.D., Maurer, J.J., Doyle, M.P., 2002. Animal sources of salmonellosis in humans. J. Am. Vet. Med. Assoc. 221, 492–497.

Sanchis, V., Viñas, I., Jimenez, M., Calvo, M.A., Hernandez, E., 1982. Mycotoxin-producing fungi isolated from bin-stored corn. *Mycopathologia* 80 (2), 89–93. doi: 10.1007/BF02993849.

Scudamore, K.A., Hetmanski, M.T., Nawaz, S., Naylor, J., Rainbird, S., 1997. Determination of mycotoxins in pet foods sold for domestic pets and wild birds using linked-column immunoassay clean-up and HPLC. Food Addit. Contam. 14, 175–186.

Segev, G., Baneth, G., Levitin, B., Shlosberg, A., Aroch, I., 2004. Accidental poisoning of 17 dogs with lasalocid. Vet. Rec. 155 (6), 174–176.

Seto, C.T., Whitesides, G.M., 1993. Molecular self-assembly through hydrogen bonding: supramolecular aggregates based on the cyanuric-acid-melamine lattice. J. Am. Chem. Soc. 115, 905–916.

Sharma, M., Marquez, C., 2001. Determination of aflatoxins in domestic pet food (dog and cat) using immunoaffinity column and HACCP. Anim. Feed Sci. Tech. 93 (1), 109–114.

Songsermsakul, P., Razzazi, E., Böhm, J., Zentek, J., 2007. Occurrence of deoxynivalenol (DON) and ochratoxin A (OTA) in dog food. Mycotoxin. Res. 23 (2), 65–67. <http://refhub.elsevier.com/B978-0-12-381504-0.00015-9/sbref48>

Sporano, V., et al., 1998. Clenbuterol residues in non-liver containing meat as a cause of collective food poisoning. Vet. Hum. Toxicol. 40 (3), 141–143.

Stenske, K.A., Smith, J.R., Newman, S.J., Newman, L.B., Kirk, C.A., 2006. Aflatoxicosis in dogs and dealing with suspected contaminated commercial foods. J. Am. Vet. Med. Assoc. 228 (11), 1686–1691.

Szczech, G.M., Carlton, W.W., Tuite, J., 1973. Ochratoxicosis in beagle dogs: ochratoxicosis in beagle dogs: 1. Clinical and clinicopathological features. Vet. Path. 10, 135–154.

Van der Linde-Sipman, J.S., van den Ingh, T.S., van Nes, J.J., Verhagen, H., Kersten, J.G., Beynen, A.C., et al., 1999. Salinomycin-induced polyneuropathy in cats: morphologic and epidemiologic data. Vet. Pathol. 36, 152–156.

Van Immerseel, F., Pasmans, F., De Buck, J., Rychlik, I., Hradecka, H., Collard, J.M., et al., 2004. Cats as a risk for transmission of antimicrobial drug-resistant *Salmonella*. Emer. Infect. Dis. 10 (12), 2169–2174. Available online at: <http://www.cdc.gov/NCIDOD/EID/vol10no12/04-0904.htm> (accessed 25.05.2013.).

Voetsch, A.C., Van Gilder, T.J., Angulo, F.J., Farley, M.M., Shallow, S., Marcus, R., et al., 2004. FoodNet estimate of the burden of illness caused by nontyphoidal *Salmonella* infections in the United States. Clin. Infect. Dis. 38 (Suppl. 3), S127–S134.

Wareing, P., Fernandes, R., 2007. Micro-Facts: The Working Companion for Food Microbiologists. Leatherhead International Limited.

WHO. 2008. Analytical Methods Available for Detecting and Quantifying Melamine and Cyanuric Acid in Food. <http://www.who.int/foodsafety/fs_management/Melamine_methods.pdf> (accessed December 2008) and Overall Conclusions and Recommendations. Expert Meeting to review toxicological aspects of melamine and cyanuric acid. Ottawa, Canada, 1–4 December 2008, <http://www.who.int/foodsafety/fs_management/infosan_events/en/> (accessed December 2008).

CHAPTER

16

Food Contact Materials

Carlos de la Cruz Garcia[1], *Gloria Sánchez Moragas*[2] *and David Nordqvist*[1]

[1]Nestec SA, Nestlé System Technology Center, Orbe, Switzerland, [2]Institute of Agrochemistry and Food Technology, Paterna, Valencia, Spain

OUTLINE

INTRODUCTION

And last, but not least, food contact materials. This sentence is usually heard in seminars and symposiums on food safety because, in principle, food contact materials (FCM) were previously not considered a source of food safety issues. But this perception is changing. Twenty

Food Safety Management.
DOI: http://dx.doi.org/10.1016/B978-0-12-381504-0.00016-0

years ago, if we asked consumers about their worries regarding packaging materials they would probably have said that waste was their main concern. Waste is certainly a concern that remains today but there is an increasing concern about the inertness of FCM following the issues of recent years. The move of perception from "source of waste" to "source of chemical contamination" is reflected in discussions with consumer associations or simply with friends, where it is quite common to hear consumers using terms that were the "reserve" of experts: migration, phthalates, set-off, benzophenone, functional barrier and bisphenol A (BPA). A simple exercise with the internet using the words "packaging waste" and "packaging migration" could give an indication of this change of perception. The "Eurobarometer survey report on risk perception in the EU" published in November 2010 shows that chemical contamination is the first thing that comes to consumers' minds when they are questioned about possible risks related to food. For the first time, FCM appear in the report where 59% of the European population admit to be worried about substances contained in materials coming into contact with food.

If we look back over the last 10 years we find the reasons why there is an improvement in consumers' vocabulary. Several crises regarding FCM were hitting food and packaging industries and damaging consumers' confidence. In 2003 the semicarbazide (SEM) and epoxidized soyabean oil (ESBO) issues impacted the metal closure industry. Two years later the isopropyl thioxanthone (ITX) case put on the table the issue of set-off in printed bricks. The migration of certain phthalates from recycled cardboard or that of BPA from polycarbonate was again showing that FCM are not always as inert as we think. In all the cases mentioned the industry was identifying gaps and authorities were setting new directives or regulations.

As the time of writing mineral oils from printed paperboard are under the close watch of authorities, consumers and industry, showing that there is room for improvement in the way FCM are handled. The food chain and especially the food safety professionals now have the challenge of returning the confidence on FCMs to consumers.

DEFINITIONS

Food contact material (FCM): From a food safety perspective, all bodies that could transfer their constituents to food under the intended conditions of use (considering expected mishandling and misuse). Includes raw material packaging, processing lines, food packaging (having direct or indirect contact), auxiliary items, some parts of vending machines and food dispensers (e.g. coffee dispensers, ice cream dispensers), among others.

Direct contact: Intimate contact with the foodstuff–food contact layer (physically or in contact with the headspace).

Indirect contact: Corresponds to all layers placed between the food contact layer and a functional barrier. There is no intimate contact but during the contact period there is a potential transfer of constituents into the food.

No contact: The potential of transferring material constituents to food is excluded (it could be proved).

Migration: Migration is the transfer of constituents from the given material or article into the food. It is a time-based process but highly dependent on temperature. It is important to keep in mind the time of contact (e.g. primary packaging from filling to consumers' last serving) and the temperature in the process (e.g. hot filling, retorting, microwaving...).

Overall migration (OM): This is a measure of the inertness of the material and prevents an unacceptable change in the composition of the foodstuffs. It is the sum of all molecules migrating.

Specific migration (SM): Applies to individual molecules. Limits are different depending on the toxicological information of the molecule.

Set-off: This normally refers to transfer of ink/lacquer constituents from the no-contact side to the contact side of the material during storage/transport of the finished article (e.g. stack, reel).

Functional barrier (FB): May be considered to be a barrier consisting of one or more layers which either reduces the migration of authorized constituents below the specific migration limit or reduces the migration of non-authorized substances into foods or food simulants to a "not detectable" level.

Declaration of compliance (DoC): A document delivered by the supplier stating the conformity of the finished article with the applicable laws. This document is a legal requirement in some countries (e.g. European Union member states).

Certificate of analysis (CoA): A document accompanying the DoC proving with data what is stated on the DoC.

Not intentionally added substances (NIAS): Impurities originating from the manufacturing or extraction process of substances used in the manufacture of plastic materials or articles.

Active materials: Materials that are intended to extend the shelf-life of or to maintain or improve the condition of packaged food; they are designed to deliberately incorporate components that would release or absorb substances into or from the packaged food or the environment surrounding the food.

Intelligent materials: Materials and articles which monitor the condition of packaged food or the environment surrounding the food.

Auxiliary items: Items that are intended for food contact and/or mouth contact and are used for food consumption, e.g. teats, measuring spoons, on-pack straws, ice cream sticks, etc.

Promotional items: Objects not necessary for food consumption placed in or on the package, e.g. toys, gadgets, cards, etc.

Recycled material: Material reprocessed in a production process of the waste materials for the original purpose or for other purposes, excluding energy recovery (direct incineration).

Reworked material: A special case of recycled material (high quality) where the cuts and scrap of the virgin material is added to the same production process without leaving the production area.

Reused material: A material that has been conceived and designed to accomplish within its life cycle a minimum number of trips or rotations, and is refilled or used for the same purpose for which it was conceived, with or without the support of auxiliary products present on the market enabling the packaging to be refilled; such reused packaging will become packaging waste when no longer subject to reuse.

CLASSIFICATION OF MATERIALS

There are different ways to approach the classification of materials used in the food industry. It is important to consider different angles because of the combination of factors that could determine the risk of use, e.g. quality of material for a given use (time of contact, type of contact). A classification based on type of contact, type of material and function is shown below.

Type of Contact

Not all materials in a processing line or in a packaging material are in contact with food. It is thus important to distinguish which ones have the potential of transferring their constituents to food from those that have no contact. There are many adjectives to define the type of contact but not always with the same understanding. The key ones are:

- **Direct contact:** Intimate physical contact with the foodstuff. The surface in contact constitutes the food contact layer.
- **Indirect contact:** Corresponds to all layers placed between the food contact layer and a functional barrier. There is no intimate contact but during the contact period there is a potential transfer of constituents into the food. Transfer of volatiles via headspace is also considered indirect contact.
- **No contact:** The potential of transferring material constituents to food is excluded.

There are other adjectives that are less used. The definitions here could be used as guidance:

- **Incidental contact:** The material design could not exclude the potential direct contact with the food for a short period (e.g. splashes, consumer foreseeable misuse).
- **Not intended contact:** The potential of transferring material constituents to food is avoided by design but could not be excluded.

Type of Material

There are many materials that could be used in the food industry. Here we group them by families adding some specific comments on food safety.

- **Metals and alloys:** These are normally used in processing equipment and household utensils. Metals are rarely used individually but the main part of the equipment is made from alloys. There are many different metals that could be used in contact with food: aluminum, chromium, copper, iron, manganese, nickel, silver, tin, titanium and zinc. They are normally present as components of alloys like stainless steel (iron–chromium), bronze (copper–tin), brass (copper–zinc) or German silver (copper–nickel–zinc).

  The main restrictions applying to metals are related to heavy metal content. It could be specified in terms of content in the material or on migration/leaching. Special attention should be paid to welding (e.g. sieve reparation) in order to avoid the introduction of lead. Alternative welding materials with a mix of tin and silver are available.

Because of the extended use of stainless steel in processing and the number of references used, it could be useful to list them for clarification. A table is available as an annex to this chapter.

- **Glass:** The composition of glass is based on sand, soda, lime and glass from recycling, so called "cullet." Modifying the minor ingredients gives an array of different colors (e.g. flint, half white, sky, sapphire, royal blue, Georgian green, light green, emerald, champagne green, dark green, antique green, feuille morte, light amber, amber or red amber). In order to facilitate production, filling and handling an external coating of polyethylene wax is applied to render the container slippery and more resistant to scratches.

  Glass is perceived by the consumer as a high-quality packaging material but from the food safety perspective glass is one of the major concerns because of the potential formation of foreign bodies.
- **Wood:** Wood is widely used for vegetable and fruit boxes but also in toothpicks, chopsticks and ice cream sticks. Pine, bamboo, birch or beech is normally used for these purposes.

  Pest infestations or the growth of molds and fungus could present a food safety issue (e.g. presence of mycotoxins). In order to avoid this kind of issue, wood is normally treated with pesticides or fungicides. A well-known issue arises from the use of 2,4,6-trichlorophenol and 2,4,6-tribromophenol during surface treatment of wooden or wood-based materials. Wine lovers, and thus the wine industry, can suffer from the musty or moldy off-odor associated with these molecules, which is perceptible from as low as 4 ng/liter (WHO, 2005). Checks for the presence of residual levels of these chemicals should then be performed.
- **Cork:** A key application in the food industry is stoppers of bottled wines and spirits. Following a Council of Europe definition, cork stoppers should contain at least 51% of cork and could be made of different pieces bound together by means of glues, adhesives or any other means.

  As in wood materials, checks for residues of fungicides and pesticides should be made.
- **Paper and paperboard:** These are made almost exclusively from cellulose fiber derived from wood. The main difference is their grammage and following international standards it could be considered that material weighing less than 250 g/m^2 is paper and the rest paperboard (ISO, 1995). Common types of paper and typical applications are mentioned here. *Newsprint* is normally used in cheap pocket-books. *Commercial* is used for higher quality articles. *Grease-proof paper* is used when contacting fatty food or food with fats on their surface. A typical application is pet food bags. *Natural Kraft* paper is the strongest type and is extensively used in carrier bags. *Bleached Kraft* is used when appearance is important while keeping the strength, e.g. sugar and flour bags. *Tissue* paper is applied to any light paper and can be used as a laminated component of packaging or stand-alone as kitchen paper towels. *Solid bleached sulfate paper* (SBS) is normally used in water-resistant applications like freezer boxes or wet food contact. When a superior printing surface is needed, the *clay-coated SBS* is the correct option. *Corrugated liner and medium* are, together with adhesives, the components of so-called corrugated boxes. The inner side of the liners is rougher to allow the adhesive gluing the medium or flute to both sides. The structure

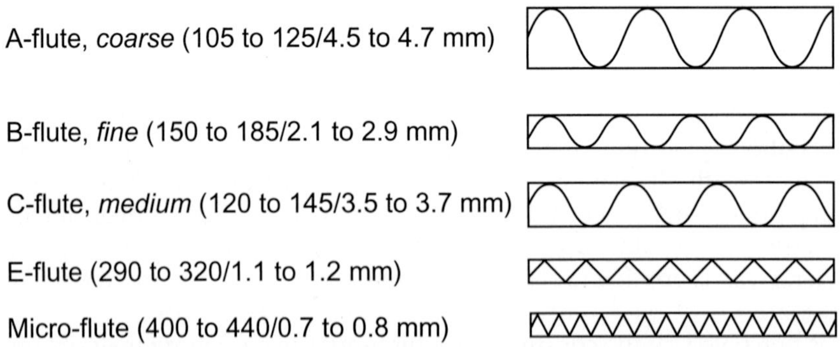

FIGURE 16.1 Comparative of corrugated board grades (approximate number of flutes per meter/flute heights).

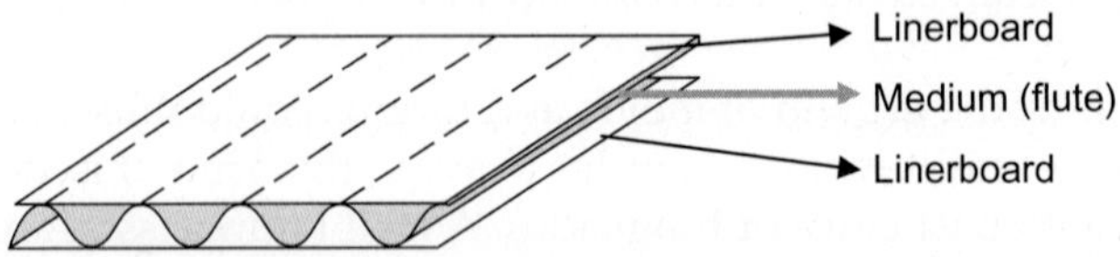

FIGURE 16.2 Structure of corrugated board.

could be doubled or tripled. Their sizes do not follow a logical alphabetical order, going from bigger to smaller A, C, B, E and micro flute. Figures 16.1 and 16.2 show the structure of the corrugated board and a description of the different types. *Chipboard* is 100% recycled paperboard and is the cheapest, with an appearance from light gray to brown. *Newsboard* is made mostly from recycled newspaper. Both are used for rigid boxes where appearance is not critical. Chipboard could be lined with virgin or high-quality recycled liner that improves the appearance. Clay coating is another option when appearance is important.

In the case of printed and/or recycled material the risk of migrating chemicals should be carefully evaluated. In many cases an intermediate barrier (e.g. plastic liner, bag in box) will be necessary. Greaseproof paper could also transfer its components to foodstuff (e.g. perfluoro compounds).

- **Regenerated cellulose:** Commonly called cellophane, the European Commission describes regenerated cellulose film as a thin sheet material obtained from a refined cellulose derived from unrecycled wood or cotton. To meet technical requirements, suitable substances may be added either in the mass or on the surface (Directive 93/10/EEC). Regenerated cellulose film may be coated on one or both sides. It is widely used in food packaging to protect baked goods and candies, and also has applications with oily products.
- **Ceramic:** Ceramic articles are manufactured from a mixture of inorganic materials with a generally high argillaceous or silicate content to which small quantities of organic materials may have been added. These articles are first shaped and the shape thus obtained is permanently fixed by firing. They may be glazed, enameled and/or decorated (Directive 84/500/EEC).

Potential migration of lead and cadmium is the main concern of food contact ceramics, especially when contacting acidic foods. In Europe, it is mandatory that a declaration of compliance (Directive 2005/31/EC) accompanies the ceramic article.

- **Plastic:** There are several polymers used for food contact. The main ones are (name/abbreviation/recycling number): polyethylene terephthalate/PET/1, high density polyethylene/HDPE/2, polyvinyl chloride/PVC/3, low density polyethylene/LDPE/4, polypropylene/PP/5 and polystyrene/PS/6. The recycling number 7 corresponds to all other resins (e.g. polyamide, polycarbonate). They could be processed by injection molding, blow molding, thermoforming or lamination. Thermoplastic rubbers are integrated into the rubber classification. These polymers incorporate additives to control or improve certain properties. As constituents of the material, antistatics, antioxidants, slip agents or UV stabilizers could migrate into food.

  The restrictions on plastic materials are usually related to their chemical composition (starting substances) and to the amount of those migrating into food (individually (SML) and globally (OML)).
- **Silicones:** Silicones constitute a group of polymeric substances and preparations, all containing polysiloxanes (characterized by Si-O-Si and Si-C bonds). Copolymers and polymer blends of polysiloxanes with organic polymers are also covered by the term "silicones," provided siloxane monomer units predominate by weight over each of the other monomer units present (CoE ResAP, 2004). We could differentiate three types of silicones based on their physical properties: oils and pastes (e.g. lubricants or release agents), resins (e.g. heat-resistant coatings) and elastomers (e.g. sealants).

  The restrictions follow the same approach as applied to plastic materials, restricting the starting substances and the amount of those migrating into food.
- **Rubbers and elastomers:** This category designates a family of materials having properties of high elasticity. In an unaged state, rubber can be substantially deformed under stress, but recovers almost to its original stage when the stress is removed. Rubber is usually made from a mixture of materials (solid and/or liquid) and can be subjected to a curing process, which changes its nature. There is also another group of rubbers, the thermoplastic rubber. This is a polymer or blend of polymers that does not require vulcanization or cross-linking during processing, yet has properties, at its service temperature, similar to those of vulcanized rubber. These properties disappear at processing temperature, so that further processing is possible, but return when the material is returned to its service temperature.

  The special properties of rubbers make this type of material ubiquitous. It can be found in food transportation (conveyer belts, hoses and tubing), food handling (gloves), food netting, pipework components (seals, gaskets, flexible connectors and diaphragm/butterfly valves), pumping systems (progressive cavity pumps stators, diaphragm pumps), plate heat exchangers (gaskets), general seals and gaskets (used in machinery and storage vessels), can sealants, bottle seals and closures or feeding teats and breast caps (nipple shields).

  Nitrosamines should be prevented, especially in sensitive applications like feeding teats and breast caps. Migration of plasticizers is another point to control since its migration could reach high levels, so much so that the use of certain plasticizers, e.g. phthalates, is a worldwide concern.

- **Resins for ion exchange and absorption:** These are synthetic organic macromolecular compounds which can be used in the processing of foodstuffs to bring about exchange of ions or adsorption of foodstuff constituents. They do not include, however, cellulosic ion exchangers (CoE ResAP(2004)3).
- **Coatings:** Coatings are finished materials prepared mainly from organic materials applied to form a layer/film on a substrate in such a way as to create a protective layer and/or to impart certain technical performance. Lacquers and varnishes are part of the coating family. Depending on their composition, there are plastic, water-based, UV cured and conventional epoxy phenolic coatings.

  Migration of chemicals due to insufficiently cured lacquers could become a food safety issue. The composition of the coating should fit food contact requirements, e.g. absence of bisphenol F diglycidyl-ether and novolac glycidyl ethers, also known as BFDGE and NOGE, in epoxy phenolic coating.
- **Adhesives:** These are complex systems. They are composed of basic raw materials (binders) which determine their adhesiveness (adhesion) and internal strength (cohesion), and of additives which determine particular end use and processing characteristics. Binders are mainly high polymers. An adhesive formulation typically consists of a binder (polymer) and one or more of the following additives: water or organic solvent carrier, plasticizers, biocides and fungicides – for natural product adhesives, paper and board adhesives, catalysts, emulsifiers, antioxidants, etc. (Bonell and Lawson, 1999).

  Chemicals migrating from adhesives (ingredients or reaction products) could cross the different layers of a plastic laminate and reach the food. An appropriate selection of the adhesive and the layers in between the food and the adhesive is needed.
- **Inks:** This category comprises complex mixes of binders, colorants, pigments, plasticizers, solvents and other additives. In their final state inks are thin layers that are dried or hardened on the material surface. Food packaging inks should not be mistaken for direct food contact printing, where food additives are used (e.g. food colorants). Inks must only be applied to the external part of the packaging material and must not be in contact with food at any stage.

  The risk of using inks in FCM use comes from an insufficient curing (e.g. high level of residual solvents, contamination of internal face by set-off) or from migration through the base material (e.g. migration of oils from printing paper inks through paperboard and plastic liner). The right quality of inks should be selected for a given application. Special attention should be given to heavy metal content of pigments (e.g. lead), photoinitiator migration of UV-cured inks (e.g. migration test of benzophenone) and to saturated and aromatic hydrocarbons from mineral oil used in paperboard printing inks (e.g. ink solvent composition).
- **Lubricants:** Lubricants are oily substances used for reducing friction, especially in the working parts of production lines. Food grade lubricants must correspond to the former USDA H1 classification, which means that the lubricated part may have incidental food contact not exceeding 10 mg/kg.

  Two factors to reduce the risk of contamination from lubricants are the right mapping of the processing line identifying the parts with incidental food contact and the right dosing of lubricants.

Further information on the basics of packaging materials can be found in the references (e.g. Soroka, 1996).

Function of Material

The tendency is to assign FCM to primary packaging of the final product, but this is quite a narrow vision of FCM. Packaging materials are already present in the transport of raw food products and ingredients (e.g. wood boxes for fruits from the producer to the fruit transformer) or in intermediate storage (e.g. transport of dried fruit pieces from the fruit supplier to the yogurt producer). The classification below intends to extend this vision into FCM.

- **Packaging materials:** These are present from the farm to the fork and could be single use (plastic wrapper for a chocolate bar) or repeated use (microwavable plastic tray for lasagna where the consumer washes and reuses it).
- **Processing materials:** Typically these are materials used by the food industry to transform the food ingredients into the finished product (ovens, vacuum dryers, mixers, extruders, etc.).
- **Auxiliary items:** Materials that are normally sold with the finished product and intended for food contact and/or mouth contact, e.g. teats, measuring spoons, on-pack straws, ice cream sticks, etc.
- **Vending machines and dispensers of prepared foods:** Typically these are beverage or ice cream machines, where the containers of ingredients and fluid parts have the potential to transfer their constituents to the food product. Special attention should be given to the hot parts (e.g. tubing after the heating block).
- **Promotional items:** Items that are sold together with the finished product and are placed in or on the package (e.g. toys, gadgets, cards included in a breakfast cereal box that are not separated by a functional barrier).

HAZARD IDENTIFICATION

Physical Hazards

Food contact materials are a potential source of physical hazards. From a food safety perspective, physical hazards are the main concern when using FCM. We could distinguish two different types, foreign bodies (such as small pieces of plastic from a badly cut container) and the finish of the food contact material itself (e.g. sharp edges on a spoon).

The safety risk of foreign bodies could due to size, leading to choking hazard, or shape, hard or sharp foreign bodies. It is not always evident to identify the potential source of foreign bodies if you are not familiar with them. If you were requested to label one type of material from the classification shown in this chapter (metal, glass, wood, paper, etc.) as a high potential of generating foreign bodies, would you select metal? Contrary to our first thought, it is the hardest materials that create the highest number of consumer complaints. The main part of the surfaces contacting food or primary packaging during production is made of metal. Food contact materials have to be carefully designed to avoid metal-to-metal friction and potential generation of particles. An appropriate lubrication is also a key factor to reduce this risk.

In evaluating the risk of physical hazards, the first hazard which may come to mind is glass. In fact, when auditing a production site against good manufacturing practices (GMP), the risk of creating glass foreign bodies is always covered (e.g. from broken lamps on the ceiling). Glass materials are of special risk since glass pieces are normally sharp and could produce injuries when touched or eaten.

But foreign bodies could be already present in the raw materials used in production or be introduced when equipment is opened (e.g. during cleaning). Correct control (e.g. sieving incoming food powder ingredient) and design of the equipment should lower the occurrence to correct levels. Once a foreign body is found it is not easy to trace its origin. Heathcock and Gibson (1990) proposed a rapid non-destructive procedure to identify the nature of glass and, in many cases, the origin of the contamination.

The finish of the FCM is another potential source of injuries. A typical case is the pieces of rigid plastic made by molding. When the pieces of the molds are not tight enough the melted plastic flows into cavities generating fine strips that could cause wounds, punctures or cuts.

Less obvious cases could arise from the selection of inappropriate materials, e.g. not-tempered glass for a tea mug, or labeling with inadequate explanations, e.g. how to open easily and correctly. The first case could result in a burn while the second in a small cut or a broken nail. How many times have we agonized when trying to open a bottle with too high a torque force? In most of these kinds of cases the potential physical hazard can cause inconvenience and should be controlled. A clearer example is the use of shelf-ready secondary/tertiary packaging. The number of accidents caused when placing articles on the shelves due to the use of cutters could be as much as 50% of total accidents at the retail stage. Even if not directly linked with food safety this illustrates the impact that an appropriate packaging design could have on the safety of workers.

Foreign bodies can be detected by human inspection, metal detection, magnetic traps, machine vision, ferrous-in-foil detection and X-ray detection. In the case of empty glass bottles electronic bottle inspection (EBI) could be a good option. In order to select the best detection system the magnitude of the problem must be studied and balanced with equipment ability.

Biological Hazards

Food contact materials are normally not considered to be a source of biological hazards. However, it is important to prevent contamination from pests (e.g. rodents during warehouse storage), dust, manure, contaminated water or raw materials as these may be a source of contamination with several pathogens (*Listeria*, *Salmonella*, leptospira, lassa virus, etc.). Published data demonstrate that the presence of pathogens in the vicinity of unprotected product in processing lines represents a significant risk of recontamination (Reij et al., 2004). Moreover, under certain circumstances some microorganisms can grow in FCM, increasing their numbers and forming biofilm; both harmless microorganisms and human pathogenic bacteria can form biofilms. Biofilms can develop on wet FCM such as those made of stainless steel and they are difficult to remove. Microorganisms in biofilms are usually protected against sanitizers due to the limited ability of the latter to penetrate the protective layer of microbial polymers in the biofilm. The poor hygienic design of equipment is often the cause of these problems. The correct hygienic design and proper maintenance of equipment as GMP are crucial to avoid recontamination through, for example, dripping condensation water or accumulating residues, cracks, micro-holes, etc.

In the last decade, FCM containing antimicrobials have been introduced to the market as a new concept to improve hygiene, by contributing to reduce the risk of cross-contamination (Moretro and Langsrud, 2011).

Raw material may also contaminate FCM, e.g. *Salmonella* in pet food factories. Therefore, process flow must ensure that raw materials move through the facility from input, where there can be high levels of contamination, to output, where levels of contamination are controlled (levels below given limits).

Microorganisms can also be carried on water droplets throughout the packing and storage areas. For instance, *Listeria* can survive in aerosols for up to 3 hours, and therefore spread throughout FCM. For this reason, water used for washing and cleaning equipment and processing lines that comes into direct contact with food must be of a high microbiological quality.

Workers can carry pathogens on their hands and in their digestive systems despite being free of symptoms of illness. In addition, workers with open sores, boils or open wounds are also a potential source of microorganisms. Unless workers understand and follow hygienic measures, they may unintentionally contaminate FCM and thereby create the opportunity to transmit pathogens.

Chemical Hazards

In general terms, chemical hazards migrating from FCM are not considered to create health issues but, as presented in the introduction, consumers' perception is changing. It is known that migration of compounds from FCM to food occurs during handling, production, storage and distribution. The majority of the potential migrants are known, coupled with their potential safety risk associated with the toxicological information available. An illustration of this is found in Table 16.1. In 2003, Laurence Castle pointed to the molecules in the table as "risk priorities." Five years later, two sound issues shook the food packaging and food industry: BPA in 2008 and benzophenone in 2009.

There are positive lists of starting substances, negative lists of non-authorized substances and lists mentioning restrictions (e.g. specific migration limit or maximum concentration in the material). These lists are sometimes in the form of law (e.g. European Regulation EC 10/2011 or USA List of Indirect Additives Used in Food Contact Substances) and sometimes in the form of recommendations or industry guidelines. Both types are normally open documents available by request or directly downloadable from the internet. The site http://www.foodcontactmaterials.com/ is a good source of these texts.

Control of migration by analysis is not an easy task (Pinalli et al., 2011) and it could be expensive and time consuming. Migration is a process that ensures that the same material could be safe and not just depend on the conditions of contact (time, temperature, etc.). The way the material is used should be considered to validate a certain application, but

TABLE 16.1 Food Contact Materials, Risk Priorities Based on Report FD 03/3 Dec2003 from Laurence Castle (Principal Scientist, The Food and Environment Research Agency, UK)

Material	Chemical
Epoxy resins/coatings and PC	Bisphenol A
PVC film	DEHA
PS	Styrene
Printed cardboard	Benzophenone
Grease-resistant paper and board, kitchenware	PTFE
PVC	Crotonic acid

foreseeable misuse should also be included. An example is the use of kitchen paper towels to absorb oil after frying foods. Even if these towels are not directly intended for food contact the fact is that consumers use them quite often. Because of this, the contents of certain plasticizers in the paper (e.g. phthalates) were readjusted.

Food contact materials may contain thousands of different molecules and not all of them have a validated method to measure their migration. Some substances like heavy metals are well known and restrictions apply to all FCM. There are materials more susceptible to containing heavy metals like inks or metals, but the main part of the components of FCM are not that toxic. Substances that are CMR (carcinogens, mutagens and substances toxic to reproduction) must not be used in the composition of FCM. The toxicity of the substances migrating is normally based on a lifetime exposure (e.g. TDI) and it is regulated via specific as well as overall migration limits (overall migration and specific migration).

An important food safety risk of packaging materials is normally linked to the use of fungicides and antimicrobials in wood, cork and paper. Active materials could also be a source of antimicrobials in a different way and should not be confused with wood treatment.

The applicable controls may come from the dosing of fungicides, as requested by the Agence française de sécurité sanitaire des aliments, and/or the residues of mycotoxins and pesticides, as recommended by the Council of Europe (ResAP(2004)2).

Two technologies are used to give antimicrobial properties to FCMs: the use of releasing molecules and the immobilization of active molecules in the FCM surface. In both cases a correct surveillance is necessary to warranty food safety (e.g. migration test).

Ultimately, FCM manufacturers and suppliers need to show evidence that articles placed on the market are safe and compliant. In order to satisfy the due diligence checks, manufacturers send their products for analysis but, before sending the samples they need to answer "the question": What should I check for? In the best case a Certificate of Compliance listing the molecules having restrictions (safety/quality) is available, simplifying the response. An issue arises when the information available is scarce. It is the aim of this book to be a practical guide so I tried to answer "the question" myself, in a context of no information available. I am not a visionary and Table 16.2 is by no means exhaustive, but following my experience, it gives some useful tips on what to focus on.

TABLE 16.2 Guidance on "What to Look for" when Information is Scarce

Material	What to Focus on (Not Exhaustive)	Comment
Active packaging	Releasing technologies (active components migrating into foodstuff or headspace)	Antimicrobials and nanotechnologies should be carefully evaluated.
Coatings for metal packaging	– Bisphenol A – Migrants below 1000 daltons	– Consumer perception to be considered – Identification and check with available lists (e.g. Council of Europe Resolution or USA-FDA)
Grease-resistant paper and board	Fluoro-based compounds	Special care when used in oven (e.g. popcorn bags, pizza boxes)
Ceramic articles	Heavy metals (cadmium, chromium VI, lead, mercury)	Especially if vitrified decoration is applied
Cork, wood and paperboard	Phenols and derivative products	The famous "cork taste" in wines could be also found in food coming from wood (e.g. pallets treated). Sensory test is highly recommended since human threshold is at low part per trillion level
Metal closure gaskets	Gasket: plasticizers (overall migration could be high) and blowing agent (Europe)	Phthalates were traditionally used; it is highly recommended to obtain information on identity of plasticizers. Azodicarbonamide is not allowed in Europe
Mineral hydrocarbons/waxes	Mineral waxes	It is important to know its composition and purity. Allowed only for contact application with dry foods
Packaging inks	– Pigments – UV printing – Inkjet printing	– Swiss positive list could be a reference. – Photoinitiators and acrylates – Methanol and ethanol residual content Exclusion lists (e.g. CEPE, Japan)
Polyacrylonitrile	Acrylonitrile and polyacrylonitrile residues	
Polystyrene	Styrene, styrene oligomers and polystyrene residues	– Max 500 mg/kg in polystyrene – Max. 0.3 mg/kg in food – The residual oligomer content in PS must be documented – Must not be used in oven application
Polyvinylchloride	– Vinyl chloride residues – Plasticized PVC	Special focus on plasticizers (phthalates are still used in a high variety and quantity)
Polyvinylidene chloride	Vinylidene chloride residues	

Allergen Hazards

The risk of allergenic reactions is due to wrong labeling (e.g. undeclared ingredients, mixed labels) or potential cross-contaminations rather than from food contact materials themselves.

FCM are not a source of allergenic hazards but there are a few exceptions. Natural rubber latex (NRL) could be considered as one of these and needs a special focus. Officially, there are 13 latex allergens listed by the World Health Organization. Depending on the NRL manufacturing processes some of these proteins lose their allergenic properties. The prevalence of latex allergy in the general population is less than 1%, but in the general pediatric population, latex sensitization is not rare when young infants have a family history of latex allergy. NRL is commonly used in gloves, cold seal adhesives but also in nipples, baby bottles or pacifiers. To minimize the risk, the quality of latex must be controlled. As a reference, a cut-off level of 0.15 μg/g of material was proposed by Palosuo et al. in 2007 as a limit below which NLR can be considered as low allergenic. Cold seal adhesives based on NRL are a common solution when sealing flow wrapped articles (Topping, 2006). In these cases the exposed cold seal surface must be kept to a minimum.

Alternative materials such as vinyl or nitrile could replace NRL in gloves. In the same way, thermoplastics (TPE) or silicone could replace NRL in many other applications. In the case of newborns or prematures this replacement is not yet possible since these materials are not flexible enough. This lack of flexibility entails a risk of low nutrition because of the higher effort needed to suck (infants could fall asleep before finishing the recommended serving).

MANAGEMENT OF SAFETY OF FOOD CONTACT MATERIALS

The Codex Alimentarius includes food contact materials in their General Principles of Food Hygiene (Recommended International Code of Practice, 2003). Section 4, dealing with design and facilities, states that surfaces and materials, in particular those in contact with food, must be non-toxic in intended use and, where necessary, suitably durable, and easy to maintain and clean. The reference to "toxic" could be linked to the correct selection of materials to avoid chemical or allergenic hazards. The reference to "easy to clean materials" could be linked to microbiological hazards. Physical hazards seem not to be a focus unless we look into the definition of contaminant: "any biological or chemical agent, *foreign matter*, or other substances not intentionally added to food which may compromise food safety or suitability."

As presented in Chapter 31 in this book, prior to application of HACCP to any sector of the food chain, that sector should have in place prerequisite programs.

For the food manufacturing and processing industry, managing the suppliers of FCM is an important prerequisite program. This should include providing clear specifications and auditing suppliers for their practices. In principle, if the supplier of the FCM has an effective food safety management system and takes adequate measures to ensure safety of materials (i.e. respecting the regulations and applying good manufacturing practices), the packaging

TABLE 16.3 Comparative of Standards Applicable for Food Contact Materials

	BRC-IoP 3 Global Standard for Packaging Materials	EN 15593 Packaging – Management of Hygiene in the Production of Packaging for Foodstuffs – Requirements	ISO 22000:2005 Food Safety Management System – Requirements for any Organization in the Food Chain
Focus	Packaging manufacturer (GMP)	Packaging manufacturer (GHP)	Food industry but includes FCM producers (hazard analysis)
Packaging materials	Fully dedicated	Fully dedicated	Considered as part of FCM
Processing line materials	Considered under chemical risk (cleaning and lubrication)	Considered under chemical risk (cleaning and lubrication)	Considered as part of FCM (requiring DoC as a PRP)

material should not present a problem and the amount of chemical migrating into the food, if any, will not be such so as to present a health risk for consumers. In such conditions, in applying the HACCP system to food manufacturing, chemicals are often not considered as a significant hazard and the hazard analysis will be as follows:

- Is the presence of the potential hazard in the food contact material probable? The answer is normally YES.
- Is an unacceptable level of this hazard in the product probable? The answer would be NO.

However, this does not preclude the processing or manufacturing industry to have a monitoring program and to verify that the prerequisite program is indeed effective. To this end the products need to be periodically tested for the chemicals which may potentially migrate to the product to confirm compliance.

For FCM, the suppliers' food safety management system could be audited against different standards. Among the different standards available, three are taken here as reference: the ISO 22000:2005, the BRC-IoP 3 and the EN 15593. Table 16.3 highlights the main characteristics of these standards.

These standards are designed to look into how the product is *manufactured*, e.g. GMP, but there is less of a focus on how the material is *designed*, e.g. chemical composition of materials or migration. Today, this is a gap in the food chain and there is a need for improvement. The last food packaging issues were pointing to this gap, showing the need of increasing the knowledge on the material composition, the lack of surveillance and the lack of partnership along the food chain. A better flow of information is needed and there are excellent tools available to develop this area (audits, specifications, declarations of compliance (DoC) and certificates of analysis (CoA)).

Some case studies about PRP and HACCP are presented in the final section of this chapter. Two of them are directly related to what was discussed above: "Extrusion of retortable and microwavable plastic bottles" and "Printing of multi-material paperboard bricks."

Regulatory Aspects

There is a link between FCM regulations and food safety. The main part of the regulations and recommendations made by authorities has positive and/or negative lists of ingredients and starting substances. These lists are based on experience (e.g. substances not used to manufacture FCM are not listed or removed from existing lists) but also on toxicological data. As an example, the specific migration limits to some molecules set by the EU Commission are based on their admissible daily intake (ADI). Making the assumption that an average consumer weighs 60 kg and eats 1 kg of packed food per day, the SML is the result of multiplying the ADI by 60.

The FCM must be compliant with the applicable regulations in the country where they are used/sold. The problem is that regulatory status of FCM around the world is quite heterogeneous. There are countries without any specific regulation on FCM and countries where national and supranational regulations apply. In the first case the industry takes the food regulation as reference, where almost all countries have a general statement like "the food placed on the market must be safe for consumption." Here, proving compliance could result in ambiguity. In the second case, compliance could become a complex task requiring a deep knowledge of the different regulations applicable.

The many differences found in the level of regulatory development by countries are similar to the ones found if we look into the different classification of FCM. From the "type of materials" perspective, plastic materials are one extreme, being highly regulated in many countries. On the contrary, metal and alloys or inks used in the FCM have almost no dedicated regulations around the world. This contrast is the same when we look into the functions of these materials. There are countries that regulate the materials used in food packaging but they make no reference to the materials used during processing.

Two regulations are generally taken as reference in the food chain, the European Regulation on Food Contact Materials and the US Code of Federal Regulations for Food and Drugs (US Department of Health and Human Services). There are countries and supranational regulations that are directly inspired by these two regulations (e.g. Mercosur and EU regulations). The approach from the EU Commission and US FDA is different. The EU shares the responsibility of FCM compliance all along the food chain while the USA centers the responsibility of FCM safety on the producers of the article (e.g. plastic bottle manufacturer or cereal extruder manufacturer). The ways to control compliance are also adapted to this approach. In the EU it is the responsibility of the supplier to deliver a compliant product and the responsibility of the customer to verify it. This cascade of responsibilities is not followed by the USA where each new packaging application is directly validated by FDA and the customer has almost no information on the composition of the material (e.g. starting substances).

Ensuring compliance of FCM is a way to ensure the safety of the food in contact. As a result of this, the industry and authorities are developing models of Declaration of Compliance. The EU Commission developed one for plastic materials in 2007 and the French Association of Food Industries (ANIA), together with the French Industry of Food Processing Equipment (FIM) and the European Hygienic Engineering & Design Group (EHEDG), developed a common document to declare compliance of all the materials used in the processing equipment. The trend is to have these models for all different FCM as reflected by national regulations of some European countries (e.g. Italy, Denmark).

RECYCLING AND REUSE

Needless to say, the quality of a product is directly linked to the quality of the ingredients. This principle is also applicable to FCM. Using virgin, recycled or a mix of both could impact the risk of introducing chemical hazards.

The risk related to the use of recycled materials for food contact could be minimized through a good selection of recycled materials (e.g. selecting waste material from the first recycling step – virgin fiber) and through an adequate recycling process (e.g. capable of removing contaminants from waste material). Plastic and paper materials have a long history of recycling and provide examples of both cases. In the first example, the recycling process used for paper was not able to remove the adhesive used for gluing the paper boxes and their components were entering the paper fibers. In this way molecules that were not expected in paper material were migrating into the food (e.g. dibuthylphthalate – DBP). The solution for this issue came from the adhesive industry by replacing the DBP with other additive. In the second example, the origin of the waste resulted also in food contamination. Waste paper from offices was used as part of the waste material used for recycling. In 1994 the first cases of food packaging samples contaminated with diisopropylnaphthalene (DIPN) were detected. The origin of this contamination was the carbonless copy paper coming from this portion of the total waste (Zhang, 2008).

THE POTENTIAL ENVIRONMENTAL IMPACT

Besides the obvious risk linked to human health with using additives irresponsibly, there is also another risk that should be addressed when it comes to additives in food packaging material. Certain additives that may not induce a risk to direct human health can have a negative environmental impact (hence, a potential "boomerang" chain effect for humans). Laws and regulations regarding the environmental impact during fabrication, usage and disposal exist in many countries (e.g. European Union, the USA, Japan) (Zweifel, 2009). Also, specific toxicological analyses on the physicochemical properties are required (e.g. acute toxicity test, skin sensitization, repeated-dose toxicity, mutagenicity testing) as part of these regulations. These tests look mostly at direct impacts on the health and environment during usage and to some extent on disposal as well.

Another aspect regarding the environmental impact from additives is the upstream, e.g. inks. Standard petroleum-based inks have a higher impact than soy-based inks. The manufacturing process of traditional inks produces a lot more pollutant in the form of VOCs (volatile organic compounds) compared to soy ink. In addition to this the recycling process of paper/cardboard of soy-based inks is easier as the de-inking process is less energy intensive and more cost efficient (US EPA, 1994). When evaluating materials and additives the full life cycle should be taken into account and a life cycle assessment (LCA) performed. This can be done by collecting as much information as possible about the scenarios (from cradle to grave) of interest and then looking at desired impact indicators such as carbon footprint. Table 16.4 lists a few impact categories together with examples of classification data and possible characterization factors. It should be pointed out that LCA is a very complex task to perform and evaluate. Today there is, however, some guidance available, e.g. ISO 14040.

TABLE 16.4 A Few Possible Impact Categories that can be Used to Assess the Environmental Impact when Performing a Life Cycle Assessment (Info from EPA's Website)

Impact Category	Scale	Classification (i.e. LCA data)	Characterization Factor
Global warming	Global	Carbon dioxide (CO_2) Methane (CH_4) Chlorofluorocarbons (CFCs)	Global warming potential
Eutrophication	Regional	Phosphate (PO_4) Nitrogen oxide (NO)	Eutrophication potential
Water usage	Regional Local	Water used or consumed	Water shortage potential
Acidification	Regional Local	Sulfur oxides (SOx) Hydrochloric acid (HCl)	Acidification potential

LESSONS FROM CASE STUDIES

Printing of Multi-material Paperboard Bricks

The issue: During an analysis of potential contaminants in infant milk a substance that was never detected before was detected and finally identified as isopropyl thioxanthone (ITX), an additive of printing ink used on the design of milk cartons. The issue was communicated to authorities (Italian Ministry of Health and European Food Safety Agency). Sparse information on the toxicity of this molecule was available at that time and following the precautionary principle Italian authorities recalled the concerned products (European Food Safety Agency, 2007). The recall and withdraws were progressively extended to other European countries, companies and products. The economic impact was huge and the damage to consumer confidence difficult to calculate (see also Chapter 41).

Cause: The printing technology used to print the external side of the brick cartons, ultraviolet printing, has many advantages, e.g. the absence of solvents. On the other hand, it requires a strict control of GMPs. In this case the ink was not correctly applied/cured, resulting in set-off. Regardless of the type of food that was put in contact subsequently, part of the not-reacted photoinitiator was already transferred to the food contact layer during storage and transportation. Once the brick carton was filled the ITX was migrating into the food. The quality controls at packaging supplier level were not sufficient. Customers were not sufficiently aware of the ink composition and the surveillance plans were not developed accordingly.

Learnings: GMPs must be respected and quality controls must consider worst-case scenarios (e.g. set-off). The lack of knowledge and surveillance along the food chain created a chain of gaps that allowed the contaminated products to reach the market. This issue was strongly impacting all sectors related to food safety in the food chain. Ink manufacturers developed new photoinitiators with lower migration profiles, packaging manufacturers reviewed their GMPs and quality controls, and the food industry reinforced the surveillance plans and reviewed the specifications of certain applications. The actions reached the authorities and new regulations were developed (e.g. EU 2023/2006/EC on Good Manufacturing Practices).

Bag in Box without Sufficient Barrier or Excess of Waxes in the Liner

The issue: In 2010 several consumers in North America complained to the producing company about the smell or taste of their breakfast cereals. Some of them reported nausea and vomiting. The company investigated the issue and decided to issue a voluntary nationwide recall of four types of their breakfast cereal.

Cause: Following the company statements, *a higher-than-normal amount of certain chemicals in its package liners caused the unusual smell and flavor*. An investigation revealed that elevated levels of hydrocarbons – including methylnaphthalene – in the packaging liners had leached into and tainted the product. The chemicals were migrating from the liner, through the inner bag into the cereals. The bag in the box was not offering sufficient protection (functional barrier) and the chemicals contaminated the food before the end of the shelf-life. Several causes were at the source. First, GMPs were not properly applied and amounts of waxes were present at a higher level than foreseen. Second, the quality controls in place were not able to detect this excess of chemicals. Third, the design of the packaging was not considered to have potential excess of waxes as a worst-case scenario and the selected material for the inner bag was not able to offer the necessary barrier properties. This resulted in a voluntary recall of 28 million boxes of cereal in June 2010. The results in North America were strongly impacted by the voluntary recall. "The estimated impact of the recall, including lost sales, reduced earnings per share by approximately $0.10 in the quarter." This corresponds approximately to a loss of 40 million dollars. The damage to company's image is difficult to consider.

Learnings: GMPs should be correctly applied (e.g. control of wax levels). Quality controls should cover worst-case scenarios. A correct selection of different packaging materials (primary, secondary) could reduce the risks of migration to a negligible level.

Extrusion of Retortable and Microwavable Plastic Bottles

The issue: At the customer's site the operator receiving the lot detected an unfamiliar odor smell when inspecting the truck. He sent some samples to the quality department for a sensory test (sniff test) where the lot obtained a score on the limit (just-out). Because of the intended use, retorting and microwaving, migration tests were requested (overall migration and volatile screening). The results showed an overall migration exceeding the specifications and levels of one molecule subjected to restrictions exceeding the specific migration limit (SML). The lot was returned to the supplier.

Cause: During a shift change on the day of production the incoming operator was informed of the delay in production due to several stops on the line (the extruder was blocking). In order to avoid the blocking issue the operator increased the temperature during the extrusion so the viscosity could be reduced, the blocking issue could be resolved and the line could run faster. The temperature was exceeding the levels specified for this application (retortable and microwavable bottle). The quality controls did not detect any physical defect and the lot was released.

Learnings: Increasing temperature and pressure over the specifications resulted in a polymer with shorter chains, increased level of volatile organic compounds (VOCs) and a higher migration (volatiles and non-volatiles). The positive release of the lot at supplier level should include sensory tests and the GMP awareness of the people operating the line should be improved. The sniff test at the customer level did not consider the retorting and microwaving steps that increase the release of VOCs.

Equipment Reparation

The issue: After reports of issues with the particle size of a cereal product, the breakage of a sieve in equipment was detected. The sieve was repaired and production restarted. During routine control of contaminants (releasing parameter) high levels of lead were detected in the product. The production was stopped, the lot destroyed and an investigation opened.

Cause: The sieve was repaired with an inadequate solder. The material contained lead and the high surface of the sieve produced migration to unacceptable levels.

Learnings: Change management and reparations on the equipment used for food processing need special attention when food contact surfaces are involved. The quality controls in place were able to detect the issue and the product was not liberated. The guidelines were reviewed and adapted to existing standards. Following the Council of Europe: the *use of lead in food contact materials should be abandoned or avoided. Parts made wholly or partly of lead and lead solder for repair should not be used in materials and articles intended to come into contact with foodstuffs including the use of lead in soldered cans* (CoE, Technical Document 2002).

Biological Contamination

The issue: An outbreak of salmonellosis affecting 79 people between 2006 and 2008 was associated with contaminated dry pet food. Exposure of humans occurred through handling of contaminated pet food. More than 23,000 tons of pet food were recalled. The implicated company recalled 105 brands of dry pet food and permanently closed the plant (Behravesh et al., 2010).

Cause: The outbreak strain was isolated from the flavoring room of the manufacturing plant, where dry food was sprayed with flavor enhancers before being packaged. Spraying was made after the killing step in the process (validated time and temperature conditions) and the contaminated pet food was able to reach customers.

Learnings: Process flow must ensure that raw materials move safely through the facility from input to output. The regular monitoring of the processing environment for significant pathogens is needed to ensure proper cleaning and disinfection of FCM.

ANNEX

Common references used for food contact stainless steel (based on Gazzetta Ufficiale della Repubblica Italiana, DECRETO 21 dicembre 2010, n. 258).

UNI EN 10088-1					
Designazione Numerica	Designazione Alfanumerica	AISI/ASTM		UNS	Note
1.4373	X12CrMnNiN 18-9-5	AISI	202	S20200	
1.4310	X10CrNi 18-8	AISI	301	S30100	
1.4325	X9CrNi 18-9	AISI	302	S30200	
1.4305	X8CrNiS 18-9	AISI	303	S30300	
–	–	AISI	303Se	S30323	
1.4301	X5CrNi 18-10	AISI	304	S30400	
1.4306	X2CrNi 19-11	AISI	304L	S30403	
1.4307	X2CrNi 18-9				
1.4303	X4CrNi 18-12	AISI	305	S30500	
–	–	AISI	308	S30800	
1.4401	X5CrNiMo 17-12-2	AISI	316	S31600	
1.4436	X3CrNiMo 17-13-3				
1.4404	X2CrNiMo 17-12-2	AISI	316L	S31603	
1.4432	X2CrNiMo 17-12-3				
–	–	AISI	316N	S31651	
1.4571	X6CrNiMoTi 17-12-2	ASTM T	Type 316Ti	S31635	
1.4541	X6CrNiTi 18-10	AISI	321	S32100	
1.4460	X3CrNiMoN 27-5-2	AISI	329	S32900	
1.4550	X6CrNiNb 18-10	AISI	347	S34700	
1.4006	X12Cr 13	AISI	410	S41000	
–	–	AISI	414	S41400	
1.4005	X12CrS 13	AISI	416	S41600	
1.4021	X20Cr 13				
1.4028	X30Cr 13	AISI	420	S42000	
1.4031	X39Cr 13				
1.4016	X6Cr 17	AISI	430	S43000	
1.4105	X6CrMoS 17	AISI	430F	S43020	
1.4057	X17CrNi 16-2	AISI	431	S43100	
1.4125	X105CrMo 17	AISI	440C	S44004	(*)
1.4542	X5CrNiCuNb 16-4	ASTM	Type 630	S17400	

(Continued)

(Continued)

UNI EN 10088-1					
Designazione Numerica	**Designazione Alfanumerica**	**AISI/ASTM**		**UNS**	**Note**
1.4462	X2CrNiMoN 22-5-3	–	–	S31803	(**)
1.4590	X2CrNbZr 17 – – –	–	–	–	(**)
1.4362	X2CrNiN 23-4	–	–	S32304	
		–	–	S32101	(***)
1.4510	X3CrTi 17	–	–	–	
1.4509	X2CrTiNb 18	–	–	S43940 S43932	
1.4521	X2CrMoTi 18-2	AISI	444	S44400	
		ASTM		S44500	

(*) *for materials intended for short contact at room temperature with foods that are related to simulants A and D during migration tests.*
(**) *only for materials exclusively intended for repeated use during short time at room or hot temperatures and for those for long contact at room temperature with foods that are related to simulant D during migration tests.*
(***) *for articles intended for repeated use at temperatures not higher than 70°C.*

References

Behravesh, C.B., Ferraro, A., Deasy III, M., Dato, V., Moll, M., Sandt, C., et al., 2010. Human Salmonella infections linked to contaminated dry dog and cat food, 2006–2008. Pediatrics 126, 477–483.

Bonell, A.E., Lawson, G., 1999. MAFF project FS 2223. Chemical composition and migration levels of packaging adhesives. MAFF R&D and surveillance report; 557. MAFF, London (UK).

European Food Safety Agency – EFSA, 2007. Scientific Statement of the Panel on food additives, flavourings, processing aids and materials in contact with food on a request from the Commission related to an update on the hazard assessment of 2-isopropyl thioxanthone (ITX) in food contact materials. <http://www.efsa.europa.eu/en/efsajournal/doc/1064.pdf>.

European Food Safety Agency – EFSA, 2011. 2010 Eurobarometer survey report on risk perception in the EU. <http://www.efsa.europa.eu/en/riskcommunication/riskperception.htm>.

Heathcock, J.F., Gibson, R.K., 1990. Identification and characterization of glass fragments in foods. Trends. Food Sci. Technol. 1 (2), 34–37.

ISO: International Organization for Standardization, 1995. Paper and board – Determination of grammage. <http://www.iso.org/iso/iso_catalogue/catalogue_tc/catalogue_detail.htm?csnumber=20674>.

Moretro, T., Langsrud, S., 2011. Effects of materials containing antimicrobial compounds on food hygiene. J. Food Prot. 74 (7), 1200–1211.

Pinalli, R., Croera, C., Theobald, A., Feigenbaum, A., 2011. Threshold of toxicological concern approach for the risk assessment of substances used for the manufacture of plastic food contact materials. Trends. Food Sci. Technol. 22 (9), 523–534.

Reij, M.W., Den Aantrekker, E.D., ILSI Europe Risk Analysis in Microbiology Task Force, 2004. Recontamination as a source of pathogens in processed foods. Int. J. Food Microbiol. 91 (1), 1–11.

Soroka, W., 1996. Fundamentals of Packaging Technology. The Institute of Packaging.

United States Environmental Protection Agency (EPA), 1994. Waste Reduction Evaluation of Soy-Based ink at a Sheet-Fed Offset Printer (Project Summary). <http://www.greenpressinitiative.org/documents/EPAinkStudy.pdf>.

United States Protection Agency (EPA). <http://www.epa.gov>.

US Department of Health and Human Services, Food and drug administration, Code of Federal Regulations, CFR 21. <http://www.access.gpo.gov/nara/cfr/waisidx_02/21cfrv3_02.html>.

World Health Organization, 2005. Concise International Chemical Assessment Document 66, 2,4,6-Tribromophenol and other simple brominated phenols, ISBN 92 4 153066 9, ISSN 1020-6167. <http://books.google.ch/books?id=2eXSO68aqsUC&lpg=PT14&ots=4tPy1qPvbt&dq=tribromophenol%20off%20taste&hl=en&pg=PP1#v=onepage&q=tribromophenol%20off%20taste&f=false>.

Further Reading

Codex Alimentarius, 2003. Recommended International Code of Practice General Principles of Food Hygiene, CAC/RCP 1-1969, Rev. 4-2003. <http://www.codexalimentarius.net/download/standards/23/cxp_001e.pdf>.

Council of Europe – CoE, 2002. Council of Europe's policy statements concerning materials and articles intended to come into contact with foodstuffs; policy concerning metals and alloys. Technical documents: Guidelines on metals and alloys used as food contact materials. <http://www.coe.int/t/e/social_cohesion/soc-sp/public_health/food_contact/TECH%20DOC%20GUIDELINES%20METALS%20AND%20ALLOYS.pdf> (accessed 13.02.2002.).

Council of Europe – CoE, 2004. Resolution ResAP(2004)3 on ion exchange and adsorbent resins used in the processing of foodstuffs. <http://www.coe.int/t/e/social_cohesion/soc-sp/public_health/food_contact/Resolution%20AP%20_2004_%203%20on%20ion%20exchange%20resins.pdf>.

Council of Europe – CoE, 2004. Resolution ResAP(2004)5 on silicones used for food contact applications. <http://www.coe.int/t/e/social_cohesion/soc-sp/public_health/food_contact/Resolution%20AP%20_2004_%205%20on%20silicones.pdf>.

EHEDG, ANIA and FIM. Common template for a declaration of product contact surfaces compliance. <http://www.ehedg.fr/actualites.php?id=61 and http://www.ania.net/fr/dossiers-ania/qualite-nutrition/qualite.htm>.

European Commission. DG Health and Consumers, Food Contact Materials, Legislative list. <http://ec.europa.eu/food/food/chemicalsafety/foodcontact/legisl_list_en.htm>.

European Parliament and Council Directive 94/62/EC of 20 December 1994 on packaging and packaging waste.

Food Contact Materials.com. The internet resource for materials that come into contact with food. <http://www.foodcontactmaterials.com>.

Gazzetta Ufficiale della Repubblica Italiana, DECRETO 21 Dicembre 2010, n. 258. <http://gazzette.comune.jesi.an.it/2011/28/2.htm>.

Palosuo, T., Reinikka-Railo, H., Kautiainen, H., Alenius, H., Kalkkinen, N., Kulomaa, M., et al., 2007. Latex allergy: the sum quantity of four major allergens shows the allergenic potential of medical gloves. Allergy 62 (7), 781–786.

Topping, J.R., Haines, J., Kneller, S., Patel, P., 2006. A preliminary investigation into the possible transfer of latex allergens from latex protein containing materials in contact with food. J. Sci. Food Agric. 12 (9), 1826–1832.

Zhang, K., Noonan, G.O., Begley, T.H., 2008. Determination of 2,6-diisopropylnaphthalene (DIPN) and *n*-dibutylphthalate (DBP) in food and paper packaging materials from US marketplaces. Food Addit. Contam. Part A 25 (11), 1416–1423.

Zweifel, H., Maier, R.D., Schiller, M., 2009. Plastics Additives Handbook, sixth ed. Carl Hanser Verlag, Munich, pp. 1103–1139.

SECTION II

TECHNOLOGIES AND FOOD SAFETY

CHAPTER

17

Thermal Treatment

Tibor Deak

Corvinus University, Budapest, Hungary

OUTLINE

INTRODUCTION

The food industry applies several processing techniques for the inhibition and/or inactivation of microorganisms in order to manufacture safe products with a long shelf-life. Thermal treatments (cooking, pasteurization, sterilization, cooling, freezing) and non-thermal treatments (among others drying, irradiation, high pressure and other methods) alone or in combination can be applied to this end. Heat treatment at high temperatures is used widely in food processing, and it is the most important method of preservation, in

Food Safety Management.
DOI: http://dx.doi.org/10.1016/B978-0-12-381504-0.00017-2

particular in the canning industry. Thermal processing is among the most important methods for controlling, eliminating or reducing pathogens to acceptable levels and sometimes also for degrading toxins and antinutritional factors, e.g. lectins in red beans.

HEAT PROCESSING

There are several unit operations that apply heat in food processing. The purpose of many of them is to abolish the raw state of material to prepare the appearance and structure of finished products. However, for raw materials that are likely to contain pathogens, these operations are also essential for safety. Among these are cooking, baking, frying, roasting, broiling and boiling (Fellows, 2009). A milder degree of heating is used for melting, tempering and blanching.

Heat treatments, such as cooking, boiling, frying and the like, make food more palatable and improve taste by altering texture, flavor and color, and improve digestibility. These do not achieve preservation, although they destroy a part of microorganisms, decrease their number, as well as inactivate enzymes and toxins. Cooking, frying and roasting are processing operations primarily used in the manufacture of meat products, whereas baking, cooking and boiling are mostly used in processing fruits and vegetables. These heat treatments are usually followed with pasteurization or sterilization in the case of canned products.

Blanching is also a cooking term that describes a preparatory process wherein the food, usually a vegetable or fruit, is heated in steam or hot water for a short time, and cooled by plunging into iced water or water spray to stop the cooking process. The purpose of blanching is to soften food, by cooking partly or fully, or to remove a strong taste (for example, of bacon, cabbage or onions). But more often, blanching is performed immediately preceding heat sterilization and can be applied before or after filling the containers (cans) with product. The reasons for blanching are the removal of gas from the tissues of the raw material; the shrinkage of this material; and the inhibition of enzymatic reactions, which, if not stopped, will adversely affect the color and nutritive value of the food. Depending on its severity, blanching will also destroy some microorganisms.

Another operation that applies heat is exhaustion. This is done after filling and before closing cans or jars. The purpose of exhaustion is to remove air from the contents and the headspace and to enable a vacuum to be formed when the container is cooled. In addition, it will remove oxygen, and protect color and flavor from oxidation and vitamin C from destruction. Usually, exhaustion is carried out by passing the containers through a steam box until the temperature at the center is at least 71°C (160°F). Because fruits are different to most vegetables, they are not usually heat blanched because heating would cause softening and juice loss.

In contrast to heat treatment as described above, the most characteristic finishing operation is heat preservation by pasteurization and sterilization. While heat processing operations will inactivate enzymes, coagulate proteins and to some degree also destroy microorganisms, the primary purpose of heat preservation is to achieve the destruction of microorganisms to assure lengthy shelf-life of canned products without spoilage. Sterilization means the use of high temperatures (over 100°C) for complete destruction of microorganisms (but see below regarding commercial sterility), whereas pasteurization means lower heat treatment (generally lower than 100°C) to destroy most vegetative pathogenic bacteria and to extend product shelf-life. Pasteurization is often combined with another means of preservation (concentration, acidification, refrigeration, etc.).

Before discussing the various methods of thermal treatment, first the fundamentals of heat destruction of microorganisms will be outlined, on which the processes of thermal treatment are based.

FUNDAMENTALS OF THERMAL DEATH OF MICROORGANISMS

The method of heat treatment rests upon the principles of thermal death of microorganisms according to which the death of a cell population follows the kinetic of a first order reaction:

$$dN/dt = -k.N$$

that is, the change in number of survivors (dN) in a given time (dt) is proportional to the actual number of living cells (N), where the k factor is called the death rate coefficient (with a negative sign as the cell number is decreasing). Integrating this equation between the limits of initial cell count (N_o) and surviving cell count (N_t) after t time, we arrive at the fundamental equation describing the death of microbial populations:

$$N_t = N_o e^{-kt}$$

often rewritten in logarithmic ($\log_{10}$) form which is called the equation of survival curve:

$$\log N_t/N_o = -k.t$$

When the logarithm of the surviving cell number is plotted against time, a straight (linear) line is obtained, the slope of which is related to the death rate coefficient (Figure 17.1).

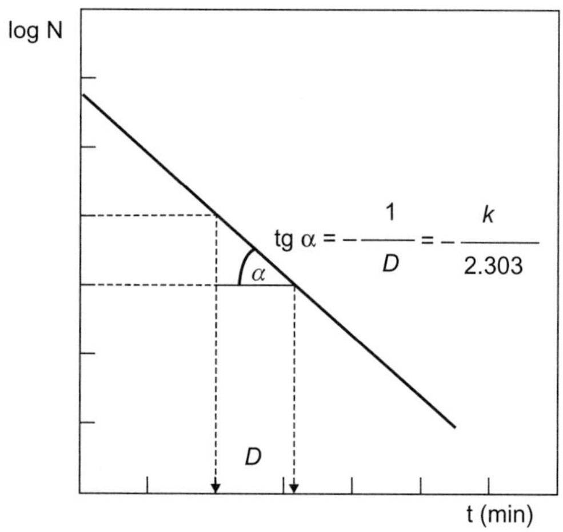

FIGURE 17.1 The survivor curve and the *D* value.

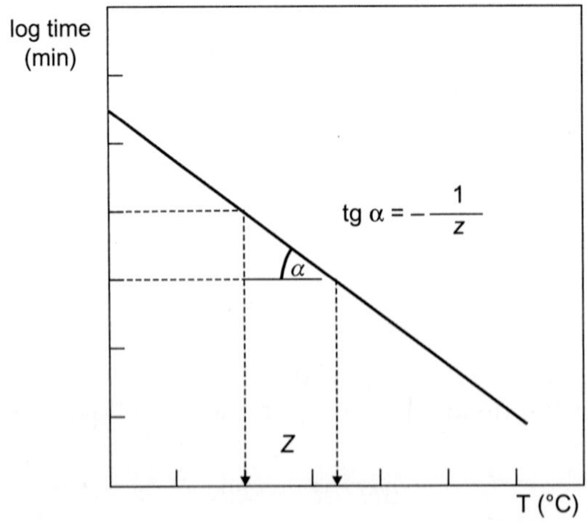

FIGURE 17.2 The thermal death curve and the z-value.

The decimal reduction time D is the time through which the number of survivors decreases to one-tenth.

$$D = t/(\log N_o - \log N_t)$$

The value of D (in minutes) is independent from the size of population but depends on the degree of temperature. Thus, the D value is also a measure of the heat resistance of a given kind (species or strain) of microorganism. The dependence of D on temperature is expressed by the value of z (in °C or °F) defined as the degrees of temperature causing a decimal change of D (Figure 17.2).

D and z are the two basic parameters defining completely the heat resistance characteristics of microorganisms.

HEAT RESISTANCE OF MICROORGANISMS

The heat resistance of microorganisms is primarily a genetically determined specific characteristic that can be modified by the environmental conditions. In general, heat resistance is in proportion to the growth temperature (Table 17.1). Psychrophilic vegetative bacteria become inactivated at about 40°C, whereas mesophiles have a decimal reduction rate of about 1 min at 55 to 60°C. Certain thermoduric bacteria (e.g. *Enterococcus, Microbacterium* species) may survive 30 min heating at 60°C, with a fairly large z-value of 15 to 20°C. Heat resistance of most vegetative pathogenic bacteria occurring in foods is similar to that of mesophiles, and they can be inactivated with the conventional pasteurizing treatments at temperatures below 100°C. The unusually high heat resistance not typical of pathogens is shown by the serotype *Salmonella* Senftenberg, approaching that of thermoduric species.

Although the vegetative cells of spore-forming bacteria are equally sensitive to heat as other bacteria are, their endospores possess high heat resistance (Table 17.2). This is attributed

TABLE 17.1 Average Heat Resistance of Vegetative Microorganisms

	D value (min) at		
Physiological Group	**40°C**	**50°C**	**60°C**
Psychrophilic bacteria	0.3	–	–
Psychrotrophic bacteria	–	1–5	–
Mesophilic bacteria	–	5–40	0.2–1
Thermoduric bacteria	–	–	1–30
Thermophilic bacteria	–	–	100
Yeasts and molds	–	1–5	0.02–0.4

Source: Tomkins and Ordal (1976).

TABLE 17.2 Thermal Resistance of Microorganisms

Microbe	***D* value (min)**	***z*-value (°C)**
Pasteurization at 65°C		
Salmonella spp.	0.02–0.25	4.4–5.5
Salmonella Seftenberg	0.80–1.00	4.4–6.7
Staphylococcus aureus	0.20–2.00	4.4–6.7
Yeasts, molds	0.50–3.00	4.4–6.7
Pasteurization at 100°C		
Alicyclobacillus acidoterrestris	3.0–8.0	6.0–8.0
Bacillus cereus	5–10	7.0–10.0
Clostridium botulinum E	15–50	5.0–8.9
Clostridium sporogenes	60–190	9.0–13.0
Sterilization at 121.1°C		
Clostridium botulinum A, B	0.10–0.20	7.8–10.0
Desulfotomaculum nigrificans	2.0–3.0	9–12
Geobacillus stearothermophilus	4.00–5.00	7.8–12.2
Clostridium thermosaccharolyticum	3.0–4.0	12–18

Data from Stumbo (1973) and Deak et al. (1980).

to the specific structure and composition of endospores, and is due essentially to the manifold layers of spore coat and the dehydrated state of spore cytoplasm. There is not much difference in the heat resistance of aerobic or facultative *Bacillus* species and anaerobic *Clostridia* in this respect; however, the thermophilic spore-forming species are remarkably more heat resistant than mesophiles. Heat resistance of mesophilic spores is characterized with $D_{121°C}$ of 0.01–0.1 min, while that of thermophiles may reach 2–5 min decimal reduction time at this temperature. From the point of food safety, *C. botulinum* is the most heat-resistant pathogenic spore-former, having 0.1–0.2 min $D_{121°C}$, in particular the strains belonging to serotypes A and B, whereas the psychrotrophic E serotype strains are less resistant, characterized by a D_{80} value of 0.3–3 min. Among the spore-formers causing spoilage in canned foods are more heat-resistant species compared to toxigenic *C. botulinum*. Spores of *G. stearothermophilus* and *C. thermosaccharolyticum* have D_{121} values of 3–5 min, and these can survive heat treatments calculated for the destruction of *C. botulinum* (see the discussion on commercial sterility below). Heat resistance of spores is also characterized with *z*-values two or three times higher than vegetative cells, in the order of 8 to 12°C, and some spores may reach 20–30°C.

The majority of yeasts and molds possess heat resistance similar to mesophilic vegetative bacteria. Heat resistance of sexual spores or asexual conidia does not surpass that of vegetative cells. However, ascospores of certain molds, such as species of *Byssochlamys*, *Neosartorya* and *Talaromyces*, have rather high heat resistance with 7–22 min *D* value at 88°C, and these can survive 30 min heat treatment at 90°C causing spoilage of pasteurized fruit juices and canned fruits.

The thermal resistance and thermal death of microorganisms are influenced by several environmental factors. Moreover, although heat resistance is a specific characteristic, it may differ between strains of a species, and may change according to the physiological state of cells. Cells in the exponential phase of growth are usually more sensitive to heat than those being in the stationary phase. For the practice of heat processing, the most important factor influencing heat resistance is the composition of the product, in particular its water activity and pH.

Decrease of water activity significantly increases thermal resistance. This often the case in foods with high sugar concentration or containing many proteins or fats. Acidic environment and low pH decrease heat resistance. Product pH is of outstanding importance for heat processing. pH 4.5 (in the USA pH 4.6) signifies a dividing line; products with pH lower than 4.5 can be pasteurized at 100°C or below, while foods of higher pH than 4.5 must be sterilized over 100°C. The fundamental safety reason for this is that the most heat-resistant pathogenic endosporic microorganism, *C. botulinum*, cannot grow or produce toxin at pH <4.6, and the spores that may survive heat treatment could not germinate either (Table 17.3, and see "Factors Determining Heat Treatment," below).

Factors affecting heat resistance are in force before, during and after heat processing. Cells surviving heat treatment become damaged and can be repaired only under optimum conditions, though not in products which may contain certain chemicals, such as preservatives and nitrite, nor in products stored at low temperature. These products, although they may contain living bacteria, are, however, not able to start growing, and such products remain in a state of "commercial sterility" without spoilage.

DETERMINATION OF HEAT PROCESS REQUIREMENT

Determination of heath process requirement and its validation is important when designing control measures in an HACCP study and determining monitoring parameter and

TABLE 17.3 Heat Processing Requirements – Dependence on Product Acidity

Acidity Class	pH Value	Food Commodity	Heat Processing Mode
Low acid	5.3–6.0	Vegetables, uncured meat, poulty, fish, soups	High temperature sterilization (115–121°C, 240–250°F)
	5.0	Tomato products	(105–115°C, 221–240°F)
Medium acid	4.5–5.3	Fruits, fruit juices	Pasteurization
			(100°C, 212°F)
Acid	3.7–4.5	Fruits	(80°C, 176°F)
High acid	3.0–3.7	Pickles, sauerkraut	(80°C, 176°F)

Source: Desrosier and Desrosier (1977).

critical limits. These need to be validated on a case-by-case basis, considering various factors which may affect the outcome such as initial bacterial load, acidity, water activity, etc. Therefore the subject is explained in detail.

The extent of microbial destruction during the process of heat treatment depends on the combined action of temperature and time. In this regard it is essential to be aware that increasing the flow rate in a pipe or the speed of a conveyor belt reduces the residence time of the product. There have been cases where an increase in the flow rate of the conveyer belt has led to an outbreak of foodborne illness (Motarjemi and Kaferstein, 1999).

In the practice of heat treatment, various degrees of temperature are applied. The values of D (or its multiple the thermal death time, TDT) at any given temperature can be obtained using a reference value, F, at a reference temperature (Figure 17.3). For the latter, 121.1°C was chosen, a temperature important in the sterilization practice (this value corresponds to a round figure, 250°F). Also, a z-value of 10°C was selected as the slope of this particular thermal death curve. With these determined points, the equation of the thermal death curve is:

$$\log(t - F) = -(T - 121.1)/z$$

This equation is used for the calculation of the thermal processing requirement and the lethality of the sterilization process. For these calculations, the lethal effect of any other temperature should be compared to the reference temperature, 121.1°C. From the above equation, the relative rate of heat destruction at various temperatures compared to that of 121.1°C can be obtained as:

$$F/t = \text{antilog}(T - 121.1)/z$$

In the practice of heat treatment (sterilization, pasteurization) the temperature is not constant but changes, increasing during warming up and decreasing during cooling. In calculating the heat process requirement, the task is to sum up the lethal effects of changing temperatures for changing time. This can be done if the thermal death time (or rate) at various temperatures is expressed in a similar manner to be integrated. As shown above, the death time of any temperature related to the reference temperature 121.1°C is expressed as

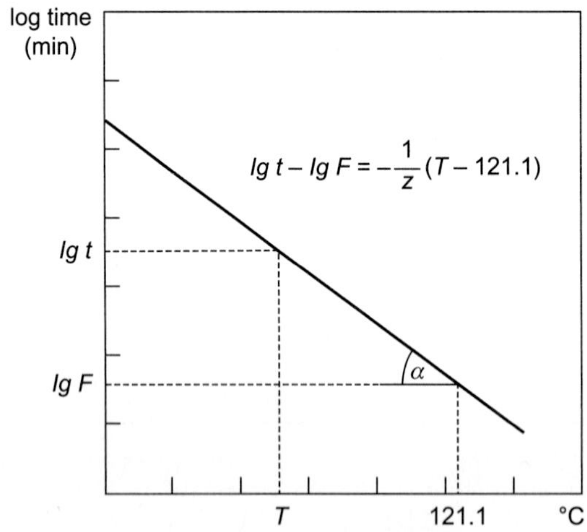

FIGURE 17.3 The reference thermal death curve.

F/t; and the integrated time-equivalent, F_i, of different temperatures in relative fractions of 121.1°C can be obtained as:

$$F_i = \int_o F/t.dt$$

e.g. if the time-equivalent of a thermal process is $F_i = 3$ min, it means that the sum of lethality of all corresponding temperature–time combinations during heat treatment will be equal to the effect of 3 min instantaneous treatment at 121.1°C. In this interpretation, the F_i value does not relate to a given kind of microorganism but to a given heat process, hence it can be used to compare the efficacy of different thermal processes.

The heat resistance of microorganisms, however, differs and changes also with temperature (as expressed by the *D* and *z*-values). For safety reasons, the minimal degree of thermal process chosen should be adequate to kill the most resistant pathogenic microbes which may occur in the practice of canning. According to common experience, it is the toxigenic *Clostridium botulinum* which constitutes the greatest health hazard and whose endospores have high heat resistance. The *D* value of the most resistant spores of *C. botulinum* at 121.1°C is 0.21 min, and its thermal dependence, the *z*-value, is 10°C. This has been chosen universally for the calculation of thermal process requirements, and the summarized lethality value of temperatures related to 121.1°C is distinctively marked as the F_o value, and called equivalent sterilization treatment. In contrast to the F_i thermal time-equivalent which may refer to any TDT curve no matter which *z*-value, the F_o value relates to a thermal process characterized with a thermal death curve of $z = 10$°C value.

The F/t values can be graphically integrated by taking the relative death rates corresponding to the different temperatures of the heat penetration curve, and plotting with time to obtain the so-called sterilization curve. The area below the sterilization curve will be equivalent with the sterilization treatment in minutes of F_o (Figure 17.4). For practical reasons the summing up of F/t values usually starts when the internal temperature reaches

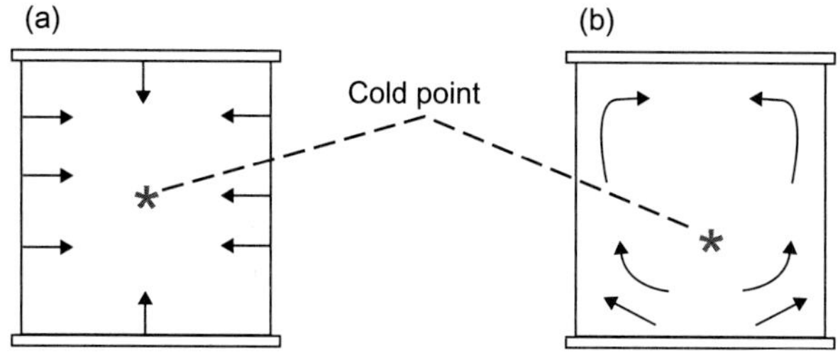

FIGURE 17.4 The sterilization curve and the sterilization equivalent value F_o.

100°C and includes also the cooling part until 100°C. The F/t values associated with temperatures below 100°C are very small and hence do not contribute significantly to the overall amount of heat treatment. On the other hand, omitting the effect of high temperatures during cooling would result in oversterilizing of the product, possibly unnecessarily resulting in additional quality losses. In recent times, with the development of computing technology, programs are available to determine thermal process requirements, and also online monitoring and controlling of the thermal process (Fellows, 2009).

Based on the sterilization equivalent F_o values, not only the efficacy of various thermal processes can be compared but also the minimal degree of heat treatment required for safety can be determined. It is a universally accepted practice to apply a heat treatment which should destroy the number of *C. botulinum* spores to 10^{-12} proportion. This is the 12D concept, which is equivalent with $12 \times 0.21 = 2.52$ min heat treatment at 121.1°C, which is the F_o for *C. botulinum* (also called "botulinum cook"). This provides high safety for heat sterilization. In modern terminology, it is referred to as "performance criterion" (see chapter 31 in this book on Hazard Analysis and Critical Control Point). Since it was introduced for commercial canning at the end of the 1920s, there is high degree of safety with industrialized canned food. Most cases of botulinum intoxications are associated with home preserved food. In the USA about 10 to 30 outbreaks of foodborne botulism are reported each year, almost all from home canning (Shapiro et al., 1998). An outbreak involving eight people, the first in 33 years, was caused by a commercially made canned food (hot chili sauce produced by a factory in Augusta, GA) (Schmit, 2008). FDA officials stepped up inspections at other canneries, and discovered botulinum spores in cans of green beans produced by a plant in Michigan. Although no illnesses were reported, the producer recalled 1.2 million cans of vegetables because of the risk.

The 12D principle of sterilization should be applied for low acid products with pH >4.5 in which *C. botulinum* can grow. In these products, however, spore-forming bacteria may occur whose heat resistance is higher than that of *C. botulinum* (Table 17.4). Although these do not present health hazards, they can survive the minimal requirement of safe heat treatment (i.e. $F_o = 2.52$ min), and can cause spoilage. For economic reasons, the spoilage ratio should be kept lower than 0.1%. When thermophilic spore-formers are to be accounted for as contaminants having D_{121} values of 3–5 min, the equivalent sterilization treatment should be much higher, sometimes reaching $F_o = 15$–20 values (Table 17.4).

TABLE 17.4 Sterilizing Time-equivalent (F_o) for Certain Canned Products

Product Type	pH	F_o (min)
Pickles	3.4–4.1	0.0002–0.004
High-acid fruits	3.2–3.8	0.002–0.007
Tomatoes	4.2–4.5	0.01–0.07
Medium-acid fruits	3.7–4.5	0.1–0.4
Medium-acid vegetables	4.0–4.5	0.1–2.0
Cooked meats	5.0–6.5	2.5–5.0
Low-acid vegetables	5.0–6.5	4.0–14.0
Ready-to-eat foods	4.5–6.5	5.0–30.0

Selected data from Richardson (2004).

On the other hand, for the heat preservation of products whose pH is lower than 4.5 (the so-called acid and high acid foods), not even the minimal requirement for botulinum cook (F_o=2.52 min) need be applied. Partly, *C. botulinum* could not grow at or below pH 4.5, and the acidic environment will decrease the heat resistance of microbes. These products can be pasteurized by heat treatment lower than 100°C.

Analogous to the calculation of the sterilization requirement and efficacy in F_o value, in the case of pasteurization the *D* and *z* parameters are to be related to fixed reference values, and the cumulative thermal destruction equivalent of changing temperatures and times is expressed in pasteurizing units (PU) or pasteurizing equivalent (*P*). The reference temperature should be marked, e.g. at 80°C the value is P_{80}.

$$P_T = \int_{T_w}^{T_c} 10^{(T-Tr)/z}.dt$$

where P_T is the pasteurization equivalent at *T* temperature of heating, integrated between the cooling temperature T_c and the warming temperature T_w, related to the reference temperature, T_r, and *t* is the time of heating.

CONVENTIONAL HEAT PRESERVATION

There are two main processes of preservation by heating: sterilization and pasteurization.

Sterilization

Sterilization means the complete destruction of microorganisms usually by temperatures over 100°C using pressurized equipment (autoclaves or retorts). Because of the resistance of bacterial spores to heat is different, sterilization frequently means a treatment of at least

121°C (250°F) of wet heat for 15 min or its equivalent to inactivate to a large extent spores of the pathogenic *C. botulinum* and most of the spore-forming spoilage microorganisms. Sterilization also means that every particle of the food must receive adequate heat treatment. Hence, the slowness of heat transfer through the food should also be considered in determining the overall heat destruction effect of the sterilizing treatment.

In practice, however, a product subjected to sterilization may not be sterile. Because the principle of exponential death rate following absolute sterility cannot be achieved (not all microorganisms will be eliminated), the probability of survival must be minimized to an acceptable degree. This has been set for a 10^{-12} part survival of *C. botulinum* spores, called 12D concept (equivalent to $F_0 = 2.5$ min heat treatment). Even in this case, some heat-resistant spore-formers, e.g. *C. thermosaccharolyticum* or *Geobacillus stearothermophilus*, may survive this and further intensive heat treatment ($F_0 = 5$ or higher values). Being of thermophilic nature, these surviving microorganisms cannot grow under the normal conditions of storage prevailing in temperate zones (at ambient temperature without refrigeration), and this condition is termed commercial sterility. In practice, most of canned or bottled products receiving the minimal 12D botulinum cook or higher exist as commercially sterile, in which some survivors of high heat-resistant spore still remain, although these will not grow under normal storage temperature at a temperate climate. However, in tropical areas an ambient temperature of 45°C may prevail long enough to cause cans to explode due to the activity of thermophiles.

Canning

Food preservation by heat treatment of products packed in containers – called canning – is a common practice of food industry. Although, from the microbial point of view, it would be ideal to employ a heat treatment that would eliminate the risk of any surviving microorganisms, most canned food products cannot be subjected to such a degree of heating because it would degrade the sensory quality and result in loss of nutritional value. Hence, in practice, a compromise is needed in order to provide a heat treatment intensive enough for the microbiological safety of the products and at the same time moderate enough for preserving product quality. Commercial sterility is a generally accepted practice of canning.

As discussed above, *Clostridium botulinum* is used as a reference organism for manufacturing safe and stable products by heat treatment with a minimum F_0 value of 2.52 min. Based on microbiological considerations and including a sufficient safety margin, most sterilized canned products should be produced with F_0 values of 4.0–5.5. The retort temperatures to be used may vary between 117 and 130°C (depending on the heat sensitivity of the individual products). It is known, however, that certain thermophilic organisms such as *G. stearothermophilus* or *C. thermosaccharolyticum* are extremely heat resistant and may survive F_0 values of 4–5.5. In the case of survival they will not grow under normal storage conditions of up to 25°C and do not pose a risk in countries with moderate temperatures. However, they may grow under tropical conditions, in particular with storage temperatures of 25°C and above. Hence, F_0 values of 12–15 have to be employed in cases having this risk (Table 17.5).

Contrary to the excellent safety record of commercially canned foods, microbial spoilage of canned products does occur, and is usually related to the following factors (Evancho et al., 2009): (1) insufficient processing, which permits survival of mesophilic microorganisms, (2) inadequate cooling after processing or high temperature storage and distribution

TABLE 17.5 Spore-forming Bacteria Causing the Spoilage of Canned Products

Type of Spoilage	pH	Products	Spoilage Bacteria	Heat Resistance (D_r min)
Flat sour	>4.5	vegetables, meat dishes	thermophiles *B. stearothermophilus*	4–5
Gaseous souring			*C. thermosaccharolyticum*	3–4
Sulfide stinker			*D. nigrificans*	2–3
Flat sour	>4.5	vegetables, canned meat	mesophiles	
			B. cereus, B. subtilis, B. brevis	0.001–0.004
Gaseous putrefaction			*C. sporogenes, C. botulinum*	0.1–0.2
			C. putrefaciens	0.001–0.01
Flat sour	<4.5	vegetables, tomato products	thermophiles	0.01–0.07
			B. coagulans	
Gaseous souring	<4.5	pickles,	mesophiles	
		tomato products	*B. polymyxa, B. macerans*	0.01–0.05
Gaseous putrefaction		canned tomato,	*C. pasteurianum*	
Butyric fermentation		canned fruits	*C. butyricum*	0.004–0.01

From Deak et al. (1980).

conditions promoting growth of thermophilic bacteria, and (3) post-processing microbial contamination due to leakage. Table 17.5 lists the possible causes and signs of spoilage in canned foods.

Insufficient processing of low-acid foods is a serious situation from the public health viewpoint because of the potential development of toxigenic spore-formers and their toxins. High-temperature (thermophilic) spoilage may occur in low-acid canned foods if growth of extremely heat-resistant spore-forming bacteria surviving the heat processing occurs. Certain ingredients (e.g. sugar and starch) may introduce excessive numbers of these organisms in the product. If thermophilic spoilage occurs, it may be caused by *Geobacillus stearothermophilus* and *B. coagulans*, so-called flat sour bacilli because they produce acid without gas and the cans do not swell. When thermophilic anaerobes such as *C. thermosaccharolyticum* cause spoilage, producing large amounts of gases (H_2 and CO_2), the cans become swollen and may even burst. The third type of thermophilic spoilage, sulfide stinker, is caused by *Desulfotomaculum nigrificans*, which produces hydrogen sulfide bound by the food or the can walls can become black. In the case of container leakage, usually a mixed spoilage population develops, consisting of lactobacilli, enterococci and other bacteria. Recontamination through faulty sealing is often due to cooling with contaminated water. Post-contamination was responsible for the largest outbreak of salmonellosis in the history of the USA affecting some 160,000 to 200,000 people in Chicago (Ryan et al., 1987).

TABLE 17.6 Comparison of Parameters of Various Methods of Pasteurization

Method	Temperature °C	Time (min, s)
Batch (vat)	65	30 min
HTST	72	15 s
Ultra pasteurization	89–100	1 s
UHT	138	2 s

From Robinson (2002).

Investigation of spoilage is important in order to determine the causes and apply control measures.

Pasteurization

Compared to sterilization, pasteurizing is a comparatively low order of heat treatment, generally at a temperature below the boiling point of water. The general objective of pasteurization is to extend product shelf-life by inactivating all non-spore-forming pathogenic bacteria and the majority of vegetative spoilage microorganisms, as well as inhibiting or stopping microbial and enzyme activity. To be effective, pasteurization is frequently combined with another means of preservation such as concentration, acidification, chemical inhibition, etc.

In pasteurizing, two types of processes can be used: slow and rapid. Slow pasteurization uses pasteurization temperatures for several minutes; e.g. typical temperature–time combinations are 63 to 65°C over 30 minutes or 75°C over 8 to 10 minutes. Rapid, high or flash pasteurization uses pasteurization temperatures of about 85 to 90°C or more for a time only in the order of seconds. Typical temperature–time combinations can be 88°C (190°F) for 1 minute; 100°C for 12 seconds; 121°C for 2 seconds (Table 17.6).

Methods of high temperatures for short time (HTST) and ultra-high temperatures (UHT) for very short holding times have been developed, replacing traditional pasteurization or sterilization processes. Such short holding times and high temperatures require special equipment to ensure uniform heat treatment, and generally are applicable for liquid products. Taking into account the short time and rapid performance of operations, this can only be achieved in a continuous process, using heat exchangers. In this process the product is heated separately, then cooled down rapidly to the temperature for filling, which will be performed in aseptic conditions in sterile receptacles.

In aseptic technology, heating is applied prior to packaging. This will cause inherently less damage to food quality. It can be applied where the food (such as liquids) can be readily distributed for rapid heat exchange. However, these methods then require packaging under aseptic conditions to prevent recontamination. On the other hand, heating within the package frequently is less costly and produces quite acceptable quality with many foods, and most canned food products are heated in the package. In-line sterilization followed by aseptic packaging is gaining in popularity for heat treatment even in the traditional canning factories.

Pasteurization is commonly associated with milk for which it is used all over the world. For the pasteurization of milk temperatures below boiling temperature are typically used since at very high temperatures casein micelles will irreversibly aggregate (or "curdle"). There are two main types of milk pasteurization used today: the conventional batch method by which the bottled milk goes through a heat treatment on a conveyor belt for the required time (e.g. at 63°C for 30 min), and the high temperature short time (HTST) method by which the milk is pasteurized at 72°C for 15 s using a continuous heat exchanger. In recent times, ultra-high temperature (UHT) is also used for milk treatment. It is in fact a sterilization process at 135°C for 2–5 s only before packaging of milk which is then filled into containers aseptically. Nowadays, batch pasteurization of milk is rarely used in large companies, but may be still used in smaller businesses and for foods other than milk (e.g. fruit juices). HTST pasteurized milk typically has a refrigerated shelf-life of 2–3 weeks, whereas UHT milk can last much longer even unrefrigerated, sometimes 6–9 months. The HTST pasteurization should achieve a 5-log reduction in the number of viable microorganisms in milk, killing almost all yeasts, mold, and common spoilage and pathogenic bacteria. UHT treatment is expected to destroy bacterial spores as well. Ultra pasteurization (UP) is a process similar to HTST pasteurization, but using slightly different equipment and higher temperatures. UP pasteurization results in a product with longer shelf-life but still requiring refrigeration. Pasteurization regimes for certain dairy products differ depending on the fat content of the product. Ice cream, dairy dessert mix, cream or processed cheese require more robust treatment, e.g. 70°C for 25–30 min or 80°C for 25 s.

Products, including dairy products, are also heated very rapidly by steam injection and cooled down by evaporation of the same amount of water as was added by the injection. This is an ultra-fast process.

Pasteurization is also widely applied to various liquid and certain viscous and particulated foods such as juices, soft drinks, beer, cider, wine, cream and processed cheese, liquid eggs, syrups, sauces, soups and some ready meals. With many products, like fruit juices and soft drinks, it is the intrinsically low pH of the product which secures a long shelf-life after the mild heat treatment. To be effective, pasteurization is frequently combined with another means of preservation such as concentration, acidification, chemical inhibition, etc.

Meat products, cured or uncured, are often subjected to pasteurization carried out at temperatures around 80°C for several minutes resulting in a limited shelf-life and the need for refrigeration. Although cooking would destroy vegetative pathogens and most spoilage bacteria, heat-resistant lactobacilli and streptococci may survive, and psychrotrophic species may cause spoilage.

Most vegetables are low-acid products with pH >4.6 and have to be sterilized, with the exception of pickles and fermented vegetables which represent high-acid products. Acidified pickled products in salt brine with 0.6–1.0% vinegar and also containing sugar are pasteurized at 80–85°C. Tomatoes are fairly acidic with a pH value around 4.6 or less, hence they can be preserved by mild heat treatment generally with pasteurization. Tomato paste is a common product which can be preserved by hot-filling at a temperature of 90–92°C without further pasteurization. Fruit products, juices and preserves have generally low pH of 3.2–3.8, and are usually pasteurized at 70–75°C. This assures a 5-log cycle reduction of vegetative form of pathogenic bacteria; however, heat resistance of yeasts can be higher. Hence, yeasts are the primary spoilage agents in fruit-based beverages and soft drinks.

Heat-resistant molds and alicyclobacilli may also survive pasteurization; however, being aerobic organisms, their spoilage potential in carbonated beverages is limited (Parish, 2006). Pasteurized fruit preserves (jams, jellies, marmalades) can be spoiled by certain fungal species with heat-resistant ascospores such as *Byssochlamys fulva*, *B. nivea*, *Neosartorya fischeri*, *Talaromyces flavus*, *T. bacillisporus* and *Eupenicillium baarnense* and some other *Eupenicillium* species (Beuchat and Pitt, 2000). Among other foods and beverages, pasteurization is widely used for beer filled and sealed in cans or bottles.

FACTORS DETERMINING HEAT TREATMENT

Since heating applied to destroy microorganisms may also exert adverse effects on the quality of foods, in practice a minimum possible heat treatment is to be used which can guarantee destruction of pathogens and toxins and give the desired storage life, but also retain the characteristic organoleptic properties of food products. This compromising requirement will determine the choice of heat treatment.

The heating (sterilization or pasteurization) process can be subdivided into three phases. By means of a heating medium (water or steam) the product temperature is increased from ambient to the required sterilization temperature (phase 1: heating phase). This temperature is maintained for a defined time (phase 2: holding phase). In phase 3 (cooling phase) the temperature in the product is decreased by introduction of cold water on the surface of the container.

In order to effectively and safely preserve foods using heat treatment, it is not enough to apply the required time–temperature combination to inactivate the most heat-resistant pathogens and spoilage organisms in a particular food. In addition to this, another factor should be also considered: the heat penetration characteristics in a particular food. In order for the heat sterilization to be efficient, the preservation processes must provide a heat treatment which will ensure that every particle of food within a container will reach a sufficient temperature, for a sufficient time, to inactivate the most resistant pathogens and the majority of spoilage organisms as well. It is usually the centermost particle, called the cold point, where the heat would penetrate least (Figure 17.5).

The course of temperature during thermal processing in a retort depends on several factors related to: (1) heating conditions (retort type, loading, time–temperature formula), (2) heating mode (still or agitated), (3) heating medium (water, steam, with/without overpressure), (4) product type (solid, liquid), and (5) container type, shape and size. The thermophysical properties of the product, in particular its consistency, will influence the mode of heat transfer, and are of utmost importance for the speed of heat penetration. Basically, heat will spread in solids by conduction and in liquids by convection; however, in real foodstuffs it is usually between the two extremes, and may change during the heating process. In the context of HACCP, determining the coldest point and monitoring temperature at this point is particularly important and from the above it can be understood that the coldest point is not always the center.

Heat penetration is extremely important, because it is the determining factor for the success of the whole operation. The most suitable and practical method to speed up thermopenetration is the movement of containers during the thermal process. Rotation of

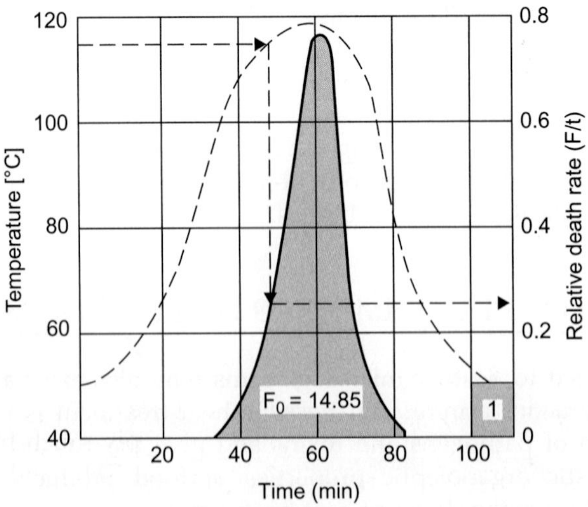

FIGURE 17.5 The slowest warming-up cold point in containers. Heat transfer is (a) by conduction, (b) by convection. *Reproduced by the authorization of the Food and Agriculture Organization of the United Nations from Heinz and Hautzinger (2007).*

containers around their axis is an efficient means to accelerate heat transfer, because this will rapidly mix the contents, enabling a more uniform heating of products, and reducing heating time and organoleptic degradation. Heat penetration is slow, especially in the case of the pasteurization of products packed in glass containers.

In addition to the composition and moisture of the food, the acidity and pH value have tremendous impact on the efficacy of heat preservation. It is customary to divide foods into groups concerning heat treatment according to their acidity. Acid foods have pH below 4.5 and low-acid foods are those with pH above 4.5. Acid foods include most fruits, and pasteurization would suffice for preservation; low-acid foods are those like meat and most vegetables, which should require sterilization treatment. Table 17.3 lists various types of foods and their pH value, together with the heat processing requirements.

NON-TRADITIONAL HEAT TREATMENT

The conventional method of heat sterilization often leads to overcooking of the food material causing unwanted loss of nutrients and organoleptic changes. Electric heating methods offer novel possibilities for sterilization providing better retention of quality attributes. Two types of electrical heating methods are known and have been practically explored: direct and indirect. In the case of the direct method electrical current is passed directly into the food (called ohmic heating, OH, or electrical resistance heating). With indirect electroheating the electric energy is first converted to electromagnetic radiation which subsequently generates heat within a product. These methods are microwave (MW) and radiofrequency (RF) heating (Figure 17.6) (Marra et al., 2009; Ramaswamy and Tang, 2008).

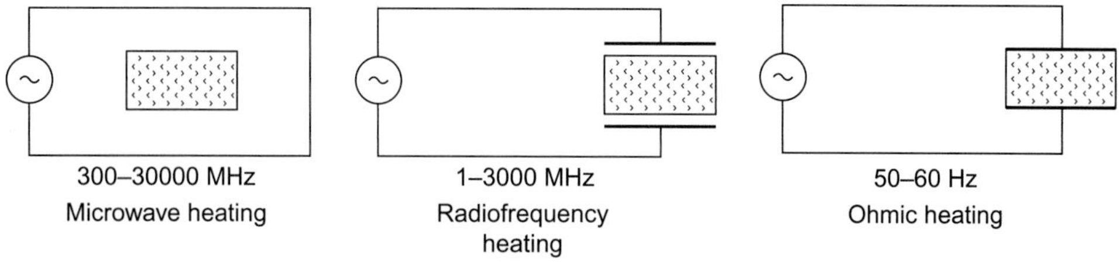

FIGURE 17.6 Schematic of electrical heating methods.

In OH the product is placed in direct contact with a pair of electrodes through which a low-frequency (50 to 60 Hz) electric current is transferred. Heat is generated due to the resistance of molecules to electrical conduction. Indirect methods apply much higher oscillating frequencies of electromagnetic waves – MW (300 MHz and 300 GHz) and RF (3 kHz–300 MHz) – which result in heating of dielectric materials by induced molecular vibration as a result of dipole rotation or ionic polarization. Changing polarity of electrical field forces oscillation of ions whose friction generates heat in the product. Movement of dipolar water molecules contributes mainly in heating due to the higher frequencies of MW. With OH an additional non-thermal effect is electroporation of cell membranes which might occur even under low frequencies (60 Hz) building up charges in the cell envelope (Lebovka et al., 2005).

OH and RF are used only for industrial purposes while MW is applied very commonly domestically and finds commercial application as well. Two frequency bands of MW are allocated in the USA. The 915 MHz band is used for industrial heating only, and the 2450 MHz band is used both in the industry and in domestic microwave ovens. Ohmic heating can be used for heating liquid foods containing large particulates, such as soups, stews, fruit slices in syrups and sauces, and heat-sensitive liquids. The technology is useful for the treatment of proteinaceous foods (e.g. liquid egg) which tend to denature and coagulate when thermally processed. Juices can be treated to inactivate enzymes with less destruction of the flavor. Other potential applications of ohmic heating include blanching, thawing, dehydration and extraction.

Due to their lower frequency levels, RF waves have a larger penetration depth than MW and hence could find better application in larger size foods. Cooking time of meat and meat products was found much shorter than conventional cooking in a water bath, and caused lower juice losses. RF radiation is also considered for post-harvest treatment and disinfestation of fruits. RF heating is also applicable for continuous flow processing of liquid foods such as fruit juices and milk.

Plastic packaging materials are transparent to microwaves. Microwaves can, therefore, be used to process prepackaged food products. Examples of in-package microwave sterilized products include different pasta dishes, pasta sauces, rice and other ready-to-eat dishes.

Electroheating has found many applications in the food processing industry, including tempering of frozen foods for further processing, precooking of meat, and finishing the drying of pasta products. In those applications, electric heating methods demonstrate significant advantages over conventional methods in reducing process time, improving food quality and reducing environmental impacts. Electric sterilization can have a major advantage over conventional retorting because of the relatively short heating time and potential to produce high-quality self-stable food products. Increasing numbers of commercial equipment appear on the market enabling more development in technology. Approaches are currently in development for continuous flow processing with aseptic packaging.

COMBINED TREATMENTS

Consumers raise increasing demand for ready-to-eat, fresh, minimally processed foods preserved by relatively mild techniques in order to minimize the loss of quality and to control microbial growth, and thus ensure product safety. To meet this demand, a hurdle approach appears to be the best method (Leistner, 2000; Alzamora et al., 2000). Hurdle technology is the term often applied when foods are preserved by a combination of processes. In the design of hurdle technology several preservation systems can be applied by the combination of factors such as temperature, water activity, pH, redox potential, preservatives and packaging. Two or more preservation methods can be applied together in smaller degrees that separately would not produce safe products. The combination can ensure not only safety but also results in better, more natural quality, and is economic by saving energy. This is because different hurdles in a food often have an additive or synergistic effect. If several hurdles are used simultaneously, a gentle preservation could be applied, which nevertheless secures stable and safe foods of high sensory and nutritional properties. Using combined technologies, moreover, the diversity of products can be increased and new types of food can be developed. Conventional and novel thermal technologies are often combined with other treatments in order to moderate the severity of doses required if applied alone (Ukuku and Geveke, 2010; Liu et al., 2011).

Examples of hurdle technology for fruit and vegetable processing are the intermediate moisture fruit product (IM, containing two hurdles as pH and a_w), a high-moisture fruit product (HMF, preserved by mild heat treatment, low pH and a_w, and/or preservative but without refrigeration), as well as a minimally processed refrigerated fruit product (MPR, treated with mild heat, then packaged and refrigerated) (Barbosa-Cánovas et al., 2003).

Several types of meat products, mainly various sausages, are processed with the combination of different preservative factors (mild cooking, low pH and a_w, nitrite or smoke curing, fermentation, refrigeration) and usually can be stored at ambient temperatures for a given time before organoleptic deterioration starts. They are called shelf-stable food (SSP), and their stability is due to the synergistic interactions of preservative effects (Kanatt et al., 2002; FSIS-USDA, 2005).

Novel types of meat products and ready-to-eat dishes can be preserved by combined treatments when packaged under vacuum, heat pasteurized at mild temperatures for long time then refrigerated. Sous-vide (French for "under vacuum") is the name of the method of cooking food sealed in airtight plastic bags in a water bath for a long time (24 to 72 hours are usually applied) at an accurately determined temperature much lower than normally used for cooking, typically around 45–60°C (111–140°F). Sealing the food in sturdy plastic bags keeps in juices and aroma that would otherwise be lost in the process. The use of temperatures much lower than for conventional cooking is an equally essential feature of sous-vide. In English these dishes are also called Repfed (refrigerated processed food of extended durability) emphasizing the third hurdle factor, i.e. low temperature storage.

These products can be stored for several weeks and are directly consumable. The mild but long heat treatment kills most vegetative bacteria. The stability and safety of sous vide products depend on the delicate balance of preservative treatments under strict hygienic conditions. Neither the cooking time nor the storage temperature provide, however accurately controlled, absolute safety for these kind of foods. In particular, spores of some strains of *C. botulinum* can survive mild cooking and grow and produce toxin at 3–5°C (Hyytiä-Trees et al., 2000). Hence, FDA suggests an additional hurdle (e.g. preservative) to be combined to ensure the safety of these foods.

A novel manufacturing technique has been suggested for the production of extended shelf-life (ESL) milk with fresh taste and prolonged stability of up to 4 weeks when distributed maintaining a cold chain. This method combines processing by microfiltration, pasteurization and subsequent refrigeration. Raw milk is separated into skimmed milk and milk fat, the former is microfiltered through caramic membranes (with pore size of 1.4 μm) and pasteurized thereafter (77°C for 30 s). The milk fat is heated at ultra-high temperature (125°C for 4 s) and then reverted to the skimmed milk. Packages should be stored at temperatures below 10°C. Various spoilage and pathogenic microorganisms (among them *B. cereus*) may survive or contaminate products post-process, hence jeopardizing the safety and stability of ESL milk (Elwele and Barbano, 2006).

References

Alzamora, S.E., Tapia, M.S., López-Malo, A., 2000. Minimally Processed Fruits and Vegetables: Fundamental Aspect and Applications. Aspen Pub. Co., Inc., Maryland, US, pp. 277–286.

Barbosa-Cánovas, G.V., Fernández-Molina, J.J., Alzamora, S.M., Tapia, M.S., López-Malo, A.L., Welti Chanes, J., 2003. Handling and preservation of fruits and vegetables by combined methods for rural areas. Techn. Man. Agric. Serv. Bull. 149, FAO, Rome.

Beuchat, L.R., Pitt, J.I., 2000. Detection and enumeration of heat resistant molds. In: Pouch-Downes, F., Ito, K. (Eds.), Compendium of Methods for the Microbiological Examination of Foods. Am. Public Health Assoc., Washington, DC, pp. 217–222.

Deak, T., Farkas, J., Incze, K., 1980. Microbiology of Canning-, Meat- and Freezing Industries (*in Hungarian*). Mezőgazdasági Kiadó, Budapest.

Desrosier, N.W., Desrosier, J.N., 1977. Technology of Food Preservation, fourth ed. AVI Publishing Co., Westport, CN.

Elwele, M.W., Barbano, D.M., 2006. Use of microfiltration to improve fluid milk quality. J. Dairy Sci. 89, E20–E30.

Evancho, G.M., Tortorelli, S., Scott, V.N., 2009. Microbiological spoilage of canned foods. In: Sperber, W.H., Doyle, M.P. (Eds.), Compendium of the Microbiological Spoilage of Foods and Beverages. Springer Sci., New York.

Fellows, P.J., 2009. Food Processing Technology. Principles and Practice, third ed. Woodhead.

FSIS-USDA, 2005. Principles of preservation of shelf-stable dried meat products. <www.fsis.usda.gov/PDF/FSRE_SS_7%205/11/05>.

Heinz, G., Hautzinger, P., 2007. Meat processing technology for small to medium scale producers. RAP Publication (FAO), no. 2009/20/FAO, Bangkok (Thailand).

Hyytiä-Trees, E., Skyttä, E., Mokkila, M., Kinnunen, A., Miia Lindström, M., Liisa Lähteenmäki Raija Ahvenainen, L., Korkeala, H., 2000. Safety evaluation of sous vide-processed products with respect to nonproteolytic *Clostridium botulinum* by use of challenge studies and predictive microbiological models. Appl. Environ. Microbiol. 66 (1), 223–229.

Kanatt, S.R., Chawla, S.P., Chander, R., Bongirwar, D.R., 2002. Shelf-stable and safe intermediate-moisture meat products using hurdle technology. J. Food Prot. 65 (10), 1628–1631.

Lebovka, N.I., Praporsic, I., Ghnimi, S., Vorobiev, E., 2005. Does electroporation occur during the ohmic heating of food? J. Food Sci. 70 (5), E308–E311.

Leistner, I., 2000. Basic aspects of food preservation by hurdle technology. Int. J. Food Microbiol. 55, 181–186.

Liu, Y., Tang, J., Mao, Z., Mah, J.-H., Jiao, S., Wang, S., 2011. Quality and mold control of enriched white bread by combined radio frequency and hot air treatment. J. Food Engn. 104, 492–498.

Marra, F., Zhang, L., Lyng, J.G., 2009. Radio frequency treatment of foods: review of recent advances. J. Food Eng. 91 (4), 497–508.

Motarjemi, Y., Kaferstein, F.K., 1999. Food safety, hazard analysis and critical control point and the increase in food-borne diseases: a paradox? Food Control 10 (1999), 325–333.

Parish, M.E., 2006. Spoilage of juices and beverages by *Alicyclobacillus* spp. In: Sapers, G.M., Gorny, J.R., Yousef, A.E. (Eds.), Microbiology of Fruits and Vegetables. Taylor and Francis, Boca Raton, pp. 159–183.

Ramaswamy, H., Tang, J., 2008. Microwave and radiofrequency heating. Food Sci. Technol. 14 (5), 423–427.

Richardson, P. (Ed.), 2004. Improving the Thermal Processing of Foods. Woodhead, Cambridge.

Robinson, R.K., 2002. The microbiology of milk and milk products. Dairy Microbiology Handbook, third ed. Wiley, New York.

Ryan, C.A., Nickels, M.K., Hargrett-Bean, N.T., Potter, M.E., Endo, T., Mayer, L., et al., 1987. Massive outbreak of antimicrobial resistant salmonellosis traced to pasteurized milk. JAMA 258 (22), 3269–3274.

Schmit, J., 2008. Case #1: management problems cited in botulism case. USA Today 6/30/08.

Shapiro, R.L., Hatheway, C., Swerdlow, D.L., 1998. Botulism in the United States: a clinical and epidemiological review. Ann. Int. Med. 129 (3), 221–228.

Stumbo, C.R., 1973. Thermobacteriology in Food Processing. Academic Press, London.

Tomkins, R.J., Ordal, Z.J., 1976. Thermal injury and inactivation of vegetative bacteria. In: Skinner, F.A., Hugo, W.B. (Eds.), Inhibition and Inactivation of Vegetative Microbes. Academic Press, London, pp. 153–190.

Ukuku, D.O., Geveke, D.J., 2010. A combined treatment of UV-light and radio frequency electric field for the inactivation of *Escherichia coli* K-12 in apple juice. Int. J. Food Microbiol. 138 (1–2), 50–55.

CHAPTER

18

Non-thermal Processing Technologies

Olga Martín-Belloso[1], *Robert Soliva-Fortuny*[1], *Pedro Elez-Martínez*[1], *A. Robert Marsellés-Fontanet*[1] *and Humberto Vega-Mercado*[2]

[1]University of Lleida, Lleida, Spain, [2]Merck & Co., West Point, PA, USA

OUTLINE

Food Safety Management.
DOI: http://dx.doi.org/10.1016/B978-0-12-381504-0.00018-4

INTRODUCTION: IDENTIFICATION OF RISKS IN NON-THERMAL PROCESSES

Every food manufacturing and processing operation has inherent risks affecting the safety of food products. Non-thermally processed foods are not exempt from those risks, which include, among others, incorrect process conditions, variability in microbial, chemical and physical characteristics of raw materials, post-processing contamination and mishandling or abuse during packaging, storage, shipping and distribution.

The biological, chemical and physical risks must be assessed for each manufacturing process, considering the type of product and the involved technologies. Therefore, the information provided in this chapter should be considered as a starting point for a more detailed analysis once a particular technology is identified to process a food product. The uncertainties regarding specific biological hazards can be clearly stated based on the technology and the product and, consequently, the selection of proper processing conditions and definition of critical process parameters can be made as well as the best way of monitoring them.

Overall Product Life Cycle

The first step in assessing the risks is to understand the overall product life cycle. Figure 18.1 shows a general life cycle of a product manufactured using in-line non-thermal processes.

Each step along the manufacturing process must be designed to prevent contamination or to reduce the extension of the assessed risks. In terms of product life cycle, such objectives require proper packaging of the raw materials, appropriate shipping and storage conditions, protection of raw materials from insects or rodents, aseptic handling of the raw materials during dispensing, appropriate cleaning and sanitization/sterilization of processing equipment, setup and operation of the equipment, controlled formulation and holding of the product prior and after non-thermal processing, packing of the product into sterile containers, integral container closures, proper storage of processed product, and proper distribution, retailing and handling of the product once in the market.

Assumption of the established good manufacturing practices (GMPs), hygienic plans and related safety tools by the different stakeholders of the food production chain are critical for the successful development and implementation of quality assurance plans. Their application is recommended or even compulsory in most countries.

Raw Materials

The chemical and physical properties of the raw materials will define the microbiological characteristics of the formulated product, and in turn the shelf-life of the unprocessed products, the minimum process conditions required to ensure microbial safety after processing, as well as suitable post-processing storage and handling conditions. Table 18.1 summarizes the chemical and physical properties typically associated with microbiological characteristics of food products.

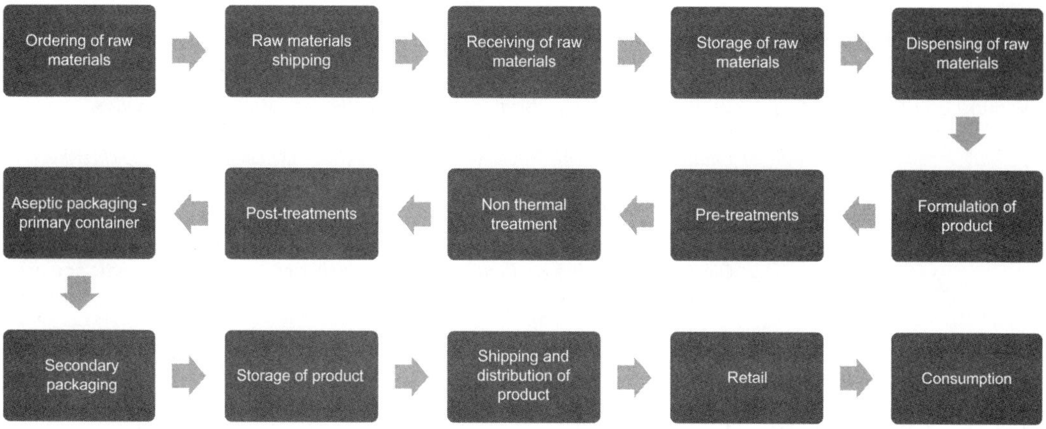

FIGURE 18.1 Overall life cycle of a product processed using non-thermal technology.

TABLE 18.1 Chemical and Physical Properties of Foods

Physical Properties	Chemical Properties
Solid	pH
Powder	Acidity and type of acid
Liquid	REDOX potential
Viscosity	Water activity
Aqueous	Protein content
Oil	Carbohydrate content
Internal structure (e.g. size distribution in emulsions)	Lipid content
	CO_2, O_2 concentrations
	Preservatives

The source and type of raw material will also have a determining influence on the kind of microorganisms that can grow in it. Table 18.2 summarizes the nature of microbial contamination related to different food products.

Non-thermal Food Processing

Food processing using non-thermal processes such as irradiation, pulsed electric fields, high-intensity pulsed light, high hydrostatic pressure, membrane filtration or a combination of any of these through a hurdle approach represents a change from the traditional heat processes that are well characterized. The technological advances associated with these processing methods and extensive scientific information demonstrating the ability of these methods for microbial inactivation provide assurance of their effectiveness in extending the shelf-life of food products. A key advantage of non-thermal processes is better retention of

TABLE 18.2 Typical Microorganisms in Food Products (ICMSF, 1998)

Food Products	Typical Microorganisms
Raw milk	Lactic acid bacteria, *Pseudomonas, Flavobacterium, Micrococcus, Bacillus, Enterobacter, Aeromonas, Alcaligenes*, viruses, *Salmonella, Yersinia enterocolitica, Campilobacter jejuni, Staphylococcus aureus, Listeria monocytogenes, Escherichia coli*
Liquid eggs	Diverse mixtures of Gram-positive and Gram-negative bacteria
Cattle and sheep carcasses	*Salmonella* spp., *C. jejuni, E. coli, Bacillus cereus, L. monocytogenes, S. aureus, Clostridium perfringens, Pseudomonas, Acinetobacter, Psychrobacter*, lactic and acetic bacteria and yeasts
Raw marine fish	*Pseudomonas, Vibrio, Acinetobacter*, Coryneform bacteria, *Flavobacterium, Micrococcus*, Enterobacteriaceae and yeast, *Anisakis simplex*
Raw vegetable products	Pseudomonads, *Erwinia carotovora*, coryneforms, spore-formers, coliforms and micrococci, several species of fungi, *B. cereus, L. monocytogenes, Clostridium botulinum, Cl. perfringens*
Raw fruits	Fungi, yeasts

TABLE 18.3 Risks Associated with Non-thermal Processes

Non-Thermal Process	Risks
Irradiation	Suboptimal irradiation dose, non-homogeneous treatment, damaged treatment containers
High hydrostatic pressure	Incorrect pressure setup or duration, too low and inhomogeneous temperatures, damaged treatment containers, contaminated treatment fluid
Pulsed electric fields	Incorrect pulse intensity strength or treatment time, too low temperature, non-homogeneous treatment, inadequate pre-decontamination of process line
High intensity pulsed light	Incorrect pulse intensity or treatment time, non-homogeneous treatment
Membrane filtration hurdle technology	Incorrect pore size, compromised membrane. Depending on the combination of technologies used

nutrients and sensory attributes close to those observed in fresh or minimally processed products. As previously mentioned, these methods have inherent risks involving the potential of microbiological contamination. Table 18.3 summarizes the risks associated with each method.

From a safety point of view, all the processing treatments mentioned above have been specifically designed to eliminate or reduce the likely occurrence of biologic hazards to an acceptable level. Therefore they must be considered critical control points (CCP) in any hazard analysis and critical control point (HACCP) program.

Packaging

Packaging of food products processed using non-thermal technologies such as pulsed electric fields, high-intensity pulse light and membrane filtration, or a combination of them, will require aseptic conditions. Aseptic packaging is not usually required for irradiated and high-pressure processed foods, which allow prepackaging of the product prior to the decontaminating step.

The use of aseptic techniques to package a product requires proper sanitization of the processing equipment. Failure to clean the equipment will result in cross-contamination or adulteration of the food product. Also, residues of the cleaning agents may represent a potential risk to food safety. Meanwhile, cleaning of the equipment may not suffice to protect the product quality. Sterilization of the processing equipment is often required to reduce or eliminate post-processing microbiological contamination. Proper sanitization or sterilization parameters must be developed and demonstrated to ensure the effectiveness in reducing or eliminating microbial contamination from the product contact surfaces.

Packaging containers must be compatible with the product and properly designed for the intended application or processing method. Containers used for products processed using pulsed electric fields, high-intensity pulsed light or membrane filtration must be sanitized or sterilized and protected from contamination prior to the filling operation. Meanwhile, containers used for products processed using irradiation or high pressure must keep their integrity throughout the processing steps and during the shelf-life of the product.

Distribution

Distribution of minimally processed food products requires proper controls to prevent spoilage or damage of the product. The main concern during distribution is the potential thermal abuse (exposure of a food product to extremely high- or low-temperature conditions). Product containers and shipping conditions are critical to protect the quality of a food. Product containers must withstand the shipping and handling process while maintaining their integrity.

NON-THERMAL TREATMENTS FOR FOOD PRESERVATION

In this section we will discuss the requirements for monitoring the manufacturing process, the critical control points and critical process parameters for each of the technologies being discussed.

Irradiation

Principles

Irradiation requires transferring energy from high energetic sources such as unstable isotopes or machine-powered irradiators to food products to inactivate/kill microorganisms so that biologic hazards are eliminated or reduced up to acceptable levels.

The isotopes used for food processing are ^{60}Co and ^{137}Cs: both irradiate γ-rays. Meanwhile, electron beams or X-rays are produced by electrical devices. All of them have the capability to randomly excite and ionize molecules, hence the name ionizing radiation (WHO, 1997):

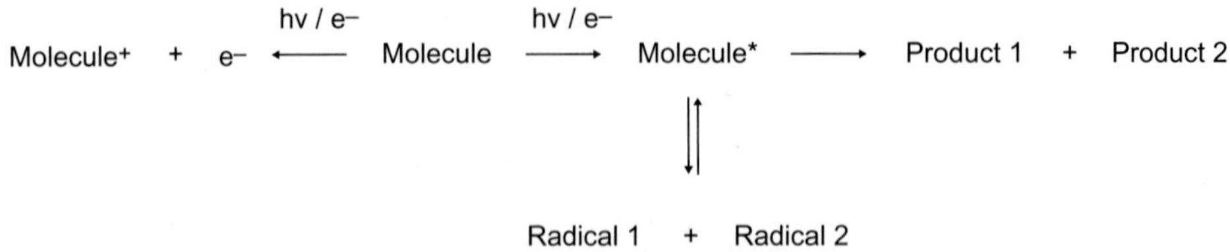

(18.1)

Being the most common component of many foodstuffs, water is a good example to show the chemistry of radiation. The products formed from water radiolysis are displayed in the following equation:

$$H_2O \xrightarrow{hv/e^-} H_3O^{\oplus} + HO^{\ominus} + H^- + e^- + H_2 + H_2O_2 \tag{18.2}$$

In summary, irradiation triggers the formation of highly unstable compounds, radicals, which are involved in several chemical transformations before yielding stable compound (Steward, 2001). These transformations are associated with the inactivation of microorganisms. The extension of chemical changes on specific molecules is proportional to both the received irradiation dose and the compound's molecular weight (Miller, 2005).

The mean-life of the radiolysis species does not exceed a few milliseconds even though the low stability of such compounds could affect biological pathways of food microorganisms disturbing the cellular homeostasis. However, the most direct effect on unwanted microorganisms occurs when ionizing radiation damages their DNA. Despite microbial repairing systems, some injuries lead to cellular death or inhibition of cellular reproduction (Dickson, 2001). Biological sensitivity to radiation can be measured as *D*-values, as occurs with thermal treatments.

Irradiation is not an intended step to reduce or eliminate physical and chemical hazards even though it should be kept in mind that some risks could come from interactions between irradiation and exogenous agents. Releasing of chemicals (e.g. leachates) from the food package or the carrier system are examples of such interactions (Arvanitoyannis, 2010).

Critical Factors and Critical Limits

The described biological effects of the irradiation treatments are related with the spatial and temporal distribution of the energy transmitted by the ionizing radiation as well as with the degree of absorption of such radiation by the food product. Consequently, the critical parameters of the irradiation treatments are the dose (*D*, Gy), and exposure time (*t*, s) at every position in the product. Exposure time should be long enough to ensure the delivery of the minimum required dose to achieve the expected effects, without reaching a dose whose adverse effects compromise the product quality or even the throughput efficiency

TABLE 18.4 Estimated D_{10}-values for Common Organisms Subjected to Ionizing Radiation Treatments (Miller, 2005; Garcia-Gonzalez et al., 2007)

Type	Organism	Medium	Temperature (°C)	D_{10}-value (kGy)
Virus	Hepatitis A	Clams, oysters	Ambient	4.8
Non-spore-forming bacteria	*Campylobacter jejuni*	Ground beef	Ambient	0.15
	Listeria monocytogenes	Poultry meat	12	0.49
	Escherichia coli	Mechanically deboned chicken meat	10	0.23
	Salmonella Enteritidis	Low-fat ground beef	Ambient	0.7
	Staphylococcus aureus	Low-fat ground beef	5	0.75
Spore-forming bacteria	*Clostridium botulinum*	Beef stew	Ambient	1.5
	Clostridium perfringens	Water	Ambient	2.1
Yeasts and molds	*Aspergillus flavus*	Growth culture	Ambient	1.0
	Trichosporon cutaneum	Fresh sausage	Ambient	1.0
Parasites	*Entamoeba hystolytica*	Water, fresh fruit and vegetables	Ambient	<0.1
	Cycsticercus bovis	Beef	Ambient	0.4
	Trichina spiralis	Pork	Ambient	0.1
Insects	Fruit fly	Fresh fruit	Ambient	0.15

of the facility. Both factors, dose and treatment time, depend on target, product and device parameters to provide safe products:

- Biological agent-dependent variables
  - Natural resistance: development stage, DNA reparation systems.
- Product-dependent variables
  - Product nature: composition, density, state, frozen or packaged goods, presence/absence of oxygen.
  - Geometry: shape and depth.
- Processing device variables
  - Source system: radioactive/electric generator device, emission energy or intensity.
  - Scanning system: single/multiple.
  - Conveyor system: continuous/batch processing equipment.

Assuming that the product receives a homogeneous dose of radiation, parasites and insect elimination require typically less than 1 kGy. Doses between 1 and 10 kGy are needed to destroy vegetative microorganisms. If sterilization is required, more than 10 kGy should be delivered. Minimum values and applications are legally defined in each country where ionizing radiation treatments are allowed for food applications (Table 18.4).

As far as maximum levels of irradiation are concerned, exhaustive studies for several decades concluded that no upper dose limit needs to be imposed, and that irradiated foods are deemed wholesome throughout the technologically useful dose range from below 10kGy to envisioned doses above 10kGy (WHO, 1997). Such studies also revealed that the compounds produced after excessive irradiation cause severe changes in the sensory characteristics of the product. Furthermore, the compounds found in such overprocessed products were very similar to those observed on products after a severe thermal treatment. This fact suggests that the maximum dose levels should comply with GMPs in countries without government maximum limits.

Monitoring

The measurement of the dose received by the product, both spatially and temporally, is the best way to monitor the irradiation process since it could change due to some factors which are more prone to change than others. Namely, the processing device variables are typically part of the design of the facility, thus well known and usually difficult to change. One paradigmatic example is the depletion of irradiation intensity with time when using isotope devices. Others, like those variables depending on the product characteristics, are commonly less homogeneous even though more easily modifiable.

Even so, there are several options to monitor the dose. Radiation dosimeters are the more direct solution and commonly used by industry. As there are several kinds of dosimeters, a way to control product dose could be by inserting a reference dosimeter on each piece of product. However, they are often very expensive and difficult to maintain, and they should usually be operated by trained people. Therefore the approach used in some facilities is to use reference dose measurements during the setup stage when a profile of the dose as a function of external dosimeters is mapped. Of course, reference dosimeters should be periodically inserted into the product or into a good product simulator in order to achieve proper data. Such information is the reference to link the external dosimeter measurements with the real dose received by products under real processing conditions. External dosimeters tend to be cheaper and more easily maintained than reference dosimeters. Moreover, they allow online data acquisition (Miller, 2005).

If the facility operates under design conditions, the described monitoring procedures should be enough. However, any product or device changes will require an exhaustive study to confirm that the minimum dose is homogeneously received prior to the commercial distribution of the product.

Supercritical Fluid Technology

Principles

This technology is grounded on the known inhibitory effect of carbon dioxide on microbial growth (Garcia-Gonzalez et al., 2007) (Table 18.5). The effect is enhanced by maintaining such gas under specific environmental conditions, known as supercritical fluid (SCF) conditions, using pressure values between 5 and 30MPa (Demazeau and Rivalain, 2011).

The mechanism of the bacteriostatic effect of CO_2 is still under discussion, even though the current working hypothesis states several steps:

- Solubilization of pressurized gas in the external liquid phase.
- Cell membrane modification.

TABLE 18.5 Examples of Susceptibility of Different Bacterial Species to Supercritical Carbon Dioxide Treatments on Several Media (Garcia-Gonzalez et al., 2007)

Target Microorganism	Solution	Process Conditions	Reduction
Saccharomyces cerevisiae	Hydrophilic filter paper disks	5 MPa, room temp., 420 min	3D
Listeria innocua	Growth medium	20.5 MPa, 34 °C, 36 min, 3 cycles	3D
		20.5 MPa, 34 °C, 36 min, 6 cycles	9D
Staphylococcus aureus	Growth medium	20.5 MPa, 34°C, 36 min, 3 cycles	3D
		20.5 MPa, 34°C, 36 min, 6 cycles	7D
Salmonella Salford	Growth medium	20.5 MPa, 34°C, 36 min, 3 cycles	3D
		20.5 MPa, 34°C, 36 min, 6 cycles	3D
Pseudomonas aeruginosa	Growth medium	20.5 MPa, 34°C, 36 min, 3 cycles	6D
Escherichia coli	Growth medium	20.5 MPa, 34°C, 30 min, 3 cycles	8D
Proteus vulgaris	Growth medium	20.5 MPa, 34°C, 36 min, 3 cycles	8D
Legionella dunnifii	Growth medium	20.5 MPa, 40°C, 90 min, 6 cycles	4D
Pseudomonas aeruginosa	Physiological saline	7.4 MPa, 38°C, 2.5 min	7D
Bacillus subtilis	Physiological saline	7.4 MPa, 38°C, 2.5 min	7D
Escherichia coli	Sterile water	20 MPa, 34°C, 10 min	2.5D
Staphylococcus aureus	Sterile water	20 MPa, 34°C, 10 min	3.5D
Saccharomyces cerevisiae	Phosphate buffer solution	7.4 MPa, 38–40°C, 10 min	5.8D
Serratia marcescens	Phosphate buffer solution	7.4 MPa, 38–40°C, 0 min	7.3D
Bacillus subtilis	Phosphate buffer solution	7.4 MPa, 38–40°C, 2.5 min	7.6D
Saccharomyces cerevisiae	Grape juice	48.3 MPa, 25°C, 5 min, 85 g/kg CO_2	5.1D
Candida stellate	Grape juice	48.3 MPa, 25°C, 5 min, 85 g/kg CO_2	5.6D
Kloeckera apiculata	Grape juice	48.3 MPa, 25°C, 5 min, 85 g/kg CO_2	3.7D

- Key enzyme inactivation and cellular metabolism inhibition due to pH reduction and direct effect of carbon dioxide or hydrogen carbonate anion such as promotion of disorders of the cellular homeostasis.

Other inert gases (N_2, N_2O, CF_2-CF_2, Ar, and mixtures between them) have been studied as a way to avoid chemical reactions promoted by carbon dioxide that could modify product characteristics. However, such gases yielded poorer results and few sensory and nutritional comparison studies have shown significant differences with unprocessed foods (Garcia-Gonzalez et al., 2007).

Critical Factors and Critical Limits

Apart from the type of gas, the most critical variables of the supercritical fluid technology are, obviously, pressure, temperature (usually 20–40°C) and treatment time. Pressure and temperature affect gas properties such as its solubility, density and therefore its diffusion into cells. Time allows controlling in part the extension of the treatment effects. Critical values of the treatments should be considered on a case-by-case basis because the inherent variability of other external factors such as the nature of the food (physical state of the food product, chemical and physical properties), the pressurizing system if an auxiliar transmitter medium is used, and the microbial susceptibility (Spilimbergo et al., 2011).

High Hydrostatic Pressure

Principles

High pressure affects biological constituents and systems (Cheftel, 1995). On foods, it has been studied as a physical agent on high hydrostatic pressure (HHP) treatments, which uses a range of pressures between 100 and 1200 MPa. The technology has the same advantage as irradiation in the sense that it allows the treatment of solids, liquids and either packaged or unpackaged goods. If packed prior to the treatment, it reduces the possibilities of microbial contamination after processing. Conversely to irradiation though, the treatment is inherently homogeneous and independent of the shape of the product because pressure changes are instantaneously and isostatically transmitted (Doona and Feeherry, 2007).

Otherwise, the HHP technology does not allow continuous processing since after the product is placed into a vessel containing the pressure transmission medium, the vessel should be kept closed up to the end of the required treatment. The main characteristics of the treatment are the come-up, holding and down times, the pressure level and the temperature of pressure processing. There exists a temperature variation (around 3–9°C) for each pressure change (100 MPa) due to the fact that work is applied in adiabatic conditions (Patazca et al., 2007).

High pressure affects microbial communities by changing cell morphology, damaging cell membranes and walls as well as by disturbing some key enzyme structures and metabolic pathways. Membrane permeabilization is considered a direct consequence of membrane thinning by compression. Vacuolar compression and ribosome dissociation have also been reported (Considine et al., 2008). The technology does not deliver enough energy to break any covalent bond, so there are very few chemical reactions. Therefore changes of enzymatic activities are related with modifications of the second or upper structural layout (Palou et al., 1999; Balny and Masson, 1993). Similar processes occur in food cells. On a liquid product such an effect should not cause any concern but on solid products such treatments are often aimed at achieving texture changes rather than ensuring food safety (Table 18.6).

The discussed critical variables (treatment time and pressure) can be easily monitored online. Temperature should also be supervised because it plays a quantifiable role, as occurs with other non-thermal technologies. Furthermore, temperature does not possess the isotropic behavior of pressure (Grauwet et al., 2010). Therefore it can be monitored selecting any of the different methods developed for thermal processing such as tracking the cold

TABLE 18.6 Bacterial Barotolerance Differences and Medium Influences (Rajkovic et al., 2010)

Pathogen	Food Product	Treatment Conditions	Log_{10} Reduction
Escherichia coli O157:H7	Apricot juice (pH 3.8)	250 MPa, 5 min, 30°C	4.85
	Orange juice (pH 3.76)	250 MPa, 5 min, 30°C	5.1
	Sour cherry juice (pH 3.3)	250 MPa, 5 min, 30°C	5.28
	Apple juice (pH 3.5)	500 MPa, 5 min, 20°C	5
	Tomato juice (pH 4.1)	500 MPa, 5 min, 20°C	5
	Orange juice (pH 3.8)	500 MPa, 5 min, 20°C	1–2
	Raw minced meat	700 MPa, 1 min, 15°C	5
	Hungarian salami	600 MPa, 6 min, 25°C	>4
Listeria monocytogenes	Human milk	400 MPa, 1.5 min, 31°C	≈6
	Turkish white cheese	600 MPa, 5 min, 25°C	4.3–4.4
	Raw milk	500 MPa, 10 min, 20°C	>4
	Fish slurry	400 MPa, 5 min, 20°C	≈3
Campylobacter jejuni	UHT whole milk	325 MPa, 10 min, 25°C	≈2.5
	UHT skim milk	325 MPa, 10 min, 25°C	≈2.5
	Soya milk	325 MPa, 10 min, 25°C	≈3
	Chicken puree	325 MPa, 10 min, 25°C	≈3.5
	Phosphate buffer	325 MPa, 10 min, 25°C	8
	Milk	300 MPa, 10 min, 20°C	0.4–1
	Broth	300 MPa, 10 min, 20°C	3–6.7
	Chicken meat slurry	200 MPa, 10 min, 20°C	0.2–2.2

spot in every treatment or performing a previous analysis to guarantee a constant effect. Package or food envelop integrity should be controlled previously rather than after the treatment if the value of the product is worth it.

Critical limits should be defined taking into account the great deal of interactions between food product and microbial susceptibility. Typical reported treatments to deal with hazardous microorganisms in vegetative form use from 300 to 600 MPa at ambient temperature for several minutes. Spores, as usual, have enhanced resistance needing more pressure and temperature. A smooth treatment followed by a harder one could first activate sporulated forms to subsequently destroy them in a vegetative state. Nevertheless the effectiveness of multiple pressurization stages is still under study. A large dependency on the processed product is being reported. A similar disparity is observed for foodborne viruses (Donsì et al., 2010).

High hydrostatic technology is currently applied in several areas of the food industry (fruit juices, ham, salsa dips) in countries that have regulated the use of this technology, e.g. USA and the EU. This fact should help to promote this technology as well as confirm that it is safe enough to be used on other food fields.

Monitoring

The measurement of the applied pressure and the treatment time are basically the main parameters monitored during HHP. Meanwhile, temperature is a secondary parameter as enhanced inactivation will be obtained with changes in temperature.

Pulse Electric Fields

Principles

The idea of using electric power to improve food safety is older than other well-established non-thermal technologies such as irradiation. It was at the end of the 19th century when the first patents of devices designed to deliver electric current to flowing food products were issued.

Technical advances have shown that using electric fields instead of electric currents yields better results. The technology is based on the fact that external electric fields with field strengths of 20–80 kV/cm of predefined duration induce an opposing membrane potential between the internal and external surfaces of the cellular membrane. Quick modifications of the external electric field such as obtained by waveform or polarity changes produce stress on the membrane as cells try to compensate the changes on the external electric field. A long enough treatment can exhaust cellular resistance, inducing pore formation in the membrane, which destroys cellular homeostasis and eventually leads to cell death.

Electric fields also induce movements of the ionic components of the foodstuffs being processed that produce friction and, consequently, heat. Thermal effects could be used if they are properly controlled, even though temperature increase is usually avoided by shortening the electric field treatment in repeated pulses in the range of microseconds (1–10 μs) as well as by using refrigerating systems. In addition, the PEF processing treatment lasts only a few milliseconds and thermal consequences for the remaining components of the food are often negligible (Soliva-Fortuny et al., 2009).

From a food safety standpoint, the technology has been studied in a variety of goods providing promising results and a feasible alternative to thermal pasteurization, especially because it is one of the few non-thermal preservation technologies that allow continuous processing of fluid foods. However, the standardization of PEF treatment conditions still under investigation. At this time, inactivation of microorganisms using PEF continues to be product microorganism specific, which hinders the use of PEF at the industrial scale (Table 18.7).

Critical Factors and Critical Control Points

The processing step within a PEF device should be considered as a biological critical control point if there is not any further biological control point. The critical parameters to keep under control can be classified as in the previous cases (device dependent, food dependent and microbial dependent factors).

TABLE 18.7 Process Parameters used for the Inactivation of Pathogenic Microorganisms in Fluid Foods by PEF Treatment (Mosqueda-Melgar et al., 2008)

Microorganism	Food	*E* (kV/cm)	n^a	τ^b (μs)	t_t^c (μs)	*F* (Hz)	*T* (°C)	Log_{10} Reductions
Listeria innocua	Orange juice	30	6	2.0	12	–	54	6.0
	Skim milk	41	63	2.5	157.5	3	37	3.9
	Liquid egg	50	32	2.0	64	3.5	36	3.4
	Whole milk	29	312	0.8	250	100	36	2.0
	Dairy cream	37.5	250	1.0	250	100	36	2.0
Listeria monocytogenes	Whole milk	30	400	1.5	600	1700	50	4.0
	Skim milk	20	10	3.25	32.5	–	35	1.0
Escherichia coli	Liquid egg	26	100	4.0	400	2.5	37	6.0
	Orange juice	30	6	2.0	12	–	54	6.0
	Liquid egg	32.89	180	0.17	30	–	20	4.7
	Milk (1,5% fat)	23	20	–	–	–	45	4.0
Escherichia coli O157:H7	Apple cider	90	10	2.0	20	–	42	5.91
	Apple juice	29	43	4.0	172	1000	42	5.0
	Skim milk	41	63	2.5	157.5	3	37	4.0
	Liquid egg	11	40	2.0	80	1	60	4.0
Escherichia coli 8739	Apple juice	29	43	4.0	172	1000	42	5.4
Bacillus cereus	Skim milk	31	20	–	6.0	–	25	0.7
Staphylococcus aureus	Raw milk	40	40	–	–	3.5	–	4.0
	Skim milk	35	124	3.7	459	250	40	3.7
	Skim milk	31	35	–	6.0	–	25	3.0
	Skim milk	35	600	4.0	2400	100	25	1.0
Salmonella Typhimurium	Orange juice	90	50	2.0	100	–	55	5.9
Salmonella Dublin	Skim milk	35	164	1.0	164	2000	50	4.0
Salmonella Enteriditis	Eggs white	35	8	–	–	900	–	3.5

[a] *Number of pulses.*
[b] *Pulse width.*
[c] *Treatment time.*

The most important factors to achieve the expected effects during PEF treatments are the electric field strength, the initial temperature and the total treatment time measured as the sum of the duration of all the delivered pulses. Typically, the longer the treatment time, the higher the microbial destruction that is achieved. Regarding the electric field strength (E) mathematical equations have been developed to predict the threshold values depending on the form factor of the specific microbial targets (Heinz et al., 2002).

Spherical shaped cells:

$$\Delta\varphi_M = -\frac{3}{2} \cdot E \cdot f(\theta) \cdot R \cdot \cos\alpha \tag{18.3}$$

Ellipsoidal shaped cells:

$$\Delta\varphi_M = -f(A) \cdot R \cdot E \tag{18.4}$$

where $\Delta\varphi_M$ is the critical membrane potential (typically $-1\,V$), R is the distance from the center of the cell, α is the angle between the cell and the applied electric field vector, and $f(\theta)$ and $f(A)$ are functions of the electrical conductivities and the semi-axes considered.

Other electrical factors such as the pulse repetition rate and the pulse shape and width are quite important as they influence how the energy is delivered to the target microorganism. However, such parameters modulate the efficiency of the process rather than modify or enhance the treatment effects. Some processing factors exert influence on the homogeneity of the treatment. The most important studied are the flow regime, the specific distribution of the electric field vector with regard to the fluid flow, the number of treatment chambers and the circulation cycles of the product through the PEF device (Pataro et al., 2011). The non-homogeneous treatment of solid goods is the main difficulty when using PEF technology on products containing solid particles of heterogeneous size.

The natural PEF resistance of the occurring microbial species and characteristics inherent to the foodstuff, such as electric conductivity and homogeneity (presence of particulate solids or bubbles) are the remaining critical parameters that should be specified before deciding the critical limits for all the described critical factors.

Monitoring

Electric field strength and total treatment time can be measured and controlled online since they come from an electrical device. Flow rate governs the homogeneity of the treatment and is managed by means of pumps that can also be controlled without effort with the currently available technology. Therefore, the whole PEF equipment can be easily linked to current computational machines. Keeping these parameters under the critical limits selected and established in the HACCP system should be enough to maintain safety of the processed product under control. The main consideration on PEF is that each system is specific in its design (e.g. number of chambers, incidence of the electric field on the food flow) and may not be modified, which requires a good understanding of the raw materials and products processed under PEF. Therefore, microbial and food characteristics must be well understood to ensure successful processing of the food product.

An important aspect that should not be forgotten is that electrodes can undergo corrosion, releasing chemicals to the product flow. All performed studies have shown that this fact should not raise any toxicological concern because of the alternating nature of the pulses (Morren et al., 2003). Nevertheless, routine inspection of the electrode status should be considered. The monitoring could include a visual inspection of the electrode thickness, a scanning of the electrode surface or the measurement of the electrical conductivity of the treated food before and after the treatment chamber.

Intense Pulsed Light

Principles

Pulsed light is considered an updated version of a treatment with continuous ultraviolet germicidal light (UV-C). In pulsed light treatments, light pulses produced by xenon lamps are released in the form of ultra-short-duration flashes of an intense broadband emission spectrum from approximately 200 to 1100 nm (López-Gómez et al., 2007).

Intense light pulses has recently received much attention as a strategy for decontaminating food, packaging, water and air (Oms-Oliu et al., 2010). Furthermore, pulsed light technology is a strong candidate for contact surface decontamination in the healthcare setting (Farrell et al., 2010).

The photochemical damage to microbial DNA, either on vegetative cells or spores, was claimed as the cause of UV-C treatment effectiveness (Guerrero-Beltrán et al., 2004). A similar inactivation mechanism is defined for pulsed light as it is rich in this kind of wavelength (200–280 nm). However, the fact that applied light also contains visible and near-infrared photons suggests the possibility of a thermal effect that is also debated. Both mechanisms could coexist because the lethal effect of the photochemical process between both modes of application should not be so different, but there will be a more reduced product processing time.

Critical Factors and Critical Control Points

The most important factors related to the treatment devices affecting the efficacy of pulsed light decontamination are the number of light pulses applied, or exposure time, and the dose received by the product, also known as fluence (J/m^2). Actually, these are the same process variables affecting irradiation because both technologies share the use of electromagnetic fields as agents to deliver energy to the product. The dose received in intense pulsed light treatments is more dependent on factors such as the lamp discharge intensity, the distance from the lamp to the treated surfaces, shading effects and product thickness than irradiation, because ultraviolet photons are less energetic than photons of either X-rays or γ-rays.

Consequently, the critical factors to be controlled are the same as those discussed in irradiation although, with current technology, dosimeters are different and pulsed light devices can be much handier.

Regarding microbial resistance, much of the aspects to be considered are the same as discussed in other non-thermal technologies. Every microorganism has a particular behavior in front of pulsed light treatments. Moreover, the photo-reactivation phenomenon, consisting on the activation of the photolyase enzyme, which is able to repair damaged DNA by light, needs to be considered. As far as food composition is concerned, a consistent decrease of effectiveness on media containing proteins and fats has been reported (Garcia-Gonzalez et al., 2007) (Table 18.8).

TABLE 18.8 Examples of Differences between Microbial Susceptibilities as well as Product Interactions using Intense Pulsed Light Technology (Martín-Belloso and Soliva-Fortuny, 2010; Rajkovic et al., 2010)

Pathogen	Food Product	Treatment Conditions	Log_{10} Reduction
Escherichia coli O157:H7	Agar	3 J/cm^2, 200 pulses, 100 ns	6.2
	Agar	7 J/cm^2, 50 pulses, 30 μs	4.7
	Agar	3 J/cm^2, 512 pulses, 1 μs	6.8
	Salmon fillets	5.6 J/cm^2, 180 pulses	1.09
	Alfalfa seeds	5.6 J/cm^2, 270 pulses	4.89
	Apple cider	1.05 J/cm^2, 12 pulses, 360 μs	3.22
	Apple juice	1.05 J/cm^2, 12 pulses, 360 μs	2.52
	Strawberries	Total energy dose 64.8 J/cm^2, 180 pulses	3.3
	Raspberries	Total energy dose 72 J/cm^2, 180 pulses	3.9
	Blueberries	Total energy dose 32.4 J/cm^2, 180 pulses	4.9
Listeria monocytogenes	Agar	3 J/cm^2, 200 pulses, 100 ns	4.4
	Agar	7 J/cm^2, 50 pulses, 30 μs	2.8
	Agar	3 J/cm^2, 512 pulses, 1 μs	6.25
	Agar	1.5 J/cm^2, 1 pulse, 300 μs	1.6
	Salmon fillets	5.6 J/cm^2, 180 pulses	1.02
	TSBYEa	7 J/cm^2, 20 pulses, 30 μs	≈1.5
Serratia marcescens	Milk	12.6 J/cm^2, 20 ns	>2.0
Staphylococcus aureus	Milk	1.27 J/cm^2, 16 pulses	7.2
Clostridium sporogenes	Honey	5.6 J/cm^2	0.89–5.46
Foodborne viruses	Phosphate buffer	1 J/cm^2	4.8–7.2
Cladosporium herbarum	Paper-polyethylene packaging material	0.977 J/cm^2, 30 pulses	2.7
Listeria innocua	Stainless steel surfaces	1.27 J/cm^2, 3 pulses	1.93–2.77

Monitoring

The measurement of the applied light intensity, number of pulses, frequency of pulsed light and overall treatment time are basically the main parameters monitored during pulsed light treatments.

Membrane Filtration

Principles

Membrane filtration allows fluid components to be separated according to their size, and in some cases this effect can be modified by interactions between the fluid components and the filtering surface. Membrane filtration technology is usually classified following the average cut-off pore diameters of the membranes. To achieve what is known as cold pasteurization, microfiltration (MF) technology, with diameter pore sizes between 0.1 and 10 μm, is used. This application is used as an alternative to thermal treatments in milk, either skimmed or not, beer, wine and fruit juice processing (Moraru and Schrader, 2009). It is possible to achieve fluid sterilization with ultra-filtration (UF) membranes that can separate even viruses, although it is only applicable to clear fluids, typically water.

The main constraint of using membrane technology is the presence of food particles and components that are retained during the filtration process. Nevertheless, there has been interest recently in combining such technology with other non-thermal preservation technologies, particularly in the dairy products area (Walkling-Ribeiro et al., 2011; Hoffmann et al., 2006; Fritsch and Moraru, 2008).

Critical Factors and Critical Control Points

From the point of view of safety assurance, the critical variables of this technology are pore size, filtration time, flow rate and trans-membrane pressure. Of course, the critical limit in this case is the size of the smallest microbial agent targeted. However, this factor is really difficult to control. Therefore, some variables related with mean pore size must be monitored in order to verify the effectiveness of the treatment:

- Permeate flow (J), measured as the permeate, which is the portion of the feed that passes through the filter, velocity (Q_p) and the effective surface area (A_e) of the membrane:

$$J = Q_p / A_e \tag{18.5}$$

- Trans-membrane pressure (TMP), which is a variable of the utmost importance, as the separation is pressure driven.
- Transmission and separation efficiency of a particular component or membrane selectivity, which can be easily monitored (Dewettinck and Le, 2011).

An indirect way to monitor the size and uniformity of pores in the membrane is through integrity testing procedures (e.g. bubble point, diffusive flow, forward flow, water intrusion). These tests are related to the ability of the membrane to retain microorganisms.

Monitoring

Filtration time, flow rate and trans-membrane pressure must be monitored throughout the filtration process. Integrity of the membrane shall be confirmed before and after the filtration step to ensure its effectiveness in removing microorganisms from the process stream.

Hurdle Technology

Principles

Briefly, the hurdle concept relies on the combination of techniques that act upon food, and process intrinsic and extrinsic factors with the aim of controlling all forms of quality deterioration (Leisner and Gould, 2002). As the worst form of quality deterioration from the human standpoint is the presence or growth of infectious or toxigenic microorganisms, the main priority is the minimization of such risks and, consequently, this section will only discuss aspects related to microbial safety. However, it should be borne in mind that hurdle technology could go further – it could even be used to better preserve quality aspects of the product such as microbial stability or sensory and nutritional food properties (Tapia de Daza et al., 1996).

The hurdle approach is not new, only the term is recent. The technology has been in use for a long time, as fermented foods and other kinds of food commodities like cured meats, fruit preserves or jams demonstrate. Actually, it is necessary for the vast majority of food products to use it to meet consumers' expectations. Other names such as "multi-target preservation," "combined methods" or even "minimal processing" have been suggested and used to describe this technology.

The main difference between traditional and novel food products is that historically the knowledge used to produce foods in a safe way was achieved following trial-and-error experimental methodology. The current knowledge on microorganisms and the ecology of food products allows not only understanding the mechanisms and factors affecting the shelf-stability of food products, but also optimizing the different hurdles for quality improvement.

There are families of food products that traditionally use a specific set of hurdles. Cured meats typically have to use a combination of preservatives and reduced water activity to be safe for long periods of time. Processed meat products such as sausages introduce mild thermal treatments and/or need to be stored under refrigeration. Recently developed food products such as ready-to-eat fruits and vegetables require a combination of chemical preservatives, low temperature and modified atmospheres to maintain product safety over the whole product shelf-life (Martín-Belloso and Soliva-Fortuny, 2010).

Consumers are currently demanding these types of convenient products, so mild non-thermal technologies such as described here are studied so that they can be applied in combination with traditional preservation methods (Soliva-Fortuny et al., 2011). Some review reports are available regarding the state-of-the-art application of hurdle technology on other important goods such as fresh meats (Zhou et al., 2010) or dairy products (Sobrino-López and Martín-Belloso, 2008), for example (Table 18.9).

Critical Factors and Critical Control Points

Critical factors of food products processed by hurdle technology depend on the combination of chosen technologies. The election of the technologies to preserve a food product is influenced basically by its nature as there are multiple hurdles that kill microorganisms and avoid microbial proliferation. The most commonly used are pH, water activity, use of preservatives, electric potential, competitive flora, physical barriers, modification of atmospheric conditions and physical treatments, either thermal or not. To add more complexity,

TABLE 18.9 Example of Combined Effect of Gamma Irradiation and Modified Atmosphere (Map) on Enterobacteriaceae Populations in Chopped Chicken Meat Stored at 4°C (Chouliara et al., 2008)

Storage Time (day)	Air Packaging (control)	Air packaging +2 kGy	MAP 1	MAP 2	MAP 2 + 2 kGy
0	2.28 ± 0.13 Aa	2.28 ± 0.13 Ab	2.28 ± 0.13 Aa	2.28 ± 0.13 Aa	2.28 ± 0.13 Aa
3	3.99 ± 0.24 Bb	<1.00	3.28 ± 0.21 Bb	2.76 ± 0.24 Ab	<1.00
6	6.15 ± 0.48 Cc	<1.00	4.85 ± 0.36 Bc	3.27 ± 0.17 Ac	<1.00
9	7.48 ± 0.51 Cd	<1.00	5.98 ± 0.39 Bd	4.08 ± 0.27 Ad	<1.00
12	ND	<1.00	6.42 ± 0.46 Bde	5.29 ± 0.51 Ae	<1.00
15	ND	1.29 ± 0.06 Aa	7.02 ± 0.52 Be	6.71 ± 0.49 Bf	<1.00
20	ND	2.88 ± 0.14 Bc	ND	ND	<1.00
25	ND	ND	ND	ND	1.91 ± 0.15 Aa

MAP 1: 30% CO_2 + 70% N_2; MAP 2: 70% CO_2 + 30% N_2; ND: not determined.
Different capital and lowercase letters in the same row and column respectively are significantly different ($p < 0.05$). Measures expressed as mean values and standard deviation in log cfu/g.

each hurdle can be modified with different agents usually in close relation with other hurdles. Thus, organic acids reduce pH as well as acting as chemical preservatives. A reduction of water activity of a product can be achieved by drying, freeze-drying, heating or by adding fat, salts or sugars to the raw product.

As already stated, hurdle technology requires an accurate case-by-case study in order to select the appropriate critical factors to be controlled. As a general rule of thumb, at least all the critical process factors of every technology used should be considered.

Regarding the decision to select critical limits for the critical parameters, there are tools that can help in such hard task due to the multiple possible combinations. Predictive microbiology attempts to provide mathematical models of microbial growth under a variety of environmental conditions (e.g. temperature, pH, a_w and the effect of preservatives). The current discussion focuses on whether the effects of a combination of hurdles are independent or interact with each other (Biesta-Peters et al., 2010). However, as prudence suggests a specific analysis for the particular combination of technologies on a certain product, the data gathered in such analysis will resolve any discussion (Figure 18.2).

If the monitored variables are the same as used for each technology the monitoring procedures should obviously be the same. Even so, during the analysis of a particular combination of technologies it would be possible to find some repetitions of the measurements. For example, in the preparation of a ready-to-use vegetable soup, it seems reasonable, *a priori*, to consider acidity a critical variable because it will prevent *a posterior* microbial growth due to cross-contamination, even though it is thermally processed before packaging. Consequently, a measure of the pH of the media is decided. The designed preparation could also require that a natural preservative and potassium sorbate are selected. The preservative concentration can be related to a pH measurement. So, both variables could be controlled by monitoring only the pH of the media at the end of the process before aseptic bottling. It is part of the

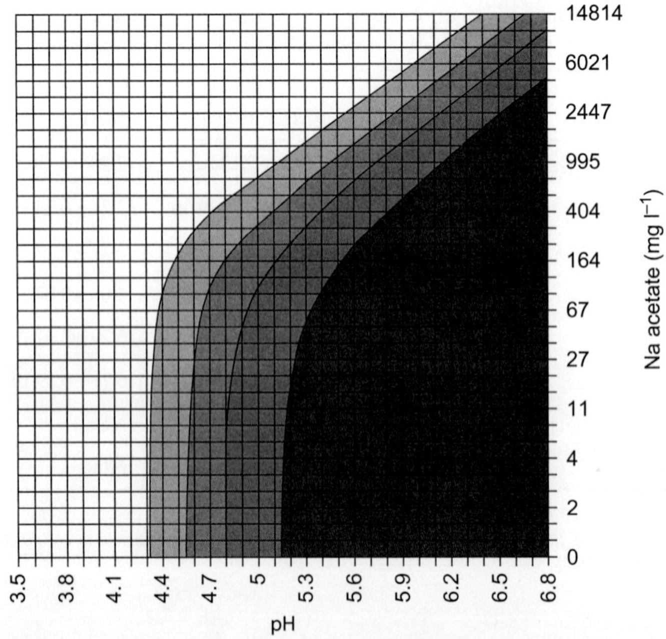

FIGURE 18.2 Example of a product stability map of total sodium acetate and pH for a cocktail of *Enterobacter sakazakii* developed using predictive microbiology (Lambert and Bidlas, 2007).

tasks of the safety management group to decide the best solution to such particular issues, as well as whether other factors should be controlled and at what stage.

VERIFICATION AND VALIDATION METHODS FOR NON-THERMAL TECHNOLOGIES

Current quality management systems are flexible enough to allow the same tools to be used to achieve both goals product safety and product quality without compromising the efficacy of the system. Actually, product safety can be considered, from a certain point of view, as the most basic aspect of product quality. Furthermore, a quality system can achieve its objectives irrespective of the area where it is applied, the processed product and the technologies used in the manufacturing process when it is properly designed and implemented. There are particular issues with specific technologies. For example, irradiation allows processing packaged products and differences between unprocessed and processed products are not easily appreciated. This point evidences the importance of the complementary measures that stakeholders should implement such as GMP, hygiene plans, preventive maintenance systems aimed at avoiding equipment failures or, in this case, an inventory system that should prevent release of non-irradiated goods.

Process verification implies a comparison of the current process variable values with the established critical limits of each critical variable of the non-thermal preservative technology. Process verification method is now highly automatic thanks to the recent advances in computers. It is quite possible to follow the manufacturing process via online physical measurements over chemical and microbiological ones. The latter methods are more suitable to confirm the relationship between the live or logged measures and the real values. This step corresponds to process validation.

In summary, the election of particular verification and validation solutions should be taken after a careful analysis of the specific requirements of the regulatory requirements as well as product quality specifications. Conversely, economic aspects should be evaluated while considering each technology to ensure a proper balance between financial goals and the production of safe products.

FINAL REMARKS

Non-thermal food preserving technologies should not add any complexity to a well-implemented quality and safety management system. Such systems are based on scientific knowledge and the critical factors of each technology have been, in most cases, already clearly determined. The main safety concern arises from the natural variability of biological systems that hinders the election of the critical limits for these variables.

The best way to solve such problems is to obtain specific data of each combination of product and process so the appropriate decision is made while defining the criticality of process variables (e.g. process limits). Meanwhile, food processors must consider the limitations associated with non-thermal processes.

Consumers are aware of the pros and cons of thermal processes and ask manufacturers for better products. This usually means more convenient foodstuffs with the added value of reduced nutritional and sensory losses because safety must be inherent in a food product. However, the novelty of non-thermal technologies and the fact that they are, by far, more specific than thermal technologies, gives consumers a feeling of insecurity. Therefore, stakeholders should provide their customers with proper information about the safety and added benefits of the products manufactured using non-thermal technologies if they want to promote their social standing (Olsen et al., 2010).

Last but not least, there is currently a concern within the scientific community, and also among food processors and legislators, that the application of sublethal stress factors could induce cross-resistance mechanisms in the surviving populations and change their virulence characteristics (Capozzi et al., 2009). This concern should motivate food industry stakeholders not only to design proper safety (quality) management systems but to implement them adequately. Such management systems provide the tools (validation) to monitor any possible deviation from what can be considered a normal microbial behavior irrespective of the preservation technology used.

References

Arvanitoyannis, I., 2010. Food packaging materials for irradiation. In: Arvanitoyannis, I. (Ed.), Irradiation of Food Commodities: Techniques, Applications, Detection, Legislation, Safety and Consumer Opinion. Elsevier, London, pp. 43–64.

Balny, C., Masson, P., 1993. Effects of high pressure on foods. Food Rev. Int. 9 (11), 611–628.

Biesta-Peters, E., Reij, M., Gorris, L., Zwietering, M., 2010. Comparing nonsynergistic Gamma models with interaction models to predict growth of emetic *Bacillus cereus* when using combinations of pH and individual undissociated acids as growth-limiting factors. Appl. Environ. Microbiol. 76 (17), 5791–5801.

Capozzi, V., Fiocco, D., Amodio, M., Gallone, A., Spano, G., 2009. Bacterial stressors in minimally processed food. Int. J. Mol. Sci. 10, 3076–3105.

Cheftel, J., 1995. High pressure, microbial inactivation and food preservation: a review. Food Sci. Technol. Int. 1 (2–3), 75–90.

Chouliara, E., Badeka, A., Sawaidis, I., Kontominas, M., 2008. Combined effect of irradiation and modified atmosphere packaging on shelf-life extension of chicken breast meat: microbiological, chemical and sensory changes. Euro. Food Res. Technol. 226 (6), 877–888.

Considine, K., Kelly, A., Fitzgerald, G., Hill, C., Sleator, R., 2008. High-pressure processing – effects on microbial food safety and food quality. FEMS Microbiol. Lett. 281, 1–9.

Demazeau, G., Rivalain, N., 2011. The development of high hydrostatic pressure processes as an alternative to other pathogen reduction methods. J. Appl. Microbiol. 110, 1359–1369.

Dewettinck, K., Le, T., 2011. Membrane separations in food processing. In: Proctor, A. (Ed.), Alternatives to Conventional Food Processing. RCS Publishing, Cambridge, UK, pp. 184–253.

Dickson, J., 2001. Radiation inactivation of microorganisms. In: Molins, R. (Ed.), Food Irradiation: Principles and Applications. John Wiley & Sons Inc., New York, pp. 23–36.

Donsì, G., Ferrari, G., Maresca, P., 2010. Pasteurization of fruit juices by means of a pulsed high pressure process. J. Food Sci. 75 (3), E169–E177.

Doona, C., Feeherry, F., 2007. High Pressure Processing of Foods. Blackwell Publishing Professional, Ames, Iowa, USA.

Farrell, H., Garvey, M., Cormican, M., Laffey, J., Rowan, N., 2010. Investigation of critical inter-related factors affecting the efficacy of pulsed light for inactivating clinically relevant bacterial pathogens. J. Appl. Microbiol. 108, 1494–1508.

Fritsch, J., Moraru, C., 2008. Development and optimization of a carbon dioxide-aided cold microfiltration process for the physical removal of microorganisms and somatic cells from skim milk. J. Dairy Sci. 91 (10), 3744–3760.

Garcia-Gonzalez, L., Geeraerd, A., Spilimbergo, S., Elst, K., Van Ginneken, L., Debevere, J., et al., 2007. High pressure carbon dioxide inactivation of microorganisms in foods: the past, the present and the future. Int. J. Food Microbiol. 117, 1–28.

Grauwet, T., Plancken, I., Venvoort, L., Hendrickx, M., van Loey, A., 2010. Mapping temperature uniformity in industrial scale HP equipment using enzymatic pressure-temperature-time indicators. J. Food Eng. 98 (1), 93–102.

Guerrero-Beltrán, J., Barbosa-Cánovas, G., 2004. Review: advantages and limitations on processing foods by UV light. Food Sci. Technol. Int. 10 (3), 137–147.

Heinz, V., Álvarez, I.A., Knorr, D., 2002. Preservation of liquid foods by high intensity pulsed electric fields – basic concepts for process design. Trends. Food Sci. Technol. 12, 103–111.

Hoffmann, W., Kiesner, C., Clawin-Rädecker, I., Martin, D., Einhoff, K., Lorenzen, P., et al., 2006. Processing of extended shelf life milk using microfiltration. Int. J. Dairy Technol. 59 (4), 229–235.

ICMSF, 1998. Microorganisms in Foods 6: Microbial Ecology of Food Commodities. Blackie Academic and Professional, London.

Lambert, R., Bidlas, E., 2007. A study of the Gamma hypothesis: predictive modelling of the growth and inhibition of Enterobacter sakazakii. Int. J. Food Microbiol. 115, 204–213.

Leisner, L., Gould, G., 2002. Hurdle Technology: Combination Treatments for Food Stability, Safety and Quality. Kluwer Academic & Plenum Publishers, New York.

López-Gómez, V., Ragaert, P., Debevere, J., Devlieghere, F., 2007. Pulsed light for food decontamination: a review. Trends. Food Sci. Technol. 18 (9), 467–473.

Martín-Belloso, O., Soliva-Fortuny, R., 2010. Advances in Fresh-Cut Fruits and Vegetables Processing. Taylor and Francis, Boca Ratón, FL, USA.

Miller, R., 2005. Electronic Irradiation of Foods: An Introduction to the Technology. Springer Science + Business Media Inc., New York.

Moraru, C., Schrader, E., 2009. In: Pabby, A., Rizvi, S., Sastre, A. (Eds.), Handbook of Membrane Separations: Chemical, Pharmaceutical, Food and Biotechnological Applications. Taylor and Francis Group, Boca-Raton, Florida, USA.

Morren, J., Roodenburg, B., Han, S.W., 2003. Electrochemical reactions and electrode corrosion in pulsed electric field (PEF) treatment chambers. Innov. Food Sci. Emerg. Technol. 4 (3), 285–295.

Mosqueda-Melgar, J., Elez-Martínez, P., Raybaudi-Massilia, R., Martín-Belloso, O., 2008. Effects of pulsed electric fields on pathogenic microorganisms of major concern in fluid foods: a review. Crit. Rev. Food Sci. Nutri. 48, 747–759.

Olsen, N., Grunert, K., Sonne, A.-M., 2010. Consumer acceptance of high-pressure processing and pulsed-electric field: a review. Trends. Food Sci. Technol. 21 (9), 464–472.

Oms-Oliu, G., Martín-Belloso, O., Soliva-Fortuny, R., 2010. Pulsed light treatments for food preservation: a review. Food Bioprocess. Technol. 3 (1), 13–23.

Palou, E., Lopez-Malo, A., Barbosa-Cánovas, G., Swanson, B., 1999. High pressure treatment in food preservation. In: Rahman, M. (Ed.), Handbook of Food Preservation. Marcel Dekker, New York, USA, pp. 533–576.

Pataro, G., Senatore, B., Donsi, G., Ferrari, G., 2011. Effect of electric and flow parameters on PEF treatment efficiency. J. Food Eng. 105 (1), 79–88.

Patazca, E., Koutchma, E., Balasubramaniam, V., 2007. Quasi-adiabatic temperature increase during high pressure processing of selected foods. J. Food Eng. 80 (1), 199–205.

Rajkovic, A., Smigic, N., Devlieghere, F., 2010. Contemporary strategies in combating microbial contamination in food chain. Int. J. Food Microbiol. 141, S29–S42.

Sobrino-López, A., Martín-Belloso, O., 2008. Use of nisin and other bacteriocins for preservation of dairy products. Int. Dairy J. 18, 329–343.

Soliva-Fortuny, R., Balasa, A., Knorr, D., Martín-Belloso, O., 2009. Effects of pulsed electric fields on bioactive compounds in foods: a review. Trends. Food Sci. Technol. 20 (11–12), 544–556.

Soliva-Fortuny, R., Grigelmo-Miguel, N., Barbosa-Cánovas, G., Martín-Belloso, O., 2011. Nonthermal processes as hurdles with selected examples. In: Zhang, Q., Barbosa-Canovas, G., Balasubramaniam, V., Dunne, P., Farkas, D., Yuan, J. (Eds.), Nonthermal Processing Technologies for Food. Wiley-Blackwell Publishing, Ames, Iowa, USA, pp. 406–427.

Spilimbergo, S., Matthews, M., Cinquemani, C., 2011. Supercritical fluid pasteurization and food safety. In: Proctor, A. (Ed.), Alternatives to Conventional Food Processing. RCS Publishing, Cambridge, UK, pp. 145–183.

Steward, E., 2001. Food irradiation chemistry. In: Molins, R. (Ed.), Food Irradiation: Principles and Applications John Wiley & Sons Inc., New York, pp. 37–76.

Tapia de Daza, M., Alzamora, S., Chanes, J., 1996. Combination of preservation factors applied to minimal processing of foods. Crit. Rev. Food Sci. Nutr. 36 (6), 629–659.

Walkling-Ribeiro, M., Rodríguez-González, O., Jayaram, S., Griffiths, M., 2011. Microbial inactivation and shelf life comparison of "cold" hurdle processing with pulsed electric fields and microfiltration, and conventional thermal pasteurisation in skim milk. Int. J. Food Microbiol. 144, 379–386.

WHO, 1997. High-dose Irradiation: Wholesomeness of Food Irradiated with Doses above 10 kGy *(Vol. WHO Technical Report Series 890)*. World Health Organization, Geneva.

Zhou, G., Xu, X., Liu, Y., 2010. Preservation technologies for fresh meat – a review. Meat. Sci. 86, 119–128.

CHAPTER

19

Acids and Fermentation

Martin R. Adams
University of Surrey, Guildford, Surrey, UK

OUTLINE

INTRODUCTION: ACIDITY AND PH

Acidity is one of the fundamental taste characteristics of food. The term itself derives from the Latin *acere*, meaning to taste sour, and consumers can readily detect acidity in foods as well as quantify it in terms of the degree of sourness.

In chemical terms, an acid, as defined by Arrhenius in 1884, is a substance that yields hydrogen ions (protons)* in solution while a base yields hydroxyl ions. This definition was later broadened independently by Brønsted and Lowry in 1923 to encompass non-ionic reactions by defining an acid as a proton donor.

When dealing with food materials we are primarily concerned with aqueous systems. In pure water a very small proportion of the water molecules dissociate into protons and hydroxyl ions:

$$H_2O \leftrightarrow H^+ + OH^-$$

The concentration of the two ionic species is equal (10^{-7} mol/L) and their product (the dissociation constant or ion product of water) is constant at 10^{-14}. If an acid is dissolved in

*Strictly speaking it is not a free hydrogen ion or proton but a solvated hydronium ion (H_3O^+).

Food Safety Management.
DOI: http://dx.doi.org/10.1016/B978-0-12-381504-0.00019-6

water then the equilibrium changes and the concentration of hydrogen ions increases, the acidity of the solution increasing with the level of hydrogen ions, while that of hydroxyl ions decreases correspondingly to maintain the ion product at 10^{-14}.

While it might be preferable in some respects to describe acidity in terms of hydrogen ion concentration, the huge range over which it can vary means that in practice a logarithmic scale is more useful. This is the pH scale, proposed in 1909 by Sørensen who was working, appropriately enough, in a food-related laboratory at the Carlsberg Brewery in Denmark. He defined pH as the negative logarithm of the hydrogen ion concentration, c_{H+}:

$$\mathrm{pH} = -\log c_{H+}$$

From the above, it follows that the pH of pure water is 7 (at 25°C). A pH below 7 indicates acidic conditions where the concentration of H^+ exceeds that of OH^-, and a pH above 7 indicates alkaline conditions where the concentration of OH^- is more than that of H^+. Thus a very acidic solution with a H^+ concentration of 1 mol/L would have a pH of 0. For most practical purposes the pH scale normally ranges between 0 and 14, although it can extend beyond that, and most foods have a pH on the acidic side of neutral ranging between 2.0 and 7.0 (Figure 19.1).

In practice pH is generally measured in the form of an electromotive force generated in an ion selective glass electrode and is a response to the activity of hydrogen ions rather than their concentration. Activity is related to concentration by a proportionality constant, the activity coefficient γ. The activity coefficient is affected by factors such as temperature and ionic strength. In very dilute solutions γ approaches 1 and activity and concentration become equal.

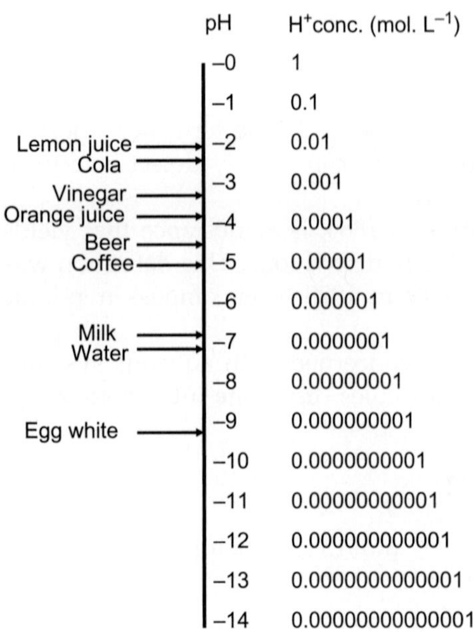

FIGURE 19.1 Most foods have a pH on the acidic side of neutral ranging between 2.0 and 7.0.

Strong acids such as hydrochloric and sulfuric dissociate completely in water to produce protons and the negatively charged counter ion but many of the acids encountered in food are described as weak acids which only partially dissociate:

$$HA \leftrightarrow H^+ + A^-$$

The extent to which acids dissociate, and hence their strength, is defined by the position of this equilibrium:

$$K_a = \frac{[H^+][A^-]}{[HA]}$$

the larger the equilibrium constant K_a, the greater the degree of dissociation and the stronger the acid.

As with hydrogen ion concentrations and pH, the range of values taken by K_a is very large so that a logarithmic scale of pK_a is used for convenience where:

$$pK_a = -\log K_a$$

Some representative values of pK_a for acids frequently encountered in foods are presented as Table 19.1.

The dissociation behavior of weak acids is described by the Henderson–Hasselbalch equation:

$$pH = pK_a + \log\frac{[A^-]}{[HA]}$$

TABLE 19.1 pK_a Values of Some Common Food Acids and Preservatives

Acid	pK_a
Acetic (ethanoic)	4.75
Propionic	4.87
Lactic	3.86
Sorbic acid	4.75
Citric	3.14, 4.77, 6.39 (tribasic)
Benzoic	4.19
Parabens	8.5
Phosphoric	2.12, 7.21, 12.67 (tribasic)
Carbonic	6.37, 10.25 (dibasic)
Nitrous	3.37
Sulfurous	1.81, 6.91 (dibasic)

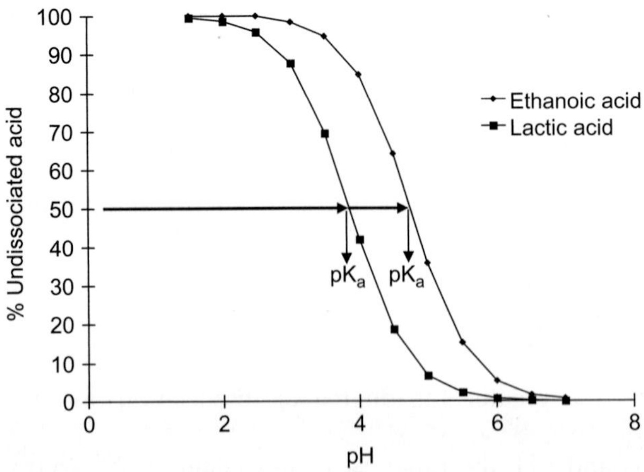

FIGURE 19.2 Dissociation of lactic acid and ethanoic acids as a function of pH.

This equation relates the strength of an acid and the pH of the solution with the relative concentration of dissociated and undissociated forms of the acid. It can be represented graphically for individual acids as shown in Figure 19.2 for ethanoic acid and lactic acid. This shows that as the pH of a solution decreases then the concentration of the undissociated acid will increase for all acids, and that for weaker acids, the undissociated proportion at any given pH will be higher than for stronger acids, i.e. those with a lower pK_a. A simple rule to remember is that when the pH is equal to an acid's pK_a, then the concentration of dissociated and undissociated forms will be equal; as the pH decreases below this then the level of undissociated acid increases.

ACIDITY AND FOODS

Acids are common components of food systems. Citric, malic and tartaric acid, among others, occur naturally in fruits and vegetables. Lactic acid produced by endogenous enzymatic activity is responsible for the postmortem acidification of meat. Adipic, citric, ethanoic, fumaric, gluconic, lactic, malic, succinic and tartaric acids are available commercially and are permitted for use as food acidulants, and acid preservatives such as benzoic acid and sorbic acid can be added to a range of food products such as jams, bread and cakes. Phosphoric acid, a relatively strong acid, is an important ingredient in soft drinks such as colas.

Lemon and lime juice, which contain citric acid, are used in products such as ceviche, some salad dressings and pickles, but traditional procedures to acidify foods and confer keeping quality and safety usually employ acids of microbial origin. The preserving power of vinegar is due to its high content of ethanoic acid. It is produced by a double fermentation process in which sugar is first converted into ethanol by yeast and a second aerobic stage in which acetic acid bacteria oxidize the ethanol to ethanoic acid.† Vinegar can be

†Edible grade acetic acid produced from petrochemical sources can also be used as an alternative to vinegar in some products. In contrast, citric acid used as a substitute for lemon juice in some circumstances is produced by a microbial fermentation.

produced with ethanoic acid concentrations in excess of 10% (most table vinegars would contain 4–5% w/v) and was the strongest acid known in antiquity. Addition of vinegar to a food material can thus considerably reduce its pH, inactivate some of its indigenous microflora and restrict the growth of those that survive.

The efficacy of acid solutions, such as vinegar, added to foods will be reduced by the diluting effect of the food's water content and by its intrinsic buffering capacity.

In many traditional products the effect of the former is mitigated by osmotic dehydration (salting) or drying of raw materials prior to pickling. While it is relatively straightforward to calculate the effect on pH of dissolving known concentrations of an acid in water and, knowing the pH, to calculate degree of dissociation of a weak acid using the Henderson–Hasselbalch equation, it is not possible to make such simple calculations of the pH resulting from addition of a weak acid such as ethanoic to the complex and ill-defined buffering system in a food. In addition to the presence of a range of buffering components such as proteins and amino acids, the issue may be further complicated by the presence of oil or fat into which the acid might partition preferentially. This would have the effect of decreasing the acidity in the aqueous phase in which microbial growth occurs and thereby the anticipated antimicrobial effect. The problem can be resolved on a purely empirical basis using test formulations but a more *a priori* approach has been described to both the problem of pH prediction and phase partitioning based on an acid titration of the food material and knowledge of the dissociation constant and the phase partition constant of the acid being used (Wilson et al., 2000).

ACIDITY AND MICROORGANISMS

The acidity of a medium will affect chemical and physicochemical reactions, the stability and activity of enzymes and other cellular components and as a consequence will affect the activity, growth and survival of microorganisms. A particular microorganism will be capable of growth over a range of pH usually spanning 2–5 pH units but will grow best over a much narrower range, typically 1–2 pH units. Different microorganisms will grow best at different pH values, but in general bacteria grow fastest in the pH range 6.0–8.0, yeasts 4.5–6.0 and filamentous fungi 3.5–4.0, although there are some notable exceptions to this.

In practical terms, the reduction of pH and an increase in acidity can have a profound effect on the microflora associated with a particular food, its shelf-life and safety. It is a commonplace observation that acidic foods such as yoghurt or fruit juices spoil as a result of the activity of more acid tolerant organisms such as yeast and molds. This effect is also seen in the greater prevalence of bacterial spoilage in vegetable products when compared to more acidic fruits where yeast and mold spoilage predominate.

Microbial susceptibility to acidity is not simply a function of the external pH but a result of the acidification of the microbial cytoplasm. Although microorganisms may tolerate external pH values that are lower than their optimum, most will strive to maintain a higher pH in their cytoplasm to ensure that the complex network of processes that comprise metabolism and growth continue to operate efficiently. The microbial cell membrane is relatively impermeable to protons and this is an important factor in maintaining the cell's intracellular pH, but microorganisms also have a battery of homeostatic mechanisms to neutralize or expel protons that enter the cell and thus help maintain a favorable internal pH.

Acids differ in their ability to acidify the cytoplasm so microbial inhibition will depend on the particular acid present as well as the pH. It is well established that weak organic acids are much more effective at inhibiting microorganisms than strong acids. This is a consequence of the physicochemical properties of weak organic acids. They only partially dissociate in solution giving a mixture of undissociated acid, protons and the acid anion. In their undissociated state they are often relatively lipophilic and can diffuse through cell membranes down a concentration gradient from high levels outside the cell to lower levels inside. Once inside the cell, where the pH is higher, the acids dissociate releasing protons and acidifying the cytoplasm. This imposes a burden on the cell's homeostatic mechanisms, and energy has to be expended to maintain the intracellular pH. At low levels of acidity, this diversion of resources results in slower growth, but the burden may become excessive in which case growth is no longer possible and the cell will eventually die.

Viruses only multiply after they have infected a susceptible cell therefore those of concern to human health will not grow in foods. They can be inactivated (killed) by low pH in a food material but are more resistant than bacteria since they are structurally much simpler and do not show the enhanced sensitivity to weak acids displayed by bacteria and other cellular microorganisms.

CONTROL OF PATHOGENS BY PH AND ACIDITY

Table 19.2 presents the minimum recorded pH values for a number of important bacterial pathogens. Because pH values below 4.5 will stop or severely curtail the growth of all the major bacterial pathogens and will, depending on the conditions, lead ultimately to their death/inactivation, food safety concerns tend to be much reduced when considering acidic foods. This is particularly true since the production of such foods often also includes a step equivalent to pasteurization such as hot bottling which is used primarily to control spoilage organisms. It is mostly where there is no pasteurization step that safety problems can arise (see later).

Acidity and pH play an important role in the safe production of shelf-stable heat-processed (commercially sterile) products such as canned foods. The major safety concern here is that spores of *C. botulinum* will survive the heat process, germinate and grow in the product during storage at ambient temperature producing the neurotoxic botulinum toxin. It has long been recognized that *C. botulinum* spores will not germinate and grow at pH values below 4.6 and this is enshrined in various codes of practice. Acidity is used as a basis of classifying canned foods since those with a pH below 4.6 will require a less severe heat process to assure safety (e.g. CAC, 1979).

Levels of acidity which do not kill pathogens or stop their growth entirely can still improve food safety. The risk from infectious pathogens such as *Salmonella* will be lower if growth and thereby numbers of the organism are restricted and, at suboptimal pH, toxigenic organisms such as *Staphylococcus aureus* may not grow to levels sufficient to produce a biologically effective concentration of toxin in the food. This is exemplified in EU regulations where food safety criteria for milk powder and some cheeses specify that only when levels of coagulase positive staphylococci exceed 10^5 cfu g^{-1} is there a requirement to test for enterotoxin.

At acidic pH levels where growth is still possible predictive models such as Combase (http://www.combase.cc/index.php/en/) or the Pathogen Modelling Programme

TABLE 19.2 Minimum Recorded pH Values for a Number of Important Bacterial Pathogens

Pathogen	Minimum Growth pH‡	Optimum Growth pH
Bacillus cereus	5.0	6.0–7.0
Campylobacter jejuni	4.9	6.5–7.5
Clostridium botulinum		
Group 1	4.6	–
Group 2	5.0	–
Clostridium perfringens	5.5–5.8	7.2
Escherichia coli	4.4	6.0–7.0
Listeria monocytogenes	4.4	7.0
Salmonella	3.8	7.0–7.5
Shigella	4.9–5.0	9.2–9.3
Staphylococcs aureus	4.0	6.0–7.0
Vibrio cholerae	5.0	7.6
Vibrio parahaemolyticus	4.8	7.8–8.6
Vibrio vulnificus	5.0	7.8
Yersinia enterocolitica	4.2	–

‡*Data from ICMSF (1996).*

(http://pmp.arserrc.gov/) can give predictions of the rate of growth under different pH conditions (as well as the effect of interaction with other factors such as temperature and water activity). These models are designed to be fail-safe so that they tend to overpredict the growth that will occur under any given set of conditions. While valuable, labor-saving tools to the food microbiologist in setting critical limits, these should be used with caution where food safety issues are concerned. Confirmation of predicted safety, particularly at the growth/no growth boundaries, may often be required using the gold standard technique of a challenge trial.

The Committee of the Mayonnaise and Condiments Sauce Industries of the EU (CIMSCEE) published formulae that predict the level of acid, salt and sugars necessary to inhibit microorganisms in cold-filled acid-preserved pickles and sauce products. One of these predicts stability, i.e. the levels of acid, salt, etc. required to inhibit the growth of spoilage organisms and confer shelf stability, while the other predicts safety. It is impossible to predict safety in any absolute sense; where this is attempted it is done on the basis of a food safety objective (FSO) that gives what is regarded as an acceptable level of protection. Ideally these should be based on some form of quantitative microbiological risk assessment which determines the level of risk and the effect of various interventions on that risk. Although not supported by such a rigorous analysis, in the CIMSCEE formulation a safe product was designated as one in which viable numbers of *E. coli* will decline by more than 3 log cycles (a factor of 10^3) in less than 72 hours at 20°C.

The formula is:

$$15.75(1-\alpha)(\%\ \text{total acetic acid}) + 3.08(\%\ \text{salt}) + (\%\ \text{hexose}) + 0.5(\%\ \text{disaccharide}) + 40(4.0 - \text{pH}) = \Sigma_s$$

where $(1-\alpha)$ is the proportion of undissociated acetic acid and α is the proportion of dissociated acid given by the Henderson–Hasselbalch equation:

$$pH = pK_a + \log\frac{\alpha}{(1-\alpha)}$$

If $\Sigma_s > 63$ then the product would be regarded as intrinsically safe, i.e. it would deliver at least the designated reduction in *E. coli*.

Though generally a very reliable guide to product formulation and the setting of critical limits, experimental studies and evidence from outbreaks have shown that survival is greater at low temperatures (chill stored products) and that higher levels of salt and sugar can also sometimes be protective (Mullan, 2009).

FERMENTED FOODS

A huge range of foods rely on the endogenous production of organic acids by lactic acid bacteria (LAB), principally lactic acid but often with lower concentrations of ethanoic acid. Although LAB produce a number of other antimicrobials such as bacteriocins which may be significant in some circumstances, the number of different species active in lactic fermentations indicates that the principal antimicrobial effect is something common to all lactic acid bacteria. This common factor is that the main mode of energy generating metabolism in these organisms is the fermentative conversion of sugars into acids, principally lactic acid. Acid production and a simultaneous reduction in pH are inevitable consequences of LAB growth and acidity levels in some fermentations can exceed 100 mM, reducing the pH to below 4.0 in weakly buffered systems (Adams, 2001).

Bacteriocins are polypeptide antimicrobials produced by bacteria which are bactericidal to other, normally closely related, organisms. Lactic acid bacteria produce a number of bacteriocins and considerable efforts have been devoted to their discovery and investigation in recent years. Despite this attention, the most effective and useful bacteriocin in food use remains nisin, a lantibiotic bacteriocin produced by some strains of *Lactococcus lactis* and first discovered as long ago as 1928. Its pre-eminence derives from its relatively broad spectrum of activity. Unlike many other bacteriocins from lactic acid bacteria, it is active against most Gram-positive bacteria and is particularly effective at inhibiting the outgrowth of bacterial endospores. In terms of its potential contribution to food safety, it can inhibit the outgrowth of spores of pathogens such as *Clostridium perfringens, Clostridium botulinum* and *Bacillus cereus* but also (at higher concentrations) has some inhibitory effect on vegetative pathogens such as *Listeria monocytogenes* and *Staphylococcus aureus*. However, where a nisin-producing strain is used in production of a fermented food its contribution to overall safety may be relatively minor. It will have no effect of Gram-negative pathogens. Vegetative Gram positives such as *Listeria monocytogenes* are known to acquire resistance to nisin quite readily and *S. aureus* is among the most inherently resistant Gram-positive species. It is more likely to be useful where there is a risk of *C. botulinum*, although usually this organism can be well controlled by efficient acid production. Nisin production may also have a detrimental effect on fermentation by inhibiting other nisin-sensitive lactic acid bacteria present and adversely affecting acid production.

For *in situ* production of lactic acid by fermentation to have a significant antimicrobial effect there must be high levels of active lactic acid bacteria present and a substantial numerical superiority over competing organisms. To ensure this, large numbers of active starter must be introduced either in the form of commercial deep frozen or freeze dried concentrates, a pre-grown starter culture or by using techniques such as back-slopping where material from a previous successful batch containing high levels of organism is retained and introduced to initiate a new fermentation. A starting level of at least 10^6 cfu g^{-1} is required to guarantee a successful fermentation. Anything that interferes with this, such as the presence of antibiotics, santizer residues or bacteriophage active against the starter will inhibit the fermentation and possibly give rise to a food safety threat.

The rate of pH drop and its final value in lactic acid fermentations depend on a number of factors such as the buffering capacity and water activity of the medium, the temperature and duration of fermentation and the activity of the lactic culture. Ideally the target pH would be around 4.5, although this in not achieved in many common fermented foods such as cheese. Even in very weakly buffered media the pH would tend to bottom out around 3.8 as lactic acid production produces a lactate buffer. Maximum effect will also be achieved if the pH drop occurs rapidly, within hours, to prevent any pathogen growth occurring, but this is less important if the raw material has been pasteurized or if other inhibitory factors are present to restrict the development of pathogens.

FOOD SAFETY PROBLEMS WITH ACIDIC FOODS

Acidic products are not immune from safety concerns; failure to achieve critical limits for pH/acidity can permit the growth/survival of pathogens and there have been some notable outbreaks of foodborne illness involving acid foods.

The most common scenario when problems arise can be viewed, in some respects, as an apparent violation of the hurdle or multiple barrier concept where two antimicrobial factors antagonize rather than supplement one another. A barrier used to slow or arrest growth – nearly always low-temperature storage – reduces the lethal effect of acidity. In mildly acidic foods chill storage will act in concert with the reduced pH to inhibit growth but some form of pasteurization may be necessary to prevent survival and assure safety. In more acidic foods where there is no pasteurization step and the acidity is potentially bactericidal, the survival time of a pathogen can be extended by low temperatures. There have been several examples of where this situation has applied over the years.

Fermented meats contain a high-risk raw material and have been associated with several outbreaks of illness. In the 1990s outbreaks associated with verotoxin-producing *E. coli* in the United States and Australia led the US Food Safety Inspection service to require any fermented sausage process to assure a 5 log reduction in the final product. Numerous studies showed that processes as applied in the USA achieve a 1–3 log reduction although this could be increased by ambient storage or by a heat processing step. This reinforced observations made after an outbreak that occurred in England in 1987/1988 when 101 people were affected by *Salmonella* Typhimurium DT 124 in a salami stick product. The subsequent investigation found that the fermentation process itself did not reduce the numbers of surviving *Salmonella* cells even though the pH dropped below 5.0. Numbers of survivors dropped during subsequent storage, the rate increasing appreciably the higher the temperature of storage. The product had a 6-month shelf-life at ambient and this is how it was generally stored elsewhere in Europe where no cases were reported. In England, however, the product was generally stored along with other cold meats at chill temperature and it was hypothesized that perhaps the lower storage temperature had allowed the salmonella to survive.

A similar scenario is apparent in outbreaks of illness associated with fruit juices: salmonellosis in unpasteurized orange juice and *E. coli* O157 in unpasteurized apple juice (USA – cider). The initial contamination originated with the fruits used to express the juices, the low pH <4.0 would generally be sufficient to inactivate the organism but low-temperature storage was essential to give the unpasteurized product a reasonable shelf-life. It also permitted the survival of the enteric pathogens sufficiently long enough to cause illness.

A number of outbreaks of salmonellosis caused by home-made mayonnaise were reported during the epidemic associated with *Salmonella* Enteritidis and poultry towards the end of the 20th century. The product was made using citric acid or vinegar, oil and eggs contaminated with *Salmonella* Enteritidis. Mindful that it is generally advised to keep ready-to-eat foods chilled, the product was refrigerated soon after production thus prolonging the survival of the *Salmonella* cells and increasing the risk of transmitting illness.

Unlike products such as fermented meats, the textural properties of cheese are not unduly affected by pasteurization of the raw material (milk), though many would claim that the flavour of raw milk cheeses is superior. Pasteurization to eliminate vegetative pathogens in the milk is therefore an important critical control point in assuring production of a safe cheese. It has, however, been associated with numerous outbreaks of foodborne illness over the years (see Table 19.3 for some examples) and in many cases inadequate pasteurization or use of/contamination with raw milk has been a factor.

A pH protective against the growth or survival of pathogens will be ineffectual in situations where a pH-stable toxin is produced elsewhere and added to the acidic product. This was the case in an outbreak of botulism caused by hazelnut yoghurt where *C. botulinum* had grown and produced toxin in the hazelnut purée used to flavor the yoghurt base (see Table 19.3). It is easy to envisage how similar situations could arise elsewhere and serves as a plangent reminder of the need for vigilance with the safety of ingredients.

Raw fish marinaded in lime juice (ceviche) was associated with the spread of pandemic cholera in South America in the early 1990s. Laboratory studies on the ability of *Vibrio cholerae* to survive in ceviche were contradictory and would clearly vary with factors such as the level of acidity, the temperature and the elapsed time between preparation and consumption. Involvement of ceviche in the pandemic could well reflect, in part, a similar variability in the method of production used.

Many parasites (helminths and protozoa) have complex life cycles involving stages that show marked resistance to adverse environmental conditions. They can occasionally be foodborne when they are acquired mainly through consumption of raw or undercooked foods. Processes such as fermentation or pickling are usually insufficient to prevent their

TABLE 19.3 Examples of Foodborne Disease Outbreaks Associated with Fermented Foods

Implicated Food	Causative Agent	Cases	Reference
Plant products			
Paste of soybeans and wax gourds	*Clostridium butyricum*	6	Meng et al. (1997)
Pruno	*C. botulinum*	5 (2 outbreaks)	Vugia (2009)
Fermented milks			
Yoghurt	*C. perfringens*	167	MOH (1993)
Hazelnut yoghurt (hazelnut purée was contaminated)	*C. botulinum*	27	O'Mahoney and Mitchell (1990)
Süzme (condensed yoghurt, Turkey)	*C. botulinum*	10	Akdeniz et al. (2007)
Fermented meats			
Semi-dry sausages	*E. coli* O111:NM	23	CDC (1995a)
Nahm (Thai fermented pork)	Trichinella	27	Khamboonruang and Nateewatana (1975)
Fermented goat (Korea)	*C. botulinum*	5	Tseng (2009)
Salami stick	*Salmonella* Typhimurium	85	Cowden et al. (1989)
Salami	*E. coli* O157	23	CDC (1995b)
Fermented fish			
Seal flipper	*C. botulinum*	1	Shaffer et al. (1990)
Salmon fish heads	*C. botulinum*	8	Shaffer et al. (1990)
Salmon eggs	*C. botulinum*	15 (7 outbreaks 1971–1984)	Hauschild and Gauvreau (1985)
Salmon eggs	*C. botulinum*	4	CCDR (2002)
Cheeses			
Soft cheese	*Salmonella* Berta	82	Ellis et al. (1998)
Cheese	*S.* Enteritidis	~700	CCDR (1999)
Goats' milk cheese	*S.* Paratyphi	273	Desenclos et al. (1996)
Soft cheese	*S.* Dublin	42	Maguire et al. (1992)
Cheddar cheese	*S.* Heidelberg	339	Fontaine et al. (1980)
Mozzarella cheese	*S.* Typhimurium	321	Altekruse et al. (1998)
Cheese	*E. coli* O157	22	The Pennington Group (1997)
Cheese (Brie, Camembert)	*E. coli* O27 H20	170	Altekruse et al. (1998)
Cheese (Brie, Camembert)	*C. botulinum*	27	Pourshafie et al. (1998)

(*Continued*)

TABLE 19.3 (Continued)

Implicated Food	Causative Agent	Cases	Reference
Mexican-style soft cheese	*Brucella melitensis*	31	Altenkruse et al. (1998)
Hand-pressed direct set cheese	*Staphylococcus aureus*	16	Altenkruse et al. (1998)
Cheese	*Shigella sonnei*	50	Sharp (1987)
Hard cheese	*S.* Typhimurium	>200	Van Duynoven et al. (2009)
Cheese	*Staphylococcus aureus*	23	Ostyn et al. (2010)

(Adapted from Motarjemi, 2001).

transmission but data on the incidence of parasitic infections associated with fermented or pickled products are sparse. There is, however, some association between, for example, salami and trichinosis and the Thai fermented fish product som-fak and *Gnathostoma*. Many parasites are susceptible to freezing and, in the absence of cooking, frozen storage is a recognized control measure to eliminate, for example, *Trichinella* in pork destined to be used in fermented meats and *Anasakis* in fish to be lightly pickled or fermented.

CONCLUSIONS

Acidity can be a potent factor in ensuring safe food. Depending on the level, it can inhibit both the growth and survival of pathogens, and there are several useful predictive tools that can help us assign critical limits for acidity. Care must be exercised, however, when there is reliance on the inactivation of contaminating pathogens by low pH and this is combined with hurdles that slow or arrest microbial metabolism such as chilling, salting or drying. In such situations unacceptable survival of a pathogen may occur.

References

Adams, M.R., 2001. Why fermented foods can be safe. In: Adams, M.R., Nout, M.J.R. (Eds.), Fermentation and Food Safety. Aspen Publishers, Gaithersburg, pp. 39–52.

Akdeniz, H., Buzgan, T., Tekin, M., Karsen, H., Karahocagil, M.K., 2007. An outbreak of botulism in a family in Eastern Anatolia associated with eating süzme yoghurt buried under soil. Scand. J. Infect. Dis. 39 (2), 108–114.

Altekruse, S.F., Babagaleh, T.B., Mowbray, J.C., Bean, N.H., Potter, M.E., 1998. Cheese-associated outbreaks of human illness in the United States, 1973 to 1992: sanitary manufacturing practices protect consumers. J. Food Prot. 61 (10), 1405–1407.

CAC, 1979. Code of hygienic practice for low and acidified low acid canned foods. CAC/RCP 23-1979. Adopted 1979, revised 1989, 1993. Editorial corrections, 2011, pp. 85.

CCDR, 1999. Salmonella enteritidis due to contaminated cheese – Newfoundland. Can. Commun. Dis. Rep. 25–30.

CCDR, 2002. Two outbreaks of botulism associated with fermented salmon roe – British Columbia. August 2001, 45–49.

CDC, 1995a. Community outbreak of hemolytic uremic syndrome attributable to *Escherichia coli* O111:NM – South Australia. Morb. Mortal. Wkly Rep. 550–551 and 557.

CDC, 1995b. *Escherichia coli* O157:H7 outbreak linked to commercially distributed dry-cured salami – Washington and California. Morb. Mortal. Wkly Rep. 44 (9), 157–160.

Cowden, J.M., O'Mahony, M., Bartlett, C.L.R., Rana, B., Smyth, B., Lynch, D., et al., 1989. A national outbreak of Salmonella typhimurium DT 124 caused by contaminated salami sticks. Epidemiol. Infect. 103, 219–225.

Desenclos, J.-C., Bouvet, P., Benz-Lemoine, E., 1996. Large outbreak of Salmonella serotype paratyphi B infection caused by goat's milk cheese, France, 1993: a case finding and epidemiological study. Br. Med. J. 312, 91–94.

Ellis, A., Preston, M., Borczyk, A., Miller, B., Stone, P., Hatton, B., et al., 1998. A community outbreak of Salmonella berta associated with a soft cheese product. Epidemiol. Infect. 120, 29–35.

Fontaine, R.E., Cohen, M.L., Martin, W.T., Vernon, T.M., 1980. Epidemic salmonellosis from cheddar cheese: surveillance and prevention. Am. J. Epidemiol. 111 (2), 247–253.

Hauschild, A.H.W., Gauvreau, L., 1985. Foodborne botulism in Canada, 1971–84. Can. Med. Assoc. J. 133, 1141–1146.

ICMSF, 1996. In: Roberts, T.A., Baird-Parker, A.C., Tompkin, R.B. (Eds.), Microorganisms in Foods 5: Characteristics of Microbial Pathogens. Blackie Academic & Professional, London.

Khamboonruang, C., Nateewatana, N., 1975. Trichinosis: a recent outbreak in Northern Thailand. Southeast Asian J. Trop. Med. Public Health 6 (1), 74–78.

Maguire, H., Cowden, J., Jacob, M., Rowe, B., Roberts, D., Bruce, J., et al., 1992. An outbreak of Salmonella dublin infection in England and Wales associated with unpasteurized cows' milk cheese. Epidemiol. Infect. 109, 389–396.

Meng, X., Karasawa, T., Zou, K., Kuang, K., Wang, X., Lu, C., et al., 1997. Characterisation of a neurotoxigenic Clostridium butyricum strain isolated from the food implicated in an outbreak of foodborne type E botulism. J. Clin. Microbiol. 35 (8), 2160–2162.

MOH, 1993. Clostridium perfringens food poisoning at Tabuk Camp. Saudi Epidemiol. Bull. 1 (1), 2.

Motarjemi, Y., Asante, A., Adams, M.R., Nout, M.J.R., 2001. Practical applications: prospects and pitfalls. In: Adams, M.R., Nout, M.J.R. (Eds.), Fermentation and Food Safety. Aspen Publishers, Gaithersburg, pp. 253–273.

Mullan, W.M.A., 2009. Comité des Industries des Mayonnaises et Sauces Condimentaires de la Communauté Économique Européenne (CIMSCEE) model for the preservation of emulsion products using acetic acid. [Online]. Available from: <http://www.dairyscience.info/calculators-models/177-cimscee-model.html> (accessed 04.06.2012.).

O'Mahony, M., Mitchell, E., 1990. An outbreak of foodborne botulism associated with contaminated hazelnut yoghurt. Epidemiol. Infect. 104, 389–395.

Ostyn, A., De Buyser, M.L., Guillier, F., Groult, J., Felix, B., Salah, S., et al., 2010. First evidence of a food poisoning outbreak due to a staphylococcus enterotoxin type E. France 2009. Eurosurveillance 15 (13), 468–474.

Pourshafie, M.R., Saifie, M., Shafiee, A., Vahdani, P., Aslani, M., Salemian, J., 1998. An outbreak of foodborne botulism associated with contaminated locally made cheese in Iran. Scan. J. Infect. Dis. 30, 92–94.

Shaffer, N., Wainwright, R.B., Middaugh, J.P., 1990. Botulism among Alaska natives – the role of changing food preparation and consumption practices. West J. Med. 153 (4), 390–393.

Sharp, J.C.M., 1987. Infections associated with milk and dairy products in Europe and North America, 1980–85. WHO Bull. 65 (3), 397–406.

The Pennington Group, 1997. Report on the circumstances leading to the 1996 outbreak of infection with *E. coli* O157 in Central Scotland, the implication for food safety and the lessons to be learned. Publications Centre, P.O. BOX 276, London, SW8 SDT, UK. Stationery Office, pp. 58.

Tseng, C.-K., Tsai, C.-H., Tseng, C.-H., Tseng, Y.-C., Lee, F.-Y., Huang, W.-S., 2009. An outbreak of foodborne botulism in Taiwan. Int. J. Hyg. Environ. Health 212, 82–86.

Van Duynoven, Y.T., Isken, L.D., Borgen, K., Besselse, M., Soethoudt, K., Haitsma, O., et al., 2009. A prolonged outbreak of Salmonella Typhimurium infection related to an uncommon vehicle: hard cheese made from raw milk. Epidemiol. Infect. 137 (11), 1548–1557.

Vugia, D.J., Mase, S.R., Cole, B., Stiles, J., Rosenberg, J., Velasquez, L., et al., 2009. Botulism from drinking pruno. Emerg. Infect. Dis. 15 (1), 69–71.

Wilson, P.D.G., Wilson, D.R., Waspe, C.R., 2000. Weak acids: dissociation in complex buffering systems and partitioning into oils. J. Sci. Food Agric. 80, 471–476.

Further Reading

Motarjemi, Y., 2001. An introduction to the Hazard Analysis and Critical Control point (HACCP) system and its application to fermented foods. In: Adams, M.R., Nout, M.J.R. (Eds.), Fermentation and Food Safety. Aspen Publishers, Gaithersburg, pp. 53–70.

USDA/FSIS. 1999. Generic HACCP Model for Not Heat Treated, Shelf Stable Meat and Poultry Products. HACCP-15, pp. 42.

CHAPTER

20

Chilling and Freezing

Stephen J. James and Christian James

The Grimsby Institute of Further & Higher Education (GIFHE), North East Lincolnshire, UK

OUTLINE

Food Safety Management.
DOI: http://dx.doi.org/10.1016/B978-0-12-381504-0.00020-2

INTRODUCTION

There is no strict definition of what constitutes a chilled food. In general, it covers any food in which the temperature of the food is reduced to, and maintained at, a temperature below that of the ambient temperature; but above the temperature where any of its water content will change from a liquid to a solid (i.e. begin to freeze). In many foods the initial freezing point will be around −1 °C; however, in food with a high salt content (such as bacon) or sugar content (such as desserts) the initial freezing point may be as low as −10 °C, or even lower. At chilled temperatures (generally between −1 °C and +14 °C) the growth of microorganisms occurs only slowly and food spoilage and deterioration reactions are inhibited to such an extent that food quality is preserved for extended periods. This can range from a few days to many weeks. However, chilled foods are perishable and they deteriorate progressively throughout their life. For many foods the maximum chilled shelf-life will be achieved at a temperature close to their initial freezing points. However, for some foods, such as bananas and other tropical fruit, low temperatures cause damage, and the optimum temperature can be as high as +14 °C.

Below the initial freezing point of a food detrimental reactions that promote food spoilage and limit quality shelf-life are significantly retarded, and in the case of microbial growth will be inhibited at temperatures below −12 °C (for the large majority of foods). Providing the food is of a safe quality prior to freezing, as long as the temperature remains below −12 °C during storage, there will be no growth of pathogenic microorganisms so the food will remain safe. Frozen storage life will be limited by physical and biochemical reactions, which although slow will continue to take place at frozen temperatures, and which ultimately affect the quality of the frozen product. The rates of these reactions are a function of temperature, so the frozen storage life will generally be longer at lower temperatures. Many of these changes will be accentuated if recommended conditions of handling, production and storage are not maintained. A frozen food has a "safe" storage and distribution life that can be measured in years when compared to the days or months of a chilled product. Once thawed, however, any microbes present can again become active, and under the right conditions will multiply to levels that can lead to foodborne illness. The production of safe frozen foods requires the same attention to good manufacturing practices (GMP) and hazard analysis and critical control point (HACCP) principles as the chilled or fresh counterpart.

The cold-chain (Figure 20.1) consists of two distinct types of operation. In processes such as primary and secondary chilling or freezing the aim is to change the average temperature of the food. In others, such as chilled or frozen storage, transport, and retail display, the prime aim is to maintain the temperature of the food. The basic requirements for the production and supply of safe chilled and frozen foods are no different to those needed for other foods. The first is that operations must be operated according to the principles of GMP or GHP (good hygiene practice). The second is the application of HACCP to assure product safety. The third is the application of all verification measures to ensure that the first two are effective. Finally, these measures should be applied in the framework of the quality management systems, such as the ISO 9000 series, to ensure that overall management complies with business excellence (see Chapter 1).

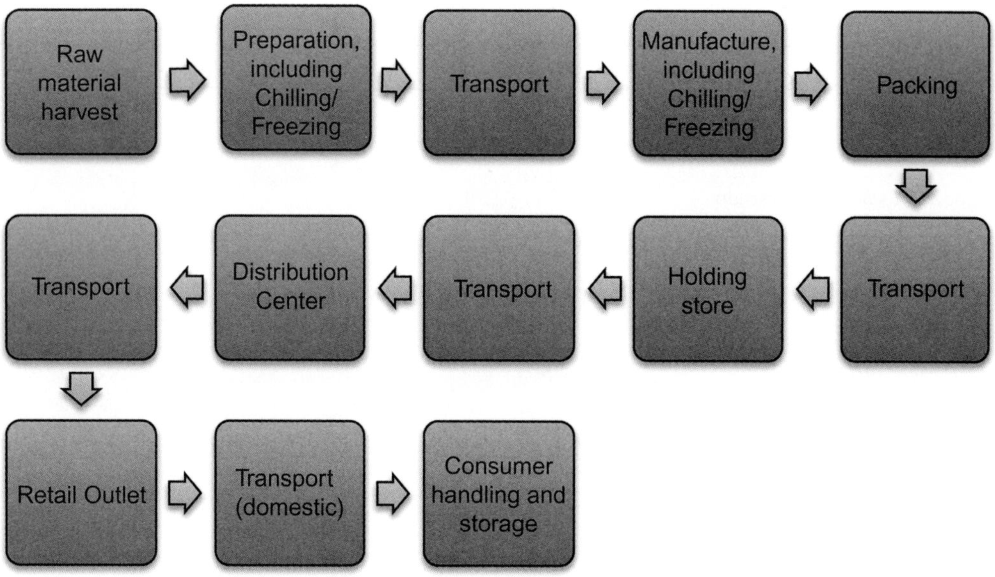

FIGURE 20.1 A typical cold chain.

EFFECT OF CHILLING ON FOOD SAFETY

The microbiological safety and rate of spoilage of chilled foods depends on what biological hazards (pathogens, etc.) are present, what other microflora are present, at what numbers they are present, whether they are on, or in, the food in question, the rate of growth of those microorganisms, the conditions of storage (temperature and gaseous atmosphere), and the characteristics (pH, a_w) of the food. Temperature is by far the most important of these factors.

The principle of chilling as a preservation process is that all biological systems are controlled by enzymatic reactions including those that control microorganisms and cause quality degradation. The rate of these reactions is directly related to temperature. Reducing temperatures below the optimum growth range of a microorganism increases its generation time. The main group of microorganisms of concern in chilled foods are psychrophiles. These organisms (such as *Pseudomonas* and *Entrococcus*) grow well at chill temperatures and cause spoilage on food at temperatures of 5 to 7°C. The optimum temperature growth range of mesophiles is 25 to 30°C and with many the minimum growth temperature is about 10°C. Since most chilled food is kept below this temperature mesophiles are not usually of concern in chilled distribution. However, some organisms (such as *Enterococcus faecalis*) can grow over a temperature range from 0 to >40°C.

Although microorganisms can grow at low temperatures, they grow more slowly as the temperature is reduced. Thus the generation time for a pseudomonad (a common form of spoilage organism) might be 1 hour at 20°C, 2.5 hours at 10°C, 5 hours at 5°C, 8 hours at

TABLE 20.1 Minimum and Optimum Growth Temperatures for Pathogens Associated with Foods

	Minimum Temperature (°C)	Optimum Temperature (°C)
Infective		
Campylobacter spp.	30	42–43
Pathogenic *Escherichia coli* strains	7	35–40
Escherichia coli O157:H7	6–7	42
Salmonella spp.	5	35–43
Aeromonas hydrophila	−0.1–1.2	15–20
Listeria monocytogenes	−1–0	30–37
Yersinia enterocolitica	−2	28–29
Toxigenic		
Clostridium perfringens	12	43–47
Clostridium botulinum proteolytic	10	35
Staphylococcus aureus	7	35–40
Bacillus cereus	4	28–35
Clostridium botulinum non-proteolytic	3	30

2°C or 11 hours at 0°C (Harrigan and Park, 1991). As temperatures are reduced below 10°C, fewer strains can grow and cause spoilage. In general, food will spoil about four times as fast at 10°C and twice as fast at 5°C, as at 0°C. Chill temperatures also have a marked effect on the type of spoilage microflora present on food by altering the microbial community. For example, raw milk stored at temperatures close to 0°C tends to putrefy because of the activity of pseudomonads, rather than to sour due to the activity of lactic acid bacteria.

Pathogenic bacteria such as *Listeria*, *Salmonella*, *Bacillus cereus* and *Yersinia* are of particular concern in chilled foods because they are capable of growth at low temperatures (Table 20.1). Many of the organisms that compete with pathogens at ambient temperatures will not grow at low temperatures, thus low temperatures may preferentially favor the growth of these pathogenic organisms. However, most will not grow, or produce toxins, below 4°C, with the exception of *Listeria* and *Yersinia* (*Yersinia* grows below 0°C).

Investigations (García de Fernando et al., 1995) into the effect of different storage atmospheres on pathogenic growth at low temperatures appear to show that carbon dioxide (CO_2)-enriched atmospheres produce the greatest inhibitory effect on psychrotrophic pathogens (*Y. enterocolitica*, *Aeromonas hydrophila* and *L. monocytogenes*).

EFFECT OF CHILLING RATE ON FOOD SAFETY

Whether "rapid" chilling offers any clear advantages to product safety will depend on what biological hazards (pathogens, etc.) are present, and at what numbers they are present, whether they are on, or in, the food in question, and how "rapid" the rate is in comparison to other rates. There is no definition of "rapid" and "slow" rates. The size of product will also have a big effect on relative rates of chilling, since conduction through the product will become the rate-limiting factor as product size increases.

Rapid chilling has been shown to be an important control measure in reducing histamine formation in at-risk fish species (pelagic species, such as mackerel, sardines, pilchards and certain tuna species) by preventing/reducing the growth of histamine-forming bacteria. High histamine levels are associated with scombroid poisoning. It is generally recommended that such fish is chilled to between 4 and 0°C in less than 12h post-harvesting.

There are instances where excessively rapid chilling rates, or too low a chilling temperature, can cause quality problems in foods. For example, a serious defect known as "woolly texture" can be produced in rapidly cooled peaches. Substantial textural problems due to a phenomenon known as "cold shortening" can occur in rapidly chilled meats (particularly beef and lamb), although electrical stimulation before rapid chilling will mitigate this problem (Chrystall and Devine, 1983).

There is little international legislation that specifies chilling rates. Legislation tends to suggest rather than define, for example the EC Regulation 852/2004 contains a requirement for the cooling of foodstuffs. Annex II, Chapter IX, 6 states: "Where foodstuffs are to be held or served at chilled temperatures they are to be cooled as quickly as possible following the heat-processing stage, or final preparation stage if no heat process is applied, to a temperature, which does not result in a risk to health."

However, there are many guidelines and recommendations, particularly for chilling cooked/pasteurized food products (Table 20.2). The aim of a pasteurization process is to ensure destruction of vegetative stages of any pathogenic microorganisms. The minimum recommended cooking temperature requirements are related to the most thermally resistant pathogen that may present a risk in such products. For many products, such as ready meals, this is *Listeria monocytogenes* and a minimum temperature of 70°C for not less than 2 minutes in the center of the food, or the equivalent, is recommended (Gaze et al., 1989). For other products, including some soups, non-proteolytic *Clostridium botulinum* or *perfringens* are of most concern. There is always the possibility that some microorganisms that produce spores will not be killed by the cooking process. Therefore the temperature of the product should be rapidly reduced between 60 and 7°C to prevent multiplication of any surviving organisms. Further reduction to 3°C is required to reduce growth of spoilage bacteria and prevent the growth/germination of any surviving pathogenic organisms/spores. Although the guidelines were produced specifically for cook–chill catering operations they are often used by the producers of chilled ready meals for retail sale.

Some examples of food poisoning outbreaks directly attributable to poor temperature control are:

1. An analysis of 1000 general outbreaks in England and Wales between 1970 and 1979 by Roberts (1982) identified that inadequate cooling was a contributory factor in 32% of outbreaks and inadequate thawing in 6%.
2. An analysis of 530 general outbreaks in England and Wales between 1992 and 1996 by Panisello et al. (2000) identified that improper storage (including foods either left at room temperature, or warm outdoor temperatures, for several hours or refrigerated in devices

TABLE 20.2 International Chilling Time Guidelines/Recommendations for the Cooling of Cooked Foods

Country	Chilling Range (°C)	Time (h)	Chilling Rate (°C/minute)	Storage Temperature (°C)	Reference
Australia	60–21	≤2	0.33	5	de Jong et al. (2004)
	21–5	≤4	0.07		
Canada	60–20	≤2	0.33	4	CFISIG (2004)
	20–4	≤4	0.07		
Codex Alimentarius	60–10	≤2	0.42	–	Codex Alimentarius Commission (1999)
Denmark	65–10	≤3	0.31	<5	Evans et al. (1996)
France	70–10	≤2	0.50	0–3	Evans et al. (1996)
Germany	80–15	≤2	0.54	2	Evans et al. (1996)
	(15–2)	≤24			
Ireland	70–3	≤2.5	0.45	3	FSAI (2004)
The Netherlands	60–7	≤5	0.18	–	de Jong et al. (2004)
	7–4	≤24			
Sweden	80–8	≤4	0.30	3	Evans et al. (1996)
UK	70–3	≤1.5	0.74	3	UK Department of Health (1989)
USA	60–5	4 to 6	0.23–0.15	–	de Jong et al. (2004)

with deficient temperature control for long periods of time) was a contributory factor in 32% of outbreaks and inadequate thawing in 2%.

3. HPA analysis of general foodborne outbreaks in England and Wales in 2011 (Health Protection Agency, 2012) showed that of the 83 general outbreaks inadequate chilling was a contributory factor in 14% (12/83), while storage of food for too long was a contributory factor in 19% (16/83).

EFFECT OF FREEZING ON FOOD SAFETY

Microorganisms vary in their ability to tolerate freezing and frozen storage. Survival is affected by the type and age of microorganism. In general, Gram-negative bacteria (which include pathogens such as *Escherichia* and *Salmonella* spp.) are more susceptible to freezing and frozen storage than Gram positives, with bacilli being more susceptible than cocci. Yeasts and molds are more resistant than bacteria, in part due their tolerance to reduced water activity (a_w). Psychrophilic and psychotropic microorganisms are generally more tolerant to freezing and frozen storage due to their ability to synthesize larger amounts of enzymes to compensate for reduced enzymic activity at low temperatures, and their reduced susceptibility to cold shock in comparison to thermopiles and mesophiles.

Spoilage microorganisms do not grow below ca. −10 to −12°C and pathogens below −1°C, thus the growth of pathogenic microorganisms is only normally relevant to handling before freezing or during thawing. In these contexts, frozen foods behave like their unfrozen counterparts, if surface temperatures are reduced rapidly during freezing this allows less time for any microorganisms to grow, although growth rates may be faster after thawing due to increased drip. Also thawing may take a long time and on large objects subjected to long uncontrolled thawing cycles, surface spoilage can occur before the center regions have fully thawed.

Repeated freeze–thaw cycles have been shown to disrupt and destroy bacteria; however, the effects of cyclic freezing on most microbial pathogens are not well documented.

Although *Salmonella*, *Staphylococci* and other potential pathogens can survive freezing and frozen storage, spoilage bacteria tend to inhibit their growth. During freezing and thawing of food, the temperature favors the growth of psychrophilic organisms, most of which are spoilage organisms. Hence, in nearly all cases, if a frozen product is mishandled, spoilage is apparent before the food becomes a health hazard.

Freezing and crust-freezing has been suggested as a means to reduce numbers of *Campylobacter* organisms on poultry carcasses. It is one of a number of measures taken to reduce the incidence of campylobacteriosis in Iceland, although the exact impact of this measure is unclear. This work in Iceland (Stern et al., 2003) has been very influential and many risk assessment models have incorporated freezing as an import factor due to this work. Freezing to ~−20°C has been reported by a number of studies to result in an initial fall in numbers of *Campylobacter* organisms, followed by a slower decline during storage. The European Food Safety Authority (EFSA) has recommended freezing as a control measure for reducing *Campylobacter*. The mechanism of damage during freezing has been attributed to mechanical damage caused by ice crystals, desiccation due to the reduced water activity, and oxidative damage.

Freezing generally has little effect on viruses. For example, the H5N1 virus (avian influenza H5N1), if present in poultry meat, is not destroyed by freezing. Food contaminated with hepatitis A is a common vehicle transmitting the virus and each year approximately 30–50,000 cases of hepatitis A-related illnesses occur in the United States (CDC, 2007). In 2012 imported frozen strawberries contaminated with hepatitis A were believed to be the source of an outbreak of food poisoning in 11,000 children in Germany (Herriman, 2012).

Freezing has been shown to be a control measure in reducing histamine formation in at-risk fish species (pelagic species such as mackerel, sardines, pilchards and certain tuna species), both by preventing the growth of histamine-forming bacteria and by reducing the activity of preformed histidine decarboxylase. However, while freezing may limit histamine formation, it has no effect on histamine that has already been formed prior to freezing.

Higher organisms, such as nematode parasites, are very susceptible to freezing and freezing is a control measure for inactivating trichinae in pork, tape worms (*Taenia saginata*) in beef and nematode parasites in seafood (particularly for lightly processed seafoods that will receive no cooking before consumption) (Archer, 2004). The USDA recommended holding times for pork to inactivate *Trichinella spiralis* range from 106 hours at −18°C to 0.5 hours at −37°C. Freezing is used as a control measure for inactivating tape worms (*Taenia saginata*) in beef carcasses with localized infections in the EU by holding at −10°C or less for 14 days or more.

Most frozen fruits and vegetables are subjected to a mild heat treatment known as blanching before freezing. Blanching is carried out to inactivate various enzymes that can lead to quality deterioration over time. Typically, blanching is done by treating the product with steam or hot water for 1–10 minutes at 75–95°C, the time–temperature combination depending on the specific product. Such treatment times and temperatures are also capable of reducing, to varying extents, the numbers of viable microorganisms on the food.

EFFECT OF FREEZING RATE ON FOOD SAFETY

Whether "rapid" freezing offers any clear advantages to product safety will depend on what biological hazards (pathogens, etc.) are present, at what numbers they are present, whether they are on, or in, the food in question, and how "rapid" the rate is in comparison to other rates. There is no definition of "rapid" and "slow" rates. Size of product will also have a big effect on relative rates of freezing, since conduction through the product will be the rate limiting factor.

CHILLING AND FREEZING PRINCIPLES

Chilling and freezing is a process of removing heat and can only be achieved by four basic mechanisms: conduction, radiation, evaporation or convection.

Conduction requires a good physical contact between the food to be chilled/frozen and the cooling medium, and this is generally achieved only with foods that can be shaped into regular shapes, such as blocks of meat or fish, etc.

Radiation does not require any physical contact but a large temperature difference is required between the surface of the food being cooled and that of surrounding surfaces to achieve significant heat flow. In primary chilling/freezing, radiation is only important in the initial stages of the process in a system where the food is not surrounded by other products. Again, in the initial stages of the chilling/freezing of cooked food products (e.g. pies and other pastry products, meat joints, baked cakes, etc.) radiant heat loss can be substantial if the products are surrounded by cold surfaces.

Evaporation from a food surface reduces yield and is not desirable in most food refrigeration operations but can be useful again in the initial cooling of cooked food products and is used in the immediate post-harvest cooling of many fruits and vegetables. However, as soon as the surface of the food is close to that of the cooling medium then any heat loss due to evaporation is minimal.

Convection is by far the most important heat transfer mechanism employed in the majority of food chilling/freezing systems. In most cases, refrigerated air is the transfer medium; however, in some cases a liquid or a cryogenic gas may be used. The rate of heat removal from the surface of a food depends on:

1. The surface area of the food available for heat flow.
2. The temperature difference between the surface of the food and the cooling medium.

3. The surface heat transfer coefficient (h). The value of h will depend on the shape of the food and its surface roughness, the type of cooling medium, the velocity of the cooling medium, and the flow regime. The higher the surface heat transfer coefficient the faster the rate of surface cooling. Air is a poorer heat transfer medium than a fluid (such as water or brine). Increasing the air flow or agitation around a food will increase the rate of surface cooling.

Heat must also be conducted from within the food to its surface before it can be removed. Since most foodstuffs are poor conductors of heat this imposes a severe limitation on attainable cooling times for either large individual items (such as meat carcasses) or small items cooled in bulk (such as a pallet of boxed product).

CHILLING/FREEZING METHODS/EQUIPMENT

There are many different types of chilling/freeing equipment, but generally all use a gas or a liquid as the cooling medium. Equipment is classified according to the method of chilling/freezing into:

- Direct methods, where the energy is extracted directly from the food into the heat transfer medium (in such cases the heat transfer medium needs to be food safe, such as air or liquid nitrogen), for example air blast chillers/freezers.
- Indirect methods, where the cooling is generated externally and then applied to the food through heat exchangers, for example plate freezers.

Chilling/freezing may be carried out as a batch (Figure 20.2) or continuous system (Figure 20.3). Generally, batch systems are used to refrigerate small quantities of food, whereas continuous systems become economic with large throughputs.

From a food safety-based approach, prepacking the food prior to chilling/freezing will lower the risk of contamination/cross-contamination during the chilling/freezing process; however, it will significantly reduce the rate of cooling, and this may allow the growth of any microorganisms present. Although there has been (and remains) a great debate regarding the virtues of "dry" air vs. spray or immersion methods for chilling products such as poultry, there is no clear scientific evidence that air chilling is hygienically better than spray or immersion (James et al., 2006). Provided the cooling medium (air, water, etc.) and refrigeration equipment used is kept clean, no one cooling method can be said to be intrinsically more hygienic than any other. The potential for the fans used in air chilling to disseminate molds and bacteria has been identified in a number of reviews but very little work has been carried out to evaluate whether this is in fact the case. A study carried out in the UK found high levels of bacterial contamination of evaporator coils in industrial chilling systems but very few pathogens (Evans et al., 1997; James et al., 1998). Further laboratory studies showed that bacteria did not grow on clean coils. It is therefore important that food refrigeration systems should be properly constructed and maintained. The design of chillers/freezers, especially drip trays, should facilitate effective cleansing and disinfection.

FIGURE 20.2 Simple batch air-cooling system for cooling trays of product.

FIGURE 20.3 Continuous air-chilling system for whole poultry.

Air Chillers/Freezers

Air is by far the most widely used method of chilling and freezing food as it is economical, hygienic and relatively non-corrosive to equipment. The big advantages of air systems are their cost and versatility, especially when there is a requirement to cool a variety of irregularly shaped products.

Three different types of air-blast (forced air) systems are used: batch (cabinets or rooms), tunnel and spiral freezers. Batch blast chillers/freezers usually consist of a room or large cabinet into which the product is loaded directly onto shelves or via trolleys that are wheeled into the chamber (Figure 20.2). These chillers/freezers are sometimes called batch-continuous systems if the trolleys are periodically removed trolley by trolley on a "first in-first out" basis. The tunnel chiller/freezer, in its simplest form, is a straight, continuous link-belt carrying product through a chamber (Figure 20.3) or tunnel. Spirals are essentially tunnel chillers/freezers in which the belt travels in a spiral (helical) motion through a near-cube-shaped room. The airflow direction in a spiral may be horizontal, vertical or some combination of these, as it flows over the product riding along on the belt. Spirals perform the same function as tunnels but require less floor space; however, they are usually more expensive. Both tunnels and spirals are best suited to small products with relatively short chilling/freezing times of less than an hour.

Operating temperatures and air speeds depend on the requirements of the product. If the risk of surface freezing is to be avoided (in the case of chilling) air temperatures will need to be above −2°C (depending on the freezing point of the product).

In general, relatively low rates of heat transfer are attained from product surfaces in air-cooled systems. In standard systems air speeds are seldom faster than $6\,ms^{-1}$, but far higher air speeds (up to $30\,ms^{-1}$) are achievable, with higher surface heat transfer rates, in impingement systems. Impingement chillers/freezers are best suited for products with high surface area to weight ratios (for example, products with one small dimension such as hamburger patties, pizzas, etc.). Testing has shown that products with a thickness less than 20 mm chill/freeze most effectively in an impingement heat transfer environment.

High heat transfer rates do not offer advantages for thick products where heat transfer within the product is the rate-limiting factor. For example, while increasing the air velocity during chilling of beef sides substantially reduces chilling times at low air velocity, the effect is smaller at higher velocities. Also the power required by the fans to move the air within a room increases with the cube of the velocity. Thus while a fourfold increase in air velocity from 0.5 to $2\,ms^{-1}$ will result in a 4.4 h (18%) reduction in chilling time for a 140 kg side, it requires a 64-fold increase in fan power. In most practical situations, where large items are being cooled it is doubtful whether an air velocity greater than $1\,ms^{-1}$ can be justified.

Even when a system has been designed to distribute the air through the product, poor management and/or poor understanding of the requirement of the plant commonly leads to uneven cooling. Products stacked or racked irregularly will leave channels around the stacks that are larger in cross-sectional area than those within the stacks and channels of differing area through the stacks. Air leaving and returning to the refrigeration coil will take the path of least resistance through the largest gaps, instead of passing evenly through or over the product.

One of the principal disadvantages of air-cooling systems is their tendency to dehydrate unwrapped products. One solution to this problem is to saturate the air with water. Wet-air cooling systems recirculate air over ice-cold water so that the air leaving the cooler is cold (0 to 1°C) and virtually saturated with water vapour (100% RH).

Immersion/Spray Chillers/Freezers

In immersion/spray systems products are either immersed in or sprayed with a cold liquid. When water is used as the heat transfer medium the process is often called "hydrocooling." Systems range in size from 2 to 3 m^3 tanks used to cool small batches of cooked products to large continuous chilling systems capable of cooling 10,000 poultry carcasses per hour. They produce high rates of heat transfer due to the intimate contact between product and cooling medium. They offer several inherent advantages over air-cooling in terms of reduced dehydration and coil frosting problems. Clearly if the food is unwrapped the heat transfer medium has to be "food safe." Cooling using ice or cryogenic substances are essentially immersion/spray processes. The freezing point of the cooling medium used dictates the temperature it can be used at. Heat transfer medium temperatures $<0°C$ necessitate the use of non-toxic salt, sugar or alcohol solutions in water, or the use of cryogens or other refrigerants. Calcium chloride solutions are capable of temperatures as low as −55°C.

Chilling with crushed ice, or an ice/water mixture, is simple, effective and commonly used for the cooling of fish (Figure 20.4), turkeys (Figure 20.5) and some fruits and vegetables. Cooling is more attributable to the contact between the fish and the cold melt water percolating through it (i.e. hydrocooling) than with the ice itself. The individual fish are packed in boxes between layers of crushed ice, which extract heat from the fish and consequently melt. Ice has the advantage of being able to deliver a large amount of refrigeration in a short time as well as maintaining a very constant temperature, 0 to −0.5°C (where sea water is present).

Solid carbon dioxide pellets or "snow" can be used in much the same way as ice for some applications, for example during sausage manufacture to remove the heat generated during chopping and mixing. Solid carbon dioxide has the advantage over ice that it rapidly sublimes to gas leaving no residue and not wetting the product.

FIGURE 20.4 Ice/water immersion cooling of whole fish.

FIGURE 20.5 Immersion cooling of turkey carcasses.

Cryogenic Freezers

The term cryogenic simply means very low temperature. Cryogenic cooling uses refrigerants, such as liquid nitrogen or solid carbon dioxide, directly. Cryogenic freezing is often treated as a specific type of freezing method on its own; however, it is essentially an immersion/spray system, depending on how the cryogen is utilized.

Although it is common in laboratory studies to freeze samples with liquid nitrogen by direct immersion, few commercial liquid nitrogen freezers employ this technique. One reason for this is that many foods will shatter and split if frozen in this way, due to rapid ice expansion; it is also inefficient. Cryogens are typically employed as sprays in tunnel, spiral or batch cabinet systems.

Cryogenic freezing is often cited as the fastest method of freezing a food. Rapid freezing in comparison to other methods is principally due to very low operating temperatures. In general, commercial cryogenic freezers do not provide substantially higher surface heat transfer coefficients between the product and medium than other refrigeration systems, unless the cryogen comes in direct contact with the product. Cryogenic systems are best suited to freezing thin products with high surface area to weight ratios in which heat conduction within the product is not rate limiting. Although running costs of cryogenic systems can be expensive, capital investment is low, with cryogenic suppliers often renting the equipment to users. Also, installation and maintenance costs are lower than mechanical refrigeration systems.

Vacuum Chillers

Vacuum cooling systems work by boiling some of the water in/on the food under vacuum conditions (typically operating at between 530 and 670 Nm^{-2}); the low pressure lowers the boiling point of water. The food cools due to the evaporation of this moisture. Evaporative cooling is quite significant, the amount of heat released through the evaporation of 1 g of water is equivalent to that released in cooling 548 g of water by 1°C (Fennema, 1975). In general terms a 5°C reduction in product temperature is achieved for every 1% of water that is evaporated. Food products that have a large surface area to volume ratio and an ability to readily release internal water are the most amenable to vacuum cooling. Suitable products, such as lettuce, can be vacuum cooled in less than 1 hour. Since vacuum cooling requires the removal of water from the product, prewetting is commonly applied to prevent the removal of water from the tissue of the product. Traditionally, this method of cooling has been relatively common for removing "field heat" of leafy vegetables immediately after harvest, but it is also suitable for many other foods, such as baked products, sauces, soups, particulate foods and meat joints (James and James, 2002; Zheng and Sun, 2005). It is particularly good for cooked fillings, stews, sauces and casseroles since pressure cooking and vacuum cooling can be combined in the same vessel, reducing both cooking and cooling times and saving space (Figure 20.6).

Vacuum cooling is rapid and economical to operate because of low labor costs, but the capital cost of the large vacuum vessels is very high and it is usually a batch process; this limits its widespread use.

FIGURE 20.6 Combined pressure cooker/vacuum cooler.

Plate Chillers/Freezers

Modern plate cooling systems differ little in principle from the first contact freezer patented in 1929 by Clarence Birdseye. Essentially product is pressed between hollow metal plates containing a circulating refrigerant. A hydraulic cylinder is used to bring the plates into pressure contact with the product. These plates can be either horizontal or vertical.

Plate coolers are more commonly used to freeze solid foods, but can be used for chilling. Contact cooling offers several advantages over air-cooling, for example better heat transfer and significant energy savings. However, the need for regularly shaped products with large flat surfaces is a major hindrance.

Belt Freezers

Belt freezers employ a similar contact method of freezing to plate freezers.

Simple belt freezers consist of an endless steel belt (around 1 mm thick), the underside of which is cooled directly with brine, glycol or cryogenic sprays, or by sliding over a stationary cold surface. Since only one side of the product is in contact with the cooling surface relatively thin products are required, such as hamburgers, fish fillets or liquid and semi-liquid products such as purées and sauces.

In double-band systems the product is frozen between two endless belts of which the top is flat and the lower belt corrugated. The product is spread into the corrugations, the top belt enclosing the exposed surface thus freezing the product as IQF (individually quick frozen) pellets. Liquids and semi-liquids are often frozen into pellets using this method.

Scraped Surface Freezers

Scraped surface, or cylindrical, freezers are designed for freezing liquid products either on the inner or the outer surface of a cooled cylinder. The layer of frozen product formed on the surface of the cylinder is continuously scraped from the cylinder surface, thus achieving high heat transfer and a rapid freezing rate. Scraped surface freezers are used for manufacturing ice creams and similar products.

Stirred Jacketed Vessels

Stirred jacketed vessels are essentially a scraped surface heat exchanger (also called swept surface heat exchanger). The blades' rotation causes an increase in the product mixing, and also leads to film removal as they continuously scrape the walls through which heat is being transferred. This type of heat exchanger is particularly useful for high viscosity products, since the heat transfer is increased by the produced turbulence.

Liquid Heat Exchangers

Liquid heat exchangers can be classified in four main categories according to construction features: tubular, plate-type, extended surface and regenerative exchangers. Tubular heat exchangers consist of one single tube or more, enclosed within a larger tube. The product flows through the smaller tubes, with the cooling medium flowing over the tubes within the larger tube.

Plate heat exchangers consist of a series of thin plates clamped together on a frame and separated by spacing gaskets. The spaces form channels in which the two fluids flow, exchanging heat through alternate plates. Suitable gaskets and channels control the flow and allow parallel or counter-current flow in any desired number of passes. The plates usually have a corrugated pattern in order to increase the available surface area for heat transfer, provide greater system support and enhance the turbulence present in the process.

Chilled Storage

Publications such as the International Institute of Refrigeration (IIR) *Recommendations for the Chilled Storage of Perishable Produce* (2000) provide data on the storage life of many foods at different temperatures.

Three factors during chilled storage – the storage temperature, the degree of fluctuation in the storage temperature and the type of wrapping/packaging in which the food is stored – are commonly believed to have the main influence on chilled storage life. The storage life of most chilled foods is limited by the growth of spoilage microorganisms. However, with unwrapped food, dehydration of the surface layers may lead to unacceptable quality changes. In general, for many foods the longest chilled storage life will be achieved by storing the food at a temperature just above its freezing point.

Most unwrapped meat, poultry, fish, fruit and vegetables and all types of wrapped foods are stored in large rooms with circulating air. To minimize energy consumption and in unwrapped foods weight loss/appearance changes associated with desiccation, air movement should be the minimum required to maintain a constant temperature. However, many storage rooms are designed and constructed with little regard to air distribution and localized velocities over products. Horizontal throw refrigeration coils are often mounted in the free space above the racks or rails of product and no attempt is made to distribute the air around the products. Using a false ceiling or other form of ducting to distribute the air throughout the storage room can substantially reduce variations in velocity and temperature.

There are some cases where maintaining a particular food at temperatures that severely limit if not completely stop chemical changes does not achieve the desired final product quality. Examples of this are in the maturing of meat, ripening of fruits and flavor development in cheese. In all these cases, the time–temperature history of the food must be carefully controlled so that periods are provided at temperatures where the desired changes can occur. However, the combination of time and temperature needs to be controlled such that undesirable and especially unsafe changes do not occur.

Controlled atmosphere storage has been developed for specialized fruit stores. In addition to the normal temperature control plant these stores also include special gas-tight seals to maintain an atmosphere that is normally lower in oxygen and higher in nitrogen and carbon dioxide than air. Additional plant is required to control the carbon dioxide concentration, generate nitrogen and consume oxygen. There is growing use of controlled atmosphere and modified atmosphere retail packs to extend the chilled storage and display life of red meats, poultry, fish and vegetables. Since the packs tend to be large and insulate the products efficient precooling before packaging is especially important if product quality is to be maintained. Provided that temperatures during chilled storage are sufficient to prevent or inhibit the growth of any pathogens present on the food in question, in general the food will spoil before unsafe pathogen growth occurs.

Frozen Storage

Publications such as the IIR *Recommendations for the Processing and Handling of Frozen Foods* (2006) provide data on the storage life of many foods at different temperatures. Storage lives for food can be as short as 3 to 4 months for individually quick frozen, polybag-packed shrimps at −18°C. On the other hand, lamb stored at −25°C can be kept for over 2.5 years.

Most of the mechanisms of quality loss in frozen foods are determined by storage temperature and are accelerated with time spent above the recommended value. They are also promoted by temperature fluctuations. Traditionally the frozen food industry was interested in two problems that were detrimental to the appearance of the frozen food: "freezer burn" and "in-package frosting" – both of which may occur during storage. Freezer burn is caused by water loss from the surface of the frozen food due to sublimation. The resulting desiccation produces a dry fibrous layer at the surface that has the appearance of a burn. It is irreversible. It only occurs in unwrapped, or poorly wrapped, foods and its development is fastest at high storage temperatures and high air movements. It occurs during storage and not during the freezing process (unless the freezing process is excessively long); it is not caused by fast freezing. In-package frosting results from a combination of water loss from the surface, loose packaging and temperature fluctuations during storage. The water lost from the surface is deposited and frozen on the inner surface of the packaging. The use of suitable packaging and good temperature control should eliminate both problems. Neither has an effect on food safety.

THAWING (DEFROSTING) AND TEMPERING SYSTEMS

Frozen raw material as supplied to the industry ranges in size and shape, although much of it is in blocks packed in boxes. Thawing (defrosting) is usually regarded as complete when the center of the block has reached 0°C, the minimum temperature at which meat or fish can be filleted or cut by hand and fruits and vegetables hand sorted. Lower temperatures (e.g. −5 to −2°C) are acceptable for produce that is destined for mechanical chopping, but such product is "tempered" rather than thawed. The two processes should not be confused because tempering only constitutes the initial phase of a complete thawing process.

Inherent in thawing is a major problem that does not occur in a freezing operation. The majority of the bacteria that cause spoilage or food poisoning are found on the surfaces of food. During the freezing operation, surface temperatures are reduced rapidly and bacterial multiplication is severely limited, with bacteria becoming completely dormant below −10°C. In the thawing operation these same surface areas are the first to rise in temperature and bacterial multiplication can recommence. On large objects subjected to long uncontrolled thawing cycles, surface spoilage can occur before the center regions have fully thawed.

Most systems supply heat to the surface and then rely on conduction to transfer that heat into the center of the foodstuff. A few use electromagnetic radiation to generate heat within the food. In selecting a thawing system for industrial use a balance must be struck between thawing time, appearance and bacteriological condition of the product, processing problems such as effluent disposal and the capital and operating costs of the respective systems.

TRANSPORTATION

Chilled/frozen foods are transported around the world and locally via a range of transportation systems. All these transportation systems are expected to maintain the temperature of the food within close limits to ensure its optimum safety and high-quality shelf-life. It is particularly important that the food is at the correct temperature before loading since the refrigeration systems used in most transport containers are not designed to extract heat from the load but to maintain the temperature of the load. In the large containers used for long distance transportation food temperatures can be kept within ±0.5°C of the set point.

Control of the oxygen and carbon dioxide levels in shipboard containers has allowed fruits and vegetables, such as apples, pears, avocados, melons, mangoes, nectarines, blueberries and asparagus, to be shipped (typically 40 days in the container) from Australia and New Zealand to markets in the USA, Europe, Middle East and Japan. Even longer shelf-lives (over 20 weeks) can now be achieved for meats, particularly beef and lamb.

Air-freighting is increasingly being used for high-value perishable products, such as strawberries, asparagus and live lobsters. Although air-freighting of foods offers a rapid method of serving distant markets, there are many problems because the product is usually unprotected by refrigeration for much of its journey. Up to 80% of the total journey time is made up of waiting on the tarmac and transport to and from the airport. During the flight the hold is normally between 15 and 20°C. Perishable cargo is usually carried in standard containers, sometimes with an insulating lining and/or dry ice but is often unprotected on aircraft pallets.

Overland transportation systems range from 12m refrigerated containers for long distance road or rail movement of bulk chilled or frozen products to small uninsulated vans supplying food to local retail outlets or even directly to the consumer. The rise in supermarket home delivery services, where there are requirements for mixed loads of products that may each require different storage temperatures, is introducing a new complexity to local overland delivery.

CHILLED RETAIL DISPLAY

The temperature of individual consumer packs, small individual items and especially thin sliced products responds very quickly to small amounts of added heat. All these products are commonly found in retail display cabinets and marketing constraints require that they have maximum visibility. Maintaining the temperature of products below set limits while they are on open display in a heated store will always be a difficult task.

The required display life and consequent environmental conditions for wrapped chilled products differ from those for unwrapped products. The desired chilled display life for wrapped meat, fish, vegetables and processed foods ranges from a few days to many weeks and is primarily limited by microbiological considerations. Retailers of unwrapped fish, meat and delicatessen products (e.g. sliced meats, pâté, cheese and prepared salads) normally require a display life of one working day. The introduction of humidification systems can significantly improve display life of unwrapped products.

Average temperatures in chilled retail display cabinets can vary considerably from cabinet to cabinet, with inlet and outlet values ranging from −6.7 to +6.0°C, and −0.3 to +7.8°C, respectively, in one survey (Lyons and Drew, 1985). The temperature performance of an individual display cabinet does not only depend on its design. Its position within a store and the way the products are positioned within the display area significantly influences product temperatures. External factors such as the store ambient temperature, the position of the cabinet and poor pretreatment and placement of products substantially affect cabinet performance. Warm and humid ambient air and loading with insufficiently cooled products can also overload the refrigeration system. Even if the food is at its correct temperature, uneven loading or too much product can disturb the airflow patterns and destroy the insulating layer of cooled air surrounding the product. One in-store survey of 299 prepackaged meat products in chilled retail displays found product temperatures in the range −8.0 to 14.0°C, with a mean of 5.3°C and 18% above 9°C (Rose, 1986). Other surveys have shown that temperatures of packs from the top of stacks were appreciably higher than those from below due to radiant heat pick-up from store and cabinet lighting. It has also been stated that products in transparent film overwrapped packs can achieve temperatures above that of the surrounding refrigerated air due to radiant heat trapped in the package by the "greenhouse" effect. However, specific investigations have failed to demonstrate this effect (Gill, 1988).

FROZEN RETAIL DISPLAY

No frozen food, with the possible exception of ice cream, should be unwrapped when in a retail display cabinet. Traditionally frozen food was displayed in a "well-type" cabinet with only the top faces of food packs being exposed. In many cases the cabinets were fitted with a see-through insulated lid to further reduce heat infiltration. There is marketing pressure to display an increasing amount of frozen food in open multi-deck display cabinets. Maintaining the temperature of products below set limits while they are on open display in a heated store will always be a difficult task. Radiant heat gain on the surfaces of exposed packs can result in the food thawing in extreme cases. During display, temperature, temperature fluctuations and packaging are the main display parameters that control quality.

Temperature fluctuations can increase the rate of weight loss from wrapped frozen food. Higher rates of dehydration have been measured in a retail cabinet operating at −15°C than in another cabinet operating at −8°C. Fluctuations in air temperature in the −15°C cabinet ranged from −5 to −21°C compared with ±1.5°C in the −8°C cabinet. Successive evaporation and condensation (as frost) caused by such a wide temperature differential resulted in exaggerated in-package dehydration.

The extent of temperature fluctuations will be dependent upon the air temperature over the product, the product packaging and the level of radiant heat. Retail display packs have a relatively small thermal mass and respond relatively quickly to external temperature changes. These can be from store and display lighting, defrost cycles and heat infiltration from the store environment. In products where air gaps exist between the packaging and the food, sublimation of ice within the product leads to condensation on the inside of the packaging, resulting in a build-up of frost. This dehydration causes small fissures in the surface of the food, allowing the ingress of any packaging gases into the food. This can aid the acceleration of oxidative rancidity within the product. Minor product temperature fluctuations are generally considered to be unimportant, especially if the product is stored below −18°C and fluctuations do not exceed 2°C.

DOMESTIC HANDLING

When removed from display cabinets the temperature of chilled and frozen foods can rise rapidly if exposed to ambient conditions. Surveys have shown that the majority of consumers do not use insulated bags or boxes to transport chilled and frozen food to their homes. Once the food has warmed during transportation it can take many hours in a domestic refrigerator and freezer for the food temperature to fall below a safe temperature. It is also common for consumers to purchase chilled products and freeze them at home. Studies have shown that it can take over 6 hours for the temperature of a chicken portion to cool from 0 to −5°C in a domestic freezer.

Generally the range of recommended refrigerator temperatures are below 8°C throughout the world, with many countries (including the UK) recommending below 5°C. The numerous surveys on the domestic storage of refrigerated foods show remarkable similarities in consumer attitudes and handling of chilled foods and the performance of their fridges. Perhaps even more remarkable is that despite numerous recommendations on handling and storage temperatures, consumer use and the performance of refrigerators remain remarkably unchanged throughout the world over the last 30 years. Numerous surveys show that mean temperatures range between 5 and 7°C, with 50 to 70% of domestic refrigerators operating at temperatures above 5°C (James et al., 2008). It is clear that many refrigerators throughout the world are running at higher than recommended temperatures. Since even these recommended temperatures are higher than the 0 to 1°C that is usually the recommended temperature range for storing fish and seafood, meat and many chilled products, the current situation is even more detrimental to maintaining the high-quality life of chilled foods. At present domestic storage of chilled foods would appear to be the weakest link in the entire chill chain.

After a frozen product has reached the operating temperature of a domestic freezer it is very unlikely that its temperature will rise above −12°C during domestic storage, unless there is an electricity cut. In a New Zealand survey (McIntyre et al., 2007) mean temperatures in domestic freezers ranged from −11.5 to −23.3°C with an overall mean value of −16.6°C. Only 28% of freezers operated at −18°C or lower, with 68% operating between −13 and −18°C. Temperature control in freezers does not appear to have improved over the last 20 years. Freezers ≤10 years old and freezers ≥21 years old had similar mean temperature values. The mean air temperature recorded in the top sections of surveyed freezers was on average 2 to 2.5°C warmer than the middle and bottom sections, respectively, which suggests that freezing could be slightly slower in the upper areas of the freezer compartment.

SPECIFYING REFRIGERATION SYSTEMS

In the authors' experience, the poor performance of new refrigeration systems used to maintain the cold chain can often be traced back to a poor, non-existent or ambiguous process specification. In older systems it is often due to a change in use that was not considered in the original specification. There are three stages in obtaining a refrigeration system that works:

1. Determining the process specification, i.e. specifying exactly the condition of the product(s) when they enter and exit the system and the amounts that have to be processed.
2. Drawing up the engineering specification, i.e. turning processing conditions into terms that a refrigeration engineer can understand, independent of the food process.
3. The procurement and commissioning of the total system, including any services or utilities.

The first task in designing a system is the preparation of a clear specification by the user of how the facility will be used at present, and in the foreseeable future. In preparing this specification the user should consult with all parties concerned: these may be officials enforcing legislation, customers, other departments within the company and engineering consultants or contractors – but the ultimate decisions in forming this specification are taken by the users alone.

The process specification must include, as a minimum, data on the food(s) to be refrigerated, in terms of size, shape and throughput. The maximum capacity must be catered for and the refrigeration system should also be specified to operate adequately and economically at all other throughputs. The range of temperature requirements for each product must also be clearly stated. If it is intended to minimize loss, it is useful to quantify at an early stage how much extra money can be spent to save a given amount of weight. All the information collected so far, and the decisions taken, will be on existing production. Another question that needs to be asked is: Will there be any changes in the use of the refrigeration system in the future?

The refrigeration system chiller, freezer, storeroom, etc. is one operation in a sequence of operations. It influences the whole production process and interacts with it. An idea must be obtained of how the system will be loaded, unloaded and cleaned, and these operations must always be intimately involved with those of the rest of the production process. There is often a conflict of interest in the usage of a chiller or freezer. In practice, a chiller/freezer can often be used as a marshaling yard for sorting orders, and as a place for storing product not sold. If it is intended that either of these operations are to take place in the chiller/freezer the design must be made much more flexible in order to cover the conditions needed in a marshaling area or a refrigerated store. In the case of a batch or semi-continuous operation, holding areas may be required at the beginning and end of the process in order to even out flows of material from adjacent processes. The time available for the process will be in part dictated by the space that is available; a slow process will take more space than a fast process, for a given throughput.

Other refrigeration loads, in addition to that caused by the input of heat from the product, also need to be specified. Many of these, such as infiltration through openings, the use of lights, machinery and people working in the refrigerated space, are all under the control of the user and must be specified so that the heat load given off by them can be incorporated in the final design. Ideally, all the loads should then be summed together on a time basis to produce a load profile. If the refrigeration process is to be incorporated with all other processes within a plant, in order to achieve an economic solution, then the load profile is important. The ambient design conditions must be specified. These are generally the temperatures in areas adjacent to the refrigerated equipment and the temperatures of the outside ambient to which heat will ultimately be rejected. In stand-alone refrigerated processes this will often be the wet and dry bulb temperatures of the outside air. If the process is to be integrated with heat reclamation then the temperature of the heat sinks must be specified. Finally, the defrost regime should also be specified. There are times in any process where it is critical that coil defrosting and its accompanying temperature rise does not take place, and that the coil is cleared of frost before commencing the specified part of the process.

Although it is common practice throughout the food industry to leave much of this specification to refrigeration contractors or engineering specialists, the end user should specify all the above requirements. The refrigeration contractors or engineering specialists are in a position to give good advice on this. However, since all the above are outside their control, the end user, using their knowledge of how well they can control their overall process, should always take the final decision.

The aim of drawing up an engineering specification is to turn the user requirements into a specification that any refrigeration engineer can then use to design a system. The first step in this process is iterative. First, a full range of time, temperature and air velocity options must be assembled for each cooling specification covering the complete range of each product. Each must then be evaluated against the user requirements. If they are not acceptable then another option is selected and the process repeated. If there are no more options available there are only two alternatives; either standards must be lowered (hence cooling specifications will not be met) or the factory operation must be altered.

A full engineering specification will typically include: the environmental conditions within the refrigerated enclosure, air temperature, air velocity and humidity; the way the air will move within the refrigerated enclosure; the size of the equipment; the refrigeration load profile; the ambient design conditions; and the defrost requirements. The final phase of the engineering specification should be to draw up a schedule for testing the engineering specification prior to handing over the equipment. This test will be in engineering and not product terms.

The specification produced should be the document that forms the basis for quotations and finally the contract between the user and his contractor, and must be stated in terms that are objectively measurable once the chiller/freezer is completed. Arguments often ensue between contractors and their clients from an unclear, ambiguous or unenforceable specification. Such lack of clarity is often expensive to all parties and should be avoided.

MANAGING/PRODUCTION PRINCIPLES FOR REFRIGERATED FOODS

Two principles control the safety and quality of chilled and frozen foods: PPP (product–process–package) factors and TTT (time–temperature–tolerance) factors.

PPP factors that need to be considered at an early stage in the production of chilled and frozen foods are:

- Product: High-quality chilled and frozen foods require high-quality raw materials and ingredients.
- Process: The speed and effectiveness of the chilling/freezing operation and the use of additional processes, e.g. heating, pasteurization.
- Package: The packaging must provide a physical/chemical barrier to protect the food; "advanced packaging," including modified atmosphere packaging.

TTT factors maintain quality and safety during storage. For different foods, different mechanisms govern the rate of quality degradation and the most successful way of determining practical storage life is to subject the food to long-term storage at different temperatures. TTT relationships should also be able to predict the effects of changing or fluctuating temperatures on high-quality shelf-life.

TEMPERATURE MEASUREMENT AND MONITORING

Temperature measurement and monitoring are integral parts of any food cold-chain management system; as well as being, in many areas of the cold chain, a legislative requirement. Monitoring the cold chain requires detailed information on food product temperatures. Temperature monitoring includes both measurement and recording. Like any food safety system, an effective temperature measurement and monitoring system should be:

- Practical to apply.
- Results oriented.
- Cost effective.
- Useful to meet applicable regulations or food safety policy.
- Applied consistently and equitably.
- Verifiable and verified.

It is important to clearly know the difference between monitoring and verification. Codex Alimentarius Commission (2008) defines them as: "Monitoring: The act of conducting a planned sequence of observations or measurements of control parameters to assess whether a control measure is under control." "Verification: The application of methods, procedures, tests and other evaluations, in addition to monitoring, to determine whether a control measure is or has been operating as intended."

When designing a temperature monitoring system there are three clear stages:

1. Identify what has to be measured and why it needs to be measured.
2. Select optimum measurement/monitoring method/system.
3. Develop a suitable method of analyzing the data gathered.

Temperature measurement can be achieved using a variety of instrumentation such as bimetal style thermometers, thermistors, thermocouples, infrared thermometers, etc. Typically, in the food industry, temperature measurement is achieved using calibrated thermocouples and data loggers, while many temperature control systems use Pt100 platinum resistance sensors. Due to the variety of available equipment, manufacturers and suppliers are best positioned to give advice to the food business on the choice of temperature measurement equipment for specific purposes and food products. Further advice can be found in publications such as the *ATP Handbook* (2012), IIR *Recommendations for the Processing and Handling of Frozen Foods* (2006) and Evans and Woolfe (2008). The following questions are a useful aide-mémoire when choosing equipment:

- What is the required temperature range and likely operating temperature range for the instrument?
- Is there a need to measure product temperatures? Ambient temperatures? Package temperatures?
- Is there a need to continuously record the temperature (temperature history), or are spot checks acceptable?
- If a temperature history is required, what sampling frequency is required?
- Does the system need to provide a permanent record of temperatures, or just act if outside limits?
- What is the required accuracy?
- What is the required response time?
- If electronic, does the battery life compromise the application?
- What shape of probe is required? For example, a flat probe to reach between packages, a sharp long robust probe to reach the deep (core) temperature of a beef side, etc.
- Is water proofing of the probe/electronics required?
- Can the temperature data be imported into commercial data analysis spreadsheets or software packages?
- Does the system allow ease of calibration?

One possible aid in the future may be the widespread use of time–temperature indicators/integrators (TTI) throughout the cold chain (Taoukis and Labuza, 2003; Evans and Woolfe, 2008; Hobday et al., 2010). Time–temperature indicators, or integrators (TTIs), are simple and potentially inexpensive devices that are capable of reporting a visual and straightforward summary of either the temperature (indicators) or time–temperature exposure history (integrators) of the product. Indicator or threshold indicators show that a product has exceeded, positively or negatively, a given temperature, while integrators monitor both time and temperature during a given period and show the cumulative effect of temperature fluctuations during the history of the product.

Recommended Temperatures

Currently, the Agreement on the International Carriage of Perishable Foodstuffs (ATP Agreement) specifies maxima for transportation of chilled and frozen foods, as shown in Table 20.3.

TABLE 20.3 Maxima for Transportation of Chilled and Frozen Foods

	Maximum Temperature (°C)	
Chilled foods:	7	Red meat and large game (other than red offal)
	6	Raw milk
	6*	Meat products, pasteurized milk, fresh dairy products (yoghurt, kefir, cream and fresh cheese), ready cooked foodstuffs (meat, fish, vegetables), ready to eat (RTE) prepared raw vegetables and vegetable products, concentrated fruit juice and fish products not listed
	4	Poultry, game (other than large game) and rabbits
	3	Red offal
	2*	Minced meat
	0**	Untreated fish, mollusks and crustaceans
Frozen foods:	−20	Ice cream
	−18	Frozen or quick (deep)-frozen fish, fish products, mollusks and crustaceans and all other quick (deep)-frozen foodstuffs
	−12	All other frozen foods (except butter)
	−10	Butter

**Or at temperature indicated on the label and/or on the transport documents.*
***At temperature of melting ice.*

These temperatures are also a good guideline to be followed during storage and retail display of such foods. It should be noted that the recommended temperature of −18°C for frozen storage is rather arbitrary and based on the historical reason that −18°C is approximately equal to 0°F. There is little evidence that frozen food is any safer at −18°C than −12°C, provided good temperature control is maintained throughout the cold chain. Data on the importance of frozen storage temperatures on shelf-life and quality are conflicting. There is a growing realization that storage lives of several foods can be less dependent on temperature than previously thought. Since research has shown that many food products, such as red meats, often produce non-linear time–temperature curves there is probably an optimum storage temperature for a particular food product. Improved packing and preservation of products can also increase storage life and may allow higher storage temperatures to be used.

Publications such as the IIR *Recommendations for the Chilled Storage of Perishable Produce* (2000) and *Recommendations for the Processing and Handling of Frozen Foods* (2006) also give indications of recommended storage life and temperatures for different foods. Specific temperatures for certain foods may also be required by national legislation.

Recommended Controls

There is much published guidance for the processing and handling of chilled and frozen foods to help ensure product safety, including many Codex texts. It is important to ensure that proper cold-chain management and control incorporates good hygienic and good manufacturing practices (GMP) and the application of the HACCP, as described in the Codex text *General Principles of Food Hygiene* (CAC/RCP 1-1969). Many Codex texts contain appropriate guidance that can be used in developing management procedures, including the *General Standard for the Labelling of Prepackaged Foods* (CODEX STAN 1-1985), codes of hygienic practice (e.g. *Code of Hygienic Practice for the Transport of Food in Bulk and Semi-packed Food* (CAC/RCP 47-2001), *Code of Hygienic Practice for Meat* (CAC/RCP 58-2005)), codes of practice (e.g. *Code of Practice for Fish and Fishery Products* (CAC/RCP 52-2003)) as well as the *Guidelines for the Validation of Food Safety Control Measures* (CAC/GL 69-2008). Guidance can also be found in the *ATP Handbook* (2012) and International Institute of Refrigeration (IIR) *Recommendations for the Chilled Storage of Perishable Produce* (2006) and *Recommendations for the Processing and Handling of Frozen Foods* (2006).

The following general factors are important in relation to achieving the necessary temperature control for chilled and frozen foods:

In raw materials selection:

- Remember that high-quality chilled and frozen foods require high-quality raw materials and ingredients.
- Stocks should be rotated to ensure that the products leave the cold store on a "first in-first out" basis or shortest durability date.

In chilled/frozen food production:

- Use product temperatures–time as parameters to monitor at the critical control points (CCP) in the HACCP plan.
- In blast-air chillers and freezers poor product loading and placement can disrupt the flow of cool air around the product adversely affecting the rate of cooling.
- Iced-up cooling coils in chillers/freezers will have an adverse effect on air flow and indicate the need for proper defrosting regimes and correct setting of thermostats.

In chilled/frozen food storage:

- The temperature of the cold store may be an essential quality provision and/or a CCP monitoring parameter to avoid a critical temperature abuse situation that may jeopardize food safety.
- Use product temperatures–time profile as a CCP monitoring parameter in the HACCP plan.
- Introducing warm products into chilled/frozen food storage rooms can cause a general temperature increase: it should be noted that storage rooms are intended only for holding and are not designed for cooling foods.
- Stocks should be rotated to ensure that the products leave the cold store on a "first in-first out" basis or shortest durability date.
- Product and environment temperatures should be closely monitored and recorded during storage. Systems available include dataloggers (both *in situ* and portable).
- Iced-up cooling coils in store rooms indicate the need for proper defrosting regimes and correct setting of thermostats.

In chilled/frozen food transport:

- The product time–temperature profile during transport and distribution may be an essential quality provision and/or CCP monitoring parameters to avoid a critical time–temperature abuse situation that may jeopardize food safety.
- The product should be at the appropriate temperature prior to loading. Unless the transportation container has been specifically designed for that purpose, distribution should never be considered a cooling operation.
- Prior cooling of the distribution vehicle container is necessary to achieve the appropriate temperature during the entire distribution process.
- Ensure that products are transferred in a continuous operation (no stopping or delays) between temperature-controlled areas, e.g. holding store to delivery truck, delivery truck to holding store.
- Product and environment temperatures should be closely monitored and recorded during the distribution process. Systems available include dataloggers (both *in situ* and portable).
- Distribution of quick frozen foods should be carried out in such a way that any rise in product temperature warmer than −18°C be kept to a minimum within, as appropriate, the limit set by competent authorities and should not in any case be warmer than −12°C in the warmest package to ensure quality of the products. After delivery, the product temperature should be reduced to −18°C as soon as possible.

In chilled/frozen food retail display:

- Introducing warm products into chilled/frozen food cabinets can cause a general temperature increase: it should be noted that retail display cabinets are intended only for holding and are not designed for cooling foods.
- Cabinets should never be stocked beyond the load line. Poor cabinet stocking and stacking arrangements and inadequate servicing can cause significant problems with maintaining low temperatures.
- Do not overload cabinets.
- Iced-up cooling coils in cabinets indicate the need for proper defrosting regimes and correct setting of thermostats.
- Interference with cabinet design can disrupt the flow of cool air through the cabinet and cause a rise in temperature.
- Cabinets should be located so that the open display area is not subject to draughts or abnormal radiant heat (e.g. direct sunlight, strong artificial light or in direct line with heat sources).
- Stocks should be rotated to ensure that the products are sold on a "first in-first out" basis or shortest durability date. In no case should products be stored beyond their specified shelf-life.

Problem Areas

Transfer points, e.g. chiller/freezer to cold store, factory to distribution vehicle, retail cabinets to consumers' refrigerators, are well-known problem areas. A useful concept is that of the "relay system," where the baton (the food product) is transferred safely from one responsible person to another, and where a signing-over system includes information on product temperature and history. Such a system necessitates thorough education and training of staff likely to come into contact with the food product.

In many air-based refrigerated systems the evaporator coils operate at temperatures below the freezing point of water in order to achieve the required air and product temperatures. During operation water vapor that is present in the air that circulates over the evaporator coil condenses and eventually freezes on the coil surface. Over time frost and ice will accumulate on the coil surface leading to a decrease both in the air flow rate and in the overall heat transfer coefficient, causing air temperatures to rise. In order to maintain satisfactory performance, evaporator coils are defrosted periodically. This is achieved by warming the coil to melt the ice. This warming can cause a brief rise in both air and product temperatures. Legislation/guidance usually allows for such brief changes, for example ATP guidance permits "a brief rise of the temperature of the surface of frozen foodstuffs of not more than 3°C in a part of the load, e.g. near the evaporator, above the appropriate temperature." Defrosting is usually controlled by a preset time cycle; however, such control may cause a number of unnecessary defrost cycles which reduces the energy efficiency of the refrigeration system as well as causing unnecessary fluctuations in air temperatures. Implementing defrosts only when they are needed, or on "demand," would reduce the number of defrost cycles, lead to savings in energy and improve product quality and safety.

References

Archer, D.L., 2004. Freezing: an underutilized food safety technology? Int. J. Food. Microbiol. 90, 127–138.

Canadian Food Inspection System Implementation Group (CFISIG), 2004. Food retail and food services code. Amended Edition, September 2004.

CDC, 2007. In: Atkinson, W., Hamborsky, J., McIntyre, L., Wolfe, S. (Eds.), Epidemiology and Prevention of Vaccine-preventable Diseases, tenth ed. Public Health Foundation, Washington, DC (Chapter 14).

Chrystall, B.B., Devine, C.E., 1983. Electrical stimulation of beef carcasses. N. Z. J. Agric. Res. 26, 89–92.

Codex Alimentarius Commission, 1999. Code of Hygienic Practice for Refrigerated Packaged Foods with Extended Shelf Life, CAC/RCP 46-(1999). Food and Agriculture Organization of the United Nations.

Codex Alimentarius Commission, 2008. Guidelines for the validation of food safety control measures (CAC/GL 69–2008). Joint FAO/WHO Food Standards Program, FAO, Rome.

De Jong, A.E.I., Rombouts, F.M., Beumer, R.R., 2004. Effect of cooling on Clostridium perfringens in pea soup. J. Food Prot. 67, 352–356.

Evans, J., Russell, S., James, S.J., 1996. Chilling of recipe dish meals to meet cook–chill guidelines. Int. J. Refrig. 19 (2), 79–86.

Evans, J.A., Woolfe, M.L., 2008. Temperature monitoring and measurement. In: Brown, M. (Ed.), Chilled Foods: A Comprehensive Guide, third ed. Woodhead Publishing Ltd, Cambridge (UK), pp. 404–444 (Chapter 15).

Evans, J.A., Corry, J.E.L. and James, S.J., 1997. Bacterial growth on food refrigeration equipment. Proceedings of the ICEF 7 Engineering & Food Section F, 25–28.

Fennema, O.R., 1975. Physical Principles of Food Preservation. Marcel Dekker Inc..

Food Safety Authority of Ireland (FSAI), 2004. Guidance note no. 15: Cook–chill systems in the food service sector. Food Safety Authority of Ireland, Dublin, Ireland.

García de Fernando, G.D., Nychas, G.J.E., Peck, M.W., Ordóñez, J.A., 1995. Growth/survival of psychrotrophic pathogen on meat packaged under modified atmospheres. Int. J. Food Microbiol. 28, 221–231.

Gaze, J.E., Brown, G.D., Gaskell, D.E., Banks, J.G., 1989. Heat resistance of *Listeria monocytogenes* in homogenates of chicken, beef steak and carrot. Food Microbiol. 6, 251–259.

Gill, J., 1988. The greenhouse effect. Food (April), 47, 49, 51.
Harrigan, W.F., Park, R.W.A., 1991. Making Safe Food. Academic Press Ltd., London (UK).
Health Protection Agency, 2012. Health Protection Report weekly report. vol. 6, No. 18–4 May 2012.
Herriman, R., 2012. Frozen strawberries likely the source of Germany's food poisoning outbreak. <Examiner.com> 5 October.
Hobday, D., Higson, S.P.J., Mena, C., 2010. Role of diagnostic packaging in food supply chain management. In: Mena, C., Stevens, G. (Eds.), Delivering Performance in Food Supply Chains. Woodhead Publishing Ltd., Cambridge (UK), pp. 345–365 (Chapter 17).
International Institute of Refrigeration, 2000. Recommendations for the Chilled Storage of Perishable Produce, fourth ed. International Institute of Refrigeration, Paris, France.
International Institute of Refrigeration, 2006. Recommendations for the Processing and Handling of Frozen Foods, fourth ed. International Institute of Refrigeration, Paris, France.
James, S.J., James, C., 2002. Meat Refrigeration. Woodhead Publishing Ltd., Cambridge, UK.
James, C., Vincent, C., de Andrade Lima, T.I., James, S.J., 2006. The primary chilling of poultry carcasses – a review. Int. J. Refrig. 29 (6), 847–862.
James, S.J., Evans, J., James, C., 2008. A review of the performance of domestic refrigeration. J. Food. Eng. 87, 2–10.
James, S.J., Evans, J.A., Russell, S.L., James, C., 1998. Bacterial contamination of food refrigeration systems. Advances in the refrigeration system, food technologies and cold chain, Proceedings of meeting of IIF-IIR Commissions B2 & C2, with D1 & D2/3, Sofia, Bulgaria 7, 544–551.
Lyons, H., Drew, K., 1985. A question of degree. Food December, 15–17.
McIntyre, L., Bayne, G., Gilbert, S., Lake, R., 2007. Domestic Food Practices in New Zealand Freezer Survey. Institute of Environmental Science & Research Limited.
Panisello, P.J., Rooney, R., Quantick, P.C., Stanwell-Smith, R., 2000. Application of foodborne disease outbreak data in the development and maintenance of HACCP systems. Int. J. Food Microbiol. 59 (3), 221–234.
Roberts, D., 1982. Factors contributing to outbreaks of food poisoning in England and Wales 1970–1979. J. Hyg. (Camb) 89, 491–498.
Rose, S.A., 1986. Microbiological and temperature observations on pre-packaged ready-to eat meats retailed from chilled display cabinets. In: Recent advances in the refrigeration of meat by chilling. Meeting of IIR Commission C2, Bristol (UK), Section 9, 463–469.
Stern, N.J., Hiett, K.L., Alfredsson, G.A., Kristinsson, K.G., Reiersen, J., Hardardottir, H., et al., 2003. *Campylobacter* spp. in Icelandic poultry operations and human disease. Epidemiol. Infect. 130, 23–32.
Taoukis, P.S., Labuza, T.P., 2003. Time–temperature indicators (TTIs). In: Ahvenainen, R. (Ed.), Novel Food Packaging Techniques. Woodhead Publishing Ltd., Cambridge (UK), pp. 103–126 (Chapter 6).
UK Department of Health, 1989. Chilled and Frozen, Guidelines on Cook–Chill and Cook–Freeze Catering Systems. Department of Health, UK.
United Nations Economic Commission for Europe Working Party on the Transport of Perishable Foodstuffs, 2012. ATP Handbook. United Nations.
Zheng, L., Sun, D.-W., 2005. Vacuum cooling of foods. In: Sun, D.-W. (Ed.), Emerging Technologies for Food Processing. Elsevier Academic Press, London, pp. 579–602.

Further Reading

ASHRAE, 2010. ASHRAE Handbook–Refrigeration. American Society of Heating, Refrigerating and Air-Conditioning Engineers, Atlanta.
Brown, M., 2008. Chilled Foods: A Comprehensive Guide, third ed. Woodhead Publishing Ltd., Cambridge, UK.
Ciobanu, A., Lascu, G., Bercescu, V., Niculescu, L., 1976. Cooling Technology in the Food Industry. Abacus Press, Tunbridge Wells, Kent (UK).
Evans, J.E., 2008. Frozen Food Science and Technology. Blackwell Publishing,
ICMSF, 1998. Microorganisms in Foods 6: Microbial Ecology of Food Commodities. Blackie Academic and Professional, London.
James, S.J., James, C., 2002. Meat Refrigeration. Woodhead Publishing Ltd., Cambridge, UK.
Kennedy, C.J., 2000. Managing Frozen Foods. Woodhead Publishing Ltd., Cambridge, UK.
Sun, D.W., 2001. Advances in Food Refrigeration. Woodhead Publishing Ltd., Cambridge, UK.

Relevant Websites

http://www.iifiir.org/
Official site for the International Institute of Refrigeration
http://www.fao.org/
Official site for the Food and Agriculture Organization of the United Nations
http://www.chilledfood.org/
Official site for the UK Chilled Food Association
http://www.ecff.net/
Official site for the European Chilled Food Federation
http://www.frperc.com
Food Refrigeration and Process Engineering Research Centre site

CHAPTER

21

Detection of Physical Hazards

Gilles Demaurex and Laurent Sallé

Nestle Product Technology Centre, Orbe, Switzerland

OUTLINE

INTRODUCTION

Foreign bodies are a major reason for consumer complaints in the food industry. They represent either a quality defect affecting company and brand reputation or a food safety hazard, due to potential injury or choking. Hard and sharp foreign bodies, named physical hazards, can lead to serious illness.

Foreign bodies are an emotive subject for consumers and should be given special attention; in particular when the product is consumed directly from the packaging (e.g. ready to drink, confectionary, ice cream) or while feeding a third person (e.g. babies or the elderly).

Food Safety Management.
DOI: http://dx.doi.org/10.1016/B978-0-12-381504-0.00021-4

The product is susceptible to contamination from raw material to final consumption. The transformation process is often responsible for introducing foreign bodies (e.g. metal particles or human hairs) but the raw material itself might also contain foreign bodies, for example fish bones or grit in mushrooms.

The application of good manufacturing practice (GMP) and hazard analysis (HACCP) through the whole food supply chain, "from plant to plate," is the most effective way to prevent and reduce contamination and thereby protect the consumer; this includes, for example, hygienic design of buildings and machinery, training of factory employees, eradication of pests or certification of raw material suppliers. In addition, separators and sorters (e.g. filters, sieves, magnets, lasers) might be placed on production lines to improve foreign body reduction.

Unfortunately, due to technical and operational limitations, the above-mentioned control measures can only mitigate the risk. This is why detection equipment, typically a metal detector or X-rays, are part of the foreign body management system, working in combination with upstream control measures to minimize the likelihood of product contamination. They act like an alarm to warn about weaknesses in control measures. However, these tools are not absolute barriers and cannot ensure "zero risk" for the consumer, and overconfidence in detection technology may create a false sense of security.

Foreign bodies might be differentiated from the product by any of their physical characteristics: magnetic or electrical conductivity, density, color, shape or dimension. Despite the emergence of new detection technologies (ultrasonic, near infrared or magnetic resonance), metal detectors and X-rays are still today the most commonly employed technology in food inspection due to reasonable cost and good detection performances on physical hazards. Both of these technologies are also supported by a mature industry providing efficient turnkey solutions, maintenance and spare parts services.

Detection equipment is a competitive market with dozens of suppliers. Significant gaps exist between manufacturers in terms of detection capabilities, product handling, reject units, human interfaces or service level. Complex integration remains the preserve of a few specialists. Given the high cost of false rejection over several years or repeated efficiency losses, it would be unreasonable to try saving money during the machine selection process.

This chapter presents the basic working principles of metal detectors and X-rays and some key rules to select the most adequate equipment according to needs. A second part introduces concepts necessary for the optimal management of the detection equipment.

SORTERS AND DETECTION EQUIPMENT (FIGURE 21.1)

Foreign body separation equipment (hereafter called sorter) and detection equipment are often incorporated into a same classification, however, there is a significant difference in their purpose:

- Sorters (e.g. sieves, filters, magnets, manual sorting) remove foreign bodies from a product identified as a known contaminate (e.g. agricultural material, nuts, cacao or green coffee beans).

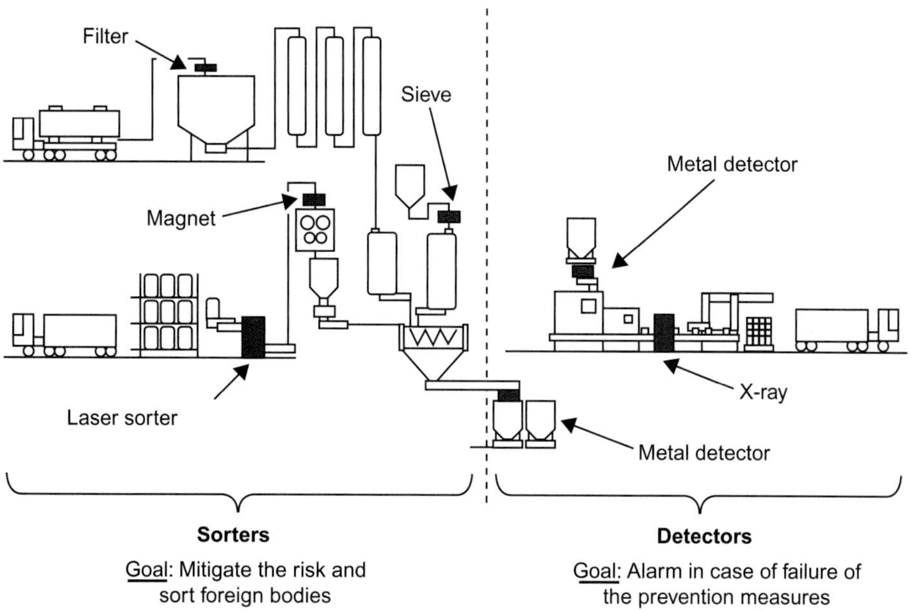

FIGURE 21.1 Sorting and detection equipment on a fictional process.

- Detection equipment (e.g. metal detector, X-ray, near infrared) is usually installed at the last stage of the process to control foreign body hazards prior to dispatch to consumers. Potential imperfections are highlighted in the upstream preventive measures.

There are also fundamental differences between the management of sorting and detection equipment. For sorters, usually no corrective actions are taken when foreign bodies are found, except in the case of abnormally high contamination or if a non-conventional type of material is discovered. However, all rejects from detection equipment should be carefully inspected and any foreign body discovered should invoke thorough investigation and batch retention (when justified) as this might be a potential failure of the preventive measures upstream (e.g. equipment maintenance, failure of sorter, hygienic design, worker's practices). In addition, most of the sorter technology does not have warning or reject mechanisms, for example a filter or a magnet will not notify when a contaminant is collected.

Sorters should be installed:

- On incoming raw materials.
- After hazardous process steps, often involving metal–metal contact, like rotary valves, grinding, mixing or cutting.
- Before some processes where foreign bodies might be broken up into smaller parts or could damage the equipment itself, generating more contamination.
- Upstream of a metal detector or X-ray as a last barrier to reduce the likelihood of an alarm.

Detection devices should be installed:

- At the end of the process, at or after packing, when no contamination is expected.
- At important intermediate process steps, for example at the filling or discharging of intermediate bulk containers.

There are countless sorting technologies, some very specific to certain types of industry. Almost all physical, chemical or biological characteristics can lead to the development of sorting techniques. Here is a brief overview of the most common:

- **Sieves and filters** are efficient, cost effective and used on free-flowing powders, granulates or liquids. The separation is based on the geometric dimensions of the contaminants. Setup might vary from a simple grid to a rotary or vibratory configuration using different mesh sizes. Sieves are often combined with magnets to capture large and small ferrous contaminants. Good build quality is vital to ensure the sieve itself does not become a source of foreign bodies. Perforated sieves are preferred to wire sieves for their hygienic design, but the clearance rate will be lower and might hinder the product flow (especially for powder). Edges must be free of sharp wires, and welds should be avoided as much as possible. For the same reason, scraping elements, such as brushes, are not recommended.
- **Magnets** attract ferromagnetic and some paramagnetic materials. They are built with permanent magnets or electromagnets. In food applications permanent rare earth magnets are the strongest available. Magnets can be configured as magnetic plates, pipeline traps, magnetic conveyors, grid or rod magnets. They are a good counterpart to remove small particles and wires that are sometimes difficult to be detected by metal detectors. They are often used upstream of equipment such as mills or cutters to prevent mechanical damage. As a drawback magnets can only remove magnetizable materials. Magnets are ideal to attract thin and flat particles; however, spherical or larger fragments may prove harder to catch.
  A point of safety: strong magnetic fields could also be a danger for people with heart pacemakers; therefore all magnets must be labeled with warning signs.
- **Optical and laser sorters** (Figure 21.2) are complex systems adapted to bulk sorting, using various types of lights and cameras to segregate contaminants according to their shape, size and color. Of the two sorts laser sorters are more complex and expensive, allowing for structure recognition. Therefore two elements with identical color can be differentiated thanks to their different outer structure. Even chlorophyll content, water content and biological characteristics may be distinguished. The product is inspected during free fall by a number of broad spectrum lasers simultaneously (infrared, ultraviolet, red, green, blue, etc.). While passing through the scan zone, the signals from the reflected lasers are evaluated. A few milliseconds later the defects are hit by timed, high-speed air jets into a reject chute.
  Both, optical and laser sorters are surface scan only. They are used on items such as vegetables, fruits, seafood or dry foods (e.g. nuts, hazelnuts).
- **Manual sorting** remains a valid solution where automatic methods are still not efficient enough or only available at prohibitive costs. This is an ideal choice for small and various operations.

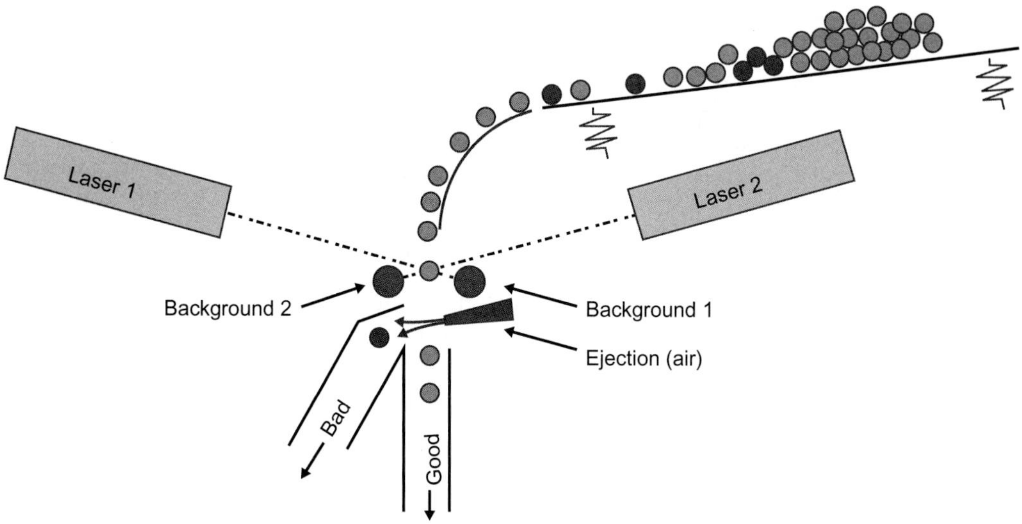

FIGURE 21.2 Laser sorter working principle.

METAL DETECTION

Working Principle

A metal detector generates a magnetic field through an emitting coil. Receiving coils constantly monitor this magnetic field for variations. Any magnetically permeable or electrically conductive materials that pass through the detector generate a disturbance of this magnetic field. Ferromagnetic metals impact on the magnetic field by both effects, which makes them easy to detect. Non-ferrous metals are non-magnetic, so more difficult to detect as the disturbance comes only from the induced magnetic field (due to the eddy current). Stainless steel is the most difficult to detect as it is usually non-magnetic and a poor conductor (usually metal detectors will catch a stainless steel piece 1.5 to 2 times larger than a ferrous piece). Note that for a metal detector the material density is not a relevant factor.

The product itself might disturb the magnetic field when passing through a metal detector; a condition called the "product effect." The amplitude of this signal depends on the conductivity of the product; this is particularly the case for products containing a high level of moisture (e.g. bread, jam or cheese) or ferrous elements (e.g. cocoa). Frozen foods are usually conductive too, but below a certain temperature (deep frozen) this electrical conductivity disappears.

Product effect is a drawback for metal detector detection capabilities. An efficient detection will therefore avoid false rejects by reducing the product effect, but without decreasing the detection capabilities: a rudimentary solution consists of moving the sensitivity threshold just higher than the product signal, reducing in the same ratio the detection sensitivity. Better results can be obtained by reducing the inspection frequency or by applying signal filtering algorithms.

Metal detector size, geometry and position of coils can be arranged in various ways: a walk-through gantry at an airport, a hand-held soil search as used by the military or a surface bar for webbing. In the food industry detectors typically have a cylindrical or rectangular aperture. Principal configurations are the horizontal flat belt (a conveyor is passed through the aperture to handle the bulk or finished product), the vertical gravity fall (used for inspection of bulk powder or product in free fall) and the pipe inspection (for pumped liquids or pastry products).

Performance between equipment and manufacturers might vary significantly; so when benchmarking performances, it is useful to remember that detection capability is proportional to the mass of the contaminant when using spheres; therefore, a detection limit of a diameter of 0.8 mm is about two times more difficult to obtain than a diameter of 1.0 mm.

How to Ensure an Efficient Detection

A high raw signal quality will always give better results than using post-processing filtering on a poor signal; some design characteristics have a significant impact on the signal quality:

- **Aperture dimensions.** Aperture dimensions should be kept to a minimum since the larger the aperture the lower the sensitivity. In practice, it is recommended to place detection equipment where the package/product is the smallest and the most homogeneous (e.g. individual pouches before grouping), rather than after grouping (e.g. shipping carton). Sensitivity is minimum at the center and maximum at the edges of the aperture.
- **Coil spacing.** Tighter coil spacings are available when geometrical constraints cannot be overcome. By bringing the coils closer together, detection quality is slightly reduced; often the shielding will also be smaller, creating more sensitivity to external electromagnetic disturbance. Such a configuration might be selected when no other can be found.
- **Operating frequencies** are referred to in kHz. The lower the detector's frequency, the lower the product effect and sensitivity. Fixed frequency detectors are built for simple applications (low product effect); a multi-frequency system offers the choice between several frequencies to optimize the inspection (often done automatically during auto-setup). Some high-end solutions generate several frequencies simultaneously and apply filters independently for each.

In addition to the above design features, some installation rules must be ensured to achieve best in class performances, in particular:

- **The metal-free zone.** This is the required area free from conductive material (especially moving parts). The dimensions for the metal-free zones are given by the suppliers and can vary from 1 to 1.5 times the largest dimension or diameter of the search head (to be confirmed by the supplier).

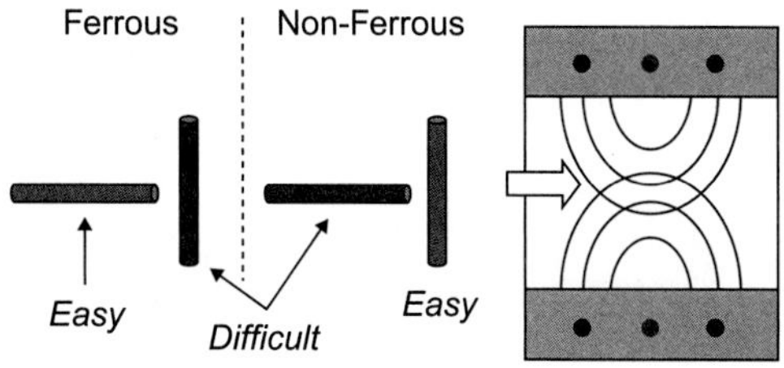

FIGURE 21.3 Orientation effect on a metal detector.

- **Power supply.** Metal detectors are sensitive to power supply quality. Use of electrical power stabilizers might be necessary in some countries.
- **Environmental conditions.** Vibrations (engines, forklift trucks) or electromagnetic interference (variable speed drives) or temperature fluctuations may affect the stability of the detection. A mechanical isolation of the detection head with anti-vibration elements is recommended. Often the reject unit itself is a source of vibration.

During operation some other aspects must be respected to minimize false rejects and ensure the best possible performance:

- **Product spacing.** Adequate spacing between products is usually calculated as half of the tunnel length plus half of the largest dimension of the aperture. If this rule is not respected the wrong pack might be detected and rejected.
- **Ensure the minimum product effect.** If the product characteristic is given, some parameters, such as temperature, may increase the product conductivity.
- **Cleaning.** Wet cleaning might disturb the metal detector if the belt or other elements are not dried correctly before restart.

Technical Limitations

As mentioned before, the product effect is the main limiting factor of metal detector capabilities; but in addition:

- The orientation of straight wire contaminants passing through the aperture might reduce the detection sensitivity as shown in Figure 21.3. Manufacturers provide equivalence tables between sphere diameter and wire length for the worst-case orientation.
- Metal detectors cannot inspect packaging containing metal parts, for example tin cans or glass jars with aluminum membranes. However, for product packed in foil (aluminum or metalized), some manufacturers propose "search in foil" technology. To do this, inspection frequencies are drastically reduced and high-power magnets might be used to increase the detectability of ferrous materials prior to inspection. This technology gives acceptable results on ferrous materials but is not very efficient for non-ferrous materials and stainless steel.

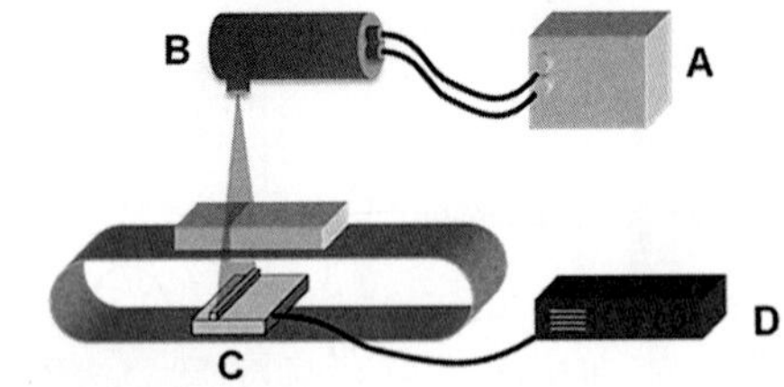

FIGURE 21.4 X-ray equipment working principle.

- Gravity fall and pipe inspection are efficient solutions but difficult to set up and test. Reject units are difficult to adjust and might generate a significant amount of rejected product to ensure the contaminant has been excluded. It is recommended to work closely with manufacturers on this topic.

X-RAY DETECTION

Working Principle

Referring to Figure 21.4, in the X-ray tube (B), electrons are accelerated between cathode and anode under strong electrical potential produced by a high-voltage generator (A). The moment the electrons hit the surface of the anode, X-rays are produced (99% of the kinetic energy of the electrons is transformed into heat, hence the importance of cooling systems). Passing through the product, X-rays are partially absorbed. The remaining radiation reaches a detector (C) where X-rays are converted into light by a scintillator and converted to a grayscale value by an array of photodiodes. A computer (D) constructs an image line by line, using the motion of the product passing in front of the detector. Some other solutions use a matrix sensor that works like a camera, taking a picture produced by a "flash X-ray," generated in milliseconds. The detection of foreign bodies is ensured through image processing software.

X-rays can penetrate all common packaging materials, which makes this method suitable for scanning glass jars, plastic bottles or tin cans. X-ray detection quality is defined by density, thickness and X-ray absorption coefficient of materials; the X-ray intensity transmitted through the material is given by the following law:

$$I = I_0 e^{-\mu \cdot x}$$

where I is the transmitted X-ray intensity, I_0 is the incident X-ray intensity, μ is the attenuation coefficient and x is the thickness of the material. The attenuation coefficient μ typically increases with material density.

Numerous image processing algorithms identify contaminants; this task becomes very complicated the more blurred the picture (some examples are shown in Figure 21.5). What might be obvious to the human eye, through the complex filtering of the brain, could be

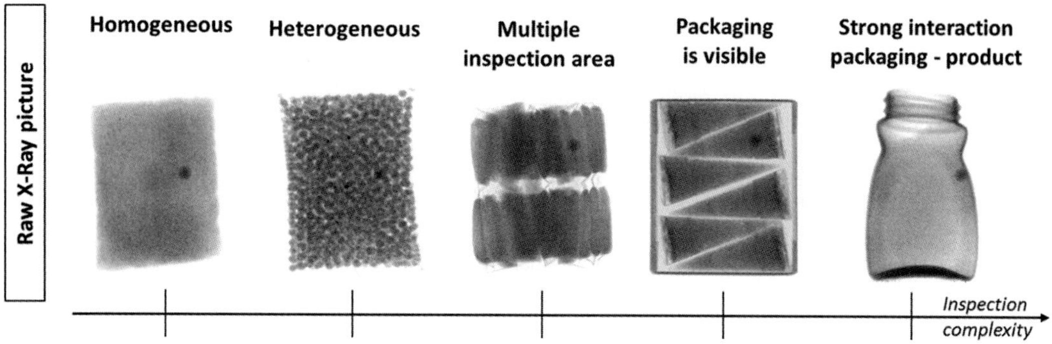

FIGURE 21.5 X-ray picture complexity.

extremely difficult, even impossile, to detect by automatic routine (which is why airports still require humans skills to analyze the X-ray scans of luggage). Basic detectors work on a simple grayscale threashold. Fixed masks/frames are then applied where pacakging shadows or a specific zone must be excluded from the inspection (e.g. promotional items). More complicated applications use active contour alogrithms to differentiate packaging from product. On high-speed applications, the processing time of complex algorithms still remains a constraint.

X-ray technology allows machine design flexibility; customized configurations might be developed according to the application by placing one or more views with different inspection angles. Flat belt is the most common application, usually a single view from the top or from the bottom crossing the conveyor belt to inspect packed, unpacked and bulk product. A similar configuration is used for the pipe inspection, scanning pumped product thought a plastic pipe transparent to X-rays. Stand-up products that cannot lie on their side are usually inspected by lateral view with the tube and the detector positioned on each side of the conveyor.

Rigid container inspection is notable because the packaging appears omnipresent and intertwined with the product. It is therefore necessary to dissociate the product from the packaging. Rigid container by single view offers good results on simple packaging shapes, but multiview X-ray (between two and four combining lateral and top views) increases significantly the volume coverage.

Glass container inspection is a typical area where multiview X-ray is widely used. Due to the density and thickness of glass, the bottom and sides of the packaging will appear blurred, limiting the inspection on a single view. Using a second view, for example crossing at 90 degrees, the areas difficult to inspect in one view are then visible in the other, as explained in Figure 21.6.

X-rays can also be used to perform further picture analysis in addition to foreign body detection, for example: filling level control, control of missing items or product, fat analysis, broken parts control or mass evaluation. However, these techniques have limitations and cannot replace dedicated tools like a checkweigher or vision system.

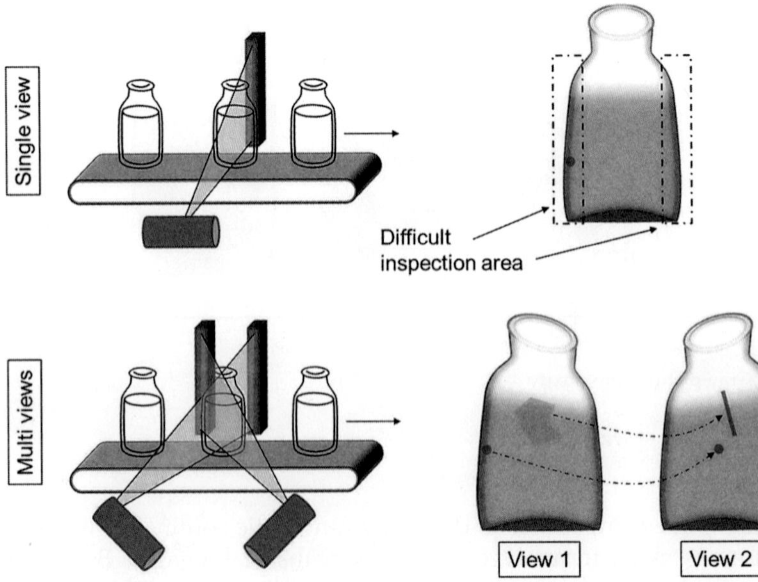

FIGURE 21.6 Double view detection advantages on rigid containers.

How to Ensure an Efficient Detection

As with metal detectors, the first priority is to ensure a superior raw signal quality (picture contrast and resolution) instead of applying ineffective post-processing on a picture of poor resolution. Therefore the following need particular attention:

- **X-ray tube power and voltage.** The voltage indicates the penetration force of the photons while amps define the image brightness. Some filters might also be applied according to the application to fine-tune the energy profile of the X-ray spectrum.
- **Focal spot size** (area where the X-rays are generated) influences the image sharpness together with the distance between tube, object and detector (as illustrated in Figure 21.7). An ideal situation is to have a small focal spot, a long distance between tube and object and a small distance between object and detector. Note that the smaller the focal point, the more important the cooling, as all the energy is concentrated in one point.
- **Scan resolution.** As known from digital cameras, the higher the number of pixels, the better the resolution. Usually for food inspection the pixel dimension should be in the range 0.4 to 1.2 mm. The resolution of the picture is the result of scan frequency in conjunction with conveyor speed. To obtain, for example, a 0.4 mm resolution at a conveyor speed of 25 m/min, a rate of 1042 scans/min is required.
- **Sufficient contrast** between the product and the contaminant is dependent upon product thickness as illustrated in Figure 21.8. As a consequence, it is recommended to inspect the smallest and thinnest possible unit. For example, detection performance will be better on a single sachet than on a shipping case.

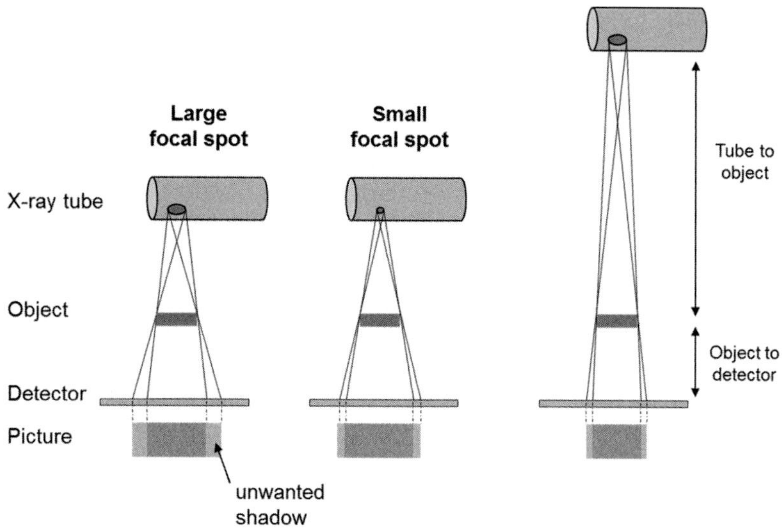

FIGURE 21.7 Scan resolution and unwanted shadow effect.

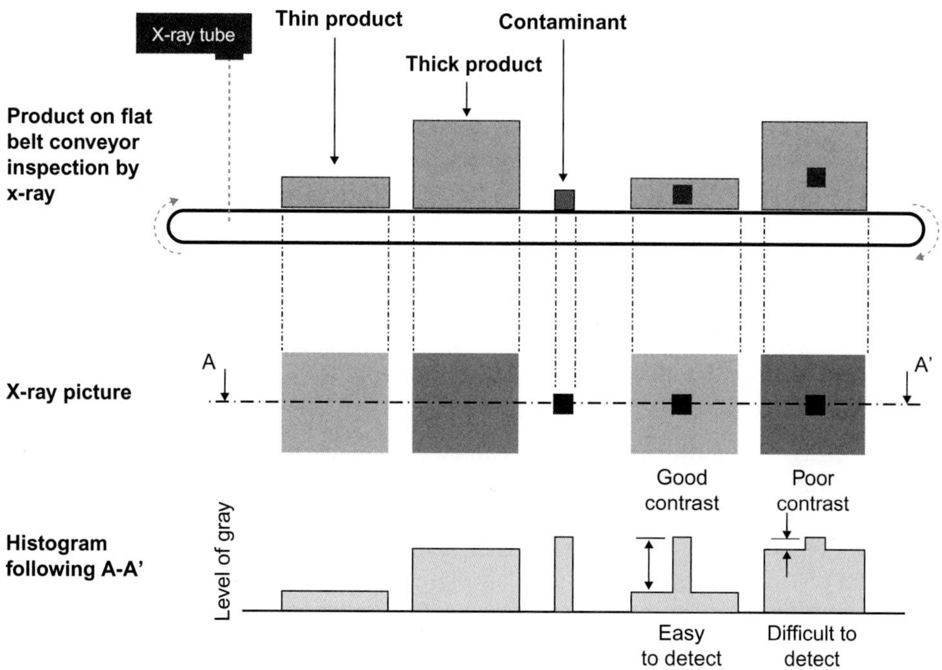

FIGURE 21.8 Product thickness influence on image contrast.

There are many rules for efficient X-ray installation and operation:

- **Precise product handling** reduces image variability and provides a greater precision of the inspection area. Protection against radiation (curtains, baffle system) can often be a source of disruption to product flow.
- **Frame accessibility.** Protection against radiation requires the use of a shield which sometimes restricts the accessibility of the transport and inspection area. Ensure the unit is easy to open and access for cleaning and maintenance is straightforward.
- **Software.** This is probably the most important aspect. The software might become extremely complex and unmanageable for a factory with limited engineering resources. Manufacturers provide remote control solutions, but this is of limited support in many situations. The software must be easy to use with a limited number of parameters to ensure manageable setup and traceability.

Technical Limitations

- **Awareness of contaminant shape and orientation** (Figure 21.9). A contaminant might appear more or less dark on the picture depending upon its orientation. This is especially limiting for flattened shapes and low density material.
  It is necessary to be vigilant when comparing the detection performance between metal detectors and X-ray. Indeed the use of a sphere or a cube is always preferable for X-rays because it concentrates the mass at a point. Multiview X-ray improves the situation by projecting the contaminant from different angles.
- **Blind area on rigid container** (Figure 21.10). Masks exclude part of the picture and therefore part of the product, where shadows generated by the packaging disturb too much of the image. This has to be considered when benchmarking X-rays on rigid containers against metal detectors before filling, where with metal detectors 100% of the product is always inspected.
- **Limitation with aluminum.** Aluminum density ($d = \sim 2.7$) is lower than other common metals, three times lower than stainless steel ($d = \sim 7.9$) and therefore more difficult to detect by X-ray. Density is not the only limiting factor; chemical elements up to atomic number 12 (magnesium) have light interaction with X-ray photons. This interaction increases gradually only from number 13 (aluminum).

Safety

Industrial X-ray for food inspection uses an electric X-ray tube, containing no radioactive sources. When turned off no residual radiation is produced. During operation the operator is protected by the lead shielding of the machine. The leakage can be measured by a gamma survey meter. There are no common international norms; each factory should refer to its local radiation protection entity to ensure compliance with local regulations.

Note that humans are constantly exposed to X-ray radiation, mainly from nature (cosmic radiation, radon in earth) and to a smaller degree from human activity (medical X-ray or air travel). For example, the average dose received by a French citizen is about 2400 microsievert per year, where a return flight from Paris to New York will generate a dose of

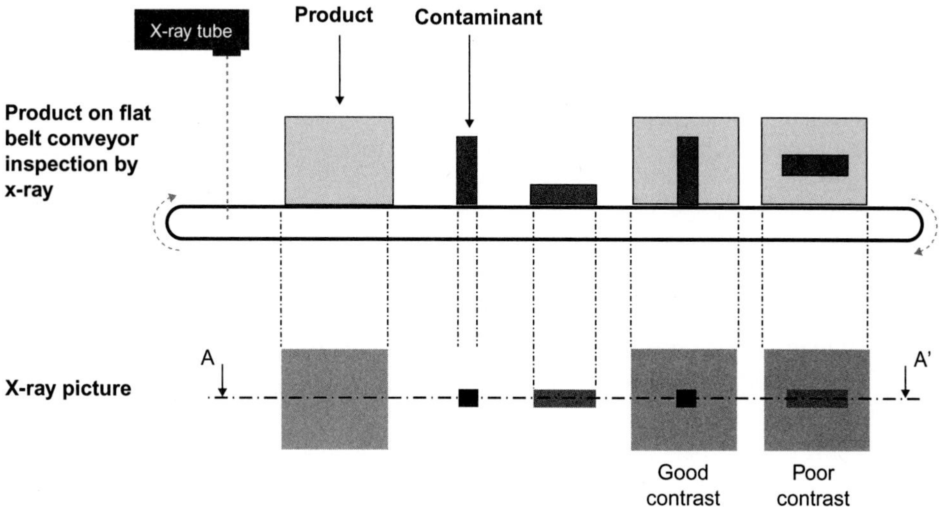

FIGURE 21.9 Contaminant orientation influence on image contrast.

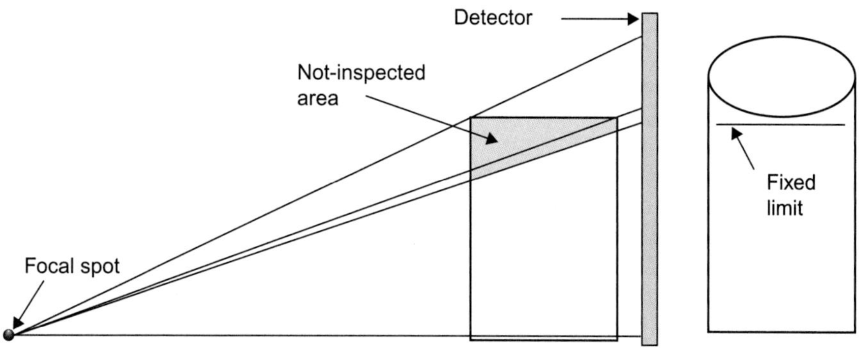

FIGURE 21.10 Limitation on rigid container.

60 microsievert and a chest X-ray about 3000 microsievert. Working in proximity to industrial X-rays might not generate any significant increased yearly irradiation.

Some recommendations:

- Leakage measurements should be carried out regularly by a radiation protection trained person on site.
- Never try to manually remove product from an X-ray device without switching off the power.
- Do not modify any machine parts; in particular do not cut the shield curtain to help product flow. If the product stream is disturbed by the curtains, another method must be discussed with the supplier, e.g. baffles or elongated tunnel.

EQUIPMENT SELECTION

There is no unique configuration and several aspects will determine the choice of detection equipment, such as contamination risk (material, shape), packaging and product characteristics, line layout, environmental conditions or process speed. No detection technology is able to detect all types of physical hazards and, according to the risks, one might install complementary technologies. Direct dialog with suppliers every so often is the best solution for professional advice. The following selection process is recommended:

1. **Determine the source and nature of physical hazards all along the production line**, based on the HACCP study and factory records. Maintenance reports or consumer complaints are an efficient way to determine the main hazards.
2. **Apply all possible prevention measures** to ensure the best reduction of foreign bodies: GMP, hygienic design, best practices and sorters according to the principle "first prevent, then detect." Working on the hygienic design of equipment with the supplier to eliminate the source of contamination in a process will deliver better results than using a metal detector and retaining the source of contamination.
3. **Select the most appropriate technology and location** according to the type of product and the packaging. This step might require modifications to the line layout. For example, the installation of a metal detector before filling instead of an X-ray device on packed product due to higher detection capabilities might require a redesign of the filler infeed or height of the building. Keep in mind that metal detectors and X-ray technology are more effective on small products (e.g. bulk before filling or individual packaging), rather than on grouped products or shipping cases.
4. **Select a shortlist of potential suppliers**, based on selection criteria, for example:
   a. Best detection performance for the specific application.
   b. Capability of the supplier to provide turnkey solutions (infeed, product handling, reject, etc.).
   c. Support level available from the supplier in the installation country. This is essential for complicated installations like multiview X-ray that might require frequent modification to settings and specific maintenance.
   d. Coherence with other equipment brands already installed in the factory to optimize maintenance and operational cost.
5. **Build the user requirement specification (URS)**, see below.
6. **Send URS together with products and contaminated test samples to the preselected suppliers to handle detection capabilities trials.** It is recommended to build your own sets of contaminated test samples to ensure a fair performance benchmarking between suppliers (with spheres and real fragments). Indicate the testing procedures to follow. Ideally the client should be present during the trials.
7. **Build a selection matrix** to make the choice, assessing the key elements of the URS. Equipment with slightly inferior performance might be chosen for its ease of cleaning and hygienic design.
8. **Handle a detailed performance qualification with the selected supplier** to assess and document the detection capabilities for each relevant foreign body type, according to the detection equipment selected.

User Requirement Specification (URS)

The URS is integral to the success of detection equipment. This is the reference document that links supplier and factory by defining all key aspects of the installation. The URS is a working document that should be developed and agreed with the supplier during the project. It should contain at least the following elements:

- **General project information**
  - Country, address, contact person
  - Project objective, type of equipment (metal detector, X-ray)
  - Confidentiality level
  - Project timeline (factory acceptance test, commissioning, start-up)
- **Product characteristics**
  - Type (powder, liquid, dry, wet, sticky, frozen, etc.)
  - General characteristics (density, flowability), particularly important for pipe or gravity fall application
  - Inspection product conditions (temperature)
- **Packaging characteristics**
  - Type and material (carton, glass, pouch, tin can, etc.)
  - Max dimensions for each format (width, height, weight)
  - Foreseen inspection orientation
  - Pictures and technical drawings
- **Production line characteristics**
  - Line speed (kg/h, m/min, pack/min)
  - Line layout and constraints (available space, accessibility, etc.)
  - Accessible power: electrical (V/Hz) and pneumatic pressure
- **Working conditions**
  - Room atmosphere (temperature, humidity) and hygiene level
  - Type of cleaning (wet cleaning, dry cleaning) and IP requirements
  - ATEX protection requirements
- **Expected performances**
  - Foreign body hazardous material foreseen on the line (material, density)
  - Maximum acceptable false reject rate
  - Expected detection limits (see Further Reading)
  - Reject unit requirements
- **Additional requirements**
  - Communication requirements (Ethernet, wireless)
  - Key items to be included in the offer: test kit, check of the rejection unit, training modules, test samples spheres and gamma survey meter
  - Training modules and documentation (language, number of copies)
  - Maintenance contract
- **Installation and commissioning procedures**
  - False reject setup procedures
  - Detection limit assessment procedure
  - Test samples definition, drawings or sketches
  - Necessary test pieces material and dimensions to be provided by supplier

TABLE 21.1 Summary of Detection Capabilities for X-ray and Metal Detector

	Material	Metal Detector	X-Ray
Metal	Ferrous	Excellent	Good
	Non-ferrous	Excellent	Good
	Stainless steel	Good	Good
	Aluminum	Excellent	Fair
Non-Metal	High density (e.g. Glass, stones)	Not possible	Fair
	Light density (e.g. Insects, wood)	Not possible	Not possible

Metal Detector or X-ray?

Testing the product on trial equipment remains the best method to evaluate the highest detection performances between a metal detector and an X-ray device. Nevertheless some basic rules will guide the choice in the early stage:

- **Non-metal hazards only:** On dense materials (e.g. glass or stone), usage of X-ray is obvious.
- **Metal hazards only:** Promote the use of a metal detector if there is no contraindication due to the packaging or product itself. While benchmarking between both technologies, remember spheres represent a favorable shape for X-ray. Therefore it is recommended, in addition to traditional spheres, to benchmark both equipments with real contaminants.
- **Aluminum hazards risk:** A metal detector is the most appropriate technology (as explained previously).

Table 21.1 gives a rough summary of material detection capacities.

In addition to the above technical considerations, investment cost, layout, maintenance complexity and difficult environment might drive the choice too, for example:

- Investment cost is usually cheaper for metal detectors. In particular, multiview X-ray costs can be prohibitive. The maintenance costs are also higher for X-ray equipment due to expensive spare parts (X-ray tube, detector, high voltage generator) with limited lifetime.
- While a metal detector has only a few parameters to adjust (amplitude, frequency and phase), X-ray machines need tens of algorithms with many parameters and are difficult to master and trace in case of modification. Factories might evaluate the complexity they are able to manage, in particular if no local support is available from the supplier.
- If the contamination risk on a tin can line is only metal particles from the process and the filler is evaluated not hazardous, a metal detector before filling to regularly check the filler integrity is a more efficient solution than a multiview X-ray alone on the tin can after filling.

Choice of Reject Unit

Reject systems are an integral part of detection equipment and must always be part of validation and monitoring processes. Faultless efficiency and synchronization with the

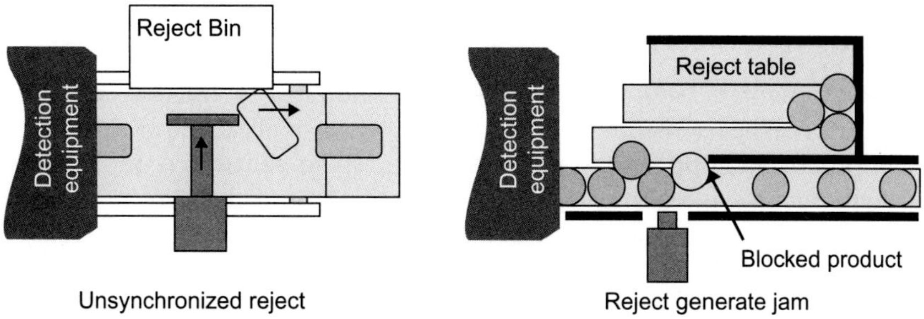

FIGURE 21.11 Reject unit malfunctioning examples.

detector must be ensured to avoid issues like mis-ejection, uncontrolled ejection or a jam (Figure 21.11).

Some key rules should be followed:

- The rejection motion should be quick enough so as not to disturb the next product and create consecutive rejections.
- If rejection is actuated by air jet, install a pressure control gauge to indicate insufficient air pressure, ensure the system has enough pressure to handle multiple rejects (air buffer tank might be needed) and carefully test the rejection unit with all products' weight and dimensions.
- Install a sensor to confirm the rejection. This control should stop the line if rejection is not effective. For flat belts this is often done by a laser barrier at the entrance of the rejection table; for gravity fall or pipe inspection it may be done by placing a second metal detector with a smaller aperture (for higher sensitivity) on the reject pipe. Install a sensor level for the rejection bin.
- Conveyor application should be equipped with an encoder to ensure automatic regulation of timing according to the conveyor speed. In the case of a long product, a synchronization photodiode should be used to detect the exact position of the product.
- For combined systems such as a checkweigher and metal detector, the rejection device should separate faulty weights from contaminated packs. The reject logic should be clear: any contaminated product should be rejected in the detection reject bin, with correct or incorrect weight.
- The line should stop automatically or an alert given in the case of a large amount (e.g. over five) of consecutive rejection during normal production.
- For rigid containers, in particular for glass jars, it is recommended to use progressive/multi-finger reject units. These rejection systems allow a smooth rejection of fragile products, like glass jars to avoid breakage that could lead to potential foreign body contamination on the line. It also allows for rejection of fallen containers or successive rejection.
- Reject units for gravity fall or pipe inspection are complex and need to be designed, installed and regulated by specialists. The relationship between the pipe size, the rejection response time and the free fall distance between the dropping point and the rejection valve must be considered. A locally made solution is not recommended.

DETECTION EQUIPMENT MANAGEMENT

Detection equipment is often the last defense against physical hazard contamination before the product reaches consumers, and warns in the case of failure of upstream measures. Detection equipment must be treated as an alarm: it has to be inactive most of the time and warn only when necessary leading to further investigation of the source of contamination. The trust placed in this equipment by the operator, engineer or quality manager will contribute to the efficiency of this alarm: in particular, a too large amount of false rejects (too sensitive settings) and insufficient level of detection are both responsible for a loss of confidence in the equipment.

This is why a detection limit cannot be decided in advance, it can only be measured once the parameters have been set for an acceptable false reject rate. Indeed, increasing the detection sensitivity results almost inevitably in an increase in the number of false rejects.

Some concepts and terminology necessary to ensure adequate detection parameter adjustment are presented below.

Product Classification

During production, the rejected product should be a contaminated product, but often the detection equipment rejects products for no apparent reason; this is the notion of **false reject** and **correct reject** as described in Figure 21.12.

False rejects are defined as products that have been ejected by the detection unit for no acceptable reason. They are mainly due to:

- Inadequate sensitivity setup. Often due to too sensitive parameters leading to the ejection of good packaging and/or good product within acceptable specification.
- Environmental disturbances (e.g. magnetic or electrical field, vibration, unstable power supply).
- Operational issues (e.g. vibration on metal detectors, wrong setup after format changeover, dust, poor cleaning, packaging with incorrect orientation/handling/spacing).
- Incorrect layout/installation (e.g. inappropriate product/packaging handling system, fallen jars on conveyor or deformed packaging, not respecting the metal-free zone area).
- Detection unit malfunctioning due to internal technical failure (electronic issues or unbalanced coil for metal detectors. Arc in X-ray tube, tube aging, software bugs or overheating of X-ray equipment.

Beside the false reject, some non-contaminated items might be rejected due to products or packaging being interpreted by the detection unit as a defect. This occurs because it is not always possible to differentiate foreign bodies from some products or packaging deviations. A metal detector might reject a product with an abnormal moisture content level or an X-ray machine might reject an overweight pack. Some typical examples are:

- Glass jar inspection by X-ray: Some metal inclusions present in the side wall of a glass jar have identical appearance to metal contamination in the product. These jars are therefore often rejected, but in this case the detection equipment should not be blamed.

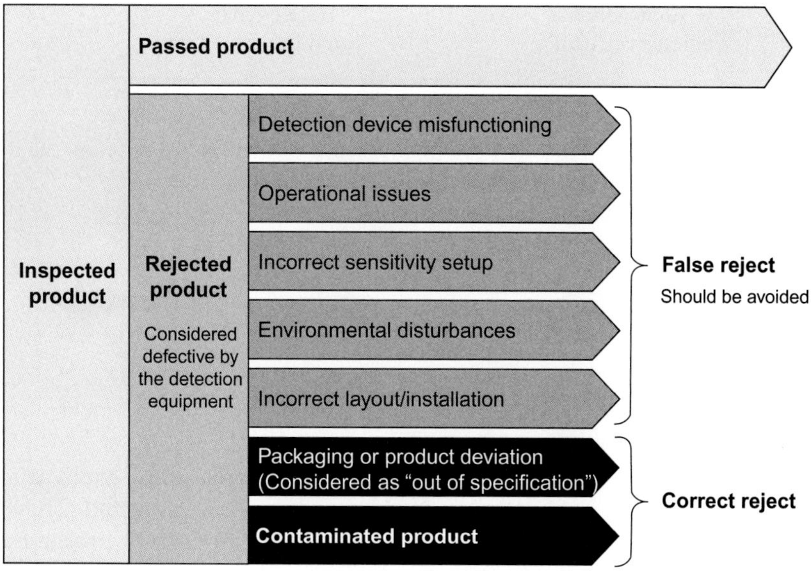

FIGURE 21.12 Reject categories.

- A tin can might have deformed during transport, and is interpreted by X-ray as not acceptable. Again this should not be considered a false reject, unless it is tolerable for the factory to produce such defects; in this case the detection sensitivity might be reduced to accept a wider packaging defect range.

The False Reject Rate (FRR)

This is the ratio of "false reject/inspected product." The FRR should be calculated on an extended period of time or quantity, after the rejected product has been sorted and classified into categories. A detailed examination of each reject is necessary to define if it is a false or correct reject.

In order to ensure a realistic workload and to maintain confidence of operators in the detection equipment, FRR should remain at an acceptable rate, ideally close to 0%. A good value recognized in industry is a maximum of 5/10,000, but it is up to each business to define the acceptable level of reject that can be manageable. However, no false reject at all (0%) might also indicate the equipment settings are not sensitive enough.

It is recommended to set the detection equipment at the higher sensitivity and to reduce it step by step once the FRR is acceptable. The sensitivity of the detection equipment is then fixed.

Representative Samples

As discussed, some product characteristics will influence the detection and generate a false reject, for example the density, humidity, temperature, mass distribution of the product or packaging tolerances.

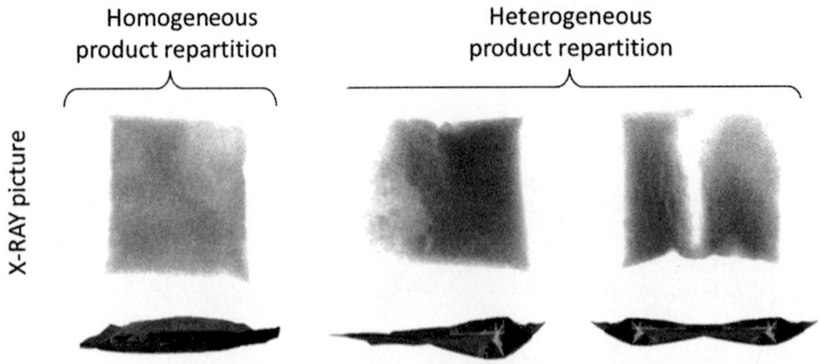

FIGURE 21.13 Heterogeneous product repartition.

Representative samples simulate the foreseen, but acceptable, industrial variations of the product and/or packaging. These samples can be hand-made or collected. They are used to set the sensitivity of the detection equipment when a large amount of product is not available. Indeed, to ensure an acceptable false rejection rate, it is more efficient to ensure that a dozen representative samples are not rejected rather than inspecting a thousand identical products.

Some examples are given below:

- For pouch inspection by X-ray, some extreme cases should be prepared, for example heterogeneous product repartition in the pack (see Figure 21.13), or over- and underweight parts.
- For conductive products inspected by a metal detector (e.g. wet, frozen, cocoa based), the standard variation range should be tested and not rejected.
- Tin cans might be artificially slightly damaged to reproduce some of the usual damage found after transport.

Probability of Detection (POD)

Foreign body occurrence is a "rare event" following Poisson's law. Detection capability is assessed by passing a contaminated test sample through the detection equipment several times. The associated probability of detection (POD) is the likelihood of a contaminant of a given size, nature and position with a defined confidence level.

In theory, a POD of 100% cannot be demonstrated by testing. Statistically, by increasing the number of passed test samples, the corresponding POD will increase. At the same time, by increasing the numbers of successfully detected test samples, the limit of detection will generally increase too. An increase in the POD is usually associated with a size increase of the corresponding test sample contaminant diameter.

A minimum number of passes must be performed to have an acceptable POD. In practice (realistic workload), a contaminated sample detected 30 times over 30 passes (POD = 30/30) ensures a statistical POD of 90.5% at 95% confidence level as per the binomial distribution law (Figure 21.14).

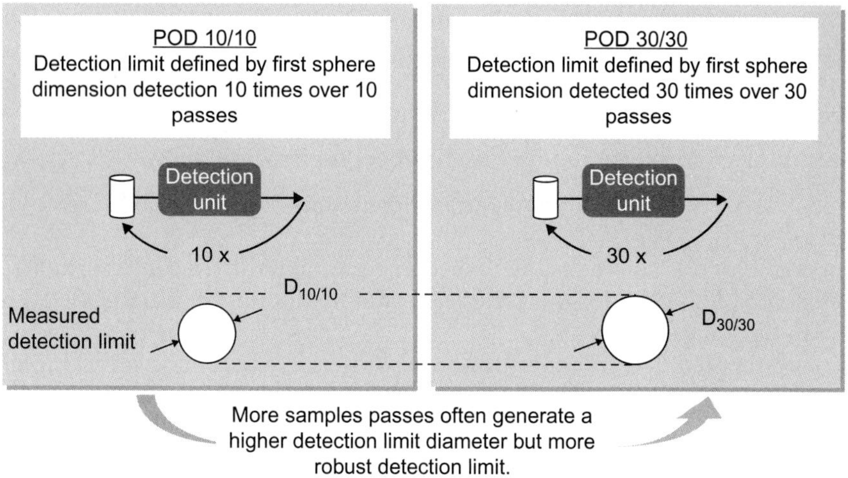

FIGURE 21.14 Impact of number of passes on detection limit.

It is important to understand that increasing the POD level will not modify the sensitivity of the equipment, but will give a more faithful representation of the real detection sensitivity.

Limit of Detection

Detection equipment performance is measured by the smallest test piece detected for a defined POD level (e.g. 30/30). For practical reasons, only spheres are used in contaminated test samples to establish the detection limit (no influence due to sphere orientation). This defined detection limit assesses, by a statistical method, the best possible machine performance in good working conditions for the spherical contaminant in a given position.

The testing procedure to establish the detection limit is:

1. Start with the equipment set up for the highest sensitivity.
2. Step by step reduce sensitivity until the FRR is not higher than expected. When a large quantity of product is not available, representative samples (see below) are used to simulate the production variability. None of them should be rejected.
3. When the settings ensure an acceptable FRR, they should be fixed, saved and not modified until the end of the procedure.
4. Pass each contaminated test sample, for each sphere dimension, the number of times required at the selected POD, starting with the smallest dimension. The detection limit is defined by the smallest dimension achieving the required POD (e.g. 30 detections on 30 passes). Always put the test pieces together with the product.
   Note: *Equipment should be designed and installed in order to ensure that several passes are easily managed. In particular, suppliers should propose test kits to ensure easy access for test sample handling on gravity fall and pipe metal detector (access/retrieval gate). This aspect might be described in the URS.*

5. Pass some test samples with real contaminants (fragments, wires, shavings, etc.) to document the detection capabilities on real contamination. These samples cannot be used to determine a detection limit because they are not reproducible, but they are essential documentation (qualitative information).

The above measured detection limit has, however, severe limitations:

- A sphere is not a true representation of usual contaminants such as wire, shavings or fragments.
- A sphere is a favorable shape for the detection equipment (mass concentration).
- Material composition used for the spherical contaminant might differ from the real hazard material composition.
- Risk of overestimated detection performance by the use of a unique test sample, even passed several times, cannot ensure the contaminant has been tested at the worst position in the product or packaging (a typical case for gravity fall or pipe inspection metal detector is where the trajectory of the contaminant cannot be controlled to pass always in the center of the aperture).

Consequently, the detection limit measured statistically at a given POD on a unique test sample (e.g. 90.5% chance of detection) is optimistic. The statistical approach always associates a risk of non-detection too. If this limit is used for frequent routine equipment checks, there is still some chance that this will not be achieved all the time. Failure in the routine check might generate operational complications (such as batch blocking, rework or waste) and loss of confidence in the equipment.

A solution to increase the POD level is to pass more samples; an alternative solution with a similar effect is to artificially counterbalance the relatively low POD by introducing higher test sample sizes (e.g. using spheres with a diameter of 0.1 or 0.2 mm larger than the detection limit) for the routine equipment checks by operators. A check of the detection limit should, however, be performed on a regular basis.

Whatever the solution selected, the effective detection performance will remain the same. Checking the equipment with a slightly larger test piece dimension than the detection limit measured during the POD test does not influence the quality of detection. Detection equipment parameters are defined during the FRR assessment and are not modified.

Detection Limitations and HACCP

Metal detectors and X-ray equipment cannot ensure removal of all foreign bodies. The detection efficiency of these two technologies is heavily influenced by factors such as material, shape, size, location and orientation of the foreign bodies. They are also limited to sharp and hard physical hazards and will not be adequate to detect common biological contaminants (for example, insects or hair).

Detection limits are measured for each material by statistical methods, using spherical contaminants and a limited number of tests. They are not absolute and only associated with a probability of detection. Therefore detection equipment cannot always guarantee 100% removal of all contaminants bigger than or equal to these detection limits.

As a consequence, detection equipment can be identified either as a control measure or a verification measure depending on local HACCP assessment, but they should not be the only measures to reduce the hazard to an acceptable level and should be used in combination with upstream control measures.

Finally, it is important to remember that all rejects must be investigated and root cause analysis and corrective actions taken when a foreign body is discovered.

Further Reading

Campden & Chorleywood Food Research Association Group, 2004. Foreign Bodies in Foods: Guidelines for Their Prevention, Control and Detection: Guideline No. 5.

Edwards, M., 2004. Detecting Foreign Bodies in Food. Woodhead Publishing Limited.

Mettler Toledo Safeline Ltd, 2007. Building an Effective X-ray Inspection Program.

Mettler Toledo Safeline Ltd, 2007. Reduction of Metal Contamination: Building an Effective Programme.

US Office of Nuclear Security and Incident Response, 2011. Access Control Systems. NUREG-1964.

Wallin, P.J., Haycock, P., 1998. Foreign Body Prevention, Detection, and Control: A Practical Approach. Blackie Academic & Professional.

Zuber, F., 2007. Détection des corps étrangers dans les produits alimentaires. Techniques de l'ingénieur.

SECTION III

FOOD SAFETY ASSURANCE SYSTEMS

CHAPTER

22

Principles and Systems for Quality and Food Safety Management

Peter Overbosch and Sarah Blanchard

Metro AG, Duesseldorf, Germany

OUTLINE

Food Safety Management.
DOI: http://dx.doi.org/10.1016/B978-0-12-381504-0.00022-6

PRINCIPLES, SYSTEMS AND SCHEMES

Background and Working Definitions

In this chapter we will first look at the scope and meaning of "quality and food safety management" and then discuss principles and their relationship with systems and schemes.

Quality and safety in food are best considered together in this context. While for some purposes (for example, HACCP) we will want to make clear distinctions between them, they also share a number of common elements, especially from a perspective of their management.

For our current purposes, therefore, we will consider the management of food in terms of the goals we aim to achieve: safety, legality, consistency and consumer acceptability.

Food businesses all around the world must build and maintain management systems around these aspects.

There are many definitions of "principles," but for our purposes we might view them as headings under which we bring together all methods, techniques and background knowledge necessary to manage a particular (sub) aspect of food quality and safety. The principles therefore all relate back to one or more of the goals listed above, and together they cover our needs. Whether the list is exhaustive will remain debatable, but it should suffice to cover our needs and relate to the "systems" (see below).

"Systems" can be understood as management tools, aimed at operationalizing the principles. Of all the methods, techniques and background knowledge associated with a principle, a system typically selects those that "must" be used to fulfill appropriate requirements in a particular context, to make them as concrete and measurable as possible and provide a defined endpoint where possible. As an example, we might take "consistency," which was mentioned above as a goal (in practice we would then need to specify targets and limits for parameters), but is also listed below as a principle (but then we need to mention related methods and techniques, and the backgrounds of statistics and metrology). 6 Sigma, for example, is a management system that operationalizes the principle of consistency: it provides steps, presents specific methods and background knowledge and defines an endpoint (in its most literal sense, that endpoint is the achievement of 6 Sigma performance for the parameter(s) in question).

We will use the term "scheme" to describe a construction where one or more systems have been enveloped in an auditing/certification format. This applies to a large degree to our current topics – management systems in the area of food quality and safety usually combine the application of various principles under one roof. The GFSI-recognized certification schemes (currently including BRC, IFS, FSSC 22000, CanadaGap, GlobalGAP, Global Red Meat Standard, Global Aquaculture Alliance Seafood Processing Standard, PrimusGFS, Safe Quality Food) are all representatives of this approach. They typically cover general quality management systems requirements, loosely based on ISO 9001, including an HACCP module (based on the Codex format), and they provide a series of "prerequisite program" requirements, which address the principles of hygiene.

These schemes provide clear advantages: they present a concrete and comprehensive format to the management of food quality and safety, their certifications can be used as the basis of acceptability within the industry and (increasingly) towards authorities, and they can be expanded and updated as necessary with the input of all stakeholders.

The disadvantages, however, relate mostly to the fact that all schemes (and their underlying systems) are compromises:

1. Specificity – in order to be widely applicable, the requirements can never be sufficiently specific to fit any particular operation and the scheme owners must strike a balance between the (commercial) scope of their scheme and the relevance of the practical guidance it provides. ISO 9001, for example, is designed to be universally applicable, and therefore contains nothing that would specifically apply to food. In contrast, one might design a standard specific to each process for each food (or raw material) product. This would lead to a totally unmanageable multitude of standards and still not be precisely applicable to every individual situation. The above-mentioned global standards have managed to strike a much more relevant balance for the food industry, but it is important to note that there will always be this balance, and it is impossible to get it "just right" for all operations and situations. The limited specificity of global standards then makes the professional judgment of the auditor the key deciding factor for the practical validity of any certificate, which represents a structural vulnerability of these approaches. The report of the Joint FAO/WHO consultation on the Role of Government Agencies in Assessing HACCP, Geneva, 2–6 June 1998 already states that it is important that the use of a checklist does not evolve into a simple "tick-box" approach where there is no critical evaluation.
2. Level of sophistication – in order to address legal and market expectations in the most developed markets, standard requirements must be set at levels that may not be easily attainable in developing markets. Any claim of global relevance then risks ending up in a compromise that is practically irrelevant everywhere. As a result, the GFSI standards now include a stepping-stone approach, the "global markets" scheme. While this clearly addresses a realistic need, it also gets us back to what are effectively local rather than global standards.

All in all, therefore, management schemes are compromises and their practical claims are critically dependent on the professional qualifications of their auditors. Still, it is probably fair to say that the world of food quality and safety management has benefited enormously from these schemes, and their development and global acceptance is likely to continue for the foreseeable future.

We will see further in this chapter the use of systems to manage food safety and quality between different elements of the value chain and between organizations, for example linking retailers with suppliers.

While IT can help structure and track processes in a system, we do not specifically tackle IT systems in this chapter. We do, however, make reference to relevant types of IT software systems, and where they are driving the adoption of overall systems to manage food safety and quality.

Conclusion

From the above it is clear that effectively and efficiently managing food safety and quality in any specific case will always have to involve the application of principles. Whether it must also involve a formal system, or a certification scheme, will depend on the need for additional structure and/or external recognition. To make certification as meaningful as possible, the choice of the appropriate scheme is important, but even more important is the professional qualification and attitude of the auditor.

Food Safety Initiatives

GFSI

The Global Food Safety Initiative was established to continuously improve food safety management systems and ensure confidence in the provision of safe food to consumers worldwide. The initiative is business driven, bringing together leading food safety experts from global organizations to collaborate through the GFSI platform. Defining food safety requirements throughout the food supply chain, benchmarking different food safety standards against the requirements, building capacity of small or less developed businesses and focusing on auditor competency are the main activities. (See http://www.mygfsi.com/.)

HACCP

Hazard analysis and critical control points is a specific food safety-oriented method that was developed in the USA. In the meantime it has become the universally recognized and accepted method for food safety assurance, part of food safety legislation in the EU, USA and many other countries. Guidelines for the application of the HACCP system have been adopted by the FAO/WHO Codex Alimentarius Commission. HACCP certification through certification bodies is available in many countries – usually through the HACCP-based ISO 22000 scheme or the GFSI-recognized FSSC 22000 scheme, which combines ISO 22000 with Prerequisite Programs as described in ISO/TS 22002-1. (See also HACCP and its validation and maintenance in this book: www.fao.org/docrep/W8088E/w8088e.htm.)

ISO 9001

ISO 9001 (the International Organization for Standardization's standard for quality management systems, currently in the 2008 version) is internationally recognized as the authoritative standard in its field, but all its requirements are generic and are intended to be applicable to all organizations, regardless of type, size and product provided. Because of the lack of specificity, an ISO 9001 certificate is normally not accepted as a sufficient proof of acceptable practices in the food industry. Still, the management system principles as outlined in 9001 have found their way into all food-specific quality standards. ISO 9001 certification is available from multiple certification bodies across the world. (See http://www.iso.org/iso/home.html.)

6 Sigma

6 Sigma is a methodology aimed at reducing product and process variability – if possible to a nominal level of 0.0003% failure rate. The method makes extensive use of applied statistics throughout its five steps: Define, Measure, Analyze, Improve, Control. Having started in the electronics industry, 6 Sigma has found applicability in many other branches. While 6 Sigma is supported by many consultants and training courses, it is not a "scheme"

as defined in this chapter. Companies can apply the method, but not be certified according to a defined 6 Sigma standard. (See http://asq.org/index.aspx.)

PRINCIPLES AND ASSOCIATED SYSTEMS

The management of quality and safety in food will require the application of the following principles.

Hygiene

Hygiene has been defined as "the practice of keeping yourself and your surroundings clean, especially in order to prevent illness or the spread of diseases." For our current purposes the concept of hygiene will be extended to include cleanliness topics that do not necessarily relate to illness, like foreign material prevention and basic housekeeping rules regarding chemicals including pesticides and lubricating oils. Taken together, they are usually covered by what is commonly referred to as "prerequisite programs" – PRPs – in the context of HACCP (see "Prevention and Risk Reduction," below), indicating that we normally do not rely on hygiene alone to prevent illness or the spread of diseases.

Hygiene will also include allergen management, which is related to illness, although not in all individuals. Nevertheless a system is required to manage cross-contamination, as a prerequisite to appropriate labeling. In many countries, there are strict laws requiring the declaration of allergens on packaging.

When designing and implementing hygiene management, we always need to consider the level of hygiene required for the specific purpose at hand. It seems obvious that a rule like "all food handlers need to wear a hairnet" would not logically apply to people harvesting lettuces by hand.

In all cases, however, we do need to consider requirements related to people, equipment, tools and materials (including anything from water to lubricating oils and pesticides) and the general environment in the context of the intrinsic vulnerability of the product and the intended later cleaning and processing.

The hairnet for lettuce harvesters would not be required because lettuces will receive a subsequent cleaning step which may include removing outer leaves and is meant to sufficiently remove more types of foreign materials than just hairs. Protecting the general lettuce-growing environment against, for example, EHEC (enterohemorrhagic *Escherichia coli*) contamination – through untreated organic fertilizers, water or direct animal activity – would be relevant because later treatment including food preparation might not eliminate the EHEC bacteria.

Hygiene management systems can therefore be understood to support HACCP systems by reducing the list of "realistic hazards" in a relatively simple and robust way. What can easily be prevented or eliminated is taken care of by the somewhat broad brush of hygiene management; the remaining hazards get individually named and treated through HACCP.

As a general source for the methods, standards and techniques of hygiene, Codex Alimentarius must be mentioned. It captures a very wide and internationally accepted array of hygiene standards. As such, it underpins practically all food quality and safety-related management systems, but it is not generally seen as a system itself, and there are no widely recognized certification schemes based on Codex (sporadically, Codex certification has been offered).

For manufacturing equipment there is, for example, EHEDG and the EU legislation.

In the context of hygiene management, the 5S methodology must be mentioned. Originally it is a Japanese workplace organization method, concentrating on the elimination of "waste" in its various forms: unnecessary tools, parts and instructions, all dirt and rejects, and all forms of disarray and untidiness. The workplace needs to be standardized, clean, clear and lean at all times. Depending on the exact format adopted locally, safety, worker satisfaction and quality can be part of the 5S program. Though it has originated outside the food industry, it is easy to see how 5S can be used in support of hygiene management and reliability, and the program is being used as a firm basis for continuous improvement programs (6 Sigma, Process Variation Reduction – PVR), as well as various HACCP-based food safety-oriented schemes (GFSI certification).

For systems approaches covering hygiene, there are the globally available and recognized GFSI certification schemes as mentioned above, and in the case of FSSC 22000 there is a specific prerequisite program standard (ISO/TS 22002-1) that was developed to augment ISO 22000 to provide an all-round food safety solution. Additionally, the AIB (American Institute of Baking) scheme must be mentioned, which heavily concentrates on hygiene topics (understandable from the perspective of AIB's bakery background, where the baking process normally provides an adequate kill step for any microbiological concerns), and a veritable multitude of local standards, which often specialize on locally relevant product categories.

Food Safety Initiatives

EHEDG

The aim of the European Hygienic Engineering & Design Group is to promote the production of safe food by ensuring the hygienic design and engineering of food manufacturing equipment. EHEDG is a consortium of equipment manufacturers, food industries, research institutes and public authorities, providing training and guidelines. Started in Europe, EHEDG has now active sections in countries on all continents and guidelines are available in many languages. (See http://www.ehedg.org/.)

PRPs

Prerequisite programs address operational conditions that must be in place if a HACCP program is to be effective. They may relate to conditions in facilities and grounds, production equipment, cleaning and sanitation, personal hygiene, control of chemicals, receiving, storage and shipping, pest control and others. Most commonly used PRPs in the food industry derive from Codex Alimentarius. ISO/TS 22002-1:2009 specifies requirements for establishing, implementing and maintaining prerequisite programs (PRPs) to assist in controlling food safety hazards, typically used in connection with ISO 22000 for certification purposes (see also Chapter 24).

PVR

Process Variation Reduction is a method for isolating and identifying sources of process variation in excess of inherent, intrinsic variation (often called common cause variation), with the intent of their removal. PVR can stand alone, or it can augment 6 Sigma. Its aim in removing assignable sources of variation is to create more laminar, less turbulent, process flow, thereby improving productivity and quality simultaneously. This is accomplished by gathering sufficient data, organized specifically to quantify common cause variation, structural variation (due to differences among parallel segments of processing) and additional sources of assignable cause variation such as those due to raw material, operator, environmental and other differences. The quantified sources of variation are accumulated into "performance" and "capability." Performance variation measures fluctuations experienced by the consumer, while capability measures the best the process can do. The difference between performance and capability can usually be set in financial terms, aiding the setting priorities for improvement projects.

5S

5S is the name of a workplace organization of Japanese origin. The method is about: (1) eliminating everything that is not needed in the workplace, (2) giving everything that remains a clear and permanent place, (3) cleaning the workplace, (4) standardizing all common elements of the workplace, and (5) sustaining and continuously improving the practice. The practices of 5S normally exceed food industry PRPs, but there can be significant synergies in the combination of the specifics of PRPs with the rigorous 5S approach. 5S is supported by many consultants and training courses, but it is not a "scheme" as defined in this chapter. Companies can apply the method, but not be certified according to a defined 5S standard. (See http://www.epa.gov/lean/environment/methods/fives.htm.)

Prevention and Risk Reduction

Prevention and risk reduction is the one principle that is generally understood to be the exclusive domain of a single system: HACCP (hazard analysis critical control points). The HACCP system was developed in the 1960s by Pillsbury and the format was given global authority by Codex in 1993. As the Codex Alimentarius series of standards is meant to form the basis of national laws and international trade, HACCP has found its way into many countries' legal systems.

ISO 22000 is the primary international standard for the certification of HACCP, but as mentioned above, it lacks a specification of necessary hygiene conditions – the PRPs – and this is what, for example, the GFSI series of schemes addresses.

At this point it is probably relevant to mention that none of the above-mentioned standards and schemes will mention any specific hazard to be "prevented or reduced to an acceptable level" (in many cases they do mention specific hygiene precautions to be taken). It therefore continues to depend on the specific expertise of those who design, verify and validate a specific HACCP system to make the appropriate choices, and on the ability of the auditor to judge the results.

Some large food manufacturers have therefore established a two-tier system, whereby for each product/processing combination category (canned pineapple, frozen vegetables, smoked sausage, etc.) highly specific process flow and core hazard identification/reduction methods are given. It is then up to the individual manufacturing site to implement the standard plan and add any locally specific additional hazards/controls as necessary. This approach is as robust as reasonably possible, but it does require the availability of high-level expertise and is only practicable where a relatively limited number of products are involved.

General claims relating to food safety "because we have HACCP" must therefore be viewed with a certain degree of skepticism.

To illustrate the point – we have seen cases where producers of canned food products did have multiple globally recognized certificates, awarded by internationally accredited certification bodies, but were unable to show records and reports of the validation of their retorting process. (In the USA, the FDA will require highly detailed information before approving a retorting operation – see, for example, http://www.fda.gov/downloads/AboutFDA/ReportsManualsForms/Forms/UCM071581.pdf.)

In practice, the HACCP system is often abused through:

- Lack of specificity: hazards are not identified individually, but collectively (for example, "microbiological pathogens"), which leads to lack of specificity of the controls.
- Identified hazards not being real hazards (product dissatisfiers, or PRPs, rather than hazards, being managed in the HACCP plan).

The status of HACCP as a legal requirement throughout the supply chain in many countries has to a certain extent facilitated its misuse, where, for example, operations handling only packed and shelf-stable foods felt compelled to establish their own CCPs.

Where large numbers of widely varying products are involved, the use of risk matrices may help to manage complexity and assure consistency. Typically, risk matrices consider a combination of inherent product risk (for example, whether or not the product in its current form would support the growth of pathogens, as indicated through a_W or pH, or the relative frequency with which a product type is associated with a certain type of chemical contamination in the EU RASFF reports) and supplier-/origin-related risk. The latter is usually more complex and may involve certification status, geographic origin (addressing questions of the type: "is this a BSE country?"), internal layout and product portfolio (allergen risks), complaints, incidents, capability history, sales turnover, etc.

Establishing and keeping a risk matrix up to date is a major undertaking if it is used to combine all current and relevant factual information about the suppliers and their products. Where the aim is to implement a risk management strategy that is both rigorous and efficient, there may not be an alternative.

Microbiological modeling is increasingly being used in support of risk matrix approaches. Growth/death kinetics of various relevant microbiological strains are being modeled under relevant and realistic conditions, including intrinsic product parameters (including pH, a_W), processing conditions (time–temperature), packaging (vacuum, modified atmosphere), logistic conditions (time, temperature, humidity), storage and retailer shelf conditions and expected treatment/preparation by the consumer. As compared to ad hoc testing of samples, this approach has many advantages, with the assessment of any individual situation being based on a body of interrelated and systematized data and trends. The sheer complexity and the efforts necessary to build these models, however, will likely preclude their general application for the near future.

In the context of prevention the due diligence defense merits discussion. The UK sees some organizations referring to their "due diligence" systems. While the defense is an integral element of UK food safety law we are seeing the principle appear in other European countries, for example in retail organizations, as an overarching guiding principle in managing food safety. The UK Food Safety Act of 1990 (since amended) established that a defense in case of court would be that a person can prove he or she took all reasonable precautions and exercised due diligence to prevent the occurrence in question. This would mean that in a food production environment there must be the appropriate risk assessment and HACCP, backed up with procedures and documentation to demonstrate all reasonable precautions have been taken. The size of the company has a bearing on the level to which these precautions extend – the larger the organization and therefore with more access to resources, the greater the expectation.

This brings us back to the beginning of this section where we mention that no scheme requires any specifically named hazard to be prevented or reduced to an acceptable level and that we depend on the expertise of those designing such a system to manage this level.

The due diligence defense is a robust way to prevent food safety incidents and puts the burden of proof on the defendant. In doing so it prevents an often seen attitude where a company that causes an offence is simply able to pay a fine – the insurance costs of which would inevitably be reflected in the cost of running this system – and continue business unabated.

The vulnerability of the food chain and the necessity for preventive action is further recognized by the signing of the FDA Food Safety Modernization Act (FSMA). While certain details remain to be finalized, the American government's intention is clear – to proactively reduce the occurrence of foodborne illnesses. This will require food and agriculture businesses to implement preventive controls, the basis of such controls being scientifically based risk assessment, which is specifically mentioned under "Hazard Analysis and Risk-based Preventive Controls" (See http://www.fda.gov/Food/GuidanceRegulation/FSMA/ucm334115.htm). We may see the introduction of such legislation having a positive effect on the adoption of certification across the supply chain – both for domestic and import suppliers.

Food Safety Initiatives

FSMA

The US Food Safety Modernization Act was signed into law on 4 January 2011 and aims to ensure the safe supply of food by preventing contamination rather than responding to it. It is said to be the most significant reform in decades. (See http://www.fda.gov/Food/FoodSafety/FSMA/default.htm.)

RASFF

The Rapid Alert System for Food and Feed (RASFF) is an online database used by European member states to quickly exchange information on consignments of food and feed where a risk to human health has been detected. The tool allows EU countries to check whether the affected product is on their market and to take necessary action, for example, to block consignments, withdraw, recall or seize. The database is also accessible by consumers and businesses. (See http://ec.europa.eu/food/food/rapidalert/index_en.htm.)

Reliability

The performance predictability of processes and equipment, reliability can be seen as another PRP underpinning food safety and consistency (see "Consistency," below). Many food scares and incidents have been caused by reliability issues, and many could have been avoided through simple – but rigorous – preventive maintenance. Carbon filters being used beyond their capacity (filtering organic chemicals or microbes out of water), in-line sieves breaking and fragmenting after a certain throughput, pumps or air conditioning devices internally accumulating dirt, are all examples of the preventable breakdown of *a priori* capable processes and equipment. The fact that preventive maintenance is still not ubiquitous is usually due to considerations of short-term versus longer-term costs (why fix it now, when it is still working?).

Another typical cause of unreliability in the food industry is *a priori* lack of capability. In one example, we have witnessed a press plate of a cocoa press, woven from steel wire, broke down and caused an avalanche of small steel particles through a system of pipes, where they were supposed to be caught by magnets. Normally the press mat would break gradually and regular inspection would enable its timely replacement, while the magnets would keep the cocoa powder from being contaminated with steel particles. When the mat suddenly broke catastrophically (though roughly around the normal period of use), the magnets were overwhelmed. Assuming capability of the magnets to deal with any eventualities, they were simply cleaned from the metal debris and heavily contaminated cocoa powder was shipped (the situation was detected by the customer through their finished product metal detection system). This episode illustrates three points: (1) replacing only parts that have already visibly started to break down is not a reliable preventive maintenance strategy, (2) a downstream process aimed to "eliminate the hazard" (HACCP terminology) must be designed to be capable of dealing with the worst-case scenario (breakdown of a complete press mat) and (3) the process must include a point that can be used as a reference to demonstrate control. (In this case, a new setup included a "final magnet" which was required to stay clean at all times. Finding any metal particle on the final magnet would be seen as an indication that metal particles would have found their way into the product.)

Organizing reliability has been the focus of TPM (Total Productive Maintenance), which aims to achieve zero unplanned equipment failures, zero product defects, and zero accidents. The first two aspects were illustrated above; the accidents aspect fits logically with the others and can also be seen in 5S programs (see "Hygiene," above). Another key element of TPM is the involvement of the operator in the maintenance program and the explicit intention to optimize and redesign the equipment, based on ongoing in-depth performance analysis. A TPM program typically starts with a complete stripdown and cleaning of the equipment, followed by initial inspection for early signs of partial deterioration.

TPM is a very time and expertise intensive program, and many companies do not feel confident to use it, but its results in terms of productivity, hygiene and consistency are potentially very significant.

Food Safety Initiatives

TPM

Total Productive Maintenance is a participative method, where production employees take responsibility for the combined preventive and corrective maintenance of the equipment with which they are working. The goals are expressed as various elements of loss reduction; downtime, out-of-spec production, planning uncertainty and the associated stock levels, safety issues, impact on morale and the costs associated with all of the above. TPM is supported by many consultants and training courses, but it is not a "scheme" as defined in this chapter. Companies can apply the method, but not be certified according to a defined TPM standard. (See http://en.wikipedia.org/wiki/Total_productive_maintenance.)

Consistency

Consistency (and variation reduction) in this context, indicates the degree to which (successive batches of) products or units processes comply with their specifications. In food this can involve any type of parameter; thermal treatment (baking, cooking, retorting), color, weight, taste characteristics, etc. While proper specification and consistent compliance is arguably as relevant in food production as it is in any other industry, it is probably fair to say that this aspect is rarely given the attention it deserves.

Proper specifications are:

- Relevant to safety, legality, customer acceptability or internal processability.
- Specific as to the exact nature of the parameter(s) and measurement methodology involved.
- Precise in defining the required targets and limits.
- Realistic in terms of the capability of the process that is being specified or that produces the product spec.

Specifications in the food industry rarely fulfill these requirements, and typical problems include:

- No target defined (which may encourage a supplier to produce in a very tight tolerance close to the economically most attractive side of the limits).
- No limits defined (in practice this means that all acceptance/rejection decisions will be arbitrary).
- Insisting on symmetrical limits around the target (for example, in a pH specification, the upper limit may reflect a very different requirement than the lower limit, and the spec may be asymmetrical).
- No regard to process capability (batches may never be in spec).
- Specifying an irrelevant parameter.
- Not specifying a relevant parameter.

In the absence of proper specifications, there can be no appropriate process control, which then may lead to noticeable issues in all relevant areas mentioned above: safety (for example, through uneven heat distribution across an oven band), legality (for example, weight or volume), customer acceptability (for example, color or texture), or internal processability (for example, the fit of exactly x cookies in a tray).

In recent years, multiple examples have demonstrated the potential of 6 Sigma approaches to increase consistency and reduce losses. At a somewhat simpler level the PVR methodology has also demonstrated the ability to improve products and processes. In its simplest form it distinguishes three types of variability: common cause (i.e. inherent in the current design of the unit operation of the process in question), structural variation (i.e. the variability related to multiple unit operations in parallel, for example multiple parallel filling heads in a bottling plant) and assignable cause variation (related to "external" influences on the process that can be identified and eliminated or reduced).

Assignable cause variation can be normally addressed without major interference with the processing conditions or equipment. Structural variation and certainly common cause variation can only be improved through rigorous maintenance and/or redesign programs, for example TPM.

Food manufacturing is often plagued by inherent losses, and often a certain level of losses can be calculated as being "unavoidable." This may involve overfilling of weight or volume, product rejections, or when equipment is jamming because of lack of fit of product into packages, or lids to tubs or any other of a multitude of variability related causes. Frequently, the role of raw material consistency on the finished product is not considered, for example in poultry production often an inordinate amount of effort is spent in cutting and sorting different sizes of chicken pieces – leg, wing, etc. – due to variability in carcass quality. A focus on the farming methods and clear specifications for suppliers would help to reduce this effort. We could see an application of a 6 Sigma approach to identify and resolve the variation.

In all those cases a careful and systematic program to reduce variability around meaningful specifications will pay significant dividends.

Traceability

Under EU law, "traceability" means the ability to track any food, feed, food-producing animal or substance that will be used for consumption, through all stages of production, processing and distribution. Traceability must be understood to apply both upstream (where does this product come from?) as well as downstream (where did this product go to?).

Legal requirements for traceability are formulated differently in different countries, but it is clearly essential in any food quality and safety management system. Claims of origin ("product of…region or…country") are the most obvious examples, but traceability also involves the coverage of products by a certificate (has this product actually been made at the manufacturing site that carried the certificate, or has it been subcontracted to a less well-controlled operation?), the assumption of *a priori* risk (did this batch of beef originate from a BSE country, did this tea shipment come from a country that has effective regulations and controls against chlorinated pesticides?), the age (production time, expiry date) or catch area of fish (fish species may be red listed in some catch areas but not in others).

In recall and withdrawal cases, traceability is essential to determine the scope (mostly downstream traceability) as well as the root cause (upstream traceability). Finding the exact root cause may again have an influence on the scope. The 2011 EHEC crisis in Germany presented a clear case in point – with recalls, withdrawals and import stops varying according to the then current root cause theory: vegetables from Spain and Holland and finally fenugreek seeds from Egypt.

A more demanding form of traceability is found in "chain of custody" requirements usually associated with laboratory testing of critical samples, environmental certification of critical commodities or criminal procedures. We might see this in certification of "GMO-free" maize and in ethically produced and labeled cocoa.

Implementing traceability involves barcode systems (a major international player in this field is GS1 – http://www.gs1.org), and/or in future RFID systems (radio-frequency identification – http://www.rfid.org/), which are potentially more capable but have so far not found wide acceptance in the industry, mainly for cost reasons. Simple maintenance and linking of documentation of incoming raw materials and outgoing finished products may suffice in smaller organizations. Inevitably as the size of an organization increases and is operating in a more developed market, the level of IT involved in managing such a system increases.

Food Safety Initiatives

GS1

An organization that provides a system of integrated global standards, GS1 aims to improve the efficiency and visibility of supply and demand chains worldwide. It works across sectors and is one of the most widely used supply chain standards globally. The GTIN (Global Trade Item Number) is one of the key elements of the GS1 system and is a unique number used to identify products and services that are made or sold at any point in the supply chain, for example warehouse, checkout. A key advantage of GS1 is the ability to secure traceability through the supply chain. (See http://www.gs1.org.)

RFID

Radio-frequency identification is a generic term used to describe wireless non-contact technologies that use radio waves to identify and track objects via a unique serial number. The most common technology is to embed an object with a tag that contains a microchip with electronically stored information and is attached to an antenna that transmits identification information to a reader. RFID tags can be attached to any object or being, for example livestock may have tags injected, allowing positive identification of the animal. RFID technology does not require contact or line of sight for communication which is the main difference to barcode technology. (See http://www.rfid.org/.)

Customer and/or Consumer Relevance

The question of customer and/or consumer relevance is normally a joint responsibility of the marketing and R&D functions (in retail and foodservice, quality assurance often assumes the R&D responsibilities). Methods range from the simplest forms of in-house product testing on the basis of externally supplied samples to more sophisticated product design methodologies and the use of professional panels.

QFD (Quality Function Deployment) is an example of a highly structured approach that has been used in the food industry for quality purposes and also for food safety purposes but not very widely. It aims to translate the "voice of the customer" (typically expressed through a prioritized list of achievables – the "whats") into the design characteristics, including types of raw materials, processing, packaging, distribution and presentation (the "hows") through successive stages.

QFD is designed to account for interactions, through the "roofs" of the "house of quality," but the complexity of the considerations have so far prevented widespread application of the methodology. We have seen successful application of QFD in large branded manufacturers who typically have greater resources at their disposal and see the investment as an essential part of their brand growth.

With products already on the market, customer response systems are widely used to track complaints, claims and comments systematically. These kinds of responses are a valuable source of information for a business, but the level of sophistication of collection, analysis and response systems varies widely.

Typical challenges include trend analysis on the basis of responses per unit sold, which may be difficult for manufacturers as they do not know exactly when products are sold to the end consumer (one approach is to try to calculate an average lag period between production and consumer use, but this is inherently imprecise) and the differences in complaint behavior between different countries. The number of complaints in Europe has traditionally been highest in the UK and much lower in countries around the Mediterranean. Globally speaking, these authors have seen a two to three order of magnitude difference in complaint rates between, for example, the USA versus Costa Rica or the Philippines for the same range of food products. It is obvious that complaint rates do not simply reflect inherent product failure rates, but they also do not reflect relative indifference to product shortcomings. In our experience, consumers in low complaining countries are – equally – less likely to buy the product again after a negative experience as customers in high complaining environments.

Quality system certification – along the lines of the ISO 9001 requirements – typically requires a company to operate a system for tracking customer satisfaction, which then should be more than a design acceptance (for example, QFD including consumer testing preproduction) and complaint registration. Ongoing comparison against relevant completion and consumer requirements is then indicated, including appropriate follow-up.

All the above-mentioned aspects of and approaches to customer/consumer relevance and acceptance are not really standardized tools and are not all explicitly required by established certification formats, although the GFSI schemes do have requirements around new product development and customer feedback. All aspects, from product design through complaint tracking, are relevant and any producer or retailer implementing a food quality and safety management system must put their own mix together.

Food Safety Initiatives

QFD

Quality Function Deployment is a method for the systematic translation of customer requirements into technical product and process specifications, through a cascading series of "what" vs. "how" matrices. Started in the field of mechanical and electronic products, its influence in the food arena has for a long time been limited. The multiple and complex interactions involved in the definition and production of food products has often made the use of the "what"/"how" matrices too complex for practical product development processes. There are, however, successful examples of QFD application in foods, and its main advantage – the explicit and transparent connection between customer requirements and technology – remains unrivaled. QFD is not a "scheme" as defined in this chapter. Companies can apply the method, but not be certified according to a defined QFD standard. (See QFD quality: http://www.mazur.net/works/Mazur%202008%20QFD%20in%20

the%20Food%20Processing%20Industry.pdf – QFD food safety: http://dspace.ucalgary.ca/bitstream/1880/48177/1/Balakrishnan_Applying_Quality_BFJ2010_postprint.pdf.)

Transparency/Accountability

Transparency/accountability refers to the integrity of the products and materials needing to be maintained and demonstrated where food is handled, anywhere in the value chain. It means that an acceptable level of transparency must be provided about all relevant parameters and conditions, related to the principles mentioned above, batch by batch. In many cases, the schemes associated with the principles are precisely designed to provide such an acceptable level of transparency. A GFSI-recognized certificate is widely accepted as a reliable reflection of an operation's quality, safety and hygiene implementation. No auditing and certification scheme can guarantee that the conditions that were in place at the moment of harvesting, production, transport, storage or sales of a particular batch of product were exactly in line with the underlying requirements of the scheme, but the audit frequency and in-built self-check requirements of the scheme should provide a reasonable assurance. Where control over individual parameters is required, however, the customer will need to see more precise information.

For batches of products moving down the food value chain, there are two extreme examples we would like to discuss here: (1) complete vertical integration and (2) open market buying.

In the case of complete vertical integration, the final seller is part of a predefined chain of custody that has in principle fulfilled all conditions for transparency: all relevant conditions and parameters are known at all times and can be recorded as part of the batch's history. Acceptance testing should not be necessary at any point, because the available records hold all relevant information in more detail than testing could ever reveal (if no chlorinated pesticides have ever been used on a crop, there is no need to test for it later). Likewise, no later "due diligence" testing should be necessary for the same reason. Where there are *a priori* uncontrollable variables (for example, mercury levels in wild caught sea fish), testing can be done only once and early in the chain.

Where products are bought on the open market, traceability is typically lost and in many cases it is even uncertain whether a certain batch has actually been produced under the same conditions – as the definition requires. Having lost traceability, acceptance testing becomes the only option to determine safety and quality, and demonstrate it to the next stage in the chain. Under these conditions we are immediately confronted with two more questions:

1. What to test for? In the absence of a known history of the batch, this may be difficult. Risk matrices (see "Prevention and Risk Reduction," above) can then be used for this purpose, but this remains an approximation of product category-typical risks and may not always give us the correct answer.
2. How to sample? When the product is not intrinsically homogeneous sampling is normally carried out through AQL (Acceptable Quality Level) sampling. The exact sampling scheme will then depend on some assumptions that need to be made about the homogeneity of the batch. In our "open market" situation, these assumptions may be problematic and an AQL sampling and testing scheme may quickly become prohibitively intensive.

Vegetable oils, bought on the world market and frequently unseen aboard a vessel, provide a useful example of how difficulties can be resolved when the typical characteristics of the trade allow. International rules require ships carrying edible oils to maintain a record of previous cargoes (providing relevant information regarding potential impurities), cleaning (reducing impurities) and acceptance testing, where sampling is relatively easy (contaminants can be assumed to be homogeneously distributed) and the use of chromatography/mass spectrometry methods allows for sufficiently rapid and comprehensive screening. In this case, the main concerns regarding lack of traceability may originate from environmental considerations (for example, sustainable palm oil).

In most cases, the various participants in the chain have to generate and communicate incomplete information, resulting in the need for some level of acceptance testing. Where possible, however, the use of aggregate internal control data (process performance, see "Consistency," above) instead of acceptance testing should be preferable. For this to be successful, a transparent view of the supply chain with information on production and processing conditions must be made available.

Transparency is essential if traceability is to be effective. The nature of today's complex food chain means that this is not always achievable and it may be in the nature of certain operators to remain guarded about their sources. For this reason we see numerous multinational organizations bundling their purchasing requirements in order to go direct to the source of supply. As well as improving traceability through a simplified supply chain (removing unnecessary middlemen and a level of complexity), this enables an organization with a brand to protect – be it a manufacturer, a retailer or food service operator – to have security over the provenance of their raw materials (not forgetting the value enhancement) and in turn provide reliable information to their customers. Hence the link between transparency and accountability becomes live.

Food Safety Initiatives

AQL

Acceptance Quality Limit is a sampling and testing methodology to determine whether a batch of products meets predefined criteria with a given level of confidence. ISO 2859-1 (See http://www.iso.org/iso/catalogue_detail.htm?csnumber=1141), on Sampling Procedures for Inspection by Attributes, describes standard methods for sampling plans and acceptance criteria.

INTEGRATED SCHEMES AND THEIR LIMITATIONS

Having to assess the specific relevance and implications of all the principles above for one's operation, the prospect of having an all-encompassing certification scheme becomes very attractive for both sides: operators and their customers. One certificate to cover the totality of all relevant requirements would be worth the effort and absolve both contract partners from their responsibilities. Scheme designers and owners have pursued this ideal for many years, and the frequent updates of their schemes testify their efforts, but the "one certificate to cover all" will probably always remain elusive.

There are five main reasons for this:

1. The balance between the need to be specific and the need to accommodate very different situations, materials and processes. The wider a standard applies, the less specifically relevant it can be. A good example is ISO 9001, which is designed to be the basis for all quality systems in all types of industries, but – for that exact reason – cannot go into any relevant detail. It is perfectly possible to design a fully adequate food quality and safety system and have it certified under ISO 9001, but that would rarely have the necessary credibility for professional partners in the industry. The GFSI set of standards therefore includes HACCP requirements and prerequisite programs – but these standards still need to cover the whole gamut of food categories and cannot be specific for canning, smoking, drying, etc. operations. Time will tell whether there will ever be a comfortable balance between practical applicability to a whole branch of industry and the need to provide specifics, but we do not expect it any time soon.
2. As mentioned above, a certification scheme cannot operate at the level of detail and focus needed for the acceptance of individual batches.
3. Linked to the above is the question of auditor competence. Globally, competent food auditors are in short supply and they can realistically be expected to be experts in a few food categories or processes only. Increasingly, auditors are qualified for specific areas only, but this then effectively reduces supply further.
4. A comprehensive certification scheme should cover the entire chain, including all logistics and storage and transfer of ownership. This is not a question of wide and shallow vs. narrow and deep, but a question of vertical integration and whether multiple independent business partners can have connected systems at the same relative level of strictness which allow for often changing business arrangements. In principle the answer must be yes, but in practice there are few examples.
5. The currently most relevant systems (the GFSI-recognized schemes) are mostly privately owned. This enables them to react more quickly to emerging needs, but it also implicates an ongoing commercial competition and a barrier to true integration.

Integrated schemes therefore are both useful and limited, and while they are constantly developed further, their basic limitations will likely remain.

Systems and the Value Chain

The value chain can be seen as the totality of the various stages of product flow (the supply chain, primary processes) and those functions that serve to support and/or innovate the supply chain. For our current purposes, we recognize:

- Product and technology development
- Primary production
- Procurement
- Logistics
- Manufacturing
- Retail and delivery
- Professional food preparation
- Human resources

All principles mentioned above (hygiene, risk prevention and reduction, reliability, consistency, traceability, consumer/customer relevance and acceptance, transparency) apply in their own specific way to each element of the value chain, but systems have typically been developed to serve the needs of one particular element of the chain and find little use in other elements. As mentioned in "Reliability," above, a reliability-oriented system like TPM was developed in the context of manufacturing and finds its application exclusively there. Also, the landscape of systems has developed at a different pace in different areas. There has been an early focus on primary production and manufacturing – inspired by Codex Alimentarius and driven by the needs of international trade. Trade groups packaged and issued most of the schemes that later were combined under GFSI, but penetration of schemes applying to retail and food service themselves has been slower.

Taking the elements of the value chain one by one:

- Product and technology development – most companies use some proprietary form of an innovation funnel, designed to administratively manage the complexities of a multitude of ideas and projects, but is not a specific food quality and safety tool. QFD has been successfully used as a specific innovation tool, also for quality and safety, and it has the capability in principle to incorporate the relevant requirements of all principles (the "whats" for hygiene, traceability, etc.) in the design phase. A comprehensive use of the methodology from this perspective in the food industry is not currently known to us. The QFD methodology could potentially be used as the basis for a comprehensive certification scheme for product and technology development.

  We have seen the use of ISO 9001 to manage new product development in food service organizations; however, this is simply a way to standardize an existing process or processes and ensure a consistent approach to product development across an organization.

  If we look at IT systems, retailers in the USA and Europe are increasingly using Product Lifecycle Management (PLM) systems to manage the NPD to launch process – while this is managed using specific software, it brings a level of discipline to a process that is often run by multiple players within organizations (retailer and supplier) with different ways of managing product development. Here we see organizations linking together using IT to increase speed of the NPD process and eventually produce a safe product that adheres to an agreed customer specification.

  Increasingly research laboratories play a role in product development, using consumer panels to test products – although laboratories accredited to ISO 17025 and able to perform effective sensory assessment are rare and costly.
- Primary production – increasingly GFSI-recognized certification schemes such as GlobalGAP, CanadaGAP and PrimusGFS are being applied in primary production, be it aquaculture, crops or livestock. However, the very nature of primary producers poses a challenge to the worldwide acceptance and hence implementation of such schemes. Frequently multinational organizations and those organizations sourcing from developed countries (although OECD countries sometimes have similar issues) struggle to implement GFSI/GlobalGAP-required standards with their suppliers. With this in mind capacity building schemes have been introduced, for example the GFSI Global Markets program to serve as a stepping-stone for those primary producers with the eventual aim

to reach full certification. The risk management and prevention principle becomes live here where the involvement of an auditor who fully understands the product and related food safety hazards can bring great value to the system. Certain retailers are going further and supplementing the certifications with additional product-based checks to ensure consistency and challenge traceability.

Simple traceability systems, often using GPS and Internet-based mapping, are appearing to link the producer with the retailer or wholesaler. We have seen this implemented successfully in China with a multinational retailer. The use of IT software and hand-held devices to manage the system of checks and audits is also proliferating. We are aware of such systems being used successfully in Europe to link fruit and vegetable growers with wholesalers and retailers. Information on the produce, for example pesticides used and quality specification parameters, is relayed up the chain and stored to allow later reporting and trending.

- Procurement – We are not aware of a scheme that specifically applies to procurement, although the food safety and quality requirements tend to be pushed back down the chain to the suppliers while being monitored by the purchasing organization. Numerous private and public sector schemes are in operation, which usually require suppliers to implement food safety management systems based on HACCP and risk assessment. The due diligence defence in the UK can be regarded as a driver of food safety in any food procurement department and we see this visibly in several public sector organizations such as schools and hospitals.

  Certification requirements would be a logical next step for suppliers including brokers. Here we see the IFS broker standard, which is applied to trading activities – brokers are obliged to ensure their suppliers have appropriate food safety and quality systems. In doing this, the customer has a certain security knowing that they do not need to carry out the checks themselves. This, however, is not a final solution and on occasion additional visits by the customer to production sites with the broker may be necessary to ensure the supplier fully understands the customer requirements.
- Storage and logistics – as well as proprietary schemes, IFS Logistics and BRC Storage and Distribution standards would apply.
- Retail and delivery – IFS Cash & Carry is gaining some ground in Europe, covering the operational aspect of retail and delivery, and again, the basis of food safety is HACCP and risk assessment. This standard is seen more in the traditional bulk sales cash and carry sector rather than in mass retail. Various proprietary standards are also in use. Taking retail on its own, procurement departments often have internal documented processes that require certain standards of suppliers, for example GFSI certification (see "Procurement," above). Such procurement departments are normally linked to their technical or quality departments and product development and here we can see the proliferation of in-house systems linking the activities, sometimes using IT to track progress. Such systems can be extended to suppliers and we have seen examples where procurement, quality and NPD work together on a web-based platform allowing capture of a full picture of the product and supplier. Elements of supplier performance (test results, speed of response, complaints, etc.) are recorded and as well as monitoring supplier performance, can be used as a basis to carry out risk assessment. Certain laboratories supply online databases with information that can contribute to the risk assessment.

Assessment of product performance may be carried out, for example, using laboratory testing of finished products (see "Transparency/Accountability," above) or using informal in-house kitchen analysis where products are "cooked up" and their performance recorded. Certain organizations require their suppliers to be present at such sessions, which is seen as a valuable way to give direct feedback to the supplier for further improvement if needed.

- Professional food preparation – we are not aware of global standards for food safety and quality in this area but there are a number of proprietary and local government standards. The focus tends more towards kitchen preparation than supplier control, although for the larger national and multinational organizations, control of the supply chain is common. Traceability and accountability are again key principles in this case.

We see the larger institutional caterers and food service organizations managing food safety and quality using ISO 9001 as a structural basis, with HACCP applied to operational food safety. Such systems enable customization to local conditions. Interestingly the integration of health and safety (H&S) systems in such environments can be seen as a way of introducing PRPs – by encouraging safe working practices that have equal relevance to food safety as personal safety. In terms of supply chain practices, it is not uncommon for such organizations to have a central procurement function that organizes ingredient specifications and certificates or audit reports from their suppliers who may be wholesalers, delivery agents or cash and carry. Product development in such organizations tends to be governed by an internal system and is usually based around informal in-house kitchen trials. The multinational branded food service organizations, however, tend to have more formal systems of product development, which may be supported by in-house R&D and external customer panels.

Food Safety Initiatives

PLM

Product Lifecycle Management (PLM) is an industry term used to describe the process of managing the full life cycle of a product from its conception, through design, development and manufacture, to launch, maintenance and disposal. IT software-supported PLM is increasingly popular in the retail industry where sharing of information and collaboration across a wide range of people with different processes is necessary for the success of a project, for example in private label development. Creation and central management of all information is the basis of PLM and while it can involve specific software, the emphasis is on PLM as a business strategy. (See http://en.wikipedia.org/wiki/Product_lifecycle_management.)

THE FUTURE OF SYSTEMS

From our overview of principles and systems it may be clear that we feel there is a bright future for systems and schemes – their demonstrable results, most notably the certificates, may increasingly be treated as international "licenses to operate." The proliferation of all kinds of overlapping and redundant schemes and standards is a concern, but there is no doubt that this rapid expansion over the last decade or so was driven by a real need and opportunity.

It may therefore be clear that we expect a number of developments:

- Systems to schemes. Where existing systems clearly have the potential to add unique value to an element of the value chain (QFD for innovation, PVR for consistency in manufacturing), we may expect a certification scheme to be developed around it. The developments around HACCP may serve here as an example – developed from a set of principles to a Codex format to an ISO standard (22000) to a GFSI-recognized scheme (FSSC 22000). Moving forward, relevant stakeholders (authorities, professional customers) may require a certified demonstration of the fact that all relevant considerations have been taken systematically into account in the design stage of a new product of technology. This may then serve as a "license to launch."
- Incorporation of additional systems into existing schemes. The current leading schemes typically do not include any mention of systematic management of reliability and/or consistency. In practice, reliability is an important factor behind many prerequisite programs as well as an economic driver. A need to establish the required level of confidence in the consistent delivery of product made under the right conditions may drive the inclusion of certain elements of TPM into mainstream certification schemes. The same holds for PVR/6 Sigma around consistency. Where legality and safety depend on strict specifications and their tolerances being upheld, a quantitative approach, linked to an ongoing improvement effort, including some of the core elements of PVR/6 Sigma underpins requirements with well-understood methodologies.
- Ongoing efforts to balance broad applicability with providing relevant detail. We may expect to see the trend set by FSSC 22000's modularity approach to PRPs – i.e. providing specific PRP packages for specific types of operations – to continue. Modularity could then also apply to the core HACCP part – specific requirements applying to canning, etc.
- Coverage of the entire value chain by schemes. As consumer awareness increases, the need to provide clear information on food safety will become imperative. Visible schemes allow consumers to make decisions and in certain environments food safety is seen as a competitive issue. We would hope that in time it would cease to do so.
- Increased attention to qualification of scheme auditors to ensure a better understanding of risk management will be proliferated throughout the value chain.

Further Reading

American Institute of Baking (AIB): https://www.aibonline.org/
Codex Alimentarius: http://www.codexalimentarius.org/.
Global Food Safety Initiative (GFSI): http://www.mygfsi.com/

- BRC Global Standards: http://www.brcglobalstandards.com/GlobalStandards/Home.aspx
- CANADAGAP: http://www.canadagap.ca/
- Food Safety System Certification 22000: http://www.fssc22000.com/en/
- Global Aquaculture Alliance Seafood Processing Standard: http://www.gaalliance.org/
- GlobalGAP: http://www.globalgap.org/

- Global Red Meat Standard: http://www.grms.org/
- International Featured Standards: http://www.ifs-certification.com/index.php/en/
- Safe Quality Food: http://www.sqfi.com/
- Primus GFS: http://www.primusgfs.com/

Report of a Joint FAO/WHO Consultation on the Role of Government Agencies in Assessing HACCP Geneva, 2–6 June 1998: http://www.who.int/foodsafety/fs_management/en/haccp98.pdf

5S: http://www.epa.gov/lean/environment/methods/fives.htm

European Hygienic Engineering & Design Group (EHEDG): http://www.ehedg.org/

Food Quality and Safety Systems – A Training Manual on Food Hygiene and the Hazard Analysis and Critical Control Point (HACCP) System, FAO 1998: www.fao.org/docrep/W8088E/w8088e.htm

The Food Safety Act 1990 – A Guide for Food Business (2009 edition): www.food.gov.uk/multimedia/pdfs/fsactguide.pdf

FDA Food Safety Modernization Act: http://www.fda.gov/Food/FoodSafety/FSMA/default.htm

TPM: http://en.wikipedia.org/wiki/Total_productive_maintenance

PVR: Chicken soup for processes, Lynne B. Hare, Quality Progress, August 2001, pp. 76–79: http://asq.org/index.aspx

6 Sigma: Leading Six Sigma, Ronald D. Snee and Roger W. Hoerl, Prentice Hall (2003); Implementing Six Sigma, Forrest W. Breyfogle III, John Wiley & Sons (1999).

Traceability: http://ec.europa.eu/food/food/foodlaw/traceability/factsheet_trace_2007_en.pdf

QFD quality: http://www.mazur.net/works/Mazur%202008%20QFD%20in%20the%20Food%20Processing%20Industry.pdf)

QFD food safety: http://dspace.ucalgary.ca/bitstream/1880/48177/1/Balakrishnan_Applying_Quality_BFJ2010_postprint.pdf

Innovation Funnel:

Groupware and teamwork in R&D: limits to learning and innovation, Claudio U. Ciborra, Gerardo Patriotta (2002): http://onlinelibrary.wiley.com/doi/10.1111/1467-9310.00080/abstract

http://www.innovationexcellence.com/blog/2011/11/29/rethinking-the-product-development-funnel/

CHAPTER

23

Hygiene in Primary Production

Gisela Kopper[1], *Slavko Mirecki*[3], *Igor S. Kljujev*[2], *Vera B. Raicevic*[2], *Blazo T. Lalevic*[2], *Jelena Petrovic-Jovicic*[2], *Stojmir Stojanovski*[4] *and Dr. Dijana Blazekovic-Dimovska*[5]

[1]Latin American and Caribbean Association of Food Science and Technology, San Jose, Costa Rica; [2]University of Belgrade, Serbia; [3]University of Montenegro, Podgorica, Montenegro; [4]Hydrobiological Institute, Ohrid, Macedonia; [5]Faculty of Biotechnical Sciences, University of Bitola, Bitola, Macedonia

OUTLINE

Food Safety Management.
DOI: http://dx.doi.org/10.1016/B978-0-12-381504-0.00023-8

INTRODUCTION

Food safety is defined as the assurance that food will not cause harm to the consumer when it is prepared or eaten according to its intended use (FAO/WHO 1997). Therefore, to achieve food safety all stakeholders in the food production chain should make every effort to reduce risks of contamination. As stated in EU Regulation 852-2004 on Hygiene Provisions, food business operators are to ensure that primary products are protected against contamination and must comply with provisions relating to the control of hazards in the primary production and associated operations. The control of hazards includes measures to control contamination arising from air, soil, water, feed, fertilizers, pesticides, veterinary drugs, etc. as well as measures relating to animal health and welfare.

Furthermore, as a result of foodborne diseases and mass outbreaks around the globe, public concern about food safety has increased dramatically recently. The World Health Organization estimates that there are hundreds of millions of people suffering from diseases resulting from contaminated food or water. In the past reporting these data was difficult, but nowadays with easier and more reliable reporting of data and occurrences, statistics show that both in developed and developing countries, foodborne diseases are rising. The WHO reported that foodborne diarrheal disease is one of the most common illnesses worldwide, estimated between 2.2 and 4 million cases per year. Every day, thousands of people die from preventable foodborne disease. In developing countries, 1.8 million children die under the age of 5 because of a diarrheal disease; up to 70% of these cases may be caused by foodborne and water pathogens (Larson, 2010; Motarjemi et al., 2012).

Countries with good reporting systems have documented significant increases in the incidence (number of cases) of foodborne diseases during the two last decades. It has been estimated that each year foodborne diseases cause approximately 76 million illnesses, 325,000 hospitalizations, 5000 deaths in the United States and 2,366,000 cases, 21,138 hospitalizations and 718 deaths in England and Wales. Data from the Netherlands indicate that out of 1.8 million cases of gastroenteritis caused by 14 infectious agents, >30% (680,000 cases) are proven to be foodborne (Motarjemi et al., 2012; Haavelar et al., 2012). From the reported number of cases, it can be assumed that the burden of foodborne disease is probably in the same order of magnitude in most countries of the Organization for Economic Cooperation and Development (OECD) (Rocourt et al., 2003).

There are numerous sources of food contamination by pathogens. In Table 23.1 some important hazards and their related epidemiological features are summarized (Käferstein et al., 2004). In OECD countries the foods most frequently involved in outbreaks are meat and meat products, poultry, eggs and egg products, with the likely implication that these foods are associated with *Salmonella* and *Campylobacter*. Case–control studies confirmed the same food sources for sporadic cases: raw and undercooked eggs, foods containing egg and poultry for salmonellosis, poultry for campylobacteriosis and raw oyster for *Vibrio* illness (Rocourt et al., 2003).

In the United States, of the total reported outbreaks and outbreak-related illnesses between 1996 and 2010, excluding meat and poultry, produce accounted for 23.3% and 42.3%, respectively. These outbreaks were associated with approximately 20 different fresh produce commodities, including sprouts, leafy greens such as lettuce and spinach, tomatoes, melons such as cantaloupe and honeydew, berries such as raspberries, blueberries,

TABLE 23.1 Some Important Foodborne Hazards and their Salient Epidemiological Features[a]

Hazards	Important Reservoir or Carrier	Transmission[b] by: Water	Food	Person to Person	Multiplication in Food	Transmission by Contact with Animals	Comments and Examples of Foods Involved
BACTERIA							
Bacillus anthracis						+	
Bacillus cereus	Soil	–	+	–	+		Cooked rice, cooked meats, vegetables, starchy puddings
Brucella spp.	Cattle, goats, sheep	–	+	–	+		Raw milk, dairy products
Campylobacter jejuni	Chickens, dogs, cats, cattle, pigs, wild birds	+	+	+	–[c]		Raw milk, poultry
Clostridium botulinum	Soil, mammals, birds, fish	–	+	–	+		Fish, meat, vegetables (home preserved), honey
Clostridium perfringens	Soil, animals, man	–	+	–	+		Cooked meat and poultry, gravy, beans
E. coli enterotoxigenic	Man	+	+	+	+		Salad, raw vegetables
E. coli enteropatogenic	Man	+	+	+	+		Milk
E. coli enteroinvasive	Man	+	+	0	+		Cheese
E. coli enterohaemorrhagic	Cattle, poultry, sheep	+	+	+	+		Undercooked meat, raw milk, cheese
Listeria monocytogens	Environment	+	+	–	+		Cheese, raw milk, coleslaw
Leptospira						+	Flooded crops, canned food
Mycobacterium bovis	Cattle	–	+	–	–		Raw milk
Salmonella Typhi and *Paratyphi*	Man	+	+	±	+		Dairy products, meat products, shellfish, vegetable salads
Salmonella Non-typhi	Man and animals	±	+	±	+		Meat, poultry, eggs, dairy products, chocolate

Shigella spp.	Man	+	+	+	+	Potato/egg salads
Staphylococcus aureus (enterotoxins)	Man	–	+	–	+	Ham, poultry and egg salads, cream-filled bakery products, ice cream, cheese,
Vibrio cholerae O1	Man, marine life	+	+	±	+	Salad, shellfish
Vibrio cholerae, non-O1	Man, marine life	+	+	±	+	Shellfish
Vibrio parahaemolyticus	Seawater, marine life	–	+	–	+	Raw fish, crabs and other shellfish
Vibrio vulnificus	Seawater, marine life	+	+	–	+	Shellfish
Yersinia enterocolitica	Water, wild animals, pigs, dogs, poultry	+	+	–	+	Milk, pork, poultry
VIRUSES						
Hepatitis A and E viruses	Man	+	+	+	–	Shellfish, raw fruit and vegetables
Calici viruses	Man	+	+	–	–	Shellfish, salad
Rotavirus	Man	+	+	+	–	0
PROTOZOA						
Cryptosporidium parvum	Man, animals	+	+	+	–	Raw milk, raw sausage (non-fermented)
Cyclospora cayetanensis	Man	+	+	0 (unlikely)	–	Raspberries
Entamoeba histolytica	Man	+	+	+	–	Vegetables, fruits
Giardia lamblia	Man, animals	+	±	+	–	Vegetables, fruit
Toxoplasma gondii	Cats, pigs	0	+	–	–	Undercooked meat, raw vegetables

(*Continued*)

TABLE 23.1 (Continued)

Hazards	Important Reservoir or Carrier	Transmission[b] by					Comments and Examples of Foods Involved
		Water	Food	Person to Person	Multiplication in Food	Transmission by Contact with Animals	
HELMINTHS							
Ascaris lumbricoides	Man	+	+	–	–		Soil-contaminated food
Clonorchis sinensis	Freshwater fish	–	+	–	–		Undercooked/raw fish
Fasciola hepatica	Cattle, goats	±	+	–	–		Watercress
Opisthorchis viverrini/ felineus	Freshwater fish	–	+	–	–		Undercooked/raw fish
Paragonimus spp.	Freshwater crabs	–	+	–	–		Undercooked/raw crabs
Taenia saginata and *T. solium*	Cattle, swine	–	+	–	–		Undercooked meat
Trichinella spiralis	Swine, carnivores	–	+	–	–		Undercooked meat
Trichuris trichiura	Man	0	+	–	–		Soil-contaminated food

+ = yes, – = no, ± = rare, 0 = no information.

[a] *Käferstein et al. (2004).*

[b] *Almost all acute enteric infections show increased transmission during the summer and/or wet months, except infections due to rotavirus and* Yersinia enterocolitica, *which show increased transmission in cooler months.*

[c] *Under certain circumstances some multiplication has been observed. The epidemiological significance of this observation is not clear.*

blackberries and strawberries, fresh herbs such as basil and parsley, and green onions as well as fresh-cut fruits and vegetables. These outbreaks involved a number of pathogens, including *E. coli* O157:H7, *E. coli* O157, *Salmonella* species, *Listeria monocytogenes, Cyclospora, Shigella sonnei* and hepatitis A (FDA, 2013b).

The emergence of new foodborne pathogens as well as the recurrence of well-known pathogens over the last decades can be explained by various factors such as world changes in society and food production systems. Globalization has undoubtedly increased world trade and travel with major consequences such as faster transfer of microorganisms from one place to another, increased opportunities for contaminations, time–temperature abuse of products and hence the risk of foodborne illness. An increasing elderly world population is now exposed to a greater number of different strains and types of pathogens. A person with a foodborne illness can expose others to a new pathogen in a location thousands of miles from the original source of infection.

Advances and changes in food production at the primary level and further processing also pose new threats to global food safety. In 1996, a new variant of Creutzfeldt–Jakob disease linked to Bovine Spongiform Encephalopathy (BSE) in cattle was diagnosed in humans. Consumption of contaminated meat products from cattle is presumed to be the cause. Modern intensive animal husbandry practices to maximize production seem to have led to the emergence and increased prevalence of *Salmonella* serovars and *Campylobacter* in herds of almost all important production animals (Rocourt et al., 2003). Excessive use of veterinary drugs in intensive animal production led also to increased awareness of the health effects of high drug residue levels in animal muscle. The use of untreated manure as organic fertilizers and contaminated irrigation waters in agriculture has been associated with unusual pathogen contamination of fresh fruits and vegetables. Infections caused by *Vibrio* spp. (*V. cholerae, V. parahaemolyticus, V. vulnifucus*) and intoxications due to naturally occurring toxins, e.g. various forms of shellfish poisoning or ciguatera, as well as trematodiasis, are common concerns with fish and fishery products. *Cyclospora* in raspberries, *E. coli* O157:H7 in apple juice and alfalfa sprouts, *Salmonella* in cantaloupes are just some of the many recent cases of foodborne outbreaks linked to practices in primary production (FDA, 2013a). The list of potential foodborne hazards and illnesses is long and the magnitude of the problem is enormous. The problem is greatest in the developing countries, although official reports of outbreaks are scarce and anecdotal. The high incidence of travelers' diarrhea in these countries is an indication of underlying food and water safety problems (Motarjemi et al., 2012). Over and above illnesses due to microbial agents, misuse of agrochemicals as well as naturally occurring toxins such as mycotoxins, in particular aflatoxins, causes major problems in the developing countries and are barriers to their export and development.

An effective prevention program must start with the prevention of food contamination in primary production, particularly considering the fact that many food products may be consumed raw and the predilection for such foods is increasing.

The present chapter addresses three main primary production systems and the challenges to reduce the threats to safety inherent to each. Good animal farming, fish health and good agricultural practices are described extensively so as to provide a clear picture of the complexity of the food production chains and the many factors that need to be under control to assess the safety of the products presented to consumers. This chapter is not intended to contain an exhaustive treaty on good agricultural, farming or aquaculture practices but to share experiences and propose different approaches to foster food safety.

PART 1: GOOD ANIMAL HUSBANDRY

INTRODUCTION

The human population is growing year by year. It is expected that by 2020, the number of people on Earth will increase from 7 billion to approximately 8 billion. In order to ensure a sufficient amount of food, it is necessary to use the most advanced agricultural technologies in plant and animal production. Although the agricultural technologies are highly sophisticated and activities are based on scientific knowledge, there are a number of negative impacts if they are not properly implemented. Those improper farm activities can cause environmental pollution and health problems to animals and farm workers, to neighboring wildlife, and to the consumers of their products.

Animals are exposed to constant activity factors such as air, water, soil and climate, and factors of an inanimate nature, but also by the presence of people, other animals, insects, microorganisms, pests and other factors of a living nature. All those factors can be carriers of agents that can directly or indirectly cause contamination of food produced on farms.

Almost every activity on a farm carries the risks of contamination from animal to animal, from animal to man, from man to animal; and, most important, from animal and man to farm products. Further, the composition of many animal products (meat, milk, eggs, etc.) is an ideal medium for the outgrowth of pathogenic microorganisms. Domestic animals may carry human pathogens which if present in food of animal origin may increase the risk of causing foodborne illness. Almost all foods have the potential to cause foodborne illness. There is also the potential of contamination of animal products with residues of veterinary drugs, hormones, pesticides and other chemical contaminants.

Therefore, implementing the proper procedures in agriculture, especially hygienic procedures, and control of animal products throughout the food chain are essential to ensure the safety and suitability of these foods.

This part will present potential microbiological risks at primary production level on animal farms, and give an overview of good practices that can prevent or solve problems caused by these risks.

POTENTIAL HEALTH RISKS ON ANIMAL FARMS

Microorganisms, viruses and parasites are the source of various animal and human diseases, and can, directly or indirectly, cause contamination of food produced on farms. They are widespread in the environment, in air, water and soil, inside or outside people, animals or insects, in barns, parlors, equipment and tools. For this reason, the "battle" against such agents is hard and must be carried out on a daily basis. In Figure 23.1, a selection of zoonoses and foodborne cases in human population in EU countries during 2010 is presented.

FOODBORNE DISEASES

Foodborne diseases are acute illnesses, usually affecting the gastrointestinal tract, brought on by consuming contaminated food or beverages. Various microbial agents

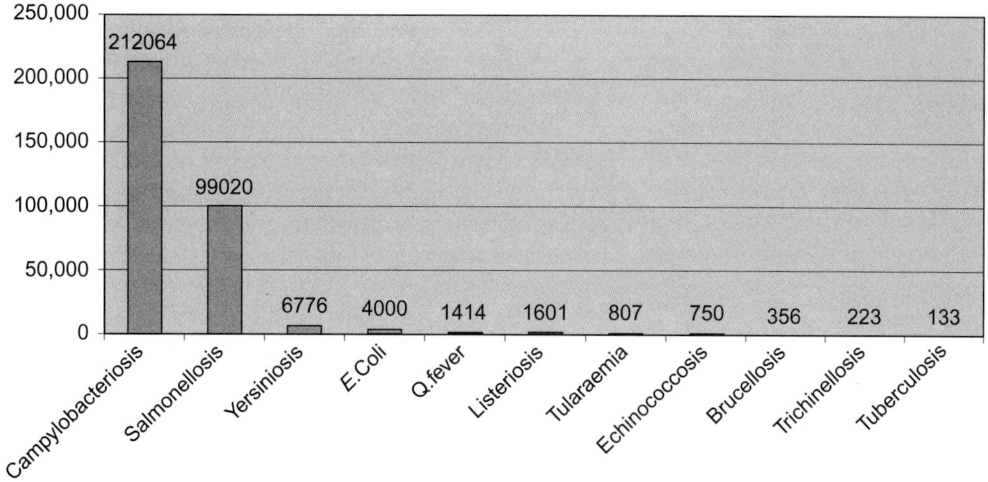

FIGURE 23.1 Number of zoonosis and foodborne cases in humans, EU countries 2010 *(data source: EFSA, 2012).*

(viruses, parasites and bacteria) can cause illness and currently more than 200 known diseases are recognized as foodborne. Contamination of food may occur at any stage in the process from "farm to fork" and can result from environmental contamination, including pollution of water, soil or air.

Despite remarkable advances in food science and technology, foodborne illness is a rising cause of morbidity in all countries and the list of potential foodborne microbial pathogens keeps increasing.

Up to 30% of the population in industrialized countries may be affected by foodborne illness each year. The global incidence is difficult to estimate, but in 1998 more than 2.2 million people, including 1.8 million children, died from foodborne diseases. In 2008, the US Center for Disease Prevention and Control (CDC) estimated that foodborne diseases caused approximately 76 million illnesses, 325,000 hospitalizations and 5,000 deaths each year. (Oliver et al., 2009).

There are significant microbiological risks associated with primary production. A wide range of agricultural products can become contaminated with microorganisms, including human pathogens. Some of these pathogenic groups come from soil and water, but for some of them, animals or humans are reservoirs from which they spread. Pathogens that live on farms are directly or indirectly recognized as risk factors in the entire commercial food chain (Tauxe, 1997). The most common foods that caused food poisoning in 2010 are presented in Figure 23.2.

EXAMPLES OF FOODBORNE PATHOGENS

Pathogens are the leading causes of foodborne morbidity and mortality. Dairy and beef cattle can harbor and shed *E. coli. Campylobacter jejuni, L. monocytogenes* and *Salmonella* spp. are carried by cattle, poultry and swine and are found in their associated farm environments

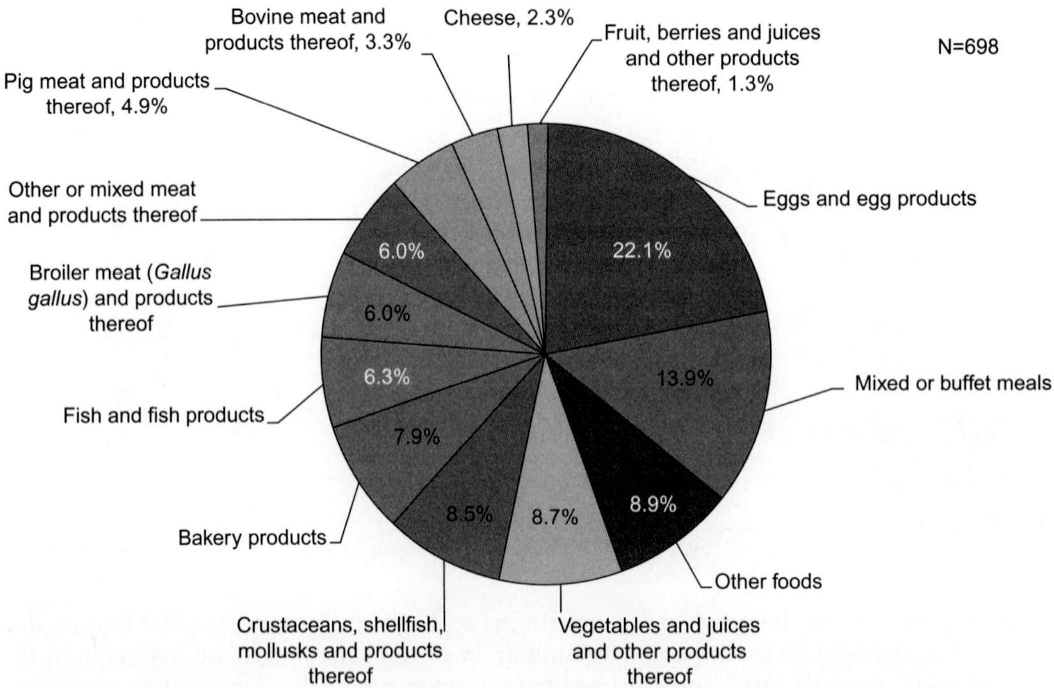

FIGURE 23.2 Distribution of outbreaks caused by food (EFSA, 2012).

(McEwen and Fedorka-Cray, 2002). *Staphylococcus aureus*, *Clostridium perfringens* and *Bacillus cereus* are also important pathogens that have origins on farms. The *Streptococcus suis* encountered in swine production is now recognized as a human pathogen. Viruses such as norovirus and hepatitis E, and parasites as *Cryptosporidium purvum* and *Toxoplasma gondii* are encountered in the farm environment and considered as human pathogens (Tauxe, 2002).

These pathogens are found in animal feces; therefore, contamination of food products by animal feces is likely to be a principal mode by which foodborne pathogens reach the consumer. Wild birds and various mammals that are common in farm environments can also be a source of these pathogens.

From the standpoint of pre-harvest food safety in general and human health in particular, *Salmonella* spp., *E. coli*, *Campylobacter jejuni* and *Listeria monocytogenes* are the most important foodborne pathogens affecting public health (Bean and Griffin, 1990).

Salmonella spp.

Salmonella spp. is the most commonly reported cause of human foodborne diseases. *Salmonella* spp. lives in the intestinal tract of various animal species and can be present on farms with absence of clinical disease. Healthy animals can become carriers and can shed *Salmonella* for long periods. Humans become infected if they consume animal products or water contaminated with feces, but direct contact with infected animals can also be a source

TABLE 23.2 Percent of Samples Positive for *Salmonella* in Various Animal Products in EU Countries during 2010

Animal Product	No. Tested Samples	% of Positive
Fresh broiler meat	21,539	4.8
RTE broiler meat	3253	0.3
Fresh turkey meat	4329	9.0
Egg and egg products	19,142	0.3
Fresh pig meat	69,005	0.9
RTE minced pig meat	11,675	0.6
Fresh bovine meat	34,236	0.2
RTE minced bovine meat	3299	0.4
Raw milk, pasteurized milk	7825	5 cases
Cheeses	34,109	0.1

Data Source: EFSA, 2012

of contamination, especially for farm families. Although a great percentage of human salmonellosis occurs through consumption of raw milk or dairy products manufactured with raw milk, human illnesses are frequently linked with consumption of poultry and pork products (Besser et al., 2000).

In beef cattle, *Salmonella* was detected in 38 of 100 feedlots (Fedorka-Cray et al., 1998). In swine farms, on 58 of 152 farms, *Salmonella* was detected in 20% of broiler carcasses and 45% of ground chicken meat (Rabsch et al., 2003). Table 23.2 shows the percent of samples positive for *Salmonella* in various animal products in EU countries during 2010.

Escherichia coli

Several strains of *E. coli* cause a variety of diseases in humans and animals. *Escherichia coli* O157:H7, also called enterohemorrhagic *E. coli*, is a type associated with a particularly severe form of human disease as hemorrhagic colitis, hemolytic uremic syndrome and thrombotic thromocytopenic purpura. The majority of human outbreaks caused by *E. coli* O157:H7 were linked to the consumption of contaminated meat and raw milk (Dorn, 1993). Sources of contamination also include: feces from infected animals, use of contaminated manure as fertilizer, fecal contamination of meat at slaughter plants, raw manure and slurry from dairy farms, and cross-contamination of other food products at farm (Tarr, 1995; Banatvala et al., 1996). Cattle are currently considered a reservoir for *E. coli* O157:H7, and cattle manure is an important vehicle for spreading contamination, but this pathogen is also detected in sheep, goats, horses, dogs, reindeer, deer, birds and rabbits (Hancock et al., 1998). In 2010, the total number of confirmed human *E. coli* cases in the EU was 4000 and almost half of the reported were serogroup O157 (41.1%). Each year in the United States, approximately 265,000 cases of *E. coli* are detected, and about 36% are O157 serotype

TABLE 23.3 Percent of Samples Positive for *Salmonella* in Various Animal Products in EU Countries during 2010

Animal Product	No. Tested Samples	% of Positive
Fresh bovine meat	21,539	4.8
Raw cow milk	3253	0.3
Fresh ovine and goat meat	4329	9.0
Milk and dairy products excluding cow milk	19,142	0.3

Data Source: EFSA, 2012

(www.cdc.gov). Table 23.3 shows the percent of samples positive for *Salmonella* in various animal products in EU countries during 2010.

Campylobacter spp.

Campylobacter jejuni and *Campylobacter coli* are the most frequently identified cause of acute infectious diarrhea in developed countries and the most commonly isolated bacterial intestinal human pathogens. Between 2 and 4 million cases of campylobacteriosis occur each year in the USA, and *Campylobacter* is associated with 120 to 360 deaths (Fahey et al., 1995). Several zoonotic sources have been identified, and *C. jejuni* has been isolated from cattle, swine, poultry, dogs, cats, birds, ferrets, hamsters, wild birds, mule deer and houseflies (Altekruse, 1994). Poultry meat products are the most common foodborne source of *Campylobacter* infection in humans (Vugia et al., 2007). Symptoms are chronic gastritis, enterocolitis and septicemia. Humans become infected by ingesting contaminated foods, untreated water or contaminated nonpasteurized or improperly pasteurized milk. In 2010, *Campylobacter* continued to be the most commonly reported gastrointestinal bacterial pathogen in humans in the EU since 2005. In 2010, 266 deaths were reported due to campylobacteriosis (reported for $N = 115{,}747$). According to the US Centers for Disease Control and Prevention, during 2008 there were 2.4 million people contaminated with *Campylobacter* (www.cdc.gov). Table 23.4 shows the percent of samples positive for *Campylobacter* in different animal products in EU countries during 2010.

Listeria Monocytogenes

Listeria is a serious foodborne illness in humans (listeriosis). It is dangerous primarily for pregnant women and their fetuses, the elderly and the immunocompromised. The biggest public health concern is that it can develop resistance to antimicrobials. *Listeria monocytogenes* is an environmental contaminant whose primary means of transmission to humans is through food, which can become contaminated during production and processing. Ready-to-eat (RTE) foods that are refrigerated before consumption and do not receive substantial treatment, such as soft cheese, RTE meats, and RTE seafoods, have been implicated in outbreaks of listeriosis (Kathariou, 2002).

TABLE 23.4 Percent of Samples Positive for *Campylobacter* in Different Animal Products in EU Countries during 2010

Animal Product	No. Tested Samples	% of Positive
Fresh poultry meat	7413	29.6
Fresh pig meat	932	0.6
Fresh bovine meat	808	0.4
RTE bovine meat product	610	1.8
RTE minced turkey	142	0.7
RTE minced pig meat	289	0.0
Cow milk	1993	1.3
Cheeses	384	1.0

Data Source: EFSA, 2012

Listeria spp. is widespread in nature, can live naturally in plants and soil environments, and grows in a wide range of temperatures and pH (Bunning et al., 1988). This adaptability enables *Listeria* to grow in refrigerated raw milk, but can also survive high-temperature, short-time pasteurization (HTST). Human contamination occurs through consumption of raw milk or products manufactured with raw milk. In dairy and beef units infection of animals occurs through ingestion of contaminated feed, especially low-quality and spoiled silage. Healthy but infected animals shed *Listeria* in feces, and fecal contamination of pastures or vegetables was also implicated as a source of contamination for humans and ruminants (Murinda et al. 2004).

Control measures should be aimed at the farm and food-processing level, in order to prevent contamination of food products. Preventive measures include providing appropriate information for consumers on how to minimize the risk of ingesting food contaminated by *Listeria*.

In 2010, there were 1601 confirmed human cases of listeriosis in EU countries. In the USA, 24 confirmed listeriosis outbreaks were reported between 1998 and 2008, resulting in 359 illnesses, 215 hospitalizations and 38 deaths (www.cdc.gov). Table 23.5 shows the percent of samples positive for *Listeria* spp. in various animal products in EU countries during 2010.

Brucella abortus

Brucella spp. is also known as "contagious abortion." It is caused by infection with the bacterium *Brucella abortus*. Brucellosis infection of cattle causes abortion or premature calving of infected animals, most often between the fifth and eighth month of pregnancy. Although most countries have federal and state regulations for controlling this disease, it is still a threat. Brucellosis is spread from the vaginal discharge of an infected cow or from an aborted fetus. Breeding bulls that are infected can also transmit the disease to cows with infected semen. Milk produced by an infected cow may also harbor the organism. Such

TABLE 23.5 Percent of Samples Positive for *Listeria* spp. in Various Animal Products in EU Countries during 2010

Animal Product	No. Tested Samples	% of Positive
RTE broiler meat	1450	1.5
RTE pig meat	22,158	2.0
RTE poultry meat	3636	1.5
Soft/semi-soft cheese, raw cow milk	1674	0.3
Soft/semi-soft cheese, raw sheep and goat milk	865	0.8
Soft/semi-soft cheese, pasteurized cow milk	5548	0.9
Soft/semi-soft cheese, pasteurized sheep and goat milk	458	0.2
Hard cheese, raw cow milk	1024	0.4
Hard cheese, raw sheep and goat milk	303	0.0
Hard cheese, pasteurized cow milk	8029	0.3
Hard cheese, pasteurized sheep and goat milk	585	0.5
RTE fish and fishery products	2938	6.0

Data Source: EFSA, 2012

infected milk is a public health hazard as this is the organism that causes undulant fever in humans. There is no treatment for brucellosis. Prevention of brucellosis is vaccination of heifer calves. Although brucellosis has been eradicated from cattle in most countries, outbreaks still appear from time to time. According to the EFSA report for 2012, 10 countries with 356 confirmed cases have been reported (EFSA, 2012). So, even though the brucellosis vaccination is mandatory, there are still cases of this disease in Europe.

Helminths

Internal parasites, as tapeworms, lungworms or liver flukes, can cause significant diseases, but also economic losses at farms. Older animals that are frequently exposed to the parasites have a certain degree of immunity, but young animals are very susceptible. Adult worms that live in the animal body produce eggs that are passed in the manure. The eggs hatch, producing larvae that develop and move up onto the pasture grasses where animals consume them. These eggs are very resistant and durable. They can survive the winter and hatch out with warm weather. The most important step in the fight against worms is deworming of animals using anthelmintics before grazing season starts. There are several approved anthelmintics available such as paste, suspensions, granules, injectable or pour-on formulations, boluses or crumbles for oral use, and drench form. However, it is very important to consult a veterinarian concerning the type to use and the timing. It is also necessary to mention one of the most common internal parasites, *Trichinella*.

Trichinella spiralis is a concern when good animal husbandry is not respected. Many animals may act as reservoirs, but the most frequently involved in cases of human infections are pigs, horses and wild boars. Infested animals harbor larvae encysted in their muscles, and consumption of raw or undercooked meat products may lead to disease. After an incubation phase of about 24–48 hours, fever and intestinal symptoms may appear. A week after infection, larvae starts invasion of the muscles, followed by muscle aches and fever. Depending on the number of viable larvae consumed, symptoms will vary from hardly any to extremely severe or even fatal.

Trichinosis prevention is based on accurate, mandatory inspection of all slaughtered pigs and horses. The number of reported trichinellosis cases in humans in 2010 was 394 with 223 of them confirmed. More than 211 million tests of pigs were provided, in which 199 were positive; 36,871 wild boars tested with 26 positive; 724,640 hunted wild boar tested with 988 positive; 9569 foxes tested with 108 positive; 589 bears tested with 28 positive; 208 raccoon dogs tested with 58 positive; and 2760 other wild animals tested with 99 positive (EFSA, 2012).

Other Animal Infections

From the perspective of animal health and welfare and/or lost productivity, a range of animal infections are of concern to the food industry. Examples are scrapie, blackleg, foot rot, infestation with ticks, lice, horn flies, face flies and stable flies, etc. Some others, such as the foot-and-mouth disease, disrupt international trade in food. Certain infections, e.g. anthrax, are also an occupational disease in humans. Infections, such as leptospirosis can be the source of contamination of the environment, e.g. water supply and food and thus an indirect source of infections of human.

Control of Pathogens on Farms

The cycle of infections in animals begins with exposure to pathogens via contaminated feed, water and other environments, followed by amplification in animal hosts and fecal dissemination in the farm. Shedding of foodborne pathogens in feces and distribution in the environment where food-producing animals live leads to animal reinfection and persistence of the pathogen on the farm. This cycle makes animals constant reservoirs of foodborne pathogens. By breaking the infection–reinfection cycle, it is possible to reduce foodborne pathogen shedding and the spread of foodborne pathogens among food-producing animals and in the farm environment (Oliver et al., 2009).

Management of manure, including feces, urine and other animal secretions or excretions such as saliva, is central for the control of contamination in food-producing animals.

GOOD FARMING PRACTICES FOR ANIMAL HUSBANDRY

Biological, chemical and physical hazards may enter food-producing animals or animal production through a wide variety of exposure points in the food chain. To address the hazards, OIE recommends practices that include general farm and animal health management; veterinary medicines and biologics; animal feeding and watering; environment and infrastructure; and animal and product handling (FAO/OIE, 2009). We will review some of these practices here.

Livestock Production

For success in livestock production farm management is essential. Very important issues for successful livestock production are (FAO/IDF, 2011):

- **Farm location** – Farms should be located in an appropriate area for good animal rearing. The environment of the farm must have minimal risks from physical, chemical and biological hazards that may affect the sanitary of animals or on-farm products. The following locations are unsuitable for animal husbandry: industrial environments, waste disposal sites, slaughterhouses, live animal markets and other farms.
- **Farm layout** – Farms should have sufficient space of suitable size, adjusted to the race and category of animals reared and designed to avoid any problem to the environment and animal health. The housing area for different categories of animals and storage area for feed and veterinary drugs should be separated and protected from pests, pets and other domestic animals that may be disease carriers. It is preferable that the farm has an open area with sufficient air flow and an appropriate pasture area with adequate shade, if necessary. Accommodation for staff and office should be located in a residential area, distinctly segregated from the rearing area.
- **Facilities** – Facilities in which the animals are bred must have enough surface area so that animals can practice their natural movements. They should be constructed using durable materials that are easy to clean and maintain. The floor should be made using non-slippery concrete and be slightly tilted with good drainage to prevent waste accumulation within the facility. For good ventilation it is necessary to elevate the roof of the facility as much as possible (e.g. for cattle at least 3 m). Appropriate and adequate water supply should be available in each facility. The facility must have sufficient daylight as well as sufficient artificial lighting for animal caring and health checks at all times. Equipment and tools for farm operation should be in good condition, adequate, easily cleaned and operated, separately stored and not cause harm to animals.
- **Feed** – Feed must be of good quality, whether produced on the farm or purchased elsewhere. Feed containers must be clean, dry, in good condition and free from contaminants. Vehicles used for feed delivery should be cleaned and dried after each use. It is necessary to check physical and chemical properties of feed, especially potential presence of molds and fungi. These microorganisms have the ability to produce mycotoxins that can be very harmful for animals and humans. Thus, such contaminated feed must be rejected and eliminated in an appropriate way. The feed should be stored in a special room that is clean, dry and free from insects, rodents and other animals.
- **Water** – Animals must have free access to fresh and clean water throughout the day and in quantities that meet their needs. Water sources should be protected from contamination by animal manure or waste water from the farm. Also, water sources should be located away from pollutants outside the farm, such as from garbage dumps, slaughterhouses or factories. If the source of water at the farm is an artesian well, it should be adequately covered and protected from atmospheric phenomena. Water for

washing and cleaning, especially water directly used to wash the animals, should be of good quality. Water containers should be clean.

- **Farm staff** – The number of farm staff should be adjusted to the size of the farm, planned daily, taking into account seasonal activities, type of housing, animal rearing system, equipment and other facilities available on the farm. All staff should have the required knowledge and skills for their tasks. Every person working on the farm should have an annual health check-up and follow good personal hygiene practices, i.e. dressing with clean clothes, washing and drying hands every time prior to any operation, and keeping hands and nails clean. It is recommended to have an assigned veterinarian responsible for animal health who can supervise animals and give proper advice on disease prevention, treatment and correct drug usage.

ANIMAL HEALTH

Prevention and Control of Diseases

To control diseases it is necessary to have means to prevent access of pathogens or their spread on the farm.

Animal Treatment

Veterinary drugs, hazardous substances and disease treatments should be under the supervision of a veterinarian. The veterinarian should recommend animal drugs and treatments with a written prescription and record all activities. Veterinarians must take into consideration the withdrawal period for drugs, defined as the interval between the time of last administration of the drug and the time when the animal can be safely slaughtered for food purposes.

BOX 23.1

PRINCIPLES FOR GOOD MANAGEMENT OF ANIMAL HEALTH

- Minimize the risk of infections through proper nutrition, grazing and housing
- Maintain hygiene of livestock, housing facilities and feed storages
- Use disinfection barriers at entrance at farm and every facility on farm
- Limit entrance to essential visitors, authorized persons and vehicles only
- Routinely test animals for specific diseases
- Carry out vaccination against specific diseases
- Carry out regular treatment for elimination of internal and external parasites
- Prevent and control pests
- Act in accordance with veterinarian advice
- Purchase, store and use only approved veterinary products
- Adequately care for injured and sick animals
- Keep accurate records of all diseases, treatments and mortality of livestock

BOX 23.2

PRINCIPLES FOR GOOD MANAGEMENT OF ANIMAL WELFARE

- Availability of feed and clean water to animals at all times
- Provide the minimum required space per animal
- Respect the maximum allowable number of animals per unit area
- Keep animals in appropriate social groups
- Avoid animal isolation, except in cases of injuries and illnesses
- Handle animals carefully and avoid use of instruments, e.g. electric scissors
- Avoid non-therapeutic and radical measures, e.g. cutting tails, beak, etc.
- Provide minimum transport and exposure to markets and animal shows

Management Procedure for Dead Animals

Infected carcasses must be destroyed in a way that prevents spreading the disease. The veterinarian should advise how to dispose of carcasses. Many countries have facilities to professionally collect, quarantine and incinerate animals that die from contagious diseases. If buried on the farm, proper disinfectants should be poured or scattered over every part of the carcass and the pit should be filled and piled up above the ground by at least 50 cm.

Animal Welfare

Animals are sensitive living beings. Their welfare must be taken into consideration (Pejanović, 2008).

PRINCIPLES OF BIOSECURITY

Biosecurity is a set of measures that are necessary to apply to keep diseases out of farms, herds and groups of animals, or to limit the spread of disease within the herd. Biosafety is one of the protective weapons for excluding pathogens from the animal's environment. The producers, herd owners and breeders are the most responsible subjects for implementing biosecurity principles on farms. The greatest risks on farms are imported new animals, farm visitors, wildlife, equipment and vehicles (David W. Snively http://www.wvu.edu/~agexten/Biosecure/Farm.pdf; Bowman and Shulaw, 2001; Gary et.al., 2001).

New Animals on a Farm

New animals on a farm present the greatest risk of introducing infectious disease. It is desirable that the breeder purchases animals from farms that have developed procedures for animal health protection. For the purpose of avoiding spreading diseases, breeders should:

- Isolate new animals and animals returning from situations where they have been exposed to other animals, such as at fairs or shows, for a minimum of 2 weeks.
- Isolate animals showing signs of disease.
- Isolate animals in a facility separate from other animals.
- If complete isolation is not possible, provide separate pen or pasture that does not permit nose-to-nose contact or use shared feed/water supplies.
- Provide parasite control and vaccination against diseases likely to be a problem on farm.
- Do all appropriate tests and treatments under the control of a veterinarian.

Farm Visitors

Any visitor that enters the farm is a potential carrier of a disease. But not all visitors present the same level of risk.

Visitors from urban areas or others who have no contact with livestock present very little risk of carrying relevant diseases. Measures that should be taken for such visitors are:

- Wear freshly laundered outerwear and clean shoes or boots.
- Ideally provide disposable plastic boots and coveralls.
- Provide disinfectant footbaths and immerse shoes or boots in disinfectant for the adequate contact time.
- Forbid visitors to enter pens or feeding areas, to contact animals, or to bring food with them.
- When visitors leave, dispose of plastic boots in a safe way and ask visitors to wash their hands.

Salespeople, delivery people, mechanics and those who routinely visit farms but have little or no contact with animals should be subjected to the above as well as the following additional procedures:

- Wear clean or disposable coveralls and boots if there is any contact with feed, animals, soil or manure.
- Ensure personal equipment and tools are cleaned and disinfected between uses if there is any contact with feed, animals, soil or manure.
- At the end of the visit, clean and disinfect dirty boots, and remove and place coveralls in a plastic bag before visitors re-enter their vehicles.

Veterinarians, livestock-owning neighbors and anyone else who has close contact with animals and their bodily discharges are visitors presenting the greatest risk. In addition to all of the above precautions, people in this group should observe the following:

- All vehicles that enter the farm should be cleaned, free of visible dirt on tires and wheel wells, and should be disinfected prior to arriving at the farm.
- Visitors should arrive with clean clothing, boots and equipment.
- Equipment and instruments that have direct contact with animals should be cleaned and disinfected before and after use.

- Any disposable disinfectable clothing, such as sleeves and gloves, should be worn whenever there is direct contact with animal discharges or tissues.
- Dirty equipment and footwear should be cleaned and disinfected with an appropriate disinfectant before leaving the farm.
- Soiled coveralls should be removed before people re-enter their vehicles.
- Hands and forearms should be washed with antibacterial soap.

Risk from Wildlife

The presence of wild animals in a farm area should not cause alarm. However, some diseases such as rabies, leptospirosis and salmonellosis can be carried and spread by some species of wildlife including rats and mice. It is necessary to take efforts to make barnyards and buildings unattractive to wildlife by:

- Cleaning up grain spills and other sources of food.
- Cleaning up old board piles or debris piles.
- Inspecting buildings for possible hiding or denning areas.
- Inspecting hay and feed storage areas for presence of animals like cats, dogs and rats.

Risk from Farm Equipment

Equipment that has been in contact with livestock or their bodily secretions can spread diseases. Also, equipment moved from one farm unit can carry heavy pathogenic contamination to another farm unit, if not thoroughly cleaned. To help minimize this risk, it is necessary to include farm equipment in the biosecurity plan and provide activities such as:

- Remove all gross organic soiling from equipment and tools because high levels of soiling reduce the efficacy of the cleaning and disinfection process.
- Do not share manure-hauling equipment between farms unless it is thoroughly cleaned and disinfected.
- Clean and disinfect front, buckets and skid steer loaders used for manure or feed handling between each use.
- Soak and scrub equipment in a tank or pressure wash with a detergent sanitizer.
- Store used equipment where it will not be recontaminated.

Risk from Vehicles

Vehicles used for livestock haulage, feed trucks and any vehicles for animal transport are excellent vectors for disease spreading. Cleaning and washing must be carefully and thoroughly done. Procedures to reduce the potential for disease transmission are:

- Only essential vehicles may enter the farm.
- If possible, vehicles visiting the farm should be kept outside the biosecurity perimeter.
- Vehicles, especially wheels, tires and wheel arches should be cleaned and disinfected upon arrival at the farm using wheel dips or sprays.
- Personnel should use foot dips, protective clothing and observe hygiene requirements prior to entry to the premises.

GOOD HYGIENE PRACTICES ON THE LIVESTOCK FARMS

Good hygiene practices can be described as a set of procedures that provide a clean, sanitary environment for the production, processing and storage of feed and animal products. In other words, good hygiene practice determines what needs to be done regarding cleaning and hygiene, as well as when and who should carry out these tasks.

Because of the great diversity in the structure of today's farms (type, number, productivity), the emphasis has to be on prevention and crisis control, but primarily it should be on hygiene at the farm. Good farm hygiene seeks to minimize noxious external stressors that lead either to acute disease or to the exacerbation of chronic disease.

During the last decades of the 20th century, there was an increase in the number of infectious diseases that evolved into an epidemic. Almost every outbreak was associated with a lack of implementation of biosecurity and/or hygiene procedures on infected farms. Since then, hygiene has become the primary tool in the health care program.

It is very important that correct hygiene measures are routine in everyday activities.

However, before a farmer starts with the application of hygiene measures, such as cleaning and disinfection of facilities, equipment and tools, it is necessary to provide conditions on the farm that will allow hygienic measures to be effective. Attention must be given to:

- Healthy soil: hygiene must start with the soils on the farm. Healthy soil means healthy and nutritious crops.
- Manure and waste management: to reduce the risk of spreading microorganisms across the farm, special attention should be paid to manure and waste originating from sick animals that should be destroyed in a way matching principles of environmental protection.
- Grazing and harvesting programs: agro/technical measures, such as tillage and planting methods, have a direct impact on soil quality.
- Plant should be selected according to the micro-environment of the farm.
- Livestock should be selected according to the environment on the farm as well as the production and management systems.
- Water supply: sufficient quantities of clean water must be accessible to animals at all times. Slop-basins should be easily cleaned and protected from any kind of direct or indirect contact with animal excrements or animal and farm waste.
- Barns must have adequate space, water supply, ventilation and light at all times when the animals use it. If animals are kept on pasture or used outlets, the farmer must also provide adequate shelter from sun, wind, rain or snow, and those shelters must be available at all times.
- Equipment and tools must be sized for the animals being worked and must be cleaned and disinfected after each use. This is especially necessary for dairy farm equipment and tools.
- Regular vaccination is necessary to conduct in accordance with State regulations, and emergency vaccination must be provided in case of uncontrolled outbreaks.

Only if these conditions are achieved can hygiene procedures can be efficient. Maintaining hygiene on the farm should be a team approach, not the sole responsibility of one worker, especially at big farms.

Cleaning

Cleaning is one of the most important activities for disease control on a farm. Equipment, facilities, machinery, tools etc., always retain an amount of feed, litter or manure on the surface after use. Beside the corrosion that these substances can cause, they are an ideal medium for microbial growth.

Therefore, the aim of cleaning must be complete removal of manure, litter or feed, by washing, scrubbing and rinsing, or pressure washing with hot water and detergent from all kinds of surfaces, done in a dedicated separate facility to avoid spread of contaminated dirt. This is difficult to accomplish in barns with wooden walls, dirt floors, open ceilings and lack of drains. This is why, when building a new barn, it is necessary to choose a design and materials that will make cleaning easier. In barns with sand or other porous floors, it is easier and more efficient to replace sand or clay than to thoroughly clean it. Cleaning can be dry and wet.

Dry cleaning is the physical removal of manure, litter, feed and other animal wastes. The disadvantage of dry cleaning is that infective material together with dust will rise and float in the air, and after some time will cover already cleaned areas. Dry cleaning is suitable only if it represents a preparatory phase before wet cleaning. Tools for dry cleaning can be numerous types of brushes, brooms, shovels, pitchforks, etc.

Wet cleaning means using cold, temperate or hot water with or without detergents. With cold or temperate, and especially with hot water, it is possible to remove almost all organic materials from surfaces, but there is still the possibility that a so-called "organic film" will remain. To remove all traces of organic matter it is necessary to use detergents in combination with temperate and hot water.

Detergents are compounds or mixtures of compounds, organic or inorganic, used for cleaning. These are compounds that should not have an adverse effect on human health and not cause corrosion of equipment, tools, walls and floors. In general, detergents can be divided into:

- Inorganic bases and their salts:
  - sodium hydroxide, sulfates, carbonates, phosphates and silicates.
- Organic and inorganic acids:
  - nitric, phosphoric, sulfamic, citric, hydroxyl acetic, gluconic and tartaric acids.

Cleaning by using products with good detergent capacity will remove soil from the walls and floors and ensure that dirty deposits will not remain on rough surfaces, e.g. concrete and wood. Detergents also reduce the time needed to clean by up to 60%, and reduce the spread of disease in washing water.

The best way for wet cleaning with detergents is to apply a mixture using a knapsack sprayer or pressure washer. The pressure washer should be set on a low pressure setting of approximately 35 bars, and a stream of mixture should fall to the surface that should be cleaned at an angle of 45 degrees. It is also important to use the appropriate application rate of applied mixture. If it is a normal liquid mixture, it should be $500\,ml/m^2$, or if it is a foam mixture it should be $250\,ml/m^2$. It is very important to start cleaning from the apex of the roof and work down the walls to the floor, paying particular attention to corners and other areas where dirt accumulates. The same procedure, from the top to the bottom, should be used to clean equipment or tools. Caked soiling should be brushed if necessary to aid removal.

Special attention should be paid to all water systems, most importantly the water supply systems for the animals. Water systems are likely to contain some bacterial contamination. This may enable diseases to pass from animal to animal or from one batch of animals to the next unless the bacterial growth is eliminated. To eliminate bacteria from water systems, it is necessary to:

- Drain
- Remove dirt from the system
- Refill the system with water
- Clean using a detergent
- Drain again
- Refill and add a disinfectant
- Leave in the system for 10 minutes
- Drain the system to remove all disinfectant
- Flush with fresh clean water
- Refill with fresh water

Disinfection

It is necessary to emphasize that dry or wet cleaning, even cleaning with detergent, can remove mechanical impurities and organic matter, but it is not possible to remove all microorganisms and their spores. In addition to cleaning and washing, to destroy microorganisms, it is necessary to implement methods of disinfection (sanitation) on farm facilities, equipment and tools. Disinfection is using specialized cleansing techniques that destroy or prevent growth of organisms capable of infection (Stojanović et al., 2003).

There are two methods of disinfection:

- Disinfection by physical means.
- Disinfection by chemical substances.

Disinfection by Physical Means

Physical means can be heat, gamma-radiation and UV radiation. Different forms of heat can be used such as wet heat (steam, hot water, boiling solution of acids and bases) and dry heat (hot air and UV radiation).

DISINFECTION BY WET HEAT

- Steam is used mainly for autoclaving and disinfection of finer equipment and milk pipes. It is suitable for application in the dairy but not for the farm. Usually the required disinfection is obtained if the temperature of the steam is 115°C at a pressure of 0.7 bar for 3–5 minutes.
- Hot water quickly destroys vegetative forms of microbes, but not their spores. A good disinfection for a dairy farm is a water temperature of 77°C for 10 minutes. A good effect can also be obtained with a CIP (Cleaning in Place) system with water of 90–95°C for 5–7 minutes. The disadvantage of hot water is that it can cause burns and may be

expensive due to the cost of the volumes of water needed to achieve disinfection and the heating of the water. On the other hand, hot water, properly used, will kill most microorganisms and is non-corrosive.
- Boiling solution of acids and bases – usage of these types of disinfectants is similar to the previous, but effects are much better, especially if the solution temperature is higher than 77°C.

DISINFECTION BY DRY HEAT

- Hot air is less used on farms. Mainly it is used in laboratories for disinfection of laboratory glassware and equipment in dry sterilizers.
- UV radiation is used to sterilize small enclosed areas such as laboratories and chambers, milking machines, milking parlors and dairies and so on. It works on surfaces that are exposed to the radiation.

Disinfection by Chemical Substances

The presence of organic material, including bedding, manure, blood and pus, interferes with the action of most disinfection methods. This is the reason that prior to implementation of the disinfection method by chemical substances, it is necessary to clean and wash the surface that should be disinfected. After cleaning and washing it is good to allow the surface to dry, if possible. Thorough cleaning, washing and drying will remove most of the contamination and allow disinfectants to contact the surfaces and kill the microorganisms.

Disinfectants

A disinfectant is a chemical or other substance that kills microorganisms but may not kill bacterial spores, which are a dormant form of some bacteria, e.g. *Clostridia* spp. They can be applied to objects, equipment or tools. Because of their potentially toxic, irritating or corrosive properties, most disinfectants cannot be applied directly to living animals or people.

There are products that are both detergent and disinfectant. Because these substances may be toxic, farmers must strictly follow the instructions on the label before use. It is very important to pay attention to:

- Dilution rate, either as a germicidal cleaner (killing microorganisms) or as a sanitizer (reducing the number of microorganisms).
- Minimum contact time, the time required to kill microorganisms.

BOX 23.3

REQUIREMENTS FOR A GOOD DISINFECTANT

- Effectively destroys microbes
- Not toxic
- Free of smell and taste
- Not harmful to human skin and materials
- Easy to use
- Efficient
- Not expensive

This information is normally stated on the label. But the dilution rate and minimum contact time depends also on:

- Presence of organic matter.
- Temperature, pH and hardness of water.
- Concentration of disinfectant.

The main group of disinfectants and their characteristics are (Stojanović et al., 2003):

ALCOHOLS

Alcohols are commonly used for cleaning equipment on the farm but do not kill bacterial spores. Isopropyl and ethanol alcohol are commonly used, mostly for disinfection of working surfaces and smaller equipment and tools.

CHLORINES/HYPOCHLORITES

Chlorine-based disinfectants have two major advantages: excellent broad-spectrum antimicrobial activity and low cost. The major disadvantages of chlorine sanitizers are that they are very corrosive to many materials of construction and they are easily inactivated by organic materials and soils. They are commonly used but care must be taken never to mix them with acids because toxic chlorine gas will be generated. They lose some of their activity above 80°C and work best in pH < 7.

CHLORHEXIDINES

Chlorhexidines are not active against all bacteria that are found on the skin or some viruses, but they can be used for washing animals or workers' hands. They are more efficient if they are used after cleaning, washing and rinsing of surfaces.

CARBOXYLIC ACID

The carboxylic acid sanitizers are also called fatty acid sanitizers. They are a combination of acidulants, such as phosphoric acid or citric acid with a fatty acid such as octanoic acid. They have the dual function. They develop acidity for rinsing and removing mineral films and killing microorganisms. They have good broad-spectrum activity and, because of their low foaming characteristics, are very good for CIP applications.

IODOPHORS/IODINE COMPLEXES

Iodine is not very soluble in water and can be inactivated by organic matter. Therefore, cleaning and washing with a detergent and rinsing with clean water is preferable before applying iodine. To improve its efficiency, a surfactant is mixed with the iodine to form a complex known as an "iodophor." A mineral acid such as phosphoric acid is added to this combination because iodine kills best at an acidic pH (pH < 7). Iodophors have very good broad-spectrum antimicrobial properties. They cannot be used at temperatures above 80°C. Because of their natural amber color, it is easy to see if iodine is present in the sanitizing solution. Iodophor solutions, shampoo or washes can be used for washing animals (udder dipping) and workers' hands.

PEROXY COMPOUNDS

Peroxy compounds are a combination of hydrogen peroxide with organic acids such as acetic acid. The resulting peracid is an excellent broad-spectrum disinfectant. An additional benefit is that it provides an acidified rinse to remove mineral films. A big advantage is its ability to kill microorganisms at temperatures as low as 4–5°C, which can be important on the farm in certain countries in the winter. Disadvantages include the fact that peroxy disinfectant loses effectiveness in water that contains iron at levels of 0.2 ppm and higher. Also, it will corrode soft metals such as brass and copper. Peroxy disinfectant is reported to be also effective against biofilms.

PHENOLS AND CRESOLS

Phenols and cresols work well in the presence of organic matter, e.g. foot baths. They work better at higher temperatures and they are most efficient at $pH < 7$. A disadvantage is that they are inhibited by hard water.

QUATERNARY AMMONIUM COMPOUNDS

Quaternary ammonium compounds are active against most bacteria, fungi and viruses but not against bacterial spores and some viruses. They work best at $pH > 7$, but are inhibited by hard water.

SODIUM HYDROXIDES (LYE)

Lye can be used as a whitewash or as a dry powder for disinfecting buildings. The usual concentration in water is 2%.

SULFATES

Sulfates are a multi-purpose disinfectant. They consist of potassium peroxymonosulfate, sodium dodecylbenzenesulfonate, sulfamic acid and inorganic buffers. They are typically used for cleaning up hazardous spills, disinfecting surfaces and rinsing equipment. The solution is used in many areas where control of pathogens is required. Sulfates have a wide spectrum of activity against viruses, some fungi and bacteria. However, they are less effective against spores and fungi than some alternative disinfectants. They are sold as tablets or powders which dissolve readily in water. They should be mixed with water to form a 1–3% solution. They are colored, which is useful to gauge the concentration during preparation. Moreover, discoloration makes it obvious when they need to be replaced. The solution disinfectant does not cause skin irritation/corrosion, but can cause eye damage and should not be used as a hand-washing liquid. They work well under all circumstances and are well known for their detergent properties.

The Cleaning and Disinfection Process

Equipment and surfaces that come into contact with food, inedible by-products and waste should be of a material that allows cleaning and disinfection. Also, it is important that the surfaces are in satisfactory condition and undamaged, because otherwise cracks and scratches can trap dirt and prevent successful disinfection.

Vehicles should also be disinfected. If they are used for the transportation of animals, disinfection should be done at the latest within 24 hours after unloading,

For disinfection of facilities, equipment and tools, especially those that are in direct contact with animals, use only approved disinfectants.

To ensure efficiency, it is important to follow strictly the directions for use of disinfectants.

Unless it is otherwise specified in the manufacturer's instructions, the cleaning and disinfection procedure consist of the following five steps:

1. Preliminary cleaning including brushing, scraping and deleting dirt and food residue and rinsing with clean water.
2. Next is the main cleaning, which consists of scrubbing the surface that was previously soaked with soapy water in order to remove dissolved residual dirt; scrubbing must be thorough.
3. Washing with water to remove the detergent and dissolved dirt and food residue.
4. Applying disinfectant to cleaned surfaces.
5. Thoroughly rinsing with water.

In order to avoid the contamination of food, animals or people, chemical products for cleaning and disinfection should be stored in a separate room used only for chemical storage. Particular attention should be paid to the cleaning equipment, which should also be regularly cleaned and disinfected; otherwise they will become a source of cross-contamination. Cleaning equipment should be kept in a separate room, which also needs to be maintained and cleaned. Each piece of cleaning equipment should be used only in certain areas, in order to prevent the spread of contamination (e.g. broom to clean floors of toilets must not be used to clean the areas in which food is treated). Marking equipment with colors is one of the simplest ways to ensure good control over the purpose and location of certain equipment. The system to follow is: one color – one room.

How to Handle Disinfectants

As we mentioned before, some disinfectants can be extremely toxic. Therefore it is very important to observe the same rules.

BOX 23.4

RULES FOR HANDLING DISINFECTANTS

- Always use eye protection (eyeglasses)
- Use protection (rubber) gloves and special clothes (working coat)
- It is mandatory to read and understand all warnings and instructions written on labels or declaration of product, before using the product
- Products should always be used in well-ventilated areas
- Use only approved disinfectants
- Follow all instructions on labels or declaration of product as to how to make solutions, how to apply, etc.
- Do not mix disinfectants with any other disinfectants or chemicals
- Disinfectants must be stored in a separate, locked room where only those who work with disinfectants can be allowed entry
- Keep disinfectants away from other personnel, especially children!

HAZARD ANALYSIS AND CRITICAL CONTROL POINTS

In the modern approach to food safety, application of the HACCP system has been recommended as a complement to prerequisite programs (in this context, good animal husbandry) (CAC, 2004).

For various reasons, application of the HACCP system, with the same degree of stringency as applied in food processing and manufacturing, may be difficult at the farm level. This should nevertheless not exclude the possibility of using of the concept, following its adaptation, for a proactive and risk-based approach to management of risks at the farm level. Considering that there are multiple sources of animal infection and numerous risks of food contamination, a risk-based approach to identify measures which need particular surveillance is even more warranted.

Small and/or less developed businesses do not always have the resources and the necessary expertise on site for the development and implementation of an effective HACCP plan. It may be possible to develop HACCP-based codes of practices that include preventive controls following a hazard analysis on risks and practices at the farm.

PART 2: GOOD AGRICULTURAL PRACTICES FOR FOOD SAFETY

INTRODUCTION

Consumption of fresh vegetables and fruits is part of a healthy diet and recommended so as to prevent illnesses. Hence, an increased consumption of fresh vegetables and fruits in the world has been documented.

The number of human outbreaks of diseases and illnesses associated with the consumption of raw vegetables, however, has increased in recent years. In the United States, available foodborne illness outbreak data document 131 outbreaks associated with 20 different contaminated commodities between 1996 and 2010, causing more than 14,000 illnesses and 34 deaths (FDA, 2013a). *Salmonella*, *Shigella*, enterotoxigenic *E. coli* and *Escherichia coli* O157:H7, hepatitis A and *Cyclospora* have been linked to fresh tomatoes, lettuce, spinach, carrots, parsley, cantaloupe, berries, seed sprouts, etc. (Ackers et al., 1998; Beuchat, 2002; FDA, 2013a).

The most likely routes of contamination of produce from growing, harvesting and on-farm post-harvest activities are associated with water, soil amendments, animals, worker health and hygiene, buildings and equipment. It is very difficult to identify primary sources for contamination of fresh vegetables. The success of the detection of human pathogen bacteria on fresh fruits and vegetables depends on the methods applied and the nature of the contamination and sporadic contamination limits effectiveness testing. According to NACMCF (National Advisory Committee on Microbiological Criteria for Food, USA), for only two out of 27 human outbreaks, contaminated fresh produce has been identified as the source (National Advisory Committee on Microbiological Criteria for Foods, 1999).

It is known that *E. coli* O157:H7 could be transported from contaminated soil and irrigation water to lettuce leaves. Also, these bacteria can migrate throughout the lettuce plant (Solomon et al., 2002; Wachtel et al., 2002; Wang & Doyle, 1998). Guo et al. (2002) detected an association of *Salmonella* spp. with stems and leaves of tomato plants that were grown hydroponically in an inoculated solution.

Knowing about microbial ecosystems on/in raw vegetables helps to understand better the nature of microbial contamination of fresh produce. Survival of human pathogen bacteria on/in fresh vegetables depends on: pH, morphology, anatomy and metabolic functions of plant organs (fruits, flowers, leaves, roots). For example, the pH of many vegetables is 4.5 or higher and this value is appropriate for growing many human pathogen bacteria. Also, differences in morphological and anatomic properties of different plant organs ensure a wide range of ecological niches, which could be colonized by different species of human pathogenic bacteria (Brandl & Mandrell, 2002; Solomon et al., 2006).

Human pathogenic bacteria as well as non-pathogenic bacteria are able to form biofilms on the surface of raw vegetables. Biofilms have been detected on leaf surfaces of lettuce, cabbage, parsley, spinach, celery, etc. (Morris et al., 1997). These biofilms represent protective environments for human pathogenic bacteria and reduce the effect of sanitizers used for washing waters. Yet, further investigation of bacterial biofilms on the surface of raw vegetables is needed.

SOURCES OF MICROBIOLOGICAL CONTAMINATIONS OF FRESH VEGETABLES

Human pathogens can contaminate fresh vegetables at any point of the production chain. They may contaminate produce in the pre-harvest and post-harvest period (see Scheme 23.1). Pre-harvest sources of contaminations implicate soil, irrigation water (Kljujev and Raicevic, 2006), water for applying pesticides, inappropriate composted manure (Fukushima et al., 1999), feces, dust, insects, wild and domestic animals, and human handling (Beuchat, 1996).

SCHEME 1 SOURCES OF MICROBIOLOGICAL CONTAMINATION (BEUCHAT, 2002)

Post-harvest sources of contamination could be: harvesting equipment, processing equipment, transport containers, transport vehicles, rinse water, ice, as well as insects, wild and domestic animals, dust, feces, and human manipulation (Burnett and Beuchat, 2001).

Non-composted and improperly composted manure can contaminate raw vegetables if it is used for fertilizing growing plants (Fukushima et.al., 1999). Many human pathogens, like *Salmonella* spp., *E. coli* O157:H7 and *Listeria monocytogenes*, can be present in animal feces. Also, these pathogen bacteria may arrive on the growing vegetables through contaminated irrigation waters.

Recent data show that pathogenic bacteria, originated from irrigation water, can contaminate vegetables (Chalmers et al., 2000; FDA, 2013a). The potential risk of infecting humans by such contamination should be seen in the context of recommendations for the microbiological quality of irrigation water. Strategies to reduce the risk of causing human illness due to pathogenic bacteria in irrigation water are needed for producing safe and healthy food.

Wachtel et al. (2002) showed contamination of cabbage root by *E. coli* when plants were irrigated with contaminated wastewater, although the edible parts of the plants had not been treated with this water. Also, Islam et al. (2004) found the presence of *Salmonella Typhimurium* on carrot and radish if water, contaminated with *S. Typhimurium*, was used. They demonstrated that *S. Typhimurium* could survive in the soil for 203 days. Lettuce plants, irrigated with *E. coli* O157:H7-contaminated water, were positive for the presence of this pathogen during the harvest period, until 30 days after the last irrigation. After 7 and 14 days, a significant increase in the number of *E. coli* O157:H7 was detected (Solomon et al., 2002).

Quantitative models of risk assessment for using wastewater for irrigation show that the risks differ between different plants. Thus, it was noted that the risk is higher with lettuce than with cucumber, broccoli and cabbage (Hamilton et al., 2006). The time interval between irrigation and harvesting has an impact on the survival of pathogenic bacteria on plants and hence on the chance that they will reach the consumer. Some investigations in the UK showed that more than 50% of producers harvest and deliver leafy vegetables within 24 hour of the last irrigation.

MICROBIOLOGICAL QUALITY OF IRRIGATION WATER

Recent investigations show high variations in total coliform counts in stream water, closed wells, public drinking fountains, underground waters and channel waters. The degree of contaminations depends on the season of the year, the location and integrity of the wells and open channel waters (Kljujev, 2012; Dulic et al., 2008).

Open channel waters or surface water pose the highest potential for contamination and the greatest variability in quality of agricultural water sources (FDA, 2013a). Three years of investigation of microbial quality of channel water through its 54 km length until it enters the Danube River show variations according to the season and the location. At the point where the city sewage wastewater enters the channel the average coliform counts were extremely high, especially close to a pig farm that drains wastewater directly into the channel; the same happens when the channel runs near settlements, industrial areas, dairy farms and meat factories (Kljujev, 2012).

PRESENCE OF PATHOGENIC BACTERIA ON FRESH VEGETABLES

It has been documented that water that is applied directly to the harvestable portion of the plant is more likely to contaminate it. The proximity to the harvestable portion and the timing of water application in produce production before consumption are important factors in determining the likelihood of contamination (FDA, 2013a).

Field experiments confirmed that pathogenic bacteria could be transported from irrigation water to edible parts of vegetables, when waste and microbiologically incorrect water was used for irrigation. Microbiological analyses of edible parts of vegetables such as carrot, parsley, celery, cabbage, spring onion, tomato, pepper and cucumber showed the presence of pathogen bacteria species. *E. coli* was found at carrot root, spring onion, tomato and pepper fruits. The bacterial strain *E. coli* O157:H7 was identified at carrot and parsley roots and tomato fruit. *Salmonella* spp. was detected at parsley root, spring onion, tomato and pepper fruits (Table 23.6) (Kljujev, 2012; Kljujev et al., 2011; Kljujev et al., 2012).

TABLE 23.6 Presence of Pathogen Bacteria Species on the Edible Parts of Vegetables

	Pathogen Bacteria Species		
Plant Species	*E. coli*	*E. coli* O157:H7	*Salmonella* spp.
Carrot	+	+	–
Parsley	–	+	+
Celery	–	–	–
Cabbage	–	–	–
Spring onion	+	–	+
Tomato	+	+	+
Pepper	+	–	+
Cucumber	–	–	–

TRANSMISSION OF PATHOGENIC BACTERIA FROM CONTAMINATED IRRIGATION WATER AND SOIL TO PLANTS

Commodity type, growth characteristics and surface properties (porosity) affect the probability and degree of contamination. Research has indicated a big potential risk if contaminated water is used for irrigation of lettuce plants. Results showed that treatments with contaminated water resulted in significant levels of *E. coli* inside roots and leaves, whereas uninoculated controls were free of detectable *E. coli* contamination. The highest number was found on the surface of roots (4.1×10^5 CFU) and the lowest was inside leaves (5.0×10^2 CFU). Quite high numbers of *E. coli* were found inside lettuce roots, 3.0×10^3 CFU. The highest number of *E. coli* was found in the soil, near the root (7.3×10^5 CFU). *E. coli* K-12 was not detected on the surface of lettuce leaves, or inside/outside roots and leaves in control plants (Kljujev, 2012).

Laser scanning microscopy confirmed the presence of *E. coli* inside roots and leaves of lettuce (Photo 23.1). *E. coli* was not found inside roots of control plants but nucleui of root cells were clearly seen (picture 1 on Photo 23.1). On picture 3, plant vascular tubes and bacteria and root cell nucleus inside them in the same layer can be seen. This suggests that bacteria could be transported though the vascular system of the plant, through the xylem to edible parts of plant – the leaves.

Also, by confocal laser scanning microscopy, *E. coli* was seen inside leaves, and microcolonies of *E. coli* were detected below the surface of leaves. *E. coli* cells were concentrated near stomata (pictures 4, 5 and 6 in Photo 23.1). This suggests that pathogenic bacteria could enter the inside of a leaf through stomata if present in irrigation water. On the surface of leaves, confocal microscopic observation did not show the presence of bacteria. Also, the obtained micrographs demonstrate the presence of these bacteria inside the leaves of lettuce and parsley plants (Photo 23.1). Observations with *Salmonella Typhimurium* LT2 showed colonization of root surface of lettuce, tomato and sweet corn plants. The same authors

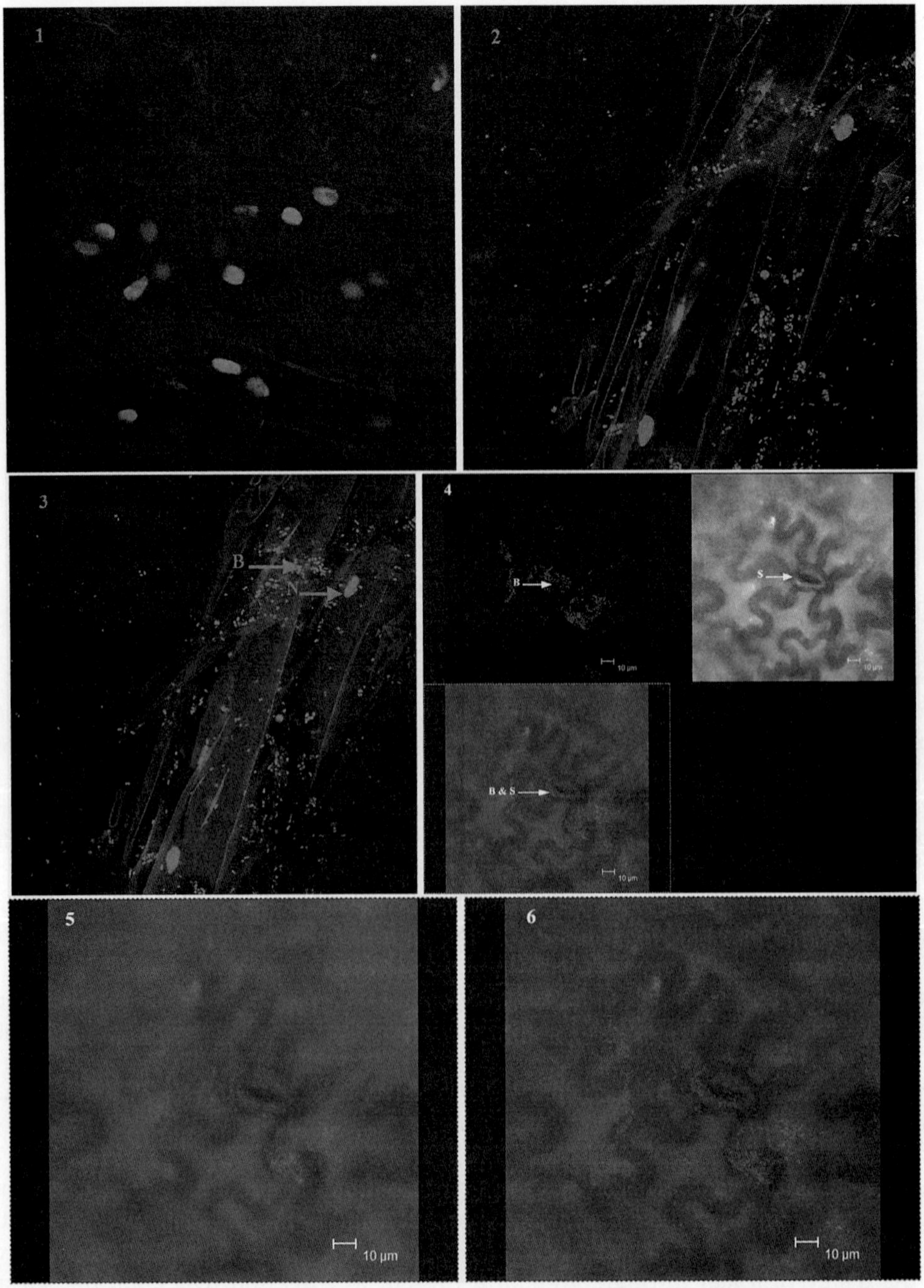

PHOTO 23.1 Microphotographs of lettuce roots and leaves, picture 1 – root of control lettuce plant irrigated with sterile water; picture 2 (the layer is 19 μm deep) and picture 3 (the layer is 20 μm deep) – root of plants irrigated with contaminated water; pictures 4, 5 and 6 (the layer is 11 μm deep) – leaves of plant irrigated with contaminated water; B – bacteria cells, N – nucleus of plant cells, S – stomata.

demonstrated that *Listeria monocytogenes* EGD-E strain has the ability for surface and endophytic colonization of carrot, parsley, celery and sweet corn root. Root colonization with *Listeria monocytogenes* SV4B strain was the most significant at herbaceous crops (lettuce and spinach). The *Listeria monocytogenes* SV4B cells were individually represented in endophytic colonization of celery and sweet corn roots (Kljujev, 2012).

GOOD AGRICULTURAL PRACTICES

The best way to control microbial, chemical and mechanical risks, but also for clear and comprehensive management strategy on farm, is to apply principles of Good Agriculture Practice (GAP). Also, practices that are directly related to monitoring and reduction or complete annulment of risks on the farm are Good Hygiene Practices, Biosecurity Principles and HACCP. These practices may be part of GAP, but also can be applied individually. Their application depends on the development of farm expertise and capabilities.

The concept of GAP is a modern agricultural management concept, which originated in the developed countries. It is an expression of the danger of ecological crisis, which seriously threatens humanity in all aspects of manufacturing activity by man. A radically different attitude to all the factors of agricultural production is required. Irrationality, inefficiency and negligence in production have resulted in increased pollution of the environment. GAP protocols were developed as a response to the increase in the number of outbreaks of foodborne diseases. Hence, GAP is both a necessity and an imperative of modern agricultural production. Key words related to GAP are knowledge, understanding, planning, measurement, control and management (Pejanović, 2008). Concept, goal and benefits of GAP are presented in Table 23.7.

Basics Principles of GAP

Basic principles of GAP include the resources, methods and practices necessary for production, which are classified into nine elements, namely: Clean Soil; Clean Landwater; Crop Production; Plant Protection; Harvesting, Processing and Storage on the Farm; Energy and Waste Management; Welfare, Health and Safety of Workers; Environment and Record Keeping.

TABLE 23.7 Concept, Goal and Benefits of Good Agriculture Practice

Concept – involves application of certain procedures in the process of agricultural production – represents the integration of the well-established work processes and well-placed controls.
Goal – to produce safe and healthy food and other agricultural products, while achieving economic values, social stability and environmental protection.
Benefits – farmers, additional value of their products and improved market access. – consumers will have safe food products – economy will make higher profits thanks to quality products – mankind will enjoy a better environment

BOX 23.5

PRINCIPLES TO ACHIEVE MINIMUM OF LOSS OF SOIL PARTICLES AND NUTRIENTS

- Production in accordance with the potential of soil fertility
- Keeping records of inputs and outputs of each parcel
- Maintenance and improvement of soil fertility using crop rotation
- Rational mechanical tillage
- Maintaining vegetation cover to reduce the soil erosion
- Using agricultural chemicals and organic and inorganic fertilizers in adequate amounts, timelines and methods that are adequate for the requirements of human health and a healthy environment

Clean Soil

It is most important for soil to be fertile and to contain no pollutants. Physical and chemical structure and biological activity of the soil determine its fertility. Maintaining and increasing soil fertility is achieved by minimizing the loss of soil particles and nutrients by applying the principles presented in Box 23.7 (Pejanović, 2008).

To avoid or minimize microbial contamination of soil, special attention should be paid to proper management of manure, animal excrements and other farm waste. That management includes (http://www.slideshare.net/dslagoriya/good-agricultural-practices):

- Orderly collection of manure and other waste from the farm.
- Prevention of wastage of manure and other waste during transport to storage places.
- Ensuring that the content of manure and other waste cannot leak or dissipate from storage places.
- Protecting the storage place from adverse weather conditions (rain, wind, sun, snow).
- Applying manure to soil, adhering to the time and limits of fertilization.
- Keeping other domestic and wild animals away from storage places.

Clean Water

Given that agriculture is one of the major water pollutants, it is necessary to carefully manage water resources on the farm and the surrounding area.

Clean water entails that all water used for washing, cooling, irrigation and processing is potable (Pejanović, 2008).

Crop Production

Beside of needs of consumers and the market, selection of plants that will be grown on the farm primarily depends on quality of the soil, availability of inputs, possibility of crop rotation, control of pests and diseases, etc. Each harvest, in fact, presents a deprivation of nutrients from the soil, so they must be replaced with new nutrients to ensure the long-term productivity of the soil (Pejanović, 2008).

BOX 23.6

PRINCIPLES FOR GOOD MANAGEMENT OF WATER SOURCES

- Use inputs of organic, inorganic and synthetic composition, in a manner that avoids contamination of water resources.
- Protect ground and surface water sources from run-off and animal contamination.
- Use underground and surface water appropriately.
- Adjust the timing and quantity of the irrigation needs of crops.
- Prevent salinization of land.
- Improve water cycles, provide permanent cover of vegetation.
- Provide an adequate, safe and clean place where animals can drink water.

BOX 23.7

GOOD AGRICULTURAL PRACTICES IN CROP PRODUCTION

- Adequate selection of species and varieties.
- Suitability of species and varieties to planting, productivity, quality, disease resistance, adaptability to soil and climate conditions, responses to fertilizer and agrochemical and market requirements.
- Inclusion of legumes in the rotation, to ensure the required amount of nitrogen.
- Use moderate amounts of organic and inorganic fertilizers.
- Inclusion of pasture land into crop rotation.
- Use of by-products of livestock production to improve soil fertility.
- Rotation of flocks on pasture to ensure natural regeneration of pasture.

Plant Protection

Plant protection must be based on a long-term strategy. All measures for plant protection, especially those that involve the use of substances that harm human health or the environment, must be conducted professionally and with the appropriate equipment (Pejanović, 2008).

Harvesting, Processing and Storage on the Farm

Product quality largely depends on how the harvest is done, on conditions of storage and on the processing of agricultural products at the farm.

Energy and Waste Management

All operations in agricultural production should be completed on time, with minimum worker downtime and with the lowest possible energy use. During the process of

BOX 23.8

PRINCIPLES FOR GOOD MANAGEMENT OF PLANT PROTECTION

- Use of species and varieties resistant to pests and diseases.
- Use of crop rotation.
- Use of production technologies that maximize biological prevention against diseases and pests.
- Application of techniques that can predict appearance of disease or pest.
- Storage and use of agrochemicals in accordance with applicable law.
- Handling and application of agrochemicals by highly trained and professional staff.
- Keeping of accurate records on the use of agrochemicals.

BOX 23.9

GOOD AGRICULTURAL PRACTICE IN HARVESTING, PROCESSING AND STORAGE

- Harvesting of agricultural products in accordance with agro-technical terms and terms of agrochemical application.
- Clean and safe manipulation during the processing.
- Use of recommended detergent and clean water for product washing.
- Storage of agricultural products in adequate hygienic and ambient conditions.
- Packaging of agricultural products in a clean container.
- Keeping of accurate records on harvesting, storage and processing.

BOX 23.10

PRINCIPLES FOR IMPROVEMENT OF ENERGY AND WASTE MANAGEMENT

- Make plans for the nutrients, energy and agrochemical inputs and outputs.
- Design objects that save energy.
- Choose adequate machinery, equipment, tools.
- Use alternative energy sources, if possible.
- Recycle organic waste and inorganic matter.
- Reduce unusable waste and dispose of it in an environmentally friendly manner.
- Store fertilizers and agrochemicals safely.
- Keep accurate records regarding energy use, storage and disposal of waste.

BOX 23.11

WELFARE, HEALTH AND SAFETY OF FARM WORKERS

- Adequate profit of agricultural households.
- Obtaining safe working conditions, with reasonable working hours.
- Training of workers for efficient and safe use of tools and machines.
- Adequate salaries, without exploitation of workers, especially women and children.

BOX 23.12

PRINCIPLES FOR ENVIRONMENT AND BIODIVERSITY PRESERVATION

- Conservation of natural habitats on the farm.
- Cultivation of as many different crops and animals on the farm.
- Minimizing the adverse impact of working operations on nature, e.g. tillage.
- Maintaining fertile agricultural land, e.g. removal of weeds, cultivating beneficial flora and fauna.
- Managing natural resources in a way that maintains biodiversity.
- Keeping garbage containers tightly closed.
- Destruction of the waste in the specified waste disposal area located separately from the rearing area.
- Removing animal manure from housing area to avoid sources of bad odor and pathogens.
- If the effluents are discharged to public water systems, an appropriate wastewater treatment device should be provided and the quality of the discharged water should meet official standards.

agricultural products, by-products can be obtained. Some of them are potential contaminants of soil, water and air. Production of harmful by-products should be minimized, while other by-products should be recycled (Pejanović, 2008).

Welfare, Health and Safety of Workers

Welfare of both the people that work on the farm and the entire community to a large extent depends on economic well-being, i.e. farm profitability. But more important issues are the health and safety of all those who are directly or indirectly involved in agricultural production (Pejanović, 2008).

Environment

Intensive agricultural production has an influence on water, soil and air pollution, and extinction of some plant and animal species because of loss of habitat. One of the major

BOX 23.13

INFORMATION THAT SHOULD BE RECORDED ON FARM

- Information on farm management, i.e. personnel information, training, health status.
- Information on pesticides and fertilizations applied to each production lot.
- Results of water analysis.
- Cleaning and sanitizing procedures.
- Information on production management, i.e. history of each animal, feed and water; farm management; animal health; production records and quality of animal products such as raw milk quality.

tasks of agricultural production is to preserve the environment and biodiversity, with simultaneous, economically justified, agricultural production.

Record Keeping

Farms should have a well-planned and established system of documentation. Important information should be archived for at least 3 years for the purposes of traceability.

PART 3: FISH HYGIENE

BACKGROUND

Statistics published by the FAO (Food and Agriculture Organization of the United Nations) show that inland capture fisheries production follows the general trend of most of the world's sea fishing areas, which have apparently reached their maximum potential, with the majority of stocks being fully exploited. In contrast, growth in aquaculture production has shown the opposite trend. While capture fisheries production has increased only very slightly, output from aquaculture (farmed fish, shellfish and algae) increased significantly from just over 13 million tonnes in 1990 to 33 million tonnes in 1999.

Despite its healthy growth, the aquaculture industry still faces problems with diseases which can affect its sustainability. Infectious diseases caused by viruses, bacteria and parasites are continuing threats to consistent industry growth. With increasing intensification, the incidence of diseases is also expected to increase proportionately.

DISEASE PREVENTION

Most diseases can be prevented through good husbandry practices and proper screening of incoming animals to the facility. When possible, bring in only eggs from a reputable supplier that can provide disease-free animals. Commercial blood test kits are available to screen fish for antibodies of several important fish pathogens. These kits normally do not determine active infections but can provide evidence of previous exposure by vaccination or live-disease-causing agents. More expensive laboratory DNA tests of fish tissue using polymerase chain reaction (PCR) can provide an even higher degree of sensitivity than antibody-based tests to identify subclinical infections or previous use of vaccinations.

Isolation, rapid removal and necropsy of dead animals will reduce the spread of disease and help to provide early diagnosis and treatment of the problem. Prophylactic external (e.g. chloramine T, etc.) treatments during and after handling procedures will prevent the start of many infections. Separate nets for each tank and iodophore disinfection baths for equipment will reduce cross-contamination problems.

Vaccines are commercially available to protect against vibriosis, furunculosis, enteric red mouth and enteric septicemia bacteria. New vaccines are in development for several commercially important viral and rickettsial fish pathogens. Specialty orders of "autogenous" vaccines can also be manufactured to protect against unique or emerging bacterial pathogens. All vaccines require that the fish be held for a period of disease-free conditions (usually 3–5 weeks) after vaccination to build up immunity before any significant exposure to infectious diseases. Vaccinations against vibriosis, enteric red mouth and enteric septicemia bacteria can be delivered to the fish by immersion. For other diseases, intraperitoneal injection is the preferred route for maximum protection and duration of immunity.

Selective breeding using quantitative genetics can be used to produce strains of fish with enhanced resistance to specific diseases. The traits for selection must be based on genes with sufficient heritability for this process to be successful. Furunculosis-resistant strains of brook trout and brown trout are examples of the great potential of these efforts.

DISEASE TREATMENT

Bacterial, parasitic and fungal diseases can all be controlled with chemo-therapeutants. Viral diseases are best prevented or eliminated by isolation and quarantine procedures. The key to successful treatment is the proper identification of the primary cause of the observed losses. For example, the observation of a single external parasite is not a reason for immediately beginning treatment without further investigation of other underlying infections or water quality problems. Consultation with a fish health pathologist or experienced veterinarian is strongly recommended before starting any treatment.

All compounds can have side effects and it is essential that caution be used in handling and use of any chemical.

Fungal infections of eggs can be treated with methylene blue or formalin. However, frequent removal of dead eggs is critical to the success of hatching survival and may limit the need for such treatments. Formalin is also useful for the treatment of external parasitic infections in juvenile or adult fish.

Damaged gills or skin can often be treated with sodium chloride to improve the osmotic balance between the water and the fish tissues. Dissolved salt treatments improve the effect of other surface compounds by removing debris and mucus from the gills and skin.

Judiciously applied, potassium permanganate can remove surface parasites and bacteria from fish in freshwater systems.

If a pathogenic bacterium is isolated, tests for antibiotic resistance should be done to select the best drug and treatment regime. Gram-negative organisms are commonly treated with oxytetracycline or sulfamethoxazole plus trimethoprim. Gram-positive bacteria are more responsive to erythromycin or doxycycline. Caution should be exercised if the biological filter media will be exposed to these compounds because water quality may quickly deteriorate if the nitrifying bacteria are lost.

Antibiotics added to the feed may not be appropriate if the affected population of fish is refusing to eat. Fish oil additives to enhance palatability or direct injection of antibiotics are alternatives to consider before applying oral drug treatments. Particular attention must be paid to local and federal regulations regarding restrictions on the use and withdrawal periods of antibiotics in food fish.

An indication of the magnitude of economic losses is illustrated by farm surveys conducted in 16 Asian countries, which show that annual losses due to disease in the region total more than USD 3 billion. Probably the most striking example of disease spread through international trade and consequential major economic loss in aquaculture is white spot disease in farmed shrimp. The disease first emerged in 1991 in a shrimp farm in an OIE member country and apparently has since spread to most other shrimp-farming countries of Asia and the Americas. This has been attributed by some experts to the uncontrolled international trade in live shrimp for aquaculture purposes and in dead shrimp for processing. Some countries with shrimp-farming activities continue to be free of the disease, almost certainly due to strict controls on imports of live shrimp and uncooked dead shrimp, in particular for use as fish bait.

The adverse social, economic and environmental consequences of uncontrolled movement of live aquatic animals and their products have increased global awareness of the need for improved health management standards. The serious impact of unrestricted international movement of aquatic animals has led to the development of health certification and risk reduction methodologies. The International Aquatic Animal Health Code and the Diagnostic Manual of Aquatic Animal Diseases are published by the OIE and provide recommendations and standards for reducing the spread of specific aquatic animal diseases considered to be of significance for international trade. They are recognized by the World Trade Organization as the international standards for trade.

The importance of containing the threat of diseases in fish production is a matter of global concern especially with increased trade and increased transboundary movements of goods which include live fish and other aquatic organisms. Due to this concern, the minimum EU measures for the control of the fish diseases are referred to in list I and II of Annex A to Council Directive 91/67/EEC (EC, concerning the animal health conditions governing the placing on the market of aquaculture animals and products, 1991). The diseases are categorized in three lists (Table 23.8).

An outbreak of a fish disease can quickly take on epizootic proportions, causing mortality and disturbances on a scale liable to reduce severely the profitability of aquaculture. Therefore it is important that control measures are taken when the presence of such a disease is suspected so that immediate and effective actions can be implemented as soon as its presence is confirmed. Such measures are aimed at preventing the spread of the disease, in particular by carefully controlling movements of fish and products liable to spread the infection.

TABLE 23.8 Listed Diseases/Pathogens of Fish, Mollusks and Crustacea (Annex A of Directive 91/67/EC)

Disease/Pathogen	Susceptible Species
LIST I	
Fish	
Infectious salmon anemia (ISA)	Atlantic salmon (*Salmo salar*)
LIST II	
Fish	
Viral hemorrhagic septicemia (VHS)	Salmonid species Grayling (*Thymallus thymallus*) Whitefish (*Coregonus* spp.) Pike (*Esox lucius*) Turbot (*Scophthalmus maximus*)
Infectious hematopoietic necrosis (IHN)	Salmonid species Pike fry (*Esox lucius*)
Mollusks	
Bonamia ostreae	Flat oyster (*Ostrea edulis*)
Marteilia refringens	Flat oyster (*Ostrea edulis*)
LIST III	
Fish	
Infectious pancreatic necrosis (IPN) Spring viremia of carp (SVC) Bacterial kidney disease (BKD) (*Renibacterium salmonidarum*) Furunculosis (*Aeromonas salmonicida*) Enteric red mouth disease (ERM) (*Yersinia ruckeri*) *Gyrodactylus salaries*	To be specified in the program referred to in Articles 12 and 13 of Directive 91/67/EC
Crustaceans	
Crayfish plague (*Aphanomyces astaci*)	

When fish on a farm are suspected of being infected with a list I disease, infectious salmon anemia (ISA), the official services in the member states must initiate official investigations to confirm or rule out the presence of the disease. No movement of fish, whether dead or alive, eggs and gametes are allowed without the authorization of the official service. When the presence of the disease is confirmed, fish infected with the disease are killed and destroyed as soon as possible to prevent the spread of the disease. Member states must have contingency plans for list I diseases.

List II diseases are important endemic diseases that should be contained and eradicated in the long term. Where fish are suspected of being infected with a list II disease, i.e. viral hemorrhagic septicemia (VHS) and infectious hematopoietic necrosis (IHN), an official investigation must be initiated to confirm or rule out the presence of the disease. Approved farms and zones will lose their status as free from the disease until it is proven that the disease is eradicated.

All farms rearing or keeping fish susceptible to list I or list II disease must be registered by the official service and keep records of mortality and the movement into and out of the farm.

Council Directive 2006/88/EC (EC 1991) on animal health requirements for aquaculture animals and products thereof, and on the prevention and control of certain diseases in aquatic animals, establishes:

- animal health requirements for the placing on the market, importation and transit of aquaculture animals (fish, mollusks and crustaceans) and their products;
- minimum measures to prevent diseases in aquaculture animals;
- minimum measures to be taken in response to suspected or established cases of certain diseases in aquatic animals.

MAJOR FISH DISEASES

Fish Viral Diseases

Prevention and control of viral diseases in fish are rather limited. Efficient chemotherapeuticals for viral diseases do not exist. Also, efficient vaccines are obtained only for a certain number of viruses. Unique practical measures for the control of viral diseases are: quarantine, control of trading of fish and their products, as well as strict disinfection and harmless removal of diseased fish.

In this respect, all reproductive centers for fish young have to be free of viral infections. Water-supplying systems in aquaculture have to be protected from the ingress of wild fish, as well as other water organisms, that might be carriers of viruses. If viruses enter an aquarium and infect the fish, it is very difficult to remove fish infected with that virus.

Viral diseases that affect the health of fish are: infectious pancreatic necrosis of salmonid fry (IPN), spring viremia of carp (SVC), pox disease of carp, viral hemorrhagic septicemia (VHS), infectious hematopoietic necrosis (IHN), infectious salmon anemia (ISA), and lymphocystis disease.

Infectious pancreatic necrosis (IPN) is an infectious, contagious disease, which attacks salmonid fry. It passes in acute stadium, characterized by a sudden explosive outbreak with high mortality. Affected fish become dark and rotate their bodies while swimming. They usually have exophthalmia and distended abdomens with the presence of a gelatinous material in the stomach and anterior intestine.

Spring viremia of carp (SVC) is an infectious, very contagious disease, which appears in acute form, and is manifested by symptoms of hemorrhagic diathesis, enteritis and peritonitis.

Pox disease of carp (*Epithelioma papillosum*) is an infectious, contagious disease, which attacks cyprinid fish species. It appears in chronic stadium, with hyperplasia of epidermis of skin and fins, such as hard gelatinous milky-white tumoroid proliferates, and in advanced cases with metabolic disorders of mineral matters. This disease has a benign character.

Viral hemorrhagic septicemia (VHS) is an infectious, contagious disease, which attacks mainly rainbow trout. It is the most serious viral disease of farmed rainbow trout, and is manifested by variable clinical symptoms: hemorrhagic syndrome, hydropsy characteristics and anemia, up to nervous disorders.

Infectious hematopoietic necrosis (IHN) of salmonid fry is manifested with hemorrhagiae and edema, accompanied by necrotic alterations of the wall of blood vessels. The hematopoietic tissues of the kidney and spleen of young fish are the most severely affected and are the first tissues to show extensive necrosis. This disease is very similar to VHS.

Lymphocystis disease is an infectious disease, with chronic progression and benign character. It is manifested with the appearance of pebble or wart-like nodules most commonly seen on the fins, skin or gills, although other tissues may be affected.

Infectious salmon anemia (ISA) is an infectious disease. It is associated with high mortalities and is of great economic significance for the Norwegian fish farming industry. Infected fish are lethargic and severely anemic. Other typical signs are ascites, petechiae in internal organs and hemorrhagic liver necrosis.

More information on viral diseases can be found in FDA (2011), Hristovski and Stojanovski (2005), OIE (2012) and Woo and Bruno (1999).

Fish Bacterial Diseases

Bacteria exist in different environments in nature. Bacteria have a major role in the circulation of matter in natural ecosystems, but certain bacteria can cause serious diseases in fish.

Some fish, which show no signs of having a certain disease, may be carriers of infective agents. But if those fish are exposed to stress factors, the disease may manifest itself and begin to excrete pathogenic microorganisms into the water, leading to repeated outbreaks of disease.

Exact identification of organisms that lead to the appearance of infective disease is particularly important, as well as determination of antimicrobial substances that successfully act against them. Different species of fish need different treatments. Usage of inappropriate antimicrobial components might create resistant lineages of bacteria.

Bacterial diseases harmful to fish health are: erythrodermatitis of carp, furunculosis of salmonids, motile aeromonas septicemia (MAS), vibriosis, yersiniosis, Edwardsiellosis, Edwardsiellosis enteritic septicemia of catfish, ulcer disease of salmonids, bacterial kidney disease, columnaris disease, bacterial cold water disease, mycobacteriosis and nokardiosis.

Erythrodermatitis of carp is an infective bacterial disease, which appears in subacute or chronic form, and is manifested with characteristic alterations of skin as erosions with progressive character and possible generalized form with hard clinical picture, where general hydropsy dominate.

Furunculosis of salmonids is an infective, very contagious bacterial disease of salmonid fish species, which appears in peracute, acute, subacute and chronic form, and is manifested with local alterations on the skin, but in certain clinical cases in the form of septicemia.

Motile aeromonas septicemia (MAS) is caused by ubiquitous *Aeromonas hydrophila* complex, and is manifested by hemorrhagic septicemia.

Vibriosis (*Erysipelosis anguillarum*) is an infective, very contagious bacterial disease of salmonid fish species, which appears in peracute, acute and chronic form, and is manifested with septicemia in acute form, and formation of abscesses and ulcers in chronic form. The losses produced by this disease are so disastrous that vibriosis caused by *V. anguillarum* has been recognized as a major obstacle for salmonid marine culture.

Yersiniosis (enteric redmouth disease) is a bacterial, subacute or acute disease of salmonids, which is manifested by hyperemia and hemorrhage on the head. It is present in a carrier state in many species of fish and remains undetected until stress.

Edwardsiellosis (*Edwardsiella septicemia*) is a serious systemic bacterial, subacute or chronic disease of warm water, rarely cold water fish species, commonly known as fish gangrene, emphysematous putrefactive disease of catfish or red disease of eels. The disease is manifested by formation of erosions, abscesses and ulcers on the skin. Infective agents can cause disease at other animals (reptiles, birds, mammals). It can cause gastroenteritis, abscesses and meningitis in humans. It sometimes produces a subclinical infection in fish intended for human consumption, where it may create problems during the cleaning process, which requires processing interruption, cleaning of equipment and disposing of infected fish.

Edwardsiellosis enteritic septicemia of catfish is a bacterial, subacute or chronic disease of cultured warm water fish species from the family Ictuliridae, which is manifested by formation of petechial hemorrhage, erosions, abscesses and ulcers on the skin. A characteristic clinical symptom is a longitudinal ulcerative lesion in the area between the eyes, which can even reach skull bones. There is no indication that *E. ictaluri* poses a health threat to aquatic animals and humans, probably due to temperature limitations under which bacteria grow.

Ulcer disease of salmonids is an infective disease, which might appear as local infection of the skin or as acute septicemia.

Bacterial kidney disease is an infective, very contagious bacterial disease of cultured and wild salmonid fish species, which appears in chronic form, and is manifested with necrotic alterations of kidneys, development of anemia and high rate of mortality.

Columnaris disease is one of the most frequent infective bacterial diseases, which appears in different fish species, and is manifested with alterations of skin and gills.

Bacterial cold water disease is a serious septicemic disease of the young of salmonid fish species bred in hatcheries. It appears at $t < 12°C$. At the start the disease has a local character, with alterations of fins, musculature, gills; later even internal organs, particularly kidneys, become affected.

Nokardiosis is a chronic, granulomatous disease of fresh- and saltwater fish, which is very similar to tuberculosis, according to clinical symptoms.

Mycobacteriosis is a bacterial, contagious, subacute to chronic, systemic, progressive disease, which appears in all fish species, and is manifested by nodules and ulcerations on the skin and tuberculous nodules in internal organs. It is not of major importance in intensive fish breeding, but is particularly harmful for breeding of aquarium fish, because they are often kept for long periods of time compared with fish raised for commercial purposes. Piscine tuberculosis is caused by three species of bacteria belonging to the genus *Mycobacterium*, which is also the causative agent of tuberculosis in humans. While the bacteria that causes this disease in fish prefers cooler temperatures than most bacteria that infect humans it is still possible for the illness to be passed on to humans. Such an infection in humans usually shows in the form of an infected nodule in the skin, although there is a chance of a more serious internal infection.

Fish pathogenic bacteria are harmful to humans and others: they are pathogenic for warm-blooded animals and humans; however, fish usually are not diseased by these pathogens, but they can be germ carriers (in internal organs, skin or gills) for some time (weeks or months). Fish pathogenic bacteria are:

1. *Salmonella*;
2. *Listeria* – salmonids might be diseased;

3. *Leptospira;*
4. *Erysipelotrix rusiopathiae* – erysipeloid diseases are found in humans who work in the fish processing industry or in fish trading. The condition known as "crayfish handler's disease" is well known in the fishing industry. It can be caused by various bacteria, but particularly *Erysipelothrix rhusiopathiae* and various species of the *Vibrio* genus. The bacteria enter the skin through abrasions, lacerations or fissures and cause a painful itching or burning sensation;
5. *Vibrio parahaemolyticus* – can cause mild disease in fish. But diseases in humans who eat fresh fish, crayfish or shellfish are frequent;
6. *Clostridium botulinum* – as with other *Clostridium* species, it is ubiquitous as well. Therefore, many kinds of food might be contaminated. Type E toxin is the most poisonous of all the toxins (A–E). It is present in sediments of open waters, near coastlines.

More information on bacterial diseases can be found in FDA (2011), Hristovski and Stojanovski (2005), OIE (2012) and Woo and Bruno (1999).

Fish Fungal Disease

Fish mycoses are considered difficult to prevent and treat, particularly in intensive freshwater systems, and are reported to be second only to bacterial disease in economic importance to aquaculture.

Ichthyophoniasis, due to infection with *Ichthyophonus hoferi*, has been known in fish since the end of the 19th century. The disease is recognized to be of economic significance, in both fish cultivation and wild fisheries, and to have a wide host and geographical distribution. Included as hosts have been various marine and freshwater crustaceans, fish (35 marine fish species and 48 freshwater species), amphibians, reptiles and piscivorous birds. *Ichthyophonus* has been recorded from many temperate and some tropical waters throughout the world. Manifested external signs include skin roughening ("sandpaper effect") and occasional ulceration. Inside the body are gross white or cream-colored nodular lesions 1–5 mm in size throughout most tissues.

Fungal infections of fish by oomycetes, commonly known as water molds, are widespread in fresh water and represent the most important fungal group affecting wild and cultured fish.

Four orders are recognized in this class and the most important are the Saprolegniales. Although eight genera have been reported in infections, namely *Saprolegnia, Achlya, Aphanomyces, Calyptratheca, Thraustotheca, Leptolegnia, Pythiopsis* and *Leptomitus*, only *Saprolegnia, Achlya* and *Aphanomyces* are significant in aquaculture.

Some species are consistently isolated from fish and generally these are assigned to a single major cluster, which form a coherent, separate taxon, *Saprolegnia parasitica* (synonym *Saprolegnia diclina* Humphrey type 1).

The Saprolegniaceae, in particular members of the genus *Saprolegnia*, are responsible for significant infections, involving both living and dead fish and eggs, particularly in aquaculture facilities. Oomycetes are classical saprophytic opportunists, multiplying on fish that are physically injured, stressed or infected. Fungal outbreaks among farmed fish stocks are frequently associated with poor water quality, injuries associated with handling and grading, temperature shock, infestation by parasites and spawning. However, there is evidence that some Saprolegniaceae act as primary pathogens.

The oomycetes are an economically important group of mycotic agents that affect salmonids and other teleosts. They are reported extensively in both wild and farmed fish and are considered ubiquitous in freshwater ecosystems. Oomycete infections have also been recorded in the marine fish species. In the marine environment, oomycetes are significant pathogens of lobsters and crayfish.

Saprolegniasis is frequently observed as a superficial and chronic infection. It may occur anywhere on the body of fish, but normally appears as a conspicuous, circular or crescent-shaped, white, cotton-like mycelium, on the integument and gills of host fish or eggs, particularly around the head and the caudal and anal fin, which may spread over the entire body surface. Most fish die due to osmotic or respiratory problems if the area of skin or gills is large.

Branchiomycosis (gill rot) is caused by two species *Branchiomyces sanguinis* and *B. demigrans*. It is primarily a problem in carp and eels. The disease occurs most commonly in ponds with abundant organic matter and high ammonia levels. Usually higher temperatures (20–25°C) bring about the disease. Affected fish usually show respiratory distress. There is prominent gill necrosis caused by thrombosis of blood vessels in the gills. Histologically the identification of nonseptated branching hyphae with an intrahyphal eosinophilic round body (apleospores) in and around blood vessels of the gill is diagnostic.

More information on fungal diseases can be found in FDA (2011), Hristovski and Stojanovski (2005), OIE (2012) and Woo and Bruno (1999).

Fish Parasitic Diseases

Causative agents of parasitic diseases in fish are different species of protozoa, helminths, leeches and crustaceans.

Protozoa: Mastigophora (flagellates), Rhizopoda (amoebae), Apicomplexa (sporozoa), Microsporidia, Myxozoa (myxosporidia), Ciliophora (ciliates).
Helminths: Trematoda, Cestoda, Nematoda and Acantocephala.

Fish parasites can be found on the skin, fins or gills – ectoparasites, or in their internal organs – endoparasites. Fish parasites appear either as direct causative agents of certain diseases or as factors leading to disorder or decrease of fish resistance, and therefore fish become sensitive to many infective diseases. Besides health problems, parasites are an important economic problem in intensive fish breeding, because their presence can cause excessive economic damages, e.g. impeding growth.

Development of parasites can occur in one or several hosts – mainly small water organisms. Higher numbers of helminth parasitize in fish as their final hosts in sexually matured (adult) form. But some of them parasitize only in their larval forms in fish, as their transitional hosts. In intensive fish culture there are outstanding conditions for the spread of substantial parasitic invasions. Usually parasites appear with simple life cycles. On the other hand, fish from open waters are found with a great number of different parasites with complex life cycles.

The following are some products that have been implicated in human parasite infection: ceviche (fish and spices marinated in lime juice); lomi lomi (salmon marinated in lemon juice, onion and tomato); poisson cru (fish marinated in citrus juice, onion, tomato and coconut milk); herring roe; sashimi (slices of raw fish); sushi (pieces of raw fish with rice and other ingredients); green herring (lightly brined herring); drunken crabs (crabs marinated in wine and pepper); cold-smoked fish; and undercooked grilled fish. Seafood-borne parasitic infections occur with sufficient frequency to recommend preventive controls during the processing of parasite-containing species of fish that are intended for raw consumption.

The process of heating raw fish sufficiently to kill bacterial pathogens is also sufficient to kill parasites.

The effectiveness of freezing to kill parasites depends on several factors, including the temperature of the freezing process, the length of time needed to freeze the fish tissue, the length of time the fish are frozen, the species and source of the fish, and the type of parasite present. For example, tapeworms are more susceptible to freezing than are roundworms. Flukes appear to be more resistant to freezing than roundworms.

Freezing and storing at an ambient temperature of −20°C or below for 7 days (total time), freezing at an ambient temperature of −35°C or below until solid and storing at an ambient temperature of −35°C or below for 15 hours, or freezing at an ambient temperature of −35°C or below until solid and storing at an ambient temperature of −20°C or below for 24 hours are sufficient to kill parasites. Note that these conditions may not be suitable for freezing particularly large fish.

Brining and pickling may reduce the parasite hazard in fish, but they do not eliminate it, nor do they minimize it to an acceptable level. Nematode larvae have been shown to survive 28 days in 21% salt by weight.

Trimming away the belly flaps of fish or candling and physically removing parasites are effective methods for reducing the numbers of parasites. However, they do not completely eliminate the hazard, nor do they minimize it to an acceptable level.

More information on parasitic diseases can be found in FDA (2011), Hristovski and Stojanovski (2005), OIE (2012) and Woo (2006).

Fish Helminth Zoonoses

Fish can also appear as carriers or act as transitional hosts of certain parasite species which attack humans. Numerous marine and freshwater fish serve as sources of medically important parasitic zoonoses. The majority of these zoonoses are found in coastal regions of the seas, big lakes and rivers, where fish and their products are consumed further. But with the increasing consumption of fish, as well as the new trend of so-called "natural cooking," the number of the registered zoonoses continuously increases. The potential danger of human infestation with certain helminths still exists, because in Europe, several helminth zoonoses have been recorded: metacercariae of trematodes *Opisthorchis felineus, Pseudamphistomum truncatum, Clinostomum complanatum, Metagonimus yokogawai, Heterophyes heterophyes, Cryptocotyle lingua, Echinochasmus perfoliatus,* plerocercoids of cestods of the genus *Diphyllobothrium* and larvae of nematodes: *Dioctophyme renale, Anisakis simplex* and *Gnathostoma hispidum*, etc. Zoonotic transmission of some bacterial diseases, such as streptococcosis or mycobacteriosis, is also possible.

In most cases fish zoonotic parasites do not lead to major health problems in fish.

Fish parasites usually cause small or moderate damages in the human body. But some of them are more frequent and are a serious threat for human health. Some show abdominal pains, diarrhea or constipation, nausea, vomiting, loss of weight, or anorexia. Hepatomegaly, eosinophilia, tetanic cramps, tremors and toxemia may also occur.

Generally, fish can be either an intermediate host of parasites involving a human as the definitive host or a carrier of larvae of animal parasites that only invade human tissues for a limited period without undergoing further development. The latter are considered incidental infections. The natural definitive hosts for parasites are usually marine mammals or birds. However, larval stages of a few fishborne parasites can mature in both animals and humans.

Fishborne trematodiasis is especially important in Southeast Asia, the Far East and regions where people are dependent on freshwater fish as the major source of protein. Infections by both large and small digenetic trematodes are common. Although the diseases are seldom fatal, they can cause morbidity and serious complications. The route of infection is by ingesting metacercariae located in muscles and subcutaneous and other tissues of fish.

There are relatively few cases of fishborne cestode infections in humans. The cestodes that mature in the small intestine of humans are not very pathogenic and the diseases are never fatal. Diphyllobothriasis is the major cestodiasis transmitted by freshwater, marine and anadromous fish.

Fishborne nematodiases are generally caused by the incidental infection of humans with nematodes whose natural definitive hosts are marine mammals, birds, pigs or other animals. Freshwater, brackish or marine fish are the second intermediate host. In most infections, the worms can only survive for a limited period after the initial invasion of the gastrointestinal tract. The method of infection is by ingesting the infective-stage larvae, which can be located in the muscles, intestine or viscera of fish. Unlike cestodiasis, some nematode infections can be fatal. In the Netherlands, since the passage of legislation against eating raw herring and requiring fish to be frozen prior to sale, anisakiasis has almost disappeared. Freezing fish for 24 hours or heating processed fish to 65°C can kill the larvae. Also, the gutting of fish soon after they are caught prevents the migration of larvae to muscles.

Theoretically, fishborne parasitic zoonoses can easily be prevented by refraining from eating raw seafood. However, in many parts of the world, such an eating habit represents an established way of life or part of the inherent culture. It cannot be easily changed, even by the implementation of a strong education program or the passage of legislation. Therefore, these diseases will remain as public health problems and there is a need to undertake regular epidemiological studies. These studies, however, cannot be carried out effectively without the development of more cost-effective, sensitive and specific diagnostic methods that can be used in large-scale screening of fish. The use of molecular biological techniques can also help to clarify species of dubious validity and to trace the source of infection. Stronger support for this neglected area of research is required.

More information on parasitic diseases can be found in FDA (2011), Hristovski and Stojanovski (2005), OIE (2012) and Woo (2006).

DISEASES OF MOLLUSCA AND CRUSTACEA

Some of the earliest records of mass mortalities of shellfish were caused by microbial disease agents, e.g. the phycomycete fungus *Ostracoblabe implexa*, responsible for "foot disease" in the European oyster (*Ostrea edulis*) and the iridoviral agent of "gill diseases" in Portuguese oysters (*Crassostrea angulata*). Increasing development of shellfish aquaculture, and recent advances in diagnostic techniques, along with diversification of cultured species, continue to provide a seemingly inexhaustible reserve of new or emerging microbial disease problems. They have also significantly broadened the scope of microbial pathogen research and are proving useful for differentiating between primary pathogens and the ubiquitous microbial fauna that surrounds shellfish in their natural environment. Note is also made of apparently non-significant pathogens, since, given the right conditions, even the most benign infectious organism may transform into a serious disease agent. Knowledge on how to distinguish between primary and opportunistic pathogens is also important for optimizing their control or treatment.

More information on diseases of Mollusca and Crustacea can be found in Hristovski and Stojanovski (2005) and Woo and Bruno (1999).

FISH TOXICITY

The discipline of toxicology involves studying the nature and mechanisms of toxic lesions, and evaluating in a quantitative manner the spectrum of biological changes produced by exposure to chemicals. It is important to realize that every chemical can be toxic to fish under certain exposure conditions. For every chemical there should be an exposure condition (i.e. dose or concentration) that is "safe" and an exposure condition that is "toxic" to fish. The range of concentrations or doses that are toxic to fish may span several orders of magnitude. It is also important to determine toxic "thresholds," that is, concentrations or doses above which toxicity occurs and below which it does not.

Until relatively recently, toxicological studies with fish focused almost exclusively on very toxic substances which produce "acutely lethal" responses, that is, mortalities in fish exposed to chemicals for only short periods. Recently, we have become concerned with substances that may produce "sublethal" responses in fish after "chronic" exposure.

There are several chemical, physical and biological factors that influence the toxicity of chemicals to fish, including the properties of the chemical in water, the water quality conditions, the route of exposure, and the species and life stage of the fish.

Chemical toxicity to fish is often affected by external factors, such as photoperiod, temperature, salinity, reproductive status, disease and exposure to other external stressors.

Factors Affecting Toxicity

Water Quality Conditions

Since fish live in water, the extent to which fish are exposed to a chemical is dependent on aqueous solubility.

The solubility of ionic chemicals, which include most salts of toxic metals and some ionic organic compounds, is usually much higher than that of non-ionic compounds.

Ions may be dissolved in water in non-toxic forms. For instance, ions may form complexes with inorganic and organic "ligands." Inorganic ligands for cations in fresh water include carbonate (CO_3^{2-}), sulfate (SO_4^{2-}), and fluoride (F^-) ions, and Cl^- is an important ligand in saline water. Complexes between cations and inorganic ligands tend to be fairly "labile" or reversible, depending on the concentration of the ion and the ligand, and the pH. However, complexes with organic ligands, such as humic acids, tend to be relatively non-labile. "Alkalinity," which is primarily the concentration of carbonate ions in solution, is an important measure of the cation-binding capacity of fresh water. Alkalinity and pH are important variables influencing the toxicity of metal ions to fish. Transformations of chemicals dissolved in water can occur by hydrolysis, photolysis and oxidation.

If we accept that the toxicity of ionic chemicals is usually dependent upon the concentrations of the free ion in solution, then various factors that affect speciation of ions will affect toxicity, including pH, alkalinity, hardness and concentrations of organic ligands.

The solubility of non-ionic chemicals, such as organic compounds and elemental forms of toxic metals (e.g. Hg), is influenced by the polarity of the compound.

The toxicity of non-ionizable chemicals, such as organic compounds, is affected to a lesser extent by water quality conditions such as pH, alkalinity and hardness. However, dissolved and particulate organic material in water can alter the toxicity of organic compounds by acting as ligands for hydrophobic substances.

For additional information on water quality conditions, see Di Giulio and Hinton (2008), Hristovski and Stojanovski (2005) and Ostrander (2000).

Biological Interactions

A chemical can be toxic to a fish in two possible ways. It may affect tissues on the surface of the organism (e.g. gill epithelium) or the chemical may enter the organism and cause toxicity. A toxic chemical must pass through cell membrane barriers to reach "target" organs or tissues. The epithelial and endothelial integument of fish is usually thickened and relatively impermeable to chemicals, except in the gill tissues, which are specialized for gas exchange, and in the gastrointestinal tract. Thus, branchial or gastrointestinal uptake routes are the most efficient mechanisms for uptake of toxic chemicals into fish.

The most prevalent route of exposure of fish for chemical agents is via gills. Fish gills have an enormous surface area, approximately 50% of the entire surface area of the fish. Gill secondary lamellae, flattened ridges protruding perpendicularly from the primary lamellae, provide an effective and extensive surface for gas exchange. Although designed to facilitate diffusion of respiratory gases, fish gills also provide routes for other molecules to be accumulated by fish. Small hydrophilic molecules (e.g. NH_3, CO_2 and urea) can pass through small aqueous pores or gaps between cells in the gill epithelium. Larger neutral hydrophobic molecules, including many drugs and toxic organic chemicals, readily diffuse across the gill epithelium into the vascular space. Diffusion or uptake efficiency of these chemicals by the gills depends primarily on their hydrophobicity and molecular size. In addition, free metal ions can bind to negatively charged sites on fish gills. Once bound to the gill epithelium many metals use existing ion transfer mechanisms, such as calcium channels or protein-mediated endocytosis for entry into the gill.

The next route of exposure of fish for chemical agents is with food. Systemic absorption of the ingested chemical is relatively rapid. Significant accumulation of radiolabeled methyltestosterone was detected in fish tissues 2 hours after feeding sprayed chow. Tissue levels of testosterone appeared to reach equilibrium concentrations 24 hours after feeding testosterone sprayed on food to coho salmon (*Oncorhynchus kisutch*). In a similar experiment with carp (*Cyprinus carpio*), 4 days' feeding was required for testosterone to reach equilibrium concentrations in fish tissues. Chemical absorption from food depends on the rate of chemical dissolution from the food, its absorption efficiency in the stomach and intestine and influences of other factors, such as chemical and microbial degradation in the gut and binding to tissues. Also, physiological or metabolic differences between the two fish species may have caused differences in absorption, distribution and metabolism of the testosterone. Thus absorption efficiencies and time needed to attain equilibrium concentrations in fish will vary, depending on chemical properties and the fish species.

The most efficient method of administering a drug to fish is by injection. Ideally the chemical agent is dissolved directly in physiological saline. Chemical agents may be injected directly into veins or arteries (intravascular), into the peritoneal cavity (intraperitoneal) or into the muscle (intramuscular) of adult fish. Fish eggs or embryos may be injected into the perivitelline space or yolk sac with a micro syringe. In general, injection techniques provide a high internal dose with rapid distribution to the tissues.

Uptake of chemicals by fish can be influenced by both the lipophilicity and molecular size of the chemical.

For some chemicals, the rate of uptake is strongly influenced by the physiology of the fish. Fish species differ widely in their sensitivity to the toxic effects of chemicals.

Gill ventilation rates and dietary intake are governed by the metabolic rates of fish. There is considerable variation in the metabolism of fish; from fast-swimming pelagic predators to slow-swimming benthivores, so toxicity thresholds may vary considerably, depending on the fish species tested. In poikilothermic organisms such as fish, metabolism changes with the water temperature, so temperature may be an important factor influencing toxicity. Similarly, dissolved oxygen concentrations may influence gill ventilatory rates. Early life stages of fish tend to have higher rates of metabolism than later life stages. Therefore, the most sensitive period for chemically induced toxicity in fish is the embryolarval or early juvenile stages.

"Bioaccumulation" of chemicals represents the uptake and retention of chemical from the environment into fish via any pathway (e.g. food, water), whereas "bio-concentration" represents uptake and retention of a chemical directly from water into fish.

Fish possess metabolic pathways capable of transforming chemicals, such as oxidation and binding of chemicals to proteins or other large biomolecules (i.e. conjugation).

Although chemical contamination of our environment is often associated with human activities, plants and animals have evolved in an environment that has included continuous exposure to toxic materials. Basic mechanisms for resisting toxicity probably evolved with early life and are likely to be highly conserved in nature. Because of the large number and wide distribution of novel anthropogenic compounds introduced into the modern world, these mechanisms have become increasingly essential for survival. Organisms surviving environments heavily contaminated with anthropogenic chemicals demonstrate a diversity of mechanisms to tolerate or resist toxic effects.

Resistance or tolerance can be defined as the relative ability to function or survive during toxicant exposures that are harmful or lethal to susceptible individuals and populations. Fish and other organisms appear to develop tolerance through a variety of short-term and long-term processes. Physiological acclimation and genetic adaptation are general terms for short-term, transitory responses and long-term, heritable responses, respectively. Physiological acclimations occur in direct response to toxic exposures and likely involve temporary alterations in the levels of expression of proteins and enzymes involved in chemical defense. Following chemical exposure, protein expression returns to normal and the state of physiological acclimation declines. Genetic adaptation or evolved tolerance occurs when the genetic basis for advantageous responses is passed on to progeny. In genetic adaptation, tolerance is retained through successive generations, even when progeny are not exposed to chemicals.

The terms physiological acclimation and genetic adaptation have been used frequently for categorizing mechanisms of chemical tolerance in fish; however, other processes and conditions may contribute to tolerance as well. For example, abundant evidence indicates that some forms of chemically induced cancer represent adaptations to harsh chemical environments, providing survival value to individuals especially during the early stages of cancer. Cancer resulting from chemically induced mutations in somatic cells and concomitant alterations in protein expression would be considered to be a genetic but nonheritable adaptation. In addition to cancer, epigenetic alterations, such as hypermethylation of promoter regions of DNA, may affect responsiveness to drug and chemical exposures. Although gene silencing due to hypermethylation has been widely studied in mammalian cancer research, it has only recently being investigated as a tolerance mechanism in fish. Finally, nongenetic but heritable factors involving maternal transfer of toxicant from an exposed parent to offspring could contribute to tolerance in offspring. In cases involving maternal transfer, tolerance may appear to have a genetic basis (i.e. tolerant field-collected parents and their progeny) but is in fact physiologically based and related to direct exposure of offspring to toxicant. Because of the possibility of maternal transfer, genetic adaptation is established only when tolerance is maintained for two or more generations.

In populations inhabiting severely contaminated sites, multiple processes likely contribute to resistance during the lives of individuals; for example, physiological acclimations could provide individuals with the ability to survive and reproduce in moderately contaminated sites. As chemical contamination at a particular site increases with time, selection of progeny that carry genetic traits with adaptive significance could become a more dominant factor. Genetically adapted individuals may rely to some extent on epigenetic mechanisms or may respond to periodic pulses of contaminants through physiological responses. Tumor-bearing individuals may also exhibit features characteristic of genetic adaptation, physiological acclimation or epigenetic alteration.

Chemical resistance in fish has been observed in response to diverse environmental contaminants, including pesticides, dioxin-like compounds, polycyclic aromatic hydrocarbons (and other compounds associated with creosote) and metals (Table 23.9).

For more information on biological interactions, see Di Giulio and Hinton (2008), FDA (2011), Hristovski and Stojanovski (2005), Ostrander (2000) and Treves-Brown (2000).

Natural Toxins

Contamination of fish with natural toxins from the harvest area can cause consumer illness. Most of these toxins are produced by species of naturally occurring marine algae (phytoplankton). They accumulate in fish when they feed on the algae or on other fish that have fed on the algae. There are also a few natural toxins that are normal constituents of certain species of fish.

TABLE 23.9 Classes and Sources of Environmental Toxicants Addressed in Toxicity Resistance Studies

Toxicant	Source
Organochlorine pesticides	Includes DDT, the first modern highly toxic pesticide, followed by toxaphene, chlordane, aldrin, dieldrin, heptachlor, mirex, and kepone. Very persistent compounds that accumulate in fatty tissues and sediments, they are toxic to fish, wildlife and humans and are banned in the United States. Some of the earliest records of toxicity resistance involve DDT.
Dioxin-like compounds (DLCs)	Includes polychlorinated biphenyls (PCBs), polychlorinated dibenzodioxins (PCDDs) and other persistent polyhalogenated aromatic hydrocarbons (PHAHs). PCBs were valuable industrial materials used in capacitors, transformers and other products. PCDDs are produced inadvertently during a variety of processes (e.g. pesticide manufacture, chlorine bleaching of pulp). The most notorious DLC is 2,3,7,8-tetrachlorodibenzo-*p*-dioxin (TCDD), often referred to simply as dioxin.
Polycyclic aromatic hydrocarbons (PAHs)	Complex mixtures of compounds produced during combustion of organic materials, especially fossil fuels; also, natural components of petroleum. PAHs are widely studied because of their abundance in the environment and because of the mutagenic and carcinogenic properties of some members (e.g. benzo(*a*)pyrene).
Creosote	Abundant pesticide mixture used to protect wood pilings, telephone poles, etc., from microbial decay. Creosote is composed primarily of PAHs; nitrogen-, sulfur- and oxygen-heterocyclic compounds; and phenols.
Metals	Naturally occurring elements (e.g. mercury, lead, cadmium, chromium), including some biologically essential elements (copper, iron, zinc). Human activities alter the environmental loading, availability and toxicity of metals through a variety of activities such as strip mining, fossil fuel combustion, smelting, and industrial processes.

For fish products in the United States there are six recognized fish poisoning syndromes that can occur from the consumption of fish or fishery products contaminated with natural toxins: paralytic shellfish poisoning (PSP), neurotoxic shellfish poisoning (NSP), diarrheic shellfish poisoning (DSP), amnesic shellfish poisoning (ASP), ciguatera fish poisoning (CFP) and azaspiracid shellfish poisoning (AZP).

1. Paralytic shellfish poisoning (saxitoxin) is generally associated with the consumption of molluscan shellfish from environments ranging from tropical to temperate waters. Certain gastropods (e.g. conch, snails and whelk) are also known to accumulate PSP toxins.
   The effects of PSP are primarily neurological with respiratory paralysis. PSP toxin is an extremely potent toxin with a high mortality rate.
2. Neurotoxic shellfish poisoning (from brevetoxin) is generally associated with the consumption of molluscan shellfish from the Atlantic coast of the USA, New Zealand, and there are some suggestions of occurrence elsewhere. NSP is characterized by gastrointestinal and neurological symptoms. There are few, if any, after-effects and there have been no reported fatalities.
3. Diarrheic shellfish poisoning (from okadaic acid and dinophysis toxins) is generally associated with the consumption of molluscan shellfish. Outbreaks have been

documented in Japan, Southeast Asia, Scandinavia, Western Europe, Chile, New Zealand, the USA and eastern Canada. DSP is characterized by gastrointestinal symptoms, including: nausea, vomiting, diarrhea, abdominal pain, headache and fever. DSP is generally not considered life threatening but complications could occur as a result of severe dehydration in some patients.

4. Amnesic shellfish poisoning (from domoic acid) is generally associated with the consumption of molluscan shellfish from the northeast and northwest coasts of North America. In these regions, domoic acid has been identified in the viscera of Dungeness, tanner and red rock crab. Domoic acid has also been identified in several fish species including anchovies, Pacific sanddab, chub mackerel, albacore tuna, jack smelt and market squid. ASP is characterized by gastrointestinal symptoms.
5. Ciguatera fish poisoning (from ciguatoxin (CTX)) is associated with consumption of toxin-contaminated subtropical and tropical predatory reef fish. The toxin is introduced to the marine food chain by microscopic algae and moves up the food chain as small plant-eating reef fish eat the toxic algae and are then eaten by larger reef fish. CFP is characterized by gastrointestinal symptoms, followed by neurological and cardiovascular symptoms.
6. Azaspiracid shellfish poisoning (AZP) is caused by the consumption of molluscan shellfish contaminated with azaspiracids (AZA). AZP was first recognized following a 1995 outbreak in the Netherlands, linked to consumption of mussels harvested in Ireland. Since then, several outbreaks of AZP have been reported in various regions in Europe. AZP is characterized by severe gastrointestinal disorders. There have been no reported fatalities.

More information on natural toxins can be found in Di Giulio and Hinton (2008), FDA (2011) and Hristovski and Stojanovski (2005).

Scombrotoxin (Histamine) Formation

Certain bacteria produce the enzyme histidine decarboxylase during growth. This enzyme reacts with histidine, a naturally occurring amino acid that is present in larger quantities in some fish than in others. The result is the formation of scombrotoxin (histamine). Scombrotoxin (histamine) formation is a result of time and temperature abuse of certain species of fish, and can cause consumer illness. Histamine is more commonly the result of high-temperature spoilage than of long-term, relatively low-temperature spoilage. Freezing may inactivate some of the enzyme-forming bacteria. Both the enzyme and the bacteria can be inactivated by cooking. However, once histamine is produced, it cannot be eliminated by heat (including retorting) or freezing. Rapid chilling of scombrotoxin-forming fish immediately after death is the most important element in any strategy for preventing the formation of scombrotoxin (histamine).

The illness is closely linked to the development of histamine in these fish. In most cases, histamine levels in illness-causing fish have been above 200 ppm, often above 500 ppm. A guidance level has been set at 50 ppm histamine in the edible portion of fish. If 50 ppm is found in one section of a fish or lot, there is the possibility that other sections may exceed 500 ppm.

However, there is some evidence that other chemicals (e.g. biogenic amines such as putrescine and cadaverine) may also play a role in the illness.

Symptoms of scombrotoxin poisoning include tingling or burning in or around the mouth or throat; rash or hives on the upper body; drop in blood pressure; headache; dizziness; itching of the skin; nausea; vomiting; diarrhea; asthmatic-like constriction of the air passage; heart palpitation; and respiratory distress. Symptoms usually occur within a few minutes to a few hours of consumption and last from 12 hours to a few days.

For more information, see Di Giulio and Hinton (2008) and FDA (2011).

Environmental Chemical Contaminants and Pesticides

Fish can be harvested from waters that are contaminated by varying amounts of industrial chemicals, including heavy metals and pesticides. These contaminants may accumulate in fish at levels that can cause human health problems (e.g. carcinogenic and mutagenic effects). The hazard is most commonly associated with exposure over a prolonged period of time (chronic exposure). Illnesses related to a single exposure (one meal) are very rare. Concern for these contaminants primarily focuses on fish harvested from aquaculture ponds, freshwater bodies, estuaries and near-shore coastal waters (e.g. areas subject to shoreside contaminant discharges), rather than from the open ocean. Environmental chemicals and pesticides may also accumulate in aquacultured fish through contaminated feed ingredients. Certain pesticides are applied directly to the water in aquaculture ponds to control weeds and algae and to eliminate fish and invertebrates.

Although some pesticides have not been produced or used for many years (e.g. dichlorodiphenyl-trichloroethane (DDT) and polychlorinated biphenyls (PCBs)), many are very persistent and tend to accumulate in soil and sediments. Once pesticides are introduced into the environment, they may travel beyond their point of application or discharge.

Many contaminants accumulate in the edible fatty tissues of fish. Concentrations of these contaminants can vary considerably in individual fish of the same species from the same location, depending on factors such as their fat content, size, age and gender.

In the case of components or extracts of whole fish (e.g. dietary supplements, dietary ingredients and flavors), the component or extract may contain higher or lower concentrations of environmental chemical contaminants and pesticides than the whole fish from which it was derived. For example, organochlorine contaminants, such as PCBs, are oil soluble. When producing fish oil and fish meal, any PCBs present will become more concentrated in the oil fraction and less concentrated in the water fraction, as compared with the levels in the whole fish.

Maximum residue levels (MRLs) are: (1) the maximum concentration of residue accepted by the European Union (EU) in a food product obtained from an animal that has received a veterinary medicine; (2) the upper legal levels of a concentration for pesticide residues in or on food or feed based on good agricultural practices and to ensure the lowest possible consumer exposure. The assessment for the safety of residues is carried out by the Committee for Medicinal Products for Veterinary Use (CVMP). In the United States, MRLs are established by the Environment Protection Agency (EPA) and the Food and Drug Administration (FDA).

Methylmercury

Mercury occurs naturally in the environment and can also be released into the air through industrial pollution. Mercury falls from the air and can accumulate in streams and oceans and is turned by bacteria into methylmercury in the water. Fish absorb the methylmercury as they feed in these waters and so it builds up in them. Nearly all fish and shellfish contain traces of methylmercury. However, larger fish (swordfish, shark, king mackerel, tuna and tilefish) that have lived longer have the highest levels of methylmercury because they have had more time to accumulate it. It is the type of mercury that can be harmful to young people. The FDA action level of methylmercury is 1.0 ppm in the edible portion of fish.

Aquaculture Drugs

A range of veterinary drugs including antimicrobial, antiparasitical and growth promoters (hormones) may be used in aquaculture to control bacterial, fungal and parasitic diseases and to control reproduction of fish. Farmers may also use a range of vitamins, immunostimulants, disinfectants and other chemotherapeutants and employ chemicals for pond soil and water treatment.

Abuse of veterinary drugs, non-respect of the withdrawal period or application of illegal drugs constitutes a potential food safety problem. The health consequences of excessive use of antimicrobial drug residues include allergies, toxic effects, changes in colonization patterns of human-gut flora and acquisition of drug resistance in pathogens. The establishment of appropriate withdrawal periods ensures that no harmful residues remain in edible tissues after use of a drug. Since fish are poikilotherms, their metabolic rate is determined by environmental temperatures. As a result, withdrawal periods are based on time and temperature, i.e. degree-days: for example, 10 days at 5°C equals 150 degree-days. Where the legislation or implementation of regulation is poor, the risk of non-compliance is greater.

Additionally, the impacts of many of these chemicals on the environment are unknown and their release into the environment is likely to have a negative effect. They can also affect microbial biodiversity and contribute to the development of antimicrobial drug resistance. Application of vaccines has been instrumental to reduce use of drugs in the farmed salmon industry.

Compliance with MRLs for products from aquaculture is beginning to be enforced. For instance, the European Union is in the course of implementing a monitoring program in which fish muscle tissue will be routinely sampled for the presence of a range of veterinary drug residues. Such monitoring programs help provide assurance that no unacceptable human health risk is posed by veterinary drug residues in products from aquaculture. Unfortunately, some countries implement monitoring programs for export products but do not offer the same assurance for domestic markets.

Reasons for the use of drugs in aquaculture include the need to (1) treat and prevent disease, (2) control parasites, (3) affect reproduction and growth, and (4) provide tranquilization (e.g. for weighing). Relatively few drugs have been approved for aquaculture. Use of unapproved drugs or misuse of approved drugs in aquacultured fish poses a potential human health hazard. These substances may be toxic, allergenic or carcinogenic, and/or may cause antibiotic resistance in pathogens that affect humans.

More details on environmental chemical contaminants and pesticides can be found in Di Giulio and Hinton (2008), FDA (2011), Hristovski and Stojanovski (2005) and Treves-Brown (2000).

PATHOGENIC BACTERIAL GROWTH AND TOXIN FORMATION

Time and Temperature Abuse

Pathogenic bacteria growth and toxin formation as a result of time and temperature abuse of fish and fishery products can cause consumer illness. This hazard is limited to bacterial pathogens since viral pathogens (viruses) are not able to grow in food. Of particular concern in seafood are the pathogenic forms of *Listeria monocytogenes, Vibrio vulnificus, Vibrio parahaemolyticus, Vibrio cholera, Escherichia coli, Salmonella* spp., *Shigella* spp., *Staphylococcus aureus, Clostridium perfringens, Bacillus cereus, Campylobacter jejuni* and *Yersinia enterocolitica.*

Pathogenic bacteria can enter the process on raw materials. They can also be introduced into foods during processing from the air, unclean hands, insanitary utensils and equipment, contaminated water or sewage and through cross-contamination between raw and cooked product. The primary method for control is to reduce levels through cooking or other treatments when feasible, and minimize the potential for recontamination and to maintain products at temperatures that do not support growth of pathogenic bacteria.

Growth rates of pathogens are highly temperature dependent. Ordinarily, pathogenic bacteria growth is relatively slow at temperatures below 20°C.

Time and temperature abuse occurs when a product is allowed to remain at temperatures favorable to pathogenic bacteria growth for sufficient time to result in unsafe levels of pathogenic bacteria or their toxins in the raw fish and fishery products (e.g. raw molluscan shellfish). Certain pathogenic bacteria grow well, and others do not. Those that grow well in time and temperature-abused raw fish include: *V. vulnificus, V. parahaemolyticus, V. cholerae* and *L. monocytogenes*. Others may grow if the natural condition of the raw fish is changed, such as through salting or reduced oxygen packaging. Those that ordinarily do not grow well, because they compete poorly with the normal spoilage bacteria, include: *C. jejuni*, pathogenic strains of *E. coli, Salmonella* spp., *Shigella* spp., *S. aureus, C. perfringens, B. cereus* and *Y. enterocolitica*.

Most pathogenic bacteria will grow well in temperature-abused cooked fish if their growth is not controlled by means such as drying, salting or acidification, because competing bacteria are destroyed by the cooking process.

Certain pathogenic bacteria are associated with specific food sources, and it may not be necessary to assume that they will be present in other foods unless introduced from a contaminated source. For example, *V. vulnificus, V. parahaemolyticus* and *V. cholerae* non-O1 and non-O139 are generally associated with marine and estuarine species of fish and not with freshwater species or non-fishery ingredients.

The infective dose or toxic dose is the total number of a pathogen, or the total amount of a toxin, that is necessary to produce human illness. The dose often varies considerably for a single pathogen based on the health of the consumer and the virulence (infective capacity) of the particular strain of the pathogen.

In humans, usually, gastrointestinal symptoms are included: nausea, vomiting, abdominal pain, abdominal cramps, diarrhea, dehydration, electrolyte imbalance, high body fluid acidity, fever, headache, muscle pain, malaise and general discomfort. Septicemia rarely appears. Symptoms usually start from 1 or few hours – 1 or few days after consumption and usually last from 1–10 days. Everyone is susceptible to pathogenic bacteria food poisoning, but it is more common in infants, the young, the elderly, the infirm, those with underlying chronic disease, with reduced stomach acidity or malnutrition, and the immunocompromised.

Strategies for Control of Pathogenic Bacteria

Management of time and temperature of product exposure is important to produce a safe product. There are a number of strategies for the control of pathogenic bacteria in fish and fishery products. They include:

- Managing the amount of time that food is exposed to temperatures that are favorable for pathogen growth and toxin production.
- Killing pathogenic bacteria by cooking, pasteurization or by retorting.
- Killing pathogenic bacteria by processes that retain the raw product characteristics.
- Controlling the amount of moisture that is available for pathogen growth (water activity) in the product by drying or formulation.
- Controlling the amount of salt or preservatives, such as sodium nitrite, in the product.
- Controlling the level of acidity (pH) in the product.
- Controlling the introduction of pathogenic bacteria after the pasteurization process.
- Controlling the source of molluscan shellfish and the time from exposure to air (e.g. by harvest or receding tide) to refrigeration to control pathogens from the harvest area.

Inadequate Drying

Dried products are usually considered shelf stable and are, therefore, often stored and distributed unrefrigerated. Examples of shelf-stable dried fish products are salmon jerky, octopus chips, dried shrimp, stock fish and shark cartilage. The characteristic of dried foods that makes them shelf stable is their low water activity (a_w). Water activity is the measure of the amount of water in a food that is available for the growth of microorganisms, including pathogenic bacteria. Pathogenic bacteria growth and toxin formation in the finished product as a result of inadequate drying of fishery products can cause consumer illness. A water activity of 0.85 or below will prevent the growth and toxin production of all pathogenic bacteria, including primary pathogens *S. aureus* and *C. botulinum*, and is critical for the safety of a shelf-stable dried product. *S. aureus* grows at a lower water activity than other pathogenic bacteria, and should, therefore, be considered the target pathogen for drying for shelf-stable products.

Cooking or Pasteurization

The survival of pathogenic bacteria through cooking or pasteurization can cause consumer illness. The primary pathogens of concern are *Clostridium botulinum, Listeria monocytogenes, Campylobacter jejuni,* pathogenic strains of *Escherichia coli, Salmonella* spp., *Shigella* spp., *Yersinia enterocolitica, Staphylococcus aureus, Vibrio cholera, V. vulnificus* and *V. parahaemolyticus.*

In addition to eliminating bacterial pathogens, cooking and pasteurization also greatly reduce the number of spoilage bacteria present in the fishery product. These bacteria normally restrict the growth of pathogens through competition. Elimination of spoilage bacteria allows rapid growth of newly introduced pathogenic bacteria. Pathogenic bacteria that may be introduced after cooking or pasteurization are, therefore, a concern. This is especially true for pasteurization, because that process can significantly extend the shelf-life of the fishery product, providing more time for pathogenic bacteria growth and toxin formation.

Retorting is a heat treatment that eliminates all foodborne pathogens and produces a product that is shelf stable.

There is a potential that *C. botulinum* type E or non-proteolytic types B and F will survive the pasteurization process and grow under normal storage conditions or moderate abuse conditions.

If the product is not reduced oxygen packaged, or contains a barrier that is sufficient to prevent the growth and toxin formation by *C. botulinum* type E or non-proteolytic types B and F, or is equipped with a time and temperature integrator, or is distributed frozen, then selection of another target pathogen may be appropriate. *L. monocytogenes* may be selected as the target pathogen for pasteurization of this type of product because it is the most resistant bacterial pathogen of public health concern that is reasonably likely to be present. Generally, *L. monocytogenes* is regarded as the most heat-tolerant, foodborne bacterial pathogen that does not form spores.

It is not practical to target viral pathogens in cooking or pasteurization processes because of their extreme heat resistance. Viral pathogens should be controlled through a rigorous sanitation regime as part of a prerequisite program or as part of hazard analysis critical control point (HACCP) itself.

Processes Designed to Retain Raw Product Characteristics

Some processes are designed to reduce specific pathogens to acceptable levels while retaining the sensory qualities (appearance, taste and texture) of the raw product. These processes are particularly useful in addressing the hazard associated with the target pathogen in raw products such as raw molluscan shellfish (i.e. oysters, clams, mussels and whole and roe-on scallops) that are intended for the raw ready-to-eat market. Because these processes do not eliminate all pathogens of public health concern, they are not considered cooking or pasteurization processes.

Examples of processes designed to retain raw product characteristics include:

- High hydrostatic pressure processing (HPP);
- Individual quick freezing (IQF) with extended frozen storage;
- Mild heat processing;
- Irradiation (ionizing radiation).

The survival of pathogenic bacteria through processes designed to retain raw product characteristics can cause consumer illness. The primary pathogens of concern are *Vibrio vulnificus* and *Vibrio parahaemolyticus*. *V. vulnificus* and *V. parahaemolyticus* are naturally occurring pathogens (i.e. not associated with human or animal sources) that may be present in fish and fishery products, and in particular raw molluscan shellfish.

Cross-Contamination of Fish and Fish Products

With fishery products, pasteurization is usually performed after the product is placed in the hermetically sealed finished product container. It is applied to fishery products that are distributed either refrigerated or frozen. Examples of pasteurized fishery products are: pasteurized crabmeat, pasteurized surimi-based analog products and pasteurized lobster meat. Because these products are cooked before they are packaged, they are at risk for recontamination between cooking and packaging. The risk of this recontamination may be minimized by filling directly from the cook kettle using a sanitary, automated, continuous-filling system (designed to minimize the risk of recontamination) while the product is still hot (i.e. hot filling). This control strategy may not be suitable for products such as crabmeat, lobster meat or crayfish meat.

There are three primary causes of recontamination after pasteurization and cooking performed immediately before reduced oxygen packaging:

- Defective container closures;
- Contaminated container cooling water;
- Recontamination between cooking and reduced oxygen packaging.

The introduction of pathogenic bacteria after pasteurization and certain specialized cooking processes can cause consumer illness. The primary pathogens of concern are *Clostridium botulinum, Listeria monocytogenes, Campylobacter jejuni*, pathogenic strains of *Escherichia coli, Salmonella* spp., *Shigella* spp., *Yersinia enterocolitica, Staphylococcus aureus, Vibrio cholerae, V. vulnificus* and *V. parahaemolyticus*.

For more on the growth of and toxin production by pathogenic microbes see FDA (2011) and Hristovski and Stojanovski (2005).

References

Ackers, M.L., Mahon, B.E., Leahy, E., Goode, B., Damrow, T., Hayes, P.S., et al., 1998. An outbreak of *Escherichia coli* O157:H7 infections associated with leaf lettuce consumption. J. Infect. Dis. 177, 1588–1593.

Altekruse, S.F., 1994. Food and animal sources of human Campylobacter jejuni infection. J. Am. Vet. Med. Assoc. 204, 57–61.

Banatvala, N.A.R., Magnano, M.L., Carter, T.J., Barret, W.F., Bibb, L.L., Vasile, P., et al., 1996. Meat grinders and molecular epidemiology: two supermarket outbreaks of *Escherichia coli* O157:H7 infection. J. Infect. Dis. 173, 480–483.

Bean, N.H., Griffin, P.M.G., 1990. Foodborne disease outbreaks in the US 1973–1987: pathogens, vehicles, and trends. J. Food Prot. 53, 804–817.

Besser, T.E., Goldoft, M., Pritchett, L.C., Khakhria, R., Hancock, D.D., Rice, D.H., et al., 2000. Multiresistant Salmonella Typhimurium DT104 infections of humans and domestic animals in the Pacific Northwest of the United States. Epidemiol. Infect. 124, 193–200.

Beuchat, L.R., 1996. Pathogenic microorganisms associated with fresh produce. J. Food Prot. 59, 204–216.

Beuchat, L.R., 1999. Survival of enterohemorrhagic *Escherichia coli* O157:H7 in bovine feces applied to lettuce and the effectiveness of chlorinated water as a disinfectant. J. Food Prot. 62, 845–849.

Beuchat, L.R., 2002. Ecological factors influencing survival and growth of human pathogens on raw fruits and vegetables. Microbes. Infect. 4, 413–423.

Bowman, G., Shulaw, W., 2001. On-Farm Biosecurity: Traffic Control and Sanitation. D.V.M. Ohio State University Extension Factsheet., 6/2001 <http://ohioline.osu.edu/vme-act/0006.html>.

Brandl, M.T., Mandrell, R.E., 2002. Fitness of Salmonella *enterica* serovar Thompson in the cilantro phyllosphere. Appl. Environ. Microbiol. 68, 3614–3621.

Bunning, V.K., Donnelly, C.W., Peter, J.T., Briggs, E.H., Bradshaw, J.G., Crawford, R.G., et al., 1988. Thermal inactivation of Listeria monocytogenes within bovine macrophages. Appl. Environ. Microbiol. 54, 364–370.

Burnett, S.L., Beuchat, L.R., 2001. Human pathogens associated with raw produce and unpasteurized juices, and difficulties in contamination. J. Indust. Microbiol. Biotechnol. 27, 104–110.

CAC. Hazard Analysis and Critical Control Point (HACCP) System and Guidelines for its Application. Annex to CAC/RCP 1-1969 (Rev. 4–2003), pp. 21–31.

Chalmers, R.M., Aird, H., Bolton, F.J., 2000. Waterborne *Escherichia coli* O157. J. Appl. Microbiol. 88, 124S–132SS.

Di Giulio, R.T., Hinton, D.E., 2008. The Toxicology of Fishes. CRC Press. Taylor & Francis Group, Boca Raton, London, New York.

Dorn, C.R., 1993. Review of food borne outbreak of *Escherichia coli* O157:H7 infection in the western United States. J. Am. Vet. Med. Assoc. 203, 1583–1587.

Dulic, Z., Kljujev, I., Raicevic, V., Zivic, I., Markovic, Z., Stankovic, M., et al., 2008. Estimation of irrigation water quality using coliform bacteria, zooplankton and zoobenthos as indicators. Arch. Biol. Sci. Belgrade 60 (1), 11P–12P.

EC. Council Directive 2006/88/EC concerning animal health requirements for aquaculture animals (and products), the placing on the market and the prevention and control of certain diseases in aquatic animals. Eureopean Commission, Brussels.

EC. Council Directive 91/67/EEC of 28 January 1991 concerning the animal health conditions governing the placing on the market of aquaculture animals and products. European Commission, Brussels.

EFSA, 2012. The European union summary report on trends and sources of zoonoses, zoonotic agents and food-borne outbreaks in 2010. EFSA J. 10 (3), 2597. [442 pp.].

Fahey, T., Morgan, D., Gunneburg, C., Adak, G.K., Majid, F., Kaczmarski, E., 1995. An outbreak of Campylobacter jejuni enteritis associated with failed milk pasteurization. J. Infect. 31, 137–143.

FAO/IDF, 2011. Guide to Good Dairy Farming Practice. Food and Agriculture Organization of United Nations and International Dairy Federation, Roma, Food and Agriculture Organization, Rome <http://www.fao.org/docrep/014/ba0027e/ba0027e00.htm>.

FAO/OIE, 2009. Guide to Good Farming Practices for Animal Production Food Safety. Food and Agriculture Organization, Rome, <http://www.oie.int/fileadmin/Home/eng/Current_Scientific_Issues/docs/pdf/eng_guide.pdf>.

FAO/WHO, 1997. Food Hygiene Basic Texts. Codex Alimentarius Commission. Joint FAO/WOHO Food Standards Programme. <ftp://ftp.fao.org/docrep/fao/005/Y1579e/Y1579e.pdf>.

FDA, 2011. Fish and Fishery Products. Hazards and Controls Guidance, fourth ed. US Department of Health and Human Services. Center for Food Safety and Applied Nutrition. Food and Drug Administration, Washington DC.

FDA, 2013a. Analysis and Evaluation of Preventive Control Measures for the Control and Reduction/Elimination of Microbiological Hazards on Fresh and Fresh Cut Produce. Food and Drug Administration, Washington DC, <http://www.fda.gov/Food/ScienceResearch/ResearchAreas/SafePracticesforFoodProcesses/ucm091270.htm>.

FDA. 2013b. Standards for the Growing, Harvesting, Packing and Holding of Produce for Human Consumption. Docket No. FDA 2011-N-0921. RIN0910-AG35. Food and Drug Administration, Washington DC. <http://www.ofr.gov/(S(mocmo3lgsnz3vp25he34mipn))/OFRUpload/OFRData/2013-03778_PI.pdf>.

Fedorka-Cray, P.J., Dargatz, D.A., Thomas, L.A., Gray, J.T., 1998. Survey of Salmonella serotypes in feedlot cattle. J. Food Prot. 61, 525–530.

Fukushima, H., Hoshina, K., Gomyoda, M., 1999. Long-term survival of Shiga toxin producing *Escherichia coli* O26, O111 and O157 in bovine feces. Appl. Environ. Microbiol. 65, 5177–5181.

Gary, L., Bowman, D.V.M., Shulaw, W.P., 2001. Disinfection in On-Farm Biosecurity Procedures. D.V.M. Ohio State University Extension Factsheet., 6/2001 <http://ohioline.osu.edu/vme-fact/0008.html>.

Guo, X., van Iersel, M.W., Chen, J., Brackett, R.E., Beuchat, L.R., 2002. Evidence of association of Salmonellae with tomato plants grown hydroponically in inoculated nutrient solution. Appl. Environ. Microbiol. 68, 3639–3643.

Hamilton, A., Stagnitti, F., Hale, G., 2006. Quantitative microbial risk assessment models for consumption of raw vegetables irrigated with reclaimed water. Appl. Environ. Microbiol. 72 (5), 3284–3290.

Hancock, D.D., Besser, T.E., Rice, D.H., Ebel, E.D., Herriot, D.E., Carpenter, L.V., 1998. Multiple sources of *Escherichia coli* O157 in feedlots and dairy farms in the Pacific Northwest. Prev. Vet. Med. 35, 11–19.

Havelaar, A.H., Haagsma, J.A., Mangen, M.J.J., et al., 2012. Disease burden of foodborne pathogens in The Netherlands, 2009. Int. J. Food Microbiol. 156, 231–238.

Hristovski, M., Stojanovski, S., 2005. Biology, Culture and Pathology of Fish (in Macedonian). University book. Cetis print. Skopje.

Islam, M., Morgan, J., Doyle, M., Phatak, S., Millner, P., Jiang, X., 2004. Fate of *salmonella enterica* serovar typhimurium on carrots and radishes grown in fields treated with contaminated manure compost or irrigation water. Appl. Environ. Microbiol. 70 (4), 2497. DOI:10.1128/AEM.70.4.2497-2502.2004.

Käferstein, F., Motarjemi, Y., Moy, G., 2004. The problems related to our food supply Food Safety – Its Role in Health and Development. UNESCO Encyclopedia of Life Science.

Kathariou, S., 2002. Listeria monocytogenes virulence and pathogenicity, a food safety perspective. J. Food Prot. 65, 1811–1829.

Kljujev, I., 2012. Contamination of plants by pathogenic bacteria from irrigation water. Doctoral Dissertation. Faculty of Agriculture, University of Belgrade.

Kljujev, I., Raicevic, V., 2006. Dynamics in Coliform Bacteria Count in Waters from the Experimental Fields of Faculty of Agriculture, Belgrade, Serbia and Montenegro. The International scientific conference BALWOIS 2006 – Conference on Water Observation and Information System for Decision Support. Ohrid, Republic of Macedonia, 23–26 May 2006 <www.balwois.org.accueil.html>.

Kljujev, I., Lalevic, B., Petrovic, J., Kikovic, D., Raicevic, V., 2011. Colonization fresh vegetables by Listeria monocytogenes. In Proceedings of the Microbiologia Balkanica 2011–Seventh Balkan Congress of Microbiology Eighth Congress of Serbian Microbiologists, Oktobar 2011. ISBN 978-86-914897-0-01. Udruženje mikrobiologa Srbije, Beograd.

Kljujev, I., Raicevic, V., Andrews, S., Jackson, R., Lalevic, B., Dorati, F., 2012. Transmission of *E. coli* from contaminated irrigation water and soil to plant tissue. J. Hyg. Eng. Design Original scientific paper, UDC 631.461:579.842.11; 628.034.3:579.842.11.

Larson, J., 2010. Ensuring global food safety – a public health priority and a global responsibility. In: Boisrobert, C., Stjepanovic, A., Oh, S., Lelieveld, H. (Eds.), Ensuring Global Food Safety Elsevier, UK.

McEwen, S.A., Fedorka-Cray, P.J., 2002. Antimicrobial use and resistance in animals. Clin. Infect. Dis. 34, S93–S106.

Morris, C.E., Monier, J.E., Jacques, M.A., 1997. Methods for observing microbial biofilms directly on leaf surfaces and recovering them for isolation of culturable microorganisms. Appl. Environ. Microbiol. 63, 1570–1576.

Motarjemi, Y., Steffen, R., Binder, H.J., 2012. Preventive strategy against infectious diarrhea—a holistic approach. Gastroenterology 143, 516–519.

Murinda, S.E., Nguyen, L.T., Ivey, S.J., Gillespie, B.E., Almeida, R.A., Oliver, S.P., 2002. Prevalence and molecular characterization of *Escherichia coli* O157:H7 in bulk tank milk and fecal samples from cull dairy cows: a 12-month survey of dairy farms in East Tennessee. J. Food Prot. 65, 752–759.

NACMF (National Advisory Committee on Microbiological Criteria for Foods), 1999. Microbiological safety evaluations and recommendations on fresh produce. Food Control 10, 117–143.

OIE, 2012. Manual of Diagnostic Tests for Aquatic Animals. World Organization for Animal Health, Paris.

Oliver, S.P., Patel, D.A., Callaway, T.R., Torrence, M.E., 2009. ASAS Centennial Paper: developments and future outlook for preharvest food safety. J. Anim. Sci. 87, 419–437.

Ostrander, G., 2000. The Laboratory Fish (Handbook of Experimental Animals). Academic Press.

Pejanović, R., 2008. Koncept dobre poljoprivredne prakse. Poslovna Politika 37 (6–7), 48–52.

Rabsch, W., Altier, C., Tschape, H., Baumler, A.J., 2003. Foodborne *Salmonella* infections Microbial Food Safety in Animal Agriculture. Blackwell Publishing Company., pp. 97–107.

Rocourt, J., Moy, G., Vierk, K., Schlundt, J., 2003. Present state of foodborne disease in OECD countries. Unpublished WHO document. WHO: Geneva. <http://www.who.int/foodsafety/publications/foodborne_disease/oecd_fbd.pdf>.

Snively, D.W. Biosecurity on the farm. Extension Service West Virginia University. <http://www.wvu.edu/~agexten/Biosecure/Farm.pdf> Accessed July 2013.

Solomon, E.B., Brandl, M.T., Mandrell, R.T., 2006. Biology of foodborne pathogens on produce. In: Matthews, K.R. (Ed.), Emerging Issues in Food Safety: Microbiology of Fresh Produce ASM Press, Washington, DC, pp. 55–83. (ARIS #179193).

Solomon, E.B., Yaron, S., Matthews, K.R., 2002. Transmission of *Escherichia coli* O157:H7 from contaminated manure and irrigation water to lettuce plant tissue and its subsequent internalization. Appl. Environ. Microbiol. 68, 397–400.

Stojanović, Z., Vukićević, Z., Hrgović, N., 2003. Zoohigijena. Zavod za udžbenike I nastavna sredstva. Beograd.

Tarr, P.I., 1995. *Escherichia coli* O157:H7: clinical, diagnostic, and epidemiological aspects of human infection. Clin. Infect. Dis. 20 (1), 1–8. quiz 9–10.

Tauxe, R.V., 1997. Emerging foodborne diseases: an evolving public health challenge. Emerg. Infect. Dis. 3 (4) National Center for Infectious Diseases, Atlanta, GA <http://www.cdc.gov/ncidod/eid/vol3no4/tauxe.htm>.

Tauxe, R.V., 2002. Emerging food borne pathogens. Int. J. Food Microbiol. 78, 31–41.

Treves-Brown, K.M., 2000. Applied Fish Pharmacology. Kluwer Academic Publishers, Dordrecht, Boston, London.

Vugia, D., Cronquist, A., Hadler, J., Tobin-D'Angelo, M., Blythe, D., Smith, K., et al., 2007. Preliminary FoodNet data on the incidence of infection with pathogens transmitted commonly through food – 10 states, 2006. MMWR 56, 336–339.

Wachtel, M.R., Whitehand, L.C., Mandrell, R.E., 2002. Association of *Escherichia coli* O157:H7 with preharvest leaf lettuce upon exposure to contaminated irrigation water. J. Food Prot. 65, 18–25.

Wang, G., Doyle, M.P., 1998. Survival of enterohemorrahagic *Escherichia coli* O157:H7 in water. J. Food Prot. 61, 662–667.

Woo, P.T.K., 2006. *Fish Diseases and Disorders*. Vol. 1: *Protozoan and Metazoan Infections*, second ed.. CAB International.

Woo, P.T.K., Bruno, D.W., 1999. *Fish Diseases and Disorders*. Vol. 3. *Viral, Bacterial and Fungal Infections*. CAB International.

CHAPTER

24

Hygiene in Food Processing and Manufacturing

John Holah

Campden BRI, UK

OUTLINE

Food Safety Management.
DOI: http://dx.doi.org/10.1016/B978-0-12-381504-0.00024-X

INTRODUCTION

The primary concern of food manufacturers is to produce a product that is both wholesome, i.e. it has all appropriate organoleptic qualities, and safe, i.e. it is free from pathogenic microorganisms and chemical and foreign body contamination. The schematic diagram shown in Figure 24.1 is a historical representation of the food industry and shows that the manufacture of safe, wholesome foods stemmed from the purchase of specified raw materials. Not all raw materials may be of the same quality, as the final product will be priced according to the intended market, though all raw materials should be safe and free from specified hazards. Figure 24.1 also shows that given specified raw materials, there are four major "building blocks" that govern the way the factory is operated to ensure that the safe, wholesome food goal is realized. Hygienic design dictates the design of the factory infrastructure and, until replaced by robots, the operatives! Hygienic practices maintain the integrity of the facility and include good hygienic practices (GHP) and good manufacturing practices (GMP). Process development enables the design of safe, validated products and processes, while process control subsequently ensures that each product in each batch on every day meets the product and process requirements.

More specifically, hygienic design is the food manufacturing infrastructure and consists of all the physical requirements necessary to manufacture the food product. Specifically, it includes the following:

- Factory site
- Factory building
- Segregation
- Food defense, biovigilance and bioterrorism
- Process lines

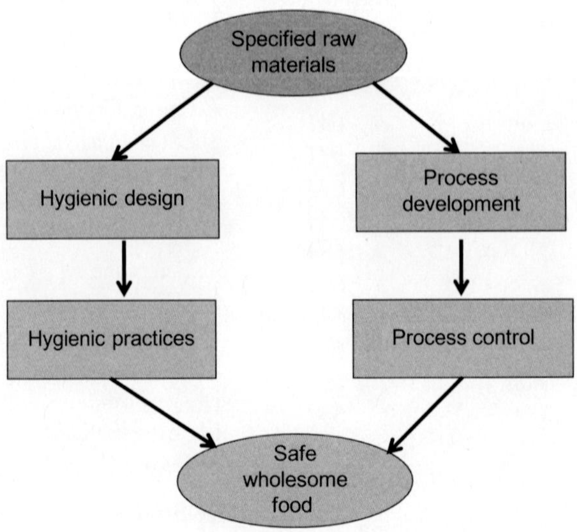

FIGURE 24.1 Schematic stages required to ensure safe, wholesome food products from a traditional basis.

- Ventilation and air flows
- Equipment
- Utensils
- Services (utilities)
- Waste disposal
- Medical screening

Hygienic practices are all the actions necessary to maintain the food manufacturing infrastructure in a hygienic manner and thus facilitate safe and wholesome food manufacture by preventing contamination to the food product, and include the following:

- Maintenance
- Housekeeping
- Cleaning and disinfection
- CIP
- Pest control
- Personal hygiene

The major change in the way that the food industry viewed food safety came with the advent of hazard analysis typified by the hazard analysis critical control point (HACCP) approach, originally developed in the late 1960s by the Pillsbury Company, the United States Army Laboratory and the National Aeronautics and Space Administration in the development of safe foods for the American Space Program. HACCP began to be practiced in the UK in the later 1980s and was published as international guidance in 1993 by Codex (Anon, 1993). Indeed, in the EU it has been a legislative requirement for food businesses since the publication of Council Regulation 852/2004 (Anon, 2004) to manage food safety with an appropriate hazard analysis-based system.

Hazard analysis via HACCP encompasses identifying the hazards that may affect the quality or safety of the food product and controlling them at all stages of the process such that product contamination is prevented. Such hazards are usually described as:

- Biological, e.g. bacteria, yeasts, molds
- Chemical, e.g. cleaning chemicals, lubricating fluids
- Physical, e.g. glass, plastic, insects or their parts, pests, metal, dust

Figure 24.2 illustrates how this affected the traditional food safety approach as identified in Figure 24.1. Hazard analysis can be viewed as an umbrella and oversees all actions related to food safety and food wholesomeness. HACCP primarily focuses on identifying hazards and controlling them via the process route, which encompasses the selection and storage of raw materials, followed by their processing and packaging and is illustrated in Figure 24.2 by the dark coloration of the "specified raw materials" box and the dark circle encircling the "product development" and "process control" boxes. HACCP recognizes, however, that the factory infrastructure, GMPs and GHPs, expressed as the "hygienic design" and "hygienic practices" boxes in Figure 24.2 are the bedrock on which the assessment and control of the hazards identified in the process are built. HACCP calls the factory infrastructure, GMPs and GHPs "HACCP prerequisites," implying that these should be in place before a hazard analysis of the process is undertaken, and these are identified by the lighter circle in Figure 24.2.

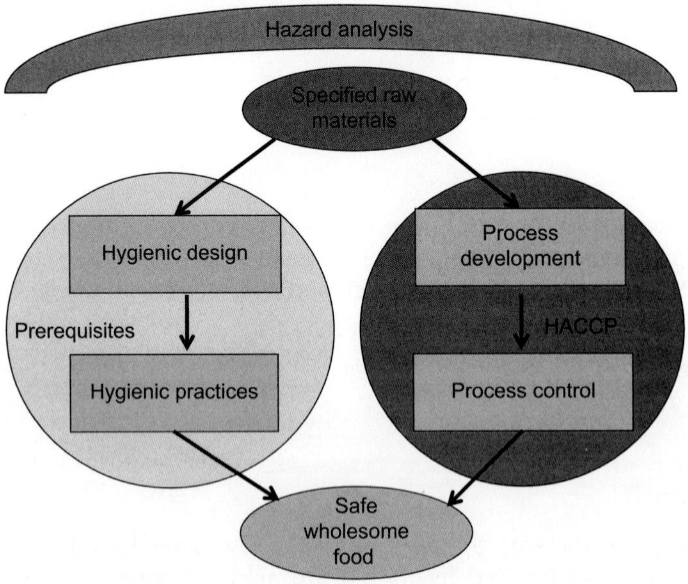

FIGURE 24.2 The advent of HACCP encouraged food manufacturers to consider the control of hazards both via the production process (specified raw materials, process development and process control) and the processing environment (hygienic design and hygienic practices) referred to as the HACCP prerequisites.

Although some principles of good hygienic practice are generic and apply to any type of business, some may be specific to a given category of food products. For instance, prerequisite programs for infant formula factories may not be the same as for coffee or fishery factories. Therefore, for the prerequisites, a hazard analysis of the product should also be undertaken at the earliest opportunity and, if possible, before the design and construction of the processing facility. This allows the design of the production facility to play a major role in reducing risks of contamination. For example, generally it is possible to identify that glass is a potential hazard and you could eliminate this hazard by designing a glass-free factory. Or more specifically in a dry foods factory, all sources of water could be eliminated from certain processing areas.

The next development in managing food safety came with the adoption of quality or business management systems. These were used as a mechanism by which food suppliers could demonstrate to their customers that they were undertaking a standardized approach to ensuring the quality of the food product. As for HACCP, the use of quality management systems began in the 1980s and was focused on BS 5750 published in 1978; this was the first published quality management system. This in turn had been developed from BS 5179 in 1969 and before that a Department of Defence standard developed to ensure the successful application of munitions. As well as providing a standard approach to managing quality, it also encouraged third party organizations to audit food companies against the standard. The adoption of BS 5750 quickly developed beyond the UK and, in 1987, this document was published as EN 29002-1987 in Europe and worldwide as ISO 9002-1987. The ISO 9000 series of standards was developed to cover the requirements of both food manufacturers and auditing bodies to ensure food manufacturers met the needs of their customers while

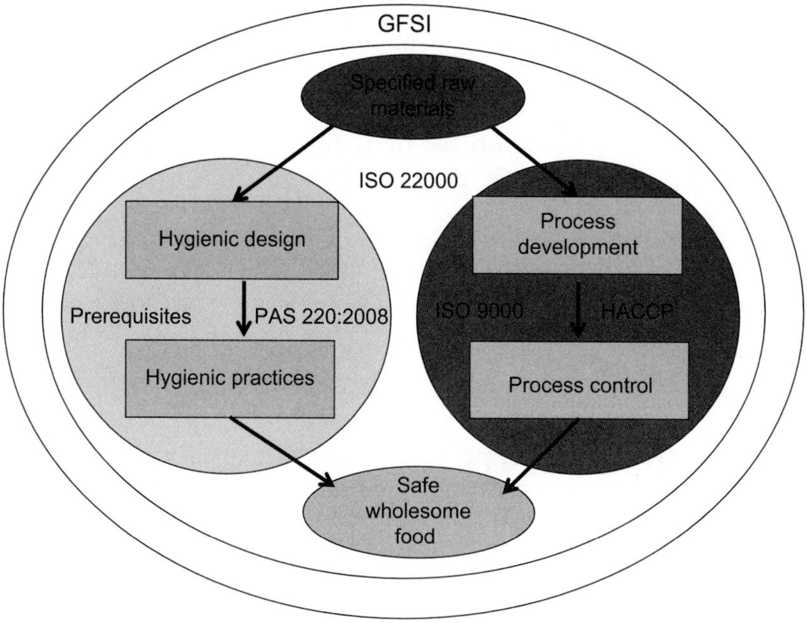

FIGURE 24.3 Current schematic food processing stages, food manufacturing standards and auditing schemes to allow food manufacturers to ensure and demonstrate safe, wholesome products.

striving to meet their expectations. Many food manufacturers still use ISO 9001:2008 (Anon, 2008a) for managing their business or quality systems, though it has little impact on prerequisites. It is noted in the dark HACCP circle in Figure 24.3.

ISO 22000 (Anon, 2005) was published in 2005 with the aim of bringing together food safety and quality management systems into one auditable standard. It supports four key elements to food safety, which it describes as interactive communication, system management, prerequisite programs and HACCP. It has also been aligned with the requirements of ISO 9001 and the HACCP principles as adopted by Codex. ISO 22000 is thus represented as a large circle in Figure 24.3 encompassing all the requirements necessary to produce safe and wholesome food products.

There was a general recognition that while there was much publically available information to support much of the four key elements in ISO 22000, particularly quality management systems and HACCP, information on prerequisite programs as a whole was lacking. A publically available specification, PAS 220:2008 (Anon, 2008b), was published by British Standards to provide additional information in this area. This document is intended to support the prerequisite program requirements of ISO 22000 (clause 7) and is noted in Figure 24.3 within the light prerequisite circle.

Finally, to complete the food safety picture, there have been complementary developments in third party auditing systems under the management of the Global Food Safety Initiative (GFSI) (www.mygfsi.org). This organization has harmonized the requirements of the major food standards organizations that produce auditable food standards such as the

British Retail Consortium (BRC, Anon, 2011a), International Food Standard (Anon, 2012) and SQF (Anon, 2008c). For example, the BRC global standard for food safety requires food manufacturers to develop and comply with the following four key elements: senior management commitment, a quality management system, prerequisite programs and an HACCP plan. Auditing standards approved by GFSI clearly then meet the requirements to audit all parts of ISO 22000 and this is represented by the circle surrounding the ISO 22000 circle in Figure 24.3.

Other than the legal requirement in some countries (e.g. in the European Union) for food manufacturers to adopt HACCP principles, the requirement for ISO 9000 or 22000 is voluntary. It may, however, be a requirement for food manufacturers to adopt these standards to comply with their customers' requirements, which may also include the need to be audited to a GFSI approved standard.

PREREQUISITE MANAGEMENT PLAN

The need to understand the role of the processing environment in how it can harbor hazards, which can be transferred on vectors to the product, has become especially important for microbial pathogen hazards in ready-to-eat (RTE) foods. While HACCP has undoubtedly improved food safety as a whole, foodborne diseases are unfortunately still highly prevalent. For instance, in both North America and Europe there have been major incidents with *Listeria* in chilled products and *Salmonella* in dry RTE products (Aavistland et al., 2001; Anon, 2008d, 2010; Centers for Disease Control and Prevention, 1998, 2008, 2009; Jackson et al., 2011). In the majority of these cases, contamination following the pathogen reduction stage (e.g. cooking or decontamination via chlorinated produce washing) is thought to be the route by which the product becomes contaminated. Post-process contamination is prevented and managed predominantly by prerequisite programs, and it can be argued that for certain products, such as RTE food products, to ensure the safety of these foodstuffs, the management of such prerequisites is as important as the management of critical control points (CCPs) where pathogens are killed, reduced to safe levels or prevented from growing.

To more effectively manage prerequisites, a prerequisite management plan (PMP) is advocated. The PMP is comprised of two elements: generic prerequisites and processing environment prerequisites. ISO 22000 and retailer audits require a number of essential hygienic practices, or prerequisites, in all food manufacturing environments. These prerequisites are generic and may be independent of the factory or the food manufacturing process it contains. Such prerequisites include those already detailed above as relating to the factory's hygienic infrastructure (e.g. factory and equipment design) and hygienic practices (e.g. maintenance, personnel hygiene and cleaning and disinfection) and should be appropriately validated, monitored and verified. Other prerequisites are recognized, e.g. in PAS 220:2008 (Anon, 2008b) management of purchased materials, rework, product recall procedures and product information/customer awareness are noted while others consider the control of foreign bodies in foods via, e.g., sieves and metal detectors as prerequisites. These prerequisites, however, relate more to quality practices or process control than to hygienic practices.

After the implementation of generic prerequisites, food manufacturers then have to consider the management of any residual hazards that may contaminate the food product during processing. Such hazards, and the transfer vectors via which they can contaminate the product, will be unique to each food manufacturer and each food manufacturing site. The assessment of hazards in the food processing environment, their risk and how they can be controlled to prevent contamination to the food product can be described as the processing environment plan (PEP).

Generic prerequisites should be undertaken to best practice standards and the following text for each of the hygienic infrastructure and practices prerequisites gives guidance as to how that can be ensured.

Factory Site

The factory site is designed to minimize the challenge of external hazards (insects, rodents, microorganisms, dust, soil, etc.) on the factory building envelope. This may be, for example, by reducing the number of pest harborage areas, controlling pest access to waste materials, reducing soil and dust, the downwind siting of effluent plants and the control of unauthorized public access (see Chapter 25 for further details).

Factory Building

The building envelope and air intakes provide a defense against external factory hazards. These could be microorganisms, pests, unauthorized human access, airborne chemical taints and airborne particulate matter. The building envelope also segregates food production from non-food production activities such as engineering workshops, boiler rooms, chemical stores, laboratories, offices, canteens, observation areas/viewing galleries, medical rooms and rest areas, etc. (see Chapter 25).

Segregation

Factories should be constructed as a series of zones and barriers that aim to limit the challenge of contaminants. The number of barriers created will be dependent on the nature of the food product, established from the HACCP study, and each barrier should reduce the challenge of a hazard to the subsequent barrier. Figure 24.4 shows that there are up to four levels of segregation that are typical for food plants. While these barriers were primarily conceived to control microbiological contamination, they are also effective at controlling many other hazards.

Level 1 represents the siting of the factory, the outer fence and the area up to the factory wall as noted above.

Level 2 represents the factory wall and other processes which should separate the factory from the external environment (e.g. rain, prevailing wind, surface run-off, delivery and dispatch vehicles, dust, odors, pests and uninvited people). Level 2 also includes all internal barriers designed to separate production stages (raw materials, intermediate product, finished product, packaged product), incompatible materials (wet, dry, chilled, frozen, allergenic, vegetarian, organic, genetically modified (GM) materials, Kosher or Halal, packaging) and non-food production areas (engineering, boiler rooms, cleaning stores, changing areas, etc.).

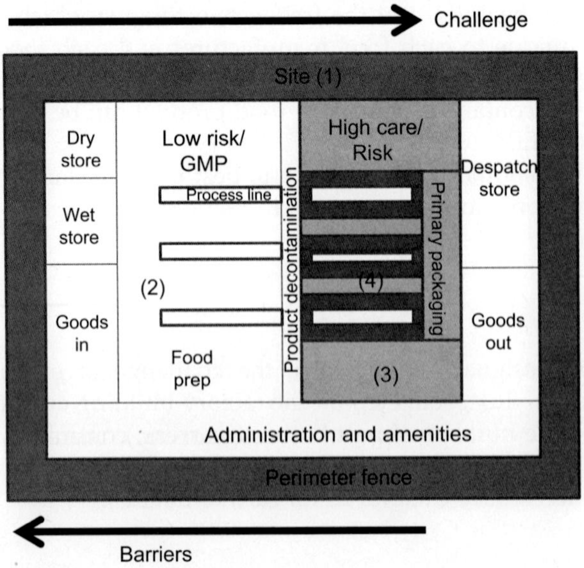

FIGURE 24.4 Schematic layout of a factory site showing 'barriers' against contamination. (1) Perimeter fence. (2) Main factory buildings. (3) Walls of high risk area. (4) Product enclosure within high risk.

Level 3 represents the internal barriers that are used to separate manufacturing processes of different microbiological risk, e.g. pre- and post-heat treatment. Product in level 3 will have a lower microbiological count than in level 2 and the microbial reduction process, incorporated as the barrier between level 2 and 3, may be a simple decontamination process (e.g. produce washing) or a pasteurization treatment (e.g. an oven, kettle or fryer). Such separation, which creates zones usually referred to as high care or risk areas, should seek to control the air, people and surfaces (e.g. the floor and drainage systems and the passage of materials and utensils across the barrier).

Level 4 represents a product enclosure zone, set within the level 3 high care/risk area. A product enclose zone could encompass true aseptic filling or "ultra clean" processing and packing areas such as glove boxes or the use of highly filtered air as a barrier around the process line (see Chapter 25).

Food Defense, Biovigilance and Bioterrorism

At all stages of production, processing and distribution, food must be protected against any contamination likely to render the food unfit for consumption and much of the protection requirements for the reduction/elimination of deliberate or accidental contamination of food products are similar, and are beneficial to both general food hygiene and bioterrorism control. The site and the production and storage areas of the factory buildings should be secured effectively by controlled access in order to prevent unauthorized entry. Site security should be reviewed and the need for fencing that fully encloses the site, close circuit television (CCTV), night lighting and/or security guards should be considered as part of a food defense program. Siting of silos, water tanks and packaging stores for raw materials, process steps or finished products outside the protection of factory buildings is not desirable as this may increase the chance of product adulteration. Bulk unloading equipment (pumps, pipes, augers, conveyors, etc.) should be owned and operated by the factory (i.e. not from the transport vehicle) and appropriately stored inside when not in use. Additional security requirements may be appropriate for any on-site laboratory, particularly if it handles pathogens or toxic chemicals (see Chapter 35).

Process Lines

Process lines are laid out so that they facilitate efficient food manufacturing, maintain product quality, reduce environmental contamination by effectively controlling the movement of raw materials and operatives, and allow space for maintenance and cleaning. The flow of ingredients and products is such that ideally raw materials enter at one end of the factory (dirty end) and are dispatched at the opposite end (clean end). There should be no backtracking or crossovers and, where there are changes in the direction of process flow, there must be adequate physical barriers. The flow of air and drainage should be away from "clean" areas towards "dirty" zones while the flow of discarded outer packaging materials should not cross or run counter to the flow of either unwrapped ingredients or finished products. Access of personnel and visitors should be controlled and the traffic pattern of personnel (and vehicles) should prevent contamination of the product (see Chapter 25).

Ventilation and Air Flows

Food factories must have sufficient means of natural or mechanical ventilation to provide fresh air for food operatives (approximately 8 liters per person per second) while not contaminating food products. Natural ventilation should be through openings which are directly connected to the outside air and so positioned in the external walls and/or roof that effective cross-ventilation is possible. Mechanical ventilation should be provided to control humidity and ambient temperatures and to effectively remove particulates, fumes, smoke, steam and vapors and microorganisms. Where there is a risk of microbiological contamination of the product by the surrounding air, the working area should be enclosed as far as possible and be maintained at a positive pressure using filtered air drawn from a clean source. The type of filters will depend on the product and process and range from dust exclusion filters to high efficiency particulate air (HEPA) filters. Air used for the transport of product must be dust filtered as a minimum and may require additional filtration, over and above that of the room air from the hygiene zone which it is in or being moved to. This is to maintain product safety and quality, as the interface between the air and the product during transport is greatly in excess of simple sedimentation to product from still air.

Equipment

Food equipment that is designed hygienically has three key advantages: food quality, reduced costs and food safety. Good hygienic design helps to ensure sustained food product quality by preventing residues remaining in the equipment that could contaminate subsequent different product batches or, in food manufacture, ensuring that product is not "held up" within the equipment where it could deteriorate and affect product quality on rejoining the main product flow. Good hygienic design reduces the time required for an item of equipment to be cleaned. This reduction of cleaning time, and the associated costs of staff training, etc., is significant over the lifetime of the equipment, which may be in excess of 25 years. Hygienically designed equipment, which may initially be more expensive (compared to similarly performing but poorly designed equipment), will be more cost effective in the long term. In addition, reduced downtime for cleaning may lead to the opportunity for increased food production. Finally, and perhaps most importantly, good hygienic design prevents the contamination of the product with substances that would adversely affect the health of the consumer. Such contamination could be microbiological (e.g. pathogens), chemical (e.g. lubricating fluids, cleaning chemicals) or physical (e.g. glass, metal swarf).

The hygienic design of equipment is a legal requirement in Europe under the auspices of Council Directive 98/37/EC (Anon, 1998a) on the approximation of the laws of member states relating to machinery. The hygienic design requirements in this directive were further developed as a European Standard and published as EN 1672-2 in 1997 (Anon, 1997a). The hygienic design of the requirements of EN 1672-2 and those of the 3-A standards group in the USA (www.3-a.org) was integrated into an international standard ISO/DIS 14159 (Anon, 2002). This standard described 11 hygienic design principles for food processing equipment, which are described below.

1. *Materials of construction* for food contact must have adequate strength over a temperature range to suit intended end-use exposure; have a reasonable life expectancy; be non-toxic, non-tainting, non-absorbent; be resistant to cracking, chipping, flaking corrosion and abrasion; prevent penetration of unwanted matter under intended use; and be easily cleaned. Stainless steel, because of its cleanability, corrosion resistance and wear resistance, usually meets all these requirements and there are various grades of stainless steel which are selected for their particular properties to meet operational requirements. For example, type 304 (AISI)/1.4301 (EN) for most general applications and type 316 (AISI)/1.4401 (EN) which contains molybdenum and has corrosion resistance are typically recommended. Elastomers and other polymers, used because of their high extensibility, particularly for seals, linings, flexible conveyors and moldings, etc., should conform to EU (Commission Regulation No. 1935/2004: Materials and articles intended to come into contact with food) or FDA (21 CFR 170) legislation.
2. *Surface finish* must be smooth enough to enable surfaces to be easily cleaned and disinfected. Surface roughness has been traditionally described by the roughness average (R_a) value which is calculated by measuring the average departure from an imaginary centerline running through the "peaks" and "valleys" of the surface profile (Anon, 1997b). For closed equipment (that used for liquid handling and usually cleaned-in-place – CIP) a surface finish of 0.8 μmR_a is recommended. For open surfaces, where more mechanical cleaning action can be applied and the effects of cleaning are more readily visible, higher R_a values may be acceptable, though again an R_a value of 0.8 μm is also appropriate.
3. *Joints*, such as those that are welded or bonded, should be smooth and continuous and free from recesses, gaps or crevices. Dismountable joints such as screwed pipe couplings must be crevice free and provide a smooth continuous surface on the product side. Flanged joints must be located with each other and be sealed with a gasket because, although metal-to-metal joints can be made leak-tight, they may still permit the ingress and harborage of microorganisms.

4. *Fasteners* such as exposed screw threads, nuts, bolts, screws and rivets must be avoided wherever possible in product contact areas. This is primarily because they have many metal-to-metal joints and crevices and are thus microbiologically uncleanable; however, fasteners in the product zone are also a foreign body hazard. When unavoidable for technical reasons, alternative methods of fastening can be used, for example where the washer used has a rubber compressible insert to form a bacteria-tight seal.
5. *Drainage* is a requirement for all pipelines and equipment surfaces because residual liquids can lead to microbial growth or, in the case of cleaning fluids, result in contamination of product. Care should be undertaken with the installation of equipment such that its drainability is not impaired.
6. *Internal angles and corners* should be well radiused, wherever possible, to facilitate cleaning. Ideally, materials should be joined away from the corner as, even if (in the case of welding) the joint is ground and polished, its 90° angle makes it difficult to access by cleaning equipment.
7. *Dead spaces* are areas within the equipment that cleaning systems are unable to reach during routine cleaning operations. In these areas, product is harbored which may cause product quality loss and a potential allergen issue. If dead spaces are unavoidable for technical reasons, they should be readily accessible for draining/ cleaning as necessary.
8. *Bearings and shaft seals* should, wherever possible, be mounted outside the product area to avoid possible contamination of product by lubricants, unless they are edible, or possible failure of the bearings due to the ingress of the product. Shaft seals must be designed so as to be easily cleaned and, if not product lubricated, then the lubricant must be edible. Where a bearing is within the product area, such as a foot bearing for an agitator shaft in a vessel, it is important that there is a groove completely through the bore of the bush, from top to bottom, to permit the passage of cleaning fluid.
9. *Instruments* must be constructed from appropriate materials and if they contain a transmitting fluid, then the fluid must be approved for food contact. Installation should avoid dead legs and in liquid handling equipment, insertion into a shortened T-piece with cleaning fluid flowing into the upstand is a favored solution.
10. *Doors, covers, panels* and door gaskets should be designed so that they prevent the entry and/or accumulation of soil. Where appropriate they should be sloped to an outside edge and should be easily removed to facilitate cleaning.
11. *Controls*, particularly those that are repeatedly touched by food handlers in normal use, should be designed to prevent contamination and should be easily cleanable. Pathogenic microorganisms have been known to be harbored in switches and be transferred to product every time they are operated.

(See Chapter 26 for further details.)

Utensils

Utensils include all tools, containers, trays and small pieces of equipment, etc. necessary to undertake the food manufacturing process. There is no specific legislation or guidance on the hygienic design of utensils and the 11 principles outlined above for food processing equipment should be followed as best practice.

Services

Services include all processing aids necessary to manufacture the food including process water, steam, electrical supplies, compressed air and other gases. Potable (hot and cold) water should be used whenever necessary to ensure that foodstuffs are not contaminated. Where appropriate, facilities for water storage, distribution and temperature control shall be adequately designed and constructed, shall be covered and shall have air vents which are insect- and rodent-proof. Plumbing shall be of adequate size and design and adequately installed to ensure potable water is not contaminated and that all hoses, taps and other similar sources of possible contamination prevent backflow or back siphonage. Where non-potable water is used, for example for fire control, steam production, refrigeration and other similar purposes, it should circulate in a separate, identified system. Non-potable water is not to connect with or allow reflux into potable systems. Steam should be generated from potable water, have approved food safe additives, be filtered, have non-return valves to prevent the drawback of product into steam lines and should have traps to ensure adequate condensate removal and elimination of foreign materials. The length of electrical cabling should be minimized and be mounted on vertical or inclined cable trays with one layer of cables per tray. Electrical connection and control boxes should be suitably protected from water and dust ingress. Compressed air or other gases mechanically introduced into food or used to clean food-contact surfaces or equipment must be dry and filtered to remove microorganisms, lubricants and particulates and ideally conform to *Food Grade Compressed Air: A Code of Practice*, British Compressed Air Society Limited (www.bcas.org.uk).

As many services as possible should be exterior to the food processing area and should be carried in false ceilings or service corridors. Pipework, suitably protected light fittings, ventilation points and other services in manufacturing areas should not run directly above food processing equipment or "open product" and should be sited (e.g. flush mounted or mounted at least 25 mm from the wall) to minimize dirt accumulation, to avoid creating recesses which are difficult to clean and to ensure that drips and condensation do not contaminate foods, raw materials or food contact surfaces. Cold water pipes and other service pipes which might be prone to condensation build-up should be insulated (see Chapter 26).

Waste Disposal

Waste disposal areas, for the storage and disposal of food waste, non-edible by-products and other refuse, must be designed and constructed so that the risk of contaminating food or the potable water supply is avoided, to reduce the attraction of pests and to minimize the potential for odor. Storage should be in a separate room within the main factory building, often part of the dispatch docking area, or in an external area that is constructed of impervious material and suitably sloped and drained. Waste disposal areas should be easily cleanable and, where necessary, suitably fly-proofed and free from animals and pests. Food waste, non-edible by-products and other refuse should be deposited in appropriately constructed, labeled, closable containers which are made of impervious material, are leak-proof and are easy to clean and disinfect. Waste containers should be specifically identifiable (e.g. by color) and be lidded and should not be moved through different hygiene zones.

Waste should be moved out of higher risk zones, ideally via existing hatches for bagged waste, though for waste collected in bins, it may be necessary to decant the waste through purpose-built, easily cleanable from high risk, waste chutes that deposit directly into waste skips (see Chapter 25).

Medical Screening

Until food factories are completely automated, food operatives are essential to the manufacture of the food product and are thus part of the hygienic infrastructure. Medical screening of food operatives is initially concerned with the requirement for medical certification of freedom of carriage of pathogenic microorganisms in prospective new employees. In addition, it involves an ongoing awareness by operatives of their own health and the health of those around them (e.g. at home), from whom they themselves may become infected and thus subsequently compromise food safety. Medical certification can be self-certifying or via a doctor or other health practitioner. In some countries this may be required by law; in others this may be voluntary. Once employment has started, any instance of potentially infectious diseases, including jaundice, fever, sore throat with fever, vomiting, stomach disorders, diarrhea, visibly infected skin lesions (such as boils, cuts) and discharge from the eyes nose or ears, must be reported to the operative's line management or directly to the first aider or medical department. This particularly applies to staff returning from areas of the world where there has been a risk of infection. If, through illness, operatives are identified as being at risk to the safety of the product, they should either be sent home or moved to other work areas or duties that do not include open food handling (see Chapter 28).

Maintenance

Effective hygienic maintenance is required both for food processing equipment and the processing environment. Maintenance should be preventive, i.e. planned such that parts and structures are changed/maintained so that they do not fail or become of a nature where they form a microbiological or physical hazard to the product. "Fire-fighting" or repairing equipment and structures after they have broken or become a hazard is not acceptable.

All materials used for maintenance and repair shall be fit for their intended use, and for replacement food contact parts (e.g. seals, gaskets, conveyor belts), certificates of conformity or other evidence shall be available to confirm their suitability for use. The traceability of replacement parts must be ensured to facilitate any potential recall of food products that could have been contaminated by defective parts. Essentially, the identification of the business from which the parts were supplied, where the parts are held in stores and the lines and equipment to which individual parts have been fitted, must be recorded.

Maintenance and repairs to food processing equipment should ideally be undertaken away from the line but if this is not possible, the worksite should be adequately screened. Similarly, repairs to the building fabric should be undertaken out of production periods but if this is not possible, screening should be in place, preferably from floor to ceiling and fully sealed. All tools, parts and materials brought to the job and/or removed should be accounted for and all parts and tools should be clean. This may necessitate the cleaning of replacement parts (and also new and second-hand equipment) prior to entering the factory. Separate toolboxes should be available for low and high risk zones. Lubricants should be food safe and ideally be preserved to prevent microbial growth and conform to ISO 21469 (Anon, 2006).

For specific hazards such as broken glass or water leaks into dry processing areas, written procedures should be available as to how to control and repair these incidents. For all maintenance procedures in food processing environments, a documented handover procedure should also be in place. Engineering staff should sign off to the fact that the area is free of tools, repair materials and engineering debris. Sanitation staff should sign off to ensure that the area is cleaned to a hygienically appropriate level and, finally, production staff should sign off that they accept the area as suitable for food processing. All such procedures should apply to both own staff and external contractor working practices.

Engineering workshops should also be recognized as a potential source of food contamination and should be kept clean and have sticky mats at their exits to help remove swarf and debris from footwear (see Chapter 26).

Housekeeping

Housekeeping is undertaken to provide a safe working environment for staff (primarily reducing slips and trip hazards), to minimize any possibilities for contamination from the food processing environment to the food product and to minimize the challenge of hazard build-up on the processing line that will need to be controlled by the end-of-production clean. Cleaning and disinfection procedures can be seen as unit operations in which, for microorganisms, a given reduction, usually 2–3 log orders for the cleaning phase and 1–2 log orders for the disinfection stage, is possible. Any increase in the starting level of microorganisms prior to the end-of-production clean can thus result in an unacceptable level of microorganisms remaining after the sanitation process. To prevent excessive microbial growth on surfaces during production periods, food operatives or cleaning staff should clean the process line prior to break periods and clean up any major product spillages. Cleaning of the processing environment during food production periods must be undertaken in a way in which contamination to the food product via cleaning aerosols is minimized (see Chapter 27).

Cleaning and Disinfection

Cleaning and disinfection, referred to together as sanitation, is undertaken primarily to remove all undesirable material (food residues, microorganisms, allergens, foreign bodies and cleaning chemicals) from surfaces in an economical manner, to a level at which any residues remaining are of minimal risk to the quality or safety of the product. The principal stages involved in a typical sanitation program are:

- *Preparation*. All product and unwanted utensils/packaging/equipment should be covered or removed from the area. Machinery should be switched off, at the machine and at the power source, and electrical and other sensitive systems protected from water/chemical ingress. Equipment should be dismantled and stored on racks.

- *Gross soil removal.* All loosely adhered or gross soil should be removed by brushing, scraping, shoveling or vacuum, etc. Wherever possible, soil on floors and walls should be picked up and placed in suitable waste containers and not washed into drains using hoses.
- *Pre-rinse.* Surfaces should be rinsed with low pressure cold water to remove loosely adhered small debris. Hot water can be used for fatty soils (approximately 60°C), but too high a temperature may coagulate proteins.
- *Cleaning.* Cleaning is a combination of mechanical or kinetic energy (physical or fluid abrasion), chemical energy (cleaning chemicals), temperature or thermal energy and cleaning time. Cleaning chemicals should break down soil-to-soil and soil-to-surface bonds and prevent the re-deposition of the dispersed soil back onto the cleaned surface. No universal cleaning chemical is available and a range of cleaning agents is used for specific food soils. Water provides the cheapest, readily available transport medium for rinsing and dispersing soils, has dissolving powers to remove ionic-soluble compounds such as salts and sugars, will help solubilize proteins below their coagulation point, emulsifies fats at temperatures above their melting point, and, in high pressure cleaning, can be used as an abrasive agent. Organic surfactants are amphipolar, are composed of a long non-polar (hydrophobic) chain or tail and a polar (hydrophilic) head and aid cleaning by reducing the surface tension of water and by emulsification of fats. Alkalis break down proteins through the action of hydroxyl ions, saponify fats and, at higher concentrations, may be bactericidal. Acids are very useful in solubilizing carbonate and mineral scales, including hard water salts and proteinaceous deposits. Sequestering agents (sequestrants or chelating agents) are employed to prevent mineral ions precipitating by forming soluble complexes with them. Cleaning and disinfection can be undertaken by hand using simple tools, e.g. brushes or cloths (manual cleaning), though as the area of open surface requiring cleaning and disinfection increases, specialist equipment becomes necessary to dispense chemicals and/or provide mechanical energy. Chemicals may be applied as low pressure mists, foams or gels, while mechanical energy is provided by high and low pressure water jets or water or electrically powered scrubbing brushes. Alternatively, dismantled equipment and production utensils may be cleaned and disinfected in soak tanks or automatically in tray or tunnel washers. Cleaning should be undertaken in a sequence so that hazards are removed from the processing environment via the drains and do not contaminate cleaned surfaces. Specifically, environment surfaces, usually in the order of drains, walls then floors are cleaned prior to food processing equipment and all food processing equipment should be cleaned at the same time.
- *Inter-rinse.* Both soil detached by cleaning operations and cleaning chemical residues should be removed from surfaces by rinsing with low pressure cold water.
- *Disinfection.* Disinfection is undertaken to remove and/or reduce the viability of remaining microorganisms to a level deemed to be of no significant risk. In exceptional circumstances and only when light soiling is to be removed, it may be appropriate to combine cleaning and disinfection by using a chemical with both cleaning and antimicrobial properties (detergent-sanitizer). Elevated temperature is the best disinfectant as it penetrates into surfaces, is non-corrosive, is non-selective to microbial types, is easily measured and leaves no residue. However, for open surfaces, the

use of hot water, steam or naked flames is uneconomic, difficult to maintain target temperatures, may bake on residues, is hazardous to operatives, and reliance is, therefore, placed on chemical biocides. Quaternary ammonium compounds (quats or QACs) and peracetic acid are universally used for open and closed surfaces, respectively; hypochlorous acid is used for high-level disinfection of, e.g., floors and drains while alcohol is used for dry disinfection. The efficacy of chemical disinfectants is generally controlled by five factors: interfering substances (primarily organic matter), pH, temperature, concentration and contact time. To demonstrate their biocidal effectiveness, in Europe, chemical disinfectants should be approved for bactericidal activity against the European Standards EN 1276 (Anon, 1997c) and EN 13697 (Anon, 2001) and for fungicidal activity against EN 1650 (Anon, 1998b). In Europe it is also acceptable to leave disinfectants on surfaces without any further rinsing prior to food production. As such, disinfectants will enter foodstuffs and must be non-toxic and non-tainting. Historically, disinfectants have been regarded as safe if their minimum acute oral toxicity (with rats) is >2000 mg/kg bodyweight. Non-toxicity is now being assessed under the auspices of European Directive 98/8/EC concerning the placing of biocidal products on the market, which seeks to produce a list of active biocidal substances that have been assessed for both their toxicological properties and also their inherent antimicrobial properties. Formulated products, sold to the final user, can then only be made by incorporating an approved active ingredient and the formulated product will then itself be assessed for its toxicological and antimicrobial properties. To assess the potential for a disinfectant to taint foods, a modification of a food container transfer test is used (Anon, 1983) in which food products are sandwiched between two sheets of stainless steel and left for 24 hours. Disinfectants can be sprayed onto the stainless steel sheets and drained off, to simulate no rinse status, or can be rinsed off prior to food contact. Control sheets are rinsed in distilled water only. The results of the test involve both a statistical assessment of any flavor differences between the control and disinfectant-treated sample and a description of any flavor changes.

- *Post-rinse.* Disinfectant residues may or may not be removed by rinsing away with low pressure cold water of known potable quality.
- *Inter-production cycle conditions.* A number of procedures may be undertaken, including the removal of excess water and/or equipment drying, to prevent the growth of microorganisms on production contact surfaces in the period up until the next production process. Alternatively, the processing area may be evacuated and whole room disinfection techniques applied, e.g. ozone, hydrogen peroxide or ultraviolet light.
- *Periodic practices.* Periodic practices increase the degree of cleaning for specific equipment or areas to return them to acceptable cleanliness levels. They include weekly acidic cleans, weekend dismantling of equipment, and cleaning and disinfection of chillers and sanitation of surfaces, fixtures and fittings above 2 meters.
- *Clean the cleaning equipment.* Following their use for cleaning, cleaning equipment should itself be cleaned and disinfected. Cleaning equipment should be visually checked for damage and any areas where microorganisms could reside, or loose parts which might become a foreign body hazard, should be replaced. Cleaning equipment should be stored in racks to dry or kept in disinfectant solution until their use is required.

Sanitation is managed via a series of plans or schedules. The cleaning plan should list all the cleaning and disinfection tasks that need to be undertaken across the food manufacturing site and their frequency (daily, weekly, monthly, etc.). Cleaning schedules are the written work instructions that detail precisely how the cleaning and disinfection procedures for each task should be undertaken and can also be used as the work instruction against which cleaning operatives can be formally trained. The "whole room" plan details all the requirements for the practical management of the cleaning and disinfection operation and includes manpower, any specialist engineering support, equipment, chemicals and their dosing, health and safety, room preparation, protection of any food production operations, how cleaned surfaces are protected from recontamination, how the room is prepared for subsequent production and how the cleaning equipment itself is cleaned and maintained. End-of-production cleaning plans detail the specific requirements of single or groups of equipment and include any specific dismantling or health and safety instructions. Periodic cleaning schedules detailing, e.g. weekend or monthly cleaning and site decontamination plans, detailing a planned decontamination of the processing areas following a potential pathogen contamination incident, may be required, particularly for ready-to-eat food (see Chapter 27).

Cleaning-In-Place (CIP)

CIP is the cleaning of complete items of equipment or pipeline circuits, *in situ*, without dismantling and with little or no manual involvement. The prime consideration of CIP cleaning is that the process and CIP equipment must be designed hygienically; otherwise cleaning will not be acceptable. Separate CIP sets for raw and post-heat-treated product lines are also required. Much of the science of CIP cleaning is the same as for open surfaces, though the relative inputs for each of the four cleaning factors can be very different. Mechanical or kinetic energy is generally limited and is provided in pipelines by turbulent flow of cleaning solutions and in vessels by either falling films or spray impingement of cleaning solutions. As the system to be cleaned is enclosed, the concentration of cleaning solutions and their circulation temperatures can thus exceed those as used for open plant cleaning. Typically, caustic detergents (e.g. sodium hydroxide) have an in-use concentration of 0.3 to 2.0% and are circulated in excess of 70°C whereas acidic detergents (e.g. phosphoric or nitric acid) are used at 0.2 to 1.0% acidity and circulated at 50 to 60°C. Methods of thermal disinfection include hot water at temperatures between 70 and 80°C and maintained for 15 minutes.

The actual cleaning mechanism within the CIP circuit is divided into the cleaning of pipelines (and other items with total submersion in the cleaning fluids) and the cleaning of vessels. The cleaning of pipelines is undertaken by circulating the cleaning fluids at a velocity of approximately 1.5 m/s throughout the whole of the pipeline system. The cleaning of vessels is undertaken via spraying devices that produce a "falling film" of cleaning fluid or an impingement jet of cleaning fluid over the whole of the vessel surface. Fixed (static) spray balls, via the use of holes drilled in specific patterns, direct cleaning fluids to cover the vessel surface. Approximately 30 to 50 liters of cleaning fluid per minute per meter of the vessel circumference, at 1.5 to 3 bar, are generated to maintain a continuous liquid film on the tank wall which falls due to gravity to provide some mechanical action. Low pressure rotating (dynamic) sprayheads can also be used which have lower water consumptions. High pressure rotating (dynamic) impingement sprayheads can be water or mechanically driven to provide a (geared) rotation around the vertical and horizontal axis such that all points on the vessel surface are impacted over a defined time period. Once this time period has been established, cleaning must always meet or exceed this time to ensure full surface coverage. The vessel must be designed to ensure adequate drainage of solution and prevent pooling of chemicals and soil in the bottom of the vessel, which is controlled by ensuring that the (scavenge) pump used to remove fluids from the vessels is operating at a flow rate in excess of the supply pump.

It is absolutely critical to prevent the cleaning and disinfection chemicals from contaminating the food product, and as such the CIP circuit must always be separated from the food product. Traditionally, the potential for contamination was minimized by bringing product and CIP lines to a flow selector plate at which specific short lengths of pipe were manually removed and reconnected to ensure that product and cleaning fluids were separated. The use of single valves between the chemicals and the product is never sufficient to separate the systems and modern circuits use, e.g., block and bleed valves to assure adequate separation (see Chapter 27).

Pest Control

Pests are attracted to food processing environments primarily for water, food and shelter. Their presence, however, can lead to consumption and/or damage of the food product and packaging, physical contamination of foodstuffs by, e.g., rodent droppings, insect parts or other foreign bodies, contamination with disease-causing agents, e.g. microbial pathogens and intestinal worms are carried in the guts and/or on the external surfaces of pests and direct damage to the building and its services. Pests can therefore be described as any animal at any stage of its life cycle that may reasonably cause biological/physical contamination to food or its presence will be detrimental to its wholesomeness, and include small mammals (e.g. black and brown rats and mice), crawling insects (e.g. cockroaches, ants, beetles), flying insects (e.g. houseflies, blowflies, fermentation flies, moths, bees, wasps) and birds (e.g. seagulls and pigeons).

Pests can be environmental, i.e. can enter the factory from the external environment, or can be associated with raw materials, often termed stored product pests. The pests relevant to a particular food processing environment can be determined by history and previous findings, inspections by pest control technicians and sightings by company staff. As company staff are regularly on the premises, and if trained in the detection of the signs of pest infestation, they are the best source of information on the presence of pest hazards and a Pest Sightings Register should be maintained at all sites which details the date, pest, location, reported by and subsequent actions taken by the food manufacturer and pest contractor.

Pest control has to be undertaken by both the food manufacture (and its raw material suppliers) and the pest control contractor (which may be an in-house function). As such, it is commonly referred to as integrated pest management (IPM) and comprises four key elements; neighboring activities; the factory environment; raw material quality control; and pest management. The activities of neighboring properties, the movement of materials on and off the factory site and any environmental features such as water courses should be considered as to their ability to attract pests. The factory site and building envelope should then be designed and maintained to minimize pest harborage, visualize pests and restrict pest entry. At the same time, however, the factory must practice good GMPs such as good housekeeping, waste handling and cleaning and disinfectant practices to limit the presence of water, food sources or harborage sites. All raw materials entering the factory which could be susceptible to stored product pests should be inspected, sampled and positively released to the food manufacturing operation. Strict stock rotation should be implemented and, as infestation can potentially remain undetected, it is important to establish batch-to-batch freedom from pest contamination.

While some food manufacturers employ their own pest control teams, most contract this service out to a pest control company. Contracts usually consist of two parts, a practical part consisting of a scope of works and an administrative part consisting of a logbook or service report. The scope of works may include the pests covered, the frequency of the services offered (technician/field biologist visits), agreed methods of pest capture or treatment (e.g. small mammal and insect traps, electric fly killers, pesticides, bird scarers, stored product fumigation and heat or freezing treatments), reporting structures, reviews and trend analysis and contractual warranties and insurances. The logbook may contain general information (pest company details, qualification certificates or licenses for pest control technician, name of daily contact person); materials and resources applied (applied pesticides overview, material safety data sheets, applied pesticide labels); monitoring devices (registration lists, maps with location indicator, capture statistics and trend analysis); and notification of pest activity, reports, advice and corrective actions (see Chapter 29).

Personal Hygiene

All food manufacturers should have in place, for all employees, visitors and contractors entering food processing areas, a personnel hygiene policy. This should cover:

- Medical screening (referred to above)
- Induction training
- Personal hygiene measures
- Company requirements – hand washing/clothing
- Monitoring and verification auditing

Induction training, provided in as many different languages as necessary for the whole work force, is essential to ensure that food operatives recognize that they may act as potential routes of food product contamination and that they should undertake personal hygiene procedures in a way that minimizes such risks. Of particular importance is the demonstration of an appropriate, validated hand hygiene procedure as operatives do not inherently know how to wash hands to maximize microbial cleanliness. This is best undertaken using a kit combining a UV-sensitive dye and a small, portable UV lamp, e.g. GloGerm System, Deb Ltd (www.debgroup.com). The dye is applied to the hands prior to hand washing and, following hand washing, the hands are placed under the UV lamp to indicate areas that have been "missed." Individuals' training records should be kept and reviewed as appropriate.

Food operatives should be encouraged to follow basic hygiene procedures at home and in the workplace to minimize their risk to foodstuffs. Such procedures cover the control of personal habits (e.g. nose picking, spitting, nail biting) and activities (e.g. eating, smoking and drinking), the wearing of make-up and jewelry and the coverage of any wounds with, e.g., blue, metal detectable plasters.

Protective clothing is provided by the food manufacturer primarily to protect the food from microorganisms released from the body and includes hair nets, hats, masks, beard snoods, overalls, coats, gloves, wrist and forearm sleeves, trousers and footwear. Personal protective equipment (PPE), which includes hard hats, gloves, safety spectacles, ear defenders, aprons, overalls and footwear with non-slip soles and metal toe caps, is provided to protect the operator from the food processing environment (cold, water, food products, etc.) and specific safety hazards as appropriate (e.g. detergents and disinfectants, falling objects, knives). Consequently the type of material used and the design of protective clothing will depend upon its prime function. Factory clothing should be hygienically designed so that it does not shed foreign bodies directly (e.g. buttons or lint) or indirectly (e.g. having outside pockets from which objects can fall out) and is often of different colors to delineate either operatives working in different risk areas or specific categories of people, e.g. engineers, cleaning staff, first aiders and management. A laundry policy should also be in place to clean and maintain such protective clothing.

Hands need to be washed before embarking upon food handling procedures and after any operation that may lead to the hands becoming contaminated, which could include: visiting the toilet; handling raw food; handling waste and chemicals; blowing noses; sneezing into hands; touching body parts; carrying out cleaning duties; removing and changing gloves; picking items off the floor and touching non-food contact surfaces, e.g. machine adjustment, power switches, buttons, etc.

Hand washing on entry to food processing areas can be combined with the donning of factory clothing in a manner that limits the transfer of hazards into food handling areas. A suggested procedure is as follows:

1. Remove outer clothing and place in personal locker.
2. Remove jewelry and watches in line with personal hygiene policy and place in personal locker.
3. Put on hair net/snood.
4. Remove shoes and place in locker.
5. Step over a barrier into the food handling area.
6. Wash and dry hands.
7. Put on clean, dry food handling area footwear.
8. Put on clean food handling area clothing.
9. Use a hand disinfectant immediately before handling food.

Good personnel hygiene and hand wash compliance can be monitored by visual assessment by line supervisors and auditing staff, or via the use of CCT cameras. It is also possible to install, e.g., turnstiles at the entrance to food processing areas such that the turnstile will only open when a recognized hand washing trigger has been activated, e.g. the application of an alcohol handrub (see Chapter 28).

RECOMMENDED PROCEDURE FOR DEVELOPING A PROCESSING ENVIRONMENT PLAN

It is recognized that as a new, additional, component of food safety, the Processing Environment Plan (PEP) is likely to be approached with a degree of suspicion by the food factory staff that have to practically implement it, on top of the requirement for the HACCP plan. To make this task easier to undertake, and from practical experience in the factory, two things have been recognized that should facilitate implementation. First, and wherever possible, the same terminology has been used in the PEP as in the HACCP plan. Second, the PEP should generally only be undertaken once, to encompass the majority of food products manufactured. Only if the products to be manufactured in the processing area of concern use different processing equipment or different ingredient transfer systems are sources and vectors of contamination likely to change.

The undertaking of the PEP follows the 14 principles of the HACCP plan as defined by Gaze (2009) and is a recently proposed food safety initiative (Holah et al., 2011, 2012).

Pathogenic microorganisms can enter food processing areas from four main routes: the external environment, raw materials, infected food operatives and visitors and laboratories undertaking pathogen testing. Microorganisms from the external environment are controlled by the design of the factory building and its segregation. Microorganisms in raw materials are controlled by the HACCP plan, while infected food handlers are managed by best practice personnel hygiene prerequisites. Finally, pathogenic microorganisms in food laboratories, typically used as media positive controls, are controlled by the complete isolation of the laboratory from the factory, including air systems, drainage systems, waste removal and the movement of laboratory staff.

The environmental plan usually starts, therefore, within the food processing environment and especially where product is most susceptible to contamination. In effect, it works in the reverse to the HACCP plan (which starts at the raw materials and moves to the finished product) as it starts from the finished product and works backwards until any contamination routes are effectively controlled. The nomenclature of the Gaze (2009) 14 HACCP principles is used as follows to define the plan, and readers should access the original document for a full description of these widely established principles.

1. Obtain Management Commitment

Senior management must be committed to providing the necessary resource for the study to be planned, undertaken, implemented and periodically reviewed. In many cases this may be implicit as the PEP and PMP can be seen as part of the HACCP plan which, for some countries, may be a legal requirement. For RTE manufacturers, management must also be committed as the outputs of the environmental plan will be critical in controlling contamination and may be significant controls for the business and the protection of its brands. Senior management should also appoint a manager and/or team leader to take responsibility for the plan's development and implementation.

2. Define the Scope or the Terms of Reference

The processing area(s) for the study should first be determined. Will the study investigate the contamination of hazards into the food manufacturing building envelope, consider the processing environment from raw materials storage up to any hazard decontamination step, or just the post-decontamination step processing environment, after which any microbiological hazards entering the product may not be controlled prior to the foods consumption?

Second, what hazards will be considered? Will all biological, chemical or physical hazards be considered and, in addition to product safety, will any effects on product quality be considered. If the study is to focus on microbiological pathogens, the specific pathogen must be noted. This is because different pathogens may require different environmental niches to survive, grow or become established; for example, in chilled environments, *Listeria* spp. may dominate in low temperatures and moisture (e.g. evaporative condenser trays), whereas *Escherichia coli* requires higher temperatures (e.g. surrounding motor drive shafts where friction creates higher local temperatures).

Lastly, the types of potential sources and routes (vectors) of environmental contamination transfer may need defining, particularly if these have already been considered at the generic prerequisite stage. For example, and for most countries' climates, if the compressed air supply is dried to a dew point of −40°C, it may not be necessary to consider the use of compressed air in a processing area as a source or vector of microbiological contamination.

3. Select the Processing Environmental Plan Assessment (PEP) Team

As the study will assess hazard sources and vectors within a given process area, many activities and events may occur in this area at different times of the day, week, year, etc. and the selection of team members should reflect all of these activities. Engineers will be required who understand the building's construction, ceiling, wall and floor finishes, food production equipment, service provisions and all maintenance activities. Production staff representing all products produced in the area, together with sanitation managers who have to clean and disinfect the area following such production, will be essential. Hazard specialists such as microbiologists, chemists, pest controllers, etc. will be required dependent on the hazards assessed in the study. Finally, HACCP, technical and quality staff, who can advise on the planning and undertaking of the study, and then will need to implement the study's findings, will also be required.

As with all projects, good teamwork is essential and the team members and details of their specific skills, qualifications and responsibilities should be recorded. A scribe may be useful in helping the team leader manage meetings and to record findings. Particularly for small companies, consultants can be used for their technical knowledge, but they should not write the plan. The PEP should be owned, written, implanted and managed by the food manufacturer.

4. Describe the Environment

All physical and operational parameters of the processing environment under study should be recorded and/or measured with due regard to activities in adjacent processing areas, beside, below or above the area of study. The physical properties will include the size and layout of the processing area; any zones of segregation; entrance barriers into the area; services flowing through or above the area; air flows, temperatures and humidity; personnel flows; transport flows for product and packaging; and solid and liquid waste streams. Operational activities will include products processed, production lengths and seasonality; housekeeping, end-of-production and periodic cleaning and disinfection practices; maintenance activities; and shutdown periods.

With respect to the hazards of concern, any historical data from previous routine sampling or observational studies (e.g. routine environmental microbiological sampling, pest control records, glass and hard plastic records) should be recovered and reviewed.

5. Identify Intended Product Use

The intended use of the product should be established with respect to the fate of any hazards entering the product in the processing area of study. First, will there be any further treatment of the product or controls of the process line that might affect the removal, reduction or growth of any hazards entering the product directly or from the food production equipment? Second, as for a classic HACCP study and particularly for RTE foods, if there is no removal or reduction of the hazard, how will these hazards affect the target consumer group?

6. Construct Flow Diagram

All information collected during principle 4 should be recorded in the form of physical maps or diagrams of the processing area. Ideally this should start with a map of the processing area with the layout of food processing equipment and services. Overlaying this map can be specific diagrams of, e.g., alternative production equipment set-ups, air flows, personnel flows, transport flows and waste flows. The diagrams present both a record of the plant construction and activities taking place in the processing area at the time of the study, together with a vehicle that can be used for entering the position of any subsequently identified hazard sources and contamination vectors.

7. On-site Confirmation of Flow Diagram

The PEP team should audit the processing area at all processing, sanitation, maintenance and down times to ensure that the flow diagrams produced are accurate and representative. The flow diagrams can then be signed off as a true record of the processing area.

8. List all Potential Hazards, Conduct a Hazard Analysis and Consider any Measures to Control the Identified Hazards

Within this step the PEP team conducts a thorough investigation of the processing environment to identify any hazard sources and any mechanisms or vectors via which these hazards could enter the food product directly or via food processing equipment. Step 8 according to Gaze (2009) equates to step 1 of the seven HACCP principles as defined by the Codex Alimentarius Commission (Anon, 1993) and subsequently, steps 9–14 relate to Codex steps 2–7. The investigation of sources and vectors in the following text is illustrated for microbial pathogens as the hazard, but is equally applicable to the analysis and control of other hazards. The identifications of hazard sources and contamination vectors can be undertaken at each food processing step within the processing environment (as for a traditional HACCP study) or can be undertaken at an environmental level, as illustrated in Figure 24.5.

Once within the processing environment, pathogens can be sporadic visitors, being present until they lose viability or are removed via cleaning and disinfection procedures or more persistent, surviving in harborage sites or growth niches to form sources of contamination. Harborage sites are physical areas in which pathogens can lodge and be protected from external forces such as cleaning and disinfection actions, e.g. poor hygienic design features of processing equipment or damaged areas of the plant's building structure. Growth niches are also harborage sites, but which also provide an environment for growth, e.g. nutrients, temperature, oxygen, water or humidity and lack of competition from other microbial flora.

Pathogens from harborage sites and growth niches and, perhaps less frequently, from the general environment as sporadic contaminants, can be transported to food products via three prime vectors. These are physical contact with a solid surface, physical contact with a liquid or settlement and/or impingement from the air (or other gases). The difference between solid contact and liquid contact is that the liquid may be absorbed into the food product which may increase the transfer of microorganisms to the food (towards 100%). For contact between solid surfaces, microorganisms will partition to and from the two surfaces, dependent on the physical properties of the microorganisms and surfaces. Smith and Holah (2007) demonstrated that the transfer of microorganisms from one contact surface to another can be approximated to 50% for practical purposes. For stationary air, transfer of microorganisms from the air is via sedimentation, which has defined rates for particles of given size (Stokes law, cited in Lamb, 1994), and the number of microorganisms transferred is dependent on the microbiological loading of the air and the exposure time. When product is transported via air, or when air is forced into the product for cooling or drying purposes, microorganisms can enter the product via impingement in addition to sedimentation, and the number of microorganisms transferred may be related to the microbial loading and volume of air that the product is exposed to.

Contamination usually occurs as a contamination event, in which a number of vectors may be involved. For example, entering a food product stream to help clear an obstructing product may have vectors of the operator's hand (or glove), the operator's sleeve, the tool to be used for cleaning away product debris and the air. In some instances the contamination event could have only a single vector, e.g. contaminated water droplets from a compressed air line.

The determination of potential pathogen sources and contamination vectors in a processing plant is a combination of physical examinations and microbiological sampling. Sources can be determined by dismantling process equipment to identify potential harborage sites and niches, together with physical inspection of the building structures and finishes. The potential presence of pathogens in such harborage sites and niches can be determined by

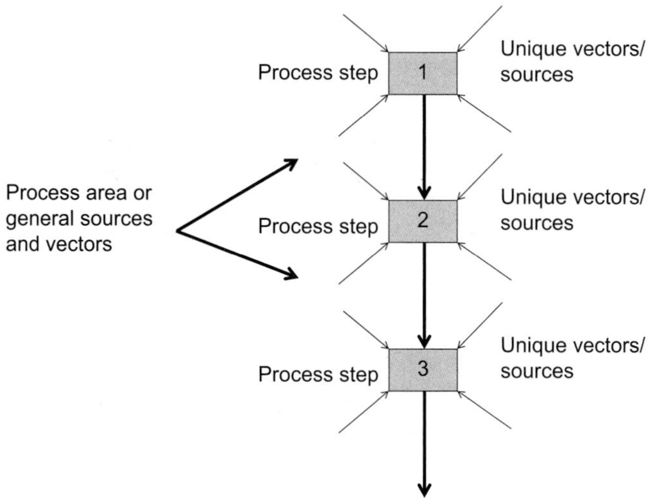

FIGURE 24.5 Hazard sources and vectors can be identified at each process step or within the general processing environment, which may affect many steps.

microbiological sampling and over prolonged periods (e.g. via environmental microbiological sampling records) an indication can be gained as to the likelihood of pathogens being present in these sites. The observation of all potential sources should be recorded as a record of the environmental survey, for example in a tabulated form as shown in Table 24.1. In these examples from the author's experience in a factory that is no longer in existence, meat residues were seen inside a meat slicer on/off switch in a high risk area of a cooked meats factory and fluid was seen oozing from underneath the meat slicer foot support plate.

Pathogens may transfer from sources directly to the food product, on product vectors, or indirectly to other parts of the processing environment via environmental vectors (Figure 24.6). If observational and microbiological data identify likely pathogen sources, all potential environmental contamination routes from this source should be determined to identify the potential for secondary or temporary sources. Using the equipment foot support plate example in Table 24.1, liquid oozing from under the plate was transferred throughout the process area on operatives' shoes and on equipment wheels and was re-deposited at random sites on the floor to act as potential temporary or short-term sources.

Contamination vectors can be identified by inspection of all of the activities associated with the production line and processing environment. Inspections should reflect all operating conditions including process type, product type, time of day or batch process, cleaning and disinfection, maintenance procedures, QC procedures, production downtimes and any seasonal events. Observations of contamination vectors should be made independently of known or likely pathogen sources, because contamination could arise from temporary sites, and is best observed from the process itself – i.e. the identification of potential transfer vectors to the process line, observed from the process line.

It is unlikely that microbiological sampling of vectors would be helpful, as the likelihood of observing a pathogen on a potential vector would be very small. Observational data for

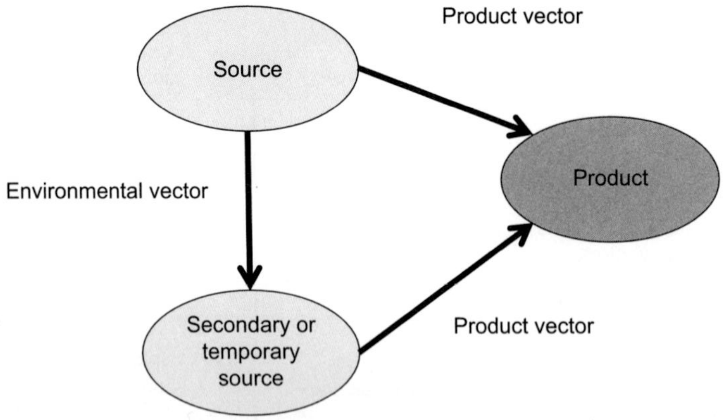

FIGURE 24.6 Transfer of pathogens from likely sources directly to food products via product vectors or indirectly to secondary or temporary sources.

vectors should also be recorded as indicated in Table 24.2, again from the author's experience in a factory that has been subsequently refurbished, showing the personnel hygiene of operatives and spray dryer interventions in a milk spray drying operation.

When observing and identifying potential contamination sources and vectors with the PEP team, any current direct controls of observed sources and vectors should be recorded as illustrated in Tables 24.1 and 24.2. For vectors, subsequent controls within the food process may have an effect on the hazard that could be transferred by the contamination event, and these should also be recorded.

When undertaking a process environment study, many potential sources and contamination vectors could be observed, though the degree of control necessary for each source and vector will depend on their potential risk to food product contamination. Hazard analysis is a fundamental aspect of HACCP studies, and a familiar approach to risk analysis is to consider the likelihood and severity of a hazard in a food as a three-point scale, or low, medium and high risk. A risk analysis for a contamination source is similar and can be described as the risk of a pathogen being present at the potential source and the ability of the pathogen to be transferred from this source via an environmental and/or product vector.

A risk assessment for a contamination transfer vector is a little more complex as it involves three factors: the potential for a pathogen being present on the product vector, the frequency of the vector and the severity of the impact of the hazard to the consumer of the product.

If a risk ranking of 1, 2 and 3 were used as a substitute for low, medium and high risk, respectively, a multiplication range can then be used to determine a risk score, which can help describe the significance of the contamination source or vector. For a source, the risk score would be in the range of 1 (low potential presence multiplied by low potential spread) to 9 (high times high). For a vector the risk score would be in the range of 1 (low potential presence multiplied by low frequency multiplied by low severity) to 27 (high times high times high).

TABLE 24.1 Potential Sources of *Listeria* Contamination Detected around a Meat Slicer in a High Risk Food Production Area that is no Longer in Existence

			Source Hazard Analysis without Controls				Source Hazard Analysis with Controls		
Process Step or Environment	**Observation**	**Likely Hazard**	**Microbial Source Presence LMH**	**Potential to Spread Via Environmental Vectors LMH**	**Risk Score**	**Current or Intended Control**	**Microbial Source Presence LMH**	**Potential to Spread Via Environmental Vectors LMH**	**Risk Score**
Meat slicer	Meat residues were seen on the inside of a switch that operated the meat slicer. When the switch was pressed in to start the machine slicing, the movement of the switch into its housing extruded meat residues onto the food operative's finger. If *Listeria* were present in the switch (which has occurred in previous installations) it could be transferred to the meat by contact with the operative's finger. Routine microbiological sampling of the switch for *Listeria* was always negative	*Listeria*	Medium 2	Low 1	2	Switches routinely cleaned as part of the end-of-production sanitation program	Low 1	Low 1	1
Meat slicer	Fluids were seen oozing out from below the foot plates supporting the legs of the meat slicer. The immediate area surrounding the foot plates is a heavy traffic area for both operatives and wheeled containers. Environmental microbiological sampling occasionally detects *Listeria* spp. from around the support plate	*Listeria*	High 3	High 3	9	Following cleaning during the end-of-production sanitation program, the foot plates are sprayed with 1% sodium hypochlorite	Medium 2	High 3	6

TABLE 24.2 Potential Vectors of *Salmonella* in a Milk Spray Drying Operation that has been Subsequently Refurbished

				Product Vector Analysis without Controls						Product Vector Analysis with Controls			
Process Step or Environment	Observation	Likely Hazard	Contamination Event Vector	Potential Presence on Vector LMH	Frequency of Vector LMH	Severity of Hazard LMH	Risk Score	Subsequent Control Step	Current or Intended Controls	Potential Presence on Vector LMH	Frequency of Vector LMH	Severity of Hazard LMH	Risk Score
Personnel hygiene	Food operatives are able to enter toilets without first removing their protective cleaning	*Salmonella*	Clothing touching food or food contact surfaces during food handling	M 2	H 3	H 3	18	None	Staff are responsible for the laundering of their own protective clothing	M 2	H 3	H 3	18
Personnel hygiene	Food operatives are able to enter toilets without first removing their protective cleaning	*Salmonella*	Clothing touching food or food contact surfaces during food handling	M 2	H 3	H 3	18	None	Staff are provided with professionally laundered clothing every day which is worn only within food production areas	L 1	H 3	H 3	9
Milk spray drying	Guillotines are inserted into the dryer to separate, e.g., the dryer from the baghouse during CIP cleaning of the dryer	*Salmonella*	Contamination on the reverse side of the guillotine entering the dryer on insertion and/or removal	M 2	H 3	H 3	18	Prior to start-up, the dryer surfaces are subjected to 200°C for 2 hours	General GMPs are practiced				
Milk spray drying	Approximately once per shift, milk injectors are removed from the dryer, cleaned, disinfected and reinserted	*Salmonella*	Dryer nozzles touch hands (gloves) clothing, tools and the dryer prior to entry	H 3	H 3	H 3	27	None	Gloves are worn and nozzles, tools and dryer contact surfaces are decontaminated with alcohol prior to dryer entry	L 1	H 3	H 3	9
Milk spray drying		*Salmonella*	Air can enter the dryer during the nozzle removal and reinsertion process	L 1	H 3	H 3	9	None	None				

Risk ranking of sources and vectors should be recorded as illustrated in Tables 24.1 and 24.2, both before and after controls are applied. Undertaking a risk analysis before and after the application of any controls can help identify whether controls are necessary and/or whether current or intended controls are sufficient to reduce the risk of the source or contamination event. As a minimum, this allows consideration of the adoption of controls for the uncontrolled sources and vectors that the environmental study has identified, which may have an immediate impact on improved food safety. In the case of current controls not being sufficient to adequately control the hazard risk, additional controls are required. To illustrate this and using the clothing vector as described in Table 24.2, a better control would be to implement a policy where protective clothing was only worn inside the food processing area and had to be removed before staff could leave this area (Table 24.2). In comparing the two controls in Table 24.2, it can be seen that the severity of the potential contamination to the food product with *Salmonella* remains the same, as does the frequency of the wearing of the uniform and its potential contact with a food product. However, as the uniform now no longer leaves the processing area, the potential presence of *Salmonella* on the uniform is significantly reduced, lowering the overall risk.

Subsequent controls should also be considered when assessing the risk of a contamination event. In the example in Table 24.2, operatives had to insert a stainless steel guillotine into the powder line to prevent CIP fluids entering sensitive areas during the dryer CIP program, e.g. the bag house or the gas burners to heat the incoming air. Any microbial contamination entering the dryer, particularly during the removal of the guillotines, would then be subjected to the dryer start-up procedure which could include the circulation of heated air for several hours (e.g. 204°C/400°F for 2 hours).

In the second dryer intervention example in Table 24.2, the removal, cleaning and insertion of the milk spray nozzles occurred every day, while CIP cleaning was undertaken every 3 weeks. Any microorganisms entering the dryer during these potential contamination events would not be subjected to a process control step. In this example, it is possible to do a risk assessment on the contamination event or, and particularly if the contamination event results in a high risk score, individual vectors related to the event to determine which of the vectors are important to control. In this case the entry of air has been chosen as an example of one of the vectors and the risk assessment for the air indicates that other vectors associated with the contamination event may be more important.

A high risk score if controls are not implemented can also be used to justify capital expenditure. For example, in Table 24.1, the capital required to stop production, remove the poorly designed foot plate, decontaminate the floor and fit and seal an improved hygienically designed foot plate could have been justified as the existing controls were inadequate and the risk of the potential for any contamination to be spread from this point remained high as the slicing machine was in a high personnel and vehicular traffic area.

9. Determine Operational Prerequisites

Control of food product contamination is a combination of reducing the number of possible hazard harborage sites and niches, controlling those that microbiological sampling has previously identified that may be a known risk, removing all unnecessary contamination vectors and controlling those that remain or are intrinsic to the food production process.

However, the control of some sources or vectors (prerequisites) may be more critical to the safety of the food product than others. The concept of a ranking system for prerequisites has been addressed by ISO 22000 (Anon, 2005), which differentiates operational prerequisites from prerequisites and defines them as being *identified by the hazard analysis as essential in order to control the likelihood of introducing food safety hazards to and/or the contamination or proliferation of food safety hazards in the product(s) or processing environment*. ISO 22000 thus suggests that a hazard analysis may identify that there may be some sources or routes of contamination that are so important to the safety of the food product that their control is essential and are thus elevated as a higher classification of prerequisite, i.e. operational prerequisites. Other definitions of operational prerequisites exist, however, for example Gaze (2009) defines operational prerequisites as control *measures associated with a particular process step and which manage specific significant hazards identified during hazard analysis but are not otherwise managed by CCPs* and *with a frequency of monitoring/checking of the control measure that is not sufficient to enable immediate corrective action*.

There is no agreed definition, therefore, of operational prerequisites and neither ISO 22000 nor Gaze (2009) give guidance as to the hazard analysis steps to be undertaken to identify an operational prerequisite. Similarly, there may be some confusion in terminology between individual operational prerequisites (OPs), which are single, and operational prerequisite programs (OPRPs), which are a collection of operational prerequisites.

The hazard analysis as described in Tables 24.1 and 24.2 for sources and contamination event vectors, respectively, can further be developed by considering the risk scores for the sources and vectors without controls. For the maximum risk scores associated with the meat slicer equipment floor plates (Table 24.1) or the removal, cleaning and reinstallation of the spray nozzles (Table 24.2), theses scores indicate that if these sources or contamination events were uncontrolled, or more practically, if the required controls failed, there would be a significant risk of pathogens being present in the processing environment (meat slicer foot plate) or product (spray nozzles). The control of these sources and vectors is thus critical to the safety of the product and, based on this risk assessment approach, such controls could be described as operational prerequisites.

10. Establish Control or Operating Limits

Wherever possible, control or operating limits should be identified for each OP. These may be defined in legislation, codes of practice and other guidance documents, though the majority are likely to be determined from collection of experimental data during trials, e.g. cleaning validation data, or from the advice of experts. In some cases there may be lower and upper control limits, together with a target limit. In rare cases the control limits may be critical limits as defined in HACCP (Gaze, 2009), though this would be unusual. For example, in water treatment systems to allow water reuse, disinfectant levels such as a minimal chlorine level of 0.5 ppm may be deemed critical to the safe disinfection of the water.

The specific control limits for each OP must be a measurable (e.g. ATP or protein levels after cleaning, disinfectant levels, flow rates, pHs, temperatures, pressures, contact times) or an observable parameter related to the control option. Measurements are preferred but where control limits are based on subjective data (e.g. visual observations) the food processor needs to provide clear guidance on requirements for compliance with practices or procedures or pictorial examples of what is acceptable (e.g. photographs to define clean surfaces or appropriate wearing of protective clothing). For the example in Table 24.3 an operating limit could be applied to a rapid assessment of the cleanliness of the wands, nozzles and tools by ATP or protein testing prior to entry. The PEP team should record details of how the control limit was determined, including relevant sources of information or experimental/validation trial data.

11. Establish a Monitoring System

Monitoring systems describe the methods by which the food processor ensures that the OPs are operating within their defined control or operating limits and are thus "in control" and, as a corollary of this, produces an accurate performance record which can be used for process verification (Stage 13). The monitoring system must be able to detect loss of control at the OP in a timeframe to provide corrective action to regain control of the OP.

Monitoring systems should ideally be on-line and could include air and gas pressure, humidity, temperature, chemical concentration, redox, conductivity or pH probes; UV intensity, flow rate; and rapid hygiene checks such as ATP, allergen and protein tests. Some on-line monitoring systems have a direct feedback system with the ability to directly control (and record) any drift in the control limit, and these are preferred. For these analytical methods, the PEP team should establish whether there are any required national reference methods for the parameters to be monitored. Microbiological sampling of source and vector controls would not be considered as a monitoring option as it may take 24–48 hours to enumerate samples; too long a time to maintain effective control.

Other monitoring checks may be visible and could include an assessment of cleanliness, an assessment of a personnel clothing changing procedure or whether a procedure is correctly being followed. For the example in Table 24.3, during the nozzle removal procedure, observations could be made to ensure that the procedure was being undertaken correctly and that there were no extrinsic factors which could act as additional contamination vectors.

The PEP team should record the job title or name of the individual(s) responsible for monitoring and ensure that they have the knowledge, competence and authority to take appropriate and stated corrective actions (Stage 12). Records of the necessary training and competence of these individuals must be signed and retained. The PEP team should also ensure that detailed specifications, procedures or work instructions to enable the monitoring to be effectively undertaken are added to the company's quality system.

12. Establish a Corrective Action Plan

Practical and achievable corrective actions to be undertaken when the results of monitoring at an OP detect a situation where a control limit has not been met (deviation) or when a treatment system is drifting out of control should be specified by the PEP team. Responsibilities for corrective actions should be clearly defined and all relevant personnel should be trained and competent. The relevant person(s) should have the authority to undertake the stated corrective actions. For the example in Table 24.3, corrective actions would review the training of the staff against removal and reinstallation procedures and the effectiveness and validation of the tools and cleaning equipment decontamination programs.

TABLE 24.3 Operational Prerequisite Management Table as Adapted from Classical HACCP CCP Management

Process Step or Area	Likely Hazard	Source or Contamination Event Vector	Control Measure(s)	Operating Limit(s)	Verification(s)	Corrective Action(s)	Records
Milk spray drying	*Salmonella*	Removal, cleaning and reinsertion of milk spraying nozzles	(1) Spray dryer processing is air filtered to 95% removal of 1.0 µm particles (2) Gloves are worn by operatives to remove and replace nozzles (3) Nozzles and support wands are removed and replaced by an alcohol-decontaminated blanking plate (4) Nozzles cleaned and covered with plastic bag until reinsertion (5) Gloves and plastic sleeve change by operatives prior to reinsertion (6) ATP assessment of cleaned nozzles. If RLU value <150, nozzles inserted (otherwise re-cleaned) (7) Use of dedicated tools (8) Alcohol decontamination of gloves, sleeves, nozzles, wands, tools and spray dryer contact surfaces (9) Blanking plate removal and nozzle reinsertion (10) Tamperproof tag installed	ATP <150RLU	(1) ATP assessment of the cleaned wand and nozzle (2) Visual assessment of the removal and reinsertion procedure (3) Occasional microbiological verification of wand, nozzle, tool and spray dryer contact surface cleaning	(1) Staff retraining (2) Revue of wand, nozzle, tool and spray dryer contact surface cleaning	(1) Dryer intervention record including correct observation of removal and reinsertion procedure (2) Post-decontamination RLU values (3) Post-decontamination microbiological values (4) Tamperproof identity tag number

Any product that could have been contaminated through any loss of control should be placed on hold following company quarantine procedures to allow authorized personnel to determine its fate. It is unlikely that a product recall would be instigated as the frequency of monitoring of the OP should be sufficient to prevent unsafe foodstuffs reaching the consumer. The cause of the deviation should then be investigated and appropriate remedial action taken, such that the OP will be returned to control. Further steps must then be taken to ensure that the same issue cannot occur in the future and the company should confirm that remedial actions have been undertaken and that they will be effective.

13. Verification

The verification stage is concerned with three activities: validation, verification and review. The objective of the validation stage is to ensure that all sources and contamination vectors for hazards that could be present in the processing environment have been considered and that the controls put in place to reduce or eliminate them are technically sound and effective. The first stage of the validation is a desktop activity to review the identification, selection and/or exclusion of hazards, the risk analysis of identified hazards, the appropriateness of the selected controls, the designation of controls as OPs, the suitability of their control limits and monitoring/verification methods and the adequateness of the corrective actions. This can be undertaken by the PEP team, but may be improved by the input of additional, independent experts, and the environmental plan must be signed off by the person ultimately responsible for product safety management in the food operation.

The second stage of the validation process is the validation of the identified control actions, as appropriate. In the example in Table 24.3, the efficacy of the tool and nozzle cleaning and disinfection process can be validated by undertaking the cleaning exercise a number of times and recording the level of cleanliness achieved as an ATP relative light unit (RLU) count. On each cleaning occasion everything should be undertaken to the best possible standards with defined chemicals at the correct concentration and temperature, the appropriate cleaning staff and techniques and for the scheduled times. The average ATP value after these cleans is thus the minimum level that could be obtained for these nozzles and tools, following their particular use and using the cleaning and disinfection method adopted. The target ATP level to be reached on each occasion may thus be this value plus a small margin for error.

Verification of the PEP gathers information from routine analytical tests that are used to demonstrate the effectiveness of the hazard controls and OPs in a timeframe beyond that of monitoring (Stage 11). For example, while microbiological sampling of the tools and nozzles in Table 24.3 is not acceptable for monitoring of a control measure, it could be undertaken for verification purposes. As with monitoring, the PEP team should record the job title or name of the individual(s) responsible for verification and ensure that they have the knowledge, competency and authority to take appropriate and stated corrective actions (Stage 12). There should also be detailed verification specifications, procedures or work instructions added to the company's quality system. Signed records of all verification activities must be retained to provide evidence that the PEP has been correctly implemented and the controls are working effectively.

Verification is also a desktop and audit exercise to examine the entire PEP and examples of such activities include: internal auditing of OPs to establish, e.g., that personnel are following the stated procedures/work instructions; external auditing programs (supplier audits, third party audits); analysis of customer complaints; trending of monitoring and verification results and a review of any deviations, corrective actions and any resulting foodstuff disposal.

In accordance with the general principles of food safety management, the safety of the Environmental Plan has to be reviewed on a regular basis and at least annually. The review should demonstrate that the plan is still relevant and that controls are working effectively. A review of the plan should also be initiated following any significant change to the food production process or the processing environment, for example (see Chapters 1 and 31):

- Change in the production process which affects its management from the processing environment, e.g. transport flows, service routes.
- Change in factory environment, e.g. building work.
- Changes in cleaning and disinfection practices.
- Changes to production equipment and maintenance schedules.
- Changes in legislation or codes of practice relating to, e.g., control limits or methods of analysis.

14. Establish Documentation and Record Keeping

Accurate and efficient record keeping is essential to the successful application of the PEP. Records should be accurate, timed and dated, include the actual as well as any calculated results, and be signed by the individual responsible for the assessment and by a delegated supervisor/manager who reviews the results. All records should be retained for at least the shelf-life of any foodstuffs and be sufficient to enable records to be available to support a defense of due diligence. In the example in Table 24.3, records would be kept of all interventions into the spray dryer, whether nozzle removal and reinstallation procedures had been correctly followed, ATP and microbiological counts following nozzle and tool cleaning and the use of any tamperproof tag numbers.

FUTURE STUDIES

The concept of the PMP and the PEP, which contain the identification of sources and vectors of contamination, their risk assessment to determine their necessary controls and the management of operational prerequisites in a similar fashion as critical control points, is a developing study. Together with the process of assessing all hazards of concern to the food product and implementing appropriate prerequisites as required by the HACCP plan, this overall concept can be represented by the prerequisite management plan pyramid as illustrated in Figure 24.7. By elevating the control of some contamination sources and vectors to the level of operational prerequisites and giving them the same management status as CCPs, this concept has aided a number of pioneering food manufacturers to focus their attention on the control of what are thought to be the highest risked contamination events to

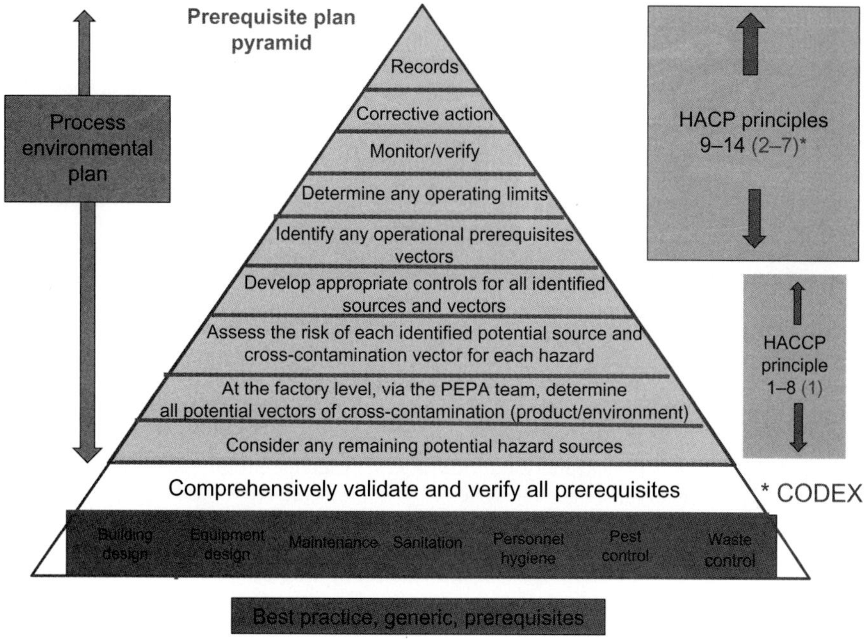

FIGURE 24.7 Prerequisite Management Plan pyramid.

the RTE product in their manufacturing process and as such has enhanced their food safety management plans.

The PMP may be constructed as a separate document with any procedures and work instructions integrated into the company business management system (BMS) or as part of the prerequisite section of the HACCP plan. What is clear, however, is that appropriate attention has to be applied to the control and management of the manufacturing process and the processing environment, via the HACCP plan, PMP and BMS to ensure that a combined food safety plan is truly effective. Additional studies are required to establish how this concept can be further improved.

References

Aavistland, P., Alvseike, O., Guérin, P.J., Stavnes, T.L., 2001. International outbreak of *Salmonella* Typhimurium DT104 – update from Norway. Eurosurveillance 5 (33), 1701.

Anon, 1983. Testing of container materials and containers for food products. DIN 10955. Deutsches Institut für Normung e.V., Berlin.

Anon, 1993. Codex Alimentarius Commission. Food hygiene basic texts.

Anon, 1995. Food handlers: fitness to work. Guidance for food businesses, enforcement officers and health professionals. Department of Health, London.

Anon, 1997a. EN 1672-2:1997 Food processing machinery – Safety and hygiene requirements – Basic concepts – Part 2: Hygiene requirements.

Anon, 1997b. Geometrical product specifications (GPS) – Surface texture: profile method – terms, definitions, and surface texture parameters. ISO 4287:1997.

Anon, 1997c. EN 1276:1997 Chemical disinfectants and antiseptics – Quantitative suspension test for the evaluation of bactericidal activity of chemical disinfectants and antiseptics used in food, industrial, domestic, and institutional areas – Test method and requirements (phase 2, step 1).

Anon, 1998a. Directive 98/37/EC of the European Parliament and of the Council of 22 June 1998 on the approximation of the laws of the Member States relating to machinery. Official Journal of the European Communities, L207.

Anon, 1998b. EN 1650:1998 Chemical disinfectants and antiseptics – Quantitative suspension test for the evaluation of fungicidal activity of chemical disinfectants and antiseptics used in food, industrial, domestic, and institutional areas – Test method and requirements (phase 2, step 1).

Anon, 2001. EN 13697:2001 Chemical disinfectants and antiseptics – Quantitative non-porous surface test for the evaluation of bactericidal and/or fungicidal activity of chemical disinfectants and antiseptics used in food, industrial, domestic, and institutional areas – Test method and requirements (phase 2, step 2).

Anon, 2002. ISO 14159:2002 Safety of machinery – Hygiene requirements for the design of machinery.

Anon, 2004. European Union Regulation (EC) No. 852/2004 of the European Parliament and of the Council on the hygiene of foodstuffs.

Anon, 2005. ISO 22000:2005 Food safety management systems – Requirements for any organisation in the food chain. <www.iso.org>.

Anon, 2006. ISO 21469:2006 Safety of machinery. Lubricants with incidental product contact. Hygiene requirements.

Anon, 2008a. ISO 9001:2008 Quality management systems – Requirements. <www.iso.org>.

Anon, 2008b. PAS 220:2008 Prerequisite programmes on food safety for food manufacturing. <www.bsigroup.com>.

Anon, 2008c. SQF 2000 Code Safe Quality Food Institute. A HACCP-based Supplier Assurance Code for the Food Manufacturing and Distributing Industries, sixth ed. <http://www.sqfi.com/wp-content/uploads/SQF-2000-Code.pdf>.

Anon, 2008d. Listeriosis outbreak in Ontario epidemiological summary. Ministry of Health and Long Term Care, Ontario, Canada. <http://www.health.gov.on.ca/english/public/pub/disease/listeria/listeriosis_outbreak_epi_sum.pdf>.

Anon, 2011a. British Retail Consortium (BRC) Global standard for food safety, Issue 6. <http://www.brcglobalstandards.com/GlobalStandards/Standards/Food.aspx>.

Anon, 2012. International Food Standard (IFS) IFS Food, Version 6. <http://www.ifs-certification.com/index.php/en/ifs-certified-companies-en/ifs-standards/ifs-food>.

Centers for Disease Control and Prevention (CDC), 1998. Multistate outbreaks of *Salmonella* serotype Agona infections linked to toasted oats cereal – United States, April–May, 1998. Morb. Mortal. Wkly Rep. 47, 462–464.

Centers for Disease Control and Prevention, 2008. Investigation of outbreak of infections caused by *Salmonella* Agona. Available online: <www.cdc.gov/print.do?url=http%3A//www.cdc.gov/salmonella/agona>.

Centers for Disease Control and Prevention (CDC), 2009. Multistate outbreaks of *Salmonella* infections associated with peanut butter and peanut butter-containing products – United States, 2008–2009. Morb. Mortal. Wkly Rep. 58, 85–90.

Gaze, 2009. HACCP: A Practical Guide, fourth ed. Guideline no. 42, Campden BRI, Chipping Campden, Gloucestershire, GL55 6LD, UK.

Holah, J., Margas, E.T., Hagberg, R., Warren, B., Fraser-Heaps, J., Mortimore, S., 2011. HACCP – the rise of the prerequisites. New Food 14, 11–15.

Holah, J., Margas, E.T., Hagberg, R., Warren, B., Fraser-Heaps, J., Mortimore, S., 2012. Identifying and controlling microbiological cross-contamination: a risk-based approach to understanding the critical role of prerequisite programs. Food Saf. Mag. February/March. <http://www.foodsafetymagazine.com/article.asp?id=4459&sub=sub1>.

Jackson, K.A., Biggerstaff, M., Tobin-D'Angelo, M., Sweat, D., Klos, R., Nosari, J., et al., 2011. Multistate outbreak of Listeria monocytogenes associated with Mexican-style cheese made from pasteurized milk among pregnant, Hispanic women. J. Food Prot. 74 (6), 949–953.

Lamb, H., 1994. Hydrodynamics, sixth ed. Cambridge University Press.

Smith, D., Holah, J., 2007. Ranking of cross-contamination vectors of ready-to-eat foods: a practical approach. Guideline no. 54, Campden BRI, Chipping Campden, Gloucestershire, GL55 6LD.

Further Reading

3-A (www.3-a.org) hygienic design guidance documents.
A code of practice for pest management in the food industry in Australia and New Zealand. Australian Environmental Pest Managers Association/Pest Management Association of New Zealand.
Anon, 2010. Six people die from Austrian Listeria contaminated cheese. USDA Foreign Agricultural Service. <http://gain.fas.usda.gov/Recent%20GAIN%20Publications/Six%20People%20Died%20from%20Austrian%20Listeria%20Contaminated%20Cheese_Vienna_Austria_2-22-2010.pdf>.
Arrowsmith, H., Brown, H., 2009. Validation of cleaning to remove food allergens. Guideline no. 59, Campden BRI, Chipping Campden, Gloucestershire, GL55 6LD, UK.
Brown, K., 2005. Guidelines on air quality standards for the food industry. Second ed. Guideline no. 12, Campden BRI, Chipping Campden, Gloucestershire, GL55 6LD, UK.
Anon, 1995. Department of Health Guideline: Food handlers: fitness to work.
European Hygienic Engineering Design Group (www.ehedg.org) hygienic design guidance documents.
Holah, J., Lelieveld, H.L.M., (Eds.), 2011. Hygienic Design of Food Factories. Woodhead Publishing, Cambridge, UK.
Middleton and Holah, 2007. Cleaning and disinfection of food factories: a practical guide. Guideline no. 55. Campden BRI, Chipping Campden, Gloucestershire, GL55 6LD.
Mortimore, S., Wallace, C., 1998. HACCP: A Practical Approach. Kluwer Academic/Plenum Publishers.
National Sanitation Foundation (www.nsf.org) hygienic design guidance documents.
Pest control procedures in the food industry. Chartered Institute of Environmental Health.
Smith, D., 2009. Hand hygiene: guidelines for best practice. Guideline no. 62. Campden BRI, Chipping Campden, Gloucestershire, GL55 6LD.
Tamime, A.Y., 2008. Cleaning-in-Place: Dairy, Food and Beverage Operations. Third ed. Society of Dairy Technology, Blackwell Publishing (Clause 4.11.6).
The Food Defense section of the American Meat Institute website. <http://www.meatami.com>.
WHO. Foodborne disease: a focus on health Education, World Health Organization, Geneva.

CHAPTER

25

Site Selection, Site Layout, Building Design

Huub Lelieveld

Formerly Unilever, Vlaardingen, The Netherlands

OUTLINE

Food Safety Management.
DOI: http://dx.doi.org/10.1016/B978-0-12-381504-0.00025-1

INTRODUCTION

The integrity of the building influences the access of pests (rodents and other small crawling animals, birds, insects), microorganisms, dust and polluted air to the products that are produced. The chances of such contaminations depend on the environment of the factory and therefore it is important to pay attention to the site. The higher the concentration of any type of contamination in the environment, the more difficult it will be to ensure that the production area will be suitable for the production of safe food products, and consequently the more expensive it will be to meet the food safety requirements.

Reconstruction and maintenance works often are done while production is continued in other areas of the same building. At such times, the safety of the food processing operation may be severely challenged. Adequate measures to prevent loss of integrity of the operation must be taken before such works start, otherwise the food operations must be interrupted until these works are finished and inspection shows that the plant is clean and ready for resuming production.

REGULATORY REQUIREMENTS

To protect the consumer, many countries currently have strict requirements with respect to food safety. Enforcement, however, may be insufficient, because in many countries legislation is weak or there is insufficient inspection capacity to ensure that regulatory requirements are met. Traditionally, this has to do with governmental budgets and priorities. Subsequent to well-publicized food scares the budget will temporarily be higher. Although the food safety requirements may be fairly, but certainly not completely (Boisrobert et al., 2010), similar between countries, it may be obvious that there are differences in requirements with respect to the environment and buildings. This is because the burden of hazards varies between regions. Factors like local climate (particularly temperature and humidity), domestic pests, husbandry (use of manure), degree of air and soil pollution and geological conditions may lead to differences in the concentration of undesirable chemicals and microorganisms in the factory environment. Some countries are prone to earthquakes, others to flooding and some to both. Consequently, risks of food safety incidents may differ too. Although in many countries food safety regulations are in place, the situation is rather different for environmental regulations. In some countries such regulations are non-existent. The consequence is that the environment of a factory may unexpectedly change and cause tremendous problems. Authorities may decide that a certain location is the best one to dump municipal wastes, because if it is disposed of closer to the population it will affect the opinion of the electorate. Nevertheless, in many countries the law requires that the premises (buildings) for food handling and processing are hygienic. These laws, however, do not specify how this must be done and hold the company fully responsible for ensuring that the premises are hygienic. An organization that does provide guidance on meeting hygiene requirement is EHEDG (see www.ehedg.org).

RETAILER'S REQUIREMENTS

Retailers are the first to deal with complaints of consumers; their reputation may be at stake, in particular where they have their own labeled goods. Therefore, retailers have good reasons to have their own requirements with respect to the hygienic condition of the factories from which they obtain their products. Even in countries where food safety regulations are adequate, retailers increasingly inspect and certify their suppliers for the simple reason that the regulators usually fail to do so or do so effectively.

SITE SELECTION

The site influences the design of the building, in order to cope with local conditions that may influence food safety. Examples are the quality of water and air, local pests (insects, birds), farms, water treatment plants, etc.

If the site is in an area with a more than average concentration of airborne microorganisms, insects and birds, there is also an above average chance of contamination of unprotected raw materials (such is often the case with fresh produce and meat) during off-loading. Sites near waste treatment plants and farms, in particular if downwind from them, may have to cope with severe problems, because untreated waste water and manure are likely to contain high concentrations of pathogenic bacteria, including *Vibrio*, *Salmonella*, *Escherichia coli*, *Campylobacter* and *Yersinia* species, and *Shigella* in addition to protozoa and viruses. These microbes may become airborne, depending on the design of the wastewater treatment system and at times when farmers spread manure to fertilize the land. The microbes will challenge the factory's air system. In addition, every time an entrance is open, anything airborne may successfully attempt to get into the factory. The building must therefore be provided with loading and off-loading bays that reduce this risk to an acceptable minimum. For the same reasons, any entrance for people, materials and air will need additional measures to keep the contamination risks sufficiently low. This is an important aspect in the selection of a suitable site. It is better not to have the factory near a sewage treatment plant, and also to make certain that such a plant will not be situated near the factory in the future. The same applies to legal waste disposal facilities and landfills. Local zoning plans will need checking and written confirmation will be needed to ensure such plants will not be built a certain distance from the factory. Nearby chemical industries may produce potentially toxic substances, which may contaminate not only the air but also the soil (with, e.g., heavy metals or chlorinated hydrocarbons), which is particularly important if well water is used, as discussed below. Be aware that the site under consideration may have been polluted, e.g. as a result of mining activities, chemical industries or (legal or illegal) waste disposal. The presence of pollution in the soil should be carefully checked.

Another requirement is the quality of water available at the site. There are large areas in the world where safe water is readily available, but there are also large areas where it is not. If available, but of unacceptable quality, an in-house water treatment facility must be installed and maintained. If the availability of water cannot be guaranteed, a well may have to be drilled. Such measures will add to the final product costs.

To operate a factory, energy is required. The energy supply may be unreliable, in particular with respect to electricity. This may severely undermine food safety management and cause incidents, because to maintain conditions to prevent the ingress of contaminated air, such as maintaining pressure differences between the various zones in the factory, electricity is essential. The same holds for cooling and freezing and for the operation of measuring, controlling and registration equipment. If the electricity fails, so will the control of processing temperatures and flow rates through pipelines. Hence, if interruptions are likely to take place, adequate back-up systems (e.g. oil-powered electricity generators) must be installed, the capacity depending on how long interruptions may last. It is important to obtain a guarantee from the local electricity supplier and to be certain also to ask for advice from a reliable local consultant.

SITE LAYOUT

For the same reasons as discussed above, the layout of the site must prevent access of pests to the factory. To keep animals at bay, there must be fences that are high enough to prevent dogs and cats from entering the area, but at the same time deep enough to prevent burrowing animals (rats, rabbits) to gain access. The fences must be such that they do not allow animals (including monkeys) to climb over them. Any unpaved surface must be covered by grass that is kept short to avoid breeding of small animals. For similar reasons, there should be no shrubs or trees or they should be remote from the factory wall and particularly its entrances and air inlets. Because they provide places for microbes and insects to breed, there should be no ponds or any other possibilities for stagnant water or mud. For these reasons, pavements should be horizontal or slightly sloping towards drain pits.

External lighting should always be away from the factory walls and entrances, luring insects away from the building instead of attracting them to it.

Every effort must be made to prevent insects, small animals and microbes from multiplying and worsening the environment of the factory: waste disposal areas too must be such that they do not allow ingress of insects and animals. Moreover, to limit undue growth of molds and bacteria, solid waste should be kept dry. This means that, although the disposal area must be outside the factory, it should nevertheless be covered to cope with precipitation. Doors to the area must be rodent resistant and be insect-proof. The doors should preferably be self-closing, otherwise an alarm should sound if the door has been open for any length of time (minutes, rather than hours).

Access to Production Areas

The entrance of the production area must be equipped with hand-washing facilities such that everybody entering the area must pass these facilities. Restrooms (toilets, washrooms, lavatories) must not be directly connected to the production area and must be easy to clean. Depending on the type of products handled in a certain area, it may be necessary to minimize the risk of transfer of contamination from the outside into that area by using a change room, where garments can be exchanged for special production room garments. The room should have a step-over barrier to leave shoes and boots that are worn outside on one side, and on the other side to put on footwear to be used exclusively in the production area. The step-over must be sealed to prevent contamination on the floor from moving to the production area. There must be a means for cleaning, disinfection and drying of hands.

Processing and packaging areas should not be used as a passageway to canteens or other amenities. Therefore, the layout of the building should take into account that cafeterias, kitchens, offices, laboratories, workshops, chemical stores, etc. are not connected to the production and packaging areas. Anybody, including laboratory staff, directors and important visitors, who needs to be in the production area should pass the change room or hand-washing facilities, as applicable, and use them as per the personal hygiene instructions. Similarly, any vehicle that is needed to transport raw materials, ingredients or finished products should have designated routes and should not be used for anything else.

BUILDING DESIGN

Supporting Structure, Foundation, External Walls and Roofs

The supporting structure for the factory should ensure that the floor is at an elevated level, so that there is no risk of rain, mud or other precipitation entering the factory. In regions with potentially heavy rainfall, the floor level should be higher than any area in the environment, to avoid the factory being flooded.

Apart from their function to protect the factory from bad weather conditions and sometimes the sun, external walls, roofs and the foundations are the first and most important barriers to the ingress of pests, in particular rodents, geckos, birds and insects. Hence, the design must be such that there are no openings that allow animals and insects to enter the building. In addition, it must be ensured that the structure is such that rodents and, in some areas of the world, termites, cannot gain access. Any areas between the ceiling and the roof must be entirely closed off to avoid them being used for nesting by birds, mice or rats. Bird droppings contain high concentrations of microorganisms, including pathogens such as *Salmonella*. Special attention should be paid to effectively seal the connections between roof and walls. Similarly, the connection between the foundation and the wall must be rodent-proof.

To avoid attracting vermin on the roof, it should be kept dry and also be sloped. External walls and roofs should be easy to clean and maintain. The outer walls should not have ridges or other protrusions that allow birds to settle and breed. Adequate measures should discourage birds from settling.

Entry and Exit Points

Entry points should be designed such that they allow passage of personnel and goods but prevent entry of pests as effectively as possible. This requires automatically closing doors that are rodent resistant and in some locations also termite resistant. The doors should close such that even small insects like ants cannot gain access.

It is important to build the factory in such a way that the openings are downwind, at least as much as possible, so that wind is not blowing any undesirable matter into the factory. Doors and windows that normally are not used are places where insects may breed in the narrow space between the frames and the doors and windows. Hence, care must be taken that such crevices be sealed with a good quality tape that is resistant to cleaning chemicals. Also, windows and doors must still be able to be used in case of emergencies.

Internal Walls, Floors and Ceilings

Walls must be non-absorbent and well cleanable, and should not have recesses or cracks that can harbor insects. The wall must be able to withstand damage that may result in product contamination. For the same reason, any paint used should be of a quality that does not flake off. The use of strong, slightly elastic wall coatings is recommended. Corners may have to be protected by metal reinforcements. The same applies to the lower part of the walls, where bumper constructions should prevent damage when fork trucks may accidentally hit the wall.

The floor often plays an important role in product contamination incidents. Improperly designed floors may accumulate moisture and nutrients for insects and bacteria. Movement (e.g. of people, vehicles) over such floors then causes aerosols that carry microorganisms. Insects full of microbes may crawl out and enter the product, visibly or not. Hence, floors, unless in areas that will never be wet, must be watertight and slope towards drains, so that any liquid spilled can be easily removed. The floor must also withstand cleaning chemicals and the temperature of hot water that may be needed for cleaning. Floors must be able to withstand damage by personnel and moving of equipment. Since the 1970s, composite floor materials (epoxy, meta-acrylate, polyester, polyurethane) have become popular, in particular in new or refurbished factories, but often with disappointing results. The materials are not as strong mechanically as good quality tiles. Forklift trucks, containers with raw materials or intermediate products and waste containers can easily damage such floors. The floors must also withstand the installation of machinery. If machines are installed on an intact epoxy floor, the integrity of that floor may be affected and moisture may penetrate to the supporting structure. The lively matter developing under these floors cannot be removed without removing the affected area of the floor. If the floor is to be subjected to heavy loads and to the movement of forklift trucks or equivalent, tiles are to be recommended. Care must be taken, however, that the tiles are grouted such that no moisture is absorbed, creating undesirable circumstances. Hence, the grouting must be resilient and water repellent. It is also easier to repair a damaged tiled floor than a damaged composite floor. Although floors should be easy to clean and hence be smooth, they should not be so smooth that they will be too slippery to walk on. Despite the drawbacks, there may be applications where composite floors provide the best solution, taking into account the prevailing operation conditions.

To allow effective cleaning, the transition between floor and walls must be rounded. For tiled floors, special tiles are available.

To avoid dust falling down on exposed product or product contact surfaces, ceilings must be tight and hence false ceilings should not be used in food processing plants. Suspended service ceilings, however, are fully acceptable and provide the advantage of mounting cables, service ducts, etc. above the process area from where they can be extended downwards to where they are needed. This way, the risk of contamination of product by dust that accumulates on these provisions is drastically reduced. The construction of these service floors must self-evidently be such that it does not allow any dust to pass from the area above the ceiling to the production area, which means that the passages of the ducts, etc. must be effectively sealed.

Lighting

There must be enough light in the factory firstly for the personnel to do their job properly and efficiently and secondly so that dust, dirt and vermin or their traces can be detected. To avoid the risk of contamination of product with glass, glass windows are generally avoided, but modern types of glass are very strong and are available in shatterproof quality. If well mounted they will not easily break and if they do, they will not splinter into small pieces, even if accidentally struck with any great force.

Lamps for artificial illumination must be covered by shatterproof covers that are tight, so that an exploding lamp cannot contaminate product. To avoid accumulation of insects and dust, lighting should preferably be mounted in the suspended service ceiling, with the underside flush with the ceiling. Besides preventing the collection of dust and insects, it also enables the replacement of the lamps from the top, without interfering with the products. In the absence of a suspended ceiling or if the ceiling is too high above the surfaces to be illuminated, the topside of the housing of the lamps must slope at an angle of approximately 45° ($\pi/4$) to prevent anything from settling.

Temperature Control

For the comfort of the personnel in the factory, air conditioning may be needed. Where food is processed under chilled conditions, cooling units are used. These units have trays underneath to prevent condensate from dripping onto personnel and product. What is not realized but is very important from a food safety point of view is that these trays are perfect places for the selective cultivation of psychrotrophic bacteria, specifically *Listeria monocytogenes*, a pathogen that may cause listeriosis. It is a fairly selective process because at low temperatures, *Listeria monocytogenes* grows faster than most other microorganisms. The fans of the cooling unit complete the food safety risk because the circulation of air helps to spread the contaminated condensate over the product. Therefore, collectors of condensate should always slope to one side, from where the condensate is led to a hygienic drain pit. Moreover, collectors need to be easily accessible for regular inspection, cleaning and disinfection.

To make the control of the temperature efficient and affordable, as well as to prevent condensation on walls, ceilings and windows, thermal insulation of walls and roof (and/or service ceiling) and double-glass windows are needed. Care must be taken that the installation of insulation panels does not create a space for the breeding of insects between the wall and the panel. The panels should be smooth for ease of cleaning and their surface should be strong enough to prevent damage under the applicable conditions.

Noise Control

Machinery may produce more noise than is desirable or legally acceptable. Noise-absorbing panels may and have been used on a large scale. Nevertheless, because they must be cleanable and hence their surface must be smooth, these hygienically acceptable panels are not very effective. Increasingly, food processing machinery is designed to produce significantly less noise than a few decades ago and the best solution for noise reduction may be encasing the noisy parts using panels with the sound-absorbing surface at the inside and having a smooth outer surface. There may also be situations where it is more attractive and efficient to use noise-canceling earphones for personnel working near noisy machinery.

Sewers, Gutters and Drains

Sewers can be a serious means of contamination of the interior of a building. The design may be such that even large animals like rats gain access, unless appropriate measures are taken to prevent this. The design must also be such that there cannot be pressure differences large enough to cause gases to enter the interior of the factory in any way. It is highly recommendable to keep the sewer system physically separated from other waste water systems, which are connected to floor drains throughout the factory. Gutters tend to be covered by perforated covers and cleaning the gutters and covers is usually troublesome. Moreover, water, often contaminated with spilled product, tends to be stagnant in most parts of the gutters, allowing microbial growth and nesting of cockroaches and other insects. Gutters therefore should be avoided and hygienic floor drains should be used instead, while the floors should slope towards these drains. The drains, however, can also become breeding places for insects and bacteria. They therefore should be of a design that can be disinfected and in high-care areas should preferably be of a kind that can hold disinfecting substances.

Internal Zoning, Ventilation and Air Conditioning

An adequate air supply is needed to ensure a sufficient supply of oxygen and to control the temperature in the production environment. Depending on the temperature required and the amount of heat produced by the machinery or the processing of the product, the amount of air needed may vary greatly between factories and hence also the dimensions and design of air ducts and exhausts. To make the risk of airborne contamination of exposed product or food contact surfaces as small as possible, the air should flow from the exposed final product area, through areas where such contamination is less important, to the area where materials arrive. This requirement influences the differences in pressure needed between the various zones in the building. The air supplied to the cleanest areas should self-evidently be adequately filtered, to the degree needed to meet the product safety requirements. HEPA (high-efficiency particulate air) filters will be required where microbiologically vulnerable products are exposed to the air. Between pre-filters, intended to remove course particles and insects, and the fine (HEPA) filters, intended to remove microorganisms, dehumidifiers must be installed to ensure that the fine filters remain dry and to prevent condensation in the production area. The combination of differences in temperature and humidity may result in condensation and hence wet spots in the process area, resulting in the growth of bacteria and fungi, which may become airborne and contaminate the product. Hence, care must be taken that either such condensation cannot take place or takes place under control, at easily accessible locations, enabling inspection, cleaning and disinfection. Air inlets should be positioned at a distance from the air outlets of the factory to prevent contaminated air to unnecessarily burden the air inlet filters and dehumidifiers. Both inlets and outlets must be provided with screens to prevent entry of flying animals and insects.

If in the factory materials are processed that contain allergens, special attention should be paid to air flows to avoid the air from the area where allergenic products are processed passing to areas where products are processed or packed which must be or are supposed to be free from these allergens.

Zoning of food production premises is important to prevent (re)contamination of exposed food and also proportionate use of protective measures and verifications (e.g. environmental monitoring). Details on zoning can be found in Holah and Lelieveld (2011). It may be difficult to realize appropriate zoning in very small premises (Todd et al., 2010). Nevertheless, if vulnerable products are made in such premises, the zoning rules should be met or such products should not be produced.

Walkways and Stairways

Footwear, also the special footwear that has been put on before entering the production hall of the factory, collects dust and dirt and hence may shed dirt again when the wearer moves around. It would be safest if levels of contamination on footwear of any person who needs to be in the factory during production would stay below those of any exposed food product and any food contact surface. There are circumstances, however, that necessitate staff to cross over such areas and hence stairways and bridges are needed. It may be required that during production, parts of machinery that are positioned high on a machine must be adjusted or replaced. In such a case, an elevated walkway is needed. Stairs, bridges and walkways must be designed and constructed such that no dirt from footwear can contaminate the food and food contact surfaces. Consequently, open structures are not acceptable and there must be sides that are high enough to prevent any dust or dirt from falling down.

Process Support and Utility Systems

To operate the food processing and packaging equipment, product, ingredients, cooling or heating media (steam, water), air, electricity, signal cables and packing material all have to be brought to where they are needed. This requires pipes, cables, conveyor systems and support structures. Together this can become a nightmare from a hygiene point of view. When combined, which is unavoidable, because they all need to reach the same machine, these items form ideal places for insects and other pests to hide and breed. They collect dust and dirt and are a source of dead insects that may fall down to contaminate product underneath. Moreover, they are very difficult if not impossible to clean.

Cables (electric power supply, signal transfer) and small pipes (compressed air, nitrogen, lubricants) should best be grouped together in ducts, which can be larger pipes or special designs. To prevent ingress of insects, these ducts must be effectively sealed at any entry or exit point of a cable or pipe, at least at the processing side. When building a new factory, probably the best way would be to have a service area below the production floor, provided that the space will be high enough for access of service personnel. From that area, service ducts may rise to the machine and the cables and pipes stay largely below the exposed product level. If a service floor is not possible, a suspended service ceiling can be a good solution. The passage of the ducts through the floor or ceiling must be such that nothing else, not even air, can pass through.

Pipes for transport of product or product ingredients may have to pass through walls or ceilings between various processing departments. Care must be taken that such passages are either tight, not allowing anything passing around them, or large enough to allow cleaning and inspection of the passage. If tight, the construction must be such that they remain tight with time and hence the passage can absorb vibrations caused by machinery to which the pipes are connected and the changes in length and diameter of the pipes as a result of thermal expansion.

Where product or packaging material must pass through walls or ceilings using conveyor belts, chains or slides, care must be taken that the passage itself is cleanable and accessible for cleaning and inspection.

Food Storage Rooms

Food storage must be designed to make certain that insects and other pests cannot reach the food, even if the food is packed. It must be possible to control the humidity to ensure that the area is always dry. Temperature control and monitoring is essential for storage of perishable products. Entrance of insects and small animals can be prevented by building the storage room on an elevated level, but such that also the entrance is higher than the outside pavement. Further, the storage room should meet the general requirements that also hold for the processing area to ensure that the space can be cleaned: smooth walls and ceilings, no ridges, no surface cracks and other crevices where insects may hide. Walls must be watertight to avoid wet surfaces on the inside. It is important that the lighting is sufficient for inspection to spot any traces of vermin. There must be enough space between the wall and the stored products for inspection.

Storage of Grain

Large qualities of grains (rice, wheat, corn, etc.) are usually stored in silos. Self-evidently, the silos must be sealed to avoid the entrance of vermin. Chances are, however, that there are insects already in the grain. Measures should be effective in ensuring that there will not be any larger animals in the silos. The insects should be prevented from multiplying by keeping the product dry. Insects, like other animals and people, need water to survive and the absence of water may perhaps not kill all insects, but at least the survivors would be dormant. Another, equally important reason to make certain that the grain remains dry is to prevent mold from growing. Molds produce metabolites that are toxic (mycotoxins) for humans in very low concentrations. For instance, *Aspergillus* species produce a variety of aflatoxins. The EU regulations require that the concentration in grains for human consumption of all aflatoxins together is below 4 μg/kg. Ochratoxins are produced by some *Aspergillus* as well as *Penicillium* species. The maximum concentration in grain for human consumption of Ochratoxin A, the most important one, is 3 μg/kg (these are parts per billion, $1{:}10^9$). Mycotoxins are also harmful to animals (e.g. horses) and hence the above also applies to feed.

The problem with large silos is that it is difficult to ensure that the temperature is the same everywhere in the product. Temperature differences, however, will cause transport of moisture to the colder spots, which subsequently may become moldy and thereby toxic. It is therefore recommended to use thermally insulated silos.

Storage of Oils

The solubility of water in oil is strongly temperature dependent. At 20°C the solubility in sunflower oil is approximately 75 mg/kg; at 40°C it is about 50% higher. The consequence is that in oil that contains more water than is soluble at the lowest temperature in the storage tanks, water will separate. In addition, if the tank is not well insulated, water may condense at the inner wall of the tank. Because of its higher density, all water will sink to the bottom of the tank and where there is water and nutrients, microbes will grow and therefore the oil becomes contaminated with microorganisms that in turn may produce potentially toxic substances. The message is that it is important to control the temperature of the room for storage of oils and to prevent oil from cooling down. In other words, letting the temperature drop in wintertime to save energy is not a good idea from a microbiological safety point of view.

Storage of Chilled Food

Self-evidently, chilled food storage rooms need adequate temperature control. It is important to take into account that lighting and ventilators produce heat and that as a consequence, despite the temperature control, there are temperature differences in the cold room. Similar to chilled rooms for food processing, condensate trays underneath cooling units should slope towards a drain, from where it is led to a hygienic drain pit. The trays need regular inspection, cleaning and disinfection.

Storage of Packing Material

Some packing material, in particular carton and paper and the increasingly popular biodegradable materials, are substrates for microorganisms and should therefore be kept dry to prevent microbial growth. Other materials, like glass, metals and non-biodegradable polymers, can be stored in areas without temperature and humidity control, unless the humidity at the location can be extreme. Switching from non-biodegradable to biodegradable materials will probably need measures to prevent microbiological problems.

Storage of Chemicals and Lubricants

The design and location of the store for chemicals for cleaning and sanitation must be such that any risk of contamination of product and packing material with chemicals is avoided. The store must be provided with a lock and there should be no direct connection between product areas and the chemicals store. The same holds in principle for lubricants, glues and inks needed in the process and packaging areas, unless the lubricants, glues and ink comply with the requirements for food contact material or they are used only in areas where the product is packed in well-sealed containers.

Storage of Refuse and Waste Materials

Waste materials, such as used cartons and boxes, must have adequate space for storage, enabling "good housekeeping" in the entire factory. Without such spaces, pests (including rodents, cockroaches) will find places to harbor and breed.

References

Boisrobert, C., Stjepanovic, A., Oh, S., Lelieveld, H. (Eds.), 2010. Ensuring global food safety – exploring global harmonization Academic Press

Todd, E.C.D., Michaels, B.S., Greig, J.D., Smith, D., Holah, J., Bartleson, C.A., 2010. Outbreaks where food workers have been implicated in the spread of foodborne disease. Part 7. Barriers to reduce contamination of food by workers. J. Food Prot. 73, 1552–1565.

Holah, J., Lelieveld, H.L.M. (Eds.), 2011. Hygienic control in the design, construction and renovation of food processing factories Woodhead Publishing

Further Reading

This chapter provides limited information on the issues that need consideration when deciding to build a new or improve an existing food factory. When it has been decided to build or refurbish a factory, much more detailed information will be needed to be able to develop a plan that covers all aspects. Such information can be found in a recently published book by Holah and Lelieveld (2011).

CHAPTER

26

Hygienic Design and Maintenance of Equipment

Frank Moerman[1] *and Jacques Kastelein*[2]

[1]Catholic University of Leuven - KU Leuven, Leuven, Belgium, [2]TNO, Zeist, The Netherlands

OUTLINE

Food Safety Management.
DOI: http://dx.doi.org/10.1016/B978-0-12-381504-0.00026-3

INTRODUCTION

There is a global trend in the food industry towards minimal food processing and preservation. Consumer demand for "fresh-like" additive-free foods, which maintain their nutritional and sensorial properties during preparation, conservation, packaging, storage and finally consumption, is increasing. But the general tendency to apply mild processing and conservation techniques to achieve that purpose often shortens the shelf life of food, may put foods at risk and may compromise consumer health. Therefore, more than ever, good hygienic engineering and design practice is one of the tools to reduce or exclude microbial (e.g. pathogens), chemical (e.g. lubricating fluids, cleaning chemicals) or physical (e.g. glass, wood) contamination of food. Good hygienic design also may eliminate product "held-up" within the process equipment where it could deteriorate and affect product quality on rejoining the main product flow. As such, good hygienic design may prevent one batch cross-contaminating a subsequent batch. Good hygienic design also reduces the downtime required for an item of process equipment to be cleaned, while at the same time allowing an increase in the time to produce. Therefore, although initially more expensive than poorly designed equipment, hygienically designed equipment will be more cost effective in the long term.

To reduce and eliminate food product recalls, lost production and site closure, due to contamination arising from poorly designed equipment, this chapter intends to inform food safety professionals and inspectors/auditors about the risks associated with poor hygienic design. With typical examples of poor hygienic design, the necessary technical and practical guidance will be given to identify and control equipment-related food safety hazards. As such, this chapter may help the food manufacturer to select the most suitable food processing equipment, to construct a food production line that meets all current and future hygienic requirements, and to set up an appropriate food safety management plan (e.g. HACCP) to eliminate or control all food safety hazards along the food chain.

In the first section, an overview is given on the current legislation and standards dealing with the hygienic design of food processing equipment. A second section lists the basic hygienic requirements that food processing equipment must meet to produce microbiologically safe food products. The third section describes the hygienic and food grade materials that can be used in the manufacturing of food processing equipment; followed by a section that outlines the requirements to the food contact surface finish. The next two sections make recommendations with respect to the hygienic design of respectively open and closed equipment for processing of food. The seventh section considers the hygienic installation of food processing equipment in the food factory. The last section deals with hygienic practices during process equipment maintenance operations in the food industry.

LEGISLATION

Many countries around the globe have developed legislation on the production of food, requiring that microbiologically safe food shall be produced by means of process equipment that minimizes the risk of contamination and that is easily cleanable. Hence, food producers are encouraged to purchase hygienically designed food processing equipment that aims to meet these criteria. In response to this demand and because they are also forced by national and/or international legislation, manufacturers of food processing equipment have developed process equipment that is hygienically designed and easily cleanable. An overview on existing legislation and standards describing the hygienic requirements applicable to food machinery is given in Hauser (2008a), van der Meulen (2010) and Moerman (2011a).

BASIC HYGIENIC REQUIREMENTS

Processing equipment intended to produce safe food should at least meet the following basic hygienic requirements (Holah, 2000; Lelieveld et al., 2003):

- Materials of construction used for equipment must be completely compatible with the food product, environment, cleaning chemicals and disinfectants, and the methods of cleaning and disinfection.
- Product contact surfaces (including the welds in the product contact area) should have a smooth surface finish to enable them to be cleaned easily.
- Food equipment should be designed to prevent bacterial ingress, survival, growth and reproduction on both product and non-product contact surfaces of the equipment. The food processing equipment must be constructed to ensure effective and efficient cleaning over the lifetime of the equipment.

FIGURE 26.1 The pits, cracks, crevices, recesses, open seams, gaps, lap seams, bolts and threads will accumulate dirt and will make this equipment not cleanable. Moreover, galvanic corrosion can be observed. *Courtesy of John Butts, Land O'Frost.*

- Welding or continuous bonding are preferred over fastenings. Exposed screw threads, nuts, bolts, screws and rivets must be avoided whenever possible in product contact areas. Alternative methods of fastening can be used where the washer used has a rubber compressible insert to form a bacteria-tight seal.
- To make permanent pipe joints, welding is the preferred method of joining. These welds must be continuous and smooth. Screwed pipe couplings must be crevice free and provide a smooth continuous surface on the product side. Flanged joints must be sealed with a gasket to avoid ingress of microorganisms.
- In design, construction, installation and maintenance, hollow areas of equipment such as frames and rollers must be eliminated or they shall be hermetically sealed. As such, bolts, studs, mounting plates, brackets, junction boxes, nameplates, end caps, sleeves and other such items must be continuously welded to the surface, and shall not be attached via drilled and tapped holes.
- Niches such as pits, cracks, crevices, open seams, gaps, lap seams, inside threads that accumulate dirt and hamper the cleanability of the process equipment are not allowed (Figure 26.1).

FIGURE 26.2 All surfaces in the product zone are designed to be self-draining for liquid food, cleaning and disinfection solutions, and rinsing water. *Courtesy of Krones AG.*

- Dead areas, dead ends, pockets or other conditions which may trap food, harbor contamination, prevent effective cleaning and disinfection, and allow cross-contamination shall be avoided.
- All inaccessible horizontal flat areas, ledges, projections, protrusions, recesses, edges, etc. where product rests can accumulate should be eliminated.
- For the same reason and to facilitate cleaning, internal angles and corners should be well radiused.
- The exterior of non-product contact surfaces should be so arranged that harboring of contamination in and on the equipment itself, as well as in its contact with other equipment, floors, walls or hanging supports, is prevented.
- All pipelines and equipment surfaces in the product zone must be so arranged that they are self-draining (Figure 26.2) to minimize contamination and corrosion risks when liquid food, cleaning and disinfection solutions, and rinsing water are retained during idle periods. Microbes can flourish in stagnant pools of water, when supported by nutrients which are trapped in the internal pockets. Moreover accumulated and pooling cleaning and disinfection solutions may contaminate food products.
- Certain equipment surfaces operate at or below the natural dew point of water vapor. Equipment design, therefore, should not permit the formation of condensate that may enter the food zone and contaminate product or product-contact surfaces.
- All parts of the equipment shall be readily accessible for inspection. Because potential contaminants on representative surfaces throughout the product contact zone must be readily detectable, all surfaces in the product zone must be immediately visible for inspection, or the design of the equipment shall allow readily dismantling without the use of tools for such inspection. Equipment surfaces must be readily accessible for manual cleaning and disinfection (Figure 26.3), unless it can be demonstrated that the

FIGURE 26.3 Product contact surfaces of this equipment are not readily accessible for manual cleaning and disinfection. Moreover, the dome screw with drive slot and washer creates gaps and crevices where debris collects. *Courtesy of Joe Stout, American Meat Institute.*

result of in-place cleaning and disinfection procedures without dismantling is equivalent to the result of dismantled and manual cleaning procedures. All potential obstructions to cleaning, disinfection and maintenance should be avoided or minimized.

- Instruments not only must be hygienically designed, but also hygienically installed.
- Equipment design also must ensure hygienic compatibility with other equipment and systems, such as electrical, hydraulics, steam, air and water.
- Maintenance equipment enclosures and human machine interfaces such as push buttons, valve handles, switches and touchscreens must be designed to ensure food product, water or product liquid does not penetrate or accumulate in and on the enclosure or interface. Also, physical design of the enclosures should be sloped or pitched to an outside edge to avoid use as storage area. Doors, covers and panels should be designed so that they prevent entry and/or accumulation of soil. To facilitate cleaning, they should be easy to remove.
- Bearings should be mounted outside the product area to avoid contamination of food products by lubricants and to exclude the ingress of bacteria. When the bearing is within the product area, its design should allow the passage of cleaning fluid.
- Food grade oil should be used, and leaking of oil onto food product has to be excluded. A drip pan which protects the product zone should be used, or motors driving equipment components such as agitators, belt drives, etc. should be placed outside the product area. If they are within the splash area, they should be protected by a removable cover.

MATERIALS OF CONSTRUCTION

General Recommendations

Materials of construction for food processing equipment, process piping and utilities should be homogeneous, hygienic (smooth, non-porous, non-absorbent, non-toxic, easily cleanable, impervious and non-mold supporting), inert (non-reactive to oil, fat, salt, etc.; may not adulterate the food by imparting deleterious substances to it, nor affect its organoleptic characteristics), chemical resistant (corrosion-proof; non-degrading and maintaining its original surface finish after sustained contact with product, process chemicals, cleaning agents and disinfectants), physically durable and mechanically stable (resistant to steam, moisture, cold, heat, the actions of cleaning and sanitizing agents; resistant to impact, stress and fatigue; resistant to wear, abrasion, erosion and chipping; not prone to cracks, crevices, scratches and pits, unbreakable) and easy to maintain, in agreement with the guidance described in EHEDG guidelines No. 8 and No. 32. Additional requirements could be availability, welding ability, machinability and capability of being shaped. Notice that materials which are worked (for instance, bent, cut, sheared, extruded or drawn) during manufacture may require additional treatment (such as surface finishing) following fabrication in order to render them corrosion resistant. Hence, materials should be selected that are suitable for surface treatment (Hauser et al., 2004a).

Product contact surfaces – all the surfaces exposed to direct contact with the product as well as indirectly impacted surfaces from which splashed product, condensate, liquid or solid particles may run off, drop off or may fall into the product – should be constructed of materials that meet the highest hygienic requirements, while materials used in the construction of components located in the non-food contact area may be of a lower grade.

Use of Metals and Alloys

Carbon steel cannot be used in the food contact area due to its corrosion sensitivity, especially by salt and chlorine-containing bleach. To retard its corrosion, it is often galvanized (zinc plated) but, with time, galvanized steel becomes damaged when the zinc coating peels off. The only permitted application of galvanized steel is in contact with dry and non-acidic foodstuffs. Painted steel never shall be used in the neighborhood of food because paints often contain zinc, lead, cadmium and phenolics. Moreover, paint can crack or flake, and some cleaning agents rupture the physical integrity of paints. Paint that peels off can fall onto the product, creating a health risk. Paint surfaces used in non-product contact areas may crack or flake and should be repainted immediately.

The austenitic chrome–nickel or chrome–nickel–molybdenum steels are mainly used for the construction of equipment and machining in the food industry. Stainless steel AISI SS 304(L) can be used for the construction of food processing equipment and food processing support systems in applications with low chloride levels (up to 50 mg/l [ppm]), near neutral pH (between 6.5 and 8) and at low temperatures (up to 25°C). However, stainless steel AISI SS 304 is sensitive to sodium hypochlorite and to salt that is usually present in food in high content. In these less appropriate circumstances, stainless steel can still be used for exterior equipment surfaces, motor and electrical cabinets, etc. Because cheaper grade AISI SS 304/304(L) will suffer some corrosion over a long time period, the small additional cost of using AISI SS 316/316L rather than AISI SS 304/304L almost certainly will be worthwhile in terms of trouble-free operation. Stainless steel AISI SS 316(L) is commonly used as construction material for food processing equipment. However, as temperatures approach 150°C, even AISI SS 316 stainless steels may suffer from stress-corrosion cracking in regions of high stress and exposure to high levels of chloride. Therefore, other stainless steel types were developed to overcome that problem (e.g. duplex steel and nickel alloys) (Hauser et al., 2004a).

The best known application of copper is vessels, traditionally used in many breweries and distilleries. Copper does not really constitute a food safety problem but it is recommended to avoid direct food contact with copper utensils, as they can cause unacceptable organoleptic effects. Moreover, copper can be quickly and severely affected by strong alkaline detergents, sodium hypochlorite, acidic and salty food, making it unsuitable in the food contact zone. The copper alloys brass (60–70% copper, 30–40% zinc) and bronze (80–95% copper, 5–20% tin) are more prone to corrosion by alkaline and acidic detergents, salty and acidic food than the ferrous steels. They become quickly porous, especially brass that undergoes de-zincification by acid and steam.

Because aluminum is attacked by alkaline detergents, sodium hypochlorite and acidic food, the use of uncoated aluminum utensils should be limited. Anodized aluminum is acceptable in the food contact area. Exposure to aluminum is usually not harmful, but its intake should be limited.

Lead, cadmium and mercury in food contact materials must be avoided. Notice, however, that these components are largely present in electrical and electronic components. In 2003, the EU adopted the *Directive on the restriction of the use of certain hazardous substances in electrical and electronic equipment* (RoHS) Directive (2002/95/EC). Alloys for food contact may only contain aluminum, chromium, copper, gold, iron, magnesium, manganese, molybdenum, nickel, platinum, silicon, silver, tin, titanium, zinc, cobalt, vanadium and carbon.

Use of Plastics

Plastic materials may be used to preclude metal-to-metal contact (e.g. for bearing surfaces), as guides and covers, or for hoses because of their plasticity and corrosion resistance. These plastics should be odorless, non-porous, smooth and free from cracks, crevices, scratches and pits which can harbor and retain soil and/or microorganisms after cleaning. They must not absorb product constituents and microorganisms, must have high mechanical strength (resistant to ageing, creep, brittleness, fatigue, etc.) and good wear/abrasion resistance, and must be resistant to heat, cold flow, hydrolysis, electrostatic charging, etc. Further, no migration of plasticizers, monomers or additives into the food product must occur.

When using a plastic material (belts, gaskets, electric cables, etc.), it is of utmost importance to ensure that the material is able to withstand all temperatures from −50°C to temperatures as high as 121°C (steam sterilization) without cracking or breaking. Moreover, the plastic material must be chemically resistant to solvents, acid, alkaline, reducing and oxidizing agents, cleaning and disinfection agents and corrosive food gases at these temperatures. The equipment manufacturer should test the chemical and temperature resistance of the plastic material (Partington et al., 2005; Moerman, 2011a).

Use of Rubbers

Elastomers must be chemically resistant to fat, cleaning agents and disinfectants; they must not show expansion and shrinking under the influence of temperature changes or chemical fluids; they must be abrasion resistant (e.g. rotary shaft seals, or seals in static

TABLE 26.1 Resistance Characteristics of Different Rubber Materials (Plett and Graßhoff, 2006)

Contact Medium	Natural Rubber	Acrylonitrile Butadiene Rubber	Silicone Rubber	Ethylene Propylene Diene Monomer	Chloroprene	Fluor Elastomer
Temperature range	−60 to 80°C	−35 to 120°C	−70 to 200°C	−60 to 135°C	−40 to 230°C	−30 to 180°C
Hot water (120°C)	–	+++	+++	+++	+++	+++
Hot water (145°C)	–	–	+++	+++	–	+++
NaOH (5%; 90°C)	++	+++	+++	+++	++	+++
NaOH (5%; 140°C)	–	–	–	++	+	+++
H_3PO_4 (2%; 90°C)	–	+++	+++	+++	+++	+++
H_3PO_4 (2%; 140°C)	–	–	–	++	–	+++
HNO_3 (1%; 70°C)	– –	–	++	++	++	++

+++ = *unlimited resistance*; ++ = *limited resistance*; + = *only short contact*; – = *non-resistant*; – – = *absolutely non-resistant.*

applications that are subjected to abrasion from dry material product); and they must retain their surface and conformational characteristics (no loss of elasticity, no embrittlement, no rubbed-off parts, no crevices, etc.). However, elastomers can be degraded by product, by cleaning agents, by disinfectants and by thermal and mechanical stress much quicker than metal components, with the following results: leakage of lubricants, loss of bacteria tightness, increased adherence and retention of dirt and bacteria in crevices leading to permanent product and process contamination, insufficient cleaning and problematic disinfection. Partly destroyed sealings allow ingress of liquids containing chlorides under gaskets and seals, so that a high chloride concentration may subsist between damaged sealings and adjacent metal, which favors crevice corrosion even in stainless steel. Therefore, gaskets and seals preferably should be of a removable type. Appropriate rubber materials are fluoro elastomers, natural rubber, silicone, neoprene, EPDM, nitrile and nitrile/butyl rubber. Their resistant characteristics can be found in Table 26.1 (Partington et al., 2005; Plett and Graßhoff, 2006).

Other Materials

Wood and certain types of insulation are not allowed within the product contact area (exceptions are butcher's blocks; wooden barrels, etc.). To avoid their exposure to the outside, they must be permanently and tightly sealed off from the product zone.

Glass may be used as a food contact surface, but its application is not recommended due to the potential for breakage. Specially formulated glass materials such as Pyrex® have proven successful. When glass is used, it must be durable, robust and heat resistant. Some applications where glass is used are light and sight openings into vessels, and to a very limited extent glass piping. Replacement by transparent alternatives like Perspex® or polycarbonate is recommended (Hauser et al., 2004b).

Ceramics are very resistant to acids and sufficiently resistant against lye. They are very hard and can withstand pressures of 100–400 MPa. They are used in the coating of other stable materials, in the production of ceramic membranes, and in the construction of pipes or processing equipment for very sensitive products. The main drawbacks of ceramics are their brittleness and porosity. To be food safe, all ceramic surfaces in direct contact with food must have smooth, unbroken and lead-free glassy surfaces, entirely free of crazing (small hairline cracks) and blemishes. Although not many bacteria may hide in a crack, in contact with food those few may be become a large culture.

The use of nanomaterials in the food industry may present potential risks, requiring the need for risk assessments to identify and quantify these risks. Some nanoparticles have been found to exhibit negative effects on tissues such as inflammation, oxidative stress and signs of early tumor formation (Stone et al., 2009; FAO/WHO, 2010; Becker et al., 2011). Because nanoparticles may become wasted in surface waters along with cleaning solutions, experimental evidence is needed to demonstrate that these nanoparticles can be removed from this surface water if it is used as a source of drinking water.

The European Hygienic Engineering & Design Group clearly states that materials which have been modified with antimicrobial chemicals may not be considered as a substitute for hygienic design. Microorganisms may build up resistance against such chemicals over a period of time, and antimicrobial chemicals are only effective if the microorganisms are in intimate contact with them.

SURFACE FINISH

Product contact surfaces must be finished to a degree of surface roughness that is smooth enough to enable them to be easily cleaned and disinfected. The surface finish must be such that there are no cracks, pits or cavities where water or soil might remain. In the pharmaceutical industry, a surface finish of roughness $R_a \leq 0.4\,\mu m$ is often used, while a surface finish of roughness $R_a \leq 0.8\,\mu m$ is considered acceptable for the food industry. Surface roughness, R_a, of enclosures in hygienic production areas should not exceed $2.5\,\mu m$. Surfaces will deteriorate making cleaning more difficult (Hauser et al., 2004a).

The technique used for achieving the appropriate surface finish is of great importance. Although with different surface finish techniques (glass blasting, ceramic beats blasting, electro polishing, pickling) a surface roughness of R_a $0.8\,\mu m$ can be achieved, the topography/structure of the surface can differ immensely, which gives different cleaning results.

HYGIENIC DESIGN OF OPEN EQUIPMENT FOR PROCESSING OF FOOD

Permanent and Dismountable Joints

Permanent Joints

It is better to use permanent joints rather than dismountable joints, because the latter type of joints may give rise to projections, protrusions, edges, recesses, metal-to-metal

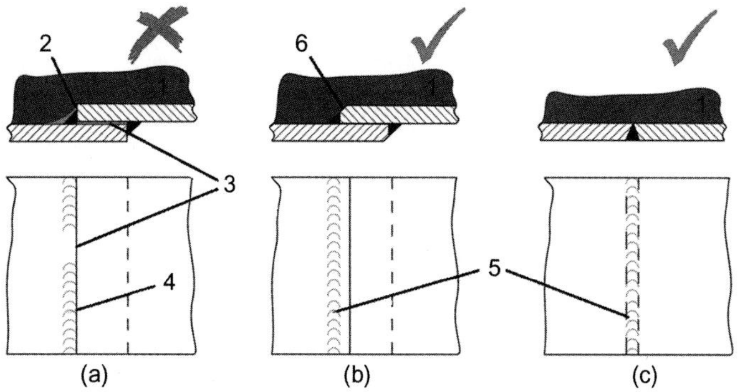

FIGURE 26.4 (a, b) In the product contact area (1), product debris may become trapped at step (2) and in the crevices and metal-to-metal contact areas between the seams (3), if overlapped sheets of metal are intermittently welded (4) instead of continuously welded. (b) Overlapped sheets of metal must have continuous welds (5) and sloped edges (6) for easy cleaning. (c) However, it is still better to avoid overlapping sheets of metal, and to give preference to smooth continuously welded sheets (Lelieveld et al., 2003; Hauser et al., 2004b).

contact, etc. In this way, welded joints are preferred over mechanical fixings, such as bolted or screwed joints.

Permanent joints of equipment should preferably be welded, but notice that several types of common defects may arise in welded joints (e.g. misalignment, cracking, porosity, inclusions), which can act as a source of microbiological problems. All welds in the product contact area are recommended to be continuously welded and with sufficient weld seam protection (inert shield-gas protection at both sides) in agreement with EHEDG guidelines No. 9 and No. 35. Higher alloyed filler metal in comparison to the welded material may reduce the risk for corrosion. When necessary welds must be polished to have the same surface finish ($R_a \leq 0.8\,\mu m$), appearance, etc. as the surrounding materials. They should be inspected for any discoloration and defects (Hauser et al., 1993; Kopitzke et al., 2006).

To avoid crevices at metal-to-metal interfaces where product debris may become trapped, intermittent or spot welds are not acceptable (all welds should be continuous or filled) and overlapping must not be used (Figure 26.4a). If overlapping is unavoidable due to the need for added strength at the weld location, reliable draining and cleaning conditions of shadow areas must also be taken into consideration. In the case of thick sheets, the edge of the upper plate must be sloped to avoid areas at the overlap edge which can retain soil and be difficult to clean (Figure 26.4b). However, it is still better to avoid overlapping sheets of metal, and to give preference to smooth continuously welded sheets (Figure 26.4c) (Hauser et al., 2004b).

Sharp corners (≤90°) and welding in sharp corners of equipment (Figure 26.5a and b) must be avoided. Radiused corners (sloped sides) and welding seams away from corners and preferably made at the non-product contact side are recommended (Figure 26.5c). Weld fillets in the food area should have a minimum radius of 6 mm. If the material is less than 4 mm thick, the minimum radius should be 3 mm. Where a corner cannot have a radius of greater than 3 mm, its cleanability should be demonstrated by testing.

Use of adhesives on metal-to-metal joints should be avoided. If adhesives are used for permanent joints they must be compatible with materials, products and cleaning/disinfecting agents with which they are in contact. All bonds should be continuous and mechanically

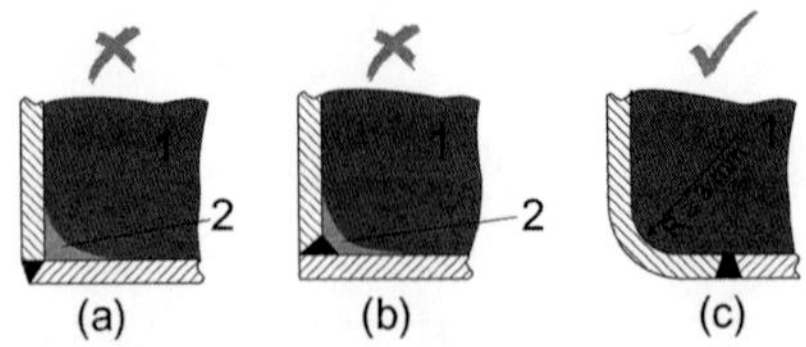

FIGURE 26.5 (a, b) Welded seams in ≤90° corners of receptacles containing food product (1) will create uncleanable areas where residual soil (2) will accumulate. (c) Well-rounded corners (radius $R \geq 3$ mm) and correctly welded seams in the plain area away from corners and preferably made at the non-product contact side avoid any hygiene risk (Lelieveld et al., 2003; Hauser et al., 2004b).

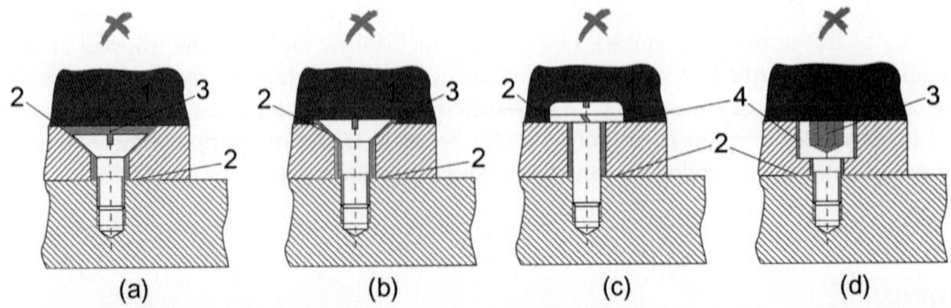

FIGURE 26.6 Screws may not be exposed to food product (1) because debris collects in the screw drive, because they give rise to metal-to-metal contact corrosion (2), and because they create gaps, dead areas (3) and crevices (4). Countersunk screws with slots or other drive configurations are not recommended for the reasons mentioned, and incorrect machining of the countersunk hole may cause the screw to either (a) form a pocket in which debris collects or (b) to protrude into the product flow giving rise to circumferential crevices where debris may become trapped. (c) Pan, dome, round and truss screws are not suitable because they protrude in the product flow. (d) Socket head cap screws are not allowed in the food area because debris accumulates in the recess or socket to fit an Allen wrench for turning. In addition, the use of counterbores is not recommended for all the reasons mentioned earlier (CFPRA, 1983; Lelieveld et al., 2003; Hauser et al., 2004b).

sound so that the adhesives do not separate from the base materials to which they are bonded.

Dismountable Joints

Dismountable joints (e.g. of plates or appendages) fixed by fasteners (e.g. screws or bolts) must only be used if dismantling is unavoidable. Joining components with hexagon nut-and-bolt pairs which protrude in the product zone or with screws exposed to product is not allowed. Besides, crevices, screws, bolts and nuts also give rise to metal-to-metal contact corrosion, and create gaps, dead areas and/or exposed threads (Figures 26.6 and 26.7).

Wing nuts and pop rivets are also not allowed on the product side. It is recommended to have a plain or domed bolt head sited on the product side, to cover exposed threads with domed nuts, and to use solid rivets instead of pop rivets (Figure 26.8). But overall, the use of welded butt joints that are ground and polished instead of fastenings is preferred (CFPRA, 1983).

Correct design of bolt heads and their effective sealing with metal-backed elastomer gaskets (Figure 26.9) can render them hygienic. The head of the hexagon headed bolts will be plain or domed. Domed nuts can be used to cover exposed threads. Sealing the

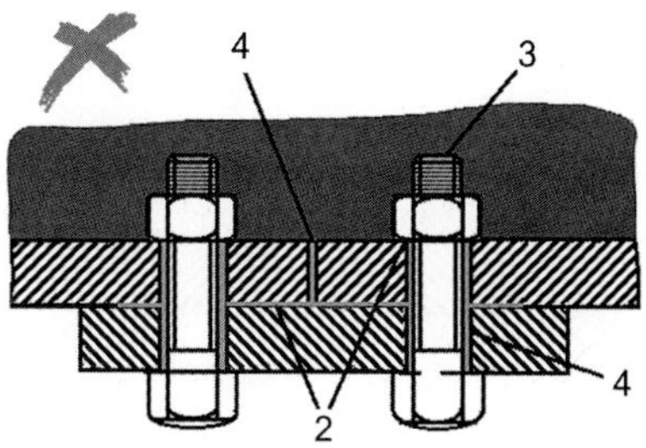

FIGURE 26.7 Exposed bolt ends and nuts in the product zone (1) are not allowed because they give rise to metal-to-metal contact corrosion (2), exposed threads (3) and crevices (4). Debris also tends to adhere to and around fixings and provides nutrients for microbial slime growth. Exposed threads should be cut to the correct length or preferably domed nuts should be used (Lelieveld et al., 2003; Hauser et al., 2004b).

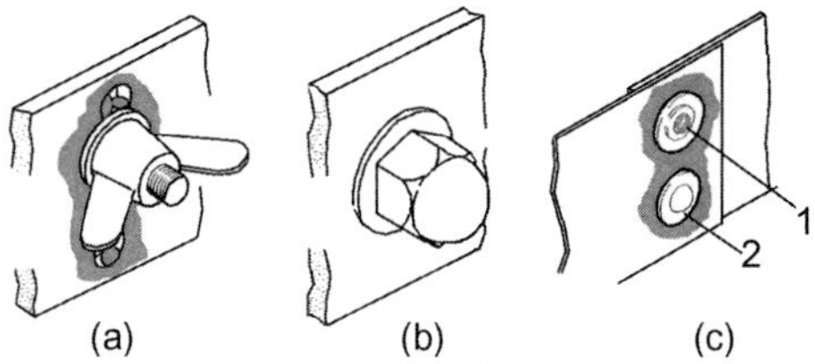

FIGURE 26.8 (a) Wing nuts are often used where adjustment is required but debris collects around and in the exposed portion of the slot behind the nut. (b) It is recommended to cover exposed threads with domed nuts. (c) Pop rivets (1) are not recommended where construction necessitates this type of fabrication. Solid rivets (2) should be used instead of open rivets (CFPRA, 1983).

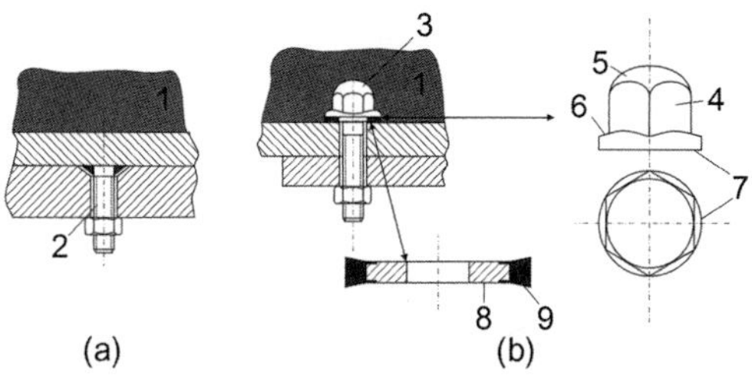

FIGURE 26.9 (a) To prevent crevices at the product side (1), screws, pins or a stud welded on the non-product side (2) should be used. (b) A bold head (3) that is hexagonal (4), domed (5) and provided with a sloped circular collar (6, 7) is easily cleanable, and the metal-backed (8) elastomer gasket (9) is used to seal the thread (Lelieveld et al., 2003; Hauser et al., 2004b).

crevice between the bolt head and the food contact surface will protect the annular clearance between the shaft of the bolt and the hole through which it passes.

Dismountable joints must be crevice free and provide a smooth continuous surface on the product side. Further, metal-to-metal contact should be avoided. Therefore, where components butt against one another in the product area, the crevice between them should also be sealed by means of an elastomer. Compression of the seal can be controlled by means of screws and interference-fit location pins on the reverse side to the product (Figure 26.10a).

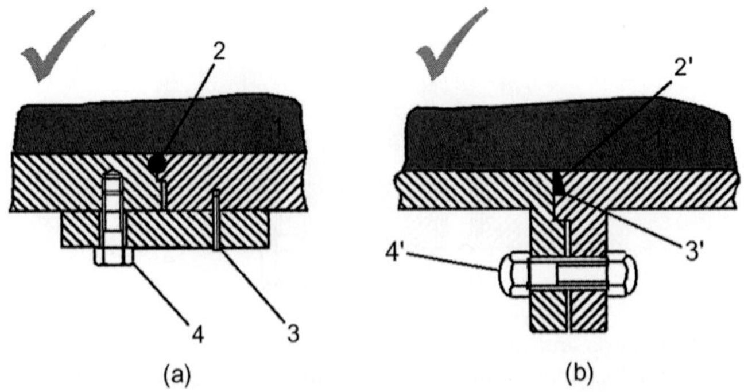

FIGURE 26.10 Where components butt against one another in the product area (1), the crevice between them should also be sealed by means of an elastomer (2). (a) Compression of the seal can be controlled by means of interference-fit location pins (3) and screws (4) on the reverse side to the product. (b) A flange (4′)-like connection can control compression (2′) and the design of the groove should allow space for expansion (3′) of the seal (Lelieveld et al., 2003; Hauser et al., 2004b).

A flange-like connection can control compression (Figure 26.10b). The design of the groove for the seal must allow space for expansion in order to avoid extension of seal material into the product area during heating.

Split pins, self-tapping screws, staples, spring tension pins, bushings, etc. which may be loose and cause damage to other equipment and physical danger to the consumer are unsuitable fastenings. Tape, rubber bands and wire should not be used to permanently modify equipment. A designer also must avoid very small fastenings, and fixings in plastics which cannot be identified by metal detectors. Stainless steel or dull-nickel-plated fixings should be used as specified in the fixings and fastenings handbook. Finally, one must allow for sufficient space around fixings for cleaning (min 25 mm).

Hygienic Design of Process Vessels, Containers, Bins, etc.

Interior and Exterior Design of Process Vessels, Containers, Bins, etc.

Appropriately designed and installed process vessels shall meet the following recommendations:

- Equipment without bottom outlets must be pivoted (Figure 26.11) for fully discharging of product and cleaning solution. Materials or contaminants from the exterior of the vessel must not gain access to the food product being discharged. Besides full drainability, the vessel tipped for discharge also should be designed for improved cleanability (e.g. vessel corners should be well rounded; hinges must allow for maximum cleanability).
- For good drainability and cleanability, food containing equipment (tanks, vessels, troughs, reservoirs, bins, etc.) shall have their discharge outlet at the lowest level; their bottom shall be sloped (more than 3° towards the outlet); and their corners shall be well rounded. These corners should preferably have a radius equal to or larger than 3 mm. Sharp corners (≤90°) must be avoided (Figure 26.12).

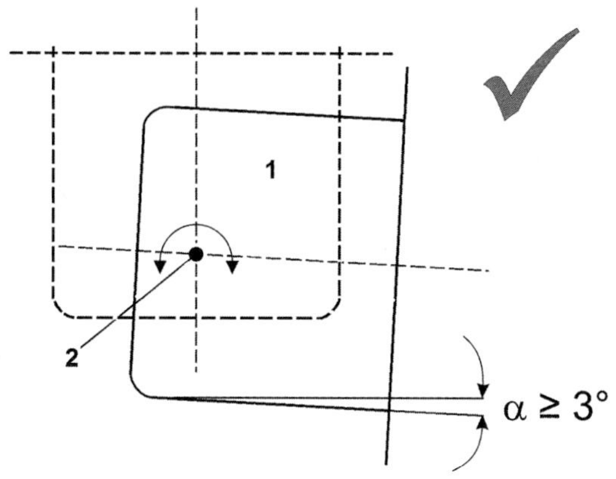

FIGURE 26.11 To fully empty containers without bottom outlet, they must tip over an angle of at least 93°. The interior and exterior of the container must be designed to exclude any contamination of the food product when it is drained. Vessel should have well-rounded bottom corners, with hinges designed for maximum cleanability (Lelieveld et al., 2003; Hauser et al., 2004b).

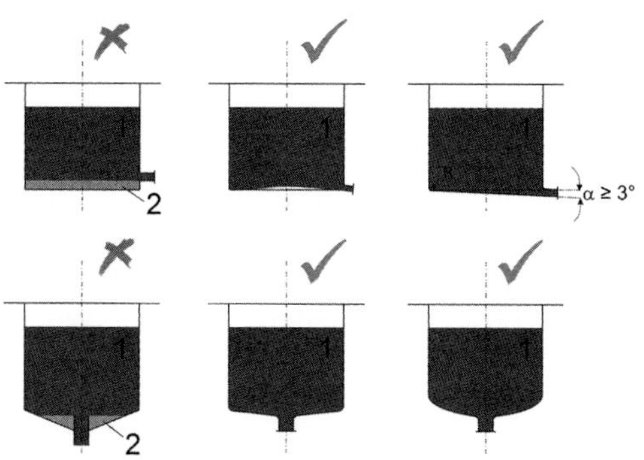

FIGURE 26.12 For good drainability and cleanability, equipment (tanks, vessels, troughs, reservoirs, bins, etc.) used in the processing of food (1) shall have their discharge outlet at the lowest level; their bottom shall be sloped (more than 3° towards the outlet), and their corners shall be well rounded. Where food product and cleaning solutions are not allowed to drain, residual soil (2) will be left (Lelieveld et al., 2003; Hauser et al., 2004b).

- The design of the top rims of product containing equipment (e.g. open tanks, chutes, boxes) must avoid ledges where product can lodge and which are difficult to clean (Figure 26.13a). Open top rim designs must be rounded and sloped for drainage (Figure 26.13b). If the top rim is welded to the wall, the weld must be flush and polished to provide a smooth surface and the rim must be totally closed. Any holes, therefore, must be sealed by welding or by fitting sealed caps (Figure 26.13b).
- Lids are used (e.g. for process vessels, tanks, bins,) to avoid contamination of product from the environment during processing or storage. They can be completely detachable for cleaning, but if they are non-removable they must be sloped for drainage. If hinged covers are used the hinge must be designed in such a way that it can be cleaned easily

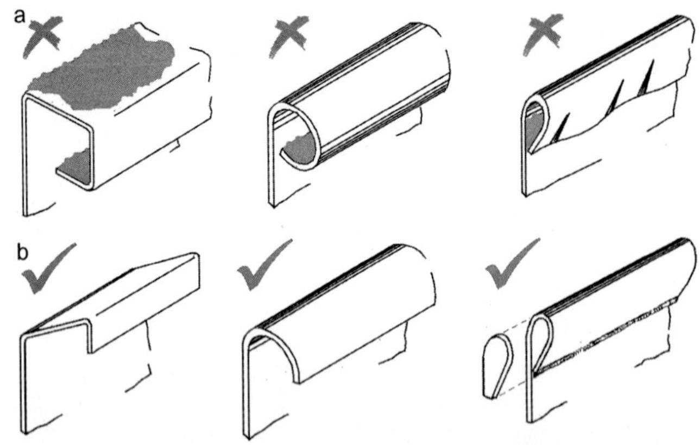

FIGURE 26.13 Top rims may impart rigidity to the construction. (a) However, a rim with an upper horizontal part provides a surface where debris may collect. When the rolled-over part of the rim is badly designed, it may provide a ledge where product debris can lodge. This soil can indirectly affect the product. (b) Open top rims must be rounded in a way that at one side the product drains back in the bulk of the product, while the more exterior part of the rim must allow drainage to the outside. Where preference is given to closed top rims, the top rim should be welded correctly to the wall over its full length. The weld must be flush and polished to provide a smooth surface and the rim must be totally closed. Any holes, therefore, must be sealed by welding or by fitting sealed caps (CFPRA, 1983).

and that accumulation of product, dust and foreign bodies (including insects, etc.) is avoided. When the vessel is covered, no sharp corner at the top should be created when the lid is placed on the vessel. Flat lids provide a horizontal surface where dirt may accumulate. Moreover a sharp corner is created at the top near the seal. Preference should be given to domed lids with a sloped top that collects less dirt and allows for proper drainage of liquids (Figure 26.14).

- Elastomers can be deformed, but the volume cannot be reduced! This means that when a flat gasket is compressed so that the thickness is reduced by say 20%, the width of the gasket is increased by 25%, assuming that the length can be kept constant. As a consequence, a considerable amount of movement takes place at the edges of the gasket. In view of the inconsistency of the friction between stainless steel and elastomers it is most uncertain how the deformation of the gasket will take place. Overcompression of the flat gasket (Figure 26.15a) may affect the hygienic characteristics of equipment in two ways. First, overcompression may lead to destruction of the gasket, particularly if it is heated (such as during hot cleaning and/or sterilization). The gasket may exceed the maximum of compression caused by thermal expansion and become brittle and fail to perform, while particles of it may break off and contaminate the product. Second, overcompression may lead to protrusion of the gasket into the product flow, thereby impeding cleaning and draining. Undercompression (Figure 26.15b) is also highly undesirable as it may lead to both indentations and crevices and failure to provide a reliable seal. Even when it is not visibly leaking, the seal may permit the ingress of microorganisms. It is good practice to slope the groove that receives the gasket (Figure 26.15c) in a way that space for expansion is provided at the non-product side while controlled compression of the gasket is possible

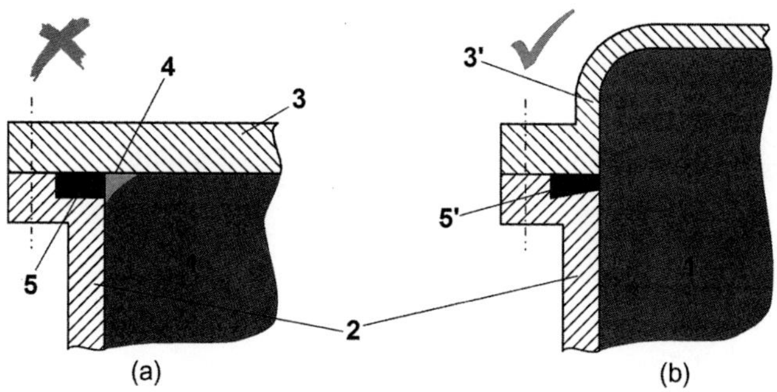

FIGURE 26.14 (a) Covers are used (e.g. for process vessels, tanks, bins, etc.) to avoid contamination of food product (1) from the environment during processing or storage. When the vessel (2) is covered with a flat lid (3), a horizontal surface is provided where dirt may accumulate. Moreover a sharp corner (4) is created at the top near the seal. This seal (5) is not very appropriate because overcompression may lead to protrusion of the seal in the product area, thereby impeding cleaning; while undercompression may lead to both indentations and crevices and failure to provide a reliable seal. Even when it is not visibly leaking, the seal may permit the ingress of microorganisms. (b) Preference should be given to domed lids (3′) with a sloped top that collect less dirt and allow for proper drainage of liquids. The present gasket groove allows for controlled compression of the gasket (5′) at the product side (Lelieveld et al., 2003; Hauser et al., 2007).

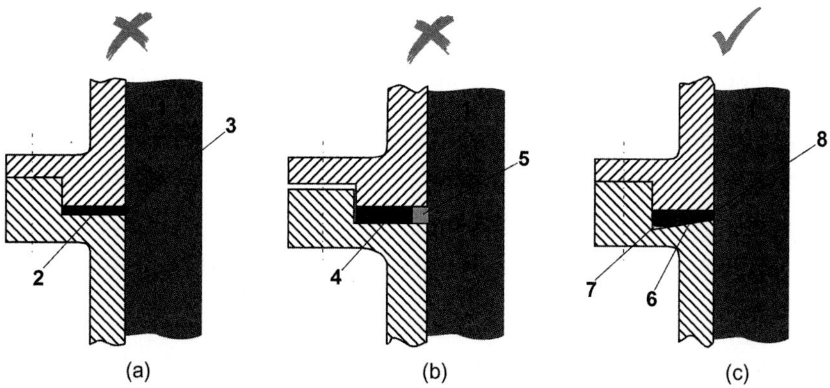

FIGURE 26.15 (a) Overcompression of the gasket (2) may lead to protrusion of the gasket (3) into the product area (1), thereby impeding cleaning and draining. Moreover, because the gasket may exceed the maximum allowable limit of compression during thermal expansion, it may become brittle and fail. (b) Insufficient compression of the gasket (4) may give rise to a crevice (5) between the two flanges and the possibility of leakage. (c) It is good practice to slope (6) the groove that receives the gasket in a way that space for expansion (7) is provided at the non-product side while controlled compression of the gasket is possible at the product side (8). A further possibility to reduce the effects of friction is to avoid even compression of the gasket by using gaskets with a profiled section which "involutes" along the sealing faces rather than sliding under compression (Lelieveld et al., 2003; Hauser et al., 2007).

at the product side. Such a design also allows reduction of the area of the gasket in direct contact with the food product. A further possibility to reduce the effects of friction is to avoid even compression of the gasket by using gaskets with a profiled section which "involutes" along the sealing faces rather than sliding under compression.

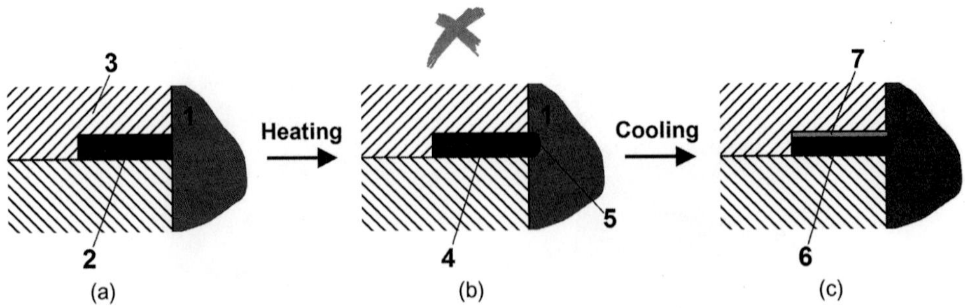

FIGURE 26.16 (a) A non-resilient flat PTFE gasket (2) is installed in a rectangular groove and compressed between the sealing faces of two stainless steel surfaces (3) to separate the product area (1) from the outside. (b) Because of the large difference in thermal expansion coefficient of both PTFE gasket and stainless steel, a heat treatment changes the shape of the PTFE gasket (4). Protrusion of the gasket (5) takes place. (c) Because of the lack of resilience, the PTFE gasket may become irreversibly deformed (6). Hence, after cooling down, the gasket will not return to its original form and proportions, and as such a crevice (7) will be generated (Lelieveld et al., 2003; Hauser et al., 2007).

- As with metals, not all polymeric materials and elastomers exhibit the same coefficients of thermal expansion (Figure 26.16). Therefore, not only the dimensions of the metal components but also those of the seal must be correct, ensuring adequate compression at the product side, under all conditions of intended use. Attention must be given to thermal expansion at high temperatures (e.g. during hot cleaning and sterilization) and to loss of resilience at low temperatures (e.g. during the manufacture of ice cream). To ensure a smooth durable surface with sufficient temperature and corrosion resistance, equipment manufacturers tend to use polytetrafluorethylene (PTFE) as gasket material in food processing equipment. However, PTFE has insufficient resilience and expands significantly more than stainless steel (expansion coefficient for PTFE is approximately $100 \times 10^{-6}/°C$, compared to approx. $16 \times 10^{-6}/°C$ for stainless steel). Due to this large difference in thermal expansion coefficient, a heat treatment changes the shape of the PTFE gasket (gasket protrusion occurs) and after cooling down a crevice occurs. For a gasket of 5 mm thickness and a temperature change from 20 to 120°C and back, the crevice may be 36 μm wide if there is no resilience at all (in practice the gap will be slightly smaller). Therefore, seals made from non-resilient materials should not be used.
- Conventionally designed right-angled grooves containing O-rings invariably create gaps and crevices that are impossible to clean in-place and/or to sterilize in-line (Figure 26.17). One cause is that the elastomer material of the O-ring has a significantly higher thermal expansion coefficient than steel. During heating the seal will expand to cover an increasingly larger surface of steel, protecting microorganisms trapped between the O-ring and the steel surface against contact with hot water, chemical solution or steam. Although the seal contact surface will usually reach the correct temperature during treatment with hot water or steam, the water activity in the grooves will be too low for the destruction of most microorganisms at the temperature and time applied. After cooling down and shrinkage of the seal, the surviving microorganisms may be released and will multiply and contaminate the product. Additionally, repeated thermal expansion of the seal into the product flow may result in it suffering damage which will

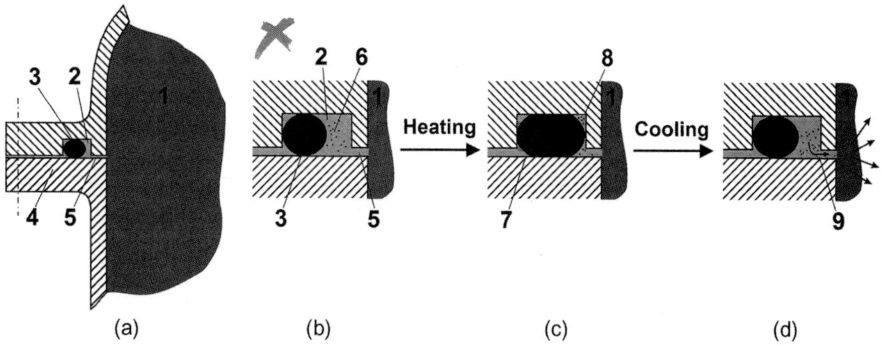

FIGURE 26.17 (a) A conventionally designed right-angled groove (2) contains an O-ring (3) that is compressed between the sealing faces of two stainless steel surfaces (4) to separate the product area (1) from the outside. (b) Such a rectangular groove-O-ring design invariably creates gaps and crevices (5) that are impossible to clean in-place and/or to sterilize in-place. The groove provides sufficient space for microorganisms (6) to enter via the crevice. (c) During heating, due to the difference in thermal expansion between metals and elastomers, the O-ring will expand (7) to cover an increasingly larger surface of steel, protecting microorganisms (8) trapped between the O-ring and the steel surface against contact with hot water, chemical solution or steam. (d) After cooling down and shrinkage of the seal, the surviving microorganisms may be released (9) and will multiply and contaminate the product (Lelieveld et al., 2003; Hauser et al., 2007).

not only contaminate the product but may also progressively reduce its ability to seal again upon re-cooling.

Installation of Agitators in Open Vessels (e.g. Kettles)

Equipment like stirrers, homogenizers or mixers should preferably be arranged in such a way that the need to seal shafts into the product is avoided. Where mounting of the equipment outside the product zone is possible, the mixer used to mix open product should be fixed beside the equipment, not only to prevent the contamination of the product with dripping oil, but also to avoid the introduction of soil, and concomitantly spoiling microorganisms and pathogens into the product along with overhanging electrical cabling (Figure 26.18).

Permanently Mounted Agitators in Closed Vessels

Top entering agitators with shaft seals are typically mounted to a vessel using a flanged or hygienic clamp connection, with hygienic O-rings or gaskets to seal between the mating surfaces. The selected mounting arrangement must support the agitator mounting design loads while achieving an appropriate seal. The upstand for the top mounting of the agitator should have limited length L because of the difficulty of cleaning of the annular space in-place. The annular space between the agitator shaft and agitator nozzle shall, for cleaning purposes, have the target maximum L/A ratio of 2:1. At least a 25 mm gap is required to facilitate CIP spray coverage (Figure 26.19) (CFCRA, 1997; BISSC, 2003; ASME, 2009).

Agitator motors should be equipped with permanently lubricated bearings. Where lubrication is required, the design and construction shall be such that lubrication cannot leak, drip or be forced into the product zone. Self-lubricating agitator shaft (packing) seals shall be provided with convenient means for adjustment to prevent leakage and to allow

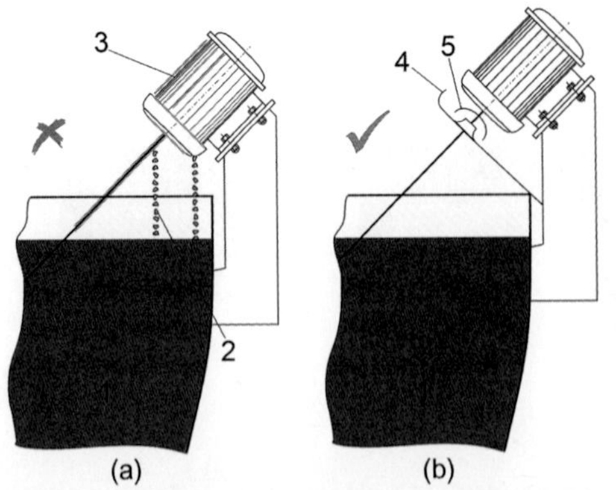

FIGURE 26.18 (a) A motor and cabling mounted over any exposed product (1) can contaminate it by soil, condensate or lubricants (2). (b) The motor drive (3) and power line should be placed beside the equipment. A self-draining protection sheet with "upstand" (4) in combination with a cowl (5) on the shaft must exclude any food safety risk. The bottom side of the thrower ring (cowl) should be made inspectable (Lelieveld et al., 2003; Hauser et al., 2004b).

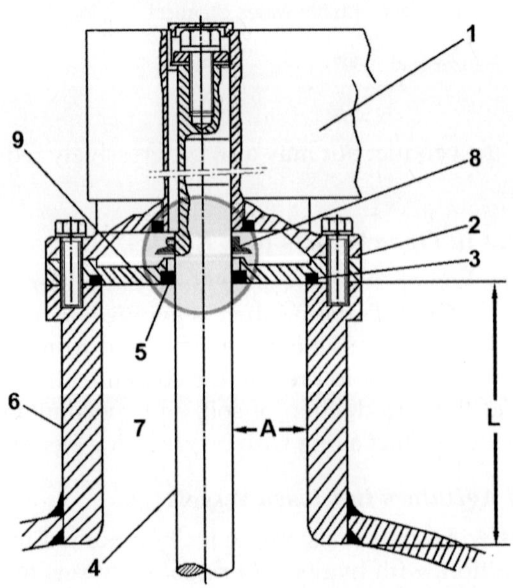

FIGURE 26.19 The top entering agitator with motor (1) is mounted to a vessel using a flanged or hygienic clamp connection (2), with hygienic O-rings or gaskets (3) to seal between the mating surfaces. A retained gasket having limited compression is more hygienic than an O-ring in the face for sealing the joint. The agitator shaft (4) passes through the mounting flange via a seal (5). The upstand (6) for the top mounting of the agitator should have limited length *L* because of the difficulty of cleaning the annular space (7) in-place. The annular space between the agitator shaft (4) and agitator nozzle (6) shall, for cleaning purposes, have the target maximum L/A ratio of 2:1. Agitator motors (1) should be equipped with permanently lubricated bearings. Where lubrication is required, the design and construction shall be such that lubrication cannot leak, drip, or be forced into the product zone. Self-lubricating agitator shaft (packing) seals (8) shall be provided with convenient means for adjustment to prevent leakage and to allow for complete drainage to the exterior. In that way, accumulations of foreign material in the event that leakage does occur can be avoided. Further, a drip protection plate (9) can be provided to prevent lubricant from entering the product zone (CFCRA, 1997; BISSC, 2003; ASME, 2009).

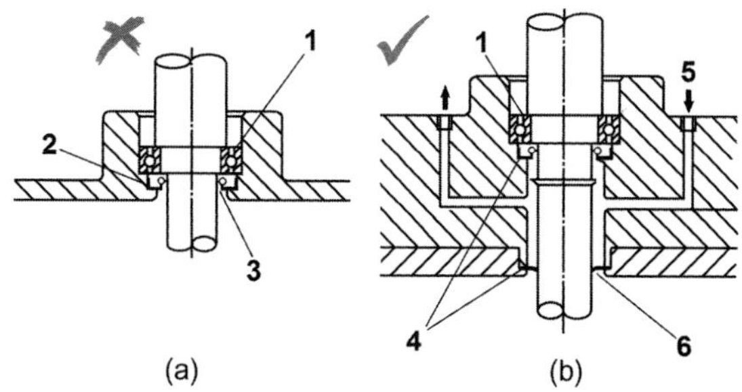

FIGURE 26.20 Rotary shafts running at a high number of revolutions are held in place in an adapter sleeve with a radial roller bearing (1). (a) Single dynamic seals (2) are lubricated by a lubricant (top mounted agitator) or the product (bottom mounted agitator) which may be transported past the seal and back again, further contaminating the product. They may be easy to clean if properly designed but they will not prevent the passage of microorganisms, and hence they are not suitable in aseptic process equipment. There is also a narrow annular space (3) at the product side in the proximity of the seal, which makes cleaning very difficult. (b) A double seal arrangement (4) allows the use of a barrier medium (5) such as steam, hot water, condensate or a disinfectant solution which makes it well suited from a microbiological standpoint. The volume of the annular gap around the shaft is increased (6), improving the cleanability of the seal and its proximity (Holah, 2000).

for complete drainage to the exterior (Figure 26.19). In that way, accumulations of foreign material in the event that leakage does occur can be avoided. Further, drip protection is commonly provided to prevent lubrication from entering the product zone. All surfaces of shaft seal ring assemblies passing through a bowl or cover shall be accessible, removable or retractable to permit cleaning of all product zone surfaces.

Rotary shafts running at a high number of revolutions are held in place in an adapter sleeve with a radial roller bearing. Single dynamic seals (Figure 26.20a) will not prevent the passage of microorganisms. If properly designed, they may be easy to clean but not bacteria tight because rotating shafts may exhibit some axial mobility. This makes single dynamic seals unsuitable for aseptic equipment. A narrow annular space at the product side in the proximity of the seal such as shown in Figure 26.20a must be avoided because it is difficult to clean. The space around the seal should be as wide as possible. Rotary shafts with a double seal arrangement allow the use of a barrier medium, and have been shown to be well suited from a microbiological standpoint. In Figure 26.20b, one seal is seated rigidly in the housing (longitudinal shading), while the other moves with the shaft. The sealing surface between the two seals must be lubricated. If the shaft opening has product flowing through it, which could be the case with agitators having a shaft entry from the bottom of vessels, the product itself can be directly used as lubricant. The product flowing through can be carried away by the barrier medium, which could be steam, hot water, condensate or a disinfectant solution (e.g. alcohol). The sterile fluid may scavenge the microorganisms that enter the space between the seals, maintaining absolutely sterile conditions. Which flushing fluid should be used will depend on the product and the process but both the barrier medium and lubricant chosen must be product compatible. To avoid transfer of microorganisms from the outside of the equipment to the inside, without an adequately long exposure to antimicrobial fluid the distance between the two seals must always be sufficiently large (Lelieveld et al., 2003; Hauser et al., 2007).

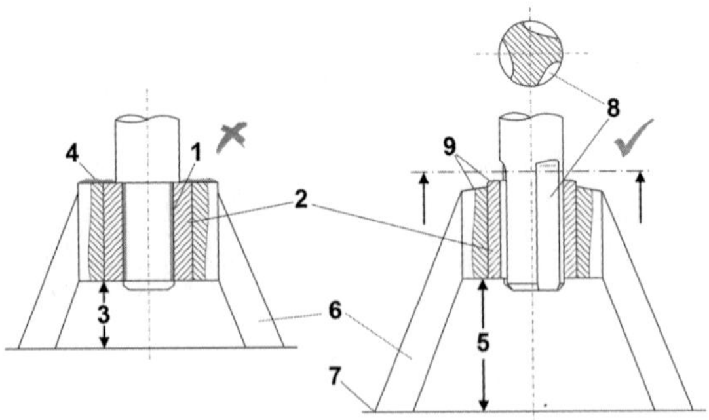

FIGURE 26.21 (a) Cleaning may be impeded due to too tight clearance (1) in the foot bearing itself (2), and due to too little clearance between it and the base (3). Horizontal ledges (4) where product may accumulate or where liquids are not allowed to drain must be avoided. (b) The foot bearing is now mounted clear of the bottom of the vessel (5), allowing free flow of product and cleaning solution around it. Bearing pedestal support members (6) should preferably be made of solid construction. Hollow constructions are not recommended, but if used, they shall be of sealed (welded) construction, inspected for integrity. Round legs are preferred over flat members, even if the latter are radiused. The legs should be flush welded in place to the tank bottom (7). All welds must be ground and polished to blend smoothly with the adjacent surfaces. The agitator shaft is provided with grooves (8) in the bearing area to facilitate both lubrication by fluid products and cleaning. Sloped and radiused surfaces (9) reduce the probability of debris getting lodged on the top of the foot bearing and allow for proper drainage of liquids (e.g. cleaning solution) (CFCRA, 1997; Lelieveld et al., 2003; Hauser et al., 2004b; ASME, 2009).

Bearings in the product area should be avoided but an application may mandate the use of foot bearings. As an example, if the shaft of a top entry agitator is very long, a foot bearing may be required at the bottom of the vessel to steady it. It shall be of a packless bearing type. The foot bearing must be mounted well clear of the base so as not to impede free draining of product and also to allow easy cleaning of their supports. Design features and/or procedures required to ensure cleanability are: drain holes, spray ball and/or wand additions, increased CIP flow, and operating the steady bearing immersed in CIP fluid. The arrangement of wear surfaces (bushing, shaft or shaft sleeve) shall facilitate drainage. A longitudinal or helical groove may be cut in either the bush or the shaft. It should be deep enough to allow access into the bearing of either the product as a lubricant or the detergent for cleaning (Figure 26.21). Sealed bearings should not be used in the product area because they can cause hygiene risks at their seals. If, however, their use is unavoidable, their lubricants should be specified as being allowed with the food contact.

Hygienic Design of Agitators

Agitators and agitator shaft assemblies passing through the seals shall be designed and constructed to be smooth, with all surfaces meeting all the hygienic design criteria applicable to a product contact area. Agitator shaft assemblies shall be readily accessible to allow all surfaces to be effectively cleaned via spray, directed flow, immersion or cleaning-in-place. Agitator ends shall have surfaces of minimum area immediately adjacent to the recipient ends and no longer than necessary to ensure proper incorporation of ingredients into a mix.

The design of agitator product contact parts should minimize the occurrence of crevices, void spaces and dead spaces in grooves. All voids should be closed by either fabrication

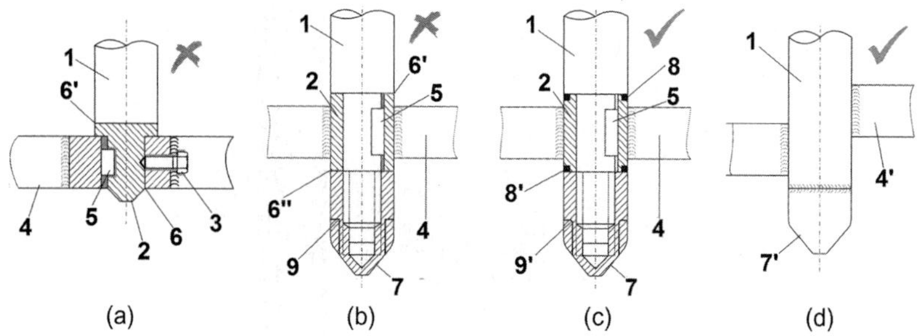

FIGURE 26.22 (a) The hub (2) is secured to the shaft (1) by means of a screw (3), which is exposed to product that may collect in and around the screw head. The hub-to-shaft connection gives rise to a metal-to-metal joint (6′) that may permit the ingress of product and bacteria. Agitator blades (4) should be welded to the hub, although screw connections are sometimes observed. These exposed screw heads (even bolts with dome head nuts and washers of suitable food grade material) again will create a food safety hazard, and the blade-to-hub connection gives rise to a new metal-to-metal joint (6). To avoid the latter problem, the joint between the blade and the lug on the hub can be sealed by a thin gasket. Keyways (5) exposed to product are not recommended, because product and microorganisms may be retained in the keyway. Keyways may require additional design and/or cleaning practice to ensure drainage and cleanability, e.g. spray ball and wand additions, increased CIP flow and adjusted spray coverage. (b) Once the hub (2) is secured to the shaft (1), an end cap (impeller nut, 7) is screwed on the interior male thread end of the shaft. The non-welded impeller hub-to-shaft and hub-to-end cap connections give rise to crevices and metal-to-metal joints (respectively 6′ and 6″) that may allow the ingress of product and bacteria. In that way, the keyway (5) also may retain product and microorganisms. The sharp corners of the spanner flats (9) on the end cap may be difficult to clean. (c) Food quality gaskets under controlled compression respectively may seal the propeller hub to the shaft (8) and to the end cap (8′). Keyways (5), where employed due to mechanical design considerations, shall have edge radii not less than 3 mm. The corners of the spanner flats on the end cap have been radiused (9′). (d) An all-welded impeller assembly (e.g. hubs, blades, end cap) is still preferred. Impeller hubs welded to the shaft are preferred over removable hubs. The designer may omit the hub and immediately attach the blades to the shaft by welding (4′). Finally, the end cap can be welded to the shaft (7′) (CFCRA, 1997; Lelieveld et al., 2003; Hauser et al., 2004b; ASME, 2009).

(welding) or approved sealing techniques (O-rings, seals, etc.) to give surfaces ground flush and free of crevices at points of metal-to-metal contact. Metal-to-metal joints (e.g. keyways, hub-to-shaft joint, hub-to-end cap joint, etc.) may allow ingress and accumulation of product and/or microorganisms (Figures 26.22a and b and 26.23).

Food quality gaskets under controlled compression may seal the propeller hub to the shaft and to the impeller nut (end cap) that secures the end of the agitator shaft (Figure 26.22c). Alternatively, the hub should be welded to the shaft and the end cap (Figure 26.22d). Because debris may collect on exposed screw threads, the hub shall not be fastened to the shaft by means of a screw. To avoid any screwed joints (even bolts with dome head nuts and washers of suitable food grade material), the blades of appendages (stirrers, homogenizers, mixers, etc.) should be welded to the hub. As an alternative to hub-to-shaft and subsequent impeller blade-to-hub attachment, blades can be attached to shafts by welding. All welds used in the assembly of agitator parts should be grounded and polished.

Permanently joined metal surfaces with a total included internal angle less than 135° on agitators (e.g. at hubs and nuts) shall have a radius of not less than 3 mm tangential to both adjacent surfaces. Corners (e.g. at hubs, nuts, spanner flats, etc.) must be radiused to facilitate cleaning, and horizontal areas must be sloped to prevent debris from becoming lodged on the surfaces and to allow for maximum drainabilty. Machined transitions such as shaft

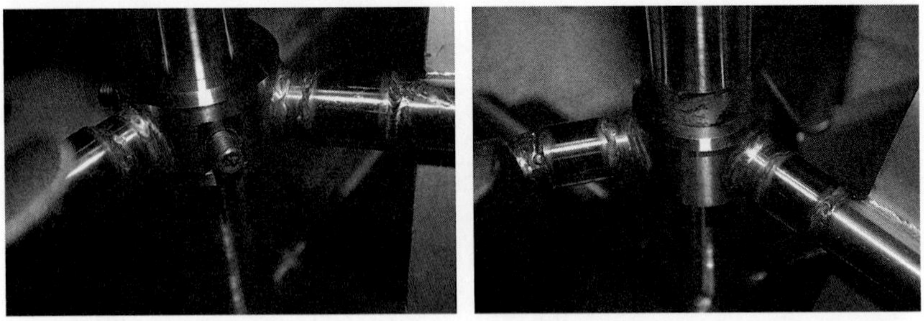

FIGURE 26.23 The hub is secured to the shaft by means of bolts with dome head nuts, which are exposed to product that may collect in and around the screw head. This non-welded hub-to-shaft joint also lacks a food grade gasket that could seal the dead spaces in the groove and avoid crevices at points of metal-to-metal contact. Ingress and accumulation of product and/or microorganisms at the inside are observed. Welds also have a high degree of roughness. *Courtesy of Burggraaf & Partners B.V., www.burggraaf.cc.*

steps, coupling surfaces, spanner flats, etc. should have 15 to 45° sloped surfaces. Impellers with flat, horizontal surfaces (e.g. flat-blade disc turbines, concave-blade disc turbines) may require additional design and/or cleaning practice to ensure drainage and cleanability, e.g. drain holes, spray ball and/or wand additions, increased CIP flow, adjusted spray coverage, and faster impeller rotation.

Agitators permanently mounted are not required to be removable if they are readily accessible and do not interfere with drainage from the tank. Where permanently installed agitators are equipped with an outer frame to which rubber, plastic or other similar scraping edges are attached, these scrapers shall be readily removable from the agitator. In kettles, however, it is recommended that the entire unit shall be constructed so that it can be tilted or lifted out of the kettle.

Welded in-tank shaft connections are preferred, although in-tank threaded shaft connections (Figure 26.24f) and in-tank shaft couplings (Figure 26.24a–e) are allowed if they are of acceptable hygienic design. Threaded shaft connections are preferred over in-tank shaft couplings, although shaft rotation of the first is limited to a single direction to avoid the shaft sections separating. The designer must ensure that the use of a threaded shaft connection is appropriate for the selected shaft diameter and design loads. To avoided exposure of the threads to the product, O-rings or flat gaskets (preference for the first mentioned) should be used to seal mating surfaces (Figure 26.24f). Hygienic bolted coupling construction may be used where appropriate for the particular application. The preferred location for fastening hardware is on the underside of couplings, and the fasteners typically used should be hex-head cap screws, acorn-head cap screws and threaded studs with acorn nuts (Figure 26.24d). These fastener heads shall be free of raised or engraved markings that might inhibit cleanability. Again O-rings or flat gaskets (preference for the first mentioned) should be used to seal coupling mating surfaces. Elastomer seal washers (Figure 26.24b–d) must avoid metal-to-metal contact.

Good Insulation Practices

Non-chloride-releasing insulation material should be used. For thermal insulation of vessels, appropriate qualities of rock wool are acceptable. However, for piping, Styrofoam,

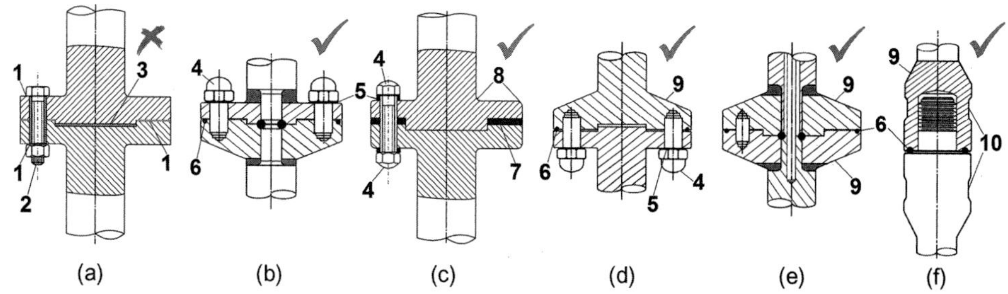

FIGURE 26.24 (a) Bolted agitator couplings with flat hexagon head screws without elastomer gasket under the bolt head and the nut give rise to metal-to-metal crevices (1) that may allow the ingress of food product and bacteria. Moreover, debris may lodge in and around the bolt thread (2). The absence of a circumferential O-ring or flat gasket gives rise to another metal-to-metal crevice, and product and microorganisms may be retained in the cavity (3). (b, c) Agitator couplings made by means of domed hexagon bolt heads and nuts (4) provided with an elastomer gasket (5) under the bolt head and the nut allow for a crevice-free joint without metal-to-metal contact. Due to the presence of a circumferential O-ring (6) or flat gasket (7), no product or microorganisms can enter inside the agitator coupling. Corners are radiused (8). However, there is still a horizontal flat surface at the upper side of the agitator coupling where debris may lodge. (d, e) Aseptic applications require for fastening hardware at the bottom side of the agitator coupling, and the upper parts of the coupling should be sloped to a minimum of 15–45° (9) to prevent debris from collecting at these places and to allow for maximum drainability. (f) The most optimal agitator coupling in an aseptic environment is a threaded shaft connection with a O-rings or flat gasket (preference for the first mentioned) (6) to seal the mating surfaces to avoid exposure of the interior thread. The corners of the spanner flats on the end cap have been radiused (10) (CFCRA, 1997; Hauser et al., 2004b; ASME, 2009).

foam glass or another rigid foam are better choices over fibrous materials. The problem with fiberglass batting is that this material has already proven to be an excellent harborage of dust, insects and rodents, and a clean-up and maintenance nightmare if not properly installed and maintained. Therefore, it is highly recommended to install fully welded, vapor-tight, aluminum or stainless steel cladding of appropriate thickness that resists tear and abrasion. The exterior of the insulation protection should be smooth, properly sealed to avoid ingress of dust, liquor, air and moisture, and should be installed in a correct way with joints facing downwards. Such ingress could promote corrosion between the walls, assisted by possible microbial growth. Damaged or wet insulation should be repaired or immediately replaced (Figure 26.25). Insulated lines should be kept high overhead where there is less chance for food products to contact the insulation. Pipes that are frequently soiled by food products or require periodic disassembly may be left uninsulated. Insulation is also often omitted around steam pipes inside cleanrooms, to preserve a clean exterior surface.

Equipment Framework

The number of support legs and cross bracings should be reduced but shall be of sufficient number and strength and so spaced that the process equipment will be adequately supported. Cross bracers should be fitted in a diamond configuration. Solid cross members as structural members are preferred over hollow section members. Although for use in the horizontal plane and to minimize horizontal ledges and crevices, completely sealed hollow section members are still preferable over open profile angle or channel sections (Figure 26.25a).

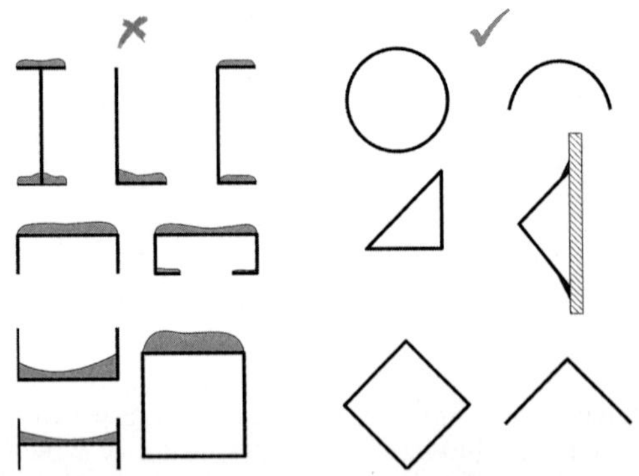

FIGURE 26.25 (a) Prevent unnecessary flat open and closed horizontal support members on which debris can lodge. (b) Round section, square section members turned through 45° and open profile members provide sloping surfaces (Lelieveld et al., 2003; Hauser et al., 2004b).

Round section members or square section members turned through 45° that provide sloping surfaces are recommended (Figure 26.25b).

For the design of framework that will be exposed to continuous vibrations (e.g. drying towers, etc.) the use of open profile construction should be considered. Small fatigue cracks can arise from vibration, allowing penetration of moisture, soil and microorganisms in closed profiles. For vertical parts of frames all the cross-sections shown in Figure 26.25 can be used when legs and supports are designed with open profiles; the folding should be turned outward for easy cleaning, or alternatively as completely closed pipes.

Rolled hollow sections must be sealed by welding, should be filled and made drainable away from the product zone. Plastic plugs are less recommended. Tubular sections shall not be penetrated, e.g. with fasteners, and hence drilled and tapped holes are not allowed. Preference should be given to welded plugs when fastening to hollow sections. Welded studs and tapping plates are not recommended.

Feet

Feet begin at the point where they attach to the leg of the body of the equipment and end at the support point on the floor. These feet are non-product contact surfaces but have a hygienic significance because they may become a harborage of soil and create a source of secondary contamination to the products (e.g. during pressure cleaning, dirt present on the feet may splash on the food contact surfaces).

Use a minimum number of support legs/floor mountings, because they are important obstacles for cleaning and service personnel. However, feet must be sufficient in number and strength and so spaced that the equipment will be adequately supported. The general rule is to minimize the floor contact area, but the contact face of the foot must be sufficient to absorb the pressure. If the equipment is heavy and requires leg pads to distribute the load, such pads or bases shall be fastened to the floor. The manner in which feet are fastened

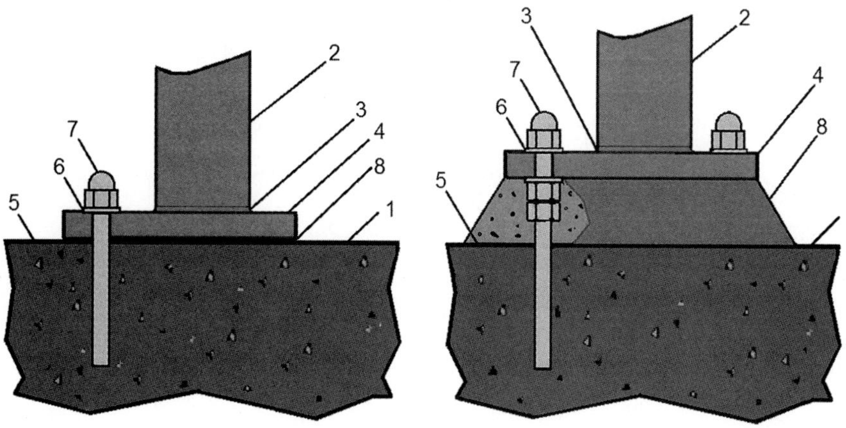

FIGURE 26.26 If the equipment is heavy, the contact face of the foot (2) with the floor (1) must be sufficient to absorb the pressure. To distribute the load, feet should be provided with leg pads or bases (4) welded to the foot leg (3). The foot may be fastened to the floor (5) by means of stainless steel anchor bolts which must have seal washer(s) (6) and dome nut(s) (7) fitted. When the equipment is bolted to the floor, pads or bases shall be sealed (figure left, 8) or grouted (figure right, 8) to the floor.

to the floor depends on the type of floor and the presence of equipment (e.g. machinery producing heat, etc.) or services (e.g. electricity, etc.) immediately below the surface. Fastening to the floor may occur by bolting, but chemical anchors without bolting (fixing to floors by means of a polymer seal) are recommended. If the equipment must be bolted to the floor, pads or bases shall be sealed or grouted to the floor (Figure 26.26). Care must be taken during installation to assure that the foot pad does not span over cracks, grout lines or other floor imperfections. Whenever anchor bolts have been drilled into the floor, the holes must be sealed with epoxy or similar materials, dependent on the floor, so that water and dirt are not allowed to leak into the hole. Floor fixings should be of stainless steel, and have dome nuts fitted.

Fixed feet should be radiused, free of sharp corners and crevices at the fixing point. Feet ends may have a foot base with flat (not recommended) or sloped surfaces (recommended), or may consist of a pivot-socket arrangement where the pivot-end of the spindle may freely swivel in the socket or internal cavity of a separate load-bearing foot base (Figure 26.27). This type of connection allows relative inclination of the foot stem and foot base as in an articulated bone joint, and is optimal to allow equipment to be repositioned or moved to uneven surfaces without loss of stability. Due to the non-rigid nature of the foot leg–foot base transition and because the load of the supported equipment is more evenly distributed about the surface of the socket, articulated support feet can better cope with the vibratory or oscillatory movements of the process equipment.

Ball feet are not recommended because they leave uncleanable crevices between the floor and the foot. Moreover, mechanically, they will almost destroy the floor, because – due to their very small contact surface with the floor – they exert locally a very high pressure. If the process equipment is heavy and prone to vibration, the floor will break up very quickly.

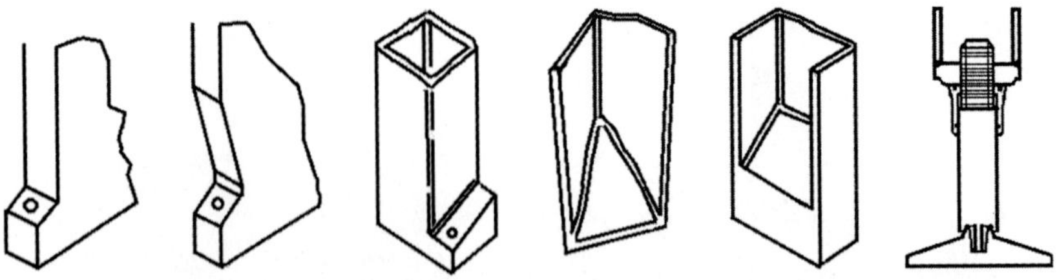

FIGURE 26.27 Feet ends may have a foot base with flat (not recommended) or sloped surfaces (recommended), or may consist of a pivot-socket arrangement where the pivot-end of the spindle may freely swivel in the socket or internal cavity of a separate load-bearing foot base. Feet ends with horizontal flat surfaces are not recommended. For maximum drainability, all surfaces of the feet not in contact with the floor should be sloped, with rounded corners and smooth welds (APV Baker, 2001).

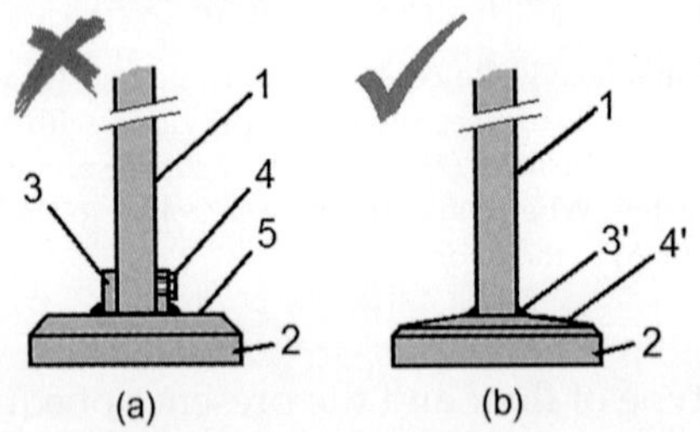

FIGURE 26.28 (a) Foot spindle (1) is inserted into a bush (3) which is welded to a foot base (2). The foot spindle is rigidly fastened to the foot base by means of a screw (4). This foot is not hygienically designed because the upper part of the foot base forms a non-drainable flat surface (5). Debris and water may collect into the crevice formed between the inner surface of the bush and the inserted part of the spindle, and around the fastening (4). (b) The foot spindle (1) is all-around flush welded (3′) to the foot base (2) from which the upper surface parts (4′) are now sloped to make them drainable (Hauser, 2008b).

All exposed surfaces shall have a smooth finish such that soil may be cleaned from the surface using manual cleaning techniques, and be free of pits, folds, cracks, crevices and other imperfections in the final fabricated form, when installed on the machinery and within the specified load conditions. Hence, feet may not create dirt traps, and further they must be self-draining which means that they shall not have pockets which retain liquids (Figure 26.28b). Feet with fixing holes should be provided only if bolting to the floor is necessary, but avoid the use of extra brackets. Figure 26.29 shows some examples of hygienically designed feet.

Equipment should be adequately located in position, with all its feet having a contact face that is even so as to ensure complete contact with or to allow fixation to the floor. For proper installation on uneven or inclined floors, the use of improvised shimming to level food processing equipment is not allowed. Equipment feet adjustable by min ±75 mm should be used. When adjustable feet with threads are used for this purpose, the threaded spindle for leveling should be completely concealed in closed profiles/pipes or enclosed so as not to cause accumulation of dirt and contaminants in the thread.

The load-bearing foot may also include a rubber layer underneath or rubber can be embedded in the load-bearing foot. The elastomeric material may dampen the vibrations of the operating equipment and may prevent slipping of the foot on the support surface. The

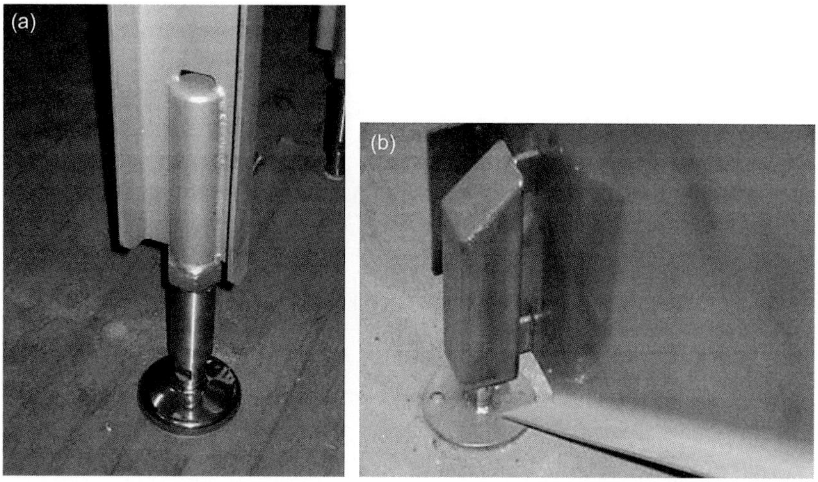

FIGURE 26.29 (a) Pivotal machine leveling mount from which the threaded spindle is completely concealed in a closed pipe that is in-welded in a sheet metal leg. (b) Stair riser legs, totally sealed, with sloped top and set off the riser. *(a) courtesy of Den Rustfri Stålindustris Kompetencecenter. (b) courtesy of Joe Stout.*

rubber used shall be of sufficient low Durometer to provide a tight continuous seal with the flooring material.

Castors

Castors are applicable in those places where equipment has to be made mobile in order to facilitate inspection and cleaning of equipment and process rooms. Transportable equipment (e.g. conveyors) also allows the layout of process lines to be changed so that products can be altered to suit demand (e.g. frozen vegetable industry). However, a castor assembly must not be used in the product zone. As an example, containers designed for elevated dumping shall not be equipped with attached castors if, when raised, the castors are over the product zone.

Castors should be made of a material that suits the floor quality, the expected loading and the frequency of movement. If underspecified castors are used, the body of their wheels can break up due to being overloaded. In general, the heavier the load, the larger the wheel required for the castor. Large wheels roll more easily, are generally more maneuverable and ride better over obstructions and floor cracks, tracks and ruts than smaller wheels. Large wheels also provide sufficient clearance between the lowest part of the equipment and the floor for easy cleaning and inspection.

Although cast iron wheels are virtually indestructible and are able to withstand the highest loads, their use in the food industry is not recommended (not acceptable), because they are prone to general corrosion and can damage floor surfaces. Castors manufactured from zinc-plated mild steel should be avoided, because the coating on the wheel may wear away, resulting in corrosion and increased friction between the wheels and the castor forks (horns). Paint shall not be used as a coating. Castors manufactured from zinc-plated mild

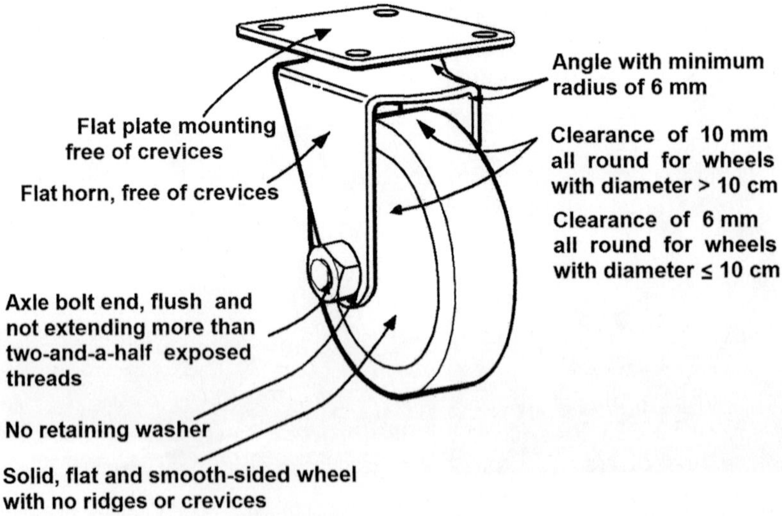

FIGURE 26.30 Hygiene design requirements that castors in the food industry must meet (APV Baker, 2001).

steel require the swivel bearing and wheel axle to remain lubricated to prevent them from corroding. Hence, lubrication of the swivel bearings and wheel/axle surfaces should be regularly and properly done, especially because lubricant can be washed away by regular cleaning. Castors (body, mounting plate, etc.) manufactured from stainless steel with stainless steel swivel bearings need no lubrication to prevent corrosion. Stainless steel axles in combination with an outer PTFE bushing provide self-lubrication of the wheel/axle surfaces. However, worn wheels and PTFE bushings still will need periodical replacement.

Thermosetting plastics, particularly phenolics, are widely used in the food industry because they can withstand high temperatures and carry high loads. However, they can become damaged by poor quality flooring and by defects in floors, such as concrete joints and ridges. Phenolic-wheeled castor types are often worn to a flatter profile or their tread can break up causing spalling. Thermoplastic wheels have better impact resistance than phenolic-wheeled castors, but they have poor resistance to higher temperatures. However, these wheels do not need bearings. Where possible, the wheel should have a color (e.g. blue). High temperature rubber-wheeled castors have a high temperature thermoplastic center with a bonded high temperature rubber tire. They will wear and may be damaged by poor or abrasive surfaces, acids, oils, chemicals and other substances that may be harmful to rubber. These soft tread wheels, however, may ride more easily over bumps, level changes, joints, drainage gullies, etc., and are less destructive to tiled, linoleum, etc. floors.

Swivel castors (Figure 26.30) only function well when they are securely mounted to a rigid frame so the swivel bearing kingpin axis remains vertical at all times. Rigid castors must be mounted (welded, sealed or readily removable) in a way that their axis and wheels are in alignment. All structural members (mounting plate and horn) shall have a minimum of horizontal flat surfaces. The plate mounting shall be constructed to have a flat top surface. The angle between the top surface and the edge of the plate shall be 90° or less.

Mounting holes and other devices provided for installation shall be so designed as to prevent the formation of pockets or areas difficult to clean. The horn assembly or fork shall be constructed so that the surface facing the wheel has no concave surface except that part joining the horn plate. Included angles between all surfaces should have a minimum radius of 6 mm. Kingpin assemblies, which have the nuts or rivets at the bottom, shall have suitable caps covering the ends. The minimum clearance between horn assembly and wheel for wheels having a diameter should be 6 mm all round, while the minimum clearance should be <10 cm all round for wheels with diameter >10 cm. Brakes and locking devices should comply with the hygienic requirements mentioned above.

Preference should be given to single-wheeled castors because dual-wheeled castors are more sensitive to contamination, and are more difficult to inspect and to clean. Castor wheels should be constructed so as to have no concave surfaces facing the horn assembly except that part which joins the hub. The included angles between all vertical and horizontal surfaces shall have a radius of not less than 6 mm. Wheels should have solid webs, smooth sided, without ridges or crevices, and their tread face should be smooth and flat. Rubber-wheeled castors should have a tire from which tread and shoulder are free of lugs, voids and indentations wherein foreign matter can penetrate. If bolted, axle bolt ends should be flush and should not extend more than two-and-a-half exposed threads beyond the retaining nut. Excess threads should be cut off and covered with a "dome"-type nut. The use on the axle of cotter pins or castellated nuts to keep the wheel attached to the horn assembly is not acceptable. Two PTFE washers (combination seals) can be fitted, one either side of the wheel, to prevent direct contact (e.g. metal-to-metal contact) between the wheel and the castor body. Although it is expected that the life of these washers should almost be as long as that of the wheel, the washers can become worn and must be replaced immediately. In general, washers (retaining washer under a nut) should not be used between the horn of the castor and the axle retaining nut, because there they are more exposed to impact from the outside.

Roller or ball bearings should be used. Roller bearings can carry heavier loads, while ball bearing wheels roll more easily but carry lighter loads. All bearing arrangements must ensure that no crevices or dead areas are present which could adversely affect cleanability and/or functional life. If no self-lubricating bearings (stainless steel with PTFE bushing) are used, they should be lubricated every 6 months. In corrosive environments, lubrication of bearings should occur once a month. In the food industry where the lubricant is washed away by daily cleaning, lubrication is sometimes required after each washing. Bearings in castors (wheels and swivel horns) should preferably be of the sealed type. These seals used to contain the lubricant oil or the grease in the bearings will wear, ultimately allowing leakage. Their integrity must be regularly checked and they should be replaced at defined maintenance intervals. If "open" ball-race bearings are used, they must be cleanable and, when required, capable of being disinfected and re-packed with food grade grease as necessary.

Belt Conveyor

Conveyor frames should have an open structure (Figure 26.31) with a minimum of hidden areas/surfaces. But guards are required in places where a drive station, a pulley, rollers or the conveyor belt may cause injury. The guards, however, should be easy to dismount to allow for complete cleaning. Solid cross-members as structural members are preferred over

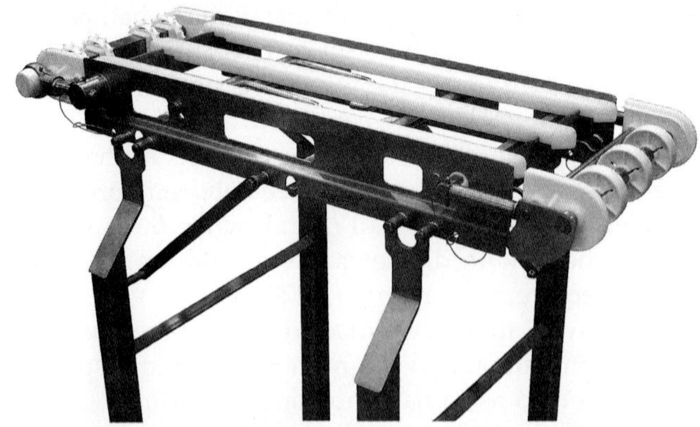

FIGURE 26.31 Conveyor frames should have an open structure without horizontal surfaces and with a minimum of hidden areas/surfaces. At the outside, the framework consists of vertical plate members positioned longitudinally, which also serve as a lateral belt guide. The conveyor frame is an all welded construction with solid round cross-members, welded at the outside framework. The use of bolts and nuts for fastenings is reduced to a minimum. The cross-members not only act as structural frame members, but also as belt supports. The weld-on flat cross-members are provided with gaps to accommodate the freely located plastic wear strips that help to support the conveyor belt. No bolts, holes or nuts were used for fastening the ultra-high-molecular-weight (UHMW) polyethylene wear strips. To minimize cleaning time, these belt supports are easily lifted out of the frame by means of a quick tension-release arrangement and without manual tools. The cut-outs in the frame allow spraying and cleaning of the inside of the conveyor without lifting the belt. The conveyor shown is provided with a swivel-mounted roller that releases tension, providing improved access to the space between belt and bearing strips for cleaning and disinfection. The frame member closest to the point where the belt runs onto the drive roller sprocket also serves as a guard. Stand-off legs keep fasteners out of the food zone. *Courtesy of Dorner Conveyors.*

hollow section members, although completely sealed hollow section members are still more preferable over open profile angle or channel sections, to minimize horizontal ledges and crevices. Hollow sections should be sealed by welding.

Conveying surfaces shall be supported by a minimum amount of carrying surface or bed as required (Figure 26.32b). The use of solid plate that expands the whole top surface of the conveyor table to provide support to a belt is likely to increase contamination problems and cause excessive wear of the belt (Figure 26.32a). Non-removable bearing surfaces for belts cannot be cleaned easily. Rollers shall be used where practical, or line supports that are easily removable for cleaning. The conveyor belt should have minimal debris retention, and running under a turned-over section of side cladding (overhanging belt edges) is not allowed because the whole surface of the belt cannot be cleaned, and the belt cannot be lifted up to allow cleaning and inspection of internal surfaces and support members. Also, pivoted covers cannot be cleaned easily. The use of fixed hinges is not recommended because of the great difficulty of removing debris and microbial slime from between the hinge segments (Figure 26.32a). Side guides used to contain product should be capable of being removed. However, removable guides may cause problems because of the possibility of the fastening system working loose. The conveyor frame must be designed so that the sides of the belt are turned up to form an integral guide to the belt. Besides this guide cladding can be made removable allowing for effective cleaning (Figure 26.32b).

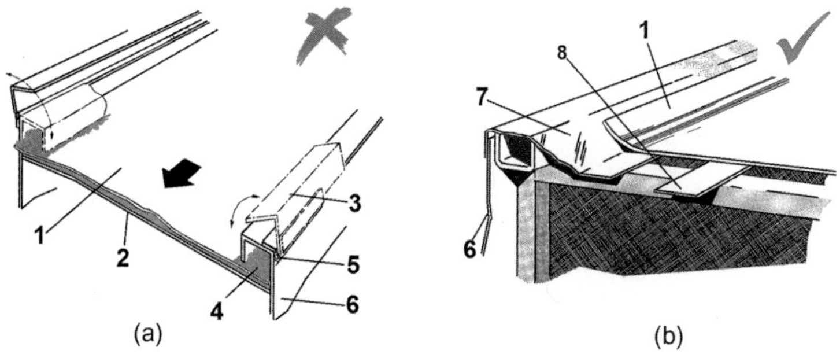

FIGURE 26.32 (a) The use of solid plate (2) expanded over the whole top area of the conveyor table to provide support to a belt (1) is likely to increase contamination problems and cause excessive wear of the belt. The non-removable bearing surface (2) for belts cannot be cleaned easily. The conveyor belt should have minimal debris retention, and hence running under the turned-over section of side cladding (overhanging belt edges) is not permitted as the whole surface of the belt cannot be cleaned (4), and the belt cannot be lifted up to allow cleaning and inspection of internal surface and support members. Side guides used to contain product should be capable of being removed. But removable guides also may cause problems because of the possibility of the fastening system working loose. Pivoted covers (3) cannot be cleaned easily, and the use of fixed hinges is not recommended because of the great difficulty of removing debris and microbial slime from between the hinge segments (5). (b) The conveyor frame (6) must be designed so that the sides of the belt are turned up to form an integral guide to the belt (7). Besides, this guide cladding can be made removable allowing for effective cleaning. The conveyor belt shall be supported by a minimum amount of carrying surface or bed (8) as required. Rods, slats, rollers or similar supports shall be used where practical (CFPRA, 1983; Hauser et al., 2004b).

The drive motor of the belt conveyor should not be positioned over the product flow, as this may result in contamination of the product by lubricants discharged from the drive system. Otherwise, an adequately sized drip tray should be fitted. However, motors should rather be located below the line of the product flow because the exposed motor may have a fan that will scatter dust and dustborne microbes. The motor, gears and the chain must be covered to avoid any contamination of food product (e.g. enclosed in a hygienically designed and hermetically sealed housing). However, a chain guard (essential from an occupational safety point of view), when open, may provide a place where product may accumulate, allowing microbes to multiply to large numbers and so posing a contamination risk for the food product on the belt (Figure 26.33).

Also notice that drive motors installed below food products are quickly splashed and difficult to keep clean. The motor is also often of a type that cannot be washed with a high pressure hose using water and cleaning agents. In that case, if installed below the line of the product flow, the gears and motors of belt drives must be covered. Alternatively, cleanable and sealed motors (wash down or easy clean motors), which do not require ventilation or housings, can be used. Where needed, the motor, gears and the chain should be enclosed in a hygienically designed enclosure or hermetically sealed housing (Figure 26.34a). IP55/54/67 motors can be easily cleaned and drained of water around the motor, if they are provided with enough air space for cleaning and disinfection, maintenance and repair. Where possible, use drum motors (motorized pulleys) (Figure 26.34b) that are fully closed, non-ventilated, conveyor belt drives where motor and gearwheels are inside, submerged in a bath of food

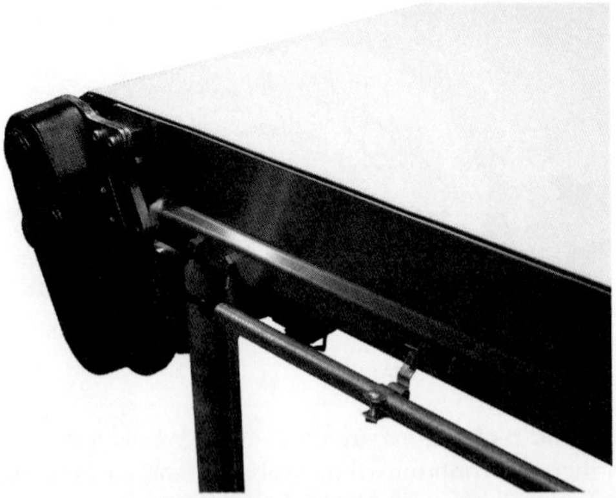

FIGURE 26.33 Motors should rather be located below the line of the product flow. Gears, chains and motors of belt drives must be covered to avoid any contamination of product. However, a chain guard (essential from an occupational safety point of view), when open, may provide a place where product may accumulate, allowing microbes to multiply to large numbers and so posing a contamination risk for the food product on the belt. *Courtesy of Dorner Conveyors.*

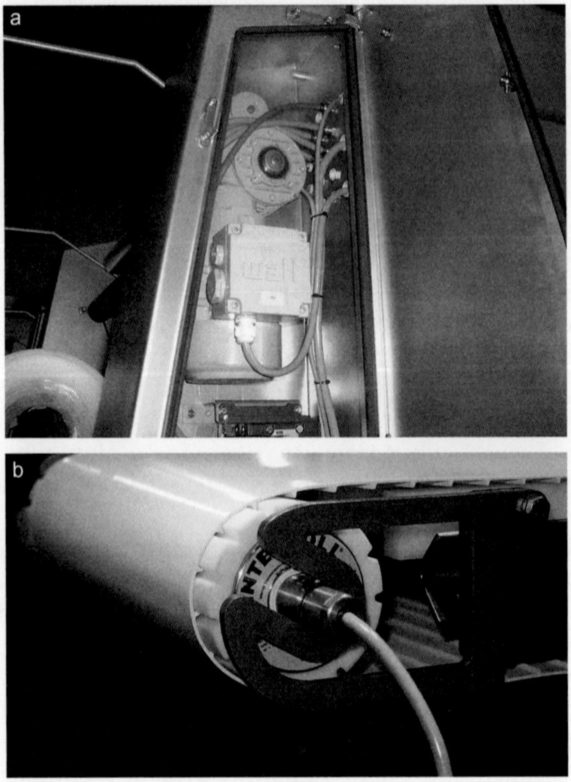

FIGURE 26.34 (a) Where possible, the motor, gears and the chain should be enclosed in a hygienically designed enclosure or hermetically sealed housing. (b) An even better solution is applying a direct-driven (drum motor) instead of a chain-driven system. *(a) Courtesy of Den Rustfri Stålindustris Kompetencecenter. (b) Courtesy of Interroll.*

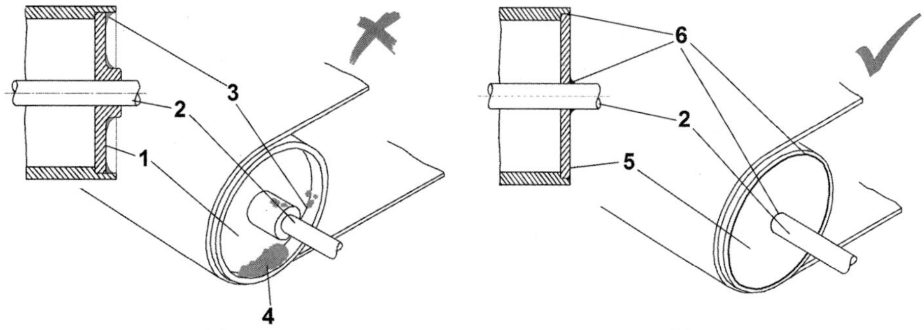

FIGURE 26.35 (a) Pressed-in roller ends (1) create dead areas and crevices (3), where residues of product and soil (4) may accumulate. (b) Flush roller ends (5) which are properly welded (6) to the roller and to the shaft (2) avoid any hazard and can be cleaned easily (CFPRA, 1983; Hauser et al., 2004b).

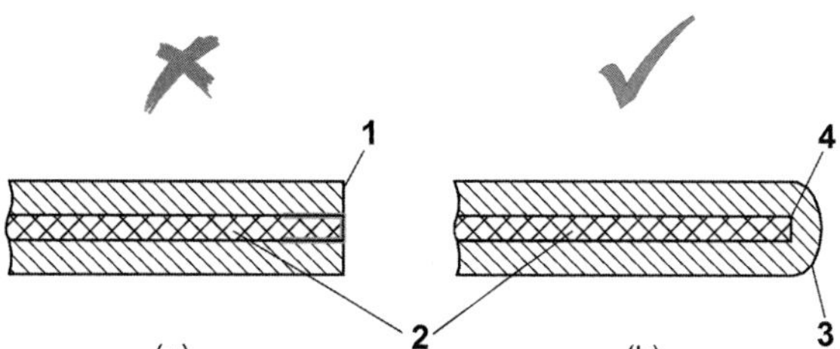

FIGURE 26.36 (a) Cut edges of belts (1) which incorporate reinforcing materials (2) are prone to penetration of liquids into the interior by wicking (capillary action). (b) Therefore, embedded reinforcements, as well as fabric backing materials (2) in conveyor belts, must be covered to avoid contact with the product. The edge should be suitably sealed and covered in a way that the covered edge (4) is shaped like a round rim (Lelieveld et al., 2003; Hauser et al., 2004b).

grade lubricant, providing at the same time lubrication and cooling. Drum motors make gears and chains redundant. (Den Rustfri Stålindustris Kompetencecenter, 2006a).

The design of rollers, pulleys and sprockets shall be free of end recesses and shall be closed if hollow. A welded construction should be preferred to a sealed design (Figure 26.35).

Embedded reinforcements, as well as fabric backing materials in conveyor belts, must be covered to avoid contact with the product. Cut edges of belts which incorporate reinforcing materials must be sealed to prevent penetration by wicking (capillary action) of liquids into the interior (Figure 26.36).

Covers and Guards

It is difficult to obtain motors, gearboxes, etc. that meet the recommendation of EN 1672-2. Protecting any of these items by means of covers or guards is recommended. These guards must also protect the food product from contact with drive parts such as lubricated chains, sprocket wheels, etc. The requirement of guarding machinery to ensure safety in operation

FIGURE 26.37 Example of a hygienically designed guard. *Courtesy of P.T. Group.*

may easily conflict with hygiene requirements unless considerable care is taken in its design, construction, installation and maintenance. However, the housings or guards should be removable to provide access for cleaning. From a hygienic and safety point of view, totally removable covers, guards or cladding should be avoided. They may not be put back, creating a hazard for the operators in the environment of the process equipment and exposing the food product to risk. Covers and guards also may become damaged during removal. Bars, perforated/punched sheet and weld mesh (Figure 26.37) stainless steel guards with an open area of 40–50% give good protection from moving equipment parts, and permit access for cleaning and disinfection by spray nozzles or hosing down procedures. For good drainability, covers should always have an angle and should be free of panel joints.

Where possible, hinged covers and guards that pivot outboard should be used. But use as few hinges as possible, with the least number of parts. In view of cleaning and disinfection, continuous and piano hinges are not allowed. Block or pin hinges are a possible option, but should have removable hinge pins or be the lift-off type. Finally, the exterior of enclosures is easier to clean if internal hinges are used.

Maintenance Enclosures

Maintenance enclosures (e.g. electric control panels, junction boxes, pneumatic/hydraulic enclosures) must be designed, constructed and maintainable to ensure that the product water or product liquid does not penetrate into, or accumulate in or on, the enclosure. The cabinet and operator panel are mounted where they will be least exposed to splashes. Electrical control cabinets mounted on the exterior of the equipment shall be watertight and sealed to the supporting member with food standard silicon seal, or spaced sufficiently away from the member to permit cleaning of all surfaces. A minimum of 20 mm between the control and supporting member shall be provided. Electrical enclosures can also be sealed to a wall (with food

FIGURE 26.38 Electrical enclosures can also be sealed to a wall (with food standard silicon seal), or shall be spaced at least 30 mm away or at a distance equal to one-fifth of the shortest dimension of the electrical enclosure parallel to that wall, to prevent a soil trap being created at the rear of the enclosure and to allow for adequate access for cleaning. Suspending members should be constructed of a solid steel round tubing to prevent the formation of a flat horizontal surface whereupon dirt may collect. *Courtesy of Rittal.*

standard silicone seal), or spaced away at least 30 mm or at a distance equal to one-fifth of the shortest dimension of the electrical enclosure parallel to that wall. The distance between the cabinet base and the floor should be no less than 0.3 m. Horizontal surfaces should be minimized or avoided by installing a top roof with a minimum 30° inclination towards the front to allow water to run off and prevent tools being placed on the top. The front edge of the inclining cabinet top should reach beyond the front door and the seal (Figure 26.38). To prevent condensate dripping from the field box into the product, field boxes should not be placed in or above the contact area. Furthermore, field boxes should be located such that easy access for maintenance and cleaning is practicable. All connections (e.g. cable ladders or wire trays, trunking, conduit, cable, etc.) to cabinets or field boxes should be made via the bottom side of the cabinet. Connections of cables and wires to housings must be sealed (Moerman, 2011a).

The control and indicator devices must be constructed of durable and mechanically stable (unbreakable, resistant to steam, moisture and the actions of cleaning and sanitizing agents, abrasion and corrosion resistant) material. Commonly used food grade plastics for the construction of control devices and indicator lights are polyamide (PA), polycarbonate (PC), polyoxymethylene (POM), silicone and acrylonitrile butadiene styrene (ABS). Control devices and indicator lights in contact with food should be shaped so as to avoid the accumulation of dirt and bacteria, and to facilitate cleaning (Figure 26.39). The device heads must have smooth and crevice-free surfaces that are easy to clean. Device head to front panel transitions must be smooth, without corners and edges. Push buttons, when touched, should not penetrate deeply in the front panel far beyond a (protruding) frame edge surrounding the button. Connections must be conceived in such a way that protruding parts, strips and concealed corners are restricted to a minimum. The connections of inside

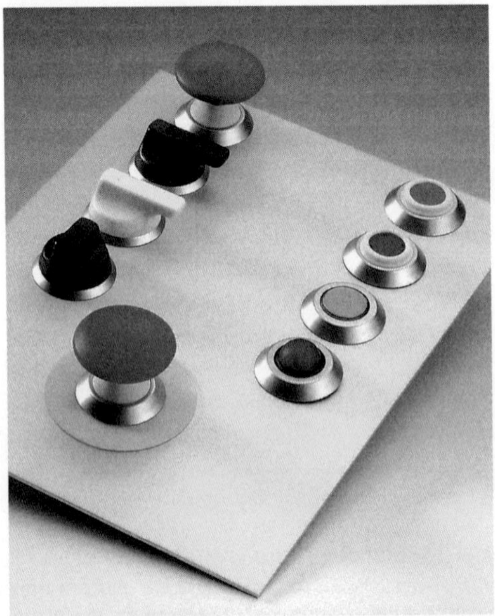

FIGURE 26.39 Control panel with hygienic control and indicator devices. Seals should fill the gaps between the fixed and moving device parts, to avoid the ingress of product residues, lubricants and organic materials. A perfect, hermetic seal is also required to prevent the ingress of moisture, dust and dirt within the control panel. Adequate space should be provided between control and indicator devices for easy cleaning. *Courtesy of Elan-Schmersal.*

surfaces must be made with curves of sufficient diameter. Seals should fill the gaps between the fixed and moving device parts, to avoid the ingress of product residues, lubricants and organic materials. A perfect, hermetic seal is also required to prevent the ingress of moisture, dust and dirt within the control panel. An IP67 or IP67K ingress protection rating for control panel enclosures is highly recommended. Control panels with control and indicator devices shoud be installed in vertical or declining position, such that fluids (splashed food and cleaning solutions) are able to flow from the control panel. Adequate space should be provided between control and indicator devices for easy cleaning. Further hygienic alternatives to control panels with push buttons and selection switches are membrane panels with a ≥2% inclination or touch screen displays.

HYGIENIC DESIGN CLOSED EQUIPMENT FOR PROCESSING OF LIQUID FOOD

Process and Utility Lines

Hygienic Design of Process and Utility Lines

To avoid the formation of standing "pools" of liquid that can support the growth of microorganisms, process and utility piping runs should be sloped to at least 3% in the direction of flow and should be properly supported to prevent sagging (Figures 26.40 and 26.41).

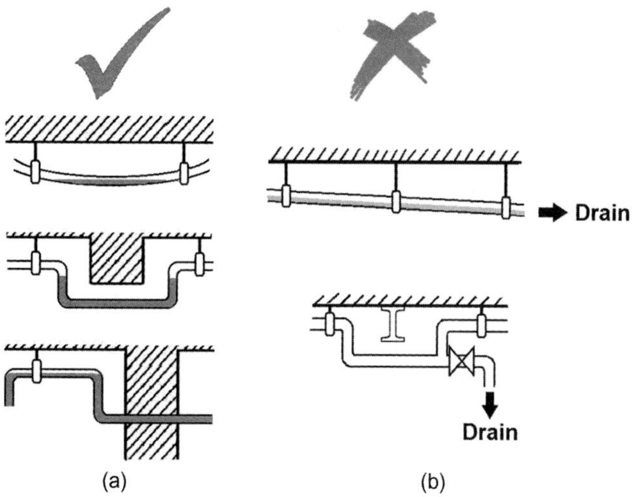

FIGURE 26.40 (a) Sagging of piping must be avoided because standing "pools" of liquid can support the growth of microorganisms. Changes in the level of horizontal runs of pipelines should be avoided otherwise there will be an undrainable section. Horizontal runs of pipe which are routed vertically up and then down to by-pass beams, doorways or other obstructions will allow air to collect in the raised section. (b) Process and utility piping runs should be sloped to at least 3% in the direction of flow. Piping must be installed in a way that air does not collect in the raised section. While automatic air release valves can be installed (on top of elevated horizontal pipe sections) to remove trapped air, the resulting dead leg may cause contamination and/or cleaning problems. Where liquid collects in a lower horizontal pipe section, fitting a valve in a shortened tee allows that liquid to be drained (CFCRA, 1997).

FIGURE 26.41 Non-drainable pipe (Hauser et al., 2007).

Blanked-off tees should be avoided where possible as they constitute a potential hazard. A dead space, being an area outside the product flow, where liquid or gas can become stagnant and where water is not exchanged during flushing, is formed. An air pocket may be present if the branch of a blanked-off tee is pointing vertically upwards (Figure 26.42a). Hence it will prevent liquids (cleaning solutions, disinfectant solutions or hot water) from reaching all surfaces to be treated, with the result that cleaning-in-place and decontamination processes will be unsatisfactory. Drain points pointing downwards that act as a dead

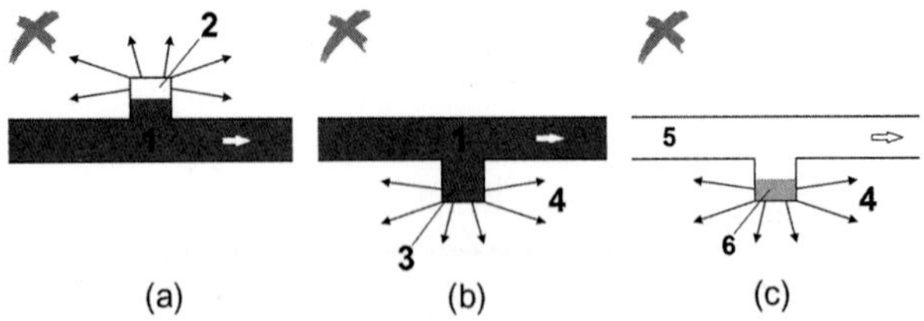

FIGURE 26.42 (a) When cleaning and disinfection solutions (1) flow through the piping, an air pocket (2) will be formed if the branch of a blanked-off tee is pointing vertically upwards. This will prevent the solutions from wetting the surface in the dead leg. (b) Drain points pointing downwards (3) again act as a dead leg, providing an area of entrapment which may not be reached by cleaning or sterilizing procedures, and hence they lead to contamination of the product. Moreover, during a hot water treatment, the hot water also will stagnate in the downwards pointing pocket, so that the temperature of the surfaces in the dead area may be lower than required as the consequence of heat loss (4). (c) A downwards pointing dead area also will collect condensate (6) due to heat loss (4) during steam sterilization (5), with the result that again the temperature of the surfaces in the dead area may be lower than required. Blanked-off tees should be positioned such that they are a few degrees above the horizontal (Lelieveld et al., 2003; Hauser et al., 2007).

leg (Figure 26.42b) are not acceptable because they provide an area of entrapment which may not be reached by cleaning or sterilizing procedures, and hence they lead to contamination of the product. During a hot water treatment, the hot water will also stagnate in the downwards pointing pocket, so that the temperature of the surfaces in the dead area may be lower than required as the consequence of heat loss. A downwards pointing dead area will also collect condensate during steam sterilization (Figure 26.42c), resulting again in the temperature of the surfaces in the dead area being lower than required.

The direction of the flow of food product has a significant influence on the residence time in the dead leg. When the food product flows in the direction as indicated in Figure 26.43a, b and c, part of the product will stand still in the dead leg, especially if the length or depth of the T-section is too long. If the length of the T-section is equivalent to the diameter of the main pipe, a flow velocity of 2 m/s in the main pipe already results in a reduced velocity of 0.3 m/s in the T-section. This decrease in flow velocity provides a relatively stable pocket or dead leg in which product residues can accumulate and microorganisms begin to multiply. Long T-sections outside of the main flow of cleaning solutions are also very difficult to clean. During cleaning there is much less transfer of thermal (heat), chemical (detergent and disinfectant chemicals) and mechanical energy (action of turbulent flow) to the food residues in the zones and T-sections which are further outside the main flow of cleaning liquids than to the soil in the main flow. Notice that flow away from the dead leg (Figure 26.43a and c) gives rise to more contamination problems and problematic cleaning, as velocities in these dead legs are even much lower.

A properly designed food processing line should not have unnecessary dead legs, and where they cannot be excluded, they should be in the correct position for the selected cleaning and decontamination process and should be as short as possible. For pipe diameters of 25 mm or larger, T-sections should have a depth/length preferably under 28 mm, while for smaller pipe diameters this length should be smaller than the diameter. Blanked-off tees should be positioned such that they are a few degrees above the horizontal. The dead leg will then be drainable but not necessarily cleanable even if made as short as possible. If a sensor must be installed in a process line, it should be installed in a bend on a shortened tee in a position so that the flow of cleaning fluid is directed into the tee (Figure 26.43e and f). Where

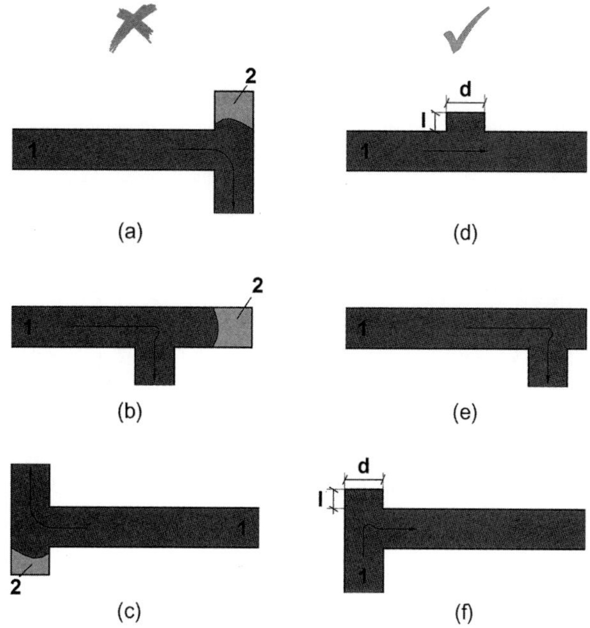

FIGURE 26.43 When the food product flows in the direction as indicated in (a), (b) and (c), part of the product will stand still in the dead leg, especially if the length or depth of the T-section is too long. Long T-sections outside of the main flow of cleaning solutions are also very difficult to clean. For most liquids, the dead leg should be positioned as shown in (d), (e) and (f). The configuration in (f) is quite acceptable if $l \leq d$, because the flow directed into the short dead leg provides sufficiently high velocities for proper cleaning. If the dead leg is very short ($l \leq d$), configuration (d) is acceptable, although flow across a dead leg results in much lower velocities within it and thus only provides moderate cleaning. Configuration (e) may not be suitable, if products contain any particulate matter that may accumulate in the dead leg (CFCRA, 1997; Lelieveld et al., 2003; Hauser et al., 2007).

an angle valve is installed in the process piping circuit, this valve must also be mounted in a shortened tee so that no or a minimum of annular space above the side branch is formed. Again the flow of cleaning solution must be directed into the tee.

For most liquids, the dead leg should be positioned as shown in Figure 26.43e, d and f. The configuration in Figure 26.43f is quite acceptable, because the flow directed into the short dead leg provides sufficiently high velocities for proper cleaning. If the dead leg is very short, configuration Figure 26.43d is acceptable, although flow across a dead leg results in much lower velocities within it and thus only provides moderate cleaning. The configuration in Figure 26.43e may not be suitable if products contain any particulate matter, which may accumulate in the dead leg. In all cases, the cleaning procedure must take the presence of the dead leg into account.

Flow diversion should not be done in a way that would cause part of the product to stand still in a dead leg. The two-valve system for flow diversion (Figure 26.44a) creates a dead leg towards the closed valve. The correct type of valve is shown in Figure 26.44b.

For horizontal piping, eccentric reducers should be used instead of concentric reducers, because the latter provides a dead spot where condensate and dirt may collect (Figure 26.45).

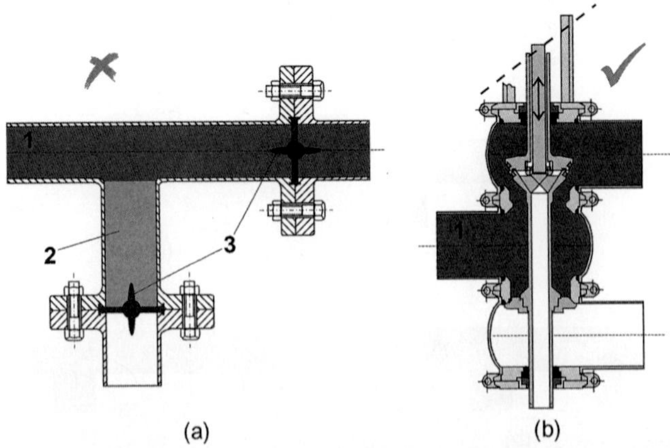

FIGURE 26.44 (a) Flow diversion should not cause part of the product (1) to stagnate in a dead area (2). The system of two butterfly valves (3) for flow diversion creates a dead area (2) towards the closed valve. (b) The correct type of valve is shown on the right (Lelieveld et al., 2003; Hauser et al., 2007).

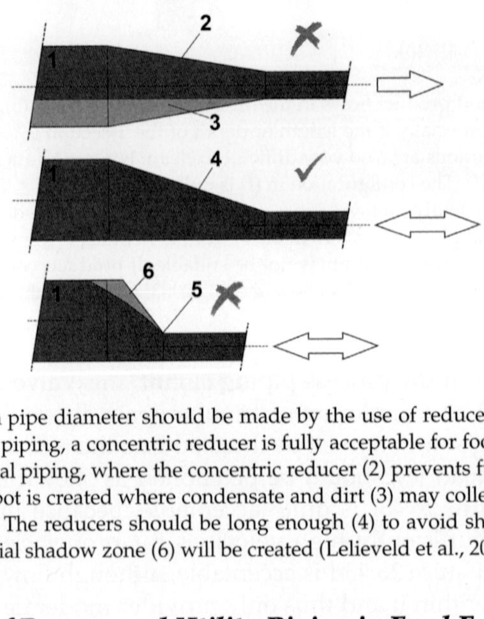

FIGURE 26.45 Changes in pipe diameter should be made by the use of reducers to ensure a smooth transition of the product flow. In vertical piping, a concentric reducer is fully acceptable for food product (1) to flow. However, this is not the case for horizontal piping, where the concentric reducer (2) prevents full drainage if product flow is in the wrong direction. A dead spot is created where condensate and dirt (3) may collect. For horizontal piping, eccentric reducers (4) are preferred. The reducers should be long enough (4) to avoid shadow zones. If a short eccentric reducer (5) is applied, a potential shadow zone (6) will be created (Lelieveld et al., 2003; Hauser et al., 2007).

Hygienic Integration of Process and Utility Piping in Food Factories

Welding of attachments on food processing support piping is not recommended. They can cause stress on the pipe and the part of the supporting anchoring structure. All hangers and supports have to be designed in such a way that they either move together with the pipe (roll or slide) or they can swing without exposing any stress either on the pipe or on the part of the supporting anchoring structure.

All process and utility piping should be grouped together in pipe trains whenever possible. All these process and utility piping should preferably be positioned in a way that all exterior surfaces are readily accessible, to allow cleaning from all sides. The points of

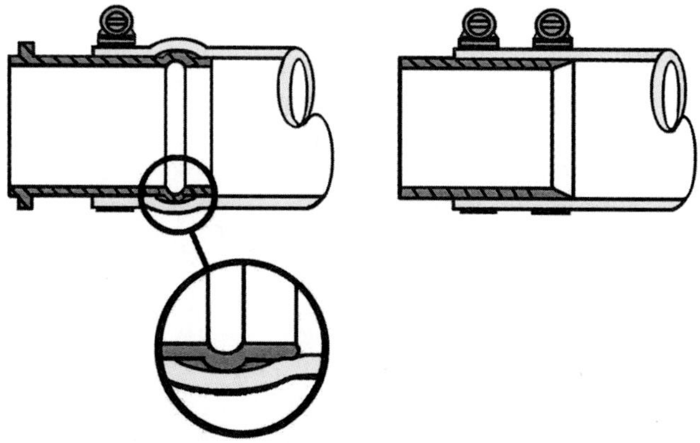

FIGURE 26.46 (a) Incorrect and correct installation of hoses on fixed pipes. (b) Hoses attached to stainless steel pipes should be clamped at the very end of the pipe to minimize the amount of dead space between the clamped portion and the end of the pipe. *Courtesy of Huub Lelieveld, personal communication.*

use should also be grouped in an attempt to minimize individual ceiling drops. Vertical entrance of piping into the equipment is more hygienic than horizontal piping runs. Running of process and utility piping over open equipment in food preparation areas is unacceptable, and nesting of ductwork should be avoided.

Hoses

The use of hoses is not recommended, because failure of hoses can occur due to over-stretching, kinking, rough handling, mechanical impact, ageing, fatigue, abrasion, corrosive atmospheres, etc., and because the chance that leakage of liquid occurs is much higher than when fixed piping is used. Therefore, hoses need regular inspection for damage, deterioration and cleanliness. They should be cleaned and maintained in good mechanical condition. Braided (woven wire or fabric) covers on hoses should not be used.

Out of service hoses must be hanging without touching the floor, and must never hang over open process equipment. Hoses attached to stainless steel pipes should be clamped at the very end of the pipe to minimize the amount of dead space between the clamped portion and the end of the pipe (Figure 26.46). Hoses should not exceed 3 meters in length. When not in use, the ends of the hoses should be covered or capped to maintain proper hygienic conditions.

Pipe Joints

Welded Pipe Joints

It is strongly recommended that the number of joints, whether welded or detachable, is minimized. Cold bending of pipes is highly preferable to the use of prefabricated bends which have to be installed using joints. Although more hygienic, this is still true for welded joints as they are the weaker places in a process system.

Welding is the preferred method of joining, provided that it is done correctly. Stainless steel sanitary tubing joints should be made by automatic orbital welding (Figure 26.47) where possible

FIGURE 26.47 Stainless steel sanitary tubing joints should be made by automatical orbital welding where possible (Kopitzke et al., 2006).

and hand welding in those places that are difficult to access. However, those welds that are difficult to access should wherever possible be completed in the workshop prior to installation in the plant. The applied materials should be easily weldable, and a higher alloyed filler metal in comparison to the welded material should be used to improve the corrosion resistance. Piping with the correct interior diameters should be applied because any mismatch in diameters or thickness may result in misalignment introducing a step in the wall or bore. If the diameters of the pipes to be joined are not the same, then the smaller pipe should be expanded to match the larger. Misalignment can also be due to incorrect fitting up (missed coincidence between the axes of the two coupled components) prior to welding. Alignment and clamping tools are available to ensure accurate alignment. Misalignment tolerance must be limited to less than 20% of the wall thickness.

For proper welding, the parts to be welded should be adequately prepared. Cutting should be done with a mechanical mill or saw to ensure that the cut face is exactly at right angles to the longitudinal axis of the pipe. Any burrs must be removed with either a file or emery paper. Care must be taken not to remove the corner edges of the pipe, as this can give rise to problems with fusion of the root of the weld. The pipe surface 25 mm either side of the weld should be roughened up with a stainless steel wire brush, or emery paper. Then both pipe ends and roughened surface area should be degreased with a solvent and cleaned of contaminants. Any organic substances remaining on the metal surface are vaporized during the welding process and form bubbles (porosity) in the weld metal that may trap product.

After two deburred pipe ends are aligned and butted together to a gap of less than 0.25 mm between both pipe faces, a butt weld joint is made by fusing together the two stainless steel edges with the aid of filler material. If the gap during the joint preparation is too wide, a crack running along the weld metal itself may result (centerline cracking). Full penetration welds should be used whenever possible to avoid pockets where volumes of gas or contaminants can be trapped. Single pass welds should be utilized instead of multi-pass welds to avoid trapped volumes. The weld metal should exactly fill the joint and remain flush with the surface. Underpenetration leaves a crevice at the joint, while excessive overpenetration can give rise to hold-up of product in pipework once taken into service. The weld metal in the joint must be fully fused to the parent, otherwise a crevice will form at the interface between weld and plate. Weld zones should be continuous, smooth and flush with the parent metal. Welding should always occur with sufficient weld seam protection, because insufficient inert gas shielding or no internal purge will result in roughened welds of lower corrosion resistance that are prone to increased adhesion of soiling and difficult to clean. Typically, where inert gas shielding was inadequate, significant discoloration or carbonization in the heat-affected zone is observed.

Weld slag and debris generated within the pipe must be removed from the inside and outside of the weld by proper maintenance and cleaning practice with an alkaline detergent solution prior to the start of the production process. This is followed by rinsing with water of good microbiological quality, usually chlorinated water to 2ppm available chlorine maximum. After draining, the access points should be covered and sealed. In some circumstances there is an additional requirement to passivate the weld area on the product contact side. The welds may be mechanically polished (outside) or electro-polished (inside and outside), but air leakage should be monitored after the polishing procedure.

Finally, weld seams should be visually inspected for any discoloration and surface-breaking defects, usually by endoscopy and aided by dye penetrant tests that highlight these defects. Inspection personnel should be trained and act with caution to avoid internal surface damages while handling endoscopic tools (Hauser et al., 1993; Kopitzke et al., 2006).

Detachable Pipe Joints

Pipework may be designed for rapid regular dismantling to permit cleaning, or the plant may be designed for cleaning-in-place (CIP) or sterilizing-in-place (SIP) without dismantling the plant. In such equipment it is important to avoid crevices and gaps where product residues can accumulate and potentially begin to decompose. Therefore, from a hygiene point of view, the use of threaded piping is not recommended, because it provides crevices and areas where bacteria can adhere and proliferate. To make detachable joints the use of conventional O-ring grooves is not recommended, because these groove designs leave a considerable free space in the groove. Other hygienic requirements for detachable joints include coaxial alignment of the two mating bores, axial stop for controlled compression of the seal, room for thermal expansion of the seal and avoidance of sharp edges such that seals are not damaged. Where there are depressions and steps of more than 0.2mm in the pipework, the flow of cleaning fluid may not thoroughly wash the surface and proper draining of the piping will be hampered. Hence, when making bolted flange fittings, a lot of care should be taken to avoid offsets, gaps, penetrations and voids. A further aspect to be considered is that the seal material must be compatible with both the system product and the cleaning fluids which may be at a much higher temperature.

A number of specific pipe couplings and seal arrangements have been developed for hygienic applications. Some types are covered by national, international or internal company standards, but many of these have been in use for some considerable time and are not considered to be compatible with current requirements in some areas of the food and drink industry (Table 26.2).

TABLE 26.2 Several Well-Established Couplings have been Assessed for Applications in the Food Industry (CFCRA, 1997; Lelieveld et al., 2003; Hauser, 2008b)

Type	Hygiene Characteristics	Application
3-A coupling – ground seat 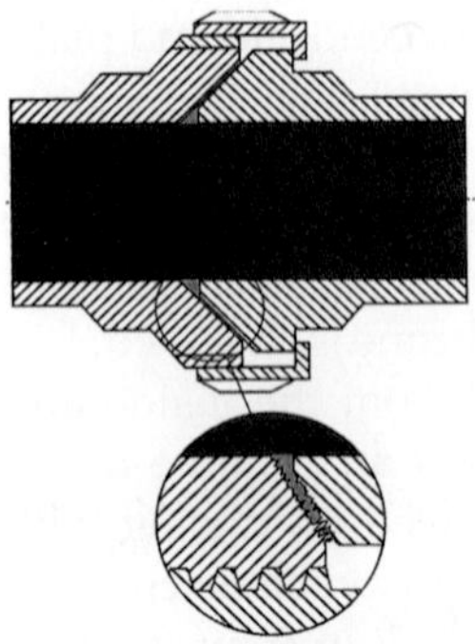	When these surfaces become permanently damaged, it becomes more difficult to obtain a tight seal after every disconnection. The metal-to-metal seat does not prevent the partial penetration of low viscosity liquids or the ingress of microorganisms. Even if the joints are not visibly leaking, the ingress of microorganisms is possible. Furthermore, the seal obtained is very unlikely to be continuous at the interface with the product. More likely, the actual seal follows an irregular line between the inside and outside. The resulting annular crevice will trap product.	Not recommended for use in hygienic plant pipelines and CIP installations, because the internal annular crevice may retain product during production and/or after cleaning in-place. It is widely used in situations where a gasket is unacceptable.
3-A coupling – gasket seat 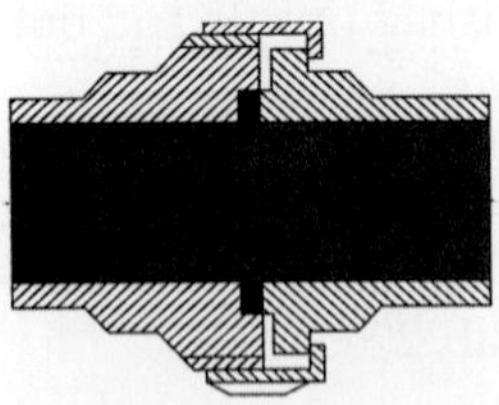	When correctly fitted and assembled a smooth, crevice-free internal surface is obtained.	Suitable for handling most products and for cleaning in-place.
Dairy coupling DIN 11851 – standard gasket 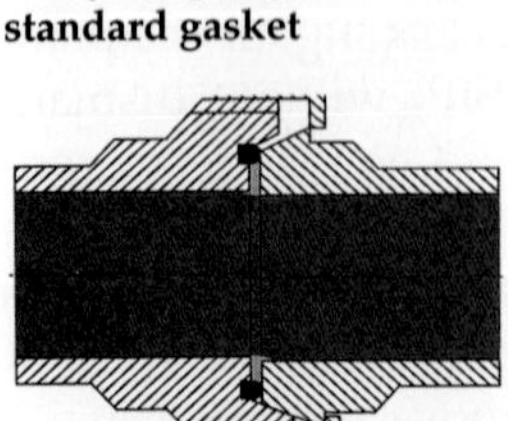	There is an internal annular crevice between the ends of the coupling parts and the bore of the gasket. Product may be retained during production and/or after CIP. An additional potential problem with the design of this fitting is that it has a clearance on the cone fitting; as a consequence the two pipes are not automatically aligned. This could give rise to a potential step in the pipe joint. Does not comply with 3-A or EHEDG sanitary design criteria.	Often found in the food industry (pipes and tanks) due to the fact that it is reasonably priced. Not considered as suitable for CIP, which means that the fitting should only be used where the pipework is manually cleaned.
Dairy coupling DIN 11851 – non-standard collared gasket 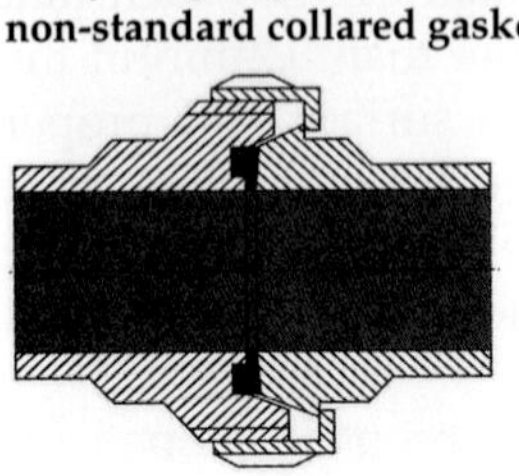	It provides a smooth crevice-free internal surface when correctly fitted and assembled. However, because of the mobility of this type of coupling and of the alternating expansion and contraction of the gasket, this gasket may be damaged by shear. Does not comply with 3-A or EHEDG sanitary design criteria.	Not recommended for use in hygienic plant process lines and CIP installations. Expensive and does not fulfill standard hygienic design criteria.

(*Continued*)

TABLE 26.2 (Continued)

Type	Hygiene Characteristics	Application
Dairy coupling DIN 11851 – alternative gasket with SKS ring 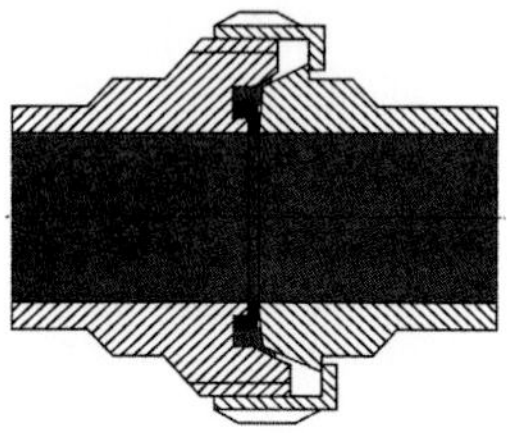	With support of the steel ring the gasket remains flush with the surface. The special designed gasket fills all dead areas in the coupling and will expand to the outside in cases of high temperature. At elevated temperatures, expansion of the seal to the inside is limited. This solution takes all critical points of a DIN 11851 coupling away. Complies with 3-A or EHEDG sanitary design criteria.	Stainless steel center ring and a gasket is an easy solution to upgrade a DIN 11851 coupling to a hygienic status. As a smooth surface gives excellent cleanability.
IDF coupling ISO 2853 with L-gasket 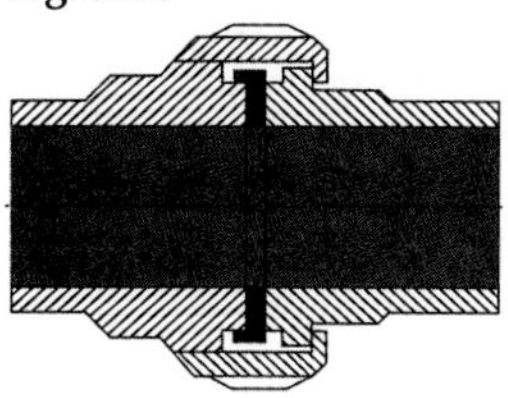	When the coupling is correctly fitted and assembled, a smooth continuous bore and internal surface without crevice is obtained, so that cleaning may be performed without any problems.	This coupling is recommended for applications where CIP is normally practiced. Widely used for pasteurized circuits where dismantling is infrequent.
IDF coupling ISO 2853 with non-standard T-shaped gasket 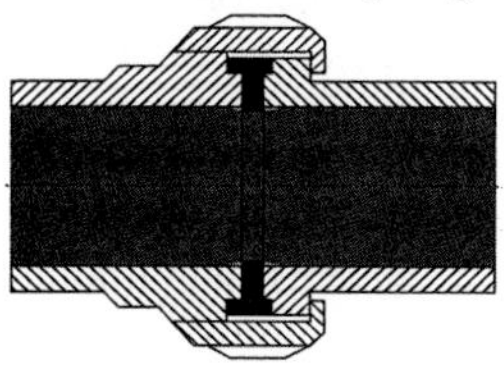	When properly made up, the joint is crevice free and has a smooth bore, flush with the pipe walls. If overtightened, the gasket may expand into the bore of the pipe, which creates a step where product can become trapped. Unless the nut is tightened correctly, the coupling will not be bacteria tight.	Most suitable for permanent or semi-permanent installations that are going to be cleaned in-place. If the seal material is suitable, then it can be sterilized.
IDF coupling ISO 2853 with metal-backed T-shaped gasket 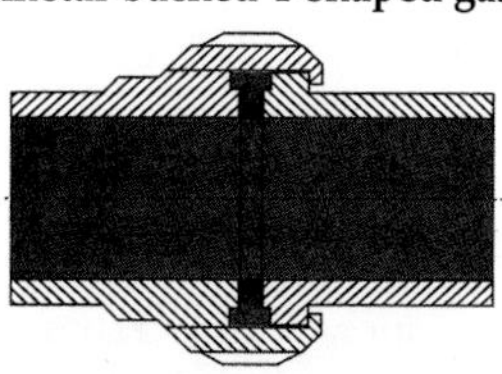	By supporting the seal with a stainless steel ring, both axial stop and centering can be achieved, allowing the connection to meet the requirements of hygienic design. The rubber is specifically shaped to give a flush interior joint when the union is tightened.	Most suitable for permanent or semi-permanent installations that are going to be cleaned in-place. If the seal material is suitable, then it can be sterilized.
Recessed ring joint type (RJT) screwed coupling 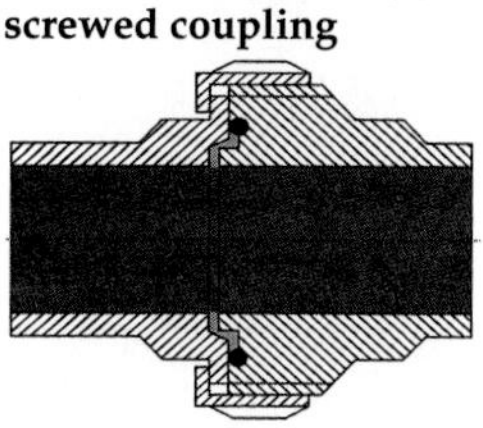	There is an internal annular crevice between the liner and the male part and the bore of the joint ring. Hence, product may be trapped and retained between the two metal components during production and could cause problems if certain products are handled. Does not comply with 3-A or EHEDG sanitary design criteria.	This type of coupling is recommended for use where piping systems are frequently dismantled, but is not suitable for CIP.

(*Continued*)

TABLE 26.2 (Continued)

Type	Hygiene Characteristics	Application
		It is used in the brewing and dairy industry in applications where pipework is manually cleaned. Excellent for flow plates, owing to wide dimensional tolerance on mating bends.
Coupling DIN 11864 form A 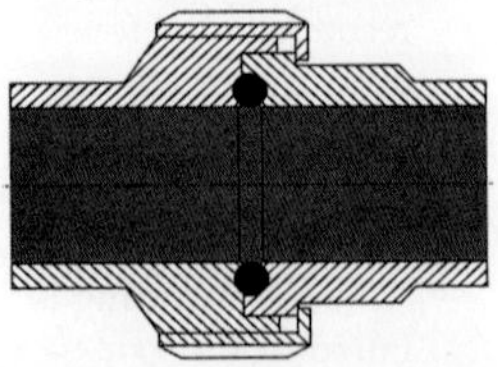	A smooth interface within the pipe work while simultaneously achieving a metal-to-metal seat behind the joint. A sufficient gap is created between the seal and the product space to facilitate rinsing in cleaning processes. This gap also serves as an expansion space that can accommodate volume expansions in the material as a result of heat or the influence of media without forces that can result in shearing. The groove is designed to minimize protrusion of the O-ring into the pipe bore. Complies with EHEDG and 3-A design criteria.	Optimal for aseptic operations because they are successfully tested for CIP-ability, steam sterilizability and bacteria tightness.
Coupling DIN 11864 form B 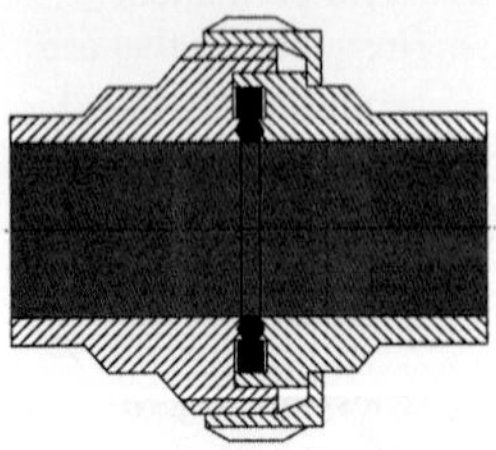	The volume of the functional part of the gasket (diamond section) is minimal to limit the effects of thermal expansion. A small area of the gasket is exposed to the product. The width of the gasket is only 1 mm. The block of elastomer behind the seal will accommodate the thermal expansion, relieve stress build-up on the sealing faces and limit expansion into the product stream to a minimum. The small functional part of the gasket can expand in two directions. To prevent air from being trapped between the gasket shoulder and the male part groove small slits are provided on the outside, acting as vents.	Optimal for aseptic operations because they are successfully tested for CIP-ability, steam sterilizability and bacteria tightness.
Standard SMS 1145 coupling 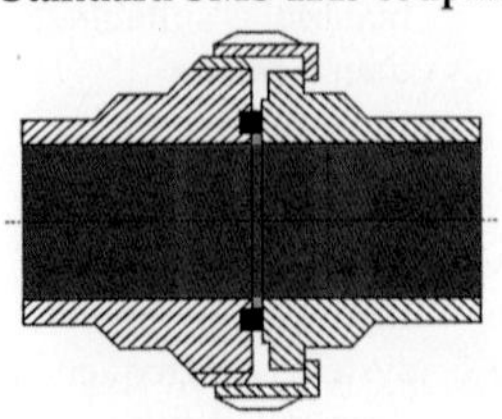 (DS coupling is similar to this coupling)	Standard SMS couplings are not hygienic because an internal annular crevice is formed in which product may be retained during production and/or after cleaning-in-place. The bore of the gasket may retain product. L-profile gasket is available but does not provide self-centering. A later version when correctly fitted and assembled provides a smooth crevice-free internal surface. Does not comply with 3-A or EHEDG sanitary design criteria	Only the latter version is suitable for handling viscous products and for in-place cleaning.

(Continued)

TABLE 26.2 (Continued)

Type	Hygiene Characteristics	Application
SMS 1145 coupling – alternative gasket 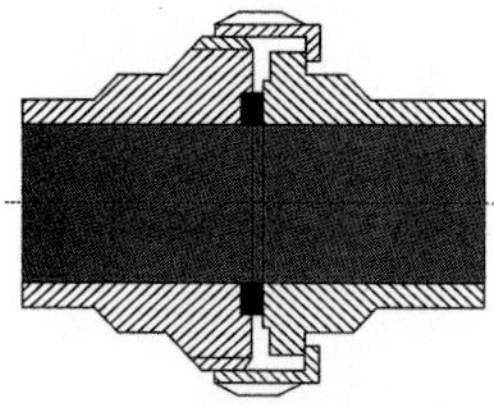	When correctly fitted and assembled gives a smooth crevice-free internal surface.	Suitable for handling viscous products and for in-place cleaning.
Clamp coupling ISO 2852 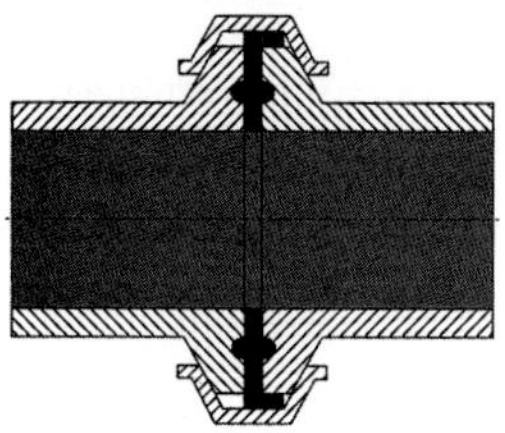	The seal is considered to form a smooth crevice-free joint between the liners, which makes clamp-type couplings suitable for CIP duties. Some users have indicated a preference for clamp fittings rather than screw-type couplings because in the event of a spill, screw threads cannot be decontaminated effectively. Clamp-type couplings are perceived to have the advantage that in the event of a product spillage at the fitting there is no thread to become filled with product that may be difficult to clean.	Often found in the food and pharmaceutical industry (pipes and tanks). Not considered as suitable for CIP.
Varivent® flange coupling 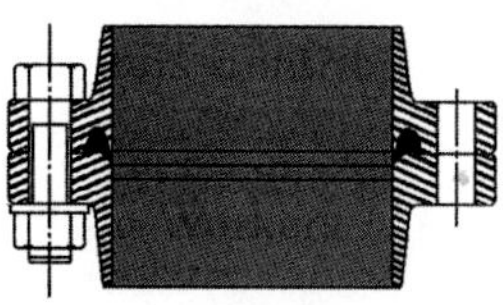	Varivent® flange coupling ensures a smooth transition, free of dead space. It complies with EHEDG and 3-A design criteria.	Successfully tested for CIP-ability. Suitable for aseptic processes.
Neumo Bioconnect® 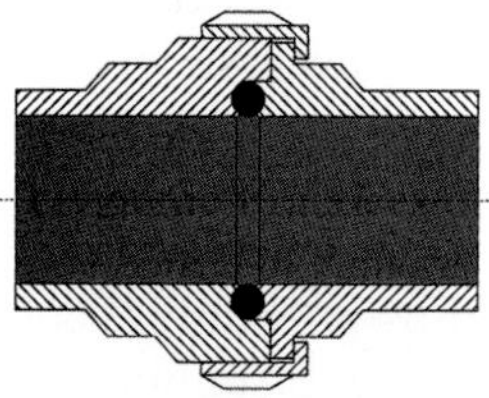	The seal is almost completely encapsulated. The highest press-on power is found at the transitions to wetted areas, preventing dirt and germs from penetrating into the sealing space behind the sealing element. Dead volume is minimized. Complies with EHEDG and 3-A design criteria. Successfully tested for CIP-ability.	Optimal for aseptic operations because it has been successfully tested for CIP-ability, steam sterilizability and bacteria tightness.

Hygienic Design of Pumps

Hygienic Design of Centrifugal Pumps

While it is often convenient for the arrangement of pipework to orientate the casing of a centrifugal pump so that the outlet port is pointing vertically up, this will result in the pump casing retaining liquid up to the level of the inlet port. The pump casing is drainable through the outlet port if the pump's outlet is arranged to point horizontally at the

FIGURE 26.48 (a) The pump casing is drainable through the outlet port if the pump's outlet is arranged to point horizontally at the bottom. (b) Now the centrifugal pump is installed in a vertical position, and hence fully drainable through its suction port. *(a) Courtesy of Patrick Wouter, Unilever. (b) Courtesy of Hilge.*

bottom (Figure 26.48a), or the pump casing can be made drainable through its suction port if installed in vertical execution (Figure 26.48b).

Hygienic Design of Rotary Lobe Pumps

Rotary lobe pumps having unhygienic design features can only be cleaned effectively after dismantling. To avoid any introduction of contaminants into food product and to allow for CIP without dismantling, rotary lobe pumps should be hygienically designed. Metal-to-metal joints should be eliminated by hygienic application of O-rings; O-ring groove design should be improved and O-rings should be positioned more appropriately, or alternatively gaskets having controlled compression should be used; sharp corners must be rounded to a minimum radius of 3mm; the length of the annular space within the mechanical seals should be reduced by changing the design of these mechanical seals (e.g. the elements of the mechanical seal should be reversed and the radial distance increased); any exposed threads (e.g. threads of the rotor shafts, Figure 26.49a) should be covered by crevice-free domed retainer nuts; or even better, the rotors and shafts should be designed as an integral construction so that rotor retaining nuts and associated metal-to-metal joints can be eliminated, so that the inside of the front cover can be made completely flat and free of space holes for rotor retainers.

Some types of rotary lobe pumps are traditionally positioned in such a way that draining is impossible without dismantling but the same type of pumps can also be designed for installation in a drainable position. As an example, the inlet and outlet ports of rotary lobe pumps have been arranged traditionally in the horizontal position as this has again been convenient for connecting the pipework. This results in the retention of liquid in the casing up to the level of the inlet and outlet ports. Nowadays there are well-designed hygienic rotary lobe pumps available with the ports arranged in the vertical plane (Figures 26.49b and 26.50) so that it is possible to drain the casing.

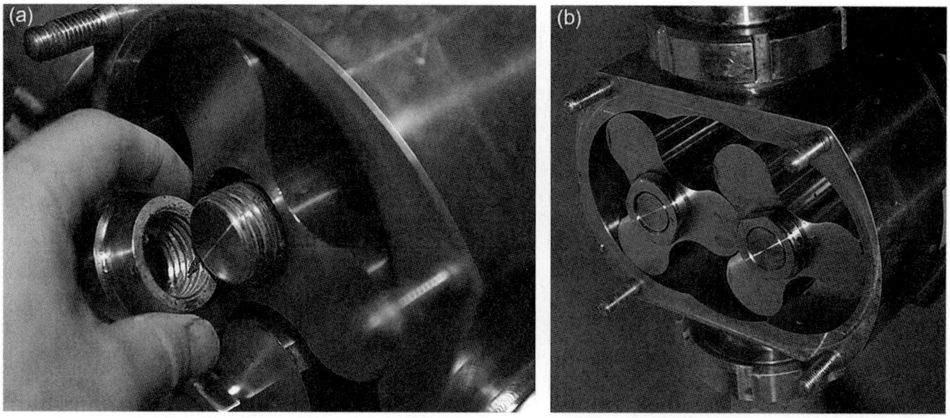

FIGURE 26.49 (a) Ingress and retention of product and/or microorganisms in the threads of the rotor retaining nuts should be avoided by making use of crevice-free domed retainer nuts and by application of O-rings. (b) In an improved version, the rotors and shaft should be designed as an integral construction. With the ports arranged in the vertical plane, it is possible to drain the lobe pump casing. *Courtesy of Burggraaf & Partners, www.burggraaf.cc.*

FIGURE 26.50 Nowadays there are hygienically designed rotary lobe pumps available with the ports arranged in the vertical plane. *SPX Flow Technology – brand Johnson Pump.*

Sensors and Instrumentation

Incorrect mounting of sensors in process lines will result in large dead areas which are unacceptable (Figure 26.51). Instrument branches, which could become a dead leg when not properly installed, should be installed vertically upwards to keep condensates, debris, suspended solid particles, flakes, etc. from collecting in the sensor or from falling into the sensor and the measurement system. However, the length of the dead area must be as short as possible and its cleanability must be demonstrated. For all pipe diameters the length of the upstand should be smaller than its diameter ($l \leq d$).

FIGURE 26.51 The pressure gauge is mounted on too long a tee branch such that an unacceptably large dead area is created. *Courtesy of Huub Lelieveld, personal communication.*

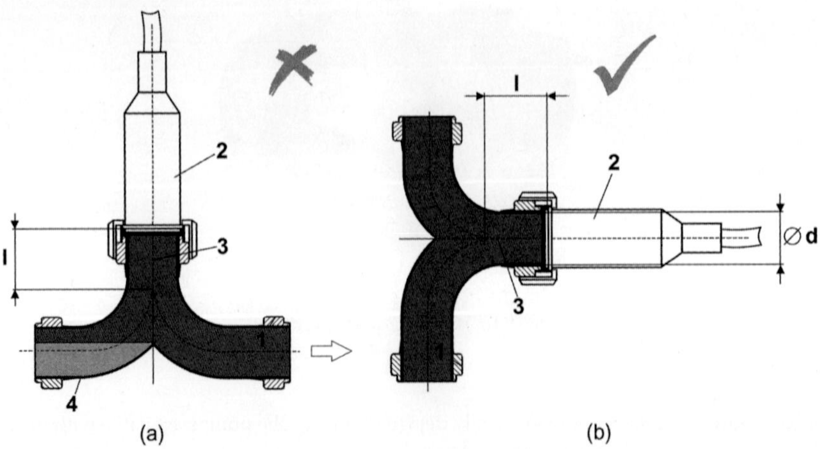

FIGURE 26.52 Incorrect mounting of sensors (2) in process lines (1) may give rise to tees with closed ends (3) that if too long will result in large dead areas. (a) But a swept tee if mounted in a horizontal pipeline may impede drainage (4). (b) Swept tees should be mounted in a vertical pipeline. Dimension l must be as short as possible relative to dimension d, maximum $l = d$ (Lelieveld et al., 2003; Hauser et al., 2007).

It is possible to avoid such dead areas by mounting, e.g., the pressure transmitter on a swept tee (Figure 26.52). However, swept tees must be used with caution, as a swept tee in a horizontal pipeline could hamper draining. Swept tees should be mounted in a vertical pipeline. Dimension l must be as short as possible relative to dimension d, maximum $l = d$. Alternatively, pressure transmitters with tubular membranes, with the same inner diameter as the adjacent pipelines, can be installed in standard spherical valve bodies welded into the

FIGURE 26.53 The pressure transmitter with tubular membranes, having the same inner diameter as the adjacent pipeline, can be integrated into the process, installed by means of a clamp fitting in a standard spherical valve body welded into the piping. The stainless steel diaphragm is sealed by O-rings fitted into grooves such that there is no metal-to-metal joint on the product side. This way of mounting a pressure transmitter provides a dead space-free, flush transition from the process line to the pressure transmitter. *Courtesy of WIKA.*

piping by means of clamp fittings. The stainless steel diaphragms are sealed by O-rings fitted into grooves such that there is no metal-to-metal joint on the product side (Figure 26.53). This way of mounting of pressure transmitters provides a dead space-free, flush transition from the process line to the pressure transmitters.

Temperature measurement is usually based on electronic detection of a change in resistance. The actual temperature sensor elements used integrate either platinum thin film resistors (Pt100, etc.), or employ other sensing elements with a varying electrical resistance against temperature (NTC or PTC resistors). Also semiconductor devices are common. The temperature sensor element itself is covered by a protective sleeve, a highly polished, closed tube typically made of stainless steel. Only one surface of the thermowell has fluid contact, the sensor being installed inside. For these temperature sensors, a close thermal and mechanical contact to the liquid to be measured is needed. Therefore, often a paste with high thermal conductivity is used inside thermowells.

Temperature sensors should not be mounted on too long a tee branch because an unacceptable large dead area is then created. Thermowells with flanged process connection (Figure 26.54) can be integrated into the process, installed by means of clamp fittings in standard spherical valve bodies welded into the piping. The sheath of the probe is welded into one of two blanks which are sealed to the spherical valve body by O-rings fitted into grooves such that there is no metal-to-metal joint on the product side. This way of mounting of a temperature sensor provides a dead space-free, flush transition from the process line to the blank containing the thermowell.

A surface probe with the inner diameter of its pipe the same as that of the adjacent piping is, from a hygiene point of view, an excellent choice. However, the thermowell can also be directly fitted via an orbital welded pocket (Figure 26.55). Attention should be given to the quality of the weld, which must be smooth and continuous. Furthermore, to avoid shadow areas, the direction of the flow must be as indicated.

For temperature measurement in tanks and larger vessels, the thermowells can be continuously welded to the tanks with welding balls or welding collars, after which the inner

FIGURE 26.54 This thermowell, having the same inner diameter as the adjacent pipeline, is integrated into the process, installed by means of a clamp fitting in a standard spherical valve body welded into the piping. The sheath of the probe is welded into one of the two blanks which are sealed to the spherical valve body by O-rings fitted into grooves such that there is no metal-to-metal joint on the product side. This way of mounting a temperature sensor provides a dead space-free, flush transition from the process line to the blank containing the thermowell. *Courtesy of WIKA.*

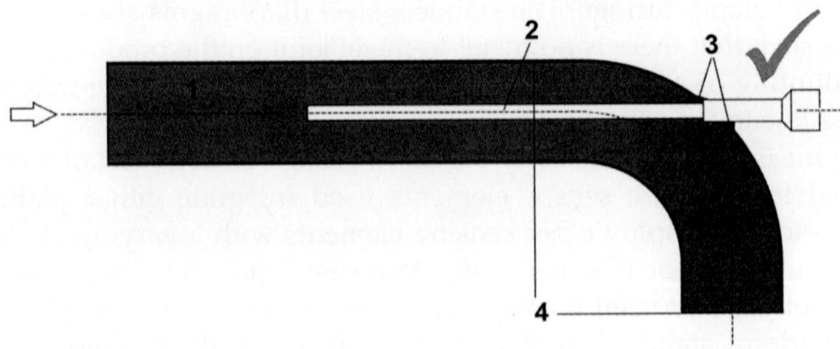

FIGURE 26.55 To avoid dead areas, the pocket for the temperature probe (2) may be welded in the product flow (1) through the pipeline. Attention should be given to the quality of the weld (3), which must be smooth and continuous. Welding of the temperature probe into the bend may be done off-line, after which the bend can be built permanently (by welding) or with dismountable joints into the piping system. In the latter case, the bend section is detachable (4) (Lelieveld et al., 2003; Hauser et al., 2007).

welding seam is polished and passivated after welding. Sensors also can be installed via a hygienic process connection sandwiched (detachable seal joints such as O-rings) into the pipeline (Figure 26.56). The dimensions of the O-ring and the design of the groove to be used for mounting sensors are critical to achieving controlled compression of the seal. The O-ring needs periodic maintenance with an inspection of the O-ring upon dismantling. Used O-rings should not be reinstalled.

Valves are used to change the direction of the flow of product or cleaning solutions (selection of the product routing) to regulate the flow and pressure to protect a process system against overpressure. The cleanability of a valve is largely determined by its internal geometry, the way in which the inlet and outlet connections are made, and the seal between the fluid and the external environment. The seals may be under a static load or dynamic with linear or rotary motion. Valves must have the following properties:

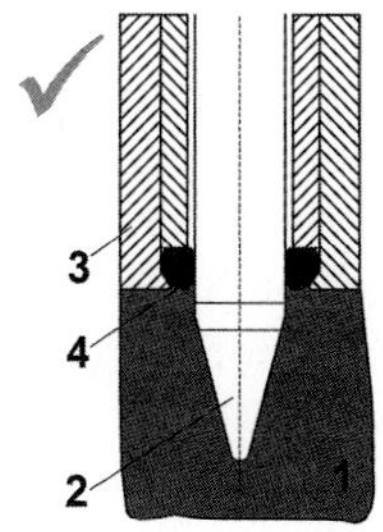

FIGURE 26.56 In the product area (1), a sensor (2) can be installed via a weld-in adapter (3) and a hygienic process connection sandwich into the pipeline. The detachable seal joint (e.g. O-ring, 4) is almost completely enclosed with the surrounding metal protecting the non-product side from the product contact area (Lelieveld et al., 2003; Hauser et al., 2007).

- Be fully drainable, without the need to dismantle;
- Be resistant to wear and easy to maintain;
- Have the minimum number of seals, positively retained and flush with adjacent surfaces;
- Dynamic seals on valve shafts in contact with product must provide an absolute barrier between the product and the environment to prevent microbial recontamination;
- Where unavoidable, springs in contact with product should have minimum surface contact area;
- Allow rapid visual detection of internal leakage.

The following are hygienic requirements for different types of valves (CFCRA, 1997; Schonrock, 2005):

- Diaphragm valves used as back pressure valves need visual detection of leakage (usually there are leakage holes in the valve bonnet), because damage to the diaphragm can result in product leaking through into the non-product side. Such an event may give rise to contamination, and cleaning and disinfection will become nearly impossible. To avoid premature rupture, they should be replaced at regular intervals depending upon the operating conditions. Diaphragm valves must be installed for full drainability.
- Butterfly valves comprise a disc and a rubber seal clamped between the halves of the body providing both a seat for the disc to close on and a seal for the disc spindles. If properly designed, they are hygienic low-cost valves, with the properties of low resistance to flow and their appropriateness to be automated and cleaned in-place. Butterfly valves with a streamlined disk free of external ribs are hygienic. However, product containing fibrous material may build up on the leading edge of the disc, and butterfly valves are suitable as long as the seals are not worn. Seals can wear and break down after a period of time due to the frequent opening and closing of the butterfly valve. Product can also migrate along the shafts due to product pressures in the system. Therefore, butterfly valves should preferably be disassembled for manual cleaning. If butterfly valves are in use, appropriate cleaning and maintenance schedules must be implemented.
- Traditional ball valves are considered unsuitable for process installations that are cleaned in-place. Due to the presence of crevices in their internal construction, the area between

ball, housing and seal face is uncleanable. Food product is transferred in the annular dead space when the valve is operated from its open to its closed position. When the ball valve is then rotated back from its closed to its open position to allow CIP, the food product trapped in the annular space between the sphere and the housing will not be removed by cleaning-in-place. Moreover, ball valves may retain condensate in their internal cavities. Often the design incorporates cavity fillers or encapsulating seals to prevent product flow around the exterior of the ball but product may still find its way under the seat surface and become an area for bacterial growth. Ball valves in existing installations must be disassembled completely for manual cleaning. However, the design and construction of a ball valve are such that it is not easily dismantled for cleaning. Certain ball valves with improved design allow for cleaning-in-place, especially in a half open position. For some applications, connections have been made to the housing so that the annular space may be continuously purged with steam throughout production.

- Plug valves are unsuitable for CIP, because product is carried around the clearance between the plug and the body during the rotation of the plug. Three-way plug cock valves allow 90° changes in flow direction of both food product and cleaning solutions. They have the disadvantage that they neither can be automated or cleaned in-place. However, plug valves can be easily manually cleaned after dismantling, which – due to their simple design – can be done very easily.
- Pressure relief valves are valves where the valve head is lifted off its seat when the product pressure exceeds that at which the valve has been set. Product then may be discharged to drain through the discharge port. To flush the inside of the valve body and the discharge port during cleaning-in-place, the valve must be opened by moving the lever through 90°. The valve body must be installed in a position so that it is fully drainable to the outlet side, and should be mounted on a short tee to avoid a large dead leg in which product will be retained throughout the production.
- Check valves with springs, hinges and flappers should be avoided as they quickly become contaminated and could give rise to cleaning problems. When spring-loaded check valves are used, the coil springs having product contact surfaces shall have at least 2 mm openings between coils, including the ends when the spring is in a free position. Spring-loaded check valves must be fully disassembled for manual cleaning. The use of ball-type check valves is the preferred practice. Springless floating ball check valves have a streamlined internal design which may reduce the potential for material to clog or hang up. Check valves must be installed in a position that allows full drainage of the check valve.
- Tank outlet valves should be installed as close as possible to the product vessel to reduce the dead leg formed by the stub pipe that connects the bottom valve with the vessel. They may be manually or mechanically operated and cleaned depending upon their design features.
- Mixproof valves are an essential part of automated processing, not only separating two different products but also preventing product contamination from cleaning fluids during mechanical cleaning. The valve uses double seats that can be operated independently, separated by a self-draining opening to the atmosphere between the valve seats. The vent space must also be cleanable and avoid a pressure build-up in case of a leak from a seal. The outlet from the vent line must be visible so leakage can be easily detected. A steam or sterile barrier may also be applied in the atmospheric opening (vent) to prevent ingress of microorganisms.

- Linear plug and stem valves may incorporate a lip seal to limit microbial contamination via the reciprocating shaft. This seal is easily cleanable but will not prevent the ingress of microorganisms. A hole is required to detect product leakage when the lip seal wear becomes excessive. Arrangements incorporating an O-ring seal are less hygienic because product can enter the clearance around the stem and become trapped in the O-ring groove from which it cannot be removed by cleaning in-place. For aseptic processing applications where ingress of microorganisms must be prevented, the shaft may be sealed by means of a diaphragm and bellows. In the case of the diaphragm type, the diaphragm must be replaced at regular intervals and a leakage hole must be provided that indicates failure of the diaphragm. With respect to the bellows sealed linear plug and stem valve, the bellows will rupture after a period of service and needs to be replaced at regular intervals. Moreover, if the product contains particulates, there may be a cleaning problem because particulate material may become trapped in the convolutions of the bellows. A steam barrier between the atmospheric and product sides of the valve stem is another method of preventing ingress of microorganisms.

INSTALLATION OF THE FOOD PROCESSING EQUIPMENT IN THE FOOD FACTORY

Clearance with Respect to the Floor, Walls and Adjacent Equipment

There should be enough clearance under the machine to allow for adequate cleaning and inspection to be carried out effectively. With that purpose, the process equipment should be installed as high off the ground as possible. The minimum height should be a function of the depth of the bottom surface above the floor (indicative: 150–300 mm). For large sized equipment, greater distances apply (at least 0.5 m from walls), as it is necessary to be able to walk around such equipment with at least enough room to facilitate cleaning. If the equipment is sealed against the mounting surface, care must be taken to avoid gaps, cracks or crevices where insects or microorganisms can remain/survive after cleaning.

Installation of large equipment (e.g. freezing equipment, meat curing chambers, etc.) on feet is technically not always possible. An alternative is sealing the equipment onto the factory floor. Proper sealing of the perimeter between the equipment and the subfloor must prevent water from accidentally getting into this space. But sealing, especially with silicone, has not always proven to be successful in excluding wet and unhygienic conditions.

Equipment must not be mounted beneath tanks or vessels so that maintenance and cleaning are impeded but must be easily accessible. Increased elevation of tanks and vessels facilitates cleaning and maintenance operations beneath them but water and condensation running down their sides may allow microbial growth and certainly must not fall onto exposed product.

Raised Walkways and Stairs

Raised walkways or stairs (Figure 26.57) over exposed product should be avoided because dirt may be transferred from clothing or footwear onto product lines beneath. The use of

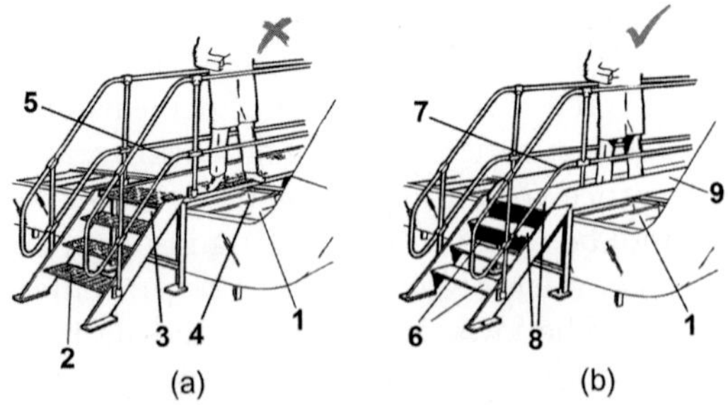

FIGURE 26.57 (a) If not appropriately designed, walkways and stairs over open product (1) may contaminate it. Open-mesh steps (2) that are not enclosed by vertical risers (3), the absence of a cover over the product area (4) and the handrail and its mountings hanging (5) over product area put the open food product at risk. (b) Now, the steps are enclosed (6), the handrail is mounted inside the walkway (7), solid anti-slip steps and floor plates are used (8), and fully welded, continuous kick plates are in place to prevent the open product from getting contaminated (Hauser, 2004b).

covers and hygienically designed walkways should be both considered. The decking of platforms and steps (crossovers on conveyor systems) should be constructed from solid plates containing a raised anti-slip material as deck. The steps can be given a small inclination for improved drainability. Mesh must be avoided to prevent soil from being transferred into the product. Further, fully welded continuous kick plates should be in place, designed as a one-piece construction. Platforms and stairs should have generous radii in the corners of kick plates, etc., to allow cleaning and disinfection. Handrails should not overhang the walkway and must be attached to the inside of the walkway. Risers of staircases must be enclosed and the steps should be constructed of the same anti-slip material as the deck.

HYGIENE PRACTICES DURING MAINTENANCE OPERATIONS IN THE FOOD INDUSTRY

Maintenance and Repair, a Necessary Evil

Physical equipment in any field or in any plant or industry is susceptible to failure through breakdown, deterioration in performance owing to wear and tear over time and to obsolescence due to improvement in technologies. Therefore, machinery should be regularly checked with respect to its performance. Equipment maintenance checks should include an assessment of the equipment's overall condition and integrity (e.g. is it working properly?), the sources of physical contaminants (e.g. damaged, lost or worn parts, rust, loose/flaking paint, broken parts such as needles and blades, loose parts on equipment prone to vibration, polymeric deposits, friction, fatigue, chemical reaction, etc.), the microorganism harborage sites (e.g. worn or frayed hoses, gaskets or belts, porous welds, product contact surfaces). Increase in noise, lubricant consumption, temperature rise or increased leakage is usually the consequence of failure of equipment and its components. Worn parts should be replaced as soon as practical, not only to ensure that production is maintained but also to prevent debris from worn or broken parts entering the product or contaminating the production line.

The operator must ensure equipment used for critical measurements is calibrated and uniquely identifiable. It must be used within its design and capacity (e.g. accuracy, calibration range, conditions of use). Items requiring calibration could include thermometers, temperature recorders, scales, test weights, metal detectors, gas analyzers, pressure or heat sensors, chemical assessment equipment, flow meters, etc.

Scheduled Preventive Maintenance

Scheduled preventive maintenance should be preferred over inefficient "breakdown" maintenance and repetitive repair. No longer does the maintenance department have the luxury of extended periods of available equipment downtime in order to carry out maintenance. Instead the maintenance function is moving toward a more predictive approach. If the failure characteristics of the equipment are known, predictive maintenance can detect the failure well in advance and appropriate actions can be taken in a planned and organized manner. Predictive maintenance makes use of a group of emerging scientific technologies that can be employed to detect potential failures: vibration analysis, thermal imaging, ultrasonic measurement and oil analysis. The maintenance technicians should be skilled to use these diagnostic tools, and they must have detailed knowledge of the operating characteristics of the equipment to make the correct failure diagnosis. By means of a risk analysis, the manufacturer may define which parts of the system are critical and allowing the necessary treatment (which interval, which time point, and which measures). The maintenance schedule should be frequently reviewed during the initial operating period of an installation to establish the optimum maintenance frequency (Jha, 2006).

Proper *a Priori* Design, Installation and Working Practices that May Reduce the Occurrence of Unhygienic Conditions during Maintenance and Repairs

Proper design and installation of the processing equipment and utility services, and common-sense measures create the appropriate conditions to keep up a sanitary process environment during maintenance and repairs (Moerman, 2011b):

- Equipment should be of such a design that the need for physical entry into the system is minimized. Enough space and clearance should be provided so that all equipment parts and components are readily and easily accessible for inspection, maintenance and troubleshooting.
- Mechanical, electrical, pneumatic, hydraulic and electronic components, together with distribution conduits, valves, pumps, pressure reducers, gas cylinders, vacuum sources, compressors, etc., should be relocated to a technical room or technical corridor adjacent to the production room, so that maintenance personnel can access the technical area without special gowning or disruption of the cleanliness of the high hygiene space below.
- Lamps with high light output should be used so that the factory staff can perform inspections of the food processing equipment and the process environment more easily and profoundly, enhancing the detection of grease, leaking oil, failures, maintenance residues, etc. Torches to light dark places with process equipment should be resistant to breakage.

- Maintenance managers and supervisors should implement "Maintenance Best Practice," eliminating the sources of breakdown and contamination that cause downtime, quality holds and lost profits.
- Correct maintenance attitudes must help to ensure that the production area and products are kept free from contamination by undesirable microorganisms, filth, debris or machine parts. Regular audits should be done to verify if the maintenance staff or contractors have adopted the correct hygienic practices during maintenance operations.

Maintenance and Repair Operations according to the Principles of Hygienic Design

Maintenance and repairs should occur according to the principles of proper hygienic design to ensure that safe food is produced once production is resumed. The following recommendations should be followed (Moerman and Degraer, 2003; Den Rustfri Stålindustris Kompetencecenter, 2006b; Moerman, 2011b):

- The construction materials used during maintenance and repair must be compatible with the food product or process aid they contain, and may not introduce contaminants that would present a risk to food safety. Piping and components should be constructed out of the same materials to prevent contact corrosion between dissimilar metals.
- Work in black steel and stainless steel must always be kept separated. Spare parts should be pre-packed in plastic, stored segregated from other non-stainless steel products.
- Stainless steel equipment components should be (shrink) wrapped with plastic film to protect them against corrosion in contact with black steel (particles), and their inlet and outlet connections should be fitted with protective caps to prevent ingress of impurities, insects and small animals (Figure 26.58).
- Prior to use, process equipment and components should be examined for debris, oil or grease; and if necessary should be cleaned.
- The body and internal parts must be handled carefully to ensure that the machined surfaces are not damaged.
- Use as much piping as possible with the same internal and external diameter over the whole factory, in particular to avoid misalignment (missed coincidence between the axes of two coupled pipe components) prior to welding.
- Reassemble piping and equipment components using a new seal, and check for leaks and retighten as necessary.
- All fastening devices should be secured firmly.
- If old insulation containing asbestos has to be removed, all precautions should be taken to avoid the spreading of asbestos fibers in the food processing environment.
- For insulation work, preference should be given to rigid foam rather than fibrous materials that have already proven to be an excellent harborage of dust, insects and rodents. Afterwards, the insulation should be covered with properly sealed cladding of appropriate thickness that resists tear and abrasion.
- When a new cable has to be installed, it should not be supported from a previously installed cable because a hygienically unacceptable entangled cable bundle may be formed. The cables should be fastened individually at a distance no less than 25 mm from each other to allow for proper cleaning.

FIGURE 26.58 Stainless steel equipment components should be (shrink) wrapped with plastic film to protect them against corrosion in contact with black steel (particles), and their inlet and outlet connections should be fitted with protective caps to prevent ingress of impurities, insects and small animals. *Courtesy of Zhejiang Jugang Pipe Co., Ltd.*

- The use of temporary devices, such as tape, wire, string, etc., should be avoided. If strips are the only option, they should preferably be of a stainless steel type that can be detected by means of a metal detector. Alternatively, a plastic strip of a color that is not omnipresent in the food product and food factory could be used. Temporary fixes should be replaced in a timely manner by permanent repairs.
- Always determine the correct installation situation and direction of fluid flow. Install for maximum cleanability and drainability.
- Calibrated equipment that is non-conforming (i.e. broken, expired calibration period) must be identified as non-conforming, and further recalibrated, repaired or replaced.

Personal Hygiene Practices During Maintenance Operations in the Food Industry

Before the onset of maintenance and repair operations, all maintenance workers shall comply with the requirements for personal hygiene appropriate to the area where maintenance and repairs will be executed (Holah and Taylor, 2003; Smith and Keeler, 2007; NZFSA, 2009):

- Both the food manufacturer's own maintenance staff and contractors should follow the food manufacturer's guidance with respect to personal safety and hygiene.
- It is recommended to encourage the maintenance staff or contractors to fill out a health questionnaire before allowing them to enter the food production area. The food manufacturer must restrict access of any person with obvious health problems such as flu, colds, skin lesions, uncovered sores or wounds, etc. All personnel are in fact responsible for reporting any such condition to their supervisor before beginning or continuing work.
- The use of cosmetics, medical substances (ointments, plasters or Band-Aids for wound healing, safety pins) or other chemicals (suntan products, etc.) on the skin are not allowed.

- Eating, drinking, chewing (gum, toothpicks, straws, etc.) and smoking are not allowed during maintenance operations.
- Maintenance staff or contractors are not allowed to enter the food production area with their casual clothes. These should be stored away from the production area. Protective clothing shall be worn, not only to safeguard the person's casual clothes during work but also to protect the food product. In order to avoid contamination of work surfaces, maintenance personnel should wear clean coveralls.
- Maintenance workers who work in a less clean area which has high microbiological activities (raw materials) must change their garments prior to entering a cleaner area where sensitive food products (e.g. finished products) are produced. Hair nets, headbands, caps, bump hat, hard hats, beard nets or other devices must be worn to control hair lost in the food, onto food surfaces and into packaging.
- All piercings, jewelry and watches should be removed.
- Hands should be washed thoroughly, including in between fingers, before entering a food processing area and after eating, drinking, smoking or using the restroom. The use of gloves may be advisable. Gloves are to be maintained in a clean, sanitary and intact condition. Gloves used in less hygienic (raw material) areas of the plant must not be used in more hygienic areas.
- Footwear should be clean. If it is necessary to stand on or over machinery, the process equipment shall be covered to prevent footwear dirt and debris from contaminating the surface. It is also recommended to cover footwear with overshoes just prior to walking on the process equipment.
- Maintenance staff or contractors must remove all unsecured objects, such as pens, pocket notebooks, small screwdrivers, non-attached earplugs, nuts and bolts in shirt pockets, etc., which could fall into the product. These items must be stored in the tool box or the carrier used to bring parts to the work site.

Hygiene Practices during Maintenance Operations in the Food Industry

Recommended Hygiene Practices to be Taken before the Onset of Maintenance and Repair Operations

The following measures and actions will create the appropriate hygienic conditions to execute maintenance and repair without compromising the safety of the food produced with that equipment when production resumes (Jha, 2006; Smith and Keeler, 2007; NZFSA, 2009):

- Work such as drilling or welding will inevitably produce debris and dust. Where possible, production operators should remove food processing equipment from the processing room before repairs are made. Coverings such as tarps or plastic sheeting (polyethylene or equivalent film) can be draped over equipment to reduce contamination.
- Maintenance could be done in a separate room outside the food processing area.
- If entry in process equipment is required, a plastic cover film must be laid down on the bottom of the process equipment.
- Where practical, maintenance tools should be dedicated for use in specific areas of their operation to avoid cross-contamination.

- Tools used for repairs and maintenance must not come in contact with, or compromise the hygienic status of, any product or packaging material. The maintenance tools must be free of rust, peeling paint, niches and threads, and not have wooden handles or knurling soft rubber grips. They should be non-corrosive, easy to clean and inspect, with smooth finish and hard plastic grips, and with fitted heads for equipment longevity. They must be designed in a way that they cannot damage the process equipment.
- The maintenance tools must be clean and used with care so that they cannot be left in the production equipment.
- Maintenance equipment and tools must not transfer microorganisms from a hygienic area into a less hygienic area.
- Ordinary steel wool or steel brushes should never be used on stainless steel surfaces as particles of steel may become embedded in stainless steel surfaces and rust.
- Debris from engineering workshops (such as swarf and other unwanted materials) must be prevented from entering processing or support areas. This is especially important where engineering workshops have access ways (e.g. doorways) that lead into processing or support areas. This may be achieved by keeping doors closed, using swarf mats, boot washes, etc.

Recommended Hygiene Practices during Maintenance and Repair

The following hygiene practices should be followed during maintenance and repair (Smith and Keeler, 2007; NZFSA, 2009):

- During maintenance operations, light sources used to provide the necessary light for proper maintenance and repair should not be placed above open process equipment, or the lamp should be housed in a shatter-resistant fixture to avoid shattered glass falling into the open processing equipment during its maintenance. By using a protective PTFE coating, one may also maintain the integrity of the lamp in the event of breakage. Light sources used during maintenance operations should not contain mercury.
- Opening the distribution system will expose it to particles from the outside environment. The contamination risk can be minimized by using strict specifications on how to conduct activities, such as cutting pipework, and handling pipes and components before the actual installation. Precautions should be taken to prevent the distribution of any contamination residues or mechanical damage residues in the surroundings. Vacuum cleaners should be applied to extract maintenance debris at the place where the maintenance takes place, and drip pans should be used to collect oil, etc. Equipment openings must be protected to maintain the interior of the process equipment and components free from any external contamination.
- Equipment components subjected to maintenance, spare parts and tools should not be placed on the ground or walking surface (e.g. deck), but on a plastic pallet, in a receptacle, a box, a carrier or a trolley provided with a plastic cover. In the food processing area, no wooden pallets should be used to store new or replaced equipment components.
- Whenever parts and tools are stored in the production area, they should preferably be kept in rooms or lockers reserved for such use.
- Equipment components in service should be clearly indicated and/or placed in quarantine.

- Care must be exercised not to lose nuts, bolts, etc. when removing them from machinery. Because small parts easily can be misplaced, loose bolts, nuts, screws, rivets, washers, etc. should be stored in maintenance receptacles.
- Bolts, nuts, screws, etc. of a lower alloy composition may not be left behind on stainless steel because they may induce corrosion.
- Maintenance personnel should not walk on the cladding of insulated piping so as not to damage it.
- Food grade maintenance chemicals (lubricants, heat transfer liquids, etc.) that do not provoke corrosion should be used.
- Personnel must be trained and suitably skilled in the correct access, handling and use of approved maintenance compounds, or have access to documented directions.
- Maintenance products (oils, greases, lubricants, ammonia, glues, chemical products, etc.) should not be left in the food processing environment when maintenance operations have ceased (e.g. during the night, during weekends, during collective holidays, etc.). They shall be stored separately from food products in clearly labeled (identifying the maintenance compound) and closed containers (e.g. bulk supply) in dedicated secure storage facilities.
- Maintenance compounds that are "in use" or for "immediate use" may be stored in processing and support areas, but only in quantities necessary for immediate use. When transferred from their original container (e.g. bulk supply) to a new container (e.g. "in use" or for "immediate use"), the latter must be labeled with the name of the maintenance compound.
- Empty maintenance compound containers must not be reused in a way that food product could be contaminated. All containers/implements should be labeled "for chemical use only."
- Excessive lubricant and grease should be removed to prevent them coming into contact with the product or food contact areas (Figure 26.59).
- Avoid placing dirty, greasy, oily hands on any surface with which the product comes into contact.

Recommended Hygiene Practices after Maintenance and Repair

After maintenance and repair operations, the following practices should be followed (Smith and Keeler, 2007; NZFSA, 2009):

- Maintenance tools or machinery must be removed or returned to storage without delay once maintenance or repair work is completed. Therefore, maintenance technicians must verify that all maintenance tools and components are removed after maintenance and repair to ensure nothing is left where it may enter the product or damaged equipment. An inventory can be made of all tools prior to maintenance.
- Any maintenance waste and other refuse (e.g. packaging materials, broken components, failed parts, dirt, dust, spilled oil) must be regularly removed immediately to a suitable storage area.
- Equipment that could be a source of contamination must be physically isolated from processing lines and product, or removed from processing areas. Damaged or decommissioned equipment that remains in processing areas must be clearly identified as

FIGURE 26.59 Avoid over-lubrication. Excessive lubricant and grease should be removed to prevent them from coming into contact with the product. *Courtesy of John Butts, Land O'Frost.*

such, to ensure that it is not used. Decommissioned equipment may be stored outdoors, but should be placed on a hard standing (e.g. concrete, sealed or paved area) and covered.

- If emergency repairs were required during production, any product that may have been left sitting for long periods of time or become contaminated during repairs should be disposed of.
- The operator must have a procedure to ensure that equipment returned to use (e.g. after repairs and maintenance, recommissioning or having previously been idle) is not a source of contamination to product because of bad maintenance or repair, because repair does not conform to rules of appropriate hygienic design, or because maintenance debris remains.
- Maintenance debris (e.g. abraded particles, swarf) must be flushed from the system after maintenance and repairs.
- When it was necessary to "break in" to the system for maintenance or inspection, equipment should be thoroughly cleaned when maintenance or repairs of any type are performed in a food processing facility. The equipment and area should be cleaned with solutions of detergents and disinfectants in the right concentration, then rinsed and finally dried prior to resuming production.

Evaluation of the Quality of Maintenance Work Done and Record Keeping

Before production resumes, the food manufacturer must evaluate whether finished maintenance operations and repairs meet the expectations with respect to the quality of the maintenance and repairs. In this perspective, the following practices should be followed:

- Equipment must be subjected to a pre-operational check before processing recommences. Are all technical problems solved? Are maintenance and repairs done in a way that the process equipment allows to produce safe food products once production resumes?
- Equipment operating under validated conditions must be revalidated if the repairs and maintenance activity will affect its validated status (e.g. replacing temperature probes/sensors in ovens/freezers).

- Maintenance records or job sheets (including when and how the defect/breakdown was repaired, who conducted the work, who has signed off that it was completed and that appropriate equipment return to use procedures have been followed) must be provided. Comprehensive maintenance records will assist the operator to verify that the repairs and maintenance program are working correctly.

Acknowledgment

We would like to thank our colleagues at the European Hygienic Engineering & Design Group for their permission to use their knowledge and photographic material generated in several EHEDG guidelines, in particular EHEDG documents Nos. 8, 10, 13, 14, 16, 20, 23, 25, 32, 35. I would like to recommend the reader to consult these documents in which the content of this chapter is approached in a more detailed way.

References

A.P.V. Baker, 2001. Hygienic design handbook, second ed. Peterborough, United Kingdom, pp. 53.

ASME, 2009. Bioprocessing Equipment, ASME BPE-2009 International Standard. New York, United States, pp. 213.

Becker, H., Herzberg, F., Schulte, A., Kolossa-Gehring, M., 2011. The carcinogenic potential of nanomaterials, their release from products and options for regulating them. Int. J. Hyg. Environ. Health 214, 231–238.

Baking Industry Sanitation Standards Committee (BISSC) of the American Society of Baking, 2003. American National Standards for Baking Equipment – Sanitation Standard, ANSI/BISSC/Z50.2-2003. Chicago, United States, pp. 35.

Campden Food Preservation Research Association (CFPRA), 1983. Hygienic Design of Food Processing Equipment. In: Dudley, K. (Ed.), Report Prepared by the Working Party on Hygienic Design of the Heat Preserved Foods Panel in Conjunction with the Research Association, Technical Manual No. 7. Chipping Campden, Gloucestershire, United Kingdom, pp. 93.

Campden Food & Chorleywood Research Association (CFCRA), 1997. Hygienic Design of Liquid Handling Equipment for the Food Industry. In: Timperley, A.W. (Ed.), Technical Manual No. 17. Chipping Campden, Gloucestershire, United Kingdom, pp. 1–204.

Den Rustfri Stålindustris Kompetencecenter, 2006a. Conveyors, with a focus on hygiene, Guideline No. 3, version 1.0. Danish Technological Institute, Kolding, Denmark, pp. 60.

Den Rustfri Stålindustris Kompetencecenter, 2006b. Design of piping systems for the food processing industry, Guideline No. 5, version 1.0. Danish Technological Institute, Kolding, Denmark, pp. 29.

FAO/WHO, 2010. FAO/WHO Expert meeting on the application of nanotechnologies in the food and agricultural sectors: potential food safety applications. Meeting Report, Food and Agriculture Organization of the United Nations (FAO), Rome, Italy – World Health Organization (WHO). Geneva, Switzerland, pp. 109.

Hauser, G., Eastwood, C.A., Woodall, D.L., Timperley, D.A., Curiel, G.J., Peschel, P., 1993. Welding stainless steel to meet hygienic requirements. EHEDG Guideline No. 9. first ed. EHEDG subgroup "Design Principles." EHEDG, Frankfurt, Germany, pp. 1–21.

Hauser, G., Curiel, G.J., Bellin, H.-W., Cnossen, H.J., Hofmann, J., Kastelein, J., et al., 2004a. Hygienic Equipment Design Criteria. EHEDG Guideline No. 8. second ed. EHEDG subgroup "Design Principles." EHEDG, Frankfurt, Germany, pp. 1–16.

Hauser, G., Curiel, G.J., Bellin, H.-W., Cnossen, H.J., Hofmann, J., Kastelein, J., et al., 2004b. Hygienic Design of Open Equipment for Processing of Food. EHEDG Guideline No. 13. second ed. EHEDG subgroup "Design Principles." EHEDG, Frankfurt, Germany, pp. 1–24.

Hauser, G., Curiel, G.J., Bellin, H.-W., Cnossen, H.J., Hofmann, J., Kastelein, J., et al., 2007. Hygienic Design of Closed Equipment for Processing of Food. EHEDG Guideline No. 10. second ed. EHEDG subgroup "Design Principles." EHEDG, Frankfurt, Germany, pp. 1–22.

Hauser, G., 2008a. Hygienische Produktionstechnology. Wiley-VCH GmbH & Co. KGaA, Weinheim, Germany, pp. 540.

Hauser, G., 2008b. Hygienegerechte Apparate und Anlagen für die Lebensmittel-, Pharma- und Kosmetikindustrie. Wiley-VCH GmbH & Co. KGaA, Weinheim, Germany, pp. 861.

Holah, J., 2000. Food processing equipment design and cleanability. In: Gormley, R. (Ed.), Flair-flow technical manual 377A/00. National Food Centre, Dublin, pp. 39.

Holah, J.T., Taylor, J., 2003. Personal hygiene. In: Lelieveld, H.L.M., Mostert, M.A., Holah, J., White, B. (Eds.), Hygiene in food processing, first ed. Woodhead Publishing, Cambridge, England, pp. 288–309 (Ch. 15, Book No. 88).

Jha, S.N., 2006. Dairy and Food Processing Plant Maintenance: Theory and Practice, first ed. International Book Distribution Co., Charbagh, Lucknow, India, pp. 140.

Kopitzke, T., Barnickel, M., Gasparetti, M., Merhof, P., Wahlers, J., 2006. Hygienic Welding of Stainless Steel Tubing in the Food Processing Industry. EHEDG Guideline No. 35. first ed. EHEDG subgroup "Welding." EHEDG, Frankfurt, Germany, pp. 1–29.

Lelieveld, H.L.M., Mostert, M.A., Curiel, G.J., 2003. Hygienic equipment design. In: Lelieveld, H.L.M., Mostert, M.A., Holah, J., White, B. (Eds.), Hygiene in food processing, first ed. Woodhead Publishing, Cambridge, England, pp. 122–166 (Ch. 8, Book No. 88).

Moerman, F., Degraer, J., 2003. Guidance for proper sanitary welding of stainless steel. Lecture in Dutch, EHEDG Belgium – the Nederlands Workshop "Welding and surface treatment of stainless steel in the hygienic design of process equipment for the food and pharmaceutical industry," Welding Week 2003 fair, 16 October 2003. Bouwcentrum, Antwerp, Belgium, pp. 27.

Moerman, F., 2011a. Hygienic supply of electricity in food factories. In: Lelieveld, H.L.M., Holah, J. (Eds.), Hygienic Design of Food Factories. Woodhead Publishing, Cambridge, United Kingdom, pp. 369–411 (Ch. 19, Book No. 216).

Moerman, F., 2011b. Hygienic design of piping for food processing support systems in food factories. In: Lelieveld, H.L.M., Holah, J. (Eds.), Hygienic design of food factories. Woodhead Publishing, Cambridge, England, pp. 471–493 (Ch. 21, Book No. 216).

NZFSA, 2009. Code of Practice – Processing of Poultry, Part 2: Good Manufacturing Practice, Chapter 3: Hygiene and Sanitation. New Zealand Food Safety Authority: Wellington, New Zealand, pp. 39.

Partington, E., Besuchet, P., Godwin, A., Hall, K., Holah, J., Holland, P., et al., 2005. Materials of construction for equipment in contact with food. EHEDG Guideline No. 32. EHEDG subgroup "Materials of Construction." EHEDG, Frankfurt, Germany, pp. 1–48.

Plett, E.A., Graßhoff, A., 2006. Cleaning and sanitation. In: Heldman, D.R., Lund, D.B. (Eds.), Handbook of food engineering. CRC Press/Taylor & Francis, Boca Raton, Florida, United States, pp. 929–975 (Ch. 14).

Schonrock, F.T., 2005. Improving the hygienic design of valves. In: Lelieveld, H.L.M., Mostert, M.A., Holah, J. (Eds.), Handbook of hygiene control in the food industry, first ed. Woodhead Publishing, Cambridge, England, pp. 263–272. (Ch. 16, Book No. 116).

Smith, D.A., Keeler, L.J., 2007. Maintenance in a Food Manufacturing Facility – Keeping a Sanitary Process Environment during Repairs. Food Processing for Entrepreneurs Series, NebGuide G1815, University of Nebraska – Lincoln Extension, Institute of Agriculture and Natural Resources, pp. 2.

Stone et al., 2009. Engineered Nanoparticles: Review of Health and Environmental Safety, Final Report of the European 7th Framework Programme. ENRHES, Edinburgh Napier University, United Kingdom, pp. 408.

van der Meulen, B.M.J., 2010. Development of legislation around the world. In: Boisrobert, C.E., Stjepanovic, A., Oh, S., Lelieveld, H.L.M. (Eds.), Ensuring global food safety – exploring global harmonization, first ed. Academic Press, Elsevier, San Diego, California, United States, pp. 5–70 (Ch. 2).

CHAPTER

27

Development of a Comprehensive Cleaning and Sanitizing Program for Food Production Facilities

Robert Ryther

ECOLAB, Eagan, MN, USA

OUTLINE

Food Safety Management.
DOI: http://dx.doi.org/10.1016/B978-0-12-381504-0.00027-5

INTRODUCTION: CLEANING AND SANITIZING OPERATIONS IN FOOD PROCESSING FACILITIES

Effective cleaning and sanitizing, whether automated or manual, requires an understanding of these operations and how to properly validate cleaning and sanitizing procedures to ensure a safe food environment. Primarily, procedures must focus on where soil and microbial contamination can reside in a food processing system. Cleaning and sanitizing procedures that fail to remove soil from food contact surfaces can lead to build-up of microbial agents and be a potential source of contamination of subsequent food production. Prevention of such contamination can be accomplished by food processors focusing on development of efficient Sanitation Standard Operating Procedures (SSOPs) by implementation of HACCP and ISO 22000-type standards (Arvanitoyannis, 2009).

Sanitation Standard Operating Procedure Development

SSOPs are documented procedures for the cleaning and sanitizing of a given piece of equipment or area in a food production facility. A verification procedure ensuring that a cleaning and sanitizing operation was actually completed should be documented in the SSOP. Each SSOP, once written, should be "validated" or proven to actually function as required. An example of a SSOP structure can be found in the United States Department of Agriculture (USDA) Code of

TABLE 27.1 Ten Principles of Sanitary Design

1. Cleanable
2. Made of Compatible Materials
3. Accessible for Inspection, Maintenance, Cleaning and Sanitation
4. No Liquid Collection
5. Hollow Areas Eliminated or Sealed to Avoid Liquid or Soil Collection
6. No Niches (also to Avoid Liquid or Soil Collection)
7. Sanitary Operational Performance (Demonstrated Ability to Execute All Aspects of Sanitation Procedures)
8. Validate Cleaning and Sanitizing Protocols
9. Separate Processes Wherever Possible
10. Meet Personnel Hygiene and Sanitation Requirements

Federal Regulations, Title 9 Part 416 (http://www.usda.gov/wps/portal/usda/usdahome). A more detailed discussion on developing a validation protocol can be found below.

Food Production Facility Cleaning Based on Sanitary Design Principles

Despite the very broad range of food systems and very specialized equipment developed for many food production and packaging operations, cleaning and sanitizing systems should be designed using standard principles. An example of these principles adapted from the American Meat Institute is shown in Table 27.1 (AMI, 2002).

Cleaning methods, manual or automated, cannot overcome poorly designed production equipment and facilities. For example, even sealed hollow areas in support structures or walls have been known to develop cracks and become microbial harborage points.

Hygienic design standards for food processing equipment and facilities, and even "hygienic zoning" concepts (designing facilities to provide "hurdles" to microbial or allergenic contamination transfer from raw food areas to post-processed food areas), should be used for all new equipment, new buildings or new sections to existing structures of food processing facilities to minimize food safety-related recalls (Lelieveld et al., 2003, 2005; Holah and Lelieveld, 2011).

Types of Cleaning and Sanitizing Systems: CIP, COP and Environmental

Cleaning systems for food plants are generally separated into three categories, clean-in-place (CIP), clean-out-of-place (COP) and environmental cleaning:

1. CIP is the automated cleaning of equipment with minimal dismantling of food production equipment prior to the cleaning and sanitizing operation.
2. COP is the removal of food production equipment or portions of the equipment as well as related food production tools to an external area for cleaning, sanitizing and drying prior to reassembly.
3. Environmental surfaces are those external to food processing equipment within the food production facility. Cleaning and sanitizing of all environmental systems is generally accomplished manually but in some cases automated cleaning systems have been utilized.

In all three categories, cleaning is usually followed by sanitizing although sanitizer chemistry and procedures will differ based on regional regulatory requirements. The methods used for cleaning and sanitizing can also vary significantly depending on the food type, food additives and processing temperature used to make the food.

Cleaning Factors

Four factors are generally accepted as being important to ensure effective cleaning and sanitizing. Cleaning time, temperature, chemical activity and mechanical energy all need to be defined for all cleaning and sanitizing programs as described below:

1. **Time to clean and sanitize** is often misunderstood, especially when chemical cleaning is involved. Optimizing the time for a cleaning operation to ensure effective soil dissolution and emulsification (tying up soil in solution to avoid redeposition) is generally a high priority for food producers. Rushing a cleaning operation can result in poor cleaning and the potential for food contamination. Improper use of cleaning chemistry, temperature or mechanical action can result in an inordinately long cleaning time.
2. **Temperature effects on cleaning and sanitizing** will vary depending on soil type and water quality. A rule of thumb is that for every 10°C increase, cleaning chemical activity doubles resulting in fatty soils, sugars and starches and many other types of food soils being more easily removed with increased temperature. High temperatures (>145°F/65°C) will kill microbes but if used properly, lower temperature cleaning and sanitizing programs can be used to achieve effective microbial kill. Increasing cleaning temperature in some cases will precipitate proteins or hardness ions (calcium or magnesium) and create difficult to remove scale deposits.
3. **Chemical activity** is important as cleaning chemistry is built to dissolve soils from the surfaces to be cleaned and emulsify these soils to avoid redeposition. A sanitizing step will kill or inhibit microbial contamination that remains after the cleaning step. Chemical activity is impeded when:
   a. Cleaning or sanitizing solutions do not reach the soils due to lack of solution flow (dead zones).
   b. Chemical concentrations are too low (cannot dissolve soils) or too high (precipitate out or react with soils).
   c. Inappropriate chemical systems are used and are not effective at cleaning or sanitizing the food processing equipment.
4. **Mechanical action** is required to move soils away from a surface. In the absence of manual cleaning, automated cleaning systems generally rely on pressurized water or air to provide mechanical force for soil removal. The need for mechanical force can be minimized if temperature, time and/or chemical activity can be optimized to permit better soil dissolution but some force is always required to move the soil away from a surface to prevent soil redeposition.

CIP BACKGROUND (FIGURE 27.1)

In practice, a standard CIP system will recirculate cleaning solution automatically through enclosed food processing equipment such as tanks, ovens, fryers, conveyors and cooling systems and the associated food transfer piping. Recirculating a cleaning solution

IMAGE COURTESY OF ECOLAB

FIGURE 27.1 CIP cleaning solution and sanitizing solution tanks: this single-use CIP system includes two medium-sized tanks for detergent and rinse/sanitizer, a steam heat exchanger, and programmable logic controllers to automate the cleaning process.

permits measurement and control of temperature, flow rate and chemical concentrations. Such a CIP operation, run for a time period sufficient to ensure all surfaces are free of contaminants, is generally completely automated to:

1. Ensure consistent cleaning results.
2. Minimize labor compared to manual cleaning.
3. Provide electronic documentation that a cleaning program was run as desired (Jennings et al., 1957; Gibson et al., 1999; Schmidt, 1997).

For a CIP system, the mechanical energy is provided by circulation systems. Liquid impingement on surfaces or turbulent flow through piping generally cannot match the mechanical energy provided by manual scraping and scrubbing. To overcome this mechanical energy deficit, the other cleaning factors – temperature, chemical activity and time – must be emphasized. As an automated and enclosed CIP system does not expose operators to chemical mixtures, stronger chemical activity and higher temperatures can be safely used in cleaning and sanitizing. Cleaning times can also be lengthened as labor can be distributed to other tasks while an automated CIP is in operation.

Food CIP systems and beverage CIP systems (such as milk, beer or soft drinks) use essentially the same cleaning and sanitizing solution transfer and spray technology but can vary greatly in the required chemical strength, water temperature and cleaning time for removing light versus heavy, aged and/or burnt on food soils. Use of automated cleaning and sanitizing CIP systems has advantages over manual cleaning in the following ways (Lowry, 2010):

1. Reduces the amount of time and labor spent on sanitation operations (increasing food production run times).

2. Decreases the impact of sanitation operations on water consumption, energy utilization and wastewater processing.
3. Reduces overall wear on process equipment due to decreased manual cleaning.

CIP – Line Circuit Cleaning

The line circuit portion of CIP cleaning is focused on ensuring proper turbulent flow at rates to provide mechanical "scrubbing" by the cleaning solution. Critical to the success of cleaning line circuits is removal of all "dead zones" where cleaning solution cannot flow and residual food product can build up and result in microbial contamination. CIP circuits can often be extremely complicated, especially when using a single CIP system to clean numerous circuits (sometimes simultaneously). Careful design of each CIP piping circuit, especially when new food processes are added to existing equipment is suggested to ensure elimination of these dead zones.

Optimizing time, temperature and chemical concentration is important after ensuring optimum mechanical action based on solution flow rates. Flow rates will need to increase with increasing pipe diameter (see Table 27.2). Care must be taken where line circuits contain multiple pipe sizes as pressure/velocity drops will occur going from smaller to larger diameter piping and inadequate cleaning of the larger piping can result. It is common for tank and lines to be cleaned in the same CIP operation and optimizing both types of CIP programs in conjunction is required.

CIP – Tank Circuit Cleaning

The tank circuit CIP program relies on a spray device or spray ball to achieve mechanical action. At a minimum a tank cleaning spray device must be designed to ensure all tank surfaces are reached by the cleaning solution. Fixtures within tanks such as mixing blades or drain valves on tank surfaces can block spray from reaching soiled surfaces. Multiple overlapping spray devices are often required to overcome such blockages. Flow rates out of a

TABLE 27.2 Flow Rates in Feet/Second (ft/sec) and Gallons/Minute (GPM) for Different CIP Line Sizes

Line Size*	Desired Velocity	Minimum Flow Rate	Drain Capacity**
1″	5 ft/sec	15 GPM	22 GPM
1.5″	5 ft/sec	24 GPM	40 GPM
2″	5 ft/sec	43 GPM	75 GPM
2.5″	5 ft/sec	69 GPM	115 GPM
3″	8 ft/sec	163 GPM	190 GPM
4″	8 ft/sec	288 GPM	350 GPM

**- Assumes Standard Sanitary Pipe.*
***- Maximum drainage through the pipe.*
(Courtesy of Ecolab Inc.)

spray device typically are on the order of 3 gallons per minute per foot of circumference of a cylindrical tank.

Single Versus Multi-use CIP Designs

CIP circuits can be designed to be (1) single use with cleaning solution dumped directly to drain after completion of the CIP cycle or (2) reuse or multi-use systems. Multi-use CIP systems are often designed to recover final rinse water for use to make up subsequent cleaning solutions. Some or all of the cleaning solution itself can also be saved and reused to minimize chemical usage. In all CIP designs, any final rinse or sanitizing rinse would not be reused in that function but fresh final steps would generally be required by food processing regulations in most regions.

Single use systems, where all cleaning, sanitizing and rinse solutions are used once, are simple to design and result in the highest level of sanitation. These single use systems would be required for food production equipment having very high soil loads or allergenic material and thereby avoid the potential for cross-contamination. Facilities with very limited space for additional equipment would also benefit from such a system.

Recovery of rinse water is a popular choice for CIP as there is little chance of soil redeposition resulting in system contamination. Multi-use systems where the cleaning chemicals are reused require a fairly complicated CIP design and a high level of understanding of the CIP circuit to avoid soil redeposition. There are many of these systems in use, some of which have automated make-up water and cleaning chemical additions to achieve equilibrium in a cleaning solution and can run weeks or months without the need to dump the cleaning solution.

COP CLEANING

Cleaning out-of-place (COP) is the cleaning of removable parts of food processing equipment after disassembly or any ancillary food production tools. Typically COP systems are open tanks where a given cleaning solution can be heated and recirculated. As with CIP systems, automation is possible with a recirculation step (permitting the monitoring of solution temperature, chemical concentration and flow rate) so that "push button" COP systems are common. The use of numerous distribution headers in COP tanks are used to create turbulence to aid in soil removal via mechanical action.

COP cleaning and sanitizing programs require:

1. Appropriate tank design:
   a. Sized for application.
   b. That can contain shelves or hangers for parts to maximize number of parts loaded while still maintaining separation between parts to ensure full access to cleaning solutions.
   c. With a recirculation system and associated headers sized to ensure tank turbulence.
2. Cleaning solution chemistry, temperature and cleaning time balanced to ensure full cleaning of the toughest to remove soils.
3. A rinse step to ensure removal of residual soil and cleaning solution.

4. A sanitizer step to kill off any residual microbial contamination.
5. Proper parts storage program to ensure complete drying and avoidance of cross-contamination prior to reassembly.

ENVIRONMENTAL CLEANING

Cleaning and sanitizing environmental surfaces in a food production facility is a critical part of a full food safety program. Poor design of a cleaning program for environmental surfaces cleaning (sometimes termed open plant cleaning) can leave microbial contamination that can migrate into food product (Samelis et al., 2001).

Environmental cleaning in a food processing environment is the cleaning of equipment's external surfaces, walls, floors, ceilings, elevated walkways, drains, piping and conduit in addition to ancillary equipment (such as motors, electrical boxes, etc.) that generally cannot be cleaned by CIP or COP methods.

Typically, cleaning environmental areas of a food production facility is done manually by first removing food debris followed by wet cleaning and sanitizing steps. Emphasis is on removing as much soil as possible before a wet cleaning operation to limit the biological load on the wastewater treatment system.

Environmental cleaning programs for food processing facilities should remove as much water as possible after completion of cleaning and final rinse steps and return to as dry a state as possible prior to resuming food production. Focusing even wet cleaned areas on maintaining as dry a state as possible will help limit microbial growth in the environment, thereby reducing the potential for microbial cross-contamination into food product.

Systems used to support environmental cleaning include the following.

Foaming or Gelling Systems

Cleaning and sanitizing with foam or gel-based chemistry increases the dwell time of chemical on the surface to help dissolve soils for cleaners or provide increased microbial kill efficacy for sanitizers. Foam generators mix a chemical solution with air to create a foam or gel. This solution is then sprayed under pressure onto the surface to be cleaned. Often entire rooms and all accessible surfaces of the food production equipment are foamed (with chemically sensitive equipment being wrapped prior to foaming).

High Pressure Cleaning Systems

High pressure air, water, steam or particle blasting systems can be used for cleaning the exterior parts of equipment, floors and some building surfaces when mechanical action is required for soil removal. Soil types dictate which type of mechanical action will be most effective. For example, particle blasting works best on brittle food soils while hot steam can clean by liquefying soils. Steam can also be used to sanitize surfaces (care must be taken that all surfaces are heated to an appropriate temperature for a reproducible time period to ensure the desired microbial kill).

Pressurized air or steam injection systems generally operate with nozzle pressures between 60 and 170 psi. Cleaning effectiveness is dependent largely upon the force of the cleaning system against the surface and will be a function of both operating pressure and nozzle design. It must be emphasized that high pressure air or water systems (especially centralized systems) must be free of microbial contamination. Additionally, care must be taken that these high pressure cleaning systems do not scatter soil (which may contain microbial or allergenic contamination) into adjacent food production systems.

Ancillary Cleaning Equipment

Programs should be developed for cleaning ancillary food production tools as well as ensuring that cleaning tools themselves have appropriate cleaning procedures to avoid the potential for microbial or allergenic cross-contamination. Strategies often involve using color-coded cleaning and waste removal equipment specific to either food contact or non-food contact surfaces. Procedures to clean, dry and store equipment after a cleaning operation is completed and isolation of equipment are used in areas where raw food products are stored separately from areas where finished food products are stored.

Master Sanitation Schedule

Development and implementation of a master sanitation schedule is highly recommended for all food production facilities. A facility's environmental areas that are not normally cleaned on a daily basis should have a strict cleaning schedule at some regular time frequency to ensure these areas are not overlooked and become sources of food product contamination. Not only cleaning activities but maintenance of equipment, sealing of roofs, wall and floor cracks where water may enter a facility as well as identification and elimination of any microbial harborage points or pest activity is recommended on a regularly scheduled basis.

CLEANING OF ALLERGENS

Food allergies affect as many as 6% of young children and 3 to 4% of adults (Sicherer and Sampson, 2006). Exposure of some individuals to very small amounts of allergenic proteins can be life-threatening. Moreover, allergen contamination makes up a significant portion of all food recalls. Control of allergens in any food facility that produces both allergen- and non-allergen-containing foods must be accompanied by a stringent allergen control program. Potential for cross-contamination of allergens into allergen-free foods can include: (1) mislabeled raw material entering the food process facility, (2) mislabeling of materials by facility staff, (3) cross-contamination from allergen-containing food remaining on food production equipment or tools and (4) mislabeling of a final product containing allergens. There are many manufacturers who have separate production equipment or even separate manufacturing facilities to avoid the potential for such cross-contamination to occur (Vierk et al., 2002).

CLEANING OF DRY OR LOW MOISTURE FOODS

Low moisture or dry food production areas are usually cleaned without water for the purpose of minimizing growth of microbial pathogens. Some of these low moisture foods include milled grains, bakery goods, cereals, chocolate, dry dairy, nuts, spices and fried or baked chips. Food facilities that have dry processing areas (in some cases the entire production facility will be dry) have been historically more concerned about pest issues than microbial contamination.

Unfortunately, recent food recalls involving food contaminated in dry food production facilities have increased the need by these food producers to identify ways to provide sanitation breaks in their facilities. *Salmonella* is the main concern in these environments due to its persistence on dry foods and in food manufacturing facilities. *E. coli* O157:H7 and *Listeria* also remain of major concern in these areas.

Developing dry cleaning and sanitizing protocols using little or no water becomes a challenge as there are few government regulations addressing such cleaning and sanitizing methods. For most dry food facilities, dry raw material delivery, storage and internal facility transport systems are rarely cleaned completely and almost never sanitized (relying on dry environment to inhibit any microbial growth).

Some examples of such dry cleaning include traditional sweeping, scraping and vacuuming coupled with sanitizing with quick drying alcohol solution applied directly to surfaces or using sanitizing wipes. Sanitizing with steam and heat also provides low water alternatives to water-based sanitizers. More unique methods include "rinsing" with the new lot of food product itself to remove any traces of older product. Non-allergenic food material such as rice or salt might also be used to rinse out systems and then tested to demonstrate removal of microbial contamination. The final "rinse" would consist of food product from a new production lot (to be discarded until all traces of a rinse raw material are removed).

CLEANING CHEMISTRY

Chemical cleaners are required where dissolving and emulsifying soils is more efficient at cleaning food processing equipment than manual cleaning would be. Food soils of concern to most food processing facilities will vary depending on food type and additives, temperature and type of food processing as well the condition of the water used in food production and cleaning. To effectively clean these soils requires an understanding of the functions of different chemical components in a cleaning system. An overview of the properties of chemicals used in cleaning procedures is provided below. A more comprehensive review of these technologies can be found in the work by Stenga (Stenga, 2010).

The pH of a cleaning solution is defined by the relative level of hydrogen cation (acidity) or hydroxyl anion (alkalinity) in a solution and is a key factor in the cleaning ability for most food soils. The chemical structures of food soils such as fats and oils, proteins, sugars and starches as well as minerals all have some ionic features under some or all conditions. A cleaning solution must be built from components that maximize the breakdown of soil residues on food facility equipment and associated environmental surfaces while minimizing the amount of chemical used for cleaning. The goal of an optimized chemical cleaning system is to use the cleaning solution to wet the soil, dislodge it from a surface to be cleaned and then emulsify it (hold the soil in solution) so it will not redeposit.

When choosing a chemical cleaning system for a given food soil the cleaning time, cost, compatibility with equipment, safety for facility operators as well as regulatory requirements for wastewater discharge and environmental sustainability are all factors that must be considered to ensure an optimum system.

Personal Protective Equipment and Safety Programs for Chemical Usage

The use of personal protective equipment (PPE) when using any chemical cleaning or sanitizing system is highly recommended. A proper sanitation program should include operator training on safety issues at the core of its training program. Continuous monitoring of compliance with all safety procedures for all employees and a safety review of all new procedures should be a part of any new SSOP development program. All chemical materials brought into a food production facility should have an easily accessible material safety data sheet (MSDS) available. These MSDS documents describe how to safely handle and dispose of the chemical as well as the appropriate PPE to be worn by any employee or contractor using the chemical.

Environmental Issues with Chemical Cleaners

Understanding regional and local regulations regarding use and discharge of chemical cleaning and sanitizing systems is important for proper operation and long-term viability of a food production facility. As regulations are constantly changing, care needs to be taken by a business to be aware of regulatory changes that might affect business practices.

Typically, a facility will be most concerned with local wastewater treatment facility requirements as chemical discharge can have almost immediate effects on a water treatment program, and discharge limits on phosphorus or nitrogen, for example, or sanitizing chemistry discharges that wipe out a bacterial culture used to break down organic waste at a water treatment facility can result in significant fines for a food production facility. Use of non-compliant chemical systems can also result in economic issues for a facility as well as environmental damage from such chemical systems.

Alkalinity

Alkaline cleaners use alkalinity to break down and solubilize fats, proteins and starches. Alkaline salts provide hydroxyl anions to an aqueous cleaning solution and can be strong alkaline (high pH) or buffered to alkalinities down to a pH below 10 (where neutral chemical components start to provide the major cleaning effects).

Sodium or potassium hydroxides are the common caustic additives that provide high strength, high pH cleaning effects. These solutions will turn fats into soaps (saponification) and are commonly used for dissolution of protein soils, especially proteins that have been denatured, precipitated or polymerized by heat. If used alone, without added buffering or threshold sequestrant components, the pH of alkaline solutions can change rapidly due to dilution by rinsing. Such pH swings for an alkaline solution can result in extremely rapid soil redeposition or scale depositing onto food production equipment.

There are many other alkaline sources effective at cleaning including alcohol amines such as monoethanolamine (MEA), silicates, used for safe cleaning of soft metals and phosphates and polyphosphates that act simultaneously as an alkalinity source, a sequestrant and anti-redeposition agent.

Reuse of alkaline solutions is common in order to save water and chemical costs in CIP systems as described above.

Acidity

Acidic cleaning products provide hydrogen cations to a cleaning solution primarily for the dissolution of inorganic scale deposits. Scale forms from precipitating metal salts such as calcium, magnesium and other multivalent metal ions and can be a problem especially in heated solutions, as calcium and magnesium salts become less soluble with increasing temperature.

Scale can result from minerals precipitating from foods (such as calcium phosphate or milk stone and calcium oxalate or beer stone), from hard water used in food production or from the cleaning step itself. Care needs to be taken in all these processes to ensure scale build-up is carefully controlled as some scales, if they become too thick, may be extremely difficult to remove. Silicate scale from food or plant water sources, for example, can be especially difficult to remove if not controlled and in some cases require the quite hazardous hydrofluoric acid to dissolve.

Certain anions interacting with hardness ions can induce scale precipitation. Stearate, oleate and laurate from fatty acids, oxalate from vegetable sources and inorganic anions such as phosphate, sulfate, fluoride and carbonate from food or water sources can induce scale under certain conditions and must be controlled to ensure equipment can be properly maintained.

It is important to note that strong acid washes can still be responsible for scaling food processing equipment. This happens when using hard water for rinsing the acid. When the acid concentration in the rinse falls below a level where the scale ion concentration is soluble, these ions will precipitate. In order to avoid such scale precipitation during an acid wash, threshold inhibitors are often added to acid cleaning products.

Acids generally are poor detergents as they tend to make soils hydrophobic and, therefore, difficult to wet and dissolve. Surfactants are often added to acidic cleaning formulation to provide some wetting ability. Surfactants can also be formulated into acid cleaners to provide visible foam for environmental cleaning.

A common cleaning practice involves washing with an alkaline cleaning solution followed by an acidic cleaning solution to ensure removal of organic soils and inorganic scale, respectively.

Strong acid cleaners (phosphoric, nitric and sulfuric blends) are the most cost-effective cleaners but have to be used carefully as they may damage many soft metals and some plastic or rubber materials. Organic acid-based cleaners (such as citric, oxalic, lactic, etc.) are more expensive but are more ecologically sound (by not contributing phosphorus or nitrogen to wastewater as well as being biodegradable) and they are generally safer to use on soft metals.

Chelants and Sequestrants

Chelants or sequestrants are chemical compounds designed to bind dissolved metal salts. Sequestrants are added to cleaning formulation for two basic functions:

1. For dissolving scale on equipment surfaces; and
2. To keep hardness ions such as calcium and magnesium from precipitating out of solution.

While acids act to dissolve scale as a separate step, as discussed above, neutral and alkaline cleaning formulations often need to be built with sequestrant components for one-step cleaning applications as well as to ensure hardness ions do not precipitate during cleaning.

Some sequestrants can work by ensuring each hardness ion is complexed (or kept bound) by a sequestrant molecule (stoichiometric sequestrant). Stoichiometric sequestrants are required for many cleaning systems, especially when cleaning solutions are reused, ensuring long-term stability of the solution against the precipitation of scale.

Other sequestrants can also work as threshold agents. A threshold sequestrant molecule effectively impedes the growth of a scale crystal in solution so that a small number of sequestrant molecules can stabilize a large number of scale forming ions. A threshold agent is added to alkaline or acid formulations to guard against scale formulation during a rinsing operation where hard water is used or where the cleaning solution has solubilized a great deal of scale during cleaning.

The strongest sequestrants, such as ethylene diamine tetraacetic acid (EDTA), or organic acids, such as gluconate or citrate, do not act as threshold agents. Tripolyphosphate, phosphonates and carboxylate polymers can act as both stoichiometric and threshold agents and can be used individually or with strong sequestrants depending on the specific cleaning requirements.

Surfactant and Solvent Systems

Surface modifying additives to cleaning formulations are also generally known as surfactants. Solvents used in food cleaning systems are essentially low molecular weight surfactants and are effectively used for the same purpose: to dissolve food soils and emulsify them into the cleaning solution. All surfactant molecules have a "water-loving" hydrophilic portion that is water soluble and solubilize ionic soils. They also contain an oily hydrophobic portion that will dissolve oil- or fat-based soils. Surfactants are added to cleaning formulations to provide:

1. Soil wetting or solvating capabilities that assist the delivery of other cleaning components (such as acidity, alkalinity or sequestrants) into the soil.
2. Emulsification of oils in order to keep oily soils from redepositing during cleaning.
3. Modification of cleaning solutions by introduction of foaming or defoaming features as required by different cleaning methods (see CIP and COP discussions above).

Proper use of surfactants can significantly reduce the requirements for other cleaning components in a formulation. Often very small amounts of surfactants can have a very large effect on a cleaning operation as these molecules only need to interact with soils on the interface or surface of the soil to be effective and not dissolve the entire soil. Solvents generally require much higher concentrations than surfactants to have an effect on cleaning but can be much better overall cleaning solutions at those concentrations.

Caustic-Oxidizer

Oxidizing agents are generally used with alkaline cleaning solutions to break down soils, such as protein, much more effectively than without the oxidants. Chlorine and hydrogen peroxide are the most common oxidizer additives in alkaline cleaners and act by breaking apart and solubilizing food protein molecules. Protein soils, especially heat-deposited proteins, can be extremely difficult to remove and can build up in food production equipment over time if the equipment is not properly cleaned and leading to harborage sites for microbial contamination.

Oxidizers such as nitric acid and hydrogen peroxide are also commonly used for stainless steel passivation which keeps equipment surfaces from oxidation damage (ASTM A380-06).

Enzymes

Enzymes in biological systems are extremely efficient at breaking down biological molecules at relatively low body temperatures. Outside of biological systems, enzyme-based cleaners are designed to break down very specific soils: proteases for proteins, lipases for fats and amylases for starches, for example. These cleaners are generally limited to lower temperature cleaning and are designed specifically for a given soil type. Enzymes are easily denatured by the wrong pH range, high temperature or various contaminants. Enzymes are especially useful in cleaning equipment such as membranes that are sensitive to temperature and many cleaning chemicals.

Cleaner-Sanitizers

In applications and/or regions where a cleaner-sanitizer single step meets regulatory requirements, such an operation can be used to clean soils as well as remove microbial contamination. In some cases a cleaner-sanitizer can be formulated to provide both functions or a wash step can be followed by heat sanitizing.

COMMON CLEANING PROBLEMS IN FOOD PROCESS ENVIRONMENTS

Issues at food production facilities involving difficult to remove soils vary significantly depending on the food being produced. Proteins, starches, fats/oils and metal ion scales can all be difficult to remove depending on the food production method, processing temperature, time between cleanings and accessibility to food soils. Effective cleaning methods depend in part on whether the equipment was built to be cleaned (which is not always the case).

Protein Cleaning Problems

Protein soils can form difficult to remove soils especially if left behind after other components (such as fats, starches and inorganic scale) are removed by the cleaning operation. Usually, protein soils from such foods as milk, eggs or meats that undergo heat treatment or are exposed for a significant amount of time to air oxidation will denature, precipitate or polymerize to form soils that can build up over time and create harborage sites for bacteria and associated biofilms.

High cleaning temperatures and poor choice of chemistry (inappropriate alkaline, oxidizer, chelant or acid levels) can precipitate or denature a protein soil and make it more difficult to remove. Each protein source will have different optimized cleaning requirements due to the wide range of protein types found in food products.

Protein cleaning is usually best using alkaline oxidizing chemistry at temperatures that support cleaning without precipitating the protein from the cleaning solution.

Fats and Oils

Oily soil from fats or vegetable oils generally will clean better with increasing temperature and use of appropriate surfactant emulsifying cleaning solutions. If unsaturated oils polymerize due to heating in the food production process, these oils can create varnish-like coatings on food equipment surfaces. Dissolving these coatings can be very difficult and can require fairly concentrated chemistries or high levels of manual labor to remove.

Cleaning Starches and Polysaccharides

Starches from food sources, gums, pectins and other thickeners are usually water soluble but when heated can dry out and become very difficult to rewet. These polymeric soils can also be held together in some cases by inorganic mineral scale. Built alkaline cleaners with oxidizers (hydrogen peroxide or chlorine) are often required to remove these soils effectively and, in some cases, pretreatment with acid followed by alkaline cleaning may result in more efficient cleaning.

Scale Removal Problems

Different scale types might require a specialized cleaning program to ensure complete removal of food production equipment scale. Examples of some important food-produced scales include:

1. Calcium oxalate or beer stone is formed from foods containing tannins, seeds, fruits or vegetables. Tea, beer, tomatoes and especially vegetables treated in a blanching process can form these difficult to remove calcium oxalate scales. Removal methods include

alkaline peroxide-EDTA-based cleaners followed by nitric acid washes to dissolve. (EDTA is often the best chelant system but in some regions must be replaced by less optimal chelants due to regulatory restrictions.)

2. Calcium phosphate or milk stone is formed from high phosphorus-containing foods such as milk. Alkaline or alkaline peroxide, EDTA or other chelants can be used as cleaners followed by a separate nitric acid wash would be a typical cleaning system.
3. Calcium soaps can form by reaction of fats with calcium from milk, meats or other processed fat or seed oil-containing foods. Acid washes of these hydrophobic organic scales are much less effective than with inorganic scales. Strong chelant systems such as EDTA and alkaline peroxide-type cleaners will be required for cleaning these soils.

Cleaning Sensitive Equipment

The food industry has a broad range of customized and unique production and packaging equipment systems all of which should be designed with sanitary principles in mind. Unfortunately, there are many applications that require complicated equipment designs that can be extremely difficult to properly clean or contain materials of construction that can be damaged by conventional cleaning chemistries.

Corrosion issues on metals and plastic degradation can occur as a result of strong acid or alkaline chemistries or other high salt (especially chloride-containing) solutions. Surfactants or solvents can damage specialized equipment, especially plastic parts or rubber gaskets. Membrane separation systems, such as reverse osmosis (RO) or ultra-filtration (UF) membranes used, for example, for process water treatment and concentrating whey and milk in the dairy industry, are very effective production systems but can be easily damaged by many standard cleaning and sanitizing chemistries.

It is highly recommended that, prior to building or purchasing new equipment or systems, facility engineering work with quality groups to evaluate the special cleaning needs required for those devices is carried out. Many facility problems can be avoided by designing equipment so it can be effectively cleaned and does not contain materials that are incompatible with the cleaning process.

SANITIZING CHEMISTRY

The US EPA defines antimicrobial agents as substances or mixtures of substances used to destroy or suppress the growth of harmful microorganisms whether bacteria, viruses or fungi on inanimate objects and surfaces. Definitions of each type of antimicrobial agent are described in the EPA Fact Sheet on their website (EPA Fact Sheet):

Sterilizers (also sporicides) "will destroy or eliminate all forms of microbial life including fungi, viruses, and all forms of bacteria and their spores."
Disinfectants are used "on hard inanimate surfaces and objects to destroy or irreversibly inactivate infectious fungi and bacteria but not necessarily their spores" (these require a final rinse if used on food contact surfaces).

Sanitizers will "reduce, but not necessarily eliminate, microorganisms from the inanimate environment to levels considered safe as determined by public health codes or regulations." Sanitizers used in food processing plants in the US generally are non-rinse agents, safe for food contact surfaces when used according to the product label requirements.

It is important to understand that for different food processing facilities, government regulations will differ based on region. An operator using sanitizers and disinfectants for direct and indirect food contact surfaces needs to follow regulations for both sanitizer and disinfectant products. For example, in the USA, any antimicrobial or chemical sanitizer used on food or for treating food contact or other surfaces in a food production facility must be registered at the EPA. The US Food and Drug Administration (FDA) regulate the use of antimicrobial agents used on food or food contact surfaces. All approved no-rinse food contact surface formulations and associated usage levels used in the USA are listed in the FDA 21 CFR 178.1010 Code of Federal Regulations. In some regions applications may require a final rinse for all sanitizer chemistry applications.

Sanitizing Systems

Chemical sanitizers fall into a number of categories based on whether they are oxidative or non-oxidative. These chemical sanitizers can be delivered as liquids, gases or in vapor/mist form depending on the application. A comparison of the different chemical sanitizers for stability, foaming, corrosivity, pH stability and efficacy against microorganisms can be found in a review by Richter and Cords (Richter and Cords, 2001).

For food processing operations, a sanitizing step in a cleaning program can be accomplished using chemical, non-chemical and combinations thereof to achieve the level of sanitization required for a given process.

A brief summary of these systems is described below.

Thermal Sanitizing

While not a chemical system, thermal sanitizing is used in many operations and often in conjunction with chemical sanitizers. A thermal sanitation process must follow some basic principles which, if not followed, can result not only in the potential for a food contamination incident but also in a thermally tolerant microbial biofilm that affects all food production through a contaminated system.

Bacterial contamination can generally be eliminated by a pasteurization step (>160°F or 70°C) for 15 seconds. Spores generally will require steam to induce the opening of the spore and temperatures on the order of >250°F or 122°C for several minutes to have acceptable kill levels (5 log or 99.999% reduction of the spores). Depending on the organism of concern and the level of microbial kill required for a given process, heat sanitizing functions need to ensure correct time, temperature as well as humidity levels (if sanitized with hot air rather than hot water) to obtain acceptable microbial kill. Equipment where there exists the potential that not all surfaces will be heated equally can leave cooler portions of the system improperly sanitized. Such areas could be dead head pipes with minimum flow for an aqueous thermal process or thicker or more intricate internal structures with high heat capacity if steam or hot air is used to sanitize.

Oxidative Sanitizers

Commonly used oxidative sanitizers include chlorine, iodine, hydrogen peroxide, peroxycarboxylic acids and chlorine dioxide.

Chlorine

Chlorine is one of the earliest sanitizers used in food processing plants and is used in sanitizing solutions in the form of aqueous sodium hypochlorite or hypochlorous acid. Chlorine can be purchased as a gas, in a stable liquid form (bleach), as a chlorine producing solid (calcium or magnesium hypochlorite as well as in the trichloroisocyanuric acid form) or produced from the chloride salt on site in an electrolytic cell (White, 2010).

As a sanitizer chlorine has the advantage of being generally inexpensive and broadly effective against all types of microorganisms and is thought to act primarily through disruption of cellular proteins and enzyme activity. It is most desirable to sanitize with chlorine in the pH 6–7.5 range as it becomes ineffective as a sanitizer above pH 9 and will evolve dangerous chlorine gas when combined with acid. Disadvantages include loss of activity in the presence of an organic load as well as being highly corrosive to stainless steel and elastomers (especially when not completely rinsed and permitted to dry onto these surfaces). Chlorine also will react with organics to form carcinogenic trihalomethanes (THM) and is restricted for use in some regions.

Iodine

Iodine-based sanitizers (iodine stabilized with surfactant iodophors), like chlorine, have broad spectrum kill of microorganisms and are effective even against difficult to kill bacterial spores. Iodine sanitizers can be used at acidic pHs which is important for scale removal function in dairy and brewery applications, for example. They are effective at lower concentrations than chlorine and can generally work at a higher organic load. Iodine can be somewhat expensive in practical use and it cannot be used hot as iodine will not remain in aqueous solution above 115°F (45°C). Corrosion can be a problem, especially at higher temperature, and staining of equipment and some starchy foods are common complaints in the use of iodine compounds (Gottardi, 2001).

Chlorine Dioxide

Chlorine dioxide is a gas at room temperature with only slight solubility in water but it is a very effective sanitizer even at low concentrations. The main advantage of a chlorine dioxide solution over chlorine is that chlorine dioxide will work effectively against a broad spectrum of microbial contaminants even under a high organic load and is one of the most active antimicrobial systems against biofilms. Chlorine dioxide (ClO_2) is considered to be more environmentally friendly than chlorine as it does not form THM compounds. ClO_2 has 2.5 times the oxidizing power of chlorine and, thus, less chemical is required. Typical use concentrations range from 1 to 10 ppm.

Care must be taken when using chlorine dioxide as some corrosion issues have been reported for some metals and electronics although the formation or co-delivery of salts at the metal surfaces contacted is thought to be responsible for a significant portion of the corrosion effect (proper water rinsing to remove salts should minimize this risk if rinsing is possible). Chlorine dioxide is explosive in nature, degrades rapidly above temperatures of 122°F (50°C) and when exposed to light. Therefore, chlorine dioxide is usually produced on-site, used as a gas in very carefully controlled conditions or charged into water as a sanitizer or disinfectant, also under carefully controlled conditions. As chlorine dioxide has very low solubility in water, care must be taken to avoid unsafe atmospheric concentrations for workers due to off-gassing (EPA Guidance Manual).

Acidified Sodium Chlorite

Acidified sodium chlorite (ASC) is an oxychlorine mixture formed by acidification of sodium chlorite to form a chlorous acid intermediate chemical species. The chlorous acid supports active concentrations of strong oxidants (i.e. chlorate, chlorite, chlorine dioxide). ASC has broad antimicrobial efficacy and is able to be used, as is iodine, at acidic pH which is valuable for scale removal. ASC has US FDA approval for most equipment and environmental sanitizing applications. ASC solutions have also been approved by the US FDA as a "secondary direct food additive permitted in food for human consumption" permitting antimicrobial surface treatment of red meat, poultry, seafood and raw agricultural commodities (Allende et al., 2009).

ASC is most popular in food tissue spray applications where it can be used to provide antimicrobial reduction compared to a normal water wash step. Corrosion can be a problem as with any halide salt-containing system. Care needs to be taken to ensure proper equipment rinsing as a dried acidic salt on metal surfaces can result in pitting corrosion if not properly rinsed.

Peroxides

Hydrogen peroxide and peroxycarboxylic acid-based sanitizers are oxidizing sanitizers that have an advantage over halide oxidizers with similar effects against bacteria and viruses but are more effective in the presence of organic loads (especially proteins). These sanitizers break down into non-hazardous by-products. Hydrogen peroxide and the peroxycarboxylic acids are non-corrosive to stainless steel and the standard elastomers commonly used in food processing facilities. While hydrogen peroxide is odorless, it generally requires a fairly high concentration to deliver acceptable sanitizing results in a food process environment. Peroxyacetic acid, the most common peroxycarboxylic acid, has an odor associated with it and can have a higher use cost but is effective at a much lower concentration than hydrogen peroxide.

Other peroxycarboxylic acids, such as peroxyoctanoic acid, provide additional efficacy as sanitizers over peroxyacetic acid alone and due to the lower vapor pressure will have less associated odor in use (Fatemi and Frank, 1999).

Non-oxidizing Sanitizers

Non-oxidizing sanitizers typically used in food processing facilities include quaternary ammonium compounds (quats), acid anionic surfactant systems and fatty acids.

Quaternary Ammonium Compound

Quaternary ammonium compounds (Quats) are surfactants containing a positive charge which will be effective at binding at the negative charge of a bacterial cell wall and thereby delivering its antimicrobial effect. Quats are very stable compounds, are tasteless and odorless in solution, non-corrosive to common food processing facility surfaces, are non-irritating to skin and can leave a bacteriostatic coating on surfaces to inhibit microbial growth after treatment. Quats can also have activity against viruses through interaction with the negatively charged lipids on virus envelopes covering their protein capsids.

Quat compounds most used in recent years have consisted of blending multiple quats for increased efficacy as well as using polymeric quats for decreased toxicity. These quats generally have increased hard water intolerance and lower sensitivity to anionic compounds as was the case for older quat structures.

Disadvantages of quats are the need for a relatively high concentration to obtain a germicidal or bacteriostatic effect; they have limited activity against Gram-negative bacteria and can have limited activity in the presence of water hardness or anionic surfactants common in detergent systems and relatively slow biodegradation rate.

Fatty Acid Sanitizers

Fatty acid sanitizers are popular due to their low environmental impact and use at an acidic pH (also effective for scale removal). They generally require higher active concentrations than other sanitizer systems which can affect the organic wastewater limitations of some food processing facilities (Marriott and Gravani, 2006).

Acid Anionic Sanitizers

Acid anionic surfactants are negatively charged surfactants which have antimicrobial properties similar to Quats. They have a negative charge but are used at such a low pH that the bacterial cell surface switches to a positive charge resulting in binding and antimicrobial activity.

Alcohol Sanitizers

The use of alcohols for sanitizing and disinfecting in a food processing facility is usually done only as a manual, spot sanitizing, step as the alcohol concentration (the most popular being ethanol and isopropanol) is usually on the order of 60% or greater to obtain complete efficacy.

Miscellaneous Sanitizing Systems

Other unique sanitizing systems such as UV radiation, filtration, cold plasma, high pressure and pulsed electric field can provide alternatives to heat and chemical sanitizer strategies for providing microbial reduction (Gachovska et al., 2008). Careful research into proper system design and rigorous validations of desired sanitizing efficacy for these systems is recommended. As with any sanitizing system, but especially with automated applications relying on complex technology, a proof of delivery for each application in conjunction with a good verification program is recommended to demonstrate continued efficacy.

APPLICATION OF SANITIZERS IN FOOD PROCESSING FACILITIES

A sanitizing operation is generally performed after a thorough cleaning operation for CIP and COP systems as well as for all environmental areas. Automated CIP/COP sanitizing would use a non-foaming sanitizer while a foaming sanitizer would be used for environmental surfaces (discussed above) as foam provides contact time for vertical surfaces as well as a visual confirmation of sanitizer application. (In the case of sanitizers that are approved for no-rinse applications, such foam would break and leave no visible residue.)

Any manual sanitizer application should not rely on mechanical action or scrubbing to assist the sanitizing operation (using the cleaning operation for such removal). If applicators, such as fabric or mop systems, are used for sanitizer application, care must be taken that these applicators are themselves thoroughly cleaned, sanitized and stored separately from equipment used for cleaning operations to avoid cross-contamination. It is usually recommended that cleaning equipment be color coded and stored in dry areas zoned against the possible contamination of food product by cleaning equipment.

CLEANING VALIDATION AND VERIFICATION TECHNOLOGY

There are three main components to developing a cleaning program for any food production area:

1. A **cleaning and sanitizing protocol** needs to be developed that is based on the specific legal and safety requirements for the food processing business.
2. **Validation** of that cleaning and sanitizing protocol requires development of tests for every point in the cleaning process to prove that the process can meet regulatory standards.
3. A **verification** program needs to be instituted (usually consisting of a subset of the validation test methods) in order to demonstrate the implemented cleaning and sanitizing programs are effective.

Too little focus has been given in most food production facilities to validation of cleaning and sanitizing protocols. Without proper validation of a cleaning and sanitizing protocol, there can be no assurance that a cleaning program will be effective in providing contamination-free, safe food product. An example validation program description that can be used in a food processing facility is shown in Table 27.3.

When a cleaning and sanitizing program is identified, the following steps are needed to ensure a successful program:

1. The validation program must determine the points in the process that will be the most difficult to clean and sanitize (Scipioni et al., 2002).
2. Cleaning and sanitizing methods should be evaluated specifically for the ability to effectively remove all soil types from these tough to clean and/or sanitizer areas.
3. A set of verification tests must be agreed upon to prove soil removal actually occurs in the difficult to clean areas. (These validation tests are generally much more rigorous than the final verification portion of the final SSOP document to ensure the program successfully cleans and sanitizes all potential failure points.)

TABLE 27.3 Components of a Typical Food Processing Validation Program

SCOPE OF VALIDATION PROTOCOL
Equipment and Area to be cleaned
Makeup of Validation Team
Critical Requirements for validation success
Timeline for completion
RECOMMENDED REPRESENTATIVES FOR VALIDATION TEAM
Sanitation
Quality
Maintenance
Engineering
SANITATION STANDARD OPERATION PROCEDURE (SSOP)
Contains cleaning and sanitizing program for system under review
Basis for validation design
SSOP reviewed and edited by Validation Team on completion of validation
DEFINE SANITATION REQUIREMENTS
Cleaning and Sanitizing equipment/chemicals
Personnel required for sanitation operation
Timeframe to complete sanitation
DEFINE QUALITY PROGRAM
Visual inspection program
Microbiological and/or allergenic tests to be used
Collection program of representative samples confirming removal of soil and/or microbial contamination
Acceptance Criteria
Response to failure of acceptance criteria
Addresses short and long term cleaning, sanitizing and repair requirements (Master Sanitation Plan)
IDENTIFY TRAINING REQUIREMENTS FOR SSOP
Sanitation team meetings for each sanitation event in addition to higher level monthly/yearly training
Includes safety training, chemical usage associated training on personal protection equipment
DEVIATIONS AND INVESTIGATIONS
Unforseen SSOP issues requiring changes in validation protocol
Equipment or facility repair required to meet validation acceptance criteria
FINAL REPORT
Contains data demonstrating effectiveness of SSOP
Sign off from validation team representatives and plant management
Insure report is accessible to other validation teams

4. Multiple cleaning trials must be evaluated using the developed validation protocol to prove that soil and microbial contamination are removed from the most difficult to clean areas.
5. If this is a new procedure for an old system, a baseline study using the old cleaning and sanitizing protocol and the validation test procedures is recommended to understand whether the new protocol is actually providing a better result than the current protocol.
6. The validation team should evaluate data from the trials, optimize the cleaning and sanitizing method and then determine verification tests and testing frequency required to ensure long-term success of the validated protocol.

Allergen Validation: Prototype for Validation of Food Cleaning and Sanitizing Operations

Validation for cleaning of food allergens can be used to define best practices for cleaning in food production facilities due to the similarity in goals between removal of allergen and removal of microbial contamination (Jackson et al., 2008). As with microbial contamination, allergen contamination can be distributed uniformly throughout food production equipment or it may be inhomogeneously distributed in hard to reach, inaccessible areas in the equipment (examples would be nut particles for allergens or biofilms for microbial contamination). Like allergens, a biofilm contamination may slough off equipment into food products resulting in large but difficult to detect product contamination.

As described above, allergen issues in the general population have required development of new cleaning techniques to ensure the safe removal of allergen proteins. Rigorous validation programs were required in order for the food industry to meet the regulatory requirements (such as the US Food Allergen Labeling and Consumer Protection Act of 2004 and similar strict labeling regulations in many other countries). Current guidelines require certain proteins to be removed to less than a 2 ppm residual – a level that is not always measurable by available protein test methods for some allergens.

When a food manufacturer produces an allergen-containing food, a fully validated sanitation procedure needs to be implemented when switching back to production of non-allergen food products. Such a validation of allergen removal, cleaning and sanitizing protocols not only has to take into account uniform allergen residue but also inhomogeneous pieces of an allergen that may not be detected by a general allergen swabbing technique. Standard cleaning methods often fail to remove detectable levels of allergens and special procedures must be developed to completely eliminate residual allergens.

Allergen validations must include evaluation of the complete allergen program. For example, if allergen cleaning is effective but the raw material handling program can permit unintended allergen contamination into a food product (due to labeling issues, employee error, etc.) the work of an allergen validation can be negated.

As with unintended microbial cross-contamination discussed above, a full allergen protection program in a facility requires a carefully planned program to place "hurdles" in front of a contamination source to significantly lower the probability of a contamination incident. In many cases, manufacturers have opted to build separate facilities and developed highly sophisticated raw material labeling and shipping programs to avoid any possibility for contamination.

Setting up a validation protocol to ensure effective cleaning and sanitizing methods for allergen systems usually will require careful planning and a cross-functional team that includes engineering, maintenance and raw material handling personnel in addition to the food safety and quality teams to ensure all potential allergen contamination sources are addressed.

It is important to point out that standard cleaning and sanitizing methods can often fail to completely remove allergen soils and that a purging of a system with a non-allergenic food product can often be the only effective way to remove all traces of a contaminant.

Validation of a Cleaning and Sanitizing Protocol

Validation of cleaning and sanitizing protocols for general food production systems should be done with the same level of scrutiny as allergen systems to ensure microbial contamination is not an issue. A full HACCP program is recommended that includes a validation program with appropriate verification tests.

It is important to understand the validation process is, at its core, an experimental process:

1. Validation of a cleaning and sanitizing protocol must identify both the soil to be removed and the pathogen of concern.
2. The validation team needs to make experimental conjectures about what specific cleaning and sanitizing methods will successfully provide hurdles to pathogen contamination of food product.
3. The validation team then needs to ensure that the verification tests represent the most appropriate tests.
4. The areas to be tested for microbial contamination should represent the most difficult areas in the process to clean.

Additionally, validation may require proof that a certain level of microbial kill is delivered throughout the food production area targeted. If a pathogen were to enter accidently into a food production process, would the validated method destroy that pathogen at all process points down to an undetectable level? Such validation testing could use the actual microbial pathogen in the testing but in many cases such testing would be too hazardous to a facility and its personnel and a surrogate system can be used in the validation protocol.

Use of Surrogates in a Sanitizing Validation Protocol

To represent pathogenic contamination, the use of surrogates in a validation study is a common way to ensure a validation protocol will meet regulatory requirements for a cleaning and sanitizing procedure (Kvenberg and Schwalm, 2000). Surrogates are chosen to provide features of a bacterial contamination without having the hazards associated with that pathogenic species introduced into a food production facility. Often food grade microbial species such as yeasts or probiotic bacteria can be directly used to demonstrate a microbial kill step. Other, non-microbial methods can include the use of food dyes, food particles or food grade inorganic particles (calcium carbonate, for example) that are easily detectable visually or can be detected by sensitive chemical or electronic detection methods thus proving contamination can be removed using the cleaning and sanitizing protocol being tested.

Surrogate validation tests need to be well thought out as microbial species should have similar temperature and pH stability to a target microbial species. Non-microbial methods should have a similar dissolution or removal rate as the food soil in the chosen cleaning and sanitizing system being tested. Delivery of the surrogate can be through manual application or added to a food product during manufacture.

Dry Food Production Cleaning Validation

As discussed above, food produced in a dry process environment provides an especially difficult challenge to food producers globally wanting to ensure pathogen-free food product. A sanitation break defines a specific time where food processing equipment has been completely cleaned and sanitized. For a dry food process, a sanitation break has not always been considered to be a required step in a food production process as the lack of water in these environments inhibits microbial growth. Many dry ingredient storage and delivery systems in these food production facilities have gone for years without sanitation breaks (with the focus instead being on elimination of pests). As a result of these practices there is no distinguishable raw material lot that can be used to determine an appropriate recall date for food produced with that raw material.

Validation of a dry sanitation break is critical to permit lot definition for food production (GMA, 2009). A dry process cleaning method, using little or no water and resulting in a sanitation break, would be focused on removal of microbial contamination and not necessarily all evidence of a dry food soil. Validation of such a dry process cleaning method would effectively involve the same validation steps as with wet cleaning. Verification testing on food contact surfaces and environmental surfaces needs to be rigorous and focus on areas most likely to harbor microbial contaminants. The special emphasis in dry facilities needs to be on ensuring water contamination does not occur and that effective sanitizing programs exist when water inadvertently is introduced into these dry environments. The sources of water contamination include unintended sources such as roofs, walls and floor leaks, water piping leaks, condensation inside and outside food production equipment as well as water intentionally brought into a facility by standard food production practices.

To ensure dry process cleaning and sanitizing methods eliminate microbial contamination in the food production equipment itself, surrogates for pathogens (as discussed above) can be especially valuable. For difficult to disassemble areas, sampling of these surfaces during a validation project can be used to prove the surrogate can be removed with a dry cleaning program or deactivated with a dry sanitizing step. Many food grade materials can be spiked into food products, placed in difficult to clean sites and subsequently detected by various methods at high sensitivity (such as proteins, inorganics or flavors) and biologically active but safe food grade yeast and bacterial cultures can be spiked into a food process and detected by standard plating techniques. The use of surrogates can, therefore, provide a safe way to demonstrate that a dry food cleaning and sanitizing process has been effective.

Cleaning Verification Tests

There are numerous verification tests to determine allergen and microbial contamination on food production equipment (Brown, 2009). The Association of Analytical Communities

AOAC Research Institute provides a wide array of validated methods for both allergens and pathogens (AOAC International). These verification tests or similar validated tests should be used in a validation program to ensure elimination of the desired contaminant.

Sampling for contaminations is typically done using swabbing techniques on hard surfaces and analyzing the swabs using the various applicable allergenic or microbial detection methods. Other techniques include testing rinse water or using air sampling techniques.

Some allergen verification tests include highly sensitive tests that can detect peanuts, gluten and related grain proteins (gliadin, secalins, hordeins), mycotoxins such as deoxynivalenol (DON) in cereals and aflatoxins common in cereals, spices, tree nuts and oilseeds (corn, peanuts and raw almonds).

The common bacterial tests include measurements of general bacterial populations such as total plate count (TPC), total viable count, standard plate count (SPC), aerobic bacteria (APC), Gram-positive cultures, thermoduric count, coliforms, Gram-negative bacilli as well as more specific tests such as *E. coli* O157:H7, *Salmonella*, *Listeria*, *Bacillus anthrasis* (anthrax), Enterobacteriaceae, *Staphylococcus* and Campylobacters. In addition yeast and mold tests are commonly used in food processing verification, especially for air quality assessments.

Verification tests can be general or specific. Common general tests demonstrate the presence of soil in areas of food production facilities and can be very sensitive. Use of general verification tests are based on the understanding that if no soil is detected, no contamination is present. Such sensitive tests include total organic carbon (TOC) which will detect any type of organic soil. Adenosine triphosphate or ATP-based detection technology will detect any type of food or microbial contaminations (from cell ATP content) but is not specific to the sources of the ATP. The ATP from cells (living or dead) reacting with the luciferase enzyme is the basis of swabbing tests that are quick and semi-quantitative, providing a light density-based numeric output that can be used for comparative purposes for hard surface soils.

The sensitivity of these ATP detection tests can vary widely, with some extremely sensitive, and used to detect very low levels of proteins (especially valuable to ensure a low level of food soil is removed for allergen cleaning verifications). The ATP tests cannot be used to confirm the presence of a specific allergen or pathogen, merely the presence or absence of food or microbial-based soil down to the sensitivity of the specific swab.

Numerous other fluorescence technologies have been employed for identifying pathogens, living from dead cells, at high sensitivity. Enzyme-linked immunoassays (ELISA) and enzyme-linked fluorescent (ELFA) techniques have been commonly used for automated detection of specific pathogens. There are numerous techniques for automated testing for specific pathogens and the technologies are getting better and more rapid (Fung, 2002). The ultimate goal for food processing facilities is obtaining test results rapidly enough to permit release of food product from a production facility with minimal need for keeping food product on hold.

CONCLUSIONS

This chapter has reviewed the different aspects of food production facility cleaning and sanitizing programs, SSOP development as well as chemical and non-chemical systems used for cleaning and sanitizing. Automated or manual cleaning and sanitizing requires an understanding of where soil and microbial contamination can reside in a food processing system. CIP, COP and environmental cleaning systems and the associated cleaning and sanitizing chemistry and procedures for food production facilities are discussed. Additionally, common problems encountered in food production facility cleaning and sanitizing programs as well as validation and verification programs are reviewed. Special topics include cleaning and sanitizing considerations and associated validation programs for allergen issues and dry food environments.

References

Allende, A., McEvoy, J., Tao, Y., Luo, Y., 2009. Antimicrobial effect of acidified sodium chlorite, sodium chlorite, sodium hypochlorite, and citric acid on *Escherichia coli* O157:H7 and natural microflora of fresh-cut cilantro. Food Control 20 (3), 230–234.

AMI (American Meat Institute), 2002. Sanitary Equipment Design. Fact sheet published by the associates on their website: <www.meatami.com>.

AOAC International. Performance Tested Method – Validated Methods. Publication published on their website: <www.aoac.org/testkits/testedmethods.html>.

Arvanitoyannis, I.S., 2009. HACCP and ISO 22000: application to foods of animal origin. Wiley-Blackwell Publishing Ltd, Hoboken, NJ.

ASTM A380-06, International Standard Method: Standard Practice for Cleaning, Descaling, and Passivation of Stainless Steel Parts, Equipment, and Systems. ASTM International, West Conshohocken, PA.

Brown, H.M., 2009. Validation of cleaning and cross-contact. In: Coutts, J., Fielder, R. (Eds.), Management of food allergens. Wiley-Blackwell, Oxford.

EPA Fact Sheet (United States Environmental Protection Agency). What are antimicrobial pesticides? Fact Sheet published on their website: <www.epa.gov/oppad001/ad_info.htm>.

EPA Guidance Manual (United States Environmental Protection Agency). EPA Guidance Manual, Chapter 4: Chlorine dioxide. Fact Sheet published on their website: <www.epa.gov/OGWDW/mdbp/pdf/alter/chapt_4.pdf>.

Fatemi, P., Frank, J.F., 1999. Inactivation of Listeria monocytogenes/Pseudomonas biofilms by peracid sanitizers. J. Food Prot. 62 (7), 761–765.

Fung, D.Y.C., 2002. Predictions for rapid methods and automation in food microbiology. J. AOAC Int. 85 (4), 1000–1002.

Gachovska, T.K., Kumar, S., Thippareddi, H., Subbiah, J., Williams, F., 2008. Ultraviolet and pulsed electric field treatments have additive effect on inactivation of *E. coli* in apple juice. J. Food Sci. 73 (9), M412–M417.

Gibson, H., Taylor, J.H., Hall, K.E., Holah, J.T., 1999. Effectiveness of cleaning techniques used in the food industry in terms of the removal of bacterial biofilms. J. Appl. Microbiol. 87 (1), 41–48.

GMA (Grocery Manufacturers Association), 2009. Control of *Salmonella* in Low-moisture Foods. Publication of the Grocery Manufacturers Associates published on their website: <www.gmaonline.org/downloads/wygwam/SalmonellaControlGuidance.pdf>.

Gottardi, W., 2001. Iodine and iodine compounds. In: Block, S. (Ed.), Disinfection, sterilization and preservation, fifth ed. Lippincott Williams & Wilkins, Philadelphia, pp. 159–184.

Holah, J., Lelieveld, H.L.M., 2011. Hygienic design of food factories. Woodhead Publishing Series in Food Science, Technology and Nutrition (No. 216).

Jackson, L.S., Al-Taher, F.M., Moorman, M., DeVries, J.W., Tippett, R., Swanson, K.M., et al., 2008. Cleaning and other control and validation strategies to prevent allergen cross-contact in food-processing operations. J. Food Prot. 71 (2), 445–458.

Jennings, W.G., McKillop, A.A., Luick, J.K., 1957. Circulation cleaning. J. Dairy Sci. 40, 1471–1479.

Kvenberg, J.E., Schwalm, D.J., 2000. Use of microbial data for hazard analysis and critical control point verification – Food and Drug Administration perspective. J. Food Prot. 63 (6), 810–814.

Lelieveld, H.L.M., Mostert, M.A., Holah, J., 2005. Handbook of hygiene control in the food industry. Woodhead Publishing Series in Food Science, Technology and Nutrition (No. 116).

Lelieveld, H.L.M., Mostert, M.A., White, B., Holah, J., 2003. Hygiene in food processing: principles and practice. Woodhead Publishing Series in Food Science, Technology and Nutrition (No. 88).

Lowry, D., 2010. Advances in cleaning and sanitation. Aust. J. Dairy Technol. 65 (2), 106–112.

Marriott, N.G., Gravani, R.B., 2006. Principles of food sanitation, fifth ed. Springer Science + Business Media, Inc..

Richter, F.L., Cords, B.R., 2001. Disinfection, sterilization, and preservation. In: Block, S.S. (Ed.), Formulation of sanitizers and disinfectants, fifth ed. Lippincott Williams & Wilkins, Philidelphia (Chapter 23).

Samelis, J., Sofos, J.N., Kendall, P.A., Smith, G.C., 2001. Fate of *Escherichia coli* O157:H7, *Salmonella* Typhimurium DT 104, and Listeria monocytogenes in fresh meat decontamination fluids at 4 and 10°C. J. Food Prot. 64 (7), 950–957.

Schmidt, R.H., 1997. Basic Elements of Equipment Cleaning and Sanitizing in Food Processing and Handling Operations, Fact sheet FS14 published by the University of Florida on their website: <edis.ifas.ufl.edu>.

Scipioni, A., Saccarola, G., Centazzo, A., Arena, F., 2002. FMEA methodology design, implementation and integration with HACCP system in a food company. Food Control 13, 8.

Sicherer, S.H., Sampson, H.A., 2006. Food allergy. J. Allergy. Clin. Immunol. 117 (2, Suppl 2), S470–S475.

Stenga, M., 2010. Sanitation: cleaning and disinfection in the food industry. Wiley-VCH Verlag GmbH & Co., KGaA, Weinheim.

USDA (United States Department of Agriculture), 2010. Food Safety Inspection Service, Sanitation Standard Operating Procedures 9 CFR 416.11 through 416.17. Published on their website: <www.fsis.usda.gov>.

Vierk, K., Falci, K., Wolyniaka, C., Klontz, K.C., 2002. Recalls of foods containing undeclared allergens reported to the US Food and Drug Administration, fiscal year 1999. J. Allergy. Clin. Immunol. 109 (6), 1022–1026.

White, G.C., 2010. White's handbook of chlorination and alternative disinfectants, fifth ed. Black & Veatch/John Wiley & Sons, Hoboken.

CHAPTER

28

Personal Hygiene and Health

Ewen C.D. Todd

Ewen Todd Consulting LLC, Okemos, MI, USA

OUTLINE

Food Safety Management.
DOI: http://dx.doi.org/10.1016/B978-0-12-381504-0.00028-7

RISKS OF OUTBREAKS ASSOCIATED FROM INFECTED FOOD WORKERS

Outbreaks Contributed by Food Workers

A food employee or worker is someone who works with unpackaged food, food equipment, kitchen utensils or food contact surfaces. A food handler may be perceived as someone who simply handles food but for this chapter all these terms are synonymous. Outbreaks involving infected food workers in many foodservice settings have been widely reported, with some resulting in many cases and deaths. In fact, food workers have probably been implicated in spreading foodborne disease ever since food was prepared and served. The case of Mary Mallon (Typhoid Mary), who was a chronic carrier of *Salmonella* Typhi, is an extreme example of the risks of colonized food workers infecting others. She is attributed to causing 47 illnesses and three deaths in the New York area between 1906 and 1915. During Mallon's temporary confinement from 1907 to 1909, health officials had analyzed stool samples from her approximately once a week. The samples came back with *S.* Typhi in 120 of 163 samples tested. In the early 1920s, Tony Labella, a food handler in the New York farming community, may have actually caused more illnesses and deaths, since he was also a chronic *S.* Typhi carrier. These were the years when typhoid was endemic in the USA, and there were probably many other undocumented carriers who spread typhoid fever.

A review of foodborne disease outbreaks worldwide shows that food workers have been responsible for many of these outbreaks (Greig et al., 2007; Todd et al., 2007a, b, 2008a, b, 2009). These authors collected 816 outbreak reports comprising 80,682 cases from events spanning 1927 to the first quarter of 2006, with most of these occurring in the last three decades. These outbreaks were caused by 14 agents: norovirus, Norwalk-like viruses or probable norovirus (338), *Salmonella enterica* (151), hepatitis A virus (84), *Staphylococcus aureus* (53), *Shigella* spp. (33), *Streptococcus* Lancefield groups A and G (17), and the parasites *Cyclospora*, *Giardia*, and *Cryptosporidium* (23). It appears from these data that the frequency of streptococcal, staphylococcal and typhoid outbreaks diminished over time, those involving hepatitis A virus (HAV) saw little change, but those with norovirus and maybe non-typhoidal *Salmonella* were increasing. The terminology of norovirus has changed over time from Norwalk or Norwalk-like or small round structured viruses and may in the past have included other similar viruses. Today norovirus (NoV) is the organism of most concern, because it is frequently spread person to person in the community, has a low infectious dose and has been implicated in many foodborne outbreaks where food workers have been found to be infected.

In some cases, the worker may have been a victim rather than the cause of the infection, becoming ill at the same time as the customers or later. In other situations, the worker blatantly disregarded normal hygienic practices, which may have been a result of inappropriate individual actions or the accepted way of doing business in the establishment. Practices leading to these actions have previously been documented, such as workers being asymptomatic carriers and excreting the pathogen unknowingly while working, or they continued to prepare food when it is obvious to them, and sometimes others, that they were ill with a high probability of contaminating food. Some of the outbreaks were very large; 11 involved more than 1000 persons, four with more than 3000 ill. The larger outbreaks tended to be extended over several days with a continuing source of infections, such as at festivals, resorts and community events, or the contaminated product had been accessed by a large number of customers. There were five outbreaks with more than 100 persons hospitalized, with attack rates ranging from 9.9 to 100%. However, overall, the hospitalization rate was low (1.4%), and deaths were rare (0.11% of the documented 80,682 cases). Many of the deaths were associated with high-risk persons (i.e. those who had underlying diseases, malnutrition, or both, as in a refugee camp, or young children), but a few occurred with apparently healthy adults. Unfortunately, and there is no indication that worker-related illnesses are diminishing. For instance, a study by the Indian Public Health Association in 2012 found *E. coli* on the hands of nearly 11.2% of those who handle food in five star hotels, 47% of chefs and waiting staff in smaller restaurants, and 84.7% of food vendors in roadside eateries (Narayan, 2012). In addition, 11.2% of these roadside vendors carried amebic cysts that can cause amebiasis, the third most common cause of death in India from parasitic infections.

Food Operations and Foods Implicated

An analysis of the settings for the food worker-related events in the above-mentioned study showed that most of the outbreaks came from foodservice facilities (46.1%), followed by catered events (15.4%), the home (10.2%), schools and child care centers (6.0%), and healthcare institutions (5.3%). The single most frequently reported setting was restaurants. Case numbers in outbreaks in homes or at community events are probably underestimated because they are less likely to be reported than those involving commercial establishments.

Large outbreaks frequently occurred because of the continual exposure of large groups to a pathogen, either because the source had not been identified soon enough or because control measures had been insufficient to eliminate the agent, such as at refugee camps or large outdoor events. However, in several other large outbreaks, the amount of contaminated food was so great that thousands of persons were exposed by the same batch of food; this occurred with frosting on cakes, imported raspberries used in a variety of dessert dishes, and items served at large receptions or commissaries. The agents in large outbreaks also tended to be highly infectious, such as *Shigella* or NoV. Because ready-to-eat (RTE) foods are not further processed or cooked, subsequent contamination by infected food workers frequently led to outbreaks. These included produce and baked goods, as well as beverages that would not normally allow the growth of pathogens. However, many of these were of viral origin with sufficient particles to cause an infection without further multiplication. Foodservice outlets, such as restaurants, catering companies and schools that served large numbers of patrons, were the most frequent settings implicated. However, episodes linked to bakeries, hospitals, camps, homes and church meals highlighted the necessity for those who prepare and serve meals in these operations to excuse themselves from food preparation if they are ill or exposed to infected individuals. There were 18 outbreaks associated with commercial travel in air flights, trains and cruise ships over several decades, although only the last seems to be a major concern today. There are several outbreaks every year from cruise ships; often with >2000 passengers per ship, who could be more likely the source of an enteric infection than the crew; staff typically ask illness questions before embarkation and squirt an antiseptic on hands of cruisers before they eat. How effective these practices are is uncertain but they do alert passengers of the risks of spread of foodborne infections in close quarters.

FACTORS CONTRIBUTING TO OUTBREAKS

The most frequently reported factor associated with the involvement of the infected worker was bare hand contact with the food, followed by failure to properly wash hands, inadequate cleaning of processing or preparation equipment or utensils, cross-contamination of RTE foods by contaminated raw ingredients, and (for bacterial pathogens) temperature abuse. Many of the workers were asymptomatic shedders or had infected family members and/or used improper hygienic practices. Outbreaks were sorted into categories based on how many workers were implicated, the origin of the infective agent (outbreak setting or off-site), the degree of certainty that the worker(s) were the cause or were victims, whether or not the workers denied illness, the ability of the agent to grow in the food, whether only the workers and not the patrons were ill, and whether patrons were more responsible for their illnesses than the workers. The most frequent scenarios were (1) a single worker causing an outbreak by directly infecting customers; (2) an infected worker contaminating foods from feces, typically after toileting, that were then temperature abused, leading to an outbreak; and (3) multiple workers linked to an outbreak, but with no clear initiating source.

EXAMPLES OF OUTBREAKS CAUSED BY FOOD WORKERS

A few examples suffice to indicate the type of food worker contamination that can lead to outbreaks (Todd et al., 2007a). In 2000, 37 students at a Minnesota college developed gastrointestinal symptoms from 25 April to 1 May, with most on 26–27 April. Illness was associated with consuming any cold salad bar items from the dining service at the college cafeteria on 25–27 April. The index case was a food preparer who reported developing vomiting and diarrhea on 23 April after being exposed to children who also were vomiting and had diarrhea on 22 April. This person called in sick on 24 April and symptoms resolved later that day. The employee then returned to work on 25 April and worked the remainder of the week in the salad bar section of the dining service, with extensive bare-hand contact of salad items during preparation and stocking of the salad bar with lettuce, salad toppings and cut fruit. Additional cases, with onset after the weekend of 29–30 April were likely due to secondary spread of viral infection within dormitories and other campus settings. The Department of Health categorized the agent as viral based on the epidemiological information available, but no specific agent was isolated. An ill call-in log was useful in determining dates that employees were ill and to ascertain the responsible employee.

In Los Angeles County in the same year, an increase in *Salmonella* Thompson infections was noted with most cases dining at a restaurant chain before developing illness (Kimura et al., 2005). A case–control study implicated burgers eaten by 23 individuals at the fast food restaurants. In addition, hamburger buns were also served at a catered luncheon and at three other restaurants where a further 15 *S.* Thompson cases were reported. The index case was a burger bun packer at a bakery that supplied buns for the chain, but she had not eaten at the restaurant chain. This full-time employee was responsible for removing freshly baked bread and buns from the cooling rack, feeding them through an automatic slicer and packaging them for distribution. She did not wear gloves and handled every individual bread item (notably hamburger buns) at least twice with her bare hands. She worked from the day of illness onset on 13 July until she required overnight hospitalization on 17 July. She resumed work after hospital discharge on 18 July and continued working until termination of employment on 23 July. Although stool specimens were taken during her hospitalization, the results were not reported until 31 July, 2 weeks after onset of illness. The patient's brother, also employed at the bakery, became ill on 17 July, and continued to work while ill until he was removed from work on 3 August. Presumably she infected him either through contact or through consumption of buns handled by her. He was mainly responsible for mixing dough but did some rotation of duties that would allow contamination of bread items. The bakery did not offer any formal training on safe food-handling practices. Furthermore, although many of the employees spoke only Spanish, the procedure manuals were written in English. As indicated above, low water activity does not allow bacterial growth in most baked goods or the ingredients of cake frosting but there were a surprising number of outbreaks, some very large, associated with contaminated icing or frosting on cakes and glazing on pastries. Direct contact between contaminated hands or arms and the ingredients was enough to transfer NoV, HAV, *Salmonella*, *S. aureus* and rotavirus to the products in sufficient quantity to cause illness. Examples of these are: (1) 414 people were ill with NoV after eating pastries served in a Winnipeg hotel; (2) 68 cases of HAV were associated with eating buns and pickles handled by a worker in a Chattanooga fast food restaurant and who was an intravenous drug user; and (3) 12 HAV cases derived from an infected cook who contaminated cream while preparing pastries in a Glasgow restaurant (Greig et al., 2007).

Some outbreaks involving food workers are international in scope (Todd et al., 2007a). For example, *Salmonella* Brandenburg was responsible for illnesses in 232 passengers, 27 cabin staff and 31 aircrew in 45 flights originating in Paris for many parts of the world, including the United States, Canada, the Caribbean, Egypt, Senegal, Japan, Venezuela, Brazil, Russia and eight other European countries in April 1976. The illnesses occurred from 6 to 11 April, and an alert was only triggered when an aircrew on a 9 April flight was concerned about the ill people. However, despite this alert, meals continued to be prepared and served until 11 April, and many more passengers are thought to have become ill than the 290 eventually reported to the authorities. The organism was isolated from a variety of cold foods, primarily fish. Only one of the 200 employees of the catering firm in Paris tested for the pathogen had *S.* Brandenburg isolated from the stools but this person had prepared these cold dishes. Unfortunately, this employee was not recognized by the regular inspection and testing program in the establishment despite the fact that their surveillance program had resulted in 14 suspensions of staff who had infections over the previous 2 years.

More recently, outbreaks from June to September 2005 of norovirus infection in Denmark were linked to frozen raspberries imported from Poland. All outbreaks occurred in institutions or commercial catering settings. A cold dessert dish prepared from frozen raspberries, which had not been heated, had been served one day before the start of each outbreak. In the first five outbreaks, frozen raspberry pieces had been used, which could be traced to the same large batch imported to Denmark from Poland in the spring of 2005. In the last outbreak in September, the frozen raspberries had been supplied by a different Polish producer to a different Danish importer and made into a traditional Danish dessert of buttermilk, fromage frais, sugar, vanilla and raspberries. With 1143 cases in total, these raspberries caused the largest number of foodborne infections attributable to a single vehicle in Denmark in many years. Delays in the implementation of a recall after the first large outbreak of 450 people in a hospital resulted in a second large outbreak among elderly clients of a meals-on-wheels service in early June. An estimated 400 mainly elderly people were affected and at least 23 required hospitalization. Surprisingly, three different genotypes of norovirus were found in the six outbreaks. As Polish frozen raspberries were exported to several European countries, outbreaks due to frozen raspberries would be expected beyond Denmark but none of Polish origin was reported. Contamination with norovirus may have occurred at farm level by fecally contaminated irrigation water, during harvesting by infected farm workers and/or during processing and freezing by infected workers at company level. The hypothesis was that several independent contamination events took place, thus explaining the heterogeneous distribution of norovirus strains in the Danish shipments. Infected workers in the harvesting or processing of the raspberries in Poland were a likely but not proven source.

PATHOGENS CARRIED BY FOOD WORKERS

Sources of Pathogens

The human body has several means of transmitting infections from body orifices, primarily fecal, nasal and skin sources (though urine can be a transmitter of HAV), because these are exposed to the external contaminated environment through air, water, food and contact with other humans and animals. Sources of pathogens, therefore, include vomitus, diarrhea, nasopharyngeal or oropharyngeal secretions, often being transmitted to food or food contact surfaces. The likelihood that workers cause illness in patrons and fellow workers depends on several factors: the numbers of organisms required to initiate an infection, the site of colonization and the length of their carriage in infected persons. Pathogens of nose, throat, skin or fecal origin are most likely to be transmitted by the hands (particularly fingertips and palms), as hands are the parts of the body that frequently touch the mouth, skin and anal areas. The pathogens most likely to be transmitted by food workers are NoV, HAV, *Salmonella*, *Shigella* and *Staphylococcus aureus*. Unfortunately, such pathogens can be in high numbers in or on the body during an infection. This is particularly true for intestinal infections where levels can reach 10^{11} infectious cells or particles per gram of feces, although 10^5–10^9/g is more frequent. Some pathogens appear to be able to infect at doses as low as 1 to 100 units, including viruses, parasites and some bacteria. Although parasitic foodborne episodes of illness are rare, the dose for *Cyclospora*, *Cryptosporidium* and *Entamoeba* may be as low as one cyst/oocyst. Based on outbreak data and other infectious disease studies, other pathogens with low minimum infections doses are *Campylobacter*, *E. coli* O157:H7, *Salmonella* Typhi (and a few other *Salmonella* serotypes), *Shigella dysenteriae*, HAV, NoV and rotavirus (<100 CFU or particles). Interestingly, only rarely have *Campylobacter*, *E. coli* O157:H7 and other *E. coli* serotypes been implicated in food worker-associated outbreaks.

Incubation Periods

For ill persons, these can range from a few hours, e.g. *S. aureus* enterotoxin, to many weeks, e.g. HAV and *S.* Typhi. The longer the incubation period, the more opportunities there are that an infected person will excrete the pathogen. This is equally important for a worker's contact persons, mostly likely the family or fellow workers, who may be the persons initially infected and excreting. The duration of illnesses is important too. Gastroenteritis symptoms may last many days or even weeks or months, e.g. chronic diarrhea, as in cases of infection with *Salmonella* Typhi, *Shigella* spp., HAV and the protozoan parasites. Since employees want to return to work quickly after illness, and they do not usually receive paid sick leave, they may work while continuing to be ill or only having mild symptoms like occasional diarrhea, without reporting their conditions to management. Post-symptomatic long-term shedding can also occur with *Campylobacter, Salmonella, Shigella, V. cholerae, Yersinia,* enteric viruses and parasites.

Fecal Contamination of Hands

Lack of hand hygiene compliance means that fecal contamination of fingers and hands is not uncommon (Todd et al., 2010e). Maybe it is not generally understood that the fecal–oral route is the main source of enteric infection from pathogens present in the feces of ill, convalescent or otherwise colonized persons because no visible feces is present on fingers or hands after normal defecation. It is difficult for managers of food operations to identify food workers who may be excreting pathogens, even when these workers voluntarily report their illnesses (whether there is a policy for reporting illnesses or not), because workers can shed pathogens during the prodrome phase of illness or can be long-term excretors or asymptomatic carriers. Some convalescing individuals have excreted *Salmonella* for 102 days, and most individuals infected with viruses remain asymptomatic while discharging infective particles into their surroundings. Fecal contamination on the hands is linked to the limited effectiveness of toilet paper use. The fecal mass on fingers after defecation has been estimated from $10^{-5.6}$ g to >1 mg (Todd et al., 2008b). This means that there could be 10^{3-4} fecal coliforms present on hands when initially contaminated, and pathogens if present could be in substantial numbers (Table 28.1). There is limited information on carriage rates before, during and after illness and in asymptomatic individuals. Carriage rates range between <1 and 36%, and shedding can occur many days before symptoms appear, making exclusion of excreting employees from the food environment very difficult. The soil matrix, relative humidity and temperature all influence pathogen survival. Declines can be rapid on hands, but most pathogens that cause foodborne illness survive long enough on hands and contact surfaces to allow some transfer to food or fellow workers during a shift. *Salmonella* can survive for several hours on fingertips if they are not washed. Non-enveloped viruses such as NoV, rhinovirus and enterovirus are more stable on skin than are viruses with envelopes, such as the influenza virus.

HYGIENIC PRACTICES OF FOOD WORKERS

Spread of Pathogens in the Food Industry

Personal hygiene is critical to reduce the opportunities for pathogen transmission. This is of particular concern in the food and healthcare industries where food is served and

TABLE 28.1 Levels of Pathogens in Clinical Specimens and Body Excretions

Pathogen	Source of Contamination	Contamination (level/g or ml)
Salmonella	Feces, while ill or early convalescence Feces, in late excretion period Infants excrete longer than adults	10^5–10^7 CFU 10^0–10^3 CFU
	Feces, while in convalescence	6×10^3 CFU 15 days after illness $5 \times 10^{2-6} \times 10^7$ CFU (median, 6.0×10^6 CFU) <10 days after illness 1.3×10^2–1.6×10^9 CFU (median, 1.0×10^5 CFU) 10–19 days after illness <$10^0 \times 3.5 \times 10^6$ CFU (median, 2.5×10^4 CFU) 20–25 days after illness 7.0×10^1–1.8×10^5 CFU (median , 1.4×10^2 CFU) 6–35 days after illness 2.0×10^0–3.5×10^4 CFU (median, 5.5×10^3 CFU) 42–50 days after illness <$10^{0-6} \times 10^4$ CFU (median, 2.5×10^4 CFU) 69–102 days after illness
S. aureus and streptococci	Pus in infected lesions	10^7–10^8 CFU (median) for intra-abdominal and anorectal and soft tissue infections (one sample with almost exclusively *S. aureus* and two with beta hemolytic streptococci)
S. pyogenes	Saliva in a sneeze from carriers	Typical person infected with streptococci: saliva 10^0 to 10^6 CFU; <100 CFU/154 sq. cm. 1.5–9.5 feet from sneeze source One carrier sneezed twice (Day 1 and Day 6): saliva 3.2–7.5×10^6 CFU; 23–500 CFU/154 sq. cm. 1.5–9.5 feet from sneeze source; 50% of the streptococci remained in the air 10–15 min. after the sneeze (10–16 min after sneezing 10^1–10^3 streptococci were cultured 9.5 feet away). A nose blow from a carrier into a handkerchief yielded 2.4×10^7 CFU material from posterior pharynx compared with 3.8×10^4 CFU saliva
S. pyogenes	Saliva in a cough from carriers	10^3–10^6 CFU (1/20 persons infected with streptococci coughed 6 CFU streptococci/154 sq. cm. 9.5 feet from cough source; most of the other 19 did not cough any streptococci)
Enteroviruses (coxsackie virus, echovirus, polio virus, etc.)		10^3–$10^{7.5}$ infectious particles, $10^{8.2}$ infectious particles
Hepatitis A virus	Feces, highest numbers before symptoms begin	10^9 virions 10^8 infectious particles

(*Continued*)

TABLE 28.1 (Continued)

Pathogen	Source of Contamination	Contamination (level/g or ml)
Norovirus Group G-I	Feces, while ill[a]	2.2×10^4 to 2.9×10^{10} copies/g fecal specimen, median = 8.4×10^5
Norovirus Group G-II		2.5×10^4 to 7.7×10^{10} copies/g fecal specimen, median = 3.0×10^8
Norovirus Group G-I		GI 2.79×10^7 copies/g of stool
Norovirus Group G-I/4		GI/4 2.02×10^8 copies/g of stool
Norovirus Group G-II		GII, 3.81×10^8 copies/g of stool
Norovirus Group G-II/4		GII/4 7.96×10^9 copies/g of stool
Rotavirus	Feces and vomitus while ill	10^{11} particles excreted but only 10^6–10^7 infectious 100 times more virus in vomitus than feces 8×10^9–10×10^9 infectious particles $>10^{12}$ infectious particles
		10^{10}–10^{12} in feces
Cryptosporidium spp.		10^8–10^9 oocysts in a single bowel movement
		10^6–10^7 oocysts 3×10^9 oocysts/day $<10^9$ cysts daily in stools
Giardia lamblia/intestinalis		1–5 $\times 10^6$ cysts

Information from Todd et al. (2008b); CFU = colony-forming units.
[a]*levels in vomitus not known but assumed to be similar to those in stools.*

people and patients touched. An extensive series of reviews of factors contributing to outbreaks by food workers and how they spread diseases revealed a composite list of problems uncovered during the investigations (Greig et al., 2007; Todd et al., 2010d). The major concerns focused on (1) hand washing, (2) sanitation of food contact surfaces, (3) facility-wide hygiene education and training, (4) incentives for workers to report their illnesses, (5) surveillance of the work force by management, and (6) regular professional screening of employees for illness, including nasal and stool samples obtained from staff returning from overseas travel. Food workers have been implicated in outbreaks of foodborne illness, and hands contaminated by human or animal feces are a well-recognized mode of pathogen transfer; sneezes, coughs, infected skin lesions and vomitus also have transmitted pathogens from workers to food, patrons and fellow workers. Physical barriers such as food shields (sneeze guards), utensils and appropriate protective clothing have value but are insufficient to completely prevent contamination of food or food contact surfaces by body secretions. By far the most important action to avoid contamination of food is the cleanliness of the hands.

Widespread lack of attention to any kind of major promotional campaign of hand hygiene was reinforced when the US Food and Drug Administration's (FDA) National

TABLE 28.2 FDA Report on the Occurrence of Foodborne Illness Risk Factors in Selected Institutional Foodservice, Restaurant and Retail Food Store Facility Types (US FDA, 2009b)

Facility Type	Proper, Adequate Hand Washing (%)	Good Hygienic Practices (%)	Prevention of Contamination from Hands (%)	Hand-washing Facility, Convenient/ AccessIble (%)	Hand-washing Facility, Cleanser/ Drying Device (%)
RESTAURANTS					
Fast food	38.8	22.5	26.3	18.4	15.5
Full service	75.8	24.2	46.3	29.2	29.2
RETAIL STORES					
Delicatessens	52.0	13.3	6.3	17.4	13.3
Meat and poultry	18.4	1.1	0.0	6.1	6.1
Seafood	21.6	5.3	2.9	7.3	7.3
Produce	24.6	9.9	6.3	15.6	17.7
INSTITUTIONAL					
Hospitals	35.6	13.3	9.0	23.3	4.4
Nursing homes	34.4	12.1	12.5	12.9	8.6
Elementary schools	27.5	11.8	8.6	21.5	5.4
Total	**18.4–75.8**	**1.1–24.2**	**2.9–46.3**	**6.1–29.2**	**4.4–29.2**

Retail Food Team conducted the third phase of a three-phase,10-year study to measure the occurrence of practices and behaviors commonly identified by the US Centers for Disease Control and Prevention (CDC) as contributing factors in foodborne illness outbreaks: food from unsafe sources; poor personal hygiene; inadequate cooking; improper holding/time and temperature; and contaminated equipment/protection from contamination (US FDA, 2009b). FDA Regional Retail Food Specialists collected data during site visits of over 800 establishments representing nine distinct facility types (Table 28.2). Direct observations, supplemented with information gained from discussions with management and food employees, were used to document the establishments' compliance status. The operational areas most in need of improvement were employee hand washing, time–temperature control for safety foods, date marking of RTE foods and cleaning and sanitizing of food contact surfaces. Unfortunately, lack of compliance of personal hygienic practices has changed little over time (Todd et al., 2010e). Auxiliary factors contributing to the lack of proper hand washing were the lack of convenient hand-washing facilities and/or supplies of hand cleanser/drying devices; temporary placement of mobile equipment in front of a hand sink and the use of hand-washing facilities for other purposes. Avoidance of employees eating, drinking and smoking in food preparation areas and working while experiencing persistent coughing and sneezing are critical to prevent pathogen transmission to foods and food

TABLE 28.3 FDA Report on the Occurrence of Foodborne Illness Risk Factors in Selected Institutional Foodservice, Restaurant and Retail Food Store Facility Types; Effect of Manager Certification on Compliance of Personal Hygiene (US FDA, 2009b)

Facility Type	Total Observed	% In-Compliance	Total Observed	% In-Compliance	Difference (% In-Compliance)
Hospitals	65	72.3	384	84.6	12.3
Nursing homes	165	85.5	290	83.1	−2.4
Elementary schools	168	86.3	295	84.4	−1.9
Fast food restaurants	186	72.0	319	78.1	6.0
Full service restaurants	**237**	**50.2**	**240**	**67.9**	**17.7**
Delicatessens	**153**	**69.9**	**335**	**83.9**	**14.0**
Meat and poultry	156	91.7	273	94.1	2.5
Seafood	125	86.4	257	93.4	7.00
Produce	127	75.6	265	89.4	13.8

NOTE: Bold facility types had overall in-compliance percentages that were significantly higher in establishments with a Certified Food Protection Manager.

contact surfaces. Improper employee actions could be attributed to a lack of knowledge and/or commitment to proper sanitation and controlling microbial growth.

Reinforcing the importance of hand washing should be supported by a management system that includes proper employee training and monitoring of the frequency and effectiveness of hand-washing practices as well as provision of physical infrastructure to facilitate hand washing. Alwood et al. (2004) conducted a survey of retail food establishments to investigate the effect of hand-washing training, availability of hand-washing facilities and the ability of the person in charge (PIC) to describe hand washing according to the Minnesota Food Code on workers' ability to demonstrate food code-compliant hand washing. Only 52% of the PICs could describe the hand-washing procedure outlined in the food code, and only 48% of workers could demonstrate code-compliant hand washing. The most common problems observed were failure to wash for 20 seconds and failure to use a fingernail brush. There was a strong positive association between the PIC being a certified food manager and being able to describe the food code hand-washing procedure, and there was an even stronger association between the PIC being able to describe hand washing and workers being able to demonstrate code-compliant hand washing. Significant associations were detected among correct hand-washing demonstration, physical infrastructure for hand washing and the hand-washing training methods used by the establishment. However, the principal determinant of successful hand-washing demonstration was the PIC's ability to describe proper hand-washing procedure. This is consistent with the results of the FDA studies that the presence of a certified food protection manager is positively correlated to the overall in-compliance percentages in certain facility types, especially in delis, full service restaurants, seafood departments and produce departments (Table 28.3).

Poor personal hygiene, improper holding/time and temperature and contaminated equipment/protection from contamination appear to be the risk factors for which the presence of a certified manager had the most positive correlation. This indicates that management can have a positive influence on personal hygiene and other factors relating to food safety in foodservice establishments. Foodservice and retail food store operators must ensure that their management systems are designed to achieve active managerial control over the risk factors, and regulators must ensure that their inspection, education and enforcement efforts are geared toward the control of the risk factors commonly found to be out of compliance.

Specifically, in the 2009 FDA report of a US national survey of 850 US food establishments for a variety of risk factors reported that observations of poor personal hygiene were extensive but varied across facility types: restaurants (fast food, 24.2%; full service, 40.9%); retail food establishments (delicatessen departments, 20.5%; meat and poultry departments, 6.8%; seafood departments, 8.9%; produce departments/markets, 15.1%); institutional foodservice facilities (nursing homes, 16.0%; hospitals, 17.1%; elementary schools, 14.9%) (US FDA, 2009b).

The types of poor hygienic practices were categorized into five areas. These were: (1) proper, adequate hand washing (hands are clean and properly washed when and as required); (2) good hygienic practices (food employees eat, drink and use tobacco only in designated areas/do not use a utensil more than once to taste food that is sold or served/ do not handle or care for animals present; food employees experiencing persistent sneezing, coughing or runny nose do not work with exposed food, clean equipment, utensils, linens, unwrapped single-service or single-use articles); (3) prevention of contamination from hands (employees do not contact exposed RTE food with their bare hands); (4) convenient and accessible hand-washing facilities (hand-washing facilities conveniently located and accessible for employees); and (5) hand-washing facility, cleanser/drying device (hand-washing facilities supplied with hand cleanser/sanitary towels/hand-drying devices). Table 28.2 indicates the out-of-compliance actions related to these. All these were critical issues to avoid transmission of pathogens, and the levels of out-of-compliance observations were mostly too high. By far the most important was inadequate, improper hand washing (18.4–75.8%); these were followed by inconvenient or inaccessible hand-washing facilities, contact of RTE food with bare hands and lack of good hygienic practices.

The availability of hand soap and sanitary towels/hand-drying devices, though not directly linked to human illness, is an essential component of the management system needed to ensure proper hand washing. Also, this study shows the difference in attitudes to personal hygiene in the different food settings. Unexpectedly, full service restaurants came off worst compared with other facilities and institutions with over 75% no or poorly observed hand washing, almost 50% seen with bare hand contact with RTE food, almost 30% for hand-washing facilities for employees inconveniently located, inaccessible or not supplied with hand cleanser/sanitary towels or hand-drying devices, and almost 25% for inadequate good hygienic practices by employees in general. Even within retail stores some departments demonstrated better hygienic practices than others; the number of observed inadequate hand washing was twice that for delicatessens than in the other departments; this may reflect the attitude of employees that slicing, handling and displaying deli meats carried less of a risk than working with raw meats, seafood and produce.

PRACTICAL ASPECTS OF HAND HYGIENE

Rationale for Hand Washing to Avoid Transmission of Pathogens

Hand washing removes dead skin cells, sweat, sebaceous secretions, associated resident bacteria, transient microorganisms and any organic material adhering to the hands. The transients are the more important to remove as they include the pathogenic bacteria, viruses and parasites obtained through contact with other persons (and their own body fluids), the environment (including water, sewage and animals), unprocessed food or ingredients and food contact surfaces. An effective hand-wash method should remove most of these, facilitated by the use of soaps, detergents and antimicrobial compounds (Todd et al., 2010c). However, hand washing never achieves sterility because of the presence of the resident skin flora; and hands can become recontaminated with transients immediately after the washing and drying process. The efficacy of microbial removal depends on the type and level of microbial and organic matter contamination present, the use of potable versus non-potable water, the wash time, the type and volume of antiseptic (soap/detergent/alcohol gel) used, the extent to which the fingers, palms, backs of hands, subungual area beneath the nails, and wrists are exposed to the washing process, and the amount and vigor of the rubbing of fingers and palms during rinsing.

Diarrheal and respiratory infections can be reduced where hand hygiene programs focus on frequent washing with soap and water and/or use of alcohol antiseptics. Thumbs, palms, spaces between fingers and fingertips (including the fingernail area) are areas where contamination is most likely to remain. A hygienic water source, typically potable water from a piped system or deep well, is vital for effective hand washing. Even in developing countries with limited resources, spread of disease can be limited by proper hygiene. In regions where sanitary waste disposable systems are limited, safe stool disposal (a primary barrier to transmission) may be more important than hand washing before eating, which constitutes a secondary barrier. Additional information and recommendations for hand hygiene are available from Boyce and Pittet (2002), although these apply to healthcare employees and are not for food processors, preparers or servers.

Removal of Soil

The first step in hand washing is removal of the bioburden, typically visible dirt or contamination with proteinaceous material, blood, other body fluids (e.g. fecal material or urine) or food (e.g. meat protein). Water soluble material is easier to remove than fat, oil or grease, but soaps can facilitate the removal of these lipid substances. Hand hygiene practices of food workers are dependent on the type of work involved and the type and nature of the soil. The contamination level of hands after toilet use, changing diapers or handling contaminated raw foods and food packing material can all contribute to soil containing up to 1 million enteric bacteria per hand. Most surface microorganisms are easily flushed off with washing, but some remain in cracks, crevices, skin folds and nail regions.

To reduce the potential for bacterial transfer, food workers may need to wash their hands for longer than 15 seconds or may need to wash more often. Thorough rinsing is important because this action also removes potential skin irritants and contact sensitizers originating in food, soaps, metals and facility disinfectants that could lead to dermatitis. Rinsing with hot water (>120°F/49°C) may cause scalding, irritation, pain, removal of the protective fatty layer, cracking, fissures and possible pathogen colonization, which can discourage future hand washing and result in subsequent increases in microbial counts on hand surfaces.

Food employees should clean their hands in a designated hand-washing sink or approved automatic hand-washing facility and not use other wash facilities such as a sink used for food preparation, janitorial purposes or ware washing. Employees may use sinks for hand washing in toilets common to both patrons and employees if approved by the health authority.

Hand Hygiene Antiseptic Products

Hand hygiene is a general term that applies to hand washing, an antiseptic hand wash, an antiseptic hand rub or surgical hand antisepsis. An antiseptic agent is an antimicrobial substance applied to the skin to reduce the microbial flora or inhibit the growth of microorganisms (Todd et al., 2010d). Examples include alcohols, chlorhexidine gluconate, chlorine derivatives, iodine, parachlorometaxylenol, chloroxylenol, quaternary ammonium compounds and triclosan. Antiseptics were formerly called sanitizers in some settings, and the term is still in use today. Soaps loosen dirt and remove microorganisms from hands in the home, the healthcare environment and food processing and foodservice operations. Soap acts as an emulsifier, suspending oil and dirt and allowing them to be washed off; it decreases water surface tension and binds to dirt, oil and bacteria. Hard water reduces the effectiveness of soaps. Detergents (surfactants) are compounds that possess a cleaning action. They are composed of hydrophilic and lipophilic parts and can be divided into four groups: anionic, cationic, amphoteric and nonionic. Detergents are often referred to as soaps in everyday language. Plain soap is a detergent that does not contain antimicrobial agents or that contains very low concentrations of antimicrobial agents that are effective solely as preservatives. Strong detergents are more effective than soaps for cleaning with hard water because these detergents contain a synthetic surfactant and other chemicals that may improve the cleaning ability. Such detergents are not usually used for hand cleaning. Milder detergents are the most frequently used agents for hand washing and are typically called soaps.

Antimicrobial soap (or detergent) contains an antiseptic agent at a concentration sufficient to reduce or inhibit the growth of microorganisms. These inactivate pathogens more effectively than does soap alone. Triclosan, triclocarban-trichlorocarbamide and parachlorometaxylenol-chloroxylenol are commonly used for their antibacterial and deodorant activities in consumer cleansing products. However, removal of transient microorganisms with either plain soap or soap with an antibacterial compound is not significantly different. However, soaps that include a disinfectant are additionally effective at lowering the resident organism population.

Adequate exposure time is also important for soaps with antimicrobial compounds to be effective. For instance, after repeated use over several days, the residual effect of chlorhexidine gluconate (CHG) substantially reduced the normal skin microflora.

When frequent hand washing is needed, a gentle product is required for acceptance by personnel. Soaps should have good lathering ability, acceptable scent and consistency, and should not contain components that will cause skin irritation or dryness. The effectiveness of these nonantimicrobial soaps can be improved with longer wash time and greater soap volume. Moisturizers and emollients are materials added to hand creams to improve their performance and the feel of the skin. Moisturizers add moisture to the skin, and emollients provide a softening or soothing effect, smoothing dry and scaly skin areas.

Waterless antiseptic agents do not require the use of exogenous water. The term includes different types of hand rubs (liquid formulations, gels, foams, leaflets, towelettes and wipes). An alcohol-based hand rub contains alcohol (in a lotion, rinse, gel or foam) and is designed for application to the hands to reduce the growth of microorganisms. After application, the individual rubs the hands together until the agent has dried and by this process reduces the number of viable microorganisms on the hands. Such preparations may contain one or more types of alcohol with excipients (inactive substances used as carriers for the active ingredients of a medication), other active ingredients and humectants (emollients or moisturizers, e.g. propylene glycol).

Effect of Friction during Hand Washing

Friction is well known as one of the most important elements in hand washing, dislodging microflora from the skin surface during both the washing and rinsing stages. Unfortunately, any aspect of the hand-washing process that decreases friction (e.g. soft water versus hard water, soft bristle brushes versus coarse bristle brushes) and any type of soap by its nature will reduce the mechanical removal of any microflora, particularly when hands are soiled. Wipes also decrease friction.

Cleaning Long and Artificial Fingernails

Outbreaks have been linked to workers with long or artificial fingernails, which are very difficult to clean even with appropriate soaps, hand rubs or gels. The greatest reduction of inoculated microbial populations is obtained by washing with liquid soap plus a nailbrush, and the least reduction was obtained by rubbing hands with the alcohol gel. Because nail polish or varnish can chip off and fall into food, and any cosmetic item that can be brushed or fall off, there should be a policy that no staff are allowed to wear nail polish, varnish, fake fingernails or false eyelashes in food processing and preparation areas. Another reason for banning fingernail polish is because chipped fingernail polish or fingernail polish worn for more than 4 days fosters increased bacterial numbers on the nails. These results indicate that best practices for fingernail sanitation by food workers include maintaining short fingernails and scrubbing them with soap and a nailbrush while washing hands; nails should not be polished. Artificial nail use by food and healthcare workers should be prevented, and the FDA Food Code (US FDA, 2009a) prohibits the use of artificial nails by food workers unless gloves are worn.

Duration and Frequency of Hand Washing

Hand-washing efficiency is affected by two aspects of hand washing: how well (soaps, friction and duration) and how often (frequency) it is done. Both aspects are important for limiting hand contamination. The duration of the hand-washing process is a critical factor for removing microorganisms, and hand-washing times of 15 to 30 seconds have been recommended by different agencies around the world with slightly different emphases. The 2009 version of the Food Code states that hands and arms should be washed for at least 20 seconds, with 10 to 15 seconds of vigorous scrubbing, and that individuals must use a paper towel or other barrier when touching surfaces to prevent recontamination of hands after washing. The World Health Organization states that 40 to 60 seconds total should be used for washing hands with soap and water, rinsing and drying them, and 20 to 30 seconds for disinfecting the hands with an alcohol-based formulation. Hand-wash times are sometimes encouraged by suggesting that everyone recite the alphabet or sing "Happy Birthday" or a similar-length ditty during washing to obtain maximum pathogen removal. The American Society for Testing and Materials (ASTM) recommends wetting hands under warm water at 100 to 108°F (38 to 42°C), applying 3 ml of hand-washing product and rubbing all hand surfaces vigorously concentrating on interdigital parts (Guzewich and Ross, 1989). Washing for too long may damage the skin, bringing the skin resident flora to the surface, increasing the number of microorganisms recovered from hands and damaging the epithelial layer. Unfortunately, observations of workers in different settings have revealed that less time is spent on actual washing than has been recommended, sometimes as low as 9–11 seconds, and soap was not always used.

Hand-washing Water Temperature

Common sense suggests that hand-washing water temperatures should be as hot as is comfortable (between 110 and 120°F/43 and 49°C). However, water temperature has been shown not to be influential in hand hygiene efficacy when plain or antimicrobial soaps are used. No significant differences in bacterial reductions of either resident or transient bacteria were found for any of the washing and rinsing temperatures during normal hand washing with a non-antimicrobial soap. Vigorous friction during washing is more effective for removal of bacteria than is the type of soap, the length of the wash time or the temperature of the water. However, washing and rinsing hands at excessively low temperatures, equivalent to those found in a refrigerated cutting room, is uncomfortable and also may result in poor hand-washing compliance. The 2001 FDA Food Code amended the 1999 version by decreasing the recommended water temperature for hand washing to 100°F/37.7°C, and has remained ever since (US FDA, 2009a). Thus, the temperature of hand-washing water should be comfortable, preferably warm but not hot.

Double Hand Washing

Double hand washing is meant to address residual fecal finger contamination, including entrapment of feces in the subungual region of the nails after defecation or contact with toilet facilities including toilet seats and door knobs or handles. In this procedure, a nailbrush is used to produce lather on fingertips, hands and arm surfaces during initial hand washing. The hands are then rinsed and relathered, without using the nailbrush, by vigorously rubbing hand and arm surfaces, thoroughly rinsing and then drying with disposable paper towels. A double wash has been recommended when employees begin a shift and after they use the toilet. Although this sequential approach has been considered to enhance the efficacy of hand washing, research has shown only a slight gain in cleanliness with the second washing, but it may be valuable as an alternative to gloving because of the high degree of enteric bacteria removal. However, if a nailbrush is used about 10 times, enough organic material will accumulate in the brush storage sanitizer solution that bacteria could begin to grow, allowing the storage solution to become a source of bacterial contamination for workers' hands. This possible contamination is the reason why nailbrushes are not recommended for use in high-care food handling facilities in Europe (Todd et al., 2010c).

Issues at Hand-washing Stations

A worker can become contaminated from hands and clothing at hygiene stations and automated hand-washing machines from organisms deposited by a previous user on water faucet handles, sink counter tops, door handles and soap dispenser buttons (Todd et al., 2010d). Whereas an ideal hand-washing station has faucets that operate automatically or through use of a knee, foot or elbow, in most restrooms and many food preparation facilities, these types of faucets are not available, increasing the risk of cross-contamination through use of contaminated faucet handles. Thus, inadequate hand washing may actually further spread microbial contaminants, particularly environmentally resistant enteric viruses. One solution to prevent recontamination of cleaned hands is to use disposable paper towels for turning off faucets and opening restroom doors. Another possible source of contamination is from water droplet sprays and aerosols dispersed from the water flow of taps or nozzles and the action of the hands during hand washing. Managers of food preparation operations should be encouraged to check for water droplet transfer by observing the station wetness after use, and modify the faucet system appropriately.

Hand-washing machines have been used as a way to improve hand-washing effectiveness and compliance since the washing time is controlled. Some units are also designed for glove washing. Automated cleansing systems have been considered to reduce variability in hand-washing effectiveness. However, manual hand washing is sometimes preferred by many employees, and less observed use of the automatic sinks by employees would decrease overall hand-washing compliance. There have also been instances of water contaminated with potential pathogens, and some designs of hand-washing machines make contamination of sleeves and already washed hands possible. Nevertheless, the FDA authorizes the use of approved automatic hand-washing facilities.

Drying of Hands

Moist hands transfer microorganisms more readily than dry hands and, therefore, drying is an important step in the hand-washing process, but is often ignored by individuals who do a quick rinse with or without soap followed by a dab on a towel or simply shake the large droplets off. Effective hand drying includes enough time to remove moisture through absorption, and microorganisms on surface skin layers through friction on towels. If pathogens are deposited onto reusable towels, they can survive long enough so that successive users' hands become contaminated. Thus, single-use paper towels are generally considered to be more hygienic than cloth towels. The main issue is that the time taken to dry hands and wrists thoroughly is considered too long. Electric air dryers are increasingly available in both food facilities and public areas, and today's models are much faster at drying. The concern that microorganisms might accumulate in the driers and create aerosols when they are turned on has not been shown to be the case, and air drying has been shown to produce the highest reduction in the numbers of bacteria and viruses compared with cloth towels.

Alcohol-based Antiseptics and Wipes

Alcohol-based hand rubs (ABHRs) are antiseptics containing isopropanol, ethanol, n-propanol or a combination of these and are now in common use. They are more effective than many non-alcoholic products when hands are relatively clean from soil, provided that enough of the compound is used and the exposure time is not too short (Todd et al., 2010d). They also do not require thorough rubbing as do soap and water and so may reduce the risk of dermatitis. However, there are issues that prevent their complete acceptance, causing dryness of skin and stinging of cuts, and they do not act well when grease or food particles are present. Flammability may be a concern but there are very few reported incidents of the alcohol burning users. There is, however, no residual effect compared with products like CHG or triclosan, and these are sometimes added to alcohols. There can be a 3.5 $\log_{10}$ reduction of bacteria on hands after a 30-second application and 4–5 $\log_{10}$ reduction after 1 minute, but viruses are more difficult to inactivate. Alcohol at 60–95% denatures enveloped viruses but not spores, oocysts and non-enveloped viruses, such as NoV, rotavirus and HAV.

Foam sanitizers may be better than gels in removing microorganisms, but they are more expensive to use. Although they are used more and more exclusively in healthcare institutions, alcohol-based antiseptics should not replace hand-washing and drying policies in the food industry because of a greater variety of soils encountered by employees, particularly with meat, poultry and fish products. Therefore, for most operations, the hands of food workers should be washed before application of hand sanitizers.

Rinsing hands under running water and use of alcohol antiseptic followed by vigorous wiping with a paper towel is the most effective approach to removing viruses. Antiseptic wipes are widely used and typically contain benzalkonium chloride, moisturizers, wetting agents and emulsifiers. Dry tissue wiping combined with antimicrobial moist wipe use without rinsing is at least as effective as washing with soap and water. Again, wipes are most effective for removing microorganisms on clean surfaces, and not where there are many food particles and fecal matter.

Vigilance during Outbreaks

When there are reported community outbreaks, more than the usual number of diarrheal illnesses in confined locations such as army bases, cruise ships and refugee camps, or even family-associated enteric illnesses, food operation managers, employees and home carers must be especially vigilant to avoid the spread of pathogens. Certain microorganisms with low infectious doses have been known for decades to quickly infect exposed persons through direct and indirect contact. Such pathogens include some of the *E. coli* strains, *Salmonella* Typhi, *Shigella dysenteriae*, *Vibrio cholerae* and more recently NoV. Cholera and shigellosis outbreaks are good examples of this.

The 1994 cholera epidemic in Guinea-Bissau, West Africa, resulted in over 15,000 reported cases and 300 deaths. The outbreak of cholera was strongly associated with eating at a funeral with a non-disinfected corpse and with touching (i.e. transporting, washing) the body. Because a corpse will commonly leak feces, persons handling dead bodies are likely to be exposed to gastrointestinal organisms (such as *V. cholerae*). The cultural practice of serving meals to guests at funerals by the same people who prepared the body was discouraged during this outbreak, and community leaders were informed about the risk of cholera transmission during funerals, that meals should not be served at funerals and that bodies of persons dying of cholera should be disinfected. Although this was done, outbreaks continued in several villages following funerals in some regions of the country. This former widespread practice only rarely occurs today because of education and government action. However, funeral employees may be exposed through direct contact with the victim's body and soiled clothes, and transmission can occur directly through the fecal–oral route. While this should provoke caution in handling the body itself, it is also important to note that equipment used by the funeral industry such as storage facilities, vehicles and stretchers may also be contaminated. Thus, especially during a cholera or any other enteric disease epidemic, disinfection of bodies and equipment must be assiduously adhered to.

The following are six other examples where the rapid spread of disease led to large outbreaks (Todd et al., 2007b); (1) at a resort in Haiti in 1984, a worker infected with *Shigella flexneri* transmitted shigellosis to 1136 guests over a 3-week period where eating a raw or very rare hamburger and having a roommate who was ill and a younger age were significantly associated with acquiring the disease; (2) at the 1987 Rainbow outdoor gathering in North Carolina, many thousands of attendees were infected with *Shigella sonnei* through water, food and person-to-person secondary spread because sanitary facilities were very limited outdoors; (3) at a Michigan music festival in 1988, again mainly outdoors, tofu salad contaminated by infected food workers resulted in 3175 cases of shigellosis caused by *Shigella sonnei* over a 3-week period; (4) in Japan in 1989, thousands of school children ate meals contaminated with NoV prepared in a central commissary; (5) NoV outbreaks twice occurred on two consecutive cruises around the Hawaiian Islands, once in 1990 and again in 1992; and (6) in 1990, Mozambican refugees in a Malawian camp were infected with cholera via contaminated water and food for an undetermined time; hands were placed into stored household drinking water, and there was improper reheating of leftover food. All of these cases reinforce the requirement that during times where there are community outbreaks, or infections at higher levels than normal, food processing and service operators need to be extremely diligent in enforcing personal hygiene and sanitation guidelines, and where necessary have a temporary closure of operations; particular attention should be given to consumption of food where preparation and hand hygiene facilities are limited such as outdoors or in camps.

BARRIERS IN FOOD OPERATIONS TO LIMIT SPREAD OF PATHOGENS

Barriers to Contamination of Food

Physical and chemical barriers to prevent microbial contamination of food are introduced to prevent or reduce the transfer of pathogens to the foods or food contact surfaces from the hands of an employee, from raw meat, poultry, fish/shellfish, fruits or vegetables or unprocessed ingredients, or from the environment in the facility or from outdoors where dust and pests may enter. In food processing and service operations, direct contact of food by hands should be prevented by the use of well-designed barriers, especially when gloves are not worn (Todd et al., 2010a; GMA, 2009). Although many barriers have been used for decades in food processing and foodservice operations, their effectiveness is sometimes questioned or their use may be ignored. Physical barriers include properly engineered building walls and doors to minimize the flow of outside particles and pests to food storage and food preparation areas; food shields to prevent aerosol contamination of displayed food by customers and workers; work clothing designated strictly for work (these include gowns, overalls, boots and hair nets), including pathogens from infected family members; and utensils such as spoons, tongs and deli papers to prevent direct contact between hands and the food being prepared or served. Food processing buildings should be so designed that the opportunities for finished product being recontaminated by the raw ingredients are minimal; this requires a flow of food through different zones to separate those for raw materials and receiving, mixing and other precooking steps, where general good manufacturing practices apply, from cooking, packaging and storage areas. Employees should only enter the zone areas they work in, or go from a more contaminated zone to a cleaner one only after changing clothing and utilizing sanitation steps like boots in disinfectant. Unfortunately, contamination can enter rooms from air ducts, fans, eroded flooring, leaky roofs or drains, damaged and wet floors, difficult-to-clean equipment, conveyor belts and cleaning and maintenance tools such as mops and buckets.

Cash and paper money, and even credit cards, should be handled separately from any other operation involving preparation or serving of food. This is preferably done by two workers, or changing gloves and washing hands between handling money and touching food. In practice, in fast food facilities and small enterprises, this is not always done because of convenience and speed required to serve patrons.

Chemical barriers include sanitizing solutions used to remove microorganisms (including pathogens) from objects or materials used during food production and preparation and to launder uniforms, work clothes and soiled linens. Laundering, especially for highly contaminated material, e.g. with feces, blood and vomitus such as in healthcare facilities or in slaughterhouses, may create aerosols of enteric pathogens and not effectively eliminate viral pathogens.

One final point is the food that employees eat. There are no regulations or guidelines on this, but clearly if food workers consume food that is likely to contain pathogens, they are more likely to be colonized and spread these organisms at work. Thus, consumption of raw or minimally processed food items identified as high risk foods should be discouraged. Managers have no control of employees outside the work environment but they can advise them not to eat risky foods and can refuse to serve these, such as steak tartar, alfalfa sprouts and raw milk cheeses, in company cafeterias.

Effectiveness of Gloves

When worn correctly in healthcare environments, gloves have consistently reduced hospital-acquired infection rates and these were adopted later in food operations. Although utensils have hygienic value during food production and preparation to limit contact between workers and food, for many operations hands need to be in regular contact with food much of the time. Glove use has been emphasized through the widespread distribution of the US Food and Drug Administration (FDA) Food Code (2009a), and their use has increased in hospital foodservice facilities operated under hazard analysis critical control point (HACCP) systems. When gloves are worn properly, the risk of pathogen transmission can be reduced considerably, but careful inspection of gloves must be done to ensure that it is appropriate for the required tasks (Todd et al., 2010b). Hoelzer et al. (2012) consider that gloves can be a major source of *Listeria monocytogenes* contamination in retail deli operations. Their use certainly can be monitored by management and food control agencies by observing the gloves on workers and the discards in trash receptacles, and by reviewing glove purchase invoices. However, it is more difficult to determine their proper use to avoid food contamination, and glove use still is not mandatory in many jurisdictions because of conflicting data on their utility.

Although gloves give some benefit, they do not completely prevent pathogen transfer and hand-washing compliance is less when gloves are worn. Some studies show that indicators of pathogens were similarly present on gloves and hands, and found that the outside of the glove was highly contaminated at the end of a 3-hour period regardless of whether gloves had been changed or hands had been washed, and that bare hands with hourly washes and antiseptics provided a higher level of hand sanitization than did gloved hands with and without washing. Hand washing and glove use were more likely to occur in conjunction with food preparation than with other activities and when workers were not busy, and in general workers who wear gloves do not remove them and wash their hands as often as they should. Gloves are also prone to pinhole leaks or punctures by jewelry on fingers or artificial or long fingernails, as well as operational stress. Occlusion of the skin after extensive use with warmth and moisture build-up can lead to microbial growth, particularly from resident staphylococci. Therefore, gloves should be considered an adjunct to and not a replacement for hand washing for food production and preparation operations.

Arguments for glove use are (1) gloves protect the worker from foods that can cause damage to the skin, e.g. acidic ingredients, (2) gloves protect the food from direct hand contact, (3) glove use is easily observed to verify hygiene compliance, unlike assessing hand-washing frequency and thoroughness, and (4) gloves can be used to cover skin damage or infections. Arguments against glove use are (1) gloves can reduce operational dexterity and increase the risk of injury, (2) higher levels of food contamination are possible in the event of glove failure, (3) a small percentage of gloves have pinhole leaks that are not possible to detect before use, (4) gloves can be worn for longer than they should be, (5) gloves give a false sense of security as a substitute for good hand hygiene practices, and (6) gloves increase the risk of hand irritation. Thus, proper hand hygiene is essential in addition to gloving and other barriers. The best approach is to use multiple hurdles, including gloves, other barriers and appropriate hand washing, to prevent transfer of bacterial, parasitic and viral pathogens to food.

The procedure for removing disposable gloves to minimize contamination of the hands and environment can be done as follows: (1) grasp one of the gloves and pull it part way off so that the glove will turn inside out; (2) with the first glove partially on the hand remove the second glove so that the exposed bare hands or fingers do not touch the outside of either glove; (3) with the first glove over the fingers, grasp the second glove near the cuff and pull it part of the way off to turn the glove inside out, keeping the second glove partially on the hand to avoid touching the outside surface of the first glove with the bare hand; (4) pull off the two gloves at the same time, being careful to touch only the inside surfaces of the gloves with the bare hands; (5) dispose of the gloves by placing inside out in the trash; (6) wash hands thoroughly. Other important points to consider are: do not touch the face or clothing with contaminated gloves; change gloves when heavily soiled or if they are torn; do not wash or reuse disposable gloves; and, in any operation with gloves and hands, work from clean to dirty to minimize potential contamination risks.

Food Shields and Utensils as Barriers against Contamination

Despite a lack of scientific data that food shields (formerly called sneeze guards) are effective for protecting food from airborne contaminants, most food businesses with a buffet, salad bar or display of saleable RTE food use these guards. Unfortunately, food shields probably cannot protect food from highly aerosolized particles such as viruses; although a vomiting event in a foodservice area is a rare event, all exposed food must be discarded (Todd et al., 2010a). Utensils adequate for dispensing foods include spatulas, tongs, scoops, spoons, ladles, single-use dispensers and thin papers for grasping and weighing deli meats and serving bakery items. Some of these utensils should also be used for mixing foods and handling potentially contaminated foods such as raw meat, so that the hands of food workers are less likely to become contaminated. When food operations adopt a policy that includes single-use items to avoid risks of contamination, these items must never be reused. These items also must be protected from contamination until their use; specifically, they must not come into contact with food or the skin or mouth of a person. Utensils that are not single use should be thoroughly washed and sanitized before reuse. In some operations utensils and papers tend to be used inconsistently or not at all. Outbreaks attributed to contaminated utensils are most likely contaminated by a food worker. Food employees also should wear hair restraints such as hats, hair coverings or nets, beard restraints and clothing that cover body hair, which are designed and worn to effectively keep their hair from contacting exposed food, food contact surfaces including clean equipment, utensils, linens and single-use items.

Protective work clothing means any clothing provided by the employer to protect the worker from hazards in the workplace or to prevent contamination of the workplace by materials the worker may bring into it on their personal clothing. Work clothes, such as overalls, can be sources of contamination of other persons, food contact surfaces and foods themselves. In general, protective clothing of food workers should not be worn while they are eating, drinking, smoking or visiting the toilet. Diapering a sick child at home while wearing a uniform has led to a foodborne disease outbreak in a healthcare institution. Yet, there are surprisingly few instances in food codes to request the proper location of toilets, the effectiveness of washrooms and that outer clothing should be removed before any toileting activity. However, it is preferable to install toilets that will flush automatically and have sink faucets and antiseptic dispensers designed to have a minimum of hand contact.

Management should insist that separate sinks are installed for food preparation, dishwashing and cleaning, and hand washing, and they are used properly. Protective clothing, such as uniforms and overalls, should not be capable of holding anything that could become foreign matter, as well as pens, repair tools, or knives, etc., as there is a chance these items could fall out and contaminate the food items, e.g., only have pockets on the inside, or preferably no pockets at all with access to pockets only in non-work clothes. Washrooms with toilets and hand sinks should be some distance away from food processing and preparation areas but not so far as to discourage their use on a regular basis. Design the washroom facility, including installing hooks on which to hang the clothes, so that staff have to remove the outer layer of protective clothing before engaging in any toileting activity (McFoodies, 2012). Dirty uniforms, overalls and other clothing should be placed in a receptacle or movable container in order to deliver them to the laundry. If employees have to launder their own work clothes, there should be periodic checks to ensure this is done as frequently and properly as required. As previously indicated, laundering of highly infectious material (most likely in healthcare facilities but also in homes where there has been diarrhea) can cause aerosols and infect the user.

The 2009 FDA Food Code states in general that workplace contaminants means chemical or biological substances arising from workplace processes, and may include airborne contaminants or contaminants on surfaces, such as tables, benches, eating utensils, clothing or skin. The employer must ensure food is not stored or consumed in areas where the presence of these contaminants could result in a hazard to workers as a result of ingestion with food or beverages. Managers should observe facility personnel for clean outer clothing, effective hair restraints, prohibited jewelry and the condition or protection of fingernails, and if the employees regularly change their clothes in the establishment, lockers or other suitable facilities shall be used for the orderly storage of employee clothing and other possessions.

In today's society where cell phones are commonplace, they should not be allowed in food processing and preparation areas as they are a distraction from the work at hand and can also fall into open containers of food; in addition, because of frequent handling, they can be contaminated with many types of microorganisms including pathogens. Make it a policy that no staff are allowed to take mobile phones into food processing areas.

Improving Compliance

Lack of compliance for hygienic practices is notorious in both the healthcare and food industries, particularly the washing and/or sanitizing of hands (Todd et al., 2010e). This can be illustrated by a study of food workers from 29 catering businesses that produced high-risk foods in Wales (Clayton and Griffiths, 2004). Hand hygiene practices were carried out adequately on only 31% of the required occasions and were not even attempted 55% of the time. Errors included touching potentially contaminated objects or surfaces including hair and face without subsequent hand washing, improper handling of potentially contaminated foods, and failure to adequately clean potentially contaminated food contact surfaces and also frequently used objects such as telephones, cupboards and shelves, food containers, equipment and door handles. Two main hand hygiene errors were identified: (1) a failure to use soap and (2) a failure to dry hands adequately. Infrequent cleaning of such surfaces coupled with a failure to wash hands may help explain the high bacterial counts noted on these same surfaces in other studies.

Other common challenges that make it hard to clean include issues that affect employee compliance with company and heath authority hygiene guidelines and regulations: (1) lack of facilities providing sufficient warm water and hand driers in an easily accessible location; (2) lack of employee motivation; (3) lack of education and training; and (4) lack of a managerial role model. One of the most effective tools for change is the culture of the organization to encourage food safety and hand hygiene through example and pertinent information.

Hand Hygiene Occasions

Hands should be clean and properly washed in food operations before or after certain actions. These include: (1) after touching bare human body parts other than clean hands and clean, exposed portions of wrists and arms; (2) after visiting the washroom for toileting or toilet cleaning; (3) after caring for or handling any kind of animal; (4) after coughing, sneezing, using a handkerchief or disposable tissue, using tobacco, eating or drinking; (5) after handling soiled equipment or utensils, e.g., after disassembly and cleaning of processing equipment; (6) before donning gloves for working with food; (7) during food production, preparation or service, as often as necessary to remove soil and contamination and to prevent cross-contamination when changing tasks, e.g., when handling both raw and RTE food; and (8) after engaging in other activities that contaminate the hands, as specified in hygiene guidelines and management policies.

There are many guidelines on how to wash and dry hands; these differ slightly in wording for approaches and times. The basic steps are outlined as follows: (1) remove watches, bangles, all jewelry, except a simple wedding band before any operation producing or preparing food; (2) rinse hands, wrists and arms (including prosthetic devices if appropriate) under clean, warm, running water, rubbing fingers and palms to remove any visible soil and food particles; (3) apply the recommended amount of antiseptic cleaning compound, typically 1 ml but more may be required after certain operations; automatic soap dispensers are preferred to bar soap, but these have to be monitored to maintain sufficient liquid soap or foam and to prevent their contamination though employee use; (4) rub hands together vigorously for 10 to 15 seconds while ensuring that soil is removed from the palms, between fingers and backs of hands and wrists; (5) remove visible soil from under the fingernails with a fingernail brush as recommended; (6) thoroughly rinse hands under clean, running water; (7) thoroughly dry hands, wrists and arms using single-use disposable towels, a continuous towel system that supplies a new towel at each use, a heated air, hand drying device or a pressurized air blast, as approved by the regulatory authority.

Employees should be careful not to recontaminate hands during the washing and drying operations by touching faucets, sinks and contact with toilets; automatic hand washers or use of paper towels to turn off faucets and open doors reduce this risk. In smaller facilities where the only employee hand-washing facility is in the public washroom (toilet), particular care must be taken to avoid recontamination. The whole hand hygiene operation should take about 40–60 seconds, depending on how much soil is originally present on the hands. In addition, double hand washing may be recommended for employees after toileting or after operations involving hands contaminated with much soil.

EXCLUSION OF INFECTED EMPLOYEES TO WORK IN SPECIFIC FOOD OPERATIONS

Policies for Food Worker Exclusions

Enteric organisms from fecal sources are excreted during an infection, whether the individual is symptomatic or not, but the infection exists over a limited time period (usually hours to weeks). Many of these enteric pathogens are of concern to food workers, e.g. *Salmonella*, *Shigella*, HAV and NoV. These pathogens can contaminate the hands after defecation or from touching fecally-soiled clothing or surfaces. The use of toilet paper is no guarantee of preventing feces reaching uncontaminated hands, and finger contact is almost 100% certain for those individuals with diarrhea. Thorough hand hygiene following any defecation has to be ingrained as a lifelong habit. Changing diapers with loose stools, cleaning after episodes of vomiting or diarrhea by family members, washing dirty linens and contact with sick or healthy pets also are well-established risk factors. Fecal matter and vomitus can remain on clothing following home clean-ups and be transmitted to kitchen environments at work. There is no way for management to detect such a scenario and workers must take precautions before returning to work, e.g., change out of street clothing into work uniforms or overalls, thoroughly wash their hands and arms, and admit potential exposure to pathogens.

Policies do exist in many jurisdictions, however, for the exclusion of food workers infected with pathogens from working with food, but these are difficult to implement effectively. Whereas it may be possible to identify an individual showing symptoms of frank enteric disease like vomiting and diarrhea, it is much more difficult to determine if workers are excreting pathogens when they appear perfectly normal healthwise. Post-symptomatic workers may continue to excrete pathogens, but at lower rates, for days, weeks and occasionally years. Shedding duration can be measured only by a regular stool-testing regimen, and even so this approach often is ineffective, costly and not always recommended because of the intermittent pathogen excretion. Even with an apparently effective policy, an outbreak can occur; several hundred cases of salmonellosis occurred from airline meals in 1976 because the infected employee was not identified during regular inspection and testing of the establishment. The extent of asymptomatic food workers excreting pathogens can be estimated from a study in Turkey where 1.6% of 307,954 workers in military food-service operations were positive for enteric pathogens in stool specimens (Kir et al., 2006). However, the World Health Organization concluded that asymptomatic carriers of nontyphoidal *Salmonella*, *Shigella*, *V. cholerae* and enteric viruses pose only minimal risks as long as good hygiene is practiced (WHO, 1989).

Stool Testing

Stool testing and exclusion of workers has been an issue for many decades, and recommendations differ among jurisdictions, with no consistent approach for stool testing and worker exclusions. Typically, three negative stool specimens are required before an infected food employee can return to work. However, the clinically well food worker with formed stools should be allowed back at work without further examination of fecal specimens. Generally, pathogen-negative stool samples, either pre-employment or from an employee recovering from a diarrheal illness, are not necessary conditions of employment or return to work. Exceptions may be considered for typhoid and paratyphoid infections and infection caused by enterohemorrhagic *E. coli*, since these are severe diseases and infected individuals can excrete the pathogens for long periods after recovery. When *E. coli* O157:H7 infection is identified in a food worker, the worker should be excluded from work until bowel movements are normal and two negative fecal samples taken 48 hours apart have been obtained. Symptomless contacts of a person with HAV infection can continue food handling, but workers who have symptoms of hepatitis, have been in an outbreak or have been associated with family members suffering from HAV infection, or have traveled to regions where HAV is endemic would be excluded from work until they have a medical release based on laboratory testing. Those infected with norovirus should be treated the same way as for HAV.

Unfortunately, from the above information it is impossible to determine when employees are free from a pathogen since excretion can occur sporadically for many weeks or months after apparent recovery. However, it is likely the pathogen levels in their stools become increasingly low over time and even if some fecal matter remains on fingers after defecation, any organisms can be removed through vigorous hand hygienic practices. The same position should apply for asymptomatic persons with levels in stools assumed to be lower than for those with frank enteric symptoms. Thus, the best policy is to exclude those who show signs of nausea, vomiting, abdominal cramps or diarrhea (concern should also be raised for sore throat, fever and jaundice though these may have non-enteric pathogen origins) until they recover. Treat every employee as if he or she is infected and excreting at low pathogen levels, but that proper hand washing and drying and other barriers to pathogen spread are taught and enforced. Employees should report gastrointestinal illness to their supervisors but this may not happen until the symptoms are obvious to co-workers.

If suffering from an illness involving jaundice, diarrhea, vomiting, fever, sore throat with fever, discharge from ear, eye and nose or visibly infected skin lesions (such as boils, cuts) a food worker should report to his or her supervisor. The supervisor should then use discretion as to whether or not the person should be subjected to certain restrictions or suspended from food processing, handling or serving duties. Medical advice may be necessary in making this decision (WHO, 2000).

Health department clearance for some symptoms may be required by law which may or may not expect three consecutive stool specimens to be negative. Also, employees returning from a region of the world with certain endemic enteric diseases like typhoid and dysentery should be asked to be cleared by the health department. When employees or patrons give indications of enteric illness, or there are community outbreaks, managers should be particularly vigilant about preventing spread of pathogens in the food production, preparation and serving environment. Since young children tend to excrete higher levels of pathogens than adults, employees with such children should also inform their managers and take special precautions to avoid any fecal transfer.

Lack of Health Benefits

Unfortunately, although more people are employed in the food industry (>15%) than in any other sector in the United States, benefits are very limited for these workers, often at the minimum wage level, and this means there is a risk of ill employees going to work. In the EU wages for food workers are generally higher than in the United States and sick leave is likely to be more prevalent. When people go to work infected and ill, they perform less efficiently and can also infect others, which can contribute to more employees who stay at home ill and also those who decide to come to work as newly infected persons (Todd et al., 2008a). Paid sick leave policies have been shown to reduce the rate of contagious infections in the workplace by isolating sick workers at home. Failure to take time off to regain one's health actually led to longer absences because health worsened, and as an illness spread within the workplace additional workers were affected, raising the total employee absence time. For instance, in a 2012 report of the Food Chain Workers Alliance, entitled "The Hands That Feed Us: Challenges and Opportunities for Workers along the Food Chain" (Food Chain, 2012), 79% of food system workers do not have a single paid sick day, or do not know if they have paid sick days, and 58% lack health coverage. Consequently, 53% admitted to picking, processing, selling, cooking and serving food while sick for an average of at least 3 days per year. This issue of employees in the food industry being paid low wages and having few or no health benefits applies to all regions of the world, and is likely to be worse in developing countries. It will be interesting to see whether existing policies for paid sick leave will remain in place during the current economic downturn, especially in Europe. Thus, until sick leave policies are more universally accepted, the risk of a food employee transmitting pathogens at work is high.

CONCLUSION

Although hand hygiene is not a new concept for prevention of disease spread, it still remains the best strategy to reduce the opportunities for transfer of microorganisms to foods. Exclusion of employees colonized and excreting enteric pathogens is a generally failed policy to detect carriers and should only be applied to workers returning from countries with endemic diseases such as typhoid and cholera, or who have been exposed to enterohemorrhagic *E. coli*, such as on a farm, or exposed to ill persons. The asymptomatic carrier state is perhaps normal for many persons, and certainly those recovering from an enteric infection can excrete the causative pathogen for weeks and months. Also, about half the population harbors *Staphylococcus aureus* in the nasopharynx, and its presence should not be used to close an operation by health authorities because it is found in one or more workers (as has happened on many occasions, particularly in developing countries), or dismiss a carrier. Proper investigations of outbreaks can help to identify risk factors associated with food workers (and other sources) and aid in implementing appropriate prevention and control strategies (Todd et al., 2011).

To limit pathogen transfer, the use of appropriate physical barriers, e.g., building design, utensils, food shields, gloves, hair nets and an effective hand hygiene policy, is the best strategy food processing and foodservice managers can implement. Yet, employees continually forget to wash their hands or do so ineffectively, and foodborne outbreaks associated with lapses in hand hygiene in food operations, particularly foodservice facilities, occur right up to the present time. The key components affecting risk of pathogen transmission are hand hygiene compliance, hygiene efficiency and cross-contamination. Compliance reflects (1) the frequency of the cleansing process, (2) the willingness to adhere to the recommended procedures, (3) hygiene efficiency through the combined effects of washing, brushing, rinsing, drying, sanitizing, gloving, etc., (4) prevention of cross-contamination by having more hands-free operations, (5) handling less raw and more processed food, and (6) working on sanitized surfaces. The parts of the hands that are most likely to retain fecal or food contamination are the thumbs, palms, spaces between fingers and fingertips (including the fingernail area), and employees and employers need to focus on ensuring these are well cleaned before they start or resume work on RTE food.

Training and monitoring of activities alone are not sufficient. Monitoring can be accomplished by direct observation and recording of positive and negative behaviors or by some automated system of recording use of water, soaps or ABHRs. A major consideration is the ability to alter human behavior by peer pressure, such as positive deviance, or through rewards and penalties applied to both management and other employees. These issues lead into the critical impact of the cultural values of both society and workers' organizations. The climate of an institution is a key element in promoting positive change (Griffith et al., 2010). There are two components to this: the food safety culture of the organization, and the knowledge and practice of the person in charge (PIC) of the specific operations such as in a franchise. How management creates and supports the food safety culture within a business may be the most important factor for determining whether that business can avoid violations on inspection, foodborne illness of its patrons or costly recall of its products. The more confident the business is in the production and/or service of its food, the more likely it will implement proper hygienic measures and institute effective training of the staff, both managers and other employees.

The PIC of the workers on the line needs complete knowledge of food safety risks in the company's operations and why hand hygiene, including adequate washing, is necessary to avoid contamination of the food and its contact surfaces. He or she has to demonstrate making use of the different barriers, including washing hands frequently. The presence of a well-trained PIC provides a system for routine observation and feedback and for making hand hygiene easy and convenient with necessary supplies regularly stocked, putting reminders in the workplace, requesting better engineered facilities, avoiding overcrowding, understaffing and excessive workload, facilitating skin care for workers' hands, and implementing administrative sanctions and/or rewards. Collaboration and advice from local health departments and their inspectors should be encouraged, because these departments should be more involved in education than in regulation to effect change. Any change in operational practices is where vigilance should be heightened, and employees and their actions must be carefully monitored to determine whether new risks may arise.

References

Allwood, P.B., Jenkins, T., Paulus, C., Johnson, L., Hedberg, C.W., 2004. Hand washing compliance among retail food establishment workers in Minnesota. J. Food Prot. 67, 2825–2828.

Boyce, J.M., Pittet, D., 2002. Guideline for hand hygiene in healthcare settings: recommendations of the Healthcare infection control practices advisory committee and the HICPAC/SHEA/APIC/IDSA hand hygiene task force. Morb. Mortal. Wkly. Rep. 51, 1–45.

Clayton, D.A., Griffith, C.J., 2004. Observation of food safety practices in catering using notational analysis. Br. Food J. 106, 211–227.

Food Chain Worker Alliance (Food Chain), 2012. The hands that feed us: Challenges and opportunities for workers along the food chain. <http://foodchainworkers.org/wp-content/uploads/2012/06/Hands-That-Feed-Us-Report.pdf> (accessed 18.05.2013).

Greig, J.D., Todd, E.C.D., Bartleson, C.A., Michaels, B.S., 2007. Outbreaks where food workers have been implicated in the spread of foodborne disease. Part 1. Description of the problem, methods, and agents involved. J. Food Prot. 70, 1752–1761.

Griffith, C.J., Livesey, K.M., Clayton, D.A., 2010. Food safety culture: the evolution of an emerging risk factor? Br. Food J. 112, 426–438.

Grocery Manufacturers' Association (GMA), 2009. Control of *Salmonella* in low-moisture foods. 4 February 2009 (minor corrections 16 March 2009). Grocery Manufacturers' Association, Washington, DC. Available at: <http://www.gmaonline.org/downloads/wygwam/SalmonellaControlGuidance.pdf> (accessed 27.06.2012.).

Guzewich, J., Ross, M.P., 1989. Interventions to prevent or minimize risks associated with bare-hand contact with ready-to-eat foods. <http://foodsafety.ksu.edu/articles/453/rte_fd_prep_risk_eval.pdf> (accessed 18.05.2013).

Hoelzer, K.H., Oliver, F., Kohl, L.R., Hollingsworth, J., Wells, M.T., Wiedmann, M., 2012. Structured expert elicitation about *Listeria monocytogenes* cross-contamination in the environment of retail deli operations in the United States. Risk Anal. 32 (7), 1139–1156.

Kimura, A.C., Johnson, K., Palumbo, M.S., Hopkins, J., Boase, J.C., Reporter, R., et al., 2005. Multistate shigellosis outbreak and commercially prepared food, United States. Emerg. Infect. Dis. 10, 1147–1149.

Kir, T., Ucar, M., Gocgeldi, E., Kilic, S., Azal, O., 2006. Evaluation of initial and periodic examinations of food handlers in military facilities. Food Control 17, 165–170.

McFoodies, 2012. Good operating practices for the food industry. <www.hazardhub.com> (accessed 18.05.2013).

Narayan, P., 2012. Many waiters carry germs on hands: study. The Times of India, Aug 3, 2012. <http://articles.timesofindia.indiatimes.com/2012-08-03/india/33018624_1_roadside-eateries-food-handlers-food-inspectors> (accessed 18.05.13).

Todd, E.C.D., Bartleson, C.A., Guzewich, J.J., Tan, A., Lee, M., Nazarowec-White, M., 2011. Procedures to Investigate Foodborne Illness, sixth ed. International Association for Food Protection, Des Moines, Iowa.

Todd, E.C.D., Greig, J.D., Bartleson, C.A., Michaels, B.S., 2007a. Outbreaks where food workers have been implicated in the spread of foodborne disease. Part 2. Description of outbreaks by size, severity, and settings. J. Food Prot. 70, 1975–1993.

Todd, E.C.D., Greig, J.D., Bartleson, C.A., Michaels, B.S., 2007b. Outbreaks where food workers have been implicated in the spread of foodborne disease. Part 3. Factors contributing to outbreaks and description of outbreak categories. J. Food Prot. 70, 2199–2217.

Todd, E.C.D., Greig, J.D., Bartleson, C.A., Michaels, B.S., 2008a. Outbreaks where food workers have been implicated in the spread of foodborne disease. Part 4. Infective doses and pathogen carriage. J. Food Prot. 71, 2339–2373.

Todd, E.C.D., Greig, J.D., Bartleson, C.A., Michaels, B.S., 2008b. Outbreaks where food workers have been implicated in the spread of foodborne disease. Part 5. Sources of contamination and pathogen excretion from infected persons. J. Food Prot. 71, 2582–2595.

Todd, E.C.D., Greig, J.D., Bartleson, C.A., Michaels, B.S., 2009. Outbreaks where food workers have been implicated in the spread of foodborne disease. Part 6. Transmission and survival of pathogens in the food processing and preparation environment. J. Food Prot. 72, 202–219.

Todd, E.C.D., Michaels, B.S., Greig, J.D., Smith, D., Holah, J., Bartleson, C.A., 2010a. Outbreaks where food workers have been implicated in the spread of foodborne disease. Part 7. Barriers to reduce contamination of food by workers. J. Food Prot. 73, 1552–1565.

Todd, E.C.D., Michaels, B.S., Greig, J.D., Smith, D., Bartleson, C.A., 2010b. Outbreaks where food workers have been implicated in the spread of foodborne disease. Part 8. Gloves as barriers to prevent contamination of food by workers. J. Food Prot. 73, 1762–1773.

Todd, E.C.D., Michaels, B.S., Smith, D., Greig, J.D., Bartleson, C.A., 2010c. Outbreaks where food workers have been implicated in the spread of foodborne disease. Part 9. Washing and drying of hands to reduce microbial contamination. J. Food Prot. 73, 1937–1955.

Todd, E.C.D., Michaels, B.S., Holah, J., Smith, D., Greig, J.D., Bartleson, C.A., 2010d. Outbreaks where food workers have been implicated in the spread of foodborne disease. Part 10. Alcohol-based antiseptics for hand disinfection and a comparison of their effectiveness with soaps. J. Food Prot. 73, 2128–2140.

Todd, E.C.D., Greig, J.D., Michaels, B.S., Bartleson, C.A., Smith, D., Holah, J., 2010e. Outbreaks where food workers have been implicated in the spread of foodborne disease. Part 11. Use of antiseptics and sanitizers in community settings and issues of hand hygiene compliance in health care and food industries. J. Food Prot. 73, 2306–2320.

United States Food and Drug Administration (FDA), 2009a. Food Code 2009. Available at: <http://www.fda.gov/Food/FoodSafety/RetailFoodProtection/FoodCode/FoodCode2009/> (accessed 18.05.2013).

United States Food and Drug Administration (FDA), 2009b. FDA report on the occurrence of foodborne illness risk factors in selected institutional foodservice, restaurant, and retail food store facility types (2009). <http://www.fda.gov/Food/FoodSafety/RetailFoodProtection/FoodborneIllnessandRiskFactorReduction/RetailFoodRiskFactorStudies/ucm224321.htm> (accessed 18.05.2013).

World Health Organization (WHO), 1989. Health surveillance and management procedures for food-handling personnel. Technical Report Series 785. World Health Organization, Geneva.

World Health Organization (WHO), 2000. Foodborne disease: a focus for health education, ISBN: 92 4 1561963.

Further Reading

United States Food and Drug Administration (FDA), 2012. Bad Bug Book, second ed. Foodborne pathogenic microorganisms and natural toxins handbook. <http://www.fda.gov/Food/FoodborneIllnessContaminants/CausesOfIllnessBadBugBook/ucm2006773.htm> (accessed 18.05.2013).
World Health Organization (WHO), 2008. Foodborne disease outbreaks: guidelines for investigation and control. <http://www.who.int/foodsafety/publications/foodborne_disease/fdbmanual/en/index.html> (accessed 18.05.2013).
World Health Organization (WHO), 2012. Five keys to safer food. WHO/SDE/PHE/FOS/01.1.

CHAPTER

29

Pest Management

Christopher H. Bell

The Food and Environment Research Agency, Sand Hutton, York, UK

OUTLINE

Food Safety Management.

DOI: http://dx.doi.org/10.1016/B978-0-12-381504-0.00029-9

INTRODUCTION

Man is in direct competition with a variety of other species for food. These competitors not only consume the product but also contaminate the product with feces, exuviae or hairs, frass and microorganisms. They can also alter the physical properties of the product by increasing temperature and possibly moisture content, and pose health threats by acting as vectors of pathogens and parasites. The importance of implementing effective pest management strategies cannot be overemphasized as the discovery of live insect stages or contaminants such as insect fragments and exuviae, or rodent hairs and droppings, has severe financial implications. Such incidents usually lead to the recall of the entire distribution of a particular product and may result in expensive litigation procedures, but potentially the greater financial loss is the longer-term effect on consumer confidence in the product which may never be fully restored.

In food production facilities there is a constant threat of pest populations becoming established as food is always present and there are many locations and access points for pests to enter and find refuges. Insect food pests are cosmopolitan while food facilities are twice as likely to encounter rodent problems in comparison with domestic premises and are legally bound to practice high food safety standards (HGCA, 2002). Many procedures can be adopted to prevent pest access, to detect their presence on arrival and to control infestations when they occur, and these are discussed below.

PESTS OF FOOD PROCESSING AND PRODUCTION FACILITIES AND THE RISKS THEY IMPOSE

Any site where food is gathered, sorted, processed or stored is an attraction to wandering rodents, birds, insects or mites whose lives depend on the successful location of food sources. Farmers, crop storage and distribution specialists, food processors and retailers all need to take precautions to render their premises less vulnerable to exploitation by pests.

While problems from vertebrate pests can largely be addressed by exclusion strategies, the same is not true for insect and mite pests, although exclusion strategies are still an important ingredient of pest management. Incoming supplies are the primary source of these pests and many species can become established in the fabric of the building, feeding on food residues and wandering to and from harborages to locate new supplies.

Vertebrate Pests

Rats, mice, sparrows and pigeons are ubiquitous and major sources of contamination of food products in food processing facilities. They act as vectors of *Salmonella*, *Shigella* and *Leptospira* bacteria, various viruses, rickettsiae causing Q fever and other pathogens. Weil's disease caused by *Leptospirosis icterohaemorrhagiae* picked up by contact with rat excreta can be fatal, as can some cases of *Salmonella* food poisoning. Rodents also cause damage by the gnawing of wood, plastics, electric cabling and even metal water pipes, sometimes with catastrophic consequences. For birds, netting of openings and needle-matting of surfaces are well established, effective strategies to prevent ingress, but problems may still occur where continual access for transport is needed or weathering of buildings provides openings in inaccessible areas of roofing where birds such as sparrows, starlings or pigeons can gain access.

Rodent-proofing is a more complex problem as in addition to the obvious exclusion of ground-level entry points, attention needs to be paid to the drainage system as well as roofing eaves as rodents will ascend drainpipes, either internally or externally, and gain access to lofts and then through the whole building via heating ducts or piping and along electrical conduit routes. Access of rats from sewerage systems is also not an uncommon occurrence so screens and other barriers should be in place and regularly maintained. The use of rodenticides for rat control requires the involvement of trained operators and even after careful observance of regulations is still a potential risk to non-target organisms (Eason et al., 2010). Resistance has developed to anticoagulants such as warfarin and now only second generation compounds are in widespread use, difenacoum and bromadiolone for use indoors and outdoors and the more toxic brodifacoum and flocoumafen for indoor use only under carefully controlled conditions. Formulation and mixture with an appropriate food is of critical importance as baits are readily rejected. All baiting stations should be checked weekly and replaced if necessary (HGCA, 2002).

Anticoagulants have always been less effective against mice because of avoidance following small intakes of bait, and since the loss of calciferol based on vitamin D_3 no really effective bait is available. Physical traps are used to complement anticoagulant baiting strategies along with single dose agents based on alphachloralose or zinc phosphide. In addition sodium cyanide and aluminum phosphide formulations are available for fumigation treatment of rat harborages and burrows away from occupied buildings. However, none of these complementary measures can guarantee adequate control and for each facility an effective exclusion and trapping strategy is therefore a necessity.

Externally, access by rodents to buildings is prevented by clearance of all shrubbery and disused machinery from the vicinity of the exterior walls and the deployment of traps at regular intervals around the property and both inside and outside potential points of entry into buildings. A typical layout of trap deployment for a food facility is presented in Figure 29.1.

Beetle Pests

Coleoptera is the largest order of insects and provides the majority of stored product pests with over 20 species of beetle or weevil of worldwide importance in international trade. Table 29.1 lists some that are commonly associated with food processing facilities, together with their food preferences and requirements for rapid development. Many species are of tropical origin that have arrived and become established in heated premises since the advent of international trade. Others, such as the biscuit beetle and granary weevil, famous for infesting sailors' biscuits and grain supplies in the days of sailing ships, are native to temperate regions. Excavations of archeological sites have found dead specimens of the biscuit beetle in leather artifacts dating back to Roman times and in the remains of food left in tombs in ancient Egypt.

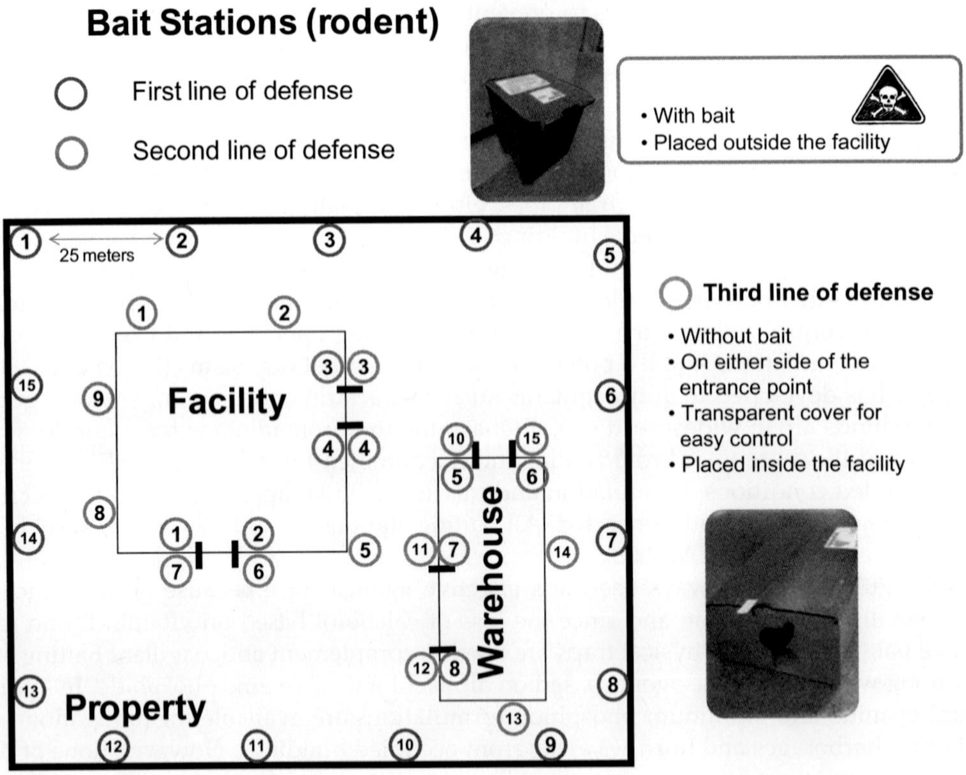

FIGURE 29.1 A typical layout of rodent traps for the protection of a food processing enterprise. *By permission of R. Stadler.*

Stored-product beetles may be divided into two categories, those developing externally on semiprocessed foods and among finely divided products, and those internal feeders developing within whole seeds such as cereal grains and legumes. The latter group includes bruchids, weevils (Curculionidae) and grain borers (Bostrichidae) which cause problems because of infested raw materials and rarely become endemic in the structure of the food production facility. They are, however, notoriously difficult to eradicate because they avoid detection and are protected from direct contact with control measures. The lesser grain borer and the granary and rice weevils occur as pests of rice and flour mills in this manner (Figure 29.2).

Those beetles feeding on semiprocessed materials or foods may again be divided into two groups, those with relatively short-lived adults (anobiids such as cigarette beetle, and dermestids such as Khapra beetle) and those whose adult stage may last a year or longer. In this latter group, including the *Tribolium* and *Cryptolestes* species which are serious flour mill pests, and the *Oryzaephilus* (and also *Tribolium*) species (Figure 29.3) occurring widely in breakfast cereal, pet food and confectionery manufacturers, both larval and adult stages actively feed on food products. It is this group that often establishes residual infestations in

TABLE 29.1 Developmental Requirements of Beetle and Moth Pests Often Found in Food Processing Facilities.

Species	Food Preferences	Developmental Range/Optimum, and Fastest Multiplication Rate
Cryptolestes ferrugineus (Stephens) Rust-red grain beetle	Grains, flour, meals, oilseeds, dried fruit and other dried vegetable materials	20–38°C, min r.h. c. 30%/32–35°C, 60-fold in 4 weeks
Cryptolestes turcicus (Grouvelle) Turkish grain beetle	Cereal products, notably wheat flour	c. 20–36°C, min r.h. 50%/28–33°C, c. 40-fold in 4 weeks
Gnatocerus cornutus (F.) Broad-horned flour beetle	Cereal products	15–35°C, min r.h. 30%/c. 30°C, c. 20-fold in 4 weeks
Lasioderma serricorne (F.) Cigarette or tobacco beetle	Cocoa, soybeans, tobacco, various cereals, spices, textiles and many other products	22–38°C, min r.h. 30%/32–35°C, 20-fold in 4 weeks
Oryzaephilus mercator (Fauvel) Merchant grain beetle	Oilseeds, dried fruit, nuts and cocoa beans	17–38°C, min 30% r.h./30–35°C, c. 30-fold in 4 weeks
Oryzaephilus surinamensis (L.) Saw-tooth grain beetle	Cereal grains, cereal products, dried fruits, nuts and some oilseeds	20–38°C, min r.h. c. 40%/31–34°C, 50-fold in 4 weeks
Rhyzopertha dominica (F.) Lesser grain borer	Cereal grains, flours, meals and macaroni	19–40°C, min r.h. 30%/32–35°C, 40-fold in 4 weeks
Sitophilus granarius (L.) Granary weevil	Cereal grains (exclusively internal grain feeder)	15–30°C, min r.h. c. 50%/25°C, 15-fold in 4 weeks
Sitophilus oryzae (L.) Rice weevil	Cereal grains (exclusively internal grain feeder)	15–34°C, min r.h. c. 40%/28–30°C, 30-fold in 4 weeks
Stegobium paniceum (L.) Biscuit or bread beetle	Cereal products and many other dried vegetable and animal products	17–32°C, min r.h. c. 60%/25–28°C, 7.5-fold in 4 weeks
Tribolium confusum J. du Val Confused flour beetle	Cereal products, copra, groundnuts, sesame and oilseeds	20–38°C, min r.h. 20%/30–32°C, 60-fold in 4 weeks
Tribolium castaneum (Herbst) Rust-red or red flour beetle	Cereal products, groundnuts, cacao, spices, dried figs and dates, copra, dried yam, palm kernels, nuts and oilseeds	22–40°C, min r.h. 20%/32–35°C, 70-fold in 4 weeks
Corcyra cephalonica (Stainton) Rice moth	Cereals, cereal products, dried fruit, seeds, cocoa and ground nuts	18–35°C, min r.h. 50%/30°C, 50-fold in 4 weeks
Ephestia cautella (Walker) Tropical warehouse moth, almond moth	Dried fruit, nuts, cereals and cereal products, cocoa beans, spices, copra, carobs, pulses and dried vegetables	17–36°C, min r.h. 25%/30–32°C, 60-fold in 4 weeks
Ephestia elutella (Hubner) Warehouse or tobacco moth	Grain, cocoa, dried vegetable products	10–30°C, min r.h. 20%/25°C, 20-fold in 4 weeks
Ephestia kuehniella Zeller The Mediterranean flour moth or mill moth	Cereals and cereal products	10–30°C, min r.h. 20%/25–28°C, 50-fold in 4 weeks
Plodia interpunctella (Hubner) Indian meal moth	Dried fruit and nuts, cereals and cereal products, cocoa, oilseeds, confectionery, citrus pulp, dried vegetables, pulses, seeds and carobs	18–36°C, min r.h. 20%/30–32°C, 50-fold in 4 weeks

Data from various sources, see Bell (2003).

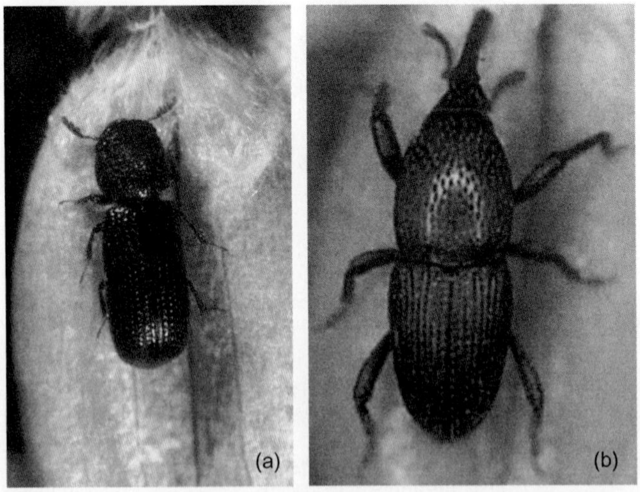

FIGURE 29.2 Two internally feeding grain beetles: A. Lesser grain borer *Rhyzopertha dominica;* B. Granary weevil *Sitophilus granarius.*

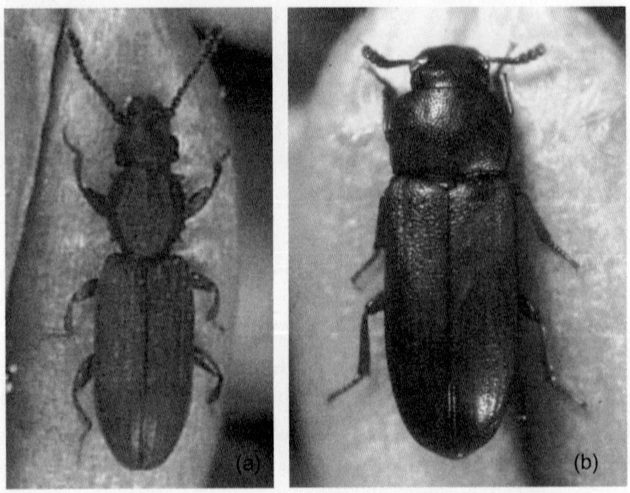

FIGURE 29.3 Two externally feeding grain beetles: A. Saw-toothed grain beetle *Oryzaephilus surinamensis;* B. Rust-red flour beetle *Tribolium castaneum.*

cracks, crevices and voids where food material escaping from processing machinery may accumulate. The long-lived adults seek out harborages from which they wander, often in a daily cycle, to scavenge for food and locate additional oviposition sites from which fresh infestations may start.

Despite their tropical origin and need of warm conditions for breeding, adults of many species of stored product beetle, both "internal" and "external" feeders, are highly cold tolerant and can readily overwinter in parts of the facility. Long-term infestation problems are revealed

FIGURE 29.4 Two moth pests of stored products: A. Brown house moth *Hofmannophila pseudospretella*; B. Almond moth *Ephestia cautella*.

if the mealworm *Tenebrio molitor* L., at 12–17 mm in length the largest of all stored product beetles, *Gnatocerus* spp. flour beetles or spider beetles (Ptinidae) are present in the facility.

Moths

Most moth pests of food processing facilities belong to the family Pyralidae, although the brown house moth *Hofmannophila pseudospretella* (Stainton) (Figure 29.4A), and the white-shouldered house moth *Endrosis sarcitrella* (L.), are commonly encountered in damper, cooler situations such as mill basements and storage areas (Bell, 2003). Adult moths do not feed and damage is caused by the larval stage which features a heavily sclerotized head capsule with biting and chewing mouthparts while the rest of the elongated body is flexible, unpigmented and unsclerotized. In addition to the consumption and contamination of food, moth larvae produce silk from glands in the mouth which builds into webbing that can obstruct machinery and slow down production lines.

Oviposition in pyralid moths occurs from dusk onwards, but is inhibited by light (Bell, 1981). The egg stage lasts a maximum of 7 days at 25°C and there are five larval instars. The duration of the larval stage is influenced by temperature, food source, humidity and whether or not a larval overwintering diapause occurs at the fifth instar. On completing their development, in preparation for pupation or diapause, larvae spin a tough cocoon which may be double-layered. The pupal stage lasts about twice the duration of the egg stage at a particular temperature and adults are short lived, females laying most of their 200–300 eggs within 3 days.

The rice moth is a serious pest of mills in hot damp climates but can become established in heated premises anywhere in the world. The tropical warehouse or almond moth *Ephestia cautella* (Figure 29.4B) is the most frequently intercepted moth pest on food imports into the developed world and a common pest of food processing facilities. The Mediterranean flour moth *E. kuehniella* is the principal moth pest of flour mills and bakeries in temperate regions of the world while the warehouse moth *E. elutella*, as its name suggests, is largely confined to warehouse storage areas where it overwinters as a diapausing larva, able to tolerate temperatures down to −10°C. The arrest is triggered by late summer day lengths of less than 14 hours (Strumpel, 1969). The Indianmeal moth *Plodia interpunctella*, perhaps the most versatile of all the pyralid species in occupying niches in the food industry, can also overwinter as a diapausing larva.

Details of the developmental limits and optima for each species are provided in Table 29.1, together with their food preferences.

Other Insects

Cockroaches, flies, ants and psocids can also cause problems in food processing facilities. The principal cockroaches belong to the genera *Periplaneta*, *Blatta* or *Blatella*. Eggs are produced in capsules, with up to 40 eggs per egg case and nymphs can mature to adults within 12 weeks. Cockroaches are disease vectors and, particularly *Blatella germanica* (L.), can cause allergenic problems, besides the obvious risks of contamination and spoilage of food (Ebeling, 1991). Most species are cryptic, hiding in refuges with access to food residues.

Many different flies are hygiene threats in industry, including house flies, blow flies, fruit flies and drain flies, each originating from different sources of hygiene failure. They can transmit many fecal and oral-borne pathogens. UV Light traps are widely employed to monitor and control flies in bakeries, restaurants and food processing plants and keep problems under control as long as adequate attention is paid to remove potential breeding sites (Taylor, 2008).

Ants have caused problems at most food production or processing premises at some time or other. Worker ants forage for food and carry it back to a central nest often at a considerable distance from the food source, leaving a chemical trail from that source. The result is that increasing numbers of workers appear in the facility, all following the same path (Beatson Campbell, 1991). Two species regularly causing problems in houses, hotels, restaurants, hospitals, warehouses and food production and processing facilities are the common black ant *Lasius niger* (L.) and the pharaoh's ant *Monomorium pharaonis* (L.). The latter can be controlled by insecticide baits based on juvenile hormone activity because, unlike the *Lasius* spp., there is usually only a single queen producing eggs in the nest.

Psocids are tiny, primitive insects feeding mainly on molds and decaying vegetable material in damp situations. They sometimes appear in huge numbers on food materials in commercial or domestic premises. The smallest opening in a food package can provide a point of entry for the minute nymphal stages. The commonest species is *Liposcelis bostrychophila* Badonnel, a rapid-moving, wingless, pale-colored insect about 1 mm long for which only females are known. Parthenogenetic multiplication can be rapid, but temperatures above 20°C and high humidity are needed for egg production (Turner, 1994).

Mites

Mites, more closely related to spiders than insects and extremely small, utilize microenvironments of moderate temperature and raised humidity. The most important family associated with food storage problems is Acaridae, though the dried fruit mite *Carpoglyphus lactis* (L.) (Carpoglyphidae) and cosmopolitan food mite *Lepidoglyphus destructor* (Schrank) (Glycyphagidae) are also common pests. The life cycle includes a brief larval stage typically followed by three nymphal stages prior to the reproductive adult stage. Their rate of increase is unparalleled by any insect, with only 14 days being needed to complete development under optimal conditions and with a single female being able to produce 555–600 eggs (Cunnington, 1965; Boczek, 1991). Eggs are cold tolerant and in some species development can proceed down to 5°C, but in all species low humidities prevent development.

The flour mite *Acarus siro* L. is able to infest any food used by humans if the local environmental conditions are suitable. The mold mite *Tyrophagus putrescentiae* (Schrank) is perhaps the most cosmopolitan mite pest of stored products, occurring in any product with a high fat or protein content. The tiniest opening permits entry of mites into packaged products and, once inside, an unpleasant taint is produced in the substrate. Many mites are also strongly allergenic.

MINIMIZING PEST OCCURRENCE IN FOOD PREMISES

It can be seen from the optimum requirements of insect pest species that the ideal environment of food processing premises should be one of low temperature, low humidity and an absence of accessible food sources. Unfortunately none of these parameters can be maintained throughout a site and so there are always tensions in striving for the right balance between production needs and pest avoidance. Most food ingredients are vulnerable to pest attack, especially those with an equilibrium relative humidity above 65%, and the continual movement of commodities to and from trade premises poses a constant threat of importing pests. In nearly every country legislation demands the highest standards for any food product destined for human consumption, so the elimination of pest contamination of food is of paramount importance for the industry.

Effective control measures carried out at the source of raw ingredients provide a vital start to the chain that leads to the final processed product. Buildings need to be designed to avoid access points from outside and doors and windows need to be precision fitted and kept closed whenever possible. Recessing of external drainpipes prevents a ready access route to the eaves for rodents and wall surfaces should have a smooth finish both inside and outside. Internally, minimization of voids, ledges, crevices or dead spaces is an important aspect at the planning stage as these provide locations where insect pests can establish refuges. New machinery and facility construction should include pest preventive design as a priority. Rigorous, systematic cleaning of processing machinery and food production areas on a regular basis helps reduce risks of infestations becoming established. Timely and appropriate removal of accumulating waste and debris by vacuum cleaning, sweeping and washing is another vital aspect to be built into management practice. Streamlining product distribution to reduce residence time in store, and avoidance of storage alongside other less secure products, are other goals in the quest to avoid infestation problems. Care should also be taken to avoid stacking products in corners or near to walls, which reduces access for cleaning and creates a harborage for pests.

Packaging can be an effective measure for reducing access of pests to food materials after processing but standard carton designs generally provide little protection against stored product insects. The spot weld glue patterns commonly used tend to leave channels through which smaller insect or mite stages can enter and does not provide a complete seal (Mullen and Mowery, 2000). Card, paper and cellophane wrappings are the least resistant to insect penetration, while polycarbonate, some polyesters, polyurethane and aluminum foil are much more resistant (Rao et al., 1972; Cline, 1978; Highland, 1984; Bowditch, 1997; Collins, 2003). All packaging is vulnerable to damage by rodents, and insects such as the lesser grain borer, biscuit beetle, cigarette beetle and larger larvae of pyralid moth species possess powerful biting mouthparts and are also able to penetrate most films. Any measures to improve packaging design by reducing the chance of an incomplete seal, and removing joins, folds and corners that are susceptible to mechanical damage or provide leverage for insect mouthparts, should be implemented. Over-wraps also improve resistance, particularly if applied as shrink-wraps fitting tightly around the package. A higher level of protection is provided by the "form-fill-seal" machines employed in modified atmosphere or vacuum packaging. A heat-molded base tray is filled with product and a flat lid is heat sealed across the top in the relevant atmosphere for the product.

All the above measures have economic implications and require there to be an adequate profit margin for the final product. Furthermore, although the presence of pests can be minimized, total elimination of pest incidence can never be guaranteed. There is therefore a need for measures to detect pests at an early stage before they locate and damage the product. Meanwhile research continues to refine methods of excluding and controlling pests (Riudavets et al., 2009; Moerman, 2010), but problems can only be avoided if vigilance is maintained and management procedures are optimized and rigorously applied.

PEST DETECTION STRATEGIES

A vital part of pest management programs is the early detection of pests. Many systems of trapping have been employed over the years, ranging from sticky papers and tapes, baited traps of various kind and thin lines of grease or food grade mineral oil around processing machinery or other vulnerable areas. The present focus is on the use of pheromones, the volatile chemicals released by the pest insects themselves that function as a means of communication between individuals (Burkholder and Ma, 1985; Campbell, 2007), and on food volatiles (Collins et al., 2007). Pheromones are particularly important for insect reproduction, both in long-range attraction of the opposite sex and short-range mate location.

The chemical structure of pheromones has been analyzed for a large number of species of concern in stored product protection (Burkholder and Ma, 1985; Phillips, 1997). A list of some of the materials that have been isolated and identified is given in Table 29.2. There are two basic types of pheromone involved in pest detection systems, sex pheromones and aggregation pheromones.

TABLE 29.2 Attractants Produced by Some Stored Product Beetles and Moths

Species	Attractant	Details
Lasioderma serricorne Cigarette beetle	Serricornin: (4,6-dimethyl-7-hydroxynonan-3-one)	Sex pheromone produced by females. Commercially available
Stegobium paniceum Biscuit beetle	Stegobinone: (2,3-dihydro-2,3,5-trimethyl-6(1-methyl-2-oxobutyl)-4H-pyran-4-one)	Sex pheromone produced by females. Commercially available
Rhyzopertha dominica Lesser grain borer	Dominicalure: (1-methylbutyl-(E)-2-methyl-2-pentenoate)	Aggregation pheromone produced by males. Commercially available
Cryptolestes ferrugineus Rust-red grain beetle	Ferrulactones I and II: [(E,E)-4-8-dimethyl-4,8-decadien-10-olide, and (3Z, 11S)-3-dodecen-11-olide, respectively]	Two-component aggregation pheromone produced by males. Commercially available
Cryptolestes turcicus Turkish grain beetle	(Z,Z)-5,8-tetradecadien-13-olide	Aggregation pheromone produced by males
Sitophilus granarius Granary weevil	Sitophilate: (1-ethylpropyl-2-methyl-3-hydroxy-pentanoate)	Aggregation pheromone produced by males. Commercially available
Sitophilus oryzae and *Sitophilus zeamais* Rice weevil and Maize weevil	Both species; Sitophinone: (5-hydroxy-4-methyl-3-heptanone, the 4S, 5R enantiomer)	Aggregation pheromone produced by males. Commercially available
Trogoderma granarium Khapra beetle	92:8 mixture of (Z)- and (E)-14-methyl-8-hexadecenal	Sex pheromone produced by females
Carpophilus hemipterus Dried fruit beetle	(2,4,6,8)E-3,5,7-trimethyl-2,4,6,8-decatetraene and related compounds	Aggregation pheromone produced by males. Commercial lure available
Carpophilus dimidiatus Corn sap beetle	(3,5,7,9)E-6,8-diethyl-4-methyl-3,5,7,9-dodecatetraene	Aggregation pheromone produced by males. Commercial lure available
Oryzaephilus mercator Merchant grain beetle	R enantiomers of Z-3-dodecen-11-olide and Z,Z-3,6-dodecadien-11-olide	Aggregation pheromone produced by males. Commercial lure available
Oryzaephilus surinamensis Saw-toothed grain beetle	R enantiomers of Z,Z-3,6-dodecadien-11-olide, Z,Z-3,6-dodecadienolide and Z,Z-5,8-tetradecadien-13-olide	Aggregation pheromone produced by males. Commercial lure available
Tenebrio molitor Yellow meal worm	4-methyl-1-nonanol	Sex pheromone produced by females
Gnatocerus cornutus Broad-horned flour beetle	(R)-acoradiene	Aggregation pheromone produced by males
Tribolium confusum, *Tribolium castaneum* Confused flour beetle, Rust-red flour beetle	Both species: 4R,8R-dimethyldecanal	Aggregation pheromone produced by males. Commercial lure available
Trogoderma spp. Warehouse beetles	(Z)-14-methyl-8-hexadecenal	Sex pheromone produced by females. Commercial lure available
Corcyra cephalonica Rice moth	Farnesal: (E,E-3,7,11-trimethyl-2,6,10-dodecatrienal)	Sex pheromone produced by males. Commercial lure available

(*Continued*)

TABLE 29.2 (Continued)

Species	Attractant	Details
Ephestia cautella Tropical warehouse or almond moth	Z-9-tetradecenyl-acetate	Sex pheromone produced by females. Commercial lure available
Ephestia elutella Warehouse moth	ZETA: (Z, *E*-9,12-tetradecadienyl-acetate), and ZETOH: (Z,E-9,12-tetradecadienol)	Sex pheromone produced by females. Commercial lure available
Ephestia kuehniella Mediterranean flour moth	ZETA: (Z, *E*-9,12-tetradecadienyl-acetate)	Sex pheromone produced by females. Commercial lure available
Plodia interpunctella Indianmeal moth	ZETOH: (Z, *E*-9,12-tetradecadienol), ZETA: (Z, *E*-9,12-tetradecadienyl-acetate), and Z, *E*-9,12-tetradecadienal	Sex pheromone produced by females. Commercial lure available

Sex Pheromones

Sex pheromones are usually emitted by females to attract males for mating. They have been reported from many moths and certain families of beetles including Anobiidae, Bruchidae and Dermestidae in which adults are relatively short-lived and feed very little or not at all (Burkholder and Ma, 1985).

Sex pheromone activity may be exclusive to a single species but commonly may be shared between several related species. Thus, the sex pheromone TDA (Z, *E*)-9,12-tetradecadienyl acetate (also known as ZETA), is active not only against *Plodia interpunctella* but also against at least four other of its pyralid relatives (Brady et al., 1971). Similarly, the anobiids *Stegobium paniceum* and *Anobium punctatum* share stegobinone (Kuwahara et al., 1978), and several *Trogoderma* spp. share (Z)-14-methyl-8-hexadecenal (Cross et al., 1976).

Aggregation Pheromones

Aggregation pheromones are usually produced by males and attract both sexes to suitable habitats and food sources where mating can then proceed. Beetles of the families Bostrichidae, Cucujidae, Curculionidae and Tenebrionidae which have adults that feed and are relatively long-lived substantially release pheromones of this type. As with the sex pheromones, aggregation pheromones may involve mixtures of materials and related species may share a common pheromone (Table 29.2). Aggregation pheromones have also been reported from mites (Kuwahara et al., 1982).

Food Volatiles

A wide range of volatiles and aromas emitted from food materials are attractive to stored product insects, notably those from groundnuts and carobs, and even plain water is effective in attracting moth species in dry conditions (Wakefield et al., 2006; Nansen et al., 2009). Food bait traps have been employed widely in food processing facilities to monitor for the presence of beetle pests with varying degrees of success. The combined use of pheromones and food attractants offers the prospect of a monitoring system for a wide range of pest species.

Pheromones as Pest Management Tools for Detection and Monitoring of Pest Populations

Pheromones are powerful attractants because of the extreme sensitivity of insects to these cues, and enable infestations to be detected at very low levels when visual or other forms of inspection are unlikely to be successful. This information is a critical input for pest management programs and associated decision-support systems in the food industry, where contamination, not only by whole insects but by fragments of them, is a major public health issue. Discovery in a laboratory oil flotation test of rodent hairs, mite or insect fragments in a product sample is the retrospective discovery of a control failure, requiring urgent action to locate the source and revise pest management procedures.

Pheromones are often complex mixtures of related compounds and their stereo-isomers can evoke vastly different responses in the species concerned; correct identification, synthesis and blending of the components is essential. Efficient delivery mechanisms for pheromones are also crucial. They must be capable of being adjusted to produce the appropriate concentration level for the species concerned, releasing the pheromone at a uniform rate, and have a capacity consistent with the particular application and operational lifetime. Trap design is important for both walking and flying insects. The distribution of traps in the treatment area is also a key factor. A vital issue after detecting the presence of insects in a facility is the accurate location of the infestation origin and to this end, precision targeting of infestation sources by spatial analysis has proved useful (Campbell, 2007; Trematerra et al., 2007), enabling control measures to get under way before other signs are evident.

PEST CONTROL STRATEGIES

Chemical Control Methods

Until recently chemicals were the mainstay for pest control in the food and agricultural industry but there has been a steady move away from reliance on biocides as a succession of adverse side effects for one compound or another have come to light. Hence the more toxic substances have largely been replaced and the use of the remaining materials is being confined to application to surfaces or areas where subsequent contact with food or packaging is unlikely, thus avoiding the problem of chemical transfer to the food (Highland et al., 1984).

Insect Growth Regulators

In recent years the focus has been on developing compounds of highly specific action, based on the physiology of the pest. In this area chemicals that act by disrupting insect life cycles have been developed. Insect growth regulators have come into use for the protection of many stored products such as grain (Oberlander et al., 1997). Methoprene, fenoxycarb and hydroprene are commercially available juvenile hormone agonists, which cause the terminal disruption of insect development but have little or no mammalian toxicity. Their use in admixture on grain or on surfaces such as fabrics can confer protection against pests for over a year.

A second group of insect growth regulators act by interfering with the molting hormone ecdysone with consequent prevention of normal metamorphosis and these are effective against Lepidoptera. A third group, effective against cockroaches, act by inhibiting the synthesis of chitin which also prevents normal molting of immature stages. Besides the very long life of the compounds, which can be an issue in international trade if residues of any added chemical are detected, another constraint for the use of insect growth regulators has been in integrated pest management programs where economically important bio-control agents may be adversely affected.

Insecticides and Repellents

The use of insecticidal sprays and dusts has been a routine measure for spot treatment of localized infestations and surface application to areas of high risk. Organophosphorus and pyrethroid compounds remain in use for this purpose though registrations on some compounds are lapsing in many countries, restricting the choice available. Much effort is being placed on the search for new insecticidal compounds of botanical origin and some such as azadirachtin from the neem tree have joined with pyrethrins as registered botanical insecticides. A more recent addition, the bacterial metabolite-based product, spinosad (Fang et al., 2002), is also available as a dust formulation. Dichlorvos space sprays have now been replaced by ULV or aerosol treatments of synergized pyrethrins or pyrethroids in food production facilities, sometimes in mixture with an insect growth regulator such as methoprene (Arthur, 2010), but are only effective against flying insects. The field of insect repellency is one still under investigation, a non-toxic, non-specific insect and mite repellent being the goal.

Fumigants

For many years fumigants have been relied upon for the whole site treatment option when infestation problems get out of control. Flour mills and chocolate factories would typically have an annual fumigation by a licensed company to have a fresh start. To be effective the fumigant had to be suitable for rapid and even distribution throughout the treatment area and in order to minimize production downtime it had to be effective against pests within 24 hours. The first fumigant in widespread use for treatment of structures, hydrogen cyanide, was replaced in the 1960s by methyl bromide, which, though less of an acute toxic risk to operators was still a highly toxic compound. It was extremely effective when used in a well-sealed structure, being an excellent penetrant of voids containing food residues and highly toxic to all pests, achieving control within 24 hours.

Methyl bromide, listed as an ozone depleting compound under the Montreal Protocol in 1992, was phased out from all but a few specialist uses in non-Article 5 (developed) countries in January 2005. Developing countries can continue using methyl bromide until 2015, beyond which their use also will be confined to a few quarantine-related circumstances (UNEP, 2006). The only other fumigant widely registered at the start of this century was phosphine, which is an excellent fumigant for commodities in store where the longer residence times permit the long exposure periods (up to 3 weeks at 15°C) required for effective control of pests. Best results are obtained by using a double layer of polythene sheets for packaged materials and treatments may only be carried by registered pest control operators who apply the aluminum or magnesium phosphide gas releasing formulations and dispose of the residues remaining at the end of the treatment according to established procedures. Access to the building in which the fumigation is carried out must be restricted and gas monitoring is required during the 24-hour aeration period after unsheeting to ensure that local atmosphere threshold limit values are not exceeded before releasing the stock for handling.

Phosphine is, however, difficult to use in food processing premises because of its corrosive properties against electronic equipment and the long exposure times required, especially at temperatures below 25°C. Although an alternative fumigant, sulfuryl fluoride (trade name Profume), has been registered for use in empty flour mills starting in Switzerland in 2003, in the UK and Italy from 2004, and now in many European countries, Australia, the USA and Canada, concerns over its global warming potential and the significance of fluoride residues have delayed registrations for use on many food materials or in structures where raw or processed food is present. Its use also requires additional heating as insect eggs are tolerant and would otherwise require long exposure times for control (Bell, 2006). With increasing pressures for the safe and effective use of chemicals, any move away from heavy reliance on them is obviously desirable.

Physical Control Methods

There are opportunities and limitations for the use of physical control methods in structures. The use of modified atmosphere (MA) techniques for space treatments, for example, is restricted to specialist chambers because whereas buildings can be sealed sufficiently for fumigation, they cannot be sealed to the much higher standard required for MA applications. Scope for use of sonic, microwave or radiation technologies is also very limited. Nevertheless several physical methods are of value in the controlling of pest outbreaks.

Heat

For the food processing industry the downtime and production loss arising from whole site treatments to combat pest problems has restricted control options to those which act most rapidly and effectively. This was the principal reason for adopting methyl bromide as the mainstay for a reliable annual whole-site treatment strategy. Heating to 47°C or above results in rapid immobilization and death of insect and mite stages within a few hours. Heat is thus one of the few options offering a similar rate of action to chemical fumigation. The principal problem for heat disinfestation, though, is the planning of heating requirements and heat source deployment to obtain a uniform heat profile throughout the structure without causing high localized temperatures which would cause damage to structural or electronic components. The temperature of air from heaters needs to be limited to 65–70°C to avoid activating sprinklers or causing expansion and cracking; and air speeds should not exceed 5 m/s to avoid dust explosions. Structural heat treatment involves raising the building temperature to 50–55°C at a rate of 5°C per hour. Sufficient heaters to ensure that 50°C is reached within 6–8 hours are required. Spot heat treatments may also be carried out where a zone of a processing facility or an item of machinery is heated to above 50°C with a forced hot air stream.

Much progress has been made using a combination of heating strategies, often in conjunction with the use of inert dusts to treat areas difficult to heat such as voids and cracks, a procedure first tested in Canada and further developed in Europe (Dowdy and Fields, 2002; Bell et al., 2004). Residual infestations in deep-seated harborages in the basement or elsewhere remain a particular problem. It must be remembered that the target temperatures for control must be reached at the point where the insects reside in the structure, a process that may take 24 hours, and that the presence of protective material such as food residues can lower the temperature experienced by the insect (Bartlett et al., 2005).

Cold

The intense periods of winter cold have long been used by millers and warehouse keepers in Canada and the northern USA for a "freeze-out" of pests and there is seldom any need for additional control methods in the first few months after treatment. Cold can also be used as a spot treatment by the injection of liquid nitrogen into confined spaces such as wall voids. However, insulation in walls can affect cold distribution, leaving protected warm spots. Also, surfaces can be stained and warping of wooden structural components may occur.

Most insects succumb to exposure at temperatures below −10°C within a few days while below 10°C insect reproduction ceases and population levels of most pests slowly decline (Fields, 1992). The stage of development of the pest is a factor in its cold resistance: eggs are more sensitive, and adults or larvae, especially those in diapause, are the most cold tolerant. Nevertheless adults of most species can survive temperatures around 4°C for many months and so can readily overwinter in buildings in temperate climates. In consequence, cold exposure requires very long holding times to be effective and this is rarely achievable in the production areas of food processing facilities. Nevertheless, the use of designated cold storage areas for incoming ingredients is a widely practiced measure in many industries in spite of the requirement for high capital investment.

Impaction

Many situations in which agricultural products are mechanically conveyed during food processing offer the opportunity for control of insects by shock, abrasion and impaction. The principle was developed over 70 years ago for use in the flour milling industry and impaction machines such as the "Entoleter" became a routine fixture in facilities such as flour mills (Pagani et al., 2006). In the Entoleter, flour falls between two rapidly spinning discs. Centrifugal force pushes the flour to the edges of the discs where it impacts a row of steel pegs mounted on the rims, and is thrown against the outer steel casing before falling into the basal receiving hopper. The material passing through the Entoleter thus encounters two major impactions and this effectively controls all free living insect stages. Impaction machines can also kill a high percentage of insects such as weevils developing inside cereal kernels (Vincent et al., 2003; Beckett, 2010).

Inert Dusts

Inert dusts cover a wide range of materials including clays, sands, ashes, diatomaceous earths (DE, fossilized remains of diatoms consisting mainly of silica with small amounts of other minerals), silica aerogels and non-silica dusts, such as phosphate and lime. Inert dusts have a long history of use for grain protection (Ebeling, 1971). Their lethal action against pests is caused by dehydration, the cuticular waxes being adsorbed by the desiccant upon prolonged contact. Abrasion of the cuticular joints in mobile stages may also be a contributory factor but recent formulations are being designed to minimize their abrasive properties to protect conveying machinery.

Inert dusts are registered in many countries for treatment of grain and pulses against insect pests and for use as sprays applied to the fabric of food premises to minimize residual infestation and migration of pests. They form a useful part of IPM strategies providing an alternative to chemical protectants for pest control (Dowdy and Fields, 2002). Some formulations are accepted as suitable for use on foods certified as "organic" in some countries. DEs are widely used as food and processing additives.

Irradiation

Irradiation from a cobalt-60 source has been used primarily as a bactericide for many years for treatment of some commodities, mainly spices, but also for dried and fresh fruit, potatoes, onions and poultry. It requires proximity to a commercial treatment source to be practical and consumer acceptance has limited its widespread use.

The methodologies for use of irradiation include exposure of a commodity by continuous flow through an irradiator or by batch treatment of cartons by pallet load, or indirect methods such as sterile male release for pest population management. A 10-MeV electron beam unit has also been in use for certain applications but the reduced safety concerns are outweighed by the very low penetrability of commodities, restricting the form in which they can be presented for treatment.

Sterile male release has given effective control of heavy field infestations of insects but apart from application in certain warehouse situations is not of relevance to the food industry where the avoidance of insect presence is the goal.

Biological Control Methods

Many organisms are known to attack, infect or parasitize stored product insects, some of which are listed in Table 29.3. The use of such organisms in food processing facilities is limited by the need to ensure that their presence does not itself lead to problems as discovery of any insect fragments in a finished product is unacceptable. Nevertheless opportunities exist for their deployment in receival facilities to deal with background pest levels in empty stores as an alternative to cold storage or fabric treatments with insecticides (Scholler et al., 1997). Pathogens are in use in conjunction with attractants to provide a control system for flying pests (Kellen and Hoffmann, 1987), and also as additives to bulk commodities such as cereals (Wakefield et al., 2010).

Use of Pheromones for Population Control

Pheromones can be used to provide the attraction agent for mass trapping to physically remove insects, by disrupting mating to prevent breeding, or by acting as an attracticide to a point where pesticides, pathogens or sterilizing agents are used as the control agent. The technique is used to reduce pest populations to manageable levels rather than eliminate them and is most suited to confined areas. It is most effective at relatively low starting population densities. Aggregation pheromones are more effective than sex pheromones

TABLE 29.3 Potential Biocontrol Agents and their Possible Target Food Pest Species

Parasite/Predator/Pathogen	Description	Host Species/Prey
Anisopteromalus calandrae (Howard)	Pteromalid wasp, endoparasite attacking larvae	*Lasioderma serricorne, Rhyzopertha dominica, Sitophilus* spp.
Choetospila elegans Westwood	Pteromalid wasp, endoparasite attacking larvae	*Lasioderma serricorne, Rhyzopertha dominica, Sitophilus* spp., *Trogoderma granarium*
Dimachus discolor (Walker)	Pteromalid wasp, endoparasite attacking larvae	*Stegobium paniceum*
Lariophagus distinguendus (Foerster)	Pteromalid wasp, endoparasite attacking larvae	*Lasioderma serricorne, Stegobium paniceum, Rhyzopertha dominica, Sitophilus granarius*
Pteromalus cerealellae Ashmead	Pteromalid wasp, endoparasite of larvae and pupae	*Lasioderma serricorne, Sitophilus* spp.
Zatropus incertus (Ashmead)	Pteromalid wasp, endoparasite attacking larvae	*Sitophilus oryzae*
Cephalonomia gallicola Ashmead	Bethylid wasp, ectoparasite attacking larvae	*Lasioderma serricorne, Stegobium paniceum*
Cephalonomia tarsalis Ashmead	Bethylid wasp, ectoparasite attacking larvae	*Oryzaephilus* spp.
Cephalonomia waterstoni Gahan	Bethylid wasp, ectoparasite attacking larvae	*Cryptolestes ferrugineus, C. turcicus*
Habrobracon brevicornis (Wesmael) and *H. hebetor* Say	Ichneumonoid (Braconid) wasps, endoparasites attacking larvae	Pyralid moths
Venturia canescens (Gravenhorst)	Ichneumonid wasp, endoparasite attacking larvae	Pyralid moths
Trichogramma cacoeciae Marschal, *T. evanescens* Westwood and *T. pretiosum* (Riley)	Trichogrammatid wasps attacking eggs	Pyralid moths
Acarophenax tribolii Newstead and Duvall	Predatory mite attacking eggs and small larvae	Tenebrionid beetles
Cheyletus eruditus (Schrank)	Predatory mite attacking eggs and small larvae	Stored product beetles and moths other than internal grain feeders
Pyemotes tritici L.-Fossat & Montagne	Predatory mite attacking eggs and small larvae	Stored product beetles and moths other than internal grain feeders
Pyemotes ventricosus (Newport)	Predatory mite attacking eggs and small larvae	Most stored product beetles and moths
Xylocoris flavipes (Reuter)	Predatory bug	All free-living stages of stored product beetles and moths
Adelina spp., *Farinocystis tribolii, Mattesia* diaspora, *M. oryzaephili, Nosema* spp.	Pathenogenic schizogregarines	*Cryptolestes ferrugineus, Oryzaephilus* spp., *Tribolium* spp., *Plodia interpunctella*
Bacillus thuringiensis and *B. cereus*	Entomopathogenic bacteria	*Lasioderma serricorne*, Pyralid moth larvae
Polyhedrosis viruses	Larval pathogens	Pyralid moths
Beauveria bassiana (Balsamo) Vuillemin	Entomopathogenic fungus	*Sitophilus* spp., *Tribolium* spp., *Oryzaephilus* spp.

because both sexes are attracted to the traps. Nevertheless mass trapping has been successfully trialed with sex pheromones against moths in flour mills to reduce pest populations to a constant low level (Trematerra and Gentile, 2010). The pheromone trap is baited with an insecticide such as cypermethrin or another quick knockdown agent or arrestant to retain the attracted moth. Alternatively a pathogen source may be incorporated to disseminate disease through the pest population.

Another approach is to use sex pheromones to disrupt mating. It is achieved by flooding the environment with the sex pheromone of the target species so that mating behavior is disrupted by false-trail following and sensory fatigue so that mate location and reproduction is minimized. The dispensers need to release adequate amounts of pheromone over a prolonged period and treatments need to be applied before emergence of the target species over a wide area for successful results.

Integrated Pest Management (IPM)

IPM is a pest risk-management approach combining a selection of the methods described above in a way that addresses socioeconomic, health and environmental risks in a sustainable manner while maintaining an acceptable level of productivity. It is highly information based, integrating knowledge about the pests with knowledge about the facility to avoid pest problems and maintain high product quality. A book edited by Heaps (2006) reviews the present status of IPM for mills and processing facilities. For successful implementation, adequate training of industry staff on the tools employed is necessary and this is a problem in some countries where there are few opportunities for formal professional education (Bartosik, 2010). In many cases pest management is contracted out by companies to a registered pest control company with specialist trained staff, but for any management strategy to work the minimum requirement is that a weekly inspection of facilities, and particularly trapping and baiting locations, is carried out and coupled with a clearly laid-out line of action if evidence of pest presence is obtained.

EMERGING THREATS FOR THE SUCCESSFUL MAINTENANCE OF PEST MANAGEMENT

The big issue regarding the continued successful use of chemicals for control of stored product pests is the development of resistance. Pests have become resistant to insecticides, insects growth regulators, fumigants such as phosphine and even to some bacteria-based sprays. The problem is often compounded by cross-resistance to other groups of compound. Resistance to phosphine was first detected more than 30 years ago and more recently occurrences of strongly resistant strains have been reported from Australia (Nayak et al., 2010) in the rusty flat grain beetle *Cryptolestes ferrugineus*. To achieve control of this strain at 20°C a concentration of 720 ppm needs to be maintained for 24 days, a far greater dosage than needed to combat previously encountered high resistance levels in the lesser grain borer *Rhyzopertha dominica*, and other pests.

Adoption of alternative strategies that avoid chemical control tends to be costly and labor intensive. This places a burden on the manufacturer that cannot always be passed on to the consumer and can result in lower standards of pest management than when chemicals were in wider use. A related effect that is often overlooked is that the reduced market for chemicals results in products being withdrawn from the market, particularly when an existing compound comes up for regulatory review on a prefixed timetable. Product registration is required in most countries for each chemical intended for use in pest control. Significant efforts have to be undertaken by commercial companies to conduct research, assemble and submit a registration package to obtain a label for legal use of a new compound or to extend the use of one that is under review. The registration process is very costly with lengthy delays and requires that the company developing the product has a high level of technically qualified personnel. Applications are often returned with requests for more data, increasing the expenditure. Where the company can only see a small market in a particular country or application, they are unlikely to proceed with registration. This can result in the disappearance of existing compounds from the market, reducing the options for pest control.

Although some problems remain, pest management standards in the food industry have never been higher and research is actively in progress to keep abreast of developments as new pests and new products and procedures come into being.

References

Arthur, F.H., 2010. Residual efficacy of aerosols to control *Tribolium castaneum* and *Tribolium confusum*. In: Proceedings of the 10th International Working Conference on Stored Product Protection, Estoril, Portugal, Julius Kühn-Archiv 425. Julius Kühn-Institut, Berlin, pp. 788–791.

Bartlett, D., Conyers, S.T., Bell, C.H., Watson, C.R. , 2005. Further development of heat-based methods for disinfesting flour mills. HGCA Project Report No. 378, Home-Grown Cereals Authority, London.

Bartosik, R., 2010. Challenges and characteristics of the South American grain and oilseed postharvest system. In: Proceedings of the 10th International Working Conference on Stored Product Protection, Estoril, Portugal, Julius Kühn-Archiv 425. Julius Kühn-Institut, Berlin, pp. 57–62.

Beatson Campbell, S.H., 1991. Ants: Formicidae, Hymenoptera. In: Gorham, J.R. (Ed.), Ecology and Management of Food Industry Pests. USA Food and Drug Administration Technical Bulletin No. 4, pp. 207–215.

Beckett, S.J., 2010. Protecting and disinfesting stored products by drying and cooling, and disinfesting stored products during handling by mechanical treatments. In: Proceedings of the 10th International Working Conference on Stored Product Protection, Estoril, Portugal, Julius Kühn-Archiv 425. Julius Kühn-Institut, Berlin, pp. 219–228.

Bell, C.H., 1981. The influence of light cycle and circadian rhythm on oviposition in five pyralid moth pests of stored products. Physiol. Ent. 6, 231–239.

Bell, C.H., 2003. Insect pests of stored products in Europe. In: Prakash, A., Rao, J., Jayas, D.S., Allotey, J. (Eds.), Insect Pests of Stored Products: A Global Scenario. AZRA, Cuttack, India, pp. 56–88.

Bell, C.H., 2006. Factors affecting the efficacy of sulphuryl fluoride as a fumigant. In: Proceedings of the 9th International Working Conference on Stored Product Protection, Campinas, Brazil. Brazilian Post-harvest Association (ABRAPOS), pp. 519–526.

Bell, C.H., Bartlett, D., Conyers, S.T., Cook, D.A., Savvidou, N., Wontner-Smith, T.J., 2004. Alternatives to methyl bromide for pest control in flour mills. HGCA Project Report No. 329, Home-Grown Cereals Authority, London, pp. 113.

Boczek, J., 1991. Mite pests in stored food. In: Gorham, J.R. (Ed.), Ecology and Management of Food Industry Pests. USA Food and Drug Administration Technical Bulletin No. 4, pp. 57–79.

Bowditch, T.G., 1997. Penetration of polyvinyl chloride and polypropylene packaging films by *Ephestia cautella* (Lepidoptera: Pyralidae) and *Plodia interpunctella* (Lepidoptera: Pyralidae) larvae, and *Tribolium confusum* (Coleoptera: Tenebrionidae) adults. J. Econ. Ent. 90, 1028–1031.

Brady, U.E., Tumlinson III, J.H., Brownlee, R.G., Silverstein, R.M., 1971. Sex pheromone of the almond moth and the Indian meal moth: cis-9, trans-12-tetradecadienyl acetate. Science 171, 801–804.

Burkholder, W.E., Ma, M., 1985. Pheromones for monitoring and control of stored-product insects. A Rev. Ent. 30, 257–272.

Campbell, J.F., 2007. Interpretation of pheromone monitoring programs for stored-product insects. Bull OILB/SROP 30, 57–62.

Cline, L.D., 1978. Penetration of seven common flexible packaging materials by larvae and adults of eleven species of stored-product insects. J. Econ. Ent. 71, 726–729.

Collins, D., 2003. Insect infestations in packaged commodities. Int. Pest Control 45, 142–144.

Collins, L.E., Bryning, G.P., Wakefield, M.E., Chambers, J., Cox, P.D., 2007. Progress towards a multi-species lure: identification of components of food volatiles as attractants for three storage beetles. J. Stored Prod. Res. 43, 53–63.

Cross, J.H., Byler, R.C., Cassidy Jr., R.F., Silverstein, R.M., Greenblatt, R.E., Burkholder, W.E., et al., 1976. Porapak-Q collection of pheromone components and isolation of (Z)- and (E)-14-methyl-8-hexadecenal, potent sex attracting components, from the frass of four species of *Trogoderma* (Coleoptera: Dermestidae). J. Chem. Ecol. 2, 457–468.

Cunnington, A.M., 1965. Physical limits for complete development of the grain mite, *Acarus siro* L. (Acarina, Acaridae) in relation to its world distribution. J. Appl. Ecol. 2, 295–306.

Dowdy, A.K., Fields, P.G., 2002. Heat combined with diatomaceous earth to control the confused flour beetle (Coleoptera: Tenebrionidae) in a flour mill. J. Stored Prod. Res. 38, 11–22.

Eason, C.T., Fagerstone, K.A., Eisemann, J.D., Humphrys, S., O'Hare, J.R., Lapidge, S.J., 2010. A review of existing and potential New World and Australasian vertebrate pesticides with a rationale for linking use patterns to registration requirements. Int. J. Pest Manage. 56, 109–125.

Ebeling, W., 1971. Sorptive dusts for pest control. A Rev. Ent. 16, 123–158.

Ebeling, W., 1991. Ecological and behavioural aspects of cockroach management. In: Gorham, J.R. (ed.), Ecology and Management of Food Industry Pests. USA Food and Drug Administration Technical Bulletin No. 4, pp. 85–119.

Fang, L., Subramanyam, Bh., Arthur, F., 2002. Effectiveness of spinosad on four classes of wheat against five stored-product insects. J. Econ. Ent. 95, 640–650.

Fields, P.G., 1992. The control of stored-product insects and mites with extreme temperatures. J. Stored Prod. Res. 28, 89–118.

Heaps, J.W., 2006. Insect Management for Food Storage and Processing, second ed. AACC International, St. Paul, Minnesota, USA.

HGCA, 2002. Rodent Control in Agriculture – A Guide. Home-Grown Cereals Authority, Caledonia House, London.

Highland, H.A., 1984. Insect infestation of packages. In: Baur, F.J. (Ed.), Insect Management for Food Storage and Processing. The American Association of Cereal Chemists Inc., St. Paul, Minnesota, pp. 311–320.

Kellen, W.R., Hoffmann, D.F., 1987. Laboratory studies on the dissemination of a granulosis virus by healthy adults of the Indian meal moth, *Plodia interpunctella* (Lepidoptera: Pyralidae). Environ. Ent. 16, 1231–1234.

Kuwahara, Y., Fukami, H., Howard, R., Ishii, S., Matsummura, F., Burkholder, W.E., 1978. Chemical studies on the Anobiidae: sex pheromone of the drugstore beetle, *Stegobium paniceum* (L.) (Coleoptera). Tetrahedron 34, 1769–1774.

Kuwahara, Y., Thi My Yen, L., Tominaga, Y., Matsumoto, K., Wada, Y., 1982. 1,3,5,7, Tetramethyldecyl formate, lardolure: aggregation pheromone of the acarid mite, *Lardoglyphus konoi* (Sasa et Asanuma). Agric Biol Chem 46, 2283.

Moerman, F., 2010. Hygienic design of food processing facilities. Food Safety Mag. 16 (5), 50–55.

Mullen, M.A., Mowery, S.V., 2000. Insect-resistant packaging. Int. Food Hyg. 11, 13–14.

Nansen, C., Davidson, D., Porter, P., 2009. Using water bottles for trapping of Indianmeal moths in stored peanuts. Ent. Exp. Appl. 133, 251–259.

Nayak, M., Holloway, J., Pavic, H., Head, M., Reid, R., Patrick, C., 2010. Developing strategies to manage highly phosphine resistant populations of flat grain beetles in large bulk storages in Australia. In: Proceedings of the 10th International Working Conference on Stored Product Protection, Estoril, Portugal, Julius Kühn-Archiv 425. Julius Kühn-Institut, Berlin, pp. 396–401.

Oberlander, H., Silhacek, D.L., Shaaya, E., Ishaaya, I., 1997. Current status and future perspectives of the use of insect growth regulators for the control of stored product insects. J. Stored Prod. Res. 33, 1–6.

Pagani, M., Cravedi, P., Fogliazza, D., 2006. Le macchine ad impatto e il controllo delle infestazioni da Artropodi in cereali e derivati nell'industria molitoria. (Impact machines and arthropod pest management in milling industry, cereals, and cereal products). Tecnica. Molitoria 57, 404–413.

Phillips, T.W., 1997. Semiochemicals of stored-product insects: research and applications. J. Stored Prod. Res. 33, 17–30.

Rao, K.M., Jacob, S.A., Mohan, M.S., 1972. Resistance of flexible packaging materials to some important pests of stored products. Indian J. Ent. 34, 94–101.

Riudavets, J., Castane, C., Alomar, O., Pons, S.J., Gabarra, R., 2009. Modified atmosphere packaging (MAP) as an alternative measure for controlling ten pests that attack processed food products. J. Stored Prod. Res. 45, 91–96.

Scholler, M., Prozell, S., Al-Kirshi, A.-G., Reichmuth, C., 1997. Towards biological control as a major component of integrated pest management in stored product protection. J. Stored Prod. Res. 33, 81–97.

Strumpel, H., 1969. Entwicklungzyklen einiger an Rohkakao schadlichen Insekten. Sonderdruck Anz Schadlingsk Pfl Schutz 42, 161–165.

Taylor, D., 2008. Cleaning up in food facilities. Pest Control Technol. 36 (6), 80–81. 76, 78.

Trematerra, P., Gentile, P., Brunetti, A., Collins, L.E., Chambers, J., 2007. Spatio-temporal analysis of trap catches of *Tribolium confusum* du Val in a semolina-mill, with a comparison of male and female distributions. J. Stored Prod. Res. 43, 315–322.

Trematerra, P., Gentile, P., 2010. Mass trapping of *Ephestia kuehniella* Zeller in a traditional flour mill. In Proceedings of the 10th International Working Conference on Stored Product Protection, Estoril, Portugal, Julius Kühn-Archiv 425. Julius Kühn-Institut, Berlin, pp. 745–752.

Turner, B.D., 1994. *Liposcelis bostrichophila* (Badonnel) (Psocoptera): a stored food pest in the UK. Int. J. Pest Manage. 40, 179–190.

UNEP, 2006. Report of the Methyl Bromide Technical Options Committee, 2006 Assessment of Alternatives to Methyl Bromide, United Nations Environment Programme, Nairobi.

Vincent, C., Hallman, G., Panneton, B., Fleurat-Lessard, F., 2003. Management of agricultural insects with physical control methods. A Rev. Ent. 48, 261–281.

Wakefield, M.E., Collins, L.E., Bryning, G.P., Chambers, J., Cox, P.D., Fennah, K., et al., 2006. Improved detection and monitoring of beetle pests in stored grain through use of a multi-species lure (LK0929). HGCA Project Report 396, HGCA, London, pp. 81.

Wakefield, M.E., Moore, D., Luke, B., Taylor, B., Storm, C.G., Collins, D.A., et al. (2010). Progress in the development of a biopesticide for the structural treatment of grain stores. In: Proceedings of the 10th International Working Conference on Stored Product Protection, Estoril, Portugal, Julius Kühn-Archiv 425. Julius Kühn-Institut, Berlin, pp. 759–764.

CHAPTER

30

Safe Handling of Food in Homes and Food Services

Patricia Desmarchelier
Food Safety Consultant, Brisbane, Australia

OUTLINE

INTRODUCTION

Preparing food in the home or food service is frequently the last link in the food chain before food is consumed and in essence it entails catering/preparing and serving of food. Safe food handling at this point is critical in preventing foodborne illness and also in maintaining the food safety measures undertaken by other supply chain participants up to this point. This is one of the most complex links within the food supply continuum and presents

Food Safety Management.
DOI: http://dx.doi.org/10.1016/B978-0-12-381504-0.00030-5

challenges in managing food safety risks. This is related to many factors such as diversity of foods prepared within a facility or at a food event; the multi-ingredient or component nature of the food handled; extremes in volume of foods handled and size of the operations; wide ranges of food worker education and communication levels and high employment turnover; and overriding socioeconomic factors enabling safe food handling practices worldwide.

The principles of safe food handling in the home and food service are not dissimilar to the basic principles in other sectors that are presented in Chapters 33–36. The technologies employed, elements of food safety assurance and food safety management in accompanying chapters are also relevant. The food safety hazards are common also although their incidences can vary in this sector. Differences at this end of the food chain are related to the nature of the food preparation practices, the potential for hazard exposure associated with those practices and the opportunity for the persistence and growth of microbial pathogens if there is poor control of food safety, particularly when there are no further controls for their elimination or reduction before consumption. While the principles of safe food handling are similar, a more flexible approach in their implementation is often necessary, tailoring food safety management to accord with local culture, economics and available infrastructure, particularly in the domestic and the informal food services sectors. In this chapter emphasis is placed on specific aspects of safe food handling in food services and the home and some of the challenges in managing food safety programs are highlighted. It is assumed the basic principles of food safety management, foodborne hazards and their control in other chapters are read in conjunction with this chapter.

Food handling in the home may simply be serving food procured outside the home or the preparation of meals from raw, partially or fully ready-to-eat food. Food services for the purpose of this chapter include the preparation of any food or meal prepared outside the home and may be either temporary or permanent, ambulatory or on a fixed site. Some examples include food preparation in vendor stalls or vehicles in streets or markets, bars and restaurants, cafeterias and canteens (e.g. in schools, workplaces, shopping centers), care facilities (e.g. child daycare and aged care centers, hospitals, institutions), catering operations, transport (boats, trains and planes) and many others. Foods available in food services such as street venders and market stalls may have been prepared in homes and small cottage industries. In poor communities, people may rely on food prepared by ambulatory vendors and in markets as they do not have the facilities for preparation at their dwelling place. In contrast, in developed countries, the increasing trend is to eat food prepared outside the home and to buy food from markets in association with busy lifestyles, income growth, health and environmental consciousness (Price, 1997; USDA, 2009). Almost one-half of every dollar spent on food in the USA is believed to be spent on food from restaurants (Jones and Angulo, 2006). Food prepared in these sectors can be for one or a few persons in a single sitting or it may be catering with thousands of meals, stored and served at a later time, resembling industrial-scale food manufacture.

EVIDENCE OF FOODBORNE ILLNESS AND CONSEQUENCES

Food safety risk managers can obtain valuable information from foodborne illness surveillance on priority hazards and their health impact, the most vulnerable populations, the foods attributed and locations where they were consumed and prepared, and factors that may have contributed to the food being hazardous. Surveillance is not universally available and most is derived from developed countries. Most data are based on outbreaks and do not include the many sporadic cases that occur. If high numbers of consumers are exposed at a common time or place there is a greater chance an outbreak will be recognized. Often locations or setting of outbreaks that are reported are places where food was consumed and not necessarily where it was prepared. Notwithstanding some of these biases, such data provide valuable insights and guidance in food safety management.

Food services and the home are frequently cited as the location for foodborne illness outbreaks. In 2007 in the USA, 40% of 503 confirmed outbreaks were reported as located in restaurants and delicatessens and 16% in private homes (CDC, 2010). In the UK in 2006, among 66 outbreaks, the setting was restaurants for 58% and homes and private locations for 18% (EFSA, 2009). In Australia in 2008, restaurants were reported as the location of 43% of confirmed and suspected outbreaks (OzFoodNet, 2009).

The health risk associated with the consumption of contaminated food depends on the level of the hazard and severity of the associated health consequences. The susceptibility of the individual consumer is an important risk factor for foodborne illness where those with a developing, declining or impaired immune status such as the very young, the elderly and the ill are most vulnerable and may experience the most severe health consequences. For example, among 17,468 laboratory-confirmed cases of enteric infection in the USA in 2009, the reported incidence was highest among children aged less than 4 years and the percentage of persons hospitalized and the case fatality rate were highest among persons aged more than 50 years (MMWR, 2010).

The number of vulnerable persons is increasing as the proportion of aged persons in the population is increasing, advancing medical intervention is extending life expectancy of the health-impaired and immune-deficient persons, and while young children remain at high risk of exposure particularly in areas of poor sanitation and hygiene. All this means safe food handling for vulnerable persons is increasingly recognized as a priority public health activity worldwide. Food consumed by vulnerable persons living in the community may be prepared domestically or purchased from the informal food sector or may be provided by a delivery service from a community food service facility. Vulnerable people may also be gathered in special care facilities (e.g. hospitals, child daycare and aged care centers) where they may be exposed to food prepared in food service operations. Authorities have taken action to recommend the avoidance of certain foods for vulnerable populations in institutions; however, adherence to these guidelines has been reported to require improvement (Dalton et al., 2010; Nelson et al., 2008).

While the public health cost of illness resulting from a food safety failure in this sector is similar to that in others, the cost to an individual business can be greater. A medium or small operator may no longer be viable following loss of a license, adverse publicity, litigation and legal costs, and loss of customers following an incident or recall. There is an increasing trend for public disclosure of regulatory hygiene inspection ratings of restaurants and similar outlets on official websites or by placing a notice on the outside of the premises. Adverse results can be quickly disseminated via the media and internet social networking. The intention of these programs, also known as "scores on doors" or "name and shame," is to allow consumers to make informed choices based on hygiene inspections about the places in which they choose to eat and from which they purchase food. It is anticipated that businesses will be encouraged to improve hygiene standards as negative publicity is damaging for trade.

At national and international levels, reports of foodborne illness are damaging for trade and tourism. Food services directly interface with travelers who are frequently compelled to eat their food while in transit or at their destination and are unable to prepare their own food. An incident of illness related to business or recreational travel has a particularly lasting impact if it is associated with loss of time and earnings or if it is incapacitating. From a public health perspective infected travelers present a specific concern as they are a potential means for the spread of exotic diseases around the world. For example, cholera and typhoid are diseases that are endemic in regions of poor sanitation and hygiene. In developed regions these diseases occur mainly in travelers returning from endemic areas. New or rare strains of biological agents can be introduced into a country via travelers as, for example, *Salmonella Enteritidis* is not endemic in Australia and of the cases of infection that occur most are among travelers from overseas (OzFoodNet, 2009).

FOOD SAFETY HAZARDS

Food safety hazards of concern may be biological, physical or chemical. In developed countries outbreaks of the following microorganisms and their toxins are most commonly reported: predominantly enteric viruses, a non-typhoidal *Salmonella enterica*, and *Campylobacter* spp., followed by staphylococcal enterotoxins, *Clostridium perfringens*, pathogenic *Escherichia coli* (particularly enterohemorrhagic pathotypes), *Bacillus cereus*, *Vibrio parahaemolyticus* and marine toxins (CDC, 2010; EFSA, 2009). Others include *Shigella* spp., other pathogenic *Vibrio* spp., and *Brucella* spp. These microorganisms may be included also in sporadic incidents where routes of transmission are not determined. Parasitic infections include *Giardia lamblia*, *Cryptosporidium* and *Cyclospora* sp. In regions of poor sanitation and hygiene and in certain endemic regions Hepatitis A virus *S. Typhi* and pathogenic *E. coli*, mycotoxigenic fungi and a broader range of parasites may be more important. Other hazards vary depending on the local features such as human and zoonotic disease epidemiology and the presence of toxic animals and plants. Physical hazards (stones, metal, plastic, insects, bone and seeds, etc.) have a greater chance of being undetected in food sold at markets and street stalls compared with packaged food screened by manufacturers. Physical hazards introduced during food handling (Band-Aids, finger-nails, broken glass, etc.) can pass unnoticed at this final stage of the food chain. They can be introduced during food preparation or by consumers when using self-service facilities.

Agricultural and veterinary chemicals, non-approved food additives and accidental contamination with chemicals used for cleaning and sanitization are important, particularly where there is limited control of or regulatory compliance with food safety in primary production and manufacturing and in implementation of safe operating procedures. Allergens are a particular concern in this sector as much of the food is unpackaged and unlabeled making it more difficult to inform the consumer of the entire ingredient list.

FOOD SAFETY RISK FACTORS

In outbreaks, foods attributed are often multi-ingredient dishes and overall most food commodity groups are included as a main component or as an ingredient. There are several food handling factors repeatedly reported following outbreak investigations. In the UK in 2006, among 66 outbreaks were included inappropriate food storage (14%), infected food handler (6%), cross-contamination (33%) and inadequate heat treatment (18%) of food (EFSA, 2009). An assessment of the foodborne illness risk factors in institutional foodservice, restaurants and retail food stores was undertaken in the USA in 2000 and 2004 with recurring risk factors identified, namely: improper holding time and temperature, poor personal hygiene and contaminated equipment (USFDA, 2004). The frequency of failure of compliance in these areas in different foodservice operations is shown in Table 30.1.

FOOD SAFETY MANAGEMENT

Most governments are moving progressively to require all food businesses, including food service operations, to be registered and to have a food safety program or plan in place.

A proactive and preventive approach is preferred and the hazard analysis critical control point (HACCP) system has been chosen by many authorities as the basis for such food safety programs although other approaches may be taken (CAC, 1969). HACCP is described in Chapter 31 and includes a series of defined steps based on scientific evidence; however, it may not be practical to apply HACCP in its entirety in this sector, for example in small and less developed businesses. To overcome this, a food safety program that is based on core HACCP principles may be more appropriate where hazards are identified and controls for their management and corrective actions are put in place, provided the approach remains based on sound knowledge of food safety (Carvalheiro and Rocha, 2008). Flexibility allowing procedures to be in proportion to the health risk, size and type of business, and capability, with emphasis on relevance and future improvement, is more important than detail and

TABLE 30.1 Foodborne Illness Risk Factors and Rates (Percent) at which Selected Institutional Foodservice, Restaurant, and Retail Food Store Facility Types Were out of Compliance in the USA in 2003. Data taken from (USFDA, 2004)

	Institutional Food Service			Restaurants		Retail		
Risk Factor	**Hospitals**	**Nursing Homes**	**Elementary Schools**	**Full Service**	**Fast Food**	**Deli**	**Meat and Poultry**	**Seafood**
Improper holding/time and temperature	40.3	30.7	30.8	63.8	41.7	64.4	29.8	42.2
Contaminated equipment/protection from contamination	18.9	20.4	22.2	37.3	21.9	23.4	24.4	20.0
Poor personal hygiene	17.5	20.2	16.3	42.7	31.2	23.5	21.4	16.8
Other/chemical	13.4	18.1	13.5	30.6	28.3	21.9	16.3	17.5
Inadequate cooking	6.3	5.0	4.5	25.8	9.1	9.2	–*	–
Food from unsafe sources	0.5	3.2	3.0	13.0	2.3	5.0	5.0	12.7

Low levels of non-compliance.

TABLE 30.2 The World Health Organization's Five Keys to Safer Food (WHO, 2006)

Keep clean
Separate raw and cooked
Cook thoroughly
Keep food at safe temperatures
Use safe water and raw materials

complexity. The challenges of universal implementation of HACCP have been recognized internationally. More detailed information is available in "Guidance to governments on the application of HACCP in small/less developed food businesses" (FAO/WHO, 2005).

Food services vary in operational organization, e.g. food may be prepared in individual units or in a centralized facility serving satellite units; food may be prepared on site or brought in partially (for assembly) or wholly prefabricated from suppliers to economize on skills, labor and equipment. Similarly food consumed in the home may be prepared there or purchased from a food vendor or food service. Food safety risk management generally lies with the owner/primary care giver and/or the person(s) responsible for the food service operation. By purchasing pre-prepared food the safety management is passed on partially; however, the food service manager is responsible for ensuring incoming products are procured from a safe and reliable source and food distributed to branch units remains safe.

Safe food handling in the home and the general community can also be based on the application of HACCP principles. Food handlers in these settings acquire their food safety knowledge through educational messages provided by their relatives and community contacts, schools, community centers, authorities and the media. These messages should be based on the identification of the major local food safety hazards together with practical and culturally appropriate safe food handling practices for their control. The *Five Keys to Safer Food* is a simple global health message developed by the World Health Organization (WHO) that is based on scientific evidence and is for use in education of all types of food handlers, including ordinary consumers. The message and training materials for adoption to different audiences are available online at http://www.who.int/foodsafety/consumer/5keys/en/. The Five Keys to Safer Food are shown in Table 30.2.

APPLICATION OF HACCP PRINCIPLES TO FOOD SERVICE AND THE HOME

The Codex Alimentarius Commission provides a sequential approach for the application of the seven HACCP principles and this approach is considered in relation to preparation of food in food services and the home (CAC, 1969).

Food service operations can range from large multinational corporations to very small and less developed businesses. The personnel similarly range from large multidisciplinary workforces to a single individual. While in the former the resources and capability to develop food safety programs may be available in-house or can be commissioned, for the latter this will most likely be lacking. For those lacking the expertise or resources expert advice and assistance in establishing an effective food safety plan may be obtained from industry and trade organizations, from regulatory authorities and hygiene inspectors, from consultants and from extension services provided by some educational institutions. Information for self-help can be obtained from published literature and from model HACCP guides and similar resources developed for various commodities and business types that can be applied following adaption to a particular business. Regulatory authorities may provide manuals, tables, forms and checklists to guide and assist in development of food safety programs that meet their requirements. The WHO (http://www.who.int/foodsafety/en/) and the Food and Agriculture Organization (FAO) (http://www.fao.org/ag/agn/agns/) provide resource material including guides to food safety risk analysis and its components of risk assessment, management and communication with guidance for developing countries and less formal food sectors such as street vended foods. Risk assessments for specific food safety hazards both biological and chemical and specific products are also available. The Codex Alimentarius Commission provides a wide range of Codes, Guidelines and Standards for general principles of food hygiene, for specific commodities and specific food handling locations. These are available in hard copy or can be downloaded from their website at http://www.codexalimentarius.net/web/index_en.jsp (cited August 2010).

PREREQUISITE PROGRAMS

A successful food safety program depends on basic hygienic and sanitary operating conditions, known as "prerequisite" programs (CAC, 1969). In undeveloped communities there may be limited understanding of basic hygiene and a lack of infrastructure available to enable basic hygiene prerequisites to be implemented (Vollaard et al., 2004). HACCP-based programs may have to be staged to allow implementation of basic hygiene first, or if the food services in question are the only source of accessible food and nutrition for the community they need to occur concurrently to maintain the food supply. In its simplest application, prerequisite requirements are not identified as separate from managing critical control points. For example, in the WHO Five Keys that can be applied in the community and with small food services such as street foods, these basic hygiene requirements are included in "keep clean" and "use safe water and raw materials" along with control points (Table 30.2). Education and training are included in prerequisite programs; however, because of their importance they are discussed under a separate heading in this chapter.

Design, Layout and Facilities

There are some unique aspects in the application of prerequisite programs in food services and homes compared with other parts of the food chain. The location, design and layout of some food services are examples as they can be temporary or permanent, mobile, ambulatory or fixed, and they may have very limited space for food preparation. Space for storage of food and for storing cleaning agents, etc. can also be limited, increasing the opportunity for cross-contamination. Those in vendor vehicles and stalls in streets and markets may lack access to services such as power, potable water, sewerage and waste removal. Affordable vehicles or stalls designed with food safety taken into account could be made or provided for rental on a community basis. The provision of designated sites for stalls and vendors with access to shared facilities and materials and conducive to the preparation of safe food also help to overcome this. Such venues provide a convenient location for authorities to communicate with and to provide education and training for the food handlers. There has been considerable study of food services and vendors of this type and the Codex Alimentarius Commission provides specific guides such as the "Revised regional guidelines for the design of control measures for street-vended foods in Africa" (CAC, 1999) as does WHO and FAO from their websites. Many local authorities provide safe food handling guidelines specifically for food services and individuals preparing food when traveling, for charities and community groups, at markets and temporary events, and in home-based businesses.

Control of Incoming Materials

Prerequisite programs include managing the quality of incoming raw materials and ensuring they are from approved suppliers and are safe and fit for the intended purpose. Food services and homes are frequently cited as the setting for illness outbreaks; however, the contamination may occur earlier in the food chain and this location is the place of consumption (Jacob and Powell, 2009). The quality of the incoming ingredients or foods is important particularly if the food is consumed with no further processing and if it is intended for high risk individuals. There have been some notable examples of contaminated raw ingredients used in food services and homes. Salmonellosis outbreaks attributed to egg and egg-related dishes prepared in food services and homes have been associated repeatedly with the purchase of cracked and dirty shell eggs and eggs not produced under approved quality assurance schemes together with inadequately cooked or uncooked eggs and egg-based dishes (Slinko et al., 2009). Pathogenic *Vibrio* spp. are a food safety risk in shellfish, particularly in oysters, that are popular in food services where they are served raw. The health risk has been closely related to pre-harvest management and post-harvest handling leading to regulatory controls (DePaola et al., 2010). Raw oysters should only be accepted from approved sites and with identification labels.

The intentional or unintentional addition of unapproved food additives or chemicals in food occurs particularly in regions where good hygiene practices and enforcement of controls are limited. The addition of melamine to powdered milk used in infant formula resulted in a nationwide outbreak of serious illness in children (FAO/WHO, 2009). Uncooked chicken pieces accidentally contaminated with ammonia during a refrigerant leak in a warehouse then prepared for a school lunch resulted in 157 persons becoming ill (Dworkin et al., 2004).

Ice, while not a food, is often used to keep food chilled. Both the ice and the water from melted ice can come into direct contact with prepared food or food to be eaten raw. Ice should be made from potable water or if this is not available at least it should be from a reliable and known source.

Personal Hygiene

The variety of food services and residential dwellings presents various opportunities for food to be exposed to the environment and pests, contact surfaces, other foods, food workers and the general public. In a review of 816 foodborne outbreaks where food workers were involved, Todd et al. (2007) reported bare hand contact was the most frequently reported risk factor followed by failure to properly wash hands, inadequate cleaning of processing or preparation equipment or utensils, cross-contamination of ready-to-eat foods by contaminated raw ingredients and (for bacterial pathogens where growth is required) temperature abuse.

Food handling in this sector remains largely manual; food is chopped, sliced, portioned, plated or assembled and embellished, sometimes repeatedly, increasing the risk of exposure to the environment, surfaces and food handlers. Food is consumed after direct contact with serving dishes and implements, plates and cutlery that are cleaned for reuse or may be intended for single use. Heat and chemicals are commonly used to ensure that the number of microorganisms on utensils and surfaces have been reduced to a level that does not compromise the safety of the food that comes in contact or cause infectious diseases. Automated washing machines are highly effective as they allow the use of high water temperatures and strong cleaning agents. Minimizing food residue, followed by manual washing and sanitizing, is also effective. In food services and homes with minimal facilities, washing with soap and clean water and drying in the sun are practical alternatives although not optimal solutions.

Food prepared in this sector is exposed to food handlers during preparation and serving and to the general public during self-service, retail purchase and in the home. Other diners can be a source of infection if infected when handling common utensils or in contact with food as can occur in self-service and smorgasbord-style operations. This is recognized in the spread of norovirus infection where vomit from an ill person can spread infection to other diners via aerosols (Boxman et al., 2009). Personal hygiene and sanitary operating conditions are important in minimizing opportunities for contamination especially when at this point food receives no further processing to eliminate contaminants. Education and training in personal hygiene, in encouraging infected workers to report illnesses, and in the provision of facilities to enable good personal hygiene is necessary and has to be ongoing especially where the workers have a high turnover and limited motivation to comply.

In these settings food can be exposed at any point through the food flow to handler's hands or excretions from lesions, sneezes and vomit, the exception being where operations are automated or enclosed. Infections reported in association with infected food handlers in food service and homes include typhoid, shigellosis, streptococcal, hepatitis A, norovirus and protozoan (*Toxoplasma, Cryptosporidium* and *Giardia*) infections, and staphylococcal intoxications. Food workers have been symptomatic or asymptomatic shedders of pathogens; they may have been exposed to infected family members or contacts; and have subsequently used improper personal hygiene practices (Todd et al., 2007). The pathogen source is not always obvious. In a salmonellosis outbreak linked to delicatessen foods the infection was believed to be transmitted via the food handlers who had contact with chickens carrying *Salmonella* at home (Hedican et al., 2010). It is not always possible to determine whether

BOX 30.1

ARGUMENTS FOR AND AGAINST THE USE OF GLOVES BY FOOD HANDLERS (TODD ET AL., 2010)

Arguments for Glove Use:	Arguments Against Glove Use:
Gloves protect the worker from foods/ingredients that can cause damage to their skin	Gloves can reduce operational dexterity and increase the risk of injury for workers
Gloves protect the food from direct hand contact	Higher levels of food contamination are possible in the event of glove failure
Glove use is more easily observed to verify hygiene compliance versus hand washing	Small numbers of gloves have pinhole leaks that are not possible to detect before use
Gloves can be used to cover worker skin damage or infections	Gloves can be worn for longer than they should be
	Gloves give a false sense of security as a substitute for good hand hygiene practices
	Gloves increase the risk of hand irritation

the food worker is also a victim having eaten some of the contaminated food while at work, whether infected by a co-worker or if they were the cause (Todd et al., 2007).

Failure to wash hands properly or lack of hand-washing supplies and facilities are frequently contributing factors in outbreaks (Todd et al., 2007). Norovirus infections are commonly linked to infected food handlers and this has been conclusively demonstrated by detection of an outbreak virus strain on the hands of the food worker preparing the implicated food (Boxman et al., 2009). Use of clean tongs, cutlery or gloves prevents bare hand contact with food. Gloves offer a barrier to hand contact and are commonly used as a hygiene measure in food service. Glove use is debated and some of the arguments for and against identified by Todd et al. (2010) are shown in Box 30.1.

While gloves can reduce the opportunity for contamination of food from bare hands they can also be a source of contamination if inappropriately used (Todd et al., 2010). Disposable gloves are used when handling ready-to-eat foods and reusable gloves may be used when handling raw food that will receive a microbiological kill step before serving, e.g. raw meats. Gloves should be clean and sanitized if reused and color coded to allow ease of detection if torn and fragmented during use. Hands should be washed before gloving and loss of integrity of the gloves should be ensured such as by limiting use of jewelry, maintaining short fingernails and avoiding punctures from sharp objects in foods or the work environment. Where gloves are used strategies will be needed to encourage food workers to comply with glove use, and management must ensure the glove supply is continuous and conveniently located (Todd et al., 2007).

Maintenance and Sanitation

Equipment used in food preparation and assembly (e.g. slicers, shredders, mixers, conveyors) have multiple parts and crevices where food residue can accumulate allowing bacteria to harbor and grow. Total disassembly of equipment, cleaning and sanitizing, or heat disinfection of the entire apparatus to reach inaccessible parts may be required for hygiene maintenance. An outbreak of salmonellosis attributed to lettuce served at a fast food chain was traced to the shredder used in its preparation (Stafford et al., 2002). The same serotype of *Salmonella* as the patients' was detected in the shredder where it had remained due to poor maintenance. The transfer of *L. monocytogenes* from meat slicers to roasted meats and fermented sausages during slicing has been demonstrated experimentally (Lin et al., 2006). *L. monocytogenes* was subsequently shown to grow on the sliced and packaged uncured roast turkey meat during storage at 4°C.

In food services there are often multiple food types prepared and handled simultaneously and the area and facilities available for food handling or storage can be minimal. Activity and demand in these operations can escalate during peak service periods such as meal times. These scenarios can provide opportunities for cross-contamination if they are not controlled with preplanning and effective prerequisite programs. The ready manner in which cross-contamination can occur has been demonstrated experimentally with *Campylobacter* spp. on raw chicken transferred to ready-to-eat salads by using the same cutting boards and knives and by unwashed hands (Van Asselt et al., 2008). Luber (2009) studied the internal contamination of eggs and external contamination of poultry with *Salmonella* and *Campylobacter*, respectively, and the associated risks. The study concluded cross-contamination from the use of the same cutting board for chicken meat and salad without cleaning in between or spreading of pathogens via the kitchen environment seemed to be of greater importance than the risk associated with undercooking of poultry meat or eggs.

In large facilities, a single direction for the flow of food from raw materials to final product, designated equipment, tools and storage/chilling units for raw and cooked foods, and restricted movement of food workers between raw and cooked food preparation areas can be engineered to prevent the opportunity for cross-contamination. In small operations this may not be possible and practical interventions may have to be devised. Examples include hand washing between handling raw and cooked foods, designated and identified equipment and utensils for use exclusively with raw or cooked foods, and protected storage of ready-to-eat and cooked food above raw food in stores or refrigerators. Products used in the preparation of allergen-free meals have to be stored and kept separate at all times.

Food residues remaining on poorly cleaned and sanitized equipment, surfaces and fixtures provide a reservoir for bacteria that can transfer to food during processing repeatedly over time. *Listeria monocytogenes*, an environmental bacterium relatively resistant to environmental conditions, is able to grow at refrigeration temperatures and has been found in many food processing environments, cold stores and refrigerators in food services and homes (FAO/WHO, 2004). The bacterium can colonize seals and surfaces in cold storage and refrigeration units, containers and pallets. Minimizing the presence of *L. monocytogenes* in equipment and premises with hygiene maintenance and standard operating procedures is essential. The numbers of *L. monocytogenes* in food can be further controlled by minimizing the time during which growth could occur; this could include labeling ready-to-eat food with the date of preparation, and determining the date by which the food can be safety consumed using predictive models.

HAZARD ANALYSIS

Product Flow

In retail and food services, products may range from a single product line (hamburger, pizza, satay, shellfish, fruit) to a diversity of products varied at regular intervals (e.g. meals in restaurants, cafeterias, institutions, transport caterers). The product may be prepared to a strictly controlled formula (e.g. fast food chains and other food service franchises) or may change opportunistically. There may be a single product and process (e.g. ice cream) or various food components and processes combined in a final complex food product (e.g. plated meal service, catering operation and restaurant). Ideally a food safety plan should be established for each product line. However, where there are multiple food lines, undertaking this for each food item would take considerable resources and is not necessary. It is more efficient to prepare plans for broadly grouped products. This is most often based on common food preparation processes and the flow of a food through the operation from raw material receipt to consumption. For example, a catering operation may prepare menu items with recipes differing in major ingredients (e.g. meat, chicken, seafood) although with common processing (e.g. cooking, holding or serving). In this example, common controls of adequate cooking and cooling would place them in the same group even though the hazards identified and the critical limits may differ.

The basis for grouping foods will vary with the type of business. Some typical process steps and examples along the product path or flow through food services or in homes for cooked food are shown in the columns in Table 30.3. These examples are not exclusive as other variations and process technologies (see Part II of this book) may be employed. The groups are formed based on common steps of cooking and further differentiated based on whether the food is held after cooking and, if so, the storage temperatures, the option of reheating and the need for final holding and holding temperature.

Another example is an approach proposed by the USA government for food services and retail establishments based on the number of times the food passes through the temperature zone where bacterial growth may occur during the flow of food through the operation (USFDA, 2006). The temperature zone referred to as the "temperature danger zone" is defined as 5°C (41°F) to 57°C (135°F). Other food flows may occur although the number of passages through the temperature danger zone is the key to the three categories of preparation processes.

Process 1: Food preparation with no cook step (no cook step to destroy pathogens).
Example flow: Receive – Store – Prepare – Hold – Serve
Process 2: Preparation for same day service (passes through the temperature danger zone once).
Example flow: Receive – Store – Prepare – Cook – Hold – Serve
Process 3: Complex food preparation (passes through the temperature danger zone two to three times always).
Example flow: Receive – Store – Prepare – Cook – Cool – Reheat – Hot Hold – Serve

TABLE 30.3 Examples of Common Processing Steps and the Flow of Food from Receipt to Consumer (Top to Bottom of Table) during Preparation in Food Service, Retail and the Home and Food Groupings

	Food Groups			
Process Steps	**Fresh – Serve**	**Cook – Serve**	**Cook – Chill (short/ extended shelf-life)**	**Cook – Freeze**
Receive	+[a]	+	+	+
Store	+	+	+	+
Prepare	+	+	+	+
Cook[c]	–[b]	+	+	+
Cool	–	–	+	+
Assemble, fill (aseptic or non-aseptic),[c] seal, label	–	+/–	+	+
Chill	+/–	–	+	+
Freeze	+/–	–	+	+
Store/distribute	–	–	+	+
Reheat	–	–	+	+
Hot hold	–	+/–	+/–	+/–
Serve/sell	+	+	+	+
Examples	Raw oysters, sashimi, green salads, cut fruits, sliced cooked meats, cheese, meat (to be cooked by customer)	Fried chicken, hamburgers, cooked eggs, hot vegetables, stir-fries, noodle dishes	Pre-cooked meals, meats, sauces, soups, entrées, desserts, pizzas	

[a]*+ Process included.*
[b]*– Process not included.*
[c]*– The product may be cooked at this step or pasteurized in packs after filling.*

Intended Use

Food items prepared for vulnerable persons should be given special attention due to their increased susceptibility to food safety hazards. Special flow diagrams for their foods may be required as additional controls and the design of a special formula or menu may be needed to minimize the risk. Foods prepared for consumers with intolerance to allergens also require individual plans to ensure freedom from contamination with the specific allergen.

Potential Hazards

The hazards that can reasonably be expected to occur in foods and be present in the food handling environment or in other inputs during food handling in this sector are as varied as the types of food involved and may be biological, physical or chemical. Evidence for some hazards commonly reported in foodborne illness outbreaks attributed to food prepared and consumed in these settings have been described above (see "Food Safety Hazards").

The efficacy, feasibility, sustainability and cost effectiveness of controlling identified hazards are important considerations especially for many small and less developed businesses at this point of the food chain. If control cannot be ensured menus should be redesigned, products or processes reformulated or foods restricted or prohibited from use. An important consideration is the intended use of the product particularly if catering for high risk groups. There have been frequent outbreak scenarios linking specific pathogens, at-risk groups and settings in this sector that have resulted in authorities developing specific regulations or guidelines to ensure safety. Neonates are susceptible to *Cronobacter sakazakii*, an environmental bacterium that may be present in powdered infant formula. If formula is reconstituted and held at temperatures allowing survival of the bacterium and for enough time to allow growth it may present a hazard for neonates and infants (FAO/WHO, 2006). This scenario could occur in hospital settings preparing formula for nursery inpatients and stringent microbiological criteria are used for these powders. A higher probability of infection with *E. coli* O157 following exposure to lower pathogen concentrations is known to occur in children and the elderly, and food services, schools, aged care and daycare facilities have been the location of outbreaks (Desmarchelier and Fegan, 2003). Foods high risk for O157 STEC should be avoided for these vulnerable groups. The elderly, immune-suppressed and pregnant women are at highest risk of listeriosis and education programs and guidelines for vulnerable persons recommending avoidance of high risks food are widely recommended or enforced by authorities (Dalton et al., 2010).

Some bacterial hazards are more frequently associated with food handling practices in this sector. For example, spore-forming bacteria (*Clostridium* spp. and *Bacillus* spp.) and toxin-producing bacteria (*Staphylococcus aureus*, *Bacillus* spp. and *C. botulinum*) that generally have to grow in food before it becomes hazardous. Suitable conditions can occur in food supporting growth with poor control of cooking, cooling and holding temperatures and lack of control of the duration the food is held at those temperatures. The extensive direct handling of foods can result in the introduction of pathogens such as norovirus, hepatitis A virus, protozoans, *S. Typhi* and *Shigella* spp., if poor personal hygiene is practiced among food workers as already discussed. This is a concern when no further treatment is applied that would result in their inactivation. Pathogenic *Vibrio* spp. and parasites in foods such as fruit and vegetables and seafood obtained from an unsafe source and served raw will similarly not be inactivated before consumption.

Just as foods can be grouped according to common processes to overcome the complexity in food services and homes, hazards that can be controlled with similar approaches may be considered collectively. Although this is not exclusive some examples are shown in Table 30.4.

Critical Control Points and Limits

Establishing and implementing control measures and critical limits in this sector is simplified if few product lines and processes are involved or if foods and hazards are grouped as described above. Food safety metrics, as described in Chapters 1, 31 and 33, can be applied although formal application will depend on the size of the operation. Some food safety controls and their critical limits may be proscribed by regulations enforced by local,

TABLE 30.4 Groups of Hazards that are of Concern in Food Services and the Home and Examples of Common Approaches for their Control. These Controls are Examples and not Meant to be Exclusive

	Approaches to Minimize or Eliminate the Hazards	
Hazard Group	**Technologies for:**	**GHP**
Non-spore forming bacteria (e.g. *Campylobacter*, *Salmonella*, *E. coli* and *Vibrio* spp.)	Inactivation or removal Controlling growth	Safe raw materials Maintain personal and equipment hygiene Avoid cross-contamination
Spore-forming bacteria (e.g. *Clostridium* and *Bacillus* spp.) and toxin-producing organisms (*S. aureus*, *Clostridium* and *Bacillus* spp.)	Controlling growth to prevent outgrowth bacterial spores and toxin production	Maintain personal and equipment hygiene Avoid cross-contamination
Fecal–oral route of transmission (enteric viruses, *S. Typhi*, *Shigella* spp., parasites)		Safe raw materials Personal hygiene Avoid cross-contamination
Physical and chemical hazards		Safe raw materials Safe chemical storage Maintain equipment hygiene and safety
Allergens		Safe raw materials Maintain equipment hygiene Avoid cross-contamination Personnel with consumer contact trained to provide consumer advice

regional or national governments. These can include specific controls, criteria and critical limits in the management of food safety practices applied at steps such as receipt, storage, processing, display, packaging, transportation and for recalls. There may also be regulations applying to the safety requirements for specific commodity groups (e.g. dairy, eggs, meat, fruit and vegetables, seafood, etc.). The person managing the food safety programs should be aware of local, regional or national regulatory requirements and must incorporate them in the food safety program.

The risk factors commonly identified in association with foodborne illness outbreaks in food services and home settings provide evidence for the most common failures in managing food safety. This information provides an indication of important food safety controls to be emphasized and re-enforced. The following lack of controls is commonly reported in different countries, the first two are most common (EFSA, 2009; USFDA, 2004).

- Inadequate heat treatment of food.
- Inappropriate storage of food, i.e. improper holding time and temperature (cold holding of potentially hazardous food and inadequate date marking of ready-to-eat food).
- Failure in control of general hygiene measures such as poor personal hygiene.
- Inadequate cleaning and sanitization of equipment leading to food contamination.
- Failure to keep raw and cooked food separated to avoid cross-contamination.

Some examples and highlights for this sector are discussed here. Cooking or alternate processing to inactivate or remove biological hazards and effective chilling and freezing to control growth of biological hazards (see Part I and II) have been presented in detail in earlier sections of this publication and should be consulted as they are not repeated.

A variety of foods prepared in food services and homes have been implicated in outbreaks where inadequate cooking was reported. *E. coli* O157 and *S. Enteritidis* are examples of particular concern as there is a higher probability of infection for at-risk groups from low numbers of these bacteria present after they have survived the cooking process (Vought and Tatini, 1998). Undercooked ground beef burgers (Desmarchelier and Fegan, 2003) and egg-based dishes have been implicated (Marcus et al., 2007) in outbreaks of infections from these bacteria, respectively. It is important that the required temperature is reached at all points in the food and for the time determined to render it safe. Determining these critical limits should be based on sound scientific evidence and cooking processes controlled. Some heating processes may be overlooked as, for example, an outbreak of campylobacteriosis was linked to garlic bread. Following laboratory investigation it was revealed the internal temperature reached 19–22°C and this was inadequate to kill the bacterium in the contaminated butter (Zhao et al., 2000).

The preparation of food before cooking can impact on the effectiveness of the cooking process. Outbreaks of salmonellosis and campylobacteriosis occur following consumption of inadequately cooked chicken (Bryan and Doyle, 1995). Frozen food such as large whole poultry carcasses and pieces of meat should be completely thawed in advance in a refrigerator or by microwaving under control. Where this is not possible due to limited facilities, thawing under potable running water under controlled conditions (e.g. at a temperature of 21°C for not more than 4 hours) can be substituted (CAC, 1999). This does not apply to smaller portions of manufactured frozen ready-to-cook products that should be cooked according to the manufacturer's cooking instructions on the package label. Improper thawing of whole frozen chickens can occur when trying to reduce time to meet unexpected demand for additional food during peak service hours or when consumer sales have been underestimated.

Food preparation in this sector takes place in advance of serving in most types of food services. It also occurs in homes with busy and time poor occupants or when food is prepared in homes for special events or for vendors. Ingredients, meal components or whole meals can be held hot, chilled or frozen until required for assembly, reheating and/or serving. As this involves food being exposed to temperature danger zones, often more than once, critical control points should be identified.

Managing the critical temperatures–time limits in these zones requires an understanding of microbial behavior as well as the dynamics of heat transfer in foods and the relationship with the structure and volume of food containers. Critical limits are chosen for individual products to meet the required performance objectives (see Chapters 31 and 33). Inexperienced food handlers may not have the level of knowledge to determine these parameters. These may be stated in food regulations that specify temperatures for hot and cold holding of ready-to-eat food, and that define the maximum duration of holding before refrigeration, before reheating or before the food should be served. Foodborne illness in the community has often been associated with events (e.g. celebrations, functions, picnics) where larger than usual quantities of food are prepared and temperature and time management is lacking by persons ignorant of hazards associated with scaling up domestic food practices (McLaughlin et al., 2006). Persons commencing cottage industries and preparation of home-based food production need to become acquainted with basic food hygiene and safety before commencing business.

The accurate way to monitor cooking and cooling is to measure the internal temperature of each product and record the duration. The practicality of temperature and time monitoring, verification and recording for every food item will vary with the size of the operation, the volume and range of food prepared. Fast food operations may cook hundreds of a particular item, e.g. hamburgers at peak times. In this situation it may be more practical to verify that the process and the equipment are consistently capable of achieving the required cooking performance and undertake less frequent but regular testing and record keeping. Implementation of controls, the use of thermometers and recording equipment, verification and monitoring can be impractical in many poor regions and in many homes. While it is an aspirational goal, some level of safety assurance may be achieved more practically in the interim by exploring local and culturally relevant practices that can be substituted. For example, recommending food or water comes to a "rolling boil" for sufficient time may be an interim compromise for cooking in areas of poor sanitation and hygiene and with minimal facilities. Identification of changes in a food's texture and taste after processes such as cooking and acidification may offer a degree of protection. Women in traditional societies after many years of experience may be able to judge when acidification is adequate by taste. Promotions of food safety for eggs and *S. Enteritidis* recommend eggs should be cooked until the yolk and white are no longer runny. However, temperature and color change in red meat and crustaceans may not reliably indicate sufficient heat has been reached to inactivate vegetative bacteria.

Producing food in this sector is dependent on multiple activities where dangerous temperature zones occur and can be cumulative, e.g. storing, thawing (raw or cooked), preparing, portioning, assembling, holding (hot or cold), reheating, displaying, packaging, transporting, etc. While having food pass through this danger zone may be unavoidable the duration in that zone can be strictly minimized so that bacterial growth is insufficient for the food to become hazardous. Bacteria may be present because they have survived cooking or other processing, they may have been introduced post-processing or the food may be unprocessed and served raw. Some important considerations include date marking and control of cold storage duration to control psychrotrophs; managing batch sizes, unit volumes and container design to enhance heat and cold transfer; and ensuring the capability and monitoring of hot or cold holding equipment (e.g. display cabinets, bain-maries, trolleys, transport modules and vehicles to meet requirements). These should be considered for both regular production volumes and take into account peak production times or unexpected events. When an operation is stretched beyond its anticipated production errors can occur as the food safety plan was not designed to accommodate these increases.

Both vegetative bacteria and those from spore-formers following vegetative cell formation, and toxin-producing bacteria, can often grow quickly in foods common in food services during cooling or reheating and have no other hurdles to inhibit growth. Errors at this stage are not reversible by cooking to eliminate bacteria as some toxins, such as staphylococcal or *B. cereus* toxins, if formed, may be heat resistant.

Inadequate cooling has often been associated with outbreaks of foodborne illness in food services and homes where the food attributed had been cooked (McLaughlin et al., 2006;

Shapiro et al., 1999) or contaminated post-cooking (Todd et al., 2007), was raw or contained raw ingredients (Mannes et al., 2010). Mannes et al. (2010) reported a salmonellosis outbreak involving 319 cases, almost a 100% attack rate, which provides an example of the extent of contamination that can occur and the consequences for public health when contaminants are present in raw ingredients and poor hygienic practices and inadequate refrigeration occur. The attributed food vehicle was raw egg mayonnaise included in chicken and pork rolls prepared in a bakery. *S. Typhimurium* isolates matching the case isolates were detected in the raw egg mayonnaise, ham, pork, chicken, pate and shell eggs, and from swabs of the preparation bench, tongs, meat slicer, floor drain and display tray, and in environmental samples taken at the source premises. A refrigeration storage breakdown and inadequate refrigeration in the display unit were observed. The count of *S. Typhimurium* in the raw egg mayonnaise sample was in excess of 1.1×10^7 colony-forming units/mL and suggested significant growth of contaminants in the raw eggs occurred in the raw egg mayonnaise during either the faulty refrigerated storage and/or after the rolls were held in a display unit with poor temperature control.

Fruits and vegetables have been linked increasingly to cases of foodborne illness in food services and home settings (Lynch et al., 2009). These foods are frequently eaten raw either alone or in salads. Viruses and parasites introduced at primary production or during food handling will not grow during storage but can persist and cause infections (CDC, 2010). Evidence is growing of the ability of bacterial pathogens (e.g. *E. coli* O157 and *L. monocytogenes*) to persist and grow in produce under certain conditions (Lynch et al., 2009). *E. coli* O157 will grow on cut or damaged surfaces of salad leaves at a more rapid rate than on intact leaf surfaces (Khalil and Frank, 2010) and growth may occur under conditions of temperature abuse during storage (Delaquis et al., 2007). Such conditions can occur in this sector.

In the home and in underdeveloped businesses, those preparing food will not determine critical control points formally. They will practice control measures they have learnt from family and community contacts, from recipes and from messages provided by the media and authorities. Public education messages can be based on sound scientific information, for example in the WHO Five Keys to Food Safety "Cook thoroughly" and "Keep food at safe temperatures" capture the important controls discussed (Table 30.2). Authorities and education groups provide numerous fact sheets on food safety controls such as cooking correctly and temperature control. Some messages are for specific consumer groups at high risk (ill, elderly, babies, pregnant women) or they may be targeted at a specific food and/or hazard (special care foods, hamburgers and *E. coli* O157, eggs and *S. Enteritidis*). Messages may be disseminated at a particular time of year when authorities know failures in food safety occur and re-enforcement is required (cooking turkey at Christmas or Thanksgiving, cooking barbecues, summer eating, emergency events).

Manufacturers may define safety control points for users of their products through the provision of instructions on labels or package inserts. Some common examples include advice on storage, e.g. refrigeration, refrigeration temperature and storage duration required for perishable or shelf-stable product after opening; on preparation, such as whether or not to wash ready-to-eat packaged fresh leafy produce, or ingredient use; on cooking or reheating, including times and temperatures and ways to measure temperature; and cooking or reheating methods such as required microwave operating capacity and use.

Illness outbreaks have been associated with manufactured foods for which consumers were required to apply critical lethality steps before consumption. Examples include

BOX 30.2

SOME EXAMPLES OF CONSUMER MISINTERPRETATIONS AND EXPECTATIONS

Consumer expectations:

- Manufactured food is safe regardless of instructions for further pathogen lethality treatments
- Methods on labels are validated for use in consumer's food preparation setting

A manufactured ready-to-cook product appearance that may cause user misinterpretation the product is fully cooked:

- Batters and crumbs that are set and have a cooked color
- Char-grill marks
- Browned pie crusts
- Plastic trays can be associated with microwave reheating rather than cooking

Instructions that can confuse lethality level of heat treatment required:

- Both microwave and cooking instructions can be confusing as microwaves are commonly used for reheating
- Consumers may have limited knowledge of microwave wattage
- Consumers are not able to calibrate temperature measuring devices

Washing fresh and fresh-cut produce in water removes all pathogens.

contaminated ready-to-cook foods such as frozen chicken entrées, breaded products (NACMCF, 2006) and pies (MacDougall et al., 2004). Such products may have been manufactured with partial cooking then chilled or frozen and may appear to the consumer to be fully cooked. The products were then reheated inadequately and consumed.

Such outbreaks have emphasized that manufacturers need to ensure the safety status of their products, how the consumer will handle their product, how accurately their package labels and instructions can be interpreted, and provide guidance on how to apply validated controls (NACMCF, 2006). Further considerations in developing cooking instructions to ensure lethality to pathogenic bacteria may include product composition, geometry, temperature before cooking and proper monitoring using thermometers. Some examples of consumer expectations and some misinterpretations when handling minimally processed and processed foods that need to be considered by manufacturers and suppliers are provided in Box 30.2.

Monitoring and Corrective Action

Managers and supervisors should have sufficient knowledge of food hygiene and safety principles and practices to be able to monitor the food safety measures and to make decisions when deviations occur and for the corrective action to be taken. Employees should be informed of their role in this process and on how to document and communicate the outcomes to management. In a small business this responsibility may lie with a single individual who should be appropriately trained and available to make decisions when the business is operating. In very small or underdeveloped businesses, street vendors, etc., the relevant authority may have to take this responsibility through regular inspection and guidance on corrective actions and solutions when necessary.

It is apparent in the salmonellosis outbreak investigation of Mannes et al. (2010) described above that failures occurred in the quality of the raw eggs, general hygiene and when storing the raw egg mayonnaise both during the refrigerated storage prior to preparation of the pork and chicken rolls and after in the display cabinet. The use of pasteurized egg, good hygienic practices, corrective action taken to discard the product after faulty refrigerated storage and monitoring display holding temperatures would all have helped to prevent the high numbers of *Salmonella*.

Validation and Verification

Validation of the effectiveness of a food safety program, verification that is operating in accordance and auditing are described in Chapters 44 and 51. The implementation of these activities and auditing of the programs should ideally be prioritized in accordance with the food safety risk of the business. For example, the risk would be significant in large institutions catering for vulnerable persons compared with a small vendor of a low risk food to low risk individuals. It is reasonable to expect that as a minimum requirement all food services should document their food safety plan, keep records relating to the procedures outlined in the plan including action taken in the event of a deviation, and review the procedures if any changes take place in the food produced or the processes used.

In some small and less developed businesses in underdeveloped regions this may be best undertaken less formally by local authorities. Large-scale commercial food service or other community facilities (e.g. schools, daycare centers, hospitals and institutional care establishments) may retain food samples for at least until the end of its shelf-life so as to be available for future possible analysis in the event of an investigation into the product's safety or quality.

EDUCATION AND TRAINING

A successful food safety program requires development of a food safety culture among management and staff. Food services present challenges as often there is a high turnover of staff and their education level can range from highly qualified chefs and food technologists to minimally educated and illiterate workers. Non-professional food work in the food service industry is attractive to poorly skilled and itinerant workers and they often receive minimal wages. These workers may include newly arrived immigrants with limited language ability and different cultural backgrounds or they may be students or others undertaking casual work. Given their employment status and the fact they may have other aspirations for their long-term life plans they may not be motivated to make a commitment to food safety and quality in the short term (Choudhury et al., 2011).

Success in establishing a sustainable food safety culture depends on commitment, leadership and support from management in making food safety a foundation value integrated with other business functions. Seaman and Eaves (2010) conducted a study among food industry managers in London, England, and found while most managers were aware of their responsibilities in training they often did not provide adequate support for practices or evaluation of their effectiveness. Poor motivation has been associated with lack of consequences for non-compliance (Niode et al., 2011).

Continuous training, education and supportive supervision are required. These should be culturally appropriate and tailored to account for language capability, literacy and education levels. Dalton et al. (2010) conducted a national case–control study of listeriosis in Australia and recommended prevention messages should be disseminated in multiple languages as they identified a lack of uptake of education messages in English among women from families where English was a second language.

The messages should be positive although the consequence of non-compliance should also be clear. Multiple approaches should be provided internally, via authorities or by consultants and some examples include demonstrations, use of multimedia communications and high profile community individuals (Powell et al., 2009). Novel approaches can be sought. Chapman et al. (2010) evaluated the use of regular topical information sheets placed in highly visible areas that had successful outcomes. Training may have to begin with basic hygiene and practical demonstrations before HACCP principles and could be provided by authorities before or at the time of business registration and licensing for less developed businesses. For domestic food handlers and food handlers in general, the WHO Keys provide basic food safety messages. Additional messages are required for at-risk groups on safe food choices and food handling. These can be provided through multiple channels such as community and industry groups, schools, public media and social networking.

When compliance is poor, understanding behaviors and barriers to adoption should be explored and alternate approaches that address overcoming the barriers developed. Niode et al. (2011) found from interviewing managers of Asian and Mexican restaurants in northern California that training would be improved if based on foods common to their cuisine and appropriate visual aids were used for employees. Barriers to be addressed in implementing food safety messages among restaurant workers have included time constraints, inconvenience, inadequate training and inadequate resources (Howells et al., 2008). Nesbitt et al. (2009) identified factors to be addressed among domestic food handlers in Canada and included demographic characteristics. They found increasing total annual household income, male sex and elderly status were associated with increases in certain high risk food behaviors.

CONCLUSIONS

Preparing food in food services and the home is the last step before food is consumed. Food production in this sector is increasing in developed countries, and in poor communities the population may be dependent on food prepared by others as their sole source of nutrition. Ensuring food safety in this sector is a critical link in the food chain continuum; however, it is also faced with specific challenges. Evidence from foodborne illness surveillance, the identification of risk factors and at-risk groups, and from understanding impediments, behaviors and cultural beliefs that influence the uptake of food safety practices are some of the activities that are helping to improve food safety in this sector.

References

Boxman, I., Dijkman, R., Verhoef, L., Maat, A., van Dijk, G., Vennema, H., et al., 2009. Norovirus on swabs taken from hands illustrate route of transmission: a case study. J. Food Prot. 72, 1753–1755.

Bryan, F.L., Doyle, M.P., 1995. Health risks and consequences of *Salmonella* and *Campylobacter jejuni* in raw poultry. J. Food Prot. 58, 326–344.

CAC, 1969. Hazard analysis and critical control point (HACCP) system and guidelines for its application. Annex to CAC/RCP 1-1969 Rev. 4-2003. Codex Alimentarius Commission.

CAC, 1999. Revised regional guidelines for the design of control measures for street-vended foods in Africa. CAC/GL-22-Rev. 1. Codex Alimentarius Commission.

Carvalheiro, R., Rocha, A., 2008. Project implementation of a food security system in an elderly center based on the HACCP methodology. Rev. de Alimentacao Hum. 14, 41–53.

CDC, OutbreakNet, Foodborne Outbreak Online Database, 2010. <http://wwwn.cdc.gov/foodborneoutbreaks/>. Cited September 2010.

Chapman, B., Eversley, T., Fillion, K., MacLaurin, T., Powell, D., 2010. Assessment of food safety practices of food service food handlers (Risk Assessment Data): testing a communication intervention (evaluation of tools). J. Food Prot. 73, 1101–1107.

Choudhury, M., Mahanta, L., Goswami, J., Mazumder, M., Pegoo, B., 2011. Socio-economic profile and food safety knowledge and practice of street food vendors in the city of Guwahati, Assam, India. Food Control. 22, 196–203.

Dalton, C.B., Merritt, T.D., Unicomb, L.E., Kirk, M.D., Stafford, R.J., Lalor, K., 2010. A national case–control study of risk factors for listeriosis in Australia. Epidemiol. Infect., 1–9.

Delaquis, P., Bach, S., Dinu, L.D., 2007. Behavior of *Escherichia coli* O157:H7 in leafy vegetables. J. Food Prot. 70, 1966–1974.

DePaola, A., Jones, J.L., Woods, J., Burkhardt III, W., Calci, K.R., Krantz, J.A., et al., 2010. Bacterial and viral pathogens in live oysters: 2007 United States market survey. Appl. Environ. Microbiol. 76, 2754–2768.

Desmarchelier, P.M., Fegan, N., 2003. *Escherichia coli*. In: Hocking, A.D. (Ed.), Foodborne Microorganisms of Public Health Significance. AIFST Inc., Sydney, pp. 267–310.

Dworkin, M.S., Patel, A., Fennell, M., Vollmer, M., Bailey, S., Bloom, J., et al., 2004. An outbreak of ammonia poisoning from chicken tenders served in a school lunch. J. Food Prot. 6, 1299–1302.

EFSA, 2009. Trends and sources of zoonoses and zoonotic agents in humans, foodstuffs, animals and feedingstuffs, United Kingdom in 2006. Zoonoses monitoring. Report referred to in Article 9 of Directive 2003/99/EC. European Food Safety Authority, Parma.

FAO/WHO, 2004. Risk assessment of *Listeria monocytogenes* in ready-to-eat foods. Interpretative Summary. Microbiological Risk Assessment Series 4, Rome.

FAO/WHO, 2005. Guidance to governments on the application of HACCP in small and/or less-developed food businesses. Food and Nutrition Paper 86, Rome.

FAO/WHO, 2006. *Enterobacter sakazakii* and *Salmonella* in powdered infant formula: Meeting Report. Microbiological Risk Assessment Series – 10, Rome.

FAO/WHO, 2009. Toxicological and health aspects of melamine and cyanuric acid. Report of a WHO Expert Meeting in collaboration with FAO supported by Health Canada, Rome.

Hedican, E., Miller, B., Ziemer, B., LeMaster, P., Jawahir, S., Leano, F., et al., 2010. Salmonellosis outbreak due to chicken contact leading to a foodborne outbreak associated with infected delicatessen workers. Foodborne Path. Dis. 7, 995–997.

Howells, A.D., Roberts, K.R., Shanklin, C.W., Pilling, V.K., Brannon, L.A., Barrett, B.B., 2008. Restaurant employees' perceptions of barriers to three food safety practices. J. Am. Diet. Assoc. 108, 1345–1349.

Jacob, C.J., Powell, D.A., 2009. Where does foodborne illness happen – in the home, at foodservice, or elsewhere – and does it matter? Foodborne Path. Dis. 6, 1121–1123.

Jones, T.F., Angulo, F.J., 2006. Eating in restaurants: a risk factor for foodborne disease? Clin. Infect. Dis. 43, 1324–1328.

Khalil, R.K., Frank, J.F., 2010. Behavior of *Escherichia coli* O157:H7 on damaged leaves of spinach, lettuce, cilantro, and parsley stored at abusive temperatures. J. Food Prot. 73, 212–220.

Lin, C.-M., Takeuchi, K., Zhang, L., Dohm, C.B., Meyer, J.D., Hall, P.A., et al., 2006. Cross-contamination between processing equipment and deli meats by *Listeria monocytogenes*. J. Food Prot. 69, 71–79.

Luber, P., 2009. Cross-contamination versus undercooking of poultry meat or eggs – which risks need to be managed first? Int. J. Food Microbiol. 134, 21–28.

Lynch, M.F., Tauxe, R.V., Hedberg, C.W., 2009. The growing burden of foodborne outbreaks due to contaminated fresh produce: risks and opportunities. Epidemiol. Infect. 137, 307–315.

MacDougall, L., Fyfe, M., McIntyre, L., Paccagnella, A., Cordner, K., Kerr, A., et al., 2004. Frozen chicken nuggets and strips – a newly identified risk factor for *Salmonella* Heidelberg infection in British Columbia, Canada. J. Food Prot. 67, 1111–1115.

Mannes, T., Gupta, L., Craig, A., Rosewell, A., Aimers McGuinness, C., Musto, J., et al., 2010. A large point-source outbreak of *Salmonella* Typhimurium phage type 9 linked to a bakery in Sydney, March 2007. Commun. Dis. Intell. 34, 41–48.

Marcus, R., Varma, J.K., Medus, C., Boothe, E.J., Anderson, B.J., Crume, T., et al., 2007. Re-assessment of risk factors for sporadic *Salmonella* serotype Enteritidis infections: a case–control study in five FoodNet Sites, 2002–2003. Epidemiol. Infect. 135, 84–92.

McLaughlin, J.B., Castrodale, L.J., Gardner, M.J., Ahmed, R., Gessner, B.D., 2006. Outbreak of multidrug-resistant *Salmonella* Typhimurium associated with ground beef served at a school potluck. J. Food Prot. 69, 666–670.

MMWR, 2010. Preliminary FoodNet data on the incidence of infection with pathogens transmitted commonly through food – 10 states, 2009. Morbid. Mort. Wkly. Rep. 59, 418–422.

NACMCF (National Advisory Committee on Microbiological Criteria for Foods), 2006. Respose to the questions posed by the Food Safety and Inspection Service regarding consumer guidelines for the safe cooking of poultry products. <http://www.fsis.usda.gov/PDF/NACMCF_Report_Safe_Cooking_Poultry_032406.pdf> cited January 2011.

Nelson, J.M., Bednarczyk, R., Nadle, J., Clogher, P., Gillespie, J., Daniels, A., et al., 2008. Foodnet survey of food use and practices in long-term care facilities. J. Food Prot. 71, 365–372.

Nesbitt, A., Majowicz, S., Finley, R., Marshall, B., Pollari, F., Sargeant, J., et al., 2009. High-risk food consumption and food safety practices in a Canadian community. J. Food Prot. 72, 2575–2586.

Niode, O., Bruhn, C., Simonne, A.H., 2011. Insight into Asian and hispanic restaurant manager needs for safe food handling. Food Control. 22, 34–42.

OzFoodNet, 2009. Monitoring the incidence and causes of diseases potentially transmitted by food in Australia: annual report of the OzFoodNet network, 2008. Commun. Dis. Intell. 33, 389–413.

Powell, D.A., Hubbell, A.L., Chapman, B., Jacob, C.J., 2009. New media for communicating food safety. Food Technol. 63, 38–43.

Price, C.C., 1997. Trends in eating out. Food Consumption. Food Review September–December. Economic Research Services. United States Department of Agriculture. <http://www.ers.usda.gov/publications/foodreview/sep1997/sept97c.pdf> cited September 2010.

Seaman, P., Eves, A., 2010. Perceptions of hygiene training amongst food handlers, managers and training providers – a qualitative study. Food Control 21, 1037–1041.

Shapiro, R., Ackers, M.-L., Lance, S., Rabbani, M., Schaefer, L., Daugherty, J., et al., 1999. *Salmonella* Thompson associated with improper handling of roast beef at a restaurant in Sioux Falls, South Dakota. J. Food Prot. 62, 118–122.

Slinko, V.G., McCall, B.J., Stafford, R.J., Bell, R.J., Hiley, L.A., Sandberg, S.M., et al., 2009. Outbreaks of *Salmonella* Typhimurium phage type 197 of multiple genotypes linked to an egg producer. Commun. Dis. Intell. 33, 419–425.

Stafford, R.J., McCall, J., Neill, A.S., Leon, D.S., Dorricott, G.J., Towner, C.D., et al., 2002. A statewide outbreak of *Salmonella* Bovismorbificans phage type 32 infection in Queensland. Commun. Dis. Intell. 26, 568–573.

Todd, E.C.D., Michaels, B.S., Greig, J.D., Smith, D., Bartleson, C.A., 2010. Outbreaks where food workers have been implicated in the spread of foodborne disease. Part 8. Gloves as barriers to prevent contamination of food by workers. J. Food Prot. 73, 1762–1773.

Todd, E.C.D., Greig, J.D., Bartleson, C.A., Michaels, B.S., 2007. Outbreaks where food workers have been implicated in the spread of foodborne disease. Part 3. Factors contributing to outbreaks and description of outbreak categories. J. Food Prot. 70, 2199–2217.

USDA, 2009. Global food markets: international consumer and retail trends. Economic Research Services. United States Department of Agriculture. <http://www.ers.usda.gov/publications/foodreview/sep1997/sept97c.pdf> cited September 2010.

USFDA, 2004. Report on the occurrence of foodborne illness risk factors in selected institutional foodservice, restaurant, and retail food store facility types. United States Food and Drug Administration. <http://www.fda.gov/Food/FoodSafety/RetailFoodProtection/FoodborneIllnessandRiskFactorReduction/RetailFoodRiskFactorStudies/ucm089696.htm> cited September 2010.

USFDA, 2006. Managing Food Safety: A manual for the voluntary use of HACCP principles for operators of food service and retail establishment. In USFDA Administration, College Park.

Van Asselt, E.D., De Jong, A.E.I., De Jonge, R., Nauta, M.J., 2008. Cross-contamination in the kitchen: estimation of transfer rates for cutting boards, hands and knives. J. Appl. Microbiol. 105, 1392–1401.

Vollaard, A.M., Ali, S., van Asten, H.A.G.H., Suhariah Ismid, I., Widjaja, S., Visser, L.G., et al., 2004. Risk factors for transmission of foodborne illness in restaurants and street vendors in Jakarta, Indonesia. Epidem. Infect. 132, 863–872.

Vought, K.J., Tatini, S.R., 1998. *Salmonella* Enteritidis contamination of ice cream associated with a 1994 multistate outbreak. J. Food Prot. 61, 5–10.

WHO, 2006. Five keys to safer food manual. World Health Organization, Geneva. <http://www.who.int/foodsafety/consumer/5keysmanual/en/index.html>.

Zhao, T., Doyle, M.P., Berg, D.E., 2000. Fate of *Campylobacter jejuni* in butter. J. Food Prot. 63, 120–122.

CHAPTER

31

Hazard Analysis and Critical Control Point System (HACCP)

Yasmine Motarjemi

Food Safety Management Consultant, Nyon, Switzerland

OUTLINE

INTRODUCTION

HACCP stands for the hazard analysis and critical control point system. Today it is known more by its acronym than its full name. HACCP as defined by the Codex Alimentarius Commission is a system that *identifies, evaluates and controls hazards which are*

Food Safety Management.
DOI: http://dx.doi.org/10.1016/B978-0-12-381504-0.00031-7

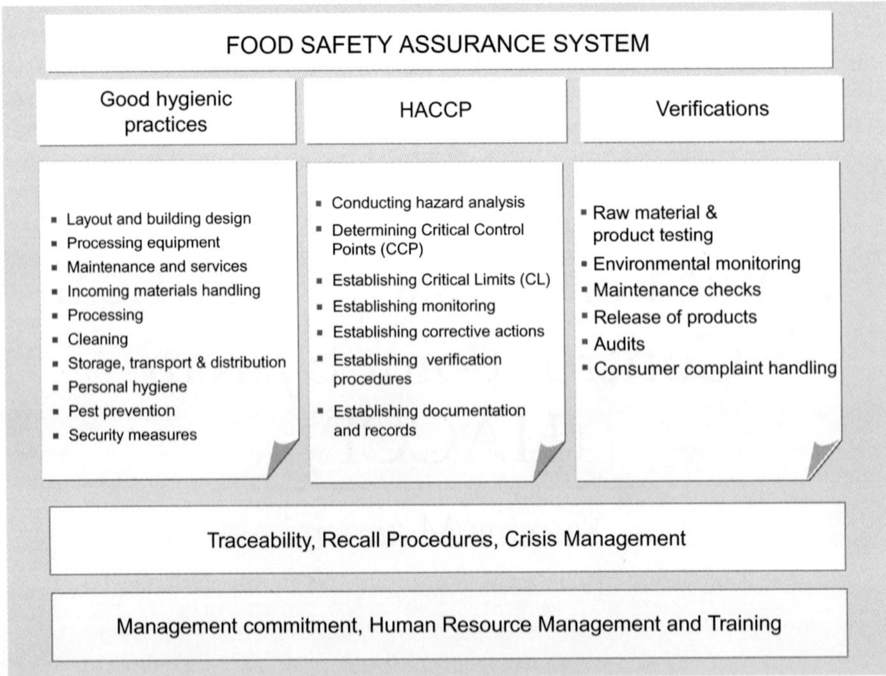

FIGURE 31.1 Overview of the food safety assurance system in the food industry.

significant for food safety. Worldwide, it is considered as the reference method for food safety assurance. However, as pointed out in Chapter 1, to be effective in ensuring food safety, HACCP has to be implemented in conjunction with a number of other programs, some of which are part of the "prerequisites programs." Examples of these are cleaning and sanitation, pest management, hygienic design, etc.; they are often grouped under the term *good hygienic practice.* Some other activities also related to food safety management, such as consumer complaint handling, environmental and pathogen monitoring, chemical contaminants monitoring and audits, are verifications which are also required in the HACCP system but are often implemented separately (Figure 31.1). It is nevertheless important that the design of these activities and programs and their outcome be carried out in coordination with the HACCP system, i.e. these different programs have to be geared into each other as the wheels of the same machinery (Figure 31.2).

Originally, the HACCP system was introduced to ensure the microbiological safety of food products. Later on, its use was extended to all types of foodborne hazards, including chemical hazards, allergens and physical hazards. The HACCP approach can also be used for auxiliary systems such as the water system.

There have been many debates and articles on the challenges in the application of the system, in particular in small and less developed businesses. Even in large food operations, the application of the system has not been without difficulties, partly because of misperception of the system and partly because of the lack of understanding and/or commitment of

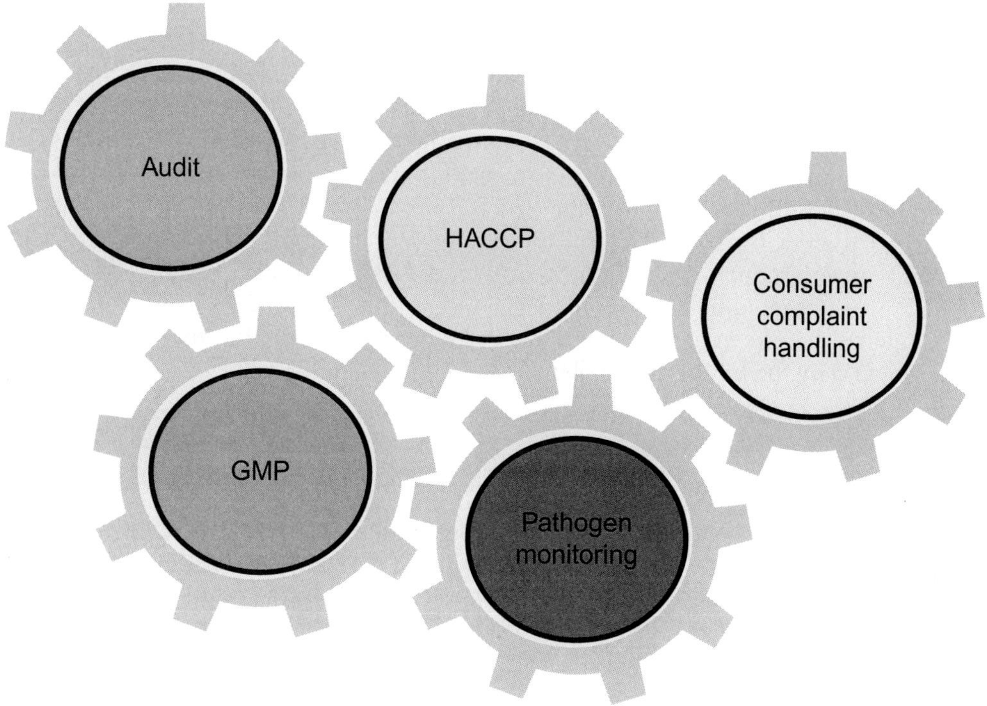

FIGURE 31.2 The figure illustrates the importance of an integrated approach to food safety management and the interaction between the different elements of the food safety assurance system.

necessary resources. The successful application of HACCP requires considerable expertise, time investment and multidisciplinary collaboration.

Considering the importance of food safety for consumers' health and for businesses, such an investment is fully justified, as without food safety there will be no business.

HISTORICAL BACKGROUND

HACCP was originally designed by the Pillsbury Company, together with the National Aeronautics and Space Administration (NASA) and the US Army Laboratories at Natick. They developed the HACCP system to ensure the safety of food for astronauts. For many years after its conception, the system was promoted by international organizations such as the World Health Organization and applied on a voluntary basis in certain food industries (Motarjemi et al., 1996). In 1993, the Codex Alimentarius Commission recognized the HACCP system as a powerful tool to improve food safety and established the Codex guidelines for the Application of the HACCP system. This has had major implications for the widespread implementation of the HACCP system. Another event in the history of food safety provided a major impetus for the promotion of HACCP. In 1995, with the establishment of the World Trade Organization and the coming into force of the Agreement on Sanitary and

Phytosanitary Measures (WTO/SPS), the work of Codex, i.e. its standards, guidelines and recommendations (including the Codex document on the Hazard Analysis and Critical Control Point system and Guidelines for its application) became the international reference or the "yardstick" for national requirements in food safety. This implied that WTO member states needed to take the work of the Codex Alimentarius into consideration and align their national legislation with the provisions of the Codex Alimentarius Commission, unless they could provide scientific evidence that the Codex Standards did not provide adequate health protection for their population. This meant that *de facto* the application of the HACCP system became an international requirement for food safety assurance. Today, the principles of HACCP are integrated in the national legislation of many countries as well as in the ISO 22000 standards, i.e. the standards defining the requirements for the management of food safety developed by the International Standard Organization (ISO, 2005).

THE NEED FOR HACCP

Up until the introduction of the HACCP system, the food safety assurance system was a reactive one, i.e. based on the implementation of directives referred to as "codes of practice" which were developed based on experience, combined with end-product testing. Such codes had to be fairly general in order to be applicable in diverse situations of food production and food processing. Therefore, they could not and did not consider hazards which where specific to a food product, i.e. its ingredients and/or the specific conditions of operation. As with globalization and changes in society, the nature and origin of raw material were becoming more and more diverse, the technology used in food production and processing more complex, and the traditional approach became increasingly inadequate for preventing and controlling hazards in foods. Also, with increased industrialization, mass production and distribution of food, the risk of large-scale foodborne disease outbreaks as has been experienced in recent years became greater. Many foodborne pathogens proved to be particularly virulent, in particular with the vulnerable group of the population, and led to severe or chronic health problems, if not death. It had become clear that end-product testing, until that time used as the main quality control method, proved to be inadequate for providing assurance, since a large number of samples would need to be tested to have a certain degree of assurance of safety; in practice, adequate end-product testing to obtain reliable information was economically not feasible, and often the results would be received after the product had been marketed and/or eaten.

Thus there was a clear need for a more effective system of food safety assurance, where hazards and risks with a given product would be identified and measures necessary for controlling these hazards would be prospectively determined and deployed. It is against this background that the HACCP system was introduced to complement the traditional approach to enhance food safety assurance. With time, it was also experienced that HACCP would be best applied if it were combined with the application of general codes of hygiene and not as a stand-alone system. In other words, the general codes of hygiene would be used as a first line of defense to have a general hygienic condition of food production, processing or any other operations (transport, distribution, preparation, etc.). HACCP would be applied as a second line of defense to have a tailor-made system of food safety assurance for the product under consideration, and end-product testing would be carried out as a last line of defense for confirmation that the preventive measures are effective (Motarjemi et al., 1996).

In summary, the benefits of the HACCP system lie in the fact that HACCP:

1. Is a proactive approach to food safety management; this means it allows conceivable and reasonably expected hazards to be identified, even when failures have not previously been experienced. It is particularly useful for new operations.
2. Is flexible, i.e. necessary control measures can be adapted to changes in operations, such as change in equipment design, in processing procedures and technological development.
3. Helps to target resources to the most critical part of the food operations.
4. Is applicable to the entire food chain, from the raw material to the end product, i.e. growing, harvesting, processing/manufacturing, transport and distribution, preparation and consumption.
5. Overcomes many of the limitations of the traditional approaches to food safety control, generally based on:
   a. snap-shot inspection, which is a rather ineffective approach in foreseeing potential problems;
   b. end-product testing, which would entail high costs for analysis and which would lead to identifying problems without understanding their cause

PRINCIPLES OF THE HACCP SYSTEM

As stipulated in the Codex guidelines on HACCP, the HACCP system is comprised of seven principles. These are as follows:

Principle 1: Conducting a hazard analysis.
Principle 2: Determining the critical control points (CCPs).
Principle 3: Establishing critical limits.
Principle 4: Establishing a system to monitor control of the CCP.
Principle 5: Establishing the corrective action to be taken when monitoring indicates that a particular CCP is not under control.
Principle 6: Establishing procedures for verification to confirm that the HACCP system is working effectively.
Principle 7: Establishing documentation concerning all procedures and records appropriate to these principles and their application.

To ensure the most effective outcome, the application of the HACCP system is carried out following a number of steps. The Codex Guidelines outline 12 steps for conducting an HACCP study and establishing an HACCP plan. To this should be added the training of different operators, implementation of the plan as well as a number of prerequisite activities.

With regard to the HACCP application, the importance of validating the elements of the HACCP system needs to be highlighted. This means that at every step in the development of the HACCP study, it is important to ensure that decisions are valid, i.e. that they are established based on a scientific and technical basis. In particular, the control measures must be effective and achieve the expected outcome (e.g. regulatory or industry limit, performance or food safety objectives). As such, validation is the assurance in the food safety assurance system. As the Codex Guidelines for HACCP are not explicit on the subject of validation, separate guidelines on the validation of control measures have been established by the Codex Alimentarius (Codex, 2008).

APPLICATION OF HACCP

Prerequisites to the Application of HACCP

The term *prerequisite* refers to all the measures and activities which need to be in place in order to support the application of the HACCP system. Very often, this term is used for technical programs that need to be in place, such as cleaning and sanitation or generally good hygienic practices. However, for the purpose of this text, the term is used in a broader sense in order to highlight the fact that certain conditions, other than technical measures, also need to be fulfilled before HACCP can be successfully implemented (Motarjemi et al., 2009):

1. Management commitment. The use of the HACCP system is very resource – and time – intensive. At times, it involves huge costs or investment as its application may underpin the need for new equipment, change in the process of production and/or quality of product, or change in the supplier of raw material. As mentioned before, a successful implementation requires high-level and multidisciplinary expertise as well as time investment. Therefore, the understanding of the management on the need and benefits of the HACCP system and its implications in terms of financial and human resources is essential and is a *conditio sine qua non* for the successful application of HACCP, short of which HACCP studies become only a paper exercise.
2. General principles of hygiene. As mentioned above, before applying the HACCP system, a certain number of programs and activities, generally considered as part of good hygienic practice, have to be implemented to ensure that products are manufactured, processed or handled in the minimum conditions of hygiene and good practices and that the generally known risks are as far as possible prevented. Failing which the number of risks to control through the HACCP will be large and the system will be difficult and costly to manage. Depending on the type of business and the stage of the food chain, these predefined rules, procedures and practices are referred to as good agriculture practice, good animal husbandry, good manufacturing practice, good transportation practice, etc.

   In practice, such codes refer to generic control measures that apply to a given sector of the food chain, regardless of its specific conditions (e.g. environment, ingredients, product formulation, production and processing). However, it is to be noted that a control measure recommended in a code, thus implemented as part of prerequisite program, can still be identified as control measure in an HACCP plan.

   The International Standardization Organization has elaborated a standard for the management of food safety in organizations, referred to as ISO 20005. This standard distinguishes between the terms "prerequisites" and "operational prerequisites"

BOX 31.1

DEFINITIONS OF PREREQUISITES ACCORDING TO ISO 22 000 STANDARDS (ISO 2005)

ISO 22000 defines the term "prerequisite" as follows:

Prerequisite program

Basic conditions and activities that are necessary to maintain a hygienic environment throughout the food chain that is suitable for the production, handling and provision of safe end products and safe food for human consumption.

Operational prerequisite program (operational PRP)

These are identified by the hazard analysis as essential in order to control the likelihood of introducing food safety hazards in, and/or the contamination or proliferation of food safety hazards in the product(s) or in, the processing environment.

(Box 31.1). The latter refers to control measures identified in the HACCP studies for controlling a hazard; without the step in the food operation being considered as critical for safety.

3. Scientific research. Scientific and technical data and know-how are fundamental to any proactive and science-based food safety assurance system, such as HACCP. The type of scientific research that is needed entails:
   a. Toxicological and epidemiological research. HACCP being a risk-based system, such toxicological and epidemiological data are needed in order to evaluate the health significance of the different compounds, the degree of risk they present and their sources. In principle, such guidance should be provided by public health authorities.
   b. Epidemiology of microorganisms. Epidemiological data can provide guidance on the type of food that is the vector of a pathogen, on risk factors, health consequences and on various information (e.g. record of previous outbreaks) necessary for hazard analysis.
   c. Ecology of microorganisms. Understanding the ecology of microorganisms is essential for evaluating the potential source, likelihood of contamination, survival or growth of a pathogen in an environment, food or during processing.
   d. Mechanism of formation of contaminants. Certain contaminants are formed during processing or manufacturing, be it in the industrial setting or in the home. Understanding the mechanism of formation of such contaminants is essential for devising control measures, e.g. designing the process to minimize their formation.
4. Validated analytical methods. Validated analytical methods will be needed to manage hazards in foods, be it for verification purposes or as control measures for certain types of hazards (e.g. chemical contaminants).

5. Data on the level of occurrence. The first principle of HACCP on hazard analysis calls for an evaluation of the risk, including the likelihood of occurrence. Data on the occurrence of hazards in food is thus fundamental for a first evaluation and hazard analysis (see also above).
6. Determination of acceptable level. A control measure is defined as "Any action and activity that can be used to prevent or eliminate a food safety hazard or reduce it to an acceptable level." This brings in the concept of "acceptable level." To be able to manage foodborne hazards, in particular when the occurrence of the agent cannot be fully prevented, there is a need to know to what level the hazard in question must be controlled and to decide on a limit of acceptability for the hazard in question. The determination of this level, preferably by regulatory authorities, or short of this by industry associations, is very important both for the industry and for consumers. Such limits, over and above providing guidance for managing the hazard in question in the context of HACCP or any other equivalent food safety assurance system, ensure a consistent approach to food safety through the food supply chain and on the global market.

   For a number of chemical hazards, such limits have been established and regulatory standards at national or international level (Codex) are available. In recent years, a similar concept has emerged for microbial agents. In principle, it is recognized that no case of foodborne illness is acceptable and ideally there should be "zero" incidents of foodborne illness. However, as for the presence of naturally occurring or environmental contaminants, realistically this is not possible since many pathogens make part of our environment and/or the microbial flora of food animals. Generally, with the presently available resources or technologies, their eradication is not feasible. Therefore, authorities are considering setting a maximum level of microbial contamination of food at the time of consumption, referred to as *food safety objective.*[1] This would ensure that the incidence of foodborne illness will be maintained within a certain accepted or tolerated level. This accepted/tolerated level of illness is referred to as the appropriate level of protection (ALOP). Food safety objectives can also be translated into performance objectives,[2] i.e. the maximum level of contamination at an early stage of the food chain, or performance criteria defining the performance of an operation in terms of growth or reduction of microorganisms. Such food safety objectives, or in the case of chemical hazards, regulatory limits, need to be decided beforehand, preferably by the regulatory authorities. In other words, regulatory norms and standards provide guidance to industry as to what level they should control the hazards in the food and how stringent should be their control measures. Such information will be necessary for their validation of their process when designing the HACCP plan.

[1]Food safety objective (FSO) is the maximum frequency and/or concentration of a hazard in a food at the time of consumption that provides or contributes to the appropriate level of protection (ALOP).
[2]The maximum frequency and/or concentration of a (microbial) hazard in a food at a specified step in the food chain before the time of consumption that still provides or contributes to the achievement of an FSO or ALOP, as applicable.

It must be remembered that against all the preconceived ideas that regulatory requirements are an unnecessary burden for the food industry, they are extremely important for the food industry in designing their food safety plan, provided that they are based on science and take into account other factors such as feasibility and costs for industry and consumers (Motarjemi and Mortimore, 2005). It is also to be noted that regulatory standards, or norms, are values which delineate between acceptable or unacceptable presence of a given contaminant or foodborne pathogen. Even though they may not always be based on a strict risk assessment, they are viewed as the safety standard by the regulatory authorities who have decided on the limits, by the consumers, and by society at large. Exceeding this limit should be considered as a violation of the safety standards of the society.

Guidance for the Application of HACCP System

Depending on the type of operations and regions of the world, there are different approaches to HACCP. For the purpose of this text, the part of work where step-by-step analysis of hazards and control measures are carried out is referred to as an HACCP study, whereas the outcome of the study where the critical control points and the measures taken at those steps of the food operation are outlined is referred to as an HACCP plan.[3] Sometimes, due to the complexity of the production, it is easier to develop different HACCP plans for different parts of the production. In such cases, it is important to ensure that a proper link between the different HACCP plans exists. It is equally important that an HACCP study also covers rework.

Application of HACCP principles is preceded by a number of activities. These preliminary steps of HACCP application set the conditions for an accurate and valid HACCP study. For instance, the validity of hazard analysis relies on the expertise of the team, the precision with which various aspects of the product are described, e.g. the raw ingredients and their source, the supplier assurance system (e.g. availability of a supplier audit report), manufacturing steps and conditions, description of packaging and other auxiliary products, intended use of product, regulatory requirement, potential use or abuse by target consumers

Step 1 – Assembling the team. Food safety is a multidisciplinary system. To design the safety of a product and foresee its safe production, there is a need for a team of experts of different background and experience. The type of expertise depends on the product and conditions of treatment or handling, and on the scope of the HACCP study, i.e. whether it covers chemical agents, microbial agents, allergens, the full food chain or part of the chain. Therefore, it should be decided on a case-by-case basis. Examples of experts who could be considered are: microbiologist, chemist, toxicologist, nutritionist, operator, veterinarian/agronomist, food technologist/food processing engineer and regulatory expert.

The importance of teamwork and of the expertise of the team cannot be emphasized enough as it is the available expertise in the team that will be the determining factor for

[3] A document prepared in accordance with the principles of HACCP to ensure control over hazards that are significant for food safety in a segment of the food chain.

the quality and accuracy of the HACCP study. What is certain is that HACCP is not a one-person job as often experienced. Usually, a team leader referred to as a "coordinator" is assigned for the team. The coordinator has the responsibility to drive the development and maintenance of the HACCP study in collaboration with the team members. The team will also decide on the scope of the HACCP study. As alluded to above, this consists in deciding which types of hazards (microbiological, chemical, physical) and which part of the food chain the study will cover. It is possible to start with one hazard (e.g. *Salmonella*) or one type of hazards (e.g. microbiological) and to extend to other types of hazards at a later stage.

Step 2 – Description of the product. A full description of the product should be drawn up, including conditions for raw material storage, transport and distribution, and preparation by the end-users. The more in-depth such a description is done, the smaller the risk of overlooking a factor that can influence the presence of a hazard. Examples of information that could be included are:

- What is the product?
- What is its formulation and composition (raw materials and ingredients, physicochemical parameters potentially influencing safety (e.g. pH, a_w)?
- What is the nature of the product, e.g. fresh, canned, dried, vacuum-packaged?
- How is the product manufactured/processed?
- What is the packaging?
- What type of storage, transport and distribution are required?
- What is the shelf-life of the product?
- Are there any other special considerations that need to be addressed, for instance a previous record of safety of the product (see "Hazard Analysis," below)?

A frequent shortcoming is that these descriptions are not detailed enough to allow an in-depth hazard analysis. In absence of such information, important hazards may be missed during hazard analysis. Not infrequently, HACCP plans are developed without fully considering the supplier's food safety assurance, and subsequently hazards that may be present in the raw material are overlooked.

Similarly, steps following manufacturing, i.e. hazards which may occur during transport, distribution, and most importantly during preparation by consumers, are frequently omitted during certain HACCP studies. For certain types of products, factors such as the conditions for storage of the product during distribution or for the target customer/consumer, or the potential mishandling of the product may be crucial for designing safety, including the necessity for providing information on the safe use of the product.

Step 3 – Identification of intended use. Information on the intended use should be based on the expected use of the product by the end-user or consumer, including the country where the product will be sold. Examples of information that need to be considered are:

- What is the intended use (home retail, foodservice, further manufacturing)?
- What preparation procedures are required by the consumer/customer?
- What is the potential for mishandling?
- Who are the target consumers (age, health status)?

The above information is important to ensure that the safety of the product is designed according to the requirements or needs of the target customer or consumers. For instance, if the product is to be sold in another country or to specific consumer groups, e.g. children, the limits applicable for the country in question or target consumers must be taken into account. The identified use will also determine which kind of information and instruction would be required on the package, as for certain products additional control measures may be at the customer or consumer end (e.g. cold storage, labeling for allergens, age of consumption or condition of consumption). As it will be explained elsewhere in the book, it is important that any assumption on the intended use of a product, e.g. age, hygienic condition of preparation and consumption, consumer practice, be validated (i.e. a reality check). For instance, in an outbreak associated with cookie dough, implicating some 70 persons in the United States in 2009, it was found that some US consumers had the practice of tasting the raw cookie dough. After the outbreak, the instruction on the package had to be revised.

Step 4 – Construct the flow diagram. The flow diagram should cover all the major steps[4] in the operation and the conditions for these, particularly steps likely to influence safety. It should, as far as possible, cover the entire food chain, including storage, warehousing, distribution and product handling downstream.

The flow diagram must reflect the real process of production, processing and manufacturing, or handling of the food product. Lack of accuracy may seriously jeopardize the quality of the HACCP study and the validity of decisions. All technical data such as temperature, time, pH, etc. should, as far as possible, be noted on the flow diagram. This can enhance visibility and understanding of the conditions of operations. It will allow an overview of the operations and identification of any possible risks associated with product design or operation.

It is also important to consider how the circulation of water and air and employee traffic can impact on the safety of the product, and, for this purpose, the flow diagram of water, air and people (or zoning plan) should also be taken into account. In this context, all building or reconstruction activities should also be considered as they may lead to the contamination of the factory environment with foodborne pathogens.

Step 5 – On-site confirmation of the flow diagram. This step in the development of the HACCP plan is intended to verify that the flow diagram reflects the real process and that no important consideration has been omitted. Very often, this step is neglected; however, a scrupulous on-site verification of the flow diagram, together with the examination of hygienic conditions by the entire HACCP team will be a strong basis for hazard analysis. To this end, it is important to check the correctness of information or whether information was overlooked. This should be checked during the period of operation and cleaning, but also during idle hours. Talking to operators working on line can also help in disclosing significant details. The on-site verification of the flow diagram is also an occasion for the members of the team to fully understand the role of the different operations units and the way the equipment works, as this insight is important for the evaluation of risks associated with operations.

[4] A point, procedure, operation or stage in the food chain, including raw materials, from primary production to final consumption. The term "step" here refers to the steps in production, manufacturing, and processing operation and not to the step in the development of the HACCP plan.

Step 6 – Hazard analysis. Hazard analysis is defined as the process of collecting and interpreting information on hazards and conditions leading to their presence to decide which are significant for food safety, and should be addressed in the HACCP plan. In practice, this consists in listing all potential hazards associated with each step of operations (from raw material to final use), and in evaluating their significance, i.e. taking into account their likelihood of occurrence and their health consequences.

For this purpose, it is imperatively important to examine the past record of safety of the product in question. This should include:

- Outbreaks, contamination or other types of adverse events (including cases of fraud) associated with the product or agent in question;
- Risk factors, e.g. level of hazard, status of the host, nature of the organism; and
- Underlying factors, i.e. errors and their root cause leading to the incident.

There have been numerous outbreaks or contamination incidents, including cases of fraud that could have been prevented if the previous record of safety of the product were examined.

Note that a hazard is defined as "a biological, chemical, or physical agent in, or condition of food, with the potential to cause an adverse health effect." The term "condition" also includes aspects related to the properties of food. Examples are its consistency and form, which may cause choking, and the nutritional composition of the food, such as infant formula, or pet food, where excess or lack of a nutrient may endanger health. The property is also very much linked to intended use of a product as some products may not be appropriate for a specific target group.

Thus, at first all potential hazards associated with the various steps will be identified; this also includes hazardous conditions of food. At each step, control measures needed to control the hazards are also determined. More than one control measure may be required to control a hazard. In deciding on the control measures, it is fundamental to understand the factors and parameters that characterize the control measures and to have full understanding of the feature of the hazards (e.g. ecology and epidemiology of microorganisms).

Very often hazards are described in general terms, e.g. "microbiological hazard" instead of specifying *Salmonella*, *S. aureus* or hepatitis A virus. Although such an approach may in some cases be practical, it is often misleading, and in regard to microbiological hazards, it may even be dangerous. The reason is that microorganisms differ in their behavior, ecology and control measures. For instance, the ecology and control measure for *S. aureus* are much different than that for *Salmonella* or viruses. Thus, unless organisms share similar ecology and epidemiology, as far as possible they should be considered specifically. Similarly, chemical contaminants, or physical hazards, should be defined individually so that valid safety limits and methods of detection or testing can be identified.

A key question in the hazard analysis is whether a hazard presents a significant risk and qualifies as a *significant hazard*. In the plethora of hazards that may theoretically be associated with a raw material or an operation, it is often difficult to decide which hazards present a real safety threat and warrant a strict control under the HACCP plan. It is clear that the risk of hazard depends on many factors, among others on the source of the raw material, i.e. quality assurance of the supplier, the hygienic and other prerequisite conditions of operations allowing contamination, survival or growth of an agent, whether during the process

TABLE 31.1 The Different Types of Data that may be Needed for Hazard Analysis

Hazards	Examples of Data
Microorgansims	Health consequences, infectious dose of the agent, epidemiological information on the prevalence of the agent in the country/raw material, thermal resistance, survival, growth characteristics Food composition and food characteristics Target consumers and their health conditions Status of Good Hygienic Practice (GHP) and other prerequisites
Mycotoxins	Legislation and safety standards Climatic conditions, droughts, insect attacks and stress factors Surveys, alerts, historical records and monitoring data Agriculture practice (harvest, transport and storage) Target consumers (infants, adults, pets)
Agrochemicals	Legislation and enforcement Agriculture practices Infections in animals or disease of plants Surveys, alerts, historical records and/or monitoring data Target consumers
Allergens	Nature of raw material Evaluation of cross-contact at the supplier or during transport Evaluation of cross-contact on the factory premises Food safety management system at the supplier Verification data/historical record Target consumers
Physical hazards	Size, shape and nature of the hazard Quality assurance at the supplier Good manufacturing/hygienic practices on the premises Target consumers and possible handling upon use

there are other steps that would remove or control the hazards to safe levels, and/or the intended use of the product and the consumer itself. For instance, opportunistic pathogens like *Cronobacter sakazakii* are a threat mainly to newborns but not to healthy adults. Thus, it can be a significant hazard in an HACCP study of infant formula and not for adult food. It is for this reason that during the hazard analysis, over and above being able to identify the potential hazards, the conditions in which food or its ingredients are produced and processed should ideally be considered from farm to fork; as far as possible, it should be ensured that good hygienic practices are implemented throughout the food chain.

At the hazard analysis step, each hazard is to be evaluated for its degree of risk and is classified as significant or not significant. Risk is defined as a function of the probability of an adverse health effect and the severity of that effect, consequential to (a) hazard(s) in food. So, in the context of HACCP and ensuring safety of products, for evaluating risks, two types of information need to be considered: (1) likelihood of occurrence of the hazard in the food and (2) the health consequences. Table 31.1 provides examples of data that will need to be considered for the analysis of different types of hazards.

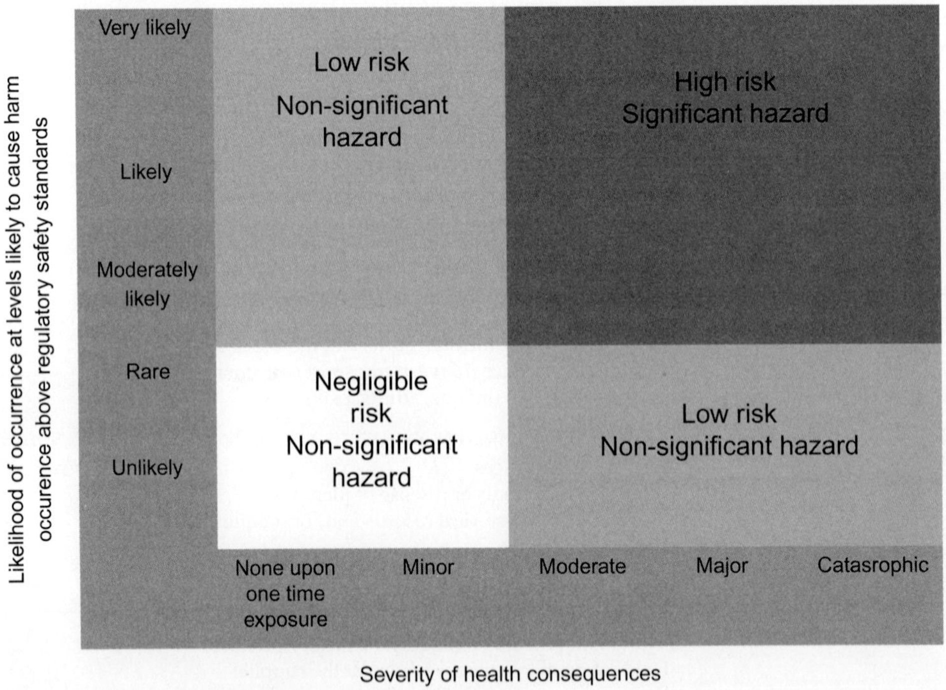

FIGURE 31.3 The evaluation of risk according to the likelihood of occurrence of the hazardous agent and the severity of health consequences. Hazards identified as high risk must be controlled at a CCP.

Hazards which are likely to occur in the raw material or during operation and present a health concern for the consumer are qualified as significant hazards and would need to be controlled through critical control points, as explained below. However, for the hazards which are viewed as non-significant on the grounds that the conditions of production, processing and/or handling will make any risk unlikely or remote, monitoring activities, referred to as *verification*, are carried out. The objective of these monitoring activities is to confirm that existing control measures are effective and that the risk of the hazard, as presumed, remains low at all times.

The evaluation of risks does not always provide a black or white answer. Even when hazards are evaluated as non-significant, there may be different degrees of residual risk (Figure 31.3). Thus, depending on the degree of risk, a different frequency of monitoring may be considered. For instance, for bolognaise sauce, where tomato is considered as a raw material, should the operator use fresh tomato from a supplier with unknown conditions of production, then *Salmonella* should be considered as a significant hazard, cooking of the sauce as the control measure and the step of cooking a critical control point (CCP), i.e. the hazard will be controlled through the HACCP plan. However, if the tomatoes are sourced with a supplier which is audited and if there is confirmation that they are produced under good agriculture practice and a good food safety assurance system, then the risk can be considered as low and the hazard as non-significant. Nevertheless, the supplier should be periodically audited and the tomatoes also periodically tested for *Salmonella* as a way to verify the hazard analysis. Clearly, if canned tomatoes are used, the canning process will remove the agent, and the risk of *Salmonella* can be considered as practicably negligible; in this case there will be no need for testing, i.e. the frequency of testing can be nil.

Step 7 – Determine the critical control points (CCPs). In general, the critical control point refers to the step in the operations at which control is essential to eliminate, reduce or maintain a hazard at an acceptable level; in other words, a step at which, if the hazard in question is not controlled, the product is likely to be unsafe. From the above we can understand that we are referring here to significant hazards. A CCP can be different steps in the operation, from raw materials to a location, process, procedure or practice, including product formulation.

The designation of a step as a CCP usually has many implications over and above the need to set up a very strict monitoring procedure. It can also imply that all the records need to be reviewed, verified and signed before the product is released. This may sometimes delay the release of the product, particularly if the monitoring is based on testing. Also, if the critical limits at the CCP have been violated and the product has accidentally reached the market, consideration must be given to product recall.

A condition for a step to be considered as a CCP is the fact that it should be possible to effectively monitor the step. Also, if a hazard is identified as significant, and no possible control measure exists, then the product or process must be modified. This principle of modification is often overlooked in practice and not infrequently, for business reasons, food establishments try to cope with a risky condition rather than to modify the process. For instance, it has occurred that a raw material is considered as a high risk because there is a doubt on the quality assurance of the supplier. Yet, instead of changing the supplier, testing of raw material is used as a means for controlling foodborne pathogens, while as mentioned before, testing for microbial agents cannot be an effective control measure and should be seen as a verification of the supplier quality assurance. In such a case, the supplier should be changed.

The determination of a CCP can be facilitated by the application of a decision tree, which supports a logical reasoning approach. Figure 31.4 provides an example of a decision tree, but other decision trees can also be used.

A lack of understanding or an inconsistent use of terms often leads to confusion. Not infrequently, the term "significant hazard" is confused with CCP, e.g. aflatoxin is referred to as CCP, or a control measure is taken for a CCP while a CCP refers to the "step" in the operation and control measure is the activity carried out at the step to prevent a food safety hazard exceeding an acceptable level.

Step 8 – Establishing critical limits. Critical limits[5] are basically limits of acceptability or unacceptability of control parameters. Therefore, they should always refer to the monitoring parameters. Depending on the type of hazards and control measures, the nature of critical limits may be different. For microorganisms or hazards where testing is not a reliable method of control and physicochemical methods are used for controlling the hazard, critical limits refer to the process parameters. They are values such as pH, a_w, temperature, time (or flow rate), salinity, level of chlorine, overpressure, etc. For hazards where testing can be

[5]A critical limit is a criterion which separates acceptability from unacceptability.

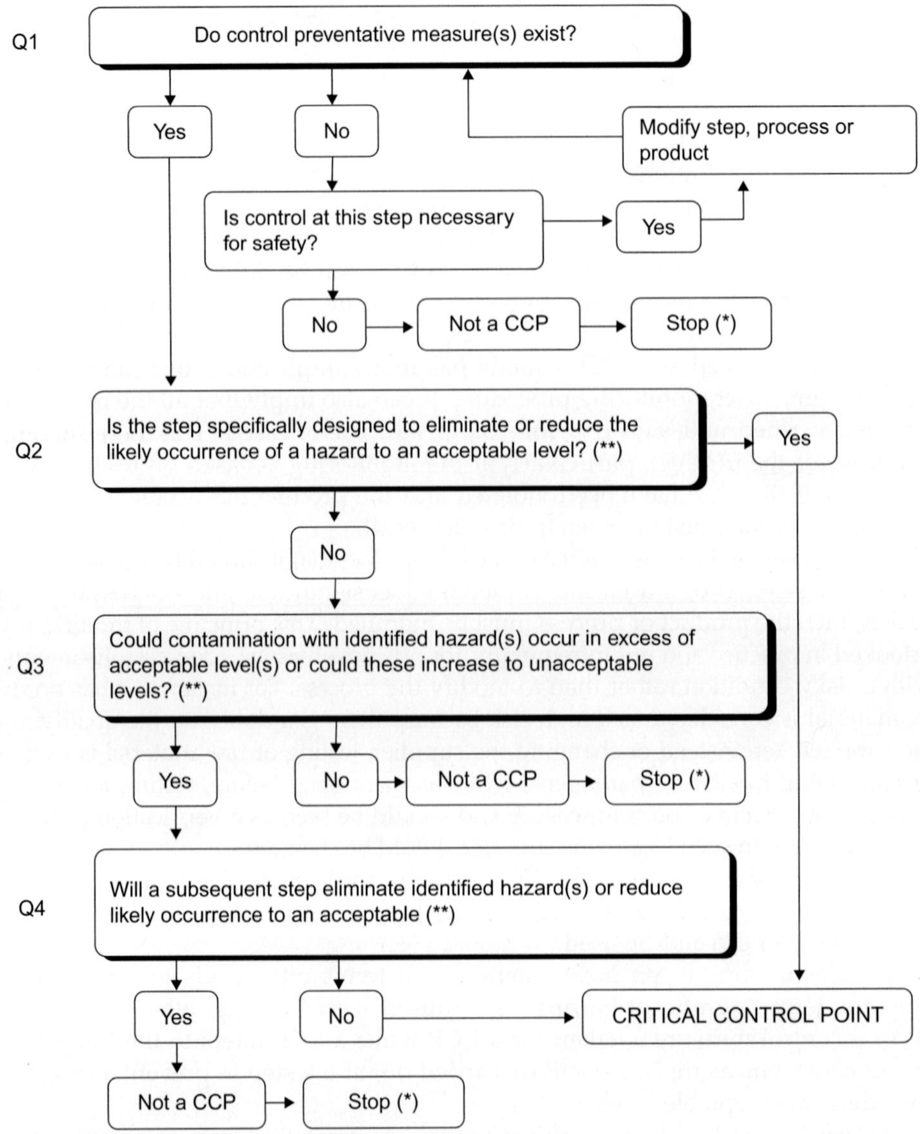

(*) Proceed to the next identified hazard in the described process.

(**) Acceptable and unacceptable levels need to be defined within the overall objectives in identifying the CCPs of HACCP plan

FIGURE 31.4 Example of a decision tree according to Codex Alimentarius Commission (CAC 2003).

used as a reliable method for control, then values such as maximum residue limits (for agrochemicals) or maximum levels (for contaminants) can be adopted.

As several parameters may be important for controlling a step (e.g. for heat treatment, both time and temperature are determining factors), more than one critical limit may be needed. Therefore, the parameters necessary to monitor a step must be carefully identified. In HACCP, critical limits are associated with the CCPs. However, even if the HACCP study does not result in specific CCPs, limits for control parameters could be defined to ensure that control measures are applied in a correct manner.

Sometimes, when the control measure takes place in homes or in foodservice establishments where tools for physical measurements do not exist, it is possible to use indirect measures such as color or texture as monitoring parameters. For instance, as a critical limit in the home environment, it could be recommended that eggs should be cooked until they become firm, or meat should be grilled until it is no longer red. Such a recommendation could be considered when formulating the instruction on the package. In food operations, visual methods of monitoring can also be used as a means to ensure that certain measures have taken place, e.g. a production line has been cleaned between two products according to the established procedure to prevent cross-contamination of allergens.

Step 9 – Establishing a monitoring system for each CCP. In the context of HACCP as defined by the Codex Alimentarius, monitoring is the act of conducting a planned sequence of observations or measurements of control parameters to assess whether a CCP is under control. Note that the monitoring which is carried out at a CCP, be it measurement or observation, is relative to the critical limits. Monitoring can be a physical measurement, visual inspection or chemical analysis.

The frequency of monitoring should be set so as to detect loss of control at the CCP and enable the necessary adjustments to control the process and prevent violation of the critical limits. Where possible, process adjustment should be made when monitoring results indicate a trend toward loss of control at a CCP. From the above, it is essential that very rigorous monitoring procedures must be established at the CCP.

Even when a step is not considered as a CCP, this should not be interpreted as that step not being important nor should it preclude monitoring of the control measure(s). Only the stringency and significance of the monitoring are different. To differentiate these two types of monitoring, for the purpose of this text we will refer to CCP monitoring versus verification monitoring. As mentioned above, the frequency of monitoring as verification can be adjusted according to the level of risk (Figure 31.3). Whether the monitoring is carried out at a CCP to ensure control, or as means for verification, it is important to have a consolidated and validated monitoring plan, indicating the type of monitoring, the frequency and time of execution, the method, the person responsible and actions in case of non-compliance.

A common failure in establishing a monitoring procedure for a CCP is omitting one or several key control parameters. For instance, not infrequently it is observed that, at the pasteurization step, which is often a CCP, only the temperature is monitored. The duration of the heat treatment, or the flow rate of the product, which is a determining factor for the duration of heat treatment, is not monitored. For water disinfection, only residual chlorine is considered and other factors such as contact time, pH of water and turbidity, which impact on the chlorination efficiency, are not considered.

It has also been experienced that some important steps in the operation have not been considered as CCPs on the grounds that they cannot be monitored "continuously" or a physical measurement method is not available. It is certainly much better to measure control parameters in an objective manner, e.g. with a temperature recorder. However, a lack of such methods for monitoring should not be a reason for not considering a step in the process as a CCP. If control at that step is important and visual observation can be effective in detecting deviations from acceptability, then the step can be considered as a CCP. The concept of continuity is also misunderstood. Monitoring a CCP should be carried out with specific and predetermined frequency. This may be every second, hour, day or defined moments as applicable. However, this has to be set in such a way that if a deviation is observed and critical or operational limits are violated, the corrective actions can be implemented in a timely manner before the product is released.

Step 10 – Establishing corrective actions. When applying the HACCP system, specific corrective actions[6] must be developed at each CCP in order to deal with deviations when monitoring shows loss of control. The actions must ensure that the CCP(s) has/have been brought under control so that only safe products reach the consumer. Strictly, this also includes possible disposition of the affected product. However, this decision should be made on a case-by-case basis. It is frequently observed that corrective actions are not well defined. There are times when they are mentioned as "see the QA manager or production manager." While it is a good practice that in times of problems the operator consult his superior, it is nevertheless important to document the specific corrective actions that have to be carried out to restore control, including ensuring that no unsafe product is released.

Corrective action should also prevent recurrence of the mishap leading to loss of control; therefore the cause of any deviation should be thoroughly investigated and the root cause of the problem determined (see Chapter 40).

Step 11 – Establishing procedures for verification. Verification refers to the application of methods, procedures, tests and other evaluations, in addition to CCP monitoring, to determine compliance with the HACCP plan. In practice, verification activities are carried out to verify the effectiveness of the HACCP plan and if it is implemented correctly, in other words, to check if what is planned is done and if the HACCP is well maintained and working as expected.

As such, verifications may include a variety of activities and collection of data to confirm that the HACCP plan is valid as well as well implemented. It can include:

1. Raw material testing and/or supplier's audit for verifying supplier's quality assurance.
2. Environmental monitoring for verifying the efficiency of cleaning and sanitation.
3. Calibration of equipment, in particular those used for monitoring of CCPs.
4. Audit of operations for confirming the adequate implementation of prerequisite programs and of the HACCP system.
5. Review of the records and of monitoring data confirming that the process parameters are kept under control and within established limits.
6. End-product testing for verifying the adequate implementation and efficiency of the system.

[6]Corrective actions are defined as the actions to be taken when the results of monitoring at the CCP indicate a loss of control.

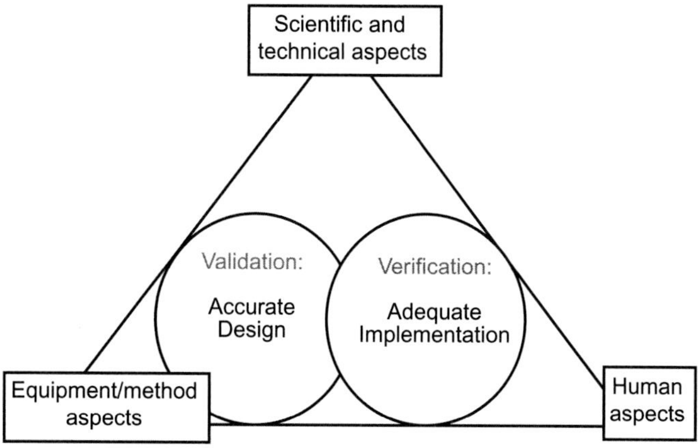

FIGURE 31.5 In validating a decision, three aspects, i.e. training and competence of employee, suitability and performance of the equipment and method, and scientific and technical data, need to be considered.

7. Verifying the consumer complaints to ensure the adequate implementation and efficacy of the food safety assurance system.
8. Verifying the training of employees involved in the implementation of prerequisite activities and of the HACCP system.

It should be noted that verification should not be ceased on the grounds that the results are negative; the results of monitoring are needed to be able to confirm that the food safety assurance system is implemented and effective at all times.

Principle 6 of HACCP also includes the concept of validation. The subject deserves some attention as validation is the assurance in the food safety assurance system. Without validation, the HACCP study has no scientific basis. According to the Codex Alimentarius Commission, validation is obtaining evidence that the elements of the HACCP plan are effective. This can be further explained as the demonstration that decisions made during the HACCP study have a scientific and/or technical basis and/or are based on accepted practices.

In validating the elements of an HACCP study, at each step it is important to consider three aspects: (1) the scientific and technical data, (2) the equipment used, i.e. if it is suitable for the intended use, and (3) the personnel who have to implement a decision (Figure 31.5). It is clear that designing a scientifically valid HACCP is of little value without the equipment used being suitable or the personnel competent to implement the decision. Validation is often confused with verification. To make a distinction between these two activities, validation is carried during the product design and the development of the HACCP plan to ensure that the plan is designed correctly, while verification is carried out as part of the implementation of the HACCP plan to ensure conformity with the plan (Figure 31.6).

While validation and verification activities are separate activities, the results of verification are of importance for (re-)validating the HACCP study as illustrated in Figure 31.7. If the results of verification show a problem, the first question that should be asked is was the HACCP plan implemented as planned? If a gap is noticed in the implementation, this needs

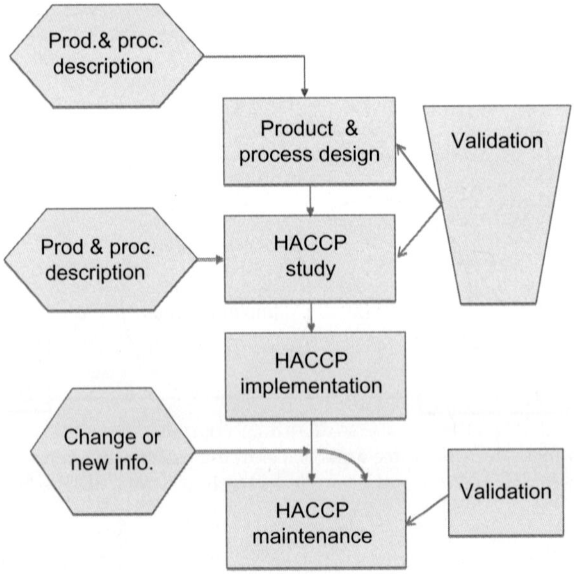

FIGURE 31.6 The process of validation and maintenance.

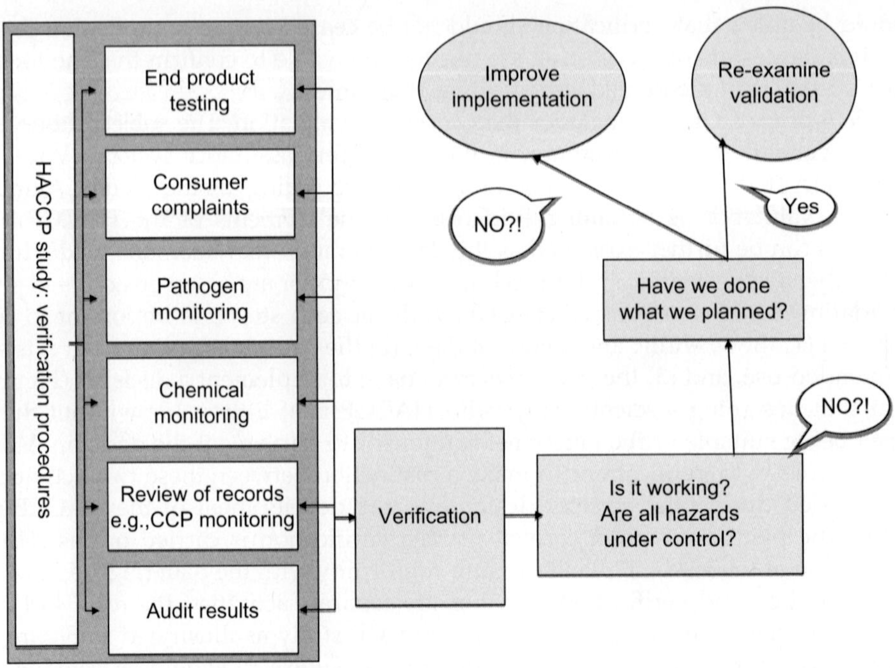

FIGURE 31.7 How verification can be used for the evaluation of HACCP implementation and efficacy.

to be corrected. However, if the investigation shows that in spite of the implementation of the HACCP system as planned, there is still an unsatisfactory situation, then the validation of the HACCP study should be questioned.

Depending on the elements that need to be validated, the exercise can be based on:

- Review of scientific literature and evaluation/extrapolation of information;
- Verification of conformance with regulatory requirements;
- Past records of verification data, surveys, or other types of historical information.
- Experimental trials, e.g. challenge tests.

Thus validation does not necessarily entail extensive and expensive studies and can be just verification of conformance with the regulatory requirements or extrapolation from studies already conducted.

Often the question is raised as to what needs to be validated. In principle, all the steps in the decision-making process need to be validated, from the hazards analysis, critical limits, monitoring parameters to corrective actions and verification procedure (Scott, 2005) (Table 31.2). The following example illustrates the importance of validating all measures: in an outbreak of *S. aureus* in Japan in the year 2000, the factory manager repasteurized a milk product which contained *S. aureus* toxin. Thus, the company applied a corrective action that was not valid, as the S. aureus toxin is heat stable. Similarly, errors can happen in the monitoring procedure if, for instance, the thermocouple used for monitoring is not set at a correct point in the food product, or the analytical method and sampling procedure for testing for a chemical are not valid. With regard to control of microorganisms, it is important to ensure that the intervention considered will ensure the performance objective or performance criterion.[7] For instance, in a canning process, the sterilization process should ensure a performance objective of <1 spore/10^{12} g *C. botulium* or a process criterion of 12 D reduction. This translates into process parameters of 2.45 min/121°C; or the process parameters of cold storage (i.e. shelf-life of a product at a given temperature) of a smoked product should ensure that there is no growth of psychrotrophic bacteria such as *Listeria monocytogenes* to above the food safety objective (e.g. 100 *Listeria*/g). Further examples of validation are provided in the Codex Guidelines for Validation of Food Safety Control (CAC 2008). While the validation of processing steps for killing and/or prevention of growth are often evident to food operations, the validation measures at the supplier, transport or consumer level are not always evident. For instance, for allergen management, to ensure correct information on the package, it is essential to verify the validation at the supplier level of their allergen control plan and of their allergen declaration. Similarly, it is important to validate the information provided to customers and consumers, i.e. that the control measure which is recommended at the consumer level, such as time and temperature of the storage of the product, is valid, and that the information is clearly communicated. For the latter, a focus group could be organized to evaluate the clarity of information. Again, in line with the principle of modification of product design where control measures are not feasible (see Q1 in Figure 31.4), if a survey

[7] Performance criterion (PC). The effect in frequency and/or concentration of a hazard in a food that must be achieved by the application of one or more control measures to provide or contribute to a PO or an FSO.

TABLE 31.2 Examples of Activities that Need to be Validated, as Applicable

Principles	Validation
Conducting a hazard analysis	Ensuring: – Appropriate expertise is used in identifying potential hazards. For instance, public health guidance or regulatory requirements are checked, scientific literature is searched, experts are consulted and previous records of contamination or outbreaks are examined. – Identified control measures are effective in achieving the food safety or performance objectives, the design of equipment is adequate to achieve these criteria and personnel are qualified for the purpose. Where hazards are considered as non-significant, ensuring that: – Prerequisite programs, including suppliers' practices, processing of food, or other conditions (e.g. regulatory measures, epidemiological situation of the country) are adequate to control the hazards to safe levels.
Determining the critical control points (CCP)	Ensuring that assumptions leading to identifications of CCPs are valid (see above), personnel are properly trained in identifying CCPs and understand the significance of a CCP.
Establishing critical limits	As applicable, ensuring that critical limits are: – Consistent with the performance or product criteria needed to meet food safety or performance objectives. – In accordance with the regulatory requirements of the country where products are marketed and/or the specifications of customers if these are stricter.
Establishing a system to monitor the CCP	Ensuring that: – All parameters needed to monitor control measures are selected correctly. – Equipment or methods used for monitoring (e.g. limit of detection or quantitation, sampling method, etc.) are appropriate, valid, calibrated and functioning correctly. – Personnel are competent, trustworthy and adequately trained in the use of equipment or the methods of monitoring (i.e. validate their training). – Personnel know what to do in case of deviation.
Establishing corrective actions	Ensuring that: – Measures foreseen for correcting a deviation are effective. – Staff are adequately trained in implementing corrective actions in case of deviation. – Traceability is working effectively.
Establishing procedures for verification	Ensuring that: – Limits (including specifications) set for product or environmental testing are valid and comply with regulatory requirements and product or process specifications. – Methods for testing and procedures are valid and carried out by competent personnel. – All verification procedures are valid; data are adequately reviewed by competent personnel and acted upon.
Establishing documentation	Ensuring that documentation and records: – Meet the regulatory and/or customer requirements. – Demonstrate the hazard analysis and HACCP plan, and include data on validation, results of CCP monitoring and verifications, as well as corrective actions in case of deviation. – Include root cause analysis in case of non-compliance (near miss or incidents). – Allow for an efficient and speedy traceability and trace-back.

of consumer/customer practice and knowledge shows that the implementation of a foreseen control measure is not feasible at the consumer and customer level, the design of the product should be modified and/or the product in question should be considered unsuitable for the customer and should not be sold.

Validation should usually take place during the design of the product or during the HACCP study, i.e. before the implementation of the HACCP study. However, if there is a change or when new information comes forth, the need for revalidation must be considered.

Step 12 – Establishing documentation and record keeping. Documentation and record keeping often give the perception that HACCP is a paper exercise. Whereas these play a pivotal role in food safety as they are an important and effective means of communication, they allow communication with other colleagues on how food safety is planned and implemented and on the information that was considered in the decision-making process. Documentation can play an important role in maintaining the HACCP plan, reviewing and if necessary revising decisions.

Documentation also provides law enforcers, auditors and customers with evidence and information on the (1) hazards analysis conducted, i.e. what hazards are considered and controlled, (2) the hazards which are considered significant and are controlled by a CCP step, and (3) evidence that the food safety assurance system has been continuously under control.

Documents are also important in case an incident needs to be investigated. They provide evidence that all appropriate actions have been taken or provide guidance on the possible source of a problem.

Examples of documentation and records that should be collected and reviewed are:

- Procedures and requirements regarding GHP (e.g. pest management plan and records, personnel health and hygiene requirements and records, etc.).
- HACCP study, including hazard analysis, determination of control measures, process parameters and critical limits, and the HACCP plan as well as the validation data.
- Product formulation (specifications) and manufacturing process.
- Specification to suppliers or any other information to stakeholders in the food chain, e.g. to transporters and distributors for further handling of food.
- Reports of audits of suppliers, transporters, distributors.
- Records of CCP monitoring and the procedures used, as well as corrective actions taken in case of deviations.
- Verification activities (see above) as well as validation data or studies.
- Records of investigation and follow-up of non-compliance and/or corrective actions.
- Records of training of personnel, the nature and scope of their training as compared to their responsibility.
- Periodical review of the HACCP study and the HACCP plan.

Tables 31.3 and 31.4 provide examples of templates that can be used to document the key decisions of the HACCP study and the HACCP plan.

Implementation of the HACCP Plan and its Maintenance

The implementation of the HACCP plan at a production site starts with the training of personnel, including all personnel involved in one or several activities of the HACCP

TABLE 31.3 Example of Template for Documenting the HACCP Study

Steps	Hazards	Control Measures	CCP Yes /No	Limits	Monitoring	Corrective Actions	Verification

TABLE 31.4 Example of Template to Record the HACCP Plan

CCPs	Hazards	Control Measures	Limits	Monitoring	Corrective Actions	Verification

system and/or the prerequisite activities (WHO, 1995; Williams et al., 2003). In doing so, the managers and employees should be informed of their tasks and responsibilities, the importance and significance of their measures, the need for informing the HACCP team or the coordinator of the team of any change or deviation from the plan or the prerequisites (e.g. non-compliance observed during audits, verification data showing unsatisfactory results, or change in the process or supplier). It is to be remembered that non-compliance with the prerequisites has a bearing on the hazard analysis done during the HACCP study and its outcome, i.e. the HACCP plan. A major outbreak of salmonellosis in Germany, affecting some 1000 children, resulted from a sudden change of the supplier. Incidents related to undeclared allergens are also frequently associated with a change of supplier without taking the necessary measures to examine the impact of the change on the operations and the product. Thus, any of the above conditions should trigger a re-examination of the HACCP study.

Also, those responsible for implementing the HACCP should fully understand the importance of compliance of CCPs and the need for thorough investigation of any non-compliance at these steps. The root cause of such near misses, up to the responsibility of management, needs to be established.

Further to the implementation of prerequisites and of the HACCP system, the HACCP plan should be maintained (Figure 31.6). This means a periodical review of the HACCP study to "verify" that the HACCP has indeed been maintained and continues to be valid in the light of the latest internal or external developments. It is to be emphasized that the periodical review is a verification of maintenance of the HACCP system and an update of the records. The maintenance of the HACCP plan itself should be done on a continual basis. Thus, contrary to the general practice, HACCP is not a one-time exercise. Also, the maintenance of HACCP does not necessarily lead to increased control but can also be beneficial, as certain changes in the environment of production or improvements in the prerequisites can result in the need for lesser control through the HACCP plan, as for instance if the raw material is changed to a product which has lesser risk of certain contaminants.

Over and above verification data, two types of information and changes should trigger a re-examination of the HACCP study and plan, and its validation.

1. Changes related to internal operations. Examples are:
   a. Change of the supplier, their practices or where the raw material is sourced
   b. Change in recipe, product formulation or packaging (including labeling)
   c. Change in the process line, equipment and material
   d. Change in personnel
   e. Change in food production environment, in particular in case of temporary maintenance work, which can lead to the exposure of the factory environment to foodborne pathogens
   f. Change in transport or distribution channel
   g. Change in target consumer and intended use and/or conditions of use (a product is intended for a younger age that previously considered)
2. Changes, external to the operations, i.e. related to the environment where the product is produced, processed or sold. Examples of these changes are:
   a. Emergence of new pathogens or chemical contaminants
   b. Changes in regulatory requirements where the product is sold
   c. Changes in technology, analytical capabilities, monitoring tools
   d. Changes in the demography and/or consumers' practices or perception
   e. Environmental contamination or social factors
   f. Report of incidents, outbreaks or errors in other operations in the world or new data on food contamination previously not known

Figure 31.8 illustrates the entire cycle of an HACCP application, from hazard analysis based on the examination of the status of prerequisite programs, to the implementation of the HACCP plan and its maintenance.

HACCP IN SMALL BUSINESSES OR LESS DEVELOPED BUSINESS

Worldwide, it is recognized that the HACCP system may be difficult to implement in small or less developed businesses due to lack of resources. An approach that some governments have taken to address this problem is to assist these businesses by carrying a generic HACCP study for a category of a product, and based on the outcome of the study, to develop an HACCP-based code of practice. Such an approach combines the requirements provided in a code as well as specific measures required for a specific product category of products.

ASSESSMENT OF HACCP

Whether as regulators, auditors or managers in a company, professionals may be brought in to evaluate the HACCP plan of a company (WHO, 1998; Motarjemi, 2000). Since HACCP requires a multidisciplinary expertise, one cannot rely on the assessment of auditors to ensure adequacy and validity of the HACCP study. However, auditors can check the understanding of the principles of HACCP and the vigilance with which these are implemented, validated

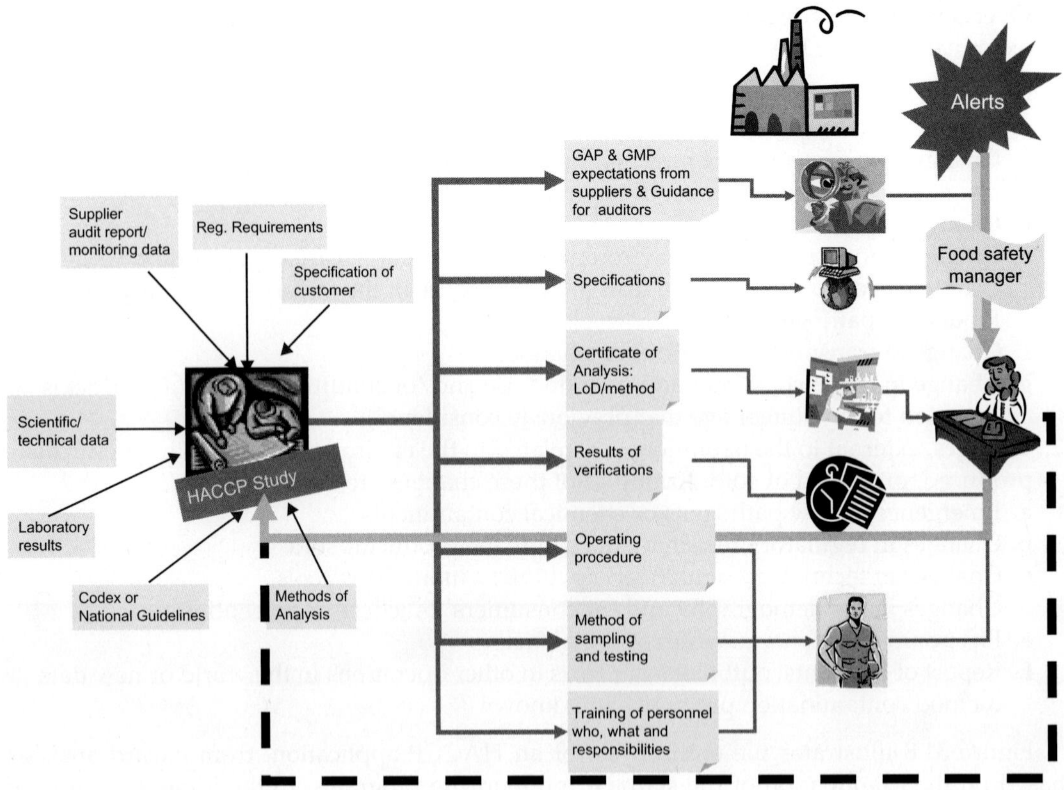

FIGURE 31.8 The process of developing an HACCP study, taking into account various information including the prerequisite situation, to implementation and maintenance.

and maintained. Auditors and inspectors are cautioned against the indiscriminate use of a checklist. While a checklist provides some benefits and ensures completeness of the assessment, it should not evolve into a tick-box approach, replacing critical evaluation. Like the development of the HACCP plan, its assessment requires a certain degree of expertise, as often audits are carried out to function as an aide-mémoire. Guidance on the regulatory assessment of HACCP as well as on internal audit is provided in Chapter 38, as well as elsewhere. (WHO, 1998; Motarjemi, 2000).

CONCLUSION

The application of the HACCP system is not a stand-alone system, but it should be seen as an element of food safety management. It complements basic good hygienic practices in food safety assurance by targeting product-specific hazards and devising control measures necessary for managing risks relevant to the product and conditions of operations.

However, it is not a magic wand and it is not a panacea for all problems. It can be a powerful tool for the management of food safety only if it is correctly understood and applied and if there is adequate commitment by the management for providing necessary resources and expertise (Motarjemi and Kaferstein, 1999). It should not be seen as a measure for regulatory authorities and/or implemented to satisfy the requirement of authorities; otherwise it becomes a bureaucratic exercise, with a massive amount of paperwork without much added value. It has to be used in the context of true commitment to food safety.

Finally, for more in-depth reading and understanding of the HACCP system, the reader is referred to a number of good books which have been written in this area; among these is Wallace et al. (2011) *Food Safety for the 21st Century: Managing HACCP and Food Safety throughout the Global Supply Chain*.

References

CAC (Codex Alimentarius Commission), 2003. Hazard analysis and critical control point (HACCP) system and guidelines for its application. In Food Hygiene Basic Texts (CAC/RCP 1-1969, Rev. 4, Recommended International Code of Practice General Principles of Food Hygiene. Food and Agriculture Organization: Rome.

CAC (Codex Alimentarius Commission), 2008. Guidelines for the Validation of Food Safety Control Measures, CAC/GL 69 – 2008 Joint FAO/WHO Food Standard Programme. Food and Agriculture Organization, Rome.

ISO, 2005. ISO 22000. Food safety management systems – requirements for any organization in the food chain. International Organization for Standardization: Geneva.

Motarjemi, Y., Käferstein, F.K., 1999. Food safety, HACCP and the increase in foodborne diseases: a paradox? Food Control 10, 325–333.

Motarjemi, Y., Käferstein, F.K., Moy, G., Miyagawa, S., Miyagishima, K., 1996. Importance of HACCP for public health and development: the role of the World Health Organization. Food Control 7, 77–85.

Motarjemi, Y., Mortimore, S., 2005. Industry's need and expectations to meet food safety. Proceedings of the Fifth International Meeting: Noordwijk Food Safety and HACCP Forum 9–10 December 2002. Food Control 16 (2005), 523–529.

Motarjemi, Y., 2000. A regulatory assessment of HACCP: a FAO/WHO consultation on the role of government agencies in assessing HACCP. Food Control, 341–344.

Motarjemi, Y., Stadler, R., Studer, A., Damiano, V., 2009. Application of the HACCP approach for the management of processing contaminants. In: Stadler, R.H., Lineback, D. (Eds.), Process Induced Food Toxicants and Health Risks John Wiley & Sons Inc., New Jersey.

Scott, V.N., 2005. How does industry validate elements of HACCP plans? Food Control 16 (6), 497–503.

WHO, 1995. Workshop on training aspects of the hazard analysis critical control point system. Geneva.

WHO, 1998. Guidance on the Regulatory Assessment of HACCP. Report of a Joint FAO/WHO Consultation on the role of governmental agencies in assessing HACCP. WHO/SF/FOS/98.5. Geneva.

Wallace, C., Sperber, W., Mortimore, S.E., 2011. Food Safety for the 21st Century: Managing HACCP and Food Safety throughout the Global Supply Chain. Wiley-Blackwell.

Williams, A.P., Smith, R.A., Gaze, R., Mortimore, S.E., Motarjemi, Y., Wallace, C.A., 2003. An international future for standards of HACCP training. Food Control 14, 111–112.

Further Reading

CAC Codex Alimentarius Commission. Guideline for the Validation of Food Safety Control Measures, Food and Agriculture Organization, Rome.

CAC Codex Alimentarius Commission. Principles and Guidelines for the Conduct of Microbiological Risk Management, Food and Agriculture Organization, Rome.

Cole, M.B., Tompkin, R.B., 2005. Microbiological performance objectives and criteria. In: Sofos, J. (Ed.), Improving the Safety of Fresh Meat. Woodhead Publishing Ltd., Cambridge, England

FAO, 2003. Assuring food safety and quality – Guidelines for strengthening national food control systems. FAO Food and Nutrition paper number 76.
Food Control, 2005. Impact of food safety objectives on microbiological food safety management. Proceedings of a workshop held on 9–11 April 2003. Marseille, France. 16(9), 775–832.
ILSI-Europe (1998). Food Safety Management tools. ISBN: 1-57881-034-5.
Motarjemi, Y., 2004a. HACCP from farm to fork. Part I. Food Beverage Int. February, iii–iv.
Motarjemi, Y., 2004b. HACCP from farm to fork. Part II. Food Beverage Int. February, vi–vii.

CHAPTER

32

HACCP Misconceptions

Yasmine Motarjemi[1], Carol Wallace[2] and Sara Mortimore[3]

[1]Food Safety Management Consultant, Nyon, Switzerland, [2]University of Central Lancashire, Preston, UK, [3]Land O'Lakes, Inc., St. Paul, MN, USA

OUTLINE

INTRODUCTION

For over two decades, HACCP has been promoted internationally, particularly in industrialized countries, as a method to enhance food safety assurance systems in the food industry. Nevertheless, in spite of the advance of HACCP as a core method for food safety management, large-scale foodborne diseases outbreaks, food contamination incidents or food recalls of major importance have continued to occur.

Food Safety Management.
DOI: http://dx.doi.org/10.1016/B978-0-12-381504-0.00032-9

Often the question has been raised as to why, despite the application of HACCP system, such incidents continue to happen? There are a number of explanations for this paradox. An in-depth review of these is given elsewhere (Motarjemi and Käferstein, 1999). Here we limit ourselves to stress only two points: (1) without any doubt, without the HACCP system, we would have experienced many more incidents and (2) considering the quantity of industrially produced food on the market, the number of incidents is relatively low.

As for any preventive measures, one cannot collect statistics and trends for events which have not occurred. Also, HACCP is not a panacea for all problems and the application of HACCP, no matter how rigorously implemented, cannot prevent all kind of incidents. Some of the major incidents, which have marked the history of food safety, could not have been prevented, with even the strictest application of the system. This is particularly the case where the problems are due to the malevolence of some individuals such as the incident of melamine with pet food in the USA in 2007, or the emergence of a new hazard or finding a new source for it, previously unknown, e.g. semicarbazide in baby food in 2003.

Having said this, when the health of consumers is at stake, any incident is one too many and all efforts should be put in place to prevent food safety incidents. What is particularly unacceptable is where by negligence an incident is repeated. This is where HACCP application should be of help, provided that it is properly understood, applied and the resource and infrastructures necessary for it are put in place, particularly so as to review and update the system in response to new knowledge. To this end, management of a company must be committed to providing the necessary resources, i.e. qualified human resources is a *conditio sine qua non* and is perhaps the most important Achilles' heel in the implementation of HACCP. Figure 32.1 shows a root cause analysis of the frequent shortcomings in the HACCP application.

This chapter highlights misconceptions and common errors in the implementation of HACCP, which have repeatedly been observed and which led to failures in the application of the system and have thus undermined the potential of the system to prevent incidents.

Some of the information presented here is mentioned in the previous chapter on Hazard Analysis and Critical Control Point System (HACCP) (Chapter 31). Nevertheless, some salient parts in relation to failures in the implementation of HACCP are highlighted here. Readers who are already conversant with the HACCP system, but would like to further improve the application of HACCP, may wish to focus on the guidance given in this chapter.

MISCONCEPTIONS

HACCP should not be Seen as a Measure for Authorities or Certification Bodies

Sometimes, HACCP is implemented mainly with the objective of satisfying the requirement of authorities, or is seen as a task that is mandatory, without the management of the business really seeing its value, understanding its principles and truly endorsing its implications.

HACCP can be helpful only if it is carried out with the specific objective of managing food safety in an effective manner, i.e. taking the right decision, ensuring that the decisions are actually effective and correctly implemented. In particular, it can help managers to identify those steps in the process that should receive the highest degree of attention and

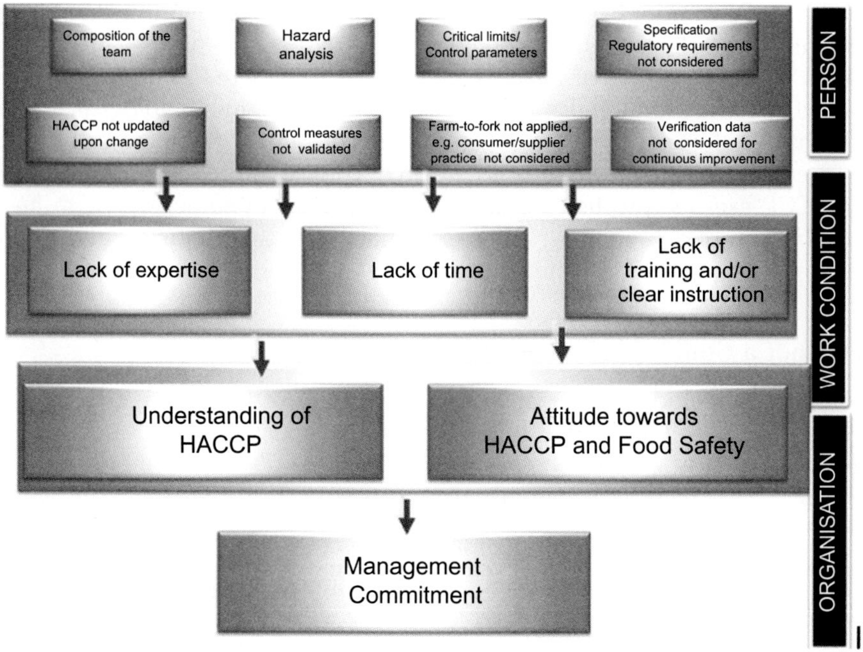

FIGURE 32.1 Root cause analysis of frequent shortcomings in the application of the HACCP system.

surveillance. When HACCP is applied only with the objective to meet the satisfaction of authorities, it will lead to a massive amount of paperwork without much added value. In this case, it has very little chance to become a meaningful exercise and there is a real risk that it will be seen as a burden by all personnel.

HACCP should not be Reduced to Simply Paperwork

Although the HACCP study includes desktop activities, HACCP is not a paper exercise. The proper application of HACCP implies scientific and technical expertise, monitoring of critical control point (CCP) parameters, verification of good manufacturing practice (GMP) on the factory floor, reviewing verification data (e.g. results of raw material monitoring, monitoring consumer complaints and calibration of key equipment), training of people, etc. HACCP should also not be equated with filling forms. In many cases, a number of forms have been created to facilitate the systematic approach and thinking process. However, this should not replace the critical thinking and scientific and technical expertise required to carry out a HACCP study – it is the quality of content that is important, not the form itself.

HACCP is not One Man's Job

As mentioned above, one of the biggest added values of HACCP is the promotion of multidisciplinary teamwork. To carry out a proper HACCP study, it is fundamental to draw on the right and adequate expertise. The importance of a multidisciplinary team is particularly high when safety is considered at the product and process development. It is at this stage that many potential hazards have to be considered and managing the safety of the product needs to be thought through. As an example, microbiologists and veterinarians usually have not been educated and trained in equipment and process line design, nor in process measurement, control and monitoring. They will therefore never be able to do a useful HACCP exercise on their own. The involvement of somebody with an adequate engineering background is in most cases essential, as is the inclusion of personnel with appropriate operations and food safety knowledge.

HACCP is not a Stand-Alone System

A major misunderstanding or error in the application of HACCP is that it is viewed as a separate system. HACCP should be seen as an approach to food safety assurance; its application draws on an array of measures such as GMP, audits, monitoring, traceability, etc. (Figure 32.2). Therefore, it is essential that as part of a HACCP study, the state of

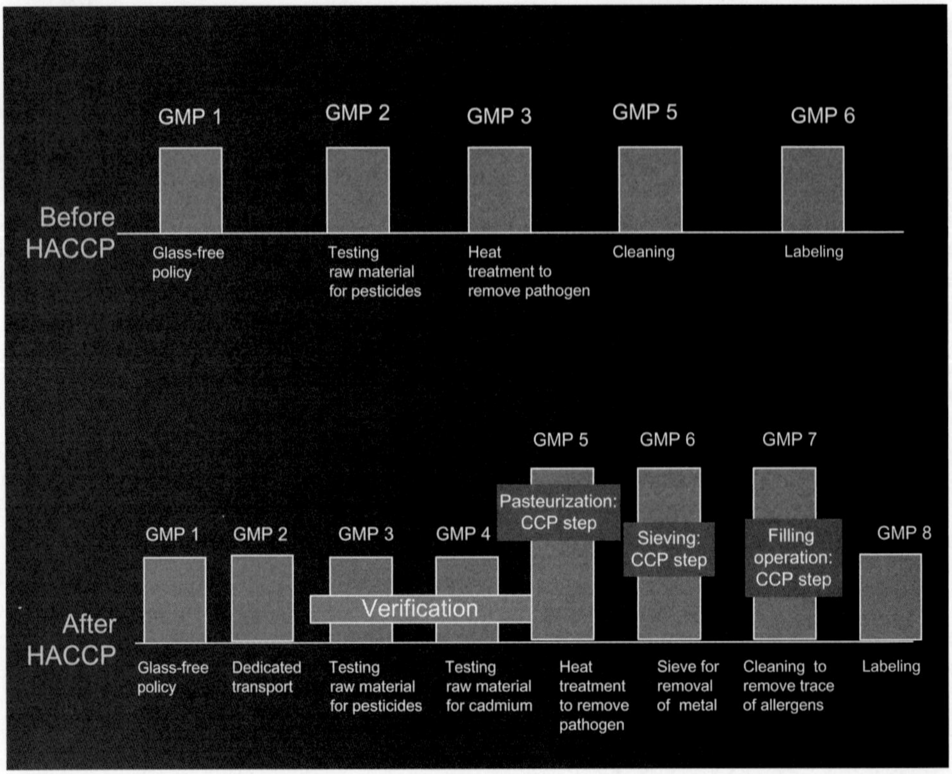

FIGURE 32.2 Schematic illustration to explain the relationship between GMP, CCP and verification activities. The figure shows that HACCP helps all necessary control measures often considered as part of GMP. It also gives weight to the control measures that are particularly important in terms of food safety.

these different measures be carefully considered. For instance, during the hazard analysis, the decision on the likelihood of the presence of a given pathogen in the ingredient will depend on its source. Knowledge about the origin of the raw material and conditions of production, previous record of safety, all audit and monitoring data of the supplier are essential for deciding on the significance of a pathogen. These require having traceability of the ingredient and information on the epidemiological situation of the country where the raw material is produced or on processing conditions of the supplier, should the ingredient be a semi-processed product. Similarly, the likelihood of recontamination of the food product is related to the cleaning procedures and the state of application of the good hygienic practices.

HACCP is not a One-off Exercise

Many see HACCP as a one-time exercise, and once the study is carried out, it is rarely reviewed or updated. First, HACCP is a tool for decision-making. As such, it should be flexible and should be part of the everyday thinking process. As the situation evolves, decisions may also need to be changed. For instance, a step in the process may temporarily need to be considered as a critical control point (CCP) until there is enough assurance that the risks are adequately under control. Or, the frequency of the verification of a supplier's products could be modified as the supplier confidence increases

Second, maintenance of a HACCP plan is as important as its development and implementation. Each time a factor related to food production, e.g. supplier, an ingredient, the process, the equipment, target consumer, etc. is changed, the consequence of this change on the hazard analysis needs to be considered and if necessary the HACCP plan needs to be changed.

Similarly, when new information becomes available, for instance when authorities communicate an alert about a new hazard or when verification data (e.g. pathogen monitoring, consumer complaints, raw material monitoring, audit reports of suppliers, industry information, etc.) indicate a new threat, such information should prompt the HACCP team to revisit their study and evaluate the adequacy of their measures. For instance, if verification data (e.g. audit report of the supplier) or raw material monitoring indicate that the supplier's food safety assurance is inadequate, this may necessitate a change of supplier or increased frequency of verification of the supplier's products.

Documentation and Record Keeping are not Bureaucratic Work

Documentation and records play a pivotal role in food safety as they are an important and effective means of communication. They allow communication with other colleagues or provide evidence to customers or regulatory authorities on how food safety is planned and implemented and on what the bases for decisions are. Documentation can play an important role in maintaining the HACCP plan, i.e. reviewing and if necessary revising decisions, providing evidence that appropriate measures have been taken in times of problems.

However, the value of the documentation lies in the quality and quantity of information that it contains. If the information is superficial or not adequate, the documentation becomes more a bureaucratic work than a tool for communication.

In case of incidents, HACCP records provide evidence that the appropriate actions have been taken and facilitate the investigation of the causes of problems.

HACCP does not Work if there is no CCP during the Food Chain from Farm to Fork

A fundamental principle of HACCP is to identify significant hazards and control measure(s) that are essential to eliminate or reduce the hazards to acceptable levels. Where such a control measure(s) is not in place, it is important to modify the production process and/or its conditions (e.g. labeling or providing information on the storage conditions or instruction of use). In many instances (particularly in the case of raw foods), significant hazards are identified, but there is no CCP during the food chain at which point the reduction of the hazard to an acceptable level can be achieved and no modification is, or can be, made to the food chain. For example, the CCP for hamburgers is at the preparation step, there is no real CCP at the slaughterhouse or meat plant level. Unless the meat is irradiated, cooking of hamburgers by consumers must become a CCP and consumers should be imperatively and adequately informed about their role. In these cases, if it is not acceptable to have a CCP at consumer level, the application of HACCP *per se* will not make food safe.

HACCP is not only Qualitative

There is a general misconception that the decisions taken within HACCP studies are qualitative. However, during the hazard analysis, the likelihood (possible, probable or likely) of contamination with a chemical contaminant, survival or growth of an organism or production of toxin and the magnitude of these events can also be estimated. Moreover, the efficiency of control measures has to be determined, i.e. how effectively a processing step can eliminate or reduce a pathogen to an acceptable level (evaluated in terms of log reduction), or what the extent of microbial growth will be during the shelf-life of a product under given conditions.

These concepts are generally addressed under the principle of validation of the control measures (CAC, 2008) and should not be mistaken for the concept of microbial risk assessment. The latter, recommended in the framework of Codex Alimentarius as one of the three elements of "risk analysis," is a governmental activity with the aim of integrating scientific facts into the decision-making process.

COMMON ERRORS OR SHORTCOMINGS IN THE APPLICATION OF HACCP

There are also a number of systematic weaknesses in the application of HACCP. The root cause of many major or international incidents can be attributed to these errors or shortcomings. Table 32.1 shows a variety of well-known food safety incidents and outbreaks and considers the likely errors and shortcomings in the application of HACCP and food safety management systems.

TABLE 32.1 Some Examples of Food Safety Incidents and Presumed Weaknesses in the Application of HACCP that Led to the Incident

Year	Place	Incident	Likely HACCP/FSMS Weakness	See Common Errors and Shortcomings Section Number
1993	Germany	*Salmonella* in paprika chips	Maintenance of the HACCP plan. i.e. Change in supplier and shortcomings in the control of the raw material	2, 13
1994	USA	*Salmonella* in ice cream	Shortcomings in the scope of the HACCP and application of perquisite programs (i.e. lack of dedicated transport, and poor GHP)	2, 7
1996	UK	*Salmonella*, cheddar cheese	Weakness in the validation of control measures, no corrective action, lack of understanding of HACCP principles	12, 11
2000	Japan	*S. aureus* with milk products	Weakness in the hazard analysis, lack of knowledge regarding risk, lack of temperature control for raw milk during power failure, non-valid corrective actions, inadequate communication during incident	1, 6, 7, 11, 12, 13
2002	Europe	Recall of honey due to Chloramphenicol	Weakness in the hazard analysis and verification of suppliers	1, 6
2002	Norway	*S. aureus* in ice cream	Poor implementation of prerequisite programmes (e.g. failure in the maintenance of the dispensing machine, cleaning and disinfection of the system, and the pasteurization of ice-cream mix). Poor verification and validation of control measures at the customer. Weakness in the scope of the HACCP	6, 12
2002	Denmark	Recall of spoiled baby food poorly sterilized	CCP was not monitored correctly Corrective actions were not valid (products released in spite of CCP violation)	9, 11
2003	Germany	Thiamine and Infant formula	Error in the validation of the composition and weakness in the verification of the product composition before its release	12
2005	International	Isopropylthioxantone (ITX) from ink of packaging	Weakness in the hazard analysis and verification	1, 6, 12
2006	International	Benzene and softdrinks	Weakness in the hazard analysis, verification and maintenance of the HACCP plan	1, 6, 12

(Continued)

TABLE 32.1 (Continued)

Year	Place	Incident	Likely HACCP/FSMS Weakness	See Common Errors and Shortcomings Section Number
2006	USA	*Salmonella* and chicken pie	Weakness in the control measures (communication with consumers on the microwave heating of products)	2, 6, 12
2006	UK	*Salmonella* in chocolate	Weaknesses in prerequisite programs, lack of knowledge and technical expertise, poor communication.	1, 5, 6, 7, 11
2007	Europe	Sunflower oil contaminated with mineral oil ex-Ukraine	Weakness in hazard analysis, maintenance of HACCP plan and verification (monitoring suppliers)	1, 6, 12, 13
2007	USA	*Salmonella* in peanut butter	Substantial weaknesses in prerequisite programs	6, 12
2009	USA	*Salmonella* in peanut butter	Substantial weaknesses in prerequisite programs, inadequate segregation between raw and roasted peanuts.	5, 7
2008	China	Melamine in infant formula[a]	Weakness in the hazard analysis, maintenance of the HACCP and failure in verification, inadequate raw material safety control; lack of knowledge of consequences of adulteration (although as a criminal offence it is possible that those involved may have gone ahead anyway)	1, 2, 6, 13
2008	USA	*E.coli* O157 in cookie dough	Weakness in the hazard analysis of the raw material and consumer practices and validation of control measures (safety instructions to consumers)	2, 6, 12
2011	Germany, France	Enteroaggregative *E. coli* O104:H4 and fenugreek	Weakness in prerequisite programs and verification of suppliers	1, 2, 12

[a]*Note this was the second well-known incident with melamine, the first occurred in 2007 in the United States as result of contaminated wheat gluten imported from China.*

1. **Expertise.** The success of a HACCP study and resulting plan depends on the expertise employed. A HACCP study is a task requiring both scientific and technical expertise (agronomy, veterinary science, food safety microbiology and chemistry, engineering and technology, consumer knowledge) and operational information and experience. When there is no access to such experts on site, the HACCP study can be reviewed by relevant experts.
2. **Past record of safety.** Analysis of foodborne disease outbreaks and incidents, even certain cases of fraud which *a priori* may seem unpredictable, are often repetition of previous events and in most cases preventable if they were better examined and taken into

account in HACCP studies. Therefore during a hazard analysis, data on the past record of the safety of the product, including any incident, case report of illness or outbreaks, epidemiological data on the event and its root causes need to be considered.

3. **Farm to fork approach.** Although the importance of an integrated approach to food safety and consideration of all steps from farm to fork has been stressed time and again in recent years, nevertheless, frequently HACCP plans are developed without fully considering this principle. For instance, HACCP plans are developed without fully considering the supplier food safety assurance, and subsequently, hazards that may be present in the raw material are overlooked.

   Similarly, those steps following manufacturing are frequently omitted during HACCP studies, for example hazards that may occur during transport, distribution and most importantly during preparation by consumers. For certain types of products, factors such as the conditions for storage of the product during distribution, target customer/consumer or the potential mishandling of the product may be crucial for designing safety, including the necessity for providing information on the safe use of the product. Shortcomings in the application of this point have been the source of numerous reported incidents: salmonellosis and chicken pie in 2006, *E. coli* infection and cookie dough 2008.

   The implication of this principle is that as far as feasible, one should investigate and understand:

   a. The source, origin, conditions of production of raw materials and ingredients;
   b. Conditions of transport and distribution; and
   c. Handling, storage and preparation practices by consumers and customers.

   This insight is essential for determining what control measures (including labeling and handling instructions) need to be considered outside the factory, e.g. at the supplier, transport, distribution and consumer/customer levels. Based on these, we should then define and communicate the:

   a. requirements to suppliers;
   b. expectations to transporters and distributors; and provide
   c. validated instructions for safe preparation and handling of products to caterers and consumers.

4. **Flow diagram.** Very often, flow diagrams used for the HACCP study do not reflect the real processing and manufacturing conditions of the product. Lack of accuracy and detail may seriously jeopardize the quality of the HACCP study and the validity of decisions.

5. **Product description.** Validity of hazard analysis relies on the precision with which various aspects of the product are described, e.g. the raw ingredients and their source, the supplier assurance system (e.g. availability of an audit report), manufacturing steps and conditions, description of packaging and other auxiliary products, potential use or abuse by target consumers. Too often these descriptions are not detailed enough to allow an in-depth hazard analysis. In absence of such information, important hazards may be missed during hazard analysis.

6. **Consideration of circulation of air and water and employee traffic.** When conducting the HACCP study, the flow diagram is often limited to the product. It is important also to consider how the circulation of water and air and employee traffic (or zoning) can impact on the safety of the product. In this context, all building or reconstruction activities should also be considered as they may often lead to increased contamination

of the environment with pathogens, as well as foreign materials. For these reasons it is useful also to consider site layout diagrams and to spend time observing the operation in practice.

7. **Hazard analysis.** Hazard analysis is a process of describing and evaluating potential hazards. Very often hazards are described in general terms, e.g. "microbiological hazard" instead of specifying *Salmonella*, *S. aureus*, hepatitis A virus, etc. Although such an approach may in some cases be practical, it is often misleading and in regard to microbiological hazards, it may even be dangerous. The reason is that microorganisms differ in their behavior, ecology and control measures. For instance, the ecology and thus control measure for *S. aureus* is much different than that for *Salmonella* or viruses. Thus, unless organisms share similar ecology and epidemiology, as far as possible they should be considered specifically. Similarly, chemical contaminants should be clearly defined so that valid safety limits and methods of testing can be identified.

   Similarly, control measures must also be defined and detailed as much as possible. For instance, instead of stating good hygiene practice (GHP), it should specify washing, disinfecting and drying hands or hand hygiene.

   The hazard analysis must include an evaluation of likelihood of occurrence and severity of effect of each hazard identified. This allows the identification of the significant hazards (see also point 8 below), which must be controlled for the food to be safe. Companies who do not take the time to do a thorough evaluation often struggle with establishment of the correct CCPs for the process concerned and their accompanying control and monitoring procedures.

   As mentioned before, hazard analysis must also be carried out taking into consideration the status and effectiveness of prerequisite activities, e.g. GMP, supplier quality assurance (review of the supplier audit report, ensuring that the audit has addressed the concern considered in the hazard analysis).

   The hazard analysis should also take into consideration relevant historical information, such as the previous safety record of a product or its ingredients or previous food safety incidents involving similar products and processes.

8. **CCPs or just good manufacturing practices.** One of the major difficulties in HACCP is the differentiation or understanding of what is at a certain production step a GMP and what is a CCP. Sometimes, operators have reported that a step, which is considered as a GMP, cannot be a CCP.

   To explain this, we need to go back to the time before HACCP. Food safety in the food processing and manufacturing industry was ensured through a number of measures referred to as good manufacturing practices (GMP). Some processing operations, today referred to as control measures, such as heat treatment (e.g. pasteurization), were then considered a good manufacturing practice. Thus, what in the past was referred to as *GMP* in HACCP terminology may today be referred to as *control measures*.

   In the context of HACCP, some of these control measures (or GMP), which play a significant role in controlling a specific hazard, received higher weight and the step at which the control measure is applied is thus now considered a CCP (Figure 32.3). In other words, a hazard analysis can actually permit to identify which good manufacturing practices are of direct relevance to food safety and if there is any additional measure which should be considered as part of GMP or with today's terminology prerequisites

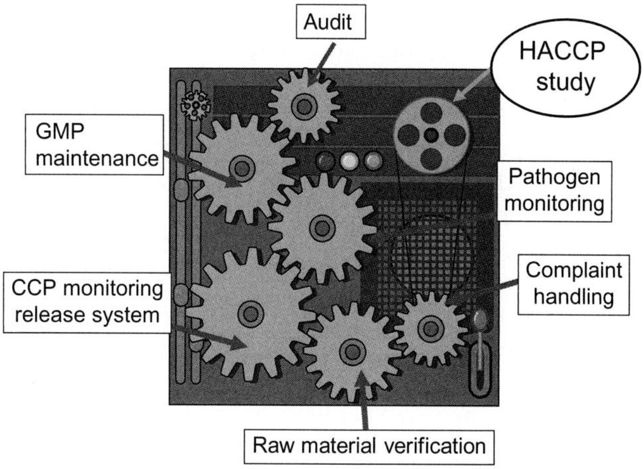

FIGURE 32.3 Schematic illustration to explain that different measures implemented as part of food safety assurance system are interrelated and need to be considered as part of the HACCP study.

programs. Therefore the question of CCP or GMP is a wrong question as both are interrelated. With the HACCP system we can strengthen the GMP to cover food safety concerns and also identify those GMP practices that are of importance for a close surveillance. This thinking process has been the basis for developing HACCP-based codes of practices for the small and developing businesses.

Part of this confusion comes from mistaking the term CCP for control measures (see below).

9. **Meaning of terms: significant hazards, CCPs and monitoring.** Lack of understanding or inconsistent use of terms often leads to confusion. Frequently, the term significant hazard is confused with CCP (e.g. aflatoxin is referred to as a CCP) or a control measure is taken for a CCP. This confusion also contributes to misconceptions mentioned above in relation to GMP versus control measure or CCP.

To be crystal clear: a CCP is a step in the food operation whereas a control measure is an intervention specifically designed to prevent, reduce or eliminate the hazard. To differentiate these two concepts, a CCP is usually a step in the flow diagram of the food production, while for the control measure this is not always the case. For instance, CIP of an installation is a *control measure* to eliminate the risk of contamination of food, but does not enter in the process of food production itself and is thus not a step or CCP. This does not preclude the fact that the processing step at which the CIP is applied can be a control measure and that the CIP is mentioned as a comment on the flow diagram.

It has also been experienced that some important steps in the operation have not been considered as CCPs on the grounds that they cannot be monitored "continuously" or a physical measurement method is not available. It is certainly much better to measure control parameters in an objective manner, e.g. using a temperature recorder. However, a lack of such methods for monitoring should not be a reason for not considering a step in the process as CCP. If control at that step is important and that visual/off-line control can be effective in detecting deviations from acceptability, then the step can be considered as

CCP. The concept of continuity is also misunderstood. The recording of a parameter leading to a line between two measurements is often mistaken for the concept of continuity. Monitoring a CCP should be carried out with specific and predetermined frequency. This may be every second, hour, day or defined moments as applicable. However, this has to be set such that if a deviation is observed and critical limits are violated, the corrective actions can be implemented in a timely manner before the product is released.

10. **Significance of CCPs.** As stated above, CCPs are steps in the food production that *must* be under control to produce a safe product. For each CCP critical limits are established to define the parameters that must be achieved to ensure safety. As such these limits define the acceptability or unacceptability of a product or a process. For food safety, CCPs and their associated critical limits are the most important steps and aspects of the operation. *Where a critical control point is violated, the product must be considered as potentially unsafe.*

    Therefore it is extremely important that:

    a. These steps are identified and controlled correctly, i.e. all parameters need to be controlled and identified and that the critical limits are validated, including consideration of any regulatory requirement.

    b. Monitoring of CCPs provides assurance that the control measures(s) are correctly implemented and are within the defined critical limits. Therefore:
    - The monitoring of CCPs must be carried out under the responsibility of trusted and well-trained operators.
    - Responsibility and the consequences of failures of CCPs must be clearly communicated to operators, including the corrective actions that must be taken in the event of a CCP failure.
    - Methods and procedures used for monitoring, be it a physical measurement, visual inspection or chemical analysis, must be up to date, and valid for the intended use (including sampling and sensitivity of method).
    - Microbial testing is verification and is generally not suitable for CCP testing. An exception is for the release of certain high-risk raw materials;
    - Equipment and materials used for monitoring must function correctly, be well maintained and calibrated. The frequency of monitoring must be set so as to ensure that if there is a deviation, corrective action is applied in time to correct the problem during production and/or to assure that unsafe product does not leave the factory.
    - Results of CCP monitoring are part of release criteria.

11. **Monitoring of CCPs**. When establishing a monitoring procedure for a CCP, care should be taken to identify all the parameters that will impact the efficiency of the control measure. For instance, it is frequently observed that at the pasteurization step, which is often a CCP, only the temperature is monitored and the residence time is ignored. Or for water disinfection, only residual chlorine is considered and other factors such as contact time, pH of water and turbidity, which impact on the chlorination efficiency, are not considered.

12. **Corrective actions**. While it is a good practice that in times of problems the operator consults his superior, it is nevertheless essential that corrective actions needed to restore control be clearly and precisely defined in the HACCP study. For instance, it should

be stated "reheating the product" or "sieving the product" rather than "call the QA or production manager."

Also, the procedures for blocking and eventually reworking or disposing of products that have not met safety or quality criteria should be carefully examined to prevent or minimize the possibility of any human error or accidental release. The efficacy of the corrective actions applied must be validated.

13. **Effectiveness of HACCP cannot be ensured without validation and verification.** One of the greatest weaknesses in the application of HACCP has been that very little attention has been given to validation and verification activities.

   a. **Validation.** A HACCP study whose elements are not valid will have limited benefits. Validation brings assurance in the design of the food safety assurance system. In absence of validation, there is no assurance that control measures will be effective in ensuring food safety and the HACCP studies may indeed become a paper exercise.

   Validation consists of proving evidence and documenting that the elements of the HACCP are effective and/or have a scientific and technical basis. It should include:

   - Identification and evaluation of hazards
   - Effectiveness of control measures (including corrective actions)
   - Correctness of CCPs
   - Critical limits
   - CCP monitoring
   - Corrective action
   - Suitability of verification activities

   Validation does not necessarily require an experimental approach or complicated tests (such as challenge tests) but may simply consist of confirming consistency with regulatory requirements, examining scientific literature, consulting experts, providing historical data and so on to substantiate the elements of the HACCP plan.

   b. **Verification**. An equally important principle of HACCP that is sometimes overlooked or carried out independently from HACCP is verification. Verification is the application of methods, procedures, tests and other evaluations, in addition to monitoring at CCPs to determine compliance with the HACCP plan. Verification provides confirmation that HACCP is implemented according to the plan, and is effective. Verification includes activities such as:

   - Environmental monitoring
   - Pathogen monitoring
   - End-product testing
   - Raw material testing for chemical contaminants
   - Audit of the factory and its food safety management systems
   - Calibration of equipment
   - Consumer complaints monitoring

   These activities, even if not part of the release procedures are essential to ensure that preventive measures are implemented correctly.

   Therefore, data collected through verification activities should be carefully examined and analyzed in terms of compliance and trends. Where a deviation from acceptable conditions is observed or the trend indicates an abnormal situation, implications for product safety should be evaluated, an investigation should be carried out as to the root cause of the problem, and the situation should be corrected. The cause may be failure in implementation or shortcomings in validation.

14. **Maintenance.** Maintenance of a HACCP is a means for addressing management of change. The environment and conditions under which food is produced is constantly changing. Examples of changes are:
   a. Emergence of new hazards and/or new information about existing hazards, e.g. knowledge about outbreaks in a sector of the food chain.
   b. Changes in the regulatory requirements.
   c. Changes in the intended use of the product and/or consumer/customer.
   d. Change and differences in the climatic conditions (where the raw material is produced and where the final product is transported and consumed).
   e. Change in the country where the product is to be exported to leading to a number of other changes as mentioned above.
   f. Changes in the source of the raw material.
   g. Changes in practices at the supplier.
   h. Changes in recipe, process or equipment.
   i. Changes in the factory environment, e.g. reconstructions, change of personnel.

   To ensure that the hazard analysis and control measures remain valid, it is important that each time a change is reported, the HACCP study is reviewed and the validation of control measures, critical limits, monitoring procedures, corrective action and verification procedures are reconfirmed. This means that each change should prompt a reflection on possible consequences for food safety and the need for amending existing control measures or setting up new measures.

   Review of the HACCP study does not necessarily mean an immediate and full revision of the HACCP study and associated plan. In many cases, it can be simply a note to document that:
   a. The change in question has been considered and control measures have been modified as follows, or
   b. It has been concluded that the change did not impact on food safety on the following grounds.

   The various notes can be consolidated during the annual review of the HACCP study.

15. **Different (Modular) HACCP plans.** Often, due to the complexity of the production, it is easier to develop different HACCP plans for different parts of the production. This is usually known as Modular HACCP plans as the process is split into "process modules" and HACCP principles are applied to each one. It is important to ensure that a proper link between the different HACCP plans exists and that errors do not occur as a result of this practice, e.g. skipping a step, omitting certain hazards.

CONCLUSIONS

The advance of HACCP has had a significant and positive impact on the management of food safety. However, to fully benefit from the advantages of HACCP, a proper understanding, application and true commitment is needed.

Some of the major shortcomings in the design and implementation of the HACCP plans have been:

- Lack of experience and expertise in the design of the HACCP plan.
- Ignoring previous records of safety of the product, e.g. incidents associated with the products.
- Failing to consider the regulatory or customer requirements.
- Overlooking implications of practices upstream, i.e. at the suppliers, or downstream (transport, distribution, handling and preparation).
- Gaps in the validation of the hazard identification and control measures and taking into consideration the state of prerequisites programs.
- Shortcomings in the review of the verification data (e.g. pathogen and environmental monitoring of food production premises) or end-product testing.
- Shortcomings in the maintenance of the HACCP plan in the light of verification data or changes.

It must be reiterated that HACCP is not a panacea to all problems and it is not a magic wand. It is a tool, among many others, to manage and enhance food safety. Its effectiveness in eliminating or reducing hazards to an acceptable level depends on how it is used.

References

Motarjemi, Y., Käferstein, F.K., 1999. Food safety, HACCP and the increase in foodborne diseases: a paradox? Food Control 10, 325–333.

CAC (Codex Alimentarius Commission), 2008. Codex Guidelines for the Validation of Food Safety Control Measures, CAC/GL 69 – 2008. Joint FAO/WHO Food Standard Programme, Codex Alimentarius Commission, the Food and Agriculture Organization of the United Nation, Rome.

Further Reading

Wallace, C.A., Sperber, W.H., Mortimore, S.E., 2011. Food Safety for the 21st Century. Wiley-Blackwell, Oxford, UK.

CHAPTER

33

Management of Microbiological Hazards: Role of Testing as Verification

Tim Jackson

Nestlé USA and Nestlé Canada

OUTLINE

Food Safety Management.
DOI: http://dx.doi.org/10.1016/B978-0-12-381504-0.00033-0

INTRODUCTION

Microbiological testing programs play an important role in the verification of the effectiveness of control measures for many food products. Such programs may include monitoring[1] of the production environment and processing equipment, and testing of raw materials,[2] in-process and finished products. The relevance and application of testing programs depend upon the design of the product and process, the hygienic status of the processing environment and the availability of other verification information about a product lot. Microorganisms are often distributed unevenly in foodstuffs and the practicality and economics of sampling make product testing ineffective as control measures. Likewise, environmental monitoring provides a snapshot of the hygienic status of the environment or processing equipment at the time and locations that samples were taken. Product and environmental testing are lagging indicators of hygienic failures as they do not directly control the root conditions that lead to contamination. As such, they are most effective when used within a system of risk-based preventive controls, such as HACCP, hygienic zoning and other prerequisite programs, and when they work together with other verification activities to assess the condition of the food safety system.

Microbiological testing may also be used to support the design and validation of control measures in a food safety management system. Testing may be used to determine initial microbial levels on raw materials or in-process product prior to the application of a microbiocidal process in order to establish the level of reduction required. Testing may also be conducted to determine the surviving levels of target microorganisms in a foodstuff after a microbiocidal process is applied in order to confirm that the desired reductions are achieved. It is often difficult to obtain quantitative information on the levels of pathogens present in a foodstuff prior to processing as levels of these organisms are often low and unevenly distributed, and information in verification testing programs and other surveys are based on analysis of presence or absence of the target organism, providing little information

[1]The term "monitoring" is used in this chapter to indicate the use of testing of products or the environment in verification programs. This is different from the use of the term "monitoring" in HACCP studies, which is the taking of measurements during processing to ensure that a critical control point is within established critical limits (Chapter 31).

[2]The term "raw materials" in this chapter refers to raw agricultural products, processed ingredients and food contact packaging materials used in the manufacture of food products.

on population levels (ILSI Europe, 2010; Jongenberger et al., 2012a,b). Where quantitative data are available for indicator organisms, it may be possible to extrapolate these data to estimate worst-case initial loads of the target pathogen. Most often validation studies are conducted using samples artificially inoculated with levels of target organisms sufficient to determine the reduction achievable by the process. Strains used in such studies are representative of those of concern in the foodstuff and are pre-conditioned to most closely approximate their physiological state prior to processing. Regulatory and industry reviews provide additional guidance on the use and application of microbiological testing in validation (ICMSF, 2011; NACMCF, 2006, 2010; Codex Alimentarius Commission, 2008b; Swanson et al., 2000; Zwietering et al., 2010).

The remainder of this chapter will focus on the use of microbiological testing as verification in food safety management systems. The role of environmental, raw material and finished product monitoring programs will be discussed as well as approaches to their development and implementation.

WHEN ARE MICROBIOLOGICAL TESTING PROGRAMS USEFUL FOR VERIFICATION?

Due to their cost and complexity, microbiological testing programs are only applied when they can provide relevant information about a product and process. Understanding the appropriate application of microbiological testing requires an understanding of the significance of microbial levels at various points in the process. The contribution of product, process and environmental factors to the ability to achieve the required microbiological limits in manufactured products is described in a conceptual equation developed by the International Commission on Microbiological Specifications for Foods (ICMSF). This equation illustrates the impact of various factors on the ability to manufacture product that does not exceed a food safety objective (FSO) or performance objective (PO; Codex Alimentarius Commission, 2007b, 2008b; ICMSF, 2002; Stringer, 2004; Van Schothorst, 2009; Motarjemi and Moy, in press):

$$H_0 - \Sigma R + \Sigma I_{(\mathrm{G} + \mathrm{RC})} \leq \mathrm{PO/FSO}$$

An FSO is the maximum frequency and/or concentration of a microbiological hazard in the foodstuff at the time of consumption necessary to achieve a public health objective such as an appropriate level of protection (ALOP). This is established by regulatory authorities as part of their risk management activities. Regulators may also define a PO, i.e. the maximum frequency and/or concentration of a hazard in a food at a specified step in the food chain before the time of consumption, in order to meet an FSO (Figure 33.1). Likewise a performance criterion (PC) may be established to communicate the required outcome for a control measure or series of control measures, such as microbiocidal or microbiostatic controls (Codex Alimentarius Commission, 2007b, 2008b; ICMSF, 2002, 2011).

A PO or PC may also be developed by the manufacturer based upon an established FSO, where one exists, or based upon the levels of relevant microbiological hazards necessary for

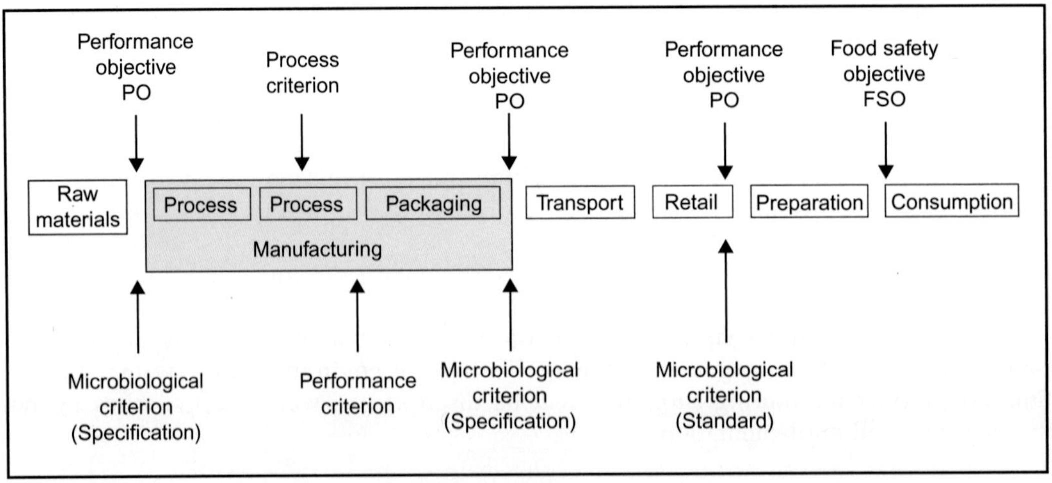

FIGURE 33.1 The role of food safety objectives, performance objectives and microbiological criteria in food safety management *(adapted from Gorris, 2005; Codex Alimentarius Commission, 2007b).*

product safety as determined in the HACCP study. The ability to produce a product that is equal to or below the PO is based upon the initial microbial levels in raw materials (H_0), the presence of conditions that could increase microbial levels during storage and processing (ΣI_G), the presence of conditions that could lead to the recontamination of the product or raw materials from equipment or the factory environment (ΣI_{RC}) and the microbial reduction achieved as a result of process controls (ΣR). Such an understanding of the contribution and interrelatedness of factors affecting microbiological quality and safety can assist those developing a product or process to determine the appropriate combination of control measures to ensure product safety. It can also be used to determine the application and nature of the corresponding monitoring programs necessary to verify the effectiveness of these controls.

Examples of the application of microbiological monitoring programs for the verification of various products are provided in Table 33.1. Testing may be useful to verify the hygienic status of raw materials that are not subject to a lethal process, either in a dry mix or assembly operation, or where they are added after the application of a lethal process. Testing may also be useful for raw materials that are exposed to conditions that would allow the outgrowth of microorganisms to levels greater than that which the applied processes are capable of inactivating. Testing is generally unnecessary for raw materials that will be subjected to a process that will inactivate the levels of pathogens or spoilage organisms present in the raw materials.

Environmental monitoring may be necessary to verify the application of environmental controls where raw materials, in-process or finished products are exposed to the production environment without a subsequent microbiocidal step. Such monitoring may not be relevant for products that are enclosed during processing and packaging unless the hygienic condition of the environment where the finished product is handled may have an impact on the ingress of microorganisms (for example, contact with cooling water or poor hygienic handling of retorted products prior to cooling).

TABLE 33.1 Examples of Testing Applied to Products and Process Controls

Product Type	Monitoring Program Applied	Parameters Evaluated	Location and Frequency
Dry mix product or other product assembled without further microbiocidal process	Raw material	Relevant pathogen and hygienic indicator organisms	Depends upon raw material risk; for example, testing of each lot of high risk material by supplier and/or receiving factory
	Environment	Relevant pathogen and hygienic indicator organisms	Areas where raw materials are handled and product is exposed prior to packaging. Weekly sampling which may involve rotation of sites
	Finished product	Relevant pathogen and hygienic indicator organisms	For new processing lines or where there is evidence of a hygienic concern for the product and process. Periodic testing according to risk
Products receiving a microbiocidal process and exposed to the environment after processing and prior to packaging	Raw material	Relevant pathogen and hygienic indicator organisms	For raw materials added after the thermal process. Depends upon raw material risk; for example, testing of each lot of high risk material by supplier and/or receiving factory
	Environment	Relevant pathogen and hygienic indicator organisms	Equipment and environment where products are exposed post-processing; interface between raw and cooked material handling areas
	Finished products	Relevant pathogen and hygienic indicator organisms	For new processing lines or where there is evidence of a hygienic concern for the product and process. Periodic testing according to risk
Products that are in-pack pasteurized	Finished product	Total plate count and total coliform or total Enterobacteriaceae	Each lot to establish history of performance; ongoing frequency based upon risk
Hot-filled products	Finished product	Total plate count and/or mold and yeast (for high-acid products)	Each lot to establish history of performance; ongoing frequency based upon risk
Product processed for commercial sterility and aseptically packaged	Finished product	Incubation testing	Representative samples from each production line, includes samples from each filler head and samples from events that could affect hygiene (start-up, stoppage, maintenance)
Low-acid products commercially sterilized in hermetically sealed containers	Finished product	Incubation testing	A small number of representative samples from each lot, includes events (start-up, stoppage, maintenance)

(Adapted from GMA, 2012b)

A finished product testing program may have a role in verifying the overall functioning of preventive control measures for products that rely on various supplier, production and environmental controls, such as products blended or assembled without a subsequent lethality control measure, or products exposed to the environment following the application of a microbiocidal process. A finished product testing program is less relevant for products that receive a process in the final package, but may play a role in verifying the application of a thermal process, evaluating the functioning of a production line over time or investigating potential process failures.

Details on the application of raw material, environmental and finished product monitoring programs are discussed in subsequent sections of this chapter.

PREREQUISITES TO THE DEVELOPMENT AND IMPLEMENTATION OF MICROBIOLOGICAL TESTING PROGRAMS

Effective product testing and environmental monitoring programs are developed and implemented only after the implementation of programs that identify and establish appropriate preventive controls:

- Hazard analysis and critical control point system;
- Hygienic design of equipment and processing environment;
- Hygienic zoning controls to prevent entry, harborage and growth of pathogens;
- A well-designed raw material selection and verification program;
- Personnel training to ensure that control measures are applied correctly.

Microbiological testing is of limited value in the absence of such preventive controls; however, microbiological testing and monitoring programs can be effective tools for verification when they are based upon a thorough understanding of the product and process as determined in these programs.

Requirements of Regulatory Agencies and Customers

Finished product requirements may be defined in customer specifications or in regulatory requirements. Regulatory requirements may be expressed as FSOs at the point of consumption or as POs for the finished product after production or for product on the market (see Chapter 31). Often, requirements are expressed indirectly as within-lot microbiological criteria for lot acceptance, increasingly following the format developed by ICMSF and adopted by the Codex Alimentarius Commission (Figure 33.1). For some products, between-lot criteria are established as an ongoing assessment of process control (ICMSF, 2002; Codex Alimentarius Commission, 1997). Some regulatory bodies have systematically established microbiological criteria for relevant categories of ready-to-eat products (e.g. Canada, European Union, Hong Kong). Other regulatory bodies have developed criteria for finished products or raw materials as needed based upon an identified risk or in response to the occurrence of public health incidents or a specific public health concern (NRC, 2003). Where such criteria exist they can be used to determine the appropriate design of products and process controls necessary to meet these criteria, and the testing programs necessary to ensure that the criteria are consistently met. Microbiological criteria should not be mistaken with FSO or PO. The former define the acceptability or unacceptability of products (in or out), the latter is used for designing the control measures, defining the expected/desired performance in verification programs and establishing expectations in contractual agreements.

In many cases there are no criteria specified for a product in regulation or in customer requirements; instead there is a general requirement for the producer to manufacture safe products. It is therefore the producer's responsibility to determine the necessary PO, PC, microbiological criteria and supporting verification programs for raw materials, processing environments and finished products. For some products, industry guidance has been developed to support manufacturers in the development of appropriate product and process criteria (Chen et al., 2009a, b; GMA, 2010; MAF/NZ, 2011; NFI/NFPA, 2002; Scott et al., 2009).

Hazard Analysis and Critical Control Point Study

Microorganisms of concern for the product and process are identified in the hazard analysis conducted as part of the HACCP study. The study will identify at what point in the process microorganisms will be introduced or multiply and will identify the type and location of control measures necessary to ensure the hazards are controlled. This will be based upon an understanding of the microbiology of raw materials, the effect of processes applied during manufacture, the exposure of the product or raw material during processing and after the application of a microbiocidal process, the behavior of the pathogen in the product (survival, growth, inactivation) and the impact of consumer preparation and reasonably expected misuse. The HACCP study will also identify the procedures, including microbiological monitoring, that are necessary to verify the ongoing functioning of the preventive controls for the identified hazards.

The HACCP study is focused on microorganisms of food safety concern; however, information from the study can also be used to evaluate the impact of product attributes, handling and distribution conditions on product spoilage and thus the relevance of non-pathogenic spoilage organisms in testing programs.

Zoning of the Factory Environment and Hygienic Design of Equipment

Hygienic zoning is the separation of factory areas based upon the risk of product contamination and the corresponding hygienic and preventive controls necessary to ensure that cross-contamination of products and raw materials does not occur (Duffy et al., 2003; Holah, 2005; Scott et al., 2009). Such controls may include physical barriers, cleaning practices, restrictions on the control of the movement of people, materials and equipment, management of tools, air flow and personnel practices required for each area and for movement between areas. Control measures include those specific to the area as well as those necessary at the entry to or transition between zones, for example between areas that must be dry cleaned and areas that are wet cleaned, or between areas where unprocessed, highly contaminated materials are handled and where products are handled that have received a microbiocidal process, or where raw materials are handled that will be used in a process without application of a microbiocidal process.

Zoning studies consider product design, process flow, equipment design, exposure of raw materials and product before and after microbiocidal processes, movement of people, materials, equipment and waste, air and utilities flow, prior history of the product type and processing facility. The studies identify sensitive areas of the process (e.g. areas where product is exposed to the environment), high risk areas and activities (handling of highly contaminated material such as raw meat, maintenance activities, management of waste) and factors that could lead to cross-contamination into sensitive areas. Good hygienic practices, structural and logistical control measures are identified including cleaning and sanitation practices necessary to ensure protection of the product from contamination. Areas of the factory are classified according to the required hygienic controls (Table 33.2).

The most stringent controls may be needed at the interface between high risk activities and areas of the factory where ingress of pathogens into processing areas can occur (through personnel or other activities), for example:

- To protect areas that must be kept dry to prevent harborage with *Salmonella* from other areas of the factory that must be wet cleaned;
- To ensure that pathogens present in materials where their presence is likely do not enter into production areas where product is exposed following a microbiocidal process;
- To ensure control of the environment where product is exposed that is intended for sensitive populations;
- To ensure the application of hygienic controls is sufficient to prevent the contamination of perishable chilled products with *Listeria monocytogenes* that may grow during storage of the product.

Sampling sites at such interfaces will be included in environmental monitoring programs to evaluate the effectiveness of control measures.

Specific examples of zoning controls are available in regulatory and industry guidance on the control of food borne pathogens (Chen et al., 2009a, Codex Alimentarius Commission, 2007a, 2008a; US FDA, 2008; USDA FSIS, 2012; GMA, 2010, 2012a; NFI/NFPA, 2002; Tompkin et al., 1999). The outcome of zoning studies is often included in zoning maps, which identify the hygienic classification of areas, and includes the movement of people, materials, equipment, waste and air. Such maps are valuable in identifying relevant sites to be included in environmental monitoring programs.

A study of the hygienic design of equipment and manufacturing environment will help to identify where potential harborage points exist in the process. These include points where food or water can collect and/or which are difficult to clean. In many cases the identification of such areas of concern in hygienic design and zoning studies, such as harborage points or hollow bodies in equipment or production areas or traffic patterns, will result in corrective actions to address the concern. Until these areas can be addressed they will be under increased scrutiny in environmental monitoring programs.

TABLE 33.2 Hygienic Zoning Classifications and Sample Prioritization

Hygiene Zone Classification	Definitions	Example	Environmental Monitoring Sites[a]
High hygiene (high care)	Area of factory where products, raw materials or equipment highly sensitive to contamination are handled, processed or stored	Infant formula dry mixing, packaging. Milk powder spray-drying. Post-oven handling/ packaging of chilled RTE foods supporting the growth of *Listeria monocytogenes*. Clean equipment and tools storage for high hygiene activities	Product contact surface/line/Z1 Z2/P1 Z3/P2
Medium hygiene (medium care)	Area where products, raw materials and equipment are exposed or stored and where they are sensitive to contamination: intended for the consumer without elevated sensitivity; where growth of microbial pathogens is not possible in the supply chain	Assembly, handling of frozen processed products. Blending, molding of confectionery products prior to packaging. Dry blending area for soup mixes. Clean equipment and tools storage for medium hygiene activities. Storage of rework, in-process products	Product contact surface/line/Z1 Z2/P1 Z3/P2 Z4/P3
Basic/low hygiene (low care)	Areas where activities will not result in the contamination of products (for example, storage of raw materials and finished products in enclosed packaging), or products or materials are handled prior to a microbiocidal process. If movement of people, material, air and water are not controlled, area could become a source of cross-contamination of sensitive processing areas	Storage of finished products. Storage area for raw material and packaging. Storage of processing equipment prior to cleaning (other than those used in high risk activities). Storage of cleaning chemicals. Storage of in-process materials or ingredients in sealed containers	Z3/P2 Z4/P3
High risk	Areas where materials are handled with a high probability of contamination with microbial pathogens	Handling and processing of raw meat and poultry, unprocessed vegetables, raw milk, raw cocoa beans and nuts, raw cereals	Z4/P3 (at interface between zones, transport equipment, etc.)
Not zoned	Areas isolated from production activities	Offices, lunch room, entry lobby, change rooms	Investigation

[a]*See sample site definitions, Table 33.4.*

MICROBIOLOGICAL MONITORING OF THE FACTORY ENVIRONMENT

Environmental monitoring programs are used as a verification of the effectiveness of control measures to prevent the ingress, harborage and multiplication of microbial pathogens in the production environment, specifically:

- Effectiveness of cleaning and sanitation procedures;
- Effectiveness of environmental controls:
  - Controls associated with hygienic zoning
  - Movement of people, equipment and materials
  - Construction and maintenance activities
- Identification of areas of ingress or harborage so that they can be eliminated;
- Investigation of the impact of adverse findings.

The application and design of environmental monitoring programs will depend upon the risks associated with the product and process. For example, an evaluation may not be needed of the processing areas for products that are processed in enclosed systems and filled aseptically or hot-filled. Likewise, products that are processed in their final package and are not exposed to the environment after processing may not require stringent sampling programs unless there is a risk of recontamination (for example, through micro-leaks in the seams of cans during cooling following a thermal process).

While an environmental monitoring program can be a valuable verification tool, it only provides a picture of the sites analyzed during the day samples are taken. However, when evaluated along with other samples taken from a production line or process environment over time, it can provide useful information regarding ongoing status or trends in hygienic control.

Selection of Pathogens and Indicator Organisms

The pathogens that are the focus of the environmental monitoring programs will be determined by the hazard analysis in the HACCP and zoning studies. Generally, *Salmonella* and *Listeria monocytogenes* are the pathogens of environmental concern, although other pathogens may be included based upon product and risk (for example, *Cronobacter* spp. in infant formula; *Staphylococcus aureus* or *Bacillus cereus* as investigative sampling in areas where outgrowth of the organism during processing is a concern). Environmental monitoring programs include the pathogen of concern; however, the infrequent or sporadic distribution of these organisms often makes them difficult to detect, even when they are present in the factory environment. A program focused solely on the isolation of the target pathogen will only identify a problem when it occurs and may not identify early enough that conditions are or have been present that would also allow the ingress or growth of the pathogen of concern. Because of this, effective environmental monitoring programs include hygienic indicators, selected according to their ability to demonstrate the presence of conditions that would lead to the presence or growth of the pathogen of concern.

Processing Environments where Wet Cleaning is Conducted

Listeria monocytogenes is the primary environmental pathogen of concern in processing environments where wet cleaning is used. The most effective indicator organisms for the presence of *L. monocytogenes* in the environment are other members of the *Listeria* genus (USDA FSIS, 2012; US FDA, 2008, 2013). Because they are very closely related to *L. monocytogenes*, the detection of non-*monocytogenes* members of the *Listeria* genus (*Listeria* spp.) indicates that conditions exist that could also lead to the presence of *L. monocytogenes*. Detection of these indicators will initiate a root cause analysis and increased investigative testing of equipment or the environment from which the isolation occurred to ensure that the root cause is investigated. Recovery of *Listeria* spp. or *L. monocytogenes* from product contact surfaces or nearby areas on equipment or the environment may also initiate or intensify finished product testing to verify that the product is not affected.

Quantitative indicators, such as Enterobacteriaceae, coliforms or total plate counts, may be useful for monitoring the effectiveness of cleaning and sanitation procedures or to assess whether conditions exist that allow multiplication of microorganisms on or around processing equipment. Because they are heat sensitive, Enterobacteriaceae and coliforms are useful for the evaluation of the hygienic status of the processing line after a thermal process.

Total plate counts (TPC) may be used to monitor that conditions are present during processing that could lead to the outgrowth of *S. aureus* or *B. cereus*. High TPC results are followed by investigative sampling of potential harborage sites in or on processing equipment or holding containers (e.g. tanks, totes, mixers), where product and moisture may be present that could lead to the growth of these organisms. Such testing is usually conducted as an investigation of out-of-specification results from finished products. Because TPC is a quantitative hygiene indicator, expected baseline levels are established through an analysis taken of clean surfaces when it is known that cleaning and sanitation were effective. Due to the broad variety of microorganisms that will be recovered for TPC analysis, it is most effectively used for the evaluation of product contact surfaces or nearby surfaces and is less useful for areas of the environment away from the processing line.

The inclusion of mold and yeast may be useful in monitoring programs for the exposed-product environment of products for which yeast and mold spoilage are a concern (such as chilled dairy products, intermediate moisture pasta, etc.).

ATP bioluminescence involves the detection of adenosine tri-phosphate (ATP) present in food material through the generation of a luminescent signal expressed in relative light units (RLU). The intensity of the signal is proportional to the level of ATP present and is an indirect indicator of the amount of biomass present. ATP bioluminescence may be used to verify the effectiveness of cleaning by measuring the presence of signals originating from residues present on product contact surfaces after cleaning (Moore et al., 2001; Powell and Atwell, 1997; Whitehead and Smith, 2008). To properly evaluate signals detected by bioluminescence equipment a baseline signal is established through the analysis of clean surfaces. An elevated signal will indicate that product residues are present, indicating that cleaning was inadequate and the surface needs to be recleaned.

Processing Environments that are Dry Cleaned or Controlled-wet Cleaned

Salmonella is the primary pathogen of concern in factory environments where dry materials are handled or low moisture products are manufactured and where dry cleaning, or in specific cases controlled-wet cleaning, is applied to ensure the absence of moisture from the environment during processing (Duffy et al., 2003). *Salmonella* can enter the environment through the movement of people, equipment and materials or through the failure of zoning controls between high risk and sensitive areas (for example, between areas where raw cocoa beans are stored and handled and roasted cocoa beans and cocoa products are exposed to the environment). When moisture is present, *Salmonella* can multiply; however, even where moisture is absent, the organism can persist for long periods up to years and multiply when moisture re-enters the environment (Scott et al., 2009). Resident strains may remain dormant only to reappear after some time due to a change in activity, such as a construction or maintenance event, or hygienic failure allowing the ingress of water.

Unlike the *Listeria* genus, there is currently no microorganism identified whose presence will closely correlate with the presence of *Salmonella. E. coli* has been used as an indicator of fecal contamination in water and as an indicator of post-process contamination in dairy products. While the monitoring of *E. coli* in the environment may be part of a monitoring program where fecal cross-contamination or growth is suspected, *E. coli* can persist in the environment and its presence in dry environments may not correlate directly with the presence of *Salmonella* (Cox et al., 1988; Kornacki and Johnson, 2001).

Salmonella is a member of the Enterobacteriaceae family and the quantitative analyses of the environment for members of this family are frequently included in pathogen monitoring programs for dry environments. Unlike *Salmonella*, which may only enter the environment rarely through a hygiene failure, many members of the Enterobacteriaceae family are likely to be present at some level even in clean environments. An evaluation of the level of Enterobacteriaceae present at a sampling site can provide information on whether conditions are or have been present that could lead to the multiplication of *Salmonella*. As *Salmonella* may or may not be present in the environment, there is not a direct correlation between the presence or population of Enterobacteriaceae and the presence of *Salmonella*; however, the use of quantitative determinations of Enterobacteriaceae in environmental monitoring programs will allow conditions that may lead to the multiplication of *Salmonella* to be identified and corrected. Enterobacteriaceae should only be included on product contact surfaces or near product contact surfaces due to the variability of levels in non-process areas of the factory without strict hygiene controls. (Figure 33.2)

For products intended for infants, *Cronobacter* spp. is a significant concern and is included in environmental monitoring programs for infant formula manufacture where ingredients, in-process or finished product are exposed. As with *Salmonella*, *Cronobacter* spp. is a member of the Enterobacteriaceae family and control measures taken to prevent *Salmonella* entry and harborage in the environment will also be effective for this organism. *Cronobacter* spp. has greater prevalence in the environment than *Salmonella*, increasing the importance of proper management of control measures and the corresponding stringency of monitoring programs. As with *Salmonella*, inclusion of quantitative Enterobacteriaceae as a hygiene indicator can help to identify the presence of conditions that could lead to *Cronobacter* spp. harborage and growth (Codex Alimentarius Commission, 2008a).

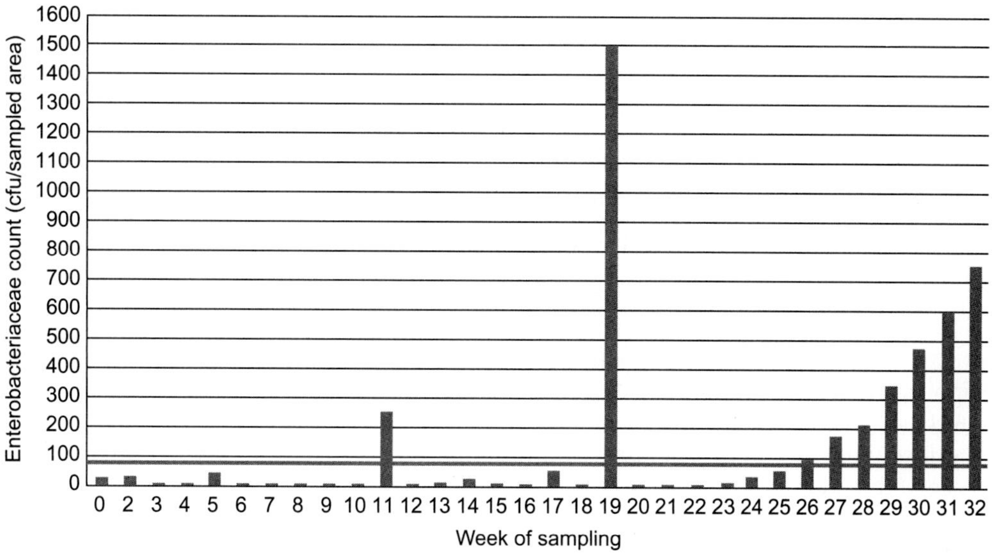

FIGURE 33.2 Enterobacteriaceae count at sample site A64 (near-product contact).

TABLE 33.3 Examples of Reaction Limits and Interpretation of Quantitative Enterobacteriaceae in Product Residue Taken from Equipment Surfaces and Process Environment in a Powdered Milk Factory

Enterobacteriaceae Level	Interpretation	Action
<100 cfu/g[a]	Acceptable (expected under good hygienic practices)	None necessary
100 to 1000 cfu/g	Marginally acceptable. Conditions may have existed to allow the increase of Enterobacteriaceae (such as ineffective dry cleaning, leakage of water into the environment, maintenance event)	Investigation conducted, potential root causes corrected
>1000 cfu/g	Unacceptable. Likely presence of water or other event has led to high levels of Enterobacteriaceae	Investigation conducted, potential root causes corrected. Increase sampling of finished product and environment for Enterobacteriaceae and pathogen

[a]*For some infant formula and milk powder factories levels <10 colony forming units (cfu)/g may be possible on product contact and near product contact sites; correspondingly reaction thresholds may need to be adjusted.*

Enterobacteriaceae is a quantitative indicator, and baseline levels and reaction limits need to be determined to facilitate the interpretation of results and corresponding corrective actions. Levels should be established with an understanding of the product and process and what is achievable under good manufacturing practices. An example of such limits for a milk powder factory is provided in Table 33.3.

ATP bioluminescence may be applied to verify the effectiveness of a periodic wet-cleaning process, but is typically not useful for environmental monitoring programs in factories manufacturing low moisture products that are dry cleaned as the presence of product particulates on equipment surfaces will interfere with the ATP signal.

Selection of Environmental Monitoring Program Sites

Sample sites for environmental monitoring programs are selected based upon risk as identified in HACCP, zoning and hygienic design review, with the primary focus on the following areas:

- Areas of equipment that are difficult to clean and could be harborage sites;
- High-traffic areas;
- Interfaces where movement occurs between hygiene zones;
- Interfaces between areas were raw, highly contaminated materials are handled and where processed materials are handled after the application of a microbiocidal process (such as cooking);
- Interfaces between wet- and dry-clean areas;
- Areas from where pathogens could be transferred into sensitive areas with exposed product or raw materials through the movement of people, equipment and materials;
- Areas where pathogens could enter into the facility from outside the factory or from higher risk areas within the factory.

Sampling sites are classified according to the potential for product contamination if the pathogen was present at that site. Some companies have classified samples according to sampling "zones" while other companies have used other terminology for sample prioritization to avoid confusion with the classification of hygiene zones (Table 33.4). In fact, there may be a variety of sites of different risk classification within a given hygienic zone (Table 33.2). The location and number of sampling sites will vary based upon the nature of the product, the complexity of the process, the degree of product and raw material exposure, the GMP practices necessary for a particular production area, and the movement of people, equipment and materials.

Examples of potential sources and harborage sites for *Salmonella* and *L. monocytogenes* in food processing facilities are available in guidance documents (Chen et al., 2009b; US FDA 2006) and can be valuable resources in establishing sites of focus for hygiene audits and environmental monitoring programs.

The weight of sampling programs is placed on the most sensitive sampling locations. This is reflected in the number of samples selected and the sample frequency, with priority given to more sensitive areas. Product contact surfaces are analyzed according to risk of product exposure and sampling history. The number of sampling sites that are included in an environmental monitoring program is based upon the nature of the product and process and the design of the processing line and factory. The number of samples taken on a given sampling day will often be weighted based upon sensitivity (for example, a proportional split of P1/Z2 60%, P2/Z2 30%, P3/Z3 10%). Greater emphasis will also be placed on historically problematic areas, or those where an investigation has identified conditions that may lead to ingress or harborage.

TABLE 55.4 Prioritization of Sampling Sites

Sampling Area	Example Classification	Definition	Examples
Product contact	Zone 1 (Z1), production line	Surfaces with direct or indirect contact with product	Surface of product conveyor. Nozzles and pipes dispensing products. Areas of product build-up. Product discharge chutes. Interior of pipes carrying product. Inside of storage totes. Inside of filler hopper. Product scrapers/utensils. Cleaning tools in contact with product or product contact surfaces. Contents of vacuum cleaners used to clean product contact surfaces. Surfaces from which product or water build-up during production could fall onto product or product contact surfaces during processing
Near product contact	Zone 2 (Z2), Priority 1 (P1)	Environmental surfaces with close proximity to product contact surfaces where contamination could easily be transferred to product contact surface	External surfaces of processing equipment. Environment near the exposed product/processing line. Floor drains near processing lines. Catwalks. Outside of tunnels. Outside of totes and fillers. Weigh scales. Outside of equipment used for dry mixing or mixing of ingredients without subsequent microbiocidal process. Cleaning tools and vacuum cleaners in contact with Z2/P1 areas
Non-product contact close to production line	Zone 3 (Z3), Priority 2 (P2)	Surfaces of equipment or the production environment in processing areas away from the production line and exposed products. The presence of pathogens could easily contaminate near-product contact sites	Hand trucks/pallet jacks used in processing areas. Forklifts used in production areas. Floors and drains in processing areas away from production lines. Wash stations. Ingredient storage areas. Traffic pathways into production areas. Wall/floor junctures. High hygiene side of shoe change area into high hygiene zone (and shoes). Interface between wet- and dry- cleaned production areas. Interface between areas handling highly contaminated materials and processing areas post-lethality
Non-product contact away from production line	Zone 4 (Z4), Priority 3 (P3)	Areas of the factory away from production	Remote locations in medium hygiene areas. Warehouses. Interface between medium and basic hygiene zones. Bathrooms. Transfer corridors

TABLE 33.5 Example of Sample Frequency Based Upon Site Prioritization and Level of Hygiene Concern

Minimum Sampling Frequencies per Sampling Site or Area			
	Level of Hygiene Concern		
Site Prioritization	**Normal/Routine**	**Elevated**	**High**
Product contact	1× week or based upon risk[a]	2× week	Investigative
Near product contact	1× week	2× week	Investigative
Non-contact near production line	1× month	1× week	Investigative
Non-contact away from production	Periodic	As needed	As needed

[a]*Proportion of samples analyzed for pathogens and hygienic indicators determined by risk and history.*

An example of sample frequency is included in Table 33.5. The frequency of sampling will vary based upon risk; however, sampling in most cases is conducted weekly and increased as a result of a finding (event potentially affecting hygiene, hygiene inspection finding, finding of pathogen or adverse trend of hygiene indicator in the environment). In some cases sampling will be conducted less frequently, for example according to a production schedule where a product or processing line is only used infrequently. "Due diligence" programs that only involve infrequent sampling occasions (such as monthly, quarterly or bi-annually) are generally not useful as they provide little information of the hygienic status of a process and do not allow a rapid correction of hygienic failures and adjustment of sampling programs when adverse results are obtained.

As their selection is based upon a risk assessment, the majority of sample sites in an environmental monitoring program are predetermined. Sampling programs should include a proportion of investigative samples, taken based upon the results of hygiene audits or of observations taken at the time of sampling. Established sampling sites may be modified based upon monitoring program findings. Where a number of sampling areas are identified, sites may be rotated with a given number of sites sampled at each sampling occasion. For example, a factory may select samples randomly using a numbering system classified by sample priority. Some factories have assigned sampling sites to alternating sample schedules, for example 1 week sampling is conducted for 50% of sites according to Schedule A, the subsequent week for the other 50% of sites according to Schedule B. In most cases the same sites are evaluated for hygiene indicators and for pathogens.

Sites are ideally identified on detailed factory maps, which may include information on hygienic zoning of areas and movement of people, equipment and materials. This facilitates the interpretation of data and also allows communication of data to factory personnel, or to corporate microbiologists or sanitarians in a different location supporting the factory in troubleshooting problems.

The stringency of an environmental monitoring program should be adaptable, increasing upon adverse findings, events or insufficient information on hygienic conditions. The increased intensity is reflected in more frequent sampling, but also may be reflected in an increased number of sampling sites and investigative sampling focused around the area of the adverse finding.

An example of program adaptation is included in Table 33.5. In this example, sample frequencies are categorized by normal/routine, elevated and high concern. The elevation of sample frequency may be applied to a specific production line where it is isolated from other lines, or to a specific processing area that is the focus of the hygienic concern.

Elevated concern could result from:

- Elevated level or adverse trend in quantitative hygiene indicator;
- Maintenance event;
- Exposure of factory area to the adverse conditions potentially impacting hygiene (for example, roof leak, sprinkler operation, burst pipe);
- Breach of hygienic zoning controls;
- Finding of pathogen in the processing area away from processing line.

High concern could result from:

- Finding of pathogen or out-of-specification hygiene indicator (presence of *Listeria* spp., elevated Enterobacteriaceae) in product, on product contact surface or the environment near the processing line.

The increased program intensity continues for a time sufficient to verify that the hygienic status of the line has returned to normal. This could vary due to the nature of the problem leading to elevated concern and the sensitivity of the product. For example, increased sampling may only be needed for a short period following a maintenance event. Sampling at elevated or high concern may be continued for a longer period of time where evidence (hygiene audits, test results) indicate a persistent problem or where investigation of a positive pathogen finding has not determined a clear root cause. Likewise heightened sampling may be conducted for several weeks or months for a new factory or production line. A line sampled under a high concern level may be placed on a sampling program for elevated concern for a period of time before returning to routine sampling.

Collection of Environmental Samples

As important as the selection of sampling locations to the success of an environmental monitoring program is the effectiveness of sample collection procedures.

Samples collected immediately after cleaning and sanitation will verify the effectiveness of this operation and the suitability of the line for the start-up of production. Samples collected during production may indirectly verify cleaning effectiveness but will also verify the effectiveness of control measures aimed at preventing contamination of processing areas or production lines, harborage or growth. Samples collected at the end of production will verify control measures but may provide additional information on microbial growth during production and can provide information on the risks associated with production, build-up of material on the line and the intervals between cleaning and sanitation activities. Samples taken towards the end of a production run are recommended; however, some sampling programs include a combination of samples taken post-sanitation and samples taken during production.

The tools selected for sampling will depend upon the nature of the site to be sampled as well as the level of residue/debris present at the site. Sterile, pre-moistened swabs may be most effective for the sampling of small cracks and crevices on equipment and the environment where moisture or product may collect, and for difficult-to-access areas. Sterile, pre-moistened sponges are more effective for sampling larger sampling areas on equipment and the environment. Sterile spatulas or scrapers may be used to sample product residue. Other tools, such as sterilized disposable dusting cloths or mop heads, may be useful tools for collecting samples from large areas in the environment during root cause investigations.

Pre-moistened swabs and sponges are available from several manufacturers, in some cases with novel features that facilitate sampling of difficult areas and aseptic transfer of sponge to neutralizing buffer or other appropriate transport medium. In cases where sponges are pre-moistened prior to use, it is important to squeeze the majority of moisture from the sponge prior to sampling. Where pre-moistened swabs or sponges are used in processing areas or processing equipment that must be dry during production, the sampled areas are dried after sampling by the technician taking the sample.

Sampling with swabs and sponges should use sufficient force to ensure that any contamination present in the sampled area is transferred to the sponge. In many cases, defined areas are sampled (e.g. 50 cm^2) which could be identified using a sterilized template, to facilitate comparison of quantitative results between sampling sites or trends in the same sampling site over time.

Agar contact plates are sometimes used to sample equipment surfaces. Sampling is conducted through direct contact with the surface being sampled. Plates are then covered and incubated until colonies develop which are then enumerated. The advantage of such methods for the analysis of quantitative indicators is that they require little or no advanced preparation and no additional preparation after sampling other than incubation. However, they are limited in their ability to transfer contamination present in cracks and crevices in equipment and may lose effectiveness on surfaces with a large build-up of soil.

Where present, the sampling of product residue or soil is preferable to the sampling of "clean surfaces" as such residue is more likely to be a source of harborage. Samples are taken with a sterilized brush, spoon, scraper or spatula, depending upon the material collected, and transferred to a whirl-pack bag. When sampling build-up of product on surfaces or the environment, care should be taken not to focus on the sampling of clean product, but instead to focus activities on areas where the build-up of product and/or moisture could lead to microbial harborage and growth.

Samples should be transported in a suitable buffer or other transport medium. In cases where the residuals of sanitation chemicals may be present, in particular when sampling following a cleaning/sanitation event, a neutralizing buffer should be used. If not analyzed immediately after collection, samples must be stored under refrigeration (0–4°C) until they are analyzed. (In some cases dry samples may be stored at room temperature if storage does not affect the survival or level of the target organism or group.) If analyzed off-site, samples must be shipped under refrigeration, with care taken to ensure that the refrigerant (such as an ice pack) does not freeze the sample. Samples need to be analyzed soon after they are taken, preferably within 36 hours (Andrews and Hammak, 2003; Evancho et al. 2002; Midura and Bryant, 2001).

In some cases, for example when sampling specific high priority sites, the same site is analyzed at each sampling event. However, many locations identified for sampling will be areas of the equipment or factory environment; for these areas specific sampling sites are varied at each sampling event during routine sampling.

When separate samples are collected from the same site, for example for *Salmonella* and total Enterobacteriaceae, care must be taken not to swab the same area for both samples at the same time; in these cases adjacent areas are sampled for the hygiene indicator or pathogen.

Investigative sampling will be conducted in the event of a pathogen finding or out-of-specification hygiene indicator, or will be conducted when issues are observed during sampling or during a hygiene audit that could impact safety. Following a pathogen finding, sampling typically involves re-examination of the location of the finding and the surrounding area and may also include strategic sampling of other areas of the factory to investigate the extent of contamination, the movement of the contaminant through the environment and/or the origin or harborage point of the contaminant. When investigating potential harborage sites in processing equipment, it may be necessary to shut down and open the processing line to allow access to sampling sites. In such cases sampling is done during a scheduled or unscheduled shutdown, or a specific shutdown is scheduled to allow a sufficient examination of the equipment. This is particularly important in an investigation of a pathogen finding.

During routine monitoring, samples are sometimes pooled (i.e. combined into one sample) for analysis. Such pooling is generally done across similar areas or sample prioritization sites (for example, samples taken from product contact surfaces) on a production line or in a production area. Pooling is not recommended across production days, between production lines or between sites of different prioritization. The advantage of pooling is greater efficiency of cost and the ability to sample more sites in the program. The disadvantage is that the source of contamination is more difficult to trace when adverse results are found (USDA FSIS, 2012). For this reason, pooling is not recommended for sampling when conducted under elevated or high concern.

Analysis and Interpretation of Environmental Monitoring Data

Monitoring data from a sampling event represents the hygienic status of processing equipment or the processing environment at the time that the samples were taken. For samples taken in response to an event (maintenance, observation of hygiene failure), such data could indicate the impact of the failure, or, if taken following corrective actions, indicate the effectiveness of those actions.

The presence of a pathogen or out-of-specification hygiene indicator are lagging indicators of a failure of hygienic controls which has led to the presence or harborage of pathogens, or to the presence of conditions that could potentially lead to the growth or multiplication of pathogens. Examples are elevated Enterobacteriaceae in dry environment, elevated coliform in product contact sample and *Listeria* spp. in "wet" processing environment. Unless observations of hygiene deviations were made at the time of sampling, additional investigation will be needed to determine the root cause of the failure and to ensure that any corrective actions taken were effective. Depending on the location of the out-of-specification sample, finished product sampling may be needed to verify that product was not affected. Where pathogens are isolated from product contact surfaces, it is assumed that corresponding product that has made contact with the surface is also positive for the pathogen. Examples of the interpretation and actions in response to findings in monitoring programs are included in Table 33.6.

TABLE 33.6 Example of Changes in Level of Concern and Resulting Actions from Pathogen and Out-of-Specification Hygiene Indicators in Various Sample Types

Location of Sample	Finding[a]	Resulting Level of Concern	Action
Product contact (production line)	Pathogen	High	Block affected product lot(s) and destroy or recondition; initiate or intensify finished product testing; increase sampling of product contact surfaces and environment to investigate; conduct investigation of harborage sites and hygienic practices; conduct raw material review
	Hygiene indicator	High	Initiate or intensify finished product testing; increase sampling of product contact surfaces and environment to investigate; conduct investigation of harborage sites and hygienic controls; conduct raw material review
Near product contact (Z2/P1)	Pathogen	High	Initiate or intensify finished product testing; increase sampling of product contact surfaces and environment to investigate; conduct investigation of harborage sites and hygienic controls; conduct raw material review
	Hygiene indicator	Elevated	Increase sampling of product contact surfaces; initiate investigative testing of the environment; if not in place implement finished product testing; conduct investigation of harborage sites and hygienic controls; conduct raw material review
Non-product contact close to production line	Pathogen	Elevated	Increase sampling of product contact surfaces; initiate investigative testing of the environment; conduct investigation of harborage sites and hygienic controls
	Hygiene indicator	Elevated	Increase sampling of product contact surfaces; initiate investigative testing of the environment; conduct investigation of harborage sites and hygienic controls
Non-product contact away from production line	Pathogen	Elevated or routine/normal[b]	Initiate investigative testing of the environment to determine root cause, impact on sensitive areas; conduct investigation of harborage sites and hygienic controls
	Hygiene indicator	Routine/normal	Increase pathogen monitoring in area; conduct investigation of harborage sites and hygienic controls
Finished product	Pathogen	High	Block affected product lot and destroy or recondition; initiate or intensify finished product testing; increase sampling of product contact surfaces and environment to investigate; conduct investigation of harborage sites and hygienic practices; conduct raw material review
	Hygiene indicator	High	Initiate or intensify finished product testing; increase sampling of product contact surfaces and environment to investigate; conduct investigation of harborage sites and hygienic controls; conduct raw material review

[a] *Presence of pathogen; presence* (Listeria spp.) *or out-of-specification (quantitative) hygiene indicator.*
[b] *Depends upon the location of the sample and the risk of contamination of sensitive production areas/equipment.*

Within-specification monitoring results, along with other verification information (hygiene audits, start-up checklist, visual evaluation of cleanliness) indicate that the environment was of acceptable hygienic status on the day samples were collected. A systematic trend analysis of environmental data over time can provide greater confidence in the hygienic status of a processing line or processing area. A periodic short-term increase in a quantitative indicator such as Enterobacteriaceae could indicate that an event occurred on or prior to the sampling day impacting the hygiene of the area and facilitate a root cause investigation. Upward trends in data could indicate a gradual loss of hygienic status and enable the problem to be identified and addressed before the underlying hygiene issue leads to harborage or cross-contamination with a pathogen.

The results of hygiene monitoring programs should be kept in a database (e.g. Excel using pivot table functionality) facilitating the evaluation of trends and correlations in data and the generation of graphical representations and reports. Results of qualitative analyses (such as presence/absence for *Salmonella, L. monocytogenes, Listeria* spp.) are often documented on a factory map to facilitate the root cause analysis.

Serotyping or genetic typing to identify strains of isolated pathogens is often useful to the root cause analysis. Serotyping is particularly useful for *Salmonella,* as there are greater than 2500 serotypes. Serotyping may also be conducted for *Listeria*; however, there are fewer serotypes identified and genetic typing, such as through pulsed field gel electrophoresis, may provide greater precision for a root cause investigation (Jadhav et al., 2012). Genetic typing has also been used for *Salmonella, Cronobacter* spp. and other pathogens. Recurrence of the same strain in multiple sampling events, on a variety of surfaces, or following cleaning usually indicates harborage in the factory environment. The detection of different strains usually indicates transient contamination due to multiple entries into the environment from one or more routes of entry (such as through raw material or from the environment external to the factory).

Data from monitoring programs communicated to the factory food safety team for review and development of corrective actions are needed. Program results may also be presented to factory personnel at operational reviews and/or through the posting of program results.

ACCEPTANCE CRITERIA AND TESTING PROGRAMS FOR FINISHED PRODUCTS AND RAW MATERIALS

Microbiological criteria may be established for finished products, raw materials and in-process products to define the conformance of a product lot or processing line to performance objectives and to define conditions of acceptance when verification testing is conducted. Criteria may be established as requirements for products on the market or at import by regulatory agencies, as a specification by a food manufacturer for finished products or raw materials or as guidance by regulators or industry groups to food manufacturers. The utility of product testing is limited when contaminants are present at low levels and unevenly distributed. The costs of product testing are often significant due to the need to hold a corresponding product lot during the time testing is conducted. Because of such limitations, food safety management systems that incorporate preventive controls including good hygienic practices and HACCP are much more effective than a reliance on finished product testing in the absence of knowledge of such controls (ICMSF, 2002; NRC, 1985).

Although statistically limited, finished product and raw material testing may be conducted where there is limited information available about the hygienic status of a product lot (for example, a regulator's analysis of imported product or a food producer's analysis of raw materials). Testing may also be used for the evaluation of the suitability of finished products or raw materials where there is information from other verification activities that indicates an increased risk of contamination.

The development and application of acceptance criteria for finished products and raw materials is discussed extensively by the ICMSF (2002). Lot acceptance criteria are expressed in sampling plans outlining the pathogen or indicator organism(s) of concern, the number of samples to be taken from a lot (n), the limits of acceptance (c, m and M) and the methodology to be used in verifying conformance. Sampling plans in specifications are most often defined as two-class attributes plans (acceptable and unacceptable) and three-class attributes plans (acceptable, marginally acceptable and unacceptable). Two-class attributes plans are defined by m, the level separating acceptable from unacceptable and c, the maximum allowable number of sample units yielding a result greater than m. For pathogens m is often set at 0, indicating an absence of the organism in the analytical unit tested. Three-class attributes plans are defined by m, the level separating acceptable from marginally acceptable, M, the level separating marginally acceptable from unacceptable, and c, the maximum allowable number of sample units yielding a result greater than m and less than M. If any sample is above M in a three-class plan the lot is rejected. Three-class plans are most often applied in criteria for quantitative hygienic indicator organisms as they account for variability in levels and allow identification and correction of trends before levels exceed criteria that would result in lot rejection.

The ICMSF (2002) has developed standardized "cases," sampling criteria with stringency based upon the relative risk of the microorganism or group to be analyzed and the effect of handling conditions on the relative product risk. The ICMSF has also developed representative criteria for specific product categories (ICMSF, 2011).

Guidance on the sampling and shipment of finished product and raw material samples for analysis is provided in industry and regulatory guidance, including APHA (Midura and Bryant, 2001), FDA (Andrews and Hammack, 2003) and Codex Almentarius Commission (2004).

MICROBIOLOGICAL MONITORING OF RAW MATERIALS

The relevance of microorganisms in raw materials is dependent upon the nature of the material, how it is processed and the material's intended use. This will be determined in the HACCP study for the raw material. Where the microbiology of the raw material is important to the finished product microbiology, or where the microbiology of the raw material is correlated to the quality of the material (for example, sensory characteristics reflected in high plate counts), microbiological criteria are established and communicated to the vendor in specifications included with the contractual agreement (Figure 33.1). Such criteria indicate how a given lot of material will perform in analyses when inspected.

Raw material analysis is statistically limited; the presence of a microbial pathogen or an out-of-specification hygienic indicator demonstrates that the lot was non-conforming, but the failure to isolate a pathogen does not necessarily indicate it is absent from the lot. As a result, raw material testing is most effective when it is part of an overall supplier management program that includes other verification activities, such as on-site audits, supplier certification, evaluation of supplier performance and other inspection (such as sensory evaluation) of incoming material.

When raw material monitoring is conducted for more than one operation, the program design will be based upon the most conservative use of the material. For example, a milk powder lot intended for a dry-mix operation where it will receive no microbiocidal control measure and for a wet-mix operation for a product that will be pasteurized will be analyzed by the manufacturer according to the sampling plan and risk level of the dry-mix operation.

Establishment of Microbiological Specifications for Raw Material

Microbiological specifications for raw materials are only established when there is a specific need relative to the use of the material. It is important that specification limits established are technically attainable by the supplier through the application of HACCP and good hygienic practice. This is determined through an understanding of the nature of the raw material and how it is processed. Unrealistic specifications can lead to the use of a material that is unsuitable for its intended use even if the supplier has agreed to the specification, or to the rejection of a raw material that by its design could not meet the specification limits.

Quantitative limits in specifications may be derived from industry guidance or regulatory standards. In the absence of such standards they are based upon an analysis of the raw material over time and from a number of operations, during normal production. Such limits must also be consistent with the expectations for finished products (as expressed in finished product specifications) and the contribution that the raw material has on the microbiological status of the finished product. Specification development should also consider those already established by the supplier; however, supplier specifications often include parameters that are not relevant to the use of the material, or do not include parameters or limits relevant to the customer need. If a raw material cannot meet expectation due to the method of manufacture of the material, it is not fit for purpose and a new material that can meet requirements should be sourced or the finished product redesigned.

The stringency of microbiological specifications is based upon the risk of the material and consequences of loss of control, and on the level of confidence needed to ensure that the raw material meets microbiological requirements.

Specifications should follow a standardized format, such as that outlined by the ICMSF (2002, 2011). Raw material specifications should be reviewed on an established frequency (e.g. annually) for relevance.

Design of a Raw Material Testing Program

The scope, frequency and location of testing are determined by the raw material risk and vendor performance. Material risk is a function of the likelihood of microbial hazards inherent the materials to be present, the severity of the hazards, and how the material is used. For example, a lower risk and thus a lower sampling frequency may be assigned to a material that has robust controls, that is to be used for a product that will be cooked by the consumer, or is from a supplier with a good history of performance. A higher risk may be assigned to a raw

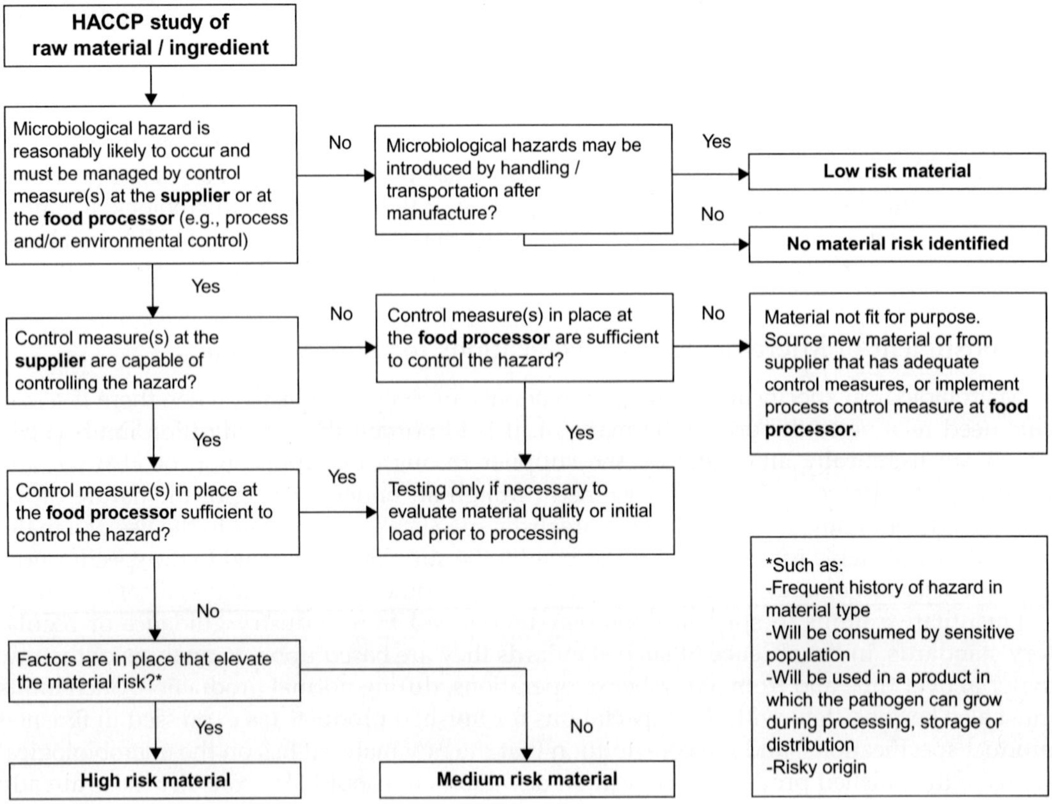

FIGURE 33.3 Example of a decision tree for categorizing raw material risk to determine verification activities. Corresponding verification testing programs are outlined in Table 33.7.

material to be used in a product without the application of a microbiocidal process by the customer that will be used by a sensitive consumer, for a material with frequent history of failure, for a material from a new supplier, or from a supplier with poor or marginal performance. An example of a decision tree to support the classification of materials according to factors affecting raw material risk is provided in Figure 33.3. The risk level may determine the stringency of criteria, frequency of testing or whether a certificate of analysis (COA) will be accepted in place of testing upon receipt by the customer. An example of a raw material verification program that is adapted to raw material risk and supplier confidence is included in Table 33.7.

Verification of the conformance of a raw material lot to specification may be conducted by the supplier and communicated in a COA, indicating through analytical testing the conformance of the specific lot to be purchased. Because testing for COA is conducted by the supplier, often at a supplier's own laboratory, customers requiring COAs from their suppliers often conduct periodic (e.g. quarterly, biannual) testing of incoming material to verify conformance of the lot, and of the COA provided by the supplier to specification requirements. Verification testing may also be conducted by the customer as pre-shipment (i.e. before the lot has left the supplier) or upon receipt at the customer site. In the latter case, the

TABLE 33.7 Example of a Raw Material Verification Program Based Upon Raw Material Risk and Supplier Confidence

Material Risk	Supplier Confidence	COA	Pre-shipment Possible	Testing upon Receipt
High	High	Each lot	Yes	Each lot
	Medium	Each lot	No	Each lot
	Low	Disqualify vendor		
Medium	High	Each lot	Yes	First 15 lots, then quarterly
	Medium	Each lot	Yes	Increase frequency (e.g. monthly)
	Low	Each lot	No	Each lot
Low	High	Quarterly	Yes	First 10 lots, then quarterly
	Medium	Each lot	Yes	Monthly
	Low	Each lot	Yes	Increase frequency

material is blocked and is not used until the results of testing are obtained and evaluated for conformance to specification.

MICROBIOLOGICAL MONITORING OF FINISHED PRODUCTS

Finished product testing may be used to verify the overall effectiveness of a food safety system. Due to statistical limitations finished product testing cannot ensure the conformance of a lot to safety requirements and is not effective as a preventive control; however, finished product testing may be useful to evaluate the conformance of a lot to specified microbiological criteria (regulatory, customer or internal), and verify the overall effectiveness of control measures.

Such testing may be conducted as within-lot or between-lot testing to demonstrate that a lot or production line is under control. Within-lot finished product testing may be conducted periodically or on each lot in response to regulatory or customer requirements. Where such testing is required as part of the contractual agreement with a customer, a COA is usually provided indicating the laboratory results. In some cases regulators may require finished product testing on a periodic frequency. Manufacturers will design control measures and conduct their own testing more frequently to ensure that their system is able to meet regulatory criteria.

The design and use of finished product monitoring is based upon a variety of factors, including:

- Sensitivity of finished product (growth, no growth, application of a lethal process);
- Exposure of product during processing (i.e. assembled, post-lethality exposed vs. in-pack pasteurization or hot fill);

TABLE 33.8 Example of the Frequency of Microbiological Testing of Finished Products in a Verification Program

Level of Hygiene Concern	Finished Product Testing	Notes
Routine/normal	Periodically based upon risk	Periodic evaluation to verify conformance to complete specification. Routine evaluation may be conducted for hygiene indicators, with evaluation against complete specification if threshold is exceeded
Elevated concern	Each production line/week	Evaluation to verify conformance to complete specification, including pathogens, hygiene indicator
High concern	Each lot	Evaluation to verify conformance to complete specification, including pathogens, hygiene indicator

- Performance objective/criteria established for the finished product;
- Results of environmental monitoring or other verification of process environment hygiene;
- Risks associated with raw materials.

In some cases it is practical to routinely examine each lot only for hygiene indicators, such as Enterobacteriaceae, coliforms or total plate count. Products that exceed a threshold on this initial examination are subject to evaluation in a detailed examination to evaluate conformance to complete criteria (including pathogens and indicators as defined in the finished product criteria).

Examples of the application of finished product testing are included in Table 33.1. The necessity and frequency of monitoring may be adapted by the level of concern of hygiene of the product and process. An example of such adjustment is included in Table 33.8.

Development of Microbiological Specifications for Finished Products

Finished product specifications take into account relevant regulatory or customer requirements, the hazards that may be present in raw materials and the environment, the nature of the product and process, and intended use of the material as determined in the HACCP study. Specifications include pathogens of concern as well as relevant indicator organisms, defined sampling plans and methodology. Sampling plans included in specifications should follow ICMSF format, with stringency based upon the severity of the pathogen of concern, the use of the product and the sensitivity of the consumer. Stringency may also be increased for new products or production lines, or where prior history of the product or process lead to a heightened concern. Sampling plan limits for *m* and *M* should be based upon an understanding of the raw materials and processes and ideally the results of testing of products manufactured under good conditions on a variety of production days.

Some regulatory authorities have established "process" criteria, which evaluate the number of positive samples as a proportion of samples collected from an operation over a period of time (ICMSF 2007). The period under evaluation is often a "moving window" of time where new results are assessed relative to a specified number of previous production days. These criteria have been applied to the analysis of pathogens in raw animal products, where control measures can reduce, but may not be able to eliminate, the presence of the pathogen of concern.

ROOT CAUSE ANALYSIS AND CORRECTIVE ACTIONS

The information collected in microbiological monitoring programs is used in conjunction with other verification activities to assess the functioning of process and environmental controls and to determine when adjustments are needed for these control measures.

A positive pathogen or out-of-specification hygiene finding in the environment, raw material or finished product is a significant event and must generate an investigation, including modification of verification activities, a determination of product impact and a root cause analysis to determine what corrective actions are needed (Table 33.6). A positive pathogen result in a product, raw material or product contact sample cannot be negated by additional sampling unless there is confirmed evidence of a sampling or analytical error. Corresponding product will need to be destroyed or reconditioned using a process sufficient to inactivate the level of pathogenic microorganisms present in the material.

A simple conclusion of a "passing contamination" with a cleaning and sanitation event followed by re-examination is not sufficient to ensure that the contamination will not recur. Root cause investigations must include a serious examination of the underlying factors and control-measure failures to ensure that appropriate corrective actions are taken and failures are not repeated. In many cases it is not possible even in an in-depth investigation to make a solid link to a specific root cause. In such situations all relevant factors that may have contributed to the contamination are addressed and monitoring programs continue with heightened stringency until there is confidence that the factors leading to the contamination have been addressed.

References

Andrews, W.H., Hammack, T.S., 2003. Food sampling and preparation of sample homogenate. Chapter 1 in Bacteriological Analytical Manual, eighth ed. United States Food and Drug Administration, Washington, DC. <http://www.fda.gov/Food/ScienceResearch/LaboratoryMethods/BacteriologicalAnalyticalManualBAM/default.htm>.

Chen, Y., Scott, V.N., Freier, T.A., Kuehm, J., Moorman, M., Meyer, J., et al., 2009a. Control of *Salmonella* in low moisture foods. II: hygiene practices to minimize *Salmonella* contamination and growth. Food Prot. Trends 29, 435–445.

Chen, Y., Scott, V.N., Freier, T.A., Kuehm, J., Moorman, M., Meyer, J., et al., 2009b. Control of *Salmonella* in low moisture foods. III: process validation and environmental monitoring. Food Prot. Trends 29, 493–508.

Codex Alimentarius Commission, 1997. Principles for the establishment and application of microbiological criteria for foods. CAC/GL 21-1997. <http://www.codexalimentarius.org/>.

Codex Alimentarius Commission, 2004. General guidelines on sampling. CAC/GL 50-2004. <http://www.codexalimentarius.org/>.

Codex Alimentarius Commission, 2007a. Guidelines on the application of general principles of food hygiene to the control of Listeria monocytogenes in foods. CAC/GL 61-2007. <http://www.codexalimentarius.org/>.

Codex Alimentarius Commission, 2007b. Principles and guidelines for the conduct of microbiological risk management (MRM). CAC/GL63-2007. <http://www.codexalimentarius.org/>.

Codex Alimentarius Commission, 2008a. Code of hygienic practice for powdered formulae for infants and young children. CAC/RCP66-2008. <http://www.codexalimentarius.org/>.

Codex Alimentarius Commission, 2008b. Guidelines for the validation of food safety control measures. CAC/GL 69-2008. <http://www.codexalimentarius.org/>.

Cox, L.J., Keller, N., van Schothorst, M., 1988. The use and misuse of quantitative determinations of Enterobacteriaceae in food microbiology. J. Appl. Bacteriol. (Suppl.), 237S–249S.

Duffy, J.L., Hauser, G., Hutten, H., Mager, K., Masters, K., Meesters, G.M.H., et al., 2003. Hygienic engineering of plants for the processing of dry particulate materials. European Hygienic Engineering Design Group. Document 26. <http://www.ehedg.org>.

Evancho, G.M., Sveum, W.H., Moberg, L.J., Frank, J.F., 2001. Microbiological monitoring of the food processing environment. In: Downes, F.P., Ito, K. (Eds.), Compendium of Methods for the Microbiological Examination of Foods American Public Health Association, Washington DC, pp. 25–35.

Grocery Manufacturers Association, 2012a. Principles for pathogen cross-contamination prevention (zoning preventative control) and zoning verification activities – environmental monitoring. GMA Submission Docket No. FDA-2011-N-0251 20 May 2011 Industry Food Safety Practices: Informing the FDA Rule-Making Process. Grocery Manufacturers Association, Washington, DC.

Grocery Manufacturers Association, 2012b. Product testing for verification of preventative controls. GMA Submission Docket No. FDA-2011-N-0251 20 May 2011 Industry Food Safety Practices: Informing the FDA Rule-making Process. Grocery Manufacturers Association, Washington, DC.

Grocery Manufacturers Association, 2010. Industry Handbook for the Safe Processing of Nuts. Grocery Manufacturers Association, Washington, DC, <http://www.gmaonline.org/downloads/technical-guidance-and-tools/Industry_Handbook_for_Safe_Processing_of_Nuts_1st_Edition_22Feb10.pdf>.

Gorris, L.G.M., 2005. Food safety objective: an integral part of food chain management. Food Control 16, 801–809.

Holah, J., 2005. Improving zoning within food processing plants. In: Lelieveld, H.L.M., Mostert, M.A., Holah, J. (Eds.), Handbook of Hygiene Control in the Food Industry Woodhead Publishing Ltd., Cambridge, England, pp. 148–167.

International Commission on Microbiological Specifications for Foods (ICMSF), 2002. In: chair Tompkin, R.B. (Ed.), Microorganisms in Foods 7. Microbiological Testing in Food Safety Management Springer, New York.

International Commission on Microbiological Specifications for Foods (ICMSF), 2011. In: chair Swanson, K.M.J. (Ed.), Microorganisms in Foods 8. Use of Data for Assessing Process Control and Product Acceptance Springer, New York.

International Life Sciences Institute Europe (ILSI), 2010. Impact of microbial distributions on food safety ILSI Europe Report Series. International Life Sciences Institute Europe, Brussels, Belgium, <http://www.ilsi.org/Europe/Publications/Microbial%20Distribution%202010.pdf>.

Jadhav, S., Bhave, M., Palombo, E.A., 2012. Methods used for the detection and subtyping of *Listeria monocytogenes*. J. Microbiol. Methods 88, 327–341.

Jongenburger, I., Bassett, J., Jackson, T., Zweitering, M.H., Jewell, K., 2012a. Impact of microbial distributions on food safety. I. Factors influencing microbial distributions and modeling aspects. Food Control 26, 601–609.

Jongenburger, I., Bassett, J., Jackson, T., Gorris, L.G.M., Jewell, K., Zweitering, M.H., 2012b. Impact of microbial distributions on food safety. II. Quantifying impacts on public health and sampling. Food Control 26, 546–554.

Kornacki, J.L., Johnson, J.L., 2001. *Enterobacteriaceae*, coliforms and *Escherichia coli* as quality and safety indicators. In: Downes, F.P., Ito, K. (Eds.), Compendium of Methods for the Microbiological Examination of Foods American Public Health Association, Washington, DC, pp. 69–82.

Midura, T.F., Bryant, R.G., 2001. Sampling plans, sample collection, shipment, and preparation for analysis. In: Downes, F.P., Ito, K. (Eds.), Compendium of Methods for the Microbiological Examination of Foods American Public Health Association, Washington, DC, pp. 13–23.

Minstry of Agriculture and Forestry, New Zealand (MAF/NZ), 2011. Guidance for the control of *Listeria monocytogenes* in ready-to-eat foods. Part 2. Good Operating Practice. July 2011 <www.foodsafety.govt.nz/elibrary/industry/good-operating-practices.pdf>.

Moore, G., Griffith, C., Felding, L., 2001. A comparison of traditional and recently developed methods for monitoring surface hygiene within the food industry: a laboratory study. Dairy Food Environ. Sanit. 21, 478–488.

Motarjemi, Y., Moy, G., Risk management: application to biological hazards. Encyclopedia of Food Safety. Elsevier (in press).

National Advisory Committee on Microbiological Criteria for Foods (NACMCF), 2006. Requisite scientific parameters for establishing the equivalence of alternative methods of pasteurization. J. Food Prot. 69, 1190–1216.
National Advisory Committee on Microbiological Criteria for Foods (NACMCF), 2010. Parameters for determining inoculated pack/challenge study protocols. J. Food Prot. 73, 140–202.
National Fisheries Institute and National Food Processors Association (NFI/NFPA), 2002. Listeria monocytogenes Control Manual, Draft 9. National Food Safety Initiative in 2000 project number 00-51110-9768. <http://foodscience.cornell.edu/research/labs/wiedmann/upload/SSWGLMManual.pdf>.
National Research Council (NRC) Subcommittee on Microbiological Criteria, Committee of Food Protection, Food and Nutrition Board, 1985. Executive summary. In: An Evaluation of the Role of Microbiological Criteria for Foods and Food Ingredients, National Academies Press, Washington, DC, pp. 1–14.
National Research Council (NRC) Committee on the Review of the Use of Scientific Criteria and Performance Standards for Safe Food, 2003. Appendix E: international microbiological criteria. In: Scientific Criteria to Ensure Safe Food. National Academies Press, Washington, DC, pp. 317–358.
Powell, S.C., Attwell, R.W., 1997. The use of ATP-bioluminescence as an objective measure of food hygiene standards. Int. J. Environ. Health Res. 7, 47–53.
Scott, V.N., Chen, Y., Freier, T.A., Kuehm, J., Moorman, M., Meyer, J., et al., 2009. Control of *Salmonella* in low moisture foods. I. Minimizing entry of *Salmonella* into a processing facility. Food Prot. Trends 29, 342–353.
Stringer, M., 2004. Food safety objectives – role in microbiological food safety management ILSI Europe Report Series. International Life Sciences Institute Europe, Brussels, Belgium, <http://www.ilsi.org/Europe/Publications/R2004Food_Safe.pdf>.
Swanson, K.M.J., Anderson, J.E., 2000. Industry perspectives on the use of microbial data for hazard analysis and critical control point validation and verification. J. Food Prot. 63, 815–818.
Tompkin, R.B., Scott, V.N., Bernard, D.T., Sveum, W.H., Gombas, K.S., 1999. Guidelines to prevent post-processing contamination from *Listeria monocytogenes*. Dairy Food Environ. Sanit. 19, 551–562.
United States Department of Agriculture, Food Safety and Inspection Service (USDA FSIS), 2012. FSIS Compliance guideline: controlling Listeria monocytogenes in post-lethality exposed ready-to-eat meat and poultry products. Sept. 2012 <http://www.fsis.usda.gov/PDF/Controlling_LM_RTE_guideline_0912.pdf>.
United States Food and Drug Administration (US FDA), 2008. Guidance for industry: control of Listeria monocytogenes in refrigerated or frozen ready-to-eat foods; draft guidance. Feb. 2008. United States Food and Drug Administration, Washington, DC. <http://www.fda.gov/food/guidancecomplianceregulatoryinformation/guidancedocuments/foodprocessinghaccp/ucm073110.htm>.
United States Food and Drug Administration (FDA), 2013. The role of testing as a verification measure in a modern food safety system. In: Current Good Manufacturing Practice and Hazard Analysis and Risk-based Preventive Controls for Human Food. Proposed rule, pp. 3812–3820. Fed Register 78, 3646–3824.
Van Schothorst, M., Zweitering, M.H., Ross, T., Buchanan, R.L., Cole, M.B., 2009. Relating microbiological criteria to food safety objectives and performance objectives. Food Control 20, 967–979.
Whitehead, K.A., Smith, L.A., 2008. The detection of food soils and cells on stainless steel using industrial methods: UV illumination and ATP bioluminescence. Int. J. Food Microbiol. 127, 121–128.
Zwietering, M.H., Stewart, C.M., Whiting, R.C., 2010. Validation of control measures in a food chain using the FSO concept. Food Control 21, 1716–1722.

Further Reading

Grocery Manufacturers Association (GMA), 2009. Control of *Salmonella* in Low-Moisture Foods, February 4, 2009. <http://www.gmaonline.org/downloads/technical-guidance-and-tools/SalmonellaControlGuidance.pdf>.
International Commission on Microbiological Specifications for Foods (ICMSF), 2002. In: chair Tompkin, R.B. (Ed.), Microorganisms in Foods 7. Microbiological Testing in Food Safety Management. Springer, New York.
International Commission on Microbiological Specifications for Foods (ICMSF), 2011. In: chair Swanson, K.M.J. (Ed.), Microorganisms in Foods 8. Use of Data for Assessing Process Control and Product Acceptance. Springer, New York.

CHAPTER

34

Management of Chemical Contaminants

Yasmine Motarjemi

Food Safety Management Consultant, Nyon, Switzerland

OUTLINE

Food Safety Management.
DOI: http://dx.doi.org/10.1016/B978-0-12-381504-0.00034-2

INTRODUCTION

Food can be the source of a broad range of chemical contaminants and residues of agrochemicals. Some may be present naturally, or they may occur as a result of contamination or processing, or they also may be applied by the agriculture or manufacturing industry for their functional properties. Sometimes, chemicals are also added for malicious reasons, e.g. economic fraud, tampering or terrorism.

Thus, considering the plethora of chemical hazards that may be present in food, a risk-based approach for the management of these is usually needed. The HACCP system, a risk-based approach to food safety assurance, was originally developed to manage the safety of microbiological hazards in the food supply. But it is recognized that the principles of the system can also be used for the management of chemical contaminants.

This chapter describes the management of food chemical contaminants, based on HACCP principles. However, it is to be noted that the application of HACCP may not be the only approach. In any case it should ideally be based on (1) a conscious and proactive analysis of potential hazards – in particular those for which there are regulatory limits, (2) the analysis of their risk, based on sound scientific evidence, (3) setting in place effective measures to prevent or control their occurrence within agreed acceptable limits, and (4) verifying that the food safety management system is effective.

NATURE OF CHEMICAL HAZARDS

Chemical hazards[1] can be broadly categorized as follows:

- **Environmental contaminants:** originate from the environment (soil, air, water), either naturally or as a result of anthropogenic activity. They are present in/on the raw material and they enter into the product in this way. Examples are toxic metals (cadmium, lead, mercury, arsenic and aluminum), polychlorinated biphenyls (PCBs), dioxins and radionuclides.
- **Naturally occurring toxins:** are produced naturally by plants, algae, fungi or marine organisms. Examples include: plant toxicants (e.g. solanine in potatoes), mycotoxins (e.g. aflatoxins), marine biotoxins (e.g. saxitoxin responsible for paralytic shellfish poisoning). Although some foodborne pathogens also produce toxins, they are often addressed in the context of microbial food safety management.
- **Processing contaminants:** are undesirable compounds that are formed during the treatment of food as the result of the interaction of its components. Examples are acrylamide, chloropropanols, furan, benzene, ethyl carbamate.
- **Packaging contaminants:** are components of packaging material or ink, which then migrate into the product. Examples are Bisphenol A diglycidyl ether (BADGE), phthalates and epoxidized soybean oil (ESBO). They are sometimes grouped under "surface contact contaminants."

[1]Codex Alimentarius considers food allergens as a chemical hazard.

- **Food additives:** certain food additives, when present in high levels in food, may present a health risk. An example is nitrate. In the scope of this text, only food additives that have an established ADI are considered as "potential hazards."
- **Agrochemicals:** include veterinary drugs and pesticides. Similar to food additives, agrochemicals are considered as a hazard if they occur at levels above regulatory limits or internally safety-based norms.

Additionally, foods may be subject to:

- **Accidental contamination** from various chemical agents used for manufacturing purposes; examples are disinfectants, cleaning agents and lubricants.
- **Adulteration**, e.g. use of unauthorized substances such as unauthorized dyes. This is often practiced for economic reasons.
- **Terrorism or sabotage.** These are often deliberately added to food for malicious reasons.

HEALTH CONSEQUENCES

It is well established that the health consequences of chemical hazards depend on three factors:

1. Nature of the agent.
2. Amount present in the food and the intake of consumers.
3. Vulnerability of consumers.

The health effects vary according to the dose. As Paracelsus (Swiss physician and chemist, 1494–1541) stated, "All things are poison, and nothing is without poison; only the dose permits something not to be poisonous," or, more concisely, "The dose makes the poison." At high doses, chemical hazards can lead to acute or fatal intoxication, or allergic reaction in the case of allergens. Upon long-term exposure at low doses, they can also cause adverse health conditions and be a risk factor for various chronic diseases. For a thorough overview of the health risks associated with chemicals, the reader is referred to Moy and Todd (in press).

FACTORS AFFECTING THE OCCURRENCE OF CHEMICAL HAZARDS

Depending on the nature and the source of chemical hazards, different factors may influence their occurrence in the raw material or during processing. Understanding these factors and their consideration in the hazard analysis is essential for evaluating the likelihood of occurrence and deciding on appropriate control measures and verification activities. Examples of such factors are:

- Agronomical
  - General farm/agricultural practices (e.g. conventional, contract, bio).
  - Disease in animals/plants.
  - Nature of soil.

- Price and availability of agrochemicals (e.g. easy access to unapproved or banned agrochemicals).
- Climatic
  - Climatic fluctuations may stress plants and promote fungal attacks, which increase the risk either of mycotoxins or of abuse when using agrochemicals.
  - Stress caused by drought or excessive rain increase the risk of pre-harvest mold growth. Droughts have also led to feeding cattle with plants not intended as feed, thus contaminating unapproved agrochemicals.
  - Insect infestations also stress/damage plant tissues and increase the risk of mold growth and subsequent mycotoxin formation.
- Environmental
  - Industrial activity and pollution can lead to contamination of soil, atmosphere and water with chemical hazards such as heavy metals or dioxins.
  - Soil may also naturally contain high levels of certain chemical agents, such as heavy metals and POPs (persistent organic pollutants). Mining activities can also increase exposure to toxic metals.
- Suppliers' practices
  - Suppliers' farm or agricultural practices.
  - Manufacturing practices and method of processing.
  - Suppliers' QA system (preventive measures, monitoring measures).
- Legislation
  - Regulatory requirements, i.e. if a country lacks appropriate legislation.
- Enforcement
  - If the authorities are not enforcing and monitoring the implementation of the legislation.

REGULATORY REQUIREMENTS AND CHALLENGES

To protect consumers' health and ensure the safety of the food supply, public health authorities establish maximum limits for various contaminants, maximum levels of use for food additives and maximum residues limits (MRLs) for veterinary drugs and pesticides. Specific migration limits are also established for various packaging contaminants.

One of the fundamental considerations in setting standards for chemicals is the health effects of chemicals, from the perspective of both short-term and long-term exposure.

At the national level, regulatory standards, or norms, are generally established based on the consideration of the health risk associated with a given chemical, but also taking into account other factors such as feasibility to comply, nutritional needs and the diet of the population. Therefore, regulatory standards are often a trade-off between the health risk of a chemical and what is achievable and appropriate for society. As such, it is a risk management decision. Nevertheless, regulatory standards established for a chemical hazard are viewed by society as a *food safety standard* and industry has the obligation to abide by these standards. With respect to chemical hazards, exceeding these standards must be seen as a violation of food safety.

At the international level, the World Health Organization (WHO) and the Food and Agriculture Organization of the United Nations (FAO) sponsor the Joint Expert Committee on Food Additives and Contaminants (JECFA) and the Joint Meeting on Pesticide Residues (JMPR), which carry out risk assessment of chemical contaminants, food additives, veterinary drugs and pesticides, respectively. Based on these risks assessments, the FAO/WHO Codex Alimentarius Commission (CAC) establishes international standards for food.

Since the establishment of the World Trade Organization (WTO) and the coming into force of the Agreement on the Sanitary and Phytosanitary Measures in 1995, the work of CAC, e.g. its standards, have become the international reference for food safety. This means that products that comply with Codex standards cannot be rejected on food safety grounds by the WTO member states unless the importing country provides scientific evidence (based on risk assessment) that the product in question is not appropriate for its population.

REGULATORY COMPLIANCE

The food industry has the obligation to comply with all the laws and regulations of the country in which they market the food. Considering that some countries may have different standards, multinational companies will have to produce foods of different standards. This raises the issue of double standards, or even that of dumping of foods of higher level of contamination in countries with lower standards. For ethical reasons, it is thus recommended that multinational companies meet the Codex Alimentarius standards as a minimum; it must be noted that the Codex standards are today recognized as the internationally agreed requirements for food safety.

Where national or international requirements are not established, the food industry still has the obligation to produce safe food; thus in some cases, internal norms may be needed on the basis of the *due diligence principle.*

MANAGEMENT OF CHEMICALS IN INDUSTRY

Prerequisites

HACCP is applied in conjunction with a number of supporting measures, which are generally referred to as prerequisites. The term is used to emphasize that the HACCP system is not a stand-alone system and that its successful implementation depends on a number of programs embedded in the food safety management system such as (1) good agricultural practice (GAP), good animal husbandry practice (GAHP) or good manufacturing practice (GMP), (2) supplier or vendor confidence level and (3) personnel training (including managers, supervisors, shop floor operators, technicians and laboratory personnel).

As most of the chemical contaminants in products come with the raw material and, once present, generally cannot be removed, supplier/vendor management is a key prerequisite in food businesses. Due to its importance, some key guidance is provided below.

Additionally, there are other "measures" or requirements which are not usually referred to as prerequisites, but which in practice are the *conditio sine qua nons* for the management of chemical hazards. Therefore, they deserve to be mentioned here. These are:

- Scientific knowledge (e.g. understanding the mechanism of formation of processing contaminants, conditions for growth of molds, impact of control measures, etc.).
- Legislation (e.g. norms, codes of practices) and enforcement.

Where these measures are not in place, the likelihood of a contaminant being present or occurring is higher. Therefore, before conducting a hazard analysis, the implementation of the above needs to be evaluated, and in case of gaps, their application needs to be improved in the first place. In the interim, the risks which may ensue from the gaps in prerequisite programs need to be considered in the hazard analysis, and the chemical in question must be considered as potentially significant. For instance, during import of a raw material from a region or country where the legislation related to the use of veterinary drugs or pesticides is not established or enforced, the likelihood of the presence of unauthorized residues in the commodity, marketed in the selling country above safe or regulatory levels, must be considered likely.

Supplier Management

Considering that many chemical hazards are introduced into products through the raw material, the importance of supplier management cannot be overemphasized.

Supplier management starts by selecting the supplier. However, before doing this, there is a need to understand the suppliers' expectations and whether they will be capable of producing the material according to specifications. Therefore, the process starts with understanding the requirements, the quality and safety objectives and formulating the specifications.

Specifications

A specification is a description of a material's properties and values (e.g. physical, chemical, sensorial, microbiological, as well as transportation and storage requirements). One may differentiate between purchasing specifications and finished product specifications.

Purchasing specifications is an important instrument to convey to suppliers the requirements in terms of food safety and quality. As such, chemical contaminants that are likely to be present in the raw material at an unacceptable level must be prescribed.

The requirements to be mentioned in the purchasing specifications must follow the hazard analysis during the HACCP study, taking into account the conditions of production or manufacturing of the raw or packaging material. In preparing the specifications, consultation of the supplier is recommended since the supplier will have specific expertise on the subject. The regulatory requirements of the country where the product is manufactured and/or sold are also important when establishing the specification.

For unauthorized compounds, the specification must indicate "absence." The minimum performance criterion of the analytical method[2] expressed as limit of detection (LoD) and/or

[2] Also referred to as the **M**inimum **R**equired **P**erformance **L**imit (MRPL) in the EU legislation.

limit of quantification (LoQ) has to be given in this case. The units of the limits should be expressed according to SI norms (e.g. mg/kg, ng/l).

Finished product specification relates to norms for the final product and is especially important for certain types of contaminants and products, such as products that constitute an important part of the diet. The finished product specification must conform to the regulatory requirements of the country where the product is sold and/or with the CAC norms, whichever is stricter. The finished product specification represents the final consolidation of all the requirements, be they regulatory, safety or quality related, and it is the key document for compliance verification.

Selection of the Supplier

When selecting and approving a supplier, consideration must be given to the supplier's ability to meet the purchasing specifications, in particular the supplier's:

- Awareness of chemical hazards associated with their products.
- Consideration of regulatory requirements in their HACCP studies.
- Raw material and management of their own supply chain.
- Traceability.
- Implementation of control measures at the CCPs.
- Practices with regard to the processing and storage of raw material and semi-finished or finished products.
- Monitoring activities and records.
- Training program for personnel as well as suppliers' laboratory capabilities and performance.

Certificate of Analysis (CoA)

As a confirmation of the suppliers' compliance with the requirements, a CoA may be required. The CoA is to be viewed as a verification of control measures at the suppliers' level. It is thus a complement to internal monitoring. However, care must be taken that the CoA is provided by a competent accredited or approved laboratory. In absence of accreditation, a periodic independent or in-house verification of the CoA is necessary.

Alternatively, suppliers may provide a certificate of compliance (CoC). This is different from a CoA. It is basically a certificate stating that the material complies with the requirements, including compliance with the regulatory requirements or recognized international standards. It is to be noted that a CoC is not based on the analytical results, and that its validity depends on the measures that the supplier puts in place to meet the set requirements. This has to be verified during audits of suppliers.

In the delivery of the certificate, the following conditions must be respected:

- The certificate must refer to an actual analysis of the lot being delivered, not to an average monthly sample, or to a previously analyzed lot. It must cover all the parameters agreed with the supplier.
- The sampling method and sampling plan must be mutually agreed upon.
- The laboratory carrying out the analysis must be clearly identifiable on the certificate.
- The report must identify the analytical methods used.
- The accuracy of results for chemical parameters must be verified periodically.

Analytical Aspects

Besides the analytical performance of a test method that is used to analyze a specific chemical hazard, the manner of reporting test results may also have an impact on the comparability and validity of analytical data. The following principles that need to be considered in the reporting of *quantitative* test results are as follows:

- **Form of the chemical hazard and unit of measurement:** A test result should be reported in the same form (active – chemical form) and with the same unit of measurement as that given in the specified requirements (e.g. local regulatory limit, Codex Alimentarius).
- **Number of significant figures:** If the requirement provides clear guidance, the same number of significant figures should be reported. Otherwise, the test result should be expressed with one significant figure more than the limit stated in the requirement. In addition, the number of significant figures depends on the uncertainty of the analytical method.
- **Correction for recovery:** Generally test results are not corrected for recovery. They may be corrected if the relative recovery is significantly different from 100% (typically <70% with good precision). In the latter case, both the measured and corrected value should be given, as well as the basis for correction. The recovery of a specific chemical hazard may vary, depending on the sample matrix.
- **Reporting limits:** Reporting limits are the LoD, which is key for banned or unauthorized chemical compounds, and the LoQ. As for the recovery, the LoD and LoQ of a specific analytical test method may vary depending on the sample matrix.
- **Uncertainty of measurement:** In accordance with the standard ISO/IEC 17025:2005, a statement on the estimated measurement of uncertainty (MU) should be included in test reports when:
  - it is relevant to the validity or application of the test results;
  - a customer's instructions so require; or
  - the uncertainty affects compliance with a specification limit.
- The final uncertainty is expressed as the interval (measured value ± expanded MU), at a 95% confidence level.
- **Uncertainty factors:** However, some factors may contribute to discrepancies in the analytical results, and should be considered in further investigation in case of a non-compliance:
  - Heterogeneity of the product batch/lot.
  - Different sampling procedures for analytical testing.
  - Different analytical testing procedures (including sample preparation, analytical method, quantification procedure, quality controls) with different performance characteristics (e.g. detection limit, measurement of uncertainty).
  - Different "rules" to assess the regulatory compliance (e.g. correction for recovery, taking into account the measurement of uncertainty).

It is important to understand the principles that regulatory authorities apply to interpret analytical test results and how they assess the compliance of a product against a requirement. The application of different principles for treating data may affect the conclusion regarding the compliance or non-compliance of a product with a requirement.

Many food business operators make use of external laboratories, or rely on the laboratories operated by suppliers or co-manufacturers. Governments also have their own control laboratories and may verify the compliance of products independently.

In order to be able to rely on the results of tests, it is best to refer to ISO accredited laboratories.

APPLICATION OF THE HACCP SYSTEM TO MANAGEMENT OF CHEMICALS

Identification of Hazards

A first step in the management of chemical contaminants consists in identifying potential hazards associated with the product and manufacturing process. The source of many chemical hazards is the raw commodity or packaging material itself. Most chemical hazards present at source, i.e. raw material, will not be eliminated through processing. Some chemical hazards may also be formed during processing or storage.

To identify potential chemical hazards, expertise is needed; hence the importance of integrating an expert on the subject into the HACCP team. As a complement, or in absence of an expert, the following sources of information can be consulted.

- Regulatory requirements (considering the requirements of the country where the product is to be sold). As products need to comply with the regulatory requirements, the contaminants that should be examined are those for which regulatory authorities have established some guidance or regulatory requirements.
- Scientific literature can provide information on the type of hazards which are associated with food, and their level of occurrence.
- Governmental and industry associations guidance material such as fact sheets, websites. The guidance provided by the International Life Science Institute (ILSI) or the Global Harmonization Database can be a source of such information.
- Reports of surveillance of governments or industry, be it monitoring of chemical contaminants in food and environment or reports of inspection of food control capabilities showing potential weaknesses in control, monitoring or analytical capabilities. A major food recall that occurred in 2001 in Europe in relation to chloramphenicol in honey could perhaps have been anticipated if the report of EU inspectors, showing lack of monitoring and of governmental laboratory capabilities in China for enforcing legislation on veterinary drugs, had been shared with industry associations.
- Portals such as the RASFF (Rapid Alert System for Food and Feed). The accessible RASFF portal database at http://ec.europa.eu//rasff enables a search for RASFF notifications on food and feed of interest or a particular hazard.

Certain chemical hazards, processing and packaging contaminants in particular, are best addressed in the design of the product. Therefore, their prevention and control must be considered during the early stage of product development and reflected in a preliminary hazard analysis using, e.g., the "Safety by Design" approach.

Analysis of the Hazards

For hazards that are identified, a decision should be taken on their degree of risk. Those which are viewed as high risk in the HACCP study are referred to as a "significant hazard." To identify which chemical hazard is significant, the following factors need to be considered:

- The likelihood of occurrence of the hazard above safety/regulatory limits. This may be estimated taking into consideration the factors influencing the occurrence of a hazard (see section on factors influencing the occurrence of hazards above) and the prerequisite programs in place. Data confirming the proper implementation of prerequisite programs must be available. Examples are audit reports of the supplier or manufacturing site and historical records such as monitoring data of the supplier.
- The severity of health consequences of the agent, taking into consideration the target consumer, the nature and the level of the chemical potentially present.

If a regulatory limit or an industry limit is not available, the decision on the significance of a hazard could be based on food safety assessment. To this end, two types of data are required:

- A reference dose: this is the dose below which exposure to that chemical can be considered as safe (e.g. ADI, TDI, PTWI).
- An estimate of exposure[3] based on food consumption data.

The degree of significance of a hazard can then be estimated by comparing the level of exposure to the particular chemical agent through a given food with the ADI or other equivalent reference dose (TDI, PTWI). If the exposure does not represent a significant proportion of the safe reference dose, the agent is not viewed as a significant food safety concern (e.g. ratio of estimated intake to TDI or ADI <1). In other words, the significance of the hazard can be evaluated based on the degree of contribution that it makes to the total exposure of the target consumer. In case that degree is negligible, the hazard is considered not to be a major food safety concern.

To calculate the level of exposure, the worst-case scenario must be considered, i.e. using the maximum consumption of the product and the maximum amount of chemical that may occur in the particular food, and based on historical records or other surveys.

Control of Hazards

Except for hazards that may occur as a result of processing or storage, for a great proportion of chemical hazards, the *control measures* are at the supplier level, i.e. the application of GAP, GAH, GMP.[4] For packaging contaminants, the design and formulation of the material as well as the application of specific GMP measures at the supplier level are the control measures. Thus, sourcing the raw material from reliable and approved suppliers is essential for preventing these types of chemical hazards. Therefore, the customer of a raw material

[3]Estimate of exposure (mg/kg bw/day) = [maximum level of agent in the finished product (mg/g) × maximum amount of food consumed (g/day)]/bodyweight/kg).
[4]In ISO 22000 these control measures are referred to as "prerequisites."

must clearly communicate its requirements (including the intended use of the raw material) to the suppliers. Purchasing specifications is an important tool for this communication.

Testing the raw materials at reception is in principle a verification activity since it confirms the suppliers' quality assurance program and compliance with the agreed specifications. However, in situations where the confidence level is low, it can be considered as a control measure, provided that it is carried out systematically on all lots of incoming materials, using a validated sampling plan. Results of the analysis will then be part of the release procedure.

For some hazards, selecting resistant varieties of raw materials can be considered as one method to control a hazard, in which case the specific variety desired must be mentioned in the raw material purchasing specification.

For processing contaminants such as acrylamide, the design and control of process parameters or the formulation of the product may constitute the main control measure. For certain types of mycotoxins, the control of storage conditions (storage time, temperature, humidity) of raw materials is the key control measure.

For lubricants, food grade quality and good maintenance practices must be considered as key control measures (this is often done as part of GMP).

Preventing accidental or cross-contamination with chemicals requires good warehouse management, e.g. separation of cleaning chemicals from food items, proper closing and labeling of chemicals, dedicated recipients, etc.

The control of hazards must at all times ensure that chemical hazards are prevented, eliminated or reduced to an *acceptable level*.

Critical Limits

The second principle of HACCP is the decision on the critical limit. This is the limit which separates the acceptability from the unacceptability of a control parameter. These limits have to be established based on the parameters that characterize a control measure. For instance, if for the application of antibiotics, the control measure has to take account of a withdrawal period, the monitoring parameter is time, and the critical limit is the number of days required for the residues of antibiotics to decrease to an acceptable level, e.g. 7 days. However, for a raw material where there is low supplier confidence or the supplier is not known, and the testing of the raw material is considered as a means for controlling the hazards, the critical limit is the regulatory standard of the country where the raw material is to be used, or preferably the Codex norms if these are more stringent. For intermarket supplies, attention must also be paid to ensuring that the finished product meets the regulatory limits of the market where the product is sold and/or the Codex norm. If this requires a more stringent norm for the raw material than the regulatory requirement of the country where the product is manufactured, then this should be stated in the requirements communicated to the supplier.

Chemicals used by producers (e.g. agrochemicals) or by food manufacturers (e.g. food additives) should not be used in food production and manufacturing if they have not been evaluated and have not been proven safe for use.

For unapproved or prohibited chemicals, some governments may apply the concept of *zero tolerance*. This concept is based on the idea that if an agent is prohibited, its mere presence at any level is an indication of violation of the legislation. However, many governments can be tolerant if the industry can demonstrate that the presence of the agent in the product was inadvertent, a due diligence measure was taken to prevent it, the contaminant can have other sources, e.g. environment as was the case with melamine, the level in the product is so minute that it does not present a harm to consumers and corrective measures are taken to prevent it in future. Under such circumstances the governments may allow the product on the market. In other cases, the governments may confiscate the food and punish the producer, but destruction of such a food considered safe would not be wise. At very low levels of contamination, there may be conflicts among the stakeholders of the presence or absence of such a substance and regulatory authorities may determine minimum required performance limits (MRPLs) of the analytical method for such substances.

For processing contaminants, the critical limits correspond to the acceptable limits of the processing parameter(s), e.g. temperature of the heat treatment. Similarly, for contaminants associated with storage, the critical limits will be the acceptable limits of storage parameters (temperature and/or humidity).

CCP Monitoring

Where the raw material is considered as a CCP, the chemical hazard must be tested on every batch and the results of testing must be made a release criterion. Correct and valid sampling is essential. Where the distribution of the hazard is heterogeneous (e.g. mycotoxins) and the raw material is considered as a CCP, the validation of the sampling method is particularly critical for food safety and must be considered as compulsory.

Similarly, in line with HACCP principles, any processing or storage step which is identified as a CCP must be monitored; the parameters and frequency of monitoring must be set so that if the critical limit is violated, corrective actions can be applied in a timely manner. For chemical agents, re-processing is generally not applicable and an infringement of the acceptable level of the agent should lead to the rejection of the raw material or product.

Personnel entrusted with the management of CCPs must be well trained, be aware of their responsibility and must completely understand the consequences of an eventual failure of the CCP.

CP Monitoring and Other Verification

In addition to CCP monitoring mentioned above, depending on the level of risk, a verification procedure must be established to confirm that control measures (preventive measures) are adequately implemented and the HACCP system is effective. For chemical hazards, verification includes activities such as:

- Audit of the supplier.
- Factory audit.
- Verification of identity of the raw material upon receipt in the factory (e.g. visual inspection) to confirm that the right variety is selected.
- Monitoring of the raw material for potential hazards according to the degree of risk as well as other factors (see "Monitoring Plans (see next section page 931),").
- Testing of the finished products.

Monitoring Plans

Frequency of Monitoring

Whether monitoring is implemented at a CCP for controlling a hazard or as a verification measure to verify that the control measures are applied correctly, the frequency of monitoring needs to be decided. In line with what has been mentioned above, this should be decided on a risk-based approach. However, frequently health risk is not a sufficient criterion since a one-time non-compliance may not present a significant health risk for the consumer or may even present no risk, but may jeopardize the reputation of the business or present economic risk in case of violation of regulations and product recall. Therefore, a two-step decision-making process is proposed here.

In a first step, the frequency of monitoring is decided taking into account:

- The likelihood of occurrence of the contaminant above acceptable levels in the raw material, e.g. taking into consideration the prerequisite conditions (e.g. availability of certificate of analysis).
- Health consequences for target consumers in case of non-compliance and the anticipated level of the contaminants in the final product.

In line with the decision tree (Figure 34.1), depending on the level of risk, the raw material is considered as a CCP or Control Point (CP)[5]. If the risk is viewed as negligible, it may

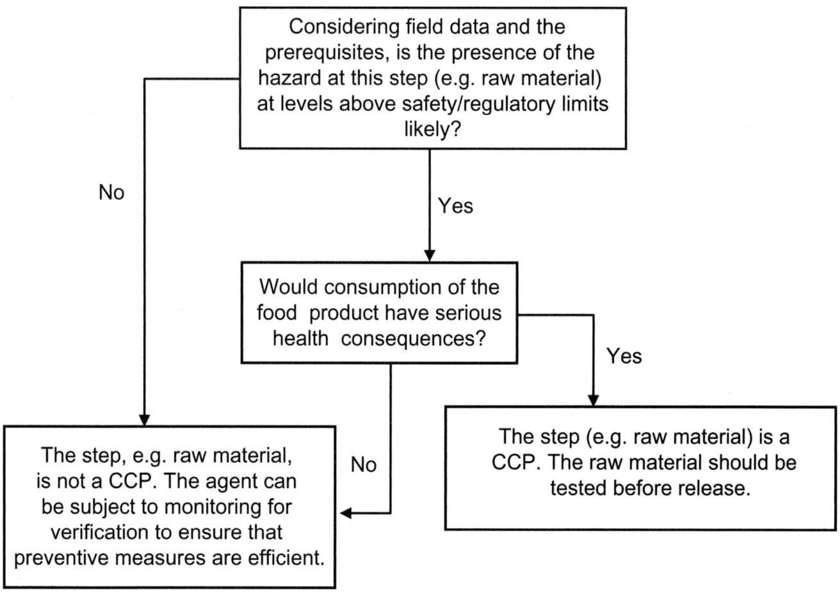

FIGURE 34.1 Simplified decision tree for hazard analysis of chemicals.

[5]Control point is a step in the operation where control measures are applied but the step is not considered as a CCP. Nevertheless, the step is checked or surveyed for verification purposes.

Health consequences

Likelihood of exceeding acceptable levels

	Minor Minor adverse health impact upon long-term exposure	Moderate Mild illness or health consequences upon long-term exposure	Major Serious illness	Catastrophic Death
Very likely			CCP monitoring I.e. 100% of lots and positive release	
Likely				
Moderately likely				
Rare		20% testing of deliveries or 2 times per year	50% testing of deliveries or 4 times per year	
Unlikely	**No testing**	10% testing of deliveries Or once per year		

FIGURE 34.2 A schematic presentation of risk-based monitoring.

also be decided that no monitoring is required for food safety reasons. The monitoring frequency decided at this step should in principle set the minimum requirement for testing (Figure 34.2).

In a second step, the frequency of monitoring can be readjusted taking into consideration "other factors and requirements"; examples are:

- Feasibility (availability of analytical method, speed of results).
- Regulatory requirements and/or public perception.
- Economic reasons: amount of raw material and/or product processes and implications in case of a product recall.
- Impact on the image and reputation.

Principles in Setting a Monitoring Plan

In setting up the monitoring plan, the following principles must be considered:

- Time of testing must be adapted to the period associated with the highest risk of contamination (e.g. monitoring of aflatoxin in cereals or nitrate in vegetables may require higher frequency in specific seasons or climatic conditions).
- Samples must be taken at the arrival of the raw material.
- A certificate of analysis can be required in place of, or in combination with, verification monitoring. However, the certificate of analysis must be obtained from an accredited or approved laboratory and be periodically verified internally.

- Except for contaminants that occur during processing and manufacturing, most of the efforts must be on the raw material. However, periodic testing of the finished products is also advised in the following cases:
  - To verify that the HACCP plan is effectively implemented. This is important for significant hazards.
  - To confirm compliance with the legislation.[6]
  - To verify that risks associated with processing or storage are under control.
  - If a certificate of analysis is required by customers.[7]
  - If a certificate of analysis is required for exportation.[8]

Corrective Action

In case of non-compliance, the following actions need to be considered:

- Non-complying raw materials must be rejected and suppliers must be advised. Depending on the situation and the gravity of the non-compliance, the need for increasing the frequency of the verification activities (i.e. frequency of auditing and/or monitoring materials for the potential hazards) or alternatively terminating the contract must be considered. Fraudulent practices must lead to an immediate termination of the contract.
- Non-complying finished products must be blocked.
- Any deviation must be immediately investigated and followed up.
- For chemical hazards occurring as a result of processing or storage conditions, a deviation from set standards must lead to a product reformulation or a change in processing or storage conditions.

Validation

The principles of the HACCP system as well as of the Codex Alimentarius Commission require validation of control measures. Validation consists in obtaining evidence that the elements of the HACCP system are effective. As such, all the decisions relating to the different principles of HACCP need to be validated to ensure that they have a scientific and/or technical basis, and/or are based on accepted practices. These include consideration of the need for validating:

- Hazards which are considered as non-significant and efficiency of control measures (operational prerequisites), e.g. suppliers' practices, monitoring and/or CoA.
- Limits and/or specifications.
- The sampling scheme and procedures.
- The analytical method (e.g. equipment, variability, sensitivity, approved and recognized method).

[6]Provided that there is scientific evidence that the risk is associated with the product or related raw materials.

[7]Idem as 5.

[8]Idem as 5.

- Frequency of monitoring activities.
- Training and competence of personnel, from operators to laboratory (e.g. accreditation).

Necessary data, records and documentation providing the basis for decision-making must be available.

Maintenance of the HACCP Plan

Results of monitoring and other verification activities (e.g. audit of supplier, preventive maintenance) as well as previous records of consumer complaints and or accidents must be the subject of a continuous review. Hazard analysis and decisions on CCPs and/or frequency of monitoring and other verification activities must be re-evaluated and the HACCP and monitoring plan must be updated in the light of these data. Examples of technical or scientific data that should prompt an update of the plan are:

- Alerts (internal or external).
- Surveys by authorities or national food institutes.
- Reports or data on previous incidents or non-compliance.
- New scientific developments (e.g. emergence of new potential hazards).
- New or change in the regulatory requirements.
- Change of supplier or suppliers' practices.
- Change in the country where the product is marketed.
- Change in the intended use, preparation method or target consumer.
- Change in the product formulation or process/storage conditions; change in factors influencing the occurrence of a hazard, such as environmental contamination or climatic changes. The latter may for instance increase the risk of mold growth and formation of mycotoxins, which in turn can lead to abuse of fungicides. Climatic changes may also increase animal infections, leading to a higher use of antibiotics.

While the maintenance of the HACCP plan must be a continuous practice, it is a good practice to periodically review collected data and their trend analysis. Types of data that should be considered during such a review are:

- Results of in-house monitoring, including out-of-norm results.
- Survey or monitoring carried out by authorities (or planned to be carried out) or third parties.
- Verification of certificates of analysis.
- Performance of suppliers (audit reports, supplier's monitoring plan) and future audit plans.
- Information on emerging chemicals.
- Reports on laboratory competences.

The results of this review must lead to an analysis of trends and decisions for:

- Enhancing preventive measures.
- Readjusting the frequency of monitoring.
- Setting up new monitoring activities or surveys.
- Communication to regulatory authorities regarding the feasibility of the legislation.
- Management of suppliers (request for audits, change in the frequency of monitoring or issuing warnings).

In case of any report by regulatory authorities or other third parties (e.g. customer, consumer organization) of non-complying products, a transparent and speedy reaction is important to maintain credibility and the confidence of authorities. The following action is recommended:

- Handling of non-compliance and corrective actions – all non-compliances must be the subject of immediate follow-up action and must lead to the adjustment of the food safety management system, including monitoring activities. Handling of non-compliance and incidents is explained in Chapter 40.

Further Reading

Motarjemi, Y., Stadler, R., Studer, A., Damiano, V., 2009. Application of the HACCP approach for the management of processing contaminants. In: Stadler, R.H., Lineback, D. (Eds.), Process Induced Food Toxicants and Health Risks. John Wiley & Sons Inc., New Jersey.

Moy, G., Todd, E. Overview of foodborne illness and public heath aspects: Part 2. Chemical, physical and other hazards. Encyclopedia of Food Safety. Elsevier (in press).

In case of any [illegible]

[illegible]

[illegible]

- [illegible] non-compliance [illegible] must be the subject to [illegible] must result in the [illegible] system [illegible] and [illegible] is explained in Chapter 20.

Further Reading

[illegible]

[illegible]

CHAPTER

35

Food Defense

Ned Mitenius[1], Shaun P. Kennedy[2] and Frank F. Busta[2]

[1]Periscope Consulting, Grapevine, TX, USA, [2]National Center for Food Protection and Defense, University of Minnesota, St. Paul, MN, USA

OUTLINE

Food Safety Management.
DOI: http://dx.doi.org/10.1016/B978-0-12-381504-0.00035-4

DESCRIPTION OF ISSUES

Food defense, as used in this chapter, concerns the *intentional* contamination of the food supply. While governments have begun to take notice of this risk only in recent years, people have been poisoning food since before the pyramids were built.

What is different today is the scale of the impact and the scope of the risk (Tumin, 2009). Modern agricultural methods, large-scale food manufacturing and efficient logistics turn what might only have been a local problem into an international crisis. A single farming region may produce a commodity for much of a nation. A single factory may manufacture food that is distributed across a continent, or produce an ingredient shipped around the world (Cavallaro et al., 2011).

At the same time, terrorist attacks that reach across the world, increased global political unrest and unprecedented global distribution of food magnify the risk. No longer must we be concerned with only personal enemies, but we must also be concerned with extremist political factions with whom we have no contact. Another potential risk is with trading partners half a world away.

DEFINITIONS FOR FOOD DEFENSE AND RELATED TERMS

Some working definitions are appropriate to frame the discussions in this chapter. As a relatively new area of focus within governments and food producers, the terms used to describe protections from *intentional contamination* are evolving. The language may change further as global partners continue to collaborate.

Food Safety

Food safety has its focus on reducing the risk of unintentional contamination in the food supply, be it natural, accidental, a result of negligence or violation of food safety principles due to technical ignorance.

Food safety is a well-established area of effort that has been around for decades. Universities worldwide have well-regarded food safety programs. Most governments have agencies dedicated to food safety. Producers increasingly have food safety management programs based on global standards. Cold-chain practices, risk analysis methods such as operation risk method (ORM) and hazard analysis critical control point (HACCP) are well understood (and highlighted prominently in reference materials such as this one).

Food safety frequently focuses on a relatively short list of well-studied pathogens, such as pathogenic strains of *Listeria, E. coli* and *Salmonella,* along with chemical contaminants and physical hazards like wood or metal fragments. Some of these contaminants are inherently found in the environment or production systems. Inspection and detection methods exist for these agents. Many food processing systems include treatments (oxidizing or heat treatments) or compositions (low water activity, pH) that inactivate pathogens.

Key points: Accidental contamination, well-established best practices, known agents.

Food Security

Food security has its focus on the availability of nutrition for a population.

Despite the large-scale agriculture, industrial food manufacturing and global distribution of the global food supply, there is still a great deal of hunger in the world. The Food and Agriculture Organization of the United Nations (FAO) estimates "that a total of 925 million people are undernourished in 2010" with "developing countries account[ing] for 98 percent of the world's undernourished people." Many populations (nations, regions or people) have an insufficient supply of food, and they are "Food Insecure." Populations with sufficient supplies of food have "Food Security." This problem is projected to increase due to population growth, climate change and related factors (FAO, 2010).

Note: Some documents may use the term *food security* to describe measures to reduce the risk of intentional contamination. If you are unsure, you should try to clarify the local context of the term.

Key points: Sufficient food for a population.

Food Defense

Food defense has its focus on the prevention of the *intentional* contamination of the food supply.

Many agents can be used for intentional contamination. In addition to the agents normally identified with food safety, these can include other chemical, biological, physical or even radiological agents. Many potential agents are highly toxic and are not prevented or inactivated by conventional food safety interventions. Most of these potential agents are difficult to detect, or at least difficult to detect when in a variety of foods.

The motivations for intentional contamination can be as varied as the agents. They can range from local grievances to economic advantage to political disruption to mass casualties. The capabilities of perpetrating an intentional contamination also have a broad range, from the limited effort of an individual to the significant capabilities of a well-organized international terrorist cell.

Key points: Intentional contamination, unknown agents, limited detection methods, motivations and capabilities vary with the perpetrator.

Food Protection

Food protection is an "umbrella term" to encompass both food safety and food defense activities.

This pays homage to the fact that it is often the food safety professionals that become responsible for food defense. This reflects various similarities and overlaps in the work of food safety and food defense. For example, both food safety and food defense benefit from control and inspection of incoming materials and from a robust recall and recovery program. Note that in terms of regulatory authorities there are different organizations involved, e.g. food control agencies for food safety versus police or homeland security agencies for intentional contamination.

Note: There are also several fundamental differences between food safety and food defense because of the intentional element. Malicious intent and intelligent adversaries must be considered in food defense. A change in thinking is required when food safety professionals also manage food defense measures. They must begin to "think like the bad guy."

Key points: Emphasizes the similarities with food safety, may diminish the differences.

Bio-terrorism, Agro-terrorism and Bio-defense

Bio-terrorism, agro-terrorism and bio-defense focus on biological attacks of several kinds. These can be attacks against crops or livestock as well as the food we eat.

The inclusion of the word "terrorism" describes crimes that are politically motivated. Food defense is concerned with intentional contamination of food that is politically motivated and motivations that are economic or revenge based.

While intentional attacks against the food we eat or the feed stocks that make our food may be considered bio-terrorism or agro-terrorism, those terms are more often used in the contexts of crops and livestock, or as an umbrella term for all biological-based terrorist attacks against food.

Key points: Attacks against any food, crop or livestock rather than a focus on the food we eat.

Summary

The terms used may be different from one country to the next or in different industry segments, and may continue to change over time. Regardless of the terms used, if you are considering the protection of your food supply from intentional contamination, this chapter is for you.

FARM TO FORK

Food defense borrows the notion from traditional food safety and cold chain to treat the entire food supply as an integrated system. The weakest link between the farm and the consumer may be the place where an intentional contamination occurs. You may also hear comparable terms such as "farm to table," or in Australia, "paddock to plate."

With food defense two factors make protecting the entire supply chain even more critical than with food safety. The first factor is the malicious intent of the perpetrator. If they have studied the food system, they may choose to attack at the location with the fewest defenses, or the location where they can do the most harm. Some vulnerability assessment methods, in fact, focus on how a perpetrator would choose a target.

The second factor is the difficulty of inactivating some agents that might be used in an intentional attack. Traditional food safety protocols heavily emphasize sanitation conditions at food producers and the critical control points (CCPs) of "kill steps" like pasteurization to reduce the risk of a food safety incident. These measures would be totally ineffective against many potential contamination agents.

Note: Several food safety incidents have illustrated the weakness of relying too heavily on the controls at the food producer. The *E. coli* outbreak in leafy greens in the United States in 2006 caused a significant health impact despite controls at the producer. Significant improvements in good agricultural practices (GAP) were instituted as a result (CDC, 2006).

Most food producers are not vertically integrated to the extent that they can control the entire food supply from "farm to fork." They must instead *influence* the defensive ability of the supply chain that is outside their control in another way. To do this they borrow techniques from traceability methods, looking one step ahead and especially one step behind as part of supply chain verification.

TYPES OF RISK AND HAZARDS

In traditional food safety we look at the various mechanisms (vectors) that could create a food safety incident. In food defense, when we talk about the types of risks and hazards we can consider the various types of *perpetrators* that might commit an intentional contamination. What kinds of people or groups are they? What are their *motivations*? What are their *capabilities*? Are there *targeted mitigations* for those specific perpetrators?

We can also consider the various *agents* that might be used. Taken in combination, these represent the *threat vectors* we need to guard against. The good news is that much of what we do to protect our processes from one potential perpetrator will also protect us from others. The bad news is that this is not entirely true, because their capabilities are different. Consider perimeter fences that will help keep out competitors or local political threats, but would do little to slow down an insider or a well-organized terrorist cell.

Perpetrators: Motivations, Capabilities and Targeted Mitigations

Owners and Managers; Economically Motivated Adulteration (EMA)

While we might first think of intentional contamination with respect to criminal or terrorist activity, often with a political motive, perhaps the most pervasive form of intentional contamination is to improve profits: economically motivated adulteration.

Economically motivated adulteration can take the form of diluting a product, substituting inferior ingredients or adulterating with potentially hazardous ingredients that might improve the apparent value (with the intention of "fooling" the current quality assurance of analytical methods used to establish value). Fake products (imitations/counterfeits) are another form of economically motivated adulteration.

A challenging factor with economically motivated adulteration is the nature of the perpetrator – the owner or key management staff. They have ready access to all systems.

They understand the process thoroughly and have the means to circumvent any physical security measures. Their workforce may be unaware of, unable or unwilling to report, their suspicious activities.

There is another interesting conundrum when considering the potential health impacts of economically motivated adulteration. The owner/perpetrator presumably does not desire people to become ill; their motivation is economic gain not human harm or political impact. So they may dilute or substitute with a material of less nutritional value though not necessarily one that is pathogenic. On the other hand, because the adulteration is more likely to go undetected for some time, the cumulative health impact could be quite great.

As an example, in 2008, about 300,000 children in China became ill and at least six babies died when the milk used for infant formula manufacturing was adulterated with melamine (*The New York Times*, 2011). In 2004, about 50 children died from fake infant formula, which provided little or no nutrition (Watts, 2004).

Special capabilities: To disguise the substitution of one ingredient for another, using the excuse of controlling proprietary information to hide the actual formulation from employees.

Limitations: The selection of agents is limited to adulterants the owner or manager believes will get around QA systems, and those that will be benign, so as to allow the adulteration to be perpetuated over time for the greatest economic gain.

Targeted mitigations: EMA agent informed analysis by the customer.

Employees and Other Insiders

A "disgruntled employee," angry with a supervisor or co-worker, may be one of the most difficult threats to guard against. This person may have access to many areas within the facility, may know the best places to contaminate without being caught and may not raise any suspicion even when away from their normal work area.

The nature of the design and operation of many food manufacturing facilities allows almost any employee to be in almost any work area. Even new or temporary service employees often have relatively unrestricted access. Color-coded hats or uniforms are often used to designate job function, like supervision or quality, and not what area the employee should be working in.

When considering employees, also consider other people who have regular access to your facility: vendors, contractors, sanitation personnel and other temporary support employees can have similar access and motivations.

Many of the physical security measures we might first think of when considering food defense, like fences and guards, are powerless against an inside threat. Even cameras might be ineffective in stopping an insider if they know they are not being monitored in real time. Enhancements of our physical security measures and additional behavioral measures are needed.

The motivation of an insider threat may be retribution for some harm or slight. As such, the goal is to do harm to the owner or company, not necessarily to create a large health impact. So the contamination may be one that becomes obvious, changing the color or composition of the product in a way that is detected before the product would make it to the consumer. There is one exception: the special case where the insider belongs to, or has been compromised by, a terrorist group that does seek to do harm.

Special capabilities: To freely bypass external security measures, and in many cases have access to food mixing operations that make a good place to introduce a contaminant.

Limitations: The agents available to employees are limited. They usually would not have access to highly toxic agents in high concentrations.

Targeted mitigations: Zoned internal security measures, "buddy" systems, surveillance. Background security screening of all personnel, not only full-time employees, but also vendor and temporary support personnel.

Competitors

Like a disgruntled employee, a competitor may wish to harm another company. A competitor likely does not have access to the facility (though this could happen in some cases of prior employment). The competitor would, however, have significant knowledge of the processes and possibly the vulnerabilities.

Special capabilities: To understand the most effective place to contaminate and recognize those processes in another operation.

Limitations: The agents available to competitors could be limited. Access may be limited as it is to other outsiders. There may be restraint on the part of a competitor if they understand that any public awareness of tainted product could damage the entire market including their own business.

Targeted mitigations: Prompt removal of access privileges for all previous employees. Industry education regarding the shared impact of a contamination event.

Local Extremists

Because the threats of local extremists are politically motivated, they can be termed terrorists. They are listed in this chapter separately from global terrorists because the capabilities of these local terrorists may be much different. (In some food defense programs, all terrorists, local and global, are considered as a single category of threat – National Standard of the People's Republic of China, 2010.)

What sets extremists apart from the previously classified groups is their willingness to cause harm to people. While they may attempt to achieve their goals through economic harm and publicity, they may prefer to make their statement by impacting public health.

There have been several intentional food contamination events by local extremists. Fortunately they have been fairly limited in scope. These events have generally been targeted at the "last mile" such as retail food establishments.

As an example, in 1984 in The Dalles, Oregon, USA, members of a religious commune deliberately contaminated salad bars at 10 restaurants. A total of 751 persons were stricken with *Salmonella* gastroenteritis associated with eating or working at these restaurants (Torok et al., 1997). This was a test run on their ability to impact voter turnout at a local election.

Special capabilities: To organize and conspire with others committed to a cause. They may research methods and agents in advance.

Limitations: The most esoteric methods and most toxic agents may not be available. They may be restricted only to simple access points.

Targeted mitigations: Special emphasis on retail food establishments. Facility access controls in manufacturing environments such as fences, guards and access badges. HR practices.

Global Terrorist Threat

Highly organized global terrorist attacks could target the food supply with the intention of causing significant illness and deaths. The impact could be more tragic than bombing a hotel or crashing an airplane.

There is evidence that global terrorist groups have at least considered attacks on the food supply. This evidence includes captured notes from a raid on an Al-Qaeda camp (Tarnak Farms Afghanistan) in 2001 that illustrated portions of the food supply and listed potential actions (Hoffman and Kennedy, 2007). There is no confirmed evidence to date that global terrorist groups have executed an *actual* attack against the food supply, although such events have aroused suspicion.

What we realize, however, is the historic vulnerability of our food supply to a potential attack. The supply chain is vast, from fields, to factories, to endless distribution routes and countless consumers. It is broadly recognized as the most complex of supply chains (*Food Safety Magazine*, 2011).

We also realize the devastating impact a significant attack could have. How many people have become sick or died in some of the major *food safety* incidents of the last decade, and how many more people could be harmed by a well-orchestrated attack on a food plant? How many more people could be harmed using highly toxic agents that are not inactivated by current processes and not detected by current methods?

Special capabilities: To organize, conduct surveillance of potential targets, practice attacks and coordinate asymmetric attacks against multiple targets. They are also able to recruit or place an insider in an organization, giving them many of the capabilities previously discussed for company insiders. Potential access to any agent desired.

Limitations: The planning activities required to organize a coordinated attack make them susceptible to detection by government authorities. Surveillance or trial exercises may make them detectable by alert plant personnel.

Targeted mitigations: Collaboration with law enforcement or national security organizations and a comprehensive and layered food defense system that includes trained and alert employees or authorities.

Agents

In food safety we concentrate on a list of established agents identified as contaminants. They are naturally occurring and/or unintentionally introduced and have been reasonably well studied. The situation is much more complex in food defense.

A broad range of toxic agents could be used for an intentional contamination, including biological, chemical, physical and radiological agents. Many of these agents are not well understood and we have little experience detecting them in the food supply or inactivating them.

It is natural to think that we should know what agent we are guarding against, but in one respect, this really is not necessary. When we "think like the bad guy," we look for an agent and a contamination point that is not inactivated by the normal manufacturing process.

If we create a food defense plan based on some specific agent, we might overlook potential vulnerabilities based on our ability to inactivate that agent, such as points upstream of a pasteurization point. If we create a food defense plan based on the idea that the agent that will be used will cause harm and will make it through our process, we can address additional vulnerabilities.

[Hint: Create your food defense plan based on an agent called "really bad stuff" that is highly toxic, difficult to detect and will not be inactivated by your process or supply chain.]

After you have created an agent-agnostic plan, there are two additional benefits to going back and looking at the specific characteristics of different agents.

1. You may be able to create some additional agent-specific mitigation that further reduces your vulnerability to specific agents or types of agents. For example, a slightly higher pasteurization temperature may inactivate a broader range of biological agent as many US dairy processors have done (Detlefsen, 2005). (Remember that contamination could occur after pasteurization, so you cannot leave those downstream systems unguarded.)
2. You may improve your ability to debunk a hoax of contamination. In the event of a hoax that claims a specific contamination of your product, you are better prepared to explain to the public how that agent would be inactivated by your process, or how the quantity added would be incapable of causing mass harm.

Much of the information on specific agents is either limited or classified as there is no "normal" reason to need it or provide information beneficial to potential perpetrators. Additional information and references on chemical and biological agents (Kennedy and Busta, 2007) of concern is, however, available.

Detection: Methods to detect the introduction of intentional contaminates are being studied. Today there are no commercially available broad-spectrum test methods available.

Summary

Multiple perpetrators and multiple agent characteristics (since agent specific available data are limited) must be considered to create a food defense plan that can reduce the risk from multiple vectors. Some mitigation measures are important for any type of risk, but there are targeted mitigation measures that can help address each individual type of risk.

METHODS OF VULNERABILITY ANALYSIS

Risk analysis for food safety focuses on known hazards that do occur with reasonable probability to analyze the impact, with a goal of creating mitigation measures to reduce the impact. The focus is on controls.

There are several methods to assess food safety risks, including the eye of an experienced practitioner. Hazard analysis critical control point (HACCP) is the global standard and is fundamentally based on operational risk management (ORM) principles.

The ORM analysis is usually displayed as a two-dimensional health risk assessment model as displayed in Figure 35.1 by the Food and Agricultural Organization of the United Nations (FAO, 1998).

For food defense, the math is very different as it focuses on vulnerability assessment rather than risk assessment. An event of intentional contamination has no normal likelihood of occurrence, and an unknown potential for occurrence but the consequences could be

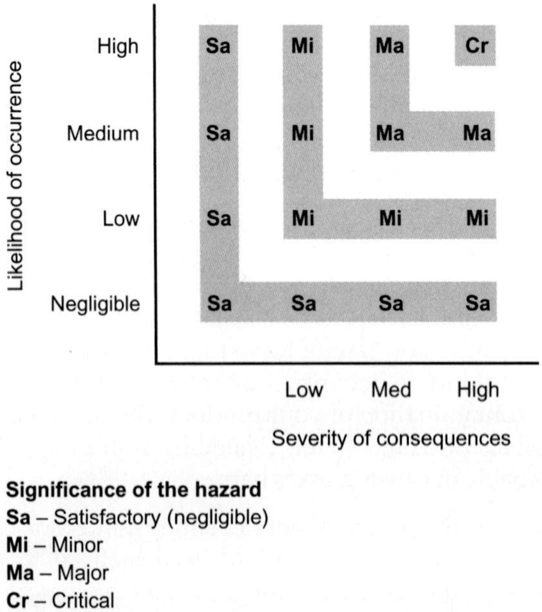

FIGURE 35.1 Example Operational Risk Management analysis, usually displayed in a two-dimensional health risk assessment model.

devastatingly off the chart. For this reason the analysis is on vulnerabilities in food defense, with a goal of creating mitigation measures to take an already low likelihood of occurrence and reduce it even further. The focus is on vulnerabilities instead of controls.

There are several methods to assess the vulnerabilities of a food production facility or system, including the eye of an experienced practitioner. CARVER + Shock, or a derivative of this method, is often used though there is as yet no apparent global standard.

CARVER + Shock

CARVER + Shock is a seven-attribute scoring method used to assess the most vulnerable operations in the food supply chain. It can be used across supply chain elements or within an individual production facility.

CARVER + Shock was adapted from a military targeting method used to select targets by giving those targets a preference score. Some targets are more hardened, others better guarded, some are easier to recognize, others easier to rebuild and some will cause more disruption.

Consider the possible military targets in Figure 35.2 and decide how you might score them low to high against Table 35.1 that lists the seven CARVER + Shock attributes.

When we use CARVER in the food industry, we score each unit operation (also known as process step) for each of the seven attributes. This results in a target preference score; if a "bad guy" were to target our operation, the higher scoring operations would make the most desirable targets.

FIGURE 35.2 Examples of possible military targets that could be scored from low to high against the seven CARVER + Shock attributes listed in Table 35.1.

Knowing *where* the best targets are helps you understand where you should focus mitigation measures. Knowing *why* some unit operations are the most preferred targets helps you understand what mitigation measures might be most effective in making those targets less desirable.

The goal of the CARVER analysis is a *ranking* of unit operations, not the actual score that is calculated, as a score is not transferable across a supply chain. The process steps with the highest ranking are the locations where you want to consider additional mitigation measures.

The CARVER analysis starts with a flow chart to describe the process flow within a portion of the supply chain such as your facility. A flow chart used for HACCP planning can usually be adopted for this purpose, but operations that would not be a food safety risk need to be added.

Scoring tables used for a CARVER analysis are often based on a catastrophic event on a national level. The scoring can be adjusted for an individual facility. The key factor is that the analysis helps you *differentiate* one unit operation from another.

Note: CARVER + Shock analysis often uses the term "nodes" for what we might call a unit operation or process step.

CARVER + Shock Software

The United States Food and Drug Administration (US FDA) has developed a software tool to assist in performing a CARVER + Shock analysis. It helps guide the user using an interview process, and does all the behind-the-scenes math that would be needed to score all of the unit operation (nodes).

TABLE 35.1 CARVER Attributes

C	Criticality	Do I hurt their economy, their health, their ability to fight?
A	Accessibility	How close? How easy to get to? Physical access?
R	Recuperability	How quickly can they rebuild?
V	Vulnerability	Can I damage the target? Is it hardened? Guarded?
E	Effect	Do they have backups or alternatives?
R	Recognizability	Can we recognize and find the target?
Plus	Shock	Psychological effects, like women and children?

Adapted from http://www.fda.gov/Food/FoodDefense/CARVER/default.htm#whatis, What is CARVER + Shock, What Does C-A-R-V-E-R + Shock Mean?

To locate and download the free software, enter the address for US FDA Food Defense, www.fda.gov/fooddefense. Use the site SEARCH tool to search for "VULNERABILITY ASSESSMENT."

The software has three main components. The first is to outline the facility using a flow diagram. The second is to respond to a series of interview questions that is generated based on the flow diagram. The final component displays the ranked scores as well as some suggested mitigation measures.

[**Hints** (based on software for manufacturing, version 2):

1. Choose a representative operation or line. Many of the insights you gain from one analysis will be appropriate for other similar lines.
2. Keep the flow chart fairly simple. The number of interview questions is proportional to the number of chart elements. Group like operations in the same area together and score for the worst (i.e. all "blending" might be described as a single chart element).
3. Rename each chart element with a preceding numeral in order of the process flow. The interview questions are asked in "alphabetical" order of the chart element names, not in the sequence of the process flow.
4. Choose a representative product and answer the interview questions based on that product. Many of the insights you gain from one product will be appropriate for similar products.]

Alternative Assessment Methods

Guidance Documents and Checklists

There are many guidance documents and checklists now available that suggest mitigation measures that a food production facility should consider. At the time of this writing, most of these are suggestive in nature, and not required by law.

Exercising one (or several) of these checklists can identify vulnerabilities when you realize a suggested mitigation measure is not in use in your facility. While not as stringent as a CARVER + Shock analysis, these checklists can still be beneficial on their own. This is because they have been created based on the results of numerous stringent vulnerability analyses, like CARVER.

Checklists are available for many industry segments and from many sources including government agencies and universities. You can search the internet for current checklists and guidance documents. There are several available from the US FDA and the United States Department of Agriculture (US DA): http://www.fda.gov/fooddefense and www.usda.gov. There is active research on new methods as well, so the tool set is evolving (Newkirk, 2010).

"Mini" CARVER + Shock

When analyzing entire industry segments and comparing one segment to another, there is some benefit in using all seven attributes in the CARVER + Shock method. Within a single organization, and certainly within a single facility, some of the attributes do not help to differentiate one unit operation from another, so they no longer add value to the analysis.

As an example, if there is a major contamination event at your facility, you either do or do not have backup manufacturing capability available (the EFFECT score). This is true regardless of what point of your operation is contaminated. So there is no benefit to you to scoring each unit operation for EFFECT; every process step would have the same score!

In fact, the second version of the US FDA CARVER + Shock software puts more emphasis on only four of the scoring criteria to simplify the interviews. This had no impact in the validity of the scores. The software scores are based primarily on CAV (Criticality, Accessibility and Vulnerability) with a little impact from R (Recuperability).

Taking this one step further within a single facility, Criticality and Recuperability will usually be scored the same or very similar. Since the goal is a *ranking* of unit operations, not a specific numerical score, using a score based only on AV (Accessibility, Vulnerability) will still provide a valid ranking of your most at risk operations.

As an example, the exercise tools used for Asian Pacific Economic Cooperation (APEC) Food Defense workshops use a simple two-attribute scoring system: Accessibility and Vulnerability (Periscope Consulting, 2010–2012).

Food AG Sector Criticality Assessment Tool (FASCAT)

The National Center for Food Protection and Defense (NCFPD) at the University of Minnesota developed an assessment tool that looks across entire systems of the food and agricultural sector. It is designed for national or state agencies to evaluate their many food operations to determine which types of operations might be more critical. It helps to retain equity in cross-sector critical system identification, using attributes like overall size and nature of distribution and the potential consequences of various threats (Hennessey, 2010; Food and Agriculture Sector-specific Plan, 2010).

A practitioner evaluating an individual factory and not a supply chain will find FASCAT of limited benefit as it is a systems analysis tool. A large producer with many facilities, especially facilities of different types, may find FASCAT or future derivative tools beneficial in determining where to emphasize their food defense efforts.

MSHARPP

MSHARPP is an Air Force targeting matrix to analyze likely terrorist targets. It is being evaluated by some countries for potential use in evaluating likely food industry targets (Air Force Antiterrorism Standards, 2005).

The scoring attributes have some similarity to CARVER, with the acronym standing for Mission, Symbolism, History, Accessibility, Recognizability, Population and Proximity.

The Eye of an Experienced Practitioner

After conducting numerous checklist-based assessments, workshops and CARVER + Shock analysis, there are some individuals who will have a very good "eye" for vulnerabilities and can develop a basic vulnerability assessment by walking through your operation. You might use an approach like this to get your food defense program started quickly.

If you use a food defense practitioner to give you a vulnerability assessment based on a tour and discussion to get your program going, you should get all recommendations in writing. It would be then prudent to consider continuing with a more detailed and systematic assessment to assist you in refining your program over time.

PREVENTIVE MEASURES

To reduce the risk of intentional contamination, the focus for industry is on mitigation measures that reduce the vulnerability of an attack (Khan et al., 2001). The focus is on *prevention*.

You may also hear the following terms used in the context of food defense: *detection*, *response* and *recovery*. These are the focus of government agencies and in some cases universities. *Detection* may be used to describe the detection of an intended attack or detection methods used to detect arcane contaminants within our foods. *Response* describes the laboratory capacity and information sharing as well as the regulatory and law enforcement work needed after a contamination event occurs. *Recovery* deals with restoring the ability of our facility or system to produce after a contamination, including the discarding of cleaning materials, decontamination of surfaces and restoration of consumer confidence.

While you may want to stay abreast of what government agencies are doing in the areas of detection, response and recovery, industry can have the most impact by helping to *prevent* a contamination. It is mitigation measures that make it more difficult to succeed that help to prevent an attack.

This section will focus on two different types of mitigation measures. First, there are *basic mitigation measures* that apply to nearly any food manufacturing facility. Second, there are additional *targeted mitigation measures* put in place after performing a vulnerability assessment (as described previously).

You can think of the basic mitigation measures like the prerequisite programs you put in place for your food safety program (GMPs, sanitation, etc.). They give you a good foundation and partially reduce your risk.

You can think of the targeted mitigation measures like the CCPs you put in place with an HACCP plan: they are specific to your particular vulnerabilities, and further reduce your risk.

Many of the mitigation measures recommended in food defense are already in place for other reasons. For example, good traceability and recall programs support food defense needs as well as food safety needs. Traceability and recall, as important as they are in food safety (*Food Safety Magazine*, 2010), are even more time critical in food defense as models show how many more lives could be harmed each day following a major intentional contamination.

But you may need to adjust these existing measures when you begin to "think like the bad guy." For example, your emergency evacuation procedures may need to address preventing others from entering your facility during the evacuation, not just getting everyone out.

Comparison with HACCP

While it is helpful to make a comparison between the *processes* of creating a HACCP plan with the *process* of creating a food defense plan, the details are quite different. In food defense, the mitigations are put in place to further reduce the *risk of occurrence* of events of exceedingly low probability, not to reduce the *impact* of events of known probability.

There is also some danger in thinking like an HACCP practitioner. Once you begin to rely on some control, like pasteurization, to reduce the *impact* of an intentional contamination, you might overlook the small risk of occurrence of a contamination *after* the pasteurizer and fail to further reduce that risk with appropriate mitigation measures. Even more easily, you might overlook the small risk of contamination upstream of the pasteurizer by a heat stable compound and fail to further reduce that risk.

Basic Mitigation Measures

Basic mitigation measures generally apply to all food establishments and should at least be considered by every practitioner of food defense. They include physical security measures like fences and door locks. They also include behavioral or procedural measures like developing employee awareness and auditing your performance. Many organizations have taken specific steps to deter intentional contamination such as tamper-evident seals on packaging.

While sophisticated physical measures can enhance your food defense program, overreliance on them can present its own danger. Even the most sophisticated locks have little value if employees intentionally jam them.

Besides being classified as physical and behavioral, basic mitigation measures are often classified in additional ways. These are artificial classifications used for administrative purposes. They organize similar topics together and help assure that a broad range of mitigation measures are included. The US FDA uses these classifications (US FDA, 2007):

- Management (systems)
- Human element – staff
- Human element – public
- Facility
- Operations

An alternative organization system (Periscope Consulting, 2010–2012) uses:

- Outside (perimeter) security
- Inside security
- Logistics, production, and storage security
- Management systems

It is beyond the scope of this chapter to provide an exhaustive list of potential mitigation measures. The food defense practitioner is encouraged to research sample plans, assessment checklists and guidance documents, including the FDA guidance document referenced above. The World Health Organization (WHO) also provides a helpful checklist (WHO, 2008).

Below, however, are a few examples of basic mitigation measures in each classification (Periscope Consulting, 2012).

Outside (Perimeter) Security

Outside (perimeter) security has to do with the walls, fences, doors, etc. that keep an attacker out of your operation. These are most effective against outsiders, but provide little protection from employees and other insiders.

There are two important considerations with perimeter security. The first consideration is the concept of "layered" security. You will see this in military history: an inner and outer city wall, for example. This means that not only is each perimeter barrier impenetrable, but that collectively they slow down an attacker, making it more likely that they are detected.

The second consideration is that you must ensure the barriers remain effective. If the fence is not inspected for damage, and underbrush is not cut away from the wall, or doors are not closed and locked, your defenses will not be effective.

Table 35.2 shows some sample mitigations to consider.

Inside Security

Inside security has to do with measures in place once an attacker has got inside, including lighting, cameras and internal access restrictions (zoning). These measures are effective against outsiders, but also provide protection from employees and other insiders.

There are two important considerations with inside security. The first consideration is the concept of detection. What can you do to make it easier to detect or recognize an attacker? Ideally this would be prior to an intentional contamination. But even detecting the contamination event has benefits, since you would not knowingly ship a contaminated product and thus would reduce the health impact.

The second consideration is to create additional *layers* of security, like the layers described in outside security. Zoning the facility, authorizing access only to individual work areas and putting in additional access controls (walled areas with locked doors) all reduce the risk of contamination. These measures do not prevent all people from accessing sensitive areas.

Table 35.3 shows some sample mitigations to consider.

Note: Closed circuit television (CCTV) is a popular security measure in many larger facilities. You should understand that cameras are primarily forensic in nature – providing information after the fact to prove what happened and capture the perpetrator. You must actively monitor any cameras and respond to unusual activities for them to be effective.

TABLE 35.2 Justifications for Exterior Defense

Mitigation Measures to Consider:	Reason for the Mitigation Measures:
Are the facility's grounds secured to prevent entry by unauthorized persons (e.g. by locked fence, wall, or other physical barriers)? Are there regular security patrols?	Physical barriers such as a fence, wall or water can be used to restrict access to the facility. Guard patrols may substitute when no physical barrier is practical or provide an additional layer of defense when used in addition to physical barriers.
Is there enough lighting outside the building to properly monitor the perimeter (fence) and the space between the perimeter and the manufacturing operation?	Good lighting at the perimeter reduces the time someone can spend getting into the facility without detection.
Are primary entrances like exterior doors and gates secured? Have the number of entrances been reduced to a minimum? For entrances that must remain open during operations, is there a procedure to secure them after hours? Are existing locks really used on a regular basis to keep the facility secure?	Having doors that lock is critical. The fewer access points (doors) the better. Just having locks available is not sufficient unless you can confirm those locks are being used.
Are operational entrances like loading dock doors locked or latched from the inside when not in use?	Some access points may be open to allow normal operations, but should be secured after hours/ weekends when the facility is not operating.

TABLE 35.3 Justifications for Interior Defense

Mitigation Measures to Consider:	Reason for the Mitigation Measures:
Is there adequate lighting throughout the facility? Is there an emergency lighting system in the facility?	Good lighting makes it more difficult to commit an intentional act without detection. It enhances the ability of CCTV (where used) to adequately record events.
Does your plant have monitored and recorded security cameras (CCTV)?	CCTV recordings are among the best ways to investigate a crime after it occurs. Recordings may help prove or disprove a threat of intentional contamination. While recorded CCTV can act as a deterrent, active monitoring of the CCTV improves the deterrent effect. Recordings should be tested from time to time so you know they are working correctly.
Is access restricted to production, storage and other sensitive areas? Are these restricted areas clearly marked? Is there a method to identify who is authorized to access these restricted areas?	Restricting access to sensitive areas provides a third layer of defense in addition to perimeter and building security. Creating zones in the facility and lists of approved persons can help you detect if only authorized persons are in an area. Color-coded uniform elements enhance this. Locked doors, keys or access cards can further protect sensitive areas.

Logistics, Production and Storage Security

Logistics, production and storage security is concerned with measures in place with materials stored in your facility as well as moving materials into and out of your facility. These measures are effective against any attacker and also help to extend protection up and down the food supply.

TABLE 35.4 Justifications for Logistics, Production and Storage Security

Mitigation Measures to Consider:	Reason for the Mitigation Measures:
When choosing suppliers for your packaging materials, labels, ingredients and raw materials, do you consider whether or not they have implemented food defense measures?	Contractual agreements, vendor/supplier surveys, audits and certification programs for your suppliers can be used to insure they have food defense measures comparable to yours.
Are trailers/trucks on the premises maintained under lock and/or tamper-evident seal when not being loaded or unloaded? (This includes during any short-term storage time before unloading or before shipping.)	Any storage outside your facility should be controlled much like you control your warehouse storage. If you temporarily hold trailers/trucks full of materials or finished product, they should be protected.
Are incoming shipments of raw materials, ingredients and finished products required to be sealed with tamper-evident or numbered seals (and documented in the shipping documents)? Are these seals verified prior to acceptance? Are suspicious alterations in the shipping documents investigated before acceptance?	Closed trailers should be sealed with tamper-evident seals to detect unauthorized access to the shipment. Seals should be numbered to reduce the risk of counterfeit. Seal numbers should be verified and the documents inspected for alteration for the same reason. Locks may be used in addition to seals to provide additional security, but locks alone do not provide tamper evidence.

There are two important considerations with logistics, production and storage security. The first consideration is the concept of "farm to fork." Because a contamination can occur anywhere in the supply chain, it is important to guard the entire supply chain. While you cannot directly control your suppliers, you can use your contractual agreements and supplier audits to help ensure that their food defense measures are comparable to your own.

The second consideration is the critical importance of ingredients. If key ingredients are contaminated on your site or at your supplier, you will inadvertently, deliberately and effectively mix those contaminated ingredients into large batches of food that result in many consumer portions affecting the health of many people. This is why extra scrutiny of suppliers, inspection and testing of incoming ingredients, and protected storage of ingredients becomes so important.

Table 35.4 shows some sample mitigations to consider.

Management Systems

Management systems are concerned with the policies, procedures and training you put in place to create and maintain a robust food defense program. These systems knit your program together into a cohesive whole and perpetuate the program over time.

There are two important considerations with management systems. The first is that your greatest risk is the people on your site. Careful selection and monitoring of all employees, contractors and temporary employees is important. Training and awareness for all employees is equally important – to promote vigilance and the reporting of unusual circumstances.

The second consideration is the importance of monitoring the program itself. Without surveys, inspections or audits, it is easy to overlook the measures you promised to implement. Without periodic review and update, your program will not improve as new best practices are established.

Table 35.5 shows some sample mitigations to consider.

TABLE 35.5 Justifications for Management Systems

Mitigation Measures to Consider:	Reason for the Mitigation Measures:
Are background checks conducted on all employees who will be working in sensitive operations?	Background checks of employment and criminal history can help reduce the risk that people with a negative history will have easy access to your product.
Are background checks conducted on all contractors (both permanent and seasonal) who will be working in sensitive operations?	If you conduct background checks of your full-time employees, you should also conduct similar checks of any other people with access to sensitive areas, whether they are part-time, temporary or contracted employees. Keep this in mind if you use contract services for work such as sanitation of your production equipment.
Do all plant employees receive training on security procedures as part of their orientation training?	All employees should have basic food defense awareness training that can be provided when they become employed. A refresher course once a year can improve their awareness.
Do you conduct regular food defense drills to test the effectiveness of your food defense measures?	Testing, inspections and audits help identify weaknesses in the plan, or shortcomings where you are not performing as your procedures indicate. This information may be used to strengthen your plan.

Targeted Mitigation Measures

In order to target mitigation measures at specific vulnerabilities, a vulnerability assessment should be completed. (See the section above on vulnerability assessments.) Based on those vulnerabilities (e.g., product access in transportation, etc.), additional targeted mitigation measures (e.g., locks and seals, etc.) can be added to your food defense plan.

Because the vulnerability assessment looks at unique characteristics of your process, the targeted mitigation measures are often specific and process related. The mitigations you choose will usually be related to the reason that location was deemed more vulnerable. For example, a blending operation is especially vulnerable because the blend tank is completely uncovered; therefore you may decide that a good mitigation measure is to add a lid. But if that measure is not practical, you could reduce the vulnerability in other ways, such as adding additional video recording or supervision of that location.

Creating zones within your facility can have special importance for the sensitive areas highlighted by your vulnerability assessment. In the case of the vulnerable blend tank just mentioned, you can install additional access controls to the specific area and designate that area only for your most trusted employees to work in.

If only your most trusted employees have access to your most sensitive work areas, you are well protected. If you want to further reduce your risk, you can consider a "buddy system" where no single employee is permitted in the area alone, as well as video recording those areas. These measures can help further reduce your risk in this special case.

Mitigation lists and databases are worth considerating. The CARVER + Shock software previously mentioned displays some sample mitigations on its results screen. The self-assessment checklists previously mentioned can be scoured for suggestions; they are available by searching online. There are also mitigation strategies databases from the US FDA and the US DA available online. These databases let you look up sample mitigation strategies based on your selection of processing operation.

Mitigation Databases

- US FDA: http://www.accessdata.fda.gov/scripts/fooddefensemitigationstrategies/Card.cfm?card=55
- US DA: http://www.fsis.usda.gov/Food_Defense_&_Emergency_Response/Risk_Mitigation_Tool/index.asp

(Alternatively these databases can be accessed by visiting www.fda.gov or www.usda.gov and searching for "food defense mitigation.")

Regulatory Requirements

There are few government regulations dealing with food defense. Much of what is published is in the form of guidance documents recommending best practices, rather than mandatory requirements.

If you operate in or export to the United States, you should familiarize yourself with the BioTerrorism Act (BT Act) and the Food Safety Modernization Act (FSMA). The BT Act requires the registration of companies supplying to and an advance notification of shipments prior to their arrival in the USA. FSMA requires the FDA to address intentional contamination; future regulations to support this law will likely require food defense plans and a consideration of mitigation measures similar to those outlined in this chapter.

In China, the Certification and Accreditation Administration (CNCA) established under the Administration for Quality Supervision, Inspection and Quarantine (AQSIQ) has put forth guidance and regulation requiring food defense plans for exporting companies.

The UK and Germany have robust food defense initiatives. Canada has its own study of vulnerability assessment methods. This is by no means an exhaustive list, but an indication of the global progress of food defense initiatives.

You should inquire about the food defense requirements in your own country. This is an area that will probably see a great deal of development in the next few years.

HOW TO MANAGE THE CASE

Case studies and table-top exercises can reinforce your knowledge and allow you to practice your skills. There are excellent case studies and exercises available at no cost.

The US FDA has created a set of five exercises: Food Related Emergency Exercise Bundle (FREE-B). They include scenarios based on both intentional and unintentional contamination events. While these are targeted towards agency use, they include the provision for the private sector to participate, and they may be instructive on their own. They are freely available at www.fda.gov/fooddefense (search for FREE-B.)

The following case studies focus on recall and were based on food safety incidents. Recall is also important to food defense, and two of the three studies have a section entitled "What if Contamination were Intentional?"

FOOD RECALL CASE STUDIES

(http://foodindustrycenter.umn.edu/EducationalResources/index.htm)

1. Westland/Hallmark 2008 Beef Recall: A Case Study by the Food Industry Center (.pdf)
   Seltzer, J., Rush, J. and Kinsey, J. The Food Industry Center, University of Minnesota. January 2010.
2. Natural Selection 2006 *E. coli* Recall of Fresh Spinach: A Case Study by the Food Industry Center (.pdf)
   Seltzer, J., Rush, J. and Kinsey, J. The Food Industry Center, University of Minnesota. October 2009.
   Natural Selection Case Study Learning Module
   Problem Set Instructions (.pdf)
   Problem Set Instructor's Guide (.pdf)
   Problem Set Data (.xls)
3. Castleberry's 2007 Botulism Recall: A Case Study by the Food Industry Center (.pdf)
   Seltzer, J., Rush, J. and Kinsey, J. The Food Industry Center, University of Minnesota. August 2008.

References

Air Force Instruction 10-245. Air Education and Training Command, Supplement 1, 15 May 2005. Air Force Antiterrorism Standards. Attachment 14.

Cavallaro, E., Date, K., Medus, C., Meyer, S., Miller, B., Kim, C., et al., 2011. *Salmonella* Typhimurium infections associated with peanut products. N. Engl. J. Med. 364, 601–610.

CDC, 2006. Ongoing Multistate Outbreak of *Escherichia coli* serotype O157:H7 Infections Associated with Consumption of Fresh Spinach – United States, September 2006. MMWR Dispatch September 26, 2006/55 (Dispatch); 1–2.

Detlefsen, C., 2005. Dairy industry vigilant in addressing food security. Cheese Market News 25(22), 1.

Food and Agricultural Organization of the United Nations (FAO), 1998. Food quality and safety systems. A training manual on food hygiene and the Hazard Analysis and Critical Control Point (HACCP) System (Section 3).

Food and Agriculture Organization of the United Nations, 2010. Global hunger declining, but still unacceptably high. Economic and Social Development Department, September 2010.

Food and Agriculture Sector-Specific Plan, 2010. An Annex to the National Infrastructure Protection Plan. <http://www.dhs.gov/files/programs/gc_1179866197607.shtm>.

Food Safety Magazine online article, October/November 2010. Product Tracing in Food Systems: Legislation versus Reality, by Jennifer Cleveland McEntire, PhD.

Food Safety Magazine, December 2011/January 2012, The Food Safety Challenge of the Global Food Supply, Gary Ades, PhD, Craig W. Henry, PhD and Faye Feldstein.

Hennessey, M., 2010. Risk Evaluation Tools and Food Defense. North Central Cheese Industry Conference, 13 October 2010.

Hoffman, J., Kennedy, S.P., 2007. International cooperation to defend the food supply chain: nations are talking; next step – action. Vanderbilt. J. Trans. Law. 40, 1169–1178.

Kennedy, S.P., Busta, F.F., 2007. Biosecurity: food protection and defense. Chapter 5. In: Doyle, M.P., Beuchat, L.R. (Eds.), Food Microbiology: Fundamentals and Frontiers, third ed. ASM Press, Washington, DC, pp. 87–102.

Khan, A.S., Swerdlow, D.L., Juranek, D.D., 2001. Precautions against biological and chemical terrorism directed at food and water supplies. Public Health Rep. 116, 3–14.

National Standard of the People's Republic of China, GB/T-27320 – 2010, Food Defense Plan and Guidelines for its Application – Food Processing Establishments.

Newkirk, R., 2010. Risk & Vulnerability Assessment Methodology for Food Systems. <http://www.orau.gov/dhssummit/2010/presentations/March10/Panel06/newkirk_ryan.pdf>.

Periscope Consulting, 2010–2012. Asian Pacific Economic Cooperative, Food Defense Workshops, Food Defense Plan Builder Tool, copyright 2010, 2011, 2012.

Periscope Consulting, 2012. The excerpts of mitigation tables are used with permission from Periscope Consulting, taken from their Food Defense Plan Builder Tool, copyright 2012.

The New York Times, 2011. Melamine – China tainted baby formula scandal. <http://topics.nytimes.com/top/reference/timestopics/subjects/m/melamine/index.html>, updated 4 March 2011.

Torok, T.J., Tauxe, R.V., Wise, R.P., Livengood, J.R., Sokolow, R., Mauvais, S., 1997. A large community outbreak of salmonellosis caused by intentional contamination of restaurant salad bars. J. Am. Med. Assoc. 278 (5), 389–395.

Tumin, T., 2009. Visualizing food safety: seeing the linkages in a networked world. A fire under embers. <http://blogs.law.harvard.edu/fireunderembers/> (accessed 21.10.11.).

United States Food and Drug Administration, 2007 Guidance for Industry: Food Producers, Processors, and Transporters: Food Security Preventive Measures Guidance. March 2003; revised October 2007.

Watts, J., 2004. Chinese baby milk blamed for 50 deaths. Guardian 20.

World Health Organization, 2008. Terrorist Threats to Food – Guidelines for Establishing and Strengthening Prevention and Response Systems (revised version – May 2008).

CHAPTER

36

Effective Leadership

Serge Imboden

Haute Ecole de Sion, Sion, Switzerland

OUTLINE

INTRODUCTION

What makes a "good boss" a good manager? This is a question that has always intrigued leadership and management theorists. According to Drucker and Maciariello (2009), managers bear responsibility for their contribution and that of their staff to the achievement of the company's objectives. Managers form an elite with social and ethical responsibility. Among the most important abilities of a manager are decisiveness, a systematic approach to decisions on human resources and communication skills. The manager of the future, according to Drucker and Maciariello, must be a good integrator whose primary goal is to acquire and to satisfy customers. As long ago as 2002 Drucker was putting the focus of modern management on the person and calling marketing and innovations the general functions of every enterprise.

For Malik (2007), management is "the transformation of resources into results." Every manager must perform five tasks: provide objectives, organize, take decisions, monitor and develop people. Managers have seven tools available to help them: meetings, written

Food Safety Management.
DOI: http://dx.doi.org/10.1016/B978-0-12-381504-0.00036-6

communication, job design and assignment control, personal working methods, budgeting, performance appraisal and "systematic waste disposal." In contrast to other authors, Malik insists that management can be learned and is the same everywhere. For him, management is communication, because the right communication brings the right solutions. Management is not a primarily economic system, but rather a system of social behavior which is about not only maximizing profits but also balancing and integrating different interests (Malik, 2008).

Porter (1985) sees the primary task of a manager as developing successful competitive strategies. He has written numerous publications on strategic management. For him, a competitive strategy is essential to a company's profitability and survival. He has developed various and now globally recognized methods, such as structural analysis and the value chain for the analysis of industries and competitors.

According to Sprenger (2002), in most companies mistrust rather than trust prevails. Trust is often vehemently claimed, but seldom lived. He is convinced that trust can be employed as a control mechanism in an organization. The first step and the key to a new relationship of trust consists in trusting others and thereby making oneself vulnerable. Sprenger (2007) also points out that the direct supervisor often exercises the greatest demotivating influence on staff. He talks of the "myth of motivation," characterized by the paradox according to which all motivation necessarily leads to demotivation, and thus praise, rewards, bribery, threats and punishment are sins of personnel management. For him, incentives have the effect of making workers not more highly motivated, but instead ever more unsatisfied. To avoid demotivation, then, a good manager should thus concentrate on intrinsic rather than extrinsic motivation.

In his book *Die Fünfte Disziplin*, Senge (2008) calls for a fundamental reorientation towards an integrated understanding of the world and of self. The orientation to linear cause–effect chains and the concentration on unconnected individual elements are inadequate for the highly complex systems that determine our modern world. At the heart of all management therefore is the learning organization, the foundation of which is systems thinking – the fifth dimension (the other four dimensions being personal mastery, mental models, a shared vision and team learning). It is oriented to interrelationships and considers individual elements relationally, i.e. it always puts them in relation with all other elements of the system. "We tend to blame outside circumstances for our problems. Systems thinking shows us there is no 'outside'; that you and the causes of your problems are part of a single system."

An interesting dimension of management is revealed by Covey (1989, 2005) in his bestseller, *The 7 Habits of Highly Effective People*. He describes seven ways to effectiveness in which moral principles such as fairness and reliability are a prerequisite for good management. He states that man is not a being conditioned by incentives, but has the freedom to have meaningful reactions. A good manager spends a lot of time discovering, planning and facilitating new possibilities. He makes the win–win paradigm a guiding principle for his action, can communicate simply and listen empathetically. Leadership personalities are thus people who understand themselves as part of a whole, synergistic and trailblazers for the next generation.

Most authors are agreed, however, that successful leadership personalities must have a particular mix and manifestation of various qualities that are specifically aligned to the expectations and objectives of the company. It is also assumed today that leadership can be learned, although putting it into practice is not always easy or comfortable. Talent is of course helpful, allowing what has been learned to be exploited to the full. It is also not disputed that good leadership cannot be exercised off the cuff but is instead a separate job description that must meet the same standards as every other job exercised professionally.

THEORIES ON THE SUBJECT OF LEADERSHIP

Various models and theories of leadership have been put forward both to explain leadership success and to deliver recommendations for action to improve leadership practice or to solve problems. Even the practice of promoting executives primarily according to their specialist knowledge, still common in some organizations today, has not proven its worth.

One of the most influential theories that has accompanied and furthered the development of leadership skills arose as long ago as the 1930s. It was the concept of leadership styles. Kurt Lewin, for instance, described the authoritarian, the cooperative and the laissez-faire style. This was followed by numerous variations, such as employee-oriented or task-oriented, participative, bureaucratic, etc. One of the best known theories to arise, which is still used in many leadership seminars today, is the "situational leadership" theory of Hersey and Blanchard (Pelz, 2004). However, there is as yet no sustainable empirical evidence that any particular leadership style is more successful in practice than another. Leadership styles have such a high degree of abstraction that they can at most be applied as a (subsequent) description of behavior.

Equally, there is as yet no empirical evidence that particular personality characteristics are associated with leadership success. The study by Harvard University (Nohira, 2003) can be cited as an example. It reveals that personal characteristics such as "visionary," "energetic," "enterprising," "passionate," "power-conscious" or "modest," "empathetic," "nurturing," "self-assurance," etc. have virtually no influence on leadership success. What is much more important is the specific observable behavior of the manager, as in the empirically validated model of the transformational leader developed by Bass and Avolio (1994). According to them, a successful manager must perform the following tasks:

- Be a role model and inspire trust in order to gain loyalty (idealized influence).
- Provide motivation with challenging, meaningful targets and thereby enhance the willingness to engage (inspirational motivation).
- Stimulate independence and creativity (intellectual stimulation).
- Encourage employees individually so that they are able to continue to develop their personal skills and strengths (individualized consideration).

In summary, it can be stated that research on the subject of leadership has so far been unable to supply a convincing concept and, despite countless publications, is still in its infancy. Many studies have determined and explained characteristics and skills of the ideal manager and, while the answers may be correct and stimulate discussion, the opinion held by many researchers that the results can be applied with general validity continues to be an illusion. Malik (2007) puts it succinctly: "The ideal managers could well be as they are presented in the studies. It is not the answers that are wrong; it is the question."

MODELS OF EFFECTIVE LEADERSHIP

It was held for a long time that leadership was a phenomenon that could not be explained; it was simply a gifting of particular "charismatic personalities" (Weber, 1922). The consequence of this was that for many decades there was no systematic development of managers in companies. It was thought that the "right" person would come through in any case (Yukl, 2006). This view, which is still sometimes put forward today, is not surprising, given that the age of industrialization to the postwar period is dominated by many charismatic leadership personalities. In recent years, experience in the theory and practice of management development has led to a new trend (Carter et al., 2005; Thomas, 2008): first, the significance of theories and models has diminished markedly, and second, attempts are being made to define leadership ability by a limited number of company-specific skills that are operationalized through behavioral descriptions that are as precise as possible (Hale, 2004). Thus managerial competence can now be summarized as the sum of company-specific behavioral expectations that may differ from one company to another and from one hierarchy level to another. The challenge associated with the search for managers is thus to arrive at the most precise possible description of these behavioral expectations. If, for instance, a technology company might seek a person who has the behavioral expectation of a "willingness to take risks," the manager should be able regularly to analyze the environment in respect of possible risks and hazards, assume responsibility for risks that actually come about or create a positive, constructive culture of a willingness to take risks.

Building models comes with the danger that circumstances are simplified so much that they no longer reflect the complexity of the reality. Krönung (2007) rightly speaks of the "management illusion" of many management "gurus" who supply simplified success models that are not appropriate to the complexity of entrepreneurial action and allow mechanistically thinking managers to believe that they can simply copy these models. Nevertheless, a model of effective leadership is described below. This model is to be understood not as a management instruction but rather as a guide to diffuse management jargon and as a basis for discussion of the definition of behavioral expectations. The model of effective leadership does not lay any claim to completeness.

The model in Figure 36.1 is read from the inside out. The core of effective leadership is formed by the areas of quality, business culture and innovation. For these areas to be developed requires skills, namely leadership, management and entrepreneurial skills. Leadership is about enhancing effectiveness (do the right things), management about enhancing efficiency (do the things right) and entrepreneurship about developing the company (do new things). There are methods, instruments and processes for each of these elements. To enhance effectiveness, for instance, one needs instruments of strategic management, to enhance efficiency controlling instruments or quality management and to develop the company innovation management or human resource management. The model is surrounded by the "Deming cycle" (plan, do, check, act), which is intended to highlight that the different areas first interact and second require constant improvement (the *kaizen* principle). As already mentioned, the purpose of this simplified model is to show that the technical occupational skills of a leadership personality should primarily be the leadership skills that can be trained and learned. The individual areas are investigated in more detail in the following sections.

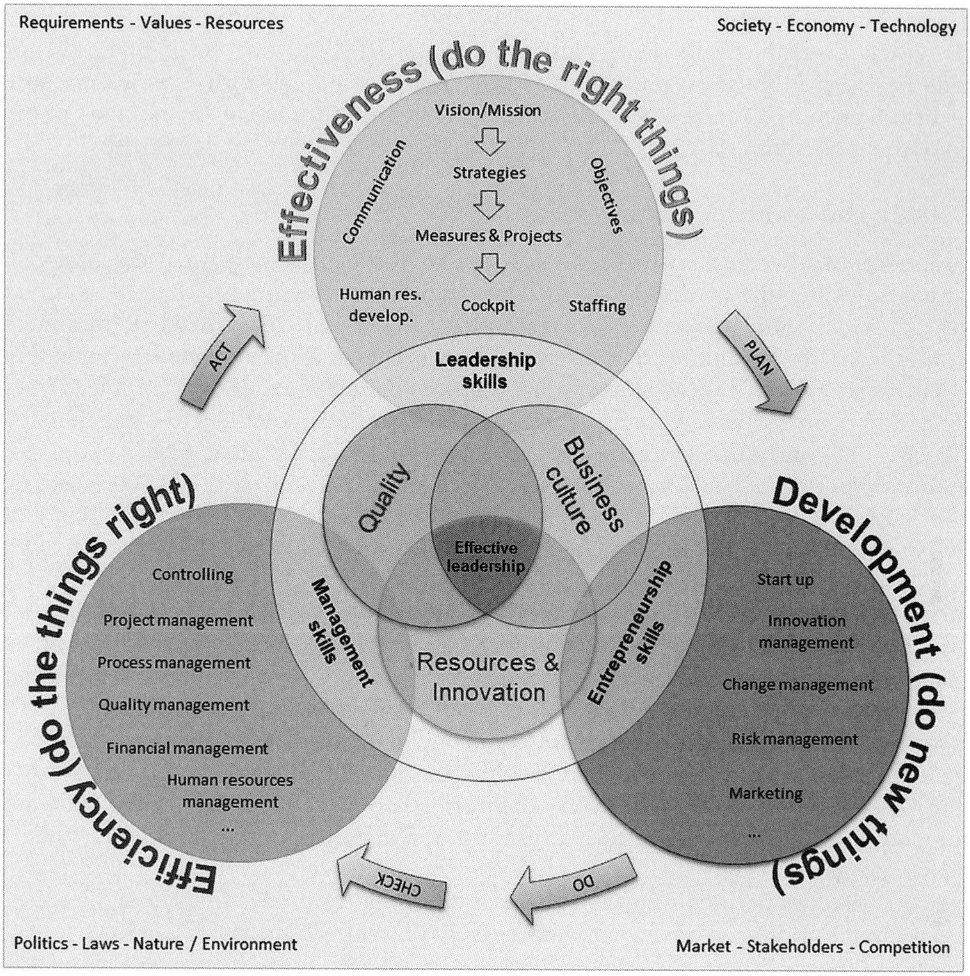

FIGURE 36.1 Model of effective leadership.

Quality, Culture, Innovation

At the heart of effective leadership are three areas: quality, the business culture and innovation. These three areas are a matter for top management and should not be delegated; they are among the fundamental leadership tasks. The first, quality, is a prerequisite if customers are to be acquired and retained in the long term. With the customer the company generates the turnover that is essential for survival. The second element, the business culture, influences the efficiency and effectiveness of the employees. Various studies have shown that there is a relationship between business culture and productivity. That is why many companies, such as Google, attach great value to a positive business culture. A lot is undertaken in order to ensure that employees feel comfortable in the work environment, because greater productivity strengthens the market position of a company. Finally, the third area, innovation, ensures the future development of the enterprise, aptly summed up by the widely disseminated slogan, "Innovate or die." It should be noted that a distinction is made between innovation and innovation management. The former is a basic attitude that is firmly anchored in the corporate culture, while the latter is the "managing" or organization of the innovation. The same applies for quality and quality management.

Leadership and Strategic Management

Leadership has become an increasingly popular term in management discussions over recent years. Leadership is often used as a synonym for management. There is as little crystal-clear differentiation between these concepts as there is a recognized definition of the individual terms. In the literature the words are sometimes used distinctively, sometimes to mean the same things. To give an example, House et al. (1999, p. 184) defines leadership as "...the ability of an individual to influence, motivate, and enable others to contribute toward the effectiveness and success of the organization...," while Drucker (2006) states tersely: "The only definition of a leader is someone who has followers." It is not the object of this chapter to open a definitional discussion of these terms. Nevertheless, a distinction is made below between leadership and management to the extent that leadership is understood to take place at strategic level (doing the right things) and management to take place at operational level (doing the things right). Covey (2005) provides an apt and interesting analogy: imagine that the objective of a company is to cross a jungle as quickly as possible. The task of the manager consists in organizing the team so that it moves forward efficiently and reaches the destination as quickly as possible and without problems. The leader, on the other hand, looks for the highest tree, climbs up and ascertains whether they are in the right jungle.

Leadership is usually associated with charismatic personalities who have forward-looking visions, whose vital energy is used almost solely to achieve the visions, who are single minded and who know how to enthuse and mobilize a group of people for their idea. Leaders are conspicuous by their farsightedness, their perseverance and their ability to establish and maintain networks. They always keep the long-term objectives in sight and know how to motivate and drive people forward even in difficult situations. There have in the past been personalities to whom considerable leadership skills have been ascribed: in the political sphere these may include people such as Mahatma Gandhi, Winston Churchill and Martin Luther King; in the scientific field names such as Albert Einstein and Robert Oppenheimer would be mentioned; and in business many entrepreneurs such as Werner von Siemens, Henry Ford, Steve Jobs and Bill Gates have shaped notions of successful leadership.

These usually very personal characteristics of a leader are difficult to describe and explain. It is hardly surprising, therefore, that these abilities have in the past been dismissed as innate talent. The strategic management approach does indeed supply possibilities of operationalization. Kaplan and Norton (2009) call for a systematic procedure according to a fixed process in which the strategy is to be developed on the basis of mission, vision and values. From the strategy are derived goals, success criteria and finally measures and projects. Instruments such as the strategy map or balanced scorecard help to maintain an overview. The authors insist both that the goals are measurable and that the operational level is separated from the strategic.

A possible systematic process with some disseminating instruments of strategic management might look something like the following (Figure 36.2).

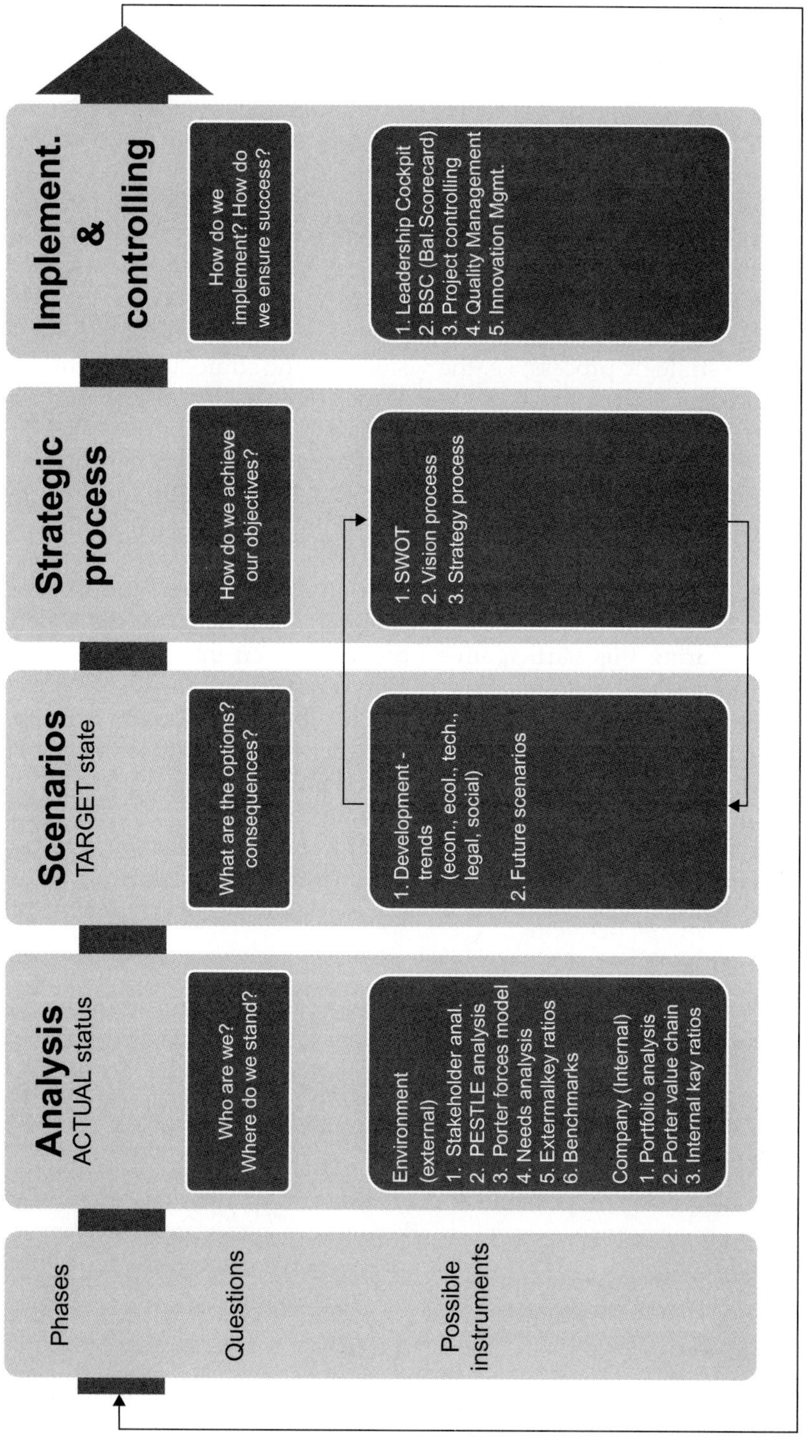

FIGURE 36.2 Stategic process.

The starting point of the strategic process is always an in-depth analysis of the current situation of the enterprise. This is where questions such as "Who are we?" and "Where do we stand?" are answered. There are numerous instruments for this phase, such as the stakeholder analysis, the PESTLE analysis, analysis of the industry structure, portfolio analysis or Porter's value chain. In the second phase the future options and their consequences for the company are assessed. This entails discussing developmental and mega trends and then elaborating possible future strategies (e.g. best case, base case and worst case). The third phase, the actual development of the strategy, begins with the SWOT analysis, which serves as a basis for formulating the vision, the mission and the strategies. Finally, in the implementation and controlling phase, the strategies are used to derive action plans and projects that go into a leadership cockpit and help managers to achieve the formulated objectives.

The results of the strategic process, i.e. the vision and the strategies, objectives and measures derived from it, form the basis of a good leader. This basis must be periodically controlled and adapted: the vision at least every 3–10 years, the strategy every 2–3 years and the measures monthly or quarterly. The next step, then, is to mobilize the employees for the set objectives. A good leader thus needs good communication skills and integrative abilities. Experience has shown that implementation critically depends on actively engaging the employees, or at least management level, throughout the whole process. This increases considerably the comprehension of the initial situation, the commitment and the identification with the elaborated objectives and hence the willingness to implement the proposed measures. Furthermore, during this participative process one can integrate team-building elements that in turn foster the development of a shared corporate culture.

The greatest danger in this phase is that the results obtained will fall victim to a bureaucratic exercise and waste away their existence on paper in some drawer somewhere. If this is to be prevented, three things are required: first, leadership strength, second, the operationalization of the measures in the form of concrete project tasks and, third, a controlling system. The controlling can be realized by means of a leadership cockpit, for instance, that integrates the balanced scorecard approach of Kaplan and Norton (1992).

Management

Management traces its origins to the organization of armies and the principle of command and obedience. Today, management is considered not as an isolated science but rather as a collection of techniques, methods and processes (Drucker, 2002). To enable better differentiation of the leadership tasks arising, the model of effective leadership (cf. Figure 36.1) distinguishes between management and leadership. It proposes that management is primarily about techniques for increasing efficiency, i.e. the question of how the tasks arising can be performed as quickly as possible and in sufficient quality. Examples of these tasks include controlling, quality management, marketing and project management. By way of elucidation, the following two sections deal briefly with two central elements, namely quality and project management.

Quality Management

The list of publications on the subject of quality management is just as long as the list on the subject of leadership. Quality management is usually understood as an integral part of leadership. In his book on quality management, Crosby (2000) notes that quality problems are a sign of poor management. He rightly goes on to point out that the introduction of quality management is a full-time job. Harvey and Green (2000) distinguish five formal and non-area-specific dimensions: quality as perfection (absence of defects); as fitness for purpose (the extent to which a product fulfills its purpose); as value for money (adequate compensation); as transformation (a qualitative change in the sense of further development); or as an exception (quality is distinctive, e.g. excellence or conformity with standards). Although such differentiations may impress, the problem with them is that they often describe quality as an entirety of characteristics of an entity or understand it as a thing, a characteristic or even an object. There is agreement, by contrast, that quality must be a part of corporate culture or, as Imboden (2004) notes, that quality is not achieved until quality is no longer talked about but is instead lived.

For quality to become part of corporate culture, quality management must be implemented as a continuous process (cf. Figure 36.3).

The starting point for any quality management is a vision that is shared by all employees and the associated strategies as well as clearly defined quality objectives. It also requires sufficient resources and tried-and-tested work processes. As already mentioned, quality cannot be delegated, but quality management can. The updating of the work processes, the organization of quality audits or the tracking of checklists and documents are among the activities that can be delegated. The initiation and maintenance of the quality spiral as shown in Figure 36.3, however, is emphatically a matter for the managing director. Only through continuous planning (plan), implementation (do), control (check) and correction (act) can a constant improvement in quality be achieved (Deming, 1986).

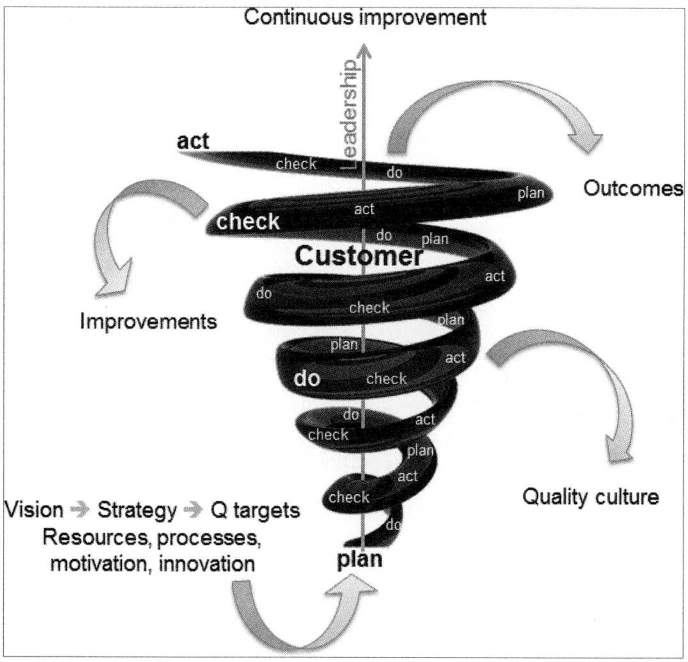

FIGURE 36.3 Quality management as a continuous process.

The significance of quality as a concept has undergone an evolution over the last 80 years. The understanding of quality has developed from the purely technical, i.e. "the machine must function," approach, through the customer perspective brought by Edwards W. Deming (1986) and Juran (1995) to excellence, where the needs of all stakeholders are to be satisfied. There is scarcely a job description at middle or senior management level that does not demand quality management skills. Good quality management knowledge and experience have now become indispensable for effective leadership. There is a great deal of literature on the subject that describes the methods, techniques, tips and tricks and the various implementation concepts. Ultimately, however, every leadership personality must gain their own experience if they are to be credible and professional leaders.

Project Management

Given the rising complexity of tasks, professional project management is increasingly becoming a factor in the success of a company. Project management may be defined with different words, depending on the source, but there is widespread unanimity in terms of its content. Figure 36.4 shows a possible subdivision of project management into five phases.

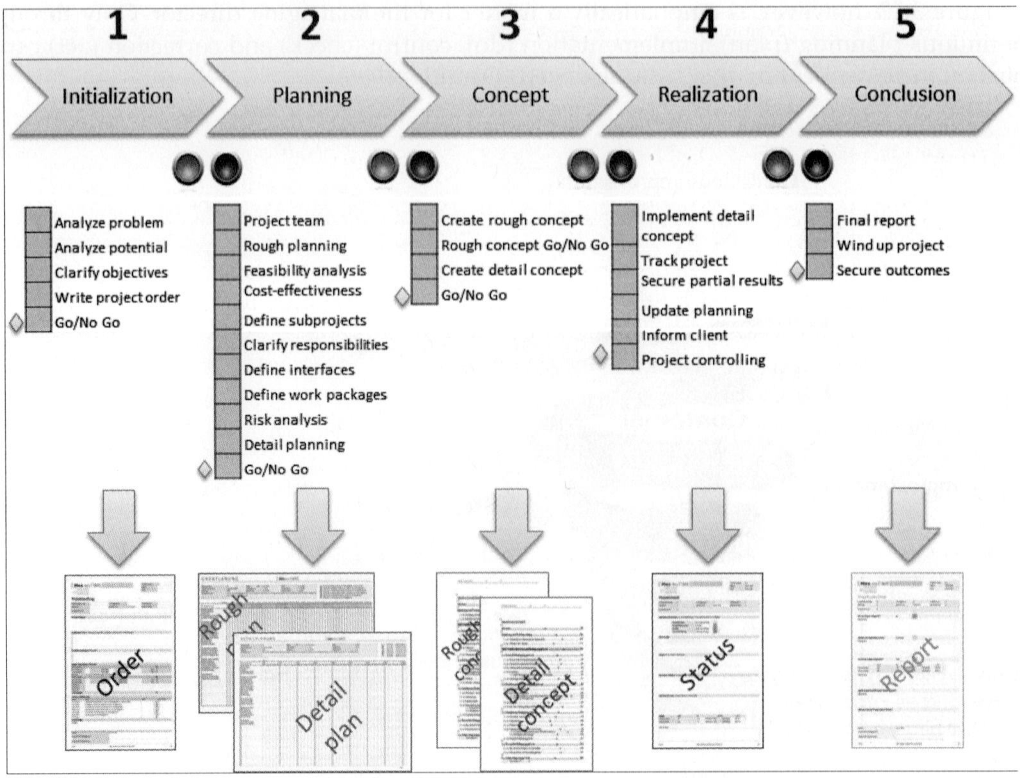

FIGURE 36.4 Project management in five phases.

In the first phase the initial situation is analyzed, the project objectives are determined and the project order is clarified with the project client. The second phase covers the planning. The project team is determined, the responsibilities clarified, the interfaces defined and the rough and detail plans created. In the next phase the rough concept and then the detail concept are created. The detail concept clarifies the feasibility, highlights the individual steps of possible implementation, sets out the resource requirement and draws attention to possible implementation problems. In the fourth phase the detail concept is implemented and finally, in the fifth phase, the concluding activities are performed, culminating in a final report.

The CHAOS study from the Standish Group (www.standishgroup.com) deals regularly with the success and failure factors in IT projects. It is among the best known and most important long-term studies in the project management field, with more than 40,000 individual projects having been scientifically examined since 1994. The Chaos Report reveals figures such as, for instance, that 70% of all IT projects fail, that the average cost overrun is 189% or that the project cancellation rate is 31.1%. Although on the one hand the high figures for the project cancelation rate and the average cost overrun are doubted by many, on the other hand the estimated number of unreported cases is high because many projects that failed according to their original definition are subsequently "sexed up," so that many project managers consider the figures to be perfectly correct from their experience. The fact is, nevertheless, that poor project management can lead to high wastage of resources for businesses and that it therefore makes absolute sense to train employees in project management.

Entrepreneurship

"Entrepreneurship is, firstly, the exploitation of business opportunities and the creative design of the business process in an organization or a phase of business change and, secondly, a scientific sub-discipline of business administration. Entrepreneurship research is concentrated primarily on scientific issues relating to the founder personality and environmental factors and on research into strategies and organizational forms that entrepreneurs make use of in order to build successful organizations" (translation from the German, *Springer Gabler Wirtschaftslexikon*, 2012). In the model of effective leadership (cf. Figure 36.1) the entrepreneurship skills are understood as the former, i.e. the exploitation of business opportunities and the creative design of innovation and human resource management processes. It deals in particular with the question of how the company can be developed further and less with the question of the foundation of the company. For a good leader, innovation, change, risk, information and human resource management (HRM) are key instruments and methods. The two central elements are innovation management and HRM. These are explored briefly below.

Innovation Management

According to the three-phase innovation process developed by Thom and Ritz (cf. Figure 36.5), innovation management is the design and control of the process, from the generation of ideas to the economic realization of an innovation. It should be noted that innovation management does not ultimately generate any innovations, but instead

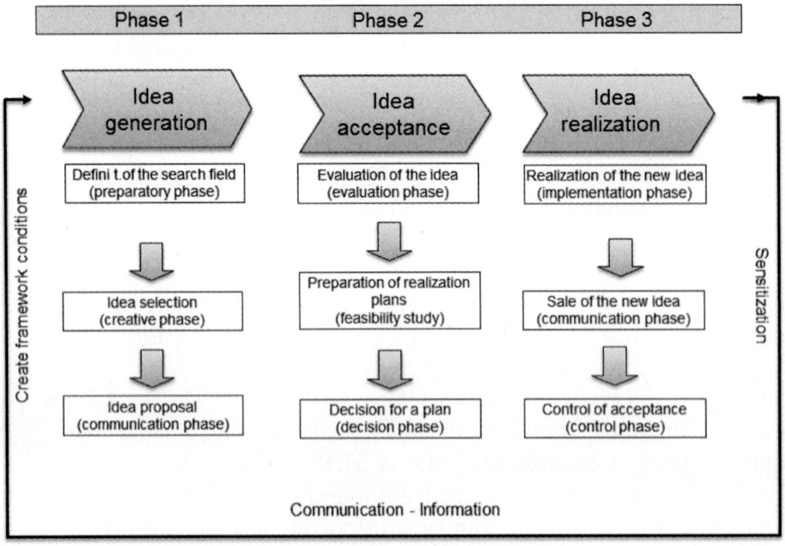

FIGURE 36.5 Innovation process according to Thom and Ritz (2002).

encourages the complex process of idea generation, idea acceptance and idea realization by providing suitable framework conditions.

The first phase, idea generation, is primarily about generating ideas in the first place. These ideas do not necessarily need to be unique and attention grabbing. The majority of innovations are in fact unspectacular optimizations of processes, products or services. The biggest challenge for leadership in this phase lies in creating an environment that allows and encourages creativity. This can also be assisted by various thinking and creativity techniques. Once an idea has been generated, the next hurdle is of a more communicative nature. Some more introverted or less articulate staff report trouble in getting their ideas across. For a good manager, the principle in this phase is that there are no bad ideas, i.e. stock phrases such as "That won't be possible" or "We don't have the budget for that" or "Unfortunately we don't have the time" should be avoided as much as possible during this stage of innovation management. It is worth taking every idea at least to the second phase, the idea acceptance phase. This is where the idea is thoroughly evaluated, realization options are discussed and a decision is taken on whether the idea will be realized. In the last phase, the idea is realized, communicated and controlled in terms of whether it exists on the market.

Experience has shown that innovation killers include disinterest or destructive feedback from the supervisor, getting neither praise nor recognition, fear of being blamed and an unwillingness to take risks. Internal resistance should not be underestimated either, because innovation also means destruction: old patterns of behavior must be abandoned in order to make space for the new. Innovation management is a complex and not to be underestimated task of a good managing director. Unfortunately, and particularly in small and medium-sized enterprises, innovation is still not managed systematically. Innovations arise almost by chance, only when problems occur or customers' wishes change.

Human Resource Management

The core task of human resource management (HRM) is to provide human resources and to use them appropriately. In the last few decades the working world has realized that the productivity, motivation and creativity of human resources are a decisive competitive advantage. This is demonstrated, for example, in the fact that in progressive businesses the HR director sits on the board and HRM takes on ever greater significance.

The core elements of HRM are summarized in Figure 36.6. Strategic human resource management is a meta-function, while HR controlling, HR marketing and the organization of human resource management are cross-departmental functions.

Before personnel are recruited, the HR requirement must first be determined. This can be done by asking the following question: How many employees with what skills will probably be required at which locations when and for how long so that the tasks in the company can be managed effectively and efficiently? Seventy-four percent of managers consider attracting good employees to be their biggest challenge (KMPG, 2001). HR recruitment consists in procuring personnel (searching for potential employees) and selecting personnel (choosing the "right" people). There are several selection instruments (e.g. tests, assessment centers, etc.) that can supply the manager with the foundation for making decisions. The final decision, however, cannot be derived from any instrument. The first working day is an important day for both the new recruit and the company, because first impressions count. It is worthwhile planning this day in minute detail.

Human resource development (HRD) is not just about bringing the requirements profiles of the posts into the best possible harmony with the skills profiles of the post-holders, but also about retaining the "marketability" of the employee. Various development instruments

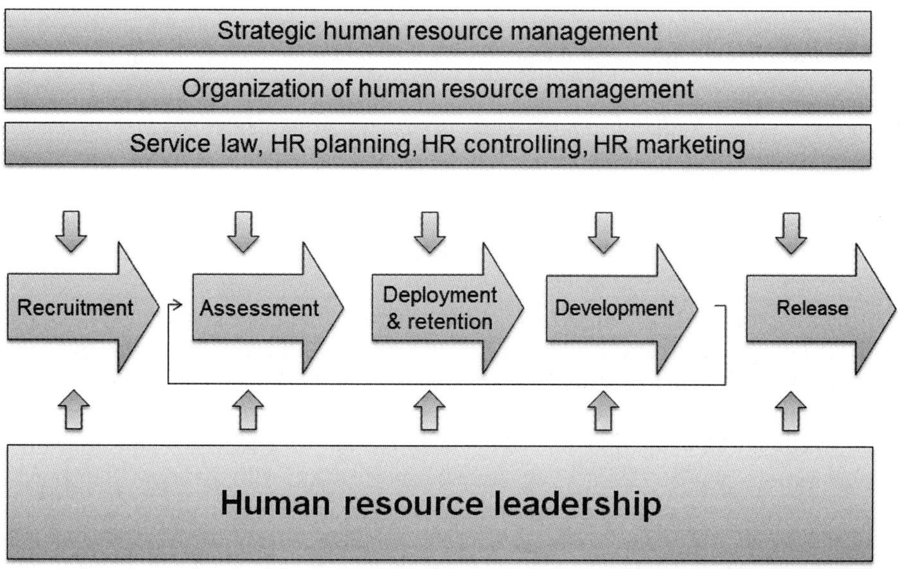

FIGURE 36.6 Human resource management, based on Thom (2001, p. 118).

are available: on-the-job measures, e.g. job enlargement, job enrichment, job rotation or coaching; off-the-job measures, such as conferences, seminars, further training; or along-the-job measures, e.g. career planning, employee appraisals and development assessments.

Retaining personnel is probably one of the most difficult leadership tasks. It requires promoting the commitment and job satisfaction of the employee. A variety of instruments are employed in an attempt to tie good workers to the institution. Motivation plays a critical role in this regard. Motivation research assumes that extrinsic incentive systems (e.g. pay, bonuses) are less effective in the long term than intrinsic motivation factors (e.g. interesting tasks, assumption of responsibility). For that reason many companies today are considering which "motivation mix" is the right one.

The release of human resources, whether voluntary or not, is another key leadership task. Normally jobs are shed when there is an excess of human resources. The problems associated with shedding personnel have been widely discussed, particularly in the last few years when markets have been liberalized and the weakness of the economy is forcing many companies to restructure. Before redundancies are announced, all available means (e.g. flexible working-time models, outplacements, management buyouts, internal transfers, short-time working, etc.) should be exhausted. As with recruitment, the last day at work is important for both the company and the person leaving, because the last impression that the employee takes with him will influence his future attitude towards the employer. Any departure must be preceded by at least one exit interview that gives the person leaving an opportunity to ask any questions that may still be relevant for him and to discuss any future plans.

"Able to go, but happy to stay" is a very apt summary of the challenge in HRM. Although employees have the opportunity to leave at any time, they are happy to remain. Behind this seemingly banal statement are many different considerations. For the working population, what is important is to remain employable. Retaining employability is a high-priority task of the supervisor, who should give the employee the opportunity to develop continuously. Since the mobility of the employee also rises as his employability on the job market improves, the associated risk of the employee going elsewhere presents the employer with a dilemma. This quandary can only be resolved with the second part of the above sentence: "...but happy to stay." The supervisor must create a working environment so that his employees are always content with their work and happy to remain.

FINAL REMARKS

Effective leadership is possible, but it places demands on leaders. This is because every leadership situation is always bounded by its context and requires creative and sensitive reactions. Effective leadership can be learned, but this entails a considerable cost. Leadership is not actually just something that can be done off the cuff; professional leadership is an independent occupation that one must spend a considerable amount of time learning and experiencing. Talent is of course helpful and allows what has been learned to be utilized. Effective leadership thus means applying the right mix of leadership, management and entrepreneurship skills in the right place and at the right time in order to achieve the set company objectives and in order to motivate and continue developing the employees.

References

Bass, B.M., Avolio, B.J. (Eds.), 1994. Improving Organizational Effectiveness through Transformational Leadership. Thousand Oaks.

Carter, L., Ulrich, D., Goldsmith, M., 2005. Best Practices in Leadership Development. San Francisco.

Covey, S.R., 1989. The 7 Habits of Highly Effective People, Powerful Lessons in Personal Change. Free Press/Simon & Schuster, Inc., NY.

Covey, S.R., 2005. Die 7 Wege zur Effektivität, Prinzipien für persönlichen und beruflichen Erfolg. Gabal.

Crosby, P.B., 2000. Qualitätsmanagement. Überreuter.

Deming, W.E., 1986. Out of the Crisis. Cambridge Massachusetts Institute of Technology Center for Advanced Engineering Study.

Drucker, P.F., 2002. Was ist Management? Das Beste aus 50 Jahren. Econ.

Drucker, P.F., 2006. The Effective Leader. New York.

Drucker, P.F., Maciariello, J.A., 2009. Management, Das Standardwerk komplett überarbeitet und erweitert. Campus.

Hale, J., 2004. Performance-based Management: What Every Manager Should Do to Get Results. San Francisco.

Harvey, L., Green, D., 2000. Qualität definieren. Fünf unterschiedliche Ansätze. In: Beiheft, S. (Ed.), Zeitschrift für Pädagogik, 41. pp. 17–39.

House, R.J., Hanges, P.J., Ruiz-Quintanilla, S.A., Dorfman, P.W., Javidan, M., Dickson, M., et al., 1999. Cultural Influences on Leadership and Organizations. Advances in Global Leadership, vol. 1. JAI Press, Stanford, CT, pp. 171–233.

Imboden, S., 2004. Excellence im Bildungsbereich – Qualitätsmanagement vor dem Hintergrund eines ganzheitlichen Führungsverständnisses. University Press, Fribourg.

Juran, J.M., 1995. A History of Managing for Quality. Milwaukee.

Kaplan, R.S., Norton, D.P., 1992. The balanced scorecard – measures that drive performance. Harv. Bus. Rev. January–February, 71–79.

Kaplan, R.S., Norton, D.P., 2009. Der effektive Strategieprozess, Erfolgreich mit dem 6-Phasen-System. Campus.

KPMG, 2001. Business Leaders in Switzerland 2001. So ticken die Schweizer Wirtschaftsführer. KPMG Fides Management AG: Zurich.

Krönung, H.-D., 2007. Die Management-Illusion, Warum Erfolg nicht kopierbar ist und was Manager daraus lernen sollten. Schäffer-Poeschel.

Malik, F., 2007. Management, Das A und O des Handwerks. Campus.

Malik, F., 2008. Strategie des Managements komplexer Systeme, Ein Beitrag zur Management-Kybernetik evolutionärer Systeme. Haupt.

Nohira, N., 2003. What really works. Harv. Bus. Rev.

Pelz, W., 2004. Kompetent führen, second ed. Wiesbaden (and on Wikipedia.org).

Porter, M.E., 1985. Competitive Advantage, Creating and Sustaining Superior Performance. Free Press.

Senge, P.M., 2008. Die fünfte Disziplin, Kunst und Praxis der lernenden Organisation. Schäffer-Poeschel.

Sprenger, R.K., 2002. Vertrauen führt, Worauf es im Unternehmen wirklich ankommt. Campus.

Sprenger, R.K., 2007. Mythos Motivation, Wege aus einer Sackgasse. Campus.

Thom, N., 2001. Personalmanagement – Überblick und Entwicklungstendenzen. In: Thom, N., Zaugg, R.J., (Ed.), Excellence durch Personal- und Organisationskompetenz, Vienna, pp. 117–131.

Thom, N., Ritz, A., 2002. Innovation, Organisation und Personal als Merkmale einer effektiven Schulführung. In: Thom, N., Ritz, A., Steiner, R. (Eds.), Effektive Schulführung. Chancen und Gefahren des Public Managements im Bildungswesen Bern, Stuttgart, Vienna, pp. 3–35.

Thomas, R.J., 2008. Crucibles of Leadership: How to Learn from Experience to Become a Great Leader. Harvard Business School Publishing, Boston.

Weber, M., 1922. Wirtschaft und Gesellschaft. Tübingen.

Yukl, G., 2006. Leadership in Organizations, sixth ed. Upper Saddle River/New Jersey.

References

CHAPTER

37

Human Factors in Food Safety Management

Yasmine Motarjemi

Food Safety Management Consultant, Nyon, Switzerland

OUTLINE

INTRODUCTION

In the past three decades, management of food safety has taken a leap forward and many systems and various technological or managerial tools have been devised to improve the safety of the food supply. The progress and development in food safety management have touched all segments of society, both in the private and public sectors. Box 37.1 shows some of the developments that have taken place at national or international level.

In the governmental sectors, these range from reconsidering the process of decision-making in risk management, strengthening of food laws and regulations, monitoring the safety of the food supply, surveillance of foodborne diseases, and promoting education of

Food Safety Management.
DOI: http://dx.doi.org/10.1016/B978-0-12-381504-0.00037-8

BOX 37.1

MAJOR DEVELOPMENTS IN FOOD SAFETY MANAGEMENT

- Increased general awareness about food safety driven by national and international media.
- Greater knowledge of pathogens, chemical contaminants and technologies and increased scientific and technical know-how.
- Development and emergence of high-performing food technologies and analytical methods.
- Increased availability of epidemiological and scientific data on foodborne pathogens and chemical contaminants.
- Improvement of the procedures for risk assessment and risk management.
- Strengthening of national legislation (standards, codes of practices), and its enforcement (inspection, monitoring).
- Strengthening of the international requirements (Codex Alimentarius, Agreement on the Application of the Sanitary and Phytosanitary Measures of the World Trade Organisation, ISO 22000).
- Increased preventive measures by the primary industry.
- Improvements in quality assurance, including application of the HACCP system.
- Strengthening of the foodborne disease and food contamination surveillance systems, alerts, traceability and incident management.
- Increased training of professionals specifically in food safety (governments, food industry and food service sector).
- Recognition of the importance of risk perception and good risk communication.
- Educational campaigns for consumers and the general public, including more informative labeling.
- Improved waste management, protection of the environment and of water and sanitation facilities.

consumers and food handlers. In the food industry, there have also been major changes, including the advance of the hazard analysis and critical control point (HACCP) system and its validation, traceability and recall, and the development of technologies and analytical methods. Consumer awareness has also increased and nowadays consumers are more actively expressing their expectations, preferences and values. Different factors have contributed to these developments; a description of these factors goes beyond the scope of this chapter and the reader is referred elsewhere (Motarjemi, in press).

In spite of the laudable efforts for improving food safety management, one area still has not received the attention which it deserves; yet, it is at the center of all the systems and tools that the society has devised to manage food safety and is a precondition for their successful implementation. That area is the role of people, from workers, managers, scientists to top management of businesses. None of the systems or controls used for the management of food safety will be effective without the proficient actions of those who have to implement them. Thus, in all sectors, from the legislator, inspectors, workers in the field or in a processing line, to the domestic or professional food handlers, management of people should be a central and integral part of food safety management. Management commitment is a *conditio sine qua non* for this. The question is then: what is management commitment and how does it impact on food safety.

The importance of management commitment is often mentioned in different text books and technical literature without going into the depth of the subject. The ISO Standard 22000 provides some insight into this subject. It explains that:

> *Top management shall provide evidence of its commitment to the development and implementation of the food safety management system and to continually improving its effectiveness by:*
> *a) showing food safety is supported by the business objectives of the organization,*
> *b) communicating to the organization the importance of meeting the requirements of this International Standard [referring to ISO 22000 standard], any statutory and regulatory requirements, as well as customer requirements relating to food safety,*
> *c) establishing the food safety policy,*
> *d) conducting management reviews, and*
> *e) ensuring the availability of resources.*

While the above requirements are a good start in explaining what is expected from management, they do not convey the day-to-day behavior that is expected from the management, as they have been developed for auditing purposes. Yet, the management of food safety relies heavily on the "quality of management" and good people management is an integral part of this.

Additionally, management of food safety requires sound judgment, objectivity in decision-making, competent managers and employees, etc., i.e. many intangible criteria which should be part of the soft skills of managers and which cannot be ticked off on a checklist. It is a question of company culture and leadership. With regard to the ISO 22000 requirements on business objectives mentioned above, there have been cases where food safety has been associated with company objectives; however, these have been less than effective, as managers have often been complacent with food safety in order to meet company production objectives and to receive their associated bonus.

Management's commitment to food safety requires first an understanding of the concept of food safety management and second, creating an organizational culture, structure and working conditions which enable and empower personnel in charge of food safety to meet their responsibilities. This is particularly important in the food business environment where there are risks with raw materials, processes, use of equipment or technologies, practices of staff and where change in any of these can impact the others. Additionally, experience from the food industry has shown that many executives and managers ignore or misperceive fundamental food safety principles, and these false perceptions can cause failures in decision-making (Table 37.1).

An analysis of organizational incidents, or "near-miss" situations, and how these take place illustrates the importance of management commitment, of an organizational culture and of the human factor and proficient food safety management. What follows is an application of the concept of Swiss cheese to food safety put forward by James Reason (1997).

TABLE 37.1 Common Misperceptions and Correction in Management of Food Businesses

Misperceptions	Correction
Food safety management is in conflict with economic interests	– There is no business without food safety – A good management of food safety can promote the business and its long-term sustainability
Food safety management is addressing food safety problems	Food safety management is about taking necessary measures to "prevent" food safety problems (including confirming that the measures are effective (validation) and are implemented correctly (verification)!
Our products are safe, as we have never had any incident	A past record of safety is no guarantee for the future, especially given the improved epidemiological tools now available
Our products are safe because the tests were negative	End-product testing is not evidence of microbial safety but merely a confirmation of the effectiveness of the food safety management system. Safety is based on the solidity of preventive measures in place.
Regulatory requirements are impediments to the business	Regulatory requirements and their enforcement will: – Facilitate fair trade and a healthy competitive environment – Ensure that all stakeholders in the food chain fulfill their role; this decreases potential risks with suppliers and their raw material – Provide guidance to businesses, in particular small and less developed businesses, on matters related to food safety, such as norms needed in designing and validating food safety assurance systems – Increase the confidence of consumers in the food supply and reassure consumers that commercial products are safe and meet the nationally and/or internationally agreed safety and quality standards

SWISS CHEESE CONCEPT

As illustrated in Figure 37.1, assuming that a potential hazard and its risk (i.e. the likelihood of its occurrence and the severity of its consequences) are known, a defense mechanism, or a series of such mechanisms, is devised to prevent the hazard from materializing. In food safety, these defenses are referred to as "control measures." In food safety assurance systems, a series of control measures are usually recommended. These can be grouped under three lines of defense: (1) basic good hygienic practices, (2) an HACCP system and (3) verification measures (Motarjemi, in press, Chapters 1, 31 and 32 in this book).

When an incident occurs, it is usually the result of a gap, or a combination of gaps, in these defenses. These gaps may be a shortcoming or an error in the design of the food safety plan or a fault in its execution. Usually, in good food safety management, a single gap in any of these defense mechanisms should not lead to an incident as the second or third line of defense should be able to detect the gap and allow for corrective actions before the contaminated or defective product reaches the market and/or consumers are exposed. Such a situation, i.e. where a failure takes place but an incident is prevented due to early detection and corrective action, is referred to as a "near-miss" situation. However, where food safety management is poor, there will be more holes in the system and the eventual alignment of these gaps increases the likelihood of an incident. Thus the more holes there are, the more

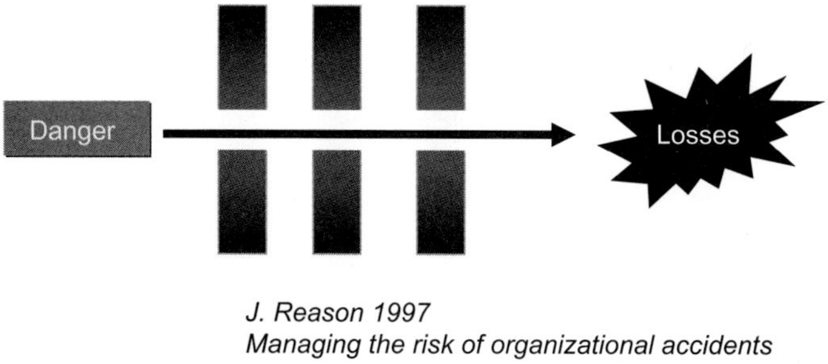

FIGURE 37.1 Swiss cheese model according to James Reason. The figure illustrates that incidents occur as a combination of a number of gaps in the food safety controls.

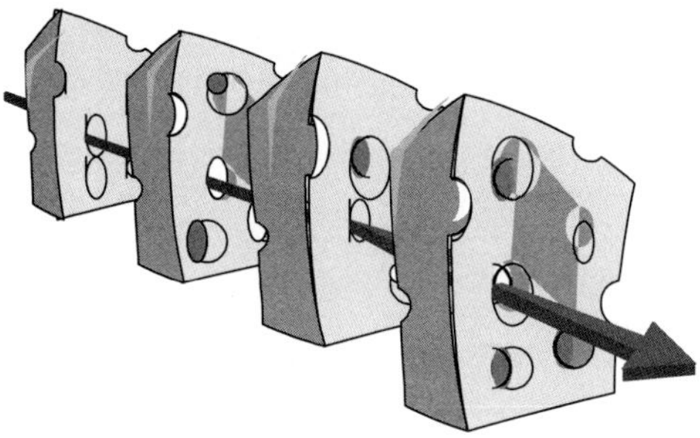

FIGURE 37.2 Figure showing that an incident occurs as a result of gaps in the defense systems. *Adapted from Reason 1997.*

the risk of an incident. The concept is referred to as the "Swiss cheese" model (Reason 1997) (Figure 37.2)

It is for this very reason that any gap in the food safety assurance system, even if *per se* not significant to produce an incident, should be addressed immediately. In other words, any detail can have its importance and should not be dismissed or neglected. Many accidents in aviation, healthcare, petrochemicals, etc., are the result of mishaps which were perceived as details. A notorious example is the case of the Concorde. The plane crashed due to a small piece of metal on the runway from which the Concorde was supposed to take off. The metal had fallen from another plane. Similar situations have occurred in the food industry. A case in point is the major outbreak that occurred in Israel and was associated with infant formula imported from Germany. The product was deficient in vitamin B1 (thiamine). Consequently, a reported 15 babies suffered from damage to their nervous system and two died. The primary failure was an error in product formulation, but a second failure was in the verification of the composition of the product before its release, which itself was due to a number of other errors. Similarly, in another incident of infant formula contaminated with the isopropylthioxanthone (ITX), a combination of gaps in the regulatory requirements, suppliers' test and practice, and customer's awareness of risks were the origin of the problem.

ROOT CAUSE OF FAILURES

A second concept that needs to be understood is the root cause of failures, which can be divided into active or latent failures (Reason 1995 and 1997) (Figure 37.3).

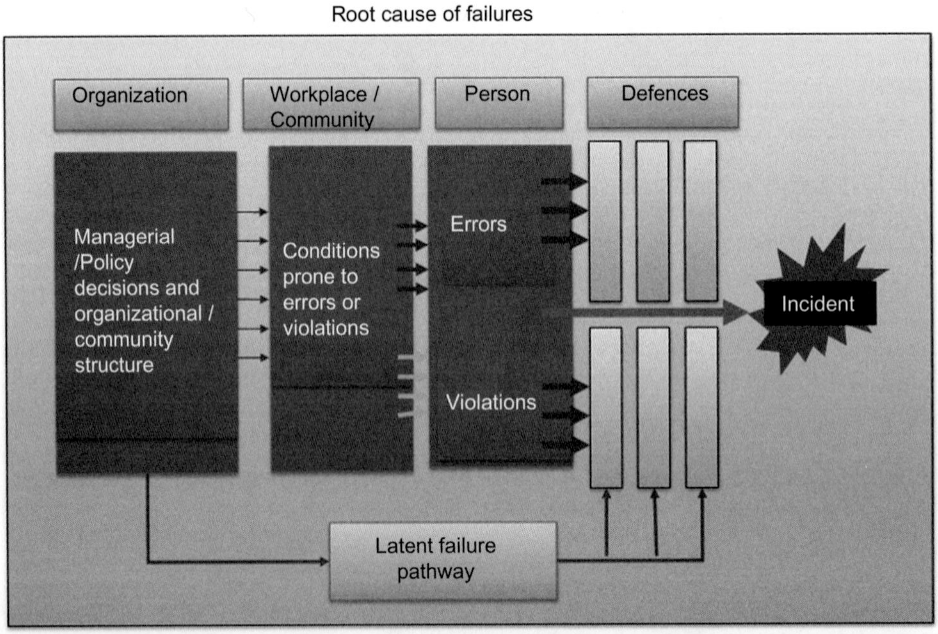

FIGURE 37.3 Figure showing the different levels and types of failures leading to an incident (Reason 1995).

Human Factors

Behind any control measure, there are people who have to implement those measures or verify that controls are implemented correctly. These can, for instance, be a worker on the line packing food, a farmer milking, an operator in the food manufacturing industry monitoring the temperature recorder, a truck driver who has to manage the temperature during transportation, a food handler who has to wash his hands before preparing food, or wash his knife and cutting board between raw and cooked foods, etc. Their failure to perform the control measures is referred to as active failures since their actions will have a direct and immediate bearing on the safety of products (Table 37.2). These are the types of failures that, in case of an incident or a "near-miss" situation, are normally investigated. In due course, the employee responsible is reprimanded, or worse fired, and the investigation ends at this point. The same process and relationship also exist between regulatory authorities and a food establishment that is incriminated for an incident. Once the vehicle of the outbreak, i.e. the implicated food item, and the error are determined, and in due course the products are recalled, generally the investigation is closed. However, it is important to pursue the investigation and understand the reason for the failure of the person(s) in charge of the control measure.

Active failures come in different forms. They may be classified by their causes or by their consequences. In the latter, the failures are described in terms of the proximal action contributing to the mishap. For instance, the consequence of an error such as the use of a wrong thermocouple may be transgressing the "critical limit," or an error in the handling of the computer system may lead to the erroneous release of a defective and blocked product.

TABLE 37.2 Frequent Active Failures in an Industrial Setting and during Food Preparation

Frequently Observed Failures in an Industrial Setting	Examples of Failures During Food Preparation
Failures in supplier management	Failure in respecting hand hygiene
Failures in design and maintenance of equipment	Food handlers handling food when suffering from a transmissible illness
Failures in hazard identification	Failure to wash utensils/equipment and allowing cross-contamination of ready-to-eat food
Failures in establishing "critical limits"	Failure to cook or refrigerate, thus allowing time–temperature abuse of food and subsequent survival and growth of microorganisms to a disease-causing level
Errors in GMP implementation	Improper use of recipients and contamination of food with chemicals/detergents
CCP monitoring failure	Failure to information consumers about essential food safety matters
Failure in applying the right corrective actions	
Human error	
Error during a change process	

Causal classification of human failures makes assumptions on the psychological mechanism implicated in the error. They are grouped by:

- *Failures in intention or "mistakes"*: These are errors in planning or problem-solving and are often related to scientific and technical misjudgment, leading to a plan which is inadequate in achieving the intended outcome. An example would be when the HACCP team decides a wrong "critical limit."
- *Execution failures such as slips, lapses*: These are failures where the plan is adequate but where the actions are not implemented as intended. There may be different psychological reasons, such as failure in attention, memory, recognition, etc.

Human errors are to be differentiated from violations. Violations are deviations from the rules, procedures and standards. They relate to deliberate action and intention. Violations also fall under different groups, depending on the incentives and reasons for violations, and range from routine or optimizing violations, such as taking a shortcut and not following the procedures, to necessary violations. The latter is when the rules and procedures are inappropriate, and violation of the rules is the only way to get the job done. Sometimes, the violation may also be the result of lack of knowledge of the rules on the part of operators. For instance, in many countries, the small or less developed businesses may simply be ignorant of the legislation and the violation of the law is not always an intentional non-compliance. Therefore, while as a principle the violation of food safety rules should not be tolerated, the cause of these should be determined and the decisions on penalties, if any, should be taken on a case-by-case basis.

Working Conditions and Environment

If root cause analysis of incidents or near misses is carried out, it can be noted that often human failures are due to the working environment and conditions. Some examples of conditions that may lead to a person committing an error or violating the rules are given in Table 37.3.

TABLE 37.3 Examples of Conditions Prone to Error or Violation

Conditions Leading to a an Error	Conditions Leading to a Violation
Unfamiliarity with the task	Misperception of the risks associated with hazards
Mismatch between the training and the education of the person with the task required	Belief that a bad outcome will not happen
Time shortage, work overload	Lack of tools, time pressure
Information overload or contradictory information	Ambiguous or apparently meaningless rules, or rules which are not applicable to the local conditions
Poor human–system or human–equipment interface	Manifest lack of organizational safety culture, or of a culture which encourages taking risks
Complex tasks or situations	Management not following the rules, or perceived lack of management's care and concern
Mental state: monotony of task or boredom, fatigue, stress	Inadequate training
Hostile environment, e.g. crowded, noisy environment	Unclear instructions
Poor instructions, procedures and definition of responsibilities	Professional attitude hostile to procedures
Poor communication or language barriers	Work conditions promoting conflict of interests
Lack of adequate scientific and technical tools or systems for performing a task	Conflicts and poor people management discouraging involvement, responsibility and ownership
Change in routine	

(Adapted from Reason 1995).

Responsibility of Management

The above-mentioned situations (Tables 37.2 and 37.3) often result from management decisions. The failure of the management to create conditions optimum for managing food safety is referred to as latent failures. The consequences of these decisions, taken at higher level in the organizational and managerial structure, may not be directly perceptible and they may not have an immediate impact on food safety, but they will create conditions favorable for non-compliance or accidents.

Latent failures have been the cause of numerous accidents in the petrochemical, transport and financial industries. While there is an abundance of reports on foodborne illnesses in the scientific literature and media, few provide in-depth information on the latent conditions leading to failures. Investigations often fail to examine these factors. A case in point is a report on an outbreak of foodborne illness associated with peanut butter in the United States. While the report provides extensive information on the epidemiological aspect of the outbreak, with regard to its cause the report is very short and gives the reader the impression that contamination of industrially produced peanut butter with *Salmonella* is to be expected. There is no information that could increase our understanding of the latent failures (CDC 2007). As a consequence, a few years later the incident repeated itself in another establishment.

Examples of the managerial decisions that lead to poor working conditions are:

1. Failing to provide the necessary policies, organizational structure and culture, adequate financial qualified human resources, or suitable equipment;
2. Appointing managers who do not have credibility or competence matching their responsibility;

3. Management behavior in contradicting, or in violating, the policies and instructions; or
4. Requiring impossible tasks which force the staff to take risky shortcuts or to violate the rules.

To alleviate the workload, but also to increase efficiency and consistency in the execution of tasks, industry has increasingly resorted to "systems," be it automation of equipment or development of checklists, for ensuring systematic coverage or handling of operations. However, it is the behavioral experts' view that while facilitating the systematic execution of tasks, automation or other systems do not fully overcome problems associated with human error or mistakes in the judgment of the scientific and technical data. Where there is too much reliance on systems, they may either inhibit critical thinking or cause boredom. As such, in the food industry, we can often see that the HACCP system is applied mechanically, without fully understanding its purpose. At times, the system is applied without adequate input from qualified food safety experts, and as a consequence the HACCP study and subsequent plan are not effective for addressing food safety.

Management style can also be the cause of failures in food safety management. A repressive or non-motivating environment may be deleterious to open and constructive working conditions and may deter employees from reporting potential problems, which helps in the early detection of gaps and in remedial action. Micro-management prevents the sense of responsibility and ownership, as well as bottom-up initiative. Unclear definition of responsibilities, expectations and procedures can create conflicts which may lead either to loss of efforts and resources, or to motivation in the execution of tasks. Therefore, the management of a company bears the ultimate responsibilities in ensuring food safety. In case of incidents, their eventual failure in providing this needs to be investigated.

Thus, *management commitment* should, among others, provide the organizational structure and culture and working conditions adequate for a professional, objective and transparent management of food safety.

The organizational culture should enable employees to openly report issues and provide them with the opportunity to see that their constraints are adequately and fairly addressed. An open and fair organizational culture is fundamental for motivation of the staff and the core of food safety management.

The organizational structure should ensure a process of decision-making based on expertise and should prevent situations of conflict of interest, for instance where audits and investigation of incidents or near-miss situations are carried out or supervised by the same person as the one who is responsible for the design and implementation of the food safety management system. It is also important that any near miss or incident be thoroughly investigated and a root cause analysis be conducted. This means that not only the primary cause of the failure, e.g. cross-contamination in a restaurant, error in a thermocouple in industry, use of contaminated water in agriculture, is examined, but the latent failures, i.e. the working conditions and management failures are also determined. In the above hypothetical cases, one may discover that the manager did not provide adequate training to the food handler or there was a lack of knives and cutting boards, the consequence of which was that the workers took a shortcut to meet the demands of the restaurant. In a food industry environment, a thermocouple may be misused because the personnel in maintenance may not appreciate the importance of the temperature for the safety of the product, or a transport company may fail to observe cold storage or the risk of cross-contamination during the transport of its product. To take an example at the agricultural level, an agriculture policy may have a direct impact on the use of safe water or fertilizers. In the case of a cholera epidemic in sub-Saharan Africa, it was seen that an increase in the price of the fertilizers led farmers to use contaminated manure for irrigation of their vegetables.

From the above it follows that proficient people management is a fundamental element of food safety management. In this respect, the factors that will influence employees in meeting their responsibilities and performing their tasks can be divided into three types of factors (WHO 2000):

1. *Predisposing factors*: These entail providing employees with the technical and scientific knowledge that they need to (a) perceive the risks associated with their job, (b) understand control measures needed to control the risks, and (c) impart the skills and competence that they need to perform their task.
2. *Enabling factors*: These consist of all infrastructures that are required for individuals to carry out the required tasks or to adopt the desired behavior. These factors can be of (a) a logistic nature such as adapted tools and equipment, having easy access to hand-washing facilities, or rapid cooling food, or (b) managerial nature such as providing the staff with the authority they need to perform their job, time for the execution of their task, etc.
3. *Reinforcing factors*: These relate to all those cultural values of the organization that encourage the individual to adopt the behavior in question, e.g. influence of peers. Management commitment and management practices are fundamental to this factor.

MANAGEMENT COMMITMENT

From the above it can be concluded that *management commitment* starts with management understanding that managers must have exemplary behavior, and stand behind the policies on a daily basis. In doing so, management must:

- Realize that any non-compliance or complacency at the higher level of management will set a bad example for the entire organization and will have serious repercussions on the entire organization, i.e. it will have a multiplying effect;
- Apply zero tolerance for leaders who violate their policies, and adapt penalties proportionate to the gravity of violations and the hierarchical level of the leaders;
- Understand what food safety means and requires in terms of measures, and realize that food safety management is not a question of the number of incidents experienced, consumer complaints received, or the results of end-product testing, but is about the measures put in place to meet the set standards, on an everyday basis;
- Ensure that their policies are not a declaration of good intentions but a description of ongoing practices, and
- Provide leadership by prioritizing food safety at levels of decision-making and ensuring that managers implement measures flawlessly.

Additionally, management commitment is founded on good people management, including:

- Ensuring that people are competent for their job and that appointments for positions are made transparently and based on merit;
- Ensuring that staff unequivocally understand their responsibility and their authority, that they have a clear job description and that there is an alignment between their responsibility and their authority;
- Providing staff necessary training and coaching according to responsibility so that they can meet what is expected from them;
- Foreseeing a succession plan and back-up positions;
- Providing necessary resources, including infrastructure, streamlined organizational structure and clear procedures so that implementation of necessary measures becomes feasible;
- Showing that the work is valued and motivating, and driving job satisfaction and providing staff with a career path;
- Creating an environment and organizational culture tolerable towards human error and allowing staff to report their impediments and the reason for their non-compliance. A fear culture and repressive style of management are banned for the benefit of early detection and the management of gaps and non-compliances;
- Being open-minded, frank with their own shortcomings or failures and investigating the incidents, or near misses, until the latent failures are determined and addressed; and
- Protecting staff who report non-compliance and whistleblowers from any retaliation.

CONCLUSIONS

A responsible food management, with a specific consideration for human factors, is central to the performance of any organization, be it a governmental institution or a business, and it is fundamental to food safety. No technological development can replace the competence and skills of managers in meeting their responsibilities. These include: good judgment in decision-making, skills in communication, training, but above all motivating and coaching staff in performing their tasks.

Management commitment is about creating an organizational culture and working conditions where the management supports the staff, so that each can become a leader in their own field and managers can exercise exemplary behavior. Good management of food safety should aspire to flawless execution; to achieve this, employees must be competent in their job, motivated and at all times vigilant.

This also means that, over and above monitoring critical limits and verifying the implementation of prerequisite programs as proposed in most food safety assurance systems, any gap or near-miss situation, such as transgression of critical limits or non-compliance practices, needs to be investigated, and their root cause, including managerial failures, determined and addressed (Figure 37.4).

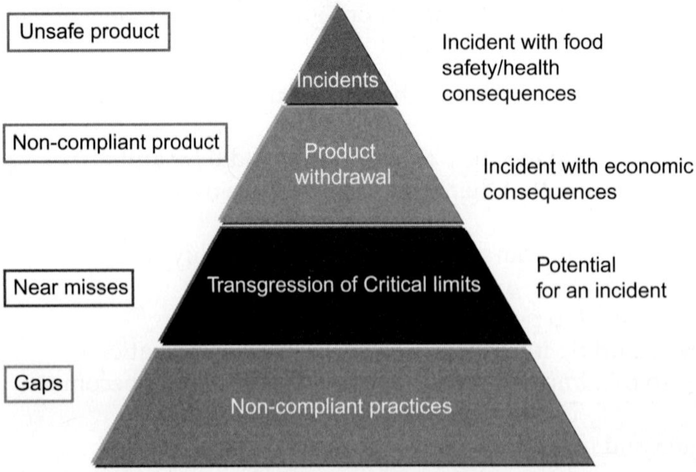

FIGURE 37.4 Every failure can be a potential for an incident. Monitoring the failures, understanding their cause, communicating and acting upon them as early as possible can reduce the risk of incidents.

References

CDC, 2007. Multistate outbreak of *salmonella* serotype tennessee infections associated with peanut butter – United States, 2006–2007. MMWR 56 (21), 521–524.

Motarjemi, Y., 2014. Modern Approach to Food Safety Management, in: Motarjemi, Y., Moy, G., Todd, E. (Eds.), Enyclopedia of Food Safety Management. Elsevier Inc, Waltham, USA.

Reason, J.T., 1997. Managing the Risks of Organizational Accidents. Ashgate, Aldershot (UK).

Reason, R., 1995. Understanding adverse events: human factors. Qual. Health Care 4, 80–89.

WHO, 2000. Foodborne Diseases: A Focus on Health Education in Food Safety. World Health Organisation, Geneva.

CHAPTER

38

Assessment of Food Safety Management Systems

Yasmine Motarjemi[1] *and Sara Mortimore*[2]

[1]Food Safety Consultant, Nyon, Switzerland, [2]Land O'Lakes, Inc., MN, USA

OUTLINE

INTRODUCTION

At first, it is important to clarify the use of the term *assessment* in this chapter. For the purpose of this book, the term assessment refers to an industry or governmental activity to verify that the food safety management system is implemented correctly and effectively, and is maintained. The primary reason for assessing a food safety management system is to establish whether a food business has the ability to consistently produce, manufacture or distribute "safe" food and to ascertain that the food safety management system provides adequate assurance.

In industry, the term is better known as *audits*, which itself is used in a variety of ways (audit of systems, processes and procedures, projects, laboratories, manufacturing,

Food Safety Management.
DOI: http://dx.doi.org/10.1016/B978-0-12-381504-0.00038-X

organizations and their management) and in different contexts such as financial, environmental or quality management.

In the framework of the enforcement of laws and regulations, governmental authorities are also led to verify the compliance[1] of industry practices with laws and regulatory requirements. This activity is usually referred to as *inspection*. In the past, inspections consisted of a snap-shot visit for checking compliance with good hygienic practice. With advances of HACCP and the development of an integrated approach to food safety management, the procedures and scope for inspection have also evolved. Today, it consists of a more comprehensive procedure similar to industry audits, referred to as regulatory audits.

Therefore, while the industry and regulatory audits differ in the purpose for which they are carried out and the authority that carries out the task, in essence they use similar processes and methods; and in both cases they are carried out with the aim of verifying compliance with a given standard. Therefore, this chapter will cover food safety audits from the perspective of both governments and industry. As will be seen later, auditing processes may also be used for reasons other than verifying compliance, such as evaluating the capability of a supplier to provide a raw material according to given safety specifications, evaluating equivalence in control measures in case of export–import certification, or evaluating the status of a factory or a business. Therefore, for the purpose of this chapter, the more neutral term of assessment is used.

In simple terms, we can distinguish different types of food safety assessments:

1. Internal assessments carried out by industry (part of self-control); and
2. External assessments carried out by either
   a. Regulatory agencies (known as inspection); or
   b. Third party assessments by customers or certification bodies.

BACKGROUND

During the past few decades, the management of food safety has greatly evolved in both the food industry sector and the governmental agencies (see Chapter 1). This change has also made an impact on the role and responsibilities of the industry versus regulatory authorities, and on the importance that is given to inspection or audits in food safety management.

While in the past the onus of safety was on governments, i.e. detecting an unsafe marketed product, with advances of the HACCP system in the last two to three decades, the responsibility for ensuring food safety and providing evidence for this has been shifted to industry. This means that the industry is to provide evidence that it is aware of the risks associated with its products and is taking the necessary preventive measures to control these.

Over and above establishing clear food safety laws, standards and regulations, the role of the regulatory authorities is to verify that the industry is complying with these; the assessment of the food safety management systems of industry is part of this verification. In industry, assessments are also used as part of self-control to verify that food safety management is effectively implemented and maintained.

[1]Compliance means that products and/or practices meet regulatory requirements.

In other words, regulatory or internal assessments are not for the purpose of controlling hazards but for confirming that control/preventive measures are implemented correctly and are effective. Governmental or supplier inspection cannot be a measure to ensure safety, but a measure to verify that the processor is implementing necessary control measures and complying with safety standards and other requirements to ensure food safety.

There are many books on the general aspects of assessments (audits or inspection) and the reader is referred to these sources, in particular to the ISO Standard ISO/TS 22003:2007 (see "Further Reading"). The objective of this chapter is not necessarily to turn the reader into a food safety assessor, but to highlight some essential points in an industry or regulatory assessment.

DEFINITION AND PURPOSE

As mentioned before, an assessment is an evaluation to verify the actual practices against set standards and codes. The purpose of an assessment may vary. It can include:

- Confirming the compliance (or identifying the divergence) with the internal rules and/or regulatory requirements. This is perhaps one of the most frequent objectives of assessments.
- Evaluating the ability of a supplier or a contractor to produce, manufacture or transport a food according to the set requirements. This can happen when choosing a supplier, a contract manufacturer or even purchasing a new business.
- Investigating violations or incidents, for example investigating a recurring CCP-related violation, employee complaints, alerts by internal whistleblowers, frequent consumer complaints or a fully fledged incident.
- Obtaining a certificate of assurance for customers that their requirements are met. This may be with customers nationally or internationally.
- Benchmarking or analyzing gaps in view of identifying the need for improvement, including the need for technical assistance, training and guidance on competences and/or improving the infrastructure (equipment, design of premises), etc. This can happen when a new factory or business is purchased, or when companies are merged. Experience has shown that small or medium-sized businesses are often not resourceful enough to know the regulations and that they often learn about these when they are visited by an inspector or assessed by a customer or the representative of a certification body. In such a situation, to avoid conflict of interest it is important that those involved in guiding the business are not the same individuals who will also assess for compliance.

SCOPE AND FREQUENCY OF ASSESSMENTS

As mentioned before, the scope and content of assessments have also evolved with time. Some years ago, depending on the stage of the food chain, such assessments were limited to verifying compliance with good fishery, agriculture, farming, manufacturing, transport or hygienic practices. Later, they were developed to include assessment of HACCP. Today, with the advance of an integrated approach to food safety management, particularly the development of ISO 22000, assessments include a variety of elements, from prerequisite programs (e.g. GMP) to HACCP, supplier management, product development, training, communication with other stakeholders of the food chain and regulatory authorities, and incident and crisis management. In this chapter, an appeal is made to give particular attention to management of people and to management commitment (see Chapter 37) since the people in a company, from the general manager to the workers on the line, play a key role in food safety management.

The decision on the scope and frequency of assessments or inspection will depend on a number of considerations, in particular whether the assessment is a first assessment or a follow-up assessment. Whether a full or partial assessment is carried out will depend on the original purpose of the assessment. For example, partial assessments might be appropriate for closing out non-compliances, for investigatory purposes after an incident or where a previous assessment has confirmed that a sound system is in place.

Classification of risks is an important criterion for prioritizing and deciding on the frequency, i.e. having more frequent assessments at higher risk premises or suppliers of high risk material. The following information can be considered in the classification of risks and in deciding on the frequency and scope of the assessments:

- The potential hazards known to be associated with the product and/or process;
- The history or level of previous compliance;
- The state of the food safety management systems and other management systems that may be in place, e.g. ISO quality management systems and certification, TQM as well as the level of in-house expertise;
- Other considerations such as processing methods, intended use and population at risk, size of operation (e.g. number of employees, volume of production, turnover), type of products and processes, complexity of operation, quantity of product affected by the raw material used, market or trade requirements.

Similarly, the following could be considered in the scope of an assessment:

- Whether it is an initial assessment or follow-up;
- Size of operation, e.g. number of employees, volume of production, turnover;
- Type of products and processes;
- Complexity of operation;
- Level of in-house expertise;
- Amount of available resources;
- Presence of management systems, e.g. ISO quality management systems, TQM;
- Results of previous assessments; and
- Population at risk.

A change in the system (process, formulation, etc.), or the aftermath of a natural accident or disaster, e.g. fire, flood etc., can also justify an assessment or an inspection. As mentioned previously, assessments may also be triggered as results of a previous food safety incident.

Subsequent frequencies for assessments and their scope can be considered in the light of the findings.

Table 38.1 presents the elements that could be the subject of a food safety assessment and presents some highlights of issues to be considered.

TABLE 38.1 Elements of Food Safety Management Systems and Some Highlights of Issues to be Assessed

Elements to be Assessed	Examples of Issues to be Considered During the Assessment
Management commitment, resources and management of people	This consists of ensuring that: – Management is aware of their responsibilities as listed below and set the example by following company policies. – A food safety policy is established and is communicated to all levels of the organization. – The food safety management system is described in an accessible language and is available at all levels. – The food safety management structure guarantees integrity and transparency. – A food safety leader is appointed with clear definition of his/her responsibilities and authority. – The food safety leader reports directly or has direct access to the top management of the company; he/she can report non-compliances without negative repercussions on his/her career. – The food safety management team is supported by an adequate infrastructure, equipment and material, and resources proportionate to its responsibilities and according to its scope of activities. – The responsibilities, interactions, reporting system and authorities are clear and mapped out. – The members of the team are knowledgeable, have integrity and are competent for their job. They receive training commensurate with their responsibilities and they are updated with development in the food safety areas, e.g. incidents, emerging risks, etc. – Periodic audits are used to verify the well functioning of the team as well as to provide an overview of the effective implementation of the food safety management system. – The system of corporate governance guarantees independence of audits and corrective actions, root cause investigation of gaps and incidents and their reporting to the higher management. – A system of whistleblowing is established and personnel grievances as related to their work are followed up, investigated objectively and corrective actions are implemented. – The management is open to providing necessary resources or investment where needed, supporting testing of products, making recalls in case of incidents. – The management gives priority to consumer health over business interests. – The crisis manager, if different from the food safety team leader, is also identified and competent for his/her job. – All important decisions, instructions, reports of non-compliance or possible risks, follow-up and closing out of issues are well recorded and documented.
Product traceability, recall and crisis management, incident management	– An effective traceability system is in place at the factory. – It is possible to trace every consumer unit. – It is possible to identify all finished products manufactured from a given consignment of incoming material, including rework. – Traceability exercises are carried out regularly, at least once a year to ensure that the traceability system is effective. – A formal written procedure for product recall is available and the system is tested periodically. – A formal early warning, crisis management procedure and crisis committee are in place and the role and responsibilities of the members are specified. – Incidents are thoroughly investigated, root causes established and lessons learned from the incidents are disseminated across the organization to prevent their recurrence. – Senior management is engaged in incident debriefs and preventive action.

(*Continued*)

TABLE 38.1 (Continued)

Elements to be Assessed	Examples of Issues to be Considered During the Assessment
Raw materials and supplier management	– Supplier management is in place: suppliers are selected and approved based on their capability to ensure safety of the raw material and are periodically audited and monitored. – Suppliers are aware of the intended use of their products and of the regulatory requirements of the country where their product will be used or these are indicated in the specifications. – Supplier auditors have the appropriate experience and training to enable them to carry out a risk-based assessment at the supplier's manufacturing location. – Raw materials (ingredients and packaging) have clear specifications and are signed to indicate agreement between the supplier and customer. – Certificates of analysis (where used) come from a laboratory that is able to competently test and samples are appropriately handled. The sensitivity of methods used are adequate for meeting the safety and regulatory requirements
Good hygienic practice (GHP)	– Incoming raw materials are inspected for their integrity authenticity and final goods are stored safely. – Warehouse management is in place, e.g. first in first out (FIFO) is respected, and raw material is properly labeled – Pest management is effective. – Adequate security measures are foreseen and visitors are screened for both health and security. – Housekeeping issues are respected, e.g. where applicable tools are labeled, chemicals are kept in a safe and secure location, identification (e.g. color coding) systems are used for tools, 5S system is used for maintaining order. – Training of employees is comprehensive and validated for effectiveness of learning: personal health and hygienic behavior protocols are observed as verification that knowledge is being reinforced in the facility. Employee facilities for hygiene, such as lockers, bench barrier entryways (where needed) and hand-washing facilities are readily available. – Basic rules of food hygiene are also respected in canteens. – Buildings (including drains) are designed to minimize risks and meet hygienic requirements. Doors and windows are appropriately closed and screened. – Zoning (including air flow and the need of a filter) and flow of people are managed to minimize risk of cross-contamination throughout the facility. – Hygienic design of equipment and maintenance programs, including calibration of equipment, are followed rigorously according to the state of the art. – Food grade lubricants are used. – Industrial services are managed to maintain a safe production environment. – Cleaning procedures are correctly laid out, are valid and the implementation is verified. – Rejected raw material or final products are correctly handled and disposed of. – In case of maintenance work, the impact of the work is considered in risk and control measures. – Foreign matters are prevented through various measures and policies, such as glass-free policy, jewelry-free policy. – Consumers', customers' and regulatory authorities' complaints are properly recorded and investigated and followed up in a timely manner. – Products are correctly labeled (content, visibility, clarity) and where consumers' practice is critical for safety, the communication of safety information is validated for accuracy and clarity. – Before launching or modifying any product, it is ensured that regulatory or safety requirements are met. – Examination of the area, such as security measures for the premises and screening of visitors and subcontractors for security as well as their health status.

(Continued)

TABLE 38.1 (Continued)

Elements to be Assessed	Examples of Issues to be Considered During the Assessment
HACCP system and implementation	– Adequacy of the hazard analysis. All potential hazards are adequately identified and evaluated, and if this evaluation indicates the risk is insignificant, is this validated by data? – Validity of control measures, i.e. the control measures eliminate or reduce the hazards to acceptable levels. – CCPs are identified, and critical limits are identified and operating within food safety and regulatory limits. Evidence should be obtained as to how these were determined, including the expertise used and any supportive document to validate these. – Evidence should be obtained that the monitoring of the critical limits indicates adequate control of the hazards. The adequacy of training in relation to personnel working at the CCPs and engaged in monitoring should also be considered, e.g. whether suitable instructions have been given to such personnel, and their role in relation to appropriate and timely actions. – An assessment as to whether the corrective actions would adequately restore control and are adequate to prevent an unsafe product from reaching the consumer. – An assessment of what, how, when and by whom the verification procedures have been undertaken, and whether these are adequate and effective. This may be indicated by an assessment of the validation data, sampling results, internal and external audit documentation as well as the frequency and thoroughness of all verification activities. The assessor should also consider whether changes, deficiencies in the HACCP plan, new emerging hazards, etc., are adequately provided for. Assessors should consider what actions are taken as a result of inadequacies in the HACCP plan or its prerequisites, or any other non-conformity[a]. – Additionally, assessors should consider whether records and documents are complete and in order. Where records indicate an issue or non-compliance, how these have been followed up. – The assessors should also evaluate the adequacy of the implementation, i.e. whether the HACCP plan and the prerequisites for HACCP have actually been implemented in the food business, maintained and are functioning correctly. – Root cause of CCP violations, or near miss investigations are carried out and short- and long-term corrective measures are in place.
Verification activities[a]	– Internal audits are carried out regularly by a competent team; they cover all levels and aspects of the operations and unsatisfactory reports are subject to an investigation and root cause analysis. – Consumer complaint handling system is valid and is working effectively, i.e. personnel are trained in what to do, how to ask questions and they have the ability to connect different sets of information to detect a pattern or a cluster of non-compliances. – Suppliers are audited according to a risk-based program by a competent team. The results of the supplier audits, including the monitoring activities of the suppliers (e.g. their end-product testing), are considered in the hazard analysis and maintenance of the HACCP plan. – The assessment of GHP and equipment maintenance programs is carried out on a regular basis. – The system for recording consumer complaints is verified and it is ensured that it is working effectively. – The pathogen and environmental monitoring as well as raw- and end-product testing are carried out effectively and results are regularly reviewed and used for the validation of the GHP program, and also for the maintenance of the HACCP system. – Laboratories carrying out chemical and microbial testing are audited for good laboratory practice and are accredited.

(Continued)

TABLE 38.1 (Continued)

Elements to be Assessed	Examples of Issues to be Considered During the Assessment
Verification activities[b]	– HACCP teams receive the results of various verification activities. – Hazard analyses are reviewed based on the verification data and where necessary, e.g. a non-compliance or a change, the HACCP plan is revised accordingly. – A release procedure is in place for finished products and raw materials. – Senior management reviews the status of the food safety program, including reports of audits, incidents, consumer complaints or other non-compliances reported by the staff. – Periodic traceability, recall and crisis management exercises, or other verifications, are carried out in an effective manner.

[a]*Conformity means that activities are carried out according to the established procedures.*
[b]*This is part of the HACCP system; however, as described in Chapter 1, to highlight the importance of verification activities for the validation of the hazard analysis and maintenance of HACCP, they are mentioned separately.*

COMPETENCE OF ASSESSORS

The validity of assessments depends to a great extent on the competencies of assessors and their integrity. Food safety being a multifaceted subject, a carefully selected team of experts will be required. The composition of this team and the expertise of the members will be all the more important as the responsibility for protecting public health is significant. In any case, for a full scope assessment, the following competences, skills and qualifications need to be considered:

- The technical competence;
- The skills in assessing and investigating (audit skills);
- The interpersonal skills and values, i.e. communications skills, diplomacy, resilience, patience, self-discipline and open mindedness. In addition they need to be curious and analytical in how they should interpret what they observe;
- Finally, a good assessment requires the cooperation and openness of the assessment entity in providing truthful information.

Other factors such as time and financial constraints and availability of documents also play a significant role. With regard to technical knowledge, the following are needed at the very least:

- Understanding the basic hygienic requirements, their relevance in supporting safe food production, and experience in assessing them;
- Knowledge of laws, regulations, standards and general codes of hygiene and/or criteria for the specific category of products;
- Knowledge of relevant industry products and processes (including past failures in the category);
- Knowledge of the HACCP system and its application, including:
    - The identification and assessment of potential hazards which may occur during food production, handling, preparation, storage and transportation, including biological, chemical and physical hazards;

- The ability to assess the effectiveness of control measures (validation) of the HACCP plan and its verification;
- Understanding the role of the human factor and of company culture in food safety.

THE PROCEDURE AND METHODOLOGY

The procedure for an assessment must be defined and carried out in accordance with a set format. Assessors should ensure that they plan the process properly, i.e. that:

- The scope of the assessment is predetermined and sufficient time is allocated;
- The required skills are available within the team;
- Tools needed are made available;
- Arrangements are communicated and agreed upon with the site being assessed.

The procedures for assessment will need to include the following stages:

1. A planning process to prioritize establishments, operations and their frequency and scope of assessments;
2. A desktop assessment;
3. An on-site assessment;
4. An evaluation process to analyze findings, determine compliance and decide corrective actions and follow-up requirements;
5. Reporting and follow-up.

The Planning Process

Initial planning is important to clarify the scope of the assessment and the approach that will be taken on-site. It helps to ensure that assessors have the necessary information and tools to complete an effective assessment. Information that will help in this planning process includes:

- Relevant company documentation;
- Previous file records, data on premises and products; and
- Results from previous visits or assessments.

The information obtained at this planning stage will also help to determine the focus of the assessment and the skills that might be necessary, particularly where assessments are carried out by a team. It also provides an opportunity to refine any checklist and protocols that might be used and, where appropriate, to communicate arrangements of the visits to the establishment. Any material such as camera, flashlight, tool kits, safety shoes, documentation and manuals can also be foreseen at this stage.

The Desktop Assessment

The assessment itself is best carried out in two steps. The first stage, desktop assessment, consists of the initial review of documentation, which may be carried out on- or off-site.

Although it is possible to carry out an assessment without a prior review of documentation, experience shows that a review of these prior to visiting the site leads to a more focused, thorough and informed assessment.

A review of the documentation allows assessors to get an idea of the standards that are relevant to examine and become familiar with the site products and processes. It will give the assessors an opportunity to carry out some research to build up knowledge of the product technology, legislative control measures and/or industry standards.

A desktop review also has the advantage of enabling assessors to plan their work, e.g. to judge how the CCPs have been established, check the personnel required for detailed discussions, review the specific questions to be asked, draw a list of priorities to focus on and/or examine areas to visit during the on-site assessment.

If the assessors find on the other hand that the document review has indicated obvious inadequacies, they may decide to stop the assessment at this point instead of proceeding to the on-site verification. Based on the findings, the assessors may decide to communicate to the company the type of measures which it needs to take.

A review of the company documentation is best carried out off-site, particularly when government agencies are concerned. In some instances, there may be some constraints that make this difficult or impractical, for example where the assessment is of an urgent investigatory nature or where it is intended to be unannounced. However, even where this can only take place on-site, it is important to review and make use of relevant documentation prior to a further physical examination of the site premises, processes and procedures. A review of the flow diagram or site plan, for example, will provide information on the nature and scale of activities carried out. This will help to target the assessment, particularly the further necessary scrutiny of records, equipment and processes.

Examples of documents to review:

- The food safety policy.
- The organigram, the responsibilities of the managers and food safety management team, and their respective technical expertise and competences.
- The operation and the type of products produced.
- The range and number of raw materials used and their origin.
- A site layout plan may give an idea of the flow of products through the site, the scale of the operation and the products produced.
- The HACCP-related documentation, including:
  - A process flow diagram and specifications relating to it;
  - The HACCP study (showing how potential hazards have been identified and on which basis they are considered as non-significant if this is the case);
  - An HACCP plan, including the monitoring plan and the validation of the control measures;
  - Records of CCP monitoring and corrective actions following the violation;
  - Verification data, e.g. consumer complaints, monitoring data for raw material, environment or end products, reports of incidents and root cause analyses.
- Training programs, e.g. the manual or other tools used for training.
- Incident and crisis management procedures.

- Records of investigation and root cause analysis of incidents (both active and latent failures) as well as evidence of follow up and corrections of gaps and dissemination of lessons learned from the incidents.
- Reports of management review of food safety and quality.

Together with the type of products and operation, such information is crucial for planning the assessment, particularly of high risk products but also to gauge if the number of personnel and their qualifications are adequate to manage the safety of the products. Many organizations use a pre-assessment survey to organize the information required for this type of desktop assessment.

On-site Assessment

The second stage is the on-site assessment. This will normally start with an initial or opening meeting to confirm, with the key people being assessed, the assessment scope, timetable, facilities and personnel required and in general to ensure cooperation. The time and location of the closing meeting could be confirmed and any additional documentation required for on-site document review could be requested at this stage.

As a regulatory authority, inspections may be carried out unannounced. This has the advantage of examining the place and practices as they are on an everyday basis and of obtaining the best picture of the real practices. However, there is also a disadvantage in that appropriate personnel may not be available to answer questions or that the inspection may disrupt the workflow, which itself can create other opportunities for mistakes leading to risk for consumers.

In an announced visit, it is helpful to prepare an agenda for the assessment program to ensure that relevant personnel are available during the assessment, and that their routine work is not disturbed more than it needs to be.

The purpose of this step of the assessment is to confirm that procedures and practices described in the food safety management system of the company or the regulatory requirements to ensure food safety are properly implemented in practice.

The scope of the assessment should have been decided during the planning stage. However, it could change depending on the findings of the on-site review of information, particularly if an off-site review (pre-assessment) was not done and the on-site assessment represents the first examination of the material. The scope of the assessment should also be changed during the assessment if serious non-compliance/deficiencies are seen. The on-site assessment will consist of a combination of activities. It should start with a review of the relevant documentation, their adequacy and accuracy.

A special focus should be put on HACCP, understanding the flow diagram, the competence of the team, the hazard analysis and validity of the decision taken in the HACCP study.

It will then move on to a physical examination of the processes, practices and records, by observation, measurement or interview to assess whether the actual operation in practice complies with the documented procedures. An important activity during this process is the evaluation of the state of prerequisite programs, including good hygienic practices according to the Codex General Principles of Hygiene or any other hygienic codes which may be applicable for the product or process in question. Such an examination will include the criteria listed earlier in Table 38.1.

During the on-site visit, specific attention should be given to HACCP implementation, including:

- Confirming the accuracy of the process flow diagram(s). This is facilitated by an initial walk through the site. The assessor will subsequently need to engage in a range of questioning and investigative activities to assess the efficacy of the HACCP system.
- Evaluating the hazard analysis taking into consideration the state of the prerequisite programs mentioned above.
- Confirming the suitability of CCPs, critical limits and corrective actions.
- Confirming that monitoring schedules are established and operating correctly.
- Confirming that persons responsible at CCPs perform activities correctly, understand the importance of the step for safety and their responsibility in case critical limits are violated. This will require specific interviews with the personnel.
- Establishing whether effective verification procedures are carried out.
- Reviewing monitoring data of raw materials, products, environment, CCPs, as well as reports of internal assessments, suppliers' assessments (inclusive of supplier monitoring programs), consumer complaints, personal reports and complaints. It is particularly important to corroborate these results with the hazard analysis (for instance, if a contaminant is considered as not significant in the raw material, this is confirmed through the monitoring carried out for verification).

During these activities, the assessors will need to keep sufficiently detailed records and to collect supporting evidence to enable conclusions to be made. Use of checklists together with a narrative, notebooks or, where appropriate, tape recorders, will assist this process. Depending on the judgment of the assessors, checks might be made on items of equipment, on-site measurements may be carried out, or product or environmental samples may be taken for subsequent laboratory analysis.

Additionally, assessors may

- Carry out tests to verify the well functioning of the traceability system.
- Check awareness of the regulatory requirements of the country where products are produced and/or marketed.
- Evaluate the knowledge and training of key personnel in food safety in relation to the job they are required to do.
- Review the handling of non-compliances (incidents of food contamination) or complaints from the regulatory authorities.
- Examine the organization's management structure to determine whether there are issues which may create conflict and undermine the reporting of non-compliances and/or investigation of incidents. The reporting system to ensure that top management is informed in a timely manner of food safety incidents or serious gaps in the company program, including managers' attitude and behavior, is a critically important element.

Evaluation Process

Where the assessment is being carried out by a team and a range of skills are being utilized, the evaluation and conclusions drawn will need to be agreed in advance of any final meeting with the site representatives. The assessor (or the team) will need to identify and analyze all information obtained during the assessment in order to draw up preliminary conclusions of deficiencies found, if any, and their effect on food safety, regulatory compliance or other trade-related concerns. Assessors should use the findings of their investigations to evaluate the effect any deficiencies may have on food safety and the speed with which they would need to be rectified.

The assessors(s) should evaluate findings based on objective evidence drawn from qualitative or quantitative information, records, statements, observations, measurements or tests which demonstrate that the prerequisites for HACCP or the HACCP system itself would not compromise food safety. Information and records gathered should be organized into a format that would support and justify the presentation of findings. It is beneficial to provide feedback on any positive findings of the assessment, where appropriate. This helps in presenting a balanced view.

At the exit meeting, the assessor will need to discuss non-compliances/deficiencies and agree on the expected corrective actions. The approach taken at this stage will depend on the purpose of the assessment, for example when the assessment had been triggered off by a serious food safety problem or where the assessment was to exclude previously identified deficiencies. However, in all circumstances, it is preferable to present any findings in a methodical manner, specifically highlighting best practices as well as areas of critical non-compliance or deficiencies.

The company should be given the opportunity to put forward its own solutions, as these may have substantial economic consequences such as capital expenditure, recruitment of new personnel, retraining of personnel or change of suppliers.

A timeframe for corrective actions should be decided according to the importance of the gaps identified. At the conclusion of any assessment, the company should be clear on any immediate remedial action required. The remedial action should be communicated to the site representatives with the appropriate responsibility. In some cases, written assessment reports might only follow more detailed off-site evaluation of the findings by the assessor. However, in all cases, it is necessary for the assessors to engage in follow-up activities to ensure that reported non-conformance is rectified.

The actions taken by government agencies where deficiencies are noted will depend on the nature of the identified deficiency, i.e. whether it is a non-conformance or a non-compliance. Some deficiencies will not have a direct impact on food safety. Assessors will need to have sufficient skills and competencies to evaluate the impact of deficiencies.

Other factors which will influence the action taken will include evidence of a repetitive pattern suggesting insufficient control that could lead to an adverse food safety problem.

Reporting and Follow-up

The format of assessor reports varies according to company policy and prior agreements with assessment bodies. However, it is essential that the results of the assessment be communicated to the management of the company and to all relevant persons within the organization (i.e. with responsibility for safety) in a timely manner.

Where an assessment report indicates critical or serious gaps, these need to be followed up rapidly, and root cause analyses of these gaps are also made to identify the latent cause of the failures (see Chapter 37).

THE DEVELOPMENT AND USE OF A CHECKLIST

Very often, to assess an operation assessors work from a checklist, i.e. a list of points to be considered during an assessment. Such a checklist is a useful tool for the assessment of the food safety management system provided that assessors are aware of its limitations and do not refrain from pursuing additional avenues of inquiry. Such a list has certain advantages and disadvantages (Table 38.2).

A checklist should be designed so that a quantitative or qualitative measure of the evaluation can be recorded. An example of qualitative evaluation would be the use of the terms: "excellent, good, medium, and poor" or "critical, serious, major, minor." Space should also be provided for written comments and objective evidence to be recorded next to each heading. The content of a checklist will depend upon the purpose of the specific assessment being undertaken and a specific checklist should be designed for each specific sector of the food chain. To facilitate their application, checklists should be supported by an assessment reference manual to guide the assessor in their correct and consistent application.

As an example, a list of commonly used questions in regard to assessment of HACCP is provided below. It does not represent a comprehensive checklist; it intends to show how a list may look and the sort of questions and activities which may lead to an effective assessment (Table 38.3).

TABLE 38.2 Advantages and Limitations of a Checklist

Advantages of Checklists	Concerns and Potential Misuse of Checklists
– Function as an aide-memoire – Help maintain the focus and objectivity of the assessment – Act as a record of the assessment itself – Ensure the completeness of the assessment – Are a useful tool in ensuring consistency of approach between different assessors – Help, together with associated reference manuals, to evaluate the comparability of different assessments, different companies or different assessors – Ensure transparency of the assessment process – Create confidence in the assessment process by all concerned, including government, industry and consumers – Enable assessment data to be more easily entered into a database which, in turn, can be used for reporting and trend analysis	– If designed or used improperly, may restrict the initiative and judgment of the assessors and discourage critical thinking and evaluation – It is important that the use of a checklist not evolve into a simple "tick-box" approach where there is no critical evaluation – A checklist may be improperly designed so that it may include unnecessary or irrelevant items, or may omit critical points

TABLE 38.3 An Example of Checklist for the Assessment of the HACCP System

Preparatory activities	What evidence is there of management commitment to HACCP use? *HACCP team* – Who was on the team? – Are all disciplines relevant to the product in question represented? – What is the likely knowledge level of the individuals (evidence of training, qualifications, experience, etc.)? – Has external expertise been sought where necessary? – What is the decision-making leverage of the HACCP team leader? *HACCP system* – How does the system fit with the overall food safety management system? – Is HACCP included in the food safety policy? – Has the scope been clearly defined? – Are previous records of safety (e.g. incidents) known to the team? – Has the product been properly described? – Are intrinsic control measures identified? *Process flow diagram (PFD)* – Is the PFD comprehensive? – How was the PFD verified for accuracy and by whom? – Are all raw materials and process/storage activities included in the flow diagram? – Are there rework opportunities and have they been included? – Is the PFD correct? – Have changes been made since the PFD was drawn up? – How is the HACCP team notified of changes to the process or product parameters? – How were the changes recorded and approved? – Were any changes discussed with the HACCP team before implementation?
Principle 1 "Conducting a hazard analysis"	How was the hazard analysis conducted? – Have all raw materials (including rework) been included? – Have all process steps been considered? – Have the potential hazards been specifically identified by type/source or have they been generalized? – How did the team assess the likelihood of occurrence? – What information sources were utilized? – Where potential hazards have been considered as insignificant have these been validated? Have appropriate control measures (CMs) been identified for each hazard? – Will the CMs control the hazards to an acceptable level and how was this validated? – Have regulatory requirements been considered in making these decisions? – Are all the CMs in place at the plant level?
Principle 2 "Determining the Critical Control Points"	How were the CCPs identified? – By expert judgment? – By the use of a decision tree (has the decision tree been used correctly?)? – By the use of consultants? – Have all necessary CCPs been identified? – Did each identified hazard undergo a systematic consideration? – How are the hazards which are not controlled by CCPs addressed?

(*Continued*)

TABLE 38.3 (Continued)

Principle 3 "Establishing Critical Limits"	How were the critical limits established? – Have critical limits been established for each CCP? – What validation exists to confirm that the critical limits control the hazards identified? – Is there evidence (experimental data, literature references, etc.)? – How do they differ from operational limits?
Principle 4 "Establishing a system to monitor the control of the CCP"	Have realistic monitoring schedules been established? – Do they cover all CCPs? – Has the reliability of monitoring procedures been assessed where appropriate? – What is the status of monitoring equipment? – Is it evidenced as being in place and calibrated appropriately? – Are the CCP log sheets being used at all CCPs? – Have CCP log sheets been filled out correctly? – Is there any evidence that procedures are not being followed consistently? – Does the frequency of monitoring adequately confirm control? – Are the sampling plans statistically valid? – Are statistical process control records being used to demonstrate that the process is in control on a day-to-day basis? – Check that records agree with stated activities. Are monitoring personnel properly identified and trained? – How was the training undertaken? – Are the monitoring records being reviewed by designated appropriate reviewers? Are violations of CCPs investigated and root cause analysis made?
Principle 5 "Establishing the corrective action to be taken when monitoring indicates that a particular CCP is not under control"	– Have the corrective actions been properly defined so that control is regained? – What evidence is there to demonstrate that this is being done in the event of a CCP deviation? – Has corrective action been recorded and how is the effectiveness being verified? – How has the authority for corrective action been assigned? – How are non-conforming products controlled and is this clearly recorded? – Are there clear disposition actions listed?
Principle 6 "Establishing procedures for verification to confirm that the HACCP system is working effectively"	– Have verification procedures been clearly and appropriately established? – How are these procedures communicated through the business? – Have responsibilities for verification procedures been allocated? – Are they being carried out effectively? – Are all CCPs covered by the verification program? – Are hazards considered as non-significant validated through verification programs? – Is there a formal system to trigger amendments? – Are control parameters being achieved? – Have process capability studies been carried out? – How are the data from HACCP being used to improve the system? – Are prerequisite support systems included within the verification program? – How is consumer complaint data being used within the verification system? – Is there a regular review of CCP failure and product dispositions?

(Continued)

TABLE 38.3 (Continued)

Principle 7 "Establishing documentation concerning all procedures and records appropriate to these principles and their application"	What format is being used to document the system? – Does the documentation cover all of the HACCP system operation, including: (1) the description of the product and its intended use, (2) the process flow diagram with the location of CCPs and related parameters available, (3) the HACCP worksheets on which are mentioned the hazards, the control measures, the CCPs, the critical limits, the monitoring procedures and the corrective actions, (4) data used for validation of hazard analysis, critical limits and monitoring parameters, corrective and verification activities, (5) the list of verification activities, (6) the results of monitoring and verification of the HACCP plan, and (7) the appropriate records necessary to ensure adequacy of prerequisite programs, particularly those used for validation of hazard analysis? – How is the documentation controlled with regard to update and issue, etc.? – Are the records accessible and are they clearly identified by unique reference numbers? – Are all documents accurate and current? – How is change control managed?
Implementation	Have the HACCP plan and the prerequisites for HACCP been implemented? – Personnel are trained in managing CCPs and know what to do when the CCPs are violated. – Personnel involved in verification activities and prerequisite activities are aware of the significance of their work for supporting the HACCP system and of the importance of reporting any non-compliance.

CONCLUSIONS

Assessment of food safety management systems is an opportunity to improve food safety management and close the gaps. It should be carried out with objectivity and integrity. An unsatisfactory audit report should not always and necessarily be a reason for reprimanding the managers; rather, over and above closing the gaps, a root cause analysis of the situation should be made and short-term or long-term corrective action should be made. Not infrequently, the root of the problem may be in the management.

Reports of audits and food incidents have shown that some of the major sources of food safety problems are:

- Raw material and supplier management.
- Failure in the design of equipment and its maintenance.
- GMP violation.
- Failure in hazard identification.
- CCP monitoring failure.
- Failure in corrective actions.
- Human negligence or error.

Acknowledgment

The authors would like to acknowledge that this chapter is based on personal experiences but also inspired from an earlier work, i.e. the FAO/WHO Expert Consultation that they convened on the subject of the Role of Government Agencies in Assessing HACCP (Geneva, 2–6 June 1998). The report of this meeting is available from the link http://www.who.int/foodsafety/fs_management/en/haccp98.pdf under the title Guidance on Regulatory Assessment of HACCP. The contribution of all experts in providing the guidance during the consultation is thankfully acknowledged. In spite of the date of publication of the report, much of the guidance is still up to date and relevant.

Further Reading

Dillon, M., Griffith, C. (Eds.), 2001. Auditing in the Food Industry. Woodhead Publishing in Food Science and Technology, Cambridge England.

Campden, B.R.I., 2009. HACCP Auditing Standard, second ed. Campden BRI, UK.

FAO/WHO Guidance on Regulatory Assessment of HACCP. Report of a Joint FAO/WHO Consultation on the role of Government Agencies in Assessing HACCP. Geneva, 2–6 June 1998, World Health Organization.

FAO/WHO. Global Forum of Food Safety Regulators, Building effective food safety systems. Proceedings of the Forum 12–14 October 2004, Bangkok, Thailand.

ISO. Food safety management systems – Requirements for bodies providing audit and certification of food safety management systems. ISO/TS 22003:2007. International Standardization Organization, 2007.

CHAPTER

39

Consumer Information and Labeling

Marjana Peterman[1] *and Tanja Pajk Žontar*[2]

[1]Slovene Consumers' Association, Ljubljana, Slovenia, [2]International Consumer Research Institute, Ljubljana, Slovenia

OUTLINE

INTRODUCTION

In effective economic markets, consumers fulfill two important roles through their purchasing decisions. First, they satisfy their own needs as individuals and second, their collective decisions ensure the competitiveness of the market-players.

American president John F. Kennedy on 15 March 1962 said that consumers by definition include us all. He added that consumers are the largest economic group, affecting and

Food Safety Management.
DOI: http://dx.doi.org/10.1016/B978-0-12-381504-0.00039-1

affected by almost every public and private economic decision. Yet, they are an important group whose views are often not heard.

President Kennedy then postulated four basic consumer rights, which are rights to safety, information, choice and representation. Some years later Consumers International (Consumers international, 2013) added four more consumer rights, which are satisfaction of basic needs, redress, consumer education and healthy environment.

In the United Nations Guidelines for consumer protection, as expanded in 1999 (United Nations Guidelines for Consumer Protection, 2013) is stated:

> The legitimate needs, which the Guidelines for consumer protection are intended to meet, are the following:
>
> a. The protection of consumers from hazards to their health and safety;
> b. The promotion and protection of the economic interests of consumers;
> c. Access of consumers to adequate information to enable them to make informed choices according to individual wishes and needs;
> d. Consumer education, including education on the environmental, social and economic impacts of consumer choice;
> e. Availability of effective consumer redress;
> f. Freedom to form consumer and other relevant groups or organizations and the opportunity of such organizations to present their views in decision-making processes affecting them;
> g. The promotion of sustainable consumption patterns.

WHO IS THE CONSUMER?

Nobody is just a consumer and consumers are not a separate group of people within society. The overwhelming majority of people are both producers and consumers during their lifetime. At some stages in an individual's life the producer role may be more important. At others – after retirement from work, for example – the consumer role may be dominant. On this basis, the individual's role as a consumer is distinct from her or his role as a producer. Put into operational terms, this concept might be rephrased as "the consumer is an individual who is offered, buys or uses goods and services, whether publicly or privately supplied, for personal or family use." In The Codex General Standard for the Labelling of Pre-packaged Foods (General standard for the labelling of prepackaged foods, 2013) consumers are defined as persons and families who purchase and receive food in order to meet their personal needs.

The word "consumer" therefore describes a person who is a buyer of goods and services as well as one who consumes goods and services and does not use these goods or services for producing and selling other goods.

In business to business communication the word "customer" means any person or business that is offered or buys goods and services for further use in the process of production and/or sale of goods and services. In the food chain a customer is any person or business that buys and sells goods and services and this includes those businesses offering catering or restaurant services and goods, including institutional catering/restaurant services to consumers. The "customer" therefore is a part of food chain business operators.

In business communication the "consumer" is sometimes referred to as "a customer, a guest, a visitor, a tourist, etc.," thus it should be prudent to use the term "consumer" in HACCP analysis as it is used in HACCP standard documents and all other legal acts.

CONSUMER PROTECTION

Governments should provide or maintain adequate infrastructure to develop, implement and monitor consumer protection policies. Special care should be taken to ensure that measures for consumer protection are implemented for the benefit of all sectors of the population, particularly the rural population and people living in poverty.

When formulating national policies and plans with regard to food, governments should take into account the need of all consumers for food safety, and should support and, as far as possible, adopt standards from the Food and Agriculture Organization of the United Nations and the World Health Organization, Codex Alimentarius or, in their absence, other generally accepted international food standards. Governments should maintain, develop or improve food safety measures, including, *inter alia*, safety criteria, food standards and dietary requirements and effective monitoring, inspection and evaluation mechanisms as well as food and health education policies and programs. Governments should also support and promote the role of consumer NGOs as consumer protection providers, since international consumer organizations on food to consumers aim to, according to the Consumers International organization food program:

- Facilitate informed and healthy choices by consumers, including vulnerable groups;
- Prevent misleading information and ensure that information can be trusted;
- Protect children from the promotion of unhealthy food;
- Ensure food sold to consumers is safe.

All enterprises should obey the relevant laws and regulations of the countries in which they do business. They should also conform to the appropriate provisions of international standards for consumer protection to which the competent authorities of the country in question have agreed, as per United Nations Guidelines (United Nations Guidelines for Consumer Protection, 2013).

GLOBAL REGULATORY MEASURES

The global or international trade in food brings to markets a wider choice of foods and at the same time provides consumers with a better choice of products. Since the establishment of the World Trade Organization (WTO) in 1995 and the WTO Agreement on the Application of Sanitary and Phytosanitary Measures (SPS agreement) the role of Codex Alimentarius standards became a legal base for food safety legislation in all countries that are members of WTO.

Codex Alimentarius General Principles of food hygiene (General principles of food hygiene CAC/RCP 1-1969, 2013) recognizes the role of consumers as: consumers should recognize their role by following relevant instructions and applying appropriate food hygiene measures.

Human and animal health and plant health protection measures are thus established and they are to be based on assessment of risk. The SPS agreement incorporates, therefore, safety aspects of foods in trade and applies the standards and related texts of the Codex Alimentarius Commission. Many countries have already incorporated HACCP (hazard analysis critical control point), Codex General Principles of food hygiene (General principles of food hygiene CAC/RCP 1-1969, 2013), into their legislation, including the European Union (Official Journal of the European Union (28 January 2002), Regulation (EC) No 178/2002, 2013).

CONSUMER CHOICE, INFORMATION AND EDUCATION

Food is a basic human need. Around the world consumers choose food for different reasons. When the choice is driven by hunger then hunger limits freedom of choice as well as concern for safety. When the choice is driven by pleasure, it means that basic human needs are satisfied. Also, there are as many drivers to food choice (besides pleasure, happiness, fun, friendship etc.) as there are consumers. Choice may also be influenced by certain experience in a given time, for example a food scare or a recall, or if some general and media information represents a threat. It is known that one bad experience may erase 10 good experiences from the brain (Maličev, 2012) and thus the freedom of choice is limited by bad experience.

Consumer choice in food may be influenced also by education, information and advice. At the early stages in life (childhood), consumers learn about food through food providers, mainly in family settings, where they also develop food preferences through experiences. This informal education continues throughout the lifespan, experiencing food in different settings and/or media exposures. Information to consumers on labels and other media used by food traders (also in advertisements) should be considered part of empowering consumers about food and its intended use, as well as safe use.

Formal education should be given by the state (health education policies and strategies) in order for the information to be autonomous and equal to all. Mostly it is given at a too early age, i.e. at elementary school, where cognitive functions are not developed yet. Consumers are then exposed to different venues of information about food and their awareness and knowledge may become biased. In this respect, information to consumers given by the food provider, such as trader, caterer or producer, is important; it is not only a legal requirement in most countries, at least the members of WTO, it is also a necessity for empowerment of the consumer to exercise an informed and safe choice. Informed choice is possible when food information provided on food or via any other means (e.g. oral, as a presentation, internet and other electronic means) to consumers is easily understood, reliable, readable, complete and not misleading.

The EU regulatory act (Official Journal of the European Union (25 October 2011), Regulation (EU) No 1169/2011, 2013) on food information to consumers is a new legal act that postulates mandatory information, taking into account that information to consumers is not only a label, but also an empowering tool (empowering also means using the information to learn how to use a product or service, which is an educational tool) and the principles are postulated as:

> Where mandatory food information is required by food information law, it shall concern information that falls, in particular, into one of the following categories:
>
> a. information on the identity and composition, properties or other characteristics of the food;
> b. information on the protection of consumers' health and the safe use of a food. In particular, it shall concern information on:
>   i. compositional attributes that may be harmful to the health of certain groups of consumers;
>   ii. durability, storage and safe use;

iii. the health impact, including the risks and consequences related to harmful and hazardous consumption of a food;

c. information on nutritional characteristics so as to enable consumers, including those with special dietary requirements, to make informed choices.

In order to achieve a high level of health protection for consumers and to guarantee their right to information, it should be ensured that consumers are appropriately informed as regards the food they consume. Consumers' choices can be influenced by, *inter alia*, health, economic, environmental, social and ethical considerations.

In order to follow a comprehensive and evolutionary approach to the information provided to consumers relating to the food they consume, there is a broad definition of food information law covering rules of a general and specific nature as well as a broad definition of food information covering information provided also by means other than the label.

The producer should consider a consumer or a vulnerable consumer group (e.g. children, pregnant women, patients) as a risk factor that is likely to occur due to poor education and empowerment thus misinterpreting the food information, when HACCP is being applied. It would be prudent also to conduct research or at-home interviews (face-to-face, not a phone question and answer exercise) or focus group discussions in order to define consumer understanding of information on the food product or need for improving understanding of the label or any other information necessary for safe use.

CLEAR AND LEGIBLE LABEL, A LEGAL REQUIREMENT

Food labels should be clear and understandable in order to assist consumers who want to make better-informed food and dietary choices. Studies show that easy legibility is an important element in maximizing the possibility for labeled information to influence its audience and that illegible product information is one of the main causes of consumer dissatisfaction with food labels. "Legibility" means the physical appearance of information, where the information is visually accessible to the general population and which is determined by various elements, *inter alia* font size, letter spacing, spacing between lines, stroke width, type color, typeface, width:height ratio of the letters, the surface of the material and significant contrast between the print and the background (Official Journal of the European Union (25 October 2011), Regulation (EU) No 1169/2011, 2013).

Figure 39.1 clearly indicates poor legibility due to a small font size and poor contrast. What type size and typeface should be used? If we consider a newspaper or a book as a benchmark, then the typeface must be of sans serif type, such as Arial or Tahoma, and as a minimum the size should be 8 pt or greater, providing that a distinct contrast (black on white) and proper spacing are used. Any producer, trader or caterer should consult individual country legislation regarding labeling and legibility, since at the time of writing there is no international agreement of the term.

PRODUCT INFORMATION WITHIN A FOOD CHAIN

It is recognized that product information is necessary not only for the final consumer, but also for anyone in the food chain in order to provide for a safe use of products and for the

FIGURE 39.1 Example of poor legibility due to a small font size and poor contrast.

purpose of tracing and traceability. Codex Alimentarius General Principles of food hygiene state that product should bear appropriate information to ensure that adequate and accessible information is available to the next person in the food chain, to enable them to handle, store, process, prepare and display the product safely and correctly. The lot or batch can be easily identified and recalled if necessary. Information for industry or trade (business to business or customer) users should be clearly distinguishable from consumer information, particularly on food labels.

Insufficient product information and/or inadequate knowledge of general food hygiene in any stage of the food chain can lead to products being mishandled at later stages in the food chain. Such mishandling can result in illness or products becoming unsuitable for consumption, even where adequate hygiene control measures have been taken earlier in the food chain.

It is generally recognized that in the catering business, where foods are used in restaurants, canteens, schools, hospitals and similar institutions and offered for immediate consumption, information to consumers about food is not customarily available. It must be recognized that caterers must be able to provide the same information to consumers as if it were a pre-packaged product.

Information about a food item, either via labeling or other means of communication, is a communication tool not only between the trader (producer, seller, caterer, etc., in short, a food business operator) and consumer, but also between producer and seller, in short, between food operators and any of the food stages within a food chain. This necessity is important in order to conduct a hazard analysis according to Codex Alimentarius Food Hygiene (General principles of food hygiene CAC/RCP 1-1969, 2013). In the process of hazard analysis, the first step of HACCP analysis requires a product description and the second step a product's intended use. These two steps can only be implemented if proper information is provided by and in between food operators in the appropriate food step.

Food chain information flow must be continuous, from farmer or food producer at the beginning of the food chain to consumer:

Farmer→distributor→processor→wholesaler→seller/caterer→consumer

CONSUMER AND RISK

A consumer addresses the hazard or risk differently from the professional or scientist. The following examples are consumer complaints received by the Slovene Consumer Association and depict certain situations and consumer understanding of food.

Example 1

A 10-year-old boy drank 1.5 to 2 liters of "ACE drink" (vitamin A, C and E-enriched drink) every day and after 2 months was admitted to hospital – he will, and most probably his family, will never consume vitamin-enriched drink again. No legal action was taken.

The producer should consider the highest possible food consumption level by a consumer of a product when designing a product and conducting HACCP, taking into consideration all groups of consumers the product might be used by (for example, small children). As an example, sugar-sweetened beverages (not juices) contributed 9 and 10% to the daily energy intake in Slovenian children (12–16 years old), respectively, which translates to an average of 650 ml of sugary drinks (not juice) consumed per day, therefore a consumption of 1.5 liters per day is possible for a boy (Fidler Mis, 2012).

Example 2

Many pre-packed products make a claim, on a front panel of a package, stating "no preservatives." A consumer with an allergy to sulfates suffered an allergic reaction, since sulfate was added as an antioxidant and not as a preservative (additives can have different functions) to this food. Although sulfate was listed in the ingredients, the consumer considered it safe to use, due to a general belief that "no preservatives" means "no additives." The consumer also stated that the ingredient list was written with such small letters that it was impossible to read the list in the store.

The producer should not have used the claim, written in large type on the front panel of the package, without putting a statement regarding allergens in the same field of vision as the claim, or should choose to omit the claim.

Example 3

A consumer bought pre-packed fresh chicken and kept it refrigerated till the end of its shelf-life (5 days). When the package was opened, the chicken had a foul smell and the consumer discarded the product. The consumer then made a complaint and sought advice from a consumer association food expert, since this was the second chicken from the same producer, bought in the same store, that had to be discarded in spite of the fact that the product was refrigerated.

The producer/packer should consider consumer behavior (for example, time and temperature of domestic refrigeration or the term "keep refrigerated") in determining the shelf-life of a microbiologically sensitive product. This may also be done by consumer behavior survey at the home. Establishing product shelf-life is the responsibility of the manufacturer or producer who needs to ensure that the safety and suitability of the food product can be retained throughout the maximum period specified, taking into consideration the potential for reasonably anticipated temperature abuse during handling by the consumer. Reasonably anticipated temperature abuse can be integrated into the shelf-life or challenge study or be taken into account by applying an appropriate safety factor. A survey or research of consumer domestic fridge temperatures and consumer practices regarding temperature and time for the food left in the fridge may be also beneficial in predicting or establishing a product shelf-life.

Example 4

A family visits a certain restaurant frequently. On one occasion, a member of the family became ill within 20 minutes after eating. When the owner/cook was asked if any ingredients had changed it was confirmed that peanut oil was now being used. The member of the family was allergic to peanuts.

The food producer, processor or provider/caterer should inform consumers of any allergens present. This example shows the importance of changing the information/label if and when there is any ingredient change.

LABELING OF ALLERGENS

The only way to avoid risks of allergic and intolerance reaction inherent in food is clear and understandable information on all ingredients present in food, either pre-packaged or served, in order for the consumer to make a safe choice.

Codex General Standard for the Labelling of Pre-packaged Foods clearly states that, when it is not possible to provide adequate information on the presence of an allergen through labeling, the food containing the allergen should not be marketed.

For example, EU Regulation (Official Journal of the European Union (25 October 2011), Regulation (EU) No 1169/2011, 2013) requires that labeling of certain substances or products causing allergies or intolerances shall meet the following requirements:

- They shall be indicated in the list of ingredients with clear reference to the name of the substance or product as listed and shall be emphasised through a typeset that clearly distinguishes it from the rest of the list of ingredients, for example, by means of the font, style or background colour.
  Some of the listed allergens are:
- Cereals containing gluten, namely: wheat, rye, barley, oats, spelt, kamut or their hybridized strains, and products thereof;

- Crustaceans and products thereof;
- Eggs and products thereof;
- Fish and products thereof;
- Peanuts and products thereof;
- Soybeans and products thereof;
- Milk and products thereof (including lactose);
- Nuts, namely: almonds (*Amygdalus communis* L.), hazelnuts (*Corylus avellana*), walnuts (*Juglans regia*), cashews (*Anacardium occidentale*), pecan nuts (*Carya illinoinensis* [Wangenh.] K. Koch), Brazil nuts (*Bertholletia excelsa*), pistachio nuts (*Pistacia vera*), macadamia or Queensland nuts (*Macadamia ternifolia*) and products thereof, except for nuts used for making alcoholic distillates including ethyl alcohol of agricultural origin;
- Celery and products thereof;
- Mustard and products thereof;
- Sesame seeds and products thereof;
- Sulfur dioxide and sulfites at concentrations of more than 10 mg/kg or 10 mg/liter in terms of the total SO_2, which are to be calculated for products as proposed ready for consumption or as reconstituted according to the instructions of the manufacturers;
- Lupin and products thereof;
- Mollusks and products thereof.

Consider an example from a consumer suffering from peanut allergy, who had had several trips to hospital (due to threatening anaphylactic shock) because of eating a food product containing "vegetable oil." Vegetable oil, although declared but not specified, was used and the consumer assumed that the food bought did not contain peanut oil. It had been a costly experience to learn which prepared foods contain peanut oil and which do not, when the label specifies "vegetable oil." It is clear that hazard analysis (in the process of HACCP) for allergen risk is a must for the producer/seller/caterer to market a safe product.

The following are applicable to the labeling of allergens:

- They should be clear, readable and understandable by any consumer;
- They should be emphasized on the label; an alert may be also used;
- Ingredients like lecithin, vegetable oil, starch, flour, whey, casein, etc. should be also labeled by the food source, to be understood by consumer;
- Formula/recipe change of a food product should be clearly indicated on the package;
- Restaurants should label allergens on the menu;
- Industry and business should take into consideration that consumers/families with hypersensitivities will avoid buying new products poorly labelled.

PRECAUTION

It is still not proven whether food colors (either azo dyes, synthetic colors or natural colors) have an effect on hyperactivity and ADHD; however, it has also not been proven that there is no effect, and effects on certain sensitive groups of children cannot be excluded. In many studies the azo dyes themselves had no effect, but the strongest effects were observed in children receiving azo dyes and benzoic acid combinations. Due to scientific uncertainty, in EU the precautionary principle was exercised by a risk manager (i.e. legislator) and the following legislative requirement is now a part of a Regulation (EC) on food additives (Official Journal of the Europe (31 December 2008), Regulation (EC) No 1333/2008, 2013) requiring, in the Annex V, that the labeling of foods include additional information, stating: *may have an adverse effect on activity and attention in children*, on foods containing one or more of the following food colors:

Sunset yellow (E 110) [*]
Quinoline yellow (E 104) [*]
Carmoisine (E 122) [*]
Allura red (E 129) [*]
Tartrazine (E 102) [*]
Ponceau 4R (E 124) [*]

[*] With the exception of foods where the color(s) has been used for the purposes of health or other marking on meat products or for stamping or decorative coloring on eggshells.

LABELING "MAY CONTAIN"

Labeling that states "may contain…an allergen" is not a precaution and should not be a substitute for good manufacturing policy or risk of legal action, but must be applied and used only if it is truthful and cannot be reasonably avoided. The statement "produced in a facility that also uses…allergen" is a statement seldom understood by the consumer and should be avoided.

CONSUMER FEEDBACK

Consumers can be information providers through a consumer complaint system. The system should not only include the process of redress in the case of a foul or non-edible food item (unsafe food or damaged product) but should also be a means of complaining by placing, for example, information on an internet page. Also, some consumer NGOs gather complaints and give advice as well as legal advice in case of damage (consumer redress procedure) to consumers. Consumer associations also, through their media, publish some consumer complaint cases in addition to publishing the results of consumer product testing. All these complaints or redress procedures can be useful to producers in evaluating the effectiveness of their food hygiene and HACCP procedures and approach to risk assessment and risk management.

This feedback information should be taken in the HACCP plan and when there are consumer complaints that indicate unsatisfactory conditions, implementation of the HACCP plan or validation of the control measures (e.g. formulation, processing, product shelf-life, etc.) needs to be re-examined.

DISCUSSION FOR THE FUTURE

Unfortunately food scandals still occur, and contaminated food can have adverse effects on consumers' health. Despite new technologies being used in food, e.g. nanotechnologies, which may also present risks, food safety will always be a key concern for consumers.

The future challenges lie in the evolving nature of risks as well as emerging risks sustainable management, taking into account climate change (ecosystems, biodiversity) and increasing global trade.

Sustainable production and consumption must become part of our lives, our decisions, our choice and behavior, the society as a whole and all stakeholders, from field to fork, in the food chain. Sustainability not only means adherence to laws and standards, it also means sustainment of four main goals: achieving a sustainable economy, ensuring a healthy and just society, living within the limits of our natural environmental and safeguarding natural resources (Our Common or Brundtland Report, 1987). In the Brundtland Report sustainability is explained as meeting the needs of the present without compromising the ability of future generations to meet their own needs.

Food chain representatives, from field to fork, need to support and manage, and strive for sustainability through improving *corporate social responsibility* and promoting *good governance* at the local, national and international level so that sustainable decisions and actions are implemented.

Corporate social responsibility and good governance should be the guiding principles used by food chain operators/businesses in order to protect consumer rights to safe food and a healthy environment as well as protecting the economy and society as a whole.

References

Consumers International, 2013. <http://www.consumersinternational.com> <www.consumersinternational.com> (19.04.13.).

General Principles of Food Hygiene, 2013. CAC/RCP 1-1969 <http://www.codexalimentarius.org/standards/listof-standards/en/?provide=standards&orderField=fullReference&sort=asc&num1=CAC/RCP> (20.05.13.).

General Standard for the Labelling of Prepackaged Foods, 1985. Labelling of Prepackaged Foods (CODEX STAN 1- <http://www.codexalimentarius.org/standards/list-of-standards/en/?provide=standards&orderField=fullReference&sort=asc&num1=CODEX> (20.05.13.).

Fidler Mis, N., 2012. Učinek pijač z dodanim sladkorjem na zdravje. In: Pavčič, M. (Ed.), Dietetikus, pp. 11–16.

Maličev, P., 2012. Nevroekonomisti bi lahko bili milni mehurček, toda nismo. In: Dobnikar Šeruga, R. (ed.), Delo, pp. 26–7, 19.5.2012.

Official Journal of the European Union, 2013. (28 January 2002), Regulation (EC) No 178/2002 of the European Parliament and of the Council of laying down the general principles and requirements of food law, establishing the European Food Safety Authority and laying down procedures in matters of food safety, <http://eur-lex.europa.eu/LexUriServ/LexUriServ.do?uri=CONSLEG:2002R0178:20031001:EN:PDF> (20.05.13.).

Official Journal of the European Union, 2013. (25 October 2011), Regulation (EU) No 1169/2011 of the European Parliament and of the Council, on the provision of food information to consumers, amending Regulations (EC) No 1924/2006 and (EC) No 1925/2006 of the European Parliament and of the Council, and repealing Commission Directive 87/250/EEC, Council Directive 90/496/EEC, Commission Directive 1999/10/EC,

Directive 2000/13/EC of the European Parliament and of the Council, Commission Directives 2002/67/EC and 2008/5/EC and Commission Regulation (EC) No 608/2004 (20.05.13.).
Our Common Future or Brundtland Report, 1987. United Nations World Commission on Environment and Development (WCED). Oxford University Press.
Official Journal of the Europe, 2013. (31 December 2008), Regulation (EC) No 1333/2008 of the European Parliament and of the Council of 16 December 2008 on food additives, <http://eur-lex.europa.eu/LexUriServ/LexUriServ.do?uri=OJ:L:2008:354:0016:0033:en:PDF> (24.05.13.).
United Nations Guidelines for Consumer Protection, 2001. As expanded in 1999, UN, New York and Geneva, <http://unctad.org/en/docs/poditcclpm21.en.pdf> (19.04.13.).

Further Reading

Griffith, C., in press. Principles and methods for health education and training in food safety. Encyclopedia on Food Safety. Elsevier.
McCrea, D., in press. Labelling and information for customers/consumers. Encyclopedia on Food Safety. Elsevier.
Motarjemi, Y., Ross, T., in press. Risk communication of biological hazards. Encyclopedia on Food Safety. Elsevier.
Motarjemi, Y., in press. Consumer education, information and risk communication. Encyclopedia on Food Safety. Elsevier.

CHAPTER

40

Incident Management and Root Cause Analysis

Yasmine Motarjemi[1] *and Carol A. Wallace*[2]

[1]Food Safety Management Consultant, Nyon, Switzerland, [2]University of Central Lancashire, Preston, UK

OUTLINE

INTRODUCTION

There are times when in spite of all efforts, some products do not meet the set standards for quality, safety or regulatory compliance. Clearly when this happens we need to take appropriate actions to protect the consumer and the brands. With proficient investigation and management of incidents, the negative consequences of these can be minimized. Over and above determining responsibilities for the mishap, it is also important to analyze incidents, investigate and understand their root cause and use the lessons learned to strengthen the food safety assurance system.

Food Safety Management.
DOI: http://dx.doi.org/10.1016/B978-0-12-381504-0.00040-8

As for management of crises, management of incidents is composed of different phases (see Chapter 41):

- Prevention of incidents;
- Reporting of incidents;
- Investigation; and
- Root cause analysis.

However, note that in this book, the terms "incident" and "crisis" are used in different manners. For the purposes of this book, an "incident" is defined as an episode resulting from:

- A deviation from the standard practice or a norm and leading to a substandard product; or
- Dissatisfaction of customers/consumers or regulatory authorities, e.g. due to injury, perceived food safety problems non-compliance with regulatory norms or sensory issues.

This compares with the definition of crisis in Chapter 41, i.e. a "crisis" is a predicted or unpredicted event that represents an immediate or future significant threat to an organization, its employees, consumers and the public at large.[1]

Incidents are often limited in time, unpredictable and lead to the need for a settlement and/or corrective actions. They are often (but not always) specific to one or several specific manufacturing sites, as opposed to wider industry crises such as avian influenza or BSE. As per the above definition, a good example of an incident is when a product is found not to meet the company or regulatory standards, or have caused injury to a consumer. What differentiates an incident from a crisis is the magnitude of the event, its consequences and the possible media attention. Frequently, incidents which are poorly managed can lead to a crisis.

Some of the principles and guidance presented in Chapter 41 also apply to the management of incidents, and the reader is invited to look into that chapter. In this chapter, the management of incidents and their root cause analysis are discussed in further detail.

PREVENTION OF INCIDENTS

Understandably, the prevention of incidents relies on a good food safety assurance system and this is addressed throughout this book. The specific focus here is the handling of non-compliances and/or *near misses* to prevent incidents. Near misses refer to situations or adverse events with the potential to cause damage and/or an injury, illness in consumers, without this actually taking place. Thus, as part of prevention, over and above a well-functioning food safety system, it is important to:

- Monitor near misses, or any unsatisfactory situation;
- Analyze their trends and their consequences; and most importantly
- Investigate their root causes so that appropriate action can be taken to prevent their recurrence.

[1] Adapted from Bartlett (1999).

Different types of data provide an indication of a potential food safety problem; these are:

- Violation of critical limits (CL). The violation of the critical limits in the HACCP system is a notorious example of a near-miss situation where the food safety standard is not met, but, with appropriate corrective actions, incidents are prevented. Monitoring unacceptable deviations of the CL and conducting a root cause analysis of the deviations is an important means for strengthening the food safety assurance system and preventing fully fledged incidents.
- Deviations in verification measures. Over and above monitoring at the critical control point (CCP), the HACCP system requires a number of verification activities. Data collected through verifications can be used as indicators to verify if the control measures at CCPs, or as part of prerequisites, are implemented as planned and are effective. Again, should verification data show an unacceptable deviation from set standards, its cause should be promptly investigated and corrected.

  Examples of verification data are:

  - Audit reports of the establishment and/or of suppliers. The reason for non-compliances reported in the audit reports must be investigated and the root cause determined.
  - Results of monitoring raw materials. These can show weaknesses in the food safety assurance system of the supplier. A non-compliance should prompt a notification to the supplier, an inquiry on the cause, and in case of repetition, possibly changing the supplier. In addition, depending on the severity of the monitoring results and status of the raw material batch concerned, there may be a need to locate and hold raw materials and/or to quarantine/withdraw/recall affected product (see "Managing an Incident," below). The scale of response needed will indicate whether the near miss is really an incident or is even progressing towards being a crisis.
  - Results of environmental monitoring. These can be an indicator that the products have been exposed to potential environmental contamination. Unsatisfactory results need to be examined and their root cause determined and followed up.
  - Results of end-product testing. Provided that substandard products have not been marketed and consumers have not been exposed, the situation can be considered as a near-miss situation. Clearly if product has been released, e.g. in the case of short shelf-life products, then an incident management or crisis response will be needed (see "Managing an Incident," below and Chapter 41).
- Reports of employees. Food establishments should be sensitive to employees' grievances or complaints about their conditions of work; they should encourage the reporting of problems, investigate these impartially, and address them in a fair manner. The importance of this point cannot be overemphasized: managing food safety is a very complex and challenging task; periodic audits and testing of products will not be sufficient to prevent incidents. Management of food safety requires the continuous vigilance of employees. Therefore, the real prevention lies in the ability to appreciate risks and to implement the control measures in a rigorous manner. The involvement and active participation of all employees in meeting this challenge is central to food safety management and this is strongly influenced by the organizational culture. A culture that intimidates or promotes fear will inevitably discourage staff from reporting problems and create an environment favorable for incidents.

BOX 40.1

In an industry context, incidents are categorized under different terms:

- Food safety incident is where consumers' health is at risk or the food safety standard has been breached.
- Regulatory incident is when a regulatory requirement is not met, without this jeopardizing the safety of the product; for instance, if there has been a mislabeling with regard to the amount of the product in the package.
- Quality incident refers to a quality defect that does not jeopardize the safety of the product, for instance when there is an agglomeration of the product.

REPORTING AN INCIDENT

The term incident refers to a situation where a non-complying product has reached the market and consumers have been exposed (Box 40.1). Not always do incidents lead to illness or injury in consumers. For instance, in some cases regulatory norms may be exceeded, but the short-term exposure of consumers to a contaminant or an ingredient may not present a significant risk for their health; nevertheless, the food safety standard has been breached and the food business has to recall its products. Over and above the economic loss, such an event can damage the reputation of the company and call into question the ability of the company to manage the safety of its products. Also, some non-compliances or defective products may not present any safety issue, but the consumers may perceive the issue otherwise. This is often the case with the spoilage issues or foreign bodies that would not meet the definition of a food safety hazard.

To prevent any adverse health effects and/or damage to consumer confidence, it is important that the business sets up a sensitive method for reporting incidents and investigating them. Examples are:

- A hotline service, preferably on a 24-hour basis.
- Information on the website on how the consumer and/or customer (e.g. retail) should contact the business.
- A clear in-house reporting system with emergency telephone numbers and a responsible person to contact in case of an incident.
- Lot coding and a traceability system are essential for the investigations and the more specific the lot coding is, i.e. providing information on the time of production, the smaller the product loss will be in case of recall.

MANAGING AN INCIDENT

Following the report of an incident, a number of measures need to be taken as first actions. These of course depend on the nature of the incident. Some of the guidance measures described below may seem self-evident to a trained or experienced food safety manager. However, experience from past incidents has shown that failures in implementing these measures have turned simple non-compliances into major crises. Therefore it is important that all relevant personnel have appropriate knowledge and training on what to do in the event of an incident. Some key measures are as follows.

The managers in charge should:

- Consider the need for blocking products; this depends on the nature of the defects and whether there is a suspicion or confirmation that the product in question is possibly implicated.
- Inquire about their eventual injury or illness in case a consumer has complained, whether directly or through a third party, e.g. regulatory authorities. In this eventuality, the manager should show empathy with the consumer's problem, whether this is an emotional affectation or an actual health injury; inform them that a thorough investigation will be initiated and that in due course the cause of the incident will be determined. Should the product be implicated, naturally consumers should be compensated and apologies be presented. As seen below, investigation and understanding the cause of the incident are also important for deciding on the follow-up actions, e.g. extension and type of recall or corrective actions.
- In case of any report by regulatory authorities or by a third party, e.g. a customer or retailers, the manager should acknowledge as soon as possible (within 24–48 hours) the receipt of the report and should assure the complainant that the issue will be investigated at once.
- Where applicable, e.g. in case of doubts on the implication of the product, reconfirm the test with an independent and accredited laboratory.
- Initiate an evaluation of the risks of the product for consumers and other consequences of the incident (regulatory violation, image).
- Depending on the nature of an incident, its consequences, e.g. an outbreak of foodborne illness, or a substantial recall[2] or a withdrawal,[3] consider the need for communicating with media (see Chapter 41).
- Decide jointly with the authorities whether a decision should be made to recall a product and whether it should be communicated internally and externally according to the circumstances (Box 40.2). Products should also be disposed of according to the regulation and in such a way that they are not at the reach of general public or employees.

In any case, a swift reaction is needed to address any ill-feelings and maintain trust. At all times, consumers' health and regulatory or customers' concerns should be the first priority, and in further discussions with the complainant, honesty, openness and transparency should be the rule of thumb. To this point, regulatory authorities or third parties should be

[2]Recall means any measure aimed at achieving the return of a product that has already been supplied or made available to consumers by the producer or distributor (adapted from EC, 2002).

[3]Withdrawal means any measure aimed at preventing the distribution, display and offer of a product to the consumer (adapted from EC, 2002).

BOX 40.2

THE DIFFERENT TYPES OF RECALLS DEPENDING ON THE TYPES OF INCIDENTS

There are different levels of recall:

- *Internal level*: products that have to be withdrawn are still within the control of the food operator, either in the factory, in transit or in company warehouses, but not at trade/retail level.
- *Trade level*: the suspected product is in the retail trade; the product is removed from the warehouses and often also from the retail shelves. This is typically done in case of regulatory (e.g. error in the name of product) or quality incidents; it is also referred to as withdrawal.
- *Public level*: recall down to the consumer level; a public recall is required when the incident is assessed to be a safety incident, whether people have been injured or not, and the public must be notified to prevent consumption or use.

provided with all the necessary data to support the findings of the investigation, if different from their report on product non-compliance.

It goes without saying that incident management, as part of food safety assurance, requires competent and well-trained and disciplined staff, as in several historical incidents blocked products were released by mistake. To this end, traceability and product recall should be part of training, yearly review and verification.

INVESTIGATION

Upon the report of non-compliance, whether or not consumers were injured or became ill, an immediate investigation should be launched. This should include:

- Examining the product for the defect and possible implication of the business. There are times when the defect may arise at another point in the food chain, e.g. at the retailer or consumer itself; for instance, many reports of glass complaints may be related to events that occurred in the home environment. The possibility of tampering should also be considered. If such is the case, over and above regulatory authorities, the police may need to be informed.
- Tracing the product to the location and time of production, processing and investigating the conditions of production, processing, transport and distribution. Depending on the nature of the incident, data on the practices of customers or consumers should also be collected.
- Examining whether the hazard was considered in the HACCP study, i.e. the hazard analysis was correctly carried out and whether the HACCP plan was accurately elaborated.

- Examining the records for any deviation in the implementation, i.e. CCP monitoring, environmental monitoring, raw material and end-product testing.
- Identifying and interviewing the operators and managers responsible for the production, eventual third parties working on the site, e.g. subcontractor for cleaning or maintenance.

The scope, method and approach for investigation can vary according to the nature of hazards. For instance, for physical hazards, an examination of the nature of the hazard can determine from which production area or equipment the agent may originate. For biological hazards, over and above the above-mentioned data, possible contamination of products by an infected or carrier employee needs to be considered. For nutritional hazards, e.g. excess or lack of vitamins, in addition to operational errors, error in product formulation needs to be considered. Chemical hazards may originate from the raw material or surface contacts (e.g. packaging, conveyor belt), and sometimes leaks from equipment (e.g. lubricants, cleaning agent residues). In case of suspicion that the raw materials may be implicated, an analysis of these will be required.

Understanding the cause of an incident is essential for determining the range and extent of products affected and the type of corrective measures that are necessary. The Perrier water crisis presented in Chapter 41 is a case in point. The fact that Perrier attributed the contamination of the mineral water to a human error in their North American facility led to a limited recall of water whereas the contamination was at the source, and a broader recall was needed. The delay in recalling the product and communicating an invalid explanation of the incident were the major reasons for the crisis. A similar situation took place with the Coca-Cola crisis in which there were many controversies regarding the cause of the problem. The faulty implication of products in the *E. coli* O104:H4 outbreak due to contaminated fenugreek in Germany in 2011, and *Salmonella* Saintpaul associated with jalapeño peppers and serrano peppers in the USA in 2008 both led to major economic losses for producers of cucumbers and tomatoes.

In case of an incident implying microbial hazards, the decision for segregating safe from unsafe products cannot be based on the testing of the product, as microbial testing alone cannot provide assurance of safety due to the likelihood of both trapping and detecting the hazard in the specific sample tested. This is of particular importance if the nature of the product or organism is such that low doses can cause a serious health effect. In a nationwide outbreak of *Salmonella* Typhimurium associated with peanut butter in 2009, the implicated company retested its products until it found negative results. Judgment of safety should be based on the confirmation that the conditions of production and processing are appropriate; microbial testing can be a further proof but should not be relied on solely. In the same line of thought, in the investigation of an outbreak, epidemiological investigation may be sufficient to render a product suspect and initiate precautionary measures, even if the microbiological testing fails to implicate the product.

ROOT CAUSE ANALYSIS

After a near-miss situation, an incident or a fully fledged crisis, a root cause analysis needs to be done and measures need to be taken to prevent recurrence of the event. While the general public may tolerate incidents caused by an unexpected event or a human error, it may not accept negligence or the repetition of incidents of the same kind, which is indicative of complacency. Most importantly, in case of an incident, it is essential to be able to demonstrate that the principle of *due diligence* has been respected and all measures have been taken to prevent future cases. In an incident where a company was implicated in a case of *E. sakazakii* in Belgium (March 2002), authorities inquired about the measures that the company had taken since a previous case to prevent recurrence.

The root cause analysis should not be confused with the investigation of the primary cause of an incident, which should be identified in the first place as part of the management of an incident. The root cause analysis is a postmortem exercise for better understanding of the underlying factors leading to the cause of the incident.

To understand the concept of root cause analysis, examining the way an incident occurs is important. This has been described by James Reason (Reason, 1997) and his approach to organizational incidents is used here.

In food safety assurance, a series of measures are foreseen to control hazards. As mentioned in Chapters 29 and 36 these can be grouped under basic good practices, HACCP and verification measures. When an incident occurs, usually it is the result of a, or rather a series of, gaps or failures in these measures. A gap or failure in any of the above-mentioned measures creates a weakness in the food safety management system and causes a threat situation which, if investigated and corrected immediately, prevents an incident from recurring. However, if a gap is not addressed, with time, combined with other gaps, it may potentially lead to an incident and, if this incident is not managed effectively, it may escalate to a crisis situation. An example of the additive effect of gaps in systems is an incident caused by vitamin B1 (thiamine)-deficient infant formula (Israel, 2003). In this incident, a reported 15 babies suffered damage to the nervous system and two died. The cause of the incident was an error in product formulation, but a second failure was in the verification of the composition of the product before its release. Similarly, in the incident of isopropylthioxanthone mentioned in Chapter 41, a combination of gaps in the regulatory requirements, suppliers' tests and practices as well as customers' awareness of risks were the origin of the problem. Such a situation where gaps of different levels and nature can combine to cause an incident is referred to as the "Swiss cheese model" (Figure 40.1).

A second concept that must be understood is the concept of active and latent failures relating to people and management (Figure 40.2). Behind any control measure, there are people who have to implement the control measures or verify that they are correctly implemented. These can be a worker on the line or in the farm, an operator monitoring the temperature recorder, a truck driver who has to manage the temperature during transportation, a food handler who has to wash his hands before preparing food, etc. Their failure to perform their work is referred to as active failures since their actions will have a direct and immediate bearing on the safety of products (Figures 40.2 and 40.3a and b). These are the types of failures that are typically investigated in case of an incident or near miss. Often, as a result of the investigation, the employee receives the blame, and may even be fired, and then the investigation ends at this point. The same process and relationship also exist between regulatory authorities and food establishments that are caught up in an incident.

However, in a root cause analysis the task is to go deeper in the investigation and understand the conditions that have led to the non-compliance of the person implicated in the incident, i.e. committing the so-called active failure. Worldwide, studies indicate that factors

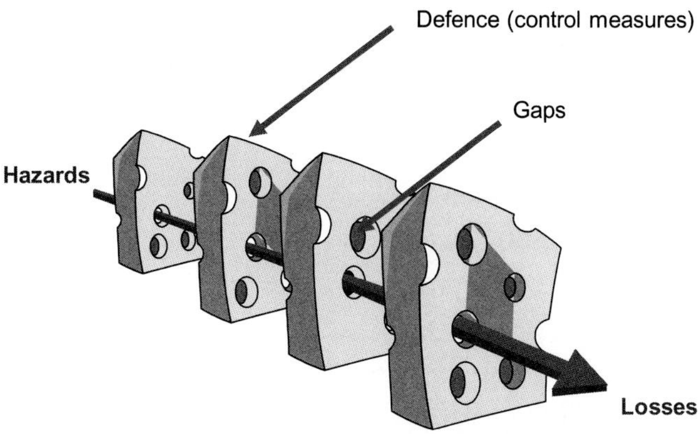

FIGURE 40.1 Swiss cheese model according to James Reason (1997).

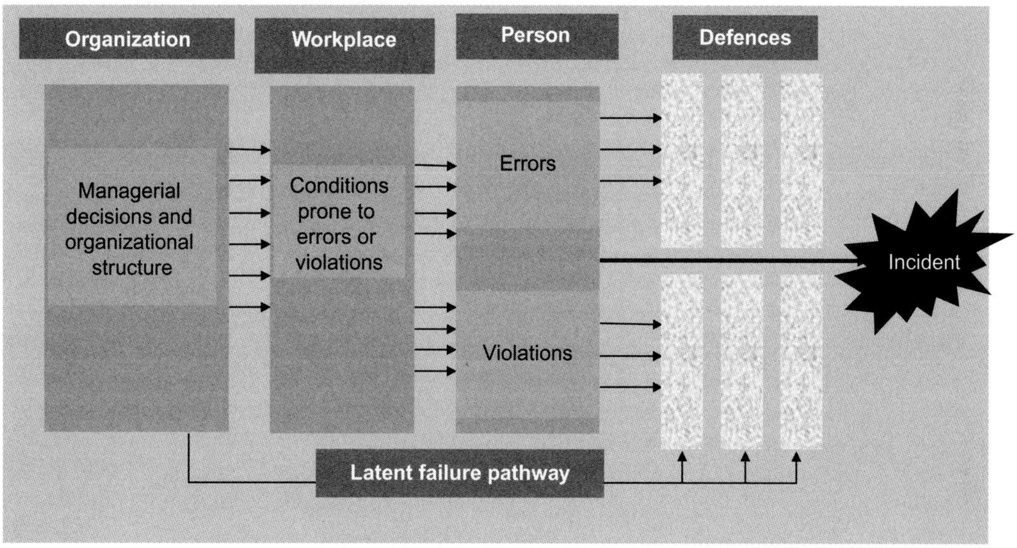

FIGURE 40.2 Levels and types of failures leading to an incident (Reason 1995).

that lead to active failures are often related to the working conditions, e.g. time constraint, lack of clear instructions, failure in defining the responsibility or authority of the person or providing adequate training and coaching, or creating a culture of fear or demonization, etc. Such situations are latent conditions which result from management decisions (Table 40.1). Thus, failures of the management in creating conditions that are optimal for managing food safety are referred to as *latent failures* (Figures 40.2, 40.3a and b). Latent failures may not have an immediate impact, but they weaken the food safety management and increase the probability of active failures, and thus of incidents. Latent failures have been the cause of numerous accidents in the petrochemical, transport and food industries and in financial institutions.

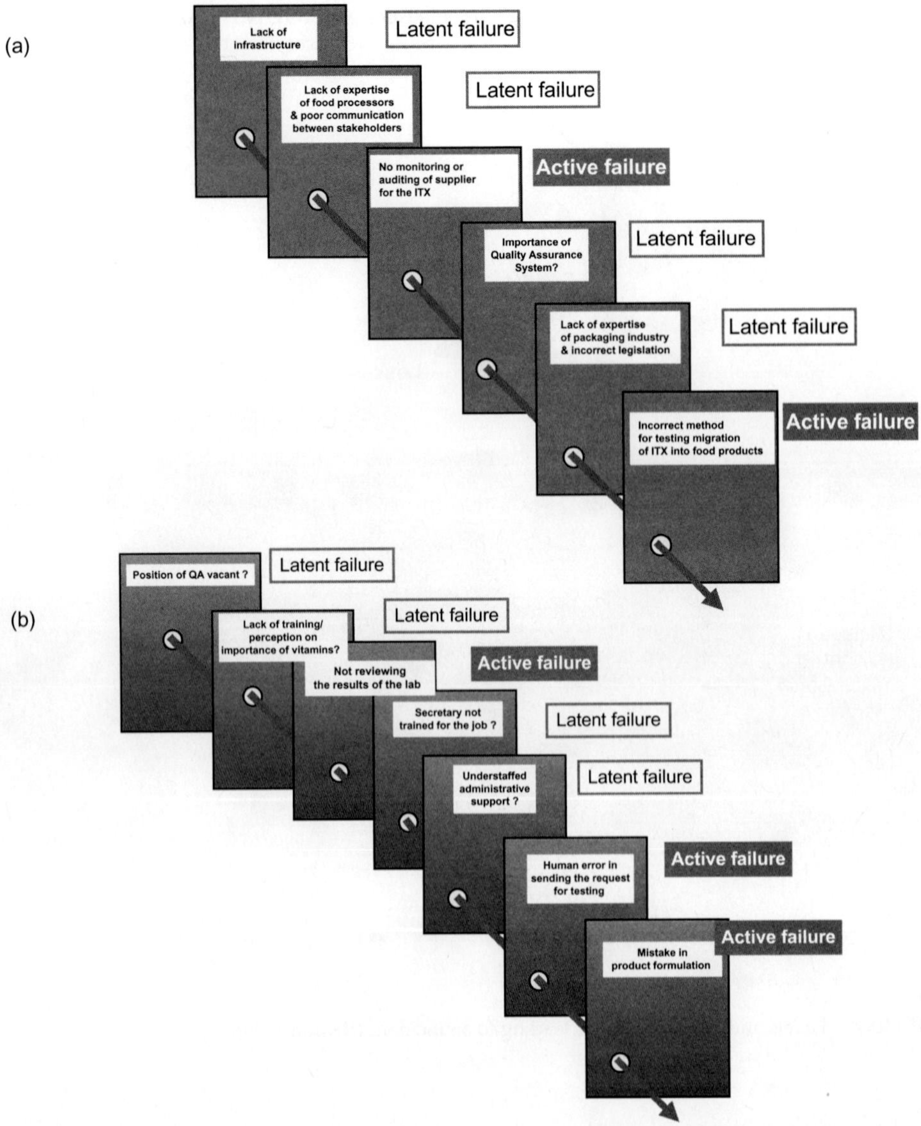

FIGURE 40.3 (a) Root cause analysis of the food safety crisis associated with infant formula contaminated with isopropylthioxantone (ITX). (b) An analysis of the root cause of an incident related to thiamine-deficient infant formula (Israel, 2003) based on information reported from unofficial sources. Some failures are hypothetical mentioned for educational purposes.

To recapitulate, a root cause analysis requires a truthful investigation of an incident at several levels, i.e. understanding:

- The primary cause of the incident: often a technical mistake, equipment failure or human error/violation. Examples are errors in the technical parameters of a product or processing, a broken sieve, or staff using a wrong thermocouple;

TABLE 40.1 Ranking of Latent Failures Preventing Efficient Implementation of HACCP

Barriers to Implementation of HACCP	Frequency of Reported Cases
Time	7
Human resources (staff)	2
Resources	5
Expertise/Knowledge in food safety and hygiene	5
Management commitment and perception of HACCP	4
Understanding of HACCP principles and systems	3
Employee motivation and attitude	3
Training	3
Weakness in regulation or enforcement	3
Lack of policies and procedures	2

(Adapted from Jevsnik, Hlebec and Raspor 2006).

- The conditions leading to the non-compliance of the person in charge of implementing the control measures, such as lack of training, time constraint, difficulty in understanding an instruction; and
- The managerial decisions that have led to those working conditions, e.g. failing to provide the necessary policies, to appoint a competent manager or personnel, to plan an optimum reporting and organizational structure, to provide adequate financial or human resources or adequate equipment, and a management behavior in contradiction or in violation with instructions, or requiring impossible tasks and forcing staff to take risky shortcuts or violate the rules. Worst would be a management that violates its own policies. This will have repercussions on the entire company.

Tools for Root Cause Analysis

Root cause analysis is used quite widely in healthcare and business settings but, as yet, it has not really been adopted by the food industry to any great extent, although the concept is identified as necessary in some food safety and quality certification standards, e.g. BRC Global Standard for Food Safety Issue 6 (BRC 2011). However, some of the tools of root cause analysis, notably structured Failure Mode and Effect Analysis (FMEA), have been used for some time in food companies for various applications. For example, Mortimore and Wallace (1994, 1998 and 2013) advocate the use of FMEA to challenge the controls within a HACCP plan before it is implemented within a food operation, the idea being that by understanding the likely causes of failure in the control systems then the controls can be strengthened further, delivering additional confidence of food safety assurance.

As discussed previously, root cause analysis needs to investigate an incident in depth to gain an understanding of all the conditions that have led to the incident occurring. It is necessary to consider all possible contributing factors and this requires both a structured approach and the ability to "think the unthinkable."

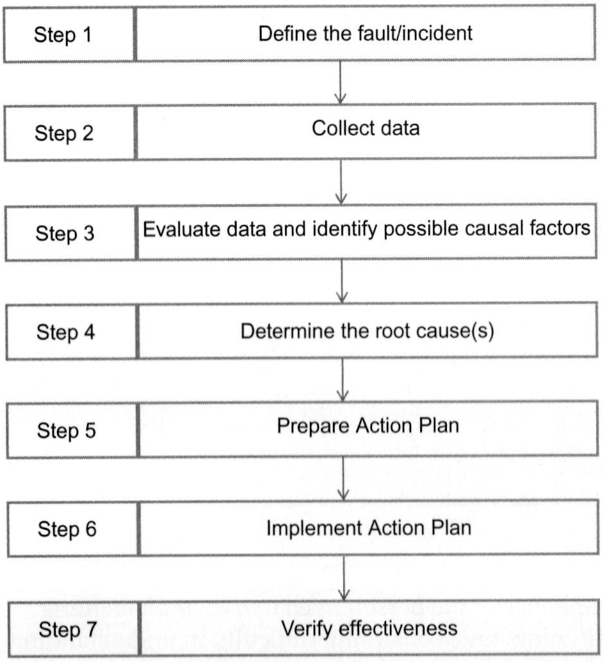

FIGURE 40.4 Root cause analysis – seven step process.

Root Cause Analysis Teams

Like many aspects of food safety management systems, root cause analysis is best performed using a team approach rather than by an individual or individuals working alone. The team needs to include personnel from within the business who have knowledge of key areas of investigation. As such, a multidisciplinary team similar to the approach used in HACCP will likely be most effective. Team members will include personnel who have knowledge and responsibility within technical/quality, manufacturing and engineering functions plus additional relevant personnel, e.g. human resources, purchasing, warehouse and transport managers and so on, depending on the nature of the incident. While the number of team members is likely to be small (four to six personnel), the team will be able to call on other personnel within the business structure to help understand what has happened and the likely contributing factors.

Structuring the Root Cause Analysis

To perform an effective route cause analysis it is important to use a stepwise approach and take the time to gain a detailed understanding at each stage before moving on. Figure 40.4 shows the steps of the structured approach to root cause analysis. Although there is general agreement on the necessary actions, various texts on root cause analysis use different numbers of steps within their route cause analysis models. We will use a seven-step process here since this covers both the analysis and the implementation and verification of corrective actions.

STEP 1 DEFINE THE FAULT/INCIDENT

Members of the root cause analysis team first need to understand what has gone wrong. At this stage it is helpful to compile as much information as possible about the fault or incident, i.e. a summary of what has gone wrong, including as much as possible on the sequence of events and what has been done so far in terms of immediate corrective action and incident management. This information is useful as background to the team allowing everyone to gain an appreciation of the incident situation.

STEP 2 COLLECT DATA

Next, it is important to collect further, more detailed information that will assist in evaluating the problem. For example, this might include:

- Product test results
- Lists of implicated products or processes
- Lists of raw materials associated with implicated products and processes
- Lists of packaging materials
- Monitoring results covering dates and times thought to be implicated. The results sample(s) should be large enough to capture all relevant data around the suspect dates, i.e. building in a margin of error.
- Corrective action records covering the dates and times thought to be implicated
- Engineering and maintenance records
- Pest management records
- Complaints records and customer contact information
- Any other relevant information, e.g. interview information from staff, etc.

Brainstorming will be a useful tool to make sure that all the necessary information sources can be identified and then members of the team can be allocated particular records to obtain and review on a preliminary basis.

STEP 3 EVALUATE DATA AND IDENTIFY POSSIBLE CAUSAL FACTORS

In order to identify the possible causal factors all the data collected so far need to be evaluated and discussed; in addition any further information required will also need to be identified. This is best done by the team working together to discuss the information found and by bringing in additional personnel as necessary to help understand the situation, e.g. factory floor staff who are familiar with the ongoing processing situation and additional experts (possibly external) who can advise on specific issues. Further discussion and brainstorming will help to elucidate the possible causal factors and these all need to be recorded by the group or its appointed secretary/scribe.

STEP 4 DETERMINE THE ROOT CAUSE(S)

The list of possible causes needs to be considered further by the team, evaluating how each one may have contributed to the problem. The use of tools from the root cause analysis toolkit (see below) will help the team understand how the possible causes may be interrelated and will assist in tracking backwards to the root cause(s). Grouping techniques such as Ishikawa Cause and Effect Analysis and questioning techniques such as the 5-Whys are particularly helpful in this context, although teams may also find some of the other tools helpful in prioritizing possible causes from their list.

The team should agree on the root cause or root causes (likely if there are distinctly different gaps or causal factors involved in an incident). The discussion can then progress onto what needs to be done to address the root cause(s). Additional tools from the root cause analysis toolkit can be helpful at this stage, such as FMEA, which considers the current controls and then identifies recommended new controls for each cause of failure.

STEP 5 PREPARE ACTION PLAN WITH TIMESCALES AND RESPONSIBILITIES

The team's recommendations for new controls, systems, personnel and infrastructure actions need to be built into an action plan with appropriate timescales for completion/implementation. Appropriate responsibility from the management hierarchy should be defined for each action point and personnel should be advised accordingly.

STEP 6 IMPLEMENT ACTION PLAN

The individual actions on the action plan all need to be implemented and signed off as complete. Depending on the nature of the actions and the timescales involved, this will need close management to make sure the plan stays on track; this can be led by members of the root cause analysis team.

STEP 7 VERIFY EFFECTIVENESS

Verification of effectiveness is the final step in the root cause analysis process and this is done to check that the necessary changes identified in the action plan are actually working in practice and are effective at addressing the root cause of the problem. It is also important to check at this stage that the changes have not introduced any other problems that were not foreseen. Verification can be done using audit techniques, and following verification it is likely that the business will wish to implement additional monitoring around the changes within the normal scheduled monitoring activities.

Root Cause Analysis Toolbox

A wide range of management problem-solving tools may be used in root cause analysis and companies will find their own preferences with the experience of trying different approaches. There are no precise rules for this; it is all about getting to understand all the possible contributing factors to gain an understanding of the likely chain(s) of events leading to the incident. This will allow prioritization of necessary changes to control systems, infrastructure and/or management practices. The following short notes are intended to help businesses understand the strengths of a selection of tools used in root cause analysis within different industries. Further, more detailed discussions on the different tools can be found in other management and problem-solving handbooks. Trial of some of these techniques within the business outside of an incident situation, perhaps as part of a business improvement project, will allow identification of preferred tools that can be used when an incident occurs.

A menu of possible tools:

- Brainstorming

  Brainstorming is an established management tool used to capture ideas from the individuals within a group. It is particularly useful because it allows for a large number of ideas to be generated in a short time and the lateral thinking involved means that initial ideas spark off other ideas and contributions from other group members. Ideas are never criticized or commented on during the brainstorming session because this may influence or even stifle subsequent suggestions. The key point is to get as many solutions down as possible for later evaluation and it is normally necessary to allocate the role of scribe to one team member in order to record the ideas effectively.

- Failure Mode and Effect Analysis

  FMEA is well known as one of the systems that helped to originate the HACCP approach to food safety management. Its method of considering the causes and potential effects of failure is useful in looking at prevention of problems but it can also be employed when investigating all the potential causes of an issue in an incident. Table 40.2 shows an example of FMEA being used to explore the causes of metal complaints due to metal detection failure.

  Some FMEA methods include a risk scoring approach although this is not often used in food manufacturing. However, it can be seen from the example in Table 40.2 that the sheer number of possible causes might mean that there is a need for prioritization of the recommended solutions/controls. This can be done using a simple likelihood of occurrence scheme, e.g. high, medium and low likelihood. Severity may also be considered, although it is likely that severity may be relatively similar in some cases, e.g. in Table 40.2 the possible causes may all result in undetected metal in product.

- 5-Whys

  The 5-Whys is a simple problem-solving technique that helps users to get to the root of the problem quickly. Made popular in the 1970s by the Toyota Production System, the strategy involves looking at any problem and asking: "Why?" and "What caused this problem?" Normally the answer to the first "why" will prompt another "why" and the answer to the second "why" will prompt another and so on. It is thought that at least five questions need to be asked to track back to the root cause, hence the name the 5-Whys strategy. In reality, there may need to be more than five questions asked depending on the complexity of the situation.

  5-Whys helps the root cause analysis team to start at the end result and work backward toward the cause by continually asking "why?" until the underlying cause of the problem becomes clear. In addition to its use in root cause analysis, it is useful at the start of a remodeling or change process and is a recognized lean manufacturing technique, challenging those working on an issue to analyze any problematic situation in a logical manner, thus enhancing change and continuous improvement.

  A number of benefits of the 5-Whys approach have been recorded:
  - Simplicity. It is easy to use and requires no advanced mathematics or tools.
  - Effectiveness. It helps to quickly separate symptoms from causes and identify the root cause of a problem.
  - Comprehensiveness. It aids in determining the relationships between various problem causes.
  - Flexibility. It works well alone and when combined with other quality improvement and troubleshooting techniques.
  - Engaging. It fosters and aids teamwork and teaming within and without the organization.
  - Inexpensive. It is a guided, team-focused exercise. There are no additional costs.

TABLE 40.2 Challenging Metal Detection Failure using Failure Mode and Effect Analysis

Issue (Outcome of Failure)	Failure	Current Control	Possible Causes of Failure[a]	Recommended Controls
Complaints of metal in product from customers. This could result in lost credibility, lost customers and bad publicity. Worse still, metal in product could cause customer injury and may result in prosecution	Failure to detect metal in products[b]	Check metal detector hourly with test pieces and record result	Metal detector breakdown Metal detector not properly calibrated Wrong sensitivity – check pieces Incorrect metal detector in use – wrong sensitivity Metal detector in wrong place in line Rejection mechanism faulty Rejection system not synchronized with detector	A range of controls will need to be considered around: – Appropriate sensitivity and calibration – Set up verification at start-up – correct sensitivity – Maintenance systems
			Rejects not controlled	Lockable receptacle needed that will accommodate all rejects
			Metal detector checks not done Metal detector checks done incorrectly Metal detector check reveals failure but this is not recorded Metal detector check reveals failure but no corrective action taken Staff not trained to perform metal detector checks Effectiveness of training not verified in terms of practice Workplace culture issues result in staff not taking responsibility for necessary checks	A range of controls will need to be considered around: – Appropriateness and coverage of training – are enough people trained and can they actually do the checks? – How can training effectiveness be verified? – What supervision is needed? – Are the checks allocated within appropriate job roles and instructions? – Management systems and commitment issues need to be investigated

[a]*This will be a brainstormed list of ideas from the root cause analysis team.*
[b]*This is likely only one failure mode associated with the issue. Other failure modes to consider will include how the metal got into the product, e.g. consideration of raw material streams and processing/equipment maintenance issues on site or possible damage of the products in distribution.*
(Adapted from Mortimore and Wallace 1998)

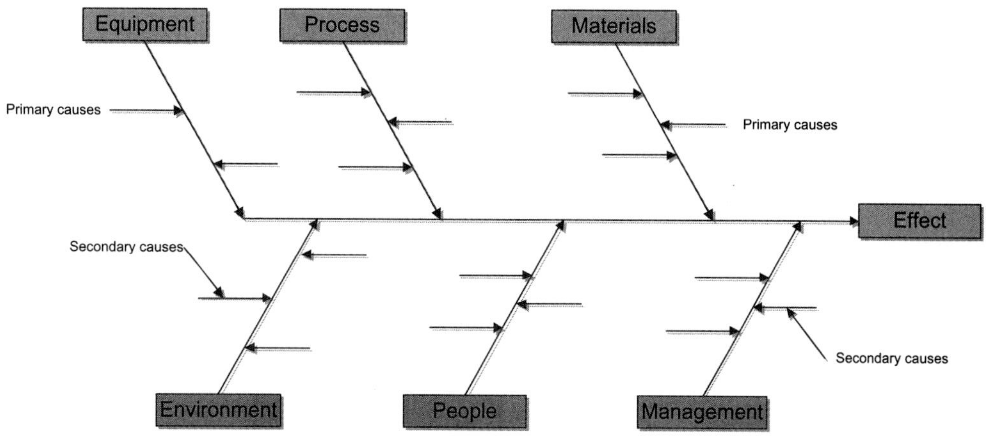

FIGURE 40.5 Example Ishikawa Cause and Effect diagram.

- Ishikawa Cause and Effect Analysis

  Ishikawa Cause and Effect Analysis (also known as fishbone diagrams) is a pictorial method (Figure 40.5) of organizing information about causes and understanding the relationships between cause and effect. It is a widely used technique in problem-solving, and seeks to understand the possible causes by asking questions such as "What happened?", "When?", "Where?", "Why?", "How?" and "What was the impact?" Ishikawa is useful in evaluating complex situations where there may be many potential causes.

  In Figure 40.5 it can be seen that causes are grouped into six categories of Equipment, Process, Materials, Environment, People and Management. These are commonly used category groupings in manufacturing situations; however, the categories in Ishikawa are not predetermined so it is possible to choose your own groupings. The diagram also shows how primary and secondary causes are portrayed and in this way the causes and causes of the causes can be identified, helping to work back to the root cause. Figure 40.6 shows how this method can be applied to an incident, based on the metal complaints issue from Table 40.2.

  This example (Figure 40.6) shows one way of grouping the possible causes identified; however, it is important to note that some causes could be grouped under more than one heading. Also in this case only the primary causes are shown; these would need to be followed up with consideration of the secondary causes and it is possible that, with further consideration, some of the points listed under the "Management" grouping might be the secondary causes affecting other groups within the diagram. There is no right or wrong way here – it is up to the team to decide how best to portray the data in their unique situation.

- HAZOP

  Hazard and Operability Studies (HAZOP) is another structured and systematic technique for examining potential faults in systems. Like HACCP, HAZOP is often

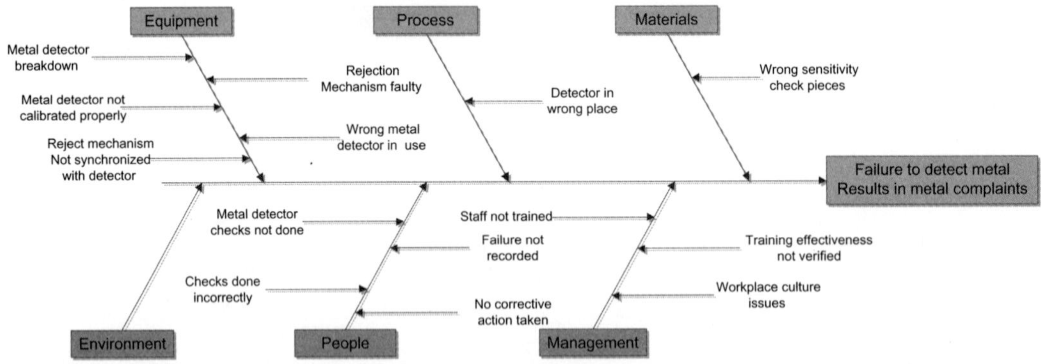

FIGURE 40.6 Ishikawa Cause and Effect diagram for metal contamination example *(after Mortimore and Wallace 1998).*

used as a technique for identifying potential hazards in a system but it also focuses on identifying operability problems that are likely to lead to nonconforming products. In HAZOP, faults or incidents are thought to be caused by deviations from design or operating intentions.

In HAZOP, The identification of deviations from the design intent is achieved by a questioning process using predetermined "guide words." The role of the guide word is to stimulate imaginative thinking, to focus the study and elicit ideas and discussion, thereby maximizing the chances of study completeness (BS IEC 2001). Further detailed guidance on how to use HAZOP, including lists of typical "guide words," can be found in the International Electro-technical Commission's guideline: Hazard and Operability Studies (HAZOP Studies) Application Guide (BS IEC 61882:2001).

- Influence diagrams

  The influence diagram approach is a further technique for visual portrayal of causal factors involved in an incident. The outcome diagram is derived in similar ways to the other tools already discussed, in that expert input, group discussion and brainstorming techniques are used. The technique differs in that it considers the possible causal factors occurring at different levels in the organization. According to Reason (1997), the levels to be considered are:
  - Influencing factor level – this includes the unsafe acts or technical failures immediately responsible for the event.
  - Performance-influencing factor level – the immediate workplace conditions that shape the occurrence of human or technical failures.
  - Implementation level – the underlying organizational factors that create the workplace performance-influencing factors.
  - Policy level – policy and regulatory factors that determine organizational processes occurring at the implementation level.
- An example influence diagram is shown in Figure 40.7. In this diagram the levels and types of failures that can result in an incident that were previously outlined in Figure 40.2 (Reason, 1995) are also highlighted on the right-hand side, indicating the practicality of application of the influence diagram approach to food safety incidents.

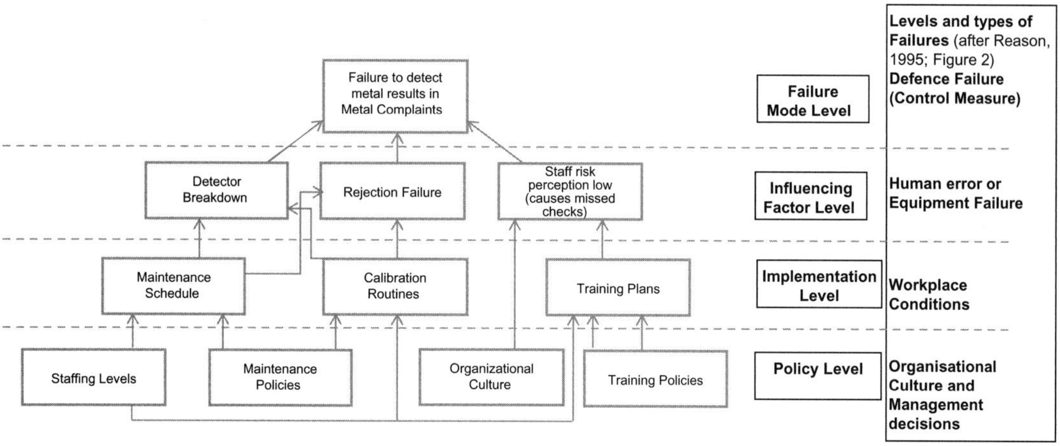

FIGURE 40.7 Example influence diagram for metal complaints (selected causes only).

- Additional specialist tools

  A variety of other tools are used for problem-solving in different industries. Root cause analysis teams may wish to consult the problem-solving, risk management and error avoidance literature to identify techniques that could be trialed for suitability in the analysis of incidents. Some of these examples are more quantitative and involve risk rating categories, which might be more difficult to apply in a food manufacturing scenario. Further tools used in other sectors include Fault Tree Analysis, the Human Error Assessment and Reduction Technique (HEART) and the Maintenance Error Decision Aid (MEDA).

As can be seen from the above, the root cause analysis toolbox contains a plethora of techniques that will assist when faced with an incident to investigate. Using any of these tools does require practice so there is no substitute for trialing chosen tools when not in the middle of an incident; the majority of these tools are also useful in preventive improvement projects, which would be a much more suitable time to try them out. A further important point is that there is no substitute for involving the correct people in root cause analysis so it is important to consider carefully who can contribute to the understanding of the incident and its causes.

CONCLUSIONS

The management of a company bears the ultimate responsibility for incidents. They are responsible for creating an organizational culture that allows employees to openly report issues and provides them with the opportunity to see that their constraints are adequately addressed. An open and fair organizational culture is fundamental for the motivation of staff and is at the core of food safety management. In case of an incident, they have not only to follow best practice in managing the ongoing incident but also be candid with analyzing the root cause of the incident, such that they can redress the situation in a fundamental way to prevent recurrence of incidents in a long-lasting manner.

As for crisis management, the lessons learned from incidents need to be reported in a final report and disseminated both internally in the organization and externally with the food safety community at large in order to prevent the recurrence of incidents in society.

References

Bartlett, R., 1999. Dancing with the Devil, Crisis Management in the Food and Drink Industry. Leatherhead Food International.

BRC, 2011. BRC Global Standard for Food Safety Issue 6. The Stationery Office, London (http://www.tsoshop.co.uk/brcbookshop/bookstore.asp?FO=1235971&DI=630504&TRACKID=004555).

EC. Directive 2001/95/EC of the European Parliament and of the Council of 3 December 2001 on general product safety. J. Eur. Communities L 31/4-17, 15.1.2002.

EU Regulation (EC) No. 178/2002 of the European Parliament and of the Council of 28th January 2002 laying down the general principles and requirements of food law, establishing the European Food Safety Authority and laying down procedures in matters of food safety. Off. J. Eur. Communities L 31/1-24, 1.2.2002.

International Electro-technical Commission, 2001. Hazard and Operability Studies (HAZOP Studies) Application Guide, BS IEC 61882:2001.

Jevšnik, M., Hlebec, V., Raspor, P., 2006. Meta-analysis as a tool for barriers identification during HACCP implementation to improve food safety. Acta Alimentaria 35 (3) September.

Mortimore, S.E., Wallace, C.A., 1994. HACCP – a practical approach, Chapman & Hall, London.

Mortimore, S.E., Wallace, C.A., 1998. HACCP: A Practical Approach, second ed. Aspen Publications, Gaithersburg.

Mortimore, S.E., Wallace, C.A., 2013. HACCP: A Practical Approach, third ed. Springer Publications.

Reason, J.T., 1997. Managing the Risks of Organizational Accidents. Ashgate, Aldershot, UK.

Further Reading

Abbott, H., 1991. *Managing Product Recall: A Comprehensive Guide to Establishing a Product Recall Plan*. Pitman, London.

BRC Global Standards, 2012. Understanding Root Cause Analysis (http://www.brcglobalstandards.com/Portals/0/media/files/Certification/BRC026%20-%20Understanding%20Root%20Cause%20Analysis.pdf EC).

FAO/WHO, 2010. FAO/WHO Framework for Developing Food Safety National Emergency Response Plans, Food and Agriculture Organisation of the United Nations and World Health Organisation (http://www.who.int/foodsafety/publications/fs_management/ERb1_E_L_101012.pdf).

FSA Task Force on Incidents, 2008. Principles for preventing and responding to food incidents. Food Standards Agency (UK) (www.food.gov.uk/multimedia/pdfs/incidentsprinciples.pdf).

Melvin, N., Kramer, M.N., Coto, D., Weidner, J.D., 2005. The science of recall. Meat Sci 71, 158–163.

CHAPTER

41

Crisis Management

Yasmine Motarjemi
Food Safety Consultant, Nyon, Switzerland

OUTLINE

INTRODUCTION

In any organization dealing with a risk-prone subject such as food, chemicals, drugs, health, transport or finance, a crisis is an almost unavoidable situation, and any organization with a professional management should be prepared for it.

Food Safety Management.
DOI: http://dx.doi.org/10.1016/B978-0-12-381504-0.00041-X

Food represents a vulnerable sector, both in terms of food safety risks and food security. Therefore, organizations operating in the food sector, whether they are food businesses or agencies responsible for overseeing food businesses, are by nature of their work exposed to such an eventuality and should have a crisis management procedure in place.

WHAT IS A CRISIS?

In food safety, a crisis is defined as a predicted or unpredicted event that represents an immediate or future significant threat to an organization, its employees, consumers and the public at large.[1]

We also know that in most crises, if not all, the media play a central role. As stated by C. Doeg (1995), "What makes a problem into a crisis is the media, or in some instances, the likelihood of media attention. Also, if a disaster strikes, it is the media's treatment of that event which determines to a great extent whether a corporation has a problem or a full-blown crisis."

Generally, a crisis reflects an acute situation requiring real-time and strategic decisions taken at high level, under harsh conditions created by time pressure, media scrutiny and often incomplete or unreliable information about the facts. A crisis is *per se* never a positive event, as it is an indication of failure in food safety management; however, a crisis which is well managed can be an opportunity for the affected organization or society to demonstrate its values and management capabilities, and for the crisis manager to demonstrate his/her leadership. It can also lead to fundamental improvement in food safety management, provided that the food safety situation and crisis management are critically reviewed, the root causes are analyzed and the gaps identified are followed up with corrective actions. In the modern history of food safety, crises have been the origin of many changes or improvements in the management of food safety, particularly in industrialized countries. Among these, the advance of the risk analysis approach to governmental decision-making processes, the restructuring of governmental organizations (e.g. creation of the European Food Safety Authority) and the strengthening of traceability are noteworthy. These developments are described in Chapter 1 of this book as well as other reference material (Motarjemi, 2014).

A crisis may occur for various reasons (Table 41.1):

- Advances in science and new scientific development or findings.
- Emergence of new hazards.
- Human error, be it scientific, managerial or operational.
- Fraud or malicious acts of sabotage, e.g. tampering, terrorism.

While the exact nature and impact of a crisis are often unpredictable, its occurrence at some point in time, in spite of all preventive measures, is to be reckoned with. However, with good management of food safety, we can minimize the likelihood of occurrence and/or impact of crises, but we cannot entirely prevent them as some of the factors leading to a crisis situation are out of the control of managers and are inherent to the nature of the subject.

[1] Adapted from Bartlett R (1999).

TABLE 41.1 Examples of Crises and their Cause

Triggered by	Examples of Crises
Advances in science and new scientific development or findings	Acrylamide, Worldwide 2002 Semicarbazide, Worldwide 2003
Emergence of new hazards	Bovine spongiform encephalopathy (BSE) and emergence of prions Worldwide, 1986 *E. coli* O157 (USA, Japan, UK, etc. 1990s) *Vibrio cholerae* (Latin America, 1993) Avian influenza (2004)
Human error: scientific, technical, managerial, operational error or violation	Bovine spongiform encephalopathy (BSE), Worldwide, 1986 *Salmonella* in chocolate (UK, 2006) *Staphylococcus aureus* (Japan, 2000) Vitamin B1-deficient infant formula, ex Germany, Israel, 2003 Isopropylthioxantone (Worldwide, 2005) *Salmonella saint paul* (USA, 2008) *Salmonella typhimurium* (USA, 2008–2009) *E. coli* O104:H4 (Germany, France, 2011)
Fraud or malicious acts of sabotage, e.g. tampering, terrorism	Lead oxide in paprika (Hungary, 1994) Dioxin in animal feed (Belgium, 1999) Sudan red in chili peppers (Europe, 2003) Wheat gluten in pet food adulterated with melamine, ex China (North America, 2007) Adulterated sunflower oil ex Ukraine (Europe, 2008) Infant formula adulterated with melamine (China, 2008)

The consequences of a crisis can be disastrous for an organisation as well as for the society. For consumers, a food safety crisis situation means that they may potentially be exposed to unsafe products, and despite risks to their health, they may also lose their trust in the food supply. For businesses, the consequences are economic and can affect their image, i.e. product recall and waste of produced food, loss of reputation, loss of market shares and loss of trust of their customers and the regulatory authorities. Additionally, they may be subject to further or more stringent regulatory measures. Loss of trust by customers may also trigger more stringent requirements, e.g. provision of a certificate of analysis. Trust of consumers, customers and the general public is one of the most important assets of a business. It takes many years to build trust, but it can be destroyed with one single incident, particularly if it is poorly managed. When lost, its impact is often long term and will take many years to rebuild.

For public health and regulatory authorities, even when a crisis is initiated in the industry, the good management of a crisis is crucial, because consumers consider their government as the guardian of the safety of the food supply and ultimately responsible for food safety.

Where governments fail to manage a food safety crisis, they may also lose their image, reputation and the trust of the general public in their capabilities to ensure safety of the food supply. In such a scenario, the trade in food can collapse. Such situations were experienced with meat and meat products in the BSE crisis in the UK and other European countries, and with fruits and vegetables in the case of *E. coli* O104: H4 in Germany (Motarjemi, 2011). Failures in managing an incident or a crisis have also been the cause of political turmoil. Following the 1999 dioxin crisis in Belgium, the Belgian Ministers of Agriculture and Health had to resign. The ruling Christian Democratic government was also voted out of office (Donal et al., 2010). In China, following the melamine crisis, the governmental officer in charge of food safety was executed!

The loss of trust of consumers in their authorities following the BSE crisis and a plethora of small- or large-scale incidents which occurred is one of the factors leading to the mistrust of consumers for new technologies like genetically modified food, food irradiation and application of agrochemicals in food production. In Europe, many consumers turned towards organically produced food.

The consequences of a crisis for an organization or a society depend on how well the organization or society is prepared for a crisis situation. As for a boat whose survival through a storm depends on its solidity, the training of the sailors and the skills of the captain, the outcome of a crisis depends on the infrastructure in place, the training of the staff and the skills of the manager.

WHAT DID WE LEARN FROM OUR CRISES?

The recent history of food safety has been interspersed with food safety crises of varying scale or degree of importance. With the globalization of the food supply and the development of rapid means of communication, many of the crises tend to take a global dimension and require fast action and, frequently, international coordination. In the following pages, a selected number of crises will be analyzed and discussed.

Perrier Mineral Water (1990)

One of the well-known and first food safety crises with international impact was the one related to Perrier (Box 41.1), which today has become a classic case study for crisis management. The Source Perrier Company tried to minimize the impact of the incident by explaining its cause before all the facts were known to them (McKoy, 2006). This led to erroneous managerial decisions, i.e. delay in worldwide recall and inconsistent communication to the general public and the media. An underlying factor for errors in the management of the crisis was weakness in international coordination and lack of leadership in an environment of global supply. Although the level of benzene present in the product did not endanger the health of consumers, it did damage the image of a product that was appreciated by consumers who perceived it as a pure product.

Animal Feed Contaminated with Dioxin (Belgium, 1999)

The first major crisis of dioxin related to animal feed occurred in Belgium. For numerous reasons, it caused tremendous outrage. First, there was a perception of an unacceptably long delay in reacting and in informing the public (Box 41.2). The difficulty in establishing a coherent list, or developing an understanding, of the range of affected products hampered

BOX 41.1

CASE OF PERRIER MINERAL WATER (1990)

In February 1990, state regulators in North Carolina found traces of benzene (at levels of 12.3–19.9 ppb) in Perrier mineral water, until that time reputed for its purity. In North America, i.e. the USA and Canada, the company immediately recalled 70 million bottles. Within 2 days, the Source Perrier Company announced that the problem was limited to North America and was caused by an employee's mistaken use of fluid containing benzene to clean the machinery in the North American bottling plant. However, further investigation revealed traces of benzene in Perrier water in other parts of the world. It was later discovered that the presence of benzene in the final product was due to failures in filtering benzene naturally present in the carbonated gas. A worldwide public recall was made. Overall, 280 million bottles were recalled and destroyed. In 1992, Nestlé bought Source Perrier Company.

the management of the crisis. For instance, despite measures taken by Belgian authorities, contaminated chicken and eggs were still on the market in late May while the incident started in late January/early February 1999. Throughout the incident, the list of affected products was constantly amended and, consequently, the decisions for product recall were haphazard, giving the impression that the government was not in control of the situation. Failures in communication and coordination with the European authorities and the European Rapid Alert System, and conflicting opinion and decisions between the European Commission and the Belgian authorities on what needed to be recalled, added to the confusion. The lack of precise scientific information or regulatory standards for the management of dioxin exacerbated the situation.

Over and above the need for strengthening the procedure for crisis management and communication, the crisis highlighted the complexity of the food chain and the need for a better traceability system, foreseeing needs for analytical capabilities and considering the farm-to-fork approach in food safety management.

Animal Feed Contaminated with Dioxin (Ireland, 2007)

A second dioxin crisis occurred in Ireland (Box 41.3). In managing this crisis, naturally the Irish authorities benefited from the experience from the previous outbreak in Belgium and the measures that had been taken since that time. In this incident, as regard to pork meat, the Irish authorities had to take drastic measures and recall all pork products as there was no traceability system for these; for beef products, because of the existing traceability system, the recall was limited to contaminated products. The rapid recall and precautionary measure increased the trust of the European Commission and consumers in the authorities and in the safety of the food supply, and it avoided conflicts with the European Commission and imposition of restrictions, as experienced in the previous incident of dioxin.

BOX 41.2

CASE OF ANIMAL FEED CONTAMINATED WITH DIOXIN (BELGIUM, 1999)

In spring 1999, it was found that some 500 tons of feed contaminated with polychlorinated biphenyls (PCBs) and dioxins were fed to farm animals in Belgium and to a lesser extent in the Netherlands, France and Germany. The source of contamination was a fat-rendering company, where transformer oil with high levels of polychlorinated biphenyls (PCBs) and dioxins was used to manufacture animal foods. Hundreds of farms were affected.

The first pathological symptoms were reported in poultry farms in February 1999. When alerted on 26 April of that year, public health authorities first took some measures to protect public health, but they omitted to inform the public. A month later, i.e. end of May, the public learned about the issue from a television report on the incident.

On 28 May, Belgian authorities ordered the withdrawal from sale of Belgian poultry and eggs from affected farms; other European countries followed. Products from farms not affected by the incident were required to be accompanied with documentation from the authorities. On 2 June, the European Community widened the ban and ordered the destruction of all food items containing >2% egg product and food containing chicken produced from 15 January to 1 June from infected farms. On 4 June, the Belgian government issued a commerce embargo of meat products (pork and beef) with a minimum of 25% fat content, not applicable for dairy products, while the European Commission extended the prohibition order to Belgian beef, pork, milk and milk products from affected holdings. Products from non-affected farms had to be specifically certified by the Belgian authorities. Belgian authorities objected to the EU restriction posed on milk, which had consequences for a large number of products, such as chocolate. A confusing situation prevailed throughout the incident. A political consequence of this incident was that the Belgian Ministers of Agriculture and Health as well as the Minister of Agriculture of the Netherlands resigned. The ruling Christian Democratic government was also voted out of office (Donal et al., 2010; Corie and Powell, 2000; Van Larebeke et al., 2001).

Another factor in the perceived successes of the management of the dioxin crisis was attributed to the leadership and transparency of the Irish authorities. Teamwork was demonstrated through open communication with the European Commission, other national food regulatory agencies and the European Food Safety Authority (EFSA).

Finally, the incident proved the importance and benefit of having a good traceability system.

Coca-Cola (1999)

In some respects, the Coca-Cola crisis in June 1999 (Box 41.4) followed the same pattern as that of the Perrier Source water mentioned above. Although the company provided a

BOX 41.3

CASE OF ANIMAL FEED CONTAMINATED WITH DIOXIN (IRELAND, 2007)

In November 2008, in the context of routine monitoring, the Irish Department of Agriculture, Fisheries and Food (DAFF) detected the presence of marker PCBs in pork meat.

Further investigation indicated that dried bread, used as an ingredient in animal feed, was also positive for marker PCBs. Considering the link between presence of PCBs and possible contamination with dioxin, further tests were carried out. They confirmed the presence of dioxin.

The source of the contaminated dried bread was identified as Millstream Power Ltd., a food recycling plant. Subsequent investigation showed that the contamination of the feed was due to recycled mineral oil used as fuel in flame drying during the processing of animal feed. In total, 10 pig farms and 38 beef farms had received the contaminated feed. In Northern Ireland, seven beef farms had also received contaminated feed.

Soon after, DAFF impounded all potentially contaminated feed on these farms, imposed restrictions on the movement of livestock and reported the problem to the Rapid Alert System for Food and Feed (RASFF) and DG SANCO.

Although the level of contamination of pork meat was in violation of the European norms, the authorities indicated that a shorter exposure did not constitute a cause of alarm or concern and would not result in adverse health effects. Nevertheless, the authorities required a full recall of pork products because the traceability system in operation for processed pork products was not capable of linking products to specific farms. For beef, the estimated exposure to dioxin was viewed as 300 times lower than that posed by pork meat and considering that as a follow-up to the BSE problem, there was a traceability system in place for bovine animals and products, it was possible to track and trace contaminated products and this allowed a partial and selective trade recall of suspected products.

public statement a week after the report of the incident and presented its regrets for the incident, its first reaction caused outrage. The company was perceived as denying responsibility and playing down the importance of the incident. It minimized the severity of the illness and claimed that the products were safe, while at the time the statement was made, the true cause of the problem was not fully known.

Besides the important role played by the media in this case, the scale of the problem was amplified by the radical measures of authorities who banned the product from the market to counteract allegations of mismanagement in the preceding crisis of animal feed contaminated with dioxin feed. This, combined with deficient communication from the company, played badly against the product, indicating to consumers that the company was not taking adequate measures. According to some analysts, the company failed to appreciate the sensitive environment and political unrest caused by the preceding dioxin crisis as well as the general climate of public mistrust created by the BSE issue.

BOX 41.4

CASE OF COCA-COLA (EUROPE, 1999)

In 1999, over 100 persons, including many children, reported feeling unwell, i.e. suffering from headache, dizziness, nausea and trembling, after drinking Coca-Cola in France and Belgium. The incident occurred right after the dioxin contamination of animal feed in Belgium. The French and Belgian governments banned the product. The company recalled some 30 million cans. Investigation of the incident revealed two unrelated failures in production: (1) contamination of carbon dioxide with carbonyl sulfide (COS) which may hydrolyze to hydrogen sulfide (H_2S) and (2) contamination of cans with a fungicide used for treating the wooden pallets. However, tests conducted of the product showed that the product could not be the cause of the illness and the symptoms – at least in most cases – were psychosomatic. Despite an aggressive marketing campaign to regain the trust of consumers, the incident caused substantial losses (an estimated $200 million) and damaged the reputation of the company (Nemery et al., 1999, 2002).

BSE I and II (Europe, 1996–2000)

The tendency for denial or playing down is often a first reflex. Such a practice can mislead decisions, i.e. allowing wishful thinking to guide decisions. It can also convey an arrogant, uncaring and unsympathetic image of an organization. This type of error was also the cause of the bovine spongiform encephalopathy (BSE) crisis, one of the most notorious crises in the history of public health and food safety (Box 41.5).

This crisis revealed various types of expectations among the general public:

- The need for transparency in decision-making and for communication of uncertainty to the general public.
- The importance of prioritizing public health over economic considerations.
- It also highlighted the importance of public perception, as the reaction of most consumers was not proportionate to the risk that they were exposed to.
- Social dimension of a food safety problem; many farmers suffered enormous economic losses and emotional distress.
- The need for better traceability and information of consumers on the source of food.
- The need for considering a farm-to-fork approach in food safety management.
- The need for functional separation between risk assessment and risk management, while maintaining an interactive communication between risk managers and risk assessors.

Many of the lessons learned from the BSE crisis were later supported by the experience in other crises. Together with other food safety concerns, BSE crises led to fundamental changes in the approach to food safety management, including the restructuring of the European institutions involved in managing food safety, as well as to the revision of the decision-making process and the rise of the risk analysis process.

BOX 41.5

CASE OF BSE (EUROPE, 1996, 2000)

In 1986, the first cases of a mysterious degenerative brain disease referred to as bovine spongiform encephalopathy (BSE), or mad cow disease, were reported in the cattle population in the United Kingdom. However, it is believed that cases may have occurred as early as the 1970s. The disease was linked to ruminant-derived meat and bone meal (MBM) fed to cattle. The disease was viewed as without risk to human health until a new variant of Creutzfeldt–Jakob disease (vCJD) was detected in 1996 in humans and was linked to the BSE epidemic in cattle. Consumption of contaminated meat and other food products from cattle was presumed to be the cause of the vCJD. In 2004, it was discovered that the disease was also transmissible to small ruminants.

In Europe, BSE caused two waves of crises. The first occurred in 1996, when the public learned about the fact that BSE may be transmissible to humans and perceived that the risk of BSE to cross the species barrier was downplayed. The second, in 2000, resulted from the mistrust of the general public in the ability of governments to contain the BSE epidemic and from the finding that the export of contaminated feed had spread BSE to other countries in the world.

To date, in the UK, which was the most severely affected by the epidemic, BSE has led to over 180,000 cases in the cattle population and to some 163 definite or probable cases of vCJD, from which 163 persons died (WHO, 2002).

Packaging Contaminants (Semicarbazide, 2003 and Isopropylthioxanthone, 2005)

In 2003 and 2005, the food industry experienced two major crises related to packaging contaminants (Box 41.6). In both cases, products were contaminated with undesirable substances, but without significant danger to health. Although the nature and source of the incident were similar, their management followed a different path.

In the case of semicarbazide (SEM), the early and transparent communication of the food industry to authorities and speedy actions on both sides increased the mutual trust and led to the smooth resolution of the crisis. A rapid risk assessment and communication by EFSA also ensured that European member states took coherent and coordinated action across Europe.

On the contrary, in the case of 2-isopropylthioxanthone (ITX), due to a conflict of opinion between different governmental agencies on the risk of the agent and the actions required, the crisis escalated in one of the European countries. This led to confiscation of Nestlé infant formula products by the police. Subsequently, to safeguard the trust of the consumers and ensure a consistent approach, Nestlé had to voluntarily recall its products from other European countries, even though an earlier decision at the European level had allowed the marketing of the product. The product recall was heavily reported in the media, who also inflated the monetary value of the recall. A succession of conflicts followed. An early risk

BOX 41.6

CASES OF SEMICARBAZIDE (EUROPE 2003) AND ITX (EUROPE 2005)

In May 2003, in the context of monitoring its products for nitrofurazone, a prohibited veterinary antibiotic in Europe, a baby food company found traces of semicarbazide (SEM), which until that time was known as a metabolite, and thus an indicator of nitrofurazone. However, in this case, it was found out that the agent originated from the breakdown of azodicarbonamide.

Azodicarbonamide is also used as a blanching agent in flour in some countries other than Europe. In this case, azodicarbonamide was used as a foaming agent in the plastic gaskets that are used to seal metal lids to glass jars. The agent had decomposed under heat treatment of the product. As azodicarbonamide was used by most of the cap manufacturers, the great majority of the baby jar products on the market worldwide were affected by this incident.

Soon after the discovery of this incident, the industry reported the case to the European and national authorities and informed them that they would take measures to change azodicarbonamide for alternative foaming agents. The European Commission requested EFSA to conduct a risk assessment and advise the commission on the risk of the agent. EFSA conducted a first evaluation in 2003, and in July 2003 stated that the risk to consumers, if any, was very small. It recognized that SEM was a very weak carcinogen in mice and had weak genotoxic activity. While it acknowledged the low safety concern, it recommended that SEM be removed from baby foods as swiftly as technological progress allowed (EFSA, 2003a, b). The European Commission banned the use of azodicarbonamide in food contact material as of August 2005, giving the industry the time to replace it. EFSA also reassured the general public that in view of the low level of SEM and low safety risk, consumers did not need to change their dietary habits and may continue to utilize all foods concerned, including baby foods.

In September 2005, a government scientist discovered traces of a 2-isopropylthioxanthone (ITX) in ready-to-feed infant formula in Nestlé products. ITX is a photo-initiator of ink used for printing on the carton packaging, contaminating the inner side of packaging during its processing, and thereafter migrating into products. Similar to semicarbazide, ITX was utilized worldwide in packaging used by Tetra Pak and other packaging companies; thus a broad range of products (e.g. milk and milk-based products, infant formula, soy beverages, fruit juices, fruit nectars and other drinks) were potentially affected on a global basis.

First on the 24 November and then on 7 December 2005, 3 months after the report of the incident and a month after the heat of the crisis, EFSA provided an opinion on the possible health risks of ITX. EFSA advised that while the presence of ITX in foods was undesirable, it did not give cause for health concern at the levels reported (EFSA, 2005).

Nestlé, the first company to be alerted by the incident, conducted a first recall of its affected products on a precautionary principle. Later, when it received confirmation of the nature of the agent, its low degree of risk and the extent of the problem, through the food industry association, it reported the

BOX 41.6 (*Contd*)

issue to the European authorities. In October 2005, a plan of action for removing ITX was agreed with the EU and communicated to the European member states through the Rapid Alert System of Food and Feed (RAFSS). Thus, contaminated products continued to stay on the market until they could be gradually replaced by other products. In November 2005, the Italian government, not agreeing with the decisions taken at the European level and the risk evaluation, seized infant formula products of Nestlé, including those in its warehouses. To maintain the trust of the consumers, the company had to remove all affected products from the European markets where the products were on sale. The seizure of the product by the police and the massive recall in Europe reflected badly on the company, who in addition to explaining the incident, had also to respond to the financial analysts about the financial consequences. Within 3 months, the food safety incident escalated first to a food safety crisis, then to a financial crisis, and thereafter to a communication crisis between the CEO of Nestlé and the Minister of Health of Italy.

assessment and risk communication on the subject by the European authorities, as in the case of SEM, could have prevented disparate and drastic measures by governments. The crisis demonstrated that an early communication by trusted sources on the nature of the event and its health implications, if any, was essential to ensure coherent measures and prevent media misreporting. It also showed the importance of coordination and clear procedures for communication among the various authorities.

The ITX incident also demonstrated a number of gaps in food safety management, which became conspicuous in the management of the crisis. First, there was an error on the part of the packaging supplier in the testing method and validation of the extent of migration of the chemical into fatty products. Regulatory standards with regard to inks were also deficient, and the official method of testing was not valid as it did not consider the fat content of some food products. Food manufacturers lacked knowledge of the risks associated with the ink. Packaging suppliers were aware of the risk but did not relay the information to food manufacturers. Subsequently, products were not monitored for this agent and the supplier was not audited for this point. Generally, a lack of expertise and infrastructure in managing this risk at all levels were the root cause of the problem. Fortunately, the chemical agent did not present a public health risk, although it was undesirable.

Melamine I and II (2007–2010)

A critical review of the two melamine incidents revealed the following weaknesses in the management of the crises (Box 41.7).

BOX 41.7

CASE OF MELAMINE (USA, 2007 AND CHINA, 2008)

At least two major incidents have occurred with regard to melamine. The first incident occurred during the period of February to March 2007, when reports of kidney failure and death were reported in pets in the United States.

Investigation traced the source of the outbreak to the pet food company Menu Foods; some other pet food companies were also implicated. Overall, more than 100 brands of wet pet food were affected. As a result, Menu Foods and other pet food companies had to recall their products. The economic impact on the pet food market has been extensive, with Menu Foods alone losing at least $42 million from the recall, not taking into account the reduced sales (Reuters, 2007).

Investigation attributed the outbreak to an adulterated wheat gluten ingredient imported from China. The investigation showed that wheat gluten was mixed with melamine, itself contaminated with cyanuric acid, to swindle buyers out of the protein content and the grade of the product.

Following this incident, it was realized that addition of melamine to animal feed was a common practice in China. Later investigation showed that rice protein concentrates were also adulterated. In the United States, the crisis culminated with the finding that the adulterated wheat gluten was recycled in feed of food animals and subsequently entered the human food chain.

A second incident of melamine occurred in July 2008, in the People's Republic of China, involving milk, infant formula and some other products. By November 2008, China reported an estimated 300,000 victims, mainly babies. As reported, six infants died from kidney stones and others suffered kidney damage; a further 860 babies were hospitalized. In this incident, the dairy company Sanlu and several other food companies were implicated. According to some sources, the issue was known a long time before the incident was revealed. However, according to some unknown sources, the regional authorities were afraid to report the problem during the Olympic Games period. The incident damaged the reputation of China and its export market. A number of criminal prosecutions occurred, and two people were executed!

Adulteration of food with melamine continued after this incident, and during the following years, Chinese authorities continued to sporadically seize contaminated food products.

Melamine 2007

- Delay of Menu Foods in reporting the potential problem: the incident was detected on 20 February, but was not reported to the USFDA until 15 March. An immediate reporting could have spared the lives of many pets.
- Lack of coordination and consistent communication between authorities and the public. This resulted on the one hand in erroneous information as to the cause of the problem

(the New York State Laboratory reported that a rodenticide was the culprit), and on the other hand a situation of a lack of information leading to various speculations and hypotheses.
- Shortcomings in the supervision of food safety. According to some sources, the Menu Foods company had never been inspected by the competent authorities. The agency was relying on local authorities to conduct inspections, whereas the central authorities had jurisdiction over all pet food manufacturing facilities. The incident showed weaknesses in procedures, regulations and inspection requirements. Having said this, it is not sure that an inspection or stronger regulation would have prevented the incident, unless there was more insight into the practices in China.
- Overlooking the disposal of adulterated wheat gluten. This resulted in some companies salvaging the adulterated wheat gluten and recycling it into animal feed, which was used for food animals.

Melamine 2008

- Shortcomings in corrective actions by the Chinese authorities and local food companies. Although few companies were implicated in the fraud, in general there was no oversight by other food companies or authorities for the prevention of such practice or of possible contamination of raw material with melamine. As a result, a number of other food companies were also marginally affected by this incident.
- The fear and the delay in reporting the incident resulted in a larger exposure of the young population and were indicative of the irresponsible and unscrupulous behavior of the companies implicated and of the local authorities.

Shiga Toxin-Producing *E. coli* (STEC) O104:H4 (Germany and France, 2011)

In the outbreak of *E. coli* O104:H4, the German authorities came under mounting criticism for their handling of the outbreak (Box 41.8). They were blamed for slow reaction and bad coordination between the states and the federal entities. Consequently, it took several weeks to contain the outbreak and identify its source. Yet hasty and invalid conclusions, which they later had to retract, and false information as to the source of the outbreak created a sense of panic in the general population (*Financial Times*, 2011c). Sale of produce, particularly that imported from Spain, was severely affected. Once again, the crisis demonstrated the importance of rapid action, coordination among authorities and the importance of giving validated information. It also showed the health and economic dimension of failures in food safety and crisis management.

Horsemeat Scandal (2013)

As this book goes to production, a new crisis is ongoing in Europe. In January 2013, the Food Safety Authority of Ireland tested a range of inexpensive frozen beef burgers in supermarkets in Ireland for the presence of DNA from other species. The test found undeclared horse DNA in over 33% of the beef burger samples and pig in 85% of them. Further investigation unraveled a huge pan-European scandal involving a vast and entangled network

BOX 41.8

CASE OF STEC O104:H4 (GERMANY AND FRANCE, 2011)

On 21 May 2011, Germany reported a dramatic outbreak of Shiga toxin-producing *Escherichia coli* bacteria (STEC), serotype O104:H4. On 24 June, France also reported a similar outbreak of *E. coli* O104:H4 in patients that participated in an event.

The investigation of the outbreak in Germany first pointed to cucumbers imported from Spain. Later, as the outbreak continued, this proved to be false and the likely source of the outbreak was attributed to sprouts of fenugreek, grown from seeds imported from Egypt (EFSA 2011). An examination of the outbreaks in France and Germany by EFSA indicated that the two outbreaks were likely to be linked. In Germany, 4321 outbreak cases, including 3469 cases of Shiga toxin-producing *E. coli* and 852 cases of the hemolytic–uremic syndrome, had been reported by 26 July 2011, when the outbreak was declared (Buchholz et al., 2011). By that time, 50 persons had died (WHO, 2011). The outbreak in Germany caused huge economic losses in many European countries.

Spain's export of cucumbers and vegetables was badly hit and export of some 150,000 tonnes of produce was affected (*Financial Times*, 2011a and 2011b).

of traders, slaughterhouses and renowned food producers. Fraudulent products have been discovered in many countries, e.g. France, Germany, Ireland, Italy, the Netherlands, Spain, Sweden, Switzerland and the UK. Additional products, e.g. frozen lasagna and spaghetti with bolognese sauce, beef ravioli and beef tortellini, as well as other abattoirs, producers, traders and supermarkets are being implicated in the scandal. There are reports indicating that in the light of the scandal, one-third of British consumers will not buy pre-prepared food products.

Although, at this stage of investigation the scandal seems not to involve a food safety issue, it revealed:

- The very complex food chain where meat sourced in one country (in the current case Romania) changed hands several times before reaching supermarkets in Europe.
- Renown brand name companies subcontract the acquisition of their raw material to other firms and who in turn subcontract to others and so on, making the present system of traceability of little value for ensuring safety, integrity and wholesomeness of food.
- Lack of ethics of managers of food businesses who would deceive consumers for whom the consumption of horsemeat is culturally and emotionally not acceptable.
- Where there are no ethics, the range of problems that can affect the food supply will be endless, unpredictable and beyond our imagination. Although the horsemeat crisis is probably not a safety issue, other cases of fraud can have food safety consequences (e.g. melamine in USA (2007) and China (2008)).

WHAT LESSONS FOR THE FUTURE?

Far from being an exhaustive list of all learnings from the various crises, the experience in the management of various crises points to the following principles:

- Speed of action and communication are vital.
- Prioritizing public health, or consumer health, over economic aspects.
- Validity of information and understanding the degree of uncertainty and/or variability, i.e. getting the facts right.
- Considering public perception, underlying science and other determinants of trust.
- Transparency, i.e. giving the full truth.
- Functional separation between risk assessment and risk management.
- Coordination between different authorities/organizations involved and consistency in communication.
- A rapid risk assessment and communication of risks by public health authorities is fundamental for preventing disparate actions.
- Maintaining the flow of communication to the general public, particularly through a trusted source.
- Monitoring the situation and consideration of the social and political context of an incident.
- Being aware of the global nature of the food supply and food safety, and of the need for international coordination.
- Analyzing other incidents and taking measures to prevent their repetition, as an incident that happens a first time can be considered a mishap; whereas the second time, it will be viewed as negligence.

A cross-cutting lesson from all crises, be it the food sector or other sectors, shows the importance of management commitment, human resource management and organizational culture in a crisis, both in terms of prevention and management of a crisis. As demonstrated in the accident that befell Concorde, in safety, any detail can be important. In this accident, a piece of another plane fell onto the runway and led to Concorde crashing plane and the demise of the company. Therefore, the collaboration of the staff, exemplary leadership and commitment of the management is needed to prevent any failure, however small this may be. Table 41.2 summarizes the key lessons from the above-mentioned and a number of other incidents and crises.

ESSENTIALS OF CRISIS MANAGEMENT

From the above, it can be understood that the importance of crisis management, and preparation for it, cannot be overemphasized. Provided that crises:

- do not occur too frequently,
- are not a consequence of obvious or gross negligence,
- do not involve unethical or malicious malpractices, and
- health and concerns of the general public are given priority.

TABLE 41.2 Summary of Lessons Learned from Various Incidents and Crises

Incidents	**Lessons Learned**
Mineral water contaminated with benzene (France, 1989)	Rapid product recall increases trust Importance of valid information for decision-making Attempts to downplay the extent of an incident will damage the reputation of a company Importance of consistent communication Importance of international coordination
BSE/vCJD (Worldwide, (Europe 1996–2000))	Importance of transparency and ability to communicate uncertainty to public Importance of prioritizing public health over economic considerations Importance of consumer/public perception (e.g. dreadful nature of disease) Need for separation of risk management from risk assessment The social dimension of food safety Importance of traceability and farm-to-fork approach Role of media Importance of risk communication by a global public health authority
S. aureus intoxication; milk (Japan, 1955)	Importance of rapid action (halt of sale, recall of products, and public apology) increases trust
S. aureus intoxication, milk powder (Japan, 2000)	Speed of action Priority to public health Communication: empathy with the victims Full transparency: any attempt of denial, or minimizing the impact (false or partial information, partial product recall) will damage the reputation more Preparation: clear procedures and training Mechanism for reporting problems to management Beware of culture of fear!
Mislabeling of beef product (January, 2001)	A good crisis management does not always help! Fraud and ethical malpractice will not be forgiven
Animal feed contaminated with dioxin (Belgium, 1999)	Importance of speed of action Demonstration of the complexity of the food chain and the need for traceability The need for farm-to-fork approach Role of media Importance of risk perception The need for resources (e.g. laboratories)
Soft drink allegedly contaminated with pesticides (Belgium, France, 1999)	Importance of communication Importance of considering the context of an incident Voluntary recall can increase trust Role of media
Packaging contaminant: semicarbazide (Worldwide, 2003)	Early and transparent communication of the food industry increases trust A rapid risk assessment and communication by trusted sources prevent escalation of a crisis and ensure coherent actions across Europe

(*Continued*)

TABLE 41.2 (Continued)

Incidents	Lessons Learned
Packaging contaminant: isopropylthioxanthone (Worldwide, 2005)	Importance of risk assessment and communication by competent authorities Alignment with government views Coordination of government agencies Importance of risk perception Financial consequences can also create a crisis Importance of documentation and records Media plays an immense role
Pet food adulterated with melamine (North America, 2007)	The need for additional resources (e.g. to handle consumer queries) Early reporting to public health authorities to minimize damage Importance of coordination and communication between authorities Importance of considering the fate of disposed products Root cause analysis of incidents and dissemination of our experience can prevent future crises
Sunflower oil adulterated with mineral oil (Europe, 2008)	International coordination to prevent dumping of contaminated food to other countries or food sectors Difference between being allowed to keep a contaminated product already in the market and being allowed to release a contaminated product
Salmonella/Peanut butter (USA, 2006 and 2008/9)	Importance of root cause, corrective actions based on understanding the underlying factors for malpractices
Melamine/Infant formula (China, 2008)	Importance of root cause, corrective actions based on understanding the underlying factors for malpractices Risks associated with fear culture Control of contaminated products and their safe disposal
Enterohemorraghic *E. coli* (EHEC O104:H4) and fenugreek (Germany, France 2011)	Importance of speedy action Validation of information before communication to the general public Coordination among authorities Impact of public fear on food market
Horsemeat (Europe, 2013)	Lack of ethics in food business Complexity of the food chain, limitation of traceability Where there is no ethic, all range of problems can happen and are unpredictable

Good crisis management can to some extent reinforce the trust of consumers and trading partners or improve the reputation of the affected organizations, or at least limit the damage. Consumers who would observe that, in case of any adverse event, the business or the government will take necessary measures to protect them will have increased trust in the businesses and governments. As mentioned above, in a crisis situation both governments and the industry may be implicated and will have to bear the consequence of a crisis; therefore, interactive communication and full transparency between the two parties is essential. Additionally, both may be asked to explain why their preventive measures failed.

While collaboration and communication between governments and industry is important to manage a crisis, it is also important that the process of decision and implementation not be biased. Therefore, the principles for risk management and the process of risk analysis developed for the management of food safety in normal times applies also in a crisis situation, except that decisions have to take place under time constraint, with incomplete data and often under media scrutiny. The more a country's food safety management is solid in terms of organization, e.g. definition of responsibilities, values, standards (existence of legislation and enforcement mechanism) and procedure, and skilled and competent managers, the better it will be prepared for managing a crisis situation. The same applies in a food business. In the dioxin case in Belgium in 1999, the lack of norms for dioxin at the European level and of traceability were gaps that, among others, negatively impacted the management of the crisis. In a similar dioxin crisis a few years later in Ireland, the existence of a norm was one of the factors that eased the decision-making process (Donal et al., 2010).

As for food safety management, crisis management requires a structured, systematic and consistent approach and includes four stages:

1. Crisis prevention
2. Crisis preparedness
3. Crisis management
4. Recovery and rebuilding after a crisis

The following principles and recommended practices for crisis management are formulated in a general manner. With some adaptation to the circumstances, they can be applied both in governments and in industry.

Crisis Prevention

It has to be recognized that a crisis situation starts when food safety management has failed. Therefore, regardless of how well it is managed, a crisis is often an indication of failures in food safety management and will have some negative consequences, particularly if crises occur too frequently or past errors are repeated.

Management of a crisis situation mobilizes many resources within the organization and disrupts the normal operations. Therefore, the repetition of a crisis situation not only will erode the trust of the trading partners, customers and the public at large, but will also undermine the routine of the operations and will wear out the staff. Subsequently, these may be more disposed to human error. A vicious cycle of vulnerability sets in. Hence the importance of preventive measures, as the better we manage food safety, the less likely we are to have a crisis. Thus, paradoxically and ironically, the best crisis management becomes its prevention.

The preventive measures are not any different than those that are necessary for food safety management; crisis management should be seen as a continuum of food safety management. For an overview of food safety management, the reader is referred to Chapter 1 and also elsewhere (Motarjemi, 2008, 2014). However, some aspects of food safety management find particular importance during a crisis situation; these are described below.

During a crisis, trust becomes a very important asset, as people operate under acute conditions requiring real-time and strategic decisions at high level. Due to the urgency of the situation, there is no time for checking the validity and thoroughness of the data as in a normal situation. Many decisions are to be taken based on the trust in the integrity and competence of staff or experts and on their respect of the values and policies of the organizations. However, trust is not built or achieved in one day; it takes years of good practice and of responsible and transparent behavior and management. Although, as mentioned above, the practices during a crisis management can enhance and reinforce trust, its foundation is to be built during normal periods, i.e. during the day-to-day operations. An organization that behaves responsibly will not have difficulty in transparency and giving the truth about the cause of its incident. Therefore, values such as open culture, transparency, prioritizing the safety of products over economic considerations should be spelled out in the policy of organizations and actively supported by the management. In other words, policies should not be a declaration of good intentions but practiced by the leaders on a daily basis.

As part of management, the definition of policies, processes, responsibilities and the provision of logistic support all are important for good management of food safety. Additionally, in a food business, the food safety assurance system, including the GMP, HACCP system and various verification measures, as well as change management, will need to be implemented in a flawless manner (see also Chapter 31 and its validation and maintenance). Regulatory authorities should also monitor the food supply for safety and have an efficient foodborne disease surveillance program to depict any problem at an early stage.

An important and integral part of the food safety management system is management of human resources, as no matter how many principles, systems and tools are innovated for managing food safety, it is finally the staff who have to implement these (see Chapter 37). Human resource management is often a neglected area in food safety management, while it is fundamental to an efficient food safety management and should be considered at the heart of the system. Experience from various crises shows that very often failures leading to an incident or a crisis are known by the staff, or could have been predicted; however, due to fear for repercussions on their career, they fail to report or do not bother to report as they do not believe in the fair evaluation of their information.

Therefore, over and above their knowledge and skills, staff need to be motivated and encouraged and, most importantly, not fear for their career or potential repercussions when reporting potential gaps or malpractices. Any gap or malpractice which is addressed at an early stage will decrease the risk of an incident and eventually of a crisis. Employees need to believe in the commitment of their management in the true sense of the word. Hence, the importance of credibility of the management and their walking the talk on a consistent and continuous basis, as any non-compliance or complacency at the higher level of management, will set a bad example and have serious repercussions on the entire organization. The importance of management commitment, having an open culture, promoting the reporting of problems, and their investigating and closing the gaps for prevention of crisis cannot be overemphasized. Naturally, governments have the leading role in protecting the right of the staff.

As part of food safety management, but of particular importance to crisis management, are of course traceability, recall procedure and the procedure for crisis management itself. The latter will be described below. With regard to traceability, without such a system, in case of a non-conforming ingredient or product, it will not be possible to make a selective recall of products, and all products suspected to be potentially affected will have to be recalled and destroyed. A case in point is the incident with dioxin in Ireland (1999) where, due to the absence of traceability, all pork products were recalled, whereas for beef, as a traceability system was established following the BSE incident, a selective recall, i.e. a recall of contaminated products, was possible. The same applies for an incident affecting a food business. The finer the traceability system, the more likely it will be possible to narrow the recall of products that are affected by a contaminant. For instance, a company who can trace a contamination to the precise time of production can limit the recall to those specific products. In absence of such a system, unless the nature of the contamination is such that it is possible to segregate affected products by testing, all the production of a day, a week or of a longer period may need to be blocked and withdrawn.

Many organizations may also benefit from an active early warning system or may even do research on potential emerging issues. Such a system is at the frontier between prevention and preparedness for crisis management. Depending on the nature and size of the organization, the system could include monitoring, surveillance and analysis of:

- Literature and scientific data;
- The regulatory development and alert networks;
- Incidents and experience of other companies and countries;
- Consumer complaints;
- Foodborne diseases, animal diseases and monitoring of contaminants;
- Post-launch of new products;
- Audit and/or inspection reports, reports of compliance of products;
- Internal account of staff reporting non-compliance or mismanagement.

Crisis Preparedness

One of the key principles for crisis management is the speed of action, be it investigation into the case or informing the general public. In all incidents where there has been delay in action, it has caused outrage and hard judgment by the public. In the dioxin incident in Belgium (1999), melamine in China (2008), *S. aureus* in Japan (2000) and *E. coli* O104:H4 in Germany (2011), one of the main failures for which the responsible authorities were severely criticized was the delay in removing products from the market or in informing the public; a similar experience was observed with Toyota (2011) and Sony who were slow to inform the public about their defective cars or their security system that was hacked. Therefore, to ensure a rapid course of action in such a situation, a certain number of actions and activities have to be carried out in advance to actively and specifically prepare for a crisis situation. To this end, it is important to consider:

1. Infrastructure and resources that may be required during a crisis. These include:
   - Developing a network of collaboration and alliance with various stakeholders, as during a crisis there will be a need for a rapid exchange of information. Examples are media, other food companies or industry associations, regulatory authorities, consumer organizations.
   - Being aware of the regulatory requirements in relation to (1) the procedure for reporting an incident (i.e. who should be informed, at which point and what kind of information needs to be provided), (2) legal requirements for withdrawing or recalling products, (3) the eventual disposal of contaminated products in a safe manner, and (4) penalties or penal actions for the responsible person, in case of consumer injuries.

- Being informed of the requirements of the Codex Alimentarius Commission and the International Health Regulation in case an incident or an outbreak takes an international dimension or affects foods entering the international trade.
- Foreseeing the scientific support (e.g. access to experts) and additional logistic support.
- Organizing a database on products and their traceability records, i.e. their destination and/or the source of the ingredients.
- Establishing a contingency plan, e.g. how the work will be delegated or an alternative source of products.
- Definition of roles and responsibilities in times of crisis and the network of people who should be informed.
- Additional administrative support and infrastructure, such as designation of a specific meeting room, extra lines for telephone, mobile phones, etc.
- Good organization (responsibilities, network) and written procedure, e.g. first minute actions.
- Red folder: data needed in case of accidents: organization chart, emergency telephone numbers, phone numbers of key partners, governmental agencies, food companies, customers, scientific experts, specialized laboratories.

2. Principles and procedures. While defining procedures and principles, it is important to also define the specific procedures for crisis management, i.e. how the early warning system should work, who decides, implements and communicates during a crisis situation. Often these may be the same as in normal circumstances; however, each organization has to give this matter specific consideration and take a conscious decision, as the same infrastructure and setup may not be suitable for all types of conditions and organizations. The types of questions that should be considered are:

- The line of reporting of information.
- The type of information (consumer complaints, regulatory actions, media, disease surveillance, food ban, etc.) that should be reported as part of early warning.
- The crisis manager and the skills required for this position.
- Composition of the core crisis management team, competence needed and responsibilities.
- Those who should possibly be informed internally and externally (regulatory authorities, medical community, consumers/general public, suppliers, customers, media, food companies, trade organizations or international organizations, police in case of tampering).
- Define the principles of decision-making and the authority, i.e. who would need to approve a decision.
- The person who will be responsible for implementing decisions and for following up.
- The spokesperson.
- The procedure for preparation of the communication.
- How the issue will be coordinated nationally or internationally.

3. Defining a crisis manager. It is to be noted that the food safety manager does not always need to be the same as the crisis manager. A crisis manager should have specific skills, such as:

- Leadership.
- Good technical knowledge (scientific, product, supply chain, regulatory information).

- Organizational management skills.
- Public relations and communication skills.
- Recognition and trust of stakeholders.
- Experience.
- Emotional intelligence (empathy).
- Pragmatism and common sense.

Finally, as part of crisis preparedness, it is important to communicate to all stakeholders, or all potentially involved parties, the flow of reporting, i.e. the communication plan, and the principles and procedures; it is also important to train the crisis team members in crisis management, and periodically perform a crisis management exercise to ensure that the procedures and principles are correctly understood, feasible and complied with. It is clear that in the light of the outcome of the exercises or in case of any new internal or external experience, the crisis management procedure is to be reviewed and improved. An important element for training is, of course, skills in crisis communication. As will be seen below for this purpose, specific skills are required.

Crisis Management

Then comes the time the house is on fire, i.e. a crisis hits. The procedure, e.g. convening the crisis management team, informing the management, is to be put in place. It is important to act swiftly but calmly. Speedy and timely decisions and actions are key but it is equally important not to take decisions in a panic mode, and as far as possible to take the decisions in consultation with the crisis management team, and/or depending on the case, with the support of the management of the organization. Where applicable, external bodies, e.g. other industries, industry associations, governments of other countries, or international organizations, may need to be consulted.

In the eye of the storm, a number of decisions are to be taken and implemented. To this end, a few principles are to be observed:

1. To get the facts right and be aware of uncertainties. Examples of information which would be required are:
   - What has happened?
   - What level of contaminants was found in the food, which method was used and its validity, sensitivity of the method, possible product variation or limitation of the analytical techniques, or the competence of the laboratory.
   - Range of products that are affected, the time the affected products have been on the market, i.e. how far back a product may need to be recalled if necessary, their expiry date, their distribution, and products which are in the warehouse at the time of the crisis.
   - What was the possible cause of the problem? This information will be essential to determine which products, up to which time, need to be pulled out of the market.
   - Who is aware of the issue?
   - Evaluating risks and possible management options. In evaluating the risks, the health consequences for consumers should be the primary concern and the priority. As part of this, different types of risks need to be considered, e.g. safety risks versus nutritional

risks. Safety risks can also entail microbial versus chemical risks. It is to be born in mind that a rapid change in a product or food consumption pattern without taking the necessary precautions may also lead to exposing consumers to new risks. Other types of factors to consider in the decision-making process are regulatory and legal aspects, potential environmental risks, reputation risk, economic and financial implications, social consequences and perception issues. Decisions are to be taken based on the above consideration and considering the pros and cons of different management options, including feasibility and possible timeframe.

To ensure that the intentions with decisions are understood and the decisions are followed, it is important to explain the basis for decisions and to keep records on the reason for them, those who participated in the decision-making process as well as the data that were considered. The implementation and the outcome of the actions are to be monitored and evaluated at all times. Where necessary, e.g. in the light of new information, the course of action or decisions may require amendments. In taking decisions, it is important to consider both short-term as well as long-term consequences, and also to think globally as today food safety is global and decisions can have broad consequences. Experience in other countries may also be beneficial. Finally, to ensure a rapid course of action, it is important to have a plan of action (including a contingency plan).

Crisis Communication

During the last two to three decades, there has been an increasing recognition of the importance of risk perception and risk communication, in particular during the period of a crisis. It has, among others, been realized that perception of the general public, although not always based on science or in line with the view of scientists, is the main driver in the acceptance of products and/or technologies and influences the food market. A huge amount of research has been carried out in recent years. It has been found that the perception of consumers is influenced by a number of factors such as:

1. Prospects of significant benefit for "me."
2. Whether the risk is voluntary (consent) or involuntary.
3. Whether the risk is familiar.
4. The "dread" factor in the risk.
5. Whether the risk and benefit are "fairly" distributed.
6. Whether the risk is part of an unethical activity.
7. Whether the risk assessor and risk manager are trustworthy.
8. Whether the risk is natural or unnatural.

Consideration of these factors is important, as much in the risk communication as in the decision-making. For instance, in the ITX incident in 2005, where infant formula was contaminated with traces of photo initiator of ink, the affected food companies decided not to recall their products on the grounds that the contamination did not represent a significant risk to health, while many consumers would not buy such a product as the contamination was an unnatural event, involuntary and the risk benefit was not equally distributed. With regard to risk assessment and risk communication, the fact that initially the risk assessment was carried out by the infant formula manufacturers and not by the authorities undermined the validity of risk assessment and communication. On the other hand, in an incident that occurred 2 years earlier in 2003, where traces of semicarbazide were found in baby food, the rapid risk assessment by the European Food Safety Authority and its communication to the general public led to a more peaceful resolution of the crisis.

Consumers and interested parties get their information in different ways. Therefore, different methods and means of communication need to be used to reach the target audience as widely as possible. These include: press release and/or press conference, TV interview, website, podcast, telephone voice messages and hot line for specific consumer queries, alert networks such as the European Rapid Alert System, WHO/FAO INFOSAN, etc. A rumor hot-line can also help people understand if any erroneous message needing correction or explanation is circulating.

Experience from a few crises shows that in the heat of the management of a crisis, a few groups of people are forgotten in the line of communication; they are mentioned here as an aide-mémoire, as depending on the situation, it is essential to keep them informed. Some may need assistance with a brief, a draft declaration or a question and answer, in case they are contacted by the public or the media. These are:

- The chief executive officer or the director general of the organization.
- Trade associations/regulatory authorities or international organizations.
- Stakeholders of the food chain, e.g. retailers.
- Employees.
- The switchboard or consumer services on possible answers to the general public or consumers.

The exact choice of the method and the mechanisms of communication depend on the case, and the strategy for communication needs to be examined very carefully so as not to create undue panic in the population, yet to inform them as needed. In the 2002 acrylamide crisis, the Swedish authorities decided to hold a press conference. According to communication experts, this method of communication created a big communication crisis and media attention, while the contaminant was not new (humans have been exposed to it as far back as the Palaeolithic period when food was cooked over fire), its risks were not yet known, and its content in food could be reduced only after extensive research, meaning that an alarming communication would only create panic without providing consumers with a solution (Löfstedt, 2003).

A few additional principles need to be considered in crisis communication:

- Speed: a communication should be made within 24 hours.
- All communications should be coordinated through one person.
- Depending on the situation, the communication could include the facts, i.e. what is known and what is not known, the decisions, actions and the basis for the decisions.
- Communications should be made in such a way as to avoid any misunderstanding by the audience, i.e. the potential for being understood in different ways than was intended. This can be done by testing the communication on a focus group or person representing the target audience.
- Confirmation of receipt of important messages.
- Full transparency and consistency, in particular being aware that any attempt to downplay or hide facts will cause more damage to the reputation.

- Having and expressing empathy with the victims and affected people.
- Frequency, mode and content of the communications should be culture specific and effective for the specific target consumers.
- Avoiding terms that would amplify the situation or unduly minimize and mislead the target audience.
- In communicating with the media, any gap in information can lead to misinformation.

Documentation and Records

Documentation and records are important means of communication and these are an equally important task in the management of food safety.

In a crisis, which is often a situation where the safety of products has gone out of control, having records of events, decisions and actions as well as supporting documentation becomes even more important. As part of this, all facts and decisions are to be recorded in a logbook. The logbook should contain records of the events, who decided what and who was informed. When meetings have taken place, it should also include minutes of the meetings. These should also record reservations made by any member of the crisis management team and should also be disseminated to all attendees and other interested parties. The preparation of a case report on the event can help in communicating the event to stakeholders in a consistent and transparent manner, and also facilitate the identification of any discrepancy and/or uncertainty that may be detrimental to the process of decision-making. It can also support the development of consensus and the later evaluation of decisions and of the crisis management.

Recovery and Rebuilding after a Crisis

Management of a crisis is often so exhausting that once the storm is over, members of the crisis team tend to return to their normal duties, without further considering the lessons from the crisis. However, the evaluation of the crisis management and determination of the root cause of the incident, its consequences, lessons learned and corrective actions are very important for preventing future incidents (see also Chapter 40). In a case when an infant had died (Belgium, 2001) where the contamination of infant formula with *Cronobacter sakazakii* (formerly known as *Enterobacter sakazakii*) was considered as a possible cause of the incident, the authorities questioned the manufacturer on the corrective actions that the company had taken since its previous *C. sakazakii* incident. Thus, the outcome of the evaluation can be instrumental in improving the food safety management system or the crisis preparedness and procedures, in order to prevent, or minimize, the impact of future problems. A final report on the case, including the root cause analysis and the lessons learned, needs to be communicated widely to prevent future similar cases. Public health authorities need to also report the results of their investigation. The repetition of several important incidents, e.g. melamine, *Salmonella* in chocolate, *Salmonella* in peanut butter, tends to indicate that the causes of incidents are not always fully investigated and the lessons learned are not widely communicated. Finally, the roles and responsibilities, including those of members of management, in the incidents need to be clarified. A major mistake would be to fire the personnel who *a priori* are viewed as responsible before the investigation is finalized and the root cause of the incident is identified. A critical review of the root cause of incidents can show that not infrequently, the failures can be traced to the management's decisions and lack of commitment.

CONCLUSIONS

In life, the unthinkable can happen. In food safety, any gap even a detail may be the occasion for an incident or mishap with the potential for causing a crisis. A proactive approach to the management of food safety can minimize the likelihood of an adverse event leading to a crisis. A crisis is never good and will cause damage, and should by all means be prevented; however, when it occurs, how it was managed can be the opportunity for demonstrating the management capabilities and the values of an organization. In the management of a crisis, the objectives should be to maintain the trust of the stakeholders, in particular the public, authorities and trading partners. Decisions should be based on facts, including consideration of the uncertainties, and putting the health of consumers as a priority. The perception of consumers is also an important consideration. Speed of action, a consistent and transparent approach and empathy for the victims are some of the key values for which an organization should be scrutinized. In a crisis situation, the media play an important role and communication with the media, or through the media with the public, is key for the management of a crisis. The values and culture that an organization promotes and actively implements is a determining factor for the early identification and management of potential issues and for the prevention of a crisis.

References

Bartlett, R., 1999. Dancing with the Devil, Crisis Management in the Food and Drink Industry. Leatherhead Food International.

Buchholz, U., 2011. German outbreak of *Escherichia coli* O104:H4 associated with sprouts. New Eng. J. Med. 365, 1763–1770.

Corie, L., Powell, D., 2000. The Belgian dioxin crisis of the summer of 1999: a case study in crisis communications and management <www.Foodsafetynetwork.ca/crisis/Belgian-dioxin-crisis>.

Doeg, C., 1995. Crisis Management in the Food and Drinks Industry: A Practical Approach, first ed. Springer, Verlag GmbH.

EFSA, 2003a. Additional advice on semicarbazide, in particular related to baby food. Ad hoc expert group meeting 9 October 2003. Parma, Italy. available at: <http://www.efsa.europa.eu/en/press/news/afc031015.htm>.

EFSA, 2003b. Background information on semicarbazide found in foods packaged in glass jars and bottles. Parma, Italy.

EFSA, 2005. Opinion on 2-isopropyl thioxanthone (ITX), adopted on 7 December 2005. Parma, Italy. available at: <http://www.efsa.europa.eu/en/scdocs/scdoc/293.htm>.

EFSA, 2011. Tracing seeds, in particular fenugreek (Trigonella foenum-graecum) seeds, in relation to the Shiga toxin-producing *E. coli* (STEC) O104:H4 2011 outbreaks in Germany and France. Techncial Report of EFSA, EFSA. Parma, Italy.

Financial Times, 2011a. Commission raises *E. coli* damages offer to Euro 210 million <http://www.ft.com/>: retrieved on 2 August 2011.

Financial Times, 2011b. Spain demand compensation over *E. coli* <http://www.ft.com/>: retrieved on 2 August 2011.

Financial Times, 2011c. Berlin criticised for *E. coli* failure <http://www.ft.com/>: retrieved on 2 August 2011.

Löfstedt, R.E., 2003. Science communication and the Swedish acrylamide alarm. J. Health Commun. 8 (5), 407–432.

McKoy, D., 2006. Business Ethic and the Neutraceutical Industry. Prospectus for an Emerging Regime. Presented at Scientific Research Council, Organization of American State Nutraceutical Symposium 2006. Kingston, Jamica, November.

Motarjemi, Y., 2008. Management of food safety in industrial setting Encyclopaedia of Biological, Physiological, and Health Sciences. UNESCO.

Motarjemi, Y., 2011. Food safety: what is the role for gastroenterologists? World Gastroenterol. News 16 (3), 3–4. 1.

Motarjemi, Y., 2014. Modern approach to food safety management. Enc. Food Saf. Academic Press, Elsevier, USA (in press).

Nemery, B., Fisher, B., Boogaerts, M., Lision, D., 1999. Dioxin, Coca Cola, and mass sociogenic illness in Beligum. Lancet 354, 77.

Nemery, B., Fisher, B., Boogaerts, M., Lision, D., Williems, J., 2002. The Coca-Cola incident in Belgium, June 1999. Food Chem. Toxicol. 40 (11), 1657–1667.

Reuters, 2007. Menu Foods Detail costs of pet food recall. Reuters, 30 May 2007. Retrieved on 05.06.2007.

Van Larebeke, N., et al., 2001. The Belgian PCB and dioxin incident of January–June 1999. Exposure data and potential impact on health. Environ. Health Perspect. 109 (39), 265–273.

WHO, 2002. Understanding the BSE Threat. World Health Organization, Geneva.

WHO/Europe, 2011. Outbreaks of *E. coli* O104:H4 infection: update 30 of 22-07-2011 <http://www.euro.who.int/en/what-we-do/health-topics/emergencies/international-health-regulations/news/news/2011/07/outbreaks-of-e.-coli-o104h4-infection-update-30>: retrieved on 2 August 2011.

Further Reading

Casey, D.K., Lawless, J.S., Wall, P.G., 2010. A tale of two crises: the Belgian and Irish dioxin contamination incidents. Brit. Food J. 112 (10), 1077–1091.

Codex, Principles and Guidelines for the exchange of information in food safety emergency situations (CAC/GL 19-1995, Rev 1-2004). Food and Agriculture Organization, Rome, Italy.

Crisis Management: Master the Skills to Prevent Disasters. Harvard Business Essentials, Harvard Business School Press, Boston Massachusetts.

EU, Modus operandi for the management of new food safety incidents with a potential for extension involving a chemical substance (Health and Consumer Protection Directorate General, 2006).

EU, Commission Decision of 29 April concerning the adoption of a general plan for food/feed crisis management (2004/478/EC) (Official Journal of the European Union).

Grant, S.E., Powell, D., Crisis Response & Communication Planning Manual Prepared for the Ontario Ministry of Agriculture, Food and Rural Affairs by: Department of Plant Agriculture, Crop Science Division, University of Guelph.

WHO, 2005. International Health Regulations, 2005. WHO, Geneva, Switzerland.

Further reading

CHAPTER

42

The Role of International, Regional and National Organizations

F. Tracy Schonrock

Schonrock Consulting Formerly with the United States Department of Agriculture (USDA)

OUTLINE

INTRODUCTION

Standards appear to be an innate, hard-wired faculty of a living brain. It is demonstrated repeatedly in even nonhuman subjects by virtue of an animal's ability to make distinctions between ripe and unripe foods, which individual will make the most suitable mate, where is the best place to build a nest or den. Creating standards appears to be the brain's preferred method to sort and classify the flood of information that it continuously confronts.

Over the course of evolution, this ability to standardize started to categorize certain sounds to mean specific things; such as food, water, danger or mate. The standardized

Food Safety Management.
DOI: http://dx.doi.org/10.1016/B978-0-12-381504-0.00042-1

sounds led to the development of language that could be understood by multiple individuals to unify a group into a society. All could learn and understand the meaning of the standardization of the various sounds. In time, the standardized sounds were further standardized into specific symbols or groups of symbols, and written language ensued.

Communication is the basis for civilization and, as such, is the lubricant that allows industry and business to flourish. Communications must include concepts, units and measurements that are commonly understood to be useful to the parties involved. The development of standards as a tool of commerce is hard work. Those actively involved with the development of standards know how difficult it can be to put concepts into words so that every reader will arrive at essentially the same interpretation. As difficult as this effort may be for physical attributes of a product, it becomes immensely more difficult when trying to describe a sensory attribute, such as a flavor or odor, in words.

The world of commerce is awash with standards. An internet search for *Food Standards* will reveal approximately 540 million citations. Every country has its own extensive list of standards for both domestic and imported products. Over time, some of those standards have been codified to become regulations. This chapter will focus on the leading organizations that develop standards that are of significance to the food processing industry. Industry members seeking to do business with a particular country or region are strongly encouraged to become familiar with their standards.

Effective standards and guidelines are developed through what is called the Consensus Process.

Wikipedia (http://wikipedia.org) defines the process as follows: "**Consensus decision-making** is a group decision-making process that seeks the consent, not necessarily the agreement of participants and the resolution of objections." Consensus is defined by Merriam-Webster as, first, general agreement, and second, group solidarity of belief or sentiment. It has its origin in a Latin word meaning literally feel together. It is used to describe both the decision and the process of reaching a decision. Consensus decision-making is thus concerned with the process of reaching a consensus decision, and the social and political effects of using this process.

All standards writing organizations using this process follow these basic principles:

- Openness
- Lack of dominance
- Balance
- Coordination and harmonization
- Notification of standards development
- Consideration of views and objections
- Consensus vote
- A process for appeals

All interested parties taking the opportunity to participate in a standard's development are assured that their comments and objections are heard, and that the work is not duplicating the work of standards already created.

Modern commerce in the food industry is both local and international in nature. Worldwide consumers are used to being able to obtain food products which are not locally grown or are not in season. They are critical of freshness, quality and safety. Even consumers in underdeveloped or developing countries that are forced to live at subsistence levels are wary of receiving cast-off or substandard products. Safe food is of primary concern to all consumers. Unsafe or low-quality food will cause a long-term loss of confidence in a supplier. This loss of confidence will result, at minimum, in a significant loss of market share and, at worst, cause a company to go out of business.

Everyone benefits from standardization. Standardization reduces production costs thus reducing consumer prices and increases the safety and desirability of products offered for sale.

LEADING INTERNATIONAL STANDARDS ORGANIZATIONS

World Health Organization

At the international level of public health oversight is the World Health Organization (WHO).[1] WHO is the directing and coordinating authority for health within the United Nations system. It is responsible for providing leadership on global health matters, shaping the health research agenda, setting norms and standards, articulating evidence-based policy options, providing technical support to countries and monitoring and assessing health trends. WHO provides services in every aspect of public health; however, its program on Food Safety (including its activities for the surveillance of foodborne diseases) is of specific importance to food safety. The Food Safety program is allied with the Consumer Protection program of the Food and Agriculture Organization of the United Nations, the host for the Codex Alimentarius Commission (see below).

WHO, as the leading authority on public health and food safety issues, has the international support and resources to conduct comprehensive risk assessment. Chemical and biological hazards in foods are a worldwide public health concern and can have major impacts on international trade. All interested parties from producers, processors to governments need to have access to reliable risk assessment, but few have the resources, expertise or funds to conduct them on the huge numbers of chemicals used in agriculture and food production. It is through the efforts of the Joint FAO/WHO Expert Committee on Food Additives (JECFA), the Joint FAO/WHO Meeting on Pesticide Residues (JMPR) and the Joint FAO/WHO Expert Meeting on Microbiological Risk Assessment (JEMRA) that the food industry benefits.

JECFA meets twice a year to conduct risk assessment and safety evaluations of food additives (including processing aids and flavorings), contaminants, naturally occurring toxicants and residues of veterinary drugs. JECFA has evaluated more than 2500 food additives, approximately 40 contaminants and naturally occurring toxicants, and residues of approximately 90 veterinary drugs since its inception. The findings of the committee are published and available from online archives:

- The WHO Technical Report Series (TRS) contains concise toxicological and chemical evaluations of each substance evaluated.
- The WHO Food Additives Series (FAS) are toxicological monographs with detailed descriptions of the biological and toxicological data from the evaluations and including intake assessments.

[1] www.who.int

- The Compendium of FAO Food Additive Specifications provides detailed specifications on the identity and purity of food additives and flavoring agents.
- The Database of FAO Veterinary Drug Residue Monographs.

JMPR meets annually to review residue and analytical data on pesticides and other contaminants. Their reviews are focused on the chemicals' rates and by-products of metabolism, effects on the environment, use patterns and to establish maximum residue levels (MRLs) and average daily intake levels (ADIs). As such, JMPR serves as the scientific advisory body for the FAO, WHO, FAO/WHO member governments, and to the Codex Alimentarius Commission via the Codex Commission on Pesticides/Residues (CCPR).

Following the Uruguay Round the World Trade Organization (WTO) succeeded the General Agreement on Tariffs and Trade (GATT) and established that scientific, risk-based standards were to be used under the Sanitary and Phytosanitary (SPS) agreement to address trade practices. This positioned the FAO/WHO, through Codex, to be the preferred source of risk assessment data.

The findings of the committee are published and available from online archives:

- Toxicological monographs summarize the data reviewed.
- Residue monographs containing information on pesticide use patterns, chemistry and composition of pesticides, methods of analysis for residues, and information on MRLs are published in the FAO Plant Production and Protection Paper series.

JEMRA was established to meet the challenges created by the emergence or re-emergence of foodborne pathogens. The risk analyses performed by JEMRA cover the entire food chain. They focus their analyses on detailed review of scientific papers with emphasis on food–pathogen combinations. They are focused on identifying what are the potential risk pathogens, what happens when the pathogen is ingested, what constitutes an infectious dose of the pathogen, who in the population are at most risk, and what measures can be employed to eliminate the chance of foods becoming infected. The primary users of the risk assessments and data they develop are the Codex Committee on Food Hygiene (CCFH) for use in the development of standards, guidelines and recommendations, and directly to WHO/FAO member countries so they can better monitor and control microbiological hazards in foods.

Of significant benefit to the food processing industry and the governmental bodies that monitor its activities are the extensive WHO efforts to develop and disseminate guidance materials on HACCP programs, the strengthening of national food safety programs, consumer education programs and international health regulations.

As noted throughout this chapter, the Food and Agriculture Organization of the United Nations (FAO)[2] is closely aligned with WHO. The two organizations support and complement each other. The mission of FAO is to provide independent scientific advice, data input for risk assessment activities and similar guidance on food safety issues to the Codex Alimentarius Commission and governmental bodies. Decisions that influence international

[2]www.fao.org

trade and food safety requirements must be based on sound, demonstrable science data. To assist this goal, FAO supports the development of member countries' resources to effectively manage food safety and quality programs by providing advice on specific scientific food safety issues. FAO provides additional guidance in food control measures through training and various publications. The following are major topic areas addressed by FAO:

- Assessment of food safety capabilities
- Establishing effective food safety programs
- Encouraging participation in Codex activities
- Development and implementation of HACCP systems
- Development and implementation of food inspection programs
- Product quality assurance programs

All of FAO's efforts are focused on achieving food security while protecting the environment and assuring sustainability of agriculture around the world.

CODEX ALIMENTARIUS
International Food Standards
World Health Organization — Food and Agriculture Organization of the United Nations

The objectives of the Codex Alimentarius Commission (CAC) are to protect the health of consumers and ensure fair practices in the food trade. Through the work of its committees, CAC develops and publishes standards and codes of practice under the auspices of WHO/FAO. Codex Alimentarius international standards, guidelines and codes of practice are recognized by the World Trade Organization's (WTO) *Agreement on Sanitary and Phytosanitary Measures (SPS Agreement)* as the reference for food safety requirements. Codex members represent 99% of the world's population. Participation offers countries a forum in which they can join the international community and assist in the development and harmonization of food standards and encourage their global adoption. Therefore, it plays a major role in international trade and in the resolution of disputes between buyers and sellers.

The WHO/FAO booklet *Understanding the Codex Alimentarius*[3] defines standards and codes of practice as follows:

> Codex standards usually relate to product characteristics and may deal with all government-regulated characteristics appropriate to the commodity, or only one characteristic.
>
> Codex Codes of Practice – including codes of hygienic practices – define the production, processing, manufacturing, transport and storage practices for individual foods or groups of foods that are considered essential to ensure the safety and suitability of food for consumption.

In many cases, Codex standards form the basis for many national standards. This provides the international trading community with a substantial foundation. Above this foundation are many standards writing organizations producing standards and guidelines.

[3]Joint FAO/WHO Food Standards Program, FAO, Rome.

These standards and guidelines are also recognized internationally. They may be specialized in various aspects of food processing and have gained wide recognition for their expertise. As with Codex, the successful organizations produce their documents through the consensus process. The consensus process assures that the documents produced have had input from all of the interested stakeholder groups; all opinions are openly discussed until a consensus is reached; the process has safeguards to protect the due diligence of the procedures; and the decisions are science based rather than on proprietary interests, arbitrary opinion or market protection.

ISO International Organization for Standardization

The largest standards writing organization in the world is the International Organization for Standardization (ISO).[4] ISO was founded in 1947 and is based in Geneva, Switzerland. Since its founding, ISO has developed more than 19,000 standards. These standards and those of its partners, the International Electrotechnical Commission (IEC) and the International Telecommunications Union (ITU), encompass the entire scope of all industry activities from agriculture, food, construction, mechanical engineering, electronics, to computer and communications technology.

ISO is a network of the national standards institutes of 164 countries. These national standards institutes may be mandated by their respective governments or they may be industry established and operated. Inclusion of standards writing organizations from both government and non-government sources provides ISO with a superior advantage in assuring that the needs of both government and industry are addressed through the consensus process. While individuals and enterprises cannot become members of ISO, they do have the opportunity to influence ISO standards and decisions by becoming active in their national ISO delegations or the delegations' member organizations. Individuals may be selected to serve on ISO technical committees as recognized subject matter experts designated by their national delegations. A principal regional partner of ISO is the European Committee for Standardization (CEN).

Support of world trade is a major activity of ISO. ISO maintains a close relationship with the World Trade Organization (WTO) which grew out of the deliberations of the Uruguay Round of 1985–94 of the General Agreement on Tariffs and Trade (GATT) as well as the specialized agencies and commissions of the United Nations. ISO has focused their technical committee to assure that all ISO standards are compatible with these international organizations.

To reach the broadest audience as possible, ISO maintains an extensive library of videos to highlight the objectives and goals of the various standards and policies. Of particular interest to the food processing industry within the ISO library are 114 standards specific to food items and processes. ISO has five current publications detailing the creation and management of food safety management systems and the certification of these systems. ISO 22000 is a food safety management system that can be applied to any organization in the food chain, farm to fork. The standard has requirements for food safety management systems processes and procedures, and requires that the organization implement prerequisite programs and HACCP.

[4] www.iso.org

LEADING REGIONAL STANDARDS ORGANIZATIONS

The European Committee for Standardization (CEN)[5] is based in Brussels. CEN was created in 1975 with goals to facilitate trade, remove trade barriers for European businesses and consumers, and promote global trading while protecting European citizens and the environment. It accomplishes these goals through the development of standards and publication of technical reports and technical specifications. CEN is the only European standards writing organization recognized by the European Union according to Directive 98/34/EC for all areas of commerce except for electrotechnology and telecommunications.

Thirty-three national members cooperate to develop and maintain European standards signified by the preface EN. The national members solicit the involvement of a vast number of technical experts, business federations, consumers and special interest groups to obtain the best and latest information when developing standards and technical specifications. CEN and ISO have a cooperative association, established by *The Vienna Agreement*, to provide representation in each group's meetings and to adopt the same text in each of their documents when applicable. Approximately 30% of CEN standards are identical to ISO standards. Once developed, the EN standards gain additional importance as they also become national standards for each of the member nations. Any existing national standards that conflict with the EN standards are withdrawn so as to provide uniformity throughout the European market. The adoption of standards and the removal of conflicting documents provide a uniform marketplace where products can travel freely without the cost burden of local or regional requirements.

CEN has established more than 400 standards as well as technical specifications and technical reports of value to the food processing and related industries. These include methods of testing of products for composition, toxins, microorganisms and allergens; materials handling equipment, food technology, metallurgy, packaging materials and systems, machinery safety, and construction materials and building designs. Of particular interest are those food and product standards which have been adopted as national standards and control the flow of products in commerce. Standards are available that also impact the food service industry.

CEN works closely with the European Commission to assure food safety:

- M/315 Standardization, Mandate in the field of method of analysis for animal feeding stuffs
- M/381 Standardization, Mandate in the field of methods of analysis of foodstuffs concerning food hygiene
- M/382 Standardization, Mandate in the field of method of analysis for animal feeding stuffs
- M/383 Standardization, Mandate in the field of method of analysis for mycotoxins in food
- M/422 Standardization, Mandate in the field of method of analysis for heavy metals and iodine in food
- M/463 Standardization, Mandate in the field of method of analysis for food contaminants

In conjunction with the European Research Center of the European Commission, CEN participates in the organization and planning of workshops for food and feeds. In this regard, *CEN Workshop 18 – Cleanability of commercial foodservice equipment used in retail and*

[5] www.cen.eu

catering sectors was created in March 2004. This workshop developed the document *CWA 15596-2006 Code of practices on cleanability of commercial food equipment used in catering sectors.* Consumers are purchasing greater quantities of ready-to-eat food items and dining away from the home more frequently. The safety of retail deli counters and salad bars in grocery stores, and restaurants ranging from food carts to fast food establishments to high-end restaurants, are all potential sources of food safety issues. Standardization of the equipment designs, processes, food handling procedures and cleanability of the equipment and facilities is of major concern to local authorities.

CEN Technical Reports are developed to complement EN standards by providing information on the technical content of the standards. These reports are informational in nature and do not place any regulatory obligation on the member nations.

Packaging is vital to the food industry. The food industry utilizes approximately 60% of all packaging materials produced. These materials include glass, paper, cardboard, metal cans, metal foils, plastics and specialized laminates. They are of concern to raw material suppliers, food processors, users and consumers, transportation firms and waste management companies. Packaging materials that have direct product contact are a food safety issue. International and smaller, local packaging material manufacturers rely on standardization and regulations to assure them that they will have broad access to food markets around the world. The CEN Technical Committee has produced approximately 200 normative documents or technical reports dealing with packaging materials.

efsa
European Food Safety Authority
Committed since 2002 to ensuring that Europe's food is safe

The European Food Safety Authority (EFSA)[6] provides independent risk assessment for food industry risk managers. EFSA has a legal obligation mandated by the European Parliament (EU Regulation 178/2002) to provide their services to EU member states. The EFSA goal is to be recognized globally as the European reference body for risk assessment, based on the highest scientific standards, on food and feed safety, animal health and welfare, plant protection and plant health.

The EFSA role is to assess and communicate risks within the entire food chain. They have a significant influence on what is available in the food retail stores and eventually on the consumer's dining table. The risk assessments conducted include adopting or revising EU legislation on food safety, the approval of regulated substances such as pesticide residues and food additives, and the development of nutritional guidelines and policy.

The primary stakeholders within the EFSA structure are the European Parliament, the EU member states, industry groups, consumer groups and non-governmental organizations (NGOs). EFSA conducts risk assessments when requested for scientific advice from any of the stakeholder groups and it may initiate activities on its own. The risk assessments are conducted by reviews of the current scientific research papers and study data available, through web-based public consultations, and may carry out research among its Scientific Panels and Directories for its key target audiences. These activities are conducted through five directories overseen by the EFSA executive director. The directorates are:

- Risk Assessment and Science Assistance
- Scientific Evaluation of Regulated Products

[6] www.efsa.europa.eu

- Science Strategy and Coordination
- Communications
- Resources and Support

LEADING NATIONAL GOVERNMENTAL ORGANIZATIONS

All national governments have regulatory and standards development or implementation programs to protect their consumers. Depending on the national structure of each country the responsibilities for these activities may be centralized or divided between multiple agencies. The following are examples of influential agencies from the United Kingdom and the United States of America.

Food Standards Agency

The Food Standards Agency (FSA)[7] is the primary standards organization for the food industry in the UK. FSA was established by the Food Standards Act of 1999 and, therefore, has a statutory obligation to protect public health and food hygiene across the UK. They accomplish this mission by working closely with local authorities in England, Scotland, Wales and Northern Ireland. FSA can commission research on food safety and hygiene issues when necessary to establish the best regulations and policies. All decisions are based on the best available science available.

FSA as the UK representative to the EU Commission and Codex assure that the concerns of the UK are considered in the policies and standards created by those agencies. The strategy of FSA is to assure:

- Foods produced or sold in the UK are safe to eat;
- Imported food is safe to eat;
- Food producers and caterers give priority to consumer interests in relation to food;
- Consumers have the information and understanding they need to make informed choices about where and what they eat;
- Regulation is effective, risk based and proportionate, is clear about the responsibilities of food business operators, and protects consumers and their interests from fraud and other risks;
- Enforcement is effective, consistent, risk-based and proportionate and is focused on improving public health.

As the primary regulatory agency for the public health aspects of food safety, FSA works with local authorities to implement and enforce the *Food Law Code of Practice*. The Code of Practice sets out instructions and criteria that local and port health authorities (food authorities) should comply with when enforcing food law. Food authorities must follow and implement the provisions of the code as they apply. Included in these regulatory activities are the establishment of a list of Approved Plants, audits of local authorities to assure uniformity of inspections, training, issuing food alerts when necessary and monitoring of food safety.

[7] www.food.gov.uk

The United States, Food and Drug Administration (FDA),[8] Center for Food Safety and Applied Nutrition (CFSAN) is the primary United States food regulatory authority except for red meats and poultry, which are regulated by the Food Safety and Inspection Service (FSIS), United States Department of Agriculture (USDA). FDA has a statutory obligation to protect the public health through the Food, Drug and Cosmetic Act of 1936 as amended. Its regulations are codified under Title 21 of the Code of Federal Regulations (CFR) and thus available to all interested parties.

The *good manufacturing practices* (GMPs) (21 CFR Part 110) and *thermally processed low-acid foods packaged in hermetically sealed containers* (21 CFR 113) are of universal interest to the food industry. CFSAN also publishes, independently from the CFR, the Pasteurized Milk Ordinance (PMO) specifically for the dairy industry, and the Food Code for the retail food industry. In addition to these regulatory and inspection guidelines, FDA also publishes Standards of Identity for most common foods. The Standards of Identity can be obtained from 21 CFR 130 through 169. FDA, like its counterparts in Europe, conducts risk assessments and approves all food additives used in the United States. On 1 January 2011, the Food Safety Management Act (FSMA) was signed into law. This act has a significant impact on both domestic and international trade within the United States.

FDA is the United States representative for interaction with Codex, the EU Commission and other national governments in regard to international food standards and international trade issues affecting food safety issues and quality.

For the food industry, FDA has interest and regulatory authority for:

- Risk assessment
- Biotechnology
- Dietary supplements
- Food defense and emergency response
- Food ingredients
- Food safety
- Retail food protection
- Guidance, compliance and regulatory information
- International activities
- Labeling and nutrition
- Animal drugs and residue monitoring
- Science and research

The United States Environmental Protection Agency (EPA)[9] is not a primary standards setting organization for the food industry. However, the EPA does conduct risk assessments for pesticide use and residues used in agriculture. These pesticide residue limits are used by the American regulatory agency for the protection of consumers. EPA risk assessments and residue limits are commonly referenced worldwide as a source of reliable data for establishing food standards limitations.

[8] www.fda.gov

[9] www.epa.gov

LEADING INDUSTRY ORGANIZATIONS

GFSI

Following a series of food safety incidents, a group of business CEOs banded together to form the Global Food Safety Initiative (GFSI).[10] This program is retailer and food service industry driven in order to maintain control over the safety of their supply chains. GFSI was formed in May 2000 under Belgium law. The stated vision of GFSI is "Safe food for consumers everywhere." Its mission further aims to "Provide continuous improvement in food safety management systems to ensure confidence in the delivery of safe food to consumers worldwide."

The Consumer Goods Forum, a global network for retail goods retailers and manufacturers worldwide, provides the direction and day-to-day management of GFSI. They recognized early on that the huge variety of products, processes, plant layouts and delivery systems was not compatible with a "one-size-fits-all" approach. Therefore, GFSI does not:

- Intervene in retailer or supplier policy;
- Make policy for standards developers;
- Undertake certification or accreditation activities of retailers.

The approach GFSI has taken is the development of a program through which the knowledge and expertise of how the various food chain participants manage food safety can be evaluated. The program will:

- Provide a method to benchmark existing food safety standards used from the farm through the processing level to the consumer;
- Compare existing standards against the requirements that have been put together by the participants in the food supply chain.

To accomplish this goal, a technical working group composed of retailers, manufacturers, certification bodies, accreditation bodies, standards developers, food service providers, food safety experts and consultants developed the GFSI Guidance Document. The current volume is the Sixth Edition Issue 3 Version 6.2 and is available for free from the GFSI website. The process outlined in this document is intended to be executed in an independent, unbiased, technically proficient and transparent manner. The Guidance Document is divided into three parts:

"Part I – The Benchmarking Process" provides the key steps developed by GFSI to rate an existing safety management system according to the key elements identified by GFSI as necessary to ensure food safety. The benchmarking process is to be done by an impartial group with full transparency so that others in the GFSI program can be assured that the benchmarking is accurate and complete. Members of the GFSI Benchmarking Process include Safe Quality Food Institute (SQF), British Retail Consortium (BRC), International Featured Standards (IFS) and Food Safety System Certification (FSSC) 2000.

"Part II – Requirements for the Management of Schemes" provides requirements necessary to the effective management and control of a food scheme. All food safety schemes require validation that they are effective and then continual verification that the scheme is working on a day-to-day basis. This part also provides the requirements for the

[10] www.mygfsi.com

competence of auditors working for the independent third party Certification Bodies that will evaluate the suitability of a food safety scheme for inclusion in the GFSI program. "Part III – Scheme Scope and Key Elements" describes and expands upon the key elements determined by the technical working group as necessary for inclusion in a food safety plan to be eligible for recognition by the GFSI program. These key elements will include the requirements for good practices which may include HACCP or HACVCP-based controls.

Firms that have successfully complied with the requirements and been certified as in conformance with the GFSI program can enjoy significant benefits as a supplier or retailer, with dealings with government agencies, recognition across borders and marketing areas, and reduction in production costs through efficiencies and reduction of duplicate inspections by multiple buyers.

International Life Sciences Institute (ILSI)[11] is based in Washington, DC. The nonprofit organization has the mission "to provide science that improves public health and well-being." They accomplish this mission by supporting and encouraging collaboration among experts, academia, government and industry on the tasks of gathering, summarizing and disseminating science. ILSI focuses its activities primarily on nutrition, health promotion, food safety, risk assessment and the environment. These are important activities that support the goal of a standards writing organization (SDO) to provide science-based standards and regulation recommendations. This, in turn, assists industry and governments to address the risks and issues that are of common concern around the world.

ILSI has identified Four Global Issues that are of common concern at the local, regional and international level, and in which nearly all ILSI entities are involved. They coordinate the scientific efforts of the programs and projects related to each Global Issue, and provide links to additional information to interested parties as available:

- Biotechnology
- Functional Foods
- Obesity
- Risk Assessment

The International Commission on Microbiological Specifications for Foods (ICMSF)[12] is a subsidiary of the International Union of Microbiological Societies (IUMS). Their mission is to be a leading source of independent, impartial scientific microbiological concepts and standards that can be adopted by government agencies and industry to reduce the incidences of pathogens in foods. ICMSF has links to WHO and Codex so its activities have worldwide availability. The commission's goals are to:

- Assemble, correlate, and evaluate data about the microbiological safety of foods;
- Consider if microbiological criteria would improve and assure the safety and quality of particular foods;

[11] www.ilsi.org
[12] www.icmsf.org

- Propose the adoption of such criteria; and
- Recommend methods of sampling and microbiological examination to assure uniformity of results worldwide.

The commission was established in response to the increasing number of foodborne diseases and the need for increased microbiological testing. The demand for international trade in foods is expected to continue to rise. Diseases caused by foodborne pathogens are a worldwide public health concern and impact on a countries' or market's food security. The food standards proposed by ICMSF are based on sound scientific principles of analysis, sampling plans and microbiological limits. Their standard's equivalency between countries is well established.

The European Food Information Council (EUFIC)[13] located in Brussels, Belgium, is a non-profit organization which communicates science-based information on nutrition and health, food safety and quality, to help consumers to be better informed when choosing a well-balanced, safe and healthful diet. EUFIC's publications are based on peer-reviewed science. Information that EUFIC publishes has been subject to a review process by members of its Scientific Advisory Board (SAB). The SAB, comprised of renowned experts from across Europe, advises EUFIC on its information and communication programs, ensuring that all information is based on scientific evidence, relevance and is factually correct. Given the broad range of subjects addressed in EUFIC's popular newsletter, Food Today, a dedicated editorial board provides additional insights and feedback for this publication.

EUFIC is supported by companies of the European food and drinks industries, but also receives some project funding from the European Commission. All members adhere to EUFIC's Transparency Statement. EUFIC's mission is to enhance the public's understanding of credible, science-based information on the nutritional quality and safety of foods and to raise consumers' awareness of the active role they play in safe food handling and choosing a well-balanced and healthy diet.

EUFIC continues to partner with a broad base of stakeholders in numerous research projects funded by the European Union. With financial support from the European Commission's Directorate General for Research, the consortia in which EUFIC participates aim to improve our knowledge about food safety and quality, and health and nutrition. Projects they participate in include CHANCE, DIETS, EATWELL, EURRECA, FLABEL, FOOD4ME, IDEFICS, NU-AGE, CONNECT4ACTION, RECAPT, INPROFOOD, and more. Additional information concerning these various programs can be obtained from the EUFIC website.

The International Organization of Consumers (IOCU)[14] is a global federation of consumer advocacy groups. As the world's consumers have become more informed about products, processes, food safety issues and the environment, they have banded together to make their voices heard by national governments and international bodies. IOCU is a non-governmental organization (NGO)

[13] www.eufic.org
[14] www.sagepublications.com

that links the activities of 170 consumer organizations residing in 60 countries. The consumer organizations support the eight following consumer rights:

1. The right to safety;
2. The right to be informed;
3. The right to choose;
4. The right to be heard;
5. The right to the satisfaction of basic needs;
6. The right to redress;
7. The right to consumer education;
8. The right to a healthy environment.

IOCU has standing committees or working groups on education, testing and development, health, transnational corporations, library and documentation, air transport, and information technology to assist their members to organize, lobby and change government regulations and policy. The organization extends the influence of the consumer by having established consultative status on several United Nations bodies. These include the UN Children's Fund (UNICEF), the World Health Organization (WHO), the UN Environment Program (UNEP), the Food and Agriculture Organization (FAO), the UN Conference of Trade and Development (UNCTAD), the UN Education and Science Organization (UNESCO), the UN Industrial Development Organization (UNIDO) and the Economic and Social Council (ECOSOC).

The European Consumer Organization (BEUC)[15] is based in Brussels, Belgium. As with other consumer advocacy organizations, their purpose is to lobby the European governments on behalf of consumers' issues. BEUC supports and reinforces the eight consumer rights promoted by the IOCU. The organization acts as an umbrella group representing 40 independent national consumer groups from 30 European countries.

BECU is duly registered with the European Parliament Lobby register which affords its representatives unlimited access to parliament facilities to meet with members of parliament and EU Commission officers. Access to the public policy developers in parliament allows BEUC to vigorously defend consumer rights and has resulted in many favorable policy decisions. Their work is to ensure that consumer policy at the EU level is sustainable for all. In BEUC terms, "sustainability" is not only the protection of the environment, including climate change, but also reduction of negative social and economic impacts. Improving well-being for all, without compromising the needs of vulnerable groups, such as children, the elderly and low income consumers, must be taken into account when designing policy.

The Grocery Manufacturers Association (GMA)[16] is located in Washington, DC. GMA is an advocate for the leading food, beverage and consumer products companies to facilitate and advance the quality of life for the consumers in the United States and around the world. GMA is active in product safety, health and nutrition, preservation of the environment, global commerce, collaboration among retailers, and providing advice and counsel to governments on consumer

[15] www.beuc.org
[16] www.gmaonline.org

products issues. They pursue their goals through a strong commitment to scientific research, testing and evaluation of consumer products and business practices. GMA assists their members as a central information resource and as a means to collaborate between members, retailers, service providers and consumers to obtain healthy, affordable, safe foods.

NSF International (NSF)[17] is based in Ann Arbor, MI, USA. Their organizational mission is "To make the world a safer place for consumers." They accomplish their mission by developing and publishing internationally recognized standards for food, water and consumer products. NSF has offices in many European countries and other international locations in the Far East, Southeast Asia, South America and Mexico.

One goal of NSF is to offer a knowledge base to support and increase legislators and regulators awareness of public health issues. The NSF Regulatory Affairs office provides information on the interpretation and application of their standards, answers to regulatory code questions, product verification to assure the products meet national standards and other requirements.

NSF offers a wide scope of programs that are of particular importance to the food and beverage industries and in auditing for conformance with international food safety initiatives, such as the Global Food Safety Standards (GFSI) (including SQF (Safe Quality Food), BRC (BRC Global Standards), Global GAP (the Global Partnership for Good Agricultural Practices), FSSC (Foundation for Food Safety Certification) and IFS (International Featured Standards)), HACCP-9000 and ISO 22000.

NSF has developed over 50 American national standards pertaining to food safety and public health under their various programs. Three standards, developed jointly with the 3-A SSI organization, are specific for meat and poultry equipment (14159-1, -2 and -3).

LEADING HYGIENIC DESIGN STANDARDS ORGANIZATIONS

The emphasis on the many product standards writing organizations often overshadows the importance of the equipment and processes used to produce safe foods. The successful conformance to product standards, as well as the reduction of production costs, is significantly enhanced when the equipment and processes are designed and fabricated to standards that will help assure that microbial and physical contamination, and the carry-over of allergens, are prevented. Multiple incidences of poorly designed processing equipment have led to problems with food safety, product quality and massive industry recalls due to an inability to properly clean and sanitize the equipment. These problems of equipment design and fabrication can be prevented by following the standardized criteria readily available.

The European Hygienic Engineering and Design Group (EHEDG),[18] based in Frankfurt, Germany, has regional offices in most European countries and international locations in the Far East, Southeast Asia, South America and Mexico. EHEDG is a consortium of equipment manufacturers, food processors, research institutes and public health authorities. It was

[17] www.NSF.org
[18] www.EHEDG.org

FIGURE 42.1 Sample EHEDG certification mark.

formed in 1989 with the mission to promote hygiene during the processing and packaging of food products. EHEDG as an organization does not develop specific standards. European legislation requires that food be processed and packaged hygienically, with hygienically designed equipment, in a hygienic facility. EHEDG takes the requirements of the legislation and presents them in a series of guidelines which are easily understood and provide guidance for specific classes of equipment and processes. Currently, EHEDG has developed 41 unique guidelines to assist the food processing industry. The scope of these guidelines cover:

- Equipment and building design and cleanability
- Equipment and building element installation
- Industrial services and utilities
- Maintenance of assets

The EHEDG certification program provides purchasers of equipment with a readily recognizable symbol (Figure 42.1) displayed on equipment that meets the guidelines' requirements. This assists purchasers to obtain equipment that will support their production of safe foods.

3-A Sanitary Standards, Inc. (3-A SSI),[19] based in McLean, VA, USA, is the premier standards writing organization for hygienic standards for food processing equipment. Their first standard was published in 1920. The historical basis for 3-A standards development has been the dairy industry. However, it has been shown over time that the fundamental principles of hygienic design and equipment cleanability encompassed by the 3-A Standards and Accepted Practices are universally applicable over most food products. The mission statement of 3-A SSI is:

> It is the mission of 3-A Sanitary Standards, Inc. to enhance product safety for consumers of food, beverages, and pharmaceutical products through the development and use of 3-A Sanitary Standards and 3-A Accepted Practices.

3-A SSI develops two types of documents. The 3-A Sanitary Standards provide criteria for specific types or classes of equipment. The 3-A Accepted Practices provide criteria

[19] www.3-A.org

for specific processing systems. There are 71 3-A Sanitary Standards and 10 3-A Accepted Practices as of 2012.

The 3-A symbol, , is a copyrighted mark to signify to buyers of the equipment that it conforms to all of the criteria of a covering 3-A Sanitary Standard.

For many years the Baking Industry Sanitation Standards Committee (BISSC) developed standards specific to baking industry equipment. As with other standards writing organizations in the United States they are accredited to develop standards using the ANSI procedures and guidelines. BISSC was formed in 1949 by representatives from six national baking industry organizations. The complete line of BISSC standards was first published as a single booklet in 1977. The current publication is the widely successful *ANSI/BISSC Z50.2-2003 Baking Equipment Sanitation Standard*.

In 2007, the board of directors of BISSC elected to become a wholly owned subsidiary of the American Institute of Baking International (AIB).

AIB International,[20] based in Manhattan, KS, USA, was established in 1919 initially as a research organization. Currently AIB offers a wide range of services to the food industry as well as the baking industry, including:

- AIB Consolidated Standards
- AIB GMP Inspections
- AIB Knowledge Center
- Analytical services
- Consulting and customized training
- Distance learning and training products
- Food defense services
- GFSI certification schemes
- HACCP accreditation
- Resource center
- Seminars, webinars and courses

CONCLUSIONS

Standards are pervasive throughout the food industry. The organizations highlighted in this chapter are a limited sampling of the most recognizable in widespread use. The information provided indicates the extremely complex interaction of all of these organizations as they work towards a common goal – food safety. Participating in standards development and the use of developed standards by manufacturers, processors, distributors and retailers provides them with the ability to assist in the creation of documents that will benefit all users through improved food safety, quality, efficiency, cost reduction and acceptance by consumers.

[20] www.aibonline.org

Further Reading

Allergen Information Manual & Auditor Guidelines, downloadable from www.aibonline.org
ANSI Essential Requirements: Due Process Requirements for American National Standards, 25 West 43rd Street, 4th Floor, New York, New York 10036.
Application of Risk Analysis to Food Standards, downloadable from www.who.int/foodsafety/micro/jemra/en/
CEN Compass: The world of European Standards, downloadable from www.cen.eu
EFSA's approach to identifying emerging risks in food and feed, downloadable from www.efsa.europa.eu/cs/Satellite
Enhancing Food Safety through Third Party Certification, downloadable from www.mygfsi.com
Food Defense Guidelines, downloadable from www.aibonline.org
GFSI Requirements on the Application of ISO/IEC 17011:2004, downloadable from www.mygfsi.com
Understanding the Codex Alimentarius, downloadable from codexalimentarius.org
WHO Global Strategy for Food Safety, downloadable from www.who.org
World Consumer, IOCU, The Hague.

SECTION IV

SUSTAINABILITY AND ETHICS

CHAPTER

43

Sustainability and Food Production

Thomas Ohlsson

Thoohls, Molndal, Sweden, Previously at SIK – The Swedish Institute for Food and Biotechnology, Gothenburg, Sweden

OUTLINE

INTRODUCTION

The chapter gives an overview of the sustainability issues facing food production today and the challenges for the future. Sustainability is a modern "buzz-word" used in many circumstances without proper consideration of what it really implies. To help amend this situation this chapter will present and discuss the concept of sustainability and its three dimensions. The interpretation of social, economic and environmental sustainability in the area of food production will be discussed with reference both to the present situation and

Food Safety Management.
DOI: http://dx.doi.org/10.1016/B978-0-12-381504-0.00043-3

also to the future. The actions taken in the food industry and by the research establishment to improve sustainability will be reviewed. The important links between food safety and sustainability will be emphasized. And finally, a number of important issues will be considered and acted on for a more sustainable global food production in the future.

SUSTAINABILITY – AN INTRODUCTION

Sustainability was highlighted in the report "Our common future" from the World Commission on Environment and Development (WCED) (Brundtland, 1987). The definition in this report stated: "Development that meets the needs of the present without compromising the ability of future generations to meet their own needs." The report also pointed out the need to assess sustainability along three pillars: Environment, Economy and Society. Sometimes the pillars are called People, Planet and Profit, or the triple P. The WCED had its roots in the 1972 Stockholm UN (United Nations) conference on the Human Environment, which was the first major international activity in the field of environment. The Stockholm meeting resulted in the establishment of many national environmental protection agencies, as well as UNEP, the UN Environmental Programme.

In the 1970s an important contribution towards improving the understanding of the global ecological situation was made by the Club of Rome. The key purpose was to perform technological forecasts taking into account the interconnectivity of ecology, economics, demography and the resource sector. Their report, entitled *The Limits to Growth* (Meadows et al., 1972), made the informed public aware that exponential economic growth has limits, set by the ecological capacity of the earth.

The Earth Summit in Rio de Janeiro in 1992 emphasized that the three pillars of sustainability should be treated in their integrity and the renowned Agenda 21 activities were derived from the principles for sustainability agreed upon during the Rio Summit (UN, 1992). At the Johannesburg World Summit on Sustainable Development in 2002 plans for implementation and sustainable development were important issues, focusing on poverty eradication, health concerns and sustainable production and consumption. From this period the word sustainability is being used extensively in documents, plan and programs, sometimes without proper consideration for what it implies. It is obvious that sustainability has and must have different meanings to different actors, organizations and countries. The consumer might have an understanding that sustainability stands for "green and healthy" with the implication that the food is deriving from production systems with positive attributes for them, like local production, animal welfare or pesticide-free production. On the industrial side the World Business Council for Sustainable Development (WBCSD) is interpreting sustainability as: "ecologically sound, economically viable and socially acceptable."

Food production and consumption is the primary requisite for a decent life and the sustainable well-being of humankind. However, human activity has the single largest global environmental impact (Smil, 2000). The majority of the global concerns for a more sustainable future are strongly related to different aspects of food production. These aspects will be presented and discussed in the following paragraphs with regard to food production and its relation to the three pillars of sustainability. Some of the major issues of sustainability of relevance to food production are outlined in Table 43.1.

TABLE 43.1 Sustainability Issues of Relevance to Food Production

Sustainability Pillar	Category	Issue
Social aspects	Human health and well-being	Food safety and nutrition
		Food security
	Human rights	Child labor
		Right of association
		No discrimination
		Rights of indigenous people
	Labour conditions	Level of wages
		Working hours
		Safety standards
Economic aspects	Corporate	Sustainable return of investment
		Corporate social responsibility and citizenship
	Fairness	Fair distribution of revenues
		Contracts and credit facilities
	Global	Food waste reduction
		Efficient use of natural resources
Environmental aspects	Global	Contribution to climate change
		Loss of biodiversity
		Land use
	Regional	Eutrophication
		Water

SOCIAL ASPECTS OF SUSTAINABILITY AND FOOD PRODUCTION

An obvious social aspect of food production is its contribution to the health and well-being of the consumer of the food. Such aspects may include an assessment of whether the characteristics of produced food are in agreement with dietary recommendations or other societal goals related to, e.g. reducing obesity or preventing cardiovascular disease. Thus, general societal changes such as increasing urbanization and a more sedentary lifestyle will form the background for these social sustainability assessments. Another important aspect will be the biological and chemical safety of food.

Food security is an obvious aspect of global food sustainability. The growing world population will require a growing food production. An often cited figure is the need to increase world food production by 70% to the year 2050 when it is projected that the world population will have increased to 9 billion from the present 7 billion people (FAO, 2009). Global food production has increased by about 2% annually in the last decades, and this increase will also be needed in the future. However, a number of doubts have cast doubt on this forecast, due the effects of climate change, the lack of arable land and to water scarcity in many regions (World Bank, 2010). In addition the UN Millennium Goal Program points out that the number of hungry people in the world (living on less than US$1 per day) continues to be around 1 billion. The elimination/reduction of poverty and hunger seems to be the goal which is most difficult to reach in the UN Millennium Goal Program (UN, 2011).

The social dimension of sustainability is also addressing the implications of the Universal Declaration of Human Rights (UN, 1948) for the people working in food production as well as others affected by the activities of food production. As an example, this may involve effects on living conditions for indigenous people affected by increased agricultural activities in their traditional living area.

Furthermore the social dimension of sustainability may consider the fairness of the working conditions for the labor force directly involved and affected by food production, assessing whether these are in agreement with international labor agreements such as the "ILO Declaration on Fundamental Principles and Right to Work." Issues such as child labor, minimum wages, working hours, freedom of association, etc. are also assessed.

Corporate social responsibility implies that the company or organization is acting as a good partner or citizen in the community and society where it is active, which is often taken as part of the assessment of social sustainability of the company or organization.

ECONOMIC ASPECTS OF SUSTAINABILITY AND FOOD PRODUCTION

The economic aspect of sustainability is most often interpreted as the ability of a commercial activity to produce a good level of return of investment (ROI) to the owners consistently over time. This assessment of economic sustainability involves looking not just at ROI but also at plans for management of economic and ecological risks and other "good management practices." This interpretation of sustainability often dominates the financial sector, where a number of assessment systems for sustainability exist, e.g. Dow Jones Sustainability Index. However, in the broader interpretation of economic sustainability, factors such as prevention of corruption and bribery are also assessed.

Many economic sustainability activities are related to ensuring fair distribution of revenues to the different actors in the food chain in order to give them the possibility of a sustainable livelihood. The focus is particularly on the small-scale farmer for whom the share of the price paid by the consumer often is less than 10%. A number of actions are taken within Sustainable Agriculture and Fair Trade programs to ensure a higher percentage being paid to the farmers (SAIPlatform, 2011; Fair Trade, 2011). Many labeling schemes among retailing and purchasing organizations also include assessments of these aspects of socio-economic sustainability.

A recent report from FAO and SIK highlights the enormous waste of food in the food chain (Gustavsson et al., 2011). The results of the study suggest that roughly one-third of food produced for human consumption is lost or wasted globally, which amounts to about

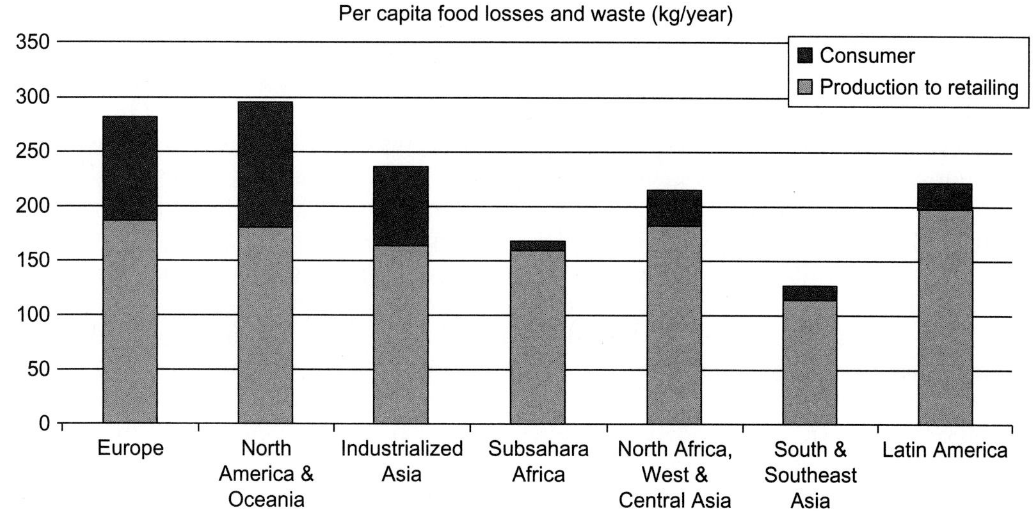

FIGURE 43.1 Per capita food losses and waste, at consumption and pre-consumptions stages, in different regions (Gustavsson et al., 2011).

1.3 billion tons per year. Food is lost or wasted throughout the supply chain, from initial agricultural production through to final household consumption. The most astonishing data come from the consumer sector in medium- and high-income countries where between 95 and 115 kg per capita per year is wasted, meaning that it is discarded even if it is still suitable for human consumption (Figure 43.1). The losses in the chain from production to retailing are substantial too, ranging from 125 to 200 kg food per capita per year. These figures should be compared to the annual production of food which per capita amounts to about 750 kg. These results indicate that a very important step to ensure global food security is to focus on reducing the food lost and wasted in the chain from the primary production to the store and eventually the consumer (Gustavsson et al., 2011).

To improve sustainability in the food sector natural resources will have to be used more efficiently, therefore an important issue on the road to sustainability is waste reduction along the chain from primary production to consumption. Whether by new preservation techniques, packaging, optimizing logistics or otherwise, new innovation must provide sustainable solutions. Furthermore the waste in industrialized countries can be reduced by raising awareness of wasteful behavior among food industries, retailers and particularly among consumers. An example of this is the UK campaign Love Food, Hate Waste.

ENVIRONMENTAL CONCERNS RELATED TO FOOD PRODUCTION

Fifty years ago the publication of Rachel Carson's *Silent Spring* (Carson, 1962) set off an alarm that the food production system used in the USA (and Western Europe) had grave ecological consequences from the spreading of toxic pesticides affecting, among other things, the birds of spring. Another early reporter on the environmental problems of food production was Georg Borgström who published and discussed much about the risk of scarcity of water for food production in many regions of the world and the related conflicts regarding freshwater availability (Borgström, 1969). The reports and books published during the 1970s and 1980s, e.g. the Club of Rome (Meadows et al., 1972), did raise awareness, both among politicians and the general public, about the ecological limits to growth including growth of food production. Among the most influential publications from this time was the series of annual *State of the World* reports published by the Worldwatch Institute, which included contributions on the environmental issues on food production (Worldwatch Institute, 2011).

A hotly debated and very visible environmental problem caused by modern food production in Western Europe in the 1980s was the leakage into lakes, rivers and seas in coastal areas of nutrients from effluents and sewage from food production causing degradation of the water quality. The most evident problem was and still is nitrogen leakage from the use of fertilizers in agriculture, which leads to eutrophication. To some extent phosphorus also contributes to the eutrophication problem. The food production system (mainly agriculture) is responsible for between 60 and 75% of the eutrophication in many industrialized countries. Eutrophication fertilizes water bodies resulting in unnaturally high rates of plant and algae production and accumulation of organic matter that degrade water and habitat quality with risks of total depletion of oxygen at the bottom of the water body. Large areas in the Baltic Sea and the Mexican Gulf, for example, are witness to this problem.

The problem of the presence of toxic compounds remains an environmental problem of food production, such as pesticides in fruits and vegetables, heavy metals (cadmium, mercury, etc.) in grains and dioxins in fish. Many of these compounds show very persistent toxicity with low rates of decline even many years after the actual source of contamination has been eliminated.

Today, awareness is growing regarding the impact on climate change from food production. It has been estimated that global food production uses about 20% of all energy used in society (Sonesson et al., 2009). It is the human activity which uses the highest amount of energy. Modern food production has an extensive dependence on fossil fuels for fertilizer manufacturing and fuel for tractors and transportation. The contribution to global warming is higher, however, estimated at about 25%. The major reason for this is the contribution to global warming of methane and nitrogen compounds from the digestive tracts of animals in meat production. About half of the global warming potential emanates from meat production according to a much discussed report from FAO (2006). The dramatic differences in global warming potential between meat and vegetable products are shown in Figure 43.2.

From a biological point of view the accelerating loss of biodiversity is a major problem caused by the methods used in food production such as mono-culture agriculture, but also many other activities in society. These losses of biodiversity will jeopardize food availability and security by reducing the "safety net" provided by the availability of alternative plants for future food production and other factors affecting human well-being. Another major impact on biodiversity comes from modern industrialized fish trawling which has led to the near depletion of a number of important fish species (UNEP, 2011). The problem of loss of biodiversity is complicated by the lack of "ownership" of the problem.

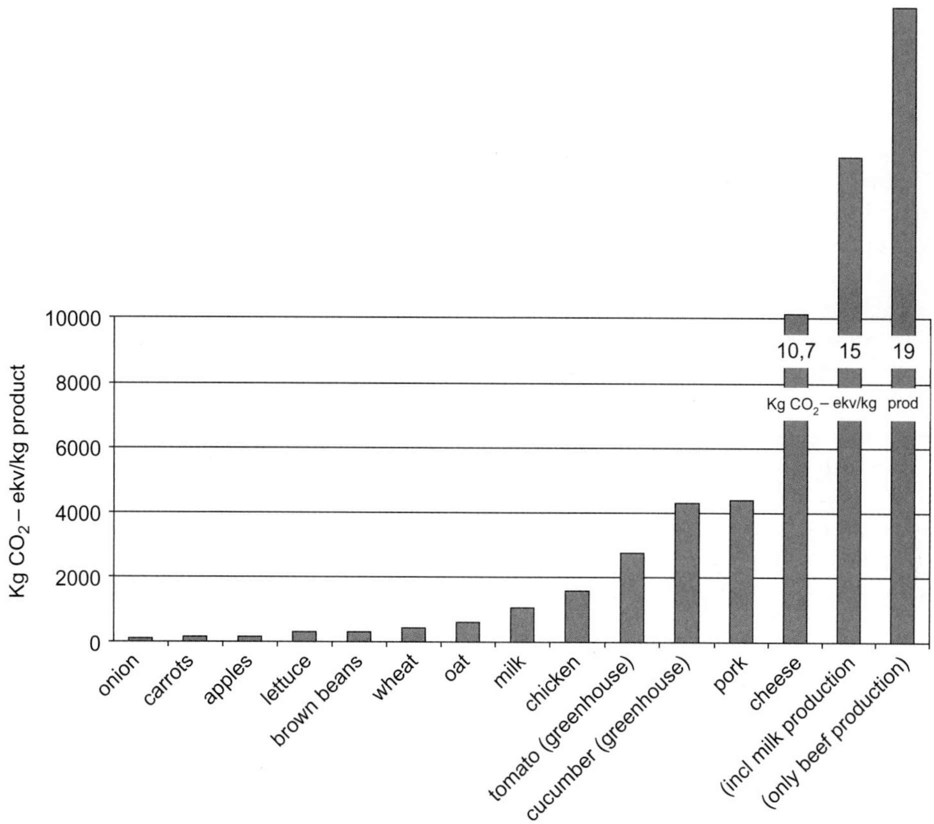

FIGURE 43.2 Global warming potential (CO_2 equivalents) for a range of foods (Sonesson, 2011).

The growing production of biofuels in many countries has brought about a debate on the most appropriate use of arable land in the world. A number of studies have pointed out that there is not much additional land available for crop production use. The need to increase food production to meet the needs of the increasing population is giving fuel to a debate on how available productive land should be used; for food, feed, fuel or forestry? The demand for more land causes more rainforests to be cut down at an alarming rate with devastating impacts on biodiversity and often limited added benefits to agricultural production (Aiking et al., 2006).

The food sector is using about 70% of the available freshwater in the world, corresponding to about 3 tons of water per capita per day. There are many problems with the present use of water in the food sector, e.g. poor water management and deteriorating water quality. Furthermore, in many parts of the world water availability is becoming an increasing problem with dropping water tables and an influx of salt into groundwater and freshwater (Ye and Van Ranst, 2009). The prediction for future climate change clearly points to the problems of availability of water, which will be much aggravated in the coming years particularly in the regions of the world where water already is a scarce resource (IPCC, 2007).

IMPROVING SUSTAINABILITY IN THE FOOD SECTOR

The first impact on the food sector of the increased concern for the environment was the requirement to reduce the biological material (BOD) in the wastewater from farms and food industries. After the Stockholm conference in 1972 many countries established "Environmental Protection Agencies" (EPAs), which often had the cleaning of effluents from industries (and municipalities) as their major plan of action. Thus the food sector started to invest in sewage treatment operations, cleaning up the effluents emanating from food production.

The food industry was also strongly affected by the new environmentally driven legislation on packaging waste imposed in many countries in Europe and Japan during the 1990s. The costs involved in these packaging waste systems (e.g. Die Grüne Punkte) drove a development to lower packaging weights and thus lower packing volumes without reducing the performance of the packaging.

In response to the growing interest and awareness in environmental and sustainability issues after the Rio 1992 conference, many countries started environmental and sustainability activities under the name of "Agenda 21." These activities often included initiatives taken by the industry and their organizations. During this period the UN Global Compact was started with partners from industry, governments, unions and NGOs committed to aligning their operations and strategies with 10 universally accepted principles in the areas of human rights, labor, the environment and anti-corruption (UN Global Compact, 2011).

After the Rio conference in 1992, the industry started the "World Business Council for Sustainable Development" (WBCSD, 2010) with the aim of providing business leadership for change toward sustainable development. Gradually the food industries also identified the importance of the issues related to sustainability, realizing the advantages of reducing the use of input resources such as energy, water and ingredients, possibly with the experiences from the reductions in packaging use as a guide. In order to demonstrate their efforts to reduce the use (and the costs) of energy, water and waste many food industries and retailers today present sustainability reports. These reports also demonstrate the companies' contribution to social and economic sustainability. This might encompass sourcing the agricultural raw materials from sustainable agriculture in developing countries with fair payment to the farmers as one of the goals, as well as better control of the use of pesticides, e.g. as demonstrated by the Roundtable on Sustainable Palm Oil (RSPO, 2011). Other food operators are participating and/or supporting sustainability related labeling schemes such as "Rainforest Alliance" or "Fairtrade," where a limited number of sustainability-related factors are assessed in order for the grower to receive the "Label."

To the general public and consumers the most obvious effort by the food and retail industry to "go green" is the marketing of "organic" or "biodynamic" food products, using a myriad of labels. However, the market segment often only captures a few percent of the total market, so the impact on global sustainability is rather limited. The advantages in terms of environmental performance of organic foods over traditionally produced foods have been much discussed and quite a number of research studies have been presented, e.g. Mattsson (1999).

In the last few decades, research on Life Cycle Assessment (LCA) of food production within the national and international research programs has provided an important contribution towards assessing the environmental sustainability of a range of food production chains and food products, particularly in Europe. Generally, LCA employs a range of environmental indicators related to the major environmental problems coupled to emissions and resource use, but the LCA method disregards the effects of temporal and spatial variation, by evaluating emissions rather than impacts.

The assessment of total sustainability of a food product or production system is not easy. A major complication is how to achieve trade-offs, where aspects of one pillar are improving at the cost of those from another pillar. Furthermore, there are dependencies through the impacts of social and economic processes on the environment, such as the ecological footprints left by international trade, which may have impacts on both ecological aspects and on fair trade relations. Therefore, it is important to continually strive for improved assessment methods for sustainability. A number of initiatives have been taken to include some of the factors of social and economic sustainability into environmental assessment methods in order to develop improved methods for assessing sustainability (UNEP, 2009).

Many other quantification methods for environmental or sustainable performance of a food product focus on one or a small number of issues, such as food miles (on transport of products) or carbon footprints (Tesco, 2007). A number of food retail companies announced in 2007–2008 that they planned to introduce carbon footprint labels on most of their food products. However, this has not materialized to any greater extent, probably due to the difficulties and complexities of how to perform the environmental analysis and the question of how to present the information in an easy-to-understand way.

FOOD SAFETY AND SUSTAINABILITY

Food production systems that lack sustainability will often also demonstrate deficiencies in food safety. The social dimension of sustainability concerns both the social conditions for the people involved in food production and the people affected by the food production. If basic human rights are not met for the workers it is likely that other basic food safety considerations are not taken into account. Moreover, there is an obvious connection between ensuring that food safety standards are met and the health and well-being aspects of social sustainability for the consumers of the food.

Inadequate risk management by food production operators has been demonstrated to jeopardize food safety with often very dramatic effects on the economic sustainability of the food production operation. Also, without adequate revenues from the operations there is a risk that the operators in the chain will shortcut the necessary food safety measures. In parts of the world where food security is a major issue, food safety might often be overlooked as the pressure on the other aspects of food security are emphasized – the availability of and access to food.

The impact on food safety from poor industrial practices resulting in persistent chemical toxins in the environment is still a major problem in many countries. Where this happens, such as fallout from nuclear power failures, the food safety risks rise dramatically and adequate risk management must be in place to minimize the effects. In natural disasters the situation may rapidly develop into a food security crisis with deteriorating food safety as a consequence. Thus environmental sustainability will require both long-term measures to reduce food safety risks and adequate risk management plans to handle sudden and unforeseen events.

SUSTAINABILITY AND FOOD PRODUCTION IN THE FUTURE

It is obvious that the dramatic climate changes predicted by the UN International Panel for Climate Change (IPCC, 2007) will greatly challenge food production in the future. The absence of effective policies to reduce emissions of carbon dioxide and other greenhouse gases is predicted to lead to significant increases in global warming and changes in precipitation patterns over the next 20–40 years (FAO, 2009). Developments of the environmental future of the world, e.g. "UN Millennium Ecosystem Assessment" and "UNEP Global Environmental Outlook," predict that the degradation of ecosystem services could grow significantly worse during the first half of the 21st century. The IPCC report also predicts decreasing agricultural production as a result of climate change. In the tropical and subtropical parts of the world temperatures will be higher and already dry areas will become even dryer, which will restrict food production. In contrast, food production in the more temperate areas of the world might benefit from a warmer and wetter climate. Furthermore, there is a risk of collapse of ecosystems due to land degradation particularly in the tropical parts of the world. The increasing frequency of extreme weather events will also negatively affect the agricultural production and the resilience of the ecosystem. Growing populations and increased food production will lead to increased demand for freshwater. Water withdrawals are expected to increase in all sectors, leading to an expansion of areas with severe water stress (IPCC, 2007; UNEP, 2011).

It is evident that food production will face many major challenges in the future to meet sustainability demands. By 2050, a world with 2 billion more people will need 70% more food (FAO, 2009). This will require enormous efforts on improving the efficiency of the food production and supply system. And this must be done with less energy and fewer water inputs than today. To reduce the contribution to global warming from food production the overall energy consumption must be reduced considerably. This will demand major contributions from all levels of society, business and individuals on all scales from local to global level. Changes will be needed all the way from more sustainable agricultural practices, better post-harvest handling and preservation, improved distribution and a less wasteful food chain including the consumption end of the chain.

Smil (2000) calculated that before large-scale application of fertilizers, the global population was effectively capped at ca. 3 billion people, less than half the present number, by nitrogen limitation. Thus we need to take advantage of man-made fertilizers in the future but they need to be used much more efficiently, as with inputs of other natural resources, not least energy and phosphorus.

On the social dimension of sustainability the UN Millennium Goal (UN, 2011) of reducing by 50% the number of really poor (and thus hungry) people in the world (those living on less than US$1 per day) has not progressed recently, due to the dramatic increase in world food prices since 2008 and the ensuing economic crisis. These "Millennium Goals" will continue to form an important part of the future demands on a sustainable future as part of human solidarity, formulated within the "Universal Declaration of Human Responsibilities," where Article 9 reads: "All people have a responsibility to make serious efforts to overcome poverty, malnutrition, ignorance, and inequality. They should promote sustainable development all over the world..." (Inter Action Council, 1997).

Scenarios for the future of food production in Europe, a part of the world not so severely affected by the climate change, predict that the changed conditions for food production in the coming 30–40 years will also have impacts on many other societal factors, related to – but not part of – food production. Climate change will induce migration to move north, away from the dry and hot parts of Europe and North Africa toward more temperate parts as quality of life deteriorates in the south (SCAR, 2007). This will also lead to strong competition for land use (and probably to more inequality). In most of the scenarios local and regional markets dominate over the global market (OECD, 2009). These scenarios predict stronger legislation to support measures to improve sustainability with more concerned and active consumers and societies as drivers (SCAR, 2007).

Animal protein products such as meat are taking an increasing share of the resources used in food production. In animals 6 kg of plant protein is required to yield 1 kg of meat protein, on average (Smil, 2000; Pimentel and Pimentel, 2003). Consequently, only 15% of protein and energy in these crops will ever reach human mouths and 85% are wasted. The inherently inefficient conversion of plant protein into animal protein makes animal protein responsible for a disproportionate share of environmental pressure of the food production system. In addition to 40% of the grain harvest, some 75% of soy is fed to livestock. The production of animal protein products accounts for over half of the global warming contribution of the overall food production (FAO, 2006).

Meat consumption has increased fivefold in the last 50 years and the increasing number of grazing animals is degrading already impoverished grassland in many parts of the world (UNEP, 2011). Globally, demand for meat and fish products is still on the rise and the projection is for a doubling of animal food products, including both meat and dairy products, until 2050 (FAO, 2006). But inevitably, the prices of meat, fish, soy and cereals will also rise and a trend towards diets containing less animal protein and more plant protein seems inevitable. Notwithstanding the environmental benefits of this move from animal to vegetable protein diets, according to many scientists, including the UK National Health Service, this would also be a trend towards a healthier human diet (NHS, 2009). As pointed out by Aiking et al. (2006), if consumers were to reduce their overall protein intake and replace animal protein products with plant-derived protein products, the majority (87–94%) of prime agricultural land currently used for feed crops might be released, with additional benefits to animal welfare and human health. Moreover, this diet transition would result in a tremendous reduction of the pressure on land and freshwater resources, and – last but most certainly not least – on biodiversity. As already pointed out, there is not much additional land that can be transferred to agriculture, in a business-as-usual scenario, without severe consequences for climate change from deforestation.

In the present situation of global economic crisis it is clear that the economic issues are at the heart of a transition towards a more sustainable future. A number of economists are questioning the possibility of reaching a more sustainable future if the present economic system based on the need for constant growth continues (Jackson, 2009). As almost all human activities mean extraction of natural resources as well as dispersing the wastes back into nature it is obvious that there is a limit to how far the growth of these activities can continue. The growth in the economy must be decoupled from the growth in the extraction of natural resources (UNEP, 2010), many of which already are past their global boundaries as discussed by Rockström et al. (2009). A number of alternative economic systems have been presented and the debate is very active among environmental economists, but the political interest in a common solution is surprisingly low. The momentum in the transition to a more sustainable development is lacking, but unity and speed are urgently required, in order to safeguard a future sustainable world with adequate, healthy and safe food for everyone. Lelieveld (2012), in his expose of the historic development of security and safety of food, also point out the importance of fair and reasonable food regulations for improving the availability of safe food globally.

References

Aiking, H., De Boer, J., Vereijken, J.M., 2006. Sustainable protein production and consumption: pigs or peas? Environment and Policy, vol. 45. Springer, Dordrecht, The Netherlands.

Borgström, G., 1969. Too Many: A Study of the Earth's Biological Limitations. Macmillan Co., London, UK.

Brundtland, G.H., 1987. Our Common Future. World Commission on Environment and Development. Oxford University Press, Oxford, UK.

Carson, R., 1962. Silent Spring. Houghton Miflin Company, Boston.

FAO, 2006. Livestock's Long Shadow. <http://www.fao.org/docrep/010/a0701e/a0701e00.HTM>.

FAO, 2009. Global Agriculture Towards 2050. <http://www.fao.org/fileadmin/templates/wsfs/docs/Issues_papers/HLEF2050_Global_Agriculture.pdf>.

Fair Trade, 2011. Fair Trade Standards. <http://www.fairtrade.net/>.

Gustavsson, J., Cederberg, C., Sonesson, U., van Otterdijk, R., Meybeck, A., 2011. Global food losses and food waste. FAO, Rome. <http://www.fao.org/fileadmin/user_upload/ags/publications/GFL_web.pdf>.

Inter Action Council, 1997. Universal Declaration of Human Responsibilities. Available at: <http://www.interactioncouncil.org/>. Inter Action Council, Tokyo, Japan.

IPCC, 2007. IPCC Fourth Assessment Report: Summary for Policymakers of the Synthesis Report. Available at: <http://www1.ipcc.ch/pdf/assessment-report/ar4/syr/ar4_syr_spm.pdf>.

Jackson, T., 2009. Prosperity without Growth: Economics for a Finite Planet. Earthscan, London, UK.

Lelieveld, H., 2012. People, planet, prosperity, the food chain and decent regulations. AgroLife Sci. J. 2012 (1), 9–17.

Mattsson, B., 1999. Environmental Life Cycle Assessment (LCA) of agricultural food production. PhD thesis, Swedish University of Agricultural Sciences, Alnarp, Sweden.

Meadows, D.H., Meadows, D.L., Randers, J., Behrens III, W.W., 1972. The Limits to Growth. Universe Books, New York, USA.

NHS, 2009. SACN Iron and Health Report. <http://www.sacn.gov.uk/reports_position_statements/reports/sacn_iron_and_health_report.html>.

OECD, 2009. OECD-FAO Agricultural Outlook 2009–2018. Available at: <http://www.oecd.org/>. Organisation for Economic Co-operation and Development, Paris, France.

Pimentel, D., Pimentel, M., 2003. Sustainability of meat-based and plant-based diets and the environment. Am. J. Clin. Nutr. 78, 660S–663S.

Rockström, J., Steffen, W., Noone, K., Persson, A., Chapin, F.S., Lambin, E.F., et al., 2009. A safe operating space for humanity. Nature 461 (7263), 472–475.

RSPO, 2011. Roundtable on Sustainable Palm Oil. <http://www.rspo.org/>.

SAIPlatform, 2011. <http://www.saiplatform.org/sustainable-agriculture/definition>.

SCAR, 2007. FFRAF Report: Foresighting Food, Rural and Agri-Futures. Standing Committee on Agricultural Research, Brussels, Belgium.

Smil, V., 2000. Feeding the World: A Challenge for the Twenty-first Century. MIT Press, Cambridge (MA), USA.

Sonesson, U., 2011. SIK Food Database. The Swedish Institute for Food and Biotechnology, Gothenburg, Sweden.

Sonesson, U., Davies, J., Ziegler, F., 2009. Food production and emission of greenhouse gases. <htpp://www.se2009.eu>.

Tesco, 2007. <http://www.tesco.com/climatechange/carbonFootprint.asp>.

UN, 1948. Universal Declaration of Human Rights. <http://www.ohchr.org/EN/UDHR/Pages/Language.aspx?LangID=eng>.

UN, 1992. Report on the UN Conference on Environment and Development, Rio de Janeiro, 3–14 June 1992. A/CONF.151/26/Rev.1, vols. I–III. United Nations, New York, USA.

UN, 2011. UN Millennium Development Goals. <http://www.un.org/millenniumgoals/>.

UN Global Compact, 2011. <http://www.unglobalcompact.org/>.

UNEP, 2009. Guidelines for Social Life Cycle Assessment of Products. <http://www.unep.fr/shared/publications/pdf/DTIx1164xPA-guidelines_sLCA.pdf>.

UNEP, 2010. Decoupling natural resource use and environmental impacts from economic growth. <http://www.unep.org/resourcepanel/Publications/Decoupling/tabid/56048/Default.aspx>.

UNEP, 2011. Keeping track of our changing environment. <http://www.unep.org/GEO/pdfs/Keeping_Track.pdf>.

WBCSD, 2010. World Business Council for Sustainable Development <http://www.wbcsd.org>.

World Bank, 2010. World Development Report 2010. <http://siteresources.worldbank.org/INTWDR2010/Resources/5287678-1226014527953/Chapter-3.pdf>.

Worldwatch Institute, 2011. State of the world 2011: innovations that nourish the planet. <http://www.worldwatch.org/>.

Ye, L., Van Ranst, E., 2009. Production scenarios and the effect of soil degradation on long-term food security in China. Glob. Environ. Change 19 (4), 464–481.

CHAPTER

44

Climatic Changes[1]

Yasmine Motarjemi
Food Safety Management Consultant, Nyon, Switzerland

OUTLINE

INTRODUCTION

Our society is subject to constant changes and these will not be without consequences on the safety of the food supply. Various factors such as

- urbanization
- demographic changes, such as population growth, migration, increase in life expectancy
- emergence of new hazards as a consequence of advances in science and technologies or biological changes
- changes in lifestyle and consumer expectations
- economic crises

will all pose new challenges to the food industry. However, perhaps, one of the most important factors that will impact food supply worldwide is climatic changes.

[1]Motarjemi, Y., 2008. Management of food safety in industrial setting *Encyclopaedia of Biological, Physiological, and Health Sciences*. UNESCO, Paris.

Food Safety Management.
DOI: http://dx.doi.org/10.1016/B978-0-12-381504-0.00044-5

IMPACT OF CLIMATE CHANGE ON FOOD SAFETY

Climatic changes can affect the safety of the food supply in a number of ways. Subsequent changes in environmental, social and economic factors may influence agricultural and livestock production, and may have a direct or indirect and unforeseeable impact on food safety and on the incidence of foodborne diseases. For instance, it is a well-known fact that certain microorganisms thrive in warmer climates and that their incidence increases during summer months. Several reports have also indicated that the incidence of cholera and diarrhea has increased under the effects of El Niño and other climatic conditions. With the increase in ambient average temperatures, it is likely that certain foodborne infections such as salmonellosis will show a longer annual peak and that their annual incidence increases.

Similarly, infections in food animals and seafood may escalate, and with this the raw material derived from these animals may be more at risk of contamination. In addition to food animals, intermediary hosts to foodborne parasites such as snails may increase in number and cause an increased contamination with parasites. Likewise, there is a heightened possibility of mycotoxin formation in plants as a consequence of the growth of fungi. A possible consequence of the above may be the excessive use of agrochemicals in animal production and agriculture. While the latter do not present a safety concern *per se* if properly carried out, the potential abuse of these chemicals to fight animal and plant diseases needs to be considered by the food industry. Climatic changes can engender extreme weather events, such as heavy rainfall, floods and droughts which may exacerbate the situation. These will contribute to the contamination of food, environment and water resources, with, among others, fecal matter; they will also stress plants and their susceptibility to diseases.

Other possible consequences of climatic disasters such as heavy rainfall or floods may be the collapse of infrastructures such as power supply, or the breakdown of water and sewage systems. These may in turn have adverse consequences on the food safety system. Several reported outbreaks of cryptosporidiosis, giardiasis and other infections have been associated with heavy rainfall. During 1997–1998, excessive flooding also caused cholera epidemics in several African countries.

Climate warming also increases coastal water temperatures and provides ideal conditions for the proliferation of microorganisms, such as *Vibrio* spp., and of planktons. Certain phytoplanktons, e.g. dinoflagellates, produce toxins and may produce biotoxin, and with algae bloom, there is a risk of increased marine biotoxin intoxications such as ciguatera. Zooplanktons are also a reservoir for *Vibrio cholerae* and facilitate the long-term survival of organisms in estuaries. A number of studies suggest a link between cholera epidemics and environmental factors such as warmer seawater and climate in general.

Weather pattern fluctuation can also lead to an increase in rodent and/or insect populations and subsequently contamination of food and water supply. For instance, it is known that rodent breeding increases during mild weather conditions whereas drought or heat may have an inverse effect. However, drought may drive rodents to seek indoor sources of water. An increase in the size of the rodent population, combined with heavy rains or floods, might lead to contamination of food with vectorborne pathogens such as *Leptospira* spp. and proliferation of infectious diseases. Several reported outbreaks of leptospirosis have attributed to such climatic conditions.

CONCLUSION

Professionals in the food industry should be vigilant with raw materials. In the context of the HACCP study of raw materials, the consequences of changes in society and climatic warming in particular should be considered in their hazard analysis, and their control and monitoring procedures adapted accordingly.

More in-depth information on climate changes and their impact on food safety can be found in the FAO report entitled "Climate change: implications for food safety" (2008).

Further Reading

FAO, 2008. Climate change: implications for food safety. Food and Agriculture Organization of the United Nations. Rome. <http://www.fao.org/docrep/010/i0195e/i0195e00.htm>.

Käferstein, F., Abdussalam, M., 1999. Food safety in the 21st century. Bull. World Health Organ. 77, 4.

Lake, I.R., Hooper, L., Abdelhamid, A., Bentham, G., Boxall, A.B.A., Draper, A., et al., 2012. Climate change and food security: health impacts in developed countries. Environ. Health. Perspect. 120, 1520–1526. <http://dx.doi.org/10.1289/ehp.1104424>. [Online 27 June 2012].

CHAPTER

45

Nutritional Trends and Health Claims

Jean-Michel Antoine

Danone, Paris, France

OUTLINE

INTRODUCTION

Consumers are expecting the food industry to deliver products that fulfill four main expectations. Two are based on immediate practical criteria: sensory characteristics, from taste, flavor and smell, to texture, noise and appearance; and services and practicality, from handling, storability, easiness to prepare and clear labeling. Two others are based on experience and intellectual criteria: safety and satisfaction of nutritional needs (see Figure 45.1).

Food is our only source of energy, and satisfaction of nutritional needs has been driving food choice for a long time. Thanks to an increase in food availability, it is becoming a stronger determinant for a growing number of consumers all over the world. Modern human nutrition science emerged in the 1950s, when it was shown that malnutrition in African children increased the risk of infectious diseases, and reciprocally that protein and energy renutrition improved some immune functions. At the same time children's malnutrition was reported in the USA and nutrition entered the political agenda. In 1968, Senator George McGovern took the lead of the first United States Committee on Nutrition and

Food Safety Management.
DOI: http://dx.doi.org/10.1016/B978-0-12-381504-0.00045-7

FIGURE 45.1 Assessment of food by the consumer. Two practical criteria: sensory and service. Two intellectual criteria: safety and satisfaction of nutritional needs.

Human Needs. Six years later he expanded the scope of the committee from reducing malnutrition to health risks associated with overeating, starting with the effect of an excess of dietary cholesterol on cardiovascular diseases, a complex relationship not yet completely deciphered after 60 years of intense research. This illustrates on the one hand the complexity of interactions between nutrition and health, the difficulty to combine observational epidemiology and experimental nutrition, and on the other hand the scientific dilemma: when is there enough scientific evidence to support a claimed benefit of a food, or a diet, on health? This is still a matter of debate today.

Diet is one of the major components of our lifestyle that we can individually control in our own environment to manage our present and future health, including the health of future generations. The double challenge is to limit the risks and to improve the benefits. Both components of this ratio are changing: foods and diets are more abundant and more energy dense, the environment and lifestyles are less demanding in terms of energy, and the world and our expectations are also changing. Our knowledge of technology and science has increased dramatically, and during the last couple of decades our focus on diet and nutrition has also changed.

HISTORICAL PERSPECTIVE

Dietary challenges have evolved during the short history of human species on earth. The first of these challenges was to select from Nature foods that can provide adequate energy and nutrients. Collecting raw foods from the environment meant ingesting a variety of foods that often contained insects and microbes; it was also essential to select foods that were first not poisonous and second nutritious. A number of explorers died during this exploratory phase of acquiring the basic knowledge of edible foods. A large part of food education was to learn which foods were safe and to train the taste buds to distinguish between edible foods and hazardous ones. This hard training may explain our fear of eating new foods – neophobia. It may explain why our reptilian brain remembers the old days when trying a new food may end in death! But how do we overcome this phobia? It is still a common practice when traveling abroad to see how local people approach their food, and then follow their example to overcome our anxieties. It transpires that people have been able either to select or to adapt and adopt certain foods. Safety was the critical point, and taste was a first warning signal: bitterness is often associated with poison in Nature, and sweetness with edibility.

The second challenge was to develop ways to store and preserve foods. Food processing must have been a key step in civilization. It dramatically improved shelf-life and storage capacity of food over a few days. Humans were able to overcome natural seasonal shortages of foods, and to explore and colonize some countries where winter reduces drastically the availability of foods. Humans have adapted the processing of local foods to local conditions and invented ways to survive in every part of the planet. Strikingly, our normal body fat contains enough energy to survive during a winter – 3 months. Increasing food storage capacity also changed the size of human groups living together: it was possible to feed a family all year around, then a larger group of people, or even a city, and nowadays the whole world benefits from improved storage of foods.

Storage also contributes to the development of agriculture and breeding on large scales. A large production of food also requires a large storage and preservation capacity to handle the harvests. This is still a challenge today as a significant part of food production is lost during storage (Gustavsson FAO, 2001).

It is quite interesting to note that food preservation used two kinds of techniques for safety: (1) killing microbial contaminants with acidic or alcoholic fermentation, and cooking; and (2) reducing the water content of foods to prevent growth of microbial contaminant. Apart from drying in a sunny wind or in a smoky cavern, some specific ingredients were used to preserve foods: salt, sugar and fat. Therefore preserved foods were high in salt, sugar or fat. It is tempting to speculate that only those who were able to cope with the necessary concentrations of salt, sugar or fat in foods survived. We can also speculate that they learned to like those concentrations: they indicated that the food was safe, and therefore survival was assured. Nowadays a too high consumption of salt, sugar or fats is identified as one of the dietary risk factors for many people. We may wonder why those nutrients are now becoming a threat. One explanation is that life expectancy is now longer, allowing metabolic disorders to develop and challenge the health of elderly consumers.

A breakthrough in preservation occurred in 1810 when Nicolas Appert published in Paris his work on the art of preserving animal and vegetable substances. He reported more than 15 years of experiments of putting foods in glass jars and boiling them in water. The same year British inventor Peter Durand patented the use of the tin can, and started a long history of canned food. The final touch came with the invention of the can opener in the USA by William Lyman using a rolling cutting wheel. Preservation can be industrialized without the constraints of salt, fats and sugars.

The third challenge was to produce enough food to feed everyone. The twentieth century was the century of industrialization of agriculture and breeding. Food production capacity increased dramatically and is still increasing. A first step in processing is the need to store this large production of raw food for very long periods of time and also during transportation all over the world to local producers. But there is still room for improvement: it is estimated that one-third of crop production is lost during storage. Obviously this huge increase in production was associated with the industrialization of preservation and storage, as well as packaging techniques to handle and distribute processed food around the globe. Packaging started another revolution: labeling can convey information for the consumer. Selecting and buying a food requires the consumer to read and understand packing information, and no longer to rely on practical experience as was the case before.

MODERN TIMES

The benefit of this increase in production is associated with two consequences: (1) seasonality is potentially redundant and it is common to find milk, apples, tomatoes, lambs, chicken, etc., all the year round in both hemispheres. Diversity of diet is diminishing accordingly, and most seasonal consumption has almost disappeared in westernized countries. On the other hand, the monotony of seasonal diet is easily diversified: the choice of processed foods in the supermarket is larger than at the local raw food market. (2) Diversity of crops is diminishing: the same cultivar is used more widely than before. For example, more and more consumers are eating the same flour: the statistical probability to have one consumer less adapted than the average population is increasing. On the other hand, a variety of cultivars are being used to make our daily bread, and we will have to adapt to new ones on a regular basis, which means that our long-term nutritional epidemiology has to deal with evolving food composition. Epidemiological correlations are stimulating but never truly demonstrative. Globally this increase in food quality and production (and improvement in politics) decreases world hunger and improves life expectancy. However, producing enough food for a still rapidly growing population remains a challenge.

With abundance and availability average dietary intakes are changing at two levels: (1) there is a reported increase in proteins, sugars and saturated fats. Spontaneously adults are not able to select a diet that fits their dietary or nutrient recommendations, when 7 years old French children were able to do so (Debry, 1980) on a 3-week basis. Simultaneously the human workload to produce food is reduced due to mechanization: far fewer human calories are required to produce food. Farmers increase their weight according to the level of mechanization (Vardavas, 2009). Man is facing a totally new challenge: how to eat less than the previous generation, when foods have never been so abundant and cheap from an energetic point of view. (2) Dietary practices are changing: surprisingly this large increase in production is correlated with a significant increase in wasted foods. A recent US survey reported that one-third of foods are thrown away: 17% of dairy products, 20% of vegetables, 15% of fruit, 33% of meat, 35% of fish and 18% of grain (NDRC, 2012). This increase in waste is a consequence of the increase of food production (Hall, 2009).

A drastic change in dietary habits and practices is happening: the tradition of eating meals at home as a family is no longer relevant: we seem now to be either eating alone or at work. Furthermore portion sizes and the composition of dishes are changing, and a re-education of our eating habits is needed.

Finally, modern times generate a brand new challenge: access to drinkable water. The amount of water on our planet is fixed and water is indefinitely recyclable. We are drinking the same water that the dinosaurs drank. In the past two processes were used to have access to drinkable water: alcohol disinfection (combining wines with water, or brewing beers), or boiling water for beverages, for example. A century ago modern technology discovered the use of chloride to reduce the microbial risk in water. Tap water became drinkable. We are still facing some challenges from microorganisms, and we also have to encounter new chemical contaminations generated by increasing amount of wastes, including drug residues and industrial processes. Pure water is both a microbial and a chemical challenge and the next dramatic test for the human race in the coming decades.

FOODS FOR HEALTH

Industrialization of food supplies increases the availability of safe food. The fight is no longer to get enough food, but to select the best foods to improve the dietary needs of consumers. Diet is important to improve a healthy lifestyle. This is more important than ever because the human species is experiencing an unprecedented event: life expectancy is increasing, from 45 years a century ago to currently 75 years. The percentage of the elderly population is increasing (e.g. from 8% in China and 12% in USA in 2005 to a projected 16% and 19% in 2030).

Due to progresses in hygiene and medicine the number of people living with diseases is increasing, generating an economic burden. And as we get older there is a physiological decrease of the different functions of the body, generating costs and/or opening new challenges for nutrition.

We are also discovering the long-term consequences of different earlier lifestyles, mainly on cardiovascular and cancer risks, and also the burgeoning epidemic of obesity.

DIET AND HEALTH

Historically, diet was one of the four elements given by the gods to humans. In Greek mythology Aesculapius had four children: two sons, Machaon (surgeon) and Podalirios (physician), and two daughters, Hygia and Panacea. Hygia used cleaning and washing to take care of health, while Panacea was in charge of identifying beneficial herbs, plants and ingredients among all the products Nature was offering. It is amazing that food risk (hygiene) and food benefits (panacea) were already under the control of women, men being physician and surgeon.

The synergism between diet and health began with the discovery that certain foods were able to cure diseases: the Andeans learned that grinding corn with lime and alkali prevented a devastating cutaneous disease, pellagra was recognized as due to a lack of vitamin B and Hippocrates used liver to prevent blindness induced by a lack of vitamin A.

The first human nutritional study was conducted by Lindt in 1747: this was a pilot study on 12 sick sailors split into six groups of two "volunteers," which suggested that lime was better than cider, vinegar, horseradish, diluted sulfuric acid and salted water to cure them. The Royal Navy was interested enough to conduct a confirmatory study on HMS *Suffolk* in 1794 during a trip from England to India. Sir Gilbert Blane, Physician Extraordinary to the Prince of Wales, gave lime to half of the crew and used the other half as a "control" group without lime. Lime prevented the experimental crew from suffering from scurvy. The first data were convincing enough for Captain James Cook, the famous explorer. He used lime to prevent the deadly disease. In 1933 vitamin C was identified by Szent-Györgyi, but the exact composition of lime is not fully known today, and lime is still more active on scurvy than its content in vitamin C indicating that some other ingredients have biological properties.

Antoine Lavoisier discovered the energy equation (energy in = energy out) in the 18th century and was used by Liebig in Germany to calculate the amount of energy needed for mine workers to extract a fair amount of coke.

Modern nutrition started after World War II, with the demonstration in the developing world of the role of energy and protein intake in two forms of malnutrition: marasmus and kwashiorkor. Adequate nutrition is also important for an efficient immune system, and adequate water and minerals intake are essential to manage acute diarrhea. These health issues are still a challenge in the developed world: a large proportion of the elderly in Europe are suffering from insufficient intake of energy, protein and water: between 50 and 75% according to a European report (So rensen et al., 2008).

The modern age of dietary science can be schematically split in two phases: the first is the deleterious effect of excessive intake of nutrients and nutritional prescriptions reduce dietary risk factors.

The increase of life expectancy was associated with an increase of new causes of death, namely cardiovascular disease, and when a US senator died of an excess of blood cholesterol, scientists explored the possible links between cholesterol in the diet and risk of cardiovascular disease. A classical correlation was published by Ancel Keys between the intake of fats and saturated fats in seven countries and their rate of cardiovascular mortality. However, changes in consumption of fats and saturated fats are not correlated with similar changes in cardiovascular mortality, and the final conclusion is still a matter of debate, even if everyone agrees on the common sense conclusion that an excess of fat (saturated or not) is not recommended.

For some time now physicians have been aware of what to ban from their patients' diets, adapting the amount of sodium to blood pressure, of protein to creatinine's clearance, of sugars to glycosylated hemoglobin, the ratio of fat to carbohydrates to the respiratory capacities, the amount of alcohol to gamma glutamyl-transferase, and so on.

After this "banning phase" came the second phase: the "recommending" phase where dietary prescriptions are about positive actions. For example, French official guidelines consist of half "reduction" and half positive guidance like: eat five servings of fruits and legumes a day. The benefits are expressed in the reduction of risk factors like high blood pressure and obesity.

Two factors contributed to starting the second phase: dieticians realized that it was necessary and more important to tell patients what to eat than what to avoid, and the majority of consumers are more interested in a healthy lifestyle and a willingness to benefit rather than associate food with diseases, even disease prevention.

The first modern use of specific foods to improve a function mimics an old tradition in Greek Olympic Games: it was a common practice to include in the diet of athletes foods associated with their sports: runners ate horse meat, weightlifters ate bear meat, fighters ate lion meat, and so on.

Modern sportsmen explored this concept and demonstrated that an improved diet was able to improve performance. This was demonstrated by Ron Hill, the Athens' marathon winner in 1969, and was supported by the success of Swedish athletes using the dissociated diet to win during the Winter Olympic Games.

The principle is to starve muscles of glucose to increase their glycogen storing capacity, then to feed them glucose and to benefit from that increased glycogen content to improve physical performance. This has been popularized by Bjorn Borg and Ivan Lendl and is becoming a mandatory habit with pasta parties before a marathon. A specific diet can improve a specific function, e.g. muscular function.

Another classical and sometimes vital specific functional food/nutrient is water: hydration is a key factor in achieving proper performance – an adequate intake of water improves physical as well as intellectual functions.

A new area in nutrition science has recently begun: Which nutrients are more specific to a given function? And equally which function can be improved by a specific nutrient, food or diet? Two European-supported programs (Fufose and Passclaim), under the umbrella of ILSI Europe, explored the concept of improvement of a function by food. This concept is based on two physiological observations:

1. Within a population every physiological function is distributed among individuals, most of the time in a Gaussian manner. Around the average value there are lower and higher functional capacities. There is room for improvement of those under the average and even those higher than average. It is a common observation that adequate training improves a functional capacity: e.g. dissociated diet improved the muscular functions of athletes.
2. Within an individual all functions oscillate around a basal value: this is chronobiology: e.g. we wake in the morning with a high level of blood cortisol and low body temperature, and sleep in the evening with a lower level of cortisol and higher body temperature. Those rhythms can be on a different time basis: day, week, season or shorter. Age is another obvious inevitable source of oscillations and decrease of functionalities. There is a possibility to reduce the duration of the low capacity period, and to prolong the period of higher capacity: e.g. coffee is able to increase the length of awareness, or reduce its evening decrease.

What are the functions that can be improved, or the reduction that can be prevented? Is it possible to prevent or slow down the aging factor? This is a challenge for each of us and our healthcare systems.

The benefit relies on scientific data and Passclaim (Aggett et al., 2005) concluded that such a benefit requires:

- an identified ingredient/food/diet ingested in an adequate amount,
- human data using commonly agreed marker(s), and
- reproducible results based on randomized controlled human trials.

One of the oldest accepted functional ingredients is wheat bran: it has been demonstrated in different human trials that a low intake of bran is associated with a long gut transit time, and that an adequate increase of wheat bran intake normalized gut transit time and increased fecal bulk and weight.

The second benefit came in 1997 from the USA when FDA agreed on the effect of oats to reduce the risk of cardiovascular disease, based on human trials.

In 1980 Japan was the first country to implement a specific regulation for this kind of food: **FO**od for **S**pecific **H**ealth **U**se FOSHU with the same scientific basis – a demonstration of the benefit through human trials. There a few hundred Japanese foods that hold a FOSHU claim, including a lot of probiotics.

In the late 20th century some European countries started to validate claims, and in 2006 the European Commission asked its European Food Safety Agency to convey scientific panels to give scientific opinions on submitted dossiers for claims related to:

- general function: based on generally accepted scientific evidence;
- new function claim: based on new scientific evidence or use of specific product or substance;

- reduction of risk factors of diseases: based on scientific evidence related to risk factors of diseases and a specific product;
- function claims in children: based on scientific evidence related to the specific children target.

EFSA used three criteria to assess the scientific evidence:

1. The food/constituent is defined and characterized.
2. The claimed effect is defined and beneficial to human health.
3. A cause and effect relationship is established between the consumption of the food and the claimed effect (for the target group under proposed conditions of use).

The scientific opinions are then reviewed by the European Commission in charge of final decision to agree or not with the proposed claim and to disseminate them to the member states.

EFSA's panels first cleared claims related to general functions, selecting those based on generally accepted nutrition science such as the benefit of vitamins and minerals. The panels were more restrictive with new function claims as the Commission asked them to use the best evidence to support their opinions, when nutrition is not the best field for providing very strong evidence due to the complexity of the domain and the variability of human physiologies.

Other panels in different countries are evaluating scientific evidence on the effect of products on health, function and risk factors. It is worth mentioning that by using similar published data different panels made different opinions on similar products.

It is interesting to note that the modern Law does not allow foods to cure nor prevent diseases, when nutrition science started by the demonstration of diseases like pellagra, scurvy or diarrhea cured or prevented by foods. This reflects, at least partly, the misconception of diseases as the consequence of an exogenous pathogenic element, often a microscopic one. This is supported by a discovery like the role of *Helicobacter pylori,* a gastric microbe responsible for gastric ulcer. The modern understanding of human physiology is that a disease is the result of an aggression by a pathogen *plus* the response of the host. Therefore the cure of a disease is based on two pillars: to destroy the pathogen (a role for a drug) and to support or enhance the host's defense (a role for diet). Nutrition is feeding the response of the host and therefore is part of the management of the disease as well as its cure. This was observed at the beginning of modern nutrition in the 1950s when the renutrition of malnourished people was able to restore immune functions and to cure infectious associated diseases.

On the other hand, a number of modern diseases result from a dysfunction of the metabolism and/or overnutrition due to inadequate dietary intake. It is logical to address the cause of a disease – malnutrition – to cure it. Science tells us that some foods and diets help to cure and prevent diseases even if they are illegal!

There is a fascinating new area for nutrition that provides adequate scientific evidence to support these new benefits. However, a difficult ethical question needs to be answered: Does the demonstration of such benefits require extensive studies due to the complexity of the nutrition and health relationship? Costs of such trials will significantly increase the cost of foods, when the consumers who need those foods the most may not have enough money to buy them: one simple human nutrition trial costs the equivalent of a million meals. When

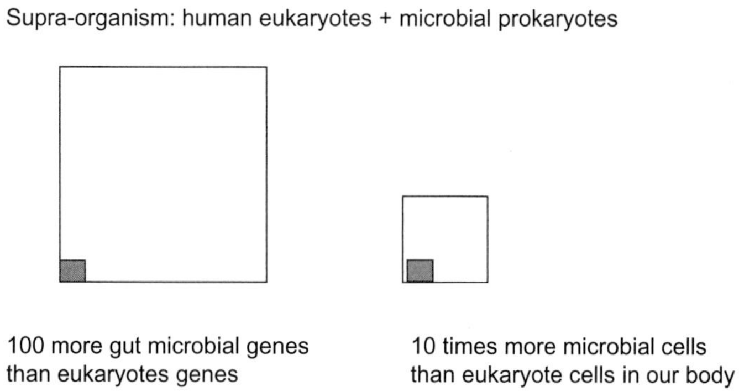

FIGURE 45.2 The ratio of microbes/host. Dark squares are eukaryotes, and the empty squares are prokaryotes.

scientific evidence are enough, they must be assessed according to the expected benefit, the potential risk and the needs of the target population.

Modern tools bring about a revolution in the physiology of nutrition and health: genomics explores the human gut through many large programs in different parts of the world and concludes that we are living with a forgotten organ – the gut microbiota. The following illustrates the importance of that new organ: it contains 10 times more microbes than we have cells in our body, it contains 100 times more genes than our cells, it weighs 1.5 to 2 kg, more than the liver, and is far more metabolically active than the liver. Edgar Lederer, a Nobel Prize winner, suggested considering this association as a "super-organism" where the major part will be the microbiota (Figure 45.2).

The importance of the microbiota can be touched upon by two correlations: J. Gordon has reported that transferring the gut microbiota of an obese mouse to a lean mouse results in a fattening of the lean mouse. In another trial he reported that obese men have a different gut microbiota profile than lean men, and when obese men lose overweight, their gut microbiota profile evolves toward a lean profile.

MetaHit reported also that the presence of a specific species, *Faecalibacterium prausnitzii,* is associated with a lower level of inflammation, a lower risk of inflammatory bowel disease and a better chance to lose weight.

Two elements of the diet have a direct impact on the gut microbiota:

1. Prebiotics: mainly indigestible carbohydrates that are fermented by some part of the gut microbiota, changing the composition and some functions of the gut microbiota.
2. Probiotics: living microorganisms that, when ingested in an adequate amount, provide health benefits to the host. Specific strains provide specific benefits. They can act directly on the host: yoghurt is able to digest lactose in the gut of lactose malabsorbers. Therefore anyone can digest yoghurt, supporting the claim "one yoghurt a day" as part of dietary recommendations, change the functioning of gut microbiota, e.g. changing the metabolism of gases and improving gut comfort (Sonnenburg, 2006), or change the functioning of digestive cells (Van Baarlen et al., 2011).

The role of microbiota on host functions within and beyond the gut offers a fascinating potential through the modulation of inflammatory status, including the aging process, or brain functions including mood and autism.

Costs and time of demonstration are not compatible with challenges, and in nutrition knowledge is still a mixture of observations supported by some data and a battlefield where different stakeholders are dependent on industry as well as on academia where the fight for official support requires one to follow the official mind set, and use the trends.

Diet is a key element of an adequate nutrition and a major contributor to health and proper functioning of our body, including gut microbiota. Is it possible to change dietary habits and to improve them?

Many education programs have been implemented in different countries using different channels, with very limited impacts.

The first element was the confusion between nutrition and diet as illustrated by the Recommended Dietary Allowances where all nutrients' needs were listed, without any dietary translation. It is impossible without a computer to build a balanced diet based on nutrient composition of foods.

Then there was the period of ban "the" culprit food, which ended in suppressing bread, or fat, or sugar, or salt, or meat, or whatever. A number of experts used that trick to get a lot of media coverage and money.

In fact it is well known that we are unable to change our diet dramatically over a long period. But we can change the serving size or the frequency of intake.

The following trend was the reverse: invest in a specific beneficial food – carrot juice, pineapple, one apple a day. But there is no perfect food, or magic food.

Moderation is the key word, and the food industry is playing a major role by adapting food contents in a progressive and unnoticed way, like the change from whole milk to half skimmed milk, or the reduction of salt in soup, or the change of fat composition in dressings, and so on.

Labeling is another dead end as there is no simple message: a food is neither good nor bad in itself, it is a matter of including it in a balanced diet – excess or deficiency must be avoided on a long-term basis. Some foods are staples like grains, legumes, fruits, dairy products and meat/fish, and some foods must be added more carefully in the diet, but none can be excluded.

Finally, we have to take into account human diversity, as illustrated by the old concept of Professor Apfelbaum's "Mangeurs inégaux." For the sake of the species we have different individual metabolic capacities and ingesting the same amount of the same fatty acid can improve or worsen blood cholesterol depending on the eater. Some of us have a high metabolic resting rate and we burn a lot of calories while some others are sparing every eaten calorie, without mentioning the role of the gut microbiota that can increase of decrease energy extraction from the diet. Some functions can be improved in some responders and not in others.

Modern tools will help us to integrate the complexity of our super-organism and hopefully identify some major crossroads, or some major clusters of responders and non-responders, and adapted functional foods.

Until we have all the information, we need to progress and use wisely scientific information. It looks like we are still in an exploratory phase where we have to be careful with risk assessment, and adventurous with benefit testing. We may not have the financial resources to wait until we know everything before incorporating some specific foods for some identified health and functional benefit. The reward of the audacious explorer will be to add years to a healthy life.

References

Aggett, P.J., Antoine, J.M., Asp, N.G., Bellisle, F., Contor, L., Cummings, J.H., et al., 2005. PASSCLAIM: consensus on criteria. Eur. J. Nutr. 44 (Suppl 1), 5–30.

Debry, G., 1980. The diet of the French child (in French). Nouv. Presse. Med. 8, 3145–3147.

Gustavsson, J., Cederberg, C., Sonesson, U., van Otterdijk, R., Meybeck, A., 2011. Global food losses and food waste. FAO Report. <fao.org/docrep/014/mb060e/mb060e00.pdf>.

Hall, K.D., Guo, M., Dore, M., Chow, C.C., 2009. The progressive increase of food waste in America and its environmental impact. PLoS ONE 4, e7940. <nrdc.org/living/eatingwell/files/foodwaste_2pgr.pdf>.

Sonnenburg, J.L., Chen, C.T.L., Gordon, J.I., 2006. Genomic and metabolic studies of the impact of probiotics on a model gut symbiont and host. PLoS Biol. 4, e413.

Sorensen, J., Kondrup, J., Prokopowicz, J., Schiesser, M., Krähenbühl, L., Meier, R., et al., 2008. EuroOOPS: an international, multicentre study to implement nutritional risk screening and evaluate clinical outcome. Clin. Nutr. 27, 340–349.

Van Baarlen, P., Troost, F., van der Meer, C., Hooiveld, G., Boekschoten, M., Brummer, R., et al., 2011. Human mucosal in vivo transcriptome responses to three lactobacilli indicate how probiotics may modulate human cellular pathways. Proc. Natl. Acad. Sci. USA 15, 4562–4569.

Vardavas, C.I., Linardakis, M.K., Hatzis, C.M., Saris, W.H., Kafatos, A.G., 2009. Prevalence of obesity and physical inactivity among farmers from Crete (Greece), four decades after the Seven Countries Study. Nutr. Metab. Cardiovasc. Dis. 19, 156–162.

CHAPTER

46

Ethics in Food Safety Management

Nina Veflen Olsen[1] *and Diána Bánáti*[2]

[1]Nofima, Ås, Norway, [2]International Life Sciences Institute (ILSI) Europe, Brussels, Belgium. Former Central Food Research Institute, Hungary

OUTLINE

INTRODUCTION

Food production is a complex matter, affecting people's lives, organizations' profits and the well-being of the whole planet. It is not always straightforward to say what is right and what is wrong when it comes to the production of food. Many ethical questions can be raised regarding the food supply chain, including agricultural production.

Climate change, animal welfare, fair trade, health and safety of employees as well as consumers, fair treatment of employees and their social rights, economic sustainability and use of natural resources are all important dimensions within the system of production, processing and trade, and where every food item often includes value conflicts. An increasing number of the Western population are becoming overweight and obese. Should the food industry limit their marketing of fat and sweet foods to reduce these problems, or is this the responsibility of consumers as far as they have an informed choice? Maybe the industry's

Food Safety Management.
DOI: http://dx.doi.org/10.1016/B978-0-12-381504-0.00046-9

aim first of all is to increase profit and secure employment? Tremendous variation in the global food production system exists. People in parts of the world are starving and to secure enough food supply is important. But is it unethical to focus on efficient food production, when this is set up against animal welfare and environmental friendly production? And is it acceptable that multinational companies have a double standard, which means that products rejected in one country can be exported to another country with less stringent safety legislation? If so, what would be the eventual consequences? Would other kinds of non-ethical practices, e.g. dumping of food, increase? The question of how far a single standard for reasons of ethics should be respected in the world was raised in an Asian country where a multinational company was required to produce its products according to the European safety standards. Is it unethical behavior when they, due to environmental contamination and local practices, were unable to have raw material meeting European requirements?

We may also ask if the short-term consequences on people are more important than the long-term consequences on the planet, and if it is all right that what is closest in time and space is most important for us. How the food industry responds to these questions will influence how consumers, citizens and other relevant stakeholders perceive them. A perception of the food industry as a large-scale sustainable production and processing system or as a business with the main purpose of making financial profit will influence consumer trust and thereby their acceptance of novel technologies, novel foods and food science and technology in general.

Food security, food safety and sustainability are considered as first priorities related to the modern developments in agricultural technologies (EGE, 2008a). In this chapter we will discuss the scope of ethics in the food industry in relation to food safety. We start out by defining ethics. Then, ethical dilemmas in food safety cases are discussed. Lastly, the precautionary principle and a framework for ethical decision-making in the food industry are highlighted.

WHAT IS ETHICS?

Ethics is defined as the philosophical study of the moral value of human conduct and of the rules and principles that ought to govern it (www.thefreedictionary.com/ethics). Simply stated, ethics refers to standards of behavior that tell us how human beings ought to act in the many situations in which they find themselves – as friends, parents, children, citizens, businesspeople, teachers, professional food producers, consumers, and so on.

Ethics is different from following the law. A good system of law incorporates many ethical standards, but law can also deviate from what is ethical. Selling tobacco to children in countries with no laws against it can be perceived as unethical. Ethics is not the same as following culturally accepted norms either. Some cultures are quite ethical, but others become corrupt – or blind to certain ethical concerns: "When in Rome, do as the Romans do" is not a satisfactory ethical standard. Ethics is not science, feelings or religions. Science can provide important data to help us make better ethical choices. But science alone does not tell us what we ought to do. Science may provide an explanation for what humans are like. But ethics provides reasons for how humans ought to act. And just because something is scientifically or technologically feasible, it may not be ethical to do it. Even though science makes it possible to breed cattle with extreme overdevelopment of muscles, we may question if it is ethical when we take animal welfare into consideration. We cannot trust our feelings either. Some people feel bad when they do something wrong, others do not. While most religions do advocate high ethical standards, they do not always address all kinds of ethical dilemmas (Velasquez et al., 2012; Baumhart, 1961).

If our ethics are not based on feelings, religion, law, accepted social practice or science, what are they based on? Some ethicists emphasize that the ethical action is the one that produces the greatest good and does the least harm for all who are affected – customers, employees, shareholders, the community and the environment (the utilitarian approach). The utilitarian approach deals with consequences; it tries both to increase the good done (e.g. ending hunger) and to reduce the harm done (e.g. environmental and social destructions). Other philosophers and ethicists suggest that the ethical action is the one that best protects and respects the moral rights of those affected (the rights approach). This approach starts from the belief that humans have a dignity based on their human nature *per se* or on their ability to choose freely what they do with their lives. On the basis of such dignity, they have a right to be treated as ends and not merely as means to other ends. They have a right to adequate food and a fundamental right to be free from hunger (FAO, 1996, Rome Declaration). The list of moral rights – including the rights to make one's own choices about what kind of life to lead, to be told the truth, not to be injured, to a degree of privacy, and so on – is widely debated; some now argue that non-humans like animals and plants have rights, too. Also, it is often said that rights imply duties – in particular, the duty to respect others' rights. Aristotle and other Greek philosophers have contributed the idea that all equals should be treated equally (the fairness or justice approach). Today we use this idea to say that ethical actions treat all human beings equally – or if unequally, then fairly based on some standard that is defensible.

The power distribution between retailers and suppliers has led many producers to state that multiple food retailers are abusing their position of power and engaging in practices that adversely affect the competitiveness of suppliers. To address these adverse effects it has been recommended that a code of practice be introduced to govern retailer–supplier relationships (Duffy et al., 2003). The Greek philosophers have also contributed the notion that life in community is a good in itself and our actions should contribute to that life (the common good approach). This approach links ethics to social responsibility and calls attention to the common conditions that are important to the welfare of everyone. Companies have a duty to be good citizens including their own workers and staff and "to do the right things" (Porter and Kramer, 2006). A very ancient approach to ethics is that ethical actions ought to be consistent with certain ideal virtues that provide for the full development of our humanity (the virtue approach). These virtues are dispositions and habits that enable us to act according to the highest potential of our character and on behalf of values like truth and beauty. Honesty, courage, compassion, generosity, tolerance, love, fidelity, integrity, fairness, self-control and prudence are all examples of virtues. Virtue ethics asks of any action, "What kind of person will I become if I do this?" or "Is this action consistent with my acting at my best?"

Each of these approaches mentioned above helps us determine what standards of behavior can be considered ethical. Different actors may not agree on the content of some of these specific approaches. They may not all agree to the same set of human and civil rights or on what constitutes the common good. They may not even agree on what is a good and what is a harm. Nonetheless, each approach gives us important information with which to determine what is ethical in a particular circumstance. And much more often than not, the different approaches do lead to similar answers (Velasquez et al., 2012).

ETHICAL ISSUES IN FOOD SAFETY

The series of food scandals and scares during the last decade resulted in a melting consumer confidence. Despite the fact that food has never been safer, it seems that consumers are considerably uncertain, anxious and increasingly critical about the safety of their food.

The safety of food products for human consumption as a precondition for their marketing must be guaranteed. But who is responsible for food safety? Are the producers and processors, or the food business operators in general responsible as stated in the European legislation (178/2022/EC), or should this responsibility be shared with consumers? Should consumers having sufficient education and knowledge be able to make informed choices? This might apply to nutrition-related choices but the safety of foods cannot be judged by consumers. Here the operators of the food supply chain are mainly kept being responsible. In food safety, the border between ethic, compliance and responsible behavior, as seen below, is not always clear cut and at times these issues are intertwined.

One of the largest food safety incidents the World Health Organization (WHO) had to deal with recently was the Chinese melamine scandal from 2008 involving milk and infant formula. By November 2008, China reported an estimated 300,000 victims, with six infants dying from kidney stones and other kidney damage, and a further 860 babies hospitalized (Branigan, 2008; see also Chapter 41). The chemical melamine appeared to have been added to milk to cause it to appear to have a higher protein content (McDonald, 2008; Macartney, 2008). A spokesman from WHO said the scale of the problem proved it was "clearly not an isolated accident, [but] a large-scale intentional activity to deceive consumers for simple, basic, short-term profits (VOA, 2008)." The issue raised concerns about food safety and political corruption in mainland China, and damaged the reputation of China's ability to manage the safety of its products. It also affected its export, with at least 11 countries stopping all imports of mainland Chinese dairy products.

This case clearly shows the catastrophic consequences when a firm puts its own goal above others. Adding a potentially fatal chemical compound to a food product is not only unethical, but also illegal and above all a criminal action. The case is not a difficult dilemma to judge in terms of ethics, legality or criminality. What was particularly unethical was that the problem was known for some time at the regional level but not divulged for fear that the scandal may have negative impact on tourism and the upcoming Olympic Games. The incident was brought to the attention of the general public after the games. An even greater negligence and source of outrage was the fact that this event was the second melamine incident; the first occurred in the USA in 2007 and was due to wheat gluten imported from China. Hundreds of dogs and cats were intoxicated (see Chapter 41). No efficient corrective action was put in place after the first incident. The products of a multinational company which had already experienced the problem in the United Stated was affected again in China and had to be recalled. Although a professional food safety management and crisis management calls for corrective actions after a first incident, it is also a matter of due diligence and ethics to be cautious on emerging risks. Consumers can forgive a first and an unexpected risk but when the same incident happens again for a second or a third time it is a matter of negligence.

Sometimes, issues are clearly illegal and reveal an underlying unethical attitude on the part of managers responsible of companies. For instance, release of a product knowing that it is contaminated or does not meet regulatory requirements, or repeating a microbial test until a negative result is obtained as evidence of safety is a sign of an unethical or immoral attitude of managers.

Other food safety issues are more problematic and less clear cut. The case of unpasteurized cheese is more controversial. Unpasteurized milk is known to be a potential source of foodborne pathogens, including *Listeria monocytogenes* (Todd, 2011). Thus, consumption of soft cheese made from unpasteurized milk is viewed as a medium to high food safety risk. *L. monocytogenes*-contaminated unpasteurized cheese has caused abortion and in a large cheese outbreak in California in 1985 one-third of 142 cases had fertile outcome (Linnan et al., 1988; Hof et al., 2003). Since the first compulsory law requiring milk from cows to be pasteurized, milk pasteurization is credited with dramatically lowering the incidence of typhoid fever, scarlet fever, diphtheria and tuberculosis. In USA today, it is forbidden to sell raw milk cheeses that have been aged for less than 60 months (US Code of Federal Regulations CFR, section 7 CFR 58.439). In Europe, the European Food Safety Authority (EFSA) states that there is a substantial risk of *Campylobacter* with milk and milk products if products are not subjected to a combination of treatments that eliminate the risk (*EFSA Journal*, 2005).

On the other hand, those in favor of unpasteurized cheese state that these cheeses, which have been made for centuries in France (e.g. Roquefort, Brie, Camembert) and other European countries (e.g. Serra da Estrela, Queijo da Ilha, mozzarella) are superior in taste. They are more complex in aroma and flavor, and they have longer lasting tastes. In addition, some believe that raw-milk bacteria and enzymes are helpful digestive aids and argue that lactose-intolerant people are able to digest raw-milk cheeses without their usual difficulty (Sheehan, 2007). Some cheese producers are also concerned with the extended ripening time for pasteurized cheese (Buffa et al., 2001). Up until the early 1900s, all cheese was made from raw milk, and raw-milk cheese defenders state that the unpasteurized cheese problems are related to hygiene problems and lack of knowledge. Raw-milk cheese makers, whether they are making young or aged cheese, must pay extra attention to the type of bacteria that develop in milk at different temperatures, and need to routinely test for bacterial counts. Food hygiene conditions must be held to the highest standards to avoid the introduction of bacteria that can develop in unheated milk. Raw milk needs to be made into cheese immediately to avoid fluctuations in temperature or possible contamination. Because the most likely source of *Campylobacter* in raw milk appears to be the feces of cows or goats, good hygienic practice (GHP) during milking is important (*EFSA Journal*, 2005). An HACCP system in place would serve as an appropriate tool for avoiding food safety outbreaks.

So what are the dilemmas? Is it all right to forbid sales of cheese made with unpasteurized milk? Here food safety is set up against people's pleasure of eating tasty cheese, traditional food cultures and the food industry's economic interest. Is unpasteurized cheese safe enough? In theory, a fully informed consumer might decide which food-related risk to take and which to avoid. But what does it mean to be fully informed? Can we expect that all consumers are able to collect the detailed information about the wide array of food safety issues and make their own decisions? Are there any good strategies for providing relevant information in such a way that consumers understand the risk? Is the correct thing to do to delegate this responsibility to responsible authorities? And what if the consumers do not trust these authorities?

The issue is even more serious when raw milk itself is sold, or worse, when raw milk is given to children. Other examples are when raw minced meat is served without information to consumers that the product may present a risk. A corollary of this situation is the subject of food irradiation. Application of this technology, evaluated as safe by the World Health Organization, can prevent a range of foodborne illnesses from foods of animal origin such as campylobacteriosis, *E. coli* O157 infection, salmonellosis and infections caused by parasites. However, its application is hampered by the fear of some misinformed consumers. Again, the question is how are consumers informed and how do they voice their view? Is it ethical that some people lose their life because of a powerful lobby against food irradiation? Or is it unethical of food scientists not to understand and pay attention to some consumers' cost of the long-term consequences of irradiation?

Another key example is when a food industry transfers responsibility of safety to consumers without providing proper warnings, crystal clear instructions on safety measures, or even worse market and promote a product in a society when it knows that consumers would not be able or have means to ensure its safety. The latter issue was raised with the question of breast milk substitute in the developing countries, until WHO established the Code of Breast-milk Substitutes that responsible companies comply with. The issue of warnings on food packaging is still not well addressed in most legislation and the clarity with which drug providers or electrical equipment or aviation companies provide information has not yet been established in the food industries. This has made some companies write an ambiguous text that does not raise the concerns of consumers, but where the company in case of an incident can decline any responsibility.

Other emerging areas where the ethics of a company is demonstrated is the validation of health claims, full consideration of safety in development of new technologies and novel foods, and in ensuring that food science and technology is not developed for the sole purpose of business interest but gives priority to consumer health.

As mentioned in Chapter 37, food safety management relies very much on the management of people, including ensuring that employees are competent in their job, have received the proper training and briefing about their responsibilities, are given the means and authority to do a professional job, and also set the right reporting structure to minimize conflict of interest in audits and investigation of incidents. Most of all a company must have a culture that fosters reporting and openly discusses problems, protects and rewards whistleblowers, and has a management that walks the talk and follows its own policies. The company culture is an area which is not legislated and enforced by law but is one of the most important aspects of the ethical practice of a company as it impacts on all aspect of operations, including safety and health and the social right of employees as well as safety of products and health of consumers.

Finally, one of the most fundamental aspects of ethics in the food industry is the commitment and "real will" of the company to build a solid food safety assurance system or just to do the minimum necessary to meet the requirements of legislation and certification bodies.

As safety of food products is the outcome of the food safety assurance system (including the professional management of its staff) of food companies, no matter how comprehensive and strong the regulatory system of a country may be, it cannot replace the everyday vigilance of managers, workers of a company who have to oversee the safety of products.

The Precautionary Principle

The precautionary principle states that if a product, an action or a policy has a suspected risk of causing harm to the public or to the environment, protective action should be supported before there is complete scientific proof of a risk. In the absence of scientific consensus, the principle implies that there is a social responsibility to protect the public from potential harm. This is a "better safe than sorry" or "caution in advance" principle that applies both to human health and to environmental protection.

One of the primary foundations of the precautionary principles, and globally accepted definitions, results from the work of the Rio Conference, or "Earth Summit" in 1992. Principle #15 of the Rio Declaration notes:

> In order to protect the environment, the precautionary approach shall be widely applied by States according to their capabilities. Where there are threats of serious or irreversible damage, lack of full scientific certainty shall not be used as a reason for postponing cost-effective measures to prevent environmental degradation.

The application of the precautionary principle has been made a statutory requirement within the European Union (Recuerda, 2006), and according to an FAO Expert Consultation report on food safety "international food safety organizations must make clear that science, while an important tool, is not sufficient in itself for food risk analysis and that it needs to operate within an ethical framework" (Kaiser, 2003). The problematic cases are the ones where there are disagreements about the value judgments made in the risk assessment. For novel food, new technologies or newly identified hazards, the answer to what is "safe" may not be the subject of a consensus. When both the likelihood of damage and the consequences of damage are unknown, then risk assessment becomes difficult and the assessors face ethical dilemmas (Kaiser, 2003; Almås, 1999). Genetically modified organisms (GMO) are one example of a new technology which has created a lot of debate. The presence of different interest groups and diverse citizens' values in different political arenas triggered a range of policy responses to GMOs in the 1990s (Vàzquez-Salat et al., 2012). While GMO was strongly supported by the scientific and biotech industry in the USA and led to a flourish of GM crops, European citizens were strongly opposed to GM plants. Consumers were skeptical and talked about "unnatural Frankenstein's food." They lacked confidence due to their governments' response to a series of food scares in the 1990s (Skogstad, 2003). Today, the GMO regulatory framework in the EU is different from the one in USA and more in line with the public's perception of risk than with the scientific definition of risk. The precautionary principle is applied in the EU and is constantly challenged.

A main ethical question which raised a controversy with regard to GM foods was a few years ago when EU countries would, as part of food aid, propose GM foods to the developing countries (Africa). In 2002 Zambia announced it would not accept GM food aid in any form. So the question was why would a technology and product rejected by the European population be good for the African population? Most African countries approach

GM technology applied to crops with caution. "Why shouldn't we be wary of this technology and its possible long-term health impacts, if the EU [European Union] is. If it is not good for them, why should it be good for us?" said Tewolde Egziabher, Ethiopia's director of the Environmental Protection Agency. Positions were polarized to a great extent after a quote from a US state department official, "Beggars can't be choosers," hit the headlines. It prompted the then president Levy Mwanawasa to say hunger was no reason for feeding his people "poison" (see the link http://www.irinnews.org/report/93991/FOOD-Rumpus-over-GM-food-aid).

The evolutionary aspect of the food system influences risk assessment of food safety and triggers some ethical dilemmas. Although consumers' food variety increases and the food industry potentials for new business grow, there are food safety issues to consider. The food market is becoming more and more global. Not only food but also food pathogens are distributed around the world. In Norway, sugar peas from Kenya led to an outbreak of dysentery in 2009, probably due to consumption of raw products. In Kenya, people boil or fry vegetables before consumption. This is not the case in Norway where unpeeled fruit and raw vegetables are consumed frequently. In many ways Norway is a food safety oasis in Europe, with a livestock population virtually free from *Salmonella* and where only one out of nine national outbreaks of infectious intestinal diseases linked to lettuce, sprouts, sugar peas and basil the last 20 years came from Norwegian produce (Røssvoll et al., 2012). The problem is that consumers' food safety habits and routines, inherited from parents, are not always adapted to handle new food scares from imported products. When food and food pathogens change, while food preparation routines stay the same, then food safety becomes an issue. The solution to this food safety problem raises ethical questions related to freedom of choice, economic prosperities for developing countries, distribution of pathogens into clean areas, etc.

ETHICAL DECISION-MAKING

Making good ethical decisions requires a trained sensitivity to ethical issues, a practiced method for exploring the ethical aspects of a decision and a weighing of the considerations that should impact our choice of a course of action. Having a method for ethical decision-making is absolutely essential. The more novel and difficult the ethical choice we face, the more we need to rely on discussion and dialogue with others about the dilemma. Only by careful exploration of the problem, aided by the insights and different perspectives of others, can we make good ethical choices in such situations.

A framework for ethical decision-making has been developed at the Markkula Center for Applied Ethics at Santa Clara University (www.scu.edu/ethics/decision). This framework for thinking ethically, which is the product of dialogue and debate among Manuel Velasquez, Dennis Moberg, Michael J. Meyer, Thomas Shanks, Margaret R. McLean, David DeCosse, Claire André and Kirk O. Hanson, is a useful method for exploring ethical dilemmas and identifying ethical courses of action. They divide ethical decision making into five blocks:

1. **Recognize an ethical issue:** Could this decision or situation be damaging to someone or to some group?

2. **Get the facts:** What are the relevant facts of the case? What facts are not known? Can I learn more about the situation? Do I know enough to make a decision?
3. **Evaluate alternative actions:** Which option will produce the most good and do the least harm? Which option best respects the rights of all who have a stake? Which option treats people equally or proportionately? Which option best serves the community as a whole, not just some members? Which option leads me to act as the sort of person I want to be?
4. **Make a decision and test it:** If I told someone I respect – or told a television audience – which option I have chosen, what would they say?
5. **Act and reflect on the outcome:** How can my decision be implemented with the greatest care and attention to the concerns of all stakeholders?

How companies respond to these questions when facing a dilemma will impact the companies' ethical image and thereby also the performance of the company. Responsibility for those concerns is shared among the players of the food supply chain, companies, decision-makers and consumers. Production, processing, storage and distribution of food and agricultural products are generally accepted as routine parts of everyday life all around the world. Therefore these activities have rarely been addressed within the realm of ethics. But food and agriculture, and the economic benefits derived from taking part in the associated system, are means to an inherently ethical end: feeding the world's population and preserving the earth's food-producing capacity and natural eco-systems for future generations. The ethical dimension of agriculture is therefore inherent to discussions on modern agricultural technologies (EGE, 2008b).

The ethics of a company and its management becomes conspicuous in times of an incident or a conflict. The ethics of a company is demonstrated by questions such as to what extent the company will:

- voluntarily acknowledge a contamination and/or if necessary recall its products;
- investigate the root cause of incidents up to the management level;
- accept loss of benefits to protect consumers;
- act transparently and reveal information on the incident and its cause; and
- take punitive actions against those who have knowingly and irresponsibly violated the policies (note: errors are different from violations and should be not the subject of punitive actions, see Chapter 37).

In the healthcare or aviation sectors, reporting of non-compliance or problems, and independence in investigations of incidents, is much more advanced and can be a model for the food sector if food safety is to be strengthened.

CONCLUSION

Consumers want a large variety of safe food choices, while producers want safe products to sell, but also less regulation. These ideas are at times in conflict. Exotic and convenience food year round and free choice of organic, raw, local and imported food create value to the producer and benefit to the consumer, but are associated with varied and sometimes serious risk to health (Todd, 2010).

New technology, which made it possible to process the food in advance, transport over long distances, display at retail and at home, made it possible for the pathogens to grow to levels capable of causing infections. Exposure to new pathogens, from imported products or due to new technologies, creates new food safety issues and raises ethical questions. Is zero risk what we aim for, or is there a level of acceptable risk?

The likelihood of becoming sick from the next meal has probably never been less than it is today, but the long-term consequences of today's food production is less known (Almås, 1999). The production process is more complex and less transparent and consumers are no longer in control of the production. They need to trust retailers and producers. Consumers are worried. Some of these worries are directly linked to the risks involved, be they real or perceived. Other worries are more linked to ethical questions related to well-being, free choice (autonomy) and fairness (justice). Availability of safe food needs to be addressed in relation to factors such as: respect for consumer choice, right to information on safety, universally affordable food, adequate income and working conditions for employees and workers, fair practice in trade, animal welfare and sustainability of biotic populations. Also, consumers are worried about new technologies and if these new techniques take into consideration their health and safety, or if they merely are developed for business interests and the benefits of producers. Some consumers wonder to what degree science is being developed impartially, if governments and public health authorities give priority to consumers' health in their opinion on risks and risk management options, and if incidents are investigated independently and transparently.

A dialogue about the ethical implications of food production, processing, policy, supply and consumption may help involved partners make better decisions. The discussion needs to be lifted to a level above what each company at any given point in time feels is best for them.

Aristotle says that identifying the good with pleasure is to prefer a life suitable for beasts:

> It is better to be a human being dissatisfied than a pig satisfied; better to be *Socrates* dissatisfied than a fool satisfied. And if the fool, or the pig, are of a different opinion, it is because they only know their own side of the question…

Every company has a social responsibility. Not behaving according to accepted norms for ethical behavior may have consequences not only for food safety, but also on a companies' image, reputation and performance. Ethics is not a question of thoughtless and slavish worship of rules, and to scrupulously check every action against a table of dos and don'ts. The fundamental question of ethics is not "What should I do?" but "What kind of person should I be?" For the food industry the question is "What image would we like for our company?" Will we accept compliances, deceive full negligence or non-compliance as long as we are not caught or will we vigilant in any condition?

"Integrity is doing the right thing, even when no one is watching" (Clive Staples Lewis, 1898–1963), and ethics is to the industry what integrity is to a person.

References

Almås, R., 1999. Food trust, ethics and safety in risk society. Sociol. Res. Online vol. 4 (3) <http://www.socresonline.org.uk/4/3/almas.html>.

Baumhart, R.C., 1961. How ethical are businessmen? Harv. Bus. Rev. 39, 6–9. (July–August).

Branigan, T., 2008. Chinese figures show fivefold rise in babies sick from contaminated milk. The Guardian (London). <http://www.guardian.co.uk/world/2008/dec/02/china> retrieved 2 April 2010.

Buffa, M., Guamis, B., Royo, C., Trujillo, A.J., 2001. Microbiological changes throughout ripening of goat cheese made from raw, pasteurized and high-pressure-treated milk. Food Microbiol. 18 (1), 45–51.

Duffy, R., Fearne, A., Hornibrook, S., 2003. Measuring distributive and procedural justice. An exploratory investigation of the fairness of retailer–supplier relationship in the UK food industry. Br. Food J. 105 (10), 682–694.

EGE, 2008a. Ethical aspects of animal cloning for food supply. Opinion No. 23 of the European Group on Ethics in Science and New Technologies to the European Commission. 16 January 2008. pp. 45.

EGE, 2008b. Ethics of modern developments in agricultural technologies. European Group on Ethics in Science and New Technologies to the European Commission. Opinion No. 24. 17/12/2008.

FAO, 1996. Report of the World Food Summit, 13–17 November 1996. Rome.

Hof, H., Nichterlein, T., Kretschmar, M., 2003. When are *Listeria* in foods a health risk? Trends. Food Sci. Technol. 5 (6), 185–190.

Kaiser, M., 2003. Ethical issues surrounding the GM-animals/GM-fish production. FAO/WHO Expert Consultation on Safety Assessment of Foods Derived from Genetically Modified Animals including Fish Meeting 17–21 November 2003.

Linnan, M.J., Mascola, L., Lou, X.D., Goulet, V., May, S., Salminen, C., et al., 1988. Epidemic listeriosis associated with Mexican-style cheese. New Engl. J. Med. 319, 823–828.

Macartney, J., 2008. China baby milk scandal spreads as sick toll rises to 13,000. The Times (London). <http://www.timesonline.co.uk/tol/news/world/asia/article4800458.ece> retrieved 2 April 2010.

McDonald, S., 2008. Nearly 53,000 Chinese children sick from milk. Associated Press. Google. Archived from the original on 21 May 2011. <http://web.archive.org/web/20110521092518/http://ap.google.com/article/ALeqM5iCL58EMBN1tqq6xujZlsaITAFpCQD93BHE880>.

Porter, M., Kramer, M.R., 2006. Strategy & Society. The link between competitive advantage and corporate social responsibility. Harv. Bus. Rev. December, 78–93.

Recuerda, M.A., 2006. Risk and Reason in the European Union Law. European Food and Feed Law Review 5. Foodstuffs (Question No. EFSA-Q-2003-081). Adopted on 27 January 2005 from the Commission related to *Campylobacter* in animals and foodstuffs.

Røssvoll, E.H., Ueland, Ø., Hagtvedt, T., Jacobsen, E., Lavik, R., Langsrud, S., 2012. Application of hazard analysis and critical control point methodology and risk-based grading to consumer food safety surveys. J. Food Prot. 75 (9), 1673–1690.

Sheehan, J.F., 2007. Testimony of John F. Sheehan. Division of Plant and Dairy Food Safety, Office of Food Safety, Center for Food Safety and Applied Nutrition, U.S. Food and Drug Administration before the Health and Government Operations Committee, Maryland House of Delegates.

Skogstad, G., 2003. Legitimacy and/or policy effectiveness? Network governance and GMO regulation in the European Union. J. Eur. Public Policy 10 (3), 321–338.

The EFSA Journal, 2005. *Campylobacter* in animals and foodstuffs. 173, pp. 1–10.

Todd, E.C.D., 2011. The international risk governance council framework and its application to Listeria monocytogenes in soft cheese made from unpasteurized milk. Food Control 22, 1513–1524.

Vàzquez-Salat, N., Salter, B., Smets, G., Houdebine, L.-M., 2012. The current state of GMO governance: are we ready for GM animals? Biotechnol. Adv. 30 (6), 1336–1343.

Velasquez, M., Andre, C., Shanks, T., Meyer, M.J., 2012. What is ethics? <http://www.scu.edu/ethics/practicing/decision/whatisethics.html>.

VOA, 2008. China's melamine milk crisis creates crisis of confidence. Archived from the original on 23 May 2011. <http://web.archive.org/web/20110523045758/http://www.voanews.com/english/2008-09-26-voa45.cfm>.

CHAPTER

47

Training and Education

Yasmine Motarjemi[1] *and Huub Lelieveld*[2]

[1]Food Safety Management Consultant, Nyon, Switzerland, [2]Global Harmonization Initiative, Bilthoven, The Netherlands

At last, few words should be said about education and training in food safety management, the purpose of this book. In this final, short, chapter we would like to draw your attention to some important points on the subject.

1. Often the terms education and training are used interchangeably and in food safety management we often use the term *training* while we may actually mean *education*. Strictly speaking there is a difference. Education is generally defined as the process of learning and acquiring information. It may be carried out for different purposes such as having a profession, a university degree, or more generally, for developing the power of reasoning and judgment. Training is the process of teaching a person (or an animal) a particular skill or type of behavior.
   A key difference between training and education is that in training, the subject may learn to practice behavior without always or necessarily knowing, or thinking of, the reason behind it. In education, the subject receives the knowledge and motivation to make informed decisions and choices. For professionals in food safety and other employees, although for simplicity we often use the term training, both education and training are essential, as managers should have the scientific knowledge and understanding to take sound decisions, but also to have the skills to perform their job.
2. It is a fallacy to automatically attribute failures in food safety management to people's incompetence and/or lack of training. As explained in Chapter 37, a failure can have different reasons, therefore, before deciding to resolve an issue with further training, there is a need for understanding the root cause of the problems and identifying the organization's needs (Figure 47.1).
3. When it has been confirmed that there is a problem with the competence of managers or employees to fulfill their responsibilities and perform their tasks, one should first determine the reasons why this is the case. Although it may be due to a lack of knowledge and skills, it may also be for a range of other reasons, such as lack of time, overload of work, weak infrastructure, lack of clear procedures or instructions, unfeasible

Food Safety Management.
DOI: http://dx.doi.org/10.1016/B978-0-12-381504-0.00047-0

FIGURE 47.1 Decision steps from identification of a company need for improvement to decisions for training. *Adapted from ISO 10015 (1999).*

policies and directives, inconsistency between responsibilities and authority, inadequate resources, lack of motivation or conflicts in values.

The design of a training program must include three types of information:

a. background knowledge and skills of trainees;
b. the work objectives;
c. the tasks that are or will be required to perform (Figure 47.2).

4. As for any management process, following the implementation of a course, the impact of that course needs to be evaluated, and based on the results of evaluation, further improvements need to be brought to its design. As part of this, it is good practice to, sometime after the course, follow up with the trainees and evaluate the impact and relevance of the course on their work performance and identify improvements needed to improve its design.

5. Often, after a course, when trainees return to their job or start their career, they may need further coaching to implement what they learned. Without that, it may be easily forgotten and thus the training will have been in vain. Involvement of a mentor would help to achieve the purpose of the training.

6. At regular times, an entire team (e.g. a department) should be given the same training together. Also, inspectors should receive the same training as professionals in the food industry and if possible also jointly. This then will enhance a common understanding of food safety management and decrease the risks of conflicts and differences in insights, which otherwise will become evident after an incident when it is too late.

7. Often we focus our training on technical people, i.e. on food safety managers or others directly involved in managing food safety, such as food safety scientists, laboratory staff, auditors, operators and line workers. There are, however, other people who may not be involved directly in food safety management but their decisions or actions can

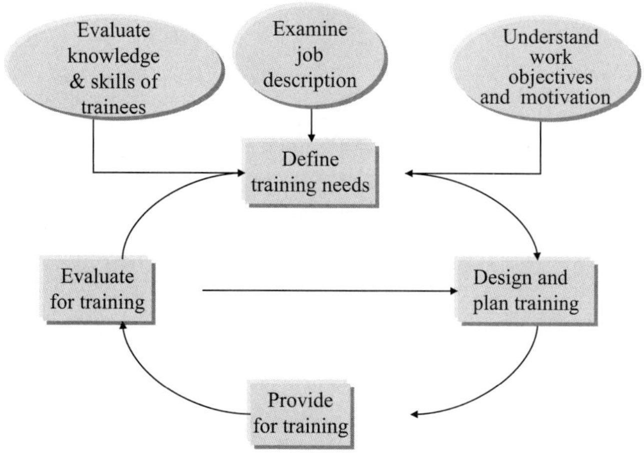

FIGURE 47.2 Steps for planning, designing and improving the training and education process. *Adapted from ISO 10015 (1999).*

have bearings on food safety. This can range from a business director to professionals working in packaging, transportation or administration. Their perception of what is a risk and what are important measures may be different and they may not be aware of the consequences of their decisions or actions for food safety. There have been numerous incidents associated with such situations. In food safety, as in other sectors (e.g. aviation), any detail is important if accidents are to be effectively prevented. The sensitization and education of these people should also be considered in food safety management; they should be made aware of the relevance of their decisions or work to food safety, in particular in an inadvertent adverse event.

8. Finally, but most importantly, the education of decision-makers, particularly the CEO of businesses, needs to be stressed.

This book has been developed with the aim of providing a common basis for training of food safety professionals working in industry and/or those who will assess food operations, to promote a common vision on food safety management and explain the underlying science. Providing training to staff, both managers and employees, is at the heart of any food safety management as without competent, knowledgeable and motivated staff, no system will be functional and effective (Figure 47.3).

Training staff to be competent for their job is not only important for consumer safety but it is also an employee right. An employee, who has not received proper training for his or her job, cannot be held responsible for incidents that are the results of mistakes. The ultimate responsibility and accountability fall on the management of the operation that failed to provide the necessary training.

Finally, a key factor in food safety management is the conscience of the staff themselves. Therefore, as final words to the practicing and future food safety professionals, and all

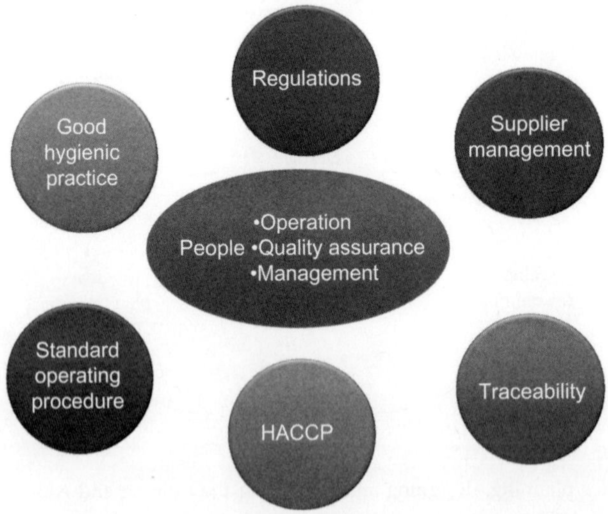

FIGURE 47.3 People are at the heart of food safety management systems and their management including their training and education is elemental in the food safety management system. *Adapted from ISO 10015 (1999).*

users of this book, we recommend that, in case of doubt, they should ask themselves three questions:

1. Would I give the food in question to my children?
2. How can I defend myself in a court of justice?
3. What did I do to prevent or minimize the risk?

Reference

ISO, 1999. ISO Quality management – Guidelines for training. ISO 10015: 1999. International Organization for Standardization, Geneva.

Index

Note: Page numbers followed by *"b," "f,"* and *"t"* refer to boxes, figures, and tables, respectively.

B

D

E

F

H

I

J

K

L

N

O

P

Q

S

T

U

V

W